OPERATIONS RESEARCH

OPERATIONS RESEARCH

Useful for:
- Honours and Postgraduate Students of Mathematics, Statistics, Commerce, Management and Engineering of all Universities
- Professional Courses like MBA, CA, ICWA, etc.

Sudhir Kumar Pundir

MSc, MPhil, NET (CSIR), JRF, SRF, PhD

Head
Department of Mathematics
SD (PG) College, Muzaffarnagar (UP)

CBSPD

CBS Publishers & Distributors Pvt Ltd

New Delhi • Bengaluru • Chennai • Kochi • Kolkata • Lucknow • Mumbai
Hyderabad • Jharkhand • Nagpur • Patna • Pune • Uttarakhand

OPERATIONS RESEARCH

ISBN: 978-93-89565-80-5

Copyright © Author and Publisher

First Edition: 2020

Reprint: 2024

Published by Satish Kumar Jain and produced by Varun Jain for

CBS Publishers & Distributors Pvt Ltd

4819/XI Prahlad Street, 24 Ansari Road, Daryaganj, New Delhi 110 002, India.
Ph: 23289259, 23266861   Website: www.cbspd.com

e-mail: delhi@cbspd.com

Corporate Office: 204 FIE, Industrial Area, Patparganj, Delhi-110092

Ph: 4934 4934   Fax: 4934 4935   e-mail: publishing@cbspd.com; publicity@cbspd.com

Branches

- **Bengaluru:** Seema House 2975, 17th Cross, K.R. Road, Banasankari 2nd Stage, Bengaluru 560 070, Karnataka, India
  Ph: +91-80-26771678/79   Fax: +91-80-26771680   e-mail: bangalore@cbspd.com
- **Chennai:** 7, Subbaraya Street, Shenoy Nagar, Chennai 600 030, Tamil Nadu, India
  Ph: +91-44-26680620, 26681266   Fax: +91-44-42032115   e-mail: chennai@cbspd.com
- **Kochi:** 42/1325, 1326, Power House Road, Opp KSEB, Ernakulam 682 018, Kochi, Kerala, India
  Ph: +91-484-4059061-67   Fax: +91-484-4059065   e-mail: kochi@cbspd.com
- **Kolkata:** 147, Hind Ceramics Compound, 1st Floor, Nilgunj Road, Belghoria, Kolkata 700 056, West Bengal, India
  Ph: +91-33-25633055/56   e-mail: kolkata@cbspd.com
- **Lucknow:** Basement, Khushnuma Complex, 7-Meerabai Marg (Behind Jawahar Bhawan), Lucknow 226 001, UP, India
  Ph: +0552-4000032   e-mail:tiwari.lucknowi@cbspd.com
- **Mumbai:** PWD Shed. Gala no. 25/26, Ramchandra Bhatt Marg, Next to JJ Hospital Gate no. 2, Opp. Union Bank of India, Noorbaug, Mumbai 400 009, Maharashtra, India
  Ph: 022-66661880/89   e-mail: mumbai@cbspd.com

Representatives

• Hyderabad	0-9885175004	• Jharkhand	0-9811541605	• Nagpur	0-8692091830
• Patna	0-9334159340	• Pune	0-9664372571	• Uttarakhand	0-9716462459

Printed at: HT Media, Greater Noida, UP

Preface

The book entitled 'OPERATIONS RESEARCH' is meant for honours and PG students of Mathematics, Management, Statistics, Commerce and Engineering of all universities on this subject. Besides, it will also be very useful for students preparing for various competitive examinations like CA, ICWA, CSIR-JRF/NET, SET, GATE and various entrance examinations for admission in M.Sc., M.Phil. and Ph.D. programme. The book covers almost the entire syllabi of various universities.

Special and conscious efforts have been made to keep the writing style simple. Students who are tired of complex concepts and abstract presentation styles, will find this book simple and straightforward. It is a collection and compilation work from various sources and has been endeavoured to include as much as information as could be possible. The book's objective is to provide a conceptual understanding of the fundamentals of Operations Research. Different concepts have been explained with the help of examples. There is plenty of scope in the form of exercise for the reader to try and solve the problem on his own.

I express my gratitude to the authors and publishers of various books, I consulted during the preparation of the book.

I wish sincerely thank **Sh. S.K. Jain** and **Sh. Varun Jain**, Managing directors, CBS publishers and Distributors, New Delhi for their encouragement and help in bringing out this publication in a present nice form.

My special thank to Sh. Y.N. Arjuna, Senior director, publishing, editorial and publicity and Smt. Ritu Chawla, publishing head, CBS publishers and distributors, New Delhi whose encouragement and unstinted support enabled me to complete my book. Sh. Sunil Dutt and Sh. Suresh Sharma, CBS publishers and distributors deserve special mention for their kind help and support. Mr. Peeyush Goel, M/s Dreamshapers also deserve special mention for nice typesetting.

I must also record my appreciation due to my wife **Dr. Rimple**, daughter **Rijuta** and son **Shrish** for their understanding and love during the long period that I have taken to complete this book. I am indebted to my colleagues and research students who generously shared their views on the need of a comprehensive book on OPERATIONS RESEARCH.

Above all I am thankful to 'The Almighty God' without whose grace nothing is possible for anyone .

Further suggestions and comments for improvement of the book will be thankfully received and duly incorporated in the next edition.

Dr. SUDHIR KUMAR PUNDIR
email : skpundir05@yahoo.co.in

Contents

Operations Research: An Introduction 1

1.1 INTRODUCTION

The term Operations Research was first coined in 1940 by McClosky and Trefthen in a small town Bowdsey of the U.K. This new science came into existence in military context. During World War II, Britain was having very limited military resources, therefore there was an urgent need to allocate resources to the various military operations, and to the activities within each operation in an effective manner. Therefore, the Britain military executives and managers called upon a team of scientists to improve the execution of various military projects. Because the team was dealing with research on military operations, this new scientific approach was called Operational Research and later it was adopted by U.S.A. as Operations Research (abbreviated as O.R.). The success of this team of scientists in Britain encouraged United States, Canada and France to start with such teams.

The apparent success of OR in the military, attracted the attention of industrial management in the new field, who were seeing solutions to the complex problems. In this way, OR began to creep into industry, business and governmental organisations.

The growth of OR has not only been limited to U.S.A. and the U.K. but has reached many countries of the world. Availability of faster and flexible computing facilities and the number of qualified professionals enhanced the acceptance and popularity of subject. Indicative of this is that the International Federation of Operations Research Societies, founded in 1959 now comprises member societies from many countries of the world. Thus the impact of OR can be felt in almost all walks of life.

1.2 THE NATURE AND DEFINITION OF OPERATIONS RESEARCH

[ROHILKHAND–1995, 98, 2000, 07; MEERUT–2006, 08, 09, 15]

As the name implies, Operations Research involves 'Research on Operations'. This says something about both the approach and the area of application of the field. Salient aspects related to definition stressed by various experts on the subject are as follows :

1. OR is a scientific method for providing executive departments with a quantitative basis for decision regarding the operations under their control.
   - Morse and Kimbal (1946)

2. OR is a scientific method of providing executive with an analytical and objective basis for decisions.
   - P.M.S. Blackett (1948)

3. OR utilizes the planned approach (updated scientific method) and an inter disciplinary team in order to represent complex functional relationships as mathematical models

for purpose of providing a quantitative basis for decision making and uncovering new problems for quantitative analysis.   - **Thieanf and Klehamp** (1975)

4. OR is the art of giving bad answers to problems to which otherwise worse answers are given.   - **T. L. Saaty** (1958)

5. OR is the application of scientific method by interdisciplinary teams to problems involving the controls of organised (man-machine) systems so as to provide solutions which best serve the purpose of the organisation as a whole.
    - **Ackoff and Saseini** (1968)

6. The term OR has hitherto been used to connate various attempts to study operations of war by scientific methods. From a more general point of view, OR can be considered to be an attempt to study those operations of modern society which involved organizations of men or of men and machines.   - **P. M. Morse** (1948)

7. OR is the application of scientific methods, techniques and tools to problems involving the operations of systems so as to provide these in control of the operations with optimum solutions to the problem.   - **Churchman, Ackoff, Amoff** (1958)

8. OR is a management activity pursued in two complementary ways – one half by the free and bold exercise of common sense untrammelled by any routine and other half by the application of a repertoire of well established precreated methods and techniques.   - **Jagjit Singh** (1968)

9. OR is an applied decision theory. It uses many scientific mathematical or logical means to attempt to cope with the problems that confront the executive when he tries to achieve a thorough going rationality in dealing with his decision problems.
    - **Miller and Starr**

10. OR is the systematic method oriented study of the basic structure characteristics, functions and relationships of an organization to provide the executive with a sound, scientific and quantitative basis for decision making.
    - **E. L. Amoff and M. J. Netzorg**

The modern definition of Operations Research can be summarized as below :

> Operations Research is the applications of the methods of mathematical science to complex problems arising in the direction and management of large system of men, machines, materials and money in industry, business, government and defence. The distinctive approach is to develop a scientific model of systems incorporating measurements of factors such as chance and risk with which to predict and compare the outcomes of alternatives, decisions, strategies or controls. The purpose is to help management determine its policy and actions scientifically.

1.3 OBJECTIVE OF OPERATIONS RESEARCH

The objective of OR is to provide a scientific basis to the decision makers for solving the problems involving the operations of system to give a solution which is in best interest of the organisation. The solution is called the optimum solution to the problem.

1.4 CHARACTERISTICS OF OPERATIONS RESEARCH   [ROHILKHAND–1996, 2000, 10]

From the evolution and development of Operations Research, the essential characteristics that have emerged are :

1. OR uses mixed team approach to find out optimum solution.
2. OR emphasis on the overall approach to the system.

3. OR is the inter-disciplinary team approach to find out the optimum return.
4. OR tries to optimize the total output by maximizing the profit and minimizing the cost.
5. OR uses scientific methods to arrive at the optimum solution.
6. OR takes into account all significant factors and evaluates them as a whole, i.e., OR considers the wholistic approach.
7. By OR techniques, we can not obtain perfect answers to our problems, but only the quality of the solution is improved from worse to bad answers.
8. The task of Operations Researcher may be said to consist of studying an operation, selecting variables to describe the operation, investigating the mathematical relationships of the variables, selecting objectives, measures and criteria of system performance and finding means to improve or optimize the system's functioning according to the selected measures and criteria.

1.5 MODELLING IN OPERATIONS RESEARCH   [MEERUT–1998, 2008; ROHILKHAND–2003]

Model plays very important role in OR. They are representations of reality. Models provide distilled and economic descriptions and explanations of the operations of the system that they represent. By experimenting on them, we can determine how the changes in the relevant system will effect its performance. Models enable us to experiment more effectively than on the system itself which is either impossible or too costly.

"A model in the sense used in OR is defined as a representation of an actual object or situation. It shows the relationships (direct or indirect) and inter-relationships of action and reaction in terms of cause and effect."

Since the model is an abstraction of reality, it thus appears to be less complete than reality itself. For a model to be complete, it must be representative of those aspects of reality that are being investigated.

> An OR model is some sort of mathematical description of various variables of a system representing some aspects of a problem on some subject of interest or inquiry. The model enables to conduct a number of experiments involving theoretical subjective manipulations to find some optimal solution of the problem.

1.5.1 ADVANTAGES OF A MODEL

Models have many advantages over a verbal description of a problem. Some of them are as follows :
1. An OR model provides a good grip over the problem.
2. It indicates the limitations and domain of the activity.
3. An OR model provides a logical, scientific and systematic approach to the problem.
4. Through OR model, one can incorporate useful tools which help in eliminating duplication of methods applied to solve specific problems.
5. OR models provide avenues for new research and improvement in the system.

1.5.2 DISADVANTAGES OF A MODEL

Models have few disadvantages also. Some of them are as follows :
1. Models are only an attempt in understanding an operation and should never be considered an absolute in any sense.
2. The validity of any model with regard to the corresponding operations can only be verified by carrying an experiment and relevant data characteristics.

☞ REMARKS

- An OR model should take into account new formulations without having any significant change in its frame.
- It should make minimum possible assumptions.
- The model should accommodate a parametric type of treatment.
- The model should be simple and coherent. The number of variables utilized by it should be less.

> Modelling is the essence of Operations Research approach. By Building Model, the complexities and uncertainties of a decision making problem can be changed to a logical structure that is amenable to formal analysis.

1.6 TYPES OF MODELS   [MEERUT–2006, 09]

There are several types of models that are commonly used in OR as well as in most sciences.

1. **Iconic Models.** Iconic models represent the system as it is but in different size. Thus, iconic models are obtained by enlarging or reducing the size of the system. In other words, they are images.

   Some common examples are photographs, drawing, model aeroplanes, ships, engines, globes, maps, etc. A toy aeroplane is an iconic model of a real one.

2. **Analogue Model.** It represents a system or an object of an inquiry by utilizing a set of properties different from what the original system processes. For example graphs are very simple analogues because distance is used to represent the properties such as time, number, percent, age, weight, and many other properties. Contour lines on a map represent the rise and fall of the heights. In general, analogues are less specific, less concrete but easier to manipulate than the iconic models.

3. **Mathematical (Symbolic) Model.** The symbolic or mathematical model is one which employs a set of mathematical symbols (*i.e.*, letters, numbers, etc.) to represent decision variables of the system. These variables are related together by means of a mathematical equation or a set of equations to describe the behaviour of the system. The solution of the problem is then obtained by applying well-developed mathematical techniques to the model.

   The symbolic model is usually the easiest to manipulate experimentally and it is the most general and abstract. Its function is more often explanatory rather than descriptive.

4. **Descriptive Models.** A descriptive model simply describes some aspects of a situation based on observation, survey, questionnaire results, or other available data. The result of an opinion poll represents a descriptive model.

5. **Predictive Model.** Such models can answer 'what if' type of question, i.e., they can make predictions regarding certain event. For example, based on the survey results, television networks such models attempt to explain and predict the election results before all the votes are actually counted.

6. **Prescriptive or Normalative Models.** Finally when a predictive model has been repeatedly successful, it can be used to prescribe a source of action. Linear programming is a normalative or prescriptive model because it prescribes what the managers ought to do.

7. **Deterministic Models.** Such models assume conditions of complete certainty and perfect knowledge. For example, linear programming, transportation and assignment models are deterministic models.

8. **Probabilistic Models.** These models handle those situations in which the consequences or pay off of managerial actions can not be predicted with certainty. However, it is possible to forecast a pattern of events, based on which managerial decisions can be made. For example, insurance companies are willing to insure against risk of fire, accidents, sickness and so on because the patterns of events have been compiled in the form of probability distributions.

9. **Dynamic Models.** In these models, time is considered as one of the important variables and admit the impact of changes generated by time. Also, in dynamic models, not only one, but a series of interdependent decisions is required during the planning horizon.

10. **Static Models.** These models do not consider the impact of changes that take place during the planning horizon, i.e., they are independent of time. Also, in static model only one decision is needed for the duration of a given time period.

1.7 GENERAL METHODS FOR SOLVING O.R. MODELS

[MEERUT–2009, 10; GARHWAL–2010; KANPUR–2012; ROHILKHAND–2003]

Solving a model consists of finding the values of the controlled variables that optimize the measure of performance, or of estimating them approximately. OR models are generally solved by the following three models :

1. **Analytic Method.** If the OR model is solved by using all the tools of classical mathematics such as differential calculus and finite differences available for this task, then such type of solutions are called analytic solutions. Solutions of various inventory models are obtained by adopting the so called analytic procedure.

2. **Numerical Method or Iterative Method.** Numerical methods concerns with the iterative or trial and error methods. Whenever the classical methods fail, we use iterative procedure. The classical methods may fail because of the complexity of the constraints or of the number of variables.

   In this procedure, we start with a trial solution and a set of rules for improving it. The trial solution is improved by the given rules and is then replaced by this improved solution. This process of improvement is repeated until no further improvement is possible or when the cost of further calculation can not be justified. Iterative procedures are divided into three groups:

   (i) After a finite number of repetitions, no further improvement will be possible.

   (ii) Although successive iterations improve the solutions, we are only guaranteed the solution as a limit of an infinite process.

   (iii) Finally, we include the trial and error method which however is likely to be lengthy, tedious and costly even if electronic computers are used.

3. **The Monte-Carlo Method.** The basis of Monte Carlo technique is random sampling of a variable's possible values. For this technique, some random numbers are required which may be converted into random variates whose behaviour is known from past experience. The main steps of Monte Carlo method are as follows :

   STEP 1. In order to have a general idea of the system, we first draw a flow diagram of the system.

   STEP 2. Then, we take correct sample observations to select some suitable model for the system. In this step, we compute the probability distributions for the variables of our interest.

   STEP 3. We, then convert the probability distributions to a cumulative distribution function.

   STEP 4. A sequence of random number is now selected with the help of random number tables.

STEP 5. Next, we determine the sequence of values of variables of interest with the sequence of random numbers obtained in step-4.

STEP 6. Finally, we construct some standard mathematical function to the values obtained in step-5.

ADVANTAGES OF MONTE-CARLO METHOD

1. These are helpful in finding solutions of complicated mathematical expressions which is not possible otherwise.
2. By these methods, difficulties of trial and error experimentation are avoided.

DISADVANTAGES OF MONTE-CARLO METHOD

1. These are costly way of getting a solution of any problem.
2. These methods do not provide optimal answers to the problems. The answers are good only when the size of the sample is sufficiently large.

1.8 PHASES OF OPERATIONS RESEARCH STUDY [ROHILKHAND–1997]

The major phases of OR study and their importance in solving the problem are described below :

1. **Identification of the Problem.** Identifying the problem correctly for the OR study is main concern of most of the organisations. It is almost impossible to obtain the 'right answer from the wrong problem'.

2. **Formulation of the Problem.** To find the solution of a problem, one must be able to find the problem and formulate it so that it is susceptible to research. Formulation of a problem for research consists in identifying, defining and specifying the measure of the components of a decision model.

   For the formulation of a problem, the following information are needed :
   (i) Who will make the decision?
   (ii) What are the objectives?
   (iii) Controllable variables and their ranges.
   (iv) The uncontrolled variables that may effect the outcomes of the available choices.
   Since, it is difficult to extract a right answer from the wrong problem, therefore, this phase should be executed with considerable care.

3. **Constructing a Mathematical Model.** After formulating the problem, we are concerned with the reformulation of the problem in an appropriate form which is convenient for analysis. The most suitable form of this purpose is to construct a mathematical model representing the system under study. It requires the identification of both static and dynamic structural elements. A mathematical model should include the following three important basic factors :
   (i) Decision variables and parameters.
   (ii) Constraints or restrictions.
   (iii) Objective function.

4. **Selecting appropriate input data.** Once an appropriate model has been constructed, the data required by the model is to be collected. This information can come from well kept records, from current tests and experiments or even from hunches based experience. Obviously, data collection is not an easy step as it can affect the output of the model significantly. It takes time to prepare these data if one wishes to reduce the possibility of data collection errors.

5. **Deriving a solution from the model.** After selecting input data, the next phase is to derive a solution from this model. Here in Operations Research, we are always in the search for an optimal solution. An optimal solution is one which maximize or minimize the objective function in the model in the best interest of the problem under consideration. Many procedures have been developed and will be discussed in detail.

6. **Testing the Model and its Solution.** After completing the model, it is once again tested as a whole for the errors if any. A model may be said to be valid if it can provide a reliable prediction of the system's performance. A good practitioner of Operations Research realises that this model be applicable for a longer time and thus he updates the model time to time by taking into account the past, present and future specifications of the problem.

7. **Establishing Controls.** Once a model and its solution is considered acceptable, controls should be placed on the solution. These controls are established to detect any significant changes in the conditions upon which the model is based. Obviously, if conditions change so much that the model is no longer an accurate representation of the system, the model is invalidated. Consequently, some monitoring procedure must be established to detect changes in the system as soon as possible to get the model revised to reflect these changes.

8. **Implementing and maintaining the solution.** The final phase of an Operations Research study is to implement the optimum solution derived by the OR team. As the conditions are constantly changing in the world, the model and the solution may not remain valid for a long time. Therefore, as the change occurs, it is to be detected as soon as possible so that the model, its solution and the resulting course of action can be modified accordingly.

1.9 SCOPE OF OPERATIONS RESEARCH [MEERUT–2007, 08, 09, 10, 14, 16; ROHILKHAND–2000]

OR has got wide scope. In general, we can say that whenever there is a problem, there is OR for help. In addition to the military, Operations Research is widely used in many organisations including business and industry. Now, we shall discuss the scope of OR in various important fields.

1. **In Agriculture.** With the increase of population and consequent shortage of food, there is a need to increase agriculture output for a country. But there are many problems faced by the agriculture department of a country, e.g., (i) climatic conditions, (ii) problems of optimal distribution of water from the resources, etc. Thus, there is a need of best policy under the given restrictions. OR techniques may be fruitful to determine the best policies. Therefore, OR has got great scope in agriculture.

2. **In Industry.** If the industry manager decides his policies (not necessarily optimum) only on the basis of his past experience (without using OR techniques) and a day comes when he gets retirement, then a heavy loss is encountered before the industry. This heavy loss can immediately be compensated by newly appointing a young specialist of OR techniques in business management. Thus, OR is useful to the industry Director in deciding optimum allocation of various limited resources such as men, machines, material, money, time, etc. to arrive at the optimum decision.

3. **In Defence.** During the World War II, OR teams of Britain and America developed the techniques and strategies that helped them to win "Air Battle of Britain", "Island Campaign in the Pacific", "Battle of North Atlantic" etc. In modern time war, the military operations are carried out by Airforce, Army and Navy. Therefore, there is a necessity to formulate optimum strategies that may give maximum benefit. OR helps

the military executives and managers to select the best strategy to win the battle. Thus, OR has got great scope in defence.

4. **In L.I.C.** OR approach is also applicable to enable the L.I.C. officers to decide :
  (i) What should be the premium rates for various modes of policies.
  (ii) How best the profits could be distributed in the cases of with profit policies, etc.

5. **In Planning.** Careful planning play an important role for the economic development of a country. OR techniques may be fruitful for such plannings. Planning Commission made use of Operations Research Techniques for planning the optimum size of the Caraveller fleet of Indian Airlines. Thus, OR has got great scope in planning also.

6. **In Marketing.** With the help of OR techniques, a marketing administrator can decide,
  (i)   where to distribute the products for sale so that the total cost of transportation etc. is minimum.
  (ii) the minimum per unit sale price.
  (iii) the size of the stock to meet the future demand.
  (iv) how to select the best advertising media with respect to time, cost, etc.
  (v)   how, when and what to purchase at the minimum possible cost.

1.10 DEVELOPMENT OF OPERATIONS RESEARCH IN INDIA

In India, Operations Research came into existence in the year 1949, with the establishment of an Operations Research unit at the regional research laboratory Hyderabad, for the prurpose of planning and organizing research. At the same time Prof. R.S. Verma (Delhi University) set up an OR team in defence science laboratory (later named Defence Science Centre) for the specific purpose of solving the problems of storing, planning and purchasing. Operations Research received a further boost with the setting up on an Operations Research team in the Indian Statistical Institute, Kolkata by Prof. P.C. Mahalanobis in 1953 for solving the problems related to national planning and survey. Later, the Operations Research Society of India (ORSI) was established in 1959. This society started publishing its journals OPSEARCH in 1964.

> Prof. Mahalanobis, first applied OR in India by formulating second five year plan with the help of OR techniques. Planning Commission made the use of OR techniques for planning the optimum size of the Caravelle Fleet of Indian airlines.

1. OR came into existence during   — World War-II
2. The first country to use OR method to solve problem is   — UK
3. The name Operations Research is first coined in the year   — 1940
4. The person(s) who coined the name Operations Research is — McClosky and Trefthen
5. OR society of India is founded in the year   — 1959

Exercise-1.1

1. What is Operations Research?
2. Give a brief account of the methods used in model formulation.
3. Explain with illustrations how Monte Carlo methods are used in OR.

REVIEW QUESTIONS

1. Define Operation Research (OR).
2. Give any three definitions of operations research and explain.
3. State the different types of models in OR. Explain briefly the general methods for solving OR models.
4. Write the main characteristics of OR.
5. Discuss the limitations of OR.
6. Discuss the various phases of OR problems.
7. Write the steps involved in the solution of an Operations Research problem.
8. Write a short note on the role of Operations Research in decision making.

MULTIPLE CHOICE QUESTIONS (CHOOSE THE MOST APPROPRIATE ONE)

1. The name 'Operations Research' is first coined in the year:
   (a) 1938   (b) 1940
   (c) 1942   (d) None of these
2. The first country to use Operations Research method to solve problems is:
   (a) UK   (b) USA
   (c) India   (d) None of these
3. OR society of India is founded in the year:
   (a) 1942   (b) 1959
   (c) 1949   (d) None of these
4. Operations Research is the outcome of:
   (a) National emergency
   (b) Political problems
   (c) Combined effects of talents of all field
   (d) None of these
5. The person who gave the name 'Operations Research' is/are:
   (a) Bellman
   (b) Newman
   (c) McClosky and Trefrhen
   (d) None of these

6. OR approach is typically based on the use of:
   (a) Physical model
   (b) Mathematical model
   (c) Descriptive model
   (d) None of these
7. Every mathematical model:
   (a) must be deteministic
   (b) represent data in numerical form
   (c) both (a) and (b) are true
   (d) None of these
8. Operations Research approach is:
   (a) scientific   (b) multidisciplinary
   (c) institution   (d) None of these
9. A model is:
   (a) an essence of reality
   (b) an approximation
   (c) both (a) and (b) are true
   (d) None of these
10. Decision variables in OR are:
    (a) controllable   (b) uncontrollable
    (c) parameters   (d) None of these

ANSWERS

1. (b)   2. (a)   3. (b)   4. (c)   5. (c)   6. (b)   7. (b)   8. (b)   9. (c)
10. (a)

ARCHIVE

1. Discuss various classification schemes of models.   [CA–1995]
2. What is meant by a mathematical model of real situations? Discuss the importance of models in the solution of OR problems.
   [AMIE–2005]
3. Explain the steps involved in the situation of an operations research problem.
   [ALLAHABAD(MBA)–1998]
4. "Whether in a private, non-profit or public organisation the most important and the distinguishing function of a

manager is problem solving. The field of quantitative methods is dedicated to aiding managers in their problem solving efforts. This is accomplished through the use of mathematical models to analyse the problem." Discuss.   [DELHI(MBA)–1998]

5. Comment on the following:

 (i) Operations Research advocates a system approach and is concerned with optimization.

 (ii) Operations Research replaces management by personality.   [DELHI(MBA)–1998]

6. Discuss in details the three types of models with special emphasis on important logical properties and the relationship the three types bear to each other and to modelled entities.   [AMIE–2004]

7. State the different types of model used in OR. Explain briefly the general methods for solving these OR models.

[INDORE–1995;

PUNJAB(Mech. Engg.)–1997, 98, 2004]

8. "In constructing any OR models, it is essential to realize that the most important purpose of modelling process is to help manager better." Keeping the purpose in mind, state any four OR models that can be of help to Charted Accountants in advising their clients.

[CA–1996]

9. Suppose you are being interviewed by the manager of a commercial firm for a job in the research department which deals with the application of quantitative techniques. Explain the scope and purpose of quantitative techniques and its usefulness in the firm. Give some examples of the applications of quantitative techniques in industry.

[AJMER(MBA)–1997]

10. "Executives at all levels in business and industry come across the problems of making decisions at every stage in their day to day activities. Operations Research provides them with various quantitative techniques for decision making and enhance their ability to make long-range plans to solve everyday problems of running a business and industry with greater efficiency, competence and confidence". Discuss the statement with examples.

[IGNOU(MBA)–1999, DELHI(MBA)–1998]

Linear Programming : Some Mathematical Preliminaries and Convex Sets

2

2.1 INTRODUCTION

Linear Programming problem (LPP) is an integral part of Operations Research which is an important branch of mathematics.

So, before discussing the basic concepts of linear programming problems, let us recall some mathematical concepts which are very useful in LPP.

2.2 MATRIX

A set of mn numbers either real or complex arranged in the form of a rectangular array in which there are m rows and n columns, is called a matrix of order $m \times n$ which is denoted by $[a_{ij}]_{m \times n}$ where $i = 1, 2, 3, ..., m$ represents the number of rows and $j = 1, 2, 3, ..., n$ represents the number of columns and thus a matrix of order $m \times n$ is usually written as

$$[a_{ij}]_{m \times n} = \begin{bmatrix} a_{11} & a_{12} & \cdots & a_{1n} \\ a_{21} & a_{22} & \cdots & a_{2n} \\ \vdots & \vdots & \vdots & \vdots \\ a_{m1} & a_{m2} & \cdots & a_{mn} \end{bmatrix}_{m \times n}$$

☞ **REMARK**

- Sometimes, a matrix is a rectangular array of numbers enclosed in double straight lines shown as '|| ||' or enclosed in parenthesis '( )'.

2.3 TYPE OF MATRICES

2.3.1 NULL MATRIX (OR ZERO MATRIX)

A matrix of order $m \times n$ is called a *null matrix* if it contains all mn elements zero. It is denoted by O and is usually written as

$$O = \begin{bmatrix} 0 & 0 & \cdots & 0 \\ 0 & 0 & \cdots & 0 \\ \vdots & \vdots & \vdots & \vdots \\ 0 & 0 & \cdots & 0 \end{bmatrix}_{m \times n}$$

2.3.2 ROW MATRIX

A matrix having only one row and n columns is called a *row matrix* of order $1 \times n$.

For example : $A = \begin{bmatrix} a_{11} & a_{12} & a_{13} & \cdots & a_{1n} \end{bmatrix}_{1 \times n}$

2.3.3 COLUMN MATRIX

A matrix having m rows and only one column is called a *column matrix* of order $m \times 1$.

For example :
$$A = \begin{bmatrix} a_{11} \\ a_{21} \\ a_{31} \\ \vdots \\ a_{m1} \end{bmatrix}_{m \times 1}$$

2.3.4 HORIZONTAL MATRIX

A matrix having more columns than the number of its rows, is called *Horizontal matrix*.

For example:
$$A = \begin{bmatrix} a_{11} & a_{12} & a_{13} \\ a_{21} & a_{22} & a_{23} \end{bmatrix}_{2 \times 3}$$

2.3.5 VERTICAL MATRIX

A matrix having more number of rows than its columns, is called *vertical matrix*.

For exmaple:
$$A = \begin{bmatrix} a_{11} & a_{12} \\ a_{21} & a_{22} \\ a_{31} & a_{32} \end{bmatrix}_{3 \times 2}$$

☛ **REMARK**

- Row matrix is also a horizontal matrix and column matrix is also a vertical matrix.

2.3.6 SQUARE MATRIX

A matrix having a number of rows equal to number of columns, is called *square matrix*.

For example :
$$A = \begin{bmatrix} a_{11} & a_{12} & a_{13} \\ a_{21} & a_{22} & a_{23} \\ a_{31} & a_{32} & a_{33} \end{bmatrix}_{3 \times 3}$$

Here, the matrix A has 3 rows and 3 columns, so it is a square matrix. Also the elements a_{11}, a_{22}, a_{33} are placed in the diagonal, so these elements are known as *diagonal elements*.

2.3.7 DIAGONAL MATRIX

A matrix of order $n \times n$ is called a *diagonal matrix* if it contains all its off diagonal elements equal to zero.

Suppose $A = [a_{ij}]_{n \times n}$ and if $a_{ij} = 0$ for all $i \neq j$, then A is a diagonal matrix. Diagonal matrix of order $n \times n$ is usually written as $\text{Diag} [a_{11} \quad a_{22} \quad a_{33} \quad \cdots \quad a_{nn}]$

For example:
$$A = \begin{bmatrix} 1 & 0 & 0 \\ 0 & 2 & 0 \\ 0 & 0 & 3 \end{bmatrix}_{3 \times 3} = \text{Diag} [1 \ 2 \ 3]$$

2.3.8 SCALAR MATRIX

A diagonal matrix whose diagonal elements are all equal but not equal to 1 is called a *scalar matrix*.

For example:
$$A = \begin{bmatrix} k & 0 & 0 \\ 0 & k & 0 \\ 0 & 0 & k \end{bmatrix}, k \neq 1$$

2.3.9 UNIT MATRIX

A square matrix of order $n \times n$ having all off-diagonal elements equal to zero and each of the diagonal elements equal to 1, is called a *unit matrix*. It is usually denoted by I_n and is written as

$$I_n = \begin{bmatrix} 1 & 0 & \ldots & 0 \\ 0 & 1 & \ldots & 0 \\ 0 & 0 & \ldots & 0 \\ \vdots & \vdots & \vdots & \vdots \\ 0 & 0 & \ldots & 1 \end{bmatrix}_{n \times n}$$

☛ REMARK
- Unit matrix can also be denoted by I.

2.3.10 TRIANGULAR MATRIX

A matrix in which the elements lying above or below principal diagonal are all zero, is called a *triangular matrix*.

There are two kinds of triangular matrix.

(a) Upper triangular matrix : A matrix of order $n \times n$ is called an *upper triangular matrix* if it contains all its elements below the diagonal elements equal to zero.

Suppose $A = [a_{ij}]_{n \times n}$ and if $a_{ij} = 0$ for all $i > j$, then A is an upper triangular matrix.

For example : $A = \begin{bmatrix} 2 & 3 & 4 \\ 0 & 1 & 5 \\ 0 & 0 & 3 \end{bmatrix}_{3 \times 3}$ is an upper triangular matrix of order 3×3.

(b) Lower triangular matrix : A matrix of order $n \times n$ is called a *lower triangular matrix* if it contains all its elements above the diagonal elements equal to zero.

Suppose $A = [a_{ij}]_{n \times n}$ and if $a_{ij} = 0$ for all $i < j$, then A is called lower triangular matrix.

For example : $A = \begin{bmatrix} 1 & 0 & 0 \\ 3 & 4 & 0 \\ 5 & 6 & 7 \end{bmatrix}_{3 \times 3}$ is a lower triangular matrix of order 3×3.

2.4 OPERATIONS ON MATRICES

2.4.1 ADDITION OF MATRICES

Suppose A and B are two matrices of same order, then the addition of these two matrices is obtained by adding corresponding elements of A and B. It is denoted by $A + B$. If the order of A and B is $m \times n$, then the order of $A + B$ will be $m \times n$.

Suppose $\qquad A = [a_{ij}]_{m \times n}$ and $B = [b_{ij}]_{m \times n}$

then $\qquad A + B = [a_{ij} + b_{ij}]_{m \times n}$

For example: If $\qquad A = \begin{bmatrix} 1 & 2 & 3 \\ 5 & 1 & 4 \\ 7 & 8 & 9 \end{bmatrix}$ and $B = \begin{bmatrix} 1 & 3 & 5 \\ 5 & 0 & 1 \\ 3 & 2 & 12 \end{bmatrix}$

then
$$A + B = \begin{bmatrix} 1 & 2 & 3 \\ 5 & 1 & 4 \\ 7 & 8 & 9 \end{bmatrix} + \begin{bmatrix} 1 & 3 & 5 \\ 5 & 0 & 1 \\ 3 & 2 & 12 \end{bmatrix}$$

$$= \begin{bmatrix} 1+1 & 2+3 & 3+5 \\ 5+5 & 1+0 & 4+1 \\ 7+3 & 8+2 & 9+12 \end{bmatrix} = \begin{bmatrix} 2 & 5 & 8 \\ 10 & 1 & 5 \\ 10 & 10 & 21 \end{bmatrix}$$

☛ **REMARK**

- If the orders of the matrices are different, then they are not conformable for addition.

2.4.2 SUBSTRACTION OF MATRICES

Suppose A and B are two matrices of same order, then the substraction of A and B, i.e., $A–B$ is obtained by substracting each element of B from the corresponding element of A. If A and B are of order $m \times n$, then $A – B$ will be of order $m \times n$.

Let $\qquad A = [a_{ij}]_{m \times n}$ and $B = [b_{ij}]_{m \times n}$

then $\qquad A - B = [a_{ij} - b_{ij}]_{m \times n}$

For example: If $\quad A = \begin{bmatrix} 1 & 2 & 3 \\ 3 & 4 & 5 \\ 5 & 6 & 7 \end{bmatrix}$ and $B = \begin{bmatrix} 0 & 5 & 2 \\ 3 & -2 & 2 \\ 5 & 7 & 8 \end{bmatrix}$

then $\qquad A - B = \begin{bmatrix} 1 & 2 & 3 \\ 3 & 4 & 5 \\ 5 & 6 & 7 \end{bmatrix} - \begin{bmatrix} 0 & 5 & 2 \\ 3 & -2 & 2 \\ 5 & 7 & 8 \end{bmatrix}$

$$= \begin{bmatrix} 1-0 & 2-5 & 3-2 \\ 3-3 & 4-(-2) & 5-2 \\ 5-5 & 6-7 & 7-8 \end{bmatrix} = \begin{bmatrix} 1 & -3 & 1 \\ 0 & 6 & 3 \\ 0 & -1 & -1 \end{bmatrix}$$

☛ **REMARK**

- If the order of matrices are different, then they are not conformable for substraction.

2.4.3 MULTIPLICATION OF A MATRIX BY A SCALAR

Suppose A is a matrix of order $m \times n$ and k is a scalar, then the multiplication of A by k, i.e. kA is obtained by multiplying each element of A by k.

Let $\qquad A = [a_{ij}]_{m \times n} \ \forall \ 1 \le i \le m$ and $1 \le j \le n$, then $kA = [ka_{ij}]_{m \times n}$

For example : If $\quad A = \begin{bmatrix} 1 & 2 & 3 \\ 4 & 5 & 6 \\ 7 & 8 & 9 \end{bmatrix}$ and $k = 3$,

then $\qquad 3A = 3 \begin{bmatrix} 1 & 2 & 3 \\ 4 & 5 & 6 \\ 7 & 8 & 9 \end{bmatrix}$

$$= \begin{bmatrix} 3 \times 1 & 3 \times 2 & 3 \times 3 \\ 3 \times 4 & 3 \times 5 & 3 \times 6 \\ 3 \times 7 & 3 \times 8 & 3 \times 9 \end{bmatrix} = \begin{bmatrix} 3 & 6 & 9 \\ 12 & 15 & 18 \\ 21 & 24 & 27 \end{bmatrix}$$

2.4.4 EQUALITY OF MATRICES

Two matrices are said to be equal if both have same order and having same corresponding elements.

For example : The matrices $A = \begin{bmatrix} 1 & 2 \\ -3 & 4 \end{bmatrix}$ and $B = \begin{bmatrix} x & y \\ z & 4 \end{bmatrix}$ are said to be equal if $x = 1$, $y = 2$ and $z = -3$.

2.5 PROPERTIES OF MATRIX ADDITION

2.5.1 COMMUTATIVE LAW

If A and B are two matrices of same order m × n, then A + B = B + A

Proof. Let $A = [a_{ij}]_{m \times n}$ and $B = [b_{ij}]_{m \times n}$ where $1 \leq i \leq m$ and $1 \leq j \leq n$. Then

$$A+B = [a_{ij}]_{m \times n} + [b_{ij}]_{m \times n}$$
$$= [a_{ij} + b_{ij}]_{m \times n} \quad \text{(By definition of addition)}$$
$$= [b_{ij} + a_{ij}]_{m \times n}$$
$$\text{(}\because \text{ Addition of Real numbers are always commutative)}$$
$$= [b_{ij}]_{m \times n} + [a_{ij}]_{m \times n}$$
$$= B + A$$

Hence, $\quad A + B = B + A$

2.5.2 ASSOCIATIVE LAW

If A ,B and C are three matrices of same order m × n, then
$$(A + B) + C = A + (B + C)$$

Proof. Let $A = [a_{ij}]_{m \times n}$ and $B = [b_{ij}]_{m \times n}$ where $1 \leq i \leq m$ and $1 \leq j \leq n$. Then

$$(A+B)+C = ([a_{ij}]_{m \times n} + [b_{ij}]_{m \times n}) + [c_{ij}]_{m \times n}$$
$$= [a_{ij} + b_{ij}]_{m \times n} + [c_{ij}]_{m \times n}$$
$$= [(a_{ij} + b_{ij}) + (c_{ij})]_{m \times n}$$
$$\text{(}\because \text{ Addition of numbers are always associative)}$$
$$= [a_{ij}]_{m \times n} + ([b_{ij} + c_{ij}]_{m \times n})$$
$$= [a_{ij}]_{m \times n} + ([b_{ij}]_{m \times n} + [c_{ij}]_{m \times n})$$
$$= A + (B + C)$$

Hence, $\quad (A+B) + C = A + (B+C)$

2.5.3 ADDITIVE IDENTITY

If A is a matrix of order m × n and O is a null matrix of the same order m ×n, then
$$A + O = A = O + A$$

Proof. Let $A = [a_{ij}]_{m \times n}$ and $O = [0]_{m \times n}$, then

$$A + O = [a_{ij}]_{m \times n} + [0]_{m \times n}$$
$$= [a_{ij} + 0]_{m \times n}$$
$$= [a_{ij}]_{m \times n} = A$$

Also $\quad O + A = [0]_{m \times n} + [a_{ij}]_{m \times n}$
$$= [0 + a_{ij}]_{m \times n}$$
$$= [a_{ij}]_{m \times n} = A$$

Hence $\qquad A + O = A = O + A$

Therefore, the null matrix O is treated as an additive identity.

2.5.4 ADDITIVE INVERSE

If A is a matrix of order $m \times n$ and $-A$ is the negative of A, so its order is also $m \times n$, then

$$-A + A = O \quad \text{(null matrix)}$$

Here, $-A$ *is the additive inverse of* A.

2.5.5 CANCELLATION LAW

If A, B and C are three matrices of order $m \times n$ then

(i) $A + B = A + C \Rightarrow B = C$ (Left cancellation law)

(ii) $B + A = C + A \Rightarrow B = C$ (Right cancellation law)

Proof.

(i) It is given that

$$A + B = A + C \qquad\qquad ...(1)$$

Adding $-A$ to the left of both sides, we get

$$-A + (A + B) = -A + (A + C)$$

$\Rightarrow \qquad\qquad (-A + A) + B = (-A + A) + C \qquad\qquad$ (By associative law)

$\Rightarrow \qquad\qquad O + B = O + C \qquad\qquad$ (By additive inverse)

$\Rightarrow \qquad\qquad B = C \qquad\qquad$ (By additive identity)

Similarly, we can prove that if $B + A = C + A$, then $B = C$.

2.6 PROPERTIES OF MULTIPLICATION OF MATRIX BY A SCALAR

(i) **Distributive law of scalar multiplication over matrix addition :** *If A and B are two matrices of order $m \times n$ and k is any scalar, then $k(A + B) = kA + kB$*

Proof. Let $A = [a_{ij}]_{m \times n}$ and $B = [b_{ij}]_{m \times n}$, then

$$
\begin{aligned}
k(A + B) &= k([a_{ij}]_{m \times n} + [b_{ij}]_{m \times n}) \\
&= k([a_{ij} + b_{ij}]_{m \times n}) \\
&= [k(a_{ij} + b_{ij})]_{m \times n} \\
&= [ka_{ij} + kb_{ij}]_{m \times n} \\
&= [ka_{ij}]_{m \times n} + [kb_{ij}]_{m \times n} \\
&= k[a_{ij}]_{m \times n} + k[b_{ij}]_{m \times n} \\
&= kA + kB
\end{aligned}
$$

Hence $\qquad k(A + B) = kA + kB.$

(ii) *If A is a matrix of order $m \times n$ and a, b are two scalars, then $(a + b)A = aA + bA$*

Proof. Let $\qquad\qquad A = [a_{ij}]_{m \times n}$, then

$$
\begin{aligned}
(a + b)A &= (a + b)[a_{ij}]_{m \times n} \\
&= [(a + b)a_{ij}]_{m \times n} \qquad \text{(By scalar multiplication)} \\
&= [aa_{ij} + ba_{ij}]_{m \times n} \qquad (\because \text{Numbers are distributive}) \\
&= [aa_{ij}]_{m \times n} + [ba_{ij}]_{m \times n}
\end{aligned}
$$

$$= a[a_{ij}]_{m \times n} + b[a_{ij}]_{m \times n}$$

$$= aA + bA$$

Hence $\qquad (a + b)A = aA + bA$

(iii) *If A is a matrix of order $m \times n$ and a, b are two scalars, then $a(bA) = (ab)A$.*

Proof. Let $A = [a_{ij}]_{m \times n}$, then

$$a(bA) = a(b[a_{ij}]_{m \times n})$$

$$= [a(ba_{ij})]_{m \times n} \qquad \text{(By scalar multiplication)}$$

$$= [(ab)a_{ij}]_{m \times n} \qquad (\because \text{ Numbers are associative})$$

$$= (ab)[a_{ij}]_{m \times n}$$

$$= (ab)\, A$$

Hence $\qquad a(bA) = (ab)A.$

(iv) *If A is a matrix of order $m \times n$ and k is any scalar, then $(-k)A = -(kA) = k(-A)$*

Proof. Let $A = [a_{ij}]_{m \times n}$, then

$$(-k)A = (-k)[a_{ij}]_{m \times n}$$

$$= [(-k)a_{ij}]_{m \times n} \qquad \text{(By scalar multiplication)}$$

$$= [-ka_{ij}]_{m \times n}$$

$$= -[ka_{ij}]_{m \times n}$$

$$= -(kA)$$

Now $\qquad (-k)A = (-k)[a_{ij}]_{m \times n}$

$$= [(-k)a_{ij}]_{m \times n}$$

$$= [k(-a_{ij})]_{m \times n}$$

$$= k[-a_{ij}]_{m \times n}$$

$$= k(-A)$$

Hence $\qquad (-k)A = -(kA) = k(-A).$

2.7 MULTIPLICATION OF MATRICES

Let A and B be two matrices of order $m \times n$ and $n \times p$ respectively. Then a matrix C of order $m \times p$ is obtained by multiplying each row of A to each column of B.

Suppose $A = [a_{ij}]_{m \times n}$, $B = [b_{jk}]_{n \times p}$, then $C = [c_{ik}]_{m \times p}$ is known as the multiplication of A and B if

$$c_{ik} = \sum_{j=1}^{n} a_{ij}b_{jk}$$

and hence we can write $\qquad C = AB$

🔖 Working Procedure

First we check whether the matrices are conformable for multiplication or not. For this we check that if the number of columns of first matrix is equal to the number of rows of the second matrix, then the matrices can be multiplied. Multiplication is operated by the rule (row × column). In this rule, we first put the first row of the first matrix next to the first column of the second matrix and the corresponding elements are now multiplied and then summed up which gives the first element of the first row of the product matrix. This process runs till the first row of the first matrix is operated to all columns of the second matrix. After that the first process is applied to the second, third etc. rows of the first matrix.

For example : If $\quad A = \begin{bmatrix} 2 & 1 & 5 \\ 6 & 2 & 3 \end{bmatrix}_{2 \times 3}$ and $B = \begin{bmatrix} 3 & 4 \\ 5 & 6 \\ 7 & 8 \end{bmatrix}_{3 \times 2}$, then

$$AB = \begin{bmatrix} 2 & 1 & 5 \\ 6 & 2 & 3 \end{bmatrix} \begin{bmatrix} 3 & 4 \\ 5 & 6 \\ 7 & 8 \end{bmatrix}$$

$$= \begin{bmatrix} 2 \times 3 + 1 \times 5 + 5 \times 7 & 2 \times 4 + 1 \times 6 + 5 \times 8 \\ 6 \times 3 + 2 \times 5 + 3 \times 7 & 6 \times 4 + 2 \times 6 + 3 \times 8 \end{bmatrix}$$

$$= \begin{bmatrix} 6 + 5 + 35 & 8 + 6 + 40 \\ 18 + 10 + 21 & 24 + 12 + 24 \end{bmatrix}$$

$$= \begin{bmatrix} 46 & 54 \\ 49 & 60 \end{bmatrix}$$

☛ REMARKS
- If the number of columns of the matrix A is equal to the number of rows of matrix B, then A and B are conformable for the multiplication AB but not for BA.
- Square matrices are always conformable for multiplication in both ways.

2.8 DETERMINANT OF A SQUARE MATRIX

Let A be a square matrix. Then the determinant which is formed by the elements of matrix A is usually denoted by $|A|$.

For example : If $\quad A = \begin{bmatrix} a_{11} & a_{12} & a_{13} \\ a_{21} & a_{22} & a_{23} \\ a_{31} & a_{32} & a_{33} \end{bmatrix}$, then its determinant is

$$A = \begin{vmatrix} a_{11} & a_{12} & a_{13} \\ a_{21} & a_{22} & a_{23} \\ a_{31} & a_{32} & a_{33} \end{vmatrix}$$

☛ REMARK
- The determinant of a matrix is reduced to a number.

2.9 PROPERTIES OF DETERMINANTS

(1) The value of a determinant is zero if all the elements of a row or column are zero.

(2) The value of a determinant remain unchanged when rows are changed into corresponding columns.

(3) If any two rows or columns of a determinant are interchanged, the sign of the determinant is changed.

(4) If any two rows or columns of a determinant are identical, then the value of the determinant is zero.

(5) If every element of same columns or row is the sum of two terms then determinant is equal to the sum of two determinants are containing only the first term and other the second term only in place of each sum.

(6) If each element of a row (or column) is multiplied by a constant k, then the value of the new determinant will be k times the value of original determinant.

(7) If each element of a row (or column) of a determinant multiplied by a constant k and then added to the corresponding elements of some other row (or column) then the value of the determinant remain same.

(8) If the elements of the determinant are the polynomial in a variable x and if by putting $x = a$, the determinant vanishes then $(x - a)$ will be a factor of determinant.

2.10 EVALUATION OF A DETERMINANT BY SARRUS DIAGRAM

$$\begin{vmatrix} a_{11} & a_{12} & a_{13} \\ a_{21} & a_{22} & a_{23} \\ a_{31} & a_{32} & a_{33} \end{vmatrix} = a_{11}(a_{22}a_{33} - a_{32}a_{23}) - a_{12}(a_{21}a_{33} - a_{31}a_{23}) + a_{13}(a_{21}a_{32} - a_{31}a_{22})$$

$$= a_{11}a_{22}a_{33} + a_{12}a_{31}a_{23} + a_{13}a_{21}a_{32} - (a_{11}a_{32}a_{23} + a_{12}a_{21}a_{33} + a_{13}a_{31}a_{22})$$

Working Procedure

Write the columns of the determinant and again write the first and second columns on the right side and draw the lines as shown in the following figure :

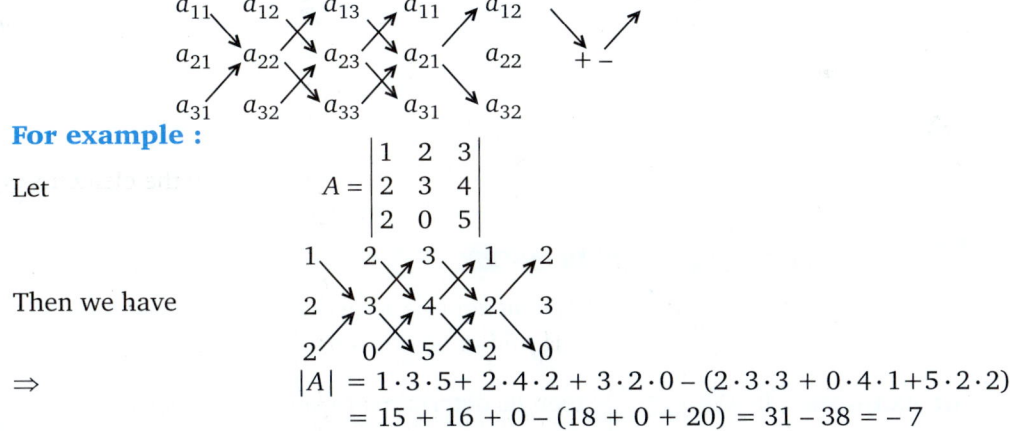

For example :

Let
$$A = \begin{vmatrix} 1 & 2 & 3 \\ 2 & 3 & 4 \\ 2 & 0 & 5 \end{vmatrix}$$

Then we have

$\Rightarrow$

$$|A| = 1 \cdot 3 \cdot 5 + 2 \cdot 4 \cdot 2 + 3 \cdot 2 \cdot 0 - (2 \cdot 3 \cdot 3 + 0 \cdot 4 \cdot 1 + 5 \cdot 2 \cdot 2)$$
$$= 15 + 16 + 0 - (18 + 0 + 20) = 31 - 38 = -7$$

2.11 MINORS AND COFACTORS

2.11.1 MINORS

In determinant, $\Delta = \begin{vmatrix} a_{11} & a_{12} & a_{13} \\ a_{21} & a_{22} & a_{23} \\ a_{31} & a_{32} & a_{33} \end{vmatrix}$ ...(1)

if we leave the row and column passing through the element a_{ij} then we obtained the second order determinant, which is called the minor of the element a_{ij}. It is denoted by M_{ij}.

Therefore, in a determinant of order 3, we may get 9 minors corresponding to the 9 elements of the determinant.

For example, in determinant (1)

$$\text{Minor of } a_{21} = \begin{vmatrix} a_{12} & a_{13} \\ a_{32} & a_{33} \end{vmatrix} = M_{21}$$

and

$$\text{Minor of } a_{32} = \begin{vmatrix} a_{11} & a_{13} \\ a_{21} & a_{23} \end{vmatrix} = M_{32}$$

If we expand the determinant along the first row, then

$$\Delta = (-1)^{1+1} a_{11} M_{11} + (-1)^{1+2} a_{12} M_{12} + (-1)^{1+3} a_{13} M_{13}$$
$$= a_{11} M_{11} - a_{12} M_{12} + a_{13} M_{13}$$

Similarly, along second column, we can write

$$\Delta = -a_{12} M_{12} + a_{22} M_{22} - a_{32} M_{32}$$

2.11.2 COFACTOR

If we multiply the minor M_{ij} by $(-1)^{i+j}$. Then resulting value is called cofactor of the element a_{ij}. If A_{ij} is the cofactor of a_{ij}, then we write

$$\text{Cofactor of } a_{ij} = A_{ij} = (-1)^{i+j} M_{ij}$$

$$\text{Cofactor of } a_{21} = A_{21} = (-1)^{2+1} M_{21} = - \begin{vmatrix} a_{12} & a_{13} \\ a_{32} & a_{33} \end{vmatrix}$$

$$\text{Cofactor of } a_{32} = A_{32} = (-1)^{3+2} M_{32} = - \begin{vmatrix} a_{11} & a_{13} \\ a_{21} & a_{23} \end{vmatrix}$$

Hence, cofactor of $a_{ij} = (-1)^{i+j}$ determinant obtained by leaving row and column passing through that element. Therefore, we can write

$$\Delta = a_{11} A_{11} + a_{12} A_{12} + a_{13} A_{13}$$
$$\Delta = a_{21} A_{21} + a_{22} A_{22} + a_{23} A_{23}$$
$$\Delta = a_{31} A_{31} + a_{32} A_{32} + a_{33} A_{33}$$

and

$$a_{11} A_{21} + a_{12} A_{22} + a_{13} A_{23} = 0$$
$$a_{11} A_{31} + a_{12} A_{32} + a_{13} A_{33} = 0$$

2.12 SINGULAR AND NON-SINGULAR MATRIX

Definition. *A matrix whose determinant value is zero, is said to be singular matrix.* If the matrix is not singular, then it is said to be non-singular.

For example : If $A = \begin{bmatrix} 2 & 3 \\ 6 & 9 \end{bmatrix}$, then its determinant value is given by

$$|A| = \begin{vmatrix} 2 & 3 \\ 6 & 9 \end{vmatrix} = 2 \times 9 - 3 \times 6 = 18 - 18 = 0$$

Thus the matrix A is singular.

2.13 TRANSPOSE OF A MATRIX

Consider a matrix $A = [a_{ij}]_{m \times n}$. Then a matrix which is obtained by interchanging the rows and columns of A is called the transpose of A. It is denoted by A' or A^T.

That is , if $A = [a_{ij}]_{m \times n}$, then $A' = [a_{ji}]_{n \times m}$.

For example : If $\quad A = \begin{bmatrix} 2 & 3 & 5 \\ 1 & 6 & 7 \end{bmatrix}_{2 \times 3}$, then its transpose is

$$A' = \begin{bmatrix} 2 & 3 & 5 \\ 1 & 6 & 7 \end{bmatrix}'$$

$$= \begin{bmatrix} 2 & 1 \\ 3 & 6 \\ 5 & 7 \end{bmatrix}_{3 \times 2}$$

☛ REMARKS
- Transpose of row matrix is a column matrix and transpose of a column matrix is a row matrix.
- If a matrix is square then its transpose will be a square matrix of same order.

2.14 PROPERTIES OF TRANSPOSE OF A MATRIX

THEOREM 1. *If A' and B' are the transpose of the matrix A and B respectively, then :*
  (i) $(A')' = A$
  (ii) $(A+B)' = A' + B'$, *here A and B must be of same order.*
  (iii) $(kA')' = kA'$, *here k is any scalar.*
  (iv) $(AB)' = B'A'$, *here AB and B'A' are conformable for multiplication.*
Proof.

(i) Let $\quad A = [a_{ij}]_{m \times n}$, then $\quad A' = [a_{ji}]_{n \times m}$

Since, $\quad (i, j)$th element in $(A')' = (j, i)$th element in A'

$$= (i, j)\text{th element in } A$$

Thus by the definition of equality of matrices, we must have $(A')' = A$.

(ii) Let $A = [a_{ij}]_{m \times n}$, $B = [b_{ij}]_{m \times n}$. So, $A' = [a_{ji}]_{n \times m}$ and $B' = [b_{ji}]_{n \times m}$, then

(i, j)th element in $(A+B)' = (j, i)$th element in $(A+B)$

$\qquad = (j, i)$th element in $A + (j, i)$th element in B

$\qquad = (i, j)$th element in $A' + (i, j)$th element in B'

$\qquad = (i, j)$th element in $(A'+B')$

Thus by the definition of equality of matrices, we get

$$(A + B)' = A' + B'$$

(iii) Let $A = [a_{ij}]_{m \times n}$ so that $\quad A' = [a_{ji}]_{n \times m}$ and k be a scalar, then

(i, j)th element in $(kA)' = (j, i)$th element in (kA)

$\qquad = (i, j)$th element in kA'

Thus by the definition of equality of matrices, we get

$$(kA)' = kA'$$

(iv) Let $A = [a_{ij}]_{m \times n}$ and $B = [b_{ij}]_{n \times p}$ then AB is conformable for multiplication and having the order $m \times p$. Therefore, the order of $(AB)'$ is $p \times m$. Since the orders of A' and B' are respectively $n \times m$ and $p \times n$ so $B'A'$ is conformable for multiplication and having the order $p \times m$.

Now $\quad (k, i)$th element in $(AB)' = (i, k)$th element in AB

$$= \sum_{j=1}^{n} a_{ij} b_{jk}$$

[By definition of multiplication of matrices]

But $\quad (k, i)$th element in $B'A' = \sum_{j=1}^{n} b_{kj} a_{ji}$

$$= \sum_{j=1}^{n} a_{ji} b_{kj}$$

$$= (i, k)\text{th element in } AB$$

$\therefore \quad (k, i)$th element in $(AB)' = (k, i)$th element in $B'A'$

Thus by the definition of equality of matrices, we must have

$$(AB)' = B'A'$$

2.15 SYMMETRIC MATRIX

A matrix A is said to be a symmetric matrix if $A' = A$, that is, the transpose of a matrix is equal to the matrix itself.

For exmaple : If $A = \begin{bmatrix} 1 & 2 & 3 \\ 2 & 4 & 5 \\ 3 & 5 & 6 \end{bmatrix}$, then $A' = \begin{bmatrix} 1 & 2 & 3 \\ 2 & 4 & 5 \\ 3 & 5 & 6 \end{bmatrix}$ so that $A' = A$

Hence, A is symmetric.

2.16 SKEW-SYMMETRIC MATRIX

A matrix A is said to be a skew-symmetric matrix if $A' = -A$.

For exmaple : If $\qquad A = \begin{bmatrix} 0 & 2 & 3 \\ -2 & 0 & 4 \\ -3 & -4 & 0 \end{bmatrix}$, then

$$A' = \begin{bmatrix} 0 & -2 & -3 \\ 2 & 0 & -4 \\ 3 & 4 & 0 \end{bmatrix}$$

$$= \begin{bmatrix} 0 & 2 & 3 \\ -2 & 0 & 4 \\ -3 & -4 & 0 \end{bmatrix} = -A$$

Hence A is skew-symmetric matrix.

2.17 RANK OF A MATRIX

Let A be a matrix of order $m \times n$, then a non-negative integer r is said to be the rank of matrix A if it possesses the following two properties :

(i) There exists at least one r-minor of A which is not equal to zero.

(ii) Every s-minor of A for all $s > r$ is zero.

We denote the rank of A by $\rho(A)$.

In other words, the rank of a matrix is the order of any highest order of a non-zero minor of the matrix.

☛ REMARKS
  • If the order of a matrix A is $m \times n$, then $\rho(A) \leq$ min. $\{m, n\}$
  • A is a null matrix iff $\rho(A) = 0$.
  • If A is any non-zero matrix, then $\rho(A) \geq 1$.
  • $\rho(A) \geq r$, if there exists a non-zero r-minor of A.
  • For any square matrix A of order n, $\rho(A) = n$ iff A is non-singular.
  • For any square matrix A of order n, $\rho(A) < n$ iff A is singular.
  • $\rho(A) \leq r$ if every s-minor of A is zero, where $s > r$.

> Every $(r+1)$- rowed minor of A can be expressed as a linear combination of its r-rowed minors, therefore if every r-minor of A is zero, then its every $(r+1)$-minor is also zero.

2.18 ECHELON FORM OF A MATRIX

A matrix A is said to be in Echelon form if :
  (i) every row of A has all its entries 0 which occurs below every row having a non-zero entry. and
  (ii) the number of zeros before the first non-zero entry in a row is less than the number of such zeros in the next row.

☛ REMARK
  • The rank of a matrix is equal to the number of non-zero rows in Echelon form of that matrix.

For example: Consider a matrix $A = \begin{bmatrix} 0 & 2 & 3 & 5 \\ 0 & 0 & 3 & 2 \\ 0 & 0 & 0 & 0 \end{bmatrix}$

Clearly, A is in Echelon form which has 2 non-zero rows, hence the rank of A is 2.

THEOREM I. *The rank of the transpose of a matrix is equal to the rank of that matrix.*

PROOF. Let A be a marix, then A' is its transpose and let $\rho(A) = r$, then there exists an r-rowed minor of A which is not equal to zero and all s-rowed minors of A are zero, where $s > r$. Let $|B|$ be a r-rowed minor of A such that $|B| \neq 0$. Since A' is the transpose of A, then $|B'|$ is the r-rowed minor of A' but $|B'| = |B| \neq 0$, therefore $\rho(A') \geq r$. Suppose there is an s-minor $|C|$ of A' such that $|C| \neq 0$, where $s > r$, then $|C'|$ will be an s-minor of A such that $|C'| = |C| \neq 0$, therefore $\rho(A) > r$ which is a contradiction, hence $\rho(A') = r$.

 Solved Examples

EXAMPLE I. ***Find the rank of the following matrices :***

  (i) $[3 \quad 0 \quad 0]$

  (ii) $\begin{bmatrix} 1 & 2 & 3 \\ 2 & 4 & 5 \end{bmatrix}$

  (iii) $\begin{bmatrix} 1 & 2 & 3 \\ 3 & 4 & 5 \\ 4 & 5 & 6 \end{bmatrix}$

  (iv) $\begin{bmatrix} 1 & 5 & 2 & 4 \\ 0 & 1 & 3 & 1 \\ 0 & 0 & 1 & 3 \end{bmatrix}$

SOLUTION. (i) Let $A = [3 \ 0 \ 0]$, then A is the non-zero rowed matrix, thereofore $\rho(A) \geq 1$.
Also A is a matrix of order 1×3, then $\rho(A) \leq 1$, hence $\rho(A) = 1$.

(ii) Let $A = \begin{bmatrix} 1 & 2 & 3 \\ 2 & 4 & 5 \end{bmatrix}$

The order of A is 2×3, then $\rho(A) \leq 2$.

Also there is a 2-minor $\begin{vmatrix} 2 & 3 \\ 4 & 5 \end{vmatrix}$ of A which is not equal to zero, then $\rho(A) \geq 2$, hence $\rho(A) = 2$.

(iii) Let $A = \begin{bmatrix} 1 & 2 & 3 \\ 3 & 4 & 5 \\ 4 & 5 & 6 \end{bmatrix}$. The order of A is 3×3, then $\rho(A) \leq 3$.

Now $|A| = \begin{vmatrix} 1 & 2 & 3 \\ 3 & 4 & 5 \\ 4 & 5 & 6 \end{vmatrix} = 1(24 - 25) - 2(18 - 20) + 3(15 - 16) = 0\,'$

$\therefore$ The only 3-minor $|A|$ of A is zero, thus $\rho(A) < 3$. Further, there is a

2-minor $\begin{vmatrix} 1 & 2 \\ 3 & 4 \end{vmatrix}$ of A which is not equal to zero, hence $\rho(A) = 2$.

(iv) Let $A = \begin{bmatrix} 1 & 5 & 2 & 4 \\ 0 & 1 & 3 & 1 \\ 0 & 0 & 1 & 3 \end{bmatrix}$

The order of A is 3×4, then $\rho(A) \leq 3$.

Now there is a 3-minor $\begin{vmatrix} 1 & 5 & 2 \\ 0 & 1 & 3 \\ 0 & 0 & 1 \end{vmatrix}$ of A which is not equal to zero, then $\rho(A) \geq 3$.

Hence $\rho(A) = 3$.

2.19 INVERSE OF A MATRIX

Let A be a non-singular matrix of order $n \times n$. Then it is said to be invertible if there exists a non-singular square matrix B of order $n \times n$ such that

$$AB = I_n = BA$$

where I_n is the unit matrix of order $n \times n$.
The matrix B is called the inverse of A and we write $B = A^{-1}$.

 Solved Examples

EXAMPLE 1. *By using elementary row-transformations find the inverse of the following matrices:*

(i) $\begin{bmatrix} 1 & 2 \\ 3 & 7 \end{bmatrix}$ (ii) $\begin{bmatrix} 1 & 2 \\ 2 & -1 \end{bmatrix}$

SOLUTION. (i) We write

$$A = I_2 A$$

or $\begin{bmatrix} 1 & 2 \\ 3 & 7 \end{bmatrix} = \begin{bmatrix} 1 & 0 \\ 0 & 1 \end{bmatrix} A$

Applying $R_2 \rightarrow R_2 - 3R_1$, we get

$$\begin{bmatrix} 1 & 2 \\ 0 & 1 \end{bmatrix} = \begin{bmatrix} 1 & 0 \\ -3 & 1 \end{bmatrix} A$$

Again applying $R_1 \rightarrow R_1 - 2R_2$, we get

$$\begin{bmatrix} 1 & 0 \\ 0 & 1 \end{bmatrix} = \begin{bmatrix} 7 & -2 \\ -3 & 1 \end{bmatrix} A$$

$\Rightarrow \qquad\qquad I_2 = BA$

$\Rightarrow \qquad\qquad A^{-1} = B = \begin{bmatrix} 7 & -2 \\ -3 & 1 \end{bmatrix}.$

(ii) We write

$$A = I_2 A$$

or $\qquad \begin{bmatrix} 1 & 2 \\ 2 & -1 \end{bmatrix} = \begin{bmatrix} 1 & 0 \\ 0 & 1 \end{bmatrix} A$

Applying $R_2 \rightarrow R_2 - 2R_1$, we get

$$\begin{bmatrix} 1 & 2 \\ 0 & -5 \end{bmatrix} = \begin{bmatrix} 1 & 0 \\ -2 & 1 \end{bmatrix} A$$

Applying $R_2 \rightarrow -\dfrac{1}{5} R_2$, we get

$$\begin{bmatrix} 1 & 2 \\ 0 & 1 \end{bmatrix} = \begin{bmatrix} 1 & 0 \\ 2/5 & -1/5 \end{bmatrix} A$$

Applying $R_1 \rightarrow R_1 - 2R_2$, we get

$$\begin{bmatrix} 1 & 0 \\ 0 & 1 \end{bmatrix} = \begin{bmatrix} 1/5 & 2/5 \\ 2/5 & -1/5 \end{bmatrix} A$$

$\Rightarrow \qquad\qquad I_2 = BA$

$\Rightarrow \qquad\qquad A^{-1} = B = \begin{bmatrix} 1/5 & 2/5 \\ 2/5 & -1/5 \end{bmatrix}$

EXAMPLE 2. *Find the inverse of the matrix*

$$A = \begin{bmatrix} 1 & 2 & 1 \\ 3 & 2 & 3 \\ 1 & 1 & 2 \end{bmatrix}$$

by using elementary row-transformation.

SOLUTION. We write $\qquad\qquad A = I_3 A$

or $\qquad \begin{bmatrix} 1 & 2 & 1 \\ 3 & 2 & 3 \\ 1 & 1 & 2 \end{bmatrix} = \begin{bmatrix} 1 & 0 & 0 \\ 0 & 1 & 0 \\ 0 & 0 & 1 \end{bmatrix} A$

Applying $R_2 \rightarrow R_2 - 3R_1$, $R_3 \rightarrow R_3 - R_1$, we get

$$\begin{bmatrix} 1 & 2 & 1 \\ 0 & -4 & 0 \\ 0 & -1 & 1 \end{bmatrix} = \begin{bmatrix} 1 & 0 & 0 \\ -3 & 1 & 0 \\ -1 & 0 & 1 \end{bmatrix} A$$

Applying $R_2 \to \dfrac{-1}{4}R_2$, we get

$$\begin{bmatrix} 1 & 2 & 1 \\ 0 & 1 & 0 \\ 0 & -1 & 1 \end{bmatrix} = \begin{bmatrix} 1 & 0 & 0 \\ 3/4 & -1/4 & 0 \\ -1 & 0 & 1 \end{bmatrix} A$$

Applying $R_3 \to R_3 + R_2$, we get

$$\begin{bmatrix} 1 & 2 & 1 \\ 0 & 1 & 0 \\ 0 & 0 & 1 \end{bmatrix} = \begin{bmatrix} 1 & 0 & 0 \\ 3/4 & -1/4 & 0 \\ -1/4 & -1/4 & 1 \end{bmatrix} A$$

Applying $R_1 \to R_1 - 2R_2$, we get

$$\begin{bmatrix} 1 & 0 & 1 \\ 0 & 1 & 0 \\ 0 & 0 & 1 \end{bmatrix} = \begin{bmatrix} -1/2 & 1/2 & 0 \\ 3/4 & -1/4 & 0 \\ -1/4 & -1/4 & 1 \end{bmatrix} A$$

Applying $R_1 \to R_1 - R_3$, we get

$$\begin{bmatrix} 1 & 0 & 0 \\ 0 & 1 & 0 \\ 0 & 0 & 1 \end{bmatrix} = \begin{bmatrix} -1/4 & 3/4 & -1 \\ 3/4 & -1/4 & 0 \\ -1/4 & -1/4 & 1 \end{bmatrix} A$$

$\Rightarrow \qquad\qquad\qquad I_3 = BA$

$\Rightarrow \qquad\qquad A^{-1} = B = \begin{bmatrix} -1/4 & 3/4 & -1 \\ 3/4 & -1/4 & 0 \\ -1/4 & -1/4 & 1 \end{bmatrix}$

Exercise-2.1

1. Are the following pairs of matrices equivalent?

(i) $\begin{bmatrix} 4 & 0 & 2 \\ 3 & 1 & 0 \\ 5 & 2 & 0 \end{bmatrix}, \begin{bmatrix} 3 & 9 & 0 & 2 \\ 7 & -2 & 0 & 1 \\ 8 & 1 & 1 & 5 \end{bmatrix}$

(ii) $\begin{bmatrix} 2 & -1 & 3 & 4 \\ 0 & 3 & 4 & 1 \\ 2 & 3 & 7 & 5 \\ 2 & 5 & 11 & 5 \end{bmatrix}, \begin{bmatrix} 1 & 0 & -5 & 6 \\ 3 & -2 & 1 & 2 \\ 5 & -2 & -9 & 14 \\ 4 & -2 & -4 & 8 \end{bmatrix}$

Determine the rank of the following matrices:

2. $\begin{bmatrix} 1 & 1 & 1 \\ 2 & 2 & 2 \\ 3 & 3 & 3 \end{bmatrix}$

3. $\begin{bmatrix} 2 & 1 & 3 \\ 4 & 7 & 13 \\ 4 & -3 & -1 \end{bmatrix}$

4. $\begin{bmatrix} 4 & 5 & 6 \\ 5 & 6 & 7 \\ 7 & 8 & 9 \end{bmatrix}$

5. $\begin{bmatrix} 1 & 2 & 3 \\ 2 & 3 & 4 \\ 3 & 5 & 7 \end{bmatrix}$

6. $\begin{bmatrix} 2 & 3 & 7 \\ 3 & -2 & 4 \\ 1 & -3 & -1 \end{bmatrix}$

7. $\begin{bmatrix} 3 & -1 & 2 \\ -6 & 2 & -4 \\ -3 & 1 & -2 \end{bmatrix}$

8. $\begin{bmatrix} 1 & 2 & 3 & 1 \\ 2 & 4 & 6 & 2 \\ 1 & 2 & 3 & 2 \end{bmatrix}$

9. $\begin{bmatrix} 1 & 3 & 4 & 3 \\ 3 & 9 & 12 & 9 \\ 1 & 3 & 4 & 1 \end{bmatrix}$

10. $\begin{bmatrix} 1 & 2 & -1 & 4 \\ 2 & 4 & 3 & 5 \\ -1 & -2 & 6 & -7 \end{bmatrix}$

11. $\begin{bmatrix} 1 & 2 & -4 & 5 \\ 2 & -1 & 3 & 6 \\ 8 & 1 & 9 & 7 \end{bmatrix}$

12. $\begin{bmatrix} 1 & -1 & 3 & 6 \\ 1 & 3 & -3 & -4 \\ 5 & 3 & 3 & 11 \end{bmatrix}$

13. $\begin{bmatrix} 1 & 2 & 3 & 0 \\ 2 & 4 & 3 & 2 \\ 3 & 2 & 1 & 3 \\ 6 & 8 & 7 & 5 \end{bmatrix}$

14. $\begin{bmatrix} 2 & 3 & -1 & -1 \\ 1 & -1 & -2 & -4 \\ 3 & 1 & 3 & -2 \\ 6 & 3 & 0 & -7 \end{bmatrix}$

15. $\begin{bmatrix} 1 & 2 & 1 & 2 \\ 1 & 3 & 2 & 2 \\ 2 & 4 & 3 & 4 \\ 3 & 7 & 4 & 6 \end{bmatrix}$

16. $\begin{bmatrix} 3 & -2 & 0 & -1 \\ 0 & 2 & 2 & 1 \\ 1 & -2 & -3 & 2 \\ 0 & 1 & 2 & 1 \end{bmatrix}$

17. $\begin{bmatrix} 0 & 1 & -3 & -1 \\ 1 & 0 & 1 & 1 \\ 3 & 1 & 0 & 2 \\ 1 & 1 & -2 & 0 \end{bmatrix}$

18. $\begin{bmatrix} 1 & 2 & -1 & 3 \\ 4 & 1 & 2 & 1 \\ 3 & -1 & 1 & 2 \\ 1 & 2 & 0 & 1 \end{bmatrix}$

19. $\begin{bmatrix} 1 & 0 & 2 & 1 \\ 0 & 1 & -2 & 1 \\ 1 & -1 & 4 & 0 \\ -2 & 2 & 8 & 0 \end{bmatrix}$

20. $\begin{bmatrix} 8 & 0 & 0 & 1 \\ 1 & 0 & 8 & 1 \\ 0 & 0 & 1 & 8 \\ 0 & 1 & 1 & 8 \end{bmatrix}$

Answers

1. (i) Not equivalent	(ii) Not equivalent	**2.** 1	**3.** 2	**4.** 2	**5.** 2
6. 1	**7.** 2	**8.** 2	**9.** 2	**10.** 3	**11.** 3	**12.** 3	**13.** 3	**14.** 3
15. 4	**16.** 2	**17.** 3	**18.** 3	**19.** 4	**20.** 2

2.20 CONVEX SET

A set of points is said to be convex if for any two points in the set, the line segment joining these points is also in the set, *i.e.*, the set is said to be convex if convex combinations of any two points in the set is also in the set. [MEERUT–2009, 10, 11, 14, 17; GORAKHPUR–2007, 10, 18; AGRA–2008]

Mathematically, for any two points x_1, x_2 in the set, if every point $x = \lambda x_1 + (1 - \lambda)x_2$, $0 \leq \lambda \leq 1$ is also in the set, then set is convex.

Following are the examples of convex sets.

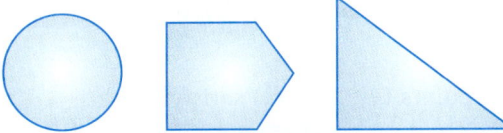

Fig. 1

Similarly some non-convex sets are given below:

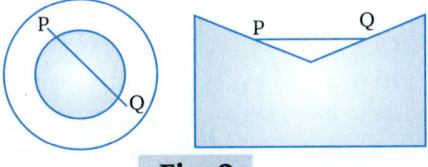

Fig. 2

☞ REMARKS
- The convex combination of any number of points in the convex set also belongs to the set.
- A set of one point is always convex.

2.21 SOME RELATED DEFINITIONS

2.21.1 POINT SET

A set whose elements are points or vectors in $\mathbf{R}^n$ is said to be point set.

For example:

1. A line $a_1x_1 + a_2x_2 = b$ represents a line in two dimensions which may be considered as a set of these points (x_1, x_2). Therefore, set of points can be written as
$$S = \{(x_1, x_2) : a_1x_1 + a_2x_2 = b\}$$

2. If we consider the set of points lying inside a circle of unit radius with centre at the origin in two dimensional space. Clearly, the points (x_1, x_2) of this set satisfy the inequality
$$x_1^2 + x_2^2 < 1$$
Therefore, set of points is given by
$$S = \{(x_1, x_2) : x_1^2 + x_2^2 < 1\}$$

2.21.2 HYPERSPHERE

In n-dimensional space, a hypersphere, with centre a and radius $r(> 0)$ is the set of points
$$X = \{x : |x - a| = r\}.$$
The equation of hypersphere in E^n (or R^n) is given by $\Sigma(x_i - a_i)^2 = r^2$

☛ REMARKS

- The set of points inside the hypersphere is the set $X = \{x : |x - a| < r\}$
- The set of points lying inside the hypersphere with centre a and radius $\epsilon > 0$ is said to be ϵ-neighbourhood about the point a.

2.21.3 LINES AND LINE SEGMENTS

Let x_1, x_2 be two distinct points in n-dimensional space E^n, then the line through the points x_1 and x_2 is defined to be the set of points given by
$$X = \{x : x = \lambda x_1 + (1 - \lambda)x_2, \text{ for } \lambda \in \mathbf{R}\}$$
and the line segment joining two points x_1 and x_2 in E^n is defined to be the set of points given by
$$X = \{x : x = \lambda x_1 + (1 - \lambda)x_2, 0 \leq \lambda \leq 1\}$$

2.21.4 HYPERPLANE

It is defined as the set of points $(x_1, x_2, ... x_n)$ satisfying
$$c_1x_1 + c_2x_2 + ... + c_nx_n = z, \text{ (not all } c_i = 0)$$
for prescribed values of $c_1, c_2, ..., c_n$ and z

We clearly observe that a hyperplane divides the whole space into three mutually disjoint sets as given below:
$$X_1 = \{x : \mathbf{cx} > \mathbf{z}\} = \{(x_1, x_2, ..., x_n) : c_1x_1 + c_2x_2 + ... + c_nx_n > z\}$$
$$X_2 = \{x : \mathbf{cx} = \mathbf{z}\} = \{(x_1, x_2, ..., x_n) : c_1x_1 + c_2x_2 + ... + c_nx_n = z\}$$
$$X_3 = \{x : \mathbf{cx} < \mathbf{z}\} = \{(x_1, x_2, ..., x_n) : c_1x_1 + c_2x_2 + ... + c_nx_n < z\}$$

The set X_1 and X_3 are known as open-half spaces and the sets $\{x : \mathbf{cx} \leq \mathbf{z}\}$ and $\{x : \mathbf{cx} \geq \mathbf{z}\}$ are known as closed-half spaces.

☛ REMARKS

- For optimum value of z, the hyperplane $\mathbf{cx} = \mathbf{z}$ is called optimal hyperplane.
- The vector $\mathbf{c}$ is known as vector normal to the hyperplane
- The value $\pm\dfrac{c}{|\mathbf{c}|}$ are called unit normals.
- The hyperplanes are always closed sets.

> In an LPP, the objective function represents a hyperplane and each constraints($\leq$ or $\geq$) is a closed-half space produced by the hyperplane given by the constraints by taking (=) sign in place of $\leq$ or $\geq$.

2.21.5 PARALLEL HYPERPLANES

Two hyperplanes $c_1x = z_1$ and $c_2x = z_2$ are said to be parallel if they have the same unit normals, *i.e.*, if $c_1 = \lambda c_2$ for some λ, $(\lambda \neq 0)$

2.21.6 CONVEX COMBINATIONS

A convex combinations of a finite number of points $x_1, x_2, ..., x_n$ is defined by the point
$$x = a_1x_1 + a_2x_2 + ... + a_nx_n, a_i \in \textbf{\textit{R}}, a_i \geq 0 \ \forall \ i \text{ and } \Sigma a_i = 1$$

In particular, the convex combination of two points x_1, x_2 be given by $x = a_1x_1 + a_2x_2$ such that $a_1, a_2 \geq 0$ and $a_1 + a_2 = 1$ which can also be written as $x = ax_1 + (1-a)x_2$, $0 \leq a \leq 1$

☞ **REMARK**
- The line segment of two points x_1 and x_2 is the set of all possible convex combinations of two points x_1 and x_2.

2.21.7 EXTREME POINT OF A CONVEX SET

A point x in a convex set C is an extreme point of C if it does not lie on the line segment of any two points, different from x in the set, *i.e.*, it can not be expressed as a convex combinations of any two distinct points x_1 and x_2 in C.

Mathematically an extreme point can be defined as follows:

A point x is said to be an extreme point of a convex set if there do not exist other points $x_1, x_2 (x_1 \neq x_2)$ in the set such that $x = \lambda x_1 + (1-\lambda)x_2$, $0 < \lambda < 1$

☞ **REMARKS**
- A convex set may also have infinite number of extreme points.
- The polygons which are convex, have the extreme points as their vertices.
- An extreme point is a boundary point of the set.
- All boundary points of a convex set are not necessarily extreme points.
- A point of a convex set C, which is not an extreme point, is said to be internal point of C.

2.22 CONVEX HULL   [MEERUT–2006]

The set of all convex combinations of set of points from the set X of points is called convex hull, *i.e.*, the intersection of all convex sets containing X in n-dimensional space is called the convex hull of X. Hence, the convex hull of a set $X \subseteq E^n$ is the smallest convex set containing X.

For example: If X is the boundary of a circle, then the convex hull $C(x)$ is the whole circle.

2.23 CONVEX FUNCTION AND CONVEX POLYHEDRON [MEERUT–2006, 12, 13; GORAKHPUR–2009, 10, 18]

Definition 1. *A function $f(x)$ is said to be strictly convex at x if for any two other distinct points x_1 and x_2*
$$f\{\lambda x_1 + (1-\lambda)x_2\} < \lambda f(x_1) + (1-\lambda) f(x_2), 0 < \lambda < 1$$

Definition 2. *The set of all convex combinations of finite number of points is said to be the convex polyhedron generated by these points.*

In other words, we can say that if the set X consist of a finite number of points, the convex hull of X is said to be the convex polyhedron with vertices at those points.

For example: The set of the area of a triangle is a convex polyhedron of its vertices.

☞ REMARK

- A function $f(\boldsymbol{x})$ is said to be strictly concave if $-f(\boldsymbol{x})$ is strictly convex.

THEOREM 1. *The hyperplane is a convex set.* [MEERUT–2007]

PROOF. Let $X = [\boldsymbol{x} : \boldsymbol{cx} = z]$ be a hyperplane and $\boldsymbol{x}_1, \boldsymbol{x}_2 \in X$ then,

$$\boldsymbol{cx}_1 = z \text{ and } \boldsymbol{cx}_2 = z \qquad \text{(By definition)}$$

Now, if $\qquad \boldsymbol{x}_3 = \lambda\boldsymbol{x}_1 + (1-\lambda)\boldsymbol{x}_2, 0 \le \lambda \le 1$

Then, $\qquad \boldsymbol{cx}_3 = \lambda\boldsymbol{c}\cdot\boldsymbol{x}_1 + (1-\lambda)\boldsymbol{cx}_2$

$$= \lambda z + (1-\lambda)z$$
$$= z$$

$\Rightarrow \qquad \boldsymbol{x}_3 = \lambda\boldsymbol{x}_1 + (1-\lambda)\boldsymbol{x}_2 \in X$

$\Rightarrow \boldsymbol{x}_3$ is also a point in X

Hence, X is a convex set.

THEOREM 2. *The closed half spaces $H_1 = \{x : cx \ge z\}$ and $H_2 = \{x : cx \le z\}$ are convex sets.*

PROOF. Let $\boldsymbol{x}_1 \in H_1$ and $\boldsymbol{x}_2 \in H_2$. Then by definition of H_1, we can write $\boldsymbol{cx}_1 \ge z : \boldsymbol{cx}_2 \ge z$

Now, if $0 \le \lambda \le 1$, then we have

$$\boldsymbol{c}[\lambda\boldsymbol{x}_1 + (1-\lambda)\boldsymbol{x}_2] = \lambda\boldsymbol{c}\cdot\boldsymbol{x}_1 + (1-\lambda)\boldsymbol{cx}_2$$
$$\ge \lambda z + (1-\lambda)z = z$$

Therefore, $\boldsymbol{x}_1, \boldsymbol{x}_2 \in H_1$ and $0 \le \lambda \le 1$ implies $\lambda\boldsymbol{x}_1 + (1-\lambda)\boldsymbol{x}_2 \in H_1$

Hence, H_1 is a convex set

Similarly, we may prove that H_2 is a convex set.

☞ REMARK

- In a similar way (as above) we may prove that the open half spaces $\{\boldsymbol{x} : \boldsymbol{cx} > z\}$ and $\{\boldsymbol{x} : \boldsymbol{cx} < z\}$ are convex sets.

THEOREM 3. *Intersection of two convex sets is also a convex set.*

[MEERUT–2007, 08, 12,15, 17, 18]

PROOF. Let X_1 and X_2 be two convex sets. We have to prove that $X_1 \cap X_2$ is also convex.

If $\boldsymbol{x}_1 \in X_1 \cap X_2 \Rightarrow \boldsymbol{x}_1 \in X_1$ and $\boldsymbol{x}_1 \in X_2$

$\boldsymbol{x}_2 \in X_1 \cap X_2 \Rightarrow \boldsymbol{x}_2 \in X_1$ and $\boldsymbol{x}_2 \in X_2$

Now, by definition of convex sets

$$\boldsymbol{x}_1, \boldsymbol{x}_2 \in X_1 \Rightarrow \lambda\boldsymbol{x}_1 + (1-\lambda)\boldsymbol{x}_2 \in X_1 ; 0 \le \lambda \le 1$$

$$\boldsymbol{x}_1, x_2 \in X_2 \Rightarrow \lambda\boldsymbol{x}_1 + (1-\lambda)\boldsymbol{x}_2 \in X_2 ; 0 \le \lambda \le 1$$

Therefore, $\lambda\boldsymbol{x}_1 + (1-\lambda)\boldsymbol{x}_2 \in X_1$ and $\lambda\boldsymbol{x}_1 + (1-\lambda)\boldsymbol{x}_2 \in X_2$

$\Rightarrow \lambda\boldsymbol{x}_1 + (1-\lambda)\boldsymbol{x}_2 \in X_1 \cap X_2$

Hence, $X_1 \cap X_2$ is a convex set.

THEOREM 4. *Finite intersection of convex sets is also a convex set.* [MEERUT–2016]

PROOF. Let $X_1, X_2, ..., X_n$ be n convex sets.

We have to prove that $X_1 \cap X_2 \cap ... \cap X_n$ is also convex.

Let $\boldsymbol{x}_1 \in X_1 \cap X_2 \cap ... \cap X_n \Rightarrow \boldsymbol{x}_1 \in X_i \forall i = 1, 2, ..., n$

$\boldsymbol{x}_2 \in X_1 \cap X_2 \cap ... \cap X_n \Rightarrow \boldsymbol{x}_2 \in X_i \forall i = 1, 2, ..., n$

Since each X_i is convex set for $i = 1, 2, ..., n$

Therefore, $x_1, x_2 \in X_i$

$\Rightarrow \lambda x_1 + (1-\lambda)x_2 \in X_i \quad \forall i = 1,2,...,n , 0 \le \lambda \le 1$

$\Rightarrow \lambda x_1 + (1-\lambda)x_2 \in X_1 \cap X_2 \cap ... \cap X_n$

$\Rightarrow x_1 \in X_1 \cap X_2 \cap ... \cap X_n$ and $x_2 \in X_1 \cap X_2 \cap ... \cap X_n$

$\Rightarrow \lambda x_1 + (1-\lambda)x_2 \in X_1 \cap X_2 \cap ... \cap X_n , 0 \le \lambda \le 1.$

Hence, $X_1 \cap X_2 \cap ... \cap X_n$ is a convex set.

☞ REMARK

- In the similar way we may extend the above result as follows:
"Arbitrary intersection of convex sets is also a convex set"

THEOREM 5. *The set of all convex combinations of a finite number of points* $x_1, x_2, ..., x_n$ *is a convex set.*

PROOF. Let us define the set X of all convex combinations as follows:

$$X = \left\{ x : x = \sum_{i=1}^{n} \lambda_i x_i, \sum_{i=1}^{n} \lambda_i = 1, \lambda_i \ge 0 \right\}$$

We have to prove that X is convex.

Let $\alpha, \beta \in X$ such that

$$\alpha = \sum_{i=1}^{n} a_i x_i ; \sum_{i=1}^{n} a_i = 1 , a_i \ge 0 \text{ and } \beta = \sum_{i=1}^{n} b_i x_i ; \sum_{i=1}^{n} b_i = 1 , b_i \ge 0$$

Let us consider

$$w = \lambda\alpha + (1-\lambda)\beta , 0 \le \lambda \le 1$$

$$= \lambda \sum_{i=1}^{n} a_i x_i + (1-\lambda) \sum_{i=1}^{n} b_i x_i = \sum_{i=1}^{n} \left\{ \lambda a_i + (1-\lambda)b_i \right\} x_i$$

$$= \sum_{i=1}^{n} c_i x_i \text{ where } c_i = \lambda a_i + (1-\lambda)b_i \qquad ...(1)$$

Now, we shall prove that

$$\sum_{i=1}^{n} c_i = \sum_{i=1}^{n} \{\lambda a_i + (1-\lambda)b_i\}$$

$$= \lambda \sum_{i=1}^{n} a_i + (1-\lambda) \sum_{i=1}^{n} b_i = \lambda \cdot 1 + (1-\lambda) \cdot 1 = 1 \qquad ...(2)$$

Further, $c_i = \lambda a_i + (1-\lambda)b_i \ge 0 \ \forall \ i$ \qquad ...(3)

Hence, from (1), (2) and (3) we conclude that $w = \sum_{i=1}^{n} c_i \cdot x_i$ is a convex combination

of $x_1, x_2, ..., x_n$. Hence X is convex.

THEOREM 6. *Let* S_1 *and* S_2 *be two convex sets in* E^n, *then for any scalar* α, β *the set* $(\alpha S_1 + \beta S_2)$ *is also convex.*

PROOF. Let S_1 and S_2 be two convex sets. We have to prove that for any two scalars α and β, $(\alpha S_1 + \beta S_2)$ is also convex.

Let $x, y \in \alpha S_1 + \beta S_2$. Then these are of the following forms:

$$x = \alpha u_1 + \beta v_1 \text{ and } y = \alpha u_2 + \beta v_2 \qquad ...(1)$$

$u_1, u_2 \in S_1$ and $v_1, v_2 \in S_2$

Now for any scalar λ, $0 \le \lambda \le 1$, we have

$$\lambda \boldsymbol{x} + (1 - \lambda)\boldsymbol{y} = \lambda(\alpha \boldsymbol{u}_1 + \beta \boldsymbol{v}_1) + (1 - \lambda)(\alpha \boldsymbol{u}_2 + \beta \boldsymbol{v}_2)$$

$$= \alpha\{\lambda \boldsymbol{u}_1 + (1 - \lambda)\boldsymbol{u}_2\} + \beta\{\lambda \boldsymbol{v}_1 + (1 - \lambda)\boldsymbol{v}_2\} \qquad \ldots(2)$$

Since, S_1 and S_2 both are convex, so by definition, we can write

$$\boldsymbol{u}_1, \boldsymbol{u}_2 \in S_1 \Rightarrow \lambda \boldsymbol{u}_1 + (1 - \lambda)\boldsymbol{u}_2 \in S_1, \, 0 \le \lambda \le 1 \qquad \ldots(3)$$

$$\boldsymbol{v}_1, \boldsymbol{v}_2 \in S_2 \Rightarrow \lambda \boldsymbol{v}_1 + (1 - \lambda)\boldsymbol{v}_2 \in S_2, \, 0 \le \lambda \le 1 \qquad \ldots(4)$$

Using (2), (3) and (4) we can write

$$\lambda \boldsymbol{x} + (1 - \lambda)\boldsymbol{y} \in \alpha S_1 + \beta S_2, \, 0 \le \lambda \le 1$$

Hence, $\alpha \cdot S_1 + \beta \cdot S_2$ is a convex set.

☞ REMARKS

- From the above theorem we may easily prove the following result:
  "The sum $(S_1 + S_2)$ and difference $(S_1 - S_2)$ of two convex sets S_1 and S_2 is again convex."
- The set of all convex combinations of a finite number of points is also convex. [MEERUT–2008, 09]
- Those convex sets which are the intersection of a finite number of closed half spaces are called 'polyhedral convex sets'.

THEOREM 7. *The set of all the internal points of a convex set is convex.*

[MEERUT–2011, 12, 15]

PROOF. Let S be the convex set and S_1 be the set of all vertices of S.

Then clearly $S - S_1$ is the set of all internal points of S.

If $\boldsymbol{u}, \boldsymbol{v} \in S - S_1$. Then $\boldsymbol{u}, \boldsymbol{v} \in S$

Suppose that $\boldsymbol{z}$ is a point on the line segment joining $\boldsymbol{u}$ and $\boldsymbol{v}$ then we can write

$$\boldsymbol{z} = \lambda \boldsymbol{u} + (1 - \lambda)\boldsymbol{v}, \, 0 \le \lambda \le 1$$

$$\in S \qquad\qquad (\because S \text{ is convex})$$

$$\Rightarrow \qquad\qquad \boldsymbol{z} \in S - S_1$$

Since $\boldsymbol{z}$ is not a vertex of S_1, therefore $S - S_1$ (set of all internal points) is convex.

Solved Examples

EXAMPLE 1. *Show that $S = \{(x_1, x_2) : 2x_1 + 3x_2 = 7\} \subset R^2$ is a convex set.*

[MEERUT–2003, 08; GORAKHPUR–2009; GARHWAL–2014]

SOLUTION. Let $\boldsymbol{u}, \boldsymbol{v} \in S$

Then we can write $\boldsymbol{u} = (u_1, u_2)$, $\boldsymbol{v} = (v_1, v_2)$

$\therefore \qquad 2u_1 + 3u_2 = 7$ and $2v_1 + 3v_2 = 7$ $\qquad\qquad \ldots(1)$

Further, suppose that $\boldsymbol{w} = (w_1, w_2)$ is a point on the line segment joining the points $\boldsymbol{u}$ and $\boldsymbol{v}$. Then we can write

$$\boldsymbol{w} = \lambda \boldsymbol{u} + (1 - \lambda)\boldsymbol{v}, \, 0 \le \lambda \le 1$$

$\therefore \qquad (w_1, w_2) = \lambda(u_1, u_2) + (1 - \lambda)(v_1, v_2)$

$$= \{\lambda u_1 + (1 - \lambda)v_1, \, \lambda u_2 + (1 - \lambda)v_2\}$$

which implies that

$$w_1 = \lambda u_1 + (1 - \lambda)v_1 \text{ and } w_2 = \lambda u_2 + (1 - \lambda)v_2$$

Consider $\quad 2w_1 + 3w_2 = 2[\lambda u_1 + (1 - \lambda)v_1] + 3[\lambda u_2 + (1 - \lambda)v_2]$

$$= \lambda[2u_1 + 3u_2] + (1 - \lambda)[2v_1 + 3v_2]$$

$$= \lambda \cdot 7 + (1 - \lambda) \cdot 7 \qquad\qquad \text{(By (1))}$$

$$= 7$$

$$\Rightarrow \qquad \boldsymbol{w} \in S$$

Hence, S is a convex set.

EXAMPLE 2. **Show that the set $S = \{(x_1, x_2, x_3) : 2x_1 - x_2 + x_3 \leq 4\} \subset R^3$ is convex.**

[MEERUT–2005, 12, 15]

SOLUTION. Let $\boldsymbol{x} = (x_1, x_2, x_3)$ and $\boldsymbol{y} = (y_1, y_2, y_3)$ be any two points of the given set S. Then by definition of S, we can write

$$2x_1 - x_2 + x_3 \leq 4 \text{ and } 2y_1 - y_2 + y_3 \leq 4 \qquad\qquad ...(1)$$

Let $\boldsymbol{w} = (w_1, w_2, w_3)$ be a point such that

$$\boldsymbol{w} = \lambda\boldsymbol{x} + (1 - \lambda)\boldsymbol{y}, \, 0 \leq \lambda \leq 1$$

which implies that

$$(w_1, w_2, w_3) = \lambda(x_1, x_2, x_3) + (1 - \lambda)(y_1, y_2, y_3)$$

$$= (\lambda x_1 + (1 - \lambda)y_1, \lambda x_2 + (1 - \lambda)y_2, \lambda x_3 + (1 - \lambda)y_3)$$

$$\Rightarrow \qquad w_1 = \lambda x_1 + (1 - \lambda)y_1$$

$$w_2 = \lambda x_2 + (1 - \lambda)y_2$$

and $\qquad w_3 = \lambda x_3 + (1 - \lambda)y_3$

Consider $2w_1 - w_2 + w_3$

$$= \lambda(2x_1 - x_2 + x_3) + (1 - \lambda)(2y_1 - y_2 + y_3)$$

$$\leq 4\lambda + 4(1 - \lambda) \qquad\qquad \text{(By (1))}$$

$$\leq 4$$

Thus, $\qquad \boldsymbol{w} = (w_1, w_2, w_3) \in S$

Hence, S is convex.

EXAMPLE 3. **Examine the convexity of the set**

$$S = \{(x_1, x_2) \in R^2 : 4x_1 + 3x_2 \leq 6, \, x_1 + x_2 \geq 1\}$$

[MEERUT–1997, 2007; 09; DELHI–2009; ASSAM–2011; PATNA–2013]

SOLUTION. We have $S = \{(x_1, x_2) \in R^2 : 4x_1 + 3x_2 \leq 6, x_1 + x_2 \geq 1\}$

Let $\boldsymbol{u} = (x_1, x_2) \in S$ and $\boldsymbol{v} = (y_1, y_2) \in S$. Then,

$$4x_1 + 3x_2 \leq 6, x_1 + x_2 \geq 1 \text{ and } 4y_1 + 3y_2 \leq 6, y_1 + y_2 \geq 1 \qquad ...(1)$$

Let $\boldsymbol{w} = (w_1, w_2)$ be a point on the line segment joining the points u and v, then

$$\boldsymbol{w} = \lambda\boldsymbol{u} + (1 - \lambda)\boldsymbol{v}$$

$$\Rightarrow \qquad (w_1, w_2) = (\lambda x_1 + (1 - \lambda)y_1, \lambda x_2 + (1 - \lambda)y_2)$$

$$\Rightarrow \qquad w_1 = \lambda x_1 + (1 - \lambda)y_1 \text{ and } w_2 = \lambda x_2 + (1 - \lambda)y_2$$

Consider, $4w_1 + 3w_2 = \lambda(4x_1 + 3x_2) + (1 - \lambda)(4y_1 + 3y_2)$

$$\leq \lambda{\cdot}6 + (1 - \lambda){\cdot}6 \qquad\qquad \text{(By (1))}$$

$$\Rightarrow \qquad 4w_1 + 3w_2 \leq 6 \qquad\qquad ...(2)$$

Also, $\qquad w_1 + w_2 = \lambda(x_1 + x_2) + (1 - \lambda)(y_1 + y_2)$

$$\geq \lambda{\cdot}1 + (1 - \lambda){\cdot}1 \qquad\qquad \text{(Again by (1))}$$

$$\Rightarrow \qquad w_1 + w_2 \geq 1 \qquad\qquad ...(3)$$

Hence, from (2) and (3) we conclude that S is a convex set.

EXAMPLE 4. **Show that the set $S = \{x : x = (x_1, x_2, x_3),\ x_1^2 + x_2^2 + x_3^2 \leq 1\}$ is a convex set.**

[MEERUT–2006]

SOLUTION. Let $x, y \in S$ be arbitrary.

Then we write

$$x = (x_1, x_2, x_3) \text{ and } y = (y_1, y_2, y_3)$$

Now by definition of S, we have

$$x_1^2 + x_2^2 + x_3^2 \leq 1 \text{ and } y_1^2 + y_2^2 + y_3^2 \leq 1 \qquad \ldots(1)$$

Further, let $z = (z_1, z_2, z_3)$ be a point on the line segment joining the points x and y then we can write

$$z = \lambda x + (1 - \lambda)y,\ 0 \leq \lambda \leq 1$$

$$\Rightarrow \quad (z_1, z_2, z_3) = \lambda(x_1, x_2, x_3) + (1 - \lambda)(y_1, y_2, y_3)$$

$$= (\lambda x_1 + (1 - \lambda)y_1, \lambda x_2 + (1 - \lambda)y_2, \lambda x_3 + (1 - \lambda)y_3)$$

which gives $\quad z_1 = \lambda x_1 + (1 - \lambda)y_1,\ z_2 = \lambda x_2 + (1 - \lambda)y_2,\ z_3 = \lambda x_3 + (1 - \lambda)y_3$

Now consider $z_1^2 + z_2^2 + z_3^2$

$$= [\lambda x_1 + (1 - \lambda)y_1]^2 + [\lambda x_2 + (1 - \lambda)y_2]^2 + [\lambda x_3 + (1 - \lambda)y_3]^2$$

$$= \lambda^2(x_1^2 + x_2^2 + x_3^2) + (1 - \lambda)^2(y_1^2 + y_2^2 + y_3^2)$$

$$+ 2\lambda(1 - \lambda)(x_1 y_1 + x_2 y_2 + x_3 y_3)$$

$$\leq \lambda^2 \cdot 1 + (1 - \lambda)^2 + 1 + 2\lambda(1 - \lambda)(x_1 y_1 + x_2 y_2 + x_3 y_3) \quad \text{(By (1))}$$

$$\ldots(2)$$

Using Lagrange's identity

$$(x_1^2 + x_2^2 + x_3^2)(y_1^2 + y_2^2 + y_3^2) - (x_1 y_1 + x_2 y_2 + x_3 y_3)^2$$

$$\equiv \Sigma(x_1 y_2 - x_2 y_1)^2 \geq 0$$

we can write

$$x_1 y_1 + x_2 y_2 + x_3 y_3 \leq \sqrt{x_1^2 + x_2^2 + x_3^2} \cdot \sqrt{y_1^2 + y_2^2 + y_3^2} \leq 1 \qquad \text{(By (1))}$$

Using these values in (2) we get

$$z_1^2 + z_2^2 + z_3^2 \leq \lambda^2 + (1 - \lambda)^2 + 2\lambda(1 - \lambda) = 1$$

$$\Rightarrow \quad\quad z = (z_1, z_2, z_3) \in S$$

Hence, the given set S is convex.

EXAMPLE 6. **Show that the set $S = \{(x_1, x_2) : 3x_1^2 + 2x_2^2 \leq 6\}$ is convex.**

SOLUTION. Let $u, v \in S$ where $u = (u_1, u_2)$ and $v = (v_1, v_2)$

Then by definition of S, we can write

$$3u_1^2 + 2u_2^2 \leq 6 \text{ and } 3v_1^2 + 2v_2^2 \leq 6 \qquad \ldots(1)$$

Let $w = (w_1, w_2)$ be a point on the line segment joining the points u and v, then

$$w = \lambda u + (1 - \lambda)v;\ 0 \leq \lambda \leq 1$$

Therefore, $\quad w = (w_1, w_2) = \lambda(u_1, u_2) + (1 - \lambda)(v_1, v_2)$

$$= (\lambda u_1 + (1 - \lambda)v_1, \lambda u_2 + (1 - \lambda)v_2)$$

$$\Rightarrow \quad\quad w_1 = \lambda u_1 + (1 - \lambda)v_1 \text{ and } w_2 = \lambda u_2 + (1 - \lambda)v_2$$

Now, consider

$$3w_1^2 + 2w_2^2 = 3\{\lambda u_1 + (1 - \lambda)v_1\}^2 + 2\{\lambda u_2 + (1 - \lambda)v_2\}^2$$

$$= \lambda^2(3u_1^2 + 2u_2^2) + (1 - \lambda)^2(3v_1^2 + 2v_2^2) + 2\lambda(1 - \lambda)(3u_1 v_1 + 2u_2 v_2)$$

$$\ldots(2)$$

But we have
$$(3u_1^2 + 2u_2^2)(3v_1^2 + 2v_2^2) - 3(u_1v_1 + 2u_2v_2)^2 = 6(u_1v_2 - u_2v_1)^2 \geq 0$$
$$\Rightarrow \quad (3u_1v_1 + 2u_2v_2)^2 \leq (3u_1^2 + 2u_2^2)(3v_1^2 + 2v_2^2) \leq 6 \times 6$$
$$\Rightarrow \quad 3u_1v_1 + 2u_2v_2 \leq 6 \qquad \qquad \text{...(3)}$$
Finally using (1) and (3) in (2) we get
$$3w_1^2 + 2w_2^2 \leq 6\lambda^2 + 6(1-\lambda)^2 + 2\lambda(1-\lambda) \times 6$$
$$\Rightarrow \quad 3w_1^2 + 2w_2^2 \leq 6$$
$$\Rightarrow \quad \boldsymbol{w} = (w_1, w_2) \in S \; \forall \; 0 \leq \lambda \leq 1$$
Hence, S is a convex set.

EXAMPLE 7. **Show that $S = \{(x_1, x_2, x_3) : 2x_1 - x_2 + x_3 \leq 4; \; x_1 + 2x_2 - x_3 \leq 1\}$ is a convex set.**

SOLUTION. Let $\boldsymbol{x}, \boldsymbol{y} \in S$ then by definition of S we can write
$$\boldsymbol{x} = (x_1, x_2, x_3) \text{ and } \boldsymbol{y} = (y_1, y_2, y_3)$$
such that
$$\left. \begin{array}{l} 2x_1 - x_2 + x_3 \leq 4; x_1 + 2x_2 - x_3 \leq 1 \\ \text{and} \quad 2y_1 - y_2 + y_3 \leq 4; y_1 + 2y_2 - y_3 \leq 1 \end{array} \right\} \qquad \text{...(1)}$$

Let $\boldsymbol{z} = (z_1, z_2, z_3)$ be such that
$$\boldsymbol{z} = \lambda\boldsymbol{x} + (1-\lambda)\boldsymbol{y}, \; 0 \leq \lambda \leq 1$$
$$\Rightarrow \quad (z_1, z_2, z_3) = \lambda(x_1, x_2, x_3) + (1-\lambda)(y_1, y_2, y_3)$$
$$= \{\lambda x_1 + (1-\lambda)y_1, \lambda x_2 + (1-\lambda)y_2, \lambda x_3 + (1-\lambda)y_3\}$$
$$\Rightarrow \quad z_1 = \lambda x_1 + (1-\lambda)y_1, z_2 = \lambda x_2 + (1-\lambda)y_2, z_3 = \lambda x_3 + (1-\lambda)y_3$$
Now, consider $2z_1 - z_2 + z_3$
$$= \lambda(2x_1 - x_2 + x_3) + (1-\lambda)(2y_1 - y_2 + y_3)$$
$$\leq 4\lambda + 4(1-\lambda) \qquad \qquad \text{(Using (1))}$$
$$= 4$$
$$\Rightarrow 2z_1 - z_2 + z_3 \leq 4$$
Similarly, $z_1 + 2z_2 - z_3 \leq 1$
Therefore, $\boldsymbol{z} = (z_1, z_2, z_3) \in S$
Hence, S is a convex set.

EXAMPLE 8. **If S_1 and S_2 be two non-empty disjoint convex sets and S be a set such that if $x_1 \in S_1, x_2 \in S_2$ then $x_1 - x_2 \in S$. Show that S is also convex and does not contain the origin.**

SOLUTION. Let us write $\boldsymbol{u} = \boldsymbol{x}_1 - \boldsymbol{x}_2$ and $\boldsymbol{v} = \boldsymbol{y}_1 - \boldsymbol{y}_2 \in S$
Then we have $\boldsymbol{x}_1, \boldsymbol{y}_1 \in S_1$ and $\boldsymbol{x}_2, \boldsymbol{y}_2 \in S_2$
If $\boldsymbol{z}$ is a point on the line segment joining $\boldsymbol{u}$ and $\boldsymbol{v}$ then
$$\boldsymbol{z} = \lambda\boldsymbol{u} + (1-\lambda)\boldsymbol{v}; \; 0 \leq \lambda \leq 1$$
$$= \lambda(\boldsymbol{x}_1 - \boldsymbol{x}_2) + (1-\lambda)(\boldsymbol{y}_1 - \boldsymbol{y}_2)$$
$$= \{\lambda\boldsymbol{x}_1 + (1-\lambda)\boldsymbol{y}_1\} - \{\lambda\boldsymbol{x}_2 + (1-\lambda)\boldsymbol{y}_2\} \qquad \text{...(1)}$$
Further, it is given that S_1 and S_2 both are convex, therefore by definition
$$\boldsymbol{x}_1, \boldsymbol{y}_1 \in S_1 \qquad \Rightarrow \qquad \lambda\boldsymbol{x}_1 + (1-\lambda)\boldsymbol{y}_1 = \boldsymbol{z}_1 \in S_1; \; (0 \leq \lambda \leq 1)$$
$$\boldsymbol{x}_2, \boldsymbol{y}_2 \in S_2 \qquad \Rightarrow \qquad \lambda\boldsymbol{x}_2 + (1-\lambda)\boldsymbol{y}_2 = \boldsymbol{z}_2 \in S_2; \; (0 \leq \lambda \leq 1)$$

Then from (1)
$$z = z_1 - z_2 \in S \ \forall \ 0 \le \lambda \le 1$$
Hence, S is convex.

Finally, let if possible $\mathbf{0} \in S$ then there exist $x_1 \in S_1$ and $x_2 \in S_2$ such that
$$\mathbf{0} = x_1 - x_2$$
$$\Rightarrow \qquad x_1 = x_2, x_1 \in S_1, x_2 \in S_2$$
$\Rightarrow$ S_1 and S_2 are not disjoint, which is a contradiction.

Hence, $\mathbf{0} \notin S$, *i.e.*, S does not contain the origin.

EXAMPLE 9.   **Examine the convexity of the following set**
$$C = \{z \in R^n : z = x + y, x \in A, y \in B\}$$
where A and B are convex sets in R^n.

SOLUTION.   Let us suppose $z_1 = x_1 + y_1$ and $z_2 = x_2 + y_2$ be any two points in the set C.
Then $x_1, x_2 \in A$ and $y_1, y_2 \in B$

If u is the point on the line segment joining the points z_1 and z_2
Then we can write
$$u = \lambda z_1 + (1 - \lambda)z_2 : 0 \le \lambda \le 1$$
$$= \lambda(x_1 + y_1) + (1 - \lambda)(x_2 + y_2)$$
$$= (\lambda x_1 + (1 - \lambda)x_2) + (\lambda y_1 + (1 - \lambda)y_2), 0 \le \lambda \le 1 \qquad ...(1)$$
Further since both sets A and B are convex, therefore, we can write
$$\left. \begin{array}{l} x_1, x_2 \in A \Rightarrow \lambda x_1 + (1 - \lambda)x_2 = u_1 \in A \ \ \forall \, 0 \le \lambda \le 1 \\ y_1, y_2 \in B \Rightarrow \lambda y_1 + (1 - \lambda)y_2 = u_2 \in B \ \ \forall \, 0 \le \lambda \le 1 \end{array} \right\} \qquad ...(2)$$
Using (2) in (1) we get
$$u = u_1 + u_2 \in C$$
$\Rightarrow$ every point of the line segment joining z_1 and z_2 of C are also in C.
Hence, C is a convex set

EXAMPLE 10.   **Express any point w inside a triangle as a convex combinations of the vertices (extreme points) x_1, x_2, x_3 of the triangle.**

SOLUTION.   Consider a triangle ABC with vertices x_1, x_2 and x_3. If P is any point w inside the triangle. Join A and P and extend this line to meet the base BC at $D(u)$, a point on the line BC.

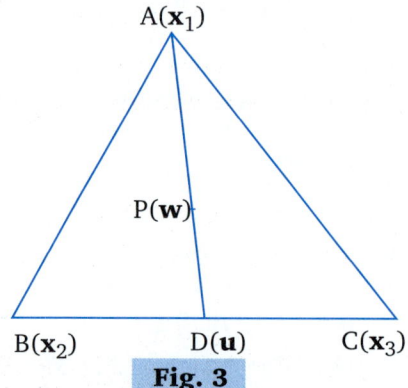

Fig. 3

$\Rightarrow$ u can be expressed as a convex combination of x_2 and x_3.
Therefore, $u = \lambda_1 x_2 + (1 - \lambda_1)x_3, 0 \le \lambda_1 \le 1$ ...(1)
Further, since P is a point on the line segment AD, therefore,
$$w = \lambda_2 x_1 + (1 - \lambda_2)u, 0 \le \lambda_2 \le 1$$

$$= \lambda_2 \, \boldsymbol{x}_1 + (1 - \lambda_2)[\lambda_1 \, \boldsymbol{x}_2 + (1 - \lambda)\boldsymbol{x}_3] \qquad \text{(Using (1))}$$
$$= \lambda_2 \, \boldsymbol{x}_1 + \lambda_1(1 - \lambda_2)\boldsymbol{x}_2 + (1 - \lambda_1)(1 - \lambda_2)\boldsymbol{x}_3$$
$$\Rightarrow \qquad \boldsymbol{w} = \mu_1 \, \boldsymbol{x}_1 + \mu_2 \, \boldsymbol{x}_2 + \mu_3 \, \boldsymbol{x}_3 \qquad \qquad \text{...(2)}$$

where $\mu_1 = \lambda_2$, $\mu_2 = \lambda_1(1 - \lambda_2)$, $\mu_3 = (1 - \lambda_1)(1 - \lambda_2)$

Clearly each μ_i lies between 0 and 1, *i.e.,* $0 \le \mu_i \le 1$, $i = 1, 2, 3$

Also, $\qquad \mu_1 + \mu_2 + \mu_3 = \lambda_2 + \lambda_1(1 - \lambda_2) + (1 - \lambda_1)(1 - \lambda_2)$
$$= \lambda_2 + \lambda_1 - \lambda_1 \lambda_2 + 1 - \lambda_1 - \lambda_2 + \lambda_1 \lambda_2 = 1$$

Thus, we conclude that $\mu_1 \, \boldsymbol{x}_1 + \mu_2 \, \boldsymbol{x}_2 + \mu_3 \, \boldsymbol{x}_3$ is a convex combination of the points $\boldsymbol{x}_1, \boldsymbol{x}_2, \boldsymbol{x}_3$.

Hence, the combination given by (2) is the required combination for the point $\boldsymbol{w}$.

EXAMPLE 11. *A hyperplane is given by the equation $3x_1 + 2x_2 + 4x_3 + 7x_4 = 8$.*
Find in which half spaces do the following points (–6, 1, 7, 2) and (1, 2, –4, 1) lie.

SOLUTION. We have

The hyperplane: $3x_1 + 2x_2 + 4x_3 + 7x_4 = 8$ \qquad ...(1)

using the values (–6, 1, 7, 2) in (1) we get

LHS = $3(-6) + 2(1) + 7(2) = 26 > 8$ = RHS

$\Rightarrow$ The point (–6, 1, 7, 2) lies in the open half spaces $3x_1 + 2x_2 + 4x_3 + 7x_4 > 8$

Now, substituting (1, 2, –4, 1) in the LHS of (1) we get

LHS = $3(1) + 2(2) + 4(-4) + 7(1) = -2 < 8$ = RHS

Hence, the point (1, 2, –4, 1) lies in the open half space $3x_1 + 2x_2 + 4x_3 + 7x_4 < 8$

EXAMPLE 12. *Sketch the convex polygon spanned by the following points in a two dimensional Euclidean space. Which of these points are vertices? Express the other as the convex linear combination of the vertices*

$(0,0), (0,1), (1,0), \left(\dfrac{1}{2}, \dfrac{1}{4}\right)$

SOLUTION. Clearly, the convex combinations of the points (0,0), (1,0) ; (0,0), (0,1) and (1,0), (0,1) give the line segments *OA*, *OB* and *AB* respectively.

Therefore, the convex combination of points (0,0), (1,0) and (0,1) is the interior of the Δ *OAB*.

Thus, the points *O*(0,0), *A*(1, 0) and *B*(0, 1) are the vertices and the point *C* is the interior point of the convex polygon spanned by the given points.

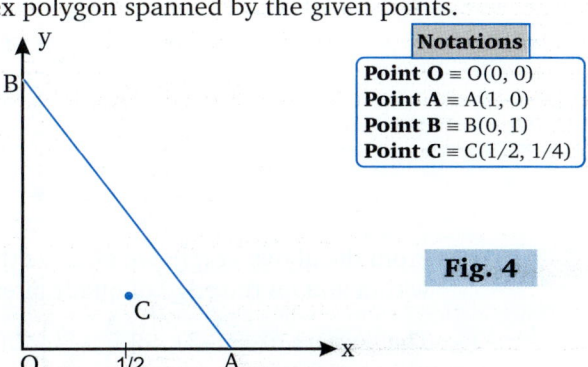

Notations
Point O $\equiv$ O(0, 0)
Point A $\equiv$ A(1, 0)
Point B $\equiv$ B(0, 1)
Point C $\equiv$ C(1/2, 1/4)

Fig. 4

Now, we have to express the point $\left(\dfrac{1}{2}, \dfrac{1}{4}\right)$ as the linear combinations of $(0,0)$, $(0, 1)$ and $(1, 0)$

Write $\left(\dfrac{1}{2}, \dfrac{1}{4}\right) = \lambda_1(0,0) + \lambda_2(0,1) + \lambda_3(1,0)$ where $\lambda_1 + \lambda_2 + \lambda_3 = 1, \lambda_i \geq 0$

$\Rightarrow \quad \left(\dfrac{1}{2}, \dfrac{1}{4}\right) = (\lambda_3, \lambda_2)$

$\Rightarrow \qquad \lambda_2 = 1/4$ and $\lambda_3 = 1/2$

Now, $\qquad \lambda_1 = 1 - \lambda_2 - \lambda_3 = = 1 - \dfrac{1}{4} - \dfrac{1}{2} = \dfrac{1}{4}$

Hence, $\left(\dfrac{1}{2}, \dfrac{1}{4}\right) = \dfrac{1}{4}(0,0) + \dfrac{1}{4}(0,1) + \dfrac{1}{2}(1,0)$

EXAMPLE 13. ***Find the extreme points of the polygonal convex set X determined by the system.***

$$2x_1 + x_2 + 9 \geq 0 \; ; \; -x_1 + 3x_2 + 6 \geq 0, \; x_1 + x_2 \leq 0, \; x_1 + 2x_2 - 3 \leq 0$$

SOLUTION. Draw the graph of the following given equations

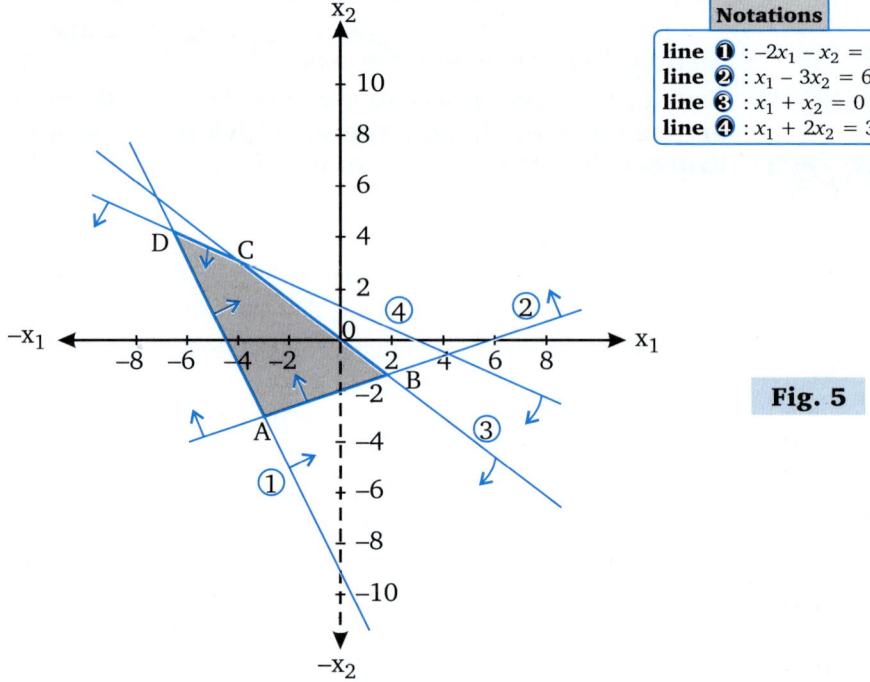

Notations
line ❶ : $-2x_1 - x_2 = 9$
line ❷ : $x_1 - 3x_2 = 6$
line ❸ : $x_1 + x_2 = 0$
line ❹ : $x_1 + 2x_2 = 3$

Fig. 5

From the above graph, we observe that the given inequalities enclose the points within and on the edge of quadrilateral *ABCD* and it is a convex set.

The corners $A(-3, -3)$, $B\left(\dfrac{3}{2}, \dfrac{-3}{2}\right)$, $C(-3, 3)$ and $D(-7, 5)$ are the extreme points of this polygonal convex set.

EXAMPLE 14. ***Find the convex hull of the set*** $A = \{(x_1, x_2) : x_1^2 + x_2^2 = 1\}$

SOLUTION. Let $\boldsymbol{x} = (x_1, x_2)$ and $\boldsymbol{y} = (y_1, y_2)$ be any two points of A

Then by definition of A, we can write

$$x_1^2 + x_2^2 = 1 \text{ and } y_1^2 + y_2^2 = 1 \qquad \qquad \text{...(1)}$$

Now, let $\boldsymbol{z}$ be the convex combination of $\boldsymbol{x}$ and $\boldsymbol{y}$

Then $\boldsymbol{z} = \lambda\boldsymbol{x} + (1 - \lambda)\boldsymbol{y} = \lambda(x_1, x_2) + (1 - \lambda)(y_1, y_2) \,; \, 0 \leq \lambda \leq 1$

$\Rightarrow \quad \boldsymbol{z} = (z_1, z_2) = (\lambda x_1 + (1 - \lambda)y_1, \, \lambda x_2 + (1 - \lambda)y_2)$

Therefore, $z_1 = \lambda x_1 + (1 - \lambda)y_1, \, z_2 = \lambda x_2 + (1 - \lambda)y_2$

Now,

$$z_1^2 + z_2^2 = (\lambda x_1 + (1 - \lambda)y_1)^2 + (\lambda x_2 + (1 - \lambda)y_2)^2$$

$$= \lambda^2(x_1^2 + x_2^2) + (1 - \lambda)^2(y_1^2 + y_2^2) + 2\lambda(1 - \lambda)(x_1 y_1 + x_2 y_2) \qquad \text{...(2)}$$

We have,

$$(x_1^2 + x_2^2)(y_1^2 + y_2^2) - (x_1 y_1 + x_2 y_2)^2 = (x_1 y_2 - x_2 y_1)^2 \geq 0$$

$$\Rightarrow \qquad (x_1 y_1 + x_2 y_2)^2 \leq (x_1^2 + x_2^2)(y_1^2 + y_2^2) = 1 \cdot 1$$

$$\Rightarrow \qquad x_1 y_1 + x_2 y_2 \leq 1 \qquad \qquad \text{...(3)}$$

Using (1) and (3) in (2) we get

$$z_1^2 + z_2^2 \leq \lambda^2 \cdot 1 + (1 - \lambda)^2 \cdot 1 + 2\lambda(1 - \lambda) \cdot 1$$

$$\Rightarrow \qquad z_1^2 + z_2^2 \leq 1$$

$$\Rightarrow \qquad \boldsymbol{z} = (z_1, z_2) \in S = \{(x_1, x_2) : x_1^2 + x_2^2 \leq 1\}$$

Hence, S is the set, containing the convex combinations of the elements of A.

Exercise-2.2

1. Show that the set $\{(x_1, x_2) : x_1 \geq 2, x_1 \leq 3)\}$ is a convex set.

2. Let A be an $m \times n$ matrix and $\boldsymbol{b}$, an m-vector, then show that $\{x \in R^n : A\boldsymbol{x} \leq \boldsymbol{b}\}$ is a convex set.

3. Show that the following sets are not convex.
   (i) $\{(x_1, x_2) : x_1 x_2 \leq 1, x_1 \geq 0, x_2 \geq 0\}$
   (ii) $\{(x_1, x_2) : x_2 - 3 \geq -x_1^2, x_1 \geq 0, x_2 \geq 0\}$

4. Show that the set $\{(x_1, x_2) : x_1^2 + x_2^2 \leq 4\}$ is a convex set. [MEERUT–2011]

5. Show that the following sets are convex
   (i) $S_1 = \left\{(x_1, x_2) : \dfrac{x_1^2}{4} + \dfrac{x_2^2}{9} \leq 1\right\}$
   (ii) $S_2 = \{(x_1, x_2) : x_1^2 + x_2^2 \leq 1, x_1 + x_2 \geq 1\}$

6. If $x_1, x_2 \in S$ implies $1/2(x_1 + x_2) \in S$. Show that S is convex.

7. Show that the set $S = \{x : |x| = 1, x \in E^n\}$ is not convex.

8. Show that union of two convex sets is not necessarily convex.

9. Show that the vector [7, 0] is a convex combination of the vectors [6, 3], [9, –6], [1, 2] and [1, –1].

10. Show that the vector [2, 1] can not be expressed as a convex combinations of [1, 1] and [–1, 2].

11. Find the extreme points of the set $\{(x, y) : |x| \leq 1, |y| \leq 1\}$.

12. Find the extreme points of the set $\{(x, y) : x^2 + y^2 \leq 25\}$.

ANSWERS

11. (1, 1), (–1, 1), (–1, –1) and (1, –1)   **12.** Every point on the circumference is an extreme point.

2.24 FEASIBLE AND BASIC FEASIBLE SOLUTIONS

Definition 1. *For a system AX = B of m equations in n unknown and if n > m and Rank(A) = Rank(A|B) = m then in order to solve the system of given equations, we set (n – m) variables to zero. Thus, a solution to resulting system of equations is said to be basic solution provided, the determinant of the coefficient of the remaining m variables is not zero.*

Definition 2. *If all the basic variables are non-negative then a basic solution is called feasible.*

Definition 3. *If at least one of the basic variables is negative then a basic solution is called infeasible.*

<u>THEOREM 1.</u> **The set of all feasible solutions (if not empty) of a LPP is a convex set.** [MEERUT-1994, 98, 2008; AGRA-1999, 2002,12; KANPUR-2008; GORAKHPUR-2007]

<u>PROOF.</u> Let X be the set of all feasible solutions of a *LPP given by*

$$Ax = b, x \geq 0 \qquad \qquad ...(1)$$

Now, we have the following cases:

Case I.

If the set X has only one element, then X is convex. In this case, theorem is true.

Case II.

If the set X has at least two elements

If $x_1, x_2 \in X$ then

$$Ax_1 = b, x_1 \geq 0 \text{ and } Ax_2 = b, x_2 \geq 0$$

Let

$$x_3 = \lambda x_1 + (1-\lambda)x_2 : 0 \leq \lambda \leq 1$$

Then

$$A x_3 = A\lambda x_1 + (1-\lambda)A x_2$$

$$= \lambda b + (1-\lambda)b = b$$

Further, since $x_1 \geq 0, x_2 \geq 0, 1-\lambda \geq 0$ $\qquad\qquad$ $(\because 0 \leq \lambda \leq 1)$

So, $\qquad\qquad x_3 = \lambda x_1 + (1-\lambda)x_2 \geq 0$

$\Rightarrow$ x_3 satisfies (1).

Therefore, $x_3 = \lambda x_1 + (1-\lambda)x_2$ is also a feasible solution and hence belongs to X.

$\Rightarrow$ Convex combination of any two points x_1 and x_2 in X belongs to X.

Hence, X is a convex set.

<u>THEOREM 2.</u> **Every basic feasible solution of the system $Ax = b, x \geq 0$ is an extreme point of the convex set of feasible solutions and conversly.** [MEERUT-1990, 94, 95, 2004]

<u>PROOF.</u> **Necessary Part:** We have to prove that every B.F.S is an extreme point of the convex set of all feasible solutions.

Let us suppose x be a B.F.S. of $Ax = b$

Also let X_B and B be the vector of m basic variables and the matrix of vectors associated to basic variables in the B.F.S. respectively, then $x = [X_B, O]$ $\qquad\qquad$ ...(1)

where O is a null vector of $(n-m)$ components

Also, $Ax = b \Rightarrow B \cdot X_B = b$ $\qquad\qquad$ ...(2)

We want to prove that x is an extreme point. Let if possible x is not an extreme point.

Since x is not an extreme point, then there exist two distinct points x_1 and x_2 of X

(the convex set of all feasible solutions of $A\boldsymbol{x} = \boldsymbol{b}$) such that $\boldsymbol{x} = \lambda \boldsymbol{x}_1 + (1-\lambda)\boldsymbol{x}_2$, $0 < \lambda < 1$ ...(3)

But we can write,

$$\boldsymbol{x}_1 = [\boldsymbol{u}_1, \boldsymbol{v}_1] \text{ and } \boldsymbol{x}_2 = [\boldsymbol{u}_2, \boldsymbol{v}_2] \qquad \text{...(4)}$$

where $\boldsymbol{u}_1$ and $\boldsymbol{u}_2$ are vectors of m components of $\boldsymbol{x}_1$ and $\boldsymbol{x}_2$ respectively and $\boldsymbol{v}_1, \boldsymbol{v}_2$ are $(n\text{-}m)$ components vectors.

Now using (1) and (4) in (3) we get

$$[X_B, \boldsymbol{0}] = \lambda[\boldsymbol{u}_1, \boldsymbol{v}_1] + (1-\lambda)[\boldsymbol{u}_2, \boldsymbol{v}_2], 0 < \lambda < 1$$
$$= [\lambda \boldsymbol{u}_1 + (1-\lambda)\boldsymbol{u}_2, \lambda \boldsymbol{v}_1 + (1-\lambda)\boldsymbol{v}_2]$$

$\Rightarrow$
$$\boldsymbol{0} = \lambda \boldsymbol{v}_1 + (1-\lambda)\boldsymbol{v}_2, \, 0 < \lambda < 1 \qquad \text{...(5)}$$

Clearly, $1 > \lambda > 0$, $1 - \lambda > 0$ and the components of $\boldsymbol{v}_1$ and $\boldsymbol{v}_2$ are greater than equal to 0. So, (5) is satisfied only when $\boldsymbol{v}_1 = 0$ and $\boldsymbol{v}_2 = 0$

So,
$$\boldsymbol{x}_1 = [\boldsymbol{u}_1, 0], \boldsymbol{x}_2 = [\boldsymbol{u}_2, 0]$$

Also, $\boldsymbol{x}_1 \in X, \boldsymbol{x}_2 \in X$, then from (2) we have

$$A\boldsymbol{x}_1 = B\boldsymbol{u}_1 = \boldsymbol{b}$$
and
$$A\boldsymbol{x}_2 = B\boldsymbol{u}_2 = \boldsymbol{b}$$
$\Rightarrow$
$$X_B = \boldsymbol{u}_1 = \boldsymbol{u}_2$$

So, $\boldsymbol{x} = x_1 = x_2$, which contradict the fact that $\boldsymbol{x}_1 \neq \boldsymbol{x}_2$

$\Rightarrow \boldsymbol{x}$ cannot be expressed as a convex combinations of any two distinct points in the set of all feasible solutions.

Hence, $\boldsymbol{x}$ is an extreme point.

Conversely, let us suppose that $\boldsymbol{x} = (x_1, x_2, ..., x_n)$ be an extreme point. We have to prove that $\boldsymbol{x}$ is a BFS.

Here, it is sufficient to prove that the vector associated with the positive elements of $\boldsymbol{x}$ are linearly independent.

Let us suppose p-components in $\boldsymbol{x}$ are non-zero and $(n\text{-}p)$ components are zero. We can assume these components as the first p-components of $\boldsymbol{x}$.

So, $\sum_{i=1}^{p} \alpha_i x_i = \boldsymbol{b}$, $x_i > 0$, $i = 1, 2, ..., p$ ...(6)

where α_i is the column vector in A associated to the ith variable in $\boldsymbol{x}$

Let if possible the column vectors $\alpha_1, \alpha_2, ..., \alpha_p$ of matrix be linearly dependent. Then by definition, there exist some scalars λ_i $(i = 1, 2, ..., p)$ with at least one of them non-zero such that $\sum_{i=1}^{p} \lambda_i \alpha_i = \boldsymbol{0}$ ...(7)

Now, from (6) and (7), for some arbitrary $\delta > 0$ we have

$$\sum_{i=1}^{p} x_i \alpha_i \pm \delta \sum_{i=1}^{p} \lambda_i \alpha_i = \boldsymbol{b}$$

$\Rightarrow$
$$\sum_{i=1}^{p} (x_i \pm \delta \lambda_i) \alpha_i = \boldsymbol{b}$$

Then, clearly, two points

$$x_1^* = [x_1 + \delta\lambda_1, x_2 + \delta\lambda_2,..., x_p + \delta\lambda_p, 0, 0,...,0 \ (n\text{-}p) \text{ in numbers}]$$

and

$$x_2^* = [x_1 - \delta\lambda_1, x_2 - \delta\lambda_2,..., x_p - \delta\lambda_p, 0, 0,...,0]$$

satisfy $Ax = b$

Further since, $x_i > 0$ therefore taking δ such that

$$0 < \delta < \min_i \left\{ \frac{x_i}{|\lambda_i|} \right\}, \ \lambda_i \neq 0, \ i = 1, 2,...,p$$

$\Rightarrow$ p component of x_1^* and x_2^* are always positive but the remaining components of x_1^* and x_2^* are zero.

$\Rightarrow$ x_1^* and x_2^* are feasible solutions different from x.

Further, $\quad x_1^* + x_2^* = 2[x_1, x_2,..., x_p, 0, 0, 0,...,0]$

$$\Rightarrow \qquad \frac{1}{2}x_1^* + \frac{1}{2}x_2^* = [x_1, x_2,..., x_p, 0, 0,...,0] = x$$

$$\Rightarrow \qquad x = \lambda x_1^* + (1-\lambda)x_2^* \qquad (\lambda = \frac{1}{2})$$

$\Rightarrow$ x can be expressed as a convex combination of two distinct feasible solutions x_1^* and x_2^*, which is a contradiction, because x is an extreme point.

$\Rightarrow$ $\alpha_1, \alpha_2, ...,\alpha_p$ are linearly independent.

$\Rightarrow$ $\alpha_1, \alpha_2, ...,\alpha_p$ can not be more than m

$\Rightarrow$ extreme point x will have atmost m non-zero variables i.e., at least $(n\text{-}m)$ variables will be zero.

$\Rightarrow$ x is a BFS.

Hence, every extreme point of the convex set of feasible solution is a BFS.

THEOREM 3. *The extreme point of the convex set of feasible solutions are finite in number.*

PROOF. Using Theorem 2 we have that there is only one extreme point for a given BFS and conversly.

$\Rightarrow$ there is one-to-one correspondance between the extreme points and the BFS in the absence of degeneracy.

Also, in case of degeneracy, corresponding to an extreme point with the number of non-zero variables less than m, we can form more than one degenerate BFS.

Hence, the number of extreme points of the feasible region is finite and it can not exceed the number of its basic feasible solution.

☞ REMARKS

- An extreme point can have atmost m positive x_i's where m is the no. of constraints.
- In an extreme point, vectors associated to the positive x_i's are linearly independent.

THEOREM 4. (Fundamental Extreme point Theorem) *If the convex set of the feasible solutions of $Ax = b$, $x \geq 0$ is a convex polyhedron then at least one of the extreme points gives an optimal solution.*

[MEERUT 2007, 08, 09, 10, 12]

PROOF. We know that the extreme points of the convex set of feasible solutions of $Ax = b$, $x \geq 0$ are finite.

Thus, we suppose that $x_1, x_2, ..., x_k$ are the extreme points of the set X of all feasible solutions of $Ax = b$, $x \geq 0$

Further, let z be the objective function such that $z = \boldsymbol{c}\boldsymbol{x}$

We have to be maximized z.

If $\boldsymbol{x}^* \in X$ is the optimal solution, then

$$\text{Max } z = \boldsymbol{c}\,\boldsymbol{x}^*$$

If $\boldsymbol{x}^*$ is the extreme point, then result is obvious.

If $\boldsymbol{x}^*$ is not an extreme point in X, then $\boldsymbol{x}^*$ can be expressed as a convex combinations of the extreme point of X.

i.e.,
$$\boldsymbol{x}^* = \lambda_1 \boldsymbol{x}_1 + \lambda_2 \boldsymbol{x}_2 + \dots + \lambda_k \boldsymbol{x}_k$$

$$= \sum_{i=1}^{n} \lambda_i \boldsymbol{x}_i \,, \ \lambda_i \geq 0 \text{ and } \Sigma\lambda_i = 1$$

$\Rightarrow$
$$z^* = \boldsymbol{c}\boldsymbol{x}^* = \boldsymbol{c}(\lambda_1 \boldsymbol{x}_1 + \lambda_2 \boldsymbol{x}_2 + \dots + \lambda_k \boldsymbol{x}_k)$$

$$= (\lambda_1 \boldsymbol{c}\,\boldsymbol{x}_1 + \lambda_2 \boldsymbol{c}\,\boldsymbol{x}_2 + \dots + \lambda_k \boldsymbol{c}\,\boldsymbol{x}_k)$$

Suppose maximum of $\boldsymbol{c}\boldsymbol{x}_i$ is $\boldsymbol{c}\boldsymbol{x}_p$, then

$$z^* \leq (\lambda_1 + \lambda_2 + \dots + \lambda_k) \cdot \boldsymbol{c}\boldsymbol{x}_p$$

$\Rightarrow$
$$z^* \leq \boldsymbol{c}\boldsymbol{x}_p$$

Since, z^* is the maximum value of z, so

$$z^* = \boldsymbol{c}\boldsymbol{x}_p$$

$\Rightarrow$
$$c\boldsymbol{x}^* = \boldsymbol{c}\boldsymbol{x}_p$$

$\Rightarrow$
$$\boldsymbol{x}^* = \boldsymbol{x}_p, \text{ one of the extreme point}$$

which shows that the optimal solution is attained at the extreme point.

THEOREM 5. *If the objective function of a LPP assumes its optimal value at more than one extreme point, then every convex combinations of these extreme points gives the optimal value of the objective function.*

PROOF. Let us consider an LPP as given below.

$$\text{Max} \cdot z = \boldsymbol{c}\,\boldsymbol{x}$$

such that
$$A\,\boldsymbol{x} = b,\ \boldsymbol{x} \geq 0$$

Let $\boldsymbol{x}_1, \boldsymbol{x}_2, \dots, \boldsymbol{x}_k$ be the extreme points of the feasible region. If the objective function z assume its optimal value z^* at the extreme points $\boldsymbol{x}_1, \boldsymbol{x}_2, \dots, \boldsymbol{x}_p$ $(p \leq k)$

Then
$$z^* = \boldsymbol{c}\,\boldsymbol{x}_1 = \boldsymbol{c}\,\boldsymbol{x}_2 = \dots = \boldsymbol{c}\,\boldsymbol{x}_p$$

If $\boldsymbol{x}_0$ is the convex combination of the extreme points $\boldsymbol{x}_1, \boldsymbol{x}_2, \dots, \boldsymbol{x}_p$

Then we have

$$\boldsymbol{x}_0 = \lambda_1 \boldsymbol{x}_1 + \lambda_2 \boldsymbol{x}_2 + \dots + \lambda_p \boldsymbol{x}_p \qquad \lambda_i \geq 0, \Sigma\lambda_i = 1$$

Thus,

$$\boldsymbol{c}\boldsymbol{x}_0 = \boldsymbol{c}(\lambda_1 \boldsymbol{x}_1 + \lambda_2 \boldsymbol{x}_2 + \dots + \lambda_p \boldsymbol{x}_p)$$

$$= \lambda_1 \boldsymbol{c}\,\boldsymbol{x}_1 + \lambda_2 \boldsymbol{c}\,\boldsymbol{x}_2 + \dots + \lambda_p \boldsymbol{c}\boldsymbol{x}_p$$

$$= \lambda_1 z^* + \lambda_2 z^* + \dots + \lambda_p z^*$$

$$= (\lambda_1 + \lambda_2 + \dots + \lambda_p) z^*$$

$$= \Sigma \lambda_i \cdot z^*$$

$$= z^* \qquad\qquad (\because \Sigma\lambda_i = 1)$$

Hence, the optimal value z^* is also attained at $\boldsymbol{x}_0$ which is the convex combination of the extreme points at which optimal value occur.

Solved Examples

EXAMPLE 1. *Show that the feasible solution $x_1 = 1$, $x_2 = 0$, $x_3 = 1$, $z = 6$ to the system*

$$x_1 + x_2 + x_3 = 2$$
$$x_1 - x_2 + x_3 = 2$$
$$2x_1 + 3x_2 + 4x_3 = z \ (minimized), x_i > 0 \ is \ not \ basic$$

SOLUTION. Here the objective funtion is given by,

minimized $z = 2x_1 + 3x_2 + 4x_3$

Here, we observe that first two equations in three variables x_1, x_2, x_3, only one variable can be assigned.

The given feasible solution is $x_1 = 1$, $x_2 = 0$, $x_3 = 1$ in which the variable x_1 and x_3 are non-zero, therefore, we shall take the vectors α_1 and α_2 associated to these variables, then

$$\alpha_1 = \text{column vector corresponding to } x_1 = \begin{bmatrix} 1 \\ 1 \end{bmatrix}$$

$$\alpha_2 = \text{column vector corresponding to } x_2 = \begin{bmatrix} 1 \\ 1 \end{bmatrix}$$

$\Rightarrow$ $\quad\quad\quad\quad \alpha_1 = \alpha_2$

$\Rightarrow$ $\quad\quad 1 \cdot \alpha_1 + (-1)\alpha_2 = 0$

$\Rightarrow$ $\quad\quad \exists$ two scalars $\lambda_1 = 1$, $\lambda_2 = -1$ such that $\lambda_1\alpha_1 + \lambda_2\alpha_2 = 0$

$\Rightarrow$ $\quad\quad \alpha_1, \alpha_2$ are linearly dependent

Hence, the given feasible solution is not basic.

EXAMPLE 2. *Find all the basic feasible solution for the system of equations*

$$2x_1 + 6x_2 + 2x_3 + x_4 = 3$$
$$6x_1 + 4x_2 + 4x_3 + 6x_4 = 2$$

and determine the associated general convex combinations of the extreme points solutions. [MEERUT–2007; KANPUR–2010; RAJASTHAN–2012]

SOLUTION. We can write the given system of equations as

$$\begin{bmatrix} 2 & 6 & 2 & 1 \\ 6 & 4 & 4 & 6 \end{bmatrix} \begin{bmatrix} x_1 \\ x_2 \\ x_3 \\ x_4 \end{bmatrix} = \begin{bmatrix} 3 \\ 2 \end{bmatrix}$$

$\Rightarrow$ $\quad\quad\quad\quad\quad AX = B$

where $A = \begin{bmatrix} 2 & 6 & 2 & 1 \\ 6 & 4 & 4 & 6 \end{bmatrix}$, $X = \begin{bmatrix} x_1 \\ x_2 \\ x_3 \\ x_4 \end{bmatrix}$, $B = \begin{bmatrix} 3 \\ 2 \end{bmatrix}$

If α_1, α_2, α_3 and α_4 are the column vectors in A then we have

$$\alpha_1 = \begin{bmatrix} 2 \\ 6 \end{bmatrix} \qquad \alpha_2 = \begin{bmatrix} 6 \\ 4 \end{bmatrix} \qquad \alpha_3 = \begin{bmatrix} 2 \\ 4 \end{bmatrix} \qquad \alpha_4 = \begin{bmatrix} 1 \\ 6 \end{bmatrix}$$

$\Rightarrow \qquad\qquad A = [\alpha_1 \quad \alpha_2 \quad \alpha_3 \quad \alpha_4]$

Here, $\qquad\qquad n$ = number of unknowns = 4

$\qquad\qquad m$ = number of equations = 2

$\Rightarrow$ There can be atmost ${}^4C_2 = \dfrac{4!}{2! \cdot 2!} = 6$ feasible solutions

Now, six set of two vectors out of α_1, α_2, α_3, α_4 are given below:

$$|B_1| = \begin{vmatrix} 2 & 6 \\ 6 & 4 \end{vmatrix} = -28 \neq 0 \;;\; |B_2| = \begin{vmatrix} 2 & 2 \\ 6 & 4 \end{vmatrix} = -4 \neq 0$$

$$|B_3| = \begin{vmatrix} 2 & 1 \\ 6 & 6 \end{vmatrix} = 6 \neq 0 \;;\; |B_4| = \begin{vmatrix} 6 & 2 \\ 4 & 4 \end{vmatrix} = 16 \neq 0$$

$$|B_5| = \begin{vmatrix} 6 & 1 \\ 4 & 6 \end{vmatrix} = 32 \neq 0 \;;\; |B_6| = \begin{vmatrix} 2 & 1 \\ 4 & 6 \end{vmatrix} = 8 \neq 0$$

$\Rightarrow$ all these set of vectors are linearly independent. Hence, all the basic solutions exist.

If $x_{B_i} : i = 1, 2, \ldots, 6$ are the vectors of corresponding basic variables respectively,

then the given system of equations reduces in the following forms.

$$B_1 x_1 = b, \qquad\qquad B_2 x_2 = b, \qquad B_3 x_3 = b$$
$$B_4 x_4 = b, \qquad\qquad B_5 x_5 = b, \qquad B_6 x_6 = b$$

which implies

$$x_{B_1} = B_1^{-1} b = -\frac{1}{28} \begin{bmatrix} 4 & 6 \\ -6 & 2 \end{bmatrix} \begin{bmatrix} 3 \\ 2 \end{bmatrix} = \begin{bmatrix} 0 \\ 1 \\ 2 \end{bmatrix}$$

$$x_{B_2} = B_2^{-1} b = -\frac{1}{4} \begin{bmatrix} 4 & -2 \\ -6 & 2 \end{bmatrix} \begin{bmatrix} 3 \\ 2 \end{bmatrix} = \begin{bmatrix} -2 \\ \frac{7}{2} \end{bmatrix}$$

$$x_{B_3} = B_3^{-1} b = \frac{1}{6} \begin{bmatrix} 6 & -1 \\ -6 & 2 \end{bmatrix} \begin{bmatrix} 3 \\ 2 \end{bmatrix} = \begin{bmatrix} \frac{8}{3} \\ \frac{-7}{3} \end{bmatrix}$$

$$x_{B_4} = B_4^{-1} b = \frac{1}{16} \begin{bmatrix} 4 & -2 \\ -4 & 6 \end{bmatrix} \begin{bmatrix} 3 \\ 2 \end{bmatrix} = \begin{bmatrix} \frac{1}{2} \\ 0 \end{bmatrix}$$

$$x_{B_5} = B_5^{-1} b = \frac{1}{32} \begin{bmatrix} 6 & -1 \\ -4 & 6 \end{bmatrix} \begin{bmatrix} 3 \\ 2 \end{bmatrix} = \begin{bmatrix} \frac{1}{2} \\ 0 \end{bmatrix}$$

and
$$x_{B_6} = B_6^{-1}\mathbf{b} = \frac{1}{8}\begin{bmatrix} 6 & -1 \\ -4 & 2 \end{bmatrix}\begin{bmatrix} 3 \\ 2 \end{bmatrix} = \begin{bmatrix} 2 \\ -1 \end{bmatrix}$$

Now, we will find the basic solutions

In the basic matrix B_1, basic vectors are α_1 and α_2 and

$$x_{B_1} = \begin{bmatrix} x_1 \\ x_2 \end{bmatrix} = \begin{bmatrix} 0 \\ \frac{1}{2} \end{bmatrix}$$

$\Rightarrow$ $x_1 = 0, x_2 = 1/2$

These two variables are the basic variables and the remaining x_3, x_4 are non-basic variables. The non-basic variables are zero.

Therefore, the basic solution associated to the basis B_1 is given by (0, 1/2, 0, 0)

In a similar way we can write all other basic solutions as follows:

$$\left(-2, 0, \frac{7}{2}, 0\right), \left(\frac{8}{3}, 0, 0, \frac{-7}{3}\right), \left(0, \frac{1}{2}, 0, 0\right), \left(0, \frac{1}{2}, 0, 0\right) \text{ and } (0, 0, 2, -1)$$

But out of these basic solutions, the BFS are $\left(0, \frac{1}{2}, 0, 0\right), \left(0, \frac{1}{2}, 0, 0\right), \left(0, \frac{1}{2}, 0, 0\right)$

Clearly, the extreme points are $\mathbf{x}_1 = \left(0, \frac{1}{2}, 0, 0\right)$, $\mathbf{x}_2 = \left(0, \frac{1}{2}, 0, 0\right)$, $\mathbf{x}_3 = \left(0, \frac{1}{2}, 0, 0\right)$

$\Rightarrow$ all the extreme points are same.

Hence, there is unique extreme point solution.

 ## Exercise-2.3

1. Find all the basic solutions for the following system of linear equations

   $x_1 + 2x_2 + x_3 = 4$

   $2x_1 + x_2 + 5x_3 = 4$

2. Find all the basic solutions of the following system of linear equations

   $x_1 + x_2 + x_3 = 4$

   $2x_1 + 5x_2 - 2x_3 = 0$

3. Show that the basic solution $x_1 = 1, x_2 = 1/2$, $x_3 = x_4 = x_5 = 0$ of the equations

   $x_1 + 2x_2 + x_3 + x_4 = 2$

   $x_1 + 2x_2 + 1/2\, x_3 + x_5 = 2$

   is not basic.

4. Find a basic feasible solution of the system of the equations

   $x_1 + 2x_3 = 3$

   $x_2 + x_3 = 4$ and $x_1, x_2, x_3 \geq 0$

5. Find all basic feasible solutions of the equations:

   $2x_1 + x_2 + 3x_3 = 3$

   $x_1 + 2x_2 + x_3 = 3$ and $x_1, x_2, x_3 \geq 0$

6. Show that if $x_1, x_2, \ldots x_k$ are k different optimal basic feasible solutions to an LPP then any convex combinations of $x_1, x_2, \ldots, x_k$ is also an optimal solution.

ANSWERS

1. $(2, 1, 0), (5, 0, -1), \left(0, \frac{5}{3}, \frac{2}{3}\right)$

2. $\left(\frac{17}{3}, \frac{-5}{3}, 0\right), \left(0, \frac{11}{7}, \frac{17}{7}\right), \left(\frac{11}{4}, 0, \frac{5}{4}\right)$

4. $x_1 = 1, x_2 = 4, x_3 = 0$

5. $(1, 1, 0), \left(0, \frac{6}{5}, \frac{3}{5}\right)$

REVIEW QUESTIONS

1. What do you mean by an extreme point of a convex set?
2. Write a short note on convex set and their applications to linear programming problem.
3. Obtain the convex hull of the boundary of a circle.
4. Prove that the convex hull of a finite number of points is a convex set.
5. Define: Hyperplane, Convex set.
6. What is meant by convex polyhedron.
7. Explain the procedure of generating extreme points solutions to a linear programming problem pointing out the assumption made. if any?

MULTIPLE CHOICE QUESTIONS (CHOOSE THE MOST APPROPRIATE ONE)

1. The number of vertices of any non empty closed bounded convex set can not be:
   (a) finite   (b) not finite
   (c) infinite   (d) None of these
2. The closed half spaces in E_n or E^n is a:
   (a) open convex set
   (b) unbounded convex set
   (c) closed convex set
   (d) no convex set
3. The set of all feasible solution (if not empty) of a L.P.P. is a:
   (a) non convex set   (b) poly convex set
   (c) convex set   (d) none of these
4. The union of two convex sets may or may not be a:
   (a) Non convex set   (b) Convex set
   (c) Poly convex set   (d) None of these
5. Every extreme point of a convex set is:
   (a) boundary value of the set
   (b) boundary point of the set
   (c) both (a) and (b)
   (d) none of these
6. A hyper plane is:
   (a) convex   (b) feasible
   (c) concave   (d) none of these
7. Let S and T are two convex sets in E^n, then $S + T, S - T$ and $\alpha S + \beta T$, where α and β are scalars, are called:
   (a) non convex   (b) convex sets
   (c) convex point   (d) none of these
8. Convex hull of set of points on the circle is:
   (a) whole circle   (b) half circle
   (c) extreme point   (d) none of these
9. A hyper plane is:
   (a) non convex   (b) convex
   (c) convex point   (d) none of these
10. In a L.P.P. the set S is polytope when S made by ... of constraints:
    (a) finite number   (b) infinite number
    (c) no number   (d) none of these
11. The vertices of the polygons when are convex sets are called:
    (a) extreme point   (b) feasible
    (c) convex   (d) none of these
12. The intersection of any finite number of convex sets is also a:
    (a) finite set   (b) convex set
    (c) infinite set   (d) non convex set
13. A set S be non empty, then it has at least:
    (a) one vertex   (b) two vertex
    (c) no vertex   (d) none of these
14. The intersection of a finite number of closed half spaces is called:
    (a) monotope   (b) polytope
    (c) nonetope   (d) polygon
15. The intersection of two convex sets is also a:
    (a) convex set   (b) non convex
    (c) extreme point   (d) none of these
16. Any points on line segment joint two points in R^n can be expressed as a convex combination of:
    (a) two points   (b) one point
    (c) variable   (d) none of these
17. A set S be closed convex bounded above/ bounded below, then S has at least:
    (a) one vertex   (b) two vertex
    (c) no vertex   (d) none of these
18. Set of all feasible solution of L.P.P. is:
    (a) convex   (b) concave
    (c) (a) and (b) both (d) none of these
19. Convex linear combination of the points x_1, x_2 is given by $x_p = \lambda_1 x_1 + \lambda_2 x_2$
    (a) $(\lambda_1 + \lambda_2)$ (b) $(\lambda_1 - \lambda_2)$
    (c) $(\lambda_2 - \lambda_1)$ (d) none of these
20. The arbitrary intersection of convex sets is also:
    (a) a convex set   (b) a non convex set
    (c) a finite set   (d) none of these

21. A point X is called a convex linear combination of points x_1 and x_2 if there exist a λ, $0 \le \lambda \le 1$ such that:
   (a) $X = \lambda x_1 + (1 - \lambda)x_2$
   (b) $X = \lambda x_1 + (\lambda - 1)x_2$
   (c) $X = \lambda x_1 - (1 - \lambda)x_2$
   (d) $X = \lambda x_1 + \lambda x_2$

22. The set of all convex linear combination of finite number of points is a:
   (a) c.l.c   (b) no c.l.c
   (c) non convex   (d) convex set

23. The set of all convex combination of a finite number of points is called the .... which generated by these points:
   (a) convex polyhedron
   (b) convex set
   (c) non convex set
   (d) none of these

24. A set C in n dimensional space is said to be convex if every point on the line segment joining any two distinct points of C lie in:
   (a) R
   (b) C
   (c) neither (a) nor (c)
   (d) none of these

25. Let S be a closed set and x is an extreme point of S then $S - \{x\}$ is:
   (a) convex
   (b) concave
   (c) may or may not be convex
   (d) none of these

26. A simplex in n dimension is a convex polyhedron having exactly:
   (a) n vertices   (b) $n + 1$ vertices
   (c) $n - 1$ vertices   (d) none of these

ANSWERS

1. (c)	2. (d)	3. (c)	4. (b)	5. (b)	6. (a)	7. (b)	8. (b)	9. (b)
10. (a)	11. (a)	12. (b)	13. (a)	14. (b)	15. (a)	16. (a)	17. (a)	18. (a)
19. (a)	20. (a)	21. (a)	22. (d)	23. (a)	24. (b)	25. (c)	26. (b)	

ARCHIVE

1. Prove that the set $\{(x_1, x_2): x_1^2 + x_2^2 \le 4\}$ is a convex set.   [MEERUT-2011]

2. Prove that the set $S = \{(x_1, x_2): 3x_1^2 + 2x_2^2 \le 6\}$ is a convex set.   [MEERUT-2006]

3. Prove that the set
   $S = \{(x_1, x_2): 2x_1 + 3x_2 = 11\} \subseteq R^2$
   is a convex set.   [MEERUT-2010]

4. If S_1 and S_2 be two non-empty disjoint convex sets and S be a set such that if $x_1 \in S_1$ and $x_2 \in S_2$ then $x_1 - x_2 \in S$. Show that S is also convex and does not contain the origin.
   [AVADH-2005, GARHWAL-2010; MEERUT 2008, 09, 12, 14]

5. Show that the set of all internal points of a convex set S is a convex set.   [MEERUT 2011, 12]

Linear Programming Problems: Formulation and Graphical Solutions

3

3.1 INTRODUCTION

Linear programming is related to the most advanced mathematics, *i.e.,* Operations Research. The word 'linear' is used to describe the relationship among two or more variables which are directly proportional while the word 'programming' means planning of activities in such a manner that achieves some optimal results with some restricted resources. Thus, linear programming (L.P.) is one of the most important optimization techniques in the field of Operations Research.

According to the obtained fact during the world war-II (1939-1945), there was a need of such an effective management that can arrange the limited military resources and the army in the most successful way. There was the need of such methods, which when adopted, gives the maximum profit and to minimize the cost. This was the birth of linear programming.

Mathematically in various situation, the problems are seen in which the number of relations is not equal to the number of variables and many of the relations are in the form of inequality ($\leq$ or $\geq$) to optimize (maximize or minimize) a linear function of the variables subject to such conditions. Therefore, we can defined the linear programming (LP) and linear programming problems (LPP) as follows:

(i) **Linear Programming (LP):** Linear programming is the most general technique, that is used for the optimization (maximization or minimization) of a function to obtain the maximum or minimum value under the certain conditions.

(ii) **Linear Programming Problems (LPP):** Linear programming problem is used to optimize a given linear function of variables, called the objective function which is subject to a certain set of linear equations or inequations (called the constraints or restrictions).

☛ REMARK

• A mathematical programming is said to be linear if both the objective function and the constraints are linear in decision variables.

3.2 BASIC TERMINOLOGY OF LINEAR PROGRAMMING

(i) **Objective function:** The function which is optimized (maximized or minimized) is known as objective function.

(ii) **Constraints:** The system of linear equations (inequations) under which the objective function is to be optimized are known as constraints.

(iii) **Decision Variables:** The decision variables refers to any activity that is competing with other activities for limited resources. In LPP, the relationship among decision variables should be linear.

3.3 BASIC REQUIREMENTS OF LPP

To define any LPP, there are some basic requirements which are given below:
  (i) The relationship between the decision variables should be linear.
 (ii) A well defined objective function must be stated which may be either to maximize or to minimize.
(iii) There must be some limitations in the form of constraints which are to be allocated among various activities.
 (iv) There must be alternative course of action to choose the best from.
  (v) All decision variables must assumes non-negative values.
 (vi) Each element of the problem is capable of being quantified by means of measurement.

☞ REMARKS
  • In LPP, both the objective function and constraints must be expressed in terms of linear equations or inequalities which can be graphically represented by a straight line.
  • Every linear programming problem has two important features in common, an objective function to be maximized or minimized and constraints or restrictions.
  • The number of activities involved in a problem should be finite, leads to a finite number of constraints in the problem.

3.4 BASIC ASSUMPTIONS OF LP MODEL

The major assumptions of LP model are given as follows:
  (i) **Certainty:** All model parameters such as availability of resources, profit contribution of a unit of decision variables and consumption of resources by a unit of decision variable must be known and may be constant i.e., the coefficient in the objective function and constraints are completely known and do not change during the period being studied.
 (ii) **Linearity:** Both the objective function and constraints must be expressed in terms of linear equations or inequations.
(iii) **Additivity:** The value of the objective function and the total amount of each resource used must be equal to the sum of the respective individual contributions by decision variables. In other words, interaction among the activities of the resources does not exists.
 (iv) **Divisibility:** Divisibility implies that solution values of decision variables and resources can take any non-negative values, i.e., fractional values of the decision variables are permitted. This, however is not always desirable.
  (v) **Non-Negativity:** It is assumed that variable will take only non-negative values.
 (vi) **Finiteness:** An optimal solution can not be evaluated in the situations where there are infinite number of alternative activities and resource restrictions.

3.5 ADVANTAGES OF LINEAR PROGRAMMING

Following are the main advantages of LP:
  (i) LP helps in attaining the optimal use of productive factors (resources). It also indicate how a decision maker can employ his productive factors effectively by selecting and

distributing these resources (factors).

(ii) LP techniques improve the quality of decision. The decision making approach of the users of this technique become more objective and less subjective.

(iii) LP techniques provide possible and practical solutions, since there might be other constraints operating outside the problem which must be taken into account.

(iv) The most significant advantage of LP technique is the highlighting of bottlenecks in the production process.

(v) LP also helps in re-evaluation of a basic plan for changing conditions. Plan can be laid for several sets of conditions to find out how to best prepare for possible future changes.

(vi) If conditions change; when the plan is partly carried out, they can be determined so as to adjust the remainder of the plans for best results.

(vii) It also contributes the development of executives through the technique of model building and corresponding interpretation.

3.6 LIMITATIONS OF LINEAR PROGRAMMING

(i) Generally, neither the objective functions nor the constraints in reality are linear related to the variable, while in LPP, all relationship among decision variables are linear.

(ii) In linear programming problem, fractional values are permitted for the decision variable. However many decision problems requires that the decision variables should be obtained in non-fractional values.

(iii) Parameters appearing in the LPP, are assumed to be constant but in real life situations they are frequently neither known nor constant.

(iv) LPP deals with only single objective function, but in real life situations, we may come across conflicting multi-objective problem.

(v) The coefficient of the basic variables can not be determined with certainity, however they can be stated only with probability.

(vi) The LPP does not take into consideration the effect of time and uncertainity.

(vii) In LPP, coefficient in the objective function and the constraints equation must be completely known and they should not change during the study period, but practically, it may not be possible to state all coefficients in the objective function and constraints with certainity.

(viii) The approximation which must be made in the complex relationship between constraints and decision variables, to reduce the problem to meaningful dimensions frequently place the final results in some doubt.

3.7 APPLICATION AREAS OF LINEAR PROGRAMMING

1. **Production Scheduling:** In production scheduling problems the technique may often be used in situations where several products can be made on each of several different machine, the problem is to decide on a programme which will maximize the profit or minimize the cost.

2. **Product Mix Problems:** In product mix problems, one make selection of the optimal product mix to make best use of machine and man hour available while maximize the firm profit.

3. **Transportation Problem:** In transportation problem, we determined the distribution system that will minimize total shipping cost from several warehouses to various market locations.

4. **Travelling Salesman Problem:** Here, we have to decide the shortest route for a salesman starting from a given city, visiting each of the specified cities and then returning to the original city of departure.

5. **Blending Problems:** Blending problem arise when a product can be made from a variety of available raw materials of various composition. In each problem we determine optimal amount of various raw materials, to be used in producing a set of products while determining the optimum quantity of each product to be produced.

6. **Investment Problems:** In investment problems, we find the amount that should be invested in a number of fixed income securities in order to maximize the return on investment.

7. **Diet Problems:** In such type of problems, we determine the amount of different feed ingradient combinations used in desired diet that will satisfy stated nutritional requirements at a minimum cost level.

8. **Media Selection Problems:** In media selection problems, we find the optimum allocation of advertisement in different effective media mix in order to maximize the audience exposure.

9. **Manufacturing Problems:** In manufacturing problems, we find the number of items of each type that should be manufactured so as to maximize the profit subject to production restrictions imposed by limitations on the use of machinery and labour.

10. **Assembly Problems:** In assembly problems, we have to determine the best combinations of basic components to produce goods according to certain specifications.

11. **Job-assigning Problems:** In such problems, we have to assign the job to the workers for maximum effectiveness and optimum results subject to the restrictions of wages and costs.

3.8 STANDARD FORM OF LINEAR PROGRAMMING PROBLEMS

In linear programming problem, we have to determine the values of n decision variables $x_1, x_2, ..., x_n$ such that the linear objective function of these variables assumes an optimal (maximum or minimum) values, when these variables are subject to the set of m linear constraints.

The general LPP with n decision variables and m constraints is stated as follows:

Optimize (maximize or minimize), $Z = c_1x_1 + c_2x_2 + ... + c_nx_n$ *(objective function) subject to the constraints*

$$\left. \begin{array}{l} a_{11}x_1 + a_{12}x_2 + ... + a_{1n}x_n (\leq, =, \geq) b_1 \\ a_{21}x_1 + a_{22}x_2 + ... + a_{2n}x_n (\leq, =, \geq) b_2 \\ \vdots \qquad\qquad\qquad \vdots \\ a_{m1}x_1 + a_{m2}x_2 + ... + a_{mn}x_n (\leq, =, \geq) b_m \end{array} \right\} \text{constraints}$$

such that $x_1, x_2, ..., x_n \geq 0$ *(non-negative restrictions)* where

(i) $x_1, x_2, ..., x_n$ are decision variables.

(ii) $c_1, c_2, ..., c_n$ are cost or profit coefficients.

(iii) $a_{ij} : i = 1, 2, ... m, j = 1, 2, ..., n$ are input output coefficients.

(iv) $b_1, b_2, ..., b_n$ represent requirements or availability of m constraints.

(v) The expression $\leq, =, \geq$ means that each constraints may assume only one of three possible forms.

(vi) The restriction, $x_i \geq 0$ implies that x_j's must be non-negative.

In compact form the general LPP can be written as follows:

$$\text{optimize (maximize or minimize) } z = \sum_{j=1}^{n} c_j x_j \qquad \text{(objective function)}$$

$$\text{subject to } \sum_{j=1}^{n} a_{ij} x_j (\leq, =, \geq) b_i : i = 1, 2, ..., m \qquad \text{(constraints)}$$

$$\text{such that } x_j \geq 0, j = 1, 2, ..., n \qquad \text{(non-negative restrictions)}$$

☛ REMARKS

• The input-output coefficients a_{ij} are also known as structural coefficients or technological coefficients which represent the exchange coefficients of the jth decision variable in the ith constraints.

• The constraints are generally expressed by "less than or equals ($\leq$)" sign in maximization problem. where as in case of minimization problem, it can be expressed by 'greater than or equal to ($\geq$) sign.

3.9 MATRIX FORM OF LPP

In matrix notation the general form of LPP can be written as follows:

$$\text{Optimize (Maximize or minimize) } Z = (c_1, c_2, ..., c_n) \begin{bmatrix} x_1 \\ x_2 \\ \vdots \\ x_n \end{bmatrix}$$

$$\text{subject to the constraints } \begin{bmatrix} a_{11} & a_{12} & \cdots & a_{1n} \\ a_{21} & a_{22} & \cdots & a_{2n} \\ \vdots & & & \\ a_{m1} & a_{m2} & \cdots & a_{mn} \end{bmatrix} \begin{bmatrix} x_1 \\ x_2 \\ \vdots \\ x_n \end{bmatrix} (\leq, =, \geq) \begin{bmatrix} b_1 \\ b_2 \\ \vdots \\ b_m \end{bmatrix}$$

such that $x_j \geq 0 \ \forall \ j = 1, 2, ..., n$

☛ REMARK

• In a general LPP, it is assumed that the number of rows of coefficient matrix is less than its number of columns.

3.10 MATHEMATICAL FORMULATION OF LINEAR PROGRAMMING PROBLEM

In order to formulate of LPP as a mathematical model we use the following procedure:

Working Procedure

STEP 1. Identify the unknown decision variables and assign symbols $x_1, x_2, \ldots$ etc to them.

STEP 2. Identify all restrictions or constraints in terms of requirement and availability of each resources and express them as linear inequations (inequalities) of decision variables.

STEP 3. Identify whether the objective function is to be maximized or minimized and then express it as a linear function of decision variables and then convert it into a linear mathematical expression in terms of decision variables multiplied by their profit or cost contributions.

☞ REMARK

• Before the formulation of linear programming problem as a mathematical model, find the key decision to be made from the study of the problem. In this connection looking for variable helps considerably.

3.11 EXTRA VARIABLE NEEDED:

1. **Slack Variables:** If a constraint has $\leq$ sign, then in order to make it an equality, we have to add something positive to the LHS.

   Definition. *The non-negative variables which are added to LHS of the constraints to convert them into equalities are known as slack variables.*

2. **Surplus Variables:** If a constraint has $\geq$ sign, then in order to make it an equality, we have to subtract something positive from its LHS.

   Definition. *The non-negative variables which are subtracted from the LHS of the constraints to convert them into equalities are called the surplus variables.*

   [MEERUT-2006, 07, 09, 11, 12, 14, 18; KANPUR-2007; GORAKHPUR-2008, 09, 10, 11]

3. **Artificial Variables:** The artificial variables are introduced for the limited purpose of obtaining an initial solution when constraints of the type $\geq$ or $=$.

   Definition. *When we use surplus variables to convert inequlities into equations, then to obtain basic matrix as identity matrix, we used artificial variable in each constraints.*

   The summary of the extra variables to be added in the given LPP to convert it into standard form is given in the following table:

Type of constraints	Extra variable	Operation	Coefficient of extra variables in the objective functions	
			Max. z	Min. z
$\leq$	slack variable	added	0	0
$\geq$	surplus variable	subtracted	0	0
	artificial variable	added	$-M$	$+M$
$=$	Artificial variable	added	$-M$	$+M$

☞ REMARKS

• Surplus variables are also known as negative slack variables.

• Surplus and slack variables carry a zero coefficient in the objective function.

Solved Examples

EXAMPLE I. **(Diet Problem)** *The objective of a diet problem is to ascertain the quantities of certain foods that should be eaten to meet certain nutritional requirement at a minimum cost. The consideration is limited to milk, green vegetables and eggs and to vitamins A, B, C. The number of milligrams of each of these vitamin contained within a unit of each food and their daily minimum requirement along with the cost of each food is given as below:*

Vitamin	Litre of milk	Vegetables (in kg)	Eggs (Dozen)	Minimum daily requirement
A	1	1	10	1 mg
B	100	10	10	50 mg
C	10	100	10	10 mg
Cost in ₹	20	10	8	

Formulate a linear programming problem for this diet problem.

[MEERUT–2004, 10, GARHWAL–2008]

SOLUTION. Let the diet contain x_1 litres of milk, x_2 kg of vegetables and x_3 dozens of eggs.
Then total cost (Z) per day in rupees is
$$Z = 20x_1 + 10x_2 + 8x_3 \qquad \text{(objective function)} \quad ...(1)$$
Further, total amount of vitamin A in daily diet is
$$(x_1 + x_2 + 10x_3)\text{mg}$$
According to question
$$x_1 + x_2 + 10x_3 \geq 1 \qquad\qquad\qquad\qquad ...(2)$$
Similarly for vitamin B and C, we must have
$$100x_1 + 10x_2 + 10x_3 \geq 50 \quad \text{(Constraints)} \quad ...(3)$$
and $\qquad\qquad 10x_1 + 100x_2 + 10x_3 \geq 10 \qquad\qquad ...(4)$

Also, the quantities of different food items to be consumed can not be negative,
so $\qquad\qquad x_1 \geq 0, x_2 \geq 0, x_3 \geq 0 \qquad$ (Non-negative restrictions)
Hence, the mathematical model of given LPP is given below:

$\quad$ *Minimize* $\qquad Z = 20x_1 + 10x_2 + 8x_3 \qquad$ *(objective function)*

$\qquad$ *subject to* $\qquad x_1 + x_2 + 10x_3 \geq 1$
$$\left.\begin{array}{l} 100x_1 + 10x_2 + 10x_3 \geq 50 \\ 10x_1 + 100x_2 + 10x_3 \geq 10 \end{array}\right\} \quad \textit{(constraints)}$$

$\qquad$ *such that* $\qquad x_1 \geq 0, x_2 \geq 0, x_3 \geq 0 \qquad$ *(Non-negative restrictions)*

EXAMPLE 2. *A furniture dealer deals in two items viz, tables and chairs. He has ₹ 10,000 to invest and a space to store almost 60 pieces. A table costs him ₹ 500 and a chair of ₹ 200. He can sell a table at profit of ₹ 50 and a chair at a profit of ₹ 15. Assume that he can sell all the items that he buys. Formulate the problem as an LPP, so that he can maximize the profit.*

SOLUTION. Let x_1 be the number of tables and x_2 be the chairs
Then, the profit on x_1 tables = ₹ $50x_1$

and, the profit on x_2 chairs = ₹ $15x_2$

$\Rightarrow$ total profit $Z = 50x_1 + 15x_2$ ...(1)

Here, we have to maximize the profit Z.

Now, the cost of x_1 tables = $500x_1$

and   the cost of x_2 chairs = $200x_2$

$\Rightarrow$ total cost of x_1 tables and x_2 chairs = $500x_1 + 200x_2$

Since, the dealer invest ₹ 10,000, so

$\qquad 500x_1 + 200x_2 \le 10000$ ...(2)

Also, $\qquad$ total items = $x_1 + x_2$

which can be atmost 60 therefore

$\qquad x_1 + x_2 \le 60$ ...(3)

Further, since number of tables and chairs cannot be negative

Therefore, $\qquad x_1 \ge 0, x_2 \ge 0$

Thus, the mathematical model of the given LPP is as follows:

$\qquad$ *Maximize* $\qquad Z = 50x_1 + 15x_2$ $\qquad$ (*objective function*)

$\qquad$ *subject to* $\qquad 5x_1 + 2x_2 \le 100$ $\qquad$ (*constraints*)

$\qquad\qquad\qquad\qquad x_1 + x_2 \le 60$

$\qquad$ *such that* $\qquad x_1 \ge 0, x_2 \ge 0$ $\qquad$ (*non-negative restrictions*)

EXAMPLE 3. ***A goldsmith manufactures necklaces and bracelets. The total number of necklaces and bracelets that he can handel per day is atmost 24. It takes one hour to make a bracelets and half an hour to make a necklace. It is assumed that he can work for a maximum of 16 hours a day. Further the profit on a bracelet is ₹300 and the profit on a necklace is ₹100. Formulate this problem as an LPP so as to maximize the profit.***

SOLUTION. Let us suppose

$\qquad$ Total number of necklaces manufactured = x_1

and Total number of bracelets manufactured = x_2

Since the profit on a necklace is ₹100 $\Rightarrow$ Profit on x_1 necklaces = $100x_1$

Similarly, the profit on a bracelet is ₹300 $\Rightarrow$ Profit on x_2 bracelets = $300x_2$

$\therefore \qquad$ Total profit $Z = 100x_1 + 300x_2$ ...(1)

To maximize the profit, we have to maximize Z.

Further, it takes one hour to make one bracelet,

$\therefore$ Total time required to make x_2 bracelets = $1 \cdot x_2 = x_2$ hours

Also, it takes half an hour to make one necklace,

$\therefore$ Total time required to make x_1 necklace = $(1/2)x_1$

So, total time required to make x_1 necklaces and x_2 bracelets = $\dfrac{x_1}{2} + x_2$

Here, total time available per day is 16 hours, so we can write $\dfrac{x_1}{2} + x_2 \le 16$

$\Rightarrow \qquad\qquad x_1 + 2x_2 \le 32$ ...(2)

It is also given that the total number of necklaces and bracelets that the goldsmith can manufacture in a day is atmost 24,

So, $\qquad\qquad x_1 + x_2 \le 24$ ...(3)

Also, the number of necklaces and bracelets manufactured can not be negative,

So, $\qquad x_1 \geq 0, x_2 \geq 0$

Hence, the mathematical model of the LPP is given by

$\qquad$ Maximize $\quad Z = 100x_1 + 300x_2$ $\qquad$ (*objective function*)

$\qquad$ subject to $\quad \left. \begin{aligned} x_1 + 2x_2 &\leq 32 \\ x_1 + x_2 &\leq 24 \end{aligned} \right\}$ $\qquad$ (*constraints*)

$\qquad$ such that $\quad x_1 \geq 0, x_2 \geq 0$ $\qquad$ (*non-negative restrictions*)

EXAMPLE 4. *A dietician decides a certain minimum intake of vitamins A, B and C for a family. The minimum daily needs of vitamins A, B, C are 30, 20, 16 units respectively. For the supply of these, the dietician depends on two types of foods X and Y. The first one gives 7, 5, 2 units per gram of vitamin A, B, C respectively.*

The first food costs ₹2 per gram and second ₹1 per gram. How many grams of each food stuff should the family buy everyday to keep the food expense at a minimum? Formulate a linear programming problem for this problem.

SOLUTION. It is often convenient to construct the following table after understanding the problem carefully.

Foods	Vitamin A	Vitamin B	Vitamin C	Cost per gram in ₹
$X(x_1)$	7	5	2	2
$Y(x_2)$	2	4	8	1
Minimum daily needs	30	20	16	

Clearly, $\qquad Z = 2x_1 + x_2$

Since, we have to keep the expenses minimum,

So we have to minimize $Z = 2x_1 + x_2$ $\qquad$ ...(1)

Also, we observe that

$\qquad 7x_1 + 2x_2 \geq 30$ $\qquad$ ...(2)

$\qquad 5x_1 + 4x_2 \geq 20$ $\qquad$ ...(3)

$\qquad 2x_1 + 8x_2 \geq 16$ $\qquad$ ...(4)

and $\qquad x_1 \geq 0, x_2 \geq 0$

Hence, the mathematical model of the LPP is given below

$\qquad$ Minimize $Z = 2x_1 + x_2$ $\qquad$ (*objective function*)

$\qquad$ subject to $\left. \begin{aligned} 7x_1 + 2x_2 &\geq 30 \\ 5x_1 + 4x_2 &\geq 20 \\ 2x_1 + 8x_2 &\geq 16 \end{aligned} \right\}$ $\qquad$ (*constraints*)

$\qquad$ such that $x_1 \geq 0, x_2 \geq 0$ $\qquad$ (*non-negative restrictions*)

EXAMPLE 5. *A toy company manufactures two type of dolls; an ordinary doll A and a deluxe doll B. Each type of doll B takes twice as long to produce as one of the type A. It is given that the company would have time to make a maximum of 2000 dolls per day if it produces* *only the ordinary version. The supply of plastic is sufficient to produce 1500 dolls per day (both A and B combined). The deluxe version requires a fancy dress of which there are only 600 pieces per day available. If the company makes profit of ₹30 and ₹50 per doll respectively on doll A and B, formulate the problem as an LPP to maximize the profit.* [MEERUT 2007, 09, 11]

SOLUTION.   Let the no. of dolls of type $A = x_1$

and the no. of dolls of type $B = x_2$

If t hours are required to produce one doll of type A then the time required to produce one doll of type B will be $2t$ hours.

Now, since the time available per day is $2000t$ hours

Therefore, $x_1 \cdot t + x_2 \cdot 2t \leq 2000t$

$\Rightarrow$ $x_1 + 2x_2 \leq 2000$ ...(1)

Further, since the supply of plastic is sufficient to produce 1500 dolls per day (both types)

So, $x_1 + x_2 \leq 1500$ ...(2)

Again, the fancy dress for deluxe version of doll B is available for 600 pieces per day only

So, $x_2 \leq 600$ ...(3)

Also, total profit on both the types of dolls per day is

$$Z = 30x_1 + 50x_2 \qquad \text{...(4)}$$

We have to maximize Z.

Also, no. of dolls can't be negative, *i.e.*, $x_1 \geq 0$, $x_2 \geq 0$

Hence, the mathematical model of LPP is given as below:

    Maximize $Z = 30x_1 + 50x_2$ (*objective function*)

    subject to $x_1 + 2x_2 \leq 2000$

                    $x_1 + x_2 \leq 1500$ (*constraints*)

                    $x_2 \leq 600$

    s.t. $x_1 \geq 0, x_2 \geq 0$ (*non-negative restrictions*)

EXAMPLE 6.   *A furniture firm manufactures chairs and tables, each requiring the use of three machines A, B and C. Production of one chair requires 2 hours on machine A, 1 hour on machine B and 1 hour on machine C. Each table requires 1 hour each on machine A and B and 3 hours on machine C. The profit realized by selling one chair is ₹300 while for a table the figure is ₹600. The total time available per week on machine A is 70 hours, on machine B is 40 hours and on machine C is 90 hours. How many chairs and tables should be made per week to maximize the profit? Formulate a mathematical model for the problem.* [MEERUT–2008, 11, 13]

SOLUTION.   Let the number of chairs manufactured by firm $= x_1$

and the number of tables manufactured by firm $= x_2$

It is often convenient to construct the table after understanding the problem carefully.

	Time required per piece on machine			Profit (in ₹)
	A	**B**	**C**	
Chair	2	1	1	300
Table	1	1	3	600
Time available per week in hours	70	40	90	

As per given, for the manufactures of x_1 chairs and x_2 tables the time required on machines A, B, C are $2x_1 + x_2$; $x_1 + x_2$ and $x_1 + 3x_2$ hours respectively.

While the time available on these machines are 70, 40 and 90 hours respectively.

Also, the profit is given by
$$Z = 300x_1 + 600x_2$$

Hence, the mathematical model of LPP is given by

Maximize $\qquad Z = 300x_1 + 600x_2$ $\qquad$ *(objective function)*

subject to $\quad \left. \begin{array}{l} 2x_1 + x_2 \leq 70 \\ x_1 + x_2 \leq 40 \\ x_1 + 3x_2 \leq 90 \end{array} \right]$ $\qquad$ *(constraints)*

s.t. $\qquad x_1 \geq 0; x_2 \geq 0$ $\qquad$ *(non-negative restrictions)*

EXAMPLE 7. *A factory produces two products A and B. Each of the product A requires 2 hours of moulding, 3 hours of grinding and 4 hours for polishing and each of the product B requires 4 hours for moulding, 2 hours for grinding and 2 hours for polishing. The moulding machine can work for 20 hours grinding machine for 24 hours and the polishing machine available for 13 hours. The profit is ₹50 per unit of A and ₹30 per unit of B. Assuming that the factory can sell all that it produces, formulate the problem as an LPP to maximize the profit.*

SOLUTION. Let total units of product A produced $= x_1$

total units of product B produced $= x_2$

After carefully understanding the problem, the given information can be systematically arranged in the form of following table

Product	Moulding time in hours	Grinding time in hours	Polishing time in hours	Profit on one unit (in ₹)
A	2	3	4	50
B	4	2	2	30
Time availability in hours	20	24	13	

Using the above information, the mathematical model of LPP is given as below:

Maximize $\qquad Z = 50x_1 + 30x_2$ $\qquad$ *(objective function)*

subject to $\quad \left. \begin{array}{l} 2x_1 + 4x_2 \leq 20 \\ 3x_1 + 2x_2 \leq 24 \\ 4x_1 + 2x_2 \leq 13 \end{array} \right|$ $\qquad$ *(constraints)*

s.t. $\qquad x_1 \geq 0, x_2 \geq 0$ $\qquad$ *(non-negative restrictions)*

EXAMPLE 8. *A diet is to contain at least 4000 units of carbohydrates 500 units of fats and 300 units of Proteins. Two foods A and B are available. Food A cost ₹20 per unit and food B cost ₹40 per unit. A unit of food A contains 10 units of carbohydrate, 20 units of fat and 15 units of protein. A unit of food B contains 25 units of carbohydrate, 10 units of fat and 20 units of protein. Formulate the problem as a LPP so as to find the minimized costs for a diet that consist of a mixture of these two foods and also meets the minimum nutritions requirements.*

SOLUTION. Let a diet consist of x_1 units of food A and x_2 units of food B. Then after carefully understanding the problem, the given information can be systematically arranged in the following table.

Food	Carbohydrates	Fat	Protein	Cost in ₹
A	10	20	15	20
B	25	10	20	40
Requirement (least)	4000	500	300	

Clearly, in a diet, carbohydrates, fats and proteins are $10x_1 + 25x_2$; $20x_1 + 10x_2$; $15x_1 + 20x_2$ units, while they are required at least 4000, 500 and 300 units respectively. Also, the profit $Z = 20x_1 + 40x_2$, has to be minimized.

Hence, the required mathematical model of LPP is given as follows:

Minimize $Z = 2x_1 + 4x_2$ *(objective function)*

subject to
$$10x_1 + 25x_2 \geq 4000$$
$$20x_1 + 10x_2 \geq 500 \qquad \text{(constraints)}$$
$$15x_1 + 20x_2 \geq 300$$

s.t. $x_1 \geq 0, x_2 \geq 0$ *(non-negative restrictions)*

Exercise-3.1

1. According to the medical experts it is necessary for an adult to consume at least 75 grams of proteins, 85 grams of fat and 300 grams of carbohydrates daily. The following table gives the analysis of the food items readily available in the market with their respective costs.

Food Type	Food value (in gms) per 100 gms			Cost in ₹ (per kg)
	Proteins	Fats	Carbohydrates	
A	18	15	—	30
B	16	4	7	40
C	4	20	2.5	20
D	5	8	40.0	15
Minimum daily requirements	75	85	300	

Formulate a linear programming problem for an optimum diet.

2. A firm manufactures two types of product A and B and sells them at a profit of ₹ 20 on type A and ₹ 30 on type B. Each product is processed on two machines E and F. Type A requires one minute of processing time on E and two minutes on F, type B requires one minute while machine F is available for 10 hours during any working day. Formulate the problem as a linear programming problem.

3. A firm can produce two products A and B during a given period of time. Each of these products requires four different operations viz. Grinding, Turning, Assembling and Testing. The requirement in hours per unit of manufacturing of these products is given below:

Products	A	B
Grinding	1	2
Turning	3	1
Assembling	4	3
Testing	5	4

The available capacities of these operations in hours for the given time period are: 30 for grinding, 60 for turning, 200 for assembling and 200 for testing. Profit on each unit of A is ₹30 and that for each unit of B is ₹20. Formulate the problem as a linear programming model to maximize the profit assuming that the firm can sell all the items that it produces at the prevailing market price.

4. A resourceful home decorator manufactures two types of lamps say A and B. Both lamps go through two technicians, first a cutter, second a finisher. Lamp A requires 2 hours of the cutter's time and 1 hour of the finisher's time. Lamp B requires 1 hour of cutters and 2 hours of finisher's time. The cutter has 104 hours and finishers has 76 hours of time available each month. Profit on the lamp A is ₹ 60 and on the lamp B is ₹ 110. Assuming that each can sell all that he produces, how many of each type of lamps should be manufacture per month to obtain the best returns? Formulate a LPP for this problem.

5. The manager of an oil refinery must decide on the optimal mix of 2 possible blending process of which the input and output per production run are as follows:

Process	Input		Output	
	Crude A	Crude B	Gasoline X	Gasoline Y
1	6	4	6	9
2	5	6	5	5

The maximum amount available of crudes A and B are 500 units and 400 units respectively. Market demand shows that at least 300 units of gasoline X and 260 units of gasoline Y must be produced. The profit per production run from process 1 and 2 are ₹400 and ₹500 respectively. Formulate the LPP for maximizing the profit.

6. A factory produces two products A and B. To manufacture one unit of product A, a machine has to work for $1\frac{1}{2}$ hours and a craftsman has to work for 2 hours. To manufacture one unit of product B, the machine has to work for 3 hours and the craftsman for one hour. In a week, the factory can avail of 80 hours of machine time and 70 hours of craftsman time. The profit on the sale of each unit of A and B is of ₹ 100 and ₹ 80 respectively. If the

manufacture can sell all the items produced; formulate the problem as a LPP.

7. A firm manufactures 3 products A, B and C. The profits are ₹30, ₹20 and ₹40 respectively. The firm has 2 machines and below the required processing time in minutes for each machine on each product.

Machine	Products		
	A	B	C
G	4	3	5
H	2	2	4

Machines G and H have 2000 and 2500 machines minutes respectively. The firm must manufactures 100 A's, 200 B's and 50 C's but not more than 150 A's. Setup a linear programming problem to maximize profit.

8. A manufacturer produces three models I, II and III of a certain product. He used two types of raw material A and B of which 4000 and 6000 units respectively are available. The raw material requirements per unit of the three models are given below:

Raw material	Requirement per unit of given model		
	I	II	III
A	2	3	5
B	4	2	7

The labour time for each unit of model I is twice that of model II and three times that of model III. The entire labour force of the factory can produce the equivalent of 2500 units of model I. A market survey indicates the minimum demand of the three models are 500, 500 and 345 units respectively. Formulate the problem as a LPP in order to determine the number of units of each product which will maximize profit.

9. A complete unit of a certain product consist of four units of component A and three units of component B. The two components (A and B) are manufactured from two different raw materials of which 100 units and 200 units respectively are available. Three departments are engaged in the production process with each department using a different method for manufacturing the components. The following table gives the raw material requirements per production run and resulting units of each component.

The objective is to determine the number of production runs for each department which will maximize the total number of component units of the final product.

Depart-ment	Input per run (units)		Output per run (units)	
	Raw Material		**Components**	
	I	**II**	**A**	**B**
1	7	5	6	4
2	4	8	5	8
3	2	7	7	3

Formulate a linear programming model to the above problem.

10. The owner of the Metro sports wishes to determine how many advertisement to place in the selected three monthly magazines A, B and C. His objective is to advertise in such a way that total exposure to principal buyers of expensive sports good is maximized. Percentage of readers for each magzine are known. Exposure in any particular magazine is the number of advertisements placed multiplied by the number of principal buyers. The following data may be used

Exposure Category	Magazine		
	A	**B**	**C**
Reader (in lakhs)	1	0.6	0.4
Principal buyers	10%	15%	7%
Cost per advertisement (in ₹)	5000	4500	4250

The budgeted amount is atmost ₹ 1 lakh for the advertisement. The owner has already decided that magazine A should have no more than 6 advertisement and that B and C each have at least two advertisement. Formulate the LPP model. [CA, NOV.-1996]

11. A city hospital has the following minimal daily requirement for nurses

Period	Clock Time (24 hr. day)	Minimum no. of nurses required
1	6 AM – 10 AM	2
2	10 AM – 2 PM	7
3	2 PM – 6 PM	15
4	6 PM – 10 PM	8
5	10 PM – 2 AM	20
6	2 AM – 6 AM	6

Nurses report to the hospital at the begining of each period and work for 8 consecutive hours. The hospital wants to determine the minimum number of nurses available for each period. Formulate this LPP.

[MEERUT–2015]

12. A tyre factory produces three types of tyres T_1, T_2, T_3. Three different type of chemicals say C_1, C_2, C_3 are required for production. One T_1 tyre needs 2 units of C_1, 3 units of C_3, one T_2 tyre needs 3 units of C_1, 2 units of C_2 and 2 units of C_3 and one T_3 tyre needs 5 units of C_2 and 4 units of C_3. The factory has only a stock of 20 units of C_1, 25 units of C_2 and 30 units of C_3. Further the profit from the sale of one tyre T_1 is ₹60, one tyre T_2 is ₹100 and one tyre of type T_3 is ₹80. Assuming that the factory can sell all that its produces. Formulate a linear programming to maximize the profit.

ANSWERS

1. Minimize $Z = 30x_1 + 40x_2 + 20x_3 + 15x_4$
   subject to the constraints
   $180x_1 + 160x_2 + 40x_3 + 50x_4 \geq 75$
   $150x_1 + 40x_2 + 200x_3 + 80x_4 \geq 85$
   $70 x_2 + 25x_3 + 400x_4 \geq 300$
   and $x_1 \geq 0; x_2 \geq 0; x_3 \geq 0; x_4 \geq 0$

2. Maximize $Z = 20x_1 + 30x_2$
   subject to the constraints
   $x_1 + x_2 \leq 400$
   $2x_1 + x_2 \leq 600$
   and $x_1 \geq 0; x_2 \geq 0$

3. Maximize $Z = 30x_1 + 20x_2$
   subject to the constraints
   $x_1 + 2x_2 \leq 30$
   $3x_1 + x_2 \leq 60$
   $4x_1 + 3x_2 \leq 200$
   $5x_1 + 4x_2 \leq 200$
   and $x_1 \geq 0; x_2 \geq 0$

4. Maximize $Z = 60x_1 + 110x_2$
   subject to the constraints
   $2x_1 + x_2 \leq 104$
   $x_1 + 2x_2 \leq 76$
   and $x_1 \geq 0; x_2 \geq 0$

5. Maximize $Z = 400x_1 + 50x_2$
   subject to the constraints
   $6x_1 + 5x_2 \leq 500$; $4x_1 + 6x_2 \leq 400$
   $6x_1 + 5x_2 \geq 300$; $9x_1 + 5x_2 \geq 260$
   and $x_1 \geq 0$; $x_2 \geq 0$

6. Maximize $Z = 100x_1 + 80x_2$
   subject to the constraints
   $1.5x_1 + 3x_2 \leq 80$
   $2x_1 + x_2 \leq 70$
   and $x_1 \geq 0$; $x_2 \geq 0$

7. Maximize $Z = 30x_1 + 20x_2 + 40x_3$
   subject to the constraints
   $4x_1 + 3x_2 + 5x_3 \leq 2000$
   $2x_1 + 2x_2 + 4x_3 \leq 2500$
   $100 \leq x_1 \leq 150$; $0 \leq x_2 \leq 200$ and $0 \leq x_3 \leq 50$

8. Maximize $Z = 60x_1 + 40x_2 + 100x_3$
   subject to the constraints
   $2x_1 + 3x_2 + 5x_3 \leq 4000$
   $4x_1 + 2x_2 + 7x_3 \leq 6000$
   $6x_1 + 3x_2 + 2x_3 \leq 15000$
   $2x_1 = 3x_2$, $5x_2 = 2x_3$
   and $x_1 \geq 500$; $x_2 \geq 500$; $x_3 \geq 375$

9. Maximize $Z = V$, where $V = \min\left[\dfrac{1}{4}(6x_1 + 5x_2 + 7x_3), \dfrac{1}{3}(4x_1 + 8x_2 + 3x_3)\right]$

   subject to the constraints
   $6x_1 + 5x_2 + 7x_3 - 4V \geq 0$; $4x_1 + 8x_2 + 3x_3 - 3V \geq 0$
   $7x_1 + 4x_2 + 2x_3 \leq 100$; $5x_1 + 8x_2 + 7x_3 \leq 200$
   and $x_1, x_2, x_3, V \geq 0$

10. Maximize $Z = 10000x_1 + 9000x_2 + 2800x_3$
    subject to the constraints
    $5000x_1 + 4500x_2 + 4250x_3 \leq 100000$
    $x_1 \leq 6$, $x_2 \geq 2$, $x_3 \geq 2$
    and $x_1, x_2, x_3 \geq 0$

11. Minimize $Z = x_1 + x_2 + x_3 + x_4 + x_5 + x_6$
    subject to the constraints
    $x_1 + x_2 \geq 7$; $x_2 + x_3 \geq 15$; $x_3 + x_4 \geq 8$;
    $x_4 + x_5 \geq 20$; $x_5 + x_6 \geq 6$; $x_6 + x_1 \geq 2$
    and $x_i \geq 0 \; \forall \; i = 1, 2, ..., 6$

12. Maximize $Z = 60x_1 + 100x_2 + 80x_3$
    subject to the constraints
    $2x_1 + 3x_2 \leq 20$
    $2x_1 + 5x_3 \leq 25$
    $3x_1 + 2x_2 + 4x_3 \leq 30$
    and $x_1 \geq 0$, $x_2 \geq 0$, $x_3 \geq 0$

3.12 SOLUTION OF LINEAR PROGRAMMING PROBLEMS

In this section we shall discuss solution method of LPP which are given below:

(1) ISO-Profit or ISO-Cost method

(2) Corner point or Extreme point solution method

Before discussing these method, let us define the following terms:

3.12.1 SOLUTION

A set of values of decision variables, x_i which satisfy all the constraints of a linear programming problem is called 'solution' of that problem.

3.12.2 FEASIBLE SOLUTION

Any solution of LPP which also satisfy the non-negative restrictions is called feasible solution, i.e., the set of values of decision variables which satisfy all the constraints and non-negative restriction of an LPP is called feasible solution.

3.12.3 INFEASIBLE SOLUTION

Any solution of an LPP which does not satisfy the non-negative restrictions, i.e., if all the values of decision variables in any solution are negative, the solution is called infeasible solution.

3.12.4 BASIC SOLUTION

For a set of m simultaneous equations in n-variables ($n > m$), a solution obtained by setting ($n - m$) variables equal to zero and solving for remaining m equations in m variables is called a basic solution.

☞ REMARK

- The $(n - m)$ variables whose value did not appear in the above solution are called non-basic variables and the remaining m variables are known as basic variables.

3.12.5 BASIC FEASIBLE SOLUTION

A feasible solution to a general LPP which is also basic solution, *i.e.*, all basic variables assume non-negative values is called basic feasible solution.

Generally, basic feasible solutions are of two types:

   (i) **Degenerate Basic Feasible solution:** A basic solution to the system of equations is called degenerate if one or more of the basic variables become equal to zero.

  (ii) **Non-degenerate Basic Feasible solution:** A basic solution is called non-degenerate if values of m basic variables are non-zero and positive.

3.12.6 OPTIMAL BASIC FEASIBLE SOLUTION

Any basic feasible solution which optimize (maximize or minimize) the objective function of a general LPP is called optimal basic feasible solution.

3.12.7 UNBOUNDED SOLUTION

A solution which can increase or decrease the value of the objective function of an LPP indefinitely is said to be an unbounded solution.

3.12.8 FEASIBLE REGION

The common region formed by all the constraints and non-negative restrictions of an LPP is called feasible region.

3.12.9 INFEASIBLE REGION

The region common to all constraints in which all the decision variables are negative is called infeasible region.

3.12.10 CONVEX REGION

If the line segment joining any two arbitrary points of the region lies entirely within the region, then this region is said to be convex region.

3.13 GRAPHICAL SOLUTION METHODS OF AN LPP

To obtain the solution of an LPP by graphical method, we shall recall the following important results:

  (1) An optimum solution of an LPP, if it exists, occurs at one of the extreme points, i.e., corner points of the convex polygon of the set of feasible solutions (Fundamental extreme point theorem)

  (2) If the optimum solution occurs at more than one extreme point the value of the objective function will be the same for all Convex Combinations of these extreme points.

  (3) The collection of all feasible solutions to an LPP constitutes a convex set whose extreme points correspond to the Basic feasible solution.

  (4) There are a finite number of basic feasible solutions within the feasible solution space.

  (5) If the convex set of all feasible solutions of the system of simultaneous equations $AX = B, X \geq 0$ is a convex polyhedron, then at least one of the extreme points gives an optimal solution.

  (6) Each corner point of the feasible region falls at the intersection of two constraints equalities.

  (7) The extreme point of the convex set give the basic feasible solution to an LPP.

Working Procedure

STEP 1. **Formulate the given problem:** If the given problem is not in mathematical form, then formulate the problem in terms of a series of mathematical constraints and an objective function.

STEP 2. **Plotting the constraints:** Each inequality in the constraints equation be written as inequality. Give any arbitrary value to one variable and get the value of other variable by solving the equation. Similarly, give another arbitrary value to the variable and find the corresponding value of the other variable. Plot these two set of values. Connect these points by a straight line. Do the same exercise for each constraints and get as many straight line as there are equations. Here each straight line representing one constraint.

STEP 3. **Identify the Feasible Region:** We have to identify the area which satisfy all the constraints simultaneously. For "greater than (>)" constraints the feasible region will be the area which lie above the constraints lines. For "less than (<)" constraints this area is generally the region below these lines. On $\geq$ or $\leq$ constraints, the feasible region includes the points on the constraints line also.

STEP 4. Select one of the following two techniques of solutions.

3.13.1 CORNER POINT METHOD

Since, we know that an optimal solution of an LPP always lie at one of the corner points of the feasible solution space. Thus, first we determined the coordinates of all corner points of the feasible region and then compute the value of the objective function at these points and then compared.

The steps of the solutions will be clear by the following working procedure.

Working Procedure

STEP 1. Identify each of the corner point of the feasible region either by inspection or method of simultaneous equations.

STEP 2. Determine the coordinate of each extreme point of the feasible region.

STEP 3. Compute the value of objective function at each extreme point and then compare.

STEP 4. Identify the extreme point that gives optimal value of the objective function.

The point where the objective function attains its optimum value, gives the optimal value of the LPP.

☞ REMARK

- If two vertices of the Convex polygon gives the same optimum value of the objective function, then all points on the line segment joining these two vertices will give the optimum value of the objective function. In this case the LPP is said to have infinite number of optimum solutions

> To determine which side of a constraint is in the feasible region, examine whether the origin (0, 0) satisfies the constraints. If it does, then all points on and below the constraint towards the origin are feasible points. If it does not, then all points on and above the constraint away from the origin are feasible points.

3.13.2 ISO-PROFIT OR ISO-COST METHOD

In this method, the optimal solution is found by using the slope of the objective function. An iso-profit line is a collection of points which designate with same value of the objective function. By assigning different values to Z, we get different profit lines. The steps of the solution by this method are as follows:

Working Procedure

STEP 1. Identify the feasible region and its extreme points

STEP 2. Assign a constant value say Z_1 to the objective function and draw the corresponding line of objective function (called iso-profit line).

STEP 3. Assign another constant value say Z_2 to the objective function and draw the corresponding line of the objective function.

STEP 4. If $Z_1 > Z_2$ and objective function Z is to be maximized then we move the line corresponding to Z_1 to the line corresponding to Z_2 parallel to itself as farthest point within the feasible region is touched by this line and any further displacement of this line takes it out of the feasible region. The coordinates of the farthest point so obtained will give the maximum value of Z.

But if we have to minimize the objective function Z, we have the line corresponding to Z_2 to the line corresponding to Z_1 and find the nearest point in the same way as we had find the farthest point. Then, coordinate of the nearest point will give minimum value of the objective function.

Since every point in the feasible region satisfies all the constraints of an LPP and there are infinitely many points so it is not easy to anyone to find a point that gives a maximum or minimum value of the objective function. To handle this situation, following results seems very useful.

Let R be the feasible region for an LPP and let $Z = ax + by$ be the objective function

(i) When Z has an optimal value, where x and y are subject to constraints describe by the line or inequation, this optimal value must occur at a corner point of the feasible region.

(ii) If R is bounded, then the objective function Z has both a maximum or minimum value at that point on R.

☛ REMARK

- The iso profit line method is also applicable if the problem has more constraints.

Solved Examples

TYPE 1. BASED ON FINDING THE SOLUTION SET OF SIMULTANEOUS LINEAR INEQUATIONS

EXAMPLE 1. ***Draw the graph of the solution set of the inequations $2x + 3y \geq 6$, $x + 4y \leq 4$, $x \geq 0$ and $y \geq 0$.***

SOLUTION. The corresponding equations of given inequations are:

$$2x + 3y = 6, x + 4y = 4, x = 0, y = 0$$

Region represented by $2x + 3y \geq 6$. The line $2x + 3y \geq 6$, meets x-axis at $A(3, 0)$ (put $y = 0$) and y-axis at $B(0, 2)$ (put $x = 0$ in it). We find that $(0, 0)$ does not satisfy the inequation $2x + 3y \geq 6$, the portion not containing the origin represented by the given inequation represents the solution set.

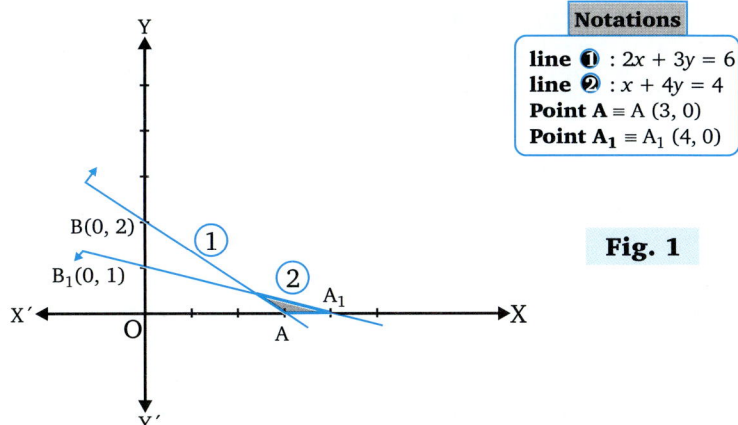

Notations

line ❶ : $2x + 3y = 6$
line ❷ : $x + 4y = 4$
Point A ≡ A (3, 0)
Point A_1 ≡ A_1 (4, 0)

Fig. 1

Region represented by $x + 4y \leq 4$. The line $x + 4y = 4$ meets x-axis at $A_1(4, 0)$ (put $y = 0$ in it) and y-axis at $B_1(0, 1)$ (put $x = 0$ in it). We find that $(0, 0)$ satisfy the inequation $x + 4y \leq 4$. So, the portion containing the origin represented by the given inequation represents the solution set.

Region represented by $x \geq 0, y \geq 0$, represent the first quadrant.

Hence, the shaded region given in the figure represents the solution set of the given linear inequations.

EXAMPLE 2. **Draw the graph of the solution set of the linear inequations.**

$$x + y \leq 5, \ 4x + y \geq 4, \ x + 5y \geq 5, \ x \leq 4, \ y \leq 3.$$

SOLUTION. The corresponding linear equations of the given linear inequations are $x + y = 5$, $4x + y = 4$, $x + 5y = 5$, $x = 4$, $y = 3$.

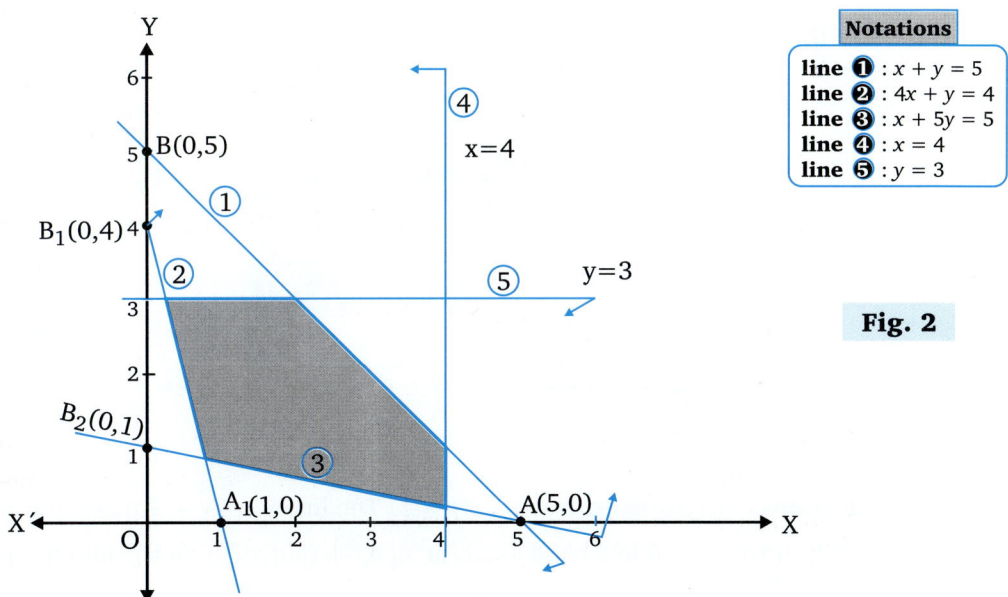

Notations

line ❶ : $x + y = 5$
line ❷ : $4x + y = 4$
line ❸ : $x + 5y = 5$
line ❹ : $x = 4$
line ❺ : $y = 3$

Fig. 2

Region represented by $x + y \leq 5$**.** The line $x + y = 5$ meets the x-axis at $A(5, 0)$ (put $y = 0$ in it) and y-axis at $B(0, 5)$ (put $x = 0$ in it). Join these points by a thick line. Clearly, $(0, 0)$ satisfy the inequation $x + y \leq 5$, because $0 \leq 5$.

So, portion containing the oritgin represents the solution set of the inequation $x + y \leq 5$.

Region represented by $4x + y \geq 4$**.** The line $4x + y = 4$ meets the x-axis at $A_1(1, 0)$ (put $y = 0$ in it) and y-axis at $B_1(0, 4)$) (put $x = 0$ in it). Join these points by a thick line. Clearly, $(0, 0)$ does not satisfy the inequation $4x + y \geq 4$. So, the portion not containing the origin is represented by the given inequation.

Region represented by $x + 5y \geq 5$**.** The line $x + 5y = 5$ meets the x-axis at $A_2(5, 0)$ (put $y = 0$ in it) and y-axis at $B_2(0, 1)$. Join these points by a thick line. Clearly, $(0, 0)$ does not satisfy the linear inequation $x + 5y \geq 5$. So, portion not containing the origin is represented by the given inequation.

Region represented by $x \leq 4$**.** Clearly, $x = 4$ is a line parallel to y-axis at a distance 4 from it to its right. Since, $(0, 0)$ satisfies the inequation $x \leq 4$. So, portion lying to the left of $x = 4$ is the shaded region.

Region represented by $y \leq 3$**.** Clearly $y = 3$ is a line parallel to x-axis at a distance 3 from it. Since, $(0, 0)$ satisfies the inequation $y \leq 3$. So, portion lying below $y = 3$ is the shaded region.

Hence, the shaded region given in figure represents the solution set of the given linear inequations.

EXAMPLE 3. ***Solve graphically the following system of inequations:***

$$x + 2y \geq 3,\ 3x + 4y \geq 12,\ x \geq 0,\ y \geq 1$$

SOLUTION. The corresponding linear equations of the given linear inequations are:

$$x + 2y = 3,\ 3x + 4y = 12,\ x = 0,\ y = 1$$

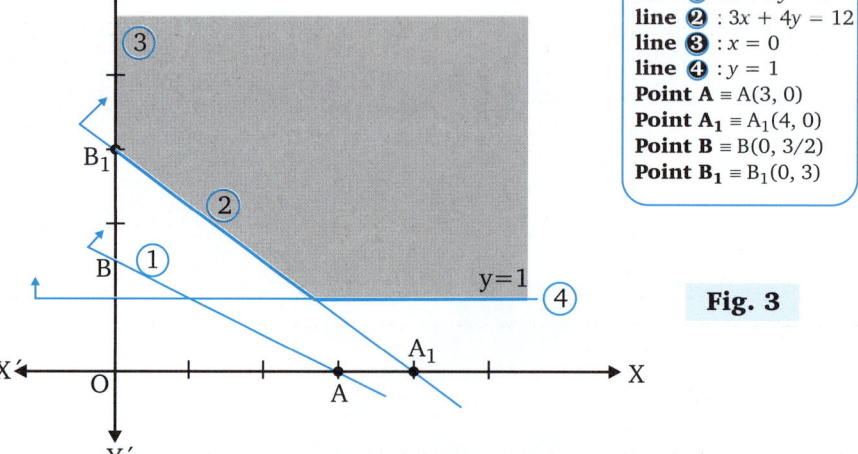

Notations
line ❶ : $x + 2y = 3$
line ❷ : $3x + 4y = 12$
line ❸ : $x = 0$
line ❹ : $y = 1$
Point A $\equiv A(3, 0)$
Point A₁ $\equiv A_1(4, 0)$
Point B $\equiv B(0, 3/2)$
Point B₁ $\equiv B_1(0, 3)$

Fig. 3

Region represented by $x + 2y \geq 3$**.** The line $x + 2y = 3$ meets the x-axis at $A(3, 0)$ (put $y = 0$ in it) and y-axis at $B\left(0, \dfrac{3}{2}\right)$ (put $x = 0$ in it). Join these points by a thick line. Clearly $(0, 0)$ does not satisfy the inequation $x + 2y \geq 3$. So, portion not containing the origin is represented by the given inequation.

Region represented by the $3x + 4y \geq 12$. The line $3x + 4y = 12$ meets the x-axis at $A_1(4, 0)$ (put $y = 0$ it it) and y-axis at $B_1(0, 3)$ (put $x = 0$ in it). Join these points by a thick line. Clearly, $(0, 0)$ does not satisfy the given inequation. So, porion not containing the origin is represented by the inequation $3x + 4y \geq 12$.

Region represented by $x \geq 0$. Clearly, $x \geq 0$ represents the right y-axis including y-axis.

Region represented by $y \geq 1$. Clearly, $y = 1$ is a line parallel to x-axis at a distance of 1 unit from it. Since, $(0, 0)$ does not satisfy $y \geq 1$. So, portion not containing the origin is represented by the given inequation.

Hence, the shaded region represents the solution set of the given inequations.

EXAMPLE 4. ***Solve the following system of inequations graphically.***
$$2x + y - 3 \geq 0$$
$$x - 2y + 1 \leq 0 \qquad\qquad (x \geq 0, y \geq 0).$$

SOLUTION. The corresponding linear equations of the given linear inequations are:
$$2x + y - 3 = 0, x - 2y + 1 = 0, x = 0, y = 0.$$

Region represented by $2x + y - 3 \geq 0$. The line $2x + y - 3 = 0$ meets the x-axis at $A\left(\dfrac{3}{2}, 0\right)$ (put $y = 0$ in it) and y-axis at $B(0, 3)$ (put $x = 0$ in it). Join these points by a thick line. Clearly, $(0, 0)$ does not satisfy the inequation $2x + y - 3 \geq 0$. So, portion not containing the origin is represented by the given inequation $2x + y - 3 \geq 0$.

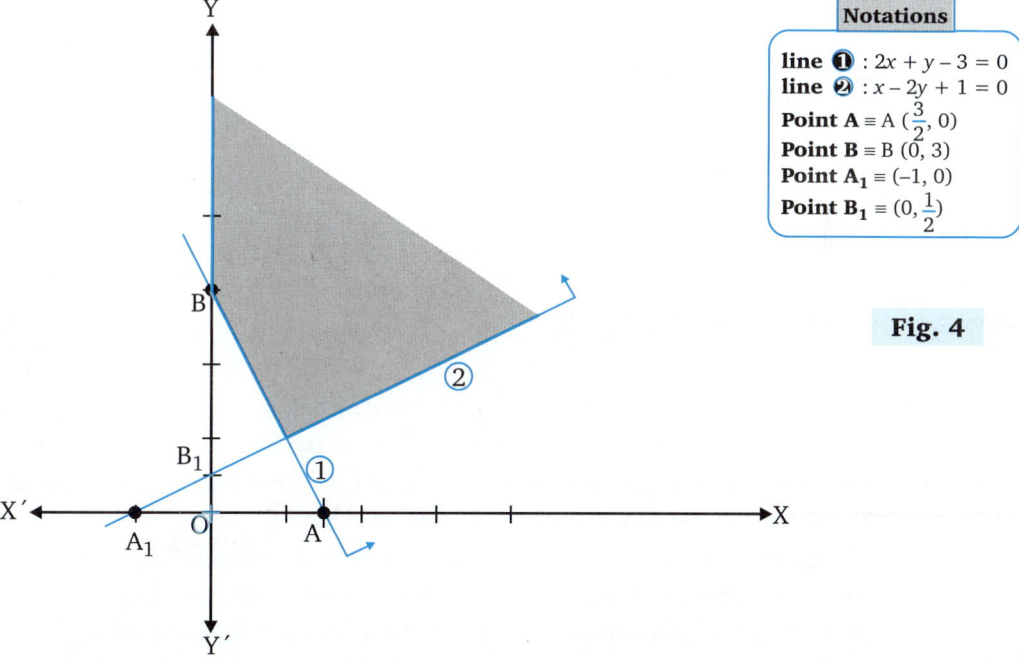

Notations

line ❶ : $2x + y - 3 = 0$
line ❷ : $x - 2y + 1 = 0$
Point A ≡ A $(\dfrac{3}{2}, 0)$
Point B ≡ B $(0, 3)$
Point A_1 ≡ $(-1, 0)$
Point B_1 ≡ $(0, \dfrac{1}{2})$

Fig. 4

Region represented by $x - 2y + 1 \leq 0$. The line $x - 2y + 1 = 0$ meets the x-axis at $A_1(-1, 0)$ and y-axis at $B_1\left(0, \dfrac{1}{2}\right)$. Join these points by a thick line. Clearly, $(0, 0)$ does not satisfy the inequation $x - 2y + 1 \leq 0$. So, portion not containing the origin is represented by the given inequation $x - 2y + 1 \leq 0$.

Region represented by $x \geq 0$, $y \geq 0$. Clearly, $x \geq 0$, $y \geq 0$ represents the first quadrant.

Hence, the shaded region given in figure represents the solution set of the given linear inequations.

TYPE 2. BASED ON FINDING THE LINEAR INEQUATIONS, WHEN THEIR SOLUTION SET IS GIVEN

EXAMPLE 1. **Find the linear inequations for which the solution set is the shaded region given in figure.**

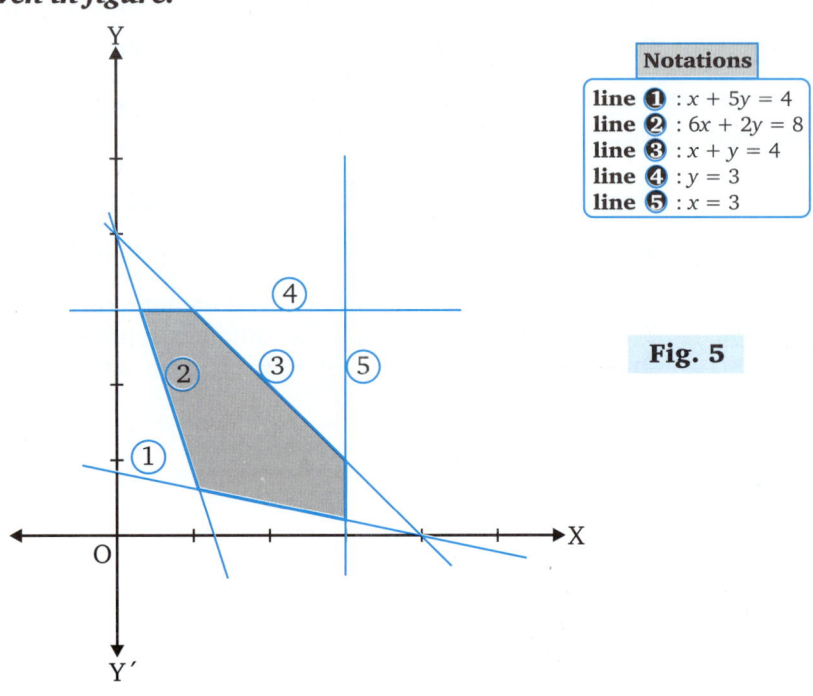

Notations

line ❶ : $x + 5y = 4$
line ❷ : $6x + 2y = 8$
line ❸ : $x + y = 4$
line ❹ : $y = 3$
line ❺ : $x = 3$

Fig. 5

SOLUTION. Consider the linear equation $x + 5y = 4$. We observe that the shaded region and the origin are on opposite side of the line $x + 5y = 4$ and $(0, 0)$ does not satisfy the linear inequation $x + 5y \geq 4$. So, we must have $x + 5y \geq 4$.

Now, consider the linear equation $6x + 2y = 8$. We observe that the shaded region and the origin are on opposite side of the line $6x + 2y = 8$ and $(0, 0)$ does not satisfy the linear inequation $6x + 2y \geq 8$. So, we must have $6x + 2y \geq 8$.

Finally, consider the line $x + y = 4$. We observe that the shaded region and the origin are on the side of the line $x + y = 4$ and $(0, 0)$ satisfy the inequation $x + y \leq 4$. So, the third linear inequation is $x + y \leq 4$. We also notice that shaded region is on the same side of the origin so, $y \leq 3$ and also $x \leq 3$.

Thus, the linear inequations corresponding to the given solution set are:

$$x + 5y \geq 4, \; 6x + 2y \geq 8, \; x + y \leq 4, \; x \leq 3, \; y \leq 3, \; x \geq 0, \; y \geq 0$$

EXAMPLE 2. **Find the linear inequations, for which the shaded region in the figure** *is the solution set.*

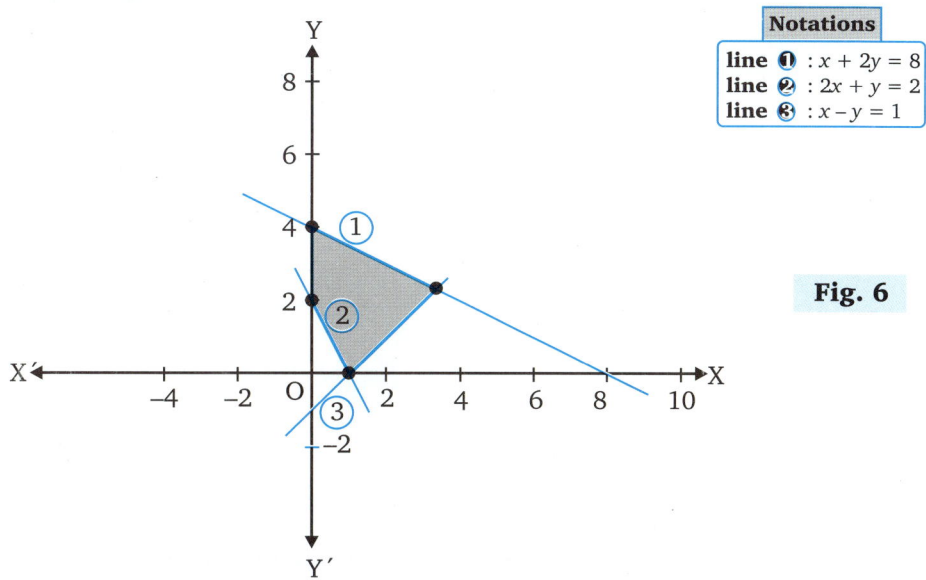

Notations

line ❶ : $x + 2y = 8$
line ❷ : $2x + y = 2$
line ❸ : $x - y = 1$

Fig. 6

SOLUTION. Consider the linear equation $x + 2y = 8$. We observe that the shaded region and the origin are on opposite side of the line $x + 2y = 8$ and $(0, 0)$ satisfy the linear inequation $x + 2y \leq 8$. So, we must have one of the linear inequations as $x + 2y \leq 8$. Now, consider the linear equation $2x + y = 2$. We find that shaded region and the origin are on opposite side of the line $2x + y = 2$ and $(0, 0)$ does not satisfy the inequation $2x + y \geq 2$. So, the second inequation is $2x + y \geq 2$. Finally, consider the line $x - y = 1$. We observe that the shaded region and the origin are on the side of the line $x - y = 1$ and $(0, 0)$ satisfy the inequation $x - y \leq 1$. So, the third inequation is $x - y \leq 1$. We also observe that the shaded region is above x-axis and is on the right side of y-axis. So, we have $x \leq 0, y > 0$.

Thus, the linear inequations are

$$x + 2y \leq 8, 2x + y \geq 2, x - y \leq 1, x \leq 0, y > 0$$

Type 3. Based on the Corner Point Method: Maximization Problems

EXAMPLE I. **Solve graphically, the following LPP**

Maximize $Z = 15x_1 + 10x_2$

subject to the constraints

$$4x_1 + 6x_2 \leq 360$$
$$3x_1 + 0x_2 \leq 180$$
$$0x_1 + 5x_2 \leq 200$$

and $\quad x_1, x_2 \geq 0$

SOLUTION. Converting the given inequalities into equations. Now find any two points that satisfy the equation and then drawing the straight line through these two points. When $x_1 = 0$ we get $6x_2 = 360 \Rightarrow x_2 = 60$. Similarly, when $x_2 = 0$, $4x_1 = 360$, i.e., $x_1 = 90$. These two points are then connected by the straight line. It is also clear that any point below the line satisfy $4x_1 + 6x_2 \leq 360$.

Similarly, the constraints $3x_1 \leq 180$ and $5x_2 \leq 200$ are plotted on the graph.

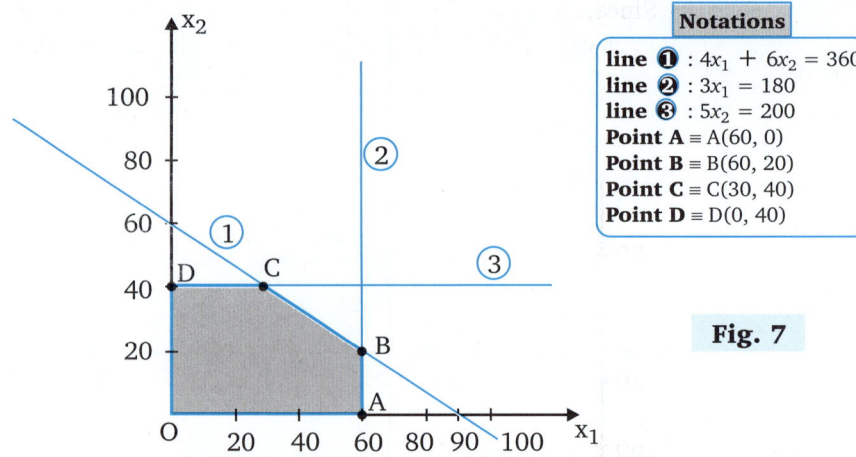

Notations

line ❶ : $4x_1 + 6x_2 = 360$
line ❷ : $3x_1 = 180$
line ❸ : $5x_2 = 200$
Point A ≡ A(60, 0)
Point B ≡ B(60, 20)
Point C ≡ C(30, 40)
Point D ≡ D(0, 40)

Fig. 7

Clearly the above shaded area $OABCDO$ is the feasible region. We know that the optimal value of the objective function occurs at one of the extreme points of the feasible region. So, we have to evaluate the value of the objective function at each extreme point of the feasible region as shown below:

Extreme point	Coordinates (x_1, x_2)	Value of the objective function $Z = 15x_1 + 10x_2$
O	(0, 0)	$15 \times 0 + 10 \times 0 = 0$
A	(60, 0)	$15 \times 60 + 10 \times 0 = 900$
B	(60, 20)	$15 \times 60 + 10 \times 20 = 1100$
C	(30, 40)	$15 \times 30 + 10 \times 40 = 850$
D	(0, 40)	$15 \times 0 + 10 \times 40 = 400$

It is clear from the above table that Z is maximum at $B(60, 20)$ and its maximum value is 1100. Hence, optimal solution to the given LPP is $x_1 = 60, x_2 = 20$ and Max $Z = 1100$.

EXAMPLE 2. *Solve the following LPP by graphical method*

Maximize Z = $4x_1 + x_2$

subject to the constraints

$x_1 + x_2 \leq 50$
$3x_1 + x_2 \leq 90$

and $x_1, x_2 \geq 0$

SOLUTION. STEP 1. First consider the constraints as equations, *i.e.,*

$x_1 + x_2 = 50$
$3x_1 + x_2 = 90$

STEP 2. In order to draw $x_1 + x_2 = 50$, put $x_2 = 0$ we get $x_1 = 50$ and by putting $x_1 = 0$, we get $x_2 = 50$

⇒ The line $x_1 + x_2 = 50$ passes through the point (50, 0) and (0, 50).

Now, plot the points (50, 0) and (0, 50) on coordinate axes and joining them by a straight line.

Similarly, we draw other line $3x_1 + x_2 = 90$.

STEP 3. Since, $x_1 + x_2 \leq 50$ and $3x_1 + x_2 \leq 90$ therefore, the region below the line $x_1 + x_2 = 50$ in the positive quadrant is the possible region and also the region below the line $3x_1 + x_2 = 90$ in the positive quadrant is the possible region. So the common region denoted by the shaded region *OABC* is the feasible region.

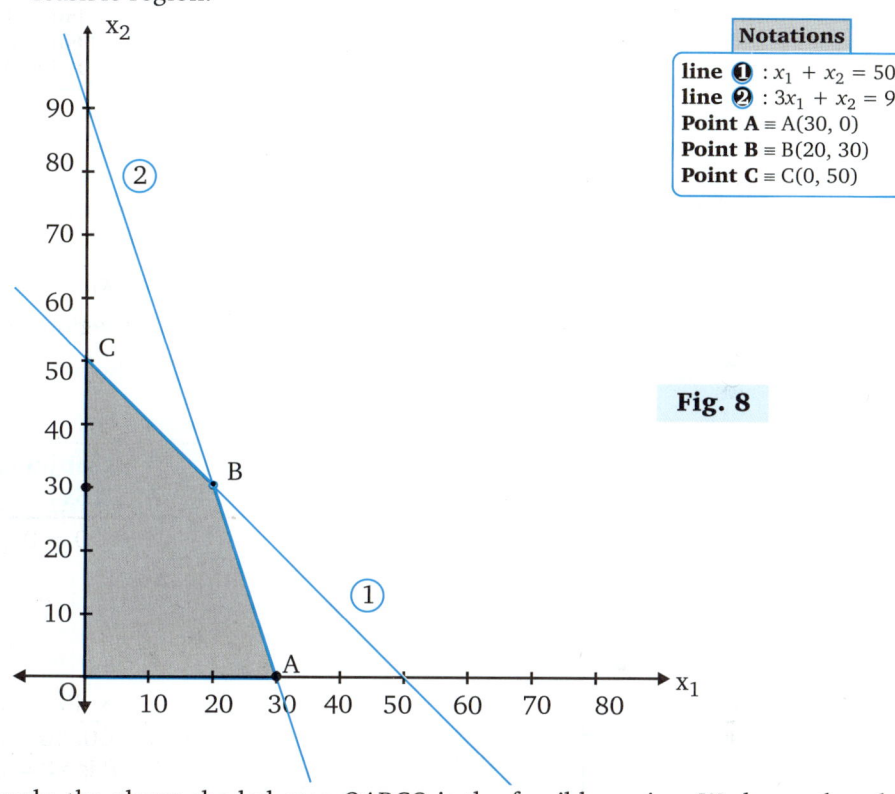

Notations
line ❶ : $x_1 + x_2 = 50$
line ❷ : $3x_1 + x_2 = 90$
Point A $\equiv$ A$(30, 0)$
Point B $\equiv$ B$(20, 30)$
Point C $\equiv$ C$(0, 50)$

Fig. 8

Clearly, the above shaded area *OABCO* is the feasible region. We know that the optimal value of the objective function occurs at one of the extreme point of the feasible region. So, we have to evaluate the value of the objective function at each extreme point of the feasible region as shown below:

Extreme point	Coordinates (x_1, x_2)	Value of the objective function $Z = 4x_1 + x_2$
O	(0, 0)	$4 \times 0 + 0 = 0$
A	(30, 0)	$4 \times 30 + 0 = 120$
B	(20, 30)	$4 \times 20 + 30 = 110$
C	(0, 50)	$4 \times 0 + 50 = 50$

It is clear from the above table that Z is maximum at $A(30, 0)$ and its maximum value is 120.

Hence, the optimal solution is given by $x_1 = 30$, $x_2 = 0$ and optimal value of Z is 120.

EXAMPLE 3. *Solve the following LPP by corner point method.*
$$\textbf{\textit{Maximize Z}} = \textbf{\textit{60x}}_1 + \textbf{\textit{40x}}_2$$

subject to the constraints

$$x_1 + 2x_2 \leq 12$$

$$2x_1 + x_2 \leq 12$$

$$x_1 + \frac{5}{4}x_2 \geq 5$$

and $\quad x_1, x_2 \geq 0$

SOLUTION. STEP 1. First consider the constraints as equations, *i.e.,*

$$x_1 + 2x_2 = 12$$

$$2x_1 + x_2 = 12$$

$$x_1 + \frac{5}{4}x_2 = 5$$

STEP 2. Clearly, the line $x_1 + 2x_2 = 12$ passes through two points $(12, 0)$ and $(0, 6)$. Plot these points on the coordinate axes and joining them by a straight line. Similarly, the line $2x_1 + x_2 = 12$ passes through two points $(6, 0)$ and $(0, 12)$. Plot these points on the coordinate axes and join them by a straight line. Also, the line $x_1 + \frac{5}{4}x_2 = 5$ passes through the point $(5, 0)$ and $(0, 4)$. Plot this line also.

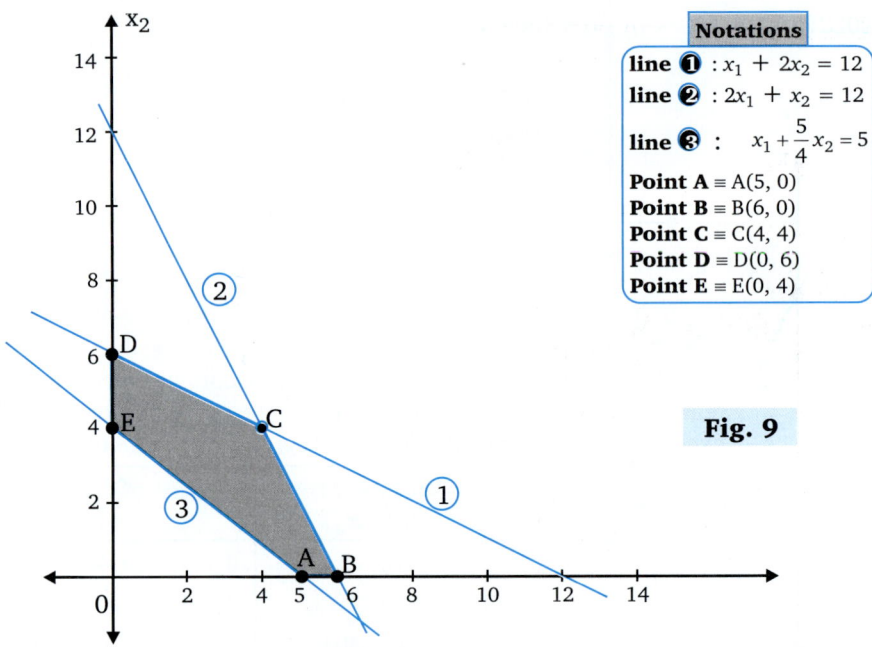

Notations

line ❶ : $x_1 + 2x_2 = 12$
line ❷ : $2x_1 + x_2 = 12$
line ❸ : $x_1 + \frac{5}{4}x_2 = 5$

Point A $\equiv$ A(5, 0)
Point B $\equiv$ B(6, 0)
Point C $\equiv$ C(4, 4)
Point D $\equiv$ D(0, 6)
Point E $\equiv$ E(0, 4)

Fig. 9

Clearly, the above shaded area *ABCDEA* is the feasible region and optimal value of the objective function at one of the extreme points of the feasible region. So, we have to evaluate the value of the objective function at each extreme point of the feasible region as shown below:

Extreme point	Coordinates (x_1, x_2)	Value of the objective function $Z = 60x_1 + 40x_2$
A	(5, 0)	$60 \times 5 + 40 \times 0 = 300$
B	(6, 0)	$60 \times 6 + 40 \times 0 = 360$
C	(4, 4)	$60 \times 4 + 40 \times 4 = 400$
D	(0, 6)	$60 \times 0 + 40 \times 6 = 240$
E	(0, 4)	$60 \times 0 + 40 \times 4 = 160$

It is clear from the above table that Z is maximum at $C(4, 4)$. Hence, the optimal solution is given by $x_1 = 4$, $x_2 = 4$ and optimal value of Z is 400.

TYPE 4. BASED ON CORNER POINT METHOD: MINIMIZATION PROBLEM:

EXAMPLE 4. *Solve the following LPP by corner point method:*

$$\text{Minimize } Z = 4x_1 + 3x_2$$

subject to the constraints

$$200x_1 + 100x_2 \geq 4000$$
$$x_1 + 2x_2 \geq 50$$
$$40x_1 + 40x_2 \geq 1400$$

and $\qquad x_1, x_2 \geq 0$

SOLUTION. STEP 1. Consider the constraints as equations, *i.e.*,

$$200x_1 + 100x_2 = 4000$$
$$x_1 + 2x_2 = 50$$
$$40x_1 + 40x_2 = 1400$$

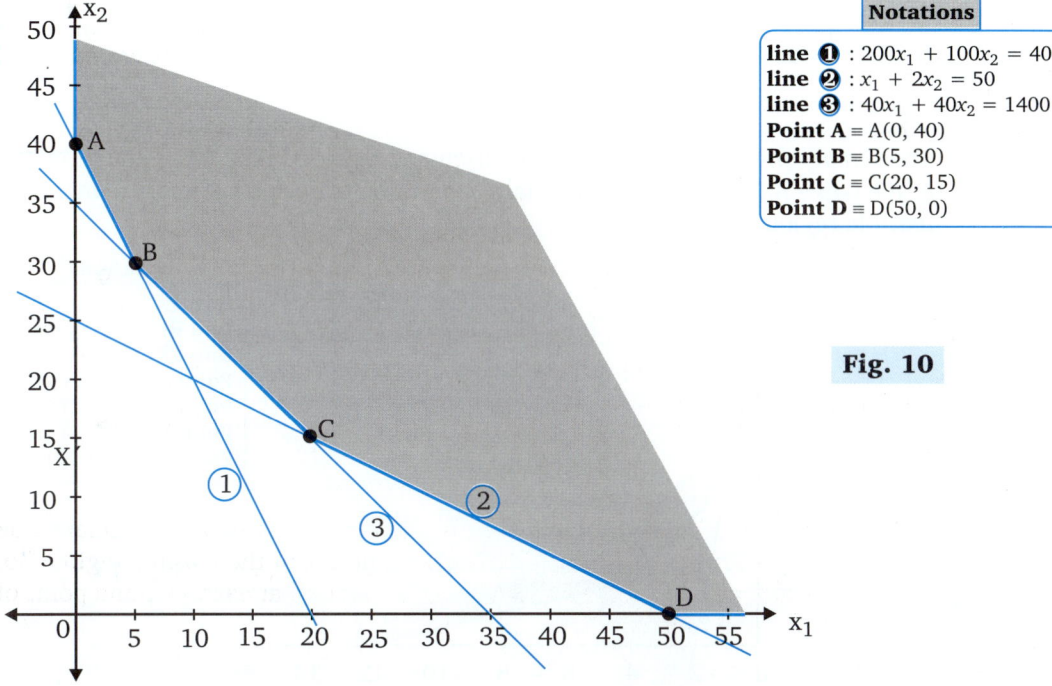

Notations

line ❶ : $200x_1 + 100x_2 = 4000$
line ❷ : $x_1 + 2x_2 = 50$
line ❸ : $40x_1 + 40x_2 = 1400$
Point A ≡ A(0, 40)
Point B ≡ B(5, 30)
Point C ≡ C(20, 15)
Point D ≡ D(50, 0)

Fig. 10

STEP 2. The line $200x_1 + 100x_2 = 4000$ passes through the point $(20, 0)$ and $(0, 40)$ and the line $x_1 + 2x_2 = 50$ passes through the point $(50, 0)$ and $(0, 25)$ while the line $40x_1 + 40x_2 = 1400$ passes through the points $(35, 0)$ and $(0, 35)$. Plot these points on the coordinate axes and join them by a straight line as shown in the adjoining graph.

Clearly, the above shaded area in the positive quadrant is the feasible region. Now, we have to evaluate the value of the objective function at each extreme point of the feasible region.

Extreme point	Coordinates (x_1, x_2)	Value of the objective function $Z = 4x_1 + 3x_2$
A	$(0, 40)$	$4 \times 0 + 3 \times 40 = 120$
B	$(5, 30)$	$4 \times 5 + 3 \times 30 = 110$
C	$(20, 15)$	$4 \times 20 + 3 \times 15 = 125$
D	$(50, 0)$	$4 \times 50 + 3 \times 0 = 200$

Clearly the point B give the minimum value of the objective function. Hence, the optimal solution of the given LPP is $x_1 = 5$ and $x_2 = 30$ and optimal value of $Z = 110$.

EXAMPLE 5. *Solve the following LPP by corner-point method.*

$$\text{Minimize } Z = 3x_1 + 2x_2$$

subject to the constraints

$$5x_1 + x_2 \geq 10, \, x_1 + x_2 \geq 6,$$
$$x_1 + 4x_2 \geq 12, \, x_1, x_2 \geq 0 \qquad \text{[MEERUT-2009]}$$

SOLUTION. Consider the constraints as equations

$$5x_1 + x_2 = 10$$
$$x_1 + x_2 = 6$$
$$x_1 + 4x_2 = 12$$
$$x_1, x_2 \geq 0$$

Then apply the usual procedure (as discussed earlier). Plot these equations on graph paper and by using the inequality condition of each constraints to mark the feasible region.

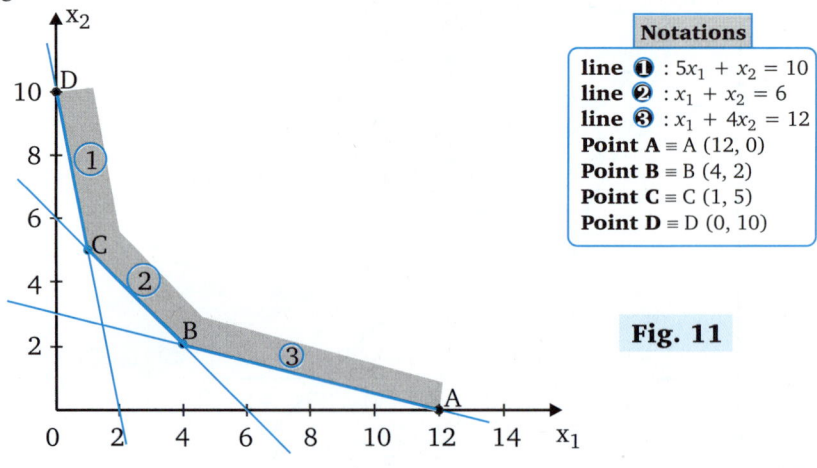

Notations

line ❶ : $5x_1 + x_2 = 10$
line ❷ : $x_1 + x_2 = 6$
line ❸ : $x_1 + 4x_2 = 12$
Point A $\equiv$ A $(12, 0)$
Point B $\equiv$ B $(4, 2)$
Point C $\equiv$ C $(1, 5)$
Point D $\equiv$ D $(0, 10)$

Fig. 11

Clearly, the above shaded region is the feasible region. Now, we have to evaluate the value of the objective function at each point (extreme) of the feasible region.

Extreme point	Coordinates (x_1, x_2)	Value of the objective function $Z = 3x_1 + 2x_2$
A	(12, 0)	$3 \times 12 + 2 \times 0 = 36$
B	(4, 2)	$3 \times 4 + 2 \times 2 = 16$
C	(1, 5)	$3 \times 1 + 2 \times 5 = 13$
D	(0, 10)	$3 \times 0 + 2 \times 10 = 20$

Clearly, the minimum value of the objective function $Z = 13$ occurs at the point $C(1, 5)$. Hence, the optimal solution to the given LPP is $x_1 = 1$, $x_2 = 5$ and Minimum $Z = 13$.

EXAMPLE 6. *Solve the following LPP by graphical method.*

$$\text{Minimize } Z = -x_1 + 2x_2$$

subject to the constraints

$$-x_1 + 3x_2 \le 10$$
$$x_1 + x_2 \le 6$$
$$x_1 - x_2 \le 2$$

and $\qquad x_1, x_2 \ge 0$

SOLUTION. First consider the constraints as equations and then plot each constraints on a graph paper. Use the inequality condition of each constraints to mark the feasible region as given below:

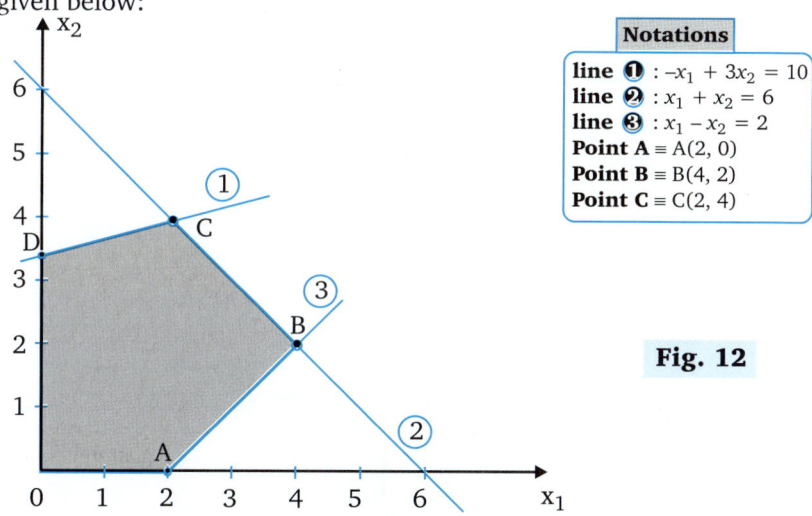

Notations

line ❶ : $-x_1 + 3x_2 = 10$
line ❷ : $x_1 + x_2 = 6$
line ❸ : $x_1 - x_2 = 2$
Point A ≡ A(2, 0)
Point B ≡ B(4, 2)
Point C ≡ C(2, 4)

Fig. 12

We observe that the area below the lines $-x_1 + 3x_2 = 10$ and $x_1 - x_2 = 2$ is not desirable, due to the reason that value of x_1 and x_2 are desired to be negative (*i.e.*, $x_1 \ge 0, x_2 \ge 0$)

Now, we have to calculate the value of the objective function at each extreme point of the above feasible region.

Extreme point	Coordinates (x_1, x_2)	Value of the objective function $Z = -x_1 + 2x_2$
O	(0, 0)	$-1 \times 0 + 2 \times 0 = 0$
A	(2, 0)	$-1 \times 2 + 2 \times 0 = -2$
B	(4, 2)	$-1 \times 4 + 2 \times 2 = 0$
C	(2, 4)	$-1 \times 2 + 2 \times 4 = 6$
D	$\left(0, \dfrac{10}{3}\right)$	$-1 \times 0 + 2 \times \dfrac{10}{3} = \dfrac{20}{3}$

Clearly, the minimum value of the objective function $Z = -2$ occur at the extreme point $A(2, 0)$. Hence, the optimal solution of the given LPP is : $x_1 = 2, x_2 = 0$ and Min $Z = -2$.

Type 5. Based on ISO-Profit Method: Maximization Problem

EXAMPLE 1. *Using, iso-profit method, find the maximum value of the function*
$$Z = 2x_1 + 3x_2$$
subject to the constraints
$$2x_1 + x_2 \le 5$$
$$x_1 - x_2 \le 1$$
$$x_2 \le 2$$
and $x_1 \ge 0, x_2 \ge 0$ [MEERUT-2008]

SOLUTION. Converting the given condition into equations and drawing these lines, the permissible region is $OPQRSO$ (shaded region) which is the set of all feasible solutions of the problem.

We have to maximize the objective function

$$Z = 2x_1 + 3x_2$$

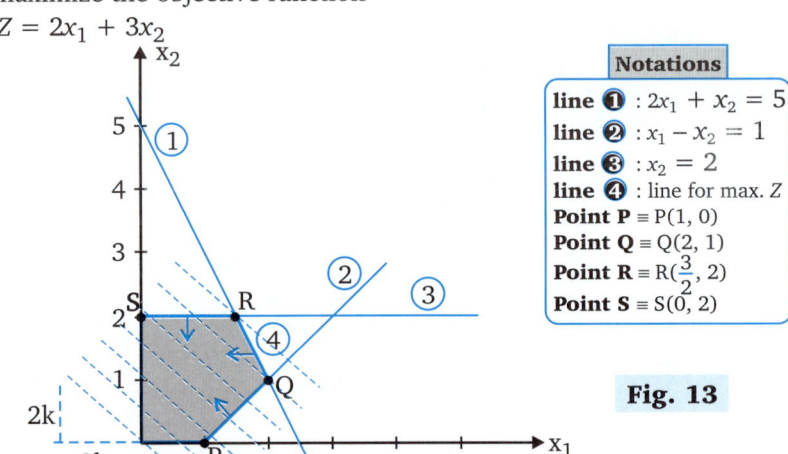

Notations

line ❶ : $2x_1 + x_2 = 5$
line ❷ : $x_1 - x_2 = 1$
line ❸ : $x_2 = 2$
line ❹ : line for max. Z
Point **P** ≡ P(1, 0)
Point **Q** ≡ Q(2, 1)
Point **R** ≡ R($\frac{3}{2}$, 2)
Point **S** ≡ S(0, 2)

Fig. 13

Taking $Z = 0$, we get $\dfrac{x_1}{x_2} = -\dfrac{3}{2}$

The line corresponding to $Z = 0$ is shown by dotted line through the origin O

which is parallel to the iso-profit line. Now draw dotted parallel lines away from the origin. The farthest line from the origin passes through $R\left(\dfrac{3}{2}, 2\right)$.

Since, the farthest line from the origin O passes through the vertex $R\left(\dfrac{3}{2}, 2\right)$.

Hence the required optimal solution is given by $x_1 = \dfrac{3}{2}, x_2 = 2$ and

$$\text{Max. } Z = 2 \times \dfrac{3}{2} + 3 \times 2 = 9.$$

EXAMPLE 2. *Solve the following L.P.P. graphically using Iso-profit method.*

$$\textbf{\textit{Maximize } } Z = 2x_1 + x_2$$

subject to the constraints

$$5x_1 + 10x_2 \leq 50$$
$$x_1 + x_2 \geq 1$$
$$x_2 \leq 4$$
$$x_1 - x_2 \leq 0$$

and $\qquad x_1 \geq 0, x_2 \geq 0$

SOLUTION. To solve the above L.P.P. we proceed as follows:

STEP 1. Making all the constraints as equations:

$$5x_1 + 10x_2 = 50$$
$$x_1 + x_2 = 1$$
$$x_2 = 4$$
$$x_1 - x_2 = 0$$

STEP 2. Draw these lines on the $x_1 x_2$ - plane as follows:

The line $5x_1 + 10x_2 = 50$ passes through the two points (10, 0) and (0, 5). Plot these points on the co-ordinate axes i.e., x_1 - axis and x_2 - axis respectively and join them by a straight line.

The line $x_1 + x_2 = 1$ passes through the points (1, 0) and (0, 1). Plot these point on the x_1 - axis and x_2 - axis respectively and join them by a straight line.

Similarly, we draw the lines $x_2 = 4$ *and* $x_1 - x_2 = 0$.

STEP 3. Since $5x_1 + 10x_2 \leq 50, x_1 + x_2 \geq 1, x_2 \leq 4, x_1 - x_2 \leq 0$. So the regions below the lines $5x_1 + 10x_2 = 50$ in the positive regions, and the regions above the line $x_1 + x_2 = 1$ and $x_1 - x_2 = 0$ in the positive quadrant is the possible region.

Now the region common to all above regions is the feasible region, which is as shown in following figure by the shaded region *ABCDEA*.

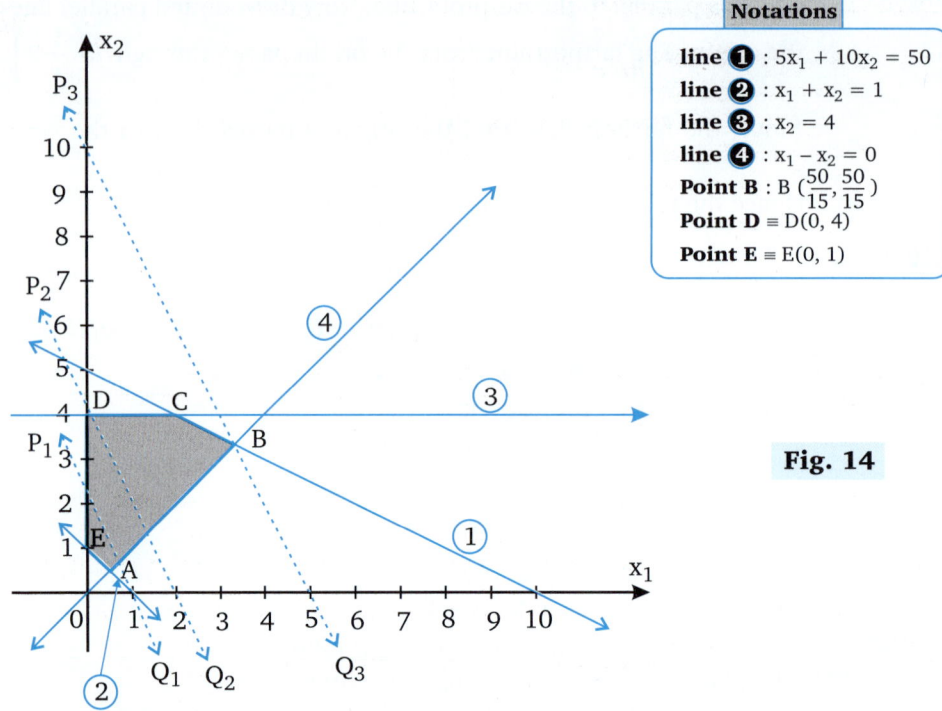

Notations

line ❶ : $5x_1 + 10x_2 = 50$
line ❷ : $x_1 + x_2 = 1$
line ❸ : $x_2 = 4$
line ❹ : $x_1 - x_2 = 0$
Point B : B $(\frac{50}{15}, \frac{50}{15})$
Point D ≡ D(0, 4)
Point E ≡ E(0, 1)

Fig. 14

STEP 4. Iso-profit Method:

(a) Assign a constant value $Z_1 = 2$ (L.C.M. of 2 and 1) to the objective function Z, we get a line

$$2x_1 + x_2 = 2 \qquad \qquad ...(1)$$

Now draw this line (1) which is as shown in figure by P_1Q_1.

(b) Again assign a constant value $Z_2 = 4$ to the objective function Z, we get a line

$$2x_1 + x_2 = 4 \qquad \qquad ...(2)$$

Now draw this line (2) which is shown in figure by P_2Q_2, clearly both the lines given by eqns. (1) and (2) and parallel to each other.

Since $Z_1 < Z_2$ and the L.P.P. is of maximization, so we move the line P_1Q_1

from P_1Q_1 to P_2Q_2 parallel to itself as far as possible, until the farthest point

$B\left(\dfrac{50}{15}, \dfrac{50}{15}\right)$ within the feasible region *ABCDEA* is touched by the line. Any

further displacement of the line takes it out of the feasible region *ABCDEA*.

Thus, the point $B\left(\dfrac{50}{15}, \dfrac{50}{15}\right)$ obtained on solving the lines $x_1 - x_2 = 0$ and

$5x_1 + 10x_2 = 50$, gives the maximum value of the objective function Z, and

the line P_3Q_3 is the line of maximum objective function.

Hence, the optimal solution is $x_1 = \dfrac{50}{15}, x_2 = \dfrac{50}{15}$

and the optimal value of $Z = 2 \times \dfrac{50}{15} + \dfrac{50}{15} = 10$

EXAMPLE 3. *Solve the following L.P.P. by graphical method:*

$$Maximize\ Z = 3x_1 + 5x_2$$

subject to the constraints

$$x_1 + x_2 \leq 1500$$
$$x_1 + 2x_2 \leq 2000$$
$$x_2 \leq 600$$

and $$x_1 \geq 0,\ x_2 \geq 0$$ (GORAKHPUR-2010)

SOLUTION. **STEP 1.** Making the given constraints as equations:

$$x_1 + x_2 = 1500$$
$$x_1 + 2x_2 = 2000$$
$$x_2 = 600$$

STEP 2. The line $x_1 + x_2 = 1500$ passes through the two points (1500, 0) and (0, 1500). Plot the points (1500, 0) and (0, 1500) on x_1- axis and x_2- axis respectively and joining them by a straight line.

The line $x_1 + 2x_2 = 2000$ passes through the points (2000, 0) and (0, 1000). Plot these point on x_1-axis and x_2- axis and joining them by a straight line.

The line $x_2 = 600$ is parallel to the x_1-axis and passing through the point (0, 600).

STEP 3. Since we have,

$$x_1 + x_2 \leq 1500,\ x_1 + 2x_2 \leq 2000,\ x_2 \leq 600\ and\ x_1 \geq 0,\ x_2 \geq 0$$

Therefore, the region below the line $x_1 + x_2 = 1500$ in the positive quadrant is the possible region, the region below the line $x_1 + 2x_2 = 2000$ in the possible region, the region below the line $x_2 = 600$ in the positive quadrant is the possible region.

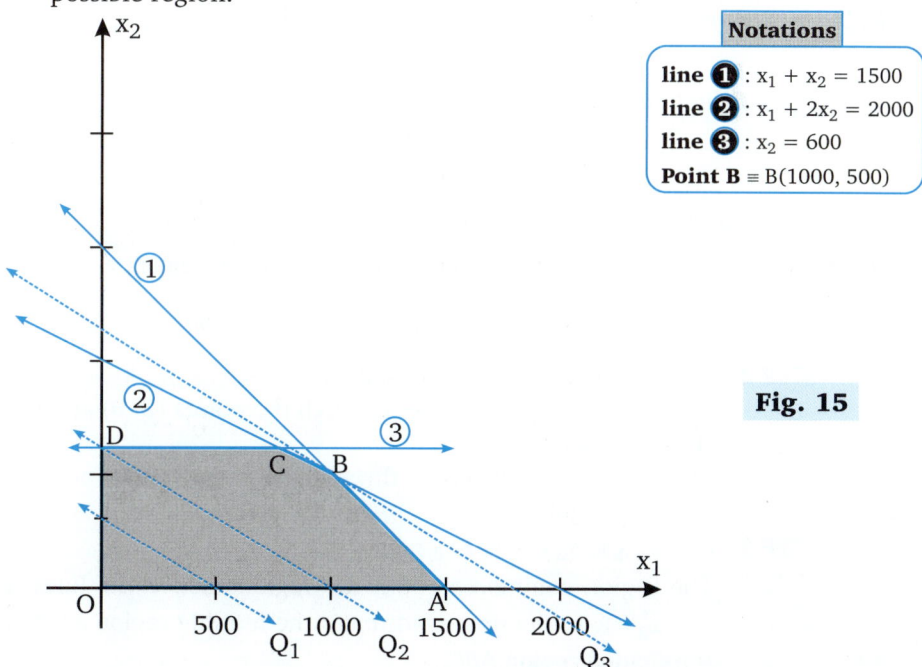

Notations

line **1** : $x_1 + x_2 = 1500$
line **2** : $x_1 + 2x_2 = 2000$
line **3** : $x_2 = 600$
Point B $\equiv$ B(1000, 500)

Fig. 15

Thus, the region common to above three regions in the positive quadrant is the feasible region, which is as shown in figure by the shaded region *OABCDO*.

STEP 4. Iso-profit Method:

(a) Assign a constant $Z_1 = 1500$ to the objective function Z, we get a line

$$3x_1 + 5x_2 = 1500 \qquad \ldots(1)$$

Now, draw this line (1), which is as shown in figure by P_1Q_1.

(b) Again assign a constant value $Z_1 = 3000$ to the objective function Z, we get a line

$$3x_1 + 5x_2 = 3000 \qquad \ldots(2)$$

Now draw this line (2), which is as shown in figure by P_2Q_2.

Clearly, both the lines P_1Q_1 and P_2Q_2 are parallel to each other.

Since $Z_1 < Z_2$ and the L.P.P is of maximization, so we move the line P_1Q_1 from P_1Q_1 to P_2Q_2 parallel to itself as far as possible, until the farthest point B (1000, 500) within the feasible region *OABCDO* is touched by the line. Any further displacement of the line takes it out of the feasible region *OABCDO*.

Thus, the point B (1000, 500) is obtained on solving the equations $x_1 + x_2 = 1500$ and $x_1 + 2x_2 = 2000$, gives the maximum value of the objective function Z, and the line P_3Q_3 is the line of maximum objective function.

Hence, the optimal solution is

$$x_1 = 1000, x_2 = 500$$

and the optimal value of $Z = 3 \times 1000 + 5 \times 500 = 5500$

TYPE 6. BASED ON ISO-PROFIT METHOD: MINIMIZATION PROBLEM

EXAMPLE 1. *Solve the following L.P.P. graphically.*

Minimize $Z = 1.5x_1 + 2.5x_2$

subject to the constraints

$$x_1 + 3x_2 \geq 3$$

$$x_1 + x_2 \geq 2$$

and $x_1 \geq 0, \ x_2 \geq 0$ [KANPUR-2010]

SOLUTION. STEP 1. Changing all the given constraints into equations:

$$x_1 + 3x_2 = 3$$

$$x_1 + x_2 = 2$$

STEP 2. Draw these lines on $x_1 x_2$-plane.

The line $x_1 + 3x_2 = 3$ passes through the points (3, 0) and (0, 1), which is as shown in figure.

The line $x_1 + x_2 = 2$ passes through the points (2, 0) and (0, 2) which is as shown in figure.

STEP 3. Since $x_1 + 3x_2 \geq 3, \ x_1 + x_2 \geq 2$ and $x_1 \geq 0, x_2 \geq 0$

The region common to both the regions above the lines $x_1 + 3x_2 = 3$ and $x_1 + x_2 = 2$ in the first quadrant is the feasible region as shown in figure by the shaded region *ABC*.

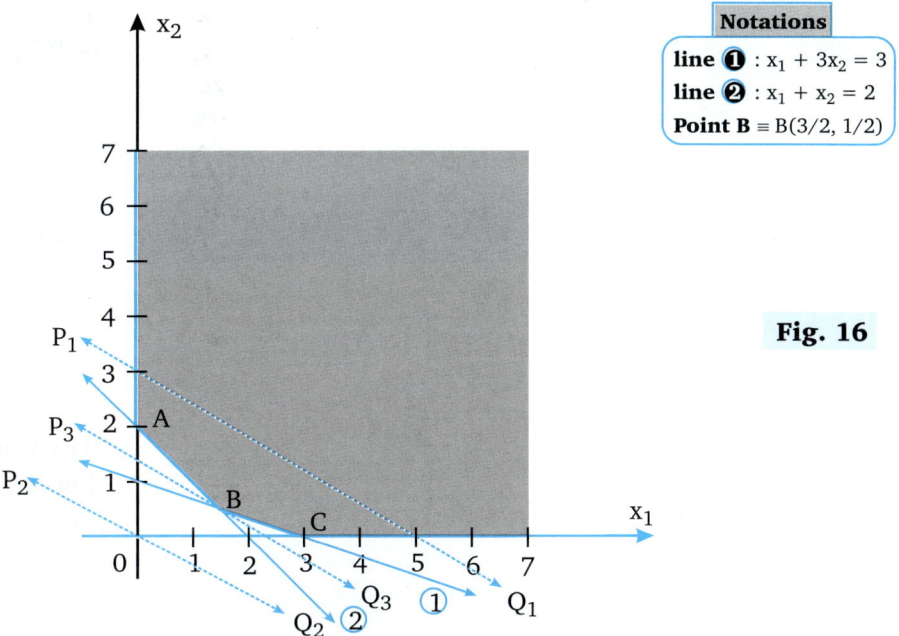

Fig. 16

STEP 4. Iso-profit Method:

(a) Assign a constant value $Z_1 = 7.5$ (LCM of 1.5 and 2.5) to the objective function Z, we get a line

$$1.5x_1 + 2.5x_2 = 7.5 \qquad \dots(1)$$

Now draw this line (1) on $x_1 x_2$-plane, which is as shown in figure by $P_1 Q_1$.

(b) Again assign a constant value $Z_2 = 10$ to the objective function Z, we get a line

$$1.5x_1 + 2.5x_2 = 10 \qquad \dots(2)$$

Now draw this line (2) on $x_1 x_2$-plane which is shown in Fig. 1.16 by $P_2 Q_2$.

Since, $Z_2 > Z_1$ so we move the line $P_1 Q_1$ from $P_1 Q_1$ to $P_2 Q_2$ parallel to itself, until a point B of the feasible region ABC is touched by the line. The point B is the intersection of the lines $x_1 + x_2 = 2$ and $x_1 + 3x_2 = 3$, so on solving these equations, we obtain the co-ordinates of B as $\left(\dfrac{3}{2}, \dfrac{1}{2}\right)$.

Thus, the point $B\left(\dfrac{3}{2}, \dfrac{1}{2}\right)$ will give the minimum value of Z.

Hence, the optimum solution of L.P.P. is $x_1 = \dfrac{3}{2}$, $x_2 = \dfrac{1}{2}$ and the optimum value of objective function $Z = 1.5 \times \dfrac{3}{2} + 2.5 \times \dfrac{1}{2} = 3.5$

TYPE 7. PROBLEMS HAVING UNBOUNDED SOLUTIONS

EXAMPLE 1. *Solve graphically the following LPP*

$$\text{Maximize } Z = 4x_1 + 5x_2$$

subject to the constraints

$$x_1 + x_2 \geq 1,$$
$$-2x_1 + x_2 \leq 1,$$
$$4x_1 - 2x_2 \leq 1,$$

and $$x_1, x_2 \geq 0$$

SOLUTION. Proceeding as usual, the permissible region is shown in the following graph by shaded region. Here, $Z = 4x_1 + 5x_2 \Rightarrow \dfrac{x_1}{x_2} = -\dfrac{5}{4}$

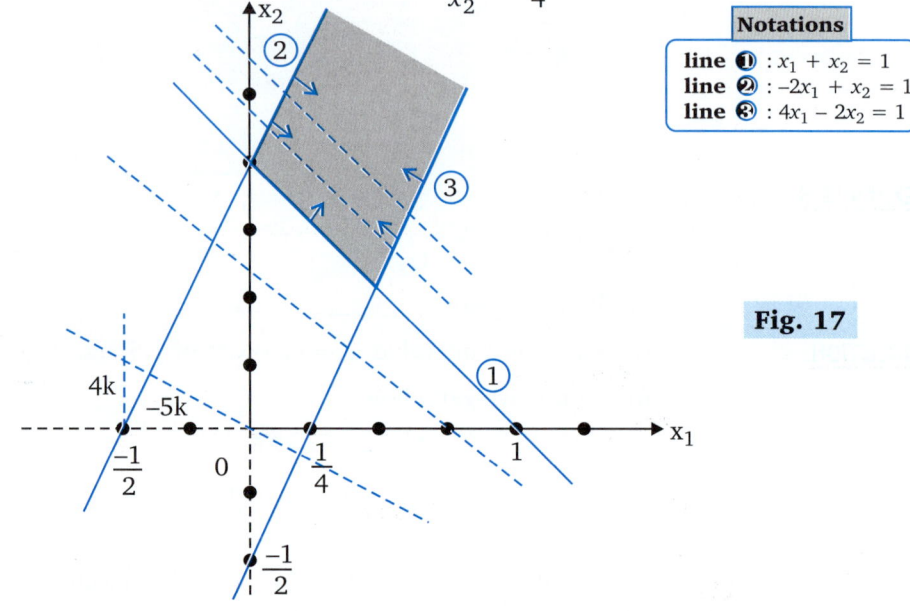

Notations

line ① : $x_1 + x_2 = 1$
line ② : $-2x_1 + x_2 = 1$
line ③ : $4x_1 - 2x_2 = 1$

Fig. 17

From the above figure, it is clear that dotted line through the origin representing $Z = 0$ can be moved parallel to itself in the direction of Z increasing and still has some point in the permissible region. Also, $Z = 4x_1 + 5x_2 = 0 \Rightarrow \dfrac{x_1}{x_2} = -\dfrac{5}{4}$

Hence, Z can be made arbitrarily large and so the problem has no finite maximum value of Z. Thus problem has unbounded solution.

EXAMPLE 2. *Solve the following LPP graphically*

$$\text{Maximize } Z = 0.75x_1 + x_2$$

subject to the constraints

$$x_1 - x_2 \geq 0, -0.5x_1 + x_2 \leq 1 \text{ and } x_1, x_2 \geq 0$$

SOLUTION. Proceeding as usual, we get the permissible region which shaded in the following figure. Hence, $Z = 0 \Rightarrow \dfrac{x_1}{x_2} = -\dfrac{4}{3}$

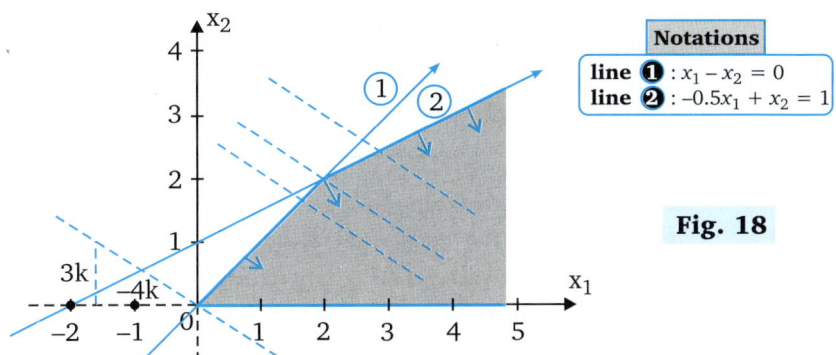

Fig. 18

From the above graph, it is clear that the dotted line passes through the origin representing $Z = 0$ can be moved parallel to itself in the direction of Z increasing and will have some points in the feasible region. Therefore, Z can be made arbitrarily large and hence the problem has no finite maximum value of Z. Thus the problem has unbounded solution.

EXAMPLE 3. ***Solve the following LPP graphically:***
$$\text{Minimize } Z = 5x_1 - 2x_2$$
subject to the constraints
$$2x_1 + 3x_2 \geq 1$$
and $\qquad\qquad x_1, x_2 \geq 0$

SOLUTION. Proceeding as usual, we get the permissible region, which is shaded in the figure given below.

Also, $Z = 0 \Rightarrow \dfrac{x_1}{x_2} = \dfrac{2}{5}$

The permissible region is the set of all the constraints and the non-negative restrictions, is unbounded. Solving simultaneously the equations of the corresponding intersecting lines we get two vertices $A\left(\dfrac{1}{2}, 0\right)$ and $B\left(0, \dfrac{1}{3}\right)$.

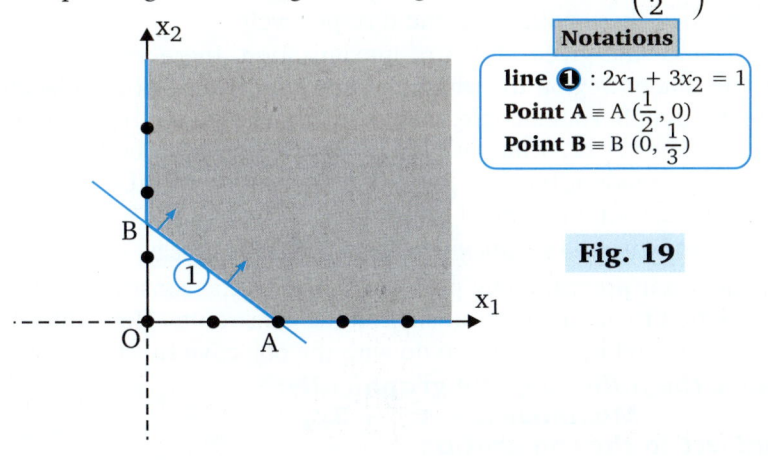

Notations

line **❶** : $2x_1 + 3x_2 = 1$
Point A $\equiv$ A $\left(\dfrac{1}{2}, 0\right)$
Point B $\equiv$ B $\left(0, \dfrac{1}{3}\right)$

Fig. 19

At A, $Z = \dfrac{5}{2} - 0 = \dfrac{5}{2}$ and at B, $Z = 0 - \dfrac{2}{3} = -\dfrac{2}{3}$

in which $Z = -\dfrac{2}{3}$ is minimum at $B\left(0, \dfrac{1}{3}\right)$

∴ The optimal solution may be $x_1 = 0$, $x_2 = \dfrac{1}{3}$ and min $Z = -\dfrac{2}{3}$. But, since the permissible region is unbounded, by taking other points in the region, the value of Z can be made small as we please.

⇒ There is no finite minimum value of Z at a point in the permissible region.

<u>EXAMPLE 4.</u>   ***Solve the following LPP graphically:***

> ***Maximize $Z = 3x_1 + 2x_2$***
>
> ***subject to the constraints $x_1 - x_2 \geq 1$, $x_1 + x_2 \geq 3$***
>
> ***and $x_1, x_2 \geq 0$***   [MEERUT-2009]

<u>SOLUTION.</u>   Apply the usual procedure, we plot the constraints on the graph and permissible region is shaded and is bounded from below and unbounded from above.

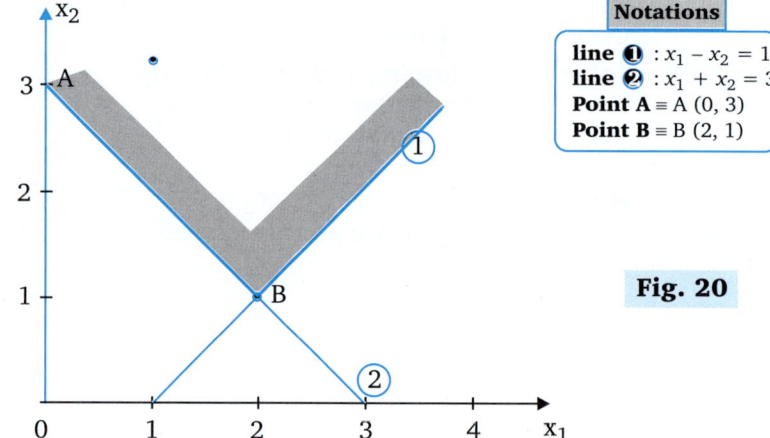

Notations

line ① : $x_1 - x_2 = 1$
line ② : $x_1 + x_2 = 3$
Point A ≡ A (0, 3)
Point B ≡ B (2, 1)

Fig. 20

Clearly, two corner points are $A(0, 3)$ and $B(2, 1)$. The value of the objective function at these points is 6 and 8 respectively.

Now, since the given LPP is of maximization, there exists a number of points in the shaded region for which the maximum value of the objective function is greater than 8. For example, the point (2, 3) lies in the region and function value at this point is 12, which is clearly more than 8. So, both the variables x_1 and x_2 can be made arbitrarily large and the value of Z also increases. Hence, the problem has an unbounded solution.

TYPE 8. PROBLEM HAVING INFEASIBLE SOLUTIONS OR MULTIPLE FEASIBLE REGION

There are some linear programming problems in which the constraints are not consistent, *i.e.*, there is no solution to an LPP that satisfies all the constraints. Here, infeasibility depends only on the constraints and has nothing to do with the objective function.

<u>EXAMPLE 1.</u>   ***Solve the following LPP graphically:***

> ***Maximize $Z = 4x_1 + 3x_2$***
>
> ***subject to the constraints***
>
> $$8x_1 + 6x_2 \leq 48$$
> $$x_1 \leq 6$$
>
> ***and*** $\qquad x_1, x_2 \geq 0$

SOLUTION. Draw the converted line on x_1x_2-plane by following the usual procedure. Clearly, the line $8x_1 + 6x_2 = 48$ passes through the point $(6, 0)$ and $(0, 8)$ and the line $x_1 = 6$ is parallel to x_2-axis and passes through the point $(6, 0)$. Also, the region below the line $8x_1 + 6x_2 = 48$ in the positive quadrant is the permissible region and the region left to the $x_1 = 6$ in the positive quadrant is the permissible region. The common to both regions is the feasible region which is shown in following figure by the shaded region $OABO$.

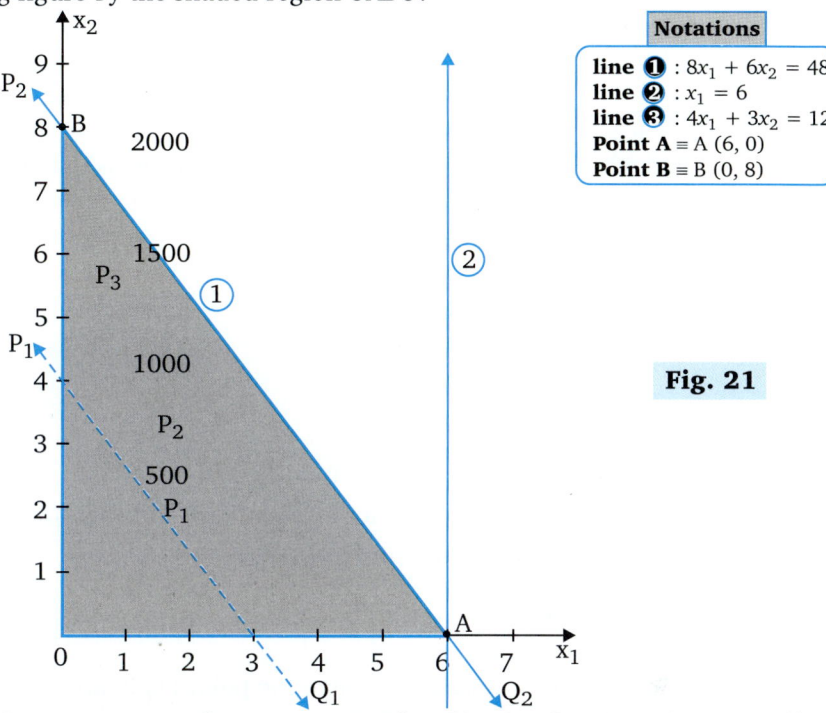

Notations
line ❶ : $8x_1 + 6x_2 = 48$
line ❷ : $x_1 = 6$
line ❸ : $4x_1 + 3x_2 = 12$
Point A $\equiv$ A $(6, 0)$
Point B $\equiv$ B $(0, 8)$

Fig. 21

Now, assign a constant value $Z_1 = 12$ to the objective function Z, we get a line $4x_1 + 3x_2 = 12$, which is shown in the above figure by P_1Q_1. Similarly, assign a constant value $Z_2 = 24$ to the objective function Z we may get a line $4x_1+3x_2=24$, shown in the above figure by the line P_2Q_2.

Now, since $Z_2 < Z_1$ and LPP is of maximization, thus we move P_1Q_1 towards P_2Q_2 parallel to itself, untill the further point of feasible region is bounded by the line. But the iso-profit line lies along the constraint $8x_1 + 6x_2 \leq 48$. Hence, the given LPP has multiple feasible solutions.

EXAMPLE 2. *Solve the following LPP graphically:*

Maximize $Z = 6x_1 - 4x_2$

subject to the constraints

$$2x_1 + 4x_2 \leq 4, \quad 4x_1 + 8x_2 \geq 16$$

and $\qquad\qquad x_1, x_2 \geq 0$

SOLUTION. Following the usual procedure, the constraints are plotted on the graph as follows:

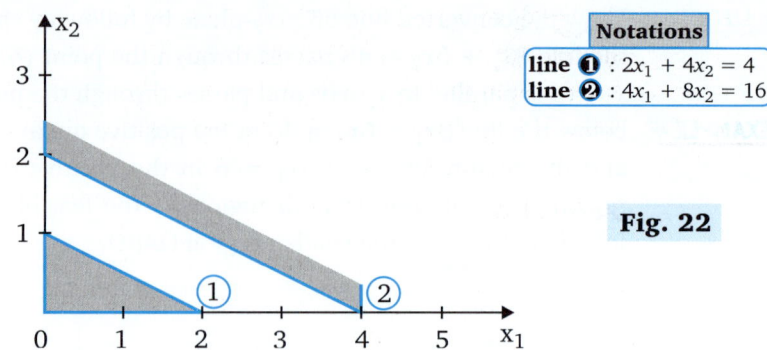

Notations

line ❶ : $2x_1 + 4x_2 = 4$
line ❷ : $4x_1 + 8x_2 = 16$

Fig. 22

Clearly, there is no unique feasible solution space to get unique set of values of variables x_1 and x_2 that satisfies all the constraints. Therefore, there is no feasible solution of this problem.

EXAMPLE 3. **Solve the following LPP graphically.**

$$\text{Minimize } Z = x_1 + x_2$$

subject to the constraints

$$x_1 + x_2 \le \frac{1}{2}$$

$$3x_1 + x_2 \ge 3$$

and $\qquad x_1, x_2 \ge 0$

SOLUTION. Changing all the constraints to equations and then draw these lines on x_1x_2-plane by following the usual procedure.

Clearly, the line $x_1 + x_2 = \frac{1}{2}$ passes through the point $\left(\frac{1}{2}, 0\right)$ and $\left(0, \frac{1}{2}\right)$ and the

line $3x_1 + x_2 = 3$ passes through the point $(1, 0)$ and $(0, 3)$.

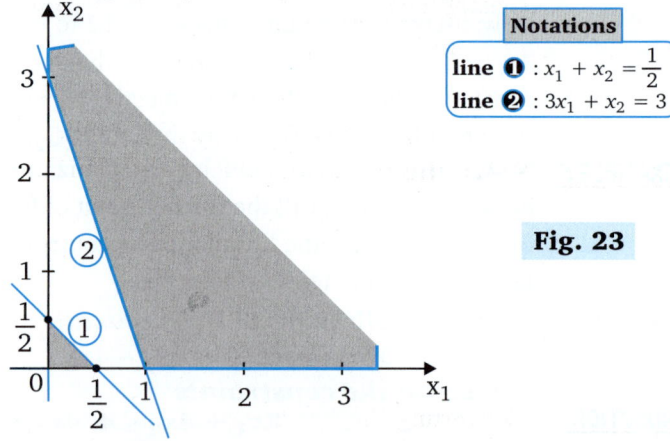

Notations

line ❶ : $x_1 + x_2 = \frac{1}{2}$
line ❷ : $3x_1 + x_2 = 3$

Fig. 23

Now, since we have $x_1 + x_2 \le \frac{1}{2}, 3x_1 + x_2 \ge 3$ and $x_1 \ge 0, x_2 \ge 0$, the region below

the line $x_1 + x_2 = 1$ is the permissible region and the region above the line

$3x_1 + x_2 = 3$ is the permissible region in the positive quadrant. The solution spaces are shaded in the above figure, which shows that there is no point satisfying both the constraints. Hence the given LPP has infeasible solution.

EXAMPLE 4. *Solve the following LPP graphically.*

$$\text{Minimize } Z = 5x_1 + 2x_2$$

subject to the constraints

$$x_1 + x_2 \le 2$$
$$3x_1 + 3x_2 \ge 12$$

and $\qquad x_1, x_2 \ge 0$

SOLUTION. Changing all the constraints to equations and draw them on x_1x_2-plane. Clearly the line $x_1 + x_2 = 2$ passes through the point (2, 0) and (0, 2) and the line $3x_1 + 3x_2 = 12$ passes through the point (4, 0) and (0, 4).

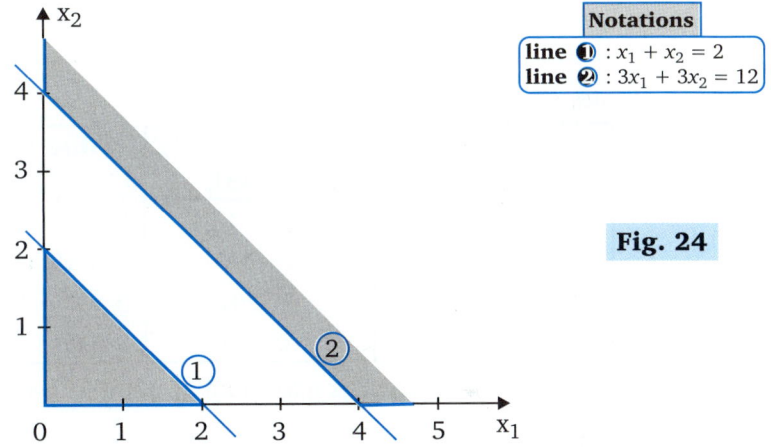

Notations
line ① : $x_1 + x_2 = 2$
line ② : $3x_1 + 3x_2 = 12$

Fig. 24

Clearly, the region below the line $x_1 + x_2 = 2$ and above the line $3x_1 + 3x_2 = 12$ is the permissible region. The solution spaces are shaded in the above figure. Clearly, there is no common part to both the region, so there is no point satisfying both the constraints. Hence, the given LPP has infeasible solutions.

EXAMPLE 5. *Solve the following LPP graphically.*

$$\text{Minimize } Z = 3x_1 + 9x_2$$

subject to the constraints

$$x_1 + 3x_2 \le 60,$$
$$x_1 + x_2 \ge 10, x_1 \le x_2$$

and $\quad x_1, x_2 \ge 0$

SOLUTION. Converting the given constraints into equations, then plot these lines on x_1x_2-plane as shown below:

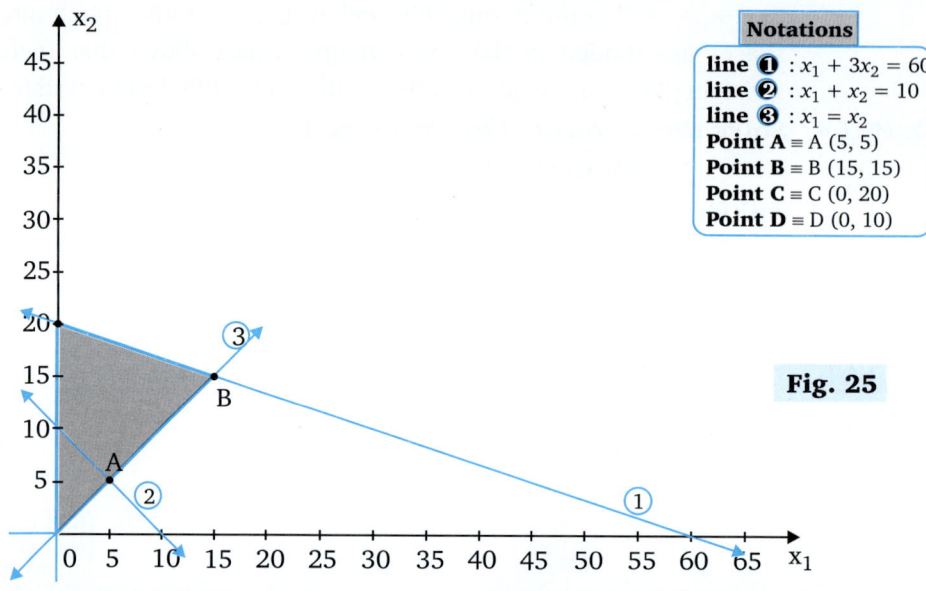

Notations
line ❶ : $x_1 + 3x_2 = 60$
line ❷ : $x_1 + x_2 = 10$
line ❸ : $x_1 = x_2$
Point A ≡ A (5, 5)
Point B ≡ B (15, 15)
Point C ≡ C (0, 20)
Point D ≡ D (0, 10)

Fig. 25

Extreme point	Coordinates (x_1, x_2)	Value of the objective function $Z = 3x_1 + 9x_2$
A	(5, 5)	$3 \times 5 + 9 \times 5 = 60$
B	(15, 15)	$3 \times 15 + 9 \times 15 = 180$
C	(0, 20)	$3 \times 0 + 9 \times 20 = 180$
D	(0, 10)	$3 \times 0 + 9 \times 10 = 90$

Here, it is clear that Z is maximum at $B(15, 15)$ and $C(0, 20)$ and its maximum value is 180. Hence, the given LPP has multiple optimal solutions at B and C.

TYPE 9. MORE PROBLEMS ON MIXED CONSTRAINTS

A redundant constraint is one that does not affect the feasible solution space and thus redundancy of any constraint does not cause any difficulty in solving LPP graphically.

EXAMPLE I.   ***Solve the following LPP graphically.***
$$\text{Minimize } Z = x_1 + x_2$$
subject to the constraints
$$5x_1 + 10x_2 \leq 50$$
$$x_1 + x_2 \geq 1$$
$$x_2 \geq 4$$
and $$x_1, x_2 \geq 0$$

SOLUTION.   Converting the given constraints into equations and drawing these lines, we have the following figure. Here, $Z = x_1 + x_2 = 0 \Rightarrow \dfrac{x_1}{x_2} = -\dfrac{1}{1}$.

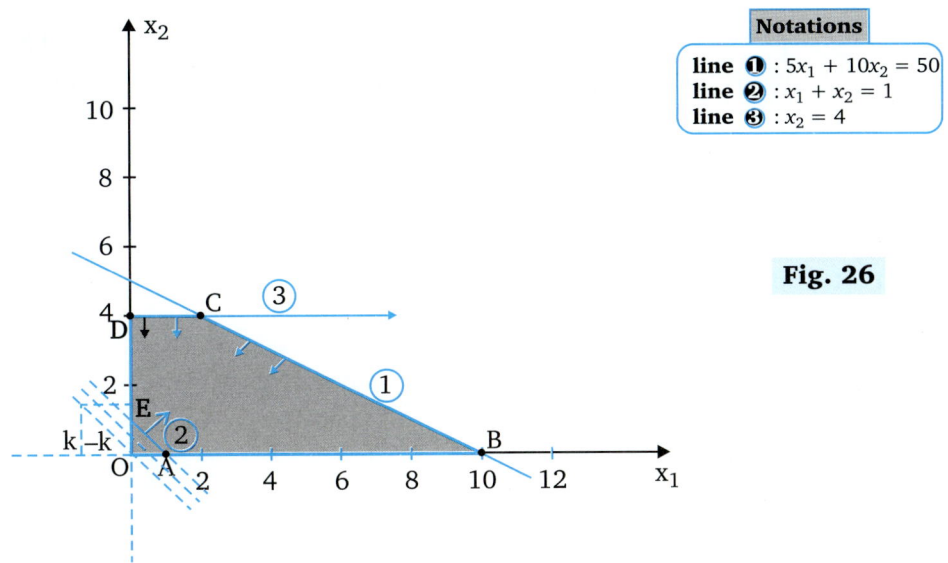

Fig. 26

Notations:
line ❶ : $5x_1 + 10x_2 = 50$
line ❷ : $x_1 + x_2 = 1$
line ❸ : $x_2 = 4$

Clearly, the permissible region is *ABCDEOA* (shaded region) which is the set of all feasible solutions of the given LPP.

Taking $Z = x_1 + x_2 = 0 \Rightarrow \dfrac{x_1}{x_2} = -\dfrac{1}{1}$

The line corresponding to $Z = 0$ is shown by dotted line through the origin *O*. Drawing lines parallel to this line, we see that a parallel dotted line passing through the vertices $A(1, 0)$ and $E(0, 1)$ of the convex polygon, which is the nearest to the origin *O*.

$\Rightarrow$ Every point of the line segment AE gives the optimal solution of the problem. Hence, the given LPP has infinite number of solutions.

EXAMPLE 2. *Solve the following LPP graphically.*

$$\text{\textit{Maximize} } Z = x_1 + \frac{1}{2}x_2$$

subject to the constraints

$$3x_1 + 2x_2 \leq 12$$
$$x_1 \leq 2$$
$$x_1 + x_2 \geq 8$$
$$-x_1 + x_2 \geq 4$$

and $\qquad\qquad x_1, x_2 \geq 0$ [MEERUT-2006]

SOLUTION. Proceeding stepwise as usual, we observe that there is no permissible region of the problem, whose points satisfy all the constraints, *i.e.*, there is no point which will satisfy all the constraints.

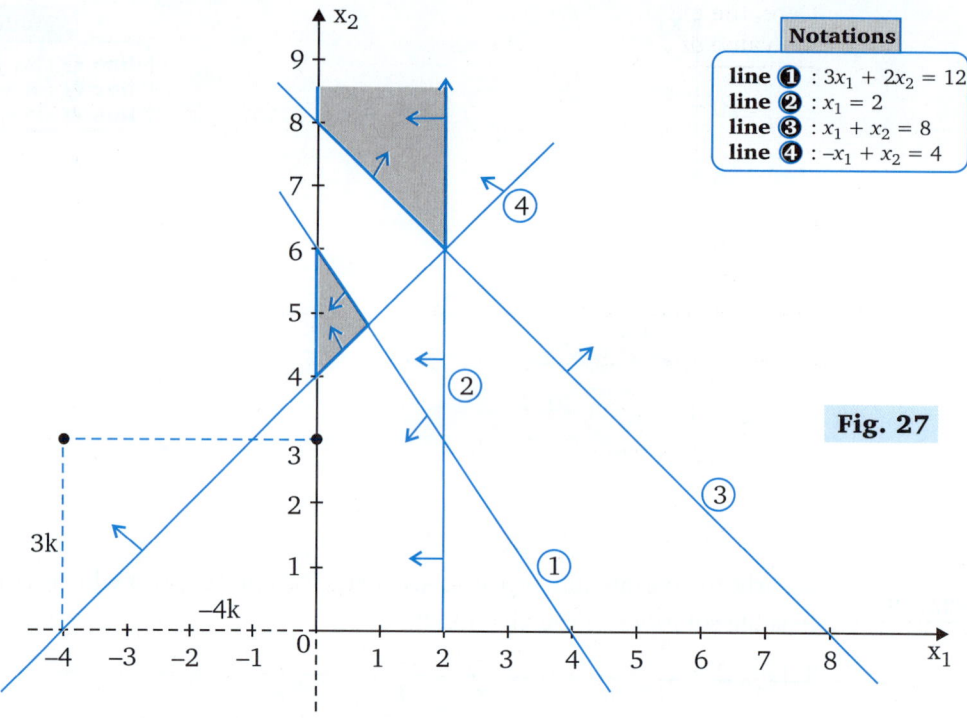

Notations

line ❶ : $3x_1 + 2x_2 = 12$
line ❷ : $x_1 = 2$
line ❸ : $x_1 + x_2 = 8$
line ❹ : $-x_1 + x_2 = 4$

Fig. 27

EXAMPLE 3. *Solve the following LPP graphically.*

Maximize $Z = 7x_1 + 3x_2$

subject to the constraints

$x_1 + 2x_2 \geq 3, x_1 + x_2 \leq 4, x_1 \leq 5/2, x_2 \leq 3/2$ and $x_1, x_2 \geq 0$

SOLUTION. Converting the given constraints into equations and then plot them by following the usual procedure and then use the inequality conditions to mark the feasible region as shown in the following figure.

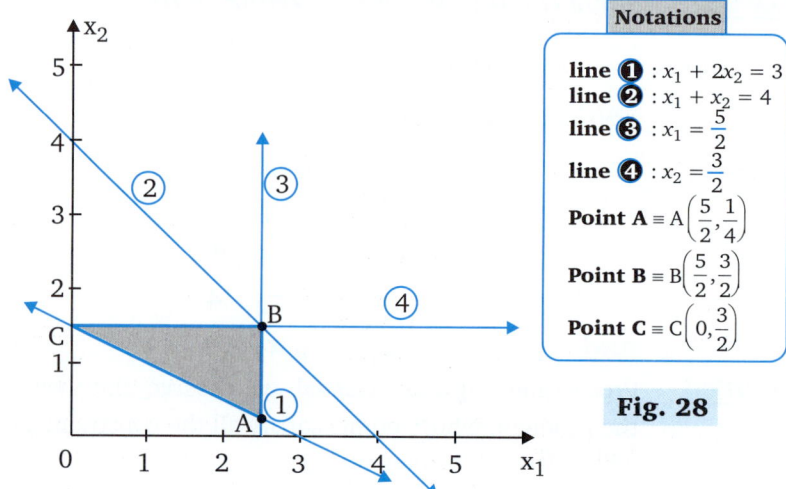

Notations

line ❶ : $x_1 + 2x_2 = 3$
line ❷ : $x_1 + x_2 = 4$
line ❸ : $x_1 = \dfrac{5}{2}$
line ❹ : $x_2 = \dfrac{3}{2}$

Point A $\equiv A\left(\dfrac{5}{2}, \dfrac{1}{4}\right)$

Point B $\equiv B\left(\dfrac{5}{2}, \dfrac{3}{2}\right)$

Point C $\equiv C\left(0, \dfrac{3}{2}\right)$

Fig. 28

Here, the extreme points of the feasible region are A, B and C. Now, we compute the value of the objective function at each of these extreme points.

Extreme point	Coordinates (x_1, x_2)	Value of the objective function $Z = 7x_1 + 3x_2$
A	$\left(\dfrac{5}{2}, \dfrac{1}{4}\right)$	$7 \times \dfrac{5}{2} + 3 \times \dfrac{1}{4} = \dfrac{73}{4}$
B	$\left(\dfrac{5}{2}, \dfrac{3}{2}\right)$	$7 \times \dfrac{5}{2} + 3 \times \dfrac{3}{2} = 22$
C	$\left(0, \dfrac{3}{2}\right)$	$0 \times 7 + 3 \times \dfrac{3}{2} = \dfrac{9}{2}$

Clearly, the maximum value of the objective function $Z = 22$ occurs at the extreme point $B\left(\dfrac{5}{2}, \dfrac{3}{2}\right)$. Hence, the optimal solution to the given LPP is $x_1 = \dfrac{5}{2}$, $x_2 = \dfrac{3}{2}$ and Max $Z = 22$.

EXAMPLE 4. *Solve the following LPP graphically.*

$$\text{Minimize } Z = 5x_1 + 6x_2$$

subject to the constraints

$$x_1 + x_2 \geq 50, \quad x_1 + 2x_2 \leq 40$$
$$3x_1 + 4x_2 \leq 100$$

and $\qquad x_1 \geq 0; \; x_2 \geq 0$

SOLUTION. Converting the given constraints into equation and then plot them by following procedure and then use the inequality conditions to mark the feasible region as shown in the following figure.

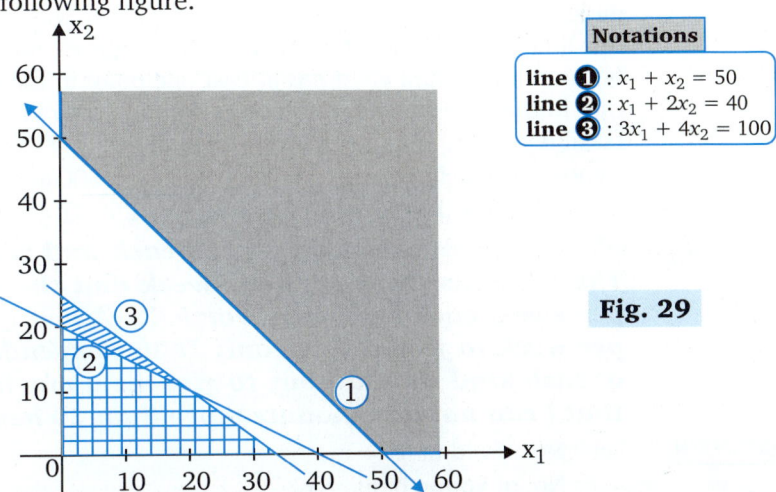

Notations

line ❶ : $x_1 + x_2 = 50$
line ❷ : $x_1 + 2x_2 = 40$
line ❸ : $3x_1 + 4x_2 = 100$

Fig. 29

Here, we observe that there exists no value of x_1 and x_2 simultaneously satisfy all the constraints and the non-negative restrictions. Hence, the given problem does not have any feasible solution.

TYPE 10. VALUE BASED QUESTIONS

EXAMPLE 1. *A toy company manufactures two types of dolls, a basic version doll A and a deluxe version doll B. Each doll of type B takes twice as long to produces as one of type A and the company would have to make a maximum 2000 per day if it produces 1500 dolls per day (both A and B combined). The deluxe version requires a fency dress of which there are only 600 pieces per day available. If the company makes a profit of ₹30 and 50 per doll respectively on doll A and B, how many of each should be produced per day in order to maximize the profit.* [MEERUT-2009]

SOLUTION. We have already formed the LPP (of Formulation of an LPP) which is given as below

Maximize $Z = 30x_1 + 50x_2$
subject to the constraints
$$x_1 + 2x_2 \leq 2000$$
$$x_1 + x_2 \leq 1500$$
$$x_2 \leq 600$$
and $x_1, x_2 \geq 0$

Now proceeding as usual, by converting the inequalities into equations and plotting them on the graph, as follows:

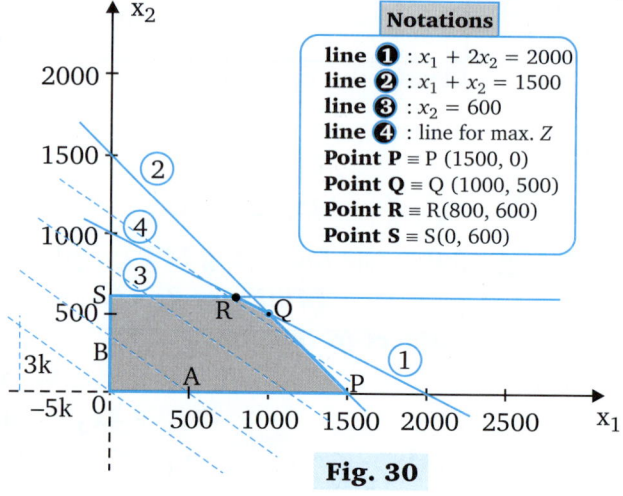

Notations

line ❶ : $x_1 + 2x_2 = 2000$
line ❷ : $x_1 + x_2 = 1500$
line ❸ : $x_2 = 600$
line ❹ : line for max. Z
Point P ≡ P (1500, 0)
Point Q ≡ Q (1000, 500)
Point R ≡ R(800, 600)
Point S ≡ S(0, 600)

Fig. 30

To find the maximum value of the objective function Z, we draw the dotted line $Z = 3x_1 + 5x_2 = 0$ passing through the origin which is parallel to the iso-profit line. We now move this line parallel to itself away from the origin so that its distance from the origin becomes maximum yet it has atleast one point in the feasible region. We observe that in this position this line passes through only one point $Q(1000, 500)$ of the feasible region. Thus Z is maximum at Q.

Hence, Z is maximum for $x_1 = 1000$ and $x_2 = 500$ and the maximum value of Z is
$$30 \times 1000 + 50 \times 500 = 30000 + 25000 = 55000$$

EXAMPLE 2. *Old hens can be bought at ₹20 each and young ones at ₹50 each. The old hens lay 3 eggs per week and the young ones lay 5 eggs per week, each egg being worth ₹3. A hen (young or old) cost ₹10 per week to feed. I have only ₹800 to spends for hens. How many of each kind should I buy to give me a maximum profit, assuming that I can not accomodate more than 20 hens.*

SOLUTION. Let   No. of old hens = x_1
           No. of young hens = x_2
Total no. of eggs I have per week = $3x_1 + 5x_2$
∴ My total income per week = $3(3x_1 + 5x_2)$
But expanses for feeding $x_1 + x_2$ hens at the rate of ₹10 per hen is = $10(x_1 + x_2)$

$\therefore$ Total profit $Z = 3(3x_1 + 5x_2) - 10(x_1 + x_2)$
$$= -x_1 + 5x_2$$

Also, $20x_1 + 50x_2 \leq 800$

and $\quad x_1 + x_2 \leq 20$

$\therefore$ The mathematical model of an LPP is given by

Max. $Z = -x_1 + 5x_2$

subject to the constraints

$$20x_1 + 50x_2 \leq 800 \Rightarrow 2x_1 + 5x_2 \leq 80$$
$$x_1 + x_2 \leq 20$$
$$x_1, x_2 \geq 0$$

Following the usual procedure, we draw the following graph:

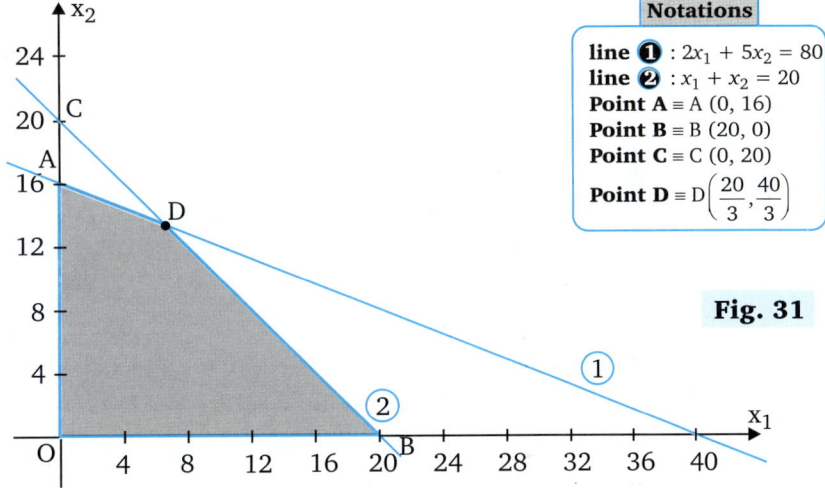

Notations

line ❶ : $2x_1 + 5x_2 = 80$
line ❷ : $x_1 + x_2 = 20$
Point A ≡ A (0, 16)
Point B ≡ B (20, 0)
Point C ≡ C (0, 20)
Point D ≡ D$\left(\dfrac{20}{3}, \dfrac{40}{3}\right)$

Fig. 31

Solving simultaneously, the equations of the corresponding intersecting lines of the feasible region, we get the coordinates of the vertices of the feasible region is

$O(0, 0), A(0, 16), B(20, 0)$ and $D = \left(\dfrac{20}{3}, \dfrac{40}{3}\right)$.

The value of the objective function at these corner points as given below:

Extreme point	Coordinates (x_1, x_2)	Value of the objective function $Z = -x_1 + 5x_2$
O	(0, 0)	$-0 + 5 \times 0 = 0$
A	(0, 16)	$-1 \times 0 + 5 \times 16 = 80$ (max.)
B	(20, 0)	$-1 \times 20 + 0 \times 5 = -20$
D	$\left(\dfrac{20}{3}, \dfrac{40}{3}\right)$	$-1 \times \dfrac{20}{3} + 5 \times \dfrac{40}{3} = 60$

Hence, the maximum value of Z is 80 at the point
$$x_1 = 0, x_2 = 16$$

EXAMPLE 3. *A soft drink plant has two bottling machines A and B. It produces and sells 8 ounce and 16 ounce bottles. The following data is available:*

Machine	8 Ounce	16 Ounce
A	100/minute	40/minute
B	60/minute	75/minute

The machines can be run 8 hours per day, 5 days per week. Weekly production of the drinks cannot exceed 300000 ounces and the market can absorb 25000 eight ounce bottles and 7000 sixteen ounce bottles per week. Profit on these bottles is 15 paise and 25 paise per bottle respectively. The planner wishes to maximize his profit subject to all the production and marketing restrictions.

Formulate the problem as an LPP and then solve graphically.

SOLUTION. Let the planner produce x_1 and x_2 numbers of bottles of 8 and 16 ounces respectively per week.

Then his profit in ₹ is $Z = 0.15x_1 + 0.25x_2$

Total production of soft drink per week to fill up these bottles $= 8x_1 + 16x_2$

Since, the production cannot exceed 300000 ounces.

Therefore, $8x_1 + 16x_2 \le 300000$

Total time taken to produce these bottles on machine A

$$= \frac{x_1}{100 \times 60} + \frac{x_2}{40 \times 60} \text{ hours}$$

On machine $B = \frac{x_1}{60 \times 60} + \frac{x_2}{75 \times 60}$ hours

But machines can run for $8 \times 5 = 40$ hours per week

$$\therefore \quad \frac{x_1}{100 \times 60} + \frac{x_2}{40 \times 60} \le 40 \text{ and } \frac{x_1}{60 \times 60} + \frac{x_2}{75 \times 60} \le 40$$

or $\qquad 2x_1 + 5x_2 \le 480000$ and $5x_1 + 4x_2 \le 720000$

Also, as per given $\quad x_1 \le 25000$ and $x_2 \le 7000$

$$x_1, x_2 \ge 0$$

Hence, the mathematical model of the given LPP as follows:

$$\text{Max } Z = 0.15x_1 + 0.25x_2$$

subject to the constraints

$$8x_1 + 16x_2 \le 300000, \, 2x_1 + 5x_2 \le 480000$$

$$5x_1 + 4x_2 \le 720000, \, x_1 \le 25000, \, x_2 \le 7000 \text{ and } x_1, x_2 \ge 0$$

Following the usual procedure, we draw the graph as follows:

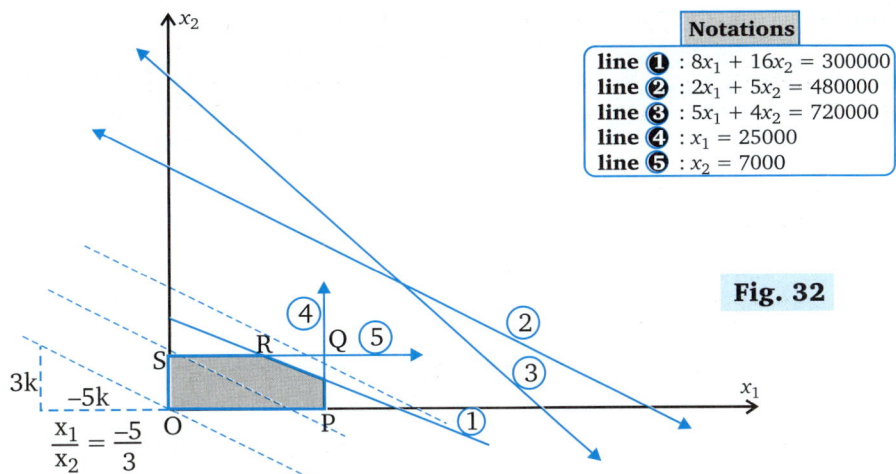

Notations

line ❶ : $8x_1 + 16x_2 = 300000$
line ❷ : $2x_1 + 5x_2 = 480000$
line ❸ : $5x_1 + 4x_2 = 720000$
line ❹ : $x_1 = 25000$
line ❺ : $x_2 = 7000$

Fig. 32

Solving simultaneously, the equations of the corresponding intersecting lines, the coordinates of the vertices of the convex polygon are $O(0, 0)$, $P(25000, 0)$, $Q(25000, 6250)$, $R(23500, 7000)$, $S(0, 7000)$.

Now, we draw dotted lines through the origin corresponding to $Z = 0$. Drawing lines parallel to this line away from the origin we see that the farthest line for maximum Z in the permissible region passes through the vertex $Q(25000, 6250)$ of the convex polygon which is the point of intersection of the lines

$$8x_1 + 16x_2 = 300000 \text{ and } x_1 = 25000.$$

Hence, the optimal solution is given by

$$x_1 = 25000, x_2 = 6250$$

and Max. $Z = 0.15 \times 25000 + 0.25 \times 6250 = ₹ 5312.50$

We can also solve this LPP by corner point method as follows:

Extreme point	Coordinates (x_1, x_2)	Value of the objective function $Z = 0.15x_1 + 0.25x_2$
O	(0, 0)	$0.15 \times 0 + 0.25 \times 0 = 0$
P	(25000, 0)	$0.15 \times 25000 + 0.25 \times 0 = 3750$
Q	(25000, 6250)	$0.15 \times 25000 + 0.25 \times 6250 = 5312.50$ (max)
R	(23500, 7000)	$0.15 \times 23500 + 0.25 \times 7000 = 5275$
S	(0, 7000)	$0 + 0.25 \times 7000 = 1750$

Therefore, $Z = ₹ 5312.50$ and is maximum when $x_1 = 25000$ and $x_2 = 6250$.

EXAMPLE 4. *A company produces two types of leather belts say type A and B. Belt A is of superior quality and B is of lower quality. Profits on the two types of belts are 40 and 30 paise per belt respectively. Each belt of type A requires twice as much time as required by a belt of type B. If all belts were of type B, the company would produces 1000 belts per day. But the supply of leather is sufficient only for 800 per day. Belt A requires a fancy buckle and 400 fancy buckles are available for this per day. For belt of type B, only 700 buckles are available per day. How should the company manufacture the two types of belts in order to have maximum overall profit.*

SOLUTION.   Firstly we shall formulate the problem as LPP.

Let the company manufacture x_1 and x_2 numbers of belt of type A and B respectively. Clearly, $x_1 \geq 0, x_2 \geq 0$.

The profit Z in ₹ is given by $Z = 0.40x_1 + 0.30x_2$

If t is the time required to produce one belt of type B and so the time required to produce one belt of type A is $2t$. So, total time required for the production of the belts of two types $= 2tx_1 + tx_2$

If belts produced are of type B then only 1000 belts can be produced and it will requires 1000t time.

Thus, $2tx_1 + tx_2 \leq 1000t \Rightarrow 2x_1 + x_2 \leq 1000$

Also, $x_1 \leq 400, x_2 \leq 700, x_1 + x_2 \leq 800$

Therefore, the mathematical formulation of the given LPP is as follows:

$$\text{Max } Z = 0.40x_1 + 0.30x_2$$

subject to the constraints $2x_1 + x_2 \leq 1000, x_1 + x_2 \leq 800, x_1 \leq 400, x_2 \leq 700$ and $x_1, x_2 \geq 0$

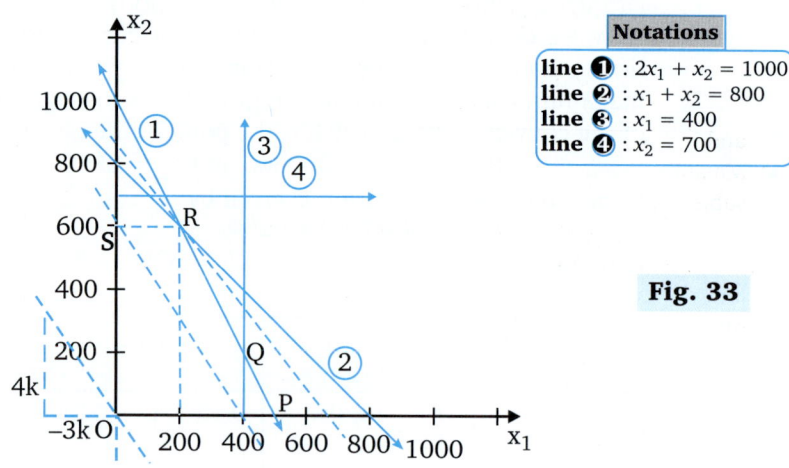

Notations
line ❶ : $2x_1 + x_2 = 1000$
line ❷ : $x_1 + x_2 = 800$
line ❸ : $x_1 = 400$
line ❹ : $x_2 = 700$

Fig. 33

To solve the above LPP, proceeding stepwise the permissible region is $OPQRSO$ and the optimal value of Z occurs at the corner $R(200, 600)$ of the convex polygon, which is clearly the point of the intersection of the lines.

$$2x_1 + x_2 = 1000$$
$$x_1 + x_2 = 800$$

Therefore, $x_1 = 200, x_2 = 600$

and Max. $Z = 0.40 \times 200 + 0.30 \times 600 = ₹ 260$

Hence, company should produce 200 belts of type A and 600 belts of type B and the maximum profit of the company will be ₹260.

Exercise-3.2

Solve the following LPP by graphical method (Qus 1-14)

1. Max. $Z = x_1 + x_2$
   subject to the constraints
   $$x_1 + 2x_2 \le 2000$$
   $$x_1 + x_2 \le 1500$$
   $$x_2 \le 600$$
   and $\quad x_1, x_2 \ge 0$

2. Minimize $Z = 20x_1 + 10x_2$
   subject to the constraints
   $$x_1 + x_2 \le 40$$
   $$3x_1 + x_2 \ge 30$$
   $$4x_1 + 3x_2 \ge 60 \quad \text{[MEERUT-2005,07,11,12;}$$
   and $\quad x_1, x_2 \ge 0 \qquad \text{KANPUR-2007]}$

3. Maximize $Z = 8x_1 + 7x_2$
   subject to the constraints
   $$3x_1 + x_2 \le 66000$$
   $$x_1 + x_2 \le 45000$$
   $$x_1 \le 20000$$
   $$x_2 \le 40000$$
   and $\quad x_1, x_2 \ge 0$

4. Maximize $Z = 6x_1 + 11x_2$
   subject to the constraints
   $$2x_1 + x_2 \le 104$$
   $$x_1 + 2x_2 \le 76$$
   and $\quad x_1, x_2 \ge 0$

5. Minimize $Z = 3x_1 + 5x_2$
   subject to the constraints
   $$-3x_1 + 4x_2 \le 12$$
   $$2x_1 - x_2 \ge -2$$
   $$2x_1 + 3x_2 \ge 12$$
   $$x_1 \le 4$$
   $$x_2 \ge 2$$
   and $\quad x_1, x_2 \ge 0$

6. Maximize $Z = 3x_1 + 4x_2$
   subject to the constraints
   $$x_1 - x_2 \le -1$$
   $$-x_1 + x_2 \le 0$$
   and $\quad x_1, x_2 \ge 0$

7. Maximize $Z = 6x_1 - 2x_2$
   subject to the constraints
   $$2x_1 - x_2 \le 2$$
   $$x_1 \le 3$$
   and $\quad x_1, x_2 \ge 0$

8. Maximize $Z = 3x_1 + 4x_2$
   subject to the constraints
   $$5x_1 + 4x_2 \le 200$$

   $$3x_1 + 5x_2 \le 150$$
   $$5x_1 + 4x_2 \ge 100$$
   $$8x_1 + 4x_2 \ge 80$$
   and $\quad x_1, x_2 \ge 0$

9. Minimize $Z = 2x_1 - 10x_2$
   subject to the constraints
   $$x_1 - x_2 \ge 0$$
   $$x_1 - 5x_2 \ge -5$$
   and $\quad x_1, x_2 \ge 0$

10. Maximize $Z = 7x_1 + 3x_2$
    subject to the constraints
    $$x_1 + 2x_2 \le 3$$
    $$x_1 + x_2 \le 4$$
    $$0 \le x_1 \le \frac{5}{2}$$
    $$0 \le x_2 \le \frac{3}{2}$$

11. Maximize $Z = 5x_1 + 3x_2$
    subject to the constraints
    $$3x_1 + 5x_2 \le 15$$
    $$5x_1 + 2x_2 \le 10$$
    and $\quad x_1, x_2 \ge 0$

12. Maximize $Z = 2x_1 + x_2$
    subject to the constraints
    $$x_1 + 2x_2 \le 10$$
    $$x_1 + x_2 \le 6$$
    $$x_1 - x_2 \le 2$$
    $$x_1 - 2x_2 \le 1$$
    and $\quad x_1, x_2 \ge 0$

13. Maximize $Z = 3x_1 + 2x_2$
    subject to the constraints
    $$x_1 - x_2 \le 1$$
    $$x_1 + x_2 \ge 3$$
    and $\quad x_1, x_2 \ge 0$

14. Maximize $Z = -3x_1 + 2x_2$
    subject to the constraints
    $$x_1 \le 3$$
    $$x_1 - x_2 \le 0$$
    and $\quad x_1, x_2 \ge 0$

15. A dietician mixes two types of food in such a way that the vitamin contents of the mixture contain at least 8 units of vitamin A and 10 units of vitamin B. Food X contains 2 units/ Kg of vitamin A and 1 unit/Kg of Vitamin C while Food Y contains 1 unit/Kg of vitamin A and 2 units/Kg of vitamin C. One Kg of Food X costs ₹5 whereas one Kg of food Y costs

₹7. Determine the minimum cost of such a mixture.

16. A farm is engaged in breeding hens. In view of the need to ensure certain nutrients (say x_1, x_2, x_3), it is necessary to buy two types of food say A and B. One unit of food A contains 36 units of x_1, 3 units of x_2 and 20 units of x_3. One unit of food B contains 6 units of x_1, 12 units of x_2 and 10 units of x_3. The minimum daily requirement of x_1, x_2 and x_3 is 108, 36 and 100 units respectively. The cost of food A is ₹20 per unit whereas food B costs ₹40 per unit. Find the minimum food cost so as to meet the minimum daily requirement of nutrients.

17. Two manager of courier company wishes to hire extra helpers during the Christmas season, because of a large increase in the volume of mail handling and number of temporary helpers must not exceed 10. According to the past experience, a man can handle 300 letters and 80 packages per day and a woman can handle 400 letters and 50 packages per day. It is believed that daily volume of the extra mail and packages will be no less than 3400 and 680 respectively. A man receives ₹25 per day and a woman receives ₹22 a day. How many men and women helpers should be hired to keep the pay roll at a minimum.

18. The *ABC* electric appliances company produces two products refrigerators and coolers. Production takes place in two separate departments. Refrigerators are produced in department-I and coolers produced in department-II. The company's two products are produced and sold on a weekly basis. The weekly production can not exceed 25 refrigerators in department-I and 35 coolers in department-II, because of the limited available facilities in these two departments. A refrigerator requires 2 man-week of labour, while a cooler requires 1 man-week of labour. A refrigerator contributes a profit of ₹60 and a cooler contributes a profit of ₹40. How many units of refrigerators and coolers should the company produces to realize maximum profit.

19. An automobile manufacture makes automobiles and trucks in a factory that is divided into two shops. Shop A, which performs the basic assembly operation, must work 5 man-days on each truck but only 2 man-days on each automobile. Shop B, which performs finishing operation must work 3 man-days for each automobile or truck that it produces. Because of men and machine limitations, shop A has 180 man-days per week available while shop B has 135 man-days per week. If the manufacturer makes a profit of ₹300 on each truck and ₹200 on each automobile, how many of each should he produce to maximize the profit.

20. A person requires 10, 12 and 12 units of chemicals A, B and C respectively for his garden. A liquid product contains 5, 2 and 1 units of A, B and C respectively per jar. A dry product contains 1, 2 and 4 units of A, B and C per carton. If the liquid product sells ₹3 per jar and the dry product sells ₹2 per carton. How many of each should be purchased to minimize the cost and meet the requirements.

ANSWERS

1. $x_1 = 1000, x_2 = 500$, Max $Z = 1500$

2. $x_1 = 6, x_2 = 12$, Min. $Z = 240$

3. $x_1 = 10500, x_2 = 34500$, Max $Z = 325000$

4. $x_1 = 44, x_2 = 16$, Max $Z = 440$

5. $x_1 = 3, x_2 = 2$, Min$(Z) = 19$ **6.** No solution

7. $x_1 = 3, x_2 = 4$, Max$(Z) = 10$

8. $x_1 = \dfrac{400}{13}, x_2 = \dfrac{150}{13}$, Max$(Z) = \dfrac{1800}{13}$

9. $x_1 = \dfrac{5}{4}, x_2 = \dfrac{5}{4}$, Min$(Z) = -10$

10. $x_1 = \dfrac{5}{2}, x_2 = \dfrac{1}{4}$, Max $Z = \dfrac{73}{4}$

11. $x_1 = \dfrac{20}{19}, x_2 = \dfrac{45}{19}$, Max $Z = \dfrac{235}{19}$

12. $x_1 = 4, x_2 = 2$, Max $Z = 10$ **13.** Unbounded solution **14.** Unbounded solution

15. ₹38 **16.** ₹160 **17.** men = 6, women = 4, ₹238

18. Refrigerators $= \dfrac{25}{2}$, Coolers = 35, ₹2150 per week

19. Automobiles = 15, Trucks = 30, maximum profit = ₹12000

20. $x_1 = 1, x_2 = 5$ and Min. $Z = 13$

REVIEW QUESTIONS

1. What is linear programming problem?
2. Define constraints, objective function and non-negative restrictions.
3. Define feasible and infeasible solution of an LPP.
4. Define optimal feasible solution and unbounded solution of an LPP.
5. Explain the graphical method of solving an LPP.
6. Define the term 'Feasible region'.
7. Define iso-profit and iso-cost lines.
8. Explain the procedure for generating extreme points solutions to an LPP.
9. How the unbounded solution be recognized in the graphical method.

MULTIPLE CHOICE QUESTIONS (CHOOSE THE MOST APPROPRIATE ONE)

1. For the inequality $4x + 3y > 24$, the point of intersection are:
   (a) $(0, 0), (0, 0)$ (b) $(0, 0), (0, 6)$
   (c) $(0, 8), (6, 0)$ (d) None of these

2. The set of values of the variable $x_1, x_2, ..., x_n$ satisfying the constraints and non negative restriction of a L.P.P. is called:
   (a) bounded Solution (b) feasible Solution
   (c) no Solution (d) none of these

3. The positive variables which are subtracted from the left hand side of the constraints to convert them into equalities are called the:
   (a) slack variables (b) surplus variable
   (c) non zero variable (d) none of the above

4. In a L.P.P. if the standard primal problem is of minimization, all the constraints involve the sign:
   (a) $\geq$ (b) $\leq$
   (c) $=$ (d) None of these

5. If the value of the objective function, Z can be increased or decreased infinitely. Such solution are called:
   (a) bounded Solution
   (b) unbounded Solution
   (c) optimal Solution
   (d) none of these

6. The system of linear inequations in an L.P.P. under which the given function is to be optimized are called as:
   (a) equations (b) inequalities
   (c) constraints (d) none of these

7. How many number of variables at least must vanish for a feasible solution to be a Basic feasible solution?

   (a) n (b) m
   (c) $n - m$ (d) $n - 1$

8. By the graphical method the solution of L.P.P. given by max $Z = 3x_1 + 2x_2$
   s.t. $x_1 - x_2 \leq 1$
   $x_1 + x_2 \geq 3$ and $x_1, x_2 \geq 0$ is:
   (a) unbounded solution
   (b) bounded solution
   (c) finite solution
   (d) None of these

9. For a L.P.P. min $Z = 2x_1 + 3x_2 + 4x_3$
   s.t. $x_1 + x_2 + x_3 = 2$
   $x_1 - x_2 + x_3 = 2$ and $x_1, x_2, x_3 \geq 0$
   the feasible solution is $x_1 = 1$, $x_2 = 0$, $x_3 = 1$ is:
   (a) a basic solution
   (b) not a basic solution
   (c) finite solution
   (d) None of these

10. How many variables in L.P.P. can be solved easily graphically?
    (a) two (b) three
    (c) four (d) None of these

11. A B.F.S. of a L.P.P. is said to be degenerate B.F.S. if:
    (a) all basic variables are zero
    (b) at least one of the basic variable is zero
    (c) none of the basic variables is zero
    (d) none of these

12. In graphical solution of solving L.P. problem to convert inequalities into equations:
    (a) use slack variables
    (b) use surplus variables
    (c) use artificial variables
    (d) simply assume them to be equations

13. Given a system of m simultaneous linear equations in n unknowns ($m < n$) the number of basic variables will be:
    (a) m (b) n
    (c) $n - m$ (d) $n + m$

14. For solution of a L.P.P. the Iso-Cost method is:
    (a) graphical Method
    (b) an analytical method
    (c) simplex method
    (d) none of these

15. In a L.P.P. :
    (a) objective function of constraints and variables are all linear
    (b) only constraints are linear
    (c) only objective function are linear
    (d) only variable are linear

16. In a L.P.P. of the standard primal problem of maximization all the constraints involve the sign:
    (a) ≥   (b) ≤
    (c) =   (d) None of these

17. Linear inequation in L.P.P. are called:
    (a) constraints   (b) F.S.
    (c) two variables   (d) none of these

18. If a L.P.P. have no feasible region then we say that the problem has:
    (a) unbounded solution
    (b) no solution
    (c) bounded solution
    (d) none of these

19. The Iso-Cost line represents:
    (a) a boundary of the feasible region
    (b) an infinite number of solution all of which yield the same profit
    (c) an infinite number of solution all of which yield the same cost
    (d) none of these

20. Solving a L.P.P. graphically, the area bounded by equation of the constraints called:
    (a) unbounded region
    (b) infeasible region
    (c) feasible region
    (d) none of these

21. Set of values of the variables $x_1, x_2, ..., x_n$ satisfying constraints and Non-negative restrictions is a:
    (a) feasible solution   (b) ^nC_m
    (c) vector   (d) None of these

22. The graphical method used in L.P.P. for:
    (a) linear equations
    (b) constraint equations
    (c) objective function equations
    (d) all of these

23. Which of the following is an assumption of an L.P. model:
    (a) additivity   (b) proportionality
    (c) divisibility   (d) all of these

24. Necessary and sufficient condition for existence and non-degeneracy of all the basic solutions of $AX = B$, is that every set of m columns of (A/B) is:
    (a) linearly independent
    (b) F.S.
    (c) vector
    (d) none of these

25. If a non-redundant constraint is removed from a L.P.P. then:
    (a) feasible region will become larger
    (b) feasible region will become smaller
    (c) solution will become infeasible
    (d) none of these

26. The variables are to be determined in L.P.P. are always:
    (a) negative   (b) non negative
    (c) positive   (d) non positive

27. Solving a L.P.P. in feasibility may be removed by:
    (a) adding another variable
    (b) adding another constraint
    (c) removing a variable
    (d) removing constraint

28. A constraint in a L.P.P. become redundant because:
    (a) the constraint is not satisfied by the solution values
    (b) two iso profit lines may be parallel to each other
    (c) the solution is unbounded
    (d) none of these

29. If a L.P.P. have B.F.S, if we drop one basis vector and introduce a Non basis vector in basis, then new solution obtained is:
    (a) a B.F.S.   (b) a non B.F.S
    (c) no solution exist   (d) none of these

30. If an iso-profit line yielding the optimal solution coincides with a constraint line, then:

(a) the solution is unbounded

(b) the solution is feasible

(c) the constraint which coincides is redundant

(d) None of these

31. Which of the following statement is true with respect to the optimal solution of an L.P. problem:

(a) every L.P. problem has an optimal solution.

(b) optimal solution of an L.P. problem always occurs at an extreme points.

(c) at optimal solution a resources are used completely.

(d) if an optimal solution exists, there will always be at least one at a corner

32. A feasible solution to an LP problem:

(a) must satisfy all the problems constraints simultaneously

(b) need not satisfy all of the constraints

(c) only some of them must be a corner point of the feasible region

(d) must optimize the value of the objective function

33. Non-Negativity condition is an important component of LP model because:

(a) variables are interrelated in terms of limited resources

(b) variable values should remain under the control of decision maker

(c) value of variables make sense and correspond to real word problems

(d) none of the above

34. Constraints in an LP model represents:

(a) limitations

(b) requirements

(c) balancing limitations and requirements

(d) all of the above

35. The distinguishing feature of an LP model is:

(a) it has single objective function and constraints

(b) value of decision variables is non-negative

(c) relationship among all variables is linear

(d) all of the above

36. A variable which has no physical meaning but is used to obtain an initial basic feasible solution to the linear programming problem is referred to as:

(a) basic variable   (b) non-basic variable

(c) artificial variable (d) basis

37. A feasible solution to the linear programming problem should:

(a) satisfy the problem constraints

(b) optimize the objective function

(c) satisfy the problem constraints and non-negativity restrictions

(d) satisfy the non-negativity restrictions

38. The solution space (region) of an LP problem is unbounded due to:

(a) objective function is unbalanced

(b) an incorrect formulation of the LP model

(c) neither (a) nor (b)

(d) both (a) and (b)

39. Constraints in LP Problem are called active if they:

(a) represent optimal solution

(b) at optimality do not consume all the available resources

(c) both (a) and (b)

(d) none of the above

40. If two constraints do not intersect in the positive quadrant of the graph, then:

(a) one of the constraint is redundant

(b) the region is unbounded

(c) the problem is infeasible

(d) none of the above

41. A constraint in an LP model becomes redundant because:

(a) two iso-profit lines may be parallel to each other

(b) the solution is unbounded

(c) this constraint is not satisfied by the solution value

(d) none of the above

42. While plotting constraints on a graph paper, terminal points on both the axes are connected by a straight line because:

(a) the constraints are linear equations or inequalities

(b) the resources are limited in supply

(c) the objective function is a linear function

(d) all of the above

43. If there is no feasible solution in a L.P.P. then we say that the problem has:

(a) infinite solution

(b) no solution

(c) unbounded solution

(d) none of these

44. An iso-profit line represents:

(a) an infinite number of solutions all of which yield the same profit

(b) an infinite number of optimum solutions

(c) an infinite number of solutions all of which yield the same cost

(d) a boundary of the feasible region

45. In a L.P.P.:

(a) number of BFS $\leq$ Number of vertices

(b) number of BFS = Number of vertices

(c) number of BFS $\geq$ Number of vertices

(d) none of these

ANSWERS

1. (c)	**2.** (b)	**3.** (b)	**4.** (a)	**5.** (b)	**6.** (c)	**7.** (c)	**8.** (a)	**9.** (b)
10. (a)	**11.** (b)	**12.** (d)	**13.** (c)	**14.** (a)	**15.** (a)	**16.** (b)	**17.** (a)	**18.** (b)
19. (c)	**20.** (c)	**21.** (a)	**22.** (d)	**23.** (d)	**24.** (a)	**25.** (a)	**26.** (b)	**27.** (d)
28. (d)	**29.** (a)	**30.** (c)	**31.** (b)	**32.** (b)	**33.** (c)	**34.** (d)	**35.** (c)	**36.** (b)
37. (c)	**38.** (c)	**39.** (b)	**40.** (a)	**41.** (b)	**42.** (b)	**43.** (b)	**44.** (d)	**45.** (c)

ARCHIVE

1. A firm manufactures three products A, B and C. Time to manufacture product A is twice that for B and thrice that for C and if the entire labour is engaged in making product A, 1600 units of this product can be produced. These products are to be produced in the ratio 3 : 4 : 5. There is demand for at least 300, 250 and 200 units of products A, B and C and the profit earned per unit is ₹ 90, ₹40 and ₹ 30 respectively. Formulate the problem as a linear programming problem.

[PT(BE)- 2001]

Raw Material	Requirement per unit of Product (kg)			Total Availability (kg)
	A	B	C	
P	6	5	2	5,000
Q	4	7	3	6,000

2. A company has two grades of inspectors 1 and 2, who are to be assigned for a quality control inspection. It is required that at least 2,000 pieces be inspected per 8-hour day. Grade 1 inspector can check pieces at the rate of 40 per hour, with an accuracy of 97 percent. Grade 2 inspector checks at the rate of 30 pieces per hour with an accuracy of 95 percent.

The wage rate of a grade 1 inspector is ₹ 5 per hour while that of a grade 2 inspector is ₹ 4 per hour. An error made by an inspector costs ₹ 3 to the company. There are only nine grade 1 inspectors and elevan grade 2 inspectors available in the company. The company wishes to assign work to the available inspectors so as to minimize the total cost of the inspection. Formulate this problem as an LP model so as to minimize daily inspection cost.

[DELHI (MBA) 2000; NIFT (MOHALI), 2000]

3. An electronic company produces three types of parts for automatic washing machines. It purchases casting of the parts from a local foundry and then finishes the part on drilling, shaping and polishing machines.

The selling prices of parts A, B and C, respectively are ₹ 8, ₹ 10 and ₹ 14. All parts made can be sold. Castings for parts A, B and C, respectively cost ₹ 5, ₹ 6 and ₹ 10.

The shop possesses only one of each type of machine. Costs per hour to run each of the three machines are ₹ 20 for drilling, ₹30 for shaping and ₹ 30 for polishing. The capacities (parts per hour) for each part of each machine are shown in the table.

Machine	Capacity Per Hour		
	Part A	Part B	Part C
Drilling	25	40	25
Shaping	25	20	20
Polishing	40	30	40

The management of the shop wants to know how many parts of each type should produce per hour in order to maximize profit for an hour's run. Formulate this problem as an LP model so as to maximize total profit to the company.   [DELHI (MBA) 1997, 2000]

4. A plant manufactures washing machines and dryers. The major manufacturing departments are the stamping deptt., motor and transmission deptt. and assembly deptt. The first two departments produce parts for both the products while the assembly lines are different for the two products. The monthly deptt. capacities are,

Stamping deptt.	1,000 washers or 1,000 dryers
Motor and transmission deptt.	1,600 washers or 7,000 dryers
Washer assembly line	9,000 washers only
Dryer assembly line	5,000 dryers only

Profits per piece of washers and dryers are ₹ 270 and ₹ 300 respectively. Formulate this problem as an LP model.

[PUNJAB (B.COM) 2005]

5. A manufacturer has five lathes and three milling machines in his workshop and produces an assembly that consists of 2 units of part A and 3 units of part B. The processing time for each part on the two types of machines is given in the table.

Part	Processing Time (in Minutes)	
	Lathe	Milling Machine
A	10	18
B	25	12

In order to maintain a uniform workload on the two types of machines, the manufacturer has framed a policy that no machine should run more than 40 minutes per day longer than the other machine. Formulate this problem as an LP model if the objective is to produce the maximum number of assemblies in any 8-hour working day.

[PT (B. Tech), 2001;
PUNJAB, BE (E AND EC), 1999]

6. *ABC* Foods Company is developing a low-calorie high-protein diet supplement called Hi-Pro. The specifications for Hi-Pro have been established by a panel of medical experts. These specifications along with calorie, protein and vitamin content of three basic foods, are given in the following table :

Nutritional Elements	Units of Nutritional Elements (Per 100 gm Serving of Basic Foods)			Basic Foods Hi-Pro specifications
	1	2	3	
Calories	350	250	200	300
Proteins	250	300	150	200
Vitamin A	100	150	75	100
Vitamin C	75	125	150	100
Cost per serving (₹)	1.50	2.00	1.20	

What quantities of foods 1, 2 and 3 should be used? Formulate this problem as an LP model to minimize cost of serving.

[DELHI, (MBA), 1995, 2000]

7. Vitamins A and B are found in foods F_1 and F_2. One unit of food F_1 contains 3 units of vitamin A and 4 units of vitamin B. One unit of food F_2 contains 6 units of vitamin A and 3 units of vitamin B. One unit of food F_1 and F_2 cost ₹ 4 and ₹ 5, respectively. The minimum daily requirement (for a person) of vitamins A and B is 80 and 100 units, respectively. Assuming that anything in excess of the daily minimum requirement of A and B is not harmful. Formulate this problem as an LP model to find out the optimum mixture of food F_1 and F_2 at the minimum cost which meets the daily minimum requirement of vitamins A and B.   [DELHI (MBA), 2000]

8. Solve the following LP problem graphically :

Maximize $Z = -x_1 + 2x_2$

subject to the constraints

$$x_1 - x_2 \leq -1;$$
$$-0.5x_1 + x_2 \leq 2$$

and $x_1, x_2 \geq 0.$

[PUNJAB(BE)–2004]

9. Use graphical method to solve the following LP problem :

Maximize $Z = 5x_1 + 4x_2$

subject to the constraints

$$x_1 - 2x_2 \geq 1;$$
$$x_1 + 2x_2 \geq 3$$

and $x_1, x_2 \geq 0.$

[PT (BE) 2001]

10. G.J. Breweries Ltd. have two bottling plants one located at 'G' and the other at 'J'. Each plant products three drinks, whisky, beer and brandy named A, B and C respectively. The number of the bottles produced per day as shown in the table :

Drink	Plant at	
	G	J
Whisky	1,500	1,500
Beer	3,000	1,000
Brandy	2,000	5,000

A market survey indicates that during the month of July, there will be a demand of 20,000 bottles of whisky, 40,000 bottles of beer and 44,000 bottles of brandy. The operating cost per day for plants at G and J are 600 and 400 monetary units. For how many days each plant be run in July so as to minimize the production cost, while still meeting the market demand? Solve graphically.   [PUNE (MBA) 2000]

HINTS AND ANSWERS

1. **Decision variables** : Let x_1, x_2 and x_3 = Number of units of products A, B and C to be manufactured, respectively.

Maximize (total profit) $Z = 90x_1 + 40x_2 + 30x_3$

Subject to the constraints

(i) Raw material P and Q required :

$P : 6x_1 + 5x_2 + 2x_3 \leq 5,000,$ $Q : 4x_1 + 7x_2 + 3x_3 \leq 6,000$

(ii) Manufacturing time required for product A, B and C :

$$x_1 + \frac{x_2}{2} + \frac{x_3}{3} \leq 1,600$$

(iii) Market demand requires : $x_1 \geq 300, x_2 \geq 250, x_3 \geq 200$

(iv) Production ratio : $x_1 : x_2 : x_3 :: 3 : 4 : 5,$ *i.e.* $\dfrac{x_1}{3} = \dfrac{x_2}{4}$ and $\dfrac{x_2}{4} = \dfrac{x_3}{5}$

or $4x_1 - 3x_2 = 0$ and $5x_2 - 4x_3 = 0$

and $x_1, x_2, x_3 \geq 0$

2. **Decision variables** : Let x_1 and x_2 = number of Grades 1 and 2 inspectors to be assigned for inspection, respectively.

The LP model : Hourly cost of each of Grades 1 and 2 inspectors can be computed as follows :

Inspector Grade 1 : ₹ (5 + 3×40×0.03) = ₹ 8.60

Inspector Grade 2 : ₹ (4 + 3×30×0.05) = ₹ 8.50

Based on the given data, the linear programming problem can be formulated as follows :

Minimize (daily inspection cost) $Z = 8(8.60x_1 + 8.50x_2)$

$$= 68.80x_1 + 68.00x_2$$

subject to the constraints

(i) Total number of pieces that must be inspected in an 8-hour day.

$$8 \times 40x_1 + 8 \times 30x_2 \geq 2,000$$

(ii) Number of inspectors of Grade 1 and 2 available

$$x_1 \leq 9; x_2 \leq 11$$

and $$x_1, x_2 \geq 0$$

3. LP Model Formulation

Let x_1, x_2 and x_3 = number of type A, B and C parts to be produced per hour, respectively.

The LP Model :

Maximize (total profit) $Z = 0.25x_1 + 1.00x_2 + 0.95x_3$

subject to the constraints

(i) Drilling machine (ii) Shaping machine (iii) Polishing machine

$$\frac{x_1}{25} + \frac{x_2}{40} + \frac{x_3}{25} \leq 1 \qquad \frac{x_1}{25} + \frac{x_2}{20} + \frac{x_3}{20} \leq 1 \qquad \frac{x_1}{40} + \frac{x_2}{30} + \frac{x_3}{40} \leq 1$$

and $x_1, x_2, x_3 \geq 0$.

4. LP Model Formulation

Decision variables :

Let x_1 and x_2 = number of washing machines and dryers to be manufactured each month, respectively.

The LP Model :

Maximize (total profit each month) $Z = 270x_1 + 300x_2$

subject to the constraints

(i) Capacity of stamping deptt. (ii) Capacity of motor and transmission deptt.

$$\frac{x_1}{1,000} + \frac{x_2}{1,000} \leq 1 \qquad \frac{x_1}{1,600} + \frac{x_2}{7,000} \leq 1$$

(iii) Capacity of washer and dryer assembly deptt.

$$x_1 \leq 9,000, x_2 \leq 5,000$$

and $x_1, x_2 \geq 0$.

5. LP Model Formulation

Decision variables :

Let x_1 and x_2 = number of units of part A and B to be produced each day that make the assembly, respectively.

The LP Model :

Maximize (number of units of the final assembly) Z = Minimum of

$$\left(\frac{x_1}{2}, \frac{x_2}{3}\right) = y \text{ (say) where } y \leq \frac{x_1}{2} \text{ or } x_1 - 2y \geq 0 \text{ and } y \leq \frac{x_2}{3} \text{ or } x_2 - 3y \geq 0$$

subject to the constraints

(i) Time for lathes machines yield

$$10x_1 + 25x_2 \leq 5 \text{ (machines)} \times 8(\text{hrs/day}) \times 60 \text{ (minutes)}$$

or $$2x_1 + 5x_2 \leq 480,$$

(ii) Time for milling machines

$$18x_1 + 12x_2 \leq 3 \times 8 \times 60 \quad \text{or} \quad 3x_1 + 2x_2 \leq 240.$$

(iii) Uniform workload

$$\left(\frac{10x_1 + 25x_2}{5}\right) - \left(\frac{18x_1 + 12x_2}{3}\right) \leq 40 \text{ or } 4x_1 - x_2 \geq 4 \text{ and } x_1, x_2, y \geq 0$$

6. Let x_1, x_2 and x_3 = quantity of foods 1, 2 and 3 to be used, respectively.

Min $Z = 1.50x_1 + 2.00x_2 + 1.20x_3$

subject to,

(i) $350x_1 + 250x_2 + 200x_3 \geq 300$;

(ii) $250x_1 + 300x_2 + 150x_3 \geq 200$;

(iii) $100x_1 + 150x_2 + 75x_3 \geq 100$;

(iv) $75x_1 + 125x_2 + 150x_3 \geq 100$;

   and $x_1, x_2, x_3 \geq 0$.

7. Let x_1 and x_2 = number of vitamin units purchased of food F_1 and F_2, respectively.

Min $Z = 4x_1 + 5x_2$

subject to,

(i) $3x_1 + 6x_2 \geq 80$;

(ii) $4x_1 + 3x_2 \geq 100$

   and $x_1, x_2 \geq 0$.

8. Given problem has multiple optimal solutions and Max. $Z = 4$.

9. Given problem has an unbounded solution.

10. The minimum value of objective function occurs at point (12, 4). Hence, the plant G should run for $x_1 = 12$ days and plant J for $x_2 = 4$ days to have a minimum production cost of ₹ 8,800.

❑❑❑❑

4 Simplex Method

4.1 INTRODUCTION

The 'simplex' is an important term in Mathematics that represents an object in n-dimensional space containing $(n + 1)$ points. Obviously, in one dimension, a simplex is a line segment connecting two points, in two dimensions, it is a triangle formed by joining these points, in three dimensions, it is a four sided pyramid having four corners. In this chapter, we shall discuss a procedure called the 'simplex method' for solving an LP model, which was developed by G.B. Dantzig in 1947. The simplex method examines the extreme points in a systematic manner repeating the same set of steps of the algorithm until an optimal solution is reached. Due to this reason, it is called an iterative method.

The simplex method is an algebraic procedure that start at a feasible extreme point of the simplex or convex set; normally the origin and systematically moves from one feasible extreme point to another until an optimal extreme point is located.

4.2 TERMINOLOGY AND NOTATIONS

The general form of an LPP is given as below:

$$\text{Max. } Z = \boldsymbol{C} \cdot \boldsymbol{x}$$

subject to $\quad A\boldsymbol{x} = \boldsymbol{b}, \boldsymbol{x} \geq 0$

where $\quad A = [a_{ij}]_{m \times N}, \boldsymbol{x} = (x_1, x_2, \ldots, x_n, \ldots, x_N)_{N \times 1}$

$\boldsymbol{C} = (c_1, c_2, \ldots, c_n, 0, 0, \ldots, 0)_{1 \times N}$

and $\quad b = [b_1, b_2, \ldots, b_m]_{m \times 1}$

Now, different symbolic representation is given in the following table:

S.No.	Name	Representation	Description
1.	Basic matrix	$B = \{\beta_1, \beta_2, \ldots, \beta_m\}$	A non-singular submatrix B of order $m \times n$ whose column vectors are m no. of linearly independent columns selected from A.
2.	Basic Variables	$x_{B_1}, x_{B_2} \ldots x_{B_m}$	The variable corresponding to $\beta_1, \beta_2, \ldots, \beta_m$ are called basic variables.
3.	Basic Feasible Solution	$\boldsymbol{X}_B = B^{-1} \cdot \boldsymbol{b}$	$\boldsymbol{X}_B = [x_{B_1}, x_{B_2}, \ldots x_{B_m}] = B^{-1} \cdot \boldsymbol{b}$
4.	Non-basic variables	$x_{B_{m_i}} i > m$	The variables, other than basic, are called non-basic variables
5.	Coefficient of basic variables	$C_{B_i} : i = 1 \text{ to } m$	Corresponding to any $\boldsymbol{x}_B$, $\boldsymbol{C}_B$ will represent the row vector containing the constants $C_{B_1}, C_{B_2} \ldots C_{B_m}$

General Representation: We shall represent column vectors by [ ] without using transpose symbol and row vector by ( ).

4.3 FUNDAMENTAL THEOREM OF LINEAR PROGRAMMING

If a LPP *Max. Z = **Cx***

*subject to A**x** = **b**, **x** ≥ 0*

where A is an m × N (N = m + n) matrix of coefficients given by A = ($\alpha_1, \alpha_2, ..., \alpha_N$) has an optimal feasible solution then at least one basic feasible solution must be optimal.

[MEERUT–2009, 10, 11, 14, 16; KANPUR–2010, 12; BANARAS–2014; DELHI–2010, 12]

Proof. Let us suppose that $x^* = [x_1, x_2, ..., x_n]$ be an optimal feasible solution of the given LPP and $Z^* = \sum\limits_{i=1}^{N} c_i x_i$ be the corresponding optimum value of the objective function.

If $k(k \le N)$ variables in x^* are non-zero and the remaining $N - k$ are zero such that

$$x^* = [x_1, x_2, ..., x_k, 0, 0, ..., 0]$$

Therefore, $\sum\limits_{i=1}^{k} x_i \alpha_i = b$ ...(1)

and $Z^* = \sum\limits_{i=1}^{k} c_i x_i$ ...(2)

Now we have the following two cases:

Case-I: If $\alpha_1, \alpha_2, ..., \alpha_k$ are linearly independent

Since, we know that any feasible solution for which the vectors $\alpha_i : i = 1, 2, ..., k$ associated with non-zero variables x_i ($i = 1, ..., k$) are linearly independent is called a basic feasible solution.

⇒ x^* is a basic feasible solution which is also optimal.

Here, it is also clear that x^* is degenerate if $k < m$ and non-degenerate if $k = m$.

Case-II: If $\alpha_1, \alpha_2, ..., \alpha_k$ are linearly dependent and $k > m$.

Let $\alpha_1, \alpha_2, ..., \alpha_k$ are linearly dependent. Therefore, we can write

$\lambda_1 \alpha_1 + \lambda_2 \alpha_2 + ... \lambda_k \alpha_k = 0$ for scalars $\lambda_1, \lambda_2 ..., \lambda_k$ such that at least one $\lambda_i \ne 0$.

⇒ $\sum\limits_{i=1}^{k} \lambda_i \alpha_i = 0$ ...(3)

Now, assume that at least one λ_i is positive (because if none is positive then we can multiply (3) by –1 and get positive λ_i)

Further suppose $V = \max\limits_{1 \le i \le k} \left(\dfrac{\lambda_i}{x_i} \right)$ ...(4)

Clearly, V is positive ($\because x_i > 0 \; \forall i$ and $\lambda_i > 0$ for at least one i)

So, multiply (3) by $\dfrac{1}{V}$ and then subtracting from (1) we get

$$\sum\limits_{i=1}^{k} \left(x_i - \frac{\lambda_i}{V} \right) \alpha_i = b$$ ...(5)

⇒ $\boldsymbol{x}' = \left[x_1 - \dfrac{\lambda_1}{V}, x_2 - \dfrac{\lambda_2}{V}, ..., x_k - \dfrac{\lambda_k}{V}, 0, 0, ..., 0 \right]$ is the new solution of $A\boldsymbol{x} = \boldsymbol{b}$.

From (4), for $i = 1, 2, ..., k$, we have

$$V \geq \frac{\lambda_i}{x_i} \text{ or } x_i \geq \frac{\lambda_i}{V} \qquad (\because x_i > 0, V > 0)$$

$$\Rightarrow \qquad x_i - \frac{\lambda_i}{V} \geq 0$$

$\Rightarrow$ all the components of x' are non-negative.

$\Rightarrow x'$ is a feasible solution of the given LPP.

Further, $V = \dfrac{\lambda_i}{x_i}$, for at least one i $(i = 1, 2, ..., k)$

$$\Rightarrow \quad V - \frac{\lambda_i}{x_i} = 0 \text{ for at least one } i.$$

$\Rightarrow x'$ can not have more than $k - 1$ non-zero variables.

Proceeding in this way, we find a new feasible solution from the given optimal feasible solution which contains less number of non-zero variables. This solution is BFS if the column vector associated to non-zero variables in this new solution are linearly independent. But if these associated vectors are not linearly independent, we shall repeat the whole process same as above. Therefore, continuing in this way for finite number of times we will find a solution in which columns corresponding to positive variables are linearly independent.

$\Rightarrow$ a BFS of the system under consideration exist.

Optimality of x^*

Let Z' be the new value of the objective function corresponding to the new solution x'. Then

$$Z' = \sum_{i=1}^{k} c_i\left(x_i - \frac{\lambda_i}{V}\right) = \sum_{i=1}^{k} c_i x_i - \frac{1}{V}\sum_{i=1}^{k} c_i\lambda_i$$

$$\Rightarrow \qquad Z' = Z^* - \frac{1}{V}\sum_{i=1}^{k} c_i\lambda_i \quad \text{(From (2))} \qquad\qquad ...(6)$$

For optimality, we must have $Z' = Z^*$, i.e., x' will be optimal if $\sum_{i=1}^{k} c_i\lambda_i = 0$ $\qquad ...(7)$

Let if possible $\sum_{i=1}^{k} c_i\lambda_i \neq 0$. Then either $\sum_{i=1}^{k} c_i\lambda_i < 0$ or $\sum_{i=1}^{k} c_i\lambda_i > 0$

In either of these two cases, we can find a real number r such that

$$r \cdot \sum_{i=1}^{k} c_i\lambda_i > 0$$

$$\Rightarrow \qquad \sum_{i=1}^{k} c_i(r\lambda_i) > 0$$

$$\Rightarrow \qquad \sum_{i=1}^{k} c_i(r\lambda_i) + \sum_{i=1}^{k} c_i x_i > \sum_{i=1}^{k} c_i x_i$$

$$\Rightarrow \qquad \sum_{i=1}^{k} c_i(x_i + r\lambda_i) > Z^* \qquad \text{(From 2)} \qquad\qquad ...(8)$$

Now, multiplying (3) by r and then adding to (1) we get

$$\sum_{i=1}^{k} x_i\alpha_i + \sum_{i=1}^{k} r\lambda_i\alpha_i = \mathbf{b}$$

$$\Rightarrow \qquad \sum_{i=1}^{k} (x_i + r\lambda_i)\alpha_i = \boldsymbol{b}$$

$$\Rightarrow \quad [x_1 + r\lambda_1, x_2 + r\lambda_2, \ldots, x_k + r\lambda_k, \underbrace{0, 0, \ldots, 0}_{(N-k)}] \qquad \ldots(9)$$

is also a solution of $A\boldsymbol{x} = \boldsymbol{b} \; \forall r$.

Let us choose r such that

$$x_i + r\lambda_i \geq 0, \; i = 1, 2, \ldots, k$$

$$\Rightarrow \qquad r\lambda_i \geq -x_i$$

$$\Rightarrow \qquad r \geq -\frac{x_i}{\lambda_i} \text{ if } \lambda_i > 0$$

$$\Rightarrow \qquad r \leq -\frac{x_i}{\lambda_i} \text{ if } \lambda_i < 0$$

and r is unrestricted if $\lambda_i = 0$.

So, $x_i + r\lambda_i \geq 0$ for $i = 1, 2, \ldots, k$ if we select r such that

$$\max_{(\lambda_i > 0)}\left(-\frac{x_i}{\lambda_i}\right) \leq r < \min_{(\lambda_i < 0)}\left(-\frac{x_i}{\lambda_i}\right) \qquad \ldots(10)$$

$$\Rightarrow \qquad \max_{i}\left(-\frac{x_i}{\lambda_i}\right) < 0 \text{ and } \min_{i}\left(-\frac{x_i}{\lambda_i}\right) > 0$$
$$\quad (\lambda_i > 0) \qquad\qquad\qquad (\lambda_i < 0)$$

It is clear that the interval given by (10) is non-empty.

$\Rightarrow$ r lies in the non-empty interval given by (10).

Thus, an infinite no. of solutions given by (9) satisfy the non-negative restriction also.

Finally, from (8) we find that $\sum_{i=1}^{k} c_i(x_i + r\lambda_i)$ gives the value of objective function which is greater than the greatest value Z^* of objective function, which is not possible.

So, we must have $\sum_{i=1}^{k} c_i\lambda_i = 0$

$$\Rightarrow Z' = Z^*$$

Hence, $\boldsymbol{x}' = \left[x_1 - \dfrac{\lambda_1}{V}, x_2 - \dfrac{\lambda_2}{V}, \ldots, x_k - \dfrac{\lambda_k}{V}, \underbrace{0, 0, \ldots, 0}_{(N-k)}\right]$ is also an optimal solution.

4.4 REDUCTION OF FEASIBLE SOLUTION TO BASIC FEASIBLE SOLUTION

<u>THEOREM I.</u> *If a linear programming problem*

$$\textbf{\textit{Max. Z = Cx}}$$

subject to Ax = b, x ≥ 0

where A = (α_1, α_2, ..., α_N) is the coefficient matrix of order m × N (N = m + n) has at least one feasible solution then it has one basic feasible solution also. [MEERUT–2005, 08]

<u>PROOF.</u> Let $\boldsymbol{x}^* = (x_1, x_2, \ldots, x_N) : x_i \geq 0$ ...(1)

be an arbitrary feasible solution of given LPP

Suppose that k ($k \leq N$) variables in $\boldsymbol{x}^*$ have positive values and the remaining $N - k$

variables are zero, which can be written as follows:

$$\boldsymbol{x}^* = [x_1, x_2, \ldots, x_k, \underbrace{0, 0, \ldots, 0}_{(N-k)}]$$

Also,
$$\sum_{i=1}^{k} x_i \alpha_i = \boldsymbol{b} \qquad \ldots(2)$$

Now, we have the following cases:

CASE I. If α_1, α_2, ..., α_k are linearly independent:

We know that any feasible solution for which the vectors $\alpha_i : i = 1, 2, \ldots, k$ associated with non-zero variables $x_i : i = 1, 2, \ldots, k$ are linearly independent is called a BFS. Therefore, clearly $\boldsymbol{x}^*$ is a BFS which is also optimal.

CASE II. If α_1, α_2, ..., α_k are linearly dependent and $k > m$

Let $\alpha_1, \alpha_2, \ldots, \alpha_k$ be the linearly dependent. Then by definition $\exists$ scalars $\lambda_1, \lambda_2, \ldots, \lambda_k$ such that

$$\lambda_1 \alpha_1 + \lambda_2 \alpha_2 + \ldots + \lambda_k \alpha_k = 0, \text{ at least one } \lambda_i \neq 0$$

$$\Rightarrow \sum_{i=1}^{k} \lambda_i \alpha_i = 0 \text{ for at least one } \lambda_i \neq 0 \qquad \ldots(3)$$

Without loss of any generality, we may assume that at least one λ_i is positive ($\because$ if none is positive then we can multiply (3) by –1 and get positive λ_i)

Further, suppose that

$$V = \max_{1 \leq i \leq k} \left(\frac{\lambda_i}{x_i} \right) \qquad \ldots(4)$$

Clearly, since $x_i > 0 \ \forall \ i$ and at least one λ_i is positive therefore, V is positive.

Now multiplying (3) by $\frac{1}{V}$ and then subtracting from (2), we get

$$\sum_{i=1}^{k} \left(x_i - \frac{\lambda_i}{V} \right) \alpha_i = \boldsymbol{b} \qquad \ldots(5)$$

Here, (5) gives a new solution of $A\boldsymbol{x} = \boldsymbol{b}$ which is given by

$$\boldsymbol{x}' = \left[x_1 - \frac{\lambda_1}{V}, x_2 - \frac{\lambda_2}{V}, \ldots, x_k - \frac{\lambda_k}{V}, 0, 0, \ldots, 0 \right]$$

Also, from (4) for $i = 1, 2, \ldots, k$ we get

$$V \geq \frac{\lambda_i}{x_i} \text{ or } x_i \geq \frac{\lambda_i}{V} \qquad (\because x_i, V \geq 0)$$

$$\Rightarrow \qquad x_i - \frac{\lambda_i}{V} \geq 0$$

$\Rightarrow$ All the components of $\boldsymbol{x}'$ are non-negative.

$\Rightarrow$ $\boldsymbol{x}'$ is a feasible solution.

Also, $V = \dfrac{\lambda_i}{x_i}$, for at least one i, $1 \leq i \leq k$

$\Rightarrow$ For this value of i, $1 \leq i \leq k, V - \dfrac{\lambda_i}{x_i} = 0$

$\Rightarrow$ New feasible solution $\boldsymbol{x}'$ cannot have more than $k - 1$ non-zero variables.

Proceeding in the same way, as above, we have derived a new feasible solution from the given optimal feasible solution which contains less number of non-zero variables. Further, this solution is BFS if the column vectors associated to non-zero variables in this new solution are linearly independent. If these associated vectors are not linearly independent we shall repeat the whole reduction process as above.

Hence, continuing in this way for a finite number of times we shall derive a solution in which column corresponding to positive variables are linearly independent, *i.e.*, we will get a basic feasible solution of the given system.

 Solved Examples

EXAMPLE 1. *If $x_1 = 2$, $x_2 = 4$ and $x_3 = 1$ be a feasible solution (FS) to the system of equations.*

$$2x_1 - x_2 + 2x_3 = 2$$

$$x_1 + 4x_2 = 18$$

Then find two basic feasible solutions(BFS). [GORAKHPUR-2007; DELHI-1995]

SOLUTION. We can write the given system of equations in matrix form as

$$\begin{bmatrix} 2 & -1 & 2 \\ 1 & 4 & 0 \end{bmatrix} \begin{bmatrix} x_1 \\ x_2 \\ x_3 \end{bmatrix} = \begin{bmatrix} 2 \\ 18 \end{bmatrix}$$

which can be written as

$$\alpha_1 x_1 + \alpha_2 x_2 + \alpha_3 x_3 = \boldsymbol{b} \qquad \ldots(1)$$

where $\alpha_1 = \begin{bmatrix} 2 \\ 1 \end{bmatrix}, \alpha_2 = \begin{bmatrix} -1 \\ 4 \end{bmatrix}, \alpha_3 = \begin{bmatrix} 2 \\ 0 \end{bmatrix}$ and $\boldsymbol{b} = \begin{bmatrix} 2 \\ 18 \end{bmatrix}$

The solution is given by

$$x_1 = 2, x_2 = 4 \text{ and } x_3 = 1$$

$$\Rightarrow \quad 2\alpha_1 + 4\alpha_2 + \alpha_3 = \boldsymbol{b} \qquad \ldots(2)$$

Clearly, we have $m = 2$, $n = 3$, therefore, the BFS have at least $n - m = 3 - 2 = 1$ zero variable.

$\Rightarrow$ BFS can not have more than two non-zero variables.

$\Rightarrow$ given FS is not BFS.

Now, we have to reduce this FS to BFS, *i.e.*, we have to make at least one variable equal to 0.

Let $\alpha_1, \alpha_2, \alpha_3$ be linearly dependent

$\Rightarrow$ $\exists$ a linear relation between them.

Let $\alpha_1 = a\alpha_2 + b\alpha_3$

$$\Rightarrow \quad \begin{bmatrix} 2 \\ 1 \end{bmatrix} = a \begin{bmatrix} -1 \\ 4 \end{bmatrix} + b \begin{bmatrix} 2 \\ 0 \end{bmatrix} = \begin{bmatrix} -a + 2b \\ 4a \end{bmatrix}$$

$$\Rightarrow \quad -a + 2b = 2, \ 4a = 1$$

$$\Rightarrow \quad a = 1/4, \ b = 9/8$$

$$\therefore \quad \alpha_1 = \frac{1}{4}\alpha_2 + \frac{9}{8}\alpha_3 \Rightarrow 8\alpha_1 - 2\alpha_2 - 9\alpha_3 = 0 \qquad \ldots(3)$$

$\Rightarrow \qquad \sum_{i=1}^{3} \lambda_i \alpha_i = \mathbf{0}$ where $\lambda_1 = 8, \lambda_2 = -2, \lambda_3 = -9$

Now, let $\qquad V = \max\left(\dfrac{\lambda_i}{x_i}\right) = \max\left(\dfrac{\lambda_1}{x_1}, \dfrac{\lambda_2}{x_2}, \dfrac{\lambda_3}{x_3}\right)$

$$= \max\left(\dfrac{8}{2}, -\dfrac{2}{4}, \dfrac{-9}{1}\right) = \dfrac{8}{2} = \dfrac{\lambda_1}{x_1}$$

So, x_1 should be zero for which the vector α_1 should be eliminated between (2) and (3), such that

$$\dfrac{2(2\alpha_2 + 9\alpha_3)}{8} + 4\alpha_2 + \alpha_3 = \mathbf{b}$$

$\Rightarrow \qquad\qquad \dfrac{9}{2}\alpha_2 + \dfrac{13}{4}\alpha_3 = \mathbf{b}$

$\therefore$ The new feasible solution is $x_1 = 0$ (non-basic), $x_2 = \dfrac{9}{2}, x_3 = \dfrac{13}{4}$ (basic)

Now, the column vectors α_2, α_3 corresponding to basic variables in this FS are linearly independent.

Since, $\ |(\alpha_2, \alpha_3)| = \begin{vmatrix} -1 & 2 \\ 4 & 0 \end{vmatrix} = -8 \neq 0$

Therefore, the feasible solution given by $x_1 = 0, x_2 = \dfrac{9}{2}, x_3 = \dfrac{13}{4}$ is a BFS.

Now, we have to find another BFS

We can rewrite equation (3) as

$$-8\alpha_1 + 2\alpha_2 + 9\alpha_3 = 0 \qquad\qquad\qquad ...(4)$$

$\Rightarrow \qquad \sum_{i=1}^{3} \lambda_i \alpha_i = \mathbf{0}$ where $\lambda_1 = -8, \lambda_2 = 2, \lambda_3 = 9$

So, $\qquad V = \max_{1 \leq i \leq 3}\left(\dfrac{\lambda_i}{x_i}\right) = \max\left(\dfrac{\lambda_1}{x_1}, \dfrac{\lambda_2}{x_2}, \dfrac{\lambda_3}{x_3}\right) = \max\left(\dfrac{-8}{2}, \dfrac{2}{4}, \dfrac{9}{1}\right) = \dfrac{9}{1} = \dfrac{\lambda_3}{x_3}$

Therefore, x_3 should be zero, for which α_3 should be eliminated between (2) and (4) such that

$$2\alpha_1 + 4\alpha_2 + (8\alpha_1 - 2\alpha_2)/9 = \mathbf{b}$$

or $\qquad \left(\dfrac{26}{9}\right)\alpha_1 + \left(\dfrac{34}{9}\right)\alpha_2 = \mathbf{b}$

So, feasible solution is given by

$$x_1 = \dfrac{26}{9}, x_2 = \dfrac{34}{9} \text{ (basic) and } x_3 = 0 \text{ (non-basic)}$$

This solution is also BFS, since the vectors α_1, α_2 corresponding to the basic variables x_1 and x_2 are linearly independent as $|(\alpha_1, \alpha_2)| = \begin{vmatrix} 2 & -1 \\ 1 & 4 \end{vmatrix} = 9 \neq 0$.

Hence, two different BFS of the given system are

$$x_1 = 0, x_2 = \dfrac{9}{2}, x_3 = \dfrac{13}{4} \text{ and } x_1 = \dfrac{26}{9}, x_2 = \dfrac{34}{9}, x_3 = 0.$$

EXAMPLE 2.   **(1, 1, 1) *is a feasible solution to the system of equations***
$$x_1 + x_2 + 2x_3 = 4$$
$$2x_1 - x_2 + x_3 = 2$$
Reduce the Feasible Solution to Basic Feasible Solution.

SOLUTION.   We can write the given system of equation in matrix form, as follows:

$$\begin{bmatrix} 1 & 1 & 2 \\ 2 & -1 & 1 \end{bmatrix} \begin{bmatrix} x_1 \\ x_2 \\ x_3 \end{bmatrix} = \begin{bmatrix} 4 \\ 2 \end{bmatrix}$$

or $\alpha_1 x_1 + \alpha_2 x_2 + \alpha_3 x_3 = \boldsymbol{b}$ ...(1)

where $\alpha_1 = \begin{bmatrix} 1 \\ 2 \end{bmatrix}, \alpha_2 = \begin{bmatrix} 1 \\ -1 \end{bmatrix}, \alpha_3 = \begin{bmatrix} 2 \\ 1 \end{bmatrix}$ and $\boldsymbol{b} = \begin{bmatrix} 4 \\ 2 \end{bmatrix}$

Here, $m = 2, n = 3 \Rightarrow n > m$

$\Rightarrow$ BFS will have at least $n - m = 3 - 2 = 1$ variable whose value is zero.

$\Rightarrow$ BFS cannot have more than two non-zero variables.

$\Rightarrow$ Given feasible solution $x_1 = 1, x_2 = 1, x_3 = 1$ is not BFS.

Now substituting the given FS in (1) we get
$$1 \cdot \alpha_1 + 1 \cdot \alpha_2 + 1 \cdot \alpha_3 = \boldsymbol{b}$$

$\Rightarrow$ $\alpha_1 + \alpha_2 + \alpha_3 = \boldsymbol{b}$ ...(2)

To reduce this FS to BFS, we have to make at least one variable zero, as follows:

Let $\alpha_1, \alpha_2, \alpha_3$ be linearly dependent. Then let
$$\alpha_1 = a\alpha_2 + b\alpha_3$$

$\Rightarrow$ $\begin{bmatrix} 1 \\ 2 \end{bmatrix} = a\begin{bmatrix} 1 \\ -1 \end{bmatrix} + b\begin{bmatrix} 2 \\ 1 \end{bmatrix} = \begin{bmatrix} a + 2b \\ -a + b \end{bmatrix}$

$\Rightarrow$ $a + 2b = 1$ and $-a + b = 2$

$\Rightarrow$ $a = -1, b = 1$

$\therefore$ $\alpha_1 = -1\alpha_2 + 1 \cdot \alpha_3$ or $\alpha_1 + \alpha_2 - \alpha_3 = 0$ ...(3)

$\Rightarrow$ $\sum\limits_{i=1}^{3} \lambda_i \alpha_i = 0$, where $\lambda_1 = 1, \lambda_2 = 1, \lambda_3 = -1$

Further, let $V = \max\limits_{1 \le i \le 3} \left(\dfrac{\lambda_i}{x_i} \right) = \max\left(\dfrac{\lambda_1}{x_1}, \dfrac{\lambda_2}{x_2}, \dfrac{\lambda_3}{x_3} \right)$

$$= \max\left(\frac{1}{1}, \frac{1}{1}, \frac{-1}{1} \right) = \frac{1}{1} = \frac{\lambda_1}{x_1} \text{ or } \frac{\lambda_2}{x_2}$$

$\Rightarrow$ Either x_1 or x_2 should be zero for which α_1 or α_2 should be eliminated.

Eliminate α_1 between (2) and (3), we get
$$(-\alpha_2 + \alpha_3) + \alpha_2 + \alpha_3 = \boldsymbol{b}$$

or $0 \cdot \alpha_2 + 2\alpha_3 = \boldsymbol{b}$

$\Rightarrow$ The new FS is $x_1 = 0$ (non-basic), $x_2 = 0, x_3 = 2$ (basic) which is also BFS, since the vectors α_2, α_3 (basic) corresponding to basic variables x_2, x_3 are linearly independent as

$$|(\alpha_2, \alpha_3)| = \begin{vmatrix} 1 & 2 \\ -1 & 1 \end{vmatrix} = 3 \ne 0$$

Hence, the new BFS is $x_1 = 0, x_2 = 0, x_3 = 2$, i.e., $(0, 0, 2)$

Now we have to find another BFS. eq^n (3) can be written as

$$-\alpha_1 - \alpha_2 + \alpha_3 = 0$$

Then $\lambda_1 = -1, \lambda_2 = -1, \lambda_3 = 1$

In this case $V = \max\limits_{1 \le i \le 3}\left(\dfrac{\lambda_i}{x_i}\right) = \dfrac{\lambda_3}{x_3}$ so variable x_3 should be zero for which the vector α_3 should be eliminated between (2) and (3).

Eliminate α_3 between (2) and (3) we get

$$2\alpha_1 + 2\alpha_2 = b$$

$\Rightarrow$ Another FS is $x_1 = 2, x_2 = 2$ (basic) and $x_3 = 0$ (non-basic)

Since, the column vector α_1, α_2 corresponding to these basic variables x_1, x_2 are linearly independent as

$$|(\alpha_1, \alpha_2)| = \begin{vmatrix} 1 & 1 \\ 2 & -1 \end{vmatrix} = -3 \ne 0$$

Hence, the FS $x_1 = 2, x_2 = 2, x_3 = 0$, i.e., $(2, 2, 0)$ is another BFS of the given system of equations.

EXAMPLE 3. *If (2, 3, 1) is a feasible solution of the linear programming problem*

$$\textbf{Max. } \textbf{\textit{Z}} = \textbf{\textit{x}}_1 + 2\textbf{\textit{x}}_2 + 4\textbf{\textit{x}}_3$$
$$\textbf{\textit{s.t.}} \quad 2\textbf{\textit{x}}_1 + \textbf{\textit{x}}_2 + 4\textbf{\textit{x}}_3 = 11$$
$$3\textbf{\textit{x}}_1 + \textbf{\textit{x}}_2 + 5\textbf{\textit{x}}_3 = 14$$
$$\textbf{\textit{x}}_1, \textbf{\textit{x}}_2, \textbf{\textit{x}}_3 \ge 0$$

Then find a basic feasible solution from the given feasible solution.

[MEERUT–1994, 95, 2006, 10]

SOLUTION. We can write the given LPP as

Max. $Z = x_1 + 2x_2 + 4x_3$

such that $A\boldsymbol{x} = \boldsymbol{b}$

or $\begin{bmatrix} 2 & 1 & 4 \\ 3 & 1 & 5 \end{bmatrix} \begin{bmatrix} x_1 \\ x_2 \\ x_3 \end{bmatrix} = \begin{bmatrix} 11 \\ 14 \end{bmatrix}$

or $\alpha_1 x_1 + \alpha_2 x_2 + \alpha_3 x_3 = \boldsymbol{b}$ ...(1)

where $\alpha_1 = \begin{bmatrix} 2 \\ 3 \end{bmatrix}, \alpha_2 = \begin{bmatrix} 1 \\ 1 \end{bmatrix}, \alpha_3 = \begin{bmatrix} 4 \\ 5 \end{bmatrix}$ and $\boldsymbol{b} = \begin{bmatrix} 11 \\ 14 \end{bmatrix}$

Since, (2, 3, 1) is a feasible solution of the given LPP, therefore from (1)

$$2\alpha_1 + 3\alpha_2 + \alpha_3 = \boldsymbol{b} \quad ...(2)$$

Clearly, the vectors $\alpha_1, \alpha_2, \alpha_3$ associated with non-zero variables x_1, x_2, x_3 will be linearly dependent if one of these vectors can be expressed as the linear combination of the remaining two vectors.

Let us suppose $\alpha_1 = a\alpha_2 + b\alpha_3$ ...(3)

$\Rightarrow \quad \begin{bmatrix} 2 \\ 3 \end{bmatrix} = a\begin{bmatrix} 1 \\ 1 \end{bmatrix} + b\begin{bmatrix} 4 \\ 5 \end{bmatrix} = \begin{bmatrix} a + 4b \\ a + 5b \end{bmatrix}$

$\Rightarrow \quad a + 4b = 2, a + 5b = 3$

$\Rightarrow \quad a = -2, b = 1$

Using these values in (3) we get

$$\alpha_1 = -2\alpha_2 + \alpha_3 \text{ or } \alpha_1 + 2\alpha_2 - \alpha_3 = 0 \qquad \text{...(4)}$$

or $\quad \sum\limits_{i=1}^{3} \lambda_i \alpha_i = 0$ which gives $\lambda_1 = 1, \lambda_2 = 2, \lambda_3 = -1$

Let $\quad V = \max\limits_{1 \le i \le 3}\left(\dfrac{\lambda_i}{x_i}\right) = \max\left(\dfrac{\lambda_1}{x_1}, \dfrac{\lambda_2}{x_2}, \dfrac{\lambda_3}{x_3}\right) = \max\left(\dfrac{1}{2}, \dfrac{2}{3}, \dfrac{-1}{1}\right) = \dfrac{2}{3} = \dfrac{\lambda_2}{x_2}$

$\Rightarrow \quad x_2$ should be 0, for which we should eliminate α_2 between (2) and (4)

Now, eliminating α_2 between (2) and (4) we get

$$2\alpha_1 - 3(\alpha_1 - \alpha_3)/2 + \alpha_3 = \boldsymbol{b}$$

$$\Rightarrow \qquad \frac{1}{2}\alpha_1 + \frac{5}{2}\alpha_3 = \boldsymbol{b}$$

$\therefore \quad$ The new FS is given by $x_1 = \dfrac{1}{2}, x_2 = 0, x_3 = \dfrac{5}{2}$

Now, $\quad |(\alpha_1, \alpha_3)| = \begin{vmatrix} 2 & 4 \\ 3 & 5 \end{vmatrix} = -2 \ne 0$

$\Rightarrow \quad \alpha_1$ and α_3 are linearly independent.

Hence, the basic feasible solution obtained from the given Feasible solution is

$x_1 = \dfrac{1}{2}, x_2 = 0, x_3 = \dfrac{5}{2}.$

EXAMPLE 4. *If (1, 1, 1, 0) is a feasible solution to the system of equations*

$$x_1 + 2x_2 + 4x_3 + x_4 = 7$$

$$2x_1 - x_2 + 3x_3 - 2x_4 = 4$$

Reduce the feasible solution to two different BFS. [MEERUT–2008]

SOLUTION. The given system of equations can be written as

$$\begin{bmatrix} 1 & 2 & 4 & 1 \\ 2 & -1 & 3 & -2 \end{bmatrix} \begin{bmatrix} x_1 \\ x_2 \\ x_3 \\ x_4 \end{bmatrix} = \begin{bmatrix} 7 \\ 4 \end{bmatrix}$$

or $\quad \alpha_1 x_1 + \alpha_2 x_2 + \alpha_3 x_3 + \alpha_4 x_4 = \boldsymbol{b} \qquad \text{...(1)}$

where $\quad \alpha_1 = \begin{bmatrix} 1 \\ 2 \end{bmatrix}, \alpha_2 = \begin{bmatrix} 2 \\ -1 \end{bmatrix}, \alpha_3 = \begin{bmatrix} 4 \\ 3 \end{bmatrix}, \alpha_4 = \begin{bmatrix} 1 \\ -2 \end{bmatrix}$ and $\boldsymbol{b} = \begin{bmatrix} 7 \\ 4 \end{bmatrix}$

Since, $x_1 = 1, x_2 = 1, x_3 = 1, x_4 = 0$ is the solution of (1) therefore,

$$\alpha_1 + \alpha_2 + \alpha_3 = \boldsymbol{b} \qquad \text{...(2)}$$

Also here $m = 2, n = 4 \quad \Rightarrow \quad m < n$

Therefore, BFS will have at least $n - m = 4 - 2 = 2$ zero variables, *i.e.*, BFS cannot have more than 2 non-zero variables. So, given FS is not BFS.

Now, to reduce FS to BFS, we have to make at least one variable zero.

We know that the column vectors $\alpha_1, \alpha_2, \alpha_3$ corresponding to non-zero variables x_1, x_2, x_3 are linearly dependent if one of them can be expressed as a linear combination of others.

So, let $\qquad \alpha_1 = a\alpha_2 + b\alpha_3$

$$\Rightarrow \qquad \begin{bmatrix} 1 \\ 2 \end{bmatrix} = a \begin{bmatrix} 2 \\ -1 \end{bmatrix} + b \begin{bmatrix} 4 \\ 3 \end{bmatrix} \qquad \Rightarrow \qquad a = -1/2, \; b = 1/2$$

So, $\alpha_1 = \left(-\dfrac{1}{2}\right)\alpha_2 + \left(\dfrac{1}{2}\right)\alpha_3$ or $2\alpha_1 + \alpha_2 - \alpha_3 = 0$ $\hspace{2cm}$ ...(3)

or $\displaystyle\sum_{i=1}^{3} \lambda_i \alpha_i = 0$ where $\lambda_1 = 2, \lambda_2 = 1, \lambda_3 = -1$

Now, $\qquad\qquad V = \max\left(\dfrac{\lambda_i}{x_i}\right) = \max\left(\dfrac{\lambda_1}{x_1}, \dfrac{\lambda_2}{x_2}, \dfrac{\lambda_3}{x_3}\right)$

$$= \max\left(\frac{2}{1}, \frac{1}{1}, \frac{-1}{1}\right) = \frac{2}{1} = \frac{\lambda_1}{x_1}$$

$\Rightarrow x_1$ should be zero for which we have to eliminate α_1 between (2) and (3) such that

$$\frac{(-\alpha_2 + \alpha_3)}{2} + \alpha_2 + \alpha_3 = \boldsymbol{b}$$

$$\Rightarrow \qquad \frac{1}{2}\alpha_2 + \frac{3}{2}\alpha_3 = \boldsymbol{b}$$

$$\Rightarrow \quad x_1 = 0 \;(\text{non-basic}), \; x_2 = \frac{1}{2}, x_3 = \frac{3}{2} \;(\text{basic})$$

Now, since $\quad |(\alpha_2, \alpha_3)| = \begin{vmatrix} 2 & 4 \\ -1 & 3 \end{vmatrix} = 10 \neq 0$

$\Rightarrow$ Column vectors α_2, α_3 corresponding to basic variables x_2, x_3 are linearly independent.

$\Rightarrow$ This feasible solution is BFS.

Further, we have to obtain another BFS. Rewrite (3) as

$\qquad -2\alpha_1 - \alpha_2 + \alpha_3 = 0$ $\hspace{5cm}$ ...(4)

$\Rightarrow \displaystyle\sum_{i=1}^{3} \lambda_i \alpha_i = 0$, where $\lambda_1 = -2, \lambda_2 = -1, \lambda_3 = 1$

$\therefore \qquad\qquad V = \max\left(\dfrac{\lambda_1}{x_1}, \dfrac{\lambda_2}{x_2}, \dfrac{\lambda_3}{x_3}\right) = \max\left(\dfrac{-2}{1}, \dfrac{-1}{1}, \dfrac{1}{1}\right) = \dfrac{1}{1} = \dfrac{\lambda_3}{x_3}$

$\Rightarrow \quad x_3$ should be zero.

For which eliminate α_3 between (2) and (4) such that

$\qquad \alpha_1 + \alpha_2 + (2\alpha_1 + \alpha_2) = 0 \quad \Rightarrow \quad 3\alpha_1 + 2\alpha_2 = 0$

$\Rightarrow \quad x_1 = 3, x_2 = 2 \;(\text{basic}), x_3 = 0, x_4 = 0 \;(\text{non-basic})$

Also, $\qquad |(\alpha_1, \alpha_2)| = \begin{vmatrix} 1 & 2 \\ 2 & -1 \end{vmatrix} = -5 \neq 0$

$\Rightarrow$ Column vectors α_1, α_2 corresponding to basic variables x_1, x_2 are linearly independent.

$\Rightarrow$ This solution is BFS.

Hence, the given FS reduces to the following two BF solutions.

$$x_1 = 0, x_2 = \frac{1}{2}, x_3 = \frac{3}{2}, x_4 = 0 \text{ and } x_1 = 3, x_2 = 2, x_3 = 0, x_4 = 0.$$

EXAMPLE 5. *If (1, 2, 1, 3) is a feasible solution of the set of equations*

$$5x_1 - 4x_2 + 3x_3 + x_4 = 3$$
$$2x_1 + x_2 + 5x_3 - 3x_4 = 0$$
$$x_1 + 6x_2 - 4x_3 + 2x_4 = 15$$
$$x_1, x_2, x_3, x_4 \geq 0$$

Find a corresponding BFS.   [MEERUT–2005, 14; GORAKHPUR–2011, 17]

SOLUTION.   We can write

$$Ax = b$$

$$\Rightarrow \begin{bmatrix} 5 & -4 & 3 & 1 \\ 2 & 1 & 5 & -3 \\ 1 & 6 & -4 & 2 \end{bmatrix} \begin{bmatrix} x_1 \\ x_2 \\ x_3 \\ x_4 \end{bmatrix} = \begin{bmatrix} 3 \\ 0 \\ 15 \end{bmatrix}$$

or $\alpha_1 x_1 + \alpha_2 x_2 + \alpha_3 x_3 + \alpha_4 x_4 = b$ ...(1)

where $\alpha_1 = \begin{bmatrix} 5 \\ 2 \\ 1 \end{bmatrix}, \alpha_2 = \begin{bmatrix} -4 \\ 1 \\ 6 \end{bmatrix}, \alpha_3 = \begin{bmatrix} 3 \\ 5 \\ -4 \end{bmatrix}, \alpha_4 = \begin{bmatrix} 1 \\ -3 \\ 2 \end{bmatrix}$ and $b = \begin{bmatrix} 3 \\ 0 \\ 15 \end{bmatrix}$

Since (1, 2, 1, 3) is a solution of (1) therefore,

$$\alpha_1 + 2\alpha_2 + \alpha_3 + 3\alpha_4 = b \quad\quad\quad ...(2)$$

The vectors $\alpha_1, \alpha_2, \alpha_3, \alpha_4$ associated with non-zero variables x_1, x_2, x_3, x_4 will be linearly dependent if one of these vectors can be expressed as the linear combination of the remaining three vectors.

Let $\alpha_1 = a\alpha_2 + b\alpha_3 + c\alpha_4$ ...(3)

$$\Rightarrow \begin{bmatrix} 5 \\ 2 \\ 1 \end{bmatrix} = a \begin{bmatrix} -4 \\ 1 \\ 6 \end{bmatrix} + b \begin{bmatrix} 3 \\ 5 \\ -4 \end{bmatrix} + c \begin{bmatrix} 1 \\ -3 \\ 2 \end{bmatrix}$$

$$\Rightarrow \begin{bmatrix} -4a + 3b + c \\ a + 5b - 3c \\ 6a - 4b + 2c \end{bmatrix} = \begin{bmatrix} 5 \\ 2 \\ 1 \end{bmatrix}$$

$\Rightarrow$ $-4a + 3b + c = 5;\ a + 5b - 3c = 2$ and $6a - 4b + 2c = 1$

On solving we get

$$a = \frac{22}{43}, b = \frac{139}{86}, c = \frac{189}{86}$$

Putting all these values in (3) we get

$$\frac{22}{43}\alpha_2 + \frac{139}{86}\alpha_3 + \frac{189}{86}\alpha_4 = \alpha_1$$

$\Rightarrow$ $86\alpha_1 - 44\alpha_2 - 139\alpha_3 - 189\alpha_4 = 0$ ...(4)

$\Rightarrow$ $\sum\limits_{i=1}^{4} \lambda_i \alpha_i = 0$ where $\lambda_1 = 86, \lambda_2 = -44, \lambda_3 = -139, \lambda_4 = -189$

Now, let $V = \max\limits_{1 \leq i \leq 4} \left(\frac{\lambda_i}{x_i} \right) = \max \left(\frac{\lambda_1}{x_1}, \frac{\lambda_2}{x_2}, \frac{\lambda_3}{x_3}, \frac{\lambda_4}{x_4} \right)$

$$= \max\left(\frac{86}{1}, \frac{-44}{2}, \frac{-139}{1}, \frac{-189}{3}\right) = \frac{86}{1} = \frac{\lambda_1}{x_1}$$

$\Rightarrow$ x_1 should be zero.

So, eliminate α_1 between (2) and (4) we get

$$\frac{1}{86}(44\alpha_2 + 139\alpha_3 + 189\alpha_4) + 2\alpha_2 + \alpha_3 + 3\alpha_4 = \boldsymbol{b}$$

$$\Rightarrow \quad 0 \cdot \alpha_1 + \frac{216}{86}\alpha_2 + \frac{225}{86}\alpha_3 + \frac{447}{86}\alpha_4 = \boldsymbol{b}$$

$$\Rightarrow \quad x_1 = 0, x_2 = \frac{216}{86}, x_3 = \frac{225}{86}, x_4 = \frac{447}{86} \text{ is the new BFS.}$$

Also, $|(\alpha_2, \alpha_3, \alpha_4)| = \begin{vmatrix} -4 & 3 & 1 \\ 1 & 5 & -3 \\ 6 & -4 & 2 \end{vmatrix} \neq 0$

$\Rightarrow$ The vectors corresponding to non-zero basic variables are linearly independent. Hence, the new FS is BFS.

EXAMPLE 6. *If (2, 1, 2, 2, 1) is a feasible solution of the set of equations*

$$2x_1 - 3x_2 + 4x_3 + 6x_4 = 21$$

$$x_1 + 2x_2 + 3x_3 - 3x_4 + 5x_5 = 9$$

Then reduce it to two different BFS.

SOLUTION.   We can write the given system of equations as

$$\begin{bmatrix} 2 & -3 & 4 & 6 & 0 \\ 1 & 2 & 3 & -3 & 5 \end{bmatrix} \begin{bmatrix} x_1 \\ x_2 \\ x_3 \\ x_4 \\ x_5 \end{bmatrix} = \begin{bmatrix} 21 \\ 9 \end{bmatrix}$$

or $\alpha_1 x_1 + \alpha_2 x_2 + \alpha_3 x_3 + \alpha_4 x_4 + \alpha_5 x_5 = \boldsymbol{b}$ ...(1)

where $\alpha_1 = \begin{bmatrix} 2 \\ 1 \end{bmatrix}, \alpha_2 = \begin{bmatrix} -3 \\ 2 \end{bmatrix}, \alpha_3 = \begin{bmatrix} 4 \\ 3 \end{bmatrix}, \alpha_4 = \begin{bmatrix} 6 \\ -3 \end{bmatrix}, \alpha_5 = \begin{bmatrix} 0 \\ 5 \end{bmatrix}$ and $\boldsymbol{b} = \begin{bmatrix} 21 \\ 9 \end{bmatrix}$

The feasible solution is given by $x_1 = 2, x_2 = 1, x_3 = 2, x_4 = 2$ and $x_5 = 1$

Then from (1)

$$2\alpha_1 + \alpha_2 + 2\alpha_3 + 2\alpha_4 + \alpha_5 = \boldsymbol{b}$$ ...(2)

Also, here $m = 2, n = 5$, i.e., $m < n$ therefore BFS will have at least $n-m=5-2=3$ zero variables, i.e., it can not have more than two non-zero variables. So, given feasible solution is not basic. To reduce FS to BFS we proceed as follows:

(1) To Reduce one of the variables to zero value.

Since, $\alpha_1, \alpha_2, \alpha_3, \alpha_4$ and α_5 are linearly dependent therefore, by inspection we have

$$2\alpha_1 + 2\alpha_2 - \alpha_3 + \alpha_4 + 0 \cdot \alpha_5 = 0$$ ...(3)

$$\Rightarrow \quad \sum_{i=1}^{5} \lambda_i \alpha_i = 0 \text{ where } \lambda_1 = 2, \lambda_2 = 2, \lambda_3 = -1, \lambda_4 = 1, \lambda_5 = 0$$

Further, let $V = \max\limits_{1 \le i \le 5}\left(\dfrac{\lambda_i}{x_i}\right) = \max\left(\dfrac{\lambda_1}{x_1}, \dfrac{\lambda_2}{x_2}, \dfrac{\lambda_3}{x_3}, \dfrac{\lambda_4}{x_4}, \dfrac{\lambda_5}{x_5}\right)$

$$= \max\left(\dfrac{2}{2}, \dfrac{2}{1}, \dfrac{-1}{2}, \dfrac{1}{2}, \dfrac{0}{1}\right) = \dfrac{2}{1} = \dfrac{\lambda_2}{x_2}$$

$\Rightarrow$ x_2 should be zero.

$\Rightarrow$ We have to eliminate α_2 between 2 and 3 as follows:

$$2\alpha_1 + \left(-\alpha_1 + \dfrac{1}{2}\alpha_3 - \dfrac{1}{2}\alpha_4\right) + 2\alpha_3 + 2\alpha_4 + \alpha_5 = \boldsymbol{b}$$

$\Rightarrow$ $\alpha_1 + \dfrac{5}{2}\alpha_3 + \dfrac{3}{2}\alpha_4 + \alpha_5 = \boldsymbol{b}$ ...(4)

$\Rightarrow x_1 = 1, x_2 = 0, \ x_3 = \dfrac{5}{2}, x_4 = \dfrac{3}{2}$ and $x_5 = 1$ is the another feasible solution of the problem, but it is again not a BFS, as it contains more than two non-zero variables.

(2) To Reduce other variables to zero value.

Since, $\alpha_1, \alpha_3, \alpha_4, \alpha_5$ are linearly dependent, so by inspection we have

$$0 \cdot \alpha_1 - 3 \cdot \alpha_3 + 2 \cdot \alpha_4 + 3\alpha_5 = \boldsymbol{0}$$ ...(5)

$\Rightarrow$ $\lambda_1 = 0, \lambda_3 = -3, \lambda_4 = 2, \lambda_5 = 3$

Further, let $V = \max\left(\dfrac{\lambda_1}{x_1}, \dfrac{\lambda_3}{x_3}, \dfrac{\lambda_4}{x_4}, \dfrac{\lambda_5}{x_5}\right)$

$$= \max\left(\dfrac{0}{1}, \dfrac{-3}{(5/2)}, \dfrac{2}{(3/2)}, \dfrac{3}{1}\right) = \dfrac{3}{1} = \dfrac{\lambda_5}{x_5}$$

So, x_5 should be zero, *i.e.*, we have to eliminate α_5 between (4) and (5) as follows:

$$\alpha_1 + \dfrac{5}{2}\alpha_3 + \dfrac{3}{2}\alpha_4 + \left(\alpha_3 - \dfrac{2}{3}\alpha_4\right) = \boldsymbol{b}$$

$\Rightarrow$ $\alpha_1 + \dfrac{7}{2}\alpha_3 + \dfrac{5}{6}\alpha_4 = \boldsymbol{b}$ ...(6)

$\Rightarrow x_1 = 1, x_2 = 0, x_3 = 7/2, x_4 = 5/6, x_5 = 0$ is another FS.

It is also not a BFS as it contains more than two non-zero variables.

(3) To Reduce one more variables to zero value.

Since, α_1, α_3 and α_4 are linearly dependent.

$\Rightarrow$ $\alpha_1 = a\alpha_3 + b\alpha_4$

$\Rightarrow$ $\begin{pmatrix} 2 \\ 1 \end{pmatrix} = a\begin{pmatrix} 4 \\ 3 \end{pmatrix} + b\begin{pmatrix} 6 \\ -3 \end{pmatrix} = \begin{pmatrix} 4a + 6b \\ 3a - 3b \end{pmatrix}$

$\Rightarrow$ $4a + 6b = 2, \ 3a - 3b = 1$

$\Rightarrow$ $a = \dfrac{2}{5}, b = \dfrac{1}{15}$

So, $\alpha_1 = \dfrac{2}{5}\alpha_3 + \dfrac{1}{15}\alpha_4 = 0$

$\Rightarrow$ $15\alpha_1 - 6\alpha_3 - \alpha_4 = 0$ ...(7)

$\Rightarrow$ $\lambda_1 = 15, \lambda_3 = -6, \lambda_4 = -1$

Further let $\quad V = \max\limits_{1 \le i \le 4}\left\{\dfrac{\lambda_i}{x_i}\right\} = \max\left\{\dfrac{\lambda_1}{x_1}, \dfrac{\lambda_3}{x_3}, \dfrac{\lambda_4}{x_4}\right\}$

$$= \max\left\{\dfrac{15}{1}, \dfrac{-6}{(7/2)}, \dfrac{-1}{(5/6)}\right\} = \dfrac{15}{1} = \dfrac{\lambda_1}{x_1}$$

$\Rightarrow$ x_1 should be zero, i.e., α_1 should be eliminated from (6) and (7) as follows:

$$\dfrac{2}{5}\alpha_3 + \left(\dfrac{1}{15}\right)\alpha_4 + \dfrac{7}{2}\alpha_3 + \dfrac{5}{6}\alpha_4 = \boldsymbol{b}$$

$\Rightarrow \qquad \dfrac{39}{10}\alpha_3 + \dfrac{9}{10}\alpha_4 = \boldsymbol{b}$

$\Rightarrow \qquad x_1 = 0, x_2 = 0, x_3 = \dfrac{39}{10}, x_4 = \dfrac{9}{10}, x_5 = 0$

which is a feasible solution with two non-zero variables.

Since, $|(\alpha_3, \alpha_4)| = \begin{vmatrix} 4 & 6 \\ 3 & -3 \end{vmatrix} = -30 \ne 0$

$\Rightarrow \qquad \alpha_3, \alpha_4$ are linearly independent.

$\Rightarrow \qquad$ Feasible solution is BFS.

Now, to get another BFS, rewrite (7) as

$$-15\alpha_1 + 6\alpha_3 + \alpha_4 = 0$$

Then $\lambda_1 = -15, \lambda_3 = 6, \lambda_4 = 1$

$\Rightarrow \quad \max\left\{\dfrac{\lambda_1}{x_1}, \dfrac{\lambda_3}{x_3}, \dfrac{\lambda_4}{x_4}\right\} = \max\left\{\dfrac{-15}{1}, \dfrac{6}{(7/2)}, \dfrac{1}{(5/6)}\right\} = \dfrac{12}{7} = \dfrac{\lambda_3}{x_3}$

$\Rightarrow x_3$ should be zero, i.e., α_3 should be eliminated between (6) and (7) as follows:

$$\dfrac{39}{4}\alpha_1 + \dfrac{1}{4}\alpha_4 = \boldsymbol{b}$$

$\therefore$ Feasible solution is $x_1 = \dfrac{39}{4}, x_2 = 0, x_3 = 0, x_4 = \dfrac{1}{4}, x_5 = 0$ which is also BFS,

because column vectors α_1, α_4 corresponding to x_1 and x_4 are linearly independent.

 Exercise-4.1

1. If $(1, 0, 1)$ is the feasible solution of the LPP given by

   Min. $Z = 2x_1 + 3x_2 + 4x_3$

   subject to

   $x_1 + x_2 + x_3 = 2$

   $x_1 - x_2 + x_3 = 0$

   $x_1, x_2, x_3 \ge 0$

   Then, show that the given feasible solution is not basic.

2. If $(2, 4, 5)$ is the feasible solution of the system of equations

   $2x_1 - x_2 + 2x_3 = 10$

   $x_1 + 4x_2 = 18$

   $x_1, x_2, x_3 \ge 0$

   Reduce this FS to BFS.

3. If $(2, 13)$ is a feasible solution of the set of equations

   $4x_1 + 2x_2 - 3x_3 = 1$

   $6x_1 + 4x_2 - 5x_3 = 1$

   Reduce this FS to BFS.

4. If $(2, 1, 3, 2, 1)$ is a Feasible solution of the system of equations

   $2x_1 - 3x_2 + 4x_3 + 6x_4 = 25$

   $x_1 + 2x_2 + 3x_3 - 3x_4 + 5x_5 = 12$

   Reduce FS to a BFS of the system.

2. $\left(0, \dfrac{9}{2}, \dfrac{29}{4}\right)$ **3.** $(1, 0, 1)$ **4.** $\left(\dfrac{147}{2}, 0, 0, \dfrac{1}{12}, 0\right)$

4.5 TO FIND THE IMPROVED BASIC FEASIBLE SOLUTION (BFS) FROM A GIVEN BFS

THEOREM 1. *Let $x_B = B^{-1}b$ be a basic feasible solution of a LPP with $Z = C_B x_B$ as the value of the objective function. If for any column α_j in A, but not in B the condition $c_j - Z_j > 0$ or < 0 holds and if at least one $y_{ij} > 0$, $i = 1, 2, ..., m$, then it is possible to find a new BFS by replacing one of the columns in B by α_j and if Z' is the new value of the objective function then $Z' \geq Z$. Further, if the initial BFS $x_B = B^{-1}b$ is non-degenerate then $Z' > Z$.*

PROOF. Let us consider an LPP given by

$$\text{Max. } Z = \boldsymbol{C}\boldsymbol{x} \text{ such that } A\boldsymbol{x} = \boldsymbol{b}, \, x \geq 0$$

where $A = (\alpha_1, \alpha_2, ..., \alpha_N)$

$$N = m + n$$

and basic matrix $B = (\beta_1, \beta_2, ..., \beta_m)$

Suppose that $x_B = [x_{B_1}, x_{B_2}, ..., x_{B_m}]$ be a basic feasible solution of the given LPP.

Now, since the vectors $\beta_1, \beta_2, ..., \beta_m$ are in the basis of A, so by definition of basis, the given vectors can be expressed as

$$\alpha_j = \sum_{i=1}^{m} y_{ij}\beta_i = y_{1j}\beta_1 + y_{2j}\beta_2 + ... + y_{mj}\beta_m \qquad ...(1)$$

Here, if $y_{rj} \neq 0$ then α_j can replace β_r in B and B is still a basis matrix.

Now, let $y_{rj} \neq 0$ then from (1) we can write

$$\beta_r = \frac{1}{y_{rj}}\alpha_j - \frac{y_{1j}}{y_{rj}}\beta_1 - ... - \frac{y_{(r-1)j}}{y_{rj}}\beta_{r-1} - \frac{y_{(r+1)j}}{y_{rj}}\beta_{r+1} - ... - \frac{y_{mj}}{y_{rj}}\beta_m$$

$$\Rightarrow \qquad \beta_r = \frac{1}{y_{rj}}\alpha_j - \sum_{\substack{i=1 \\ i \neq r}}^{m} \frac{y_{ij}}{y_{rj}}\beta_i \qquad ...(2)$$

Consider $B \cdot x_B = \boldsymbol{b}$

$$\Rightarrow \quad (\beta_1, \beta_2, ..., \beta_r, ..., \beta_m)[x_{B_1}, x_{B_2}, ..., x_{B_r}, ..., x_{B_m}] = \boldsymbol{b}$$

$$\Rightarrow \quad \beta_1 x_{B_1} + \beta_2 x_{B_2} + ... + \beta_r x_{B_r} + ... + \beta_m x_{B_m} = \boldsymbol{b}$$

$$\Rightarrow \qquad \sum_{\substack{i=1 \\ i \neq r}}^{m} \beta_i x_{B_i} + \beta_r x_{B_r} = \boldsymbol{b} \qquad ...(3)$$

Using (2) in (3) we get

$$\sum_{\substack{i=1 \\ i \neq r}}^{m} \beta_i \left[x_{B_i} - \frac{y_{ij}}{y_{rj}} x_{B_r} \right] + \frac{x_{B_r}}{y_{rj}}\alpha_j = \boldsymbol{b} \qquad ...(4)$$

$$\Rightarrow \qquad \sum_{\substack{i=1 \\ i \neq r}}^{m} x'_{B_i}\beta_i + x'_{B_r}\alpha_j = \boldsymbol{b} \qquad ...(5)$$

where $\quad x_B' = x_{B_i} - x_{B_r}\dfrac{y_{ij}}{y_{rj}}; i = 1,2,\dots,m, i \neq r$

and $\qquad x_{B_r}' = \dfrac{x_{B_r}}{y_{rj}}, i = r$ $\qquad\qquad$ ...(6)

On comparing (3) and (5), the new basic solution of $Ax = b$ is given by

$$x_B' = [x_{B_i}', x_{B_r}']\, i = 1,2,\dots,m, i \neq r$$

The above basic solution will be feasible if

$x_{B_i} - \dfrac{y_{ij}}{y_{rj}}x_{B_r} \geq 0, i = 1,2,\dots,m, i \neq r$

and $\qquad \dfrac{x_{B_r}}{y_{rj}} \geq 0, i = r$ $\qquad\qquad$ ...(7)

Now, since x_B is the IBFS, we have

$$x_{B_r} \geq 0 : r = 1,2,\dots,m$$

Then (7) holds only if $y_{rj} > 0$ and $y_{ij} \leq 0, i = 1, 2, \dots, m, i \neq r$

If $y_{rj} > 0$ and $y_{ij} > 0$ then (7) is satisfied only if

$$\frac{x_{B_i}}{y_{ij}} - \frac{x_{B_r}}{y_{rj}} \geq 0 \Rightarrow \quad \frac{x_{B_r}}{y_{rj}} \leq \frac{x_{B_i}}{y_{ij}}$$

$\Rightarrow \qquad \dfrac{x_{B_r}}{y_{rj}} = \min_i\left\{\dfrac{x_{B_i}}{y_{ij}}, y_{ij} > 0\right\} = V$ $\qquad\qquad$ ...(8)

Thus, we conclude that

> A new basic feasible solution can be obtained from the initial basic feasible solution by removing the column vector β_r of the basis matrix B by α_j is r is to be selected such that
>
> $$V = \frac{x_{B_r}}{y_{rj}} = \operatorname{Min}_i\left\{\frac{x_{B_i}}{y_{ij}}, y_{ij} > 0\right\}$$

Now, it remains to prove that $Z' \geq Z$.

Clearly, the value of the objective function for the IBFS x_B is given by

$$Z = C_B x_B$$
$$= (C_{B_1} C_{B_2} \dots C_{B_m})[x_{B_1}, x_{B_2}, \dots, x_{B_m}]$$
$$= \sum_{i=1}^{m} C_{B_i} x_{B_i}$$

Now, corresponding to new BFS, x_B', the value of the objective function is Z', therefore, we have

$$Z' = \sum_{i=1}^{m} C_{B_i}' x_{B_i}' \qquad\qquad \text{...(9)}$$

Here, C_{B_i}' are the coefficients of the basic variables $x_{B_i}' : i = 1,2,\dots,m$ in the objective function.

Clearly, $\qquad\qquad C_{B_i}' = C_{B_i} : i = 1,2,\dots,m, i \neq r$

and $\qquad C'_{B_r} = c_j$

$\Rightarrow \qquad Z' = \overset{m}{\underset{\substack{i=1 \\ i \neq r}}{\sum}} C_{B_i} x'_{B_i} + c_j C'_{B_r}$ ...(10)

Using (6) in (10) we get

$$Z' = \overset{m}{\underset{\substack{i=1 \\ i \neq r}}{\sum}} C_{B_i} \left(x_{B_i} - x_{B_r} \frac{y_{ij}}{y_{rj}} \right) + c_j \frac{x_{B_r}}{y_{rj}}$$

$$= \overset{m}{\underset{i=1}{\sum}} C_{B_i} \left(x_{B_i} - x_{B_r} \frac{y_{ij}}{y_{rj}} \right) + c_j \frac{x_{B_r}}{y_{rj}}$$

Since, $C_{B_i} \left(x_{B_i} - x_{B_r} \dfrac{y_{ij}}{y_{rj}} \right) = 0$ when $i = r$, therefore,

$$Z' = \overset{m}{\underset{i=1}{\sum}} C_{B_i} x_{B_i} - \frac{x_{B_r}}{y_{rj}} \overset{m}{\underset{i=1}{\sum}} C_{B_i} y_{ij} + \frac{x_{B_r}}{y_{rj}} c_j$$

$$= Z + \frac{x_{B_r}}{y_{rj}} (c_j - z_j) \text{ where } Z_j = \overset{m}{\underset{i=1}{\sum}} C_{B_i} y_{ij}$$

$$= Z + V(c_j - z_j) \qquad \qquad ...(11)$$

Equation (11) clearly shows that $Z' \geq Z$ if $V[c_j - z_j] \geq 0$

$\because \qquad V \geq 0$ therefore, $Z' \geq Z$ only if $c_j - z_j \geq 0$

So, we conclude that

> By choosing the vector α_j for which $c_j - z_j > 0$ and at least one $y_{ij} > 0$, we obtain a new improved value of the objective function.

☞ REMARKS

- If $V = 0$ then initial basic feasible solution is degenerate and hence new BFS is also degenerate.
- If the initial Basic Feasible solution is non-degenerate then $V > 0$ and hence $Z' > Z$.

4.6 CONDITIONS FOR THE EXISTENCE OF UNBOUNDED SOLUTIONS

We have already discussed in previous article that for any column α_j in A (not in B) the condition $c_j - z_j > 0$ holds and if at least one $y_{ij} > 0$, $i = 1, 2, ..., m$ then it is possible to obtain an improved BFS. But if for at least one α_j, all $y_{ij} \leq 0$, then we get an unbounded solution.

__THEOREM I.__ *If for any basic feasible solution $x_B = B^{-1}b$ to $Ax = b$ there is some column α_j in A but not in B for which $c_j - z_j > 0$ and $y_{ij} < 0$, $i = 1, 2, ..., m$ then if the objective function is to be maximized, then problem has an unbounded solution.*

__PROOF.__ Let $\quad$ Max. $Z = \boldsymbol{Cx}$

such that $Ax = \boldsymbol{b}, x \geq 0$

where $A = (\alpha_1, \alpha_2, ..., \alpha_N)$ $(N = m + n)$

and $\quad B = (\beta_1, \beta_2, ..., \beta_m)$

be the given LPP.

Further, suppose that $\boldsymbol{x}_B = [x_{B_1}, x_{B_2}, ..., x_{B_m}]$ be a BFS of the given LPP. Then

$$B\boldsymbol{x}_B = \boldsymbol{b} \text{ or } \overset{m}{\underset{i=1}{\sum}} x_{B_i} \beta_i = \boldsymbol{b} \qquad \qquad ...(1)$$

and $$Z = C_B x_B = \sum_{i=1}^{m} C_{B_i} x_{B_i} \qquad \qquad \dots(2)$$

Equation (1) can be written as

$$\sum_{i=1}^{m} x_{B_i} \beta_i - \lambda \alpha_j + \lambda \alpha_j = b \text{ for some scalars } \lambda. \qquad \dots(3)$$

Suppose $\alpha_j \in A$ such that $\alpha_j \notin B$

Now, since the vectors $\beta_1, \beta_2, \dots, \beta_m$ are in the basis of A, some α_j can be express as the linear combination of β's.

So, $$\alpha_j = \sum_{i=1}^{m} y_{ij} \beta_i$$

$$\Rightarrow \qquad -\lambda \alpha_j = -\lambda \sum_{i=1}^{m} y_{ij} \beta_i$$

Using the above equation in (3) we get

$$\sum_{i=1}^{m} x_{B_i} \beta_i - \lambda \sum_{i=1}^{m} y_{ij} \beta_i + \lambda \alpha_j = b$$

$$\Rightarrow \qquad \sum_{i=1}^{m} (x_{B_i} - \lambda y_{ij}) \beta_i + \lambda \alpha_j = b$$

which gives a new solution given by

$$x_B' = [x_{B_1}', x_{B_2}', \dots, x_{B_m}', \lambda] \qquad \dots(4)$$

where $x_{B_i}' = x_{B_i} - \lambda y_{ij} : i = 1, 2, \dots, m$

When $\lambda > 0$, $x_{B_i} - \lambda y_{ij} \geq 0 (\because y_{ij} \leq 0)$. Then from (4) we get a Feasible solution in which the number of positive variables is less than or equal to $(m + 1)$. It may be less than $(m + 1)$ because $x_{B_i} - \lambda y_{ij}$ may be zero for some i. If the number of positive variables in this solution is equal to $(m + 1)$ then this solution will be non-basic feasible solution.

Further, if Z' is the new value of the objective function corresponding to new solution then

$$Z' = \sum_{i=1}^{m} C_{B_i} (x_{B_i} - \lambda y_{ij}) + c_j \lambda$$

$$\Rightarrow \qquad Z' = \sum_{i=1}^{m} C_{B_i} x_{B_i} + \lambda(c_j - C_{B_i} y_{ij}) = Z + \lambda(c_j - z_j)$$

Since, $c_j - z_j > 0 \Rightarrow Z'$ can be made as large as we want by giving sufficiently large values to λ. Hence the given LPP has an unbounded solution.

☛ **REMARKS**
- An LPP has an unbounded solution if the value of the objective function can be increased or decreased arbitrarily.
- If for some α_j, $z_j - c_j > 0$ and $y_{ij} \leq 0 : i = 1, 2, \dots, m$ then the LPP has an unbounded solution if the objective function is to be minimized.

4.7 **CONDITIONS FOR IMPROVED BASIC FEASIBLE SOLUTION TO BECOME OPTIMAL**

Let us suppose $x_B = B^{-1} b$ be the basic feasible solution of an LPP given by $Z = Cx$ such that $Ax = b, x \geq 0$.

Let $Z^* = C_B x_B$ be the value of the objective function at any iteration of simplex method. If $c_j - z_j \leq 0$ for every column α_j in A but not in B, then Z^* is the optimum value of the objective function Z and x_B is an optimal basic feasible solution.

4.8 ALTERNATIVE OPTIMAL SOLUTIONS

An LPP is said to have an alternative optimal solution if the set of variables giving the optimal value of the objective function is not unique.

CONDITION FOR ALTERNATIVE OPTIMUM SOLUTION

Let us suppose there exists an optimal basic feasible solution to the given LPP and
(i) if for some α_j in A but not in B, $c_j - z_j = 0$ and $y_{ij} \leq 0$ for all $i = 1, 2, ..., m$ then a non-basic alternative optimal solution will exist.
(ii) if for some α_j in A but not in B, $c_j - z_j = 0$ and $y_{ij} > 0$ for at least one i, then an alternative basic optimal solution will exist.

4.9 INCONSISTENCY AND REDUNDANCY

Definition 1. *If the system of equations has more than enough number of constraints equations, i.e., it has more constraints equations than no. of variables. Then it is called redundancy.*
i.e., $\qquad r(A) = r(Ab) = k \leq n < m$
In this case, there will be $(m - k)$ redundant equations.

Definition 2. *The set of linear equations is said to be inconsistent if $r(A) \neq r(Ab)$.*

☛ **REMARK**
- When we solve an LPP by simplex method we should have $r(A) = r(Ab)$, *i.e.*, the constraints equations after introducing the slack and artificial variables should be consistent.

If the system $Ax = b$ involves artificial variables, then we have the following cases:
(i) If the basis B consist no artificial vector and optimality condition is satisfied, then the current solution is a Basic Feasible Solution.
(ii) If one or more artificial variable appear in B at a zero level, *i.e.*, the value of the artificial variables corresponding to aritificial vectors in B are zero and the optimality condition is satisfied, then the system is consistent. But if $y_{ij} = 0 \ \forall \ j$ and $x_{B_r} = 0$ and r corresponding to the row containing an artificial vector, then rth equation is redundant.
(iii) If at least one artificial variable appears in the basis B at a positive level, *i.e.*, the value of at least one artificial variable corresponding to artificial vector in B is non-zero and the optimality condition is satisfied then there exists no feasible solution.

4.10 PROCEDURE TO OBTAIN INITIAL BASIC FEASIBLE SOLUTION

CASE I. WHEN ALL ORIGINAL CONSTRAINTS HAVE $\leq$ SIGN

Firstly convert all the constraints into equations by introducing slack variables s_i, as given below:

$$a_{11}x_1 + a_{12}x_2 + ... + a_{1n}x_n + 1 \cdot s_1 = b_1$$
$$a_{21}x_1 + a_{22}x_2 + ... + a_{2n}x_n + 1 \cdot s_2 = b_2$$
$$\vdots$$
$$a_{m1}x_1 + a_{m2}x_2 + ... + a_{mn}x_n + 1 \cdot s_m = b_m$$

where each s_i is a slack variable.

The above system of equations can be written in matrix form as follows:

$$\begin{bmatrix} a_{11} & a_{12} & \cdots & a_{1n} & 1 & 0 & \cdots & 0 \\ a_{21} & a_{22} & \cdots & a_{2n} & 0 & 1 & \cdots & 0 \\ \vdots & & & & & & & \\ a_{m1} & a_{m2} & \cdots & a_{mn} & 0 & 0 & \cdots & 1 \end{bmatrix} \begin{bmatrix} x_1 \\ x_2 \\ \vdots \\ x_n \\ s_1 \\ s_2 \\ \vdots \\ s_m \end{bmatrix} = \begin{bmatrix} b_1 \\ b_2 \\ \vdots \\ b_m \end{bmatrix}$$

Let $B = I_m$, a unit matrix of order $m \times m$. Then initial basis matrix is given by

$$x_B = B^{-1}\boldsymbol{b} = I_m \cdot \boldsymbol{b} = \boldsymbol{b} \geq 0$$

Hence, the initial basic feasible solution (IBFS) is given by

$$s_1 = x_{B_1} = b_1, s_2 = x_{B_2} = b_2, \ldots, s_m = x_{B_m} = b_m$$

 ## Working Procedure

The initial basic feasible solution can be obtained by writing all the non-basic variables $x_1, x_2, \ldots, x_n$ equal to zero and solving the equations for remaining variables $s_1, s_2, \ldots, s_m$ (i.e., slack variables).

CASE-2 WHEN ALL ORIGINAL CONSTRAINTS HAVE $\geq$ SIGN

Firstly, convert all the constraints into equations by introducing surplus variables s_i, as given below:

$$a_{11}x_1 + a_{12}x_2 + \ldots + a_{1n}x_n - s_1 = b_1$$
$$a_{21}x_1 + a_{22}x_2 + \ldots + a_{2n}x_n - s_2 = b_2$$
$$\vdots$$
$$\vdots$$
$$a_{m1}x_1 + a_{m2}x_2 + \ldots + a_{mn}x_n - s_m = b_m$$

where each s_i is a surplus variable.

The above system of equations can be written in matrix form as follows:

$$\begin{bmatrix} a_{11} & a_{12} & \cdots & a_{1n} & -1 & 0 & \cdots & 0 \\ a_{21} & a_{22} & \cdots & a_{2n} & 0 & -1 & \cdots & 0 \\ \vdots & & & & & & & \\ a_{m1} & a_{m2} & \cdots & a_{mn} & 0 & 0 & \cdots & -1 \end{bmatrix} \begin{bmatrix} x_1 \\ x_2 \\ \vdots \\ x_n \\ s_1 \\ s_2 \\ \vdots \\ s_m \end{bmatrix} = \begin{bmatrix} b_1 \\ b_2 \\ \vdots \\ b_m \end{bmatrix}$$

If we take the initial basis matrix $B = -I_m$. Then

$$x_B = B^{-1}\boldsymbol{b} = -I_m \cdot \boldsymbol{b} = -\boldsymbol{b} \leq 0$$

which is not a basic feasible solution (BFS).

To counter this difficulty introduce artificial variables in each constraints as given below:

$$a_{11}x_1 + a_{12}x_2 + \dots + a_{1n}x_n - s_1 + a_1 = b_1$$
$$a_{21}x_1 + a_{22}x_2 + \dots + a_{2n}x_n - s_2 + a_2 = b_2$$
$$\vdots$$
$$a_{m1}x_1 + a_{m2}x_2 + \dots + a_{mn}x_n - s_m + a_m = b_m$$

which can be written in the matrix form as follows:

$$\begin{bmatrix} a_{11} & a_{12} & \dots & a_{1n} & -1 & 0 & \dots & 0 & 1 & 0 & \dots & 0 \\ a_{21} & a_{22} & \dots & a_{2n} & 0 & -1 & \dots & 0 & 0 & 1 & \dots & 0 \\ \vdots & & & & & & & & & & & \\ a_{m1} & a_{m2} & \dots & a_{mn} & 0 & 0 & \dots & -1 & 0 & 0 & \dots & 1 \end{bmatrix} \begin{bmatrix} x_1 \\ x_2 \\ \vdots \\ x_n \\ s_1 \\ s_2 \\ \vdots \\ s_m \\ a_1 \\ a_2 \\ \vdots \\ a_m \end{bmatrix} = \begin{bmatrix} b_1 \\ b_2 \\ \vdots \\ b_m \end{bmatrix}$$

Now, taking the basis matrix $B = I_m$, therefore, $x_B = B^{-1}b = I_m b = b \geq 0$ which is the required basic feasible solution.

Working Procedure

The basic feasible solution is $a_1 = x_{B_1} = b_1, a_2 = x_{B_2} = b_2, \dots, a_m = x_{B_m} = b_m$ can be obtained by writing all the non-basic variables $x_1, x_2, \dots, x_n, s_1, s_2, \dots, s_m$ equal to zero and solving the equations for the remaining basic variables (artificial variables) $a_1, a_2, \dots, a_m$.

CASE -3 WHEN ALL ORIGINAL CONSTRAINTS HAVE '$\leq$', '$\geq$' AND '$=$' SIGN

In such type of cases, we convert the constraints into equations by introducing and inserting slack, surplus and artificial variables. The basis matrix $B = I_m$ is obtained by introducing unit column vectors corresponding to the slack and artificial variables.

Working Procedure

To find the initial basic feasible solution of the given LPP, put all the non-basic variables equal to zero and solve for remaining basic variables. Then initial basic feasible solution is $x_B = b \geq 0$.

> If identity is present in coefficient matrix A without introducing the artificial variables, then there is no need to introduce the artificial variables.

4.11 SIMPLEX ALGORITHM

To find the optimal solution of the given linear programming problem, we use the following steps:

[MEERUT–2007]

STEP 1. General Steps:

(i) If the given problem is of minimization, first convert it into the maximization problem by multiplying both sides of the objective function by –1 and put $-z = z^*$. Remember that if V is the maximum value of z^* then $-V$ will be the minimum value of z.

(ii) The RHS of each of the constraints should be non-negative. If there is any constraints for which b_i is negative then multiply this constraints by –1 to convert it into positive values.

(iii) Replace each unrestricted variable (if any) with the difference of two non-negative variables, replace each non-positive variables with a new non-negative variable whose value is the negative of the original variable.

(iv) Express the given LPP into standard form by adding additional variables to the left side of each constraints and assign a zero cost coefficients to these in the objective function.

STEP 2.
Convert the given inequalities of constraints into equations. For this we introduce, slack, surplus or artificial variables (whichever is required). The coefficients of slack or surplus variables in the objective function is zero. Note that, if artificial variables are introduced then we use two-phase or Big-M method.

STEP 3. Construction of Initial Simplex Table:
Construct the initial simplex table as given below:

		$c_j \rightarrow$	c_1	c_2	...	c_n	0	0	...	0
Basic Variables B.V.	Coefficients of basic variables C_B	Value of basic variables b $(= X_B)$	x_1	x_2	...	x_n	s_1	s_2	...	s_m
s_1	C_{B_1}	$x_{B_1} = b_1$	a_{11}	a_{12}	...	a_{1n}	1	0	...	0
s_2	C_{B_2}	$x_{B_2} = b_2$	a_{21}	a_{22}	...	a_{2n}	0	1	...	0
$\vdots$	$\vdots$	$\vdots$	$\vdots$							
s_m	C_{B_m}	$x_{B_m} = b_m$	a_{m1}	a_{m2}	...	a_{mn}	0	0	...	1
$Z = \Sigma C_{B_i} x_{B_i}$		$z_j = \Sigma C_{B_i} x_j$	0	0	...	0	0	0	...	0
		$c_j - z_j$	$c_1 - z_1$	$c_2 - z_2$	...	$c_n - z_n$	0	0	...	0

In the above table we observe that

(i) The first row provide coefficients of the current basic variables in the objective function. Column headed by X_B represents the current values of the corresponding variables in the basis.

(ii) The second row indicate the coefficient c_j of variables in the objective function which remains the same in successive simplex tables. These values are used to determine the variables to be entered into the basis matrix B.

(iii) The basic matrix (identity matrix) represents the coefficients of slack variables which have been added to the constraints. Each column of the identity matrix also represents a basic variables to be listed in Column B.

(iv) The value of z_j represent the amount by which the value of the objective function z should be decreased (or increased) if one unit of given variable is added to the new solution.

Each of the value in the $c_j - z_j$ row represents the net amount of increase (or decrease) in the objective function that would occur when one unit of the variables represented by the column head is introduced into the solution.

STEP 4. **Test for optimality:**

Calculate $\Delta_j = c_j - z_j$ for all non-basic variables. Then we have the following cases:

(i) If $\Delta_j \leq 0 \ \forall \ j$, then solution under consideration is optimal.

(ii) Alternative optimal solution will exist if any Δ_j, for non-basic variables is zero.

(iii) If $\Delta_j < 0 \ \forall \ j$, corresponding to non-basic variables, then unique optimal solution exists.

(iv) If $\Delta_j > 0$ for any j, *i.e.*, if at least one Δ_j is positive then solution is not optimal and it indicates that an improvement in the value of the objective function Z is possible.

(v) If corresponding to maximum positive Δ_j, all elements of the column of basic variables are negative or zero, then solution is unbounded.

STEP 5. **Selection of incoming and outgoing vector:**

(i) The incoming vector α_k is selected corresponding to the largest positive value of Δ_j. But if maximum value of Δ_j occur for more than one α_j then we may select any of these vectors as incoming vector.

(ii) The outgoing vector β_r is selected corresponding to that value of r for which

$$\frac{x_{B_r}}{y_{rk}} = \min_i \left\{ \frac{x_{B_i}}{y_{i_k}}, y_{i_k} > 0 \right\}$$

where α_k is the incoming vector.

If the minimum value is not unique then more than one variable will vanish in the next solution and a degenerate basic feasible solution exists.

Here, it must be noted that the division by negative or zero is not permitted.

STEP 6. **Test for Feasibility.** If α_k is incoming vector and β_r is the outgoing vector then the element y_{rk} which lies at the intersection of minimum ratio arrow ($\rightarrow$) and incoming vector arrow ($\uparrow$) is called the pivot element. Put this element in the box $\square$.

STEP 7. **Finding the New Solution.** Use the following steps:

(i) If the key element is 1, then row remains unchanged in the next simplex table.

(ii) If the key element is not 1 then divide each element of the row by the key element.

(iii) The new values of the element in the remaining rows for the new simplex table can be obtained by performing elementary row operations on all rows so that all the elements except the kew element in the key column are zero.

Then, new entries in C_B and X_B columns are updated in the new simplex table of the current solution.

Therefore, we get improved basic feasible solution.

STEP 8. Test the improved BFS for optimality.

Solved Examples

EXAMPLE I. *Using simplex method, solve the following LPP:*

$$\text{Max. } Z = 16x_1 + 17x_2 + 10x_3$$

subject to the constraints

$$x_1 + x_2 + 4x_3 \leq 2000$$
$$2x_1 + x_2 + x_3 \leq 3600$$
$$x_1 + 2x_2 + 2x_3 \leq 2400$$
$$x_1 \leq 30$$

and $\qquad x_1, x_2, x_3 \geq 0$

SOLUTION. Clearly, the given LPP is in standard form.

Now introducing slack variables s_1, s_2, s_3 and s_4 to convert the given inequalities to equations. Then LP model can be written as

Max. $Z = 16x_1 + 17x_2 + 10x_3 + 0s_1 + 0s_2 + 0s_3 + 0s_4$
subject to the constraints
$$x_1 + x_2 + 4x_3 + s_1 = 2000$$
$$2x_1 + x_2 + x_3 + s_2 = 3600$$
$$x_1 + 2x_2 + 2x_3 + s_3 = 2400$$
$$x_1 + s_4 = 30$$
and $x_1, x_2, x_3, s_1, s_2, s_3, s_4 \geq 0$

Now we prepare the initial simplex table as follows:

Simplex Table-1

		$c_j \rightarrow$	16	17	10	0	0	0	0	
B.V.	C_B	X_B	x_1	x_2	x_3	s_1	s_2	s_3	s_4	Min Ratio X_B/x_2
s_1	0	2000	1	1	4	1	0	0	0	2000/1 = 2000
s_2	0	3600	2	1	1	0	1	0	0	3600/1 = 3600
s_3	0	2400	1	②	2	0	0	1	0	2400/2 = 1200 → (min)
s_4	0	30	1	0	0	0	0	0	1	—
$\Delta_j = c_j - z_j$			16	17 ↑	10	0	0	0	0	

In the above simplex table, $c_2 - z_2 = 17$ in x_2-column is the largest positive value. Now to obtain a new improved solution, by entering variable x_2 into the basis and removing variable s_3 from the basis, use the following steps:

$$R_3(\text{new}) \rightarrow R_3 \text{ (old)} \div 2$$
$$R_1 \text{ (new)} \rightarrow R_1 \text{ (old)} - R_3 \text{ (new)}$$
$$R_2 \text{ (new)} \rightarrow R_2 \text{ (old)} - R_3 \text{ (new)}$$

Then we have the following simplex table:

Simplex Table-2

		$c_j \rightarrow$	16	17	10	0	0	0	0	Min Ratio
B.V.	C_B	X_B	x_1	x_2	x_3	s_1	s_2	s_3	s_4	X_B/x_1
s_1	0	800	1/2	0	3	1	0	−1/2	0	800/(1/2) = 1600
s_2	0	2400	3/2	0	0	0	1	−1/2	0	2400/(3/2) = 1600
x_2	17	1200	1/2	1	1	0	0	1/2	0	1200/(1/2) = 2400
s_4	0	30	①	0	0	0	0	0	1	30/1 = 30 → (min)
$\Delta_j = c_j - z_j$			15/2 ↑	0	−7	0	0	−17/2	0	

The solution obtained in the above table is not optimal because $\Delta_j > 0$ in x_1-column. To improved this solution, applying the following row operations to get a new improved solution by entering variable x_1 into the basis and removing the variable s_4 from the basis.

$$R_4 \text{ (new)} \rightarrow R_4 \text{ (old)} \div \text{ (Key element)}$$

$$R_1 \text{ (new)} \rightarrow R_1 \text{ (old)} - \frac{1}{2}R_4 \text{ (new)}$$

$$R_2 \text{ (new)} \rightarrow R_2 \text{ (old)} - \frac{3}{2}R_4 \text{ (new)}$$

$$R_3 \text{ (new)} \rightarrow R_3 \text{ (old)} - \frac{1}{2}R_4 \text{ (new)}$$

Then we have the following simplex table:

Simplex Table-3

		$c_j \rightarrow$	16	17	10	0	0	0	0
B.V.	C_B	X_B	x_1	x_2	x_3	s_1	s_2	s_3	s_4
s_1	0	785	0	0	3	1	0	−1/2	−1/2
s_2	0	2355	0	0	0	0	1	−1/2	−3/2
x_2	17	1185	0	1	1	0	0	1/2	−1/2
x_1	16	30	1	0	0	0	0	0	1
$\Delta_j = c_j - z_j$			0	0	−7	0	0	−17/2	−15/2

Clearly, in the above table all $\Delta_j < 0$ corresponding to non-basic variables column. Hence, the solution is optimal and given by

$$x_1 = 30, x_2 = 1185, x_3 = 0$$

and Max. $Z = 0 \times 785 + 0 \times 2355 + 17 \times 1185 + 16 \times 30 = 20625$

EXAMPLE 2. **Solve the following LPP by simplex method**

$$\textbf{Max. } Z = 8x_1 + 11x_2$$

subject to the constraints
$$3x_1 + x_2 \le 7$$
$$x_1 + 3x_2 \le 8$$
and
$$x_1, x_2 \ge 0$$

SOLUTION. The given maximization problem is in standard form with all $b_i's$ are non-negative.

Now introduce the slack variables s_1 and s_2 to convert the given inequalities into equations. Then given problem becomes

Max. $Z = 8x_1 + 11x_2 + 0s_1 + 0s_2$

s.t.
$$3x_1 + x_2 + s_1 = 7$$
$$x_1 + 3x_2 + s_2 = 8$$
and
$$x_1, x_2, s_1, s_2 \ge 0$$

Then apply the simplex method, we construct the following table:

Simplex Table

			$c_j \to$	8	11	0	0	
B.V.	C_B	X_B	x_1	x_2	s_1	s_2		Min Ratio $X_B \mid x_2$
s_1	0	7	3	1	1	0		$7/1 = 7$
s_2	0	8	1	③	0	1		8/3 (min) →
$Z = C_B X_B = 0$		$\Delta_j =$	8	11	0	0		x_B/x_1
				↑		↓		
s_1	0	13/3	(8/3)	0	1	−1/3		13/8 (min) →
x_2	11	8/3	1/3	1	0	1/3		$8/1 = 8$
$Z = C_B X_B = 88/3$		$\Delta_j =$	13/3	0	0	−11/3		
			↑		↓			
x_1	8	13/8	1	0	3/8	−1/8		
x_2	11	17/8	0	1	−1/8	3/8		
$Z = C_B X_B = 291/8$		$\Delta_j =$	0	0	−13/8	−25/8		

Clearly, in the last table all $\Delta_j \le 0$. Thus, solution is optimal and is given by

$$x_1 = \frac{13}{8}, x_2 = \frac{17}{8} \text{ and Max. } Z = \frac{291}{8}$$

EXAMPLE 3. *Solve the following LPP by simplex method*

$$\textbf{Max. } Z = 3x_1 + 2x_2$$

subject to the constraints

$$x_1 + x_2 \le 4$$
$$x_1 - x_2 \le 2$$

and $x_1, x_2 \ge 0$ [MADURAI–1997; IAS(Main)–1992; CALICUT(B.Tech)–1990]

SOLUTION. The given LPP is in standard form. Introduce slack variables s_1 and s_2 to convert the given inequalities into equations. The given LPP becomes

Max. $Z = 3x_1 + 2x_2 + 0s_1 + 0s_2$

s.t.
$$x_1 + x_2 + s_1 = 4$$
$$x_1 - x_2 + s_2 = 2$$
and
$$x_1, x_2, s_1, s_2 \geq 0$$

Then solution to the problem using simplex method is given in the following table.

Simplex Table

B.V.	C_B	X_B	$c_j \rightarrow$ 3 x_1	2 x_2	0 s_1	0 s_2	Min Ratio $X_B	x_1$
s_1	0	4	1	1	1	0	4/1 = 4
s_2	0	2	①	−1	0	1	2/1 = 2 (min)→
$Z = C_B x_B = 0$		$\Delta_j =$	3 ↑	2	0	0 ↓	
s_1	0	2	0	②	1	−1	2/2 = 1 (min) →
x_1	3	2	1	−1	0	1	—
$Z = C_B x_B = 6$		$\Delta_j =$	0	5 ↑	0	−3 ↓	
x_2	2	1	0	1	1/2	−1/2	
x_1	3	3	1	0	1/2	1/2	
$Z = C_B x_B = 11$		$\Delta_j =$	0	0	−5/2	−1/2	

We observe that in the last table all $\Delta_j < 0$.
$\Rightarrow$ Solution is optimal.
Hence, solution is given by
$$x_1 = 3, x_2 = 1 \text{ and Max } Z = 11.$$

EXAMPLE 4. *Solve the following LPP by simplex method*

$$\textbf{Max. } Z = 3x_1 + 2x_2 - 2x_3$$

subject to the constraints

$$x_1 + 2x_2 + 2x_3 \leq 10$$
$$2x_1 + 4x_2 + 3x_3 \leq 15$$

and $$x_1, x_2, x_3 \geq 0$$

SOLUTION. Clearly, the given maximization problem is in standard form in which all $b_i's$ are non-negative.

Now, introduce the slack variables s_1 and s_2 to convert the given inequalities into equations such that
$$\text{Max. } Z = 3x_1 + 2x_2 - 2x_3 + 0s_1 + 0s_2$$
s.t.
$$x_1 + 2x_2 + 2x_3 + s_1 = 10$$
$$2x_1 + 4x_2 + 3x_3 + s_2 = 15$$
and $$x_1, x_2, x_3, s_1, s_2 \geq 0$$

Now by applying the simplex method, we have the following simplex table.

Simplex Table

B.V.	C_B	X_B	x_1	x_2	x_3	s_1	s_2	Min Ratio $X_B \mid x_1$
			3	2	–2	0	0	$c_j \rightarrow$
s_1	0	10	1	2	2	1	0	10/1 = 10
s_2	0	15	②	4	3	0	1	15/2 = (min)→
$Z = C_B X_B = 0, \Delta_j =$			3 ↑	2	–2	0	0 ↓	
s_1	0	5/2	0	0	1/2	0	–1/2	
x_1	3	15/2	1	2	3/2	0	1/2	
$Z = C_B X_B = \dfrac{45}{2}, \Delta_j =$			0	– 4	–13/2	0	–3/2	

In the last row of the above table all $\Delta_j < 0$. So, the obtained solution is optimal, which is given by

$$x_1 = \frac{15}{2}, x_2 = 0, x_3 = 0 \text{ and Max. } Z = \frac{45}{2}$$

EXAMPLE 5. **Solve the following LPP by simplex method.**

$$\textbf{Max. } Z = 3x_1 + 5x_2$$

subject to the constraints

$$3x_1 + 2x_2 \leq 18$$
$$x_1 \leq 4$$
$$x_2 \leq 6$$

and $x_1, x_2 \geq 0$

SOLUTION. Clearly, the given problem of maximization is in standard form such that all $b_i's \geq 0$.

Converting the inequalities into equations by introducing slack variables s_1, s_2, s_3 as follows:

$$\text{Max. } Z = 3x_1 + 5x_2 + 0s_1 + 0s_2 + 0s_3$$

s.t. $3x_1 + 2x_2 + s_1 = 18$
$$x_1 + s_2 = 4$$
$$x_2 + s_3 = 6$$

and $x_1, x_2, s_1, s_2, s_3 \geq 0$

Then we have the following simplex table

Simplex Table

B.V.	C_B	X_B	x_1	x_2	s_1	s_2	s_3	Min Ratio
			3	5	0	0	0	$c_j \rightarrow$
s_1	0	18	3	2	1	0	0	18/2 = 9
s_2	0	4	1	0	0	1	0	—
s_3	0	6	0	①	0	0	1	6/1 = 6 (min)→
$Z = C_B X_B = 0, \Delta_j =$			3	5 ↑	0	0	0 ↓	

s_1	0	6	③	0	1	0	-2	6/3 = 2 (min) →
s_2	0	4	1	0	0	1	0	4/1 = 4
x_2	5	6	0	1	0	0	1	—
$Z = C_B X_B = 30, \Delta_j =$			3↑	0	0↓	0	-5	
x_1	3	2	1	0	1/3	0	-2/3	
s_2	0	2	0	0	-1/3	1	2/3	
x_2	5	6	0	1	0	0	1	
$Z = C_B X_B = 36, \Delta_j =$			0	0	-1	0	-3	

In the last row of the above table all $\Delta_j < 0$, so the obtained solution is optimal and is given by $x_1 = 2$, $x_2 = 6$ and Max. $Z = 36$

EXAMPLE 6. **Solve the following LPP by simplex method**

$$\text{Max. } Z = 3x_1 + 4x_2$$

subject to the constraints

$$x_1 + 3x_2 \leq 9,\ 2x_1 - x_2 \leq 8,\ x_1 + x_2 \leq 5$$

and $$x_1,\ x_2 \geq 0$$

SOLUTION. Since, the given problem is in standard form. Converting the given inequalities into equations by introducing slack variables s_1, s_2, s_3 such that

$$\text{Max. } Z = 3x_1 + 4x_2 + 0s_1 + 0s_2 + 0s_3$$

s.t. $$x_1 + 3x_2 + s_1 = 9,\ 2x_1 - x_2 + s_2 = 8,\ x_1 + x_2 + s_3 = 5$$

and $$x_1, x_2, s_1, s_2, s_3 \geq 0$$

Now applying simplex method, we get the following table:

Simplex Table

B.V.	C_B	X_B	$c_j \rightarrow$ 3	4	0	0	0	Min Ratio
			x_1	x_2	s_1	s_2	s_3	
s_1	0	9	1	③	1	0	0	9/3 = 3 (min)→
s_2	0	8	2	-1	0	1	0	—
s_3	0	5	1	1	0	0	1	5/1 = 5
$Z = C_B X_B = 0, \Delta_j \rightarrow$			3	4↑	0↓	0	0	
x_2	4	3	1/3	1	1/3	0	0	9
s_2	0	11	7/3	0	1/3	1	0	33/7
s_3	0	2	②/3	0	-1/3	0	1	3 (min) →
$Z = C_B X_B = 12, \Delta_j \rightarrow$			5/3↑	0	-4/3	0	0↓	
x_2	4	2	0	1	1/2	0	-1/2	
s_2	0	4	0	0	3/2	1	-7/2	
x_1	3	3	1	0	-1/2	0	3/2	
$Z = C_B X_B = 17, \Delta_j \rightarrow$			0	0	-1/2	0	-5/2	

In the last row of the above table all $\Delta_j \leq 0$. Hence, the obtained solution is optimal which is given by

$$x_1 = 3, x_2 = 2 \text{ and Max. } Z = 17$$

EXAMPLE 7. **Solve the following LPP by simplex method**

$$\text{Max. } Z = 2x_1 + x_2$$

subject to the constraints

$$x_1 + 2x_2 \leq 10, x_1 + x_2 \leq 6$$

$$x_1 - x_2 \leq 2, x_1 - 2x_2 \leq 1$$

and $x_1, x_2 \geq 0$

SOLUTION. Clearly, the given LPP is in standard form.

Now, introduce slack variables s_1, s_2, s_3 and s_4 to convert the given inequalities into equations such that

$$\text{Max. } Z = 2x_1 + x_2 + 0s_1 + 0s_2 + 0s_3 + 0s_4$$

s.t. $\quad x_1 + 2x_2 + s_1 = 10, x_1 + x_2 + s_2 = 6$

$$x_1 - x_2 + s_3 = 2, x_1 - 2x_2 + s_4 = 1$$

and $\quad x_1, x_2, s_1, s_2, s_3, s_4 \geq 0$

Now we have the following simplex table

Simplex Table

B.V.	C_B	X_B	$c_j \rightarrow$ 2 x_1	1 x_2	0 s_1	0 s_2	0 s_3	0 s_4	Min Ratio
s_1	0	10	1	2	1	0	0	0	10/1 = 10
s_2	0	6	1	1	0	1	0	0	6/1 = 6
s_3	0	2	1	−1	0	0	1	0	2/1 = 2
s_4	0	1	①	−2	0	0	0	1	1/1 = 1 (min)→
$Z = 0,$		$\Delta_j \rightarrow$	2 ↑	1	0	0	0	0 ↓	
s_1	0	9	0	4	1	0	0	−1	9/4
s_2	0	5	0	3	0	1	0	−1	5/3
s_3	0	1	0	①	0	0	1	−1	1/1 = 1 (min)→
x_1	2	1	1	−2	0	0	0	1	—
$Z = 2,$		$\Delta_j \rightarrow$	0	5 ↑	0	0	0 ↓	−2	
s_1	0	5	0	0	1	0	− 4	3	5/3
s_2	0	2	0	0	0	1	−3	②	2/2 = 1 (min)→
x_2	1	1	0	1	0	0	1	−1	—
x_1	2	3	1	0	0	0	2	−1	—
$Z = 8,$		$\Delta_j \rightarrow$	0	0	0	0 ↓	−5	3 ↑	

B.V.	C_B	X_B						
s_1	0	2	0	0	1	$-3/2$	$1/2$	0
s_4	0	1	0	0	0	$1/2$	$-3/2$	1
x_2	1	2	0	1	0	$1/2$	$-1/2$	0
x_1	2	4	1	0	0	$1/2$	$1/2$	0
$Z = C_B x_B = 10, \Delta_j \rightarrow$			0	0	0	$-3/2$	$-1/2$	0

In the last row of the above table all $\Delta_j \leq 0$. So, obtained solution is optimal and is given by $x_1 = 4$, $x_2 = 2$ and Max. $Z = 10$

EXAMPLE 8. *Solve the following LPP by simplex method*

$$\textbf{Max. } \mathbf{Z = 3x_1 + 5x_2 + 4x_3}$$

subject to the constraints

$$\mathbf{2x_1 + 3x_2 \leq 8, \ 2x_2 + 5x_3 \leq 10, \ 3x_1 + 2x_2 + 4x_3 \leq 15}$$

and $\qquad \mathbf{x_1, x_2, x_3 \geq 0}$ [MEERUT–2007, 11, 14; KANPUR–2010, AVADH–2012; ALLAHABAD–2014; REWA–1993; JODHPUR1992]

SOLUTION. Clearly, the LPP is given in standard form.

Now, to convert the given inequalities into equations, we introduce the slack variables s_1, s_2, s_3 such that

$$\text{Max. } Z = 3x_1 + 5x_2 + 4x_3 + 0s_1 + 0s_2 + 0s_3$$

s.t. $\qquad 2x_1 + 3x_2 + s_1 = 8, \ 2x_2 + 5x_3 + s_2 = 10, \ 3x_1 + 2x_2 + 4x_3 + s_3 = 15$

and $\qquad x_1, x_2, x_3, s_1, s_2, s_3 \geq 0$

Now applying the usual procedure, we have the following simplex table.

Simplex Table

B.V.	C_B	X_B	$c_j \rightarrow$ 3 x_1	5 x_2	4 x_3	0 s_1	0 s_2	0 s_3	Min Ratio
s_1	0	8	2	③	0	1	0	0	$\dfrac{8}{3}$ (min)→
s_2	0	10	0	2	5	0	1	0	$10/2 = 5$
s_3	0	15	3	2	4	0	0	1	$15/2$
$Z = C_B X_B = 0, \quad \Delta_j \rightarrow$			3	5 ↑	4 ↓	0	0	0	
x_2	5	$\dfrac{8}{3}$	$\dfrac{2}{3}$	1	0	$1/3$	0	0	—
s_2	0	$\dfrac{14}{3}$	$-\dfrac{4}{3}$	0	⑤	$-2/3$	1	0	$\dfrac{14}{15}$ (min)→
s_3	0	$\dfrac{29}{3}$	$\dfrac{5}{3}$	0	4	$-2/3$	0	1	$\dfrac{29}{12}$
$Z = C_B X_B = \dfrac{40}{3}, \ \Delta_j \rightarrow$			$-\dfrac{1}{3}$	0	4 ↑	$-5/3$	0 ↓	0	

x_2	5	$\dfrac{8}{3}$	$\dfrac{2}{3}$	1	0	1/3	0	0	1
x_3	4	$\dfrac{14}{15}$	$-\dfrac{4}{15}$	0	1	−2/15	1/5	0	—
s_3	0	$\dfrac{89}{15}$	$\left(\dfrac{41}{15}\right)$	0	0	−2/15	− 4/5	1	$\dfrac{89}{41}$ (min)→
$Z = C_B X_B = \dfrac{256}{45}, \Delta_j \to$			$\dfrac{11}{15}$ ↑	0	0	−17/15	−4/5 ↓	0	
x_2	5	$\dfrac{50}{41}$	0	1	0	−15/41	8/41	−10/41	
x_3	4	$\dfrac{62}{41}$	0	0	1	−6/41	5/41	4/41	
x_1	3	$\dfrac{89}{41}$	1	0	0	−2/41	−12/41	15/41	
$Z = C_B X_B = \dfrac{765}{41}, \Delta_j \to$			0	0	0	−45/41	−14/41	−11/41	

Clearly, in the last row of the above table, all $\Delta_j \le 0$.

Hence, the optimal solution is given by

$$x_1 = \frac{89}{41}, x_2 = \frac{50}{41}, x_3 = \frac{62}{41} \text{ and Max. } Z = \frac{765}{41}$$

EXAMPLE 9. **Solve the following LPP by simplex method**

$$\textbf{Max. } Z = 3x_1 + 2x_2 + 5x_3$$

subject to the constraints

$$x_1 + 2x_2 + x_3 \le 430$$
$$3x_1 + 2x_3 \le 460$$
$$x_1 + 4x_2 \le 420$$

and $$x_1, x_2, x_3 \ge 0$$

[GARHWAL–2013; MEERUT–2008, 12; GORAKHPUR–2010, 11]

SOLUTION. Clearly, the given LPP is in standard form. Now, introduce slack variables s_1, s_2 and s_3 to convert the given inequalities into equations such that

Max. $Z = 3x_1 + 2x_2 + 5x_3 + 0s_1 + 0s_2 + 0s_3$

s.t. $x_1 + 2x_2 + x_3 + s_1 = 430$

$3x_1 + 2x_3 + s_2 = 460$

$x_1 + 4x_2 + s_3 = 420$

and $x_1, x_2, x_3, s_1, s_2, s_3 \ge 0$

Simplex Table

B.V.	C_B	X_B	$c_j \to$ 3 x_1	2 x_2	5 x_3	0 s_1	0 s_2	0 s_3	Min Ratio
s_1	0	430	1	2	1	1	0	0	430
s_2	0	460	3	0	②	0	1	0	230 (min)→
s_3	0	420	1	4	0	0	0	1	—
$Z = 0,$		$\Delta_j \to$	3	2	5↑	0	0↓	0	
s_1	0	200	$-1/2$	②	0	1	$-1/2$	0	100 (min)→
x_3	5	230	3/2	0	1	0	1/2	0	—
s_3	0	420	1	4	0	0	0	1	420/4
$Z = 1150,$		$\Delta_j \to$	$-9/2$	2↑	0	0	$-5/2$↓	0	
x_2	2	100	$-1/4$	1	0	1/2	$-1/4$	0	
x_3	5	230	3/2	0	1	0	1/2	0	
s_3	0	20	2	0	0	-2	1	1	
$Z = 1350,$		$\Delta_j \to$	-4	0	0	-1	-2	0	

Clearly in the last row of the above table all $\Delta_j \leq 0$.

$\Rightarrow$ Solution is optimal.

Hence, optimal solution is given by

$x_1 = 0, x_2 = 100, x_3 = 230$ and Max. $Z = 1350$.

EXAMPLE 10. *Solve the following LPP by simplex method*

$$\text{Max. } Z = 2x_1 + 4x_2 + 3x_3$$

subject to the constraints

$$3x_1 + 4x_2 + 2x_3 \leq 60$$
$$2x_1 + x_2 + 2x_3 \leq 40$$
$$x_1 + 3x_2 + 2x_3 \leq 80$$

and $x_1, x_2, x_3 \geq 0$

SOLUTION.   Clearly, the given LPP is in standard form. Introduce slack variables s_1, s_2 and s_3 to convert the given inequalities into equations. Then we get

Max. $Z = 2x_1 + 4x_2 + 3x_3 + 0s_1 + 0s_2 + 0s_3$

s.t. $3x_1 + 4x_2 + 2x_3 + s_1 = 60$

$2x_1 + x_2 + 2x_3 + s_2 = 40$

$x_1 + 3x_2 + 2x_3 + s_3 = 80$

and $x_1, x_2, x_3, s_1, s_2, s_3 \geq 0$

Now, apply the simplex method, we have the following simplex table:

Simplex Table

B.V.	C_B	X_B	$c_j \to$ 2 x_1	4 x_2	3 x_3	0 s_1	0 s_2	0 s_3	Min Ratio
s_1	0	60	3	④	2	1	0	0	60/4 = 15 (min)→
s_2	0	40	2	1	2	0	1	0	40/1 = 40
s_3	0	80	1	3	2	0	0	1	80/3
$Z = 0$,		$\Delta_j \to$	2 ↑	4	3 ↓	0	0	0	
x_2	4	15	3/4	1	1/2	1/4	0	0	—
s_2	0	25	5/4	0	③/②	−1/4	1	0	25/(3/2)=(50/3)(min)→
s_3	0	35	−5/4	0	1/2	−3/4	0	1	—
$Z = 60$,		$\Delta_j \to$	−1	0	1 ↑	−1 ↓	0	0	
x_2	4	20/3	1/6	1	0	1/3	−1/3	0	
x_3	3	50/3	5/6	0	1	−1/6	2/3	0	
s_3	0	80/3	−5/3	0	0	−1/12	−1/3	1	
$Z = C_B X_B = 230/3$, $\Delta_j \to$			−11/6	0	0	−5/6	−2/3	0	

Clearly in the last row of the above table all $\Delta_j \leq 0$.

⇒   Solution is optimal.

Hence, optimal solution is given by

$$x_1 = 0, x_2 = \frac{20}{3}, x_3 = \frac{50}{3} \text{ and Max. } Z = \frac{230}{3}$$

 ## Exercise-4.2

Solve the following LPP by simplex method:

1. Max. $Z = 40x_1 + 35x_2$
s.t. $2x_1 + 3x_2 \leq 60$
$4x_1 + 3x_2 \leq 96$
$x_1, x_2 \geq 0$

2. Max. $Z = 7x_1 + 5x_2$
s.t. $-x_1 - 2x_2 \geq -6$
$4x_1 + 3x_2 \leq 12$
$x_1, x_2 \geq 0$

3. Max. $Z = 5x_1 + 3x_2$
s.t. $3x_1 + 5x_2 \leq 15$
$5x_1 + 2x_2 \leq 10$
$x_1, x_2 \geq 0$

4. Max. $Z = 7x_1 + x_2 + 2x_3$
s.t. $x_1 + x_2 - 2x_3 \leq 10$
$4x_1 + x_2 + x_3 \leq 20$
$x_1, x_2, x_3 \geq 0$

5. Max. $Z = 5x_1 + 7x_2$

s.t. $x_1 + x_2 \leq 4$
$3x_1 - 8x_2 \leq 24$
$10x_1 + 7x_2 \leq 35$
$x_1, x_2 \geq 0$

6. Max. $Z = 3x_1 + 2x_2$
s.t. $2x_1 + x_2 \leq 40$
$x_1 + x_2 \leq 24$
$2x_1 + 3x_2 \leq 60$
$x_1, x_2 \geq 0$

7. Max. $Z = 2x_1 + 4x_2$
s.t. $2x_1 + 3x_2 \leq 48$
$x_1 + 3x_2 \leq 42$
$x_1 + x_2 \leq 21$
$x_1, x_2 \geq 0$

8. Max. $Z = 5x_1 + 10x_2 + 8x_3$
s.t. $3x_1 + 5x_2 + 2x_3 \leq 60$
$4x_1 + 4x_2 + 4x_3 \leq 72$
$2x_1 + 4x_2 + 5x_3 \leq 100$
$x_1, x_2, x_3 \geq 0$ [MEERUT–2009, 12]

9. Max. $Z = x_1 - x_2 + 3x_3$

s.t. $x_1 + x_2 + x_3 \leq 10$

$2x_1 - x_3 \leq 2$

$2x_1 - 2x_2 + 3x_3 \leq 0$

$x_1, x_2, x_3 \geq 0$

10. Max. $Z = 2x_1 + 5x_2 + 7x_3$

s.t. $3x_1 + 2x_2 + 4x_3 \leq 100$

$x_1 + 4x_2 + 2x_3 \leq 100$

$x_1 + x_2 + 3x_3 \leq 100$

$x_1, x_2, x_3 \geq 0$

[MEERUT–2009, 11; KANPUR–2009]

11. Max. $Z = 4x_1 + 5x_2 + 9x_3 + 11x_4$

s.t. $x_1 + x_2 + x_3 + x_4 \leq 15$

$7x_1 + 5x_2 + 3x_3 + 2x_4 \leq 120$

$3x_1 + 5x_2 + 10x_3 + 15x_4 \leq 100$

$x_1, x_2, x_3, x_4 \geq 0$

ANSWERS

1. $x_1 = 8, x_2 = 18$, Max. $Z = 1000$

2. $x_1 = 3, x_2 = 0$, Max. $Z = 21$

3. $x_1 = \dfrac{20}{19}, x_2 = \dfrac{45}{19}$, Max. $Z = \dfrac{235}{19}$

4. $x_1 = x_2 = 0, x_3 = 20$, Max. $Z = 40$

5. $x_1 = 0, x_2 = 4$, Max. $Z = 28$

6. $x_1 = 16, x_2 = 8$, Max. $Z = 64$

7. $x_1 = 6, x_2 = 12$, Max. $Z = 60$

8. $x_1 = 0, x_2 = 8, x_3 = 10$, Max. $Z = 160$

9. $x_1 = 0, x_2 = 6, x_3 = 4$, Max. $Z = 6$

10. $x_1 = 0, x_2 = \dfrac{50}{3}, x_3 = \dfrac{50}{3}$, Max. $Z = 200$

11. $x_1 = \dfrac{50}{7}, x_2 = 0, x_3 = \dfrac{55}{7}$ Max. $Z = \dfrac{695}{7}$

4.12 SIMPLEX METHOD : CASE OF MINIMIZATION

In case of minimization of an LPP, we have the following cases:

Case I. When the constraints are of the type $\leq$, *i.e.*,

$$\sum_{j=1}^{n} a_{ij}x_j \leq b_i, x_j \geq 0$$

But some right hand side constraints are negative ($b_i < 0$). Then, after adding the non-negative slack variables s_i, the initial solution so obtained will be $s_i = -b_i$ for some i. It is not the feasible solution.

Case II. When the constraints are of the $\geq$ type, *i.e.*,

$$\sum_{j=1}^{n} a_{ij}x_j \geq b_i, x_j \geq 0$$

Then convert the inequalities into equation's form, adding surplus variables such that

$$\sum_{j=1}^{n} a_{ij}x_j - s_i = b_i, x_j \geq 0, s_i \geq 0$$

Letting $x_j = 0$ we get an initial solution $-s_i = b_j$ or $s_j = -b_i$ which is also not a feasible solution. In this case we add artificial variable A_i to get an initial basic feasible solution.

If the given problem is of minimization, we multiply the objective function by –1, *i.e.*,

$$Z' = -Z = -c_i x_i.$$

4.13 ARTIFICIAL VARIABLE TECHNIQUE

There are following two methods for eliminating artificial variables from the solution:

(i) Big-*M* method

(ii) Two phase method

(i) Big-M method: The artificial variable like surplus and slack variables are imaginary variables which are introduced in the constraints with $\geq$ or $=$ sign. These are just introduced to form an identity matrix. These variables have no physical significance. So, these variables have to be removed by the simplex method to have the meaningful solution. Here a very large price say $-M$ has to be alloted to the artificial variable in the objective function. This method of solution by aritificial variables was given by Charnes and is hence known as Charnes Big-M method.

 Solved Examples

EXAMPLE 1. *Solve the following LPP by Big-M method.*

$$Min\ Z = 5x_1 + 6x_2$$

subject to the constraints

$$2x_1 + 5x_2 \geq 1500,\ 3x_1 + x_2 \geq 1200$$

and $\qquad x_1, x_2 \geq 0$

SOLUTION. The given problem is a minimization problem.

So, first convert it to maximization problem by using $Z' = Z$.

Then objective function becomes Max. $Z' = -5x_1 - 6x_2$

Now, introduce the necessary surplus variables s_1, s_2 and artificial variables A_1 and A_2, then given problem becomes

$$Max\ Z' = -5x_1 - 6x_2 + 0s_1 + 0s_2 - MA_1 - MA_2$$

s.t. $\qquad 2x_1 + 5x_2 - s_1 + A_1 = 1500$

$\qquad\qquad 3x_1 + x_2 - s_2 + A_2 = 1200$

and $\qquad x_1, x_2, s_1, s_2, A_1, A_2 \geq 0$

Now apply the simplex method, we have the following simplex table.

Simplex Table

	$c_j \rightarrow$		-5	-6	0	0	$-M$	$-M$	
B.V.	C_B	X_B	x_1	x_2	s_1	s_2	A_1	A_2	Min Ratio
A_1	$-M$	1500	2	⑤	-1	0	1	0	$1500/5 = 300$ (Min) $\rightarrow$
A_2	$-M$	1200	3	1	0	-1	0	1	$1200/1$
$Z' = C_B X_B,$ $= -2700M$		$\Delta_j \rightarrow$	$5(M-1)$	$6(M-1)$ $\uparrow$	$-M$	$-M$ $\downarrow$	0	0	
x_2	-6	300	$2/5$	1	$-1/5$	0	—	0	750
A_2	$-M$	900	⑬⁄₅	0	$1/5$	-1	—	1	$4500/13$ (Min) $\rightarrow$
$Z' = -900M$ $= -1800$		$\Delta_j \rightarrow$	$(13/5)$ $(M-1)$ $\uparrow$	0	$(1/5)$ $(M-6)$	$-M$	—	0 $\downarrow$	
x_2	-6	$2100/13$	0	1	$-3/13$	$2/13$	—	—	
x_1	-5	$4500/13$	1	0	$1/13$	$-5/13$	—	—	
$Z' = C_B X_B,$ $= -2700$		$\Delta_j \rightarrow$	0	0	-1	-1			

Computation of Δ_j

For first table:

$$\Delta_1 = c_1 - C_B x_1 = -5 - (-M, -M)(2, 3) = 5(1 + M)$$

$$\Delta_2 = c_2 - C_B x_2 = -6 - (-M, -M)(5, 1) = 6(M - 1)$$

$$\Delta_3 = c_3 - C_B s_1 = 0 - (-M, -M)(-1, 0) = -M$$

$$\Delta_4 = c_4 - C_B s_2 = 0 - (-M, -M)(0, -1) = -M$$

For second table:

$$\Delta_1 = c_1 - C_B x_1 = -5 - (-6, -M)\left(\frac{2}{5}, \frac{13}{5}\right) = \frac{13(M - 1)}{5}$$

$$\Delta_2 = 0 = \Delta_5, \Delta_3 = c_3 - C_B s_1 = 0 - (-6, -M)\left(-\frac{1}{5}, \frac{1}{5}\right) = \frac{(M - 6)}{5}$$

$$\Delta_4 = c_4 - C_B s_2 = 0 - (-6, -M)(0, -1) = -M$$

For third table:

$$\Delta_1 = 0 = \Delta_2, \Delta_3 = c_3 - C_B s_1 = 0 - (-6, -5)\left(-\frac{3}{13}, \frac{1}{13}\right) = -1$$

$$\Delta_4 = c_4 - C_B s_2 = 0 - (-6, -5)\left(\frac{2}{13}, \frac{-5}{13}\right) = -1$$

Conclusions: In the last row of the third table all $\Delta_j \leq 0$ and no artificial variables appears in the basis. Thus, the solution is optimal and is given by

$$x_1 = \frac{4500}{13}, x_2 = \frac{2100}{13}, \text{Min.} Z = -\text{Max.} Z' = 2700$$

EXAMPLE 2. *Solve the following LPP by simplex method*

$$\text{Max. } Z = 2x_1 + 4x_2$$

subject to the constraints

$$2x_1 + x_2 \leq 18$$

$$3x_1 + 2x_2 \geq 30$$

$$x_1 + 2x_2 = 26$$

and $x_1, x_2 \geq 0$ [MEERUT–2007, 10, 12; BHOPAL–2014]

SOLUTION. Clearly, the problem under consideration is of maximization and all $b_i's$ are positive.

Now, introduce slack variable s_1, surplus variable s_2 and artificial variables A_1 and A_2. Then we get the following LPP

     Max. $Z = 2x_1 + 4x_2 + 0s_1 + 0s_2 - MA_1 - MA_2$

     s.t. $2x_1 + x_2 + s_1 = 18$

            $3x_1 + 2x_2 - s_2 + A_1 = 30$

            $x_1 + 2x_2 + A_2 = 26$

     and $x_1, x_2, s_1, s_2, A_1, A_2 \geq 0$

Now apply the simplex method, we have the following simplex table.

Simplex Table

B.V.	C_B	X_B	x_1	x_2	s_1	s_2	A_1	A_2	Min Ratio
$c_j \rightarrow$			2	4	0	0	$-M$	$-M$	
s_1	0	18	2	1	1	0	0	0	18
A_1	$-M$	30	3	2	0	-1	1	0	15
A_2	$-M$	26	1	②	0	0	0	1	13 (Min) $\rightarrow$
$Z=C_BX_B=-56M, \Delta_j \rightarrow$			$(2+4M)$	$(4+4M)$ $\uparrow$	0	$-M$	0	0 $\downarrow$	
s_1	0	5	3/2	0	1	0	0	—	10/3
A_1	$-M$	4	②	0	0	-1	1	—	2 (Min) $\rightarrow$
x_2	4	13	1/2	1	0	0	0	—	26
$Z = 52-4M, \quad \Delta_j \rightarrow$			$2M$ $\uparrow$	0	0	$-M$	0	—	
s_1	0	2	0	0	1	3/4	—	—	
x_1	2	2	1	0	0	$-1/2$	—	—	
x_2	4	12	0	1	0	1/4	—	—	
$Z = 52 \qquad \Delta_j \rightarrow$			0	0	0	0			

Computation of Δ_j:

(i) For first table:

$$\Delta_1 = c_1 - C_B x_1 = 2 - (0, -M, -M)(2, 3, 1) = 2 + 4M$$
$$\Delta_2 = c_2 - C_B x_2 = 4 - (0, -M, -M)(1, 2, 2) = 4 + 4M$$
$$\Delta_3 = 0$$
$$\Delta_4 = c_4 - C_B s_2 = 0 - (0, -M, -M)(0, -1, 0) = -M$$

(ii) For second table:

$$\Delta_1 = c_1 - C_B x_1 = 2 - (0, -M, 4)\left(\frac{3}{2}, 2, \frac{1}{2}\right) = 2M$$
$$\Delta_2 = 0 = \Delta_3 = \Delta_5$$
$$\Delta_4 = c_4 - C_B s_2 = 0 - (0, -M, 4)(0, -1, 0) = -M$$

(iii) For third table:

$$\Delta_1 = 0 = \Delta_2 = \Delta_3$$
$$\Delta_4 = c_4 - C_B s_2 = 0 - (0, 2, 4)\left(\frac{3}{4}, \frac{-1}{2}, \frac{1}{4}\right) = 0$$

Conclusions.

In the last row of third table all $\Delta_j \leq 0$ and no artificial variable appears in the basis, so solution is optimal.

Hence, the optimal solution is given by

$$x_1 = 2, x_2 = 12 \text{ and Max. } Z = 52$$

EXAMPLE 3. *Solve the following LPP by Big-M method*
$$\text{Max. } Z = -2x_1 - x_2$$
subject to the constraints
$$3x_1 + x_2 = 3$$
$$4x_1 + 3x_2 \geq 6$$
$$x_1 + 2x_2 \leq 4$$
and
$$x_1, x_2 \geq 0$$

SOLUTION. The given maximization problem is in standard form. Now, introducing the surplus variables s_1 and s_2 and artificial variables A_1 and A_2, to convert the given inequalities into equations, we get,

$$\text{Max. } Z = -2x_1 - x_2 + 0s_1 + 0s_2 - MA_1 - MA_2$$
s.t.
$$3x_1 + x_2 + A_1 = 3$$
$$4x_1 + 3x_2 - s_1 + A_2 = 6$$
$$x_1 + 2x_2 + s_2 = 4$$

Apply the Big-M method, we have the following simplex table:

Simplex Table

	$c_j \rightarrow$		-2	-1	0	0	$-M$	$-M$	
B.V.	C_B	X_B	x_1	x_2	s_1	s_2	A_1	A_2	Min Ratio
A_1	$-M$	3	③	1	0	0	1	0	$3/3 = 1$ (Min) $\rightarrow$
A_2	$-M$	6	4	3	-1	0	0	1	$6/4$
s_2	0	4	1	2	0	1	0	0	$4/1$
$Z=C_BX_B=-9M,\ \Delta_j \rightarrow$			$(-2+7M)$ $\uparrow$	$(-1+4M)$	$-M$	0	0	0 $\downarrow$	
x_1	-2	1	1	$1/3$	0	0	—	0	3
A_2	$-M$	2	0	⑤/③	-1	0	—	1	$6/5$ (Min) $\rightarrow$
s_2	0	3	0	$5/3$	0	1	—	0	$9/5$
$Z=C_BX_B=-2-2M, \Delta_j \rightarrow$			0	$\left(\dfrac{-1+5M}{3}\right)$ $\uparrow$	$-M$	0	—	0 $\downarrow$	
x_1	-2	$3/5$	1	0	$1/5$	0	—	—	
x_2	-1	$6/5$	0	1	$-3/5$	0	—	—	
s_2	0	1	0	0	1	1	—	—	
$Z=C_BX_B=-12/5, \Delta_j \rightarrow$			0	0	$-1/5$	0			

Computation of Δ_j:

(i) For the first table:
$$\Delta_1 = c_1 - C_Bx_1 = -2 - (-M, -M, 0)(3, 4, 1) = -2 + 7M$$
$$\Delta_2 = c_2 - C_Bx_2 = -1 - (-M, -M, 0)(1, 3, 5) = -1 + 4M$$
$$\Delta_3 = c_3 - C_Bs_1 = 0 - (-M, -M, 0)(0, -1, 0) = -M$$

(ii) For the second table:
$$\Delta_2 = c_2 - C_Bx_2 = -1 - (-2, -M, 0)\left(\frac{1}{3}, \frac{5}{3}, \frac{5}{3}\right) = \frac{1}{3}(-1 + 5M)$$

$$\Delta_3 = c_3 - C_B s_1 = 0 - (-2, -M, 0)(0, -1, 0) = -M$$

(iii) For the third table:

$$\Delta_1 = \Delta_2 = \Delta_4 = 0$$

$$\Delta_3 = c_3 - C_B s_1 = 0 - (-2, -1, 0)\left(\frac{1}{5}, -\frac{3}{5}, 1\right) = -\frac{1}{5}$$

Conclusions. Since in the last row of the last table, all $\Delta_j \leq 0$ and no artificial variable appears in the basis, so solution is optimal.

Hence, optimal solution is given by $x_1 = \dfrac{3}{5}, x_2 = \dfrac{6}{5}, \text{Max}.Z = \dfrac{-12}{5}$

EXAMPLE 4. *Solve the following LPP by Big-M method.*

$$\textbf{Min. } \textbf{\textit{Z}} = \textbf{\textit{x}}_1 + \textbf{\textit{x}}_2 + \textbf{3\textit{x}}_3$$

subject to the constraints

$$\textbf{3\textit{x}}_1 + \textbf{2\textit{x}}_2 + \textbf{\textit{x}}_3 \leq \textbf{3}, \; \textbf{2\textit{x}}_1 + \textbf{\textit{x}}_2 + \textbf{2\textit{x}}_3 \geq \textbf{3}$$

and $\quad\quad \textbf{\textit{x}}_1, \textbf{\textit{x}}_2, \textbf{\textit{x}}_3 \geq \textbf{0}$

SOLUTION. Convert the given minimization problem to maximization such that

$$Z' = -Z = -x_1 - x_2 - 3x_3$$

Now introduce the slack variable s_1, surplus variable s_2 and artificial variable A_1. Then given problem reduces to

$$\text{Max. } Z' = -Z = -x_1 - x_2 - 3x_3 + 0x_4 + 0x_5 - MA_1$$

s.t.
$$3x_1 + 2x_2 + x_3 + s_1 = 3$$
$$2x_1 + x_2 + 2x_3 - s_2 + A_1 = 3$$

and $\quad x_1, x_2, x_3, s_1, s_2, A_1 \geq 0$

Simplex Table

B.V.	C_B	X_B	$c_j \rightarrow$ x_1	x_2	x_3	s_1	s_2	A_1	Min Ratio
			-1	-1	-3	0	0	$-M$	
s_2	0	3	③	2	1	1	0	0	$3/3 = 1$ (Min) $\rightarrow$
A_1	$-M$	3	2	1	2	0	-1	1	$3/2$
$Z' = C_B X_B,$ $= -2M$		$\Delta_j \rightarrow$	$(-1+2M)$ $\uparrow$	$(-1 + M)$	$(-3+2M)$	0	$-M$ $\downarrow$	0	
x_1	-1	1	1	$2/3$	$1/3$	$1/3$	0	0	3
A_1	$-M$	1	0	$-1/3$	④/③	$-2/3$	-1	1	$3/4$ (Min) $\rightarrow$
$Z' = C_B X_B,$ $= -1 - M$		$\Delta_j \rightarrow$	0	$-\left(\dfrac{1+M}{3}\right)$	$\dfrac{4}{3}(-2+M)$ $\uparrow$	$\dfrac{1}{3}(1-2M)$	$-M$	0 $\downarrow$	
x_1	-1	$3/4$	1	$3/4$	0	$1/2$	$1/4$	—	
x_3	-3	$3/4$	0	$-1/4$	1	$-1/2$	$-3/4$	—	
$Z' = C_B X_B,$ $= -3$		$\Delta_j \rightarrow$	0	-1	0	-1	-2		

Computation of Δ_j:

(i) For the first table:

$$\Delta_1 = c_1 - C_B x_1 = -1 - (0, -M)(3, 2) = -1 + 2M$$
$$\Delta_2 = c_2 - C_B x_2 = -1 - (0, -M)(2, 1) = -1 + M$$
$$\Delta_3 = c_3 - C_B s_1 = -3 - (0, -M)(1, 2) = -3 + 2M$$
$$\Delta_4 = 0$$
$$\Delta_5 = c_5 - C_B s_2 = 0 - (0, -M)(0, -1) = -M$$

(ii) For the second table:

$$\Delta_1 = \Delta_6 = 0$$

$$\Delta_2 = c_2 - C_B x_2 = -1 - (-1, -M)\left(\frac{2}{3}, -\frac{1}{3}\right) = -\frac{(1+M)}{3}$$

$$\Delta_3 = c_3 - C_B x_3 = -3 - (-1, -M)\left(\frac{1}{3}, \frac{4}{3}\right) = \frac{4(-2+M)}{3}$$

$$\Delta_4 = c_4 - C_B s_1 = 0 - (-1, -M)\left(\frac{1}{3}, \frac{-2}{3}\right) = \frac{(1-2M)}{3}$$

$$\Delta_5 = c_5 - C_B s_2 = 0 - (-1, -M)(0, -1) = -M$$

(iii) For the third table:

$$\Delta_1 = \Delta_3 = 0$$

$$\Delta_2 = c_2 - C_B x_2 = -1 - (-1, -3)\left(\frac{3}{4}, -\frac{1}{4}\right) = -1$$

$$\Delta_4 = c_4 - C_B s_1 = 0 - (-1, -3)\left(\frac{1}{2}, -\frac{1}{2}\right) = -1$$

$$\Delta_5 = c_5 - C_B s_2 = 0 - (-1, -3)\left(\frac{1}{4}, \frac{-3}{4}\right) = -2$$

Conclusions. Since in the last row of the last table all $\Delta_j \leq 0$, so solution is optimal and hence optimal solution is given by

$$x_1 = \frac{3}{4}, x_2 = 0, x_3 = \frac{3}{4} \text{ and } Min.Z = -Max.Z' = 3$$

EXAMPLE 5. *Solve the following LPP by Big-M method*

$$\text{Max. } Z = x_1 + 2x_2 + 3x_3 - x_4$$

subject to the constraints

$$x_1 + 2x_2 + 3x_3 = 15$$
$$2x_1 + x_2 + 5x_3 = 20$$
$$x_1 + 2x_2 + x_3 + x_4 = 10$$

and $\quad x_1, x_2, x_3, x_4 \geq 0$ [MEERUT–2009; GORAKHPUR–2008; IAS–1995;

DELHI–1997; PTU(BE(Mech.))–2006; PUNJAB(BBA)–2001]

SOLUTION. The given maximization problem is in standard form. Also, constraints are equations. We observe that, to obtain a unit matrix of order 3, we need two more unit vectors as one unit vector is formed by the coefficient of x_4.

So, introducing two artificial variables A_1 and A_2 in the first two constraints, the given problem becomes

$$\text{Max. } Z = x_1 + 2x_2 + 3x_3 - x_4 - MA_1 - MA_2$$

s.t.
$$x_1 + 2x_2 + 3x_3 + A_1 = 15$$
$$2x_1 + x_2 + 5x_3 + A_2 = 20$$
$$x_1 + 2x_2 + x_3 + x_4 = 10$$

and
$$x_1, x_2, x_3, x_4, A_1, A_2 \geq 0$$

Now we have the following simplex table:

Simplex Table

B.V.	C_B	X_B	$c_j \rightarrow$ 1 x_1	2 x_2	3 x_3	-1 x_4	$-M$ A_1	$-M$ A_2	Min Ratio
A_1	$-M$	15	1	2	3	0	1	0	5
A_2	$-M$	20	2	1	⑤	0	0	1	4 (Min) →
x_4	-1	10	1	2	1	1	0	0	10
$Z = C_B X_B,$ $\Delta_j \rightarrow$ $= -35M - 10$			$(3M + 2)$	$(3M + 4)$	$(8M + 4)$ ↑	0	0	0 ↓	
A_1	$-M$	3	$-1/5$	⑦/5	0	0	1	—	15/7(Min)→
x_3	3	4	2/5	1/5	1	0	0	—	20
x_4	-1	6	3/5	9/5	0	1	0	—	10/3
$Z = C_B X_B,$ $\Delta_j \rightarrow$ $= -3M + 6$			$\dfrac{(-M+2)}{5}$ ↑	$\dfrac{7M+16}{5}$	0	0	0	0 ↓	
x_2	2	15/7	$-1/7$	1	0	0	—	—	$-$ve
x_3	3	25/7	3/7	0	1	0	—	—	25/3
x_4	-1	15/7	⑥/7	0	0	1	—	—	5/2 (Min)→
$Z = 90/7,$ $\Delta_j \rightarrow$			6/7 ↑	0	0	0 ↓			
x_2	2	5/2	0	1	0	1/6			
x_3	3	5/2	0	0	1	$-1/2$			
x_1	1	5/2	1	0	0	7/6			
$Z = C_B X_B,$ $\Delta_j \rightarrow$ $= 15$			0	0	0	-1			

In the above table (last row) all $\Delta_j \leq 0$

Hence solution is optimal and is given by

$$x_1 = \frac{5}{2} = x_2 = x_3, x_4 = 0 \text{ and Max. } Z = 15$$

EXAMPLE 6. *Solve the following LPP by Big-M method*

$$\text{Max. } Z = 4x_1 + 5x_2 - 3x_3$$

subject to the constraints

$$x_1 + x_2 + x_3 = 10$$
$$x_1 - x_2 \geq 1$$
$$2x_1 + 3x_2 + x_3 \leq 40$$

and

$$x_1, x_2, x_3 \geq 0$$

SOLUTION. The given maximization problem is in standard form. So, introduce the surplus variable s_1, slack variable s_2 and artificial variable A_1 and A_2 in the following manner:

$$\text{Max. } Z = 4x_1 + 5x_2 - 3x_3 + 0s_1 + 0s_2 - MA_1 - MA_2$$

s.t.

$$x_1 + x_2 + x_3 + A_1 = 10$$
$$x_1 - x_2 - s_1 + A_2 = 1$$
$$2x_1 + 3x_2 + x_3 + s_2 = 40$$

and

$$x_1, x_2, x_3, s_1, s_2, A_1, A_2 \geq 0$$

Now we have the following simplex table:

Simplex Table

B.V.	C_B	X_B	$c_j \rightarrow$ x_1 4	x_2 5	x_3 −3	s_1 0	s_2 0	A_1 −M	A_2 −M	Min Ratio
A_1	−M	10	1	1	1	0	0	1	0	10/1 = 10
A_2	−M	1	①	−1	0	−1	0	0	1	1/1 = 1 (Min)→
s_2	0	40	2	3	1	0	1	0	0	40/2 = 20
$Z = -21M,$	$\Delta_j \rightarrow$		(4 + 2M) ↑	5	(−3 + M)	−M	0	0 ↓	0	
A_1	−M	9	0	②	1	1	0	1	—	9/2 (Min)→
x_1	4	1	1	−1	0	−1	0	0	—	Negative
s_2	0	38	0	5	1	2	1	0	—	38/5
$Z = -9M + 4, \Delta_j \rightarrow$			0	(9 + 2M) ↑	−3 + M	4 + M	0	0 ↓		
x_2	5	9/2	0	1	1/2	1/2	0	—	—	
x_1	4	11/2	1	0	1/2	−1/2	0	—	—	
s_2	0	31/2	0	0	−3/2	−1/2	1	—	—	
$Z = C_B X_B,$ $= 89/2$	$\Delta_j \rightarrow$		0	0	−15/2	−1/2	0			

Computation of Δ_j:

(i) For the first table:

$$\Delta_1 = c_1 - C_B x_1 = 4 - (-M, -M, 0)(1, 1, 2) = 4 + 2M$$
$$\Delta_2 = c_2 - C_B x_2 = 5 - (-M, -M, 0)(1, -1, 3) = 5$$
$$\Delta_3 = c_3 - C_B x_3 = -3 - (-M, -M, 0)(1, 0, 1) = -3 + M$$

$$\Delta_4 = c_4 - C_B s_1 = 0 - (-M, -M, 0)(0, -1, 0) = -M$$

$$\Delta_5 = \Delta_6 = \Delta_7 = 0$$

(ii) For the second table:

$$\Delta_1 = 0, \Delta_2 = c_2 - C_B x_2 = 5 - (-M, 4, 0)(2, -1, 5) = 9 + 2M$$

$$\Delta_3 = c_3 - C_B x_3 = -3 - (-M, 4, 0)(1, 0, 1) = -3 + M$$

$$\Delta_4 = c_4 - C_B s_1 = 0 - (-M, 4, 0)(1, -1, 2) = 4 + M$$

$$\Delta_5 = \Delta_6 = 0$$

(iii) For the third table:

$$\Delta_1 = \Delta_2 = \Delta_5 = 0$$

$$\Delta_3 = C_3 - C_B x_3 = -3 - (5, 4, 0)\left(\frac{1}{2}, \frac{1}{2}, -\frac{3}{2}\right) = -\frac{15}{2}$$

$$\Delta_4 = C_4 - C_B s_1 = 0 - (5, 4, 0)\left(\frac{1}{2}, \frac{-1}{2}, \frac{-1}{2}\right) = \frac{-1}{2}$$

Conclusions: In the last row of last table all $\Delta_j \leq 0$, so solution is optimal. Hence, optimal solution is given by

$$x_1 = \frac{11}{2}, x_2 = \frac{9}{2}, x_3 = 0 \text{ and Max. } Z = \frac{89}{2}$$

Exercise-4.3

Solve the following LPP by Big-M method.

1. Max. $Z = 4x_1 + 2x_2$

    s.t. $3x_1 + x_2 \leq 27$

    $x_1 + x_2 \geq 21$

    and $x_1, x_2 \geq 0$

2. Min. $Z = 2x_1 + 3x_2$

    s.t. $x_1 + x_2 \geq 5$

    $x_1 + 2x_2 \geq 6$

    and $x_1, x_2 \geq 0$

3. Max. $Z = 3x_1 - x_2$

    s.t. $2x_1 + x_2 \geq 2$

    $x_1 + 3x_2 \leq 3$

    $x_2 \leq 4$

    and $x_1, x_2 \geq 0$

4. Min. $Z = 0.60x_1 + 0.80x_2$

    s.t. $20x_1 + 30x_2 \geq 900$

    $40x_1 + 30x_2 \geq 1200$

    and $x_1, x_2 \geq 0$

5. Min. $Z = 2x_1 + 8x_2$

    s.t. $5x_1 + 10x_2 = 150$

    $x_1 \leq 20$

    $x_2 \geq 14$

    and $x_1, x_2 \geq 0$

ANSWERS

1. $x_1 = 0, x_2 = 27$, Max. $Z = 54$

2. $x_1 = 4, x_2 = 1$ and Min. $Z = 11$

3. $x_1 = 3, x_2 = 0$, Max. $Z = 9$

4. $x_1 = 15, x_2 = 20$, Min. $Z = 25$

5. $x_1 = 2, x_2 = 14$, Min. $Z = 116$

4.14 TWO PHASE METHOD

[KANPUR–2012]

This is an alternative of Big-M method. Using this method, we obtain the solution in two phases given as follows:

(1) In phase-1 all the artificial variables are eliminated from the basis.

(2) In phase-2, we use the solution from phase-I as the initial basic feasible solution and then use the simplex method to obtain the optimal solution.

Working Procedure

(A) For Phase-1

STEP 1. Convert the given LPP in the standard form.

STEP 2. Add the necessary artificial variables to the constraints as done in Big-M method to obtain an initial basic feasible solution.

STEP 3. Formulate an artificial objective function Z^* such that
$$Z^* = -A_1 - A_2 - \dots - A_n \; (= - \text{(sum of the artificial variables)})$$
by assigning -1 cost to each artificial variable A_i and zero cost to all other variables.

STEP 4. Maximize Z^* subject to the constraints of the original problem using the simplex method.
Now, we have the following cases:

CASE-I. If max. $Z^* < 0$ and at least one artificial variable appears in the optimal basis at a positive level, then given LPP will not have any feasible solution and then we will not move to phase-II.

CASE-II. If max. $Z^* = 0$ and no artificial variables appears in the optimal basis then BFS is not obtained and in order to obtain optimal BFS, we move to phase-II.

CASE-III. If max. $Z^* = 0$ and at least one artificial variable appears in the optimal basis at zero level, then a feasible solution to the auxiliary LPP is also a feasible solution of the given LPP by setting all artificial variables to zero. Finally to obtain the basic feasible solution, remove all the artificial variable from the basis matrix.

(B) For Phase-2

STEP 1. Take the basic feasible solution, which was found at the end of Phase-1 as the starting BFS for the given LPP.

STEP 2. Apply simplex method to find the optimal basic feasible solution.

 Solved Examples

EXAMPLE 1. *Apply Two-phase method, solve the following LPP.*
$$\text{Min. } Z = 40x_1 + 24x_2$$
subject to the constraints
$$20x_1 + 50x_2 \geq 4800$$
$$80x_1 + 50x_2 \geq 7200$$
and $\qquad x_1, x_2 \geq 0$ [MEERUT–2008, 12]

SOLUTION. Firstly, convert the given minimization problem into maximization by taking the objective function as
$$Z^* = -Z = -40x_1 - 24x_2$$
Now, introduce the surplus variables s_1, s_2 and artificial variables A_1, A_2 such that
$$20x_1 + 50x_2 - s_1 + A_1 = 4800$$
$$80x_1 + 50x_2 - s_2 + A_2 = 7200$$
and $\qquad x_1, x_2, s_1, s_2, A_1, A_2 \geq 0$

PHASE-1. We assign the cost -1 to the artificial variable, and 0 to all other variables. Then the new objective function of auxiliary LPP becomes

$$\text{Max. } Z^* = 0x_1 + 0x_2 + 0s_1 + 0s_2 - A_1 - A_2$$

subject to the above constraints.

Now by applying the simplex method in a usual manner, we have the following simplex table:

Simplex Table

		$c_j \rightarrow$	0	0	0	0	-1	-1	
B.V.	C_B	X_B	x_1	x_2	s_1	s_2	A_1	A_2	Min Ratio
A_1	-1	4800	20	50	-1	0	1	0	4800/20=240
A_2	-1	7200	⑧⓪	50	0	-1	0	1	7200/80=90 (Min) →
$Z^* = -12000, \Delta_j \rightarrow$			100 ↑	100	-1	-1	0	0 ↓	
A_1	-1	3000	0	⑦⑤/②	-1	1/4	1	—	80 (Min)→
x_1	0	90	1	5/8	0	$-1/30$	0	—	144
$Z^* = -2910, \Delta_j \rightarrow$			0	75/2 ↑	0	1/4	0	↓	
x_2	0	80	0	1	$-2/75$	1/50			
x_1	0	40	1	0	1/60	$-1/60$			
$Z^* = 0,$		$\Delta_j \rightarrow$	0	0	0	0			

Clearly, in the last row of the above table all $\Delta_j \leq 0$. Also, no artificial variable appears in the basis. So, phase-1 is complete.

PHASE-2. Assign the actual costs to the original variables and cost zero to the surplus variables, the objective function becomes

$$\text{Max. } Z^* = -40x_1 - 24x_2 + 0s_1 + 0s_2$$

Further, replace the c_j-row values in the final simplex table of phase-1 by the c_j-values of the original objective function and delete the artificial variables from the final simplex table of phase-1, we write the first simplex table of phase-2.

Simplex Table

		$c_j \rightarrow$	-40	-24	0	0	
B.V.	C_B	X_B	x_1	x_2	s_1	s_2	Min Ratio
x_2	-24	80	0	1	$-2/75$	1/50	−ve
x_1	-40	40	1	0	①/⑥⓪	$-1/60$	2400 (Min) →
$Z^* = -3520, \Delta_j \rightarrow$			0 ↓	0	2/75 ↑	$-38/75$	
x_2	-24	144	8/5	1	0	$-1/50$	
s_1	0	2400	60	0	1	-1	
$Z^* = -3456, \Delta_j \rightarrow$			$-8/5$	0	0	$-12/25$	

Clearly in the last row of the above table, all $\Delta_j \leq 0$.

Hence, the solution is optimal and is given by

$$x_1 = 0, x_2 = 144 \text{ and Min. } Z = -Z^* = 3456.$$

EXAMPLE 2. *Solve the following LPP by Two-phase method.*

$$\text{Max. } Z = 3x_1 - x_2$$

subject to the constraints

$$2x_1 + x_2 \geq 2$$
$$x_1 + 3x_2 \leq 2$$
$$x_2 \leq 4$$

and $$x_1, x_2 \geq 0$$

SOLUTION. Let us introduce the surplus variable s_1, slack variables s_2, s_3 and artificial variable A_1 such that the given constraints reduces to the following equations

$$2x_1 + x_2 - s_1 + A_1 = 2$$
$$x_1 + 3x_2 + s_2 = 2$$
$$x_2 + s_3 = 4$$

and $$x_1, x_2, s_1, s_2, s_3, A_1 \geq 0$$

PHASE-1. Now assigning the cost -1 to the artificial variable and cost 0 to all other variables, to get the new objective function of this auxiliary problem is given by

$$\text{Max. } Z^* = 0x_1 + 0x_2 + 0s_1 + 0s_2 + 0s_3 - A_1$$

Now, apply the simplex method to get the following simplex table.

Simplex Table

B.V.	C_B	X_B	$c_j \rightarrow$ x_1	x_2	s_1	s_2	s_3	A_1	Min Ratio
			0	0	0	0	0	-1	
A_1	-1	2	②	1	-1	0	0	1	$2/2 = 1$ (Min) $\rightarrow$
s_2	0	2	1	3	0	1	0	0	$2/1$
s_3	0	4	0	1	0	0	1	0	—
$Z^* = C_B X_B = -2, \Delta_j \rightarrow$			2 $\uparrow$	1	-1	0	0	0 $\downarrow$	
x_1	0	1	1	1/2	$-1/2$	0	0	—	
s_2	0	1	0	5/2	1/2	1	0	—	
s_3	0	4	0	1	0	0	1	—	
$Z^* = C_B X_B = 0, \Delta_j \rightarrow$			0	0	0	0	0		

In the last row of the above table $\Delta_j \leq 0$. Also, no artificial variable appears in the basis. So, we move to the phase-2.

PHASE-2. Assign the original costs to the original variables, cost 0 to slack and surplus variables, the objective function becomes

$$\text{Max. } Z = 3x_1 - x_2 + 0s_1 + 0s_2 + 0s_3$$

Further, delete the artificial columns from the last simplex table of phase-1, we write the first simplex table of phase-2.

Now apply the simplex method in a usual manner to get the following simplex table.

Simplex Table

B.V.	C_B	X_B	$c_j \rightarrow$ 3 x_1	-1 x_2	0 s_1	0 s_2	0 s_3	Min Ratio
x_1	3	1	1	1/2	$-1/2$	0	0	$-$ve
s_2	0	1	0	5/2	⟨1/2⟩	1	0	2 (Min) $\rightarrow$
s_3	0	4	0	1	0	0	1	—
$Z^*=C_BX_B=3$, $\Delta_j \rightarrow$			0	$-5/2$	3/2 $\uparrow$	0 $\downarrow$	0	
x_1	3	2	1	3	0	1	0	
s_1	0	2	0	5	1	2	0	
s_3	0	4	0	1	0	0	1	
$Z^*=C_BX_B=6$, $\Delta_j \rightarrow$			0	-10	0	-3	0	

In the last row of the above table all $\Delta_j \leq 0$, so solution is optimal and is given by

$$x_1 = 2, x_2 = 0 \text{ and Max. } Z = 6$$

EXAMPLE 3. *Solve the following LPP by two-phase method*

$$\text{Min. } Z = x_1 - 2x_2 - 3x_3$$

subject to the constraints

$$-2x_1 + x_2 + 3x_3 = 2$$
$$2x_1 + 3x_2 + 4x_3 = 1$$

and $\qquad\qquad x_1, x_2, x_3 \geq 0$

[MEERUT–2005, 09, 10; KANPUR–2012; AVADH–2008; LUCKNOW–2012]

SOLUTION. Since the given problem is of minimization, so convert it into maximization problem such that $Z = -Z^*$.

$$\text{Max. } Z^* = -x_1 + 2x_2 + 3x_3$$

Now, introduce the artificial variables A_1 and A_2 such that

$$-2x_1 + x_2 + 3x_3 + A_1 = 2$$
$$2x_1 + 3x_2 + 4x_3 + A_2 = 1$$

and $\qquad\qquad x_1, x_2, x_3, A_1, A_2 \geq 0$

PHASE-1. In a usual manner, assigning cost -1 to the artificial variables and cost 0 to all other variables, the new objective function of the auxiliary problem is:

$$\text{Max. } Z^* = 0x_1 + 0x_2 + 0x_3 - A_1 - A_2$$

Now, we have the following simplex table:

B.V.	C_B	x_B	x_1	x_2	x_3	A_1	A_2	Min Ratio
		$c_j \rightarrow$	0	0	0	−1	−1	
A_1	−1	2	−2	1	3	1	0	2/3
A_2	−1	1	2	3	④	0	1	1/4 (Min) →
$Z^* = -3,$		$\Delta_j \rightarrow$	0	4	7	0	0	
					↑		↓	
A_1	−1	5/4	−7/2	−5/4	0	1	−3/4	
x_3	0	1/4	1/2	3/4	1	0	1/4	
$Z^* = -5/4,$		$\Delta_j \rightarrow$	−7/4	−5/4	0	0	−3/4	

Clearly, all $\Delta_j's$ in the last row of above table are negative or zero, so an optimum basic feasible solution to the auxiliary problem has been attained. But on the other hand, the artificial variable A_1 appears in the basic solution at a positive level. Hence, the given LPP does not have any feasible solution.

Exercise-4.4

Solve the following LPP by Two-phase method:

1. Min. $Z = x_1 + x_2$
    subject to the constraints
$$2x_1 + x_2 \geq 4$$
$$x_1 + 7x_2 \geq 7$$
and $\qquad x_1, x_2 \geq 0$

2. Max. $Z = 5x_1 + 8x_2$
    subject to the constraints
$$3x_1 + 2x_2 \geq 3$$
$$x_1 + 4x_2 \geq 4$$
$$x_1 + x_2 \leq 5$$
and $\qquad x_1, x_2 \geq 0$

3. Max. $Z = 3x_1 + 2x_2 + x_3 + 4x_4$
    subject to the constraints
$$4x_1 + 5x_2 + x_3 + 5x_4 = 5$$
$$2x_1 - 3x_2 - 4x_3 + 5x_4 = 7$$
$$x_1 + 4x_2 + 5x_3 - 4x_4 = 6$$
and $\qquad x_1, x_2, x_3, x_4 \geq 0$

ANSWERS

1. $x_1 = \dfrac{21}{13}, x_2 = \dfrac{10}{13}$ and Min. $Z = \dfrac{31}{13}$ **2.** $x_1 = 0, x_2 = 5$ and Max. $Z = 40$

3. No feasible solution exists.

4.15 SOME SPECIAL LINEAR PROGRAMMING PROBLEMS

(1) PROBLEM OF UNBOUNDED FEASIBLE SOLUTION BUT BOUNDED OPTIMAL SOLUTION

EXAMPLE 1.   *Solve the following LPP:*
$$\textbf{Max. } Z = 6x_1 - 2x_2$$
       subject to the constraints
$$2x_1 - x_2 \leq 2$$
$$x_1 \leq 4$$
    and $x_1, x_2 \geq 0$

SOLUTION.   Introduce the slack variables s_1 and s_2 in the given LPP, then we have
$$\text{Max. } Z = 6x_1 - 2x_2 + 0s_1 + 0s_2$$

s.t. $2x_1 - x_2 + s_1 = 2$
$x_1 + s_2 = 4$
and $x_1, x_2, s_1, s_2 \geq 0$

Now, apply the simplex method in a usual manner, we have the following simplex table:

Simplex Table

		$c_j \rightarrow$	6	−2	0	0	
B.V.	C_B	X_B	x_1	x_2	s_1	s_2	Min Ratio
s_1	0	2	②	−1	1	0	1 (Min) →
s_2	0	4	1	0	0	1	4
$Z = 0,$		$\Delta_j \rightarrow$	6 ↑	−2 ↓	0	0	
x_1	6	1	1	−1/2	1/2	0	−ve
s_2	0	3	0	①/2	−1/2	1	(min) →
$Z = 6,$		$\Delta_j \rightarrow$	0	1 ↑	−3	0 ↓	
x_1	6	4	1	0	0	1	
x_2	−2	6	0	1	−1	2	
$Z = 12,$		$\Delta_j \rightarrow$	0	0	−2	−2	

Clearly, in the last row of the above table all $\Delta_j \leq 0$, so solution is optimal and is given by $x_1 = 4$, $x_2 = 6$ and Max. $Z = 12$

Now, in the first simplex table, we note that the elements of column vector x_2 are negative or zero which indicate that the feasible region is not bounded.

Hence, the given LPP has unbounded feasible solution but not bounded optimal solution.

(2) PROBLEM OF UNBOUNDED SOLUTION

It is a very well known fact that, if in any situation there is at least one Δ_j greater than zero but there is no non-negative ratios or min ratios → ∞ then we have unbounded solution. Unboundedness describes a LP problem that do not have finite solution.

EXAMPLE I.   **Solve the following LPP :**
Max. $Z = 2x_1 + x_2$
subject to the constraints
$x_1 - x_2 \leq 10$
$2x_1 - x_2 \leq 40$
and $x_1, x_2 \geq 0$

SOLUTION.   Introducing slack variables s_1 and s_2 in the given LPP. Then we have
Max. $Z = 2x_1 + x_2 + 0s_1 + 0s_2$
s.t. $x_1 - x_2 + s_1 = 10$
$2x_1 - x_2 + s_2 = 40$
and $x_1, x_2, s_1, s_2 \geq 0$

Then apply the simplex method in the usual manner, we have the following simplex table.

Simplex Table

B.V.	C_B	X_B	$c_j \rightarrow$ x_1	x_2	s_1	s_2	Min Ratio
			2	1	0	0	
s_1	0	10	1	-1	1	0	
s_2	0	40	2	-1	0	1	
$Z = C_B X_B = 0,$		$\Delta_j \rightarrow$	2	1	0	0	

Clearly in the last row of the above table, we observe that the value of $x_2 \notin B$ (the set of basic variable) and for x_2, we have $\Delta_j = 1 > 0$ and $x_{i2} \leq 0$, $i = 1, 2$. Hence, the solution is unbounded.

EXAMPLE 2.   **Solve the following LPP:**

$$\text{Max. } Z = 10x_1 + 20x_2$$

     subject to the constraints

$$2x_1 + 4x_2 \geq 16, \ x_1 + 5x_2 \geq 15$$

     and $x_1, x_2 \geq 0$

SOLUTION.   Using surplus variables s_1, s_2 and artificial variables A_1, A_2, the given LPP becomes

$$\text{Max. } Z = 10x_1 + 20x_2 + 0s_1 + 0s_2 - MA_1 - MA_2$$

     subject to $2x_1 + 4x_2 - s_1 + A_1 = 16$

                       $x_1 + 5x_2 - s_2 + A_2 = 15$

     and $x_1, x_2, s_1, s_2, A_1, A_2 \geq 0$

Now, apply the simplex method in a usual manner, we have the following simplex table:

Simplex Table

B.V.	C_B	X_B	$c_j \rightarrow$ x_1	x_2	s_1	s_2	A_1	A_2	Min Ratio
			10	20	0	0	$-M$	$-M$	
A_1	$-M$	16	2	4	-1	0	1	0	16/4
A_2	$-M$	15	1	⑤	0	-1	0	1	15/5(Min) $\rightarrow$
$Z = -31M,$		$\Delta_j \rightarrow$	$(3M+10)$	$(9M+20)$ $\uparrow$	$-M$	$-M$	0	0 $\downarrow$	
A_1	$-M$	4	⑥/⑤	0	-1	4/5	1	—	10/3(Min) $\rightarrow$
x_1	20	3	1/5	1	0	$-1/5$	0	—	15
$Z = -4M + 60, \Delta_j \rightarrow$			$\dfrac{6}{5}(M+5)$ $\uparrow$	0	$-M$	$\dfrac{4}{5}(M+5)$ $\downarrow$	0	—	
x_1	10	10/3	1	0	$-5/6$	2/3	—	—	$-$ve
x_2	20	7/3	0	1	①/⑥	$-1/3$	—	—	14 (Min) $\rightarrow$
$Z = 80,$		$\Delta_j \rightarrow$	0	0	5 $\downarrow$	0 $\uparrow$			
x_1	10	15	1	5	0	-1			
s_1	0	14	0	6	1	-2			
$Z = 150,$		$\Delta_j \rightarrow$	0	-30	0	10 $\uparrow$			

Clearly, in the last table, we observe that s_2 is the incoming vector but outgoing vector can not be found because all the elements in this column are negative. Hence, the solution is unbounded.

(3) PROBLEM HAVING MORE THAN ONE OPTIMAL SOLUTIONS

An alternative optimal solutions can be obtained by considering the $c_j - z_j$ row of the simplex table. We know that for optimal solution of maximization problem all $\Delta_j \leq 0$. But if $\Delta_j = 0$ for some non-basic variables columns in the optimal simplex table, optimal solution may exist, because if a non-basic variable corresponding to which $\Delta_j = 0$ is entered into the basis, a new solution will be arrived at but value of the objective function remains the same.

EXAMPLE 1. **Solve the following LPP:**

$$Max. \ Z = 6x_1 + 4x_2$$

subject to the constraints

$$2x_1 + 3x_2 \leq 30$$
$$3x_1 + 2x_2 \leq 24$$
$$x_1 + x_2 \geq 3$$

and $\quad x_1, x_2 \geq 0$ [JODHPUR–1993]

SOLUTION. Using slack variables s_1, s_2, surplus variable s_3 and artificial variable A_1 we have the following LPP.

$$Max. \ Z = 6x_1 + 4x_2 + 0s_1 + 0s_2 + 0s_3 - MA_1$$

s.t.
$$2x_1 + 3x_2 + s_1 = 30$$
$$3x_1 + 2x_2 + s_2 = 24$$
$$x_1 + x_2 - s_3 + A_1 = 3$$

and $\quad x_1, x_2, s_1, s_2, s_3, A_1 \geq 0$

Then apply the simplex method in a usual manner, we have the following simplex table.

Simplex Table

B.V.	C_B	X_B	$c_j \rightarrow$ 6 x_1	4 x_2	0 s_1	0 s_2	0 s_3	$-M$ A_1	Min Ratio
s_1	0	30	2	3	1	0	0	0	15
s_2	0	24	3	2	0	1	0	0	8
A_1	$-M$	3	①	1	0	0	-1	1	3 (Min) →
$Z = -3M$,		$\Delta_j \rightarrow$	(6 + M)↑	(4 + M)	0	0	$-M$	0 ↓	
s_1	0	24	0	1	1	0	2	—	12
s_2	0	15	0	-1	0	1	③	—	5 (Min) →
x_1	6	3	1	1	0	0	-1	—	—
$Z = 18$,		$\Delta_j \rightarrow$	0	-2	0	0 ↓	6 ↑		
s_1	0	14	0	⑤/3	1	$-2/3$	0	—	12/5 (Min) →
s_3	0	5	0	$-1/3$	0	$1/3$	1	—	—
x_1	6	8	1	$2/3$	0	$1/3$	0	—	24/2 = 12
$Z = 48$,		$\Delta_j \rightarrow$	0	0	0	-2	0		

We observe that in the last table all $\Delta_j \leq 0$, therefore, solution is optimal and is given by $x_1 = 8$, $x_2 = 0$, Max. $Z = 48$.

Also, corresponding to non-basic variables x_2 in the last table $\Delta_2 = 0$ and x_2 is not in the basis B, so an alternative solution will exists. For alternative optimal solution we have the following simplex table.

Simplex Table

B.V.	C_B	X_B	$c_j \rightarrow$ 6 x_1	4 x_2	0 s_1	0 s_2	0 s_3	Min Ratio
x_2	4	42/5	0	1	3/5	−2/5	0	
s_3	0	39/5	0	0	1/5	1/5	1	
x_1	6	12/5	1	0	−2/5	3/5	0	
$Z = C_B X_B = 48$,		$\Delta_j \rightarrow$	0	0	0	−2	0	

Clearly all $\Delta_j \leq 0$ implies solution is optimal and is given by

$$x_1 = \frac{12}{5}, x_2 = \frac{42}{5} \text{ and Max. } Z = 48.$$

☞ REMARK

• We know that the convex combination of BFS is also an optimal solution. Hence, if we obtain two alternative optimal solutions, then we can obtain any number of optimal solutions.

(4) LPP WITH UNRESTRICTED VARIABLES

Generally in LPP, we assume that all $x_j \geq 0$. But one or more variables can have either positive, negative or zero value, then these are called unrestricted variables.

For example, if x_r is unrestricted then we assume

$$x_r = x_r' - x_r''$$

where $x_r', x_r'' \geq 0$

EXAMPLE 1. **Solve the following LPP :**
$$\textbf{Max. } \boldsymbol{Z = x_1 + 3x_2}$$
subject to the constraints
$$\boldsymbol{x_1 + x_2 \leq 2}$$
$$\boldsymbol{-x_1 + x_2 \leq 4}$$
and $\boldsymbol{x_2 \geq 0, x_1 \text{ is unrestricted.}}$

SOLUTION. Here x_1 is unrestricted. Assume $x_1 = x_1' - x_1''$ such that $x_1', x_1'' \geq 0$. Also, introducing slack variables s_1, s_2, the given LPP becomes

$$\text{Max. } Z = (x_1' - x_1'') + 3x_2 + 0s_1 + 0s_2$$

s.t. $$x_1' - x_1'' + x_2 + s_1 = 2$$

$$-x_1' + x_1'' + x_2 + s_2 = 4$$

and $$x_1', x_1'', x_2, s_1, s_2 \geq 0$$

Now, apply the simplex method in a usual manner we have the following simplex table.

Simplex Table

B.V.	C_B	X_B	x_1'	x_1''	x_2	s_1	s_2	Min Ratio
	$c_j \rightarrow$		1	−1	3	0	0	
s_1	0	2	1	−1	①1	1	0	2 (Min) →
s_2	0	4	−1	1	1	0	1	4
$Z = 0,$		$\Delta_j \rightarrow$	1	−1	3 ↑	0 ↓	0	
x_2	3	2	1	−1	1	1	0	− 2
s_2	0	2	−2	②2	0	−1	1	1 (Min) →
$Z = 6,$		$\Delta_j \rightarrow$	−2	2 ↑	0	−3	0 ↓	
x_2	3	3	0	0	1	1/2	1/2	
x_1''	−1	1	−1	1	0	−1/2	1/2	
$Z = 8,$		$\Delta_j \rightarrow$	0	0	0	−2	−1	

Clearly, in the last row of the above table all $\Delta_j \leq 0$, so solution is optimal and is given by

$$x_1' = 0, x_1'' = 1, x_2 = 3 \text{ and Max. } Z = 8$$

$$\Rightarrow \quad x_1 = x_1' - x_1'' = -1, x_2 = 3 \text{ and Max. } Z = 8$$

EXAMPLE 2. **Solve the following LPP :**

$$\textbf{Max. } \textbf{Z = 2x}_1 + \textbf{3x}_2$$

subject to the constraints

$$\textbf{--x}_1 + \textbf{2x}_2 \leq \textbf{4}$$

$$\textbf{x}_1 + \textbf{x}_2 \leq \textbf{6}$$

$$\textbf{x}_1 + \textbf{3x}_2 \leq \textbf{9}$$

and $\quad$ **x_1, x_2 are unrestricted.**

SOLUTION. In the given LPP, x_1 and x_2 are unrestricted. So, making the transformation.

$$x_1 = x_1' - x_1'' \text{ and } x_2 = x_2' - x_2''$$

and using the slack variables s_1, s_2 and s_3. The given LPP reduces to

$$\text{Max. } Z = 2x_1' - 2x_1'' + 3x_2' - 3x_2'' + 0 \cdot s_1 + 0 \cdot s_2 + 0 \cdot s_3$$

$$\text{s.t.} \quad -x_1' + x_1'' + 2x_2' - 2x_2'' + s_1 = 4$$

$$x_1' - x_1'' + x_2' - x_2'' + s_2 = 6$$

$$x_1' - x_1'' + 3x_2' - 3x_2'' + s_3 = 9$$

and $\quad x_1', x_1'', x_2', x_2'', s_1, s_2, s_3 \geq 0$

Now apply simplex method in usual manner, we have the following simplex table.

Simplex Table

B.V.	C_B	X_B	$c_j \rightarrow$ 2 x_1'	-2 x_1''	3 x_2'	-3 x_2''	0 s_1	0 s_2	0 s_3	Min Ratio
s_1	0	4	-1	1	②2	-2	1	0	0	2 (Min) $\rightarrow$
s_2	0	6	1	-1	1	-1	0	1	0	6
s_3	0	9	1	-1	3	-3	0	0	1	3
Z= 0,		$\Delta_j \rightarrow$	2	-2	3 $\uparrow$	-3 $\downarrow$	0	0	0	
x_2'	3	2	-1/2	1/2	1	-1	1/2	0	0	-ve
s_2	0	4	3/2	-3/2	0	0	-1/2	1	0	8/3
s_3	0	3	⑤5/2	-5/2	0	0	-3/2	0	1	6/5 (Min) $\rightarrow$
Z=6,		$\Delta_j \rightarrow$	7/2 $\uparrow$	-7/2	0	0	-3/2	0 $\downarrow$	0	
x_2'	3	13/5	0	0	1	-1	1/5	0	1/5	13
s_2	0	11/5	0	0	0	0	2/5	1	-3/5	11/2(Min) $\rightarrow$
x_1'	2	6/5	1	-1	0	0	-3/5	0	2/5	-ve
Z = 51/5,		$\Delta_j \rightarrow$	0	0	0	0	3/5 $\uparrow$	0 $\downarrow$	-7/5	
x_2'	3	3/2	0	0	1	-1	0	-1/2	1/2	
s_1	0	11/2	0	0	0	0	1	5/2	-3/2	
x_1'	2	9/2	1	-1	0	0	0	3/2	-1/2	
Z = 27/2,		$\Delta_j \rightarrow$	0	0	0	0	0	-3/2	-1/2	

We observe that all $\Delta_j \leq 0$ (in the last row of the last table). Hence, solution is optimal and is given by

$$x_1' = \frac{9}{2}, x_1'' = 0, x_2' = \frac{3}{2}, x_2'' = 0, Max.Z = \frac{27}{2}$$

$$\Rightarrow \qquad x_1 = \frac{9}{2}, x_2 = \frac{3}{2} \text{ and Max. } Z = \frac{27}{2}$$

(5) LPP WITH DEGENERACY: TIE FOR ENTERING AND LEAVING BASIC VARIABLES

Sometimes, at the stage of improving the solution during simplex algorithm, minimum ratio (x_B/x_k) is determined by the last column of simplex table to find the key row. But sometimes this ratio may not be unique.

(A) In order to break the tie the selection of key column (entering variable) can be made arbitrarily by keeping in mind the following points.

   (i) If there is a tie between two decision variable, then the selection can be made arbitrarily.

   (ii) If there is a tie between decision variable and slack (or surplus) variable, then select the decision variable to enter in the basis first.

   (iii) If there is a tie between two slack (or surplus) variables, then selection can be made arbitrarily.

(iv) If there is a tie between a slack and artificial variable, preference shall be given to the artificial variable.

(B) When there is a tie between two or more basic variables for leaving the basis, *i.e.*, the minimum ratio is not unique or values of one or more variables in x_B become equal to zero; then use the following points

(i) Divide the coefficient of slack variables in the simplex table where degeneracy is detected by the corresponding positive numbers of the key column in the row, starting from left to right.

(ii) The row which contains smallest ratio comparing from left to right columnwise becomes the key row.

EXAMPLE 1. ***Solve the following LPP by simplex method***

$$\textbf{Max. } Z = 3x_1 + 9x_2$$

subject to the constraints

$$x_1 + 4x_2 \le 8$$
$$x_1 + 2x_2 \le 4$$

and $x_1, x_2 \ge 0$

SOLUTION. Introducing slack variables, s_1 and s_2, the given LPP becomes

Max. $Z = 3x_1 + 9x_2 + 0s_1 + 0s_2$

s.t. $x_1 + 4x_2 + s_1 = 8$

$x_1 + 2x_2 + s_2 = 4$

and $x_1, x_2, s_1, s_2 \ge 0$

Now apply the simplex method in a usual manner, we have the following simplex table.

Simplex Table

B.V.	C_B	X_B	x_1	x_2	s_1	s_2	Min Ratio
		$c_j \rightarrow$	3	9	0	0	
s_1	0	8	1	4	1	0	$\frac{8}{4} = 2$ Tie
s_2	0	4	1	(2)	0	1	$\frac{4}{2} = 2$
$Z = 0$,		$\Delta_j \rightarrow$	3	9 ↑	0	0	
s_1	0	0	–1	0	1	–2	
x_2	9	2	1/2	1	0	1/2	
$Z = 18$,		$\Delta_j \rightarrow$	–3/2	0	0	–9/2	

Clearly all $\Delta_j \le 0 \Rightarrow$ solution is optimal and is given by

$$x_1 = 0, x_2 = 2, \text{Max. } Z = 18$$

Explanation of Tie: In the initial simplex table, both variables s_1 and s_2 are eligible to leave the basis as the minimum ratio (*i.e.*, 2) is same, *i.e.*, there is a tie between s_1 and s_2. To avoid this, we have used the following steps:

STEP 1. Write the coefficient of slack variables as given below.

Row	Key Column	Column	
		s_1	s_2
s_1	4	1	0
s_2	2	0	1

STEP 2. Dividing the coefficients by the corresponding elements of the key column, we get the following ratio.

Row	Key Column	Column	
		s_1	s_2
s_1	4	$\dfrac{1}{4} = \dfrac{1}{4}$	$\dfrac{0}{4} = 0$
s_2	2	$\dfrac{0}{2} = 0$	$\dfrac{1}{2} = \dfrac{1}{2}$

STEP 3. Comparing the ratio of step 2 from left to right, otherwise the minimum ratio occurs for the second row.

Hence, the variable s_2 is selected to leave the basis.

Miscellaneous Exercise

1. Apply Big-M method to solve the following LPP
   Min. $Z = 4x_1 + 8x_2 + 3x_3$
   subject to the constraints
   $$x_1 + x_2 \geq 2$$
   $$2x_1 + x_3 \geq 5$$
   and $x_1, x_2, x_3 \geq 0$

2. Apply Big-M method to solve the following LPP
   Max. $Z = -x_1 - x_2$
   subject to the constraints
   $$3x_1 + 2x_2 \geq 30$$
   $$-2x_1 + 3x_2 \leq -30$$
   $$x_1 + x_2 \leq 5$$
   and $x_1, x_2 \geq 0$

3. Solve the following LPP
   Max. $Z = 6x_1 - 2x_2$
   subject to the constraints
   $$2x_1 - x_2 \leq 2$$
   $$x_1 \leq 4$$
   and $x_1, x_2 \geq 0$

4. Solve by simplex method the following LPP
   Max. $Z = 3x_1 + 4x_2$
   subject to
   $$x_1 - x_2 \leq 1$$
   $$-x_1 + x_2 \leq 2$$

   and $x_1, x_2 \geq 0$

5. Solve the following LPP
   Max. $Z = 6x_1 + 4x_2$
   subject to the constraints
   $$2x_1 + 3x_2 \leq 30$$
   $$3x_1 + 2x_2 \leq 24$$
   $$x_1 + x_2 \geq 3$$
   and $x_1, x_2 \geq 0$ [KANPUR–2011]

6. Solve the following LPP
   Max. $Z = 4x_1 + 10x_2$
   subject to the constraints
   $$2x_1 + x_2 \leq 50$$
   $$2x_1 + 5x_2 \leq 100$$
   $$2x_1 + 3x_2 \leq 90$$
   and $x_1, x_2 \geq 0$
   Also find the alternative optimal solution if exist.

7. Solve the following LPP
   Min. $Z = x_1 + x_2 + x_3$
   subject to the constraints
   $$x_1 - 3x_2 + 4x_3 = 5$$
   $$x_1 - 2x_2 \leq 3$$
   $$2x_2 - x_3 \geq 4$$
   and $x_1, x_2 \geq 0$, x_3 is unrestricted.

ANSWERS

1. $x_1 = \dfrac{5}{2}, x_2 = 0, x_3 = 0, \text{Min}.Z = 10$ **2.** No feasible solution

3. $x_1 = 4, x_2 = 6, \text{Max}. Z = 12$ **4.** Unbounded solution

5. (I) $x_1 = 8, x_2 = 0, \text{Max}. Z = 48$ (II) $x_1 = \dfrac{12}{5}, x_2 = \dfrac{42}{5}, \text{Max}.Z = 48$

6. $x_1 = \dfrac{75}{4}, x_2 = \dfrac{25}{2}, \text{Max}.Z = 200$ **7.** $x_1 = 0, x_2 = \dfrac{21}{5}, x_3 = \dfrac{22}{5}, \text{Min}.Z = \dfrac{43}{5}$

4.16 SOLUTION OF SIMULTANEOUS LINEAR EQUATIONS BY SIMPLEX METHOD

Using simplex method, we can solve the system of simultaneous linear equations. For this, we consider a dummy objective function with zero cost to those variables involved in the equations with cost –1 to each artificial variable that is introduced. After that, apply the simplex method in a usual manner.

 Solved Examples

EXAMPLE I. ***Using simplex method, solve the following system of linear equations:***

$$x_1 - x_3 + 4x_4 = 3, \quad 2x_1 - x_2 = 3, \quad 3x_1 - 2x_2 - x_4 = 1$$

and $x_1, x_2, x_3, x_4 \geq 0$ [MEERUT–2001, 06, 07, 13; PUNJAB–2011; HIMACHAL–2010]

SOLUTION. Let us introduce a dummy objective function Z, with zero cost to each given variable and cost –1, to each artificial variable. Then we can write the LPP as

Max. $Z = 0x_1 + 0x_2 + 0x_3 + 0x_4 - A_1 - A_2 - A_3$

subject to

$$x_1 - x_3 + 4x_4 + A_1 = 3, \quad 2x_1 - x_2 + A_2 = 3, \quad 3x_1 - 2x_2 - x_4 + A_3 = 1$$

and $x_1, x_2, x_3, x_4, A_1, A_2$ and $A_3 \geq 0$

Now, apply the simplex method in a usual manner, we have the following simplex table.

Simplex Table

B.V.	C_B	X_B	$c_j \to$ x_1	x_2	x_3	x_4	A_1	A_2	A_3	Min Ratio
			0	0	0	0	–1	–1	–1	
A_1	–1	3	1	0	–1	4	1	0	0	3
A_2	–1	3	2	–1	0	0	0	1	0	3/2
A_3	–1	1	③	–2	0	–1	0	0	1	1/3 (Min) →
$Z = 7,$	$\Delta_j \to$		6	–3	–1	3	0	0	0	
			↑					↓		
A_1	–1	8/3	0	2/3	–1	⑬/3	1	0	—	8/13 (Min) →
A_2	–1	7/3	0	1/3	0	2/3	0	1	—	7/2
x_1	0	1/3	1	–2/3	0	–1/3	0	0	—	–ve
$Z = -5,$	$\Delta_j \to$		0	1	–1	5	0	0		
				↑	↓					

x_4	0	8/13	0	⟨2/13⟩	−3/13	1	0	8/2=4(Min)→
A_2	−1	25/13	0	3/13	2/13	0	1	25/3
x_1	0	7/13	1	−8/13	−1/13	0	0	−ve
$Z=-25/13, \Delta_j \rightarrow$			1	3/13	2/13	0	0	
				↑	↓			
x_2	0	4	0	1	−3/2	13/2	0	−ve
A_2	−1	1	0	0	⟨1/2⟩	−3/2	1	2 (Min) →
x_1	0	3	1	0	−1	4	0	−ve
$Z = -1, \quad \Delta_j \rightarrow$			0	0	1/2	−3/2	0	
					↑			
x_2	0	7	0	1	0	2		
x_3	0	2	0	0	1	−3		
x_1	0	5	1	0	0	1		
$Z = 0, \quad \Delta_j \rightarrow$			0	0	0	0		

Since, in the last row of the above table all $\Delta_j \leq 0$. Hence, solution is optimal, and is given by $x_1 = 5, x_2 = 7, x_3 = 2, x_4 = 0$.

EXAMPLE 2. *Using simplex method, solve the following system of linear equations.*

$$3x_1 + 2x_2 = 4$$
$$4x_1 - x_2 = 6$$

SOLUTION. Clearly, to apply simplex method, all variables should be non-negative, so let us assume

$$x_1 = x_1' - x_1'', x_2 = x_2' - x_2'' \text{ such that } x_1', x_1'', x_2', x_2'' \geq 0$$

Now, introduce a dummy objective function Z with zero cost to each given variable and cost −1 to each artificial variables A_1 and A_2. Then we can write

$$\text{Max.} Z = 0x_1' + 0x_1'' + 0x_2' + 0x_2'' - 1 \cdot A_1 - 1 \cdot A_2$$

s.t. $\qquad 3x_1' - 3x_1'' + 2x_2' - 2x_2'' + A_1 = 4$

$\qquad\qquad 4x_1' - 4x_1'' - x_2' + x_2'' + A_2 = 6$

and $\qquad\qquad x_1', x_1'', x_2', x_2'', A_1, A_2 \geq 0$

Now, apply the simplex method in a usual manner, we have the following simplex table.

Simplex Table

B.V.	C_B	X_B	$c_j \to$ 0 x_1'	0 x_1''	0 x_2'	0 x_2''	-1 A_1	-1 A_2	Min Ratio
A_1	-1	4	③	-3	2	-2	1	0	4/3 (Min) $\to$
A_2	-1	6	4	-4	-1	1	0	1	6/4
$Z = -10,$	$\Delta_j \to$		7	-7	1	-1	0	0	
			$\uparrow$			$\downarrow$			
x_1'	0	4/3	1	-1	2/3	$-2/3$	—	0	$-$ve
A_2	-1	2/3	0	0	$-11/3$	⑪/3	—	1	2/11 (min)$\to$
$Z = -2,$	$\Delta_j \to$		0	0	$-11/3$	11/3		0	
						$\uparrow$		$\downarrow$	
x_1'	0	16/11	1	-1	0	0			
x_2''	0	2/11	0	0	-1	1			
$Z = 0,$	$\Delta_j \to$		0	0	0	0			

In the last row of the above table, we observe that all $\Delta_j \leq 0$.

$\Rightarrow$ solution is optimal and is given by

$$x_1' = \frac{16}{11}, x_1'' = 0, x_2' = 0, x_2'' = \frac{2}{11}$$

i.e., $\qquad x_1 = x_1' - x_1'' = \frac{16}{11}, x_2 = x_2' - x_2'' = -\frac{2}{11}$

4.17 INVERSE OF A MATRIX BY SIMPLEX METHOD

Let A be a non-singular matrix of order $n \times n$ with real entries.

Now, consider the system of equations

$$Ax = b, x \geq 0$$

where b is a dummy real matrix of order $n \times 1$.

Then, by introducing the artificial variable vector $A_i \geq 0$, form a dummy objective function Z with cost zero to the variable x and cost -1 to each artificial variable as

$$Z = 0 \cdot x - A_i$$

Then find the solution of the LPP given below by simplex method

$$\text{Max. } Z = 0 \cdot x - A_i \cdot 1$$

subject to the constraints

$$Ax + 1 \cdot A_i = b$$

and $\qquad x, A_i \geq 0$

Then use the following concepts

"When the columns of A becomes the column of unit matrix I_n, then the columns of the vectors which are in the initial basis matrix in the starting simplex table give the inverse of A."

☛ REMARK

• To find the dummy matrix, we assign any non-negative integer to each variable Z.

Solved Examples

EXAMPLE 1. *Find the inverse of the matrix* $\begin{bmatrix} 4 & 3 \\ 3 & 2 \end{bmatrix}$, *by simplex method.*

SOLUTION. Consider the system of equation [MEERUT–2008]

$$\begin{bmatrix} 4 & 3 \\ 3 & 2 \end{bmatrix} \begin{bmatrix} x_1 \\ x_2 \end{bmatrix} = \begin{bmatrix} b_1 \\ b_2 \end{bmatrix}$$

To find the dummy matrix $B = \begin{bmatrix} b_1 \\ b_2 \end{bmatrix}$, taking $x_1 = 1, x_2 = 1$, we get

$$B = \begin{bmatrix} 7 \\ 5 \end{bmatrix}$$

Now, introduce the artificial variables A_1, A_2 and the dummy objective function Z, we obtained the following form of LPP

$$\text{Max. } Z = 0 \cdot x_1 + 0 x_2 - A_1 - A_2$$

s.t. $\quad 4x_1 + 3x_2 + A_1 + 0A_2 = 7$

$\quad\quad\quad 3x_1 + 2x_2 + 0A_1 + A_2 = 5$

and $\quad\quad\quad x_1, x_2, A_1, A_2 \geq 0$

Then apply the simplex method in a usual manner, we have the following simplex table.

Simplex Table

		$c_j \rightarrow$	0	0	−1	−1	
B.V.	C_B	X_B	x_1	x_2	A_1	A_2	Min Ratio
A_1	−1	7	4	3	1	0	7/4
A_2	−1	5	③	2	0	1	5/3 (Min) $\rightarrow$
$Z = -12,$		$\Delta_j \rightarrow$	7 ↑	5	0	0	
A_1	−1	1/3	0	①/3	1	− 4/3	1 (Min) $\rightarrow$
x_1	0	5/3	1	2/3	0	1/3	5/2
x_2	0	1	0	1	3	− 4	
x_1	0	1	1	0	−2	3	
$Z = 0,$		$\Delta_j \rightarrow$	0	0	−1	−1	

In the last row of the above table $\Delta_j \leq 0 \Rightarrow$ solution is optimal.

Also, the columns of the given matrix A are converted into the columns of unit matrix I. Hence, in the final iteration, we observe that $(-2, 3)$ is corresponding to x_1 and $(3, -4)$ is that of x_2.

Therefore, inverse of A is $\begin{bmatrix} -2 & 3 \\ 3 & -4 \end{bmatrix}$.

EXAMPLE 2. *Find the inverse of the following matrix by simplex method*

$$A = \begin{bmatrix} 1 & 2 & 3 \\ 2 & 4 & 5 \\ 3 & 5 & 6 \end{bmatrix}$$

SOLUTION. Consider

$$Ax = B$$

$$\Rightarrow \quad \begin{bmatrix} 1 & 2 & 3 \\ 2 & 4 & 5 \\ 3 & 5 & 6 \end{bmatrix} \begin{bmatrix} x_1 \\ x_2 \\ x_3 \end{bmatrix} = \begin{bmatrix} b_1 \\ b_2 \\ b_3 \end{bmatrix}$$

To find the dummy matrix B, let us take $x_1 = 1, x_2 = 0, x_3 = 1$ we get

$$b_1 = 4, b_2 = 7, b_3 = 9$$

$$\Rightarrow \quad B = \begin{bmatrix} 4 \\ 7 \\ 9 \end{bmatrix}$$

Further, introducing the artificial variables A_1, A_2, A_3 and the dummy objective function Z we get the following LPP

Max. $Z = 0 \cdot x_1 + 0 \cdot x_2 + 0 \cdot x_3 - A_1 - A_2 - A_3$

subject to the constraints

$$x_1 + 2x_2 + 3x_3 + A_1 + 0A_2 + 0A_3 = 4$$
$$2x_1 + 4x_2 + 5x_3 + 0 \cdot A_1 + A_2 + 0 \cdot A_3 = 7$$
$$3x_1 + 5x_2 + 6x_3 + 0 \cdot A_1 + 0 \cdot A_2 + A_3 = 9$$

and

$$x_1, x_2, x_3, A_1, A_2, A_3 \geq 0$$

Now, apply the simplex method in a usual manner, we have the following simplex table.

Simplex Table

B.V.	C_B	X_B	x_1	x_2	x_3	A_1	A_2	A_3	Min Ratio
	$c_j \rightarrow$		0	0	0	−1	−1	−1	
A_1	−1	4	1	2	③	1	0	0	4/3 (Min) →
A_2	−1	7	2	4	5	0	1	0	7/5
A_3	−1	9	3	5	6	0	0	1	9/6
$Z = -20,$	$\Delta_j \rightarrow$		6	11	14 ↑	0	0	0	
x_3	0	4/3	1/3	2/3	1	1/3	0	0	2
A_2	−1	1/3	1/3	②/3	0	−5/3	1	0	1/2 (Min) →
A_3	−1	1	1	1	0	−2	0	1	1
$Z = -4/3,$	$\Delta_j \rightarrow$		4/3	5/3 ↑	0	−14/3 ↓	0	0	
x_3	0	1	0	0	1	2	−1	0	—
x_2	0	1/2	1/2	1	0	−5/2	3/2	0	1/2
A_3	−1	1/2	①/2	0	0	1/2	−3/2	1	1/2 (Min) →
$Z = -1/2,$	$\Delta_j \rightarrow$		1/2 ↑	0	0	−1/2	−5/2 ↓	0	
x_3	0	1	0	0	1	2	−1	0	
x_2	0	0	0	1	0	−3	3	−1	
x_1	0	1	1	0	0	1	−3	2	
$Z = 0,$	$\Delta_j \rightarrow$		0	0	0	−1	−1	−1	

We observe that in the last row of the above table all $\Delta_j \leq 0$. So, optimality condition is satisfied. Also, the columns of the given matrix A are converted into the columns of the unit matrix I_3.

Also, in the final iteration, we see that $(1, -3, 2)$ corresponds to x_1, $(-3, 3, -1)$ is that of x_2 and $(2, -1, 0)$ is that of x_3.

Hence,
$$A^{-1} = \begin{bmatrix} 1 & -3 & 2 \\ -3 & 3 & -1 \\ 2 & -1 & 0 \end{bmatrix}.$$

 ## Exercise-4.5

1. Solve the following system of linear equations by simplex method:
   (i) $x_1 + x_2 = 1$, $2x_1 + x_2 = 3$
   (ii) $x_1 + x_2 + x_3 = 6$, $x_1 + 2x_2 + 3x_3 = 14$, $x_1 + 4x_2 + 9x_3 = 36$

2. Find the inverse of the following matrix by simple method:

   (i) $\begin{bmatrix} 1 & 2 \\ 3 & 2 \end{bmatrix}$

   (ii) $\begin{bmatrix} 3 & 2 \\ 4 & -1 \end{bmatrix}$ [GORAKHPUR–2007, 09]

   (iii) $\begin{bmatrix} 4 & 1 & 2 \\ 0 & 1 & 0 \\ 8 & 4 & 5 \end{bmatrix}$

ANSWERS

1. (i) $2, -1$ (ii) $1, 3, 3$

2. (i) $\dfrac{1}{4}\begin{bmatrix} -2 & 2 \\ 3 & -1 \end{bmatrix}$ (ii) $\dfrac{1}{11}\begin{bmatrix} 1 & 2 \\ 4 & -3 \end{bmatrix}$ (iii) $\dfrac{1}{4}\begin{bmatrix} 5 & 3 & -1 \\ 0 & 4 & 0 \\ -8 & -8 & 4 \end{bmatrix}$

 ## REVIEW QUESTIONS

1. Explain various steps involved in simplex method.
2. Explain various steps involved in Big-M method.
3. Explain various steps involved in two-phase method.
4. Define optimal feasible solution and basic feasible solution.
5. Define artificial variable and its need.
6. Explain the uses of artificial variables in linear programming problems.
7. Define slack and surplus variables in linear programming problem.
8. Give outlines of the simplex method in linear programming.

MULTIPLE CHOICE QUESTIONS (CHOOSE THE MOST APPROPRIATE ONE)

1. A basic feasible solution is said to be optimal if $c_j - z_j \ (\forall \ j)$:
   (a) ≤ 0 (b) ≥ 0
   (c) 0 (d) None of these

2. If a LPP involves $\geq$ constraints, then IBFS is obtained by introducing:
   (a) decision variable   (b) artificial variable
   (c) two variables   (d) None of these

3. In Big-M method of the LPP, we take cost of slack surplus and artificial variables in the objective function respectively:
   (a) $0, 0, -M$ (b) $0, 0, M$
   (c) $-1, -1, M$ (d) None of these

4. When we solve simultaneous linear equations by simplex method we take cost of artificial variables in objective function and cost of the given variables as:
   (a) $-M, 0$ (b) $0, -1$
   (c) $-1, 0$ (d) None of these

5. To solve a LPP by simplex method all $b'_j s$ should be:
   (a) negative   (b) non-negative
   (c) positive   (d) None of these

6. Simplex method to solve LPP was developed by George Dantzig in:
   (a) 1945   (b) 1946
   (c) 1947   (d) None of these

7. Big-M method of LPP was developed by:

(a) A. Carners   (b) G.Dantizig

(c) Lagrange   (d) None of these

8. In two phase method, for solving a LPP in Phase-1 the new objective function will be max $Z' =$

(a) – (sum of artificial variables)

(b) sum of slack variables

(c) – (sum of slack variables)

(d) None of these

9. In Big-M method, M stands for:

(a) very large positive price

(b) very large negative price

(c) very small negative price

(d) None of these

10. In LPP, if in the final simplex table all $\Delta_j \leq 0$, then the optimal solution is:

(a) unique   (b) not unique

(c) does not exist   (d) None of these

Answers

1. (a)   2. (b)   3. (a)   4. (c)   5. (b)   6. (c)   7. (a)   8. (a)   9. (b)

10. (a)

Archive

1. Find all the basic feasible solution to the system of linear equations

(i) $x_1 + 2x_2 + x_3 = 4$ and

(ii) $2x_1 + x_2 + 5x_3 = 5$

Are the solutions degenerate?

[BHARTHIDASAN–1994]

2. Compute all the basic feasible solutions to the LP problem :

Maximize $Z = 2x_1 + 3x_2 + 4x_3 + 7x_4$

subject to the constraints

(i) $2x_1 + 3x_2 - x_3 - 4x_4 = 8$

(ii) $x_1 - 2x_2 + 6x_3 - 7x_4 = -3$

and $x_1, x_2, x_3, x_4 \geq 0$ [MEERUT–1994]

3. Solve the following LP problem using simplex method :

Minimize $Z = x_1 - 3x_2 + 3x_3$,

subject to the constraints

(i) $3x_1 - x_2 + 2x_3 \leq 7$, (ii) $2x_1 + 4x_2 \geq -12$

(iii) $- 4x_1 + 3x_2 + 8x_3 \leq 10$

and $x_1, x_2, x_3 \geq 0$.   [PUNJAB–2005]

4. Solve the following LP problem using simplex method :

Maximize $Z = 3x_1 - x_2$

subject to the constraints

(i) $2x_1 + x_2 \leq 2$ (ii) $x_1 + 3x_2 \geq 3$,

(iii) $x_2 \leq 4$

and $x_1, x_2 \geq 0$.

[PUNJAB (BE(Elect))1993; IAS–1992]

5. Solve the following LP problem using simplex method :

Maximize $Z = x_1 - 3x_2 + 2x_3$

subject to the constraints

(i) $3x_1 - x_2 + 3x_3 \leq 7$, (ii) $-2x_1 + 4x_2 \leq 12$

(iii) $- 4x_1 + 3x_2 + 8x_3 \leq 10$

and $x_1, x_2, x_3 \geq 0$.

[KARNATAKA(BE (M.Tech)–1992; IAS–1993; DIBRUGARH–1994]

6. Use the penalty (Big-M) method to solve the following LP problem.

Maximize $Z = 2x_1 + x_2 + 3x_3$

subject to the constraints

(i) $x_1 + x_2 + 2x_3 \geq 5$,

(ii) $2x_1 + 3x_2 + 4x_3 = 12$

and $x_1, x_2, x_3 \geq 0$.

[BHARTHIDASAN–1995; DELHI–1996]

7. Use the two-phase simplex method to solve the following LP problem.

Maximize $Z = 5x_1 - 4x_2 + 3x_3$

subject to the constraints

(i) $2x_1 + x_2 - 6x_3 = 20$,

(ii) $6x_1 + 5x_2 + 10x_3 \leq 76$,

(iii) $8x_1 - 3x_2 + 6x_3 \leq 50$

and $x_1, x_2, x_3 \geq 0$. [PUNJAB (BE (Elect.)-1994]

8. Use the two-phase simplex method to solve the following LP problem.

Maximize $Z = 3x_1 + 2x_2 + 2x_3$

subject to the constraints

(i) $5x_1 + 7x_2 + 4x_3 \leq 7$,

(ii) $-4x_1 + 7x_2 + 5x_3 \geq -2$,

(iii) $3x_1 + 4x_2 - 6x_3 \geq 29/7$

and $x_1, x_2, x_3 \geq 0$.

[PUNJAB(BE(E&C)–2006, BE(IT)–2004]

9. A transport company is considering the purchase of new vehicles for transportation between the Delhi Airport and hotels in the city. There are three vehicles under consideration: Station wagons, minibuses and large buses. The purchase price would be ₹ 1,45,000 for each station wagon, ₹ 2,50,000 for the minibus and ₹ 4,00,000 for the large bus. The board of directors has authorized a maximum amount of ₹ 50,00,000 for these purchases. Because of the heavy air travel, the new vehicles would be utilized at maximum capacity, regardless of the type of vehicles purchased. The expected net annual profit would be ₹ 15,000 for the

station wagon, ₹ 35,000 for the minibus and ₹ 45,000 for the large bus. The company has hired 30 new drivers for the new vehicles. They are qualified drivers for all three types of vehicles. The maintenance department has the capacity to handle an additional 80 station wagons. A minibus is equivalent to 1.67 station wagons and each large bus is equivalent to 2 station wagons in terms of their use of the maintenance department. Determine the number of each type of vehicle to purchase in order to maximize profit. [DELHI(MBA)–1998]

10. A farmer has a 100 acre farm. He can sell all the tomatoes, lettuce or radishes which he

produces. The price he can obtain is ₹1 per kilogram for tomatoes, ₹0.75 a head for lettuce and ₹2 per kilogram for radishes. The average yield per acre is 2,000 kilograms of tomatoes, 3,000 heads of lettuce and 1,000 kilograms of radishes. Fertilizer is available at ₹0.50 per kilogram and the amount required per acre is 100 kilogram each for tomatoes and lettuce and 50 kilogram for radishes. Labour required for sowing cultivating and harvesting per acre is 5 man-days for tomatoes and radishes and 6 man-days for lettuce. A total of 400 man-days of labour are available at ₹20 per man-day. Determine crop mix so as to maximize the farmer's total profit. [DELHI(MBA)–2000]

HINTS AND ANSWERS

1. The summary of the solutions is given below :

S.No.	Basic Vector	Basic Variables	Non-basic Variables
1.	$x_B = \begin{bmatrix} 3 \\ -1 \end{bmatrix}$; $x_1 = 3, x_3 = -1$	x_1, x_3	x_2
2.	$x_B = \begin{bmatrix} \frac{5}{3} \\ \frac{2}{3} \end{bmatrix}$; $x_2 = \frac{5}{3}, x_3 = \frac{2}{3}$	x_2, x_3	x_1
3.	$x_B = \begin{bmatrix} 2 \\ 1 \end{bmatrix}$; $x_1 = 2, x_2 = 1$	x_1, x_2	x_3

Here it may be noted that all these solutions are non-degenerate.

2. The maximum value of objective function $Z = \dfrac{144}{5}$ occurs at the basic feasible solution, $x_1 = x_2 = 0$, $x_3 = \dfrac{44}{17}, x_4 = \dfrac{45}{17}$.

3. Optimal solution is, $x_1 = \dfrac{31}{5}, x_2 = \dfrac{58}{5}, x_3 = 0$ and Min $Z = \dfrac{143}{5}$.

4. $x_1 = \dfrac{3}{5}, x_2 = \dfrac{4}{5}$ and Max. $Z = 1$.

5. Max $Z^* = -x_1 + 3x_2 - 2x_3$ where $Z^* = -Z$. $x_1 = 4, x_2 = 5, x_3 = 0$ and $Z^* = 11$.

6. $x_1 = 3, x_2 = 2$ and Max. $Z = 8$. 　　7. $x_1 = \dfrac{55}{7}, x_2 = \dfrac{30}{7}, x_3 = 0$ and Max. $Z = \dfrac{155}{7}$.

8. $x_1 = 1, x_2 = \dfrac{2}{7}, x_3 = 0$ and Max. $Z = \dfrac{25}{7}$.

9. x_1, x_2 and x_3 = number of station wagons, minibuses and large buses to be purchased, respectively.
Max. $Z = 15,000x_1 + 35,000x_2 + 45,000x_3$
subject to the constraints,
(i) $x_1 + x_2 + x_3 \le 30$; 　　(ii) $1,45,000x_1 + 2,50,000x_2 + 4,00,000x_3 \le 50,00,000$
(iii) $x_1 + 1.67x_2 + 0.5x_3 \le 80$; and $x_1, x_2, x_3 \ge 0$

10. Let x_1, x_2 and x_3 = number of units of tomatoes, lettuce and radishes to be produced, respectively.
Max. (profit) Z = Selling price-fertilizer cost-labour cost
$= (1 \times 2,000 - 0.50 \times 100 - 20 \times 5)x_1 + (0.75 \times 3,000 - 0.50 \times 100 - 20 \times 6)x_2$
$+ (2 \times 1,000 - 0.50 \times 50 - 20 \times 5)x_3$
$= 1,850x_1 + 2,080x_2 + 1,875x_3$
subject to, (i) $x_1 + x_2 + x_3 \le 100$; 　(ii) $5x_1 + 6x_2 + 5x_3 \le 400$; and $x_1, x_2, x_3 \ge 0$.

Degeneracy in Linear Programming

5.1 INTRODUCTION

In previous chapter, we have discussed that if there are m equations in n variables ($n > m$), a basic solution is obtained by solving for m variables in terms of the remaining ($n - m$) variables and setting these ($n - m$) variables equal to zero. Then remaining ($n - m$) variables are called non-basic variables, whereas the m variables are known as basic. Further we have also discussed that, in a basic feasible solution, if all m variables are positive, then it is called a non-degenerate solution. On the other hand if at least one basic variable in a basic feasible solution is zero, then it is known as a degenerate basic solution.

5.2 DEGENERACY IN LINEAR PROGRAMMING

The phenomenon of obtaining a degenerate basic feasible solution in linear programming is called the degeneracy in linear programming.

5.3 THE NECESSARY AND SUFFICIENT CONDITIONS FOR THE EXISTENCE OF NON-DEGENERACY

(1) A necessary and sufficient condition for the existence of the non-degeneracy of the basic solution of a matrix equation $Ax = b$ is that every set of m columns of the augumented matrix $[A\,|\,b]$ is linearly independent.

(2) A necessary and sufficient condition for a given basic feasible solution $x_B = B^{-1}b$ to be non-degenerate is the linear independence of b and every set of ($m - 1$) columns of B.

5.4 OCCURANCE OF DEGENERACY IN LINEAR PROGRAMMING

In a linear programming problem, the degeneracy may appear in the following two ways:

(1) The degeneracy appear in a LPP at the very first iteration when some component of vector b is zero.

(2) If none of the components of b is zero at any iteration and choice of outgoing vector at some iteration is not unique, then next solution is bound to be degenerate.

5.5 RESOLUTION OF DEGENERACY

Let x_k ($= \alpha_k$) be the incoming vector and $\min\left\{\dfrac{x_{B_i}}{x_{ik}}, x_{ik} > 0\right\}$ is not unique, i.e., same minimum ratio $\dfrac{x_{B_i}}{x_{ik}}, x_{ik} > 0$ occurs for more than one value of i.

i.e., the key element (hence the variable to leave the basis) is not uniquely determined or at the very first iteration, the value of one or more basic variables in the x_B column become equal to zero, this cause the problem of degeneracy.

However, if the minimum ratio is zero for two or more basic variables, degeneracy may result the simplex routine to cycle indefinitely. That is the solution which we have obtained in one iteration may repeat again after few iteration and therefore no optimum solution may be obtained under such circumstances. In such type of situation we use the following procedure.

Working Procedure

STEP 1. Pick up the rows for which the minimum, non-negative ratio is same (tied). To be definite, suppose such rows are first, second, third etc. for example.

STEP 2. Rearrange the columns of the usual simplex table so that the columns forming the original unit matrix come first in proper order.

STEP 3. Find the minimum ratio.

$$\left[\frac{\text{elements of the first column of unit matrix}}{\text{Corresponding elements of Key column}} \right]$$

only for the rows for which minimum ratio was not unique. That is for the rows, first, second, third, etc. as picked in step 1.

Then observe that

   (i) If the minimum is attained for the third row (say) then this row will determine the key element by intersecting the key column.

   (ii) If this minimum is also not unique then go to the next step.

STEP 4. Now compute the minimum of the ratio

$$\left[\frac{\text{elements of second column of unit matrix}}{\text{Corresponding elements of Key column}} \right]$$

only for the rows for which minimum ratio was not unique in step 3. Then

   (i) If this minimum ratio is unique for the first row (say) then this row will determine the key element by intersecting the key column.

   (ii) If this minimum is still not unique then go to next step.

STEP 5. Compute the minimum of the ratio

$$\left[\frac{\text{elements of the third column of unit matrix}}{\text{Corresponding elements of Key column}} \right]$$

only for the rows for which minimum ratio was not unique in step 4. Then

   (i) If this minimum ratio is unique for the third row (say) then this row will determine the key element by intersecting the key column.

   (ii) If this minimum is still not unique, then go on repeating the above outlined procedure till the unique minimum ratio is obtained to resolve the degeneracy.

After the resolution of this tie, simplex method is applied to obtain the optimum solution.

Solved Examples

EXAMPLE 1. **Solve the following LPP:**

$$\text{Max. } Z = 2x_1 + 3x_2 + 10x_3$$

subject to the constraints

$$x_1 - 2x_3 = 0$$
$$x_2 + x_3 = 1$$

and $\quad x_1, x_2, x_3 \geq 0$

SOLUTION. We can write the given LPP as

Max. $Z = 2x_1 + 3x_2 + 10x_3$

s.t.

$$\left. \begin{array}{l} x_1 + 0x_2 - 2x_3 = 0 \\ 0x_1 + x_2 - x_3 = 1 \end{array} \right] \qquad \ldots(1)$$

Clearly, the coefficients of x_1 and x_2 in (1) produce a unit matrix I_2 so that there is no need to introduce artificial variable.

We observe that by putting x_3 (non-basic variable) $= 0$ in (1) we get

$$x_1 = 0, x_2 = 1$$

$\Rightarrow \quad x_1 = 0, x_2 = 1, x_3 = 0$ is the starting basic feasible solution.

Now we have the following simplex table.

Simplex Table-1

B.V.	C_B	X_B	$c_j \rightarrow$ 2 x_1	3 x_2	10 x_3	Min Ratio
x_1	2	0	1	0	-2	–ve
x_2	3	1	0	1	①	1 (Min) $\rightarrow$
$Z = C_B X_B = 3,\quad \Delta_j \rightarrow$			0	0	11	
				$\downarrow$	$\uparrow$	

In the above table, using $\Delta_j = c_j - C_B x_j$, we have

$$\Delta_1 = c_1 - C_B x_1 = 2 - (2,3)\begin{pmatrix} 1 \\ 0 \end{pmatrix} = 2 - 2 = 0$$

$$\Delta_2 = c_2 - C_B x_2 = 3 - (2,3)\begin{pmatrix} 0 \\ 1 \end{pmatrix} = 3 - 3 = 0$$

$$\Delta_3 = c_3 - C_B x_3 = 10 - (2,3)\begin{pmatrix} -2 \\ 1 \end{pmatrix} = 10 - (-1) = 11$$

We observe that $\Delta_j \geq 0 \; \forall \; j$, so IBFS is not optimal.

Now, $\quad \Delta_k = \max\{\Delta_j\} = \max\{\Delta_3\} = \Delta_3$

$\Rightarrow \quad k = 3$

$\Rightarrow x_3$ is the incoming vector.

Further, $\qquad \dfrac{x_{B_r}}{x_{rk}} = \min_i \left\{ \dfrac{x_{B_i}}{x_{ik}}, x_{ik} > 0 \right\}$

$\Rightarrow \qquad \dfrac{x_{B_r}}{x_{r3}} = \min_i \left\{ \dfrac{x_{B_i}}{x_{i3}}, x_{i3} > 0 \right\} = \min \left\{ \dfrac{1}{1} \right\} = \dfrac{x_{B_2}}{x_{23}}$

$\Rightarrow \qquad\qquad r = 2$

$\Rightarrow \qquad x_2$ is the outgoing vector.

So, key element $= x_{23} = 1$

Now, entering vector x_3 in the basis matrix and remove x_2, the second simplex table is given as follows:

Simplex Table-2

B.V.	C_B	X_B	x_1	x_2	x_3	Min Ratio
		$c_j \rightarrow$	2	3	10	
x_1	2	2	1	2	0	
x_3	10	1	0	1	1	
$Z = C_B X_B = 10,\quad \Delta_j \rightarrow$			0	–11	0	

Here, in the last row of the above table, we observe that all $\Delta_j \leq 0$.

Hence, basic feasible solution $x_1 = 2$, $x_2 = 0$, $x_3 = 1$ is the non-degenerate optimal solution and the improved value of the objective function is Max. $Z = 10$.

EXAMPLE 2. **Solve the following LPP :**

$$\text{Max. } Z = 2x_1 + 3x_2 + 10x_3$$

subject to the constraints

$$x_1 + 2x_3 = 0$$
$$x_2 + x_3 = 1$$

and $\qquad x_1,\ x_2,\ x_3 \geq 0$

SOLUTION. Clearly, this is the problem of degeneracy at the initial stage because in the first constraints, we have $b_1 = 0$.

Here, the given LPP can be written as

$$\text{Max. } Z = 2x_1 + 3x_2 + 10x_3$$

s.t. $\qquad x_1 + 0x_2 + 2x_3 = 0$

$\qquad\qquad 0x_1 + x_2 + x_3 = 1$

and $\qquad\qquad x_1, x_2, x_3 \geq 0$

We observe that the coefficients of x_1 and x_2 in both constraints produce a unit matrix such that there is no need to introduce the artificial variables.

Also, by putting $x_1 = 0$ in the given constraints we get $x_1 = 0$, $x_2 = 1$.

$\Rightarrow \quad x_1 = 0$, $x_2 = 1$ and $x_3 = 0$ is the starting basic feasible solution.

Now, we have the following simplex table.

Simplex Table-1

B.V.	C_B	X_B	x_1	x_2	x_3	Min Ratio
		$c_j \rightarrow$	2	3	10	
x_1	2	0	1	0	②	0 (Min) $\rightarrow$
x_2	3	1	0	1	1	1
$Z = 3,$		$\Delta_j \rightarrow$	0 $\downarrow$	0	3 $\uparrow$	

In the above table, by using $\Delta_j = c_j - C_B x_j$ we get
$$\Delta_1 = c_1 - C_B x_1 = 2 - 2 = 0$$
$$\Delta_2 = c_2 - C_B x_2 = 3 - 3 = 0$$
$$\Delta_3 = c_3 - C_B x_3 = 10 - 7 = 3$$
Clearly, $\Delta_j \geq 0$, so solution is not optimal.

Now, $\Delta_k = $ max. $\{\Delta_j\}$ for all non-basic variables.
                        $= $ max $\{\Delta_3\} = \Delta_3$
$\Rightarrow$ $k = 3$
$\Rightarrow$ Δ_3 is the incoming vector, i.e., x_3 is the incoming vector.

Also, $\dfrac{x_{B_r}}{x_{rk}} = \min\limits_{i} \left\{ \dfrac{x_{B_i}}{x_{ik}}, x_{ik} > 0 \right\}$

$\Rightarrow$ $\dfrac{x_{B_r}}{x_{r3}} = \min\limits_{i} \left\{ \dfrac{x_{B_i}}{x_{i3}}, x_{i3} > 0 \right\} = \min \left\{ \dfrac{0}{2}, \dfrac{1}{2} \right\} = \dfrac{0}{2} = \dfrac{x_{B_1}}{x_{13}}$

$\Rightarrow$ $r = 1$
$\Rightarrow$ x_1 is the outgoing vector.

Now, entering vector x_3 in basis matrix and deleting x_1, the second simplex table is as follows:

Simplex Table-2

B.V.	C_B	X_B	x_1	x_2	x_3	Min Ratio
		$c_j \rightarrow$	2	3	10	
x_3	10	0	1/2	0	1	
x_2	3	1	−1/2	1	0	
$Z = C_B X_B = 3,$		$\Delta_j \rightarrow$	−3/2	0	0	

From the above table, we observe that all $\Delta_j \leq 0$. Hence, the BFS, $x_1 = 0$, $x_2 = 1$, $x_3 = 0$ is the degenerate BFS and Max. $Z = 3$.

☞ REMARK

• In the above example, we observe that it is possible to move from one table to next without any harm due to degeneracy.

EXAMPLE 3.   **Solve the following LPP :**
$$\textbf{Max. } Z = 2x_1 + x_2$$
subject to the constraints
$$4x_1 + 3x_2 \leq 12$$

$$4x_1 + x_2 \leq 8$$
$$4x_1 - x_2 \leq 8$$

and $\qquad\qquad x_1, x_2 \geq 0$ [KANPUR–2008, 12]

SOLUTION. Using the slack variables s_1, s_2 and s_3 in the given LPP, we get

Max. $Z = 2x_1 + x_2 + 0s_1 + 0s_2 + 0s_3$

s.t. $\quad 4x_1 + 3x_2 + s_1 + 0s_2 + 0s_3 = 12$

$\qquad 4x_1 + x_2 + 0s_1 + s_2 + 0s_3 = 8$

$\qquad 4x_1 - x_2 + 0s_1 + 0s_2 + s_3 = 8$

and $\quad x_1, x_2, s_1, s_2, s_3 \geq 0$

Now apply the simplex method in a usual manner, we have the following simplex table.

Simplex Table-1

B.V.	C_B	X_B	$\overline{s_2}$ (x_1)	$\overline{s_3}$ (x_2)	$\overline{x_1}$ (s_1)	$\overline{x_2}$ (s_2)	$\overline{s_1}$ (s_3)	Min Ratio
	$c_j \to$		2	1	0	0	0	
s_1	0	12	4	3	1	0	0	3
s_2	0	8	$4(\overline{x}_{24})$	1	$0(\overline{x}_{21})$ $1(\overline{x}_{22})$		0	2
s_3	0	8	$4(\overline{x}_{34})$	-1	$0(\overline{x}_{31})$ $0(\overline{x}_{23})$		1	2
$Z = C_B X_B = 0$,	$\Delta_j \to$		2 ↑	1	0	0	0 ↓	

In the above table, by using $\Delta_j = c_j - C_B x_j$, we have

$$\Delta_1 = c_1 - C_B x_1 = 2 - (0,0,0)\begin{bmatrix} 4 \\ 4 \\ 4 \end{bmatrix} = 2$$

$$\Delta_2 = c_2 - C_B x_2 = 1 - (0,0,0)\begin{bmatrix} 3 \\ 1 \\ -1 \end{bmatrix} = 1$$

$$\Delta_3 = 0 = \Delta_4 = \Delta_5$$

$$\text{max. } \{\Delta_j\} = \Delta_1 = \alpha_k \quad \Rightarrow \quad k = 1$$

$\Rightarrow \quad x_1$ is the incoming vector.

Now, $\dfrac{x_{B_r}}{x_{r1}}$ is not unique and minimum occur at $i = 2, 3$.

$\Rightarrow$ degeneracy exists.

Now to select the outgoing vector, proceed as follows:

Firstly renumber the columns in the starting simplex table such that

$$s_1 = \overline{x}_1, s_2 = \overline{x}_2, s_3 = \overline{s}_1, x_1 = \overline{s}_2 \text{ and } x_2 = \overline{s}_3$$

Since, minimum ratio occurs at $i = 2$ and $i = 3$

$\Rightarrow \qquad\qquad I_1 = \{2, 3\}$

$\because \quad x_1 (= \overline{s}_2)$ is the incoming vector, so we take $k = 4$. Now, for $i \in I_1$, we have

$$\min_{i \in I_1}\left\{\frac{\overline{x}_{i1}}{\overline{x}_{ik}}, x_{ik} > 0\right\} = \min_{i \in I_1}\left\{\frac{\overline{x}_{i1}}{\overline{x}_{14}}, \overline{x}_{14} > 0\right\} = \min\left\{\frac{\overline{x}_{21}}{\overline{x}_{24}}, \frac{\overline{x}_{31}}{\overline{x}_{34}}\right\} = \min\left\{\frac{0}{4}, \frac{0}{4}\right\} = \min\{0,0\}$$

$\Rightarrow$ minimum ratio is not unique, because it occurs at $i = 2, 3$.

Let $I_2 = \{2, 3\} \subseteq I_1$. Now, for $i \in I_2$, we have

$$\min_{i \in I_2}\left\{\frac{\bar{x}_{12}}{\bar{x}_{14}}, \bar{x}_{14} > 0\right\} = \min\left\{\frac{\bar{x}_{22}}{\bar{x}_{24}}, \frac{\bar{x}_{32}}{\bar{x}_{34}}\right\} = \min\left\{\frac{1}{4}, \frac{0}{4}\right\} = 0 = \frac{\bar{x}_{32}}{\bar{x}_{34}}$$

$\Rightarrow$ minimum occurs at $i = 3$, so $\bar{s}_1(= s_3)$ is the outgoing vector.

Now, we have the following simplex table.

Simplex Table-2

B.V.	C_B	X_B	$c_j \rightarrow$ 2 x_1	1 x_2	0 $s_1(x_3)$	0 $s_2(x_4)$	0 $s_3(x_5)$	Min Ratio
s_1	0	4	0	4	1	0	–1	4/4
s_2	0	0	0	②	0	1	–1	0/2 (Min) $\rightarrow$
s_3	2	2	1	–1/4	0	0	1/4	–ve
$Z = C_B X_B = 4,$		$\Delta_j \rightarrow$	0	3/2 $\uparrow$	0	0 $\downarrow$	–1/2	

In the above table, using $\Delta_j = c_j - C_B x_j$ we have

$$\Delta_1 = 0, \Delta_2 = c_2 - C_B x_2 = 1 - (0, 0, 2)\begin{pmatrix} 4 \\ 2 \\ -1/4 \end{pmatrix} = 1 + \frac{1}{2} = \frac{3}{2}$$

$$\Delta_3 = 0, \Delta_4 = 0, \Delta_5 = c_5 - C_B x_5 = 0 - (0, 0, 2)\begin{pmatrix} -1 \\ -1 \\ 1/4 \end{pmatrix} = -\frac{1}{2}$$

$\Rightarrow$ all Δ_j are not negative.

For incoming vector $\alpha_k = \max\{\Delta_2, \Delta_5\} = \max\left\{\frac{3}{2}, -\frac{1}{2}\right\} = \frac{3}{2} = \Delta_2$

Also, $$\frac{x_{B_r}}{x_{r2}} = \min_i\left\{\frac{x_{B_i}}{x_{i2}}, x_{i2} > 0\right\} = \min\left\{\frac{4}{4}, \frac{0}{2}\right\} = \frac{0}{2} = \frac{x_{B_2}}{x_{22}}$$

$\Rightarrow$ $r = 2$

$\Rightarrow$ s_2 is the outgoing vector.

$\Rightarrow$ key element $= x_{22} = 2$

So, the next simplex table is given as below.

Simplex Table-3

B.V.	C_B	X_B	$c_j \rightarrow$ 2 x_1	1 x_2	0 s_1	0 s_2	0 s_3	Min Ratio
s_1	0	4	0	0	1	–2	①	4/1 (Min) $\rightarrow$
x_2	1	0	0	1	0	1/2	–1/2	–ve
x_1	2	2	1	0	0	1/8	1/8	16
$Z = C_B X_B = 4,$		$\Delta_j \rightarrow$	0	0	0	–3/4 $\downarrow$	1/4 $\uparrow$	

In the above table, using $\Delta_j = c_j - x_B x_j$, we have
$$\Delta_1 = 0, \Delta_2 = 0, \Delta_3 = 0$$

$$\Delta_4 = c_4 - C_B x_4 = 0 - (0,1,2)\begin{pmatrix} -2 \\ 1/2 \\ 1/8 \end{pmatrix} = -\frac{3}{4}$$

$$\Delta_5 = c_5 - C_B x_5 = 0 - (0,1,2)\begin{pmatrix} 1 \\ -1/2 \\ 1/8 \end{pmatrix} = \frac{1}{4}$$

$\because$ Δ_j are non-negative, so solution is not optimal.

Now, for incoming vector, we have

$$x_k = \max(\Delta_j) = \max(\Delta_4, \Delta_5) = \max\left\{-\frac{3}{4}, \frac{1}{4}\right\} = \frac{1}{4} = \Delta_5$$

$\Rightarrow$ $\qquad k = 5$

$\Rightarrow$ $x_5(s_3)$ is the incoming vector.

Now, for outgoing vector, we have

$$\frac{x_{B_r}}{x_{r5}} = \min_i\left\{\frac{x_{B_i}}{x_{i5}}, x_{i5} > 0\right\}$$

$$= \min_i\left\{\frac{4}{1}, \frac{2}{1/8}\right\} = \frac{4}{1} = \frac{x_{B_1}}{x_{15}}$$

$\Rightarrow$ $\qquad r = 1$

$\Rightarrow$ s_1 is the outgoing vector.

$\Rightarrow$ key element $= 1$

Then proceed as usual, we have the following simplex table.

Simplex Table-4

B.V.	C_B	x_B	$c_j \rightarrow$ x_1	x_2	s_1	s_2	s_3	Min Ratio
			2	1	0	0	0	
s_3	0	4	0	0	1	-2	1	
x_2	1	2	0	1	1/2	$-1/2$	0	
x_1	2	3/2	1	0	$-1/8$	3/8	0	
$Z = C_B x_B = 5$,		$\Delta_j \rightarrow$	0	0	$-1/4$	$-1/2$	0	

We observe that in the above table all $\Delta_j \leq 0 \Rightarrow$ solution is optimal and is given by

$$x_1 = \frac{3}{2}, x_2 = 2 \text{ and Max. } Z = 5$$

EXAMPLE 4. *Solve the following LPP*

$$\textit{Max. } Z = \frac{3}{4}x_1 - 150x_2 + \frac{1}{50}x_3 - 6x_4$$

subject to the constraints

$$\frac{1}{4}x_1 - 60x_2 - \frac{1}{25}x_3 + 9x_4 \leq 0$$

$$\frac{1}{2}x_1 - 90x_2 - \frac{1}{50}x_3 + 3x_4 \leq 0$$

$$x_3 \leq 1$$

and $\qquad\qquad x_1, x_2, x_3, x_4 \geq 0$

SOLUTION. Introducing slack variables s_1, s_2, s_3, the given LPP becomes

$$\text{Max. } Z = \frac{3}{4}x_1 - 150x_2 + \frac{1}{50}x_3 - 6x_4 + 0s_1 + 0s_2 + 0s_3$$

s.t. $\quad \frac{1}{4}x_1 - 60x_2 - \frac{1}{25}x_3 + 9x_4 + s_1 + 0s_2 + 0s_3 = 0$

$$\frac{1}{2}x_1 - 90x_2 - \frac{1}{50}x_3 + 3x_4 + 0s_1 + s_2 + 0s_3 = 0$$

$$x_3 + 0s_1 + 0s_2 + s_3 = 1$$

and $\qquad\qquad x_1, x_2, x_3, s_1, s_2, s_3 \geq 0$

Firstly, putting $x_1 = 0, x_2 = 0, x_3 = 0, x_4 = 0$ in the above constraints we get

$$s_1 = 0, s_2 = 0, s_3 = 1$$

So, the starting basic feasible solution is given by

$$x_1 = x_2 = x_3 = x_4 = s_1 = s_2 = 0, s_3 = 1$$

Now, apply the simplex method in a usual manner, we obtained the following simplex table.

Simplex Table-1

		$c_j \rightarrow$	3/4	−150	1/50	−6	0	0	0	
B.V.	C_B	X_B	$x_1(\bar{x}_4)$	$x_2(\bar{s}_1)$	$x_3(\bar{s}_2)$	$x_4(\bar{s}_3)$	$s_1(\bar{x}_1)$	$s_2(\bar{x}_2)$	$s_3(\bar{x}_3)$	Min Ratio
s_1	0	0	$\frac{1}{4}(\bar{x}_{14})$	−60	−1/25	9	$1(\bar{x}_{11})$	0	0	0
s_2	0	0	$\left(\frac{1}{2}\right)(\bar{x}_{24})$	−90	−1/50	3	$0(\bar{x}_{21})$	1	0	$0 \rightarrow$
s_3	0	1	0	0	1	0	0	0	1	—
$Z = C_B X_B = 0, \Delta_j \rightarrow$			3/4	−150	1/50	−6	0	0	0	
			$\uparrow$					$\downarrow$		

In the above table, using $\Delta_j = c_j - C_B x_j$, we obtained

$$\Delta_1 = \frac{3}{4}, \Delta_2 = -150, \Delta_3 = \frac{1}{50}, \Delta_4 = -6, \Delta_5 = 0, \Delta_6 = 0, \Delta_7 = 0$$

Clearly, $\Delta_1 = \max(\Delta_j) = \alpha_k \quad \Rightarrow \quad k = 1$

$\Rightarrow \quad x_1$ is the incoming vector.

Now, for outgoing vector, we have

$$\frac{x_{B_r}}{x_{r1}} = \min_i \left\{ \frac{x_{B_i}}{x_{i1}}, x_{i1} > 0 \right\}$$

$$= \min \left\{ \frac{x_{B_1}}{x_{11}}, \frac{x_{B_2}}{x_{21}} \right\}$$

$$= \min \left\{ \frac{0}{1/4}, \frac{0}{1/2} \right\} = \min\{0, 0\}$$

$\Rightarrow \quad$ minimum is not unique as it occurs at $i = 1, 2$.

So, this is a problem of degeneracy, so for finding the outgoing vector, we use the following procedure.

First of all renumber the columns in simplex table such that

$$s_1 = \bar{x}_1, s_2 = \bar{x}_2, s_3 = \bar{x}_3, x_1 = \bar{x}_4, x_2 = \bar{s}_1, x_3 = \bar{s}_2, x_4 = \bar{s}_3$$

Here, $I_1 = \{1, 2\}$ and x_1 is the incoming vector, i.e., $x_1 = \bar{x}_4$ so we take $k = 4$ (new)

Now for $i \in I_1$, we have

$$\min_{i \in I_1}\left\{\frac{\bar{x}_{i1}}{\bar{x}_{i4}}, \bar{x}_{i4} > 0\right\} = \min\left\{\frac{\bar{x}_{11}}{\bar{x}_{14}}, \frac{\bar{x}_{21}}{\bar{x}_{24}}\right\}$$

$$= \min\left\{\frac{1}{1/4}, \frac{0}{1/2}\right\} = \min\{4, 0\} = 0 = \frac{\bar{x}_{21}}{\bar{x}_{24}}$$

This minimum is unique and it occurs at $i = 2$.

$\Rightarrow$ $\bar{x}_2(= s_2)$ is the outgoing vector.

Then by usual procedure, we have the following simplex table.

Simplex Table-2

B.V.	C_B	X_B	x_1	x_2	x_3	x_4	s_1	s_2	s_3	Min Ratio
	$c_j \rightarrow$		3/4	−150	1/50	−6	0	0	0	
s_1	0	0	0	−15	−3/100	15/2	1	−1/2	0	—
x_1	3/4	0	1	−180	−1/25	6	0	2	0	—
s_3	0	1	0	0	①	0	0	0	1	1 (Min) →
$Z = C_B X_B = 0, \Delta_j \rightarrow$			0	−15	1/20 ↑	−21/2	0	−3/2 ↓	0	
s_1	0	3/10	0	−15	0	15/2	1	−1/2	3/100	
x_1	3/4	1/25	1	−180	0	6	0	2	11/25	
x_3	1/50	1	0	0	1	0	0	0	0	
$Z = C_B X_B = 1/20, \Delta_j \rightarrow$			0	−15	0	−21/2	0	−3/2	−1/20	

In the last row of the above table all $\Delta_j \leq 0$. Hence, solution is optimal and is given by

$$x_1 = \frac{1}{25}, x_2 = 0, x_3 = 1, x_4 = 0 \text{ and Max. } Z = \frac{1}{20}.$$

EXAMPLE 5. **Solve the following LPP :**

$$\textbf{Max. } Z = 5x_1 + 3x_2$$

subject to the constraints

$$x_1 + x_2 \leq 2$$
$$5x_1 + 2x_2 \leq 10$$
$$3x_1 + 8x_2 \leq 12$$

and $$x_1, x_2 \geq 0$$

<u>SOLUTION.</u> Using the slack variables s_1, s_2 and s_3, the given LPP becomes

$$\text{Max. } Z = 5x_1 + 3x_2 + 0s_1 + 0s_2 + 0s_3$$

s.t.
$$x_1 + x_2 + s_1 = 2$$
$$5x_1 + 2x_2 + s_2 = 10$$
$$3x_1 + 8x_2 + s_3 = 12$$

and $\qquad x_1, x_2, s_1, s_2, s_3 \geq 0$

Now, apply the simplex method in a usual manner, we have the following simplex table

Simplex Table-1

B.V.	C_B	X_B	x_1 (s_2)	x_2 (s_3)	s_1 (x_1)	s_2 (x_2)	s_3 (s_1)	Min Ratio
		$c_j \rightarrow$	5	3	0	0	0	
s_1	0	2	1	1	$1(\bar{x}_{11})$	0	0	$2/1 = 2$
s_2	0	10	⑤$(\bar{x}_{24})$	2	$0(\bar{x}_{21})$	1	0	$10/5 = 2 \rightarrow$
s_3	0	12	3	8	0	0	1	$12/3 = 4$
$Z = C_B X_B = 0,$		$\Delta_j \rightarrow$	5 ↑	3 ↓	0	0	0	

In the above table, by using $\Delta_j = c_j - C_B x_j$ we obtained

$$\Delta_1 = c_1 - C_B x_1 = 5 - (0,0,0)\begin{pmatrix}1\\5\\3\end{pmatrix} = 5$$

$$\Delta_2 = c_2 - C_B x_2 = 3 - (0,0,0)\begin{pmatrix}1\\2\\8\end{pmatrix} = 3$$

$$\Delta_3 = c_3 - C_B x_3 = 0 - (0,0,0)\begin{pmatrix}1\\0\\0\end{pmatrix} = 0$$

$$\Delta_4 = c_4 - C_B x_4 = 0 - (0,0,0)\begin{pmatrix}0\\1\\0\end{pmatrix} = 0$$

and $\qquad \Delta_5 = c_5 - C_B x_5 = 0 - (0,0,0)\begin{pmatrix}0\\0\\1\end{pmatrix} = 0$

Clearly, $\Delta_j \geq 0 \ \forall \ j$, so solution is not optimal.

Now, $\qquad\qquad \alpha_k = \max\{\Delta_j\}$ for all non-basic variables.

$$= \max\{\Delta_1, \Delta_2\} = \max \{5, 3\} = 5$$

$\Rightarrow \qquad\qquad k = 1$

$\Rightarrow$ x_1 is the incoming vector.

Further, by minimum ratio column in starting simplex table, we find that the minimum is not unique and it occurs at $i = 1$ and $i = 2$. So, it is a problem of degeneracy.

To resolve the degeneracy, firstly renumber the columns in the starting table such that

$$s_1 = \bar{x}_1, s_2 = \bar{x}_2, s_3 = \bar{s}_1, x_1 = \bar{s}_2 \text{ and } x_2 = \bar{s}_3$$

Here, $I_1 = \{1, 2\}$ ($\because$ minimum ratio occurs at $i = 1$ and $i = 2$)

Now, for $i = 1, 2 \in I_1$

$$\min_{i \in I_1} \left\{ \frac{\bar{x}_{i1}}{\bar{x}_{iK}}, \bar{x}_{iK} > 0 \right\} = \min \left\{ \frac{\bar{x}_{i1}}{\bar{x}_{i4}}, \bar{x}_{14} > 0 \right\}$$

$$= \min \left\{ \frac{\bar{x}_{11}}{\bar{x}_{14}}, \frac{\bar{x}_{21}}{\bar{x}_{24}} \right\}$$

$$= \min \left\{ \frac{1}{1}, \frac{0}{5} \right\} = 0 = \frac{\bar{x}_{21}}{\bar{x}_{24}}$$

This minimum is unique and occurs at $i = 2$. Therefore, the vector s_2 ($= \bar{x}_2$) is the outgoing vector. And key element $= 5$

Now, apply the usual procedure, we have the following simplex table.

Simplex Table-2

B.V.	C_B	X_B	x_1	x_2	s_1	s_2	s_3	Min Ratio
		$c_j \rightarrow$	5	3	0	0	0	
s_1	0	0	0	③/5	1	$-1/5$	0	0 (Min) $\rightarrow$
x_1	5	2	1	2/5	0	1/5	0	5
s_3	0	6	0	34/5	0	$-3/5$	1	15/17
$Z = C_B X_B = 10,$		$\Delta_j \rightarrow$	0	1	0	-1	0	
				$\uparrow$	$\downarrow$			
x_2	3	0	0	1	5/3	$-1/3$	0	
x_1	5	2	1	0	$-2/3$	1/3	0	
s_3	0	6	0	0	$-34/3$	5/3	1	
$Z = C_B X_B = 10,$		$\Delta_j \rightarrow$	0	0	$-5/3$	$-2/3$	0	

We observe that all $\Delta_j \leq 0$. Hence, solution is optimal and is given by

$$x_1 = 2, x_2 = 0 \text{ and Max. } Z = 10.$$

Exercise-5.1

Solve the following LPP:

1. Max. $Z = x_1 + 2x_2 + x_3$
   s.t. $2x_1 + x_2 - x_3 \leq 2$
   $-2x_1 + x_2 - 5x_3 \leq -6$
   $4x_1 + x_2 + x_3 \leq 6$
   and $x_1, x_2, x_3 \geq 0$

2. Max. $Z = 3x_1 + 9x_2$
   s.t. $x_1 + 4x_2 \leq 8$
   $x_1 + 2x_2 \leq 4$
   and $x_1, x_2 \geq 0$

3. Max. $Z = 3x_1 + 5x_2 + 4x_3$
   s.t. $2x_1 + 3x_3 \leq 18$
   $2x_2 + 5x_3 \leq 18$
   $3x_1 + 2x_2 + 4x_3 \leq 25$
   and $x_1, x_2, x_3 \geq 0$

4. Max. $Z = x_1 - x_2 + 3x_3$
   s.t. $x_1 + x_2 + x_3 \leq 10$
   $2x_1 - x_3 \leq 12$
   $2x_1 - 2x_2 + 3x_3 \leq 0$
   and $x_1, x_2, x_3 \geq 0$

5. Max. $Z = 3x_1 + 5x_2$
   s.t. $x_1 + x_3 = 4$
   $x_2 + x_4 = 6$
   $3x_1 + 2x_2 + x_5 = 12$
   and $x_1, x_2, x_3, x_4, x_5 \geq 0$

6. Max. $Z = 3x_1 + 4x_2$
   s.t. $7x_1 + 5x_2 + x_3 = 40$
   $3x_1 + 4x_2 + x_4 = 20$
   $x_1 - x_2 + x_5 = 0$
   and $x_1, x_2, x_3, x_4, x_5 \geq 0$

ANSWERS

1. $x_1 = 0, x_2 = 4, x_3 = 2$, Max. $Z = 10$ 2. $x_1 = 0, x_2 = 2$, Max. $Z = 18$

3. $x_1 = \dfrac{7}{3}, x_2 = 9, x_3 = 0$, Max. $Z = 52$ 4. $x_1 = 0, x_2 = 6, x_3 = 4$, Max. $Z = 6$

5. $x_1 = 0, x_2 = 6, x_3 = 4, x_4 = 0, x_5 = 0$, Max. $Z = 30$
6. $x_1 = 0, x_2 = 5, x_3 = 15, x_4 = 0, x_5 = 5$, Max. $Z = 20$

REVIEW QUESTIONS

1. Explain the degeneracy in linear programming problem.

2. Explain the simplex method to resolve degeneracy.

MULTIPLE CHOICE QUESTIONS (CHOOSE THE MOST APPROPRIATE ONE)

1. The degeneracy may appear in a LPP at the very first iteration when some component of **b** is:
   (a) one   (b) zero
   (c) non-zero   (d) None of these

2. If one or more of the basic variables in a BFS is zero then the solution is known as:
   (a) degenerate   (b) non-degenerate
   (c) optimal   (d) None of these

3. A non-degenerate optimal solution may be obtained from a degenerate:
   (a) feasible solution

   (b) basic feasible solution
   (c) can't say
   (d) None of these

4. An optimal degenerate solution may be obtained from a/an:
   (a) optimal solution
   (b) non-degenerate solution
   (c) degenerate solution
   (d) None of these

5. If the choice of outgoing vector at any iteration in simplex method is not unique then the next solution is bound to be:

(a) degenerate   (b) non-degenerate

(c) optimal   (d) None of these

6. The procedure which prevent cycling within the simplex routine is called the resolution of:

(a) non-degeneracy   (b) degeneracy

(c) can't say   (d) None of these

7. A non-degenerate solution becomes degenerate in the next iteration if the outgoing vector is:

(a) unique   (b) not unique

(c) can't say   (d) None of these

8. If some components of vector **b** of LPP is zero, then degeneracy may occur at the first iteration and the next solution:

(a) may be degenerate

(b) may not be degenerate

(c) can't say

(d) None of these

ANSWERS

1. (b)	2. (a)	3. (b)	4. (c)	5. (a)	6. (b)	7. (b)	8. (a)

❑❑❑❑

6 Revised Simplex Method

6.1 INTRODUCTION

Revised simplex method is an another efficient method to solve an LPP. It is also developed by G.B. Dantzig. In simplex method, at each iteration, we had to calculate $c_j - z_j$ corresponding to non-basic variables. Columns to decide whether the current solution is optimal or not. In revised simplex method we need to recompute the values of B^{-1} ($\boldsymbol{B}$ is the basis matrix), x_B and Z.

In revised simplex method the word 'revised' refers to the procedure of changing or updating the simplex table. In revised simplex method, we have to calculate B^{-1} at each iteration from its previous value only when y_i is changed at each iteration for which non-basic variable is entered into the basis.

6.2 STANDARD FORMS OF REVISED SIMPLEX METHOD

There are following two standard forms:
(1) **Standard form I:** In this form, only slack and surplus variables are introduced to form basis matrix and there is no need for artificial variables, *i.e.*, an identity matrix is obtained by adding slack variables.
(2) **Standard form II:** In this form, the artificial variables are used to form basis matrix. In this case two phase method is used in a slightly different way to remove artificial variables.

6.3 REVISED SIMPLEX METHOD FOR STANDARD FORM-I

In the revised simplex method, we treat the objective function of the given LPP as additional constraints along with other constraints. So, the number of constraints in revised simplex method should be $(m + 1)$ while in simple simplex method, they were m in numbers (m is the no. of constraints in the given LPP).

Consider an LPP in the standard form:
$$\text{Max. } Z = c_1x_1 + c_2x_2 + \ldots + c_nx_n$$

subject to the constraints

$$\left.\begin{array}{l} a_{11}x_1 + a_{12}x_2 + \ldots + a_{1n}x_n \le b_1 \\ a_{21}x_1 + a_{22}x_2 + \ldots + a_{2n}x_n \le b_2 \\ \vdots \\ a_{m1}x_1 + a_{m2}x_2 + \ldots + a_{mn}x_n \le b_m \end{array}\right] \qquad \ldots(1)$$

and
$$x_1, x_2, \ldots, x_n \ge 0$$

To write the above LPP in standard form I, we proceed as follows:

Firstly consider the objective function as an additional constraints and then use slack (or surplus) variables. Then we have the following form,

$$\left.\begin{array}{c} Z - c_1 x_1 - c_2 x_2 - \ldots - c_n x_n - 0 \cdot x_{n+1} - 0 \cdot x_{n+2} - \ldots - 0 \cdot x_{n+m} = 0 \\ a_{11} x_1 + a_{12} x_2 + a_{13} x_3 + \ldots + a_{1n} x_n + x_{n+1} = b_1 \\ a_{21} x_1 + a_{22} x_2 + a_{23} x_3 + \ldots + a_{2n} x_n + x_{n+2} = b_2 \\ \vdots \\ a_{m1} x_1 + a_{m2} x_2 + a_{m3} x_3 + \ldots + x_{m+n} = b_m \end{array}\right] \qquad \ldots(2)$$

Here, $x_{n+1} (= s_1), x_{n+2} (= s_2), \ldots, x_{n+m} (= s_m)$ are slack variables.

Here the system (2) contains $(m + 1)$ simultaneous equations in $(n + m + 1)$ variables, i.e., $Z, x_1, x_2, \ldots, x_n, x_{n+1}, \ldots, x_{n+m}$.

In matrix form the system (2) can be written as

$$\begin{bmatrix} 1 & -c_1 & -c_2 & \cdots & -c_n & 0 & 0 & \cdots & 0 \\ 0 & a_{11} & a_{12} & \cdots & a_{1n} & 1 & 0 & \cdots & 0 \\ 0 & a_{21} & a_{22} & \cdots & a_{2n} & 0 & 1 & \cdots & 0 \\ \vdots & \vdots & \vdots & & \vdots & \vdots & \vdots & & \vdots \\ 0 & a_{m1} & a_{m2} & \cdots & a_{mn} & 0 & 0 & \cdots & 1 \end{bmatrix} \begin{bmatrix} Z \\ x_1 \\ x_2 \\ \vdots \\ x_{n+m} \end{bmatrix} = \begin{bmatrix} 0 \\ b_1 \\ b_2 \\ \vdots \\ b_m \end{bmatrix}$$

which can also be written as

$$\begin{bmatrix} 1 & -\boldsymbol{C} \\ \boldsymbol{0} & \boldsymbol{A} \end{bmatrix} \begin{bmatrix} Z \\ \boldsymbol{X} \end{bmatrix} = \begin{bmatrix} 0 \\ \boldsymbol{b} \end{bmatrix} \qquad \ldots(3)$$

where,
$$\boldsymbol{0} = \begin{bmatrix} 0 \\ 0 \\ \vdots \\ 0 \end{bmatrix}_{m \times 1}, -\boldsymbol{C} = \begin{bmatrix} -c_1 - c_2 - c_3 - \ldots - c_n \; 0 \ldots 0 \end{bmatrix}_{1 \times (m+n)}$$

$$\boldsymbol{A} = \begin{bmatrix} a_{11} & a_{12} & \cdots & a_{1n} & 1 & 0 & \cdots & 0 \\ a_{21} & a_{22} & \cdots & a_{2n} & 0 & 1 & \cdots & 0 \\ \vdots & \vdots & & \vdots & \vdots & \vdots & & \vdots \\ a_{m1} & a_{m2} & \cdots & a_{mn} & 0 & 0 & \cdots & 1 \end{bmatrix}_{(m+n) \times (m+n)}$$

and
$$\boldsymbol{b} = \begin{bmatrix} b_1 \\ b_2 \\ \vdots \\ b_m \end{bmatrix}$$

Now we can consider an identity matrix I_m as the basis in the simplex method.

Further $\boldsymbol{A} = (\alpha_1, \alpha_2, \ldots, \alpha_n, \alpha_{n+1}, \ldots, \alpha_{n+m})$

where each α_j is m-dimensional column vector.

Now corresponding to each α_j in $\boldsymbol{A}$, we can define a new $(m+1)$ component vector by $\alpha_j^{(1)}$

such that
$$\alpha_j^{(1)} = [-c_j \alpha_j]' ; j = 1, 2 \ldots (n + m)$$

$$= \begin{bmatrix} -c_j \\ \alpha_j \end{bmatrix} \qquad \ldots(4)$$

and corresponding to $\boldsymbol{b}$, we can define $(m + 1)$ component vector by

$$\boldsymbol{b}^{(1)} = [0, b_1, b_2, \ldots, b_m]' = [0, \boldsymbol{b}]'$$

and the column corresponding to Z is the $(m + 1)$ component unit vector denoted by e_1 such that

$$e_1 = \begin{bmatrix} 1 \\ 0 \\ 0 \\ \vdots \\ 0 \end{bmatrix}_{m \times 1}$$

Also, in revised simplex method for standard form-I, the basis matrix be denoted by $\boldsymbol{B}_1$ of order $(m + 1) \times (m + 1)$ and e_1 is the first column in it and remaining m columns are any m $\alpha_j^{(1)}$ which are linearly independent and denoted by

$$\beta_i^{(1)}; i = 1, 2, \ldots, m$$

Therefore, we can write

$$\boldsymbol{B}_1 = [e_1, \beta_1^{(1)}, \beta_2^{(1)}, \ldots, \beta_m^{(1)}]$$

$$= \begin{bmatrix} 1 & -C_{B_1} & -C_{B_2} & \cdots & \cdots & -C_{B_m} \\ 0 & \beta_{11} & \beta_{12} & \cdots & \cdots & \beta_{1m} \\ 0 & \beta_{21} & \beta_{22} & \cdots & \cdots & \beta_{2m} \\ \vdots & \vdots & \vdots & & & \vdots \\ 0 & \beta_{m1} & \beta_{m2} & \cdots & \cdots & \beta_{mm} \end{bmatrix} = \begin{bmatrix} 1 & -C_B \\ \boldsymbol{0} & \boldsymbol{B} \end{bmatrix}$$

Here, $C_B = [C_{B_1} \quad C_{B_2} \quad \cdots \quad C_{B_m}], \boldsymbol{0} = \begin{bmatrix} 0 \\ 0 \\ \vdots \\ 0 \end{bmatrix}_{m \times 1}$

and $\boldsymbol{B} = \begin{bmatrix} \beta_{11} & \beta_{12} & \cdots & \beta_{1m} \\ \beta_{21} & \beta_{22} & \cdots & \beta_{2m} \\ \vdots & \vdots & & \vdots \\ \beta_{m1} & \beta_{m2} & \cdots & \beta_{mm} \end{bmatrix} = [\beta_1, \beta_2, \ldots, \beta_m]$

Further we proceed as follows:

(i) To find $\boldsymbol{B}_1^{-1}$: We have

$$\boldsymbol{B}_1 = \begin{bmatrix} 1 & -C_B \\ \boldsymbol{0} & \boldsymbol{B} \end{bmatrix}$$

$$\Rightarrow \qquad \boldsymbol{B}_1^{-1} = \begin{bmatrix} 1 & C_B \boldsymbol{B}^{-1} \\ \boldsymbol{0} & \boldsymbol{B}^{-1} \end{bmatrix} \qquad \qquad \ldots(5)$$

But $\boldsymbol{B} = \boldsymbol{I_m}$ then $\boldsymbol{B}^{-1} = \boldsymbol{I}_m^{-1} = \boldsymbol{I_m}$

$$\Rightarrow \qquad \boldsymbol{B}_1^{-1} = \begin{bmatrix} 1 & C_B \boldsymbol{I_m} \\ 0 & \boldsymbol{I_m} \end{bmatrix} = \begin{bmatrix} 1 & C_B \\ \boldsymbol{0} & \boldsymbol{I_m} \end{bmatrix}$$

If all $b_i \geq 0$, then we use only slack variables to form the basis matrix, so we have

$$B = I_m, \quad C_B = 0$$

$$\Rightarrow \quad B_1^{-1} = \begin{bmatrix} 1 & 0 \\ 0 & I_m \end{bmatrix} = I_{m+1}$$

(ii) To find $\alpha_j^{(1)}$ (not in the basis matrix B_1):

If $\alpha_j^{(1)}$ is not in B_1, then it can be expressed as the linear combination of column vectors of B_1. Therefore,

$$\alpha_j^{(1)} = x_{0j}e_1 + x_{1j}\beta_1^{(1)} + x_{2j}\beta_2^{(1)} + \ldots + x_{mj}\beta_m^{(1)}$$

$$= [e_1, \beta_1^{(1)}, \beta_2^{(1)}, \ldots, \beta_m^{(1)}] \begin{bmatrix} x_{0j} \\ x_{1j} \\ \vdots \\ x_{mj} \end{bmatrix}$$

$$\Rightarrow \quad \alpha_j^{(1)} = B_1 x_j^{(1)}, \quad x_j^{(1)} = \begin{bmatrix} x_{0j} \\ x_{1j} \\ \vdots \\ x_{mj} \end{bmatrix}$$

$$\Rightarrow \quad x_j^{(1)} = B_1^{-1} \alpha_j^{(1)} \qquad \ldots(6)$$

$$= \begin{bmatrix} 1 & C_B B^{-1} \\ 0 & B^{-1} \end{bmatrix} \begin{bmatrix} -c_j \\ \alpha_j \end{bmatrix} \qquad \text{[By (4) and (5)]}$$

$$= \begin{bmatrix} -c_j + C_B B^{-1}\alpha_j \\ 0 + B^{-1}\alpha_j \end{bmatrix} = \begin{bmatrix} -c_j + C_B x_j \\ 0 + x_j \end{bmatrix} \qquad [\because B^{-1}\alpha_j = X_j]$$

$$= \begin{bmatrix} -c_j + z_j \\ x_j \end{bmatrix} \qquad [\because z_j = C_B x_j]$$

$$= \begin{bmatrix} -\Delta_j \\ x_j \end{bmatrix} \qquad [\because \Delta_j = c_j - z_j] \qquad \ldots(7)$$

From (6) and (7), we observed that $-\Delta_j$ is obtained when the first row of B_1^{-1} is multiplied by $\alpha_j^{(1)}$ (not in the basis B_1)

$$\Rightarrow \quad -\Delta_j = \text{(first row of } B_1^{-1}) \times \alpha_j^{-1}(\text{not in the basis } B_1)$$

The first element in $x_j^{(1)}$ is $-\Delta_j$, decides the optimality of the solution.

(iii) To find the Initial Basic Feasible Solution:

If x_B^{-1} is the basic solution corresponding to the basis matrix B_1, then

$$x_B^{-1} = B_1^{-1} b^{(1)} = \begin{bmatrix} 1 & C_B B^{-1} \\ 0 & B^{-1} \end{bmatrix} \begin{bmatrix} 0 \\ b \end{bmatrix}$$

$$= \begin{bmatrix} C_B \boldsymbol{B}^{-1} \boldsymbol{b} \\ \boldsymbol{B}^{-1} \boldsymbol{b} \end{bmatrix} = \begin{bmatrix} C_B x_B \\ x_B \end{bmatrix} \qquad [\because x_B = \boldsymbol{B}^{-1} \boldsymbol{b}]$$

$$= \begin{bmatrix} Z \\ x_B \end{bmatrix} \qquad [\because Z = C_B x_B] \qquad \ldots(8)$$

In equation (8), the first component is x_B^{-1} in Z which is the value of objective function and the second component is $\boldsymbol{x}_B$ which is the basic feasible solution of the original LPP. Now since Z can be negative so that the initial solution x_B^{-1} is not necessarily basic feasible solution.

Working Procedure

STEP 1. Write the given LPP in the standard form of maximization.

STEP 2. Write the given LPP in standard form-I of revised simplex method.

STEP 3. Write basis matrix $\boldsymbol{B}_1$ and B_1^{-1} and then find IBFS.

$$\boldsymbol{x}_B^{-1} = \boldsymbol{B}_1^{-1} \boldsymbol{b}^{(1)}$$

If for all $b_i \geq 0$ then $\boldsymbol{B}_1 = I_{m+1}$, $\boldsymbol{B}_1^{-1} = I_{m+1}$ then

$$x_B^{-1} = I_{m+1} \boldsymbol{b}^{(1)} = \boldsymbol{b}^{(1)} = \begin{bmatrix} 0 \\ b_1 \\ b_2 \\ \vdots \\ b_m \end{bmatrix}$$

STEP 4. Construct the starting revised simplex table as given below:

B.V.	x_B^{-1}	B_1^{-1}					$x_k^{-1} = \boldsymbol{B}^{-1} \alpha_k^{(1)}$	Min. Ratio $\dfrac{x_{B_i}}{x_{ik}}, x_{ik} > 0$
		e_1	$\beta_1^{(1)}$	$\beta_2^{(1)}$	$\cdots$	$\beta_m^{(1)}$		
Z	0	0	0	0	$\cdots$	0	$z_k - c_k = -\Delta_k$	
s_1	$b_1(x_{B_1})$	1	0	0	$\cdots$	0	x_{1k}	
s_2	$b_2(x_{B_2})$	0	1	0		0	x_{2k}	
s_3	$b_3(x_{B_3})$	0	0	1		0	x_{3k}	
$\vdots$	$\vdots$	$\vdots$	$\vdots$	$\vdots$		$\vdots$	$\vdots$	
s_m	$b_m(x_{B_m})$	0	0	0		1	x_{mk}	

STEP 5. Now compute Δ_j for all $\alpha_j^{(1)}$ (not in the basis $\boldsymbol{B}_1$) by using the following formula.

$$\Delta_j = -(\text{first row of } \boldsymbol{B}_1^{-1}) \times \alpha_j^{(1)}$$

and if all $\Delta_j \leq 0$ then the obtained BFS is optimal, but if atleast one of $\Delta_j > 0$, then solution is not optimal and go to next step.

STEP 6. Now we have to find incoming and outgoing vector. For this, compute

$$\max_j(\Delta_j) \forall j \text{ for which } \alpha_j^{(1)} \text{ are not in } \boldsymbol{B}_1.$$

If $\max_{j}\{\Delta_j\} = \Delta_k$, then $\alpha_k^{(1)}$ will be incoming vector.

To find outgoing vector, use the formula

$$x_k^{(1)} = B_1^{-1}\alpha_k^{(1)} = \begin{bmatrix} -\Delta_k \\ x_{1k} \\ x_{2k} \\ \vdots \\ x_{mk} \end{bmatrix}$$

Now by minimum ratio rule, we may find

$$\max_{j}\left\{\frac{X_{B_i}}{x_{ik}}, x_{ik} > 0\right\}$$

If $\min_{j}\left\{\dfrac{X_{B_i}}{x_{ik}}, x_{ik} > 0\right\} = \dfrac{X_{B_r}}{x_{rk}}$, then the vector $\beta_r^{(1)}$ will be the outgoing vector.

STEP 7. In this step, we have to find the improved solution by finding the key element as in the ordinary simplex method. If x_{rk} is the key element then bring $\alpha_k^{(1)}$ in place for $\beta_r^{(1)}$ and construct the new revised simplex table, from which one can find improved basic feasible solution.

Repeat step (6) and (7) untill we get an optimal solution of the given LPP.

 Solved Examples

EXAMPLE 1. *Solve the following LPP by revised simplex method.*

$$Max. \ Z = 6x_1 - 2x_2 + 3x_3$$

subject to the constraints

$$2x_1 - x_2 + 2x_3 \leq 2$$
$$x_1 + 4x_3 \leq 4$$

and $x_1, \ x_2, \ x_3 \geq 0$ [DELHI 1993; MADRAS(BE)–1999]

SOLUTION.   The given LPP of maximization can be written in the revised simplex form as follows:

$$Max. \ Z = 6x_1 - 2x_2 + 3x_3 + 0s_1 + 0s_2$$

$$\left. \begin{array}{r} Z - 6x_1 + 2x_2 - 3x_3 + 0s_1 + 0s_2 = 0 \\ 2x_1 - x_2 + 2x_3 + s_1 = 2 \\ x_1 + 4x_3 + s_2 = 4 \end{array} \right] \quad \ldots(1)$$

s.t.

In the matrix form, the above system can be written as

$$\begin{array}{cccccc} e_1 & \alpha_1^{(1)} & \alpha_2^{(1)} & \alpha_3^{(1)} & \alpha_4^{(1)} & \alpha_5^{(1)} \end{array}$$

$$\begin{bmatrix} 1 & -6 & 2 & -3 & 0 & 0 \\ 0 & 2 & -1 & 2 & 1 & 0 \\ 0 & 1 & 0 & 4 & 0 & 1 \end{bmatrix} \begin{bmatrix} Z \\ x_1 \\ x_2 \\ x_3 \\ s_1 \\ s_2 \end{bmatrix} = \begin{bmatrix} 0 \\ 2 \\ 4 \end{bmatrix} \quad \ldots(2)$$

$$\Rightarrow \qquad B_1 = \begin{bmatrix} 1 & 0 & 0 \\ 0 & 1 & 0 \\ 0 & 0 & 1 \end{bmatrix} = [e_1, \alpha_4^{(1)}, \alpha_5^{(1)}]$$

By hypothesis, we take $\beta_1^{(1)} = \alpha_4^{(1)}, \beta_2^{(1)} = \alpha_5^{(1)}$

Therefore, $\qquad B_1 = \begin{bmatrix} e_1 & \beta_1^{(1)} & \beta_2^{(1)} \end{bmatrix} = I_3$

$$\Rightarrow \qquad B_1^{-1} = I_3^{-1} = I_3$$

and the initial basic feasible solution is given by

$$x_B^{(1)} = B_1^{-1} \begin{bmatrix} 0 \\ 2 \\ 4 \end{bmatrix} = I_3 \begin{bmatrix} 0 \\ 2 \\ 4 \end{bmatrix} = \begin{bmatrix} 0 \\ 2 \\ 4 \end{bmatrix}$$

Now, we have the following revised simplex table.

Revised Simplex Table-1

Variables in the basis	solution $x_B^{(1)}$	B_1^{-1}			$x_1^{(1)} = B_1^{-1}\alpha_1^{(1)}$	Min Ratio
		e_1	$\beta_1^{(1)}(\alpha_4^{(1)})$	$\beta_2^{(1)}(\alpha_5^{(1)})$		
Z	0	1	0	0	-6	$\min\limits_{i}\left\{ \dfrac{X_{B_i}}{x_{i1}}, x_{i1} > 0 \right\}$
s_1	$2(x_{B_1})$	0	1	0	②(x_{11})	$2/2$ (min) $\rightarrow$
s_2	$4(x_{B_2})$	0	0	1	$1(x_{21})$	$4/1 = 4$

Since $\alpha_1^{(1)}, \alpha_2^{(1)}$ and $\alpha_3^{(1)}$ are not in the basis B_1 therefore,

$$\Delta_1 = -(\text{first row of } B_1^{-1}) \times \alpha_1^{(1)}$$

$$= -[1,0,0] \begin{bmatrix} -6 \\ 2 \\ 1 \end{bmatrix} = 6$$

$$\Delta_2 = -[1,0,0] \begin{bmatrix} 2 \\ -1 \\ 0 \end{bmatrix} = -2$$

$$\Delta_3 = -[1,0,0] \begin{bmatrix} -3 \\ 2 \\ 4 \end{bmatrix} = 3$$

Clearly, $\Delta_1, \Delta_3 > 0$ therefore, the initial basic feasible solution is not optimal.

To find incoming vector:

Since, $\qquad \Delta_1 = \max\{\Delta_1, \Delta_2, \Delta_3\} = \Delta_k$

$$\Rightarrow \qquad k = 1$$

$$\Rightarrow \qquad \alpha^{(1)} \text{ is the incoming vector.}$$

To find outgoing vector:

Using minimum ratio rule,

$$\frac{x_{B_r}}{x_{r1}} = \min_i \left\{ \frac{x_{B_i}}{x_{i1}}, x_{i1} > 0 \right\}$$

$$= \min \left\{ \frac{x_{B_1}}{x_{11}}, \frac{x_{B_2}}{x_{21}} \right\} = \min \left\{ \frac{2}{2}, \frac{4}{1} \right\} = \frac{2}{2} = \frac{x_{B_1}}{x_{11}}$$

$\Rightarrow \quad r = 1$

$\Rightarrow \quad \beta_1^{(1)} (= \alpha_4^{(1)})$ is the outgoing vector.

Hence, $x_{11} = 2$ is the required key element.

Now, apply the usual procedure, we have the following revised simplex table.

Revised Simplex Table-2

Variables in the basis	solution $x_B^{(1)}$	B_1^{-1}			$x_2^{(1)} = B_1^{-1}\alpha_2^{(1)}$	Minimum Ratio
		e_1	$\beta_1^{(1)}\left(\alpha_1^{(1)}\right)$	$\beta_2^{(1)}\left(\alpha_5^{(1)}\right)$		
Z	6	1	3	0	− 1	$\min_i \left\{ \frac{x_{B_i}}{x_{i2}}, x_{i2} > 0 \right\}$
x_1	1	0	1/2	0	$-1/2 (x_{12})$	–ve
s_2	3	0	−1/2	1	(1/2) (x_{22})	3/1 (min) →

Clearly, from the above table

$x_1 = 1, x_2 = 0$ and Max $Z = 6$ is the new improved solution.

Now, we will check the optimality of this solution.

Since $\alpha_2^{(1)}, \alpha_3^{(1)}$ and $\alpha_4^{(1)}$ are not in the basis $\boldsymbol{B}_1$ so

$$\Delta_2 = -(\text{first row of } \boldsymbol{B}_1^{-1} \text{ in second table}) \times \alpha_2^{(1)}$$

$$= -[1,3,0] \begin{bmatrix} 2 \\ -1 \\ 0 \end{bmatrix} = 1$$

$$\Delta_3 = -[1,3,0] \begin{bmatrix} -3 \\ 2 \\ 4 \end{bmatrix} = -3$$

$$\Delta_4 = -[1,3,0] \begin{bmatrix} 0 \\ 1 \\ 0 \end{bmatrix} = 3$$

Clearly, $\Delta_2 > 0 \Rightarrow$ improved solution is not optimal.

To find incoming vector:

$\because \quad \Delta_2 = \max\{\Delta_2, \Delta_3, \Delta_4\} = \Delta_k \quad \Rightarrow \quad k = 2$

$\Rightarrow \quad \alpha_2^{(1)}$ is the incoming vector.

Now, $\quad x_2^{(1)} = B_1^{-1} \alpha_2^{(1)} = \begin{bmatrix} 1 & 3 & 0 \\ 0 & 1/2 & 0 \\ 0 & -1/2 & 1 \end{bmatrix} \begin{bmatrix} 2 \\ -1 \\ 0 \end{bmatrix} = \begin{bmatrix} -\Delta_2 \\ x_{12} \\ x_{22} \end{bmatrix}$

To find the outgoing vector:
Apply minimum ratio rule, we have

$$\frac{x_{B_r}}{x_{r2}} = \min_i \left\{ \frac{x_{B_i}}{x_{i2}}, x_{i2} > 0 \right\} = \min \left\{ \frac{x_{B_2}}{x_{22}} \right\} \qquad [\because x_{12} = -1]$$

$$\Rightarrow \qquad \frac{x_{B_r}}{x_{r2}} = \frac{x_{B_2}}{x_{22}} \qquad \Rightarrow \qquad r = 2$$

Therefore, $\beta_2^{(1)} (= \alpha_5^{(1)})$ is the outgoing vector and hence $x_{22} = \dfrac{1}{2}$ is the key element.

Further to find the improved solution, we have the following simplex table:

Revised Simplex Table-3

Variables in the basis	solution $x_B^{(1)}$	e_1	B_1^{-1} $\beta_1^{(1)} \left(\alpha_1^{(1)} \right)$	$\beta_2^{(1)} \left(\alpha_2^{(1)} \right)$	$x_k^{(1)}$	Min. Ratio
Z	12	1	2	2		
x_1	4	0	0	1		
x_2	6	0	-1	2		

From the above table, we obtained
$\qquad x_1 = 4, x_2 = 6, x_3 = 0$ and Max. $Z = 12$

Now we shall check the optimality of this solution.

Clearly, $\alpha_3^{(1)}, \alpha_4^{(1)}$ and $\alpha_5^{(1)}$ are not in the basis B_1. Therefore,

$$\Delta_3 = -(\text{first row of } B_1^{-1} \text{ of third table}) \times \alpha_3^{(1)}$$

$$= -[1,2,2] \begin{bmatrix} -3 \\ 2 \\ 4 \end{bmatrix} = -9$$

$$\Delta_4 = -[1,2,2] \begin{bmatrix} 0 \\ 1 \\ 0 \end{bmatrix} = -2$$

$$\Delta_5 = -[1,2,2] \begin{bmatrix} 0 \\ 0 \\ 1 \end{bmatrix} = -2$$

We observe that $\Delta_3, \Delta_4, \Delta_5 \leq 0$. Hence, this improved solution is optimal and is given by

$$x_1 = 4, x_2 = 6, x_3 = 0 \text{ and Max. } Z = 12$$

EXAMPLE 2. *Solve the following LPP by revised simplex method:*

$$\text{Max. } Z = x_1 + 2x_2$$

subject to the constraints

$$x_1 + x_2 \le 3$$
$$x_1 + 2x_2 \le 5$$
$$3x_1 + x_2 \le 6$$

and $$x_1, x_2 \ge 0$$ [GUAHATI (MCA)–1992]

SOLUTION.   The given maximization LPP can be written in revised simplex form as follows:

$$\text{Max. } Z = x_1 + 2x_2 + 0s_1 + 0s_2 + 0s_3$$

s.t.

$$\left.\begin{array}{r} Z - x_1 - 2x_2 + 0s_1 + 0s_2 + 0s_3 = 0 \\ x_1 + x_2 + s_1 = 3 \\ x_1 + 2x_2 + s_2 = 5 \\ 3x_1 + x_2 + s_3 = 6 \end{array}\right\} \qquad \ldots(1)$$

and $$x_1, x_2, s_1, s_2, s_3 \ge 0$$

The above system can be written in matrix form as follows:

$$
\begin{matrix}
e_1 & \alpha_1^{(1)} & \alpha_2^{(1)} & \alpha_3^{(1)} & \alpha_4^{(1)} & \alpha_5^{(1)} \\
\end{matrix}
$$

$$
\begin{bmatrix}
1 & -1 & -2 & 0 & 0 & 0 \\
0 & 1 & 1 & 1 & 0 & 0 \\
0 & 1 & 2 & 0 & 1 & 0 \\
0 & 3 & 1 & 0 & 0 & 1
\end{bmatrix}
\begin{bmatrix} Z \\ x_1 \\ x_2 \\ s_1 \\ s_2 \\ s_3 \end{bmatrix}
=
\begin{bmatrix} 0 \\ 3 \\ 5 \\ 6 \end{bmatrix}
\qquad \ldots(2)
$$

Here, $$B_1 = \begin{bmatrix} 1 & 0 & 0 & 0 \\ 0 & 1 & 0 & 0 \\ 0 & 0 & 1 & 0 \\ 0 & 0 & 0 & 1 \end{bmatrix} = \begin{bmatrix} e_1 & \alpha_3^{(1)} & \alpha_4^{(1)} & \alpha_5^{(1)} \end{bmatrix}$$

Now, by hypothesis we have

$$\alpha_3^{(1)} = \beta_1^{(1)}, \alpha_4^{(1)} = \beta_2^{(1)}, \alpha_5^{(1)} = \beta_3^{(1)}$$

So, $$B_1 = \begin{bmatrix} e_1 & \beta_1^{(1)} & \beta_2^{(1)} & \beta_3^{(1)} \end{bmatrix} = I_4$$

$$\Rightarrow \quad B_1^{-1} = I_4^{-1} = I_4$$

Also, the initial solution is given by

$$x_B^{(1)} = \begin{bmatrix} 0 \\ 3 \\ 5 \\ 6 \end{bmatrix}$$

Now we have the following revised simplex table.

Revised Simplex Table-1

Variables in the basis	solution $x_B^{(1)}$	B_1^{-1}				$x_2^{(1)} = B_1^{-1}\alpha_2^{(1)}$	Min Ratio $\dfrac{x_{B_i}}{x_{i2}}, x_{i2} > 0$
		e_1	$\beta_1^{(1)}$	$\beta_2^{(1)}$	$\beta_3^{(1)}$		
Z	0	1	0	0	0	-2	$\min\limits_{i}\left\{\dfrac{x_{B_i}}{x_{i2}}, x_{i2} > 0\right\}$
s_1	$3(x_{B_1})$	0	1	0	0	$1\ (x_{12})$	$3/1$
s_2	$5(x_{B_2})$	0	0	1	0	②$\ (x_{22})$	$5/2$ (min) $\rightarrow$
s_3	$6(x_{B_3})$	0	0	0	1	$1\ (x_{32})$	$6/1$

Now, since $\alpha_1^{(1)}$ and $\alpha_2^{(1)}$ are not in the basis therefore,

$$\Delta_1 = -\begin{bmatrix} 1 & 0 & 0 & 0 \end{bmatrix}\begin{bmatrix} -1 \\ 1 \\ 1 \\ 3 \end{bmatrix} = 1$$

$$\Delta_2 = -\begin{bmatrix} 1 & 0 & 0 & 0 \end{bmatrix}\begin{bmatrix} -2 \\ 1 \\ 2 \\ 1 \end{bmatrix} = 2$$

which implies that $\Delta_1, \Delta_2 > 0$, so solution is not optimal.

To find incoming vector:

Here, $\max(\Delta_1, \Delta_2) = \Delta_2 = \Delta_k$ $\Rightarrow$ $k = 2$

$\Rightarrow$ $\alpha_2^{(1)}$ is the incoming vector.

Also, $x_k^{(1)} = x_2^{(1)} = B_1^{-1}\alpha_2^{(1)}$

$$= \begin{bmatrix} 1 & 0 & 0 & 0 \\ 0 & 1 & 0 & 0 \\ 0 & 0 & 1 & 0 \\ 0 & 0 & 0 & 1 \end{bmatrix}\begin{bmatrix} -2 \\ 1 \\ 2 \\ 1 \end{bmatrix} = \begin{bmatrix} -2 \\ 1 \\ 2 \\ 1 \end{bmatrix}$$

To find outgoing vector:

Apply the minimum ratio rule, we have

$$\frac{x_{B_r}}{x_{r2}} = \min_{i}\left\{\frac{x_{B_i}}{x_{i2}}, x_{i2} > 0\right\} = \min\left\{\frac{3}{1}, \frac{5}{2}, \frac{6}{1}\right\} = \frac{5}{2} = \frac{x_{B_2}}{x_{22}}$$

$\Rightarrow$ $r = 2$

$\Rightarrow$ $\beta_2^{(1)}(= \alpha_4^{(1)})$ is the outgoing vector.

Hence, $x_{22} = 2$ is the key element.

Now, to find improved solution, apply the usual procedure we have the following revised simplex table:

Revised Simplex Table-2

Variables in the basis	solution $x_B^{(1)}$	e_1	B_1^{-1}			$x_k^{-1} = B_1^{-1}\alpha_k^{(1)}$	Min Ratio
			$\beta_1^{(1)}\left(\alpha_3^{(1)}\right)$	$\beta_2^{(1)}\left(\alpha_2^{(1)}\right)$	$\beta_3^{(1)}\left(\alpha_5^{(1)}\right)$		
Z	5	1	0	1	0		
s_1	1/2	0	1	−1/2	0		
x_2	5/2	0	0	1/2	0		
s_3	7/2	0	0	−1/2	1		

From the above table we observe that

$$x_2 = \frac{5}{2}, s_1 = \frac{1}{2}, s_3 = \frac{7}{2} \text{ and Max. } Z = 5$$

Now, we have to check the optimality of this improved solution.

Since $\alpha_1^{(1)}$ and $\alpha_4^{(1)}$ are not in the basis B_1. Then

$$\Delta_1 = -(\text{first row of } B_1^{-1} \text{ in the second table}) \times \alpha_1^{(1)}$$

$$= -\begin{bmatrix} 1 & 0 & 1 & 0 \end{bmatrix} \begin{bmatrix} -1 \\ 1 \\ 1 \\ 3 \end{bmatrix} = 0$$

$$\Delta_4 = -\begin{bmatrix} 1 & 0 & 1 & 0 \end{bmatrix} \begin{bmatrix} 0 \\ 0 \\ 1 \\ 0 \end{bmatrix} = -1$$

We observe that $\Delta_1, \Delta_4 \le 0 \Rightarrow$ solution is optimal and is given by

$$x_1 = 0, x_2 = \frac{5}{2} \text{ and Max. } Z = 5$$

EXAMPLE 3. *Solve the following LPP by revised simplex method:*

Max. Z = 3x₁ + x₂ + 2x₃ + 7x₄

subject to the constraints

$$2x_1 + 3x_2 - x_3 + 4x_4 \le 40$$
$$- 2x_1 + 2x_2 + 5x_3 - x_4 \le 35$$
$$x_1 + x_2 - 2x_3 + 3x_4 \le 100$$
and $x_1 \ge 2, x_2 \ge 1, x_3 \ge 3, x_4 \ge 4$

SOLUTION. In the given LPP, the non-negative restrictions are given by

$$x_1 \ge 2, x_2 \ge 1, x_3 \ge 3 \text{ and } x_4 \ge 4$$

$\Rightarrow$ lower bounds of x_1, x_2, x_3 and x_4 are not zero.

Thus, introduce new variables y_1, y_2, y_3 and y_4 such that

$$x_1 = y_1 + 2, x_2 = y_2 + 1, x_3 = y_3 + 3, x_4 = y_4 + 4$$

such that $y_1 \ge 0, y_2 \ge 0, y_3 \ge 0, y_4 \ge 0$

Therefore, the given LPP reduces to
$$\text{Max. } Z = 3(y_1 + 2) + (y_2 + 1) + 2(y_3 + 3) + 7(y_4 + 4)$$
$$= 3y_1 + y_2 + 2y_3 + 7y_4 + 41$$
or
$$\text{Max. } Z' = 3y_1 + y_2 + 2y_3 + 7y_4 = Z - 41$$
s.t.
$$2y_1 + 3y_2 - y_3 + 4y_4 \leq 20$$
$$-2y_1 + 2y_2 + 5y_3 - y_4 \leq 26$$
$$y_1 + y_2 - 2y_3 + 3y_4 \leq 91$$
and
$$y_1, y_2, y_3, y_4 \geq 0$$

The above LPP is of maximization. This LPP in revised simplex form can be written as
$$\text{Max. } Z' = 3y_1 + y_2 + 2y_3 + 7y_4 + 0s_1 + 0s_2 + 0s_3$$

s.t.

$$\left. \begin{array}{l} Z' - 3y_1 - y_2 - 2y_3 - 7y_4 + 0s_1 + 0s_2 + 0s_3 = 0 \\ 2y_1 + 3y_2 - y_3 + 4y_4 + s_1 = 20 \\ -2y_1 + 2y_2 + 5y_3 - y_4 + s_2 = 26 \\ y_1 + y_2 - 2y_3 + 3y_4 + s_3 = 91 \end{array} \right] \quad \text{...(1)}$$

and
$$y_i, s_i \geq 0$$

The above system of equations can be written in matrix form as follows:

$$\begin{array}{ccccccc} e_1 & \alpha_1^{(1)} & \alpha_2^{(1)} & \alpha_3^{(1)} & \alpha_4^{(1)} & \alpha_5^{(1)} & \alpha_6^{(1)} & \alpha_7^{(1)} \end{array}$$

$$\begin{bmatrix} 1 & -3 & -1 & -2 & -7 & 0 & 0 & 0 \\ 0 & 2 & 3 & -1 & 4 & 1 & 0 & 0 \\ 0 & -2 & 2 & 5 & -1 & 0 & 1 & 0 \\ 0 & 1 & 1 & -2 & 3 & 0 & 0 & 1 \end{bmatrix} \begin{bmatrix} Z' \\ y_1 \\ y_2 \\ y_3 \\ y_4 \\ s_1 \\ s_2 \\ s_3 \end{bmatrix} = \begin{bmatrix} 0 \\ 20 \\ 26 \\ 91 \end{bmatrix} \quad \text{...(2)}$$

$$\Delta_4 = -\begin{bmatrix} 1 & 0 & 0 & 0 \end{bmatrix} \begin{bmatrix} -7 \\ 4 \\ -1 \\ 3 \end{bmatrix} = 7$$

Clearly all $\Delta_j > 0 \Rightarrow$ solution is not optimal.

To find incoming vector:

$\because \max\{\Delta_1, \Delta_2, \Delta_3, \Delta_4\} = \max\{3, 1, 2, 7\} = 7 = \Delta_4$

$\Rightarrow \qquad\qquad k = 4$

$\Rightarrow \quad \alpha_4^{(1)}$ is the incoming vector which enter in the basis matrix.

Also, $\qquad y_4^{(1)} = B_1^{-1}\alpha_4^{(1)} = I_4\alpha_4^{(1)} = \alpha_4^{(1)} = \begin{bmatrix} -7 \\ 4 \\ -1 \\ 3 \end{bmatrix} = \begin{bmatrix} -\Delta_4 \\ y_{14} \\ y_{24} \\ y_{34} \end{bmatrix}$

To find outgoing vector: Apply minimum ratio, we get

$$\frac{y_{B_r}}{y_{r4}} = \min_i \left\{ \frac{y_{B_i}}{y_{i4}}, y_{i4} > 0 \right\}$$

$$= \min \left\{ \frac{y_{B_1}}{y_{14}}, \frac{y_{B_2}}{y_{34}} \right\} = \min \left\{ \frac{20}{4}, \frac{91}{3} \right\} = \frac{20}{4} = \frac{y_{B_1}}{y_{14}}$$

$$\Rightarrow \qquad r = 1$$

So, $\beta_1^{(1)} (= \alpha_5^{(1)})$ is the outgoing vector.

$$\Rightarrow \quad y_{14} = 4 \text{ is the key element.}$$

Revised Simplex Table-1

Variables in the basis	solution $y_B^{(1)}$	e_1	$\beta_1^{(1)}(\alpha_4^{(1)})$	$\beta_2^{(1)}(\alpha_6^{(1)})$	$\beta_3^{(1)}(\alpha_7^{(1)})$	$y_3^{(1)} = B_1^{-1}\alpha_3^{(1)}$	Min Ratio
			B_1^{-1}				
Z'	35	1	7/4	0	0	–15/4	$\min_i \left\{ \dfrac{y_{B_i}}{y_{i3}}, y_{i3} > 0 \right\}$
y_4	$5(y_{B_1})$	0	1/7	0	0	–1/4 (y_{13})	–ve
s_2	$31(y_{B_2})$	0	1/4	1	0	19/4 (y_{23})	124/19 (min) →
s_3	$76(y_{B_3})$	0	–3/4	0	1	–5/4 (y_{33})	–ve
			↓				

From the above table, we observe that

$$y_4 = 5, s_2 = 31, s_3 = 76 \text{ and Max. } Z' = 35$$

From (2), the basis matrix

$$B_1 = \begin{bmatrix} 1 & 0 & 0 & 0 \\ 0 & 1 & 0 & 0 \\ 0 & 0 & 1 & 0 \\ 0 & 0 & 0 & 1 \end{bmatrix} = \begin{bmatrix} e_1 & \alpha_5^{(1)} & \alpha_6^{(1)} & \alpha_7^{(1)} \end{bmatrix}$$

By hypothesis $\alpha_5^{(1)} = \beta_1^{(1)}, \alpha_6^{(1)} = \beta_2^{(1)}, \alpha_7^{(1)} = \beta_3^{(1)}$

$$\Rightarrow \qquad B_1 = [e_1 \quad \beta_1^{(1)} \quad \beta_2^{(1)} \quad \beta_3^{(1)}] = I_4$$

$$\Rightarrow \qquad B_1^{-1} = I_4^{-1} = I_4$$

Now apply the revised simplex method in a usual manner, we have the following revised simplex table.

Revised Simplex Table-2

Variables in the basis	solution $y_B^{(1)}$	e_1	$\beta_1^{(1)}\left(\alpha_5^{(1)}\right)$	$\beta_2^{(1)}\left(\alpha_6^{(1)}\right)$	$\beta_3^{(1)}\left(\alpha_7^{(1)}\right)$	$y_4^{(1)} = B_1^{-1}\alpha_4^{(1)}$	Minimum Ratio
			$\mathbf{B}_1^{-1}$				
Z'	0	1	0	0	0	-7	$\min\limits_{i}\left\{\dfrac{y_{B_i}}{y_{i4}}, y_{i4} > 0\right\}$
s_1	$20(y_{B_1})$	0	1	0	0	④(y_{14})	$20/4 = 5(\min) \rightarrow$
s_2	$26(y_{B_2})$	0	0	1	0	$-1\ (y_{24})$	$-ve$
s_3	$91(y_{B_3})$	0	0	0	1	$3\ (y_{34})$	$91/3$

Clearly, $\alpha_1^{(1)}, \alpha_2^{(1)}, \alpha_3^{(1)}, \alpha_4^{(1)}$ are not in the basis matrix $\mathbf{B}_1$ then

$$\Delta_1 = -(\text{first row of } \mathbf{B}_1^{-1} \text{ of first table}) \times \alpha_1^{(1)}$$

$$= -[1 \quad 0 \quad 0 \quad 0]\begin{bmatrix} -3 \\ 2 \\ -2 \\ 1 \end{bmatrix} = 3$$

$$\Delta_2 = -[1 \quad 0 \quad 0 \quad 0]\begin{bmatrix} -1 \\ 3 \\ 2 \\ 1 \end{bmatrix} = 1$$

$$\Delta_3 = -[1 \quad 0 \quad 0 \quad 0]\begin{bmatrix} -2 \\ -1 \\ 5 \\ -2 \end{bmatrix} = 2$$

Now we check the optimality of this solution.

$\because$ $\alpha_1^{(1)}, \alpha_2^{(1)}, \alpha_3^{(1)}$ and $\alpha_5^{(1)}$ are not in the basis matrix $\mathbf{B}_1$.

So, now basis matrix is

$$\mathbf{B}_1 = [e_1 \quad \alpha_4^{(1)} \quad \alpha_6^{(1)} \quad \alpha_7^{(1)}]$$

Using $\Delta_j = -(\text{first row of } \mathbf{B}_1^{-1} \text{ in second table}) \times \alpha_j^{(1)}$, we get

$$\Delta_1 = -\left[1 \quad \frac{7}{4} \quad 0 \quad 0\right]\begin{bmatrix} -3 \\ 2 \\ -2 \\ 1 \end{bmatrix} = -\frac{1}{2}$$

$$\Delta_2 = -[1 \quad \frac{7}{4} \quad 0 \quad 0] \begin{bmatrix} -1 \\ 3 \\ 2 \\ 1 \end{bmatrix} = -\frac{19}{4}$$

$$\Delta_3 = -[1 \quad \frac{7}{4} \quad 0 \quad 0] \begin{bmatrix} -2 \\ -1 \\ 5 \\ -2 \end{bmatrix} = \frac{15}{4}$$

and

$$\Delta_5 = -[1 \quad \frac{7}{4} \quad 0 \quad 0] \begin{bmatrix} 0 \\ 1 \\ 0 \\ 0 \end{bmatrix} = -\frac{7}{4}$$

Clearly, $\Delta_3 > 0 \implies$ solution is not optimal.

To find incoming vector:

$\because \quad \max\{\Delta_1, \Delta_2, \Delta_3, \Delta_4\} = \max\left\{\dfrac{-1}{2}, \dfrac{-19}{4}, \dfrac{15}{4}, \dfrac{-7}{4}\right\} = \dfrac{15}{4} = \Delta_3$

$\implies \qquad\qquad k = 3$

$\implies \quad \alpha_3^{(1)}$ is the incoming vector.

and

$$y_3^{(1)} = B_1^{-1}\alpha_3^{(1)} = \begin{bmatrix} 1 & 7/4 & 0 & 0 \\ 0 & 1/4 & 0 & 0 \\ 0 & 1/4 & 1 & 0 \\ 0 & -3/4 & 0 & 1 \end{bmatrix}\begin{bmatrix} -2 \\ -1 \\ 5 \\ -2 \end{bmatrix}$$

$$= \begin{bmatrix} -15/4 \\ -1/4 \\ 19/4 \\ -5/4 \end{bmatrix} = \begin{bmatrix} -\Delta_3 \\ y_{13} \\ y_{23} \\ y_{33} \end{bmatrix}$$

To find outgoing vector: Apply minimum ratio rule, we get

$$\frac{y_{B_r}}{y_{r3}} = \min_i \left\{\frac{y_{B_i}}{y_{i3}}, y_{i3} > 0\right\} = \min\left\{\frac{y_{B_1}}{y_{13}}, \frac{y_{B_2}}{y_{23}}, \frac{y_{B_3}}{y_{33}}\right\}$$

$$= \left\{\frac{31}{19/4}\right\} = \frac{y_{B_2}}{y_{23}}$$

$\implies \qquad\qquad r = 2$

$\implies \quad \beta_2^{(1)}(= \alpha_6^{(1)})$ is the outgoing vector.

$\implies \quad y_{23} = \dfrac{19}{4}$ is the key element.

Revised Simplex Table-3

Variables in the basis	solution $y_B^{(1)}$	B_1^{-1}				$y_1^{(1)} = B_1^{-1}\alpha_1^{(1)}$	Minimum Ratio
		e_1	$\beta_1^{(1)}\left(\alpha_4^{(1)}\right)$	$\beta_2^{(1)}\left(\alpha_3^{(1)}\right)$	$\beta_3^{(1)}\left(\alpha_7^{(1)}\right)$		
Z'	1130/19	1	37/19	15/19	0	–13/19	$\min\limits_{i}\left\{\dfrac{y_{B_i}}{y_{i1}}, y_{i1} > 0\right\}$
y_4	126/19	0	5/19	1/19	0	⑧/19 (y_{11})	63/4 (min) →
y_3	124/19	0	1/19	4/19	0	–6/19 (y_{21})	–ve
s_3	1599/19	0	–13/19	5/19	1	–17/19 (y_{31})	–ve

We observe from the above table that

$$y_4 = \frac{126}{19}, y_3 = \frac{124}{19}, s_3 = \frac{1599}{19} \text{ and Max } Z = \frac{1130}{19}$$

Now, we check the optimality of this solution.

New basis $B_1 = [e_1 \quad \alpha_4^{(1)} \quad \alpha_3^{(1)} \quad \alpha_7^{(1)}]$

$\Rightarrow \quad \alpha_1^{(1)}, \alpha_2^{(1)}, \alpha_5^{(1)}$ and $\alpha_6^{(1)}$ are not in B_1.

Using $\Delta_j = -(\text{first row of } B_1^{-1} \text{ of third table}) \times \alpha_j^{(1)}, j = 1, 2, 5, 6$,

we get

$$\Delta_1 = -\begin{bmatrix} 1 & \dfrac{37}{19} & \dfrac{15}{19} & 0 \end{bmatrix}\begin{bmatrix} -3 \\ 2 \\ -2 \\ 1 \end{bmatrix} = \frac{13}{19}$$

$$\Delta_2 = -\begin{bmatrix} 1 & \dfrac{37}{19} & \dfrac{15}{19} & 0 \end{bmatrix}\begin{bmatrix} -1 \\ 3 \\ 2 \\ 1 \end{bmatrix} = -\frac{122}{19}$$

$$\Delta_5 = -\begin{bmatrix} 1 & \dfrac{37}{19} & \dfrac{15}{19} & 0 \end{bmatrix}\begin{bmatrix} 0 \\ 1 \\ 0 \\ 0 \end{bmatrix} = -\frac{37}{19}$$

$$\Delta_6 = -\begin{bmatrix} 1 & \dfrac{37}{19} & \dfrac{15}{19} & 0 \end{bmatrix}\begin{bmatrix} 0 \\ 0 \\ 1 \\ 0 \end{bmatrix} = -\frac{15}{19}$$

$\because \quad \Delta_1 > 0 \quad \Rightarrow \quad$ solution is not optimal.

To find incoming vector:

$\because \quad \max \{\Delta_1, \Delta_2, \Delta_5, \Delta_6\} = \Delta_k$

$$\Rightarrow \quad \max\left\{\frac{13}{19}, \frac{-122}{19}, \frac{-37}{19}, \frac{-15}{19}\right\} = \frac{13}{19} = \Delta_1$$

$$\Rightarrow \quad k = 1$$

$$\Rightarrow \quad \alpha_1^{(1)} \text{ is incoming vector.}$$

Also, $\quad y_1^{(1)} = B_1^{-1}\alpha_1^{(1)} = \begin{bmatrix} 1 & 37/19 & 15/19 & 0 \\ 0 & 5/19 & 1/19 & 0 \\ 0 & 1/19 & 4/19 & 0 \\ 0 & -13/19 & 5/19 & 1 \end{bmatrix}\begin{bmatrix} -3 \\ 2 \\ -2 \\ 1 \end{bmatrix}$

$$= \begin{bmatrix} -13/19 \\ 8/19 \\ -6/19 \\ -17/19 \end{bmatrix} = \begin{bmatrix} -\Delta_1 \\ y_{11} \\ y_{21} \\ y_{31} \end{bmatrix}$$

To find outgoing vector:

$$\because \qquad \frac{y_{B_r}}{y_{r1}} = \min_i\left\{\frac{y_{B_i}}{y_{i1}}, y_{i1} > 0\right\} = \min\left\{\frac{y_{B_1}}{y_{11}}, \frac{y_{B_2}}{y_{21}}, \frac{y_{B_3}}{y_{31}}\right\}$$

$$= \min\left\{\frac{126/19}{8/19}\right\} \qquad\qquad \{\because y_{21}, y_{31} < 0\}$$

$$= \frac{126}{8} = \frac{y_{B_1}}{y_{11}}$$

$$\Rightarrow \qquad r = 1$$

$$\Rightarrow \quad \beta_1^{(1)} \text{ is outgoing vector.}$$

$$\Rightarrow \quad y_{11} \text{ is the key element.}$$

Further, we have to improved this basic feasible solution. Apply the usual procedure, we have the following revised simplex table.

Revised Simplex Table-4

Variables in the basis	solution $y_B^{(1)}$	B_1^{-1}					
		e_1	$\beta_1^{(1)}\left(\alpha_1^{(1)}\right)$	$\beta_2^{(1)}\left(\alpha_3^{(1)}\right)$	$\beta_3^{(1)}\left(\alpha_7^{(1)}\right)$		
Z'	281/4	1	19/8	7/8	0		
y_1	63/4	0	5/8	1/8	0		
y_3	23/2	0	1/4	1/4	0		
s_3	393/4	0	–1/8	3/8	1		

From the above table, we observed that the improved BFS is given by

$$y_1 = \frac{63}{4}, y_2 = 0, y_3 = \frac{23}{2}, y_4 = 0 \text{ and } \text{Max } Z' = \frac{281}{4}$$

Now, we check the optimality of this solution

$$\because \quad \text{New } \boldsymbol{B}_1 = [e_1 \quad \alpha_1^{(1)} \quad \alpha_3^{(1)} \quad \alpha_7^{(1)}]$$

$\Rightarrow \quad \alpha_2^{(1)}, \alpha_4^{(1)}, \alpha_5^{(1)}$ and $\alpha_6^{(1)}$ are not in the basis $\boldsymbol{B}_1$.

Then by using $\Delta_j = -(\text{first row of } \boldsymbol{B}_1^{-1} \text{ of fourth table}) \times \alpha_j^{(1)}, j = 2, 4, 5, 6$

We get $\qquad \Delta_2 = -\begin{bmatrix} 1 & \dfrac{19}{8} & \dfrac{7}{8} & 0 \end{bmatrix} \begin{bmatrix} -1 \\ 3 \\ 2 \\ 1 \end{bmatrix} = -\dfrac{63}{8}$

$$\Delta_4 = -\begin{bmatrix} 1 & \dfrac{19}{8} & \dfrac{7}{8} & 0 \end{bmatrix} \begin{bmatrix} -7 \\ 4 \\ -1 \\ 3 \end{bmatrix} = -\dfrac{13}{8}$$

$$\Delta_5 = -\begin{bmatrix} 1 & \dfrac{19}{8} & \dfrac{7}{8} & 0 \end{bmatrix} \begin{bmatrix} 0 \\ 1 \\ 0 \\ 0 \end{bmatrix} = -\dfrac{19}{8}$$

$$\Delta_6 = -\begin{bmatrix} 1 & \dfrac{19}{8} & \dfrac{7}{8} & 0 \end{bmatrix} \begin{bmatrix} 0 \\ 0 \\ 1 \\ 0 \end{bmatrix} = -\dfrac{7}{8}$$

$\Rightarrow \quad$ all $\Delta_j < 0, j = 2, 4, 5, 6$
$\Rightarrow \quad$ solution is optimal.

Hence, the final optimal solution of the given LPP is

$$x_1 = y_1 + 2 = \dfrac{63}{4} + 2 = \dfrac{71}{4}$$
$$x_2 = y_2 + 1 = 0 + 1 = 1$$
$$x_3 = y_3 + 3 = \dfrac{23}{2} + 3 = \dfrac{29}{2}$$
$$x_4 = y_4 + 4 = 0 + 4 = 4$$

and $\quad$ Max. $Z = \text{Max}.Z' + 41 = \dfrac{281}{4} + 41 = \dfrac{445}{4}$

6.4 REVISED SIMPLEX METHOD FOR STANDARD FORM-II

We have already discussed that in the given LPP if we used artificial variables to form the basis matrix then it is treated as standard form-II. Such type of problems can be solved by two phase method. Phase-1 deals with the removal of artificial variables and an IBFS is obtained while phase-2 deals with IBFS (obtained in phase-1) to obtain optimal solution.

PHASE-I (FOR REVISED SIMPLEX METHOD)

Let $A_1, A_2, ..., A_m$ be the artificial variables and Z_a be the artificial objective function defined by

$$\text{Max. } Z_a = -A_1 - A_2 - ... - A_m$$

If the given LPP is of maximization, cost -1 is assign to each variable, but if the problem is of minimization, then we assign the cost $+1$ to each artificial variable.

Here, we have two objective functions with m constraints, so we have to consider the problem in revised simplex form with $(m + 2)$ constraints.

Thus, we can write

$$\left.\begin{array}{r}Z - c_1 x_1 - c_2 x_2 - \ldots - c_n x_n + A_1 + A_2 + \ldots + A_m = 0 \\ Z_a + 0x_1 + 0x_2 + \ldots + 0x_n + A_1 + A_2 + \ldots + A_m = 0 \\ a_{11} x_1 + a_{12} x_2 + \ldots + a_{1n} x_n + A_1 = b_1 \\ a_{21} x_1 + a_{22} x_2 + \ldots + a_{2n} x_n + A_2 = b_2 \\ \ldots\ldots\ldots\ldots\ldots\ldots\ldots\ldots\ldots\ldots\ldots\ldots\ldots\ldots\ldots\ldots\ldots \\ \ldots\ldots\ldots\ldots\ldots\ldots\ldots\ldots\ldots\ldots\ldots\ldots\ldots\ldots\ldots\ldots\ldots \\ a_{m1} x_1 + a_{m2} x_2 + \ldots + a_{mn} x_n + A_m = b_m\end{array}\right] \quad \ldots(1)$$

and $\qquad x_i, A_j \geq 0$

Further, in phase-I the problem is to maximize Z_a first subject to the constraints (1) with Z_a and Z both unrestricted in sign.

Here, we have the following possibilities:

(i) **If Max. $Z_a = 0$**, then $A_1 = A_2 = \ldots = A_m = 0$, therefore the values of $x_1, x_2, x_3, \ldots, x_n$ will give the IBFS of phase I.

(ii) **If Max. $Z_a < 0$**, then at least one artificial variable has a non-negative value. In this case, the original problem has no feasible solution.

PHASE-2

After removing all the artificial variables we get a basic feasible solution of the problem. In phase-2, we proceed to get the optimal solution by using exactly the same procedure as in revised simplex method in standard form-I.

Here, we use the following steps:

(1) Formation of Basis and its inverse:

The above system of equations (1) can be written as:

$$\begin{bmatrix} 1 & 0 & -c_1 & -c_2 & \cdots & -c_n & 0 & 0 & \cdots & 0 \\ 0 & 1 & 0 & 0 & \cdots & 0 & 1 & 1 & \cdots & 1 \\ 0 & 0 & a_{11} & a_{12} & \cdots & a_{1n} & 1 & 0 & \cdots & 0 \\ 0 & 0 & a_{21} & a_{22} & \cdots & a_{2n} & 0 & 1 & \cdots & 0 \\ \vdots & \vdots & \vdots & \vdots & & \vdots & & & & \vdots \\ 0 & 0 & a_{m1} & a_{m2} & \cdots & a_{mn} & 0 & 0 & \cdots & 1 \end{bmatrix} \begin{bmatrix} Z \\ Z_a \\ x_1 \\ x_2 \\ \vdots \\ x_n \\ A_1 \\ A_2 \\ \vdots \\ A_m \end{bmatrix} = \begin{bmatrix} 0 \\ 0 \\ b_1 \\ b_2 \\ \vdots \\ b_m \end{bmatrix} \quad \ldots(2)$$

or $\quad [e_1^{(2)} \quad e_2^{(2)} \quad \alpha_1^{(2)} \quad \alpha_2^{(2)} \quad \cdots \quad \alpha_n^{(2)} \quad \alpha_{n+1}^{(2)} \quad \alpha_{n+2}^{(2)} \quad \cdots \quad \alpha_{n+m}^{(2)}] x_B^{(2)} = b^{(2)}$

where $e_1^{(2)} = \begin{bmatrix} 1 \\ 0 \\ 0 \\ \vdots \\ 0 \end{bmatrix}_{(m+2)\times 1}$, $e_2^{(2)} = \begin{bmatrix} 0 \\ 1 \\ 0 \\ 0 \\ \vdots \\ 0 \end{bmatrix}_{(m+2)\times 1}$

and $\alpha_j^{(2)} = \begin{bmatrix} -c_j \\ 0 \\ a_{1j} \\ a_{2j} \\ \vdots \\ a_{mj} \end{bmatrix} = \begin{bmatrix} -c_j \\ 0 \\ \alpha_j \end{bmatrix}$ for $j = 1, 2, 3, \ldots, n$

where $\alpha_j = \begin{bmatrix} a_{1j} \\ a_{2j} \\ \vdots \\ a_{mj} \end{bmatrix}$ and $\alpha_j^{(2)} = \begin{bmatrix} 0 \\ 1 \\ 1 \\ 0 \\ \vdots \\ 0 \end{bmatrix} = \begin{bmatrix} 0 \\ 1 \\ e_j \end{bmatrix}$ for $j = n + 1, n + 2, \ldots, n + m$

Here, e_j is the unit vector and $\alpha_j^{(2)}$ denotes the columns of the coefficients of artificial variables $A_1, A_2, \ldots, A_m$.

and $x_B^{(2)} = \begin{bmatrix} Z \\ Z_a \\ x_1 \\ x_2 \\ \vdots \\ x_n \\ A_1 \\ A_2 \\ \vdots \\ A_m \end{bmatrix}$, $b^{(2)} = \begin{bmatrix} 0 \\ 0 \\ b_1 \\ b_2 \\ \vdots \\ b_m \end{bmatrix}$

and $e_1^{(2)}, e_2^{(2)}$ are the $(m + 2)$ component unit vector corresponding to the objective function Z and Z_a respectively.

(2) To find basis B_2:

Since, there are $(m + 2)$ constraints, so $\textbf{\textit{B}}_2$ must be of order $(m + 2) \times (m + 2)$. such that

$$B_2 = \begin{bmatrix} 1 & 0 & 0 & 0 & \cdots & 0 \\ 0 & 1 & 1 & 1 & \cdots & 1 \\ 0 & 0 & 1 & 0 & & 0 \\ 0 & 0 & 0 & 1 & & 0 \\ \vdots & \vdots & \vdots & \vdots & & \vdots \\ 0 & 0 & 0 & 0 & & 1 \end{bmatrix}_{(m+2)\times(m+2)}$$

$$= [e_1^{(2)}, e_2^{(2)}, \alpha_{n+1}^{(2)}, \alpha_{n+2}^{(2)}, \ldots \alpha_{n+m}^{(2)}]$$

$$= [e_1^{(2)}, e_2^{(2)}, \beta_1^{(2)}, \beta_2^{(2)} \ldots \beta_m^{(2)}]$$

where $\beta_j^{(2)}$ are any of $\alpha_j^{(2)}$ which are linearly independent.

Also, $\quad B_2 = \begin{bmatrix} 1 & 0 & -C_B \\ 0 & 1 & -C_{B_a} \\ \hline O & O & B \end{bmatrix} = \begin{bmatrix} 1 & 0 & -C_B \\ 0 & 1 & -C_{B_a} \\ \hline O & & B \end{bmatrix}$

where $\quad -C_B = [0 \quad 0 \quad \cdots \quad 0]_{1 \times m}$

$\qquad -C_{B_a} = [1 \quad 1 \quad \cdots \quad 1]_{1 \times m}$

$$O = \begin{bmatrix} 0 & 0 \\ 0 & 0 \\ \vdots & \vdots \\ 0 & 0 \end{bmatrix}_{m \times 2} \text{ and } B = \begin{bmatrix} 1 & 0 & 0 & \cdots & 0 \\ 0 & 1 & 0 & \cdots & 0 \\ \vdots & & & \\ 0 & 0 & 0 & \cdots & 1 \end{bmatrix} = I_m$$

Also, $\quad C_B = [C_{B_1} \quad C_{B_2} \quad \cdots \quad C_{B_m}]$ and $C_{B_a} = [C_{B_{a_1}} \quad C_{B_{a_2}} \quad \cdots \quad C_{B_{a_m}}]$

Here, c_j are the coefficient of the basic variables $x_1, x_2, \ldots, x_n$ in the given objective function Z and $C_{B_{a_j}}$ denote the coefficients of the artificial variables in artificial objective function Z_a.

(3) To find B_2^{-1} :

We have $\qquad B_2 = \begin{bmatrix} 1 & 0 & -C_B \\ 0 & 1 & -C_{B_a} \\ \hline O & & B \end{bmatrix}$

which can be written as

$$B_2 = \begin{bmatrix} I_2 & -C_B^{(2)} \\ O & B \end{bmatrix}$$

where $\qquad I_2 = \begin{bmatrix} 1 & 0 \\ 0 & 1 \end{bmatrix}, C_B^{(2)} = \begin{bmatrix} C_B \\ C_{B_a} \end{bmatrix}$

Clearly, $\qquad B_2^{-1} = \begin{bmatrix} I_2 & C_B^{(2)} B^{-1} \\ O & B^{-1} \end{bmatrix} = \begin{bmatrix} I_2 & C_B^{(2)} \\ O & I_m \end{bmatrix} \qquad [\because B = I_m^{-1} = I_m]$

(4) To find $x_j^{(2)}$ corresponding $\alpha_j^{(2)}$ not in the basis B_2:

Since $\alpha_j^{(2)}$ is not in the basis so it can be written as the linear combination of the components of B_2, so

$$\alpha_j^{(2)} = x_{0j}e_1^{(2)} + x'_{0j}e_2^{(2)} + x_{1j}\beta_1^{(2)} + x_{2j}\beta_2^{(2)} + \ldots + x_{mj}\beta_m^{(2)}$$

$$= [e_1^{(2)} \quad e_2^{(2)} \quad \beta_1^{(2)} \quad \beta_2^{(2)} \quad \cdots \quad \beta_m^{(2)}] \begin{bmatrix} x_{0j} \\ x'_{0j} \\ x_{1j} \\ x_{2j} \\ \vdots \\ x_{mj} \end{bmatrix}$$

$\Rightarrow \qquad \alpha_j^{(2)} = B_2 x_j^{(2)}$ where $x_j^{(2)} = \begin{bmatrix} x_{0j} \\ x'_{0j} \\ x_{1j} \\ x_{2j} \\ \vdots \\ x_{mj} \end{bmatrix}$

(or) $\qquad x_j^{(2)} = B_2^{-1}\alpha_j^{(2)} = \begin{bmatrix} I_2 & C_B^{(2)}B^{-1} \\ O & B^{-1} \end{bmatrix}\begin{bmatrix} -c_j \\ 0 \\ \alpha_j \end{bmatrix}$ for $j = 1, 2, 3, \ldots, n$

$$= \begin{bmatrix} 1 & 0 & C_B B^{-1} \\ 0 & 1 & C_{B_a} \cdot B^{-1} \\ O & O & B^{-1} \end{bmatrix}\begin{bmatrix} -c_j \\ 0 \\ \alpha_j \end{bmatrix} = \begin{bmatrix} -c_j + C_B B^{-1}\alpha_j \\ 0 + C_{B_a} B^{-1}\alpha_j \\ B^{-1}\alpha_j \end{bmatrix} = \begin{bmatrix} -c_j + C_B x_j \\ C_{B_a} x_j \\ B^{-1}\alpha_j \end{bmatrix}$$

$$x_j^{(2)} = \begin{bmatrix} z_j - c_j \\ z_{ja} - c_{ja} \\ x_j \end{bmatrix} \begin{bmatrix} -\Delta_j \\ -\Delta_{ja} \\ x_j \end{bmatrix}$$

Therefore,

(i) $\Delta_j = -(\text{first row of } B_2^{-1}) \times \alpha_j^{(2)}$

(ii) $\Delta_{j_a} = -(\text{second row of } B_2^{-1}) \times \alpha_j^{(2)}$

(iii) x_j is obtained when the last m rows of B_2^{-1} are multiplied with $\alpha_j^{(2)}$.

(5) To find initial BFS $x_B^{(2)}$:

Here we have

$$x_B^{(2)} = B_2^{-1} b^{(2)} = \begin{bmatrix} 1 & 0 & C_B B^{-1} \\ 0 & 1 & C_{B_a} \cdot B^{-1} \\ O & O & B^{-1} \end{bmatrix} \begin{bmatrix} 0 \\ 0 \\ b \end{bmatrix}$$

$$= \begin{bmatrix} C_B B^{-1} b \\ C_{B_a} B^{-1} b \\ B^{-1} b \end{bmatrix} = \begin{bmatrix} C_B x_B \\ C_{B_a} x_b \\ x_B \end{bmatrix} = \begin{bmatrix} Z \\ Z_a \\ x_B \end{bmatrix} \qquad [\because x_B = B^{-1} b]$$

Therefore,

(i) $Z = $ (first row of B_2^{-1}) $\times b^{(2)}$

(ii) $Z_a = $ (second row of B_2^{-1}) $\times b^{(2)}$

(iii) $X_B = $ (last m rows of B_2^{-1}) $\times b^{(2)}$

 ## Working Procedure

For Phase-1

STEP 1. Write the given LPP in revised simplex method standard form-II.

STEP 2. Form initial basis matrix B_2 and its inverse B_2^{-1}.

STEP 3. Using $x_B^{(2)} = $ (last m rows of B_2^{-1}) $\times b^{(2)} = \begin{bmatrix} 0 \\ 0 \\ x_B \end{bmatrix}$

Find the value of $x_B^{(2)}$.

Here, all $x_i \geq 0$ but no restriction on Z and Z_a.

STEP 4. Construct the initial revised simplex table for standard form-II in the following form.

Variables in the basis	solution $x_B^{(2)}$	B_2^{-1}						$x_k^{(2)} = B_2^{-1} \alpha_k^{(2)}$	Min. Ratio $\min\limits_{i} \left\{ \dfrac{x_{B_i}}{x_{ik}}, x_{ik} > 0 \right\}$
		$e_1^{(2)}$	$e_2^{(2)}$	$\beta_1^{(2)}$	$\beta_2^{(2)}$	...	$\beta_m^{(2)}$		
Z	0	1	0	...	...	...		$-\Delta_k$	
Z_a	0	0	1	...	...	...		$-\Delta_{k_a}$	
s_1	x_{B_1}	0	0						
s_2	x_{B_2}	0	0						
$\vdots$	$\vdots$	$\vdots$	$\vdots$						
s_m	x_{B_m}	0	0						

STEP 5. Maximize Z_a by using the following results.

    (i) If $Z_a = 0$ then we take $A_1 = 0, A_2 = 0, ..., A_m = 0$ and use phase II.

    (ii) If $Z_a < 0$ then calculate Δ_{ja} for all $\alpha_j^{(2)}$ not in the basis $\mathbf{B}_2$ by using the following formula

$$\Delta_{ja} = -(\text{second row of } \mathbf{B}_2^{-1}) \times \alpha_j^{(2)}$$

Here,

    (i) if all $\Delta_{ja} \leq 0$, then Z_a is maximum and no feasible solution exists.

    (ii) if at least one $\Delta_{ja} > 0$, then go to the next step.

STEP 6. **To find incoming and outgoing vectors**

If $\Delta_{ka} = \max_j\{\Delta_{ja}\} \; \forall j$ such that $\alpha_j^{(2)} \notin \mathbf{B}_2$. Then $\alpha_k^{(2)}$ will be the incoming vector.

Now calculate $x_k^{(2)} = \mathbf{B}_2^{-1}\alpha_k^{(2)}$

Then find outgoing vector by using the following formula

$$\frac{x_{B_r}}{x_{rk}} = \min_i\left\{\frac{x_{B_i}}{x_{ik}}, x_{ik} > 0\right\}$$

If minimum occur at $i = r$ then $\beta_r^{(2)}$ will be the outgoing vector.

STEP 7. Bringing $\alpha_k^{(2)}$ in place of $\beta_r^{(2)}$ to get next revised simplex table. Repeat step (1) to (7) till we get Max. $Z_a = 0 \; \forall \Delta_{ja} \leq 0$.

For Phase-2

Here, we have to maximize Z by removing Z_a row and $e_2^{(2)}$ column from the last revised simplex table of phase 1. During this process, the basis matrix $\mathbf{B}_2$ of order $(m + 2) \times (m + 2)$ is now reduced to basis matrix $\mathbf{B}_1$ of order $(m + 1) \times (m + 1)$.

Then proceed exactly the same way as in revised simplex method of standard form I.

 Solved Examples

EXAMPLE 1. *Solve the following LPP by revised simplex method:*

$$\textbf{Max. } Z = x_1 + 2x_2 + 3x_3 - x_4$$

subject to the constraints

$$x_1 + 2x_2 + 3x_3 = 15$$
$$2x_1 + x_2 + 5x_3 = 20$$
$$x_1 + 2x_2 + x_3 + x_4 = 10$$

and $x_1, x_2, x_3, x_4 \geq 0$

SOLUTION. Using artificial variables, the constraints of given LPP can be written as

$$x_1 + 2x_2 + 3x_3 + A_1 = 15$$
$$2x_1 + x_2 + 5x_3 + A_2 = 20$$
$$x_1 + 2x_2 + x_3 + x_4 + A_3 = 10$$

and $x_1, x_2, x_3, x_4, A_1, A_2, A_3 \geq 0$

Now the given LPP can be written in revised simplex standard form II as follows:

$$\left.\begin{array}{l} Z - x_1 - 2x_2 - 3x_3 + x_4 + 0A_1 + 0A_2 + 0A_3 = 0 \\ Z_a + 0x_1 + 0x_2 + 0x_3 + 0x_4 + A_1 + A_2 + A_3 = 0 \\ x_1 + 2x_2 + 3x_3 + A_1 = 15 \\ 2x_1 + x_2 + 5x_3 + A_2 = 20 \\ x_1 + 2x_2 + x_3 + x_4 + A_3 = 10 \end{array}\right] \qquad \ldots(1)$$

Phase 1: We can write the above system (1) in the following form.

$$
\begin{array}{ccccccccc}
e_1^{(2)} & e_2^{(2)} & \alpha_1^{(2)} & \alpha_2^{(2)} & \alpha_3^{(2)} & \alpha_4^{(2)} & \alpha_5^{(2)} & \alpha_6^{(2)} & \alpha_7^{(2)}
\end{array}
$$

$$
\begin{bmatrix}
1 & 0 & -1 & -2 & -3 & 1 & 0 & 0 & 0 \\
0 & 1 & 0 & 0 & 0 & 0 & 1 & 1 & 1 \\
0 & 0 & 1 & 2 & 3 & 0 & 1 & 0 & 0 \\
0 & 0 & 2 & 1 & 5 & 0 & 0 & 1 & 0 \\
0 & 0 & 1 & 2 & 1 & 1 & 0 & 0 & 1
\end{bmatrix}
\begin{bmatrix} Z \\ Z_a \\ x_1 \\ x_2 \\ x_3 \\ x_4 \\ A_1 \\ A_3 \\ A_3 \end{bmatrix}
=
\begin{bmatrix} 0 \\ 0 \\ 15 \\ 20 \\ 10 \end{bmatrix}
$$

with $x_B^{(2)}$ column and $b^{(2)}$ column.

Therefore, $\quad B_2 = \begin{bmatrix} 1 & 0 & 0 & 0 & 0 \\ 0 & 1 & 1 & 1 & 1 \\ 0 & 0 & 1 & 0 & 0 \\ 0 & 0 & 0 & 1 & 0 \\ 0 & 0 & 0 & 0 & 1 \end{bmatrix} = \begin{bmatrix} e_1^{(2)} & e_2^{(2)} & \alpha_5^{(2)} & \alpha_6^{(2)} & \alpha_7^{(2)} \end{bmatrix}$

$$= [e_1^{(2)} \quad e_2^{(2)} \quad \beta_1^{(2)} \quad \beta_2^{(2)} \quad \beta_3^{(2)}]$$

where $\alpha_5^{(2)} = \beta_1^{(2)}, \alpha_6^{(2)} = \beta_2^{(2)}, \alpha_7^{(2)} = \beta_3^{(2)}$

$\Rightarrow \quad B_2 = \begin{bmatrix} 1 & 0 & -C_B \\ 0 & 1 & -C_{B_a} \\ O & O & B \end{bmatrix}$ where $C_B = [0 \quad 0 \quad 0], C_{B_a} = [-1 \quad -1 \quad -1]$

and $\quad B = \begin{bmatrix} 1 & 0 & 0 \\ 0 & 1 & 0 \\ 0 & 0 & 1 \end{bmatrix} = I_3$

So, $\quad B^{-1} = I_3^{-1} = I_3 = \begin{bmatrix} 1 & 0 & C_B B^{-1} \\ 0 & 1 & C_{B_a} B^{-1} \\ O & O & B^{-1} \end{bmatrix} = \begin{bmatrix} 1 & 0 & C_B \\ 0 & 1 & C_{B_a} \\ O & O & I_3 \end{bmatrix}$

$$= \begin{bmatrix} 1 & 0 & 0 & 0 & 0 \\ 0 & 1 & -1 & -1 & -1 \\ 0 & 0 & 1 & 0 & 0 \\ 0 & 0 & 0 & 1 & 0 \\ 0 & 0 & 0 & 0 & 1 \end{bmatrix}$$

The initial solution $x_B^{(2)}$ is given by

$$x_B^{(2)} = \boldsymbol{B}_2^{-1} \cdot \boldsymbol{b}^{(2)} = \begin{bmatrix} 1 & 0 & 0 & 0 & 0 \\ 0 & 1 & -1 & -1 & -1 \\ 0 & 0 & 1 & 0 & 0 \\ 0 & 0 & 0 & 1 & 0 \\ 0 & 0 & 0 & 0 & 1 \end{bmatrix} \begin{bmatrix} 0 \\ 0 \\ 15 \\ 20 \\ 10 \end{bmatrix} = \begin{bmatrix} 0 \\ -45 \\ 15 \\ 20 \\ 10 \end{bmatrix}$$

Then we have the following table:

Revised Simplex Table-1

Variables in the basis	solution $x_B^{(2)}$	$e_1^{(2)}$	$e_2^{(2)}$	B_2^{-1} $\beta_1^{(2)}$ $(\alpha_5^{(2)})$	$\beta_2^{(2)}$ $(\alpha_6^{(2)})$	$\beta_3^{(2)}$ $(\alpha_7^{(2)})$	$x_3^{(2)} = B_2^{-1}\alpha_3^{(2)}$	Min. Ratio $\left\{ \dfrac{x_{B_i}}{x_{i3}}, x_{i3} > 0 \right\}$
Z	0	1	0	0	0	0	-3	
Z_a	-45	0	1	-1	-1	-1	-9	
A_1	15 (x_{B_1})	0	0	1	0	0	3 (x_{13})	$15/3 = 5$
A_2	20 (x_{B_2})	0	0	0	①	0	5 (x_{23})	$20/5 = 4$ (min) $\rightarrow$
A_3	10 (x_{B_3})	0	0	0	0	1	1 (x_{33})	$10/1 = 10$
				$\downarrow$				

Clearly, from the above table

Max. $Z_a = -45 < 0$, so we have to find $\Delta_{ja} \ \forall \ j$ for which $\alpha_j^{(2)}$ are not in the basis $\boldsymbol{B}_2$.

Since $\quad \boldsymbol{B}_2 = [e_1^{(2)} \ \ e_2^{(2)} \ \ \alpha_5^{(2)} \ \ \alpha_6^{(2)} \ \ \alpha_7^{(2)}]$

$\alpha_j^{(2)} \notin \boldsymbol{B}_2, j = 1, 2, 3, 4.$

Now, using $\quad \Delta_{ja} = -(\text{second row of } \boldsymbol{B}_2^{-1}) \times \alpha_j^{(2)}, j = 1, 2, 3, 4$ we get

$$\Delta_{1a} = -[0 \ \ 1 \ \ -1 \ \ -1 \ \ -1] \begin{bmatrix} -1 \\ 0 \\ 1 \\ 2 \\ 1 \end{bmatrix} = 4$$

$$\Delta_{2a} = -[0 \quad 1 \quad -1 \quad -1 \quad -1] \begin{bmatrix} -2 \\ 0 \\ 2 \\ 1 \\ 2 \end{bmatrix} = 5$$

$$\Delta_{3a} = -[0 \quad 1 \quad -1 \quad -1 \quad -1] \begin{bmatrix} -3 \\ 0 \\ 3 \\ 5 \\ 1 \end{bmatrix} = 9$$

$$\Delta_{4a} = -[0 \quad 1 \quad -1 \quad -1 \quad -1] \begin{bmatrix} 1 \\ 0 \\ 0 \\ 0 \\ 1 \end{bmatrix} = 1$$

Clearly $\Delta_{1a}, \Delta_{2a}, \Delta_{3a}, \Delta_{4a} > 0$. So, we have to find the improved value of Z_a.

To find incoming vector

$$\Delta_{ka} = \max_j \{\Delta_{ja}\}, j = 1, 2, 3, 4$$

$$= \max\{\Delta_{1a}, \Delta_{2a}, \Delta_{3a}, \Delta_{4a}\} = \max\{4, 5, 9, 1\} = 9 = \Delta_{3a}$$

$\Rightarrow \quad k = 3$

$\Rightarrow \quad \alpha_3^{(2)}$ is the incoming vector.

Now,
$$x_3^{(2)} = B_2^{-1}\alpha_3^{(2)} = \begin{bmatrix} 1 & 0 & 0 & 0 & 0 \\ 0 & 1 & -1 & -1 & -1 \\ 0 & 0 & 1 & 0 & 0 \\ 0 & 0 & 0 & 1 & 0 \\ 0 & 0 & 0 & 0 & 1 \end{bmatrix} \begin{bmatrix} -3 \\ 0 \\ 3 \\ 5 \\ 1 \end{bmatrix} = \begin{bmatrix} -3 \\ -9 \\ 3 \\ 5 \\ 1 \end{bmatrix} = \begin{bmatrix} -\Delta_3 \\ -\Delta_{3a} \\ x_{13} \\ x_{23} \\ x_{33} \end{bmatrix}$$

To find outgoing vector:

Using $\dfrac{x_{B_r}}{x_{rk}} = \min_i \left\{ \dfrac{x_{B_i}}{x_{ik}}, x_{ik} > 0 \right\}$ we get

$$\frac{x_{B_r}}{x_{r3}} = \min_i \left\{ \frac{x_{B_i}}{x_{i3}}, x_{i3} > 0 \right\} = \min \left\{ \frac{x_{B_1}}{x_{13}}, \frac{x_{B_2}}{x_{23}}, \frac{x_{B_3}}{x_{33}} \right\}$$

$$= \min \left\{ \frac{15}{3}, \frac{20}{5}, \frac{10}{1} \right\} = \min\{5, 4, 10\} = 4 = \frac{x_{B_2}}{x_{23}}$$

$\Rightarrow \quad r = 2$

So, $\beta_2^{(2)} (= \alpha_6^{(2)})$ is the outgoing vector.

Therefore, key element $= x_{23} = 5$

Now, to improved the value of Z_a, following the usual procedure we get the following second simplex table.

Revised Simplex Table-2

Variables in the basis	solution $x_B^{(2)}$	$e_1^{(2)}$	$e_2^{(2)}$	B_2^{-1} $\beta_1^{(2)}$ $(\alpha_5^{(2)})$	$\beta_2^{(2)}$ $(\alpha_3^{(2)})$	$\beta_3^{(2)}$ $(\alpha_7^{(2)})$	$x_2^{(2)} = B_2^{-1}\alpha_2^{(2)}$	Min. Ratio $\left\{\dfrac{x_{B_i}}{x_{i2}}, x_{i2} > 0\right\}$
Z	12	1	0	0	3/5	0	−7/5	
Z_a	− 9	0	1	−1	4/5	−1	−16/5	
A_1	3 (x_{B_1})	0	0	1	−3/5	0	⑦/5 (x_{12})	15/7 (min) →
A_2	4 (x_{B_2})	0	0	0	1/5	0	1/5 (x_{22})	20
A_3	6 (x_{B_3})	0	0	0	−1/5	1	9/5 (x_{32})	10/3

From above table, we observe that

$$\text{Max. } Z_a = -9 < 0$$

So, we find Δ_{ja} for those j for which $\alpha_j^{(2)} \notin B_2$

$\therefore$ New B_2 is given by

$$B_2 = [e_1^{(2)} \quad e_2^{(2)} \quad \alpha_5^{(2)} \quad \alpha_3^{(2)} \quad \alpha_7^{(2)}]$$

$$\Rightarrow \quad \alpha_1^{(2)}, \alpha_2^{(2)}, \alpha_4^{(2)}, \alpha_6^{(2)} \notin B_2$$

Then using $\Delta_{ja} = -$ (second row of B_2^{-1} of second table) $\times \alpha_j^{(2)}, j = 1, 2, 4, 6$

$$\Delta_{1a} = -\begin{bmatrix} 0 & 1 & -1 & \dfrac{4}{5} & -1 \end{bmatrix} \begin{bmatrix} -1 \\ 0 \\ 1 \\ 2 \\ 1 \end{bmatrix} = \dfrac{2}{5}$$

$$\Delta_{2a} = -\begin{bmatrix} 0 & 1 & -1 & \dfrac{4}{5} & -1 \end{bmatrix} \begin{bmatrix} -2 \\ 0 \\ 2 \\ 1 \\ 2 \end{bmatrix} = \dfrac{16}{5}$$

$$\Delta_{4a} = -\begin{bmatrix} 0 & 1 & -1 & \dfrac{4}{5} & -1 \end{bmatrix} \begin{bmatrix} 1 \\ 0 \\ 0 \\ 0 \\ 1 \end{bmatrix} = 1$$

$$\Delta_{6a} = -\begin{bmatrix} 0 & 1 & -1 & \dfrac{4}{5} & -1 \end{bmatrix} \begin{bmatrix} 0 \\ 1 \\ 0 \\ 1 \\ 0 \end{bmatrix} = -\dfrac{9}{5}$$

Clearly, $\Delta_{1a}, \Delta_{2a}, \Delta_{4a} > 0$

So to improve the value of Z_a, we proceed as follows.

To find incoming vector:

$$\Delta_{ka} = \max_{j}\{\Delta_{ja}\}, j = 1, 2, 4, 6$$

$$= \max\{\Delta_{1a}, \Delta_{2a}, \Delta_{4a}, \Delta_{6a}\} = \max\left\{\frac{2}{5}, \frac{16}{5}, 1, -\frac{9}{5}\right\}$$

$$= \frac{16}{5} = \Delta_{2a}$$

$$\Rightarrow \qquad k = 2$$

$$\Rightarrow \qquad \alpha_2^{(2)} \text{ is the incoming vector.}$$

Now, $\quad x_2^{(2)} = B_2^{-1}\alpha_2^{(2)} = \begin{bmatrix} 1 & 0 & 0 & 3/5 & 0 \\ 0 & 1 & -1 & 4/5 & -1 \\ 0 & 0 & 1 & -3/5 & 0 \\ 0 & 0 & 0 & 1/5 & 0 \\ 0 & 0 & 0 & -1/5 & 1 \end{bmatrix}\begin{bmatrix} -2 \\ 0 \\ 2 \\ 1 \\ 2 \end{bmatrix} = \begin{bmatrix} -7/5 \\ -16/5 \\ 7/5 \\ 1/5 \\ 9/5 \end{bmatrix} = \begin{bmatrix} -\Delta_2 \\ -\Delta_{2a} \\ x_{12} \\ x_{22} \\ x_{32} \end{bmatrix}$

To find outgoing vector:

Using $\dfrac{x_{B_r}}{x_{rk}} = \min_{i}\left\{\dfrac{x_{B_i}}{x_{ik}}, x_{ik} > 0\right\}$ we get

$$\frac{x_{B_r}}{x_{r2}} = \min_{i}\left\{\frac{x_{B_i}}{x_{i2}}, x_{i2} > 0\right\} = \min\left\{\frac{x_{B_1}}{x_{12}}, \frac{x_{B_2}}{x_{22}}, \frac{x_{B_3}}{x_{32}}\right\}$$

$$= \min\left\{\frac{3}{7/5}, \frac{4}{1/5}, \frac{6}{9/5}\right\} = \min\left\{\frac{15}{7}, 20, \frac{30}{9}\right\} = \frac{15}{7} = \frac{x_{B_1}}{x_{12}}$$

$$\Rightarrow \qquad r = 1$$

$$\Rightarrow \qquad \beta_1^{(1)}(= \alpha_5^{(2)}) \text{ is the outgoing vector.}$$

and hence, key element $= x_{12} = \dfrac{7}{5}$

Now apply the same procedure we get the following simplex table-3.

Revised Simplex Table-3

Variables in the basis	solution $x_B^{(2)}$	B_2^{-1}					$x_4^{(2)} = B_2^{-1}\alpha_2^{(2)}$	Min. Ratio $\min_{i}\left\{\dfrac{x_{B_i}}{x_{i4}}, x_{i4} > 0\right\}$
		$e_1^{(2)}$	$e_2^{(2)}$	$\beta_1^{(2)}$ $(\alpha_2^{(2)})$	$\beta_2^{(2)}$ $(\alpha_3^{(2)})$	$\beta_3^{(2)}$ $(\alpha_7^{(2)})$		
Z	15	1	0	1	0	1	1	
Z_a	$-15/17$	0	1	9/7	$-4/7$	-1	-1	
x_2	15/7 (x_{B_1})	0	0	5/7	$-3/7$	0	0 (x_{14})	—
x_3	25/7 (x_{B_2})	0	0	$-1/7$	2/7	0	0 (x_{24})	—
A_3	15/7 (x_{B_3})	0	0	$-9/7$	4/7	1	①(x_{34})	15/7 (min) →
							↑	

From the above table, we observe that

$$\text{Max.}\, Z_a = -\frac{15}{7} < 0$$

So, we again find Δ_{ja} for all j for which $\alpha_j^{(2)} \notin \boldsymbol{B}_2$.

Thus, new $\boldsymbol{B}_2$ is given by

$$\boldsymbol{B}_2 = [e_1^{(2)} \quad e_2^{(2)} \quad \alpha_2^{(2)} \quad \alpha_3^{(2)} \quad \alpha_7^{(2)}]$$

$$\Rightarrow \qquad \alpha_1^{(2)}, \alpha_4^{(2)}, \alpha_5^{(2)}, \alpha_6^{(2)} \notin \boldsymbol{B}_2$$

Now, using $\Delta_{ja} = -$ (second row of $\boldsymbol{B}_2^{-1}$ of third table) $\times\, \alpha_j^{(2)}$, for $j = 1, 4, 5, 6$, we get

$$\Delta_{1a} = -\begin{bmatrix} 0 & 1 & \dfrac{9}{7} & -\dfrac{4}{7} & -1 \end{bmatrix} \begin{bmatrix} -1 \\ 0 \\ 1 \\ 2 \\ 1 \end{bmatrix} = \frac{6}{7}; \quad \Delta_{4a} = -\begin{bmatrix} 0 & 1 & \dfrac{9}{7} & -\dfrac{4}{7} & -1 \end{bmatrix} \begin{bmatrix} 1 \\ 0 \\ 0 \\ 0 \\ 1 \end{bmatrix} = 1$$

$$\Delta_{5a} = -\begin{bmatrix} 0 & 1 & \dfrac{9}{7} & -\dfrac{4}{7} & -1 \end{bmatrix} \begin{bmatrix} 0 \\ 1 \\ 1 \\ 0 \\ 0 \end{bmatrix} = -\frac{16}{7}; \quad \Delta_{6a} = -\begin{bmatrix} 0 & 1 & \dfrac{9}{7} & -\dfrac{4}{7} & -1 \end{bmatrix} \begin{bmatrix} 0 \\ 1 \\ 0 \\ 1 \\ 0 \end{bmatrix} = -\frac{3}{7}$$

Clearly, $\Delta_{1a}, \Delta_{4a} > 0 \Rightarrow$ we can improve the value of Z_a.

To find incoming vector:

$$\Delta_{ka} = \max_j \{\Delta_{ja}\} \text{ for } j = 1, 4, 5, 6$$

$$= \max\{\Delta_{1a}, \Delta_{4a}, \Delta_{5a}, \Delta_{6a}\}$$

$$= \max\left\{\frac{6}{7}, 1, \frac{-16}{7}, \frac{-3}{7}\right\} = 1 = \Delta_{4a}$$

$$\Rightarrow \qquad k = 4$$

$$\Rightarrow \qquad \alpha_4^{(2)} \text{ is the incoming vector.}$$

Now $\qquad x_4^{(2)} = \boldsymbol{B}_2^{-1}\alpha_4^{(2)} = \begin{bmatrix} 1 & 0 & 1 & 0 & 1 \\ 0 & 1 & 9/7 & 4/7 & -1 \\ 0 & 0 & 5/7 & -3/7 & 0 \\ 0 & 0 & -1/7 & 2/7 & 0 \\ 0 & 0 & -9/7 & 4/7 & 1 \end{bmatrix} \begin{bmatrix} 1 \\ 0 \\ 0 \\ 0 \\ 1 \end{bmatrix} = \begin{bmatrix} 1 \\ -1 \\ 0 \\ 0 \\ 1 \end{bmatrix} = \begin{bmatrix} -\Delta_4 \\ -\Delta_{4a} \\ x_{14} \\ x_{24} \\ x_{34} \end{bmatrix}$

To find outgoing vector:

Using $\qquad \dfrac{x_{B_r}}{x_{rk}} = \min_i\left\{\dfrac{x_{B_i}}{x_{ik}}, x_{ik} > 0\right\}$, we get $\dfrac{x_{B_r}}{x_{r4}} = \min_i\left\{\dfrac{x_{B_i}}{x_{i4}}, x_{i4} > 0\right\}$

$$= \min \left\{ \frac{x_{B_1}}{x_{14}}, \frac{x_{B_2}}{x_{24}}, \frac{x_{B_3}}{x_{34}} \right\}$$

$$= \min \left\{ \frac{x_{B_3}}{x_{34}} \right\} \qquad (\because x_{14} = x_{24} = 0)$$

$$= \frac{x_{B_3}}{x_{34}}$$

$\Rightarrow \qquad r = 3$

$\therefore \qquad \beta_3^{(2)}(= \alpha_7^{(2)})$ is the outgoing vector.

Hence, key element $= x_{34} = 1$

Further, apply the similar procedure, we have the following table.

Revised Simplex Table-4

Variables in the basis	solution $x_B^{(2)}$	$e_1^{(2)}$	$e_2^{(2)}$	B_2^{-1} $\beta_1^{(2)}$ $\left(\alpha_2^{(2)}\right)$	$\beta_2^{(2)}$ $\left(\alpha_3^{(2)}\right)$	$\beta_3^{(2)}$ $\left(\alpha_4^{(2)}\right)$		
Z	90/7	1	0	16/7	–4/7	–1		
Z_a	0	0	1	0	0	0		
x_2	15/7	0	0	5/7	–3/7	0		
x_3	25/7	0	0	–1/7	2/7	0		
x_4	15/7	0	0	–9/7	4/7	1		

From the above table, we observe that Max. $Z_a = 0$ and all the artificial variables are removed. Hence, phase-1 is completed.

Now for phase-2, the subscript 2 is changed to 1 and the next simplex table is constructed by removing Z_a-row and $e_2^{(2)}$-column from the fourth table, we get

Revised Simplex Table-5

Variables in the basis	solution $x_B^{(1)}$	$e_1^{(1)}$	B_1^{-1} $\beta_1^{(1)}$ $\left(\alpha_2^{(1)}\right)$	$\beta_2^{(1)}$ $\left(\alpha_3^{(1)}\right)$	$\beta_3^{(1)}$ $\left(\alpha_4^{(1)}\right)$	$x_1^{(1)} = B_1^{-1}\alpha_1^{(1)}$	Min. Ratio $\min\limits_i \left\{ \frac{x_{B_i}}{x_{i1}}, x_{i1} > 0 \right\}$
Z	90/7	1	16/7	– 4/7	–1	–6/7	
x_2	15/7 (x_{B_1})	0	5/7	–3/7	0	–1/7	—
x_3	25/7 (x_{B_2})	0	–1/7	2/7	0	3/7	25/3
x_4	15/7 (x_{B_3})	0	–9/7	4/7	1	(6/7)	5/2 (min) →
						↓	

From above table, we observe that

$$x_1 = 0, x_2 = \frac{15}{7}, x_3 = \frac{25}{7}, x_4 = \frac{15}{7} \text{ and Max}.Z = \frac{90}{7}$$

is the solution of the given LPP.

Now we have to check the optimality of the solution.

Since, $\alpha_1^{(1)} \notin \boldsymbol{B}_1$

Then $\qquad \Delta_1 = - \text{ (first row of } \boldsymbol{B}_1^{-1}) \times \alpha_1^{(1)}$

$$= - \begin{bmatrix} 1 & \dfrac{16}{7} & -\dfrac{4}{7} & -1 \end{bmatrix} \begin{bmatrix} -1 \\ 1 \\ 2 \\ 1 \end{bmatrix} = \frac{6}{7}$$

$\Rightarrow \qquad\qquad \Delta_1 = \dfrac{6}{7} > 0$

$\Rightarrow$ solution is not optimal

To find incoming vector:

$$\Delta_k = \max_j\{\Delta_j\}, j = 1 = \max\{\Delta_1\} = 1 = \Delta_1$$

$\Rightarrow \qquad\qquad k = 1$

$\Rightarrow \quad \alpha_1^{(1)}$ is the incoming vector.

Now, $\qquad x_1^{(1)} = B_1^{-1}\alpha_1^{(1)} = \begin{bmatrix} 1 & 16/7 & -4/7 & -1 \\ 0 & 5/7 & -3/7 & 0 \\ 0 & -1/7 & 2/7 & 0 \\ 0 & -9/7 & 4/7 & 1 \end{bmatrix} \begin{bmatrix} -1 \\ 1 \\ 2 \\ 1 \end{bmatrix} = \begin{bmatrix} -6/7 \\ -1/7 \\ 3/7 \\ 6/7 \end{bmatrix} = \begin{bmatrix} -\Delta_1 \\ x_{11} \\ x_{21} \\ x_{31} \end{bmatrix}$

To find outgoing vector:

Using $\dfrac{x_{B_r}}{x_{rk}} = \min_i\left\{\dfrac{x_{B_i}}{x_{ik}}, x_{ik} > 0\right\}$, we get

$$\dfrac{x_{B_r}}{x_{r1}} = \min_i\left\{\dfrac{x_{B_i}}{x_{i1}}, x_{i1} > 0\right\} = \min\left\{\dfrac{x_{B_2}}{x_{21}}, \dfrac{x_{B_3}}{x_{31}}\right\} \qquad \left(\because x_{11} = -\dfrac{1}{7} < 0\right)$$

$$= \min\left\{\dfrac{25/7}{3/7}, \dfrac{15/7}{6/7}\right\} = \min\left\{\dfrac{25}{2}, \dfrac{5}{2}\right\} = \dfrac{5}{2} = \dfrac{x_{B_3}}{x_{31}}$$

$\Rightarrow \qquad\qquad r = 3$

$\Rightarrow \quad \beta_3^{(1)}(= \alpha_4^{(1)})$ is the outgoing vector and hence, key element $= x_{31} = \dfrac{6}{7}$

Now, to check the optimality, proceed in a usual manner, we get the following simplex table.

Revised Simplex Table-6

Variables in the basis	solution $x_B^{(1)}$	B_1^{-1}				$x_k^{(1)}$	Min. Ratio
		$e_1^{(1)}$	$\beta_1^{(1)}$ $\left(\alpha_2^{(1)}\right)$	$\beta_2^{(1)}$ $\left(\alpha_3^{(1)}\right)$	$\beta_3^{(1)}$ $\left(\alpha_1^{(1)}\right)$		
Z	15	1	1	0	0		
x_2	5/2	0	1/2	–1/3	1/6		
x_3	5/2	0	1/2	0	–1/2		
x_1	5/2	0	–3/2	2/3	7/6		

From the above table, we observe that, the solution is

$$x_1 = \frac{5}{2}, x_2 = \frac{5}{2}, x_3 = \frac{5}{2}, \text{Max}.Z = 15$$

To check the optimality of this solution, we proceed as follows:

$\because \qquad \alpha_4^{(1)} \notin B_1$ so,

$$\Delta_4 = -(\text{first row of } B_1^{-1}) \times \alpha_4^{(1)}$$

$$= -[1 \quad 1 \quad 0 \quad 0] \begin{bmatrix} 1 \\ 0 \\ 0 \\ 1 \end{bmatrix} = -1 < 0$$

$\Rightarrow$ solution is optimal and is given by

$$x_1 = x_2 = x_3 = \frac{5}{2} \text{ and Max. } Z = 15$$

EXAMPLE 2. *Using revised simplex method, solve the following LPP :*

$$\textbf{Min. } \textbf{\textit{Z}} = \textbf{\textit{x}}_1 + \textbf{2\textit{x}}_2$$

subject to the constraints

$$2\textbf{\textit{x}}_1 + 5\textbf{\textit{x}}_2 \geq 6$$

$$\textbf{\textit{x}}_1 + \textbf{\textit{x}}_2 \geq 2$$

and $\qquad\qquad \textbf{\textit{x}}_1, \textbf{\textit{x}}_2 \geq \textbf{0}$

SOLUTION.   The given LPP is of minimization, so first we convert it into maximization problem such that

$$\text{Max. } Z' = -Z = -x_1 - 2x_2$$

Now proceed as usual, using surplus variables s_1, s_2 and artificial variables A_1 and A_2, we have the following form of phase 1

Phase 1. The given LPP in revised simplex form-II can be written as

$$\left.\begin{array}{l} Z' + x_1 + 2x_2 + 0s_1 + 0s_2 + 0A_1 + 0A_2 = 0 \\ Z_a + 0x_1 + 0x_2 + 0s_1 + 0s_2 + A_1 + A_2 = 0 \\ 2x_1 + 5x_2 - s_1 + A_1 = 6 \\ x_1 + x_2 - s_2 + A_2 = 2 \end{array}\right] \qquad ...(1)$$

and $\ x_i, s_i \geq 0, A_i \geq 0$

Here, the artificial objective function which to be maximized in phase 1 is given by

$$\text{Max. } Z_a = -A_1 - A_2$$

In matrix form, system (1) can be written as

$$
\begin{array}{cccccccc}
e_1^{(2)} & e_2^{(2)} & \alpha_1^{(2)} & \alpha_2^{(2)} & \alpha_3^{(2)} & \alpha_4^{(2)} & \alpha_5^{(2)} & \alpha_6^{(2)}
\end{array}
\begin{array}{c}
\\
\\
\end{array}
$$

$$
\begin{bmatrix}
1 & 0 & 1 & 2 & 0 & 0 & 0 & 0 \\
0 & 1 & 0 & 0 & 0 & 0 & 1 & 1 \\
0 & 0 & 2 & 5 & -1 & 0 & 1 & 0 \\
0 & 0 & 1 & 1 & 0 & -1 & 0 & 1
\end{bmatrix}
\begin{bmatrix}
Z' \\
Z_a \\
x_1 \\
x_2 \\
s_1 \\
s_2 \\
A_1 \\
A_2
\end{bmatrix}
=
\begin{bmatrix}
0 \\
0 \\
6 \\
2
\end{bmatrix}
$$

where the column vector is $x_B^{(2)}$ and right side is $b^{(2)}$.

The basis matrix B_2 is given by

$$
B_2 = \begin{bmatrix}
1 & 0 & 0 & 0 \\
0 & 1 & 1 & 1 \\
0 & 0 & 1 & 0 \\
0 & 0 & 0 & 1
\end{bmatrix} = [e_1^{(2)} \quad e_2^{(2)} \quad \alpha_5^{(2)} \quad \alpha_6^{(2)}]
$$

$$
= [e_1^{(2)} \quad e_2^{(2)} \quad \beta_1^{(2)} \quad \beta_2^{(2)}] \text{ where } \alpha_5^{(2)} = \beta_1^{(2)}, \alpha_6^{(2)} = \beta_2^{(2)}
$$

Here, B_2 can be written as

$$
B_2 = \begin{bmatrix}
1 & 0 & -C_B \\
0 & 1 & -C_{B_a} \\
O & O & B
\end{bmatrix}
$$

where $C_B = [0 \quad 0], C_{B_a} = [-1 \quad -1], B = \begin{bmatrix} 1 & 0 \\ 0 & 1 \end{bmatrix} = I_2 \Rightarrow B_2^{-1} = I_2$

So, $B_2^{-1} = \begin{bmatrix} 1 & 0 & -C_B B^{-1} \\ 0 & 1 & -C_{B_a} B^{-1} \\ O & O & B \end{bmatrix} = \begin{bmatrix} 1 & 0 & -C_B \\ 0 & 1 & -C_{B_a} \\ O & O & B \end{bmatrix} = \begin{bmatrix} 1 & 0 & 0 & 0 \\ 0 & 1 & -1 & -1 \\ 0 & 0 & 1 & 0 \\ 0 & 0 & 0 & 1 \end{bmatrix}$

and initial solution $x_B^{(2)}$ is given by

$$x_B^{(2)} = B_2^{-1} b^{(2)}$$

$$
= \begin{bmatrix}
1 & 0 & 0 & 0 \\
0 & 1 & -1 & -1 \\
0 & 0 & 1 & 0 \\
0 & 0 & 0 & 1
\end{bmatrix}
\begin{bmatrix}
0 \\
0 \\
6 \\
2
\end{bmatrix}
=
\begin{bmatrix}
0 \\
-8 \\
6 \\
2
\end{bmatrix}
=
\begin{bmatrix}
Z' \\
Z_a \\
6 \\
2
\end{bmatrix}
$$

Now, we have the following simplex table

Revised Simplex Table-1

Variables in the basis	solution $x_B^{(1)}$	B_2^{-1}				$x_2^{(2)} = B_2^{-1}\alpha_2^{(2)}$	Min. Ratio $\min\limits_{i}\left\{\dfrac{x_{B_i}}{x_{ik}}, x_{ik} > 0\right\}$
		$e_1^{(2)}$	$e_2^{(2)}$	$\beta_1^{(2)}$ $(\alpha_5^{(2)})$	$\beta_2^{(2)}$ $(\alpha_6^{(2)})$		
Z'	0	1	0	0	0	2	
Z_a	–8	0	1	–1	–1	–6	
A_1	6 (x_{B_1})	0	0	1	0	5 (x_{12})	6/5 (min) →
A_2	2 (x_{B_2})	0	0	0	1	①(x_{22})	2/1
				↓			

$\Rightarrow$ Max. $Z_a = -8 < 0$

So, we have to find $\Delta_{ja} \ \forall \ j$ for which $\alpha_j^{(2)}$ are not in the basis.

By definition, $\alpha_1^{(2)}, \alpha_2^{(2)}, \alpha_3^{(2)}, \alpha_4^{(2)}$ are not in the basis.

Now, using $\Delta_{ja} = -(\text{second row of } B_2^{-1}) \times \alpha_j^{(2)}$ for j = 1, 2, 3, 4, we get

$$\Delta_{1a} = -[0 \quad 1 \quad -1 \quad -1]\begin{bmatrix}1\\0\\2\\1\end{bmatrix} = 3; \ \Delta_{2a} = -[0 \quad 1 \quad -1 \quad -1]\begin{bmatrix}2\\0\\5\\1\end{bmatrix} = 6$$

$$\Delta_{3a} = -[0 \quad 1 \quad -1 \quad -1]\begin{bmatrix}0\\0\\-1\\0\end{bmatrix} = -1; \ \Delta_{4a} = -[0 \quad 1 \quad -1 \quad -1]\begin{bmatrix}0\\0\\0\\-1\end{bmatrix} = -1$$

Clearly $\Delta_{1a}, \Delta_{2a} > 0 \ \Rightarrow$ The value of Z_a can be improved as follows:

To find incoming vector:

We have

$$\Delta_{ka} = \max_{j}\{\Delta_{ja}\}, j = 1, 2, 3, 4$$

$$= \max\{\Delta_{1a}, \Delta_{2a}, \Delta_{3a}, \Delta_{4a}\} = \max\{3, 6, -1, -1\} = 6 = \Delta_{2a}$$

$\Rightarrow$ k = 2

$\Rightarrow$ $\alpha_2^{(2)}$ is the incoming vector.

Also, $x_2^{(2)} = B_2^{-1}\alpha_2^{(2)} = \begin{bmatrix}1 & 0 & 0 & 0\\0 & 1 & -1 & -1\\0 & 0 & 1 & 0\\0 & 0 & 0 & 1\end{bmatrix}\begin{bmatrix}2\\0\\5\\1\end{bmatrix} = \begin{bmatrix}2\\-6\\5\\1\end{bmatrix} = \begin{bmatrix}-\Delta_2\\-\Delta_{2a}\\x_{12}\\x_{22}\end{bmatrix}$

To find outgoing vector:

Using $\dfrac{x_{B_r}}{x_{rk}} = \min\limits_{i}\left\{\dfrac{x_{B_i}}{x_{ik}}, x_{ik} > 0\right\}$ we get

$$\frac{x_{B_r}}{x_{r2}} = \min\left\{\frac{x_{B_1}}{x_{12}}, \frac{x_{B2}}{x_{22}}\right\} = \min\left\{\frac{6}{5}, \frac{2}{1}\right\} = \frac{6}{5} = \frac{x_{B_1}}{x_{12}}$$

$\Rightarrow \qquad r = 1$

$\Rightarrow \qquad \beta_1^{(2)}(= \alpha_5^{(2)})$ is the outgoing vector and key element $= x_{12} = 5$

Now, to improve the value of Z_a, proceed as usual, we have the following table.

Revised Simplex Table-2

Variables in the basis	solution $x_B^{(1)}$	B_2^{-1}				$x_1^{(2)} = B_2^{-1}\alpha_1^{(2)}$	Min. Ratio $\min\limits_{i}\left\{\dfrac{x_{B_i}}{x_{i1}}, x_{i1} > 0\right\}$
		$e_1^{(2)}$	$e_2^{(2)}$	$\beta_1^{(2)}$ $(\alpha_2^{(2)})$	$\beta_2^{(2)}$ $(\alpha_6^{(2)})$		
Z'	$-12/5$	1	0	$-2/5$	0	$1/5$	
Z_a	$-4/5$	0	1	$1/5$	-1	$-3/5$	
x_2	$6/5$ (x_{B_1})	0	0	$1/5$	0	$2/5$ (x_{11})	3
A_2	$4/5$ (x_{B_2})	0	0	$-1/5$	1	$\boxed{3/5}$ (x_{22})	$4/3$ (min) $\rightarrow$

We observe that

$$\text{Max } Z_a = -\frac{4}{5} < 0$$

So, we have to find $\Delta_{ja} \; \forall\, j$ for which $\alpha_j^{(2)}$ are not in B_2, which becomes

$$B_2 = [e_1^{(2)} \quad e_2^{(2)} \quad \alpha_2^{(2)} \quad \alpha_6^{(2)}]$$

$$\Delta_{1a} = -(\text{second row of } B_2^{-1} \text{ of second table}) \times \alpha_1^{(2)}$$

$$= -\begin{bmatrix} 0 & 1 & \dfrac{1}{5} & -1 \end{bmatrix}\begin{bmatrix} 1 \\ 0 \\ 2 \\ 1 \end{bmatrix} = \frac{3}{5}$$

$$\Delta_{3a} = -\begin{bmatrix} 0 & 1 & \dfrac{1}{5} & -1 \end{bmatrix}\begin{bmatrix} 0 \\ 0 \\ -1 \\ 0 \end{bmatrix} = \frac{1}{5}$$

$$\Delta_{4a} = -\begin{bmatrix} 0 & 1 & \dfrac{1}{5} & -1 \end{bmatrix} \begin{bmatrix} 0 \\ 0 \\ 0 \\ -1 \end{bmatrix} = -1$$

$$\Delta_{5a} = -\begin{bmatrix} 0 & 1 & \dfrac{1}{5} & -1 \end{bmatrix} \begin{bmatrix} 0 \\ 1 \\ 1 \\ 0 \end{bmatrix} = -\dfrac{6}{5}$$

$\Rightarrow$ $\Delta_{1a}, \Delta_{3a} > 0$, so we can improved the value of Z_a.

To find incoming vector:

$$\Delta_{ka} = \max_{j}\{\Delta_{ja}\}, j = 1,3,4,5$$

$$= \max\{\Delta_{1a}, \Delta_{3a}, \Delta_{4a}, \Delta_{5a}\} = \max\left\{\frac{3}{5}, \frac{1}{5}, -1, -\frac{6}{7}\right\}$$

$$= \frac{3}{5}$$

$\Rightarrow$ $k = 1$

$\Rightarrow$ $\alpha_1^{(2)}$ is the incoming vector.

Now, $x_1^{(1)} = B_2^{-1}\alpha_1^{(2)} = \begin{bmatrix} 1 & 0 & -2/5 & 0 \\ 0 & 1 & 1/5 & -1 \\ 0 & 0 & 1/5 & 0 \\ 0 & 0 & -1/5 & 1 \end{bmatrix} \begin{bmatrix} 1/5 \\ -3/5 \\ 2/5 \\ 3/5 \end{bmatrix} = \begin{bmatrix} -\Delta_1 \\ -\Delta_{1a} \\ x_{11} \\ x_{21} \end{bmatrix}$

To find outgoing vector:

Using $\dfrac{x_{B_r}}{x_{rk}} = \min_{i}\left\{\dfrac{x_{B_i}}{x_{ik}}, x_{ik} > 0\right\}$ we get

$$\frac{x_{B_r}}{x_{r1}} = \min_{i}\left\{\frac{x_{B_i}}{x_{i1}}, x_{i1} > 0\right\} = \min\left\{\frac{x_{B1}}{x_{11}}, \frac{x_{B2}}{x_{21}}\right\}$$

$$= \min\left\{\frac{6/5}{2/5}, \frac{4/5}{3/5}\right\} = \min\left\{3, \frac{4}{3}\right\} = \frac{4}{3} = \frac{x_{B_2}}{x_{21}}$$

$\Rightarrow$ $r = 2$

$\Rightarrow$ $\beta_2^{(2)}(= \alpha_6^{(2)})$ is the outgoing vector.

$\Rightarrow$ $x_{21} = \dfrac{3}{5}$ is the key element.

Now again apply the usual procedure to improve the value of Z_a, we get the following table.

Revised Simplex Table-3

Variables in the basis	solution $x_B^{(2)}$	B_2^{-1}				$x_k^{(2)}$	Min. Ratio
		$e_1^{(2)}$	$e_2^{(2)}$	$\beta_1^{(2)} \left(\alpha_2^{(2)}\right)$	$\beta_2^{(2)} \left(\alpha_1^{(2)}\right)$		
Z'	$-8/3$	1	0	$-1/3$	$-1/3$		
Z_a	0	0	1	0	0		
x_2	$2/3$	0	0	$1/3$	$-2/3$		
x_1	$4/3$	0	0	$-1/3$	$5/3$		

Clearly, from the above table, Max. $Z_a = 0$

$\Rightarrow$ Process of Phase 1 comes to an end.

Now, we enter phase 2 by making a change in subscript 2 to 1.

Phase 2. Now construct the new table by removing second row and $e_2^{(2)}$ column from third table as follows.

Revised Simplex Table-4

Variables in the basis	solution $x_B^{(1)}$	B_1^{-1}			$x_k^{(1)}$	Min. Ratio
		$e_1^{(2)}$	$\beta_1^{(2)} \left(\alpha_2^{(2)}\right)$	$\beta_2^{(2)} \left(\alpha_1^{(2)}\right)$		
Z'	$-8/5$	1	$-1/3$	$-1/3$		
x_2	$2/3$	0	$1/3$	$-2/3$		
x_1	$4/3$	0	$-1/3$	$5/3$		

The basis B_1 is given by $B_1 = [e_1^{(1)} \quad \alpha_2^{(1)} \quad \alpha_1^{(1)}] = [e_1^{(1)} \quad \beta_1^{(1)} \quad \beta_2^{(1)}]$

$\Rightarrow \quad \alpha_3^{(1)}, \alpha_4^{(1)} \notin B_1$

Now, $\quad \Delta_3 = -(\text{first row of } B_1^{-1}) \cdot \alpha_3^{(1)} = -\begin{bmatrix} 1 & -\dfrac{1}{3} & -\dfrac{1}{3} \end{bmatrix} = \begin{bmatrix} 0 \\ -1 \\ 0 \end{bmatrix} = -\dfrac{1}{3}$

$$\Delta_4 = -\begin{bmatrix} 1 & -\dfrac{1}{3} & -\dfrac{1}{3} \end{bmatrix}\begin{bmatrix} 0 \\ 0 \\ -1 \end{bmatrix} = -\dfrac{1}{3}$$

Clearly $\Delta_3, \Delta_4 < 0$

$\Rightarrow$ solution obtained in table-4 is optimal and is given by

$$x_1 = \frac{4}{3}, x_2 = \frac{2}{3} \text{ and Max.} Z' = -\frac{8}{5}, \text{ i.e., Min } Z = \frac{8}{5}$$

6.5 COMPARISON OF SIMPLEX METHOD AND REVISED SIMPLEX METHOD

When we consider an LPP with constraints $Ax = b$ where A is a matrix of order $m \times n$. Then for solving this LP by simplex method, we have to transform $(n + 1)$ columns at each iteration. Further at each iteration one variable is introduced and one is removed from the basis, so we have to compute total $n - m + 1$ columns. Also, for each of these columns we have to transforms $m + 1$ elements and for moving one iteration to another, we also need to calculate minimum ratio. Hence, we have to perform multiplication operation $(m+1)$ $(n-m+1)$ times and addition $m(n-m+1)$ times.

But in revised simplex method, there are total $(m+1)$ rows and $(m+2)$ columns. Therefore, when we move from one iteration to another we have to perform $(m+1)^2$ multiplication operation in addition to $m(n-m)$ operation for calculating $c_j - z_j$. Also, in revised simplex method, while updating the table to move from one solution to another, an additional table of original non-basic variables, not in the basis is required, which may cause of some error.

Though the revised simplex method has not met with wide acceptance for hand computation even then there are a few advantages of revised simplex method over the simplex method.

(1) The total no. of operations for the revised simplex method is approximately $mn + m^2 + 3m$, whereas in simplex method the total no. of operations is approximately $mn - m^2 + n + m$. Thus, revised simplex method deals with less computations than that of simplex method.

(2) In revised simplex method, the inverse of the current basis is automatically generated and the next BFS is also obtained, whereas it is not so in simplex method.

(3) In revised simplex method, we introduce $(m+1)(m+2)$ entries in each table while, in simplex method these are only $(n+1)(m+1)$. If $n << m$, a lot of labour and so error can be avoided.

 Exercise-6.1

Using revised simplex method, solve the following LPP:

1. Max. $Z = 2x_1 + x_2$
   s.t. $3x_1 + 4x_2 \le 6$
   $6x_1 + x_2 \le 3$
   and $x_1, x_2 \ge 0$ [DELHI-1993; MEERUT-1998]

2. Max. $Z = x_1 + x_2 + 3x_3$
   s.t. $3x_1 + 2x_2 + x_3 \le 3$
   $2x_1 + x_2 + 2x_3 \le 2$
   and $x_1, x_2, x_3 \ge 0$

3. Max. $Z = 3x_1 + 6x_2 + 2x_3$
   s.t. $3x_1 + 4x_2 + x_3 \le 2$
   $x_1 + 3x_2 + 2x_3 \le 1$
   and $x_1, x_2, x_3 \ge 0$

4. Min. $Z = 2x_1 + x_2$
   s.t. $3x_1 + x_2 \le 3$
   $4x_1 + 3x_2 \ge 6$
   $x_1 + 2x_2 \le 3$
   and $x_1, x_2 \ge 0$

5. Min. $Z = x_1 + x_2$

   s.t. $x_1 + 2x_2 \ge 7$
   $4x_1 + x_2 \ge 6$
   and $x_1, x_2 \ge 0$

6. Max. $Z = 5x_1 + 3x_2$
   s.t. $4x_1 + 5x_2 \ge 10$
   $5x_1 + 2x_2 \le 10$
   $3x_1 + 8x_2 \le 12$
   and $x_1, x_2 \ge 0$

7. Max. $Z = x_1 + 2x_2$
   s.t. $3x_1 + 2x_2 \ge 6$
   $x_1 + 6x_2 \ge 3$
   and $x_1, x_2 \ge 0$

8. Max. $Z = 2x_1 + 4x_2 + 6x_3 - 2x_4$
   s.t. $x_1 + 2x_2 + 3x_3 = 15$
   $2x_1 + x_2 + 5x_3 = 20$
   $3x_1 + 6x_2 + 3x_3 + 3x_4 = 30$
   and $x_1, x_2, x_3 \ge 0$

9. Max. $Z = -5x_2$
   s.t. $x_1 + x_2 \le 2$
   $x_1 + 5x_2 \ge 10$
   and $x_1, x_2 \ge 0$

ANSWERS

1. $x_1 = \dfrac{2}{7}, x_2 = \dfrac{9}{7}, \text{Max}.Z = \dfrac{13}{7}$

2. $x_1 = x_2 = 0, x_3 = 1, \text{Max}.Z = 3$

3. $x_1 = \dfrac{2}{5}, x_2 = \dfrac{1}{3}, x_3 = 0, \text{Max } Z = \dfrac{12}{5}$

4. $x_1 = \dfrac{3}{5}, x_2 = \dfrac{6}{5}, \text{Min}.Z = \dfrac{12}{5}$

5. $x_1 = \dfrac{5}{7}, x_2 = \dfrac{22}{7}, \text{Min}.Z = \dfrac{2}{7}$

6. $x_1 = \dfrac{28}{17}, x_2 = \dfrac{15}{17}, \text{Max}.Z = \dfrac{185}{17}$

7. $x_1 = 0, x_2 = 3, Max.Z = 6$
8. $x_1 = \dfrac{5}{2}, x_2 = \dfrac{5}{2}, x_3 = \dfrac{5}{2}, Max.Z = 30$

9. $x_1 = 0, x_2 = 2, Max.Z = -10$

REVIEW QUESTIONS

1. Explain the revised simplex method and compare it with simplex method.
2. Formulate a linear programming problem in the form of revised simplex method.
3. Explain the revised simplex method of standard form I and II.
4. Write the differences of revised simplex method and simplex method.
5. Describe the revised simplex method when artificial vectors are added to obtain an identity matrix for the initial basis matrix.

MULTIPLE CHOICE QUESTIONS (CHOOSE THE MOST APPROPRIATE ONE)

1. In LPP if FS is optimal then basic feasible solution will be:
   (a) non-zero solution
   (b) may or may not be optimal
   (c) optimal
   (d) None of these
2. In standard form-II of revised simplex method the basis matrix is denoted by:
   (a) B_1 (b) B_0
   (c) B_2 (d) None of these
3. In LPP, in simplex table, $\Delta j \leq 0$, then solution under test:
   (a) optimal   (b) minimum
   (c) unbounded   (d) none of these
4. Extreme points of $\{(x, y)/m \leq 1, \|y\| \leq 1\}$ is:
   (a) $(1, -1)$ (b) $(-1, 1)$
   (c) $(1, 1)$ (d) $(-1, -1)$
5. The number of additional constraints in standard form II of revised simplex method:
   (a) 0   (b) 2
   (c) 1   (d) None of these
6. Simplex method was developed by:
   (a) Mody   (b) Maxwell
   (c) Dantzig   (d) None of these
7. The Union of two convex sets is:
   (a) not a convex
   (b) a line segment
   (c) a convex set
   (d) may or may not be a convex set
8. In standard form II of revised simplex method we need:
   (a) slack variables
   (b) surplus variables
   (c) artificial variables
   (d) none of these

9. Matrix form of a LPP is Max. $Z = Cx$ subject to $AX = b$, $X \geq 0$, $A = [a_{ij}]_{m \times n}$, $N > m + n$. Suppose B be a submatrix of order $m \times n$ which is non-singular selected from A then $B^{-1}b$ is called:
   (a) Optimal
   (b) BFS of LPP
   (c) Non basic feasible solution of LPP
   (d) None of these
10. If a LPP Max $Z = cx$ s.t. $Ax = b, X \geq 0$, where $A = \{\lambda_1, \lambda_2, ..., \lambda_n\}$ is coefficient matrix of order $m \times N$, $N = m + n$, has at least one feasible solution then it has:
    (a) zero solution
    (b) one feasible solution but not basic
    (c) one BFS
    (d) none of these
11. In LPP, Max $Z = cx$, $AX \leq b$, then all:
    (a) $b_i \geq 0$ (b) $b_i \leq 0$
    (c) $b_i = 0$ (d) None of these
12. If the solution contains one or more artificial variable as basic and variable LPP has:
    (a) No F.S.   (b) F.S.
    (c) N.S.   (d) None of these
13. Simplex method was developed in the year:
    (a) 1947   (b) 1747
    (c) 1917   (d) None of these
14. In Final simplex table if $\Delta_j \leq 0$ then optimal solution is:
    (a) optimal   (b) unique
    (c) bounded   (d) none of these
15. The set of all the internal points of a convex set is a:
    (a) line segment   (b) convex
    (c) concave   (d) none of these

16. If max position Δj are minimum ratio $\to$ –ve or $\to \infty$ non-solution under test is:
    (a) unbounded
    (b) bounded
    (c) rational
    (d) none of these

17. In standard form-I of revised simplex method the basis matrix is denoted by:
    (a) B_0
    (b) B_1
    (c) B_2
    (d) None of these

18. In standard form-I of revised simplex method we do not need:
    (a) surplus variables
    (b) artificial variables
    (c) slack variables
    (d) none of these

19. The number of additional constraints in standard form-I of revised simplex method is:
    (a) 0
    (b) 1
    (c) 2
    (d) None of these

ANSWERS

1. (c)	**2.** (c)	**3.** (a)	**4.** (c)	**5.** (b)
10. (d)	**11.** (a)	**12.** (a)	**13.** (a)	**14.** (b)
19. (b)				

6. (c)   **7.** (c)   **8.** (c)   **9.** (b)
15. (b)   **16.** (a)   **17.** (b)   **18.** (b)

ARCHIVE

1. Use the revised simplex method to solve the following LP problem:

    Min. $Z = -4x_1 + x_2 + 2x_3$
    subject to

    (i) $2x_1 - 3x_2 + 2x_3 \le 12$
    (ii) $-5x_1 + 2x_2 + 3x_3 \ge 4$
    (iii) $3x_1 - 3x_3 = -1$
    and $x_1, x_2, x_3 \ge 0$ [PUNJAB BE(IT)–2004]

HINTS AND ANSWERS

1. $x_1 = \dfrac{59}{17}, x_2 = \dfrac{36}{17}, x_3 = \dfrac{97}{17}$ and Min. $Z = -\dfrac{6}{17}$

□□□□

7 Duality in Linear Programming

7.1 INTRODUCTION

The concept of duality is most important and useful tools in mathematics and many branches of engineering. It explains that for every LPP there is a related unique LPP involving the same data that also describes the original LPP, *i.e.*, each linear programming can be analysed in two different ways but having equivalent solutions. The original LPP is called 'primal' and other related problem is called 'dual'. In this chapter we shall discuss the duality in linear programming along with some related theorems.

7.2 RELATIONSHIP BETWEEN PRIMAL AND DUAL

The relationship between primal and dual of an LPP contains the following points:

(i) The maximization (minimization) in primal becomes the minimization (maximization) in its dual and conversely.

(ii) If the primal contains n variables and m constraints, then the dual will contains m variables and n constraints.

(iii) The coefficients $c_1, c_2, ..., c_n$ in the objective function of the primal appear in the constraints in the dual.

(iv) The constants $b_1, b_2, ..., b_m$ in the constraints of the primal appear in the objective function of the dual.

(v) If any of the two (either primal or dual) has an infeasible solution, then the value of objective function of other is unbounded.

(vi) If either the primal or dual has an unbounded solution, then the solution to the other problem is infeasible.

(vii) If some primal variables are unrestricted in sign, then these dual constraints will be equations that correspond to the said primal variables.

(viii) If a primal contains a constraint as an equation, then the variable in its dual corresponding to that equation will be unrestricted in sign.

This relationship between prime and dual can be summarized in the following table.

S.No.	Primal	Dual
1.	Objective function is of maximization	Objective function is of minimization
2.	i^{th} primal variable, x_i	i^{th} dual constraints
3.	i^{th} primal constraints	i^{th} dual variable, y_i
4.	Primal constraints $\leq$ type	Dual constraints $\geq$ type
5.	Primal variable x_i unrestricted in sign	Dual constraints i is $=$ type
6.	Primal constraints i is $=$ type	Dual variable y_i is unrestricted in sign

☞ REMARKS

- The variables in both primal and dual are non-negative.
- The value of the objective function for any feasible solution of the primal is less than the value of the objective function for any feasible solution of the dual.

7.3 SYMMETRIC PRIMAL-DUAL PROBLEMS

There are two important forms of primal and dual problems namely the standard form and symmetrical (or canonical) form.

7.3.1 STANDARD FORM OF A PRIMAL PROBLEM   [MEERUT–2007, 09, 12]

The given LPP is said to be in standard form if:

   (i)   for maximization problem, all the constraints have $\leq$ sign.

   (ii) for minimization problem, all the constraints have $\geq$ sign.

If the primal is in standard form, then we call it symmetric primal and its dual is symmetric dual.

7.3.2 FORMULATION OF DUAL LINEAR PROGRAMMING PROBLEM   [GORAKHPUR–2011]

Let us suppose the primal LPP is given in the form of

$$\text{Max. } Z_x = c_1 x_1 + c_2 x_2 + \ldots + c_n x_n$$

subject to the constraints

$$a_{11} x_1 + a_{12} x_2 + \ldots + a_{1n} x_n \leq b_1$$
$$a_{21} x_1 + a_{22} x_2 + \ldots + a_{2n} x_n \leq b_2$$
$$\vdots$$
$$a_{m1} x_1 + a_{m2} x_2 + \ldots + a_{mn} x_n \leq b_m$$

and $\qquad x_1, x_2, \ldots, x_n \geq 0$

Then to obtain the dual of the above primal, we use the following steps:

(1)   minimize the objective function instead of maximization

(2)   interchange $c_1, c_2, \ldots, c_n$ and $b_1, b_2, \ldots, b_n$, i.e., interchange the role of constant terms and the coefficients of the objective function.

(3)   Replace A (the coefficient matrix) by A' (transpose of A).

(4)   Change $\leq$ by $\geq$ sign.

Then the corresponding dual LPP is defined as

$$\text{Min. } Z_y = b_1 y_1 + b_2 y_2 + \ldots + b_m y_m$$

subject to the constraints

$$a_{11}y_1 + a_{21}y_2 + \ldots + a_{m1}y_m \geq c_1$$
$$a_{12}y_1 + a_{22}y_2 + \ldots + a_{m2}y_m \geq c_2$$
$$\vdots$$
$$a_{1n}y_1 + a_{2n}y_2 + \ldots + a_{mn}y_m \geq c_n$$

and $\qquad y_1, y_2, \ldots, y_m \geq 0$

The above conversion (primal to dual) can be expressed in the following figure.

Primal	**Dual**
Max. $Z_x = \sum\limits_{j=1}^{n} c_j x_j$	Min. $Z_y = \sum\limits_{i=1}^{m} b_i y_i$
subject to the constraints	subject to the constraints
$\sum\limits_{j=1}^{n} a_{ij} x_j \leq b_i$ $\quad i = 1, 2, \ldots, m$ $a_{ij} = a_{ji}$	$\sum\limits_{i=1}^{m} a_{ji} y_i \geq c_j \quad j = 1, 2, \ldots, n$ and $\quad y_i \geq 0, i = 1, 2, \ldots, m$
and $\quad x_j \geq 0, j = 1, 2, \ldots, n$	

7.3.3 MATRIX FORM OF PRIMAL-DUAL PROBLEM

[MEERUT–2004]

Consider the above primal-dual problem in matrix form

$$\text{Max.} Z_x = \boldsymbol{CX}$$

subject to the constraints

$$\boldsymbol{AX} \leq \boldsymbol{b}$$

and $\qquad \boldsymbol{X} \geq \boldsymbol{0}$ $\qquad\qquad$ ...(1)

where, $\qquad \boldsymbol{C} = [c_1 \quad c_2 \quad \cdots \quad c_n]_{1 \times n}$

$$\boldsymbol{X} = \begin{bmatrix} x_1 \\ x_2 \\ \vdots \\ x_n \end{bmatrix}_{n \times 1} = [x_1 \quad x_2 \quad \cdots \quad x_n]'_{1 \times n}$$

$$\boldsymbol{A} = \begin{bmatrix} a_{11} & a_{12} & \cdots & a_{1n} \\ a_{21} & a_{22} & \cdots & a_{2n} \\ \vdots & & & \\ a_{m1} & a_{m2} & \cdots & a_{mn} \end{bmatrix}_{m \times n}, \boldsymbol{b} = \begin{bmatrix} b_1 \\ b_2 \\ \vdots \\ b_m \end{bmatrix}_{m \times 1}$$

Let $\boldsymbol{O}$ be a zero matrix of order $m \times 1$. Then dual of the above problem (1) be given by

$$\text{Min.} Z_y = \boldsymbol{b}' \boldsymbol{Y}$$

subject to the constraints

$$\boldsymbol{A}' \boldsymbol{Y} \geq \boldsymbol{C}'$$

and $\qquad\qquad \boldsymbol{Y} \geq \boldsymbol{0}$

Here, $\boldsymbol{A}'$, $\boldsymbol{b}'$ and $\boldsymbol{C}'$ are the transpose of $\boldsymbol{A}$, $\boldsymbol{b}$ and $\boldsymbol{C}$ respectively.

7.3.4 Unsymmetric Primal-Dual Form

If in the given LPP, all the constraints are equations, then it is called unsymmetric primal problem.

Consider the unsymmetric primal

$$Z_x = CX$$

subject to the constraints

$$AX = b$$

and

$$X \geq 0$$

Then dual of the above primal is given by

$$Z_y = b'Y$$

subject to the constraints

$$A'Y \geq C'$$

and

$$Y \geq 0$$

Here, $'$ (dash) denote the transpose.

☞ REMARK

- In the above case, the dual variables are unrestricted in sign.

7.4 DUAL OF AN LPP WITH MIXED RESTRICTIONS

There are some situations when LPP contains a mixture of inequalities (*i.e.*, $\leq$ and $\geq$) non-negative variables and unrestricted variables. To find the dual of such case, we use the following steps:

STEP 1. If the given LP is a problem of maximization and contains some constraints of the type $\geq$, then to make all constraints $\leq$ type, multiply both sides of the such constraints by -1 and make the sign $\leq$.

STEP 2. If the given LP is a problem of minimization and contains some constraints $\leq$ type then to make all constraints $\geq$ type, multiply both sides of such constraints by -1 and make the sign $\geq$.

STEP 3. If a constraint is an equation (has $=$ sign), then replace it by two constraints involving both sign $\leq$ and $\geq$ by using the fact that $a = b \Leftrightarrow a \leq b$ or $a \geq b$.

STEP 4. If the given LPP has some unrestricted variables, then replace it by the difference of two non-negative variables.

 Solved Examples

EXAMPLE 1. *Write the dual of the following LPP:*

$$\text{Max. } Z = x_1 - x_2 + 3x_3$$

subject to the constraints

$$x_1 + x_2 + x_3 \leq 10$$
$$2x_1 - x_2 - x_3 \leq 2$$
$$2x_1 - 2x_2 - 3x_3 \leq 6$$

and $\quad x_1, x_2, x_3 \geq 0$

SOLUTION. We observe that in the given LPP there are $m = 3$ constraints and $n = 3$ variables which shows that there should be $m = 3$ dual variables and $n = 3$ constraints.

Further, the coefficient of the primal variables $c_1 = 1$, $c_2 = -1$, $c_3 = 3$ become right hand side constraints of the dual and right hand constraints $b_1 = 0$, $b_2 = 2$, $b_3 = 6$ becomes the coefficients of the dual objective function. Also, the required dual must have a minimizing objective function with all $\geq$ constraints in another variables say y_1, y_2, y_3.

Hence, the resultant dual is given by

Min. $Z = 10y_1 + 2y_2 + 6y_3$

subject to the constraints

$$y_1 + 2y_2 + 2y_3 \geq 1$$
$$y_1 - y_2 - 2y_3 \geq -1$$
$$y_1 - y_2 - 3y_3 \geq 3$$

and $\quad y_1, y_2, y_3 \geq 0$

EXAMPLE 2. *Find the dual of the following LPP:*

$$\textbf{Max. } \textbf{\textit{Z}} = \textbf{3}\textbf{\textit{x}}_\textbf{1} + \textbf{5}\textbf{\textit{x}}_\textbf{2} + \textbf{4}\textbf{\textit{x}}_\textbf{3}$$

subject to the constraints

$$\textbf{2}\textbf{\textit{x}}_\textbf{1} + \textbf{3}\textbf{\textit{x}}_\textbf{2} \leq \textbf{8}$$
$$\textbf{2}\textbf{\textit{x}}_\textbf{2} + \textbf{5}\textbf{\textit{x}}_\textbf{3} \leq \textbf{10}$$
$$\textbf{3}\textbf{\textit{x}}_\textbf{1} + \textbf{2}\textbf{\textit{x}}_\textbf{2} + \textbf{4}\textbf{\textit{x}}_\textbf{3} \leq \textbf{15}$$

and $\qquad \textbf{\textit{x}}_\textbf{1}, \textbf{\textit{x}}_\textbf{2}, \textbf{\textit{x}}_\textbf{3} \geq \textbf{0}$ [MEERUT–2012]

SOLUTION. We shall find the dual of this LPP by alternative matrix method.

Clearly, the given LPP is written in the standard primal form. In matrix form it can be written as

Max. $Z = \begin{bmatrix} 3 & 5 & 4 \end{bmatrix} \begin{bmatrix} x_1 & x_2 & x_3 \end{bmatrix} = \boldsymbol{C} \cdot \boldsymbol{x}$

s.t.

$$\begin{bmatrix} 2 & 3 & 0 \\ 0 & 2 & 5 \\ 3 & 2 & 4 \end{bmatrix} \begin{bmatrix} x_1 \\ x_2 \\ x_3 \end{bmatrix} \leq \begin{bmatrix} 8 \\ 10 \\ 15 \end{bmatrix}$$

or $\qquad \boldsymbol{A}\boldsymbol{x} \leq \boldsymbol{b}$

and $\qquad x_1, x_2, x_3 \geq 0$

Now, dual of this LPP is given by

$$\text{Min } Z_y = b' \cdot y = \begin{bmatrix} 8 & 10 & 15 \end{bmatrix} \begin{bmatrix} y_1 & y_2 & y_3 \end{bmatrix}$$

$$= 8y_1 + 10y_2 + 15y_3$$

s.t. $\qquad \boldsymbol{A}'\boldsymbol{y} \geq \boldsymbol{C}'$

or $\begin{bmatrix} 2 & 0 & 3 \\ 3 & 2 & 2 \\ 0 & 5 & 4 \end{bmatrix} \begin{bmatrix} y_1 \\ y_2 \\ y_3 \end{bmatrix} \geq \begin{bmatrix} 3 \\ 5 \\ 4 \end{bmatrix} \qquad \Rightarrow \qquad \begin{bmatrix} 2y_1 + 0y_2 + 3y_3 \\ 3y_1 + 2y_2 + 2y_3 \\ 0y_1 + 5y_2 + 4y_3 \end{bmatrix} \geq \begin{bmatrix} 3 \\ 5 \\ 4 \end{bmatrix}$

Finally the dual problem of the given LPP is given by

Min. $Z_y = 8y_1 + 10y_2 + 15y_3$

subject to the constraints

$$2y_1 + 3y_3 \geq 3$$
$$3y_1 + 2y_2 + 2y_3 \geq 5$$
$$5y_2 + 4y_3 \geq 4$$

and $\qquad y_1, y_2, y_3 \geq 0$

EXAMPLE 3. *Find the dual of the following LPP:*

$$\text{Min. } Z = 3x_1 + x_2$$

subject to the constraints

$$2x_1 + 3x_2 \geq 2$$
$$x_1 + x_2 \geq 1$$

and $\qquad x_1, x_2 \geq 0$ [KANPUR–2012]

SOLUTION. The given LPP of minimization is in its standard form. In the matrix form it can be written as

$$\text{Min.} Z = 3x_1 + x_2 = [3 \quad 1][x_1 \quad x_2] = \boldsymbol{C} \cdot \boldsymbol{x}$$

s.t. $\qquad \begin{bmatrix} 2 & 3 \\ 1 & 2 \end{bmatrix} \begin{bmatrix} x_1 \\ x_2 \end{bmatrix} \geq \begin{bmatrix} 2 \\ 1 \end{bmatrix}$

or $\qquad \boldsymbol{Ax} \geq \boldsymbol{b}$

and $\qquad x_1, x_2 \geq 0$

Now proceed same as in example 2, the dual of the given LPP is

$$\text{Max.} Z_y = \boldsymbol{b'} \cdot \boldsymbol{y} = [3 \quad 1][y_1 \quad y_2] = 3y_1 + y_2$$

such that $\qquad A'\boldsymbol{y} \leq \boldsymbol{C'}$

or $\qquad \begin{bmatrix} 2 & 1 \\ 3 & 2 \end{bmatrix} \begin{bmatrix} y_1 \\ y_2 \end{bmatrix} \leq \begin{bmatrix} 3 \\ 1 \end{bmatrix}$

$\Rightarrow \qquad \begin{bmatrix} 2y_1 + y_2 \\ 3y_1 + 2y_2 \end{bmatrix} \leq \begin{bmatrix} 3 \\ 1 \end{bmatrix}$

or $\qquad 2y_1 + y_2 \leq 3, \ 3y_1 + 2y_2 \leq 1$

and $\qquad y_1, y_2 \geq 0$

Hence, the dual of the given LPP is given by

$$\text{Max. } Z_y = 3y_1 + y_2$$

s.t. $\qquad 2y_1 + y_2 \leq 3$
$$3y_1 + 2y_2 \leq 1$$

and $\qquad y_1, y_2 \geq 0$

EXAMPLE 4. *Find the dual of the following LPP:*

$$\text{Max. } Z = 3x_1 + 4x_2$$

subject to the constraints

$$2x_1 + 6x_2 \leq 16$$
$$5x_1 + 2x_2 \geq 20$$

and $\qquad x_1, x_2 \geq 0$ [KANPUR–2007]

SOLUTION. The given LPP of maximization is not in standard form. Firstly we shall write the given LPP in the standard primal form as follows.

$$\text{Max. } Z = 3x_1 + 4x_2$$

s.t. $\qquad 2x_1 + 6x_2 \leq 16$
$$-5x_1 - 2x_2 \leq -20$$

and $\qquad x_1, x_2 \geq 0$

In matrix form it can be written as

Max. $Z = 3x_1 + 4x_2 = [3 \quad 4][x_1 \quad x_2] = \boldsymbol{C} \cdot \boldsymbol{x}$

such that

$$\begin{bmatrix} 2 & 6 \\ -5 & -2 \end{bmatrix}\begin{bmatrix} x_1 \\ x_2 \end{bmatrix} \leq \begin{bmatrix} 16 \\ -20 \end{bmatrix}, \; i.e., \; \boldsymbol{Ax} \leq \boldsymbol{b}, \; x_1, x_2 \geq 0$$

Now, the dual of the above problem is given by

$$\text{Min } Z_y = \boldsymbol{b}' \cdot \boldsymbol{y} = [16 \quad -20][y_1 \quad y_2] = 16y_1 - 20y_2$$

s.t. $\boldsymbol{A}' \cdot \boldsymbol{y} \geq \boldsymbol{C}'$

$$\Rightarrow \begin{bmatrix} 2 & -5 \\ 6 & -2 \end{bmatrix}\begin{bmatrix} y_1 \\ y_2 \end{bmatrix} \geq \begin{bmatrix} 3 \\ 4 \end{bmatrix}$$

or $\begin{bmatrix} 2y_1 - 5y_2 \\ 6y_1 - 2y_2 \end{bmatrix} \geq \begin{bmatrix} 3 \\ 4 \end{bmatrix}$

$\Rightarrow 2y_1 - 5y_2 \geq 3$ and $6y_1 - 2y_2 \geq 4$

Hence, the dual of the given LPP is

$\text{Min } Z_y = 16y_1 - 20y_2$

s.t. $2y_1 - 5y_2 \geq 3$

$6y_1 - 2y_2 \geq 4$

and $y_1, y_2 \geq 0$

EXAMPLE 5. *Find the dual of the following LPP:*

$$\textbf{Min } Z = 2x_1 + 2x_2 + 4x_3$$

subject to the constraints

$$2x_1 + 3x_2 + 5x_3 \geq 2$$

$$3x_1 + x_2 + 7x_3 \leq 3$$

$$x_1 + 4x_2 + 6x_3 \leq 5$$

and $\quad x_1, x_2, x_3 \geq 0$ [MEERUT–2011]

SOLUTION. Firstly, we shall write the given LPP in the standard primal form as follows:

$\text{Min. } Z = 2x_1 + 2x_2 + 4x_3$

s.t $\quad 2x_1 + 3x_2 + 5x_3 \geq 2$

$-3x_1 - x_2 - 7x_3 \geq -3$

$-x_1 - 4x_2 - 6x_3 \geq -5$

and $\quad x_1, x_2, x_3 \geq 0$

which can be written in matrix form as follows:

$$\text{Min } Z = 2x_1 + 2x_2 + 4x_3 = [2 \quad 2 \quad 4][x_1 \quad x_2 \quad x_3] = \boldsymbol{C} \cdot \boldsymbol{x}$$

s.t. $\begin{bmatrix} 2 & 3 & 5 \\ -3 & -1 & -7 \\ -1 & -4 & -6 \end{bmatrix}\begin{bmatrix} x_1 \\ x_2 \\ x_3 \end{bmatrix} \geq \begin{bmatrix} 2 \\ -3 \\ -5 \end{bmatrix}$

i.e., $\boldsymbol{Ax} \geq \boldsymbol{b}$ and $x_1, x_2, x_3 \geq 0$

Now, dual of the given LPP is

$$\text{Max } Z_y = \boldsymbol{b}' \cdot \boldsymbol{y} = [2 \quad -3 \quad 5][y_1 \quad y_2 \quad y_3] = 2y_1 - 3y_2 - 5y_3$$

such that $\qquad A'\boldsymbol{y} \le \boldsymbol{C}'$

or $\begin{bmatrix} 2 & -3 & -1 \\ 3 & -1 & -4 \\ 5 & -7 & -6 \end{bmatrix} \begin{bmatrix} y_1 \\ y_2 \\ y_3 \end{bmatrix} \le \begin{bmatrix} 2 \\ 2 \\ 2 \end{bmatrix} \implies \begin{bmatrix} 2y_1 - 3y_2 - y_3 \\ 3y_1 - y_2 - 4y_3 \\ 5y_1 - 7y_2 - 6y_3 \end{bmatrix} \le \begin{bmatrix} 2 \\ 2 \\ 4 \end{bmatrix}$

$\implies \qquad 2y_1 - 3y_2 - y_3 \le 2; 3y_1 - y_2 - 4y_3 \le 2; 5y_1 - 7y_2 - 6y_3 \le 4$

Hence, the required dual of the given LPP is

$\qquad$ Max. $Z_y = 2y_1 - 3y_2 + 5y_3$

$\qquad$ s.t. $\qquad 2y_1 - 3y_2 - y_3 \le 2$

$\qquad\qquad\qquad 3y_1 - y_2 - 4y_3 \le 2$

$\qquad\qquad\qquad 5y_1 - 7y_2 - 6y_3 \le 4$

$\qquad$ and $\qquad y_1, y_2, y_3 \ge 0$

EXAMPLE 6. *Find the dual of the following LPP:*

$$\text{Min. } Z = 7x_1 + 3x_2 + 8x_3$$

subject to the constraints

$$8x_1 + 2x_2 + x_3 \ge 3$$
$$3x_1 + 6x_2 + 4x_3 \ge 4$$
$$4x_1 + x_2 + 5x_3 \ge 1$$
$$x_1 + 5x_2 + 2x_3 \ge 7$$

and $\qquad\qquad x_1, x_2, x_3 \ge 0$

SOLUTION. The given LPP is in standard form. Since there are four constraints in the problem, so in the dual problem there will be four variables.

The given LPP can be written in matrix form as follows:

Min. $Z = 7x_1 + 3x_2 + 8x_3 = [7 \quad 3 \quad 8][x_1 \quad x_2 \quad x_3] = \boldsymbol{C} \cdot \boldsymbol{x}$

such that

$$\begin{bmatrix} 8 & 2 & 1 \\ 3 & 6 & 4 \\ 4 & 1 & 5 \\ 1 & 5 & 2 \end{bmatrix} \begin{bmatrix} x_1 \\ x_2 \\ x_3 \end{bmatrix} \le \begin{bmatrix} 3 \\ 4 \\ 1 \\ 7 \end{bmatrix}$$

or $\qquad A\boldsymbol{x} \ge \boldsymbol{b}, x_1, x_2, x_3 \ge 0$

Now, the dual of this LPP is given by

$\qquad$ Max $Z_y = \boldsymbol{b}' \cdot \boldsymbol{y} = [3 \quad 4 \quad 1 \quad 7][y_1 \quad y_2 \quad y_3 \quad y_4]$

$\qquad\qquad = 3y_1 + 4y_2 + y_3 + 7y_4$

s.t. $\qquad A'\boldsymbol{y} \le \boldsymbol{C}'$

$\implies \begin{bmatrix} 8 & 3 & 4 & 1 \\ 2 & 6 & 1 & 5 \\ 1 & 4 & 5 & 2 \end{bmatrix} \begin{bmatrix} y_1 \\ y_2 \\ y_3 \\ y_4 \end{bmatrix} \le \begin{bmatrix} 7 \\ 3 \\ 8 \end{bmatrix}$

$\implies \quad 8y_1 + 3y_2 + 4y_3 + y_4 \le 7;$

$$2y_1 + 6y_2 + y_3 + 5y_4 \le 3$$
$$y_1 + 4y_2 + 5y_3 + 2y_4 \le 8$$

and $\qquad y_1, y_2, y_3, y_4 \ge 0$

Hence, the dual problem of the given LPP is

Max. $Z_y = 3y_1 + 4y_2 + y_3 + 7y_4$

s.t. $8y_1 + 3y_2 + 4y_3 + y_4 \le 7$
$\qquad 2y_1 + 6y_2 + y_3 + 5y_4 \le 3$
$\qquad y_1 + 4y_2 + 5y_3 + 2y_4 \le 8$

and $\qquad y_1, y_2, y_3, y_4 \ge 0$

EXAMPLE 7. **Write the dual of the following LPP:**

$$\textbf{Min. } Z = 2x_2 + 5x_3$$

subject to the constraints

$$x_1 + x_2 \ge 2$$
$$2x_1 + x_2 + 6x_3 \le 6$$
$$x_1 - x_2 + 3x_3 = 4$$

and $\qquad x_1, x_2, x_3 \ge 0$ [MEERUT–2005]

SOLUTION. To write the given LPP in standard form, we proceed as follows:

(i) Multiply the second constraint by –1, then it becomes
$$-2x_1 - x_2 - 6x_3 \ge -6$$

(ii) The third constraint is an equality. Replace it by two constraints such that
$$x_1 - x_2 + 3x_3 \ge 4$$
and $\qquad x_1 - x_2 + 3x_3 \le 4 \quad \Rightarrow \quad -x_1 + x_2 - 3x_3 \ge -4$

Thus, the given LPP can be written as (in standard form)

Min. $Z = 0x_1 + 2x_2 + 5x_3$

s.t. $\qquad x_1 + x_2 \ge 2$
$$-2x_1 - x_2 - 6x_3 \ge -6$$
$$x_1 - x_2 + 3x_3 \ge 4$$
$$-x_1 + x_2 - 3x_3 \ge -4$$

and $\qquad x_1, x_2, x_3 \ge 0$

The above LPP can be written in matrix form as

Min. $Z = [0 \quad 2 \quad 5][x_1 \quad x_2 \quad x_3] = \textbf{C} \cdot \textbf{x}$

such that $\begin{bmatrix} 1 & 1 & 0 \\ -2 & -1 & -6 \\ 1 & -1 & 3 \\ -1 & 1 & -3 \end{bmatrix} \begin{bmatrix} x_1 \\ x_2 \\ x_3 \end{bmatrix} \ge \begin{bmatrix} 2 \\ -6 \\ 4 \\ -4 \end{bmatrix}$

$\Rightarrow \qquad A\textbf{x} \ge \textbf{b}$ and $x_1, x_2, x_3 \ge 0$

$\therefore$ Dual is

Max. $Z_y = \textbf{b}' \cdot \textbf{y} = [2 \quad -6 \quad 4 \quad -4][y_1 \quad y_2 \quad y_3' \quad y_3'']$

$\qquad = 2y_1 - 6y_2 + 4(y_3' - y_3'')$

(y_3', y_3'' are taken because the third constraint of the primal have = sign) such that $A'\mathbf{y} \le \mathbf{C}'$

$$\Rightarrow \begin{bmatrix} 1 & -2 & 1 & 1 \\ 1 & -1 & -1 & -1 \\ 0 & -6 & 3 & 3 \end{bmatrix} \begin{bmatrix} y_1 \\ y_2 \\ y_3' \\ y_3'' \end{bmatrix} \le \begin{bmatrix} 0 \\ 2 \\ 5 \end{bmatrix}$$

or $\begin{bmatrix} y_1 - 2y_2 + y_3' - y_3'' \\ y_1 - y_2 - y_3' + y_3'' \\ 0y_1 - 6y_2 + 3y_3' - 3y_3'' \end{bmatrix} \le \begin{bmatrix} 0 \\ 2 \\ 5 \end{bmatrix}$

$$y_1, y_2, y_3', y_3'' \ge 0$$

or   Max. $Z_y = 2y_1 - 6y_2 - 4(y_3' - y_3'')$

subject to the constraints

$$y_1 - 2y_2 + (y_3' - y_3'') \le 0$$
$$y_1 - y_2 - (y_3' - y_3'') \le 2$$
$$-6y_1 + 3(y_3' - y_3'') \le 5$$

and $y_1, y_2, y_3', y_3'' \ge 0$

Finally, put $y_3 = y_3' - y_3''$, the required dual be given by

    Max. $Z_y = 2y_1 - 6y_2 + 4y_3$

    subject to

$$y_1 - 2y_2 + y_3 \le 0$$
$$y_1 - y_2 - y_3 \le 2$$
$$-6y_1 + 3y_3 \le 5$$

and $y_1, y_2 \ge 0, y_3$ is unrestricted.

EXAMPLE 8.   ***Write the dual of the following LPP:***

        Max. Z = 2x₁ + 3x₂ + x₃

        subject to the constraints

            4x₁ + 3x₂ + x₃ = 6

            x₁ + 2x₂ + 5x₃ = 4

      and   ***x₁, x₂, x₃ ≥ 0***

SOLUTION.   The given primal problem is not in standard form. Following the usual procedure, the standard form of the given LPP is

      Max. $Z = 2x_1 + 3x_2 + x_3$

      s.t. $4x_1 + 3x_2 + x_3 \le 6$

            $-4x_1 - 3x_2 - x_3 \le -6$

            $x_1 + 2x_2 + 5x_3 \le 4$

            $-x_1 - 2x_2 - 5x_3 \le -4$

      and $x_1, x_2, x_3 \ge 0$

The above problem can be written in matrix form as

      Max. $Z = 2x_1 + 3x_2 + x_3 = \begin{bmatrix} 2 & 3 & 1 \end{bmatrix} \begin{bmatrix} x_1 & x_2 & x_3 \end{bmatrix} = \mathbf{C} \cdot \mathbf{x}$

subject to

$$\begin{bmatrix} 4 & 3 & 1 \\ -4 & -3 & -1 \\ 1 & 2 & 5 \\ -1 & -2 & -5 \end{bmatrix} \begin{bmatrix} x_1 \\ x_2 \\ x_3 \end{bmatrix} \leq \begin{bmatrix} 6 \\ -6 \\ 4 \\ -4 \end{bmatrix}$$

or $\quad Ax \leq b$

and $x_1, x_2, x_3 \geq 0$

Now, the required dual is

$$\text{Min } Z_y = b'y = [6 \quad -6 \quad 4 \quad -4] = [y_1' \quad y_1'' \quad y_2' \quad y_2'']$$

$$= 6(y_1' - y_1'') + 4(y_2' - y_2'')$$

such that $A'y \geq C'$

$$\Rightarrow \quad \begin{bmatrix} 4 & -4 & 1 & -1 \\ 3 & -3 & 2 & -2 \\ 1 & -1 & 5 & -5 \end{bmatrix} \begin{bmatrix} y_1' \\ y_1'' \\ y_2' \\ y_2'' \end{bmatrix} \geq \begin{bmatrix} 2 \\ 3 \\ 1 \end{bmatrix}$$

$$\Rightarrow \quad \begin{bmatrix} 4y_1' - 4y_1'' + y_2' - y_2'' \\ 3y_1' - 3y_1'' + 2y_2' - 2y_2'' \\ y_1' - y_1'' + 5y_2' - 5y_2'' \end{bmatrix} \geq \begin{bmatrix} 2 \\ 3 \\ 1 \end{bmatrix} \quad y_1', y_1'', y_2', y_2'' \geq 0$$

$$\Rightarrow \quad 4(y_1' - y_1'') + y_2' - y_2'' \geq 2$$

$$3(y_1' - y_1'') + 2(y_2' - y_2'') \geq 3$$

$$(y_1' - y_1'') + 5(y_2' - y_2'') \geq 1$$

Putting $y_1 = y_1' - y_1''$ and $y_2 = y_2' - y_2''$, we get the required dual is

$$\text{Min. } Z_y = 6y_1 + 4y_2$$

$$\text{s.t.} \quad 4y_1 + y_2 \geq 2$$

$$3y_1 + 2y_3 \geq 3$$

$$y_1 + 5y_2 \geq 0$$

and y_1, y_2 are unrestricted.

EXAMPLE 9. *Find the dual of the following LPP:*

$$\textbf{Min. } \textbf{Z} = \textbf{x}_1 - 3\textbf{x}_2 - 2\textbf{x}_3$$

subject to the constraints

$$3\textbf{x}_1 - \textbf{x}_2 + 2\textbf{x}_3 \leq 7$$

$$2\textbf{x}_1 - 4\textbf{x}_2 \geq 12$$

$$-4\textbf{x}_1 + 3\textbf{x}_2 + 8\textbf{x}_3 = 10$$

and $\textbf{x}_1, \textbf{x}_2 \geq 0$, $\textbf{x}_3$ is unrestricted.

SOLUTION. Following the usual procedure, the standard form of the given LPP is

$$\text{Min. } Z = x_1 - 3x_2 - 2(x_3' - x_3'') = x_1 - 3x_2 - 2x_3' + 2x_3''$$

$$\text{s.t.} \quad -3x_1 + x_2 - 2x_3' + 2x_3'' \geq -7$$

$$2x_1 - 4x_2 \geq 12$$
$$-4x_1 + 3x_2 + 8x_3' - 8x_3'' \geq 10$$
$$4x_1 - 3x_2 - 8x_3' + 8x_3'' \geq -10$$

and $x_1, x_2, x_3', x_3'' \geq 0$

The above LPP can be written in matrix from as follows:

$$\text{Min. } Z = [1 \quad -3 \quad -2 \quad 2][x_1 \quad x_2 \quad x_3' \quad x_3''] = \boldsymbol{C} \cdot \boldsymbol{x}$$

such that

$$\begin{bmatrix} -3 & 1 & -2 & 2 \\ 2 & -4 & 0 & 0 \\ -4 & 3 & 8 & -8 \\ 4 & -3 & -8 & 8 \end{bmatrix} \begin{bmatrix} x_1 \\ x_2 \\ x_3' \\ x_3'' \end{bmatrix} \geq \begin{bmatrix} -7 \\ 12 \\ 10 \\ -10 \end{bmatrix} \implies \boldsymbol{Ax} \geq \boldsymbol{b}$$

and $x_1, x_2, x_3', x_3'' \geq 0$

Now the dual of the above problem is given by

$$\text{Max. } Z_y = \boldsymbol{b'} \cdot \boldsymbol{y} = [-7 \quad 12 \quad 10 \quad -10][y_1 \quad y_2 \quad y_3' \quad y_3'']$$
$$= -7y_1 + 12y_2 + 10(y_3' - y_3'')$$

such that $A'\boldsymbol{y} \leq \boldsymbol{C'}$

$$\implies \begin{bmatrix} -3 & 2 & -4 & 4 \\ 1 & -4 & 3 & -3 \\ -2 & 0 & 8 & -8 \\ 2 & 0 & -8 & 8 \end{bmatrix} \begin{bmatrix} y_1 \\ y_2 \\ y_3' \\ y_3'' \end{bmatrix} \leq \begin{bmatrix} 1 \\ -3 \\ -2 \\ 2 \end{bmatrix}$$

$$\implies -3y_1 + 2y_2 - 4(y_3' - y_3'') \leq 1$$
$$y_1 - 4y_2 + 3(y_3' - y_3'') \leq -3$$
$$-2y_1 - 4y_2 + 8(y_3' - y_3'') \leq -2$$
$$2y_1 - 8(y_3' - y_3'') \leq 2$$

and $y_1, y_2, y_3', y_3'' \geq 0$

Now using $y_3 = y_3' - y_3''$, we get the required dual as

$$\text{Max. } Z_y = -7y_1 + 12y_2 + 10y_3$$

s.t. $-3y_1 + 2y_2 - 4y_3 \leq 1$
$$y_1 - 4y_2 + 3y_3 \leq -3$$
$$-2y_1 + 8y_3 \leq -2 \implies 2y_1 - 8y_3 \geq 2$$
$$2y_1 - 8y_3 \leq 2$$

and $y_1, y_2 \geq 0, y_3$ is unrestricted.

Hence, the required dual of the given LPP is

$$\text{Max. } Z_y = -7y_1 + 12y_2 + 10y_3$$

s.t. $-3y_1 + 2y_2 - 4y_3 \leq 1$
$$-y_1 + 4y_2 - 3y_3 \geq 3$$
$$2y_1 - 8y_3 = 2$$

and $y_1, y_2 \geq 0, y_3$ is unrestricted.

EXAMPLE 10. ***Write the dual of the following problem***

$$\text{Min } Z = x_1 + x_2 + x_3$$
$$\text{s.t.} \quad x_1 - 3x_2 + 4x_3 = 5$$
$$x_1 - 2x_2 \le 3$$
$$2x_2 - x_3 \ge 4$$

and $x_1, x_2 \ge 0$; x_3 ***is unrestricted.***

[MEERUT-1994, 95; GARHWAL-1997]

SOLUTION. Using $x_3 = x_3' - x_3''$, the given LPP can be written in standard form as follows.

Min. $Z = x_1 + x_2 + x_3' - x_3''$

subject to the constraints

$$-x_1 + 3x_2 - 4(x_3' - x_3'') \ge -5$$
$$x_1 - 3x_2 + 4(x_3' - x_3'') \ge 5$$
$$-x_1 + 2x_2 \ge -3$$
$$2x_2 - (x_3' - x_3'') \ge 4$$

and $\qquad x_1, x_2, x_3', x_3'' \ge 0$

The above problem can be written in matrix form as follows:

Min. $Z = [1 \quad 1 \quad 1 \quad -1][x_1 \quad x_2 \quad x_3' \quad x_3''] = \boldsymbol{C} \cdot \boldsymbol{x}$

$\Rightarrow \quad A\boldsymbol{x} \ge \boldsymbol{b}, \ x_1, x_2, x_3', x_3'' \ge 0$

Now, the dual of the given primal is

Max. $Z_y = [-5 \quad 5 \quad -3 \quad 4][y_1' \quad y_1'' \quad y_2 \quad y_3]$

$$= -5(y_1' - y_1'') - 3y_2 + 4y_3$$

subject to the constraints $A' \boldsymbol{y} \le \boldsymbol{C}'$

$$\Rightarrow \quad \begin{bmatrix} -1 & 1 & -1 & 0 \\ 3 & -3 & 2 & 2 \\ -4 & 4 & 0 & -1 \\ 4 & -4 & 0 & 1 \end{bmatrix} \begin{bmatrix} y_1' \\ y_1'' \\ y_2 \\ y_3 \end{bmatrix} \le \begin{bmatrix} 1 \\ 1 \\ 1 \\ -1 \end{bmatrix}$$

$$\Rightarrow \quad -(y_1' - y_1'') - y_2 \le 1$$
$$3(y_1' - y_1'') + 2y_2 + 2y_3 \le 1$$
$$-4(y_1' - y_1'') + y_3 \le -1$$

and $\qquad y_1', y_1'', y_2, y_3 \ge 0$

Using $y_1 = y_1' - y_1''$, we get the dual is

Max. $Z_y = -5y_1 - 3y_2 + 4y_3$

subject to

$$-y_1 - y_2 \le 1; \ 3y_1 + 2y_2 + 2y_3 \le 1$$
$$-4y_1 - y_3 \le 1, \ 4y_1 + y_3 \le -1 \ \Rightarrow \ -4y_1 - y_3 \ge 1$$
$$y_2, y_3 \ge 0 \text{ and } y_1 \text{ is unrestricted.}$$

Finally, the required dual is

Max. $Z_y = -5y_1 - 3y_2 + 4y_3$

s.t. $-y_1 - y_2 \leq 1$

$$3y_1 + 2y_2 + 2y_3 \leq 1$$

$$-4y_1 - y_3 = 1$$

and $y_2, y_3 \geq 0$, y_1 is unrestricted in sign.

 ## Exercise-7.1

Find the dual of the following LPP:

1. Max. $Z = x_1 - x_2 + 3x_3$
subject to the constraints
$$x_1 + x_2 + x_3 \leq 10$$
$$2x_1 - x_3 \leq 2$$
$$2x_1 - 2x_2 + 3x_3 \leq 3$$
and $x_1, x_2, x_3 \geq 0$

2. Min $Z = 4x_1 + 6x_2 + 18x_3$
subject to the constraints
$$x_1 + 3x_3 \geq 3$$
$$x_2 + 2x_3 \geq 5$$
and $x_1, x_2, x_3 \geq 0$

3. Min. $Z = 10x_1 + 20x_2$
subject to the constraints
$$3x_1 + 2x_2 \geq 18$$
$$2x_1 - x_2 \leq 6$$
and $x_1, x_2 \geq 0$

4. Max. $Z = 3x_1 + x_2 + 4x_3 + x_4 + 9x_5$
subject to the constraints
$$4x_1 - 5x_2 - 9x_3 + x_4 - 2x_5 \leq 6$$
$$2x_1 + 3x_2 + 4x_3 - 5x_4 + x_5 \leq 9$$
$$x_1 + x_2 - 5x_3 - 7x_4 + 11x_5 \leq 10$$
and $x_1, x_2, x_3, x_4, x_5 \geq 0$

5. Min. $Z = 3x_1 - 2x_2 + 4x_3$
subject to the constraints
$$3x_1 + 5x_2 + 4x_3 \geq 7$$
$$6x_1 + x_2 + 3x_3 \geq 4$$
$$7x_1 - 2x_2 - x_3 \leq 10$$
$$x_1 - 2x_2 + 5x_3 \geq 3$$
$$4x_1 + 7x_2 - 2x_3 \geq 2$$
and $x_1, x_2, x_3 \geq 0$
[GORAKHPUR–2009, 11]

6. Max. $Z = x_1 + 3x_2$
subject to the constraints
$$3x_1 + 2x_2 \leq 6$$
$$3x_1 + x_2 = 4$$
and $x_1, x_2 \geq 0$ [MEERUT–2017]

7. Min. $Z = x_1 + x_2 + x_3$
subject to the constraints
$$x_1 - 3x_2 + 4x_3 = 5$$
$$x_1 - 2x_2 \leq 3$$
$$2x_2 - x_3 \geq 4$$
and $x_1, x_3 \geq 0$, x_2 is unrestricted
[MEERUT–2006]

8. Max. $Z = 3x_1 + x_2 + x_3 - x_4$
subject to the constraints
$$x_1 + 5x_2 + 3x_3 + 4x_4 \leq 5$$
$$x_1 + x_2 = -1$$
$$x_3 - x_4 \geq -5$$
and $x_1, x_2, x_3, x_4 \geq 0$

9. Max. $Z = 3x_1 + 5x_2 + 7x_3$
subject to the constraints
$$x_1 + x_2 + 3x_3 \leq 10$$
$$4x_1 - x_2 + 2x_3 \geq 15$$
and $x_1, x_2 \geq 0$; x_3 is unrestricted

10. Max. $Z = 3x_1 + x_2 + 2x_3 - x_4$
subject to the constraints
$$2x_1 - x_2 + 3x_3 + x_4 = 1$$
$$x_1 + x_2 - x_3 + x_4 = 3$$
and $x_1, x_2, x_3 \geq 0$; x_4 is unrestricted.

ANSWERS

1. Max. $Z_y = 10y_1 + 2y_2 + 3y_3$
s.t.
$$y_1 + 2y_2 + 2y_3 \geq 1$$
$$y_1 - 2y_3 \geq -1$$
$$y_1 - y_2 + 3y_3 \geq 3$$
and $y_1, y_2, y_3 \geq 0$

2. Max. $Z_y = 3y_1 + 5y_2$
s.t.
$$y_1 \leq 4$$
$$y_2 \leq 6$$
$$3y_1 + 2y_2 \leq 18$$
and $y_1, y_2 \geq 0$

3. Max. $Z_y = 18y_1 + 8y_2 - 6y_3$
s.t.
$3y_1 + y_2 - 2y_3 \leq 10$
$2y_1 + 3y_2 + y_3 \leq 20$
and $y_1, y_2, y_3 \geq 0$

4. Min. $Z_y = 6y_1 + 9y_2 + 10y_3$
s.t.
$4y_1 + 2y_2 + y_3 \geq 3$
$-5y_1 + 3y_2 + y_3 \geq 1$
$-9y_1 + 4y_2 - 5y_3 \geq 4$
$y_1 - 5y_2 - 7y_3 \geq 1$
$-2y_1 + y_2 + 11y_3 \geq 9$
and $y_1, y_2, y_3 \geq 0$

5. Max. $Z_y = 7y_1 + 4y_2 - 10y_3 + 3y_4 + 2y_5$
s.t.
$3y_1 + 6y_2 - 7y_3 + y_4 + y_5 \leq 3$
$5y_1 + y_2 + 2y_3 - 2y_4 + 7y_5 \leq -2$
$4y_1 + 3y_2 + y_3 + 5y_4 - 2y_5 \leq 4$
and $y_1, y_2, y_3, y_4, y_5 \geq 0$

6. Min. $Z_y = 6y_1 + 4y_2$
s.t.
$3y_1 + 3y_2 \geq 1$
$2y_1 + y_2 \geq 3$
and $y_1 \geq 0, y_2$ is unrestricted in sign

7. Max. $Z_y = 5y_1 - 3y_2 + 4y_3$
s.t.
$y_1 - y_2 \leq 1$
$-3y_1 + 2y_2 + 2y_3 = 1$
$4y_1 - y_3 \leq 1$
and $y_2, y_3 \geq 0; y_1$ is unrestricted

8. Min. $Z_y = 5y_1 - y_2 + 5y_3$
s.t.
$y_1 + y_2 \geq 3$
$5y_1 + y_2 \geq 1$
$3y_1 - y_3 \geq 1$
$4y_1 + y_3 \geq -1$
and $y_1, y_3 \geq 0; y_2$ is unrestricted

9. Min. $Z_y = 10y_1 - 15y_2$
s.t.
$y_1 - 4y_2 \geq 3$
$y_1 + y_2 \geq 5$
$3y_1 - 2y_2 = 7$
and $y_1, y_2 \geq 0$

10. Min. $Z_y = y_1 + 3y_2$
s.t.
$2y_1 + y_2 \geq 3$
$-y_1 + y_2 \geq 1$
$3y_1 - y_2 \geq 2$
$y_1 + y_2 = -1$
and y_1, y_2 are unrestricted.

7.5 SOME RESULTS ON DUALITY

THEOREM I. *The dual of a dual of the given primal is the primal itself.*

[KANPUR–2011; MEERUT–2007, 09, 10, 14]

PROOF. Consider the primal problem

$$\text{Max.} Z_x = c_1x_1 + c_2x_2 + \ldots + c_nx_n$$

subject to the constraints

$$a_{11}x_1 + a_{12}x_2 + \ldots + a_{1n}x_n \leq b_1$$
$$a_{21}x_1 + a_{22}x_2 + \ldots + a_{2n}x_n \leq b_2$$
$$\vdots$$
$$a_{m1}x_1 + a_{m2}x_2 + \ldots + a_{mn}x_n \leq b_m$$

...(1)

and $\qquad x_1, x_2, \ldots, x_n \geq 0$

Following the usual procedure, the dual of the above primal is

$$\text{Min.} Z_w = b_1w_1 + b_2w_2 + \ldots + b_mw_m$$

subject to the constraints

$$a_{11}w_1 + a_{21}w_2 + \ldots + a_{m1}w_m \geq c_1$$
$$a_{12}w_1 + a_{22}w_2 + \ldots + a_{m2}w_m \geq c_2$$
$$\vdots$$
$$a_{1n}w_1 + a_{2n}w_2 + \ldots + a_{mn}w_m \geq c_n$$

...(2)

and $$w_1, w_2, \ldots, w_m \geq 0$$

Next to find the dual of the above dual, firstly we shall write this into standard form as follows:

The standard maximization form is

$$\text{Max.}(-Z_w) = -b_1 w_1 - b_2 w_2 - \ldots - b_m w_m$$

subject to the constraints

$$\left.\begin{array}{c} -a_{11} w_1 - a_{21} w_2 - \ldots - a_{m1} w_m \leq -c_1 \\ -a_{12} w_1 - a_{22} w_2 - \ldots - a_{m2} w_m \leq -c_2 \\ \vdots \\ -a_{1n} w_1 - a_{2n} w_2 - \ldots - a_{mn} w_m \leq -c_n \end{array}\right] \qquad \ldots(3)$$

and $$w_1, w_2, \ldots, w_m \geq 0$$

Following the usual procedure, the dual of (3) is given by

$$\text{Min.} Z_y = -c_1 y_1 - c_2 y_2 - \ldots - c_n y_n$$

subject to the constraints

$$\left.\begin{array}{c} -a_{11} y_1 - a_{12} y_2 - \ldots - a_{1n} y_n \geq -b_1 \\ -a_{21} y_1 - a_{22} y_2 - \ldots - a_{2n} y_n \geq -b_2 \\ \vdots \\ -a_{m1} y_1 - a_{m2} y_2 - \ldots - a_{mn} y_n \geq -b_m \end{array}\right] \qquad \ldots(4)$$

and $$y_1, y_2, \ldots, y_n \geq 0$$

Finally system (4) can be written in standard maximization form as given below:

$$\text{Max.} Z'_y = c_1 y_1 + c_2 y_2 + \ldots + c_n y_n \quad (\text{Here } Z'_y = -Z_y)$$

subject to the constraints

$$\left.\begin{array}{c} a_{11} y_1 + a_{12} y_2 + \ldots + a_{1n} y_n \leq b_1 \\ a_{21} y_1 + a_{22} y_2 + \ldots + a_{2n} y_n \leq b_2 \\ \vdots \\ a_{m1} y_1 + a_{m2} y_2 + \ldots + a_{mn} y_n \leq b_m \end{array}\right] \qquad \ldots(5)$$

and $$y_1, y_2, \ldots, y_n \geq 0$$

We observe that given primal system (1) and dual of dual (5) are identical.

Hence, we conclude that dual of a dual is the primal.

THEOREM 2. **(Weak LP duality Theorem)**

If x is any feasible solution to the primal given by $Z_p = Cx$ subject to $Ax \leq b$, $x \geq 0$ and w is any feasible solution to the dual problem.

$$\text{Min. } Z_D = b'\cdot w \text{ subject to } A'w \geq C', w \geq 0$$

Then $C\cdot x \leq b'w$, i.e., $Z_p \leq Z_D$

PROOF. Consider the given primal

$$\text{Max.} Z_p = C \cdot x$$

subject to the constraints

$$\left.\begin{array}{c} Ax \leq b \\ \\ x \geq 0 \end{array}\right] \qquad \ldots(1)$$

and

Following the usual procedure, the dual of primal (1) is given by

$$\text{Min.} Z_D = \boldsymbol{b}'\boldsymbol{w}$$

subject to the constraints

$$A'\boldsymbol{w} \geq \boldsymbol{C}'$$

and

$$\boldsymbol{w} \geq 0$$ ...(2)

Further, suppose that $\boldsymbol{w} = (w_1, w_2, ..., w_m)$ be any feasible solution of the dual (2). Since, $A\boldsymbol{x} \leq \boldsymbol{b}$ are the constraints of the primal (1), multiply both sides of $A\boldsymbol{x} \leq \boldsymbol{b}$ with $\boldsymbol{w}'$, we get

$$\boldsymbol{w}'(A\boldsymbol{x}) \leq \boldsymbol{w}' \cdot \boldsymbol{b}$$

$$\Rightarrow \qquad (A'\boldsymbol{w})'\boldsymbol{x} \leq (\boldsymbol{b}'\,\boldsymbol{w})' \qquad \text{...(3)}$$

Further, $\boldsymbol{x}$ is the feasible solution of the primal (1) and $A'\boldsymbol{w} \geq \boldsymbol{C}'$ denote the constraints of the dual (2). Thus,

$$\boldsymbol{x}'(A'\boldsymbol{w}) \geq \boldsymbol{x}' \cdot \boldsymbol{C}'$$

$$\Rightarrow \qquad \boldsymbol{x}'(\boldsymbol{w}'A)' \geq (\boldsymbol{C} \cdot \boldsymbol{x})' \qquad [\because \ (ab)' = b'a']$$

$$\Rightarrow \qquad [(\boldsymbol{w}'A)\boldsymbol{x}]' \geq (\boldsymbol{C}\boldsymbol{x})'$$

$$\Rightarrow \qquad (\boldsymbol{w}'A)\boldsymbol{x} \geq \boldsymbol{C} \cdot \boldsymbol{x}$$

$$\Rightarrow \qquad (A'\boldsymbol{w})' \geq \boldsymbol{C} \cdot \boldsymbol{x} \qquad \text{...(4)}$$

On combining (3) and (4), we get

$$\boldsymbol{C} \cdot \boldsymbol{x} \leq (A'\boldsymbol{w})'\boldsymbol{x} \leq (\boldsymbol{b}'\boldsymbol{w})'$$

$$\Rightarrow \qquad \boldsymbol{C} \cdot \boldsymbol{x} \leq (\boldsymbol{b}'\boldsymbol{w})'$$

$$\Rightarrow \qquad \boldsymbol{C}\boldsymbol{x} \leq \boldsymbol{b}'\boldsymbol{w}$$

Hence, $Z_p \leq Z_D$.

<u>**THEOREM 3.**</u> **(Basic Duality Theorem)** *If $\boldsymbol{x_0}$ is an optimal solution to the primal then there exists a feasible solution $\boldsymbol{w_0}$ to the dual such that $C\boldsymbol{x_0} = \boldsymbol{b}'\boldsymbol{w_0}$, $\boldsymbol{b}'$ is the transpose of $\boldsymbol{b}$.*

<u>PROOF.</u> Consider the primal problem

$$\text{Max } Z_p = \boldsymbol{C}\boldsymbol{x}$$

subject to the constraints

$$A\boldsymbol{x} \leq \boldsymbol{b}$$

and

$$\boldsymbol{x} \geq 0$$ ...(1)

Let the dual of (1) be given by

$$\text{Min } Z_d = \boldsymbol{b}'\boldsymbol{w}$$

subject to the constraints

$$A'\boldsymbol{w} \geq \boldsymbol{C}'$$

and

$$\boldsymbol{w} \geq 0$$ ...(2)

As per given, $\boldsymbol{x_0}$ is an optimal solution to the primal (1), then as in simplex method system (1), can be written as

$$\text{Max. } Z_p = \boldsymbol{C}\boldsymbol{x}$$

subject to the constraints
$$Ax + IS = b$$
where S is the vector of slack variables and I is the associated $m \times n$ identity matrix. If x_0 $(=(x_B, 0))$ is an optimal solution to the primal (1) where x_B denote the optimal basic feasible solution such that
$$x_B = B^{-1}b, B \text{ is the optimal basis of } A$$
then the optimal primal objective function is
$$Z = Cx_0 = C_B x_B, C_B \text{ is the vector containing the prices of the basic variables.}$$
Now, consider

$$z_j - c_j = C_B y_j - c_j = \begin{cases} C_B \cdot B^{-1} \alpha_j - c_j \ \forall \ \alpha_j \in A \\ C_B \cdot B^{-1} e_j - 0 \ \forall \ e_j \in I \end{cases}$$

Now, since x_0 is the optimal solution, so we must have
$$z_j - c_j \geq 0 \ \forall \ j$$
which implies $C_B B^{-1} \alpha_j \geq c_j, C_B B^{-1} e_j > 0 \ \forall \ j$

In matrix form, it can be written as
$$C_B B^{-1} A \geq C, C_B B^{-1} \geq 0$$
$$\Rightarrow \qquad A'B^{-1} C_B' \geq C'B^{-1} C_B \geq 0$$
$$\Rightarrow \qquad A'w_0 \geq C'; w_0 \geq 0 \qquad\qquad (B^{-1}C_B = w_0)$$
which shows that w_0 is a feasible solution of the dual problem and corresponding dual objective function is
$$b'w_0 = w_0' \cdot b = C_B \cdot B^{-1}b = C_B x_B = C \cdot x_0$$
Hence, we conclude that corresponding to a given optimal solution x_0 of the primal, there exists a feasible solution w_0 of dual such that
$$Cx_0 = b'w_0$$

☛ REMARK

- In a similar manner, we may prove the following result:
  "If w_0 is an optimal solution to the dual, then there exists a feasible solution x_0 to the primal such that $Cx_0 = b'w_0$."

THEOREM 4. **(The necessary and sufficient condition for an LPP and its dual to have optimal solution)**

The necessary and sufficient condition for any linear programming problem and its dual to have optimal solution is that both have feasible solution.

PROOF. Consider the primal problem

$$\left. \begin{array}{l} \text{Max.} Z_p = C \cdot x \\[2ex] Ax \leq b \\[2ex] x \geq 0 \end{array} \right] \qquad \ldots(1)$$

subject to the constraints

and

Let x be any feasible solution of (1)

The dual of (1) is given by

$$\text{Min.} Z_D = b' \cdot w$$

subject to the constraints

$$A'w \ge C'$$

and

$$w \ge 0$$

...(2)

Let $\hat{w}$ be any feasible solution of (2) such that $C \cdot \hat{x} = b \cdot \hat{w}$. We have to prove that $\hat{x}$ is the optimal solution of the primal and $\hat{w}$ is the optimal solution of the dual.

Using weak LP duality theorem, we can write

$$Z_p \le Z_D$$

$$\Rightarrow \quad Cx \le b' \cdot w'$$

$$\Rightarrow \quad Cx \le C\hat{x} \qquad\qquad [\because \ C\hat{x} = b'\hat{w} \ \text{(given)}]$$

which shows that the value of the objective function of the primal at the feasible solution $\hat{x}$ is greater than its value at any other feasible solution x.

$\Rightarrow \ \hat{x}$ is the optimal solution of the primal maximization problem.

Further, suppose that w is any feasible solution of dual (2), $\hat{x}$ is the given feasible solution of the primal (1). Then again by weak LP duality theorem.

$$C\hat{x} \le b'w$$

$$\Rightarrow \quad b'\hat{w} \le b'w \qquad\qquad (\because \ C\hat{x} = b'\hat{w}) \quad \text{(By basic duality theorem)}$$

which shows that the value of the objective function of the dual problem at the given feasible solution $\hat{w}$ is less than its value at any other feasible solution w.

$\Rightarrow \hat{w}$ is the optimal solution of the dual minimization problem.

☞ **REMARK**

- The above theorem can be restated as:

"If $\hat{x}$ is a feasible solution to the primal problem given by Max. $Z_p = C \cdot x$ subject to $Ax \le b$, $x \ge 0$ and $\hat{w}$ is a feasible solution to its dual Min. $Z_D = b'w$ subject to $A'w \ge C'$, $w \ge 0$ such that $C\hat{x} = b' \cdot \hat{w}$, then $\hat{x}$ is the optimal solution of the primal and $\hat{w}$ is the optimal solution of the dual problem."

THEOREM 5. **(Fundamental Theorem of Duality)** *If either the primal or the dual problem has a finite optimal solution, then the other problem has a finite optimal solution and the optimal values of the objective function in both the problems are the same.* [DELHI–2004; PATNA–2007]

PROOF. Consider the primal problem

$$\text{Max} \, Z_p = Cx$$

subject to the constraints

$$Ax \le b$$

and

$$x \ge 0$$

...(1)

The dual of (1) is given by

$$\text{Min } Z_d = \boldsymbol{b'w}$$

subject to the constraints

$$A'\boldsymbol{w} \geq \boldsymbol{C'}$$

...(2)

and

$$\boldsymbol{w} \geq 0$$

Let us assume that the primal has a finite optimal feasible solution $\boldsymbol{x_B}$. Now as in simplex method, introduce slack variables to each of the constraints (1) then primal (1) becomes

$$\text{Max } Z_p = \boldsymbol{Cx}$$

subject to the constraints

$$A\boldsymbol{x} + \boldsymbol{IS} = \boldsymbol{b}$$

...(3)

and

$$\boldsymbol{x} \geq 0, \boldsymbol{S} \geq 0$$

Here, $\boldsymbol{S}$ is the vector of slack variables and $\boldsymbol{I}$ is the associated $m \times n$ identity matrix. $\boldsymbol{B}$ is the basis matrix and $\boldsymbol{C_B}$ is the m-component row vector containing the prices of the basis variables.

Now, since $\boldsymbol{x_B}$ is the optimal solution of (1) therefore, we must have

$$c_j - z_j \leq 0 \ \forall \ j$$

Here, $Z_j = \boldsymbol{C_B} y_i = \boldsymbol{C_B} B^{-1} \alpha_j$

Therefore, $c_j - \boldsymbol{C_B} B^{-1} \alpha_j \leq 0 \ \forall \ \alpha_j$

or

$$\boldsymbol{C_B} B^{-1} \alpha_j \geq c_j$$

...(4)

$\Rightarrow \qquad \boldsymbol{C_B} B^{-1}(\alpha_1, \alpha_2, ..., \alpha_n) \geq (c_1, c_2, ..., c_n)$

which implies that

$$\boldsymbol{C_B} \cdot B^{-1} \cdot A \geq \boldsymbol{C}$$

...(5)

Let us take $(\hat{\boldsymbol{w}})' = \boldsymbol{C_B} \cdot B^{-1}$ where $\hat{\boldsymbol{w}} = [w_1, w_2, ..., w_m]$

Then (5) reduces to

$$(\hat{\boldsymbol{w}})A \geq \boldsymbol{C} \qquad \Rightarrow \qquad [(\hat{\boldsymbol{w}})' \cdot A]' \geq \boldsymbol{C'}$$

$\Rightarrow \qquad A'(\hat{\boldsymbol{w}}) \geq \boldsymbol{C'} \qquad \Rightarrow \qquad \hat{\boldsymbol{w}} \text{ satisfy (2)}$

$\Rightarrow \qquad \hat{\boldsymbol{w}} \text{ is the solution of (2).}$

Next, we show that $\hat{\boldsymbol{w}}$ is the optimal solution to the dual (2).

Consider, $\text{Min } Z_d = \boldsymbol{b'}\hat{\boldsymbol{w}} = [(\hat{\boldsymbol{w}})' \boldsymbol{b}]' = (\hat{\boldsymbol{w}})' \boldsymbol{b}$

$$= [\boldsymbol{C_B} B^{-1}] \boldsymbol{b} = \boldsymbol{C_B}(B^{-1} \cdot \boldsymbol{b}) = \boldsymbol{C_B} \cdot \boldsymbol{x_B} = \text{Max } Z_p$$

Here, $\hat{\boldsymbol{w}}$ and $\boldsymbol{x_B}$ are the feasible solution of the dual (2) and the primal (1) respectively and

$$\text{Min } Z_d = \text{Max. } Z_p$$

THEOREM 6. ***If the primal problem has an unbounded solution then its dual is infeasible.***

PROOF. Let if possible, the dual problem has feasible solution, then by weak duality theorem, any feasible solution to the dual would provide an upper bound on the primal objective function, which is a contradiction, because primal problem is unbounded. Hence, dual has infeasible solution.

In a similar manner, we can prove that if the primal problem has infeasible solution, then its dual has unbounded solution.

- If primal (dual) problem has unbounded optimum solution, the other problem has either no solution at all or an unbounded solution.

THEOREM 7. *If any of the constraints in the primal is a perfect equality, the corresponding dual variable is unrestricted in sign.*

PROOF. Let us suppose in the given primal, the kth constraint is an equality. Then we can write its primal in standard form as follows.

$$\text{Max. } Z = c_1 x_1 + c_2 x_2 + \ldots + c_n x_n$$

subject to the constraints

$$a_{11} x_1 + a_{12} x_2 + \ldots + a_{1n} x_n \leq b_1$$
$$a_{21} x_1 + a_{22} x_2 + \ldots + a_{2n} x_n \leq b_2$$
$$\vdots$$
$$a_{k1} x_1 + a_{k2} x_2 + \ldots + a_{kn} x_n \leq b_k$$
$$-a_{k1} x_1 - a_{k2} x_2 - \ldots - a_{kn} x_n \leq -b_k$$
$$\vdots$$
$$a_{m1} x_1 + a_{m2} x_2 + \ldots + a_{mn} x_n \leq b_m$$

and $\qquad x_1, x_2, \ldots, x_n \geq 0$

We can write the dual of the above primal as follows:

$$\text{Min. } Z_d = b_1 y_1 + b_2 y_2 + \ldots + b_k (y_k' - y_k'') + \ldots + b_m y_m$$

subject to the constraints

$$a_{11} y_1 + a_{21} y_2 + \ldots + a_{k1} (y_k' - y_k'') + \ldots + a_{m1} y_m \geq c_1$$
$$a_{12} y_1 + a_{22} y_2 + \ldots + a_{k2} (y_k' - y_k'') + \ldots + a_{m2} y_m \geq c_2$$
$$\vdots$$
$$a_{1n} y_1 + a_{2n} y_2 + \ldots + a_{kn} (y_k' - y_k'') + \ldots + a_{mn} y_m \geq c_n$$

and $\qquad y_1, y_2, \ldots, y_k', y_k'', \ldots, y_m \geq 0$

Using $y_k = y_k' - y_k''$ in the above dual we get

$$\text{Min. } Z_d = b_1 y_1 + b_2 y_2 + \ldots + b_k y_k + \ldots + b_m y_m$$

subject to the constraints

$$a_{11} y_1 + a_{21} y_2 + \ldots + a_{k1} y_k + \ldots + a_{m1} y_m \geq c_1$$
$$a_{12} y_1 + a_{22} y_2 + \ldots + a_{k2} y_k + \ldots + a_{m2} y_m \geq c_2$$
$$\vdots$$
$$a_{1n} y_1 + a_{2n} y_2 + \ldots + a_{kn} y_k + \ldots + a_{mn} y_m \geq c_n$$

and $\qquad y_1, y_2, \ldots, y_{k-1}, y_{k+1}, \ldots, y_m \geq 0$ and y_k is unrestricted in sign because

$$y_k > 0 \text{ if } y_k' > y_k'' \text{ and } y_k < 0 \text{ if } y_k' < y_k''.$$

THEOREM 8. *If any variable of the primal is unrestricted in sign, the corresponding constraints in the dual will be a strict equality.*

PROOF. Let us consider the primal problem (with k^{th} variable unrestricted) as

$$\text{Max. } Z = c_1 x_1 + c_2 x_2 + \ldots + c_k x_k + \ldots + c_n x_n$$

subject to the constraints

$$a_{11}x_1 + a_{12}x_2 + \ldots + a_{1k}x_k + \ldots + a_{1n}x_n \le b_1$$
$$a_{21}x_1 + a_{22}x_2 + \ldots + a_{2k}x_k + \ldots + a_{2n}x_n \le b_2$$
$$\vdots$$
$$a_{m1}x_1 + a_{m2}x_2 + \ldots + a_{mk}x_k + \ldots + a_{mn}x_n \le b_m$$

and $\quad x_1, x_2, \ldots, x_{k-1}, x_{k+1}, \ldots, x_m \ge 0$ and x_k is unrestricted.

Let us put $x_k = x_k' - x_k''$, $x_k', x_k'' \ge 0$ in the above primal, then we get

$$\text{Max. } Z = c_1x_1 + c_2x_2 + \ldots + c_k(x_k' - x_k'') + \ldots + c_nx_n$$

subject to the constraints

$$a_{11}x_1 + a_{12}x_2 + \ldots + a_{1k}(x_k' - x_k'') + \ldots + a_{1n}x_n \le b_1$$
$$a_{21}x_1 + a_{22}x_2 + \ldots + a_{2k}(x_k' - x_k'') + \ldots + a_{2n}x_n \le b_2$$
$$\vdots$$
$$a_{m1}x_1 + a_{m2}x_2 + \ldots + a_{mk}(x_k' - x_k'') + \ldots + a_{mn}x_n \le b_m$$

and $\quad x_1, x_2, \ldots, x_{k-1}, x_k', x_k'', x_{k+1}, \ldots, x_m \ge 0$

Now, we can write the dual of the above LPP in a usual manner as follows.

$$\text{Min. } Z_d = b_1y_1 + b_2y_2 + \ldots + b_my_m$$

subject to the constraints

$$a_{11}y_1 + a_{21}y_2 + \ldots + a_{m1}y_m \ge c_1$$
$$a_{12}y_1 + a_{22}y_2 + \ldots + a_{m2}y_m \ge c_2$$
$$\vdots$$
$$a_{1k}y_1 + a_{2k}y_2 + \ldots + a_{mk}y_m \ge c_k$$
$$-a_{1k}y_1 - a_{2k}y_2 - \ldots - a_{mk}y_m \ge -c_k$$
$$\vdots$$
$$a_{1n}y_1 + a_{2n}y_2 + \ldots + a_{mn}y_m \ge c_n$$

and $\quad y_1, y_2, \ldots, y_m \ge 0$

Clearly, the two constraints

$$a_{1k}y_1 + a_{2k}y_2 + \ldots + a_{mk}y_k \ge c_k$$

and $\quad -a_{1k}y_1 - a_{2k}y_2 - \ldots - a_{mk}y_k \ge -c_k$

will be equivalent to the single equation

$$a_{1k}y_1 + a_{2k}y_2 + \ldots + a_{mk}y_k = c_k$$

Hence, we conclude that, if the kth-variable of the primal is unrestricted, then the kth-constraints in the dual will be an equality.

THEOREM 9. **(Existence Theorem)** *There exists a bounded (finite) optimum solution to a linear programming problem if and only if there exists a feasible solution to both primal and its dual.*

PROOF. Consider the primal problem

$$\text{Max. } Z_p = \boldsymbol{Cx} \Bigg]$$

subject to the constraints

$$A\boldsymbol{x} \le \boldsymbol{b} \qquad \ldots(1)$$

and

$$\boldsymbol{x} \ge 0 \Bigg]$$

The dual of (1) can be written as

$$\text{Min.} Z_d = \boldsymbol{b'w}$$

subject to the constraints

$$\boldsymbol{A'w} \geq \boldsymbol{C'}$$

and

$$\boldsymbol{w} \geq 0$$

...(2)

Now, suppose that there exists an optimum feasible solution to the primal problem, then by fundamental theorem of duality, we can say that the dual problem has at least one feasible solution.

Conversely, let us suppose that both the primal and the dual possess feasible solution $\hat{\boldsymbol{x}}$ and $\hat{\boldsymbol{w}}$ respectively.

Then clearly $\boldsymbol{C\hat{x}}$ and $\boldsymbol{b'\hat{w}}$ both are finite and $\boldsymbol{C\hat{x}} \leq \boldsymbol{b'\hat{w}}$

$\Rightarrow \boldsymbol{b'\hat{w}}$ acts as an upper bound on $\boldsymbol{Cx}$ although not necessarily least upper bound (supremum).

Hence, the primal must have finite optimum solution.

☞ REMARK

• In the above manner, we may prove the following results:
  (i) If there does not exist any feasible solution to the dual (primal) but there exists at least one of the primal (dual) then there does not exist any finite optimum solution to the primal (dual).
  (ii) If there does not exist any finite optimum solution to the primal (dual) then there does not exists any feasible solution to the dual (primal).

THEOREM 10. **(Complimentry Slackness Theorem)** *In an optimal solution of a linear programming problem*
  (i) *if i^{th} dual variable is positive, then i^{th} constraint must be an equation.*
  (ii) *if the ith constraint in the given primal is a strict inequality, then i^{th} dual variable must be zero.*

PROOF. Consider a primal problem

$$\text{Max.} Z_p = \sum_{j=1}^{n} c_j x_j$$

subject to the constraints

$$\sum_{j=1}^{n} a_{ij} x_j \leq b_i \quad (i = 1, 2, \ldots, m)$$

and

$$x_j \geq 0 \ \forall \ j$$

...(1)

Following the usual procedure, the dual of (1) is given by

$$\text{Min.} Z_d = \sum_{i=1}^{n} b_i y_i$$

subject to the constraints

$$\sum_{i=1}^{m} a_{ij} y_i \geq c_j \quad (j = 1, 2, \ldots, n)$$

and

$$y_i \geq 0 \ \forall \ i$$

...(2)

Using weak duality theorem, we have

$$\sum_{j=1}^{n} c_j x_j \leq \sum_{i=1}^{m} \sum_{j=1}^{n} a_{ij} x_j y_i \leq \sum_{i=1}^{m} b_i y_i \qquad \ldots(3)$$

for any x_j and y_i feasible to primal and dual problem respectively.

Since, these solutions are not only feasible but also optimal to these problem, so by fundamental theorem of duality, we must have

$$\sum_{i=1}^{m} \sum_{j=1}^{n} a_{ij} x_j y_i = \sum_{i=1}^{m} b_i y_i$$

which implies

$$\sum_{i=1}^{m} \left[\sum_{j=1}^{n} a_{ij} x_j - b_i \right] y_i = 0 \qquad \ldots(4)$$

Equation (4) holds only if each of its term is equal to zero, *i.e.*,

$$\left(\sum_{j=1}^{n} a_{ij} x_j - b_i \right) y_i = 0 \qquad \ldots(5)$$

(i) Here, since the ith dual variable y_i is positive then from (5), we get

$\sum_{j=1}^{n} a_{ij} x_j = b_i$, which is the ith constraint of the given primal problem

(ii) If the ith constraint of the primal problem is a strict inequality, *i.e.*,

$$\sum_{j=1}^{n} a_{ij} x_j < b_i$$

Then using (5) we get $y_i = 0$.

Converse of (i)

If the jth primal variable is positive, then the jth constraint of the dual must be an equation. For the proof of it, we proceed as follows.

Clearly, at the optimum solution, (3) reduces to

$$\sum_{j=1}^{n} c_j x_j = \sum_{j=1}^{n} \sum_{i=1}^{m} a_{ij} y_i x_j$$

$\Rightarrow \qquad \sum_{j=1}^{n} \left[c_j - \sum_{i=1}^{m} a_{ij} y_i \right] \cdot x_j = 0$

$\Rightarrow \qquad \left[c_j - \sum_{i=1}^{m} a_{ij} y_i \right] x_j = 0 \qquad \ldots(6)$

Now since $x_j \geq 0$, therefore, for the consistency of (6) we must have

$$c_j - \sum_{i=1}^{m} a_{ij} y_i = 0$$

$\Rightarrow \qquad \sum_{i=1}^{m} a_{ij} y_i = c_j$, which is the jth constraint of the primal problem.

Converse of (ii)

If the jth constraint in dual problem is a strict inequality, then j^{th}-primal variable must be zero.

To prove it, we proceed as follows:

∵ the jth constraint in dual problem is a strict inequality, *i.e.*,

$$\sum_{i=1}^{m} a_{ij} y_i > c_j$$

Then, for the consistency of (6) we must have

$$x_j = 0$$

- The above theorem can be restated as follows:

"A necessary and sufficient condition for any pair of feasible solutions to the primal and dual to the optimal is that $y_i x_{n+i} = 0, j = 1, 2, ..., m$ where x_{n+i} is the slack variable in the primal and $x_j y_{m+j} = 0, j = 1, 2, ..., n$ where y_{m+j} is the surplus variable for the dual."

7.6 METHOD FOR OBTAINING THE SOLUTION TO THE DUAL FROM THE FINAL SIMPLEX TABLE OF THE PRIMAL AND VICE-VERSA

To obtain the solution to the dual problem from final simplex table of the primal problem we proceed as follows:

(i) Firstly we write optimal solution of the primal problem from final simplex table which gives the optimal solution to its dual problem by using duality theorem in which we have proved that

$$\text{Min } Z_p = \text{Max } Z_d$$

(ii) Next, we write the value of $\Delta_j \, (= c_j - z_j)$ corresponding to the slack (or surplus) variables from final simplex table, these value of Δ_j with different sign give the values of dual variables.

(iii) If either problem has unbounded solution, then the other will have no feasible solution.

Working Procedure

STEP 1. Apply the simplex method to the problem (primal or dual) with lessor number of constraints.

STEP 2. Read the solution of the other from the final simplex table according to the rule discussed above.

Solved Examples

EXAMPLE 1. *Find the solution of the following problem and its dual by simplex method. Also, read the solution of each problem from the final simplex table of the other*

$$\text{Max. } Z = 40x_1 + 50x_2$$

subject to the constraints

$$2x_1 + 3x_2 \le 3$$
$$8x_1 + 4x_2 \le 5$$

and $$x_1, x_2 \ge 0$$

SOLUTION. Clearly, the given LPP of maximization is in standard form. Using slack variables s_1, s_2 the problem reduces to

Max. $Z = 40x_1 + 50x_2 + 0s_1 + 0s_2$

s.t. $\quad 2x_1 + 3x_2 + s_1 = 3$

$\quad\quad\quad 8x_1 + 4x_2 + s_2 = 5$

and $\quad\quad x_1, x_2, s_1, s_2 \geq 0$

Apply the simplex method in a usual manner, we have the following simplex table.

Simplex Table

		$c_j \to$	40	50	0	0	
B.V.	C_B	X_B	x_1	x_2	s_1	s_2	Min Ratio
s_1	0	3	2	③	1	0	3/3 = 1 (min)→
s_2	0	5	8	4	0	1	5/4
$Z = C_B X_B = 0, \Delta_j \to$			40	50	0	0	
				↑	↓		
x_2	50	1	2/3	1	1/3	0	3/2
s_2	0	1	⑯/③	0	−4/3	1	3/16 (min)→
$Z = C_B X_B = 50, \Delta_j \to$			20/3	0	−50/3	0	
			↑			↓	
x_2	50	7/8	0	1	1/2	−1/8	
x_1	40	3/16	1	0	−1/4	3/16	
$Z = C_B X_B = \dfrac{205}{4}, \Delta_j \to$			0	0	−15	−5/4	

We observe that in the last row of last table, $\Delta_j \leq 0$ for all j.

$\Rightarrow$ solution is optimal and is given by

$$x_1 = \frac{3}{16}, x_2 = \frac{7}{8} \text{ and Max. } Z = \frac{205}{4}$$

Dual Problem. The dual of the given primal problem is

$$\text{Max. } Z_y = \mathbf{b}' \cdot \mathbf{y} = [3 \quad 5][y_1 \quad y_2] = 3y_1 + 5y_2$$

such that $\quad A'\mathbf{y} \geq \mathbf{C}' \quad$ or $\quad \begin{bmatrix} 2 & 8 \\ 3 & 4 \end{bmatrix}\begin{bmatrix} y_1 \\ y_2 \end{bmatrix} \geq \begin{bmatrix} 40 \\ 50 \end{bmatrix}$

$\Rightarrow \quad\quad 2y_1 + 8y_2 \geq 40$

$\quad\quad\quad\quad 3y_1 + 4y_2 \geq 50$

and $\quad\quad y_1, y_2 \geq 0$

Now, converting this dual problem to maximization problem and introducing surplus variables s_1, s_2 and artificial variables A_1, A_2, we get

$$\text{Max.} Z_y' = -Z_y = -3y_1 - 5y_2 + 0s_1 + 0s_2 - MA_1 - MA_2$$

s.t. $\quad 2y_1 + 8y_2 - s_1 + A_1 = 40$

$\quad\quad\quad 3y_1 + 4y_2 - s_2 + A_2 = 50$

and $\quad y_1, y_2, s_1, s_2, A_1, A_2 \geq 0$

Now, apply the simplex method in a usual manner, we have the following simplex table.

Simplex Table

B.V.	C_B	X_B	y_1	y_2	s_1	s_2	A_1	A_2	Min Ratio
	$c_j \rightarrow$		-3	-5	0	0	$-M$	$-M$	
A_1	$-M$	40	2	⑧	-1	0	1	0	$40/8 = 5$ (Min) $\rightarrow$
A_2	$-M$	50	3	4	0	-1	0	1	$50/4$
$Z'_y = C_B X_B,$ $\Delta_j \rightarrow$ $= -90M$			$(5M-3)$ $\uparrow$	$(12M-5)$	$-M$	$-M$ $\downarrow$	0	0	
y_2	-5	5	$1/4$	1	$-1/8$	0	—	0	20
A_2	$-M$	30	②	0	$1/2$	-1	—	1	$30/2 = 15$ (Min) $\rightarrow$
$Z'_y = -30M - 25\Delta_j \rightarrow$			$(8M-7)/4$ $\uparrow$	0	$(4M-5)/8$	$-M$	—	0 $\downarrow$	
y_2	-5	$5/4$	0	1	$-3/16$	$1/8$	—	—	
y_1	-3	15	1	0	$1/4$	$-1/2$	—	—	
$Z'_y = -205/4,$ $\Delta_j \rightarrow$			0	0	$-3/16$	$-7/8$			

Clearly, all $\Delta_j \leq 0 \Rightarrow$ solution is optimal and is given by

$$y_1 = 15, y_2 = \frac{5}{4} \text{ and Min. } Z_y = -\text{Max.} Z'_y = \frac{205}{4}$$

How to Read the solution? In the final simplex table of the primal, negative of the values of $\Delta'_j s$ for slack variables s_1 and s_2 are the values of y_1, y_2 of the dual variables.

i.e., $y_1 = -\Delta_3 = 15, y_2 = -\Delta_4 = \dfrac{5}{4}$

and in the final simplex table of the dual, negative of the values of $\Delta'_j s$ for surplus variables are the values of x_1, x_2 of the primal, *i.e.,*

$$x_1 = -\Delta_3 = \frac{3}{16}, x_2 = -\Delta_4 = \frac{7}{8}, \text{Max.} Z = \frac{205}{4} = \text{Min.} Z_y$$

EXAMPLE 2. *Write the dual of the following LPP and hence solve it.*

Max. Z = 3x₁ − 2x₂

subject to the constraints

$$x_1 \leq 4$$
$$x_2 \leq 6$$
$$x_1 + x_2 \leq 5$$
$$-x_2 \leq -1$$

and $x_1, x_2 \geq 0$ [MEERUT–2007, 08; DELHI–2011]

SOLUTION. The given LPP of maximization is in standard form.

Following the usual procedure, the dual of the given LPP is given by

$$\text{Min. } Z_d = 4y_1 + 6y_2 + 5y_3 - y_4$$
$$\text{s.t.} \qquad y_1 + y_3 \geq 3$$
$$y_2 + y_3 - y_4 \geq -2$$
$$\text{and } y_1, y_2, y_3, y_4 \geq 0$$

Changing this dual problem to maximization, introduce surplus variable s_1 and slack variable s_2, the dual problem reduces to

$$\text{Max.} Z'_d = -4y_1 - 6y_2 - 5y_3 + y_4 + 0s_1 + 0s_2$$
$$\text{s.t.} \qquad y_1 + y_3 - s_1 = 3$$
$$-y_2 - y_3 + y_4 + s_2 = 2$$
$$\text{and} \qquad y_i, s_i \geq 0$$

(Here, artificial variable is not used because in the first constraint y_1 will serve the purpose of unit matrix)

Now, apply the simplex method in a usual manner, we have the following simplex table.

Simplex Table

B.V.	C_B	X_B	$c_j \rightarrow$ y_1	-6 y_2	-5 y_3	1 y_4	0 s_1	0 s_2	Min Ratio
y_1	-4	3	1	0	1	0	-1	0	—
s_2	0	2	0	-1	-1	①	0	1	2 (Min) →
$Z'_d = C_B X_B,$ $= -12$		$\Delta_j \rightarrow$	0	-6	-1	1	-4	0	
						↑		↓	
y_1	-4	3	1	0	1	0	-1	0	
y_4	1	2	0	-1	-1	1	0	1	
$Z'_d = -10,$		$\Delta_j \rightarrow$	0	-5	0	0	-4	-1	

$\because$ in the last row of the above table, all $\Delta_j \leq 0$

$\Rightarrow$ solution is optimal for dual problem and is given by

$$y_1 = 3, y_2 = 0, y_3 = 0, y_4 = 2, \text{ Min.} Z_d = -\text{Max} Z'_d = 10$$

Solution of the primal: From the final simplex table of the dual the required solution of the primal problem is

$$x_1 = -\Delta_5 = 4; \ x_2 = -\Delta_6 = 1 \text{ and Max. } Z = -\text{ min. } Z_d = 10$$

EXAMPLE 3. *Using the principle of duality, solve the following LPP*

$$\textbf{Max. } \textbf{Z} = \textbf{2x}_1 + \textbf{x}_2$$

subject to the constraints

$$x_1 + 2x_2 \leq 10$$
$$x_1 + x_2 \leq 6$$
$$x_1 - x_2 \leq 2$$
$$x_1 - 2x_2 \leq 1$$

and $\qquad \textbf{x}_1, \textbf{x}_2 \geq \textbf{0}$ \qquad [GORAKHPUR–2007; MEERUT–2009]

SOLUTION. Since, the given LPP of maximization is in standard form, so following the usual procedure, the dual of the given primal problem is

$$\text{Min } Z_d = 10y_1 + 6y_2 + 2y_3 + y_4$$

s.t. $\quad y_1 + y_2 + y_3 + y_4 \geq 2$
$\quad\quad 2y_1 + y_2 - y_3 - 2y_4 \geq 1$
and $\quad\quad y_1, y_2, y_3, y_4 \geq 0$

Now, changing this minimization problem into standard maximization and then introduce surplus variables s_1, s_2 and artificial variables A_1 and A_2, the dual problem becomes

$$\text{Max}. Z'_d = -Z'_d = -10y_1 - 6y_2 - 2y_3 - y_4 + 0s_1 + 0s_2 - MA_1 - MA_2$$

s.t. $\quad y_1 + y_2 + y_3 + y_4 - s_1 + A_1 = 2$
$\quad\quad 2y_1 + y_2 - y_3 - 2y_4 - s_2 + A_2 = 1$

and $\quad\quad y_1, y_2, y_3, y_4, s_1, s_2, A_1, A_2 \geq 0$

Now apply the usual procedure, we have the following simplex table.

Simplex Table

B.V.	C_B	X_B	y_1	y_2	y_3	y_4	s_1	s_2	A_1	A_2	Min Ratio
		$c_j \rightarrow$	-10	-6	-2	-1	0	0	$-M$	$-M$	
A_1	$-M$	2	1	1	1	1	-1	0	1	0	$2/1 = 2$
A_2	$-M$	1	②	1	-1	-2	0	-1	0	1	$1/2$ (min)→
$Z'_d = -3M,$		$\Delta_j \rightarrow$	$(3M{-}10)\uparrow$	$(2M{-}6)$	-2	$(-M-1)$	$-M$	$-M$	0	$0\downarrow$	
A_1	$-M$	$3/2$	0	$1/2$	$3/2$	②	-1	$1/2$	1	—	$3/4$ (min)→
y_1	-10	$1/2$	1	$1/2$	$-1/2$	-1	0	$-1/2$	0	—	$-$ve
$Z'_d = -\frac{1}{2}(3M+10), \Delta_j \rightarrow$			0	$\frac{-1+M}{2}$	$\frac{-7+3M}{2}$	$(-11+2M)\uparrow$	$-M$	$\left(\frac{M}{2}-5\right)\downarrow$	0	—	
y_4	-1	$3/4$	0	$1/4$	$3/4$	1	$-1/2$	$1/4$	—	—	3
y_1	-10	$5/4$	1	③/4	$1/4$	0	$-1/2$	$-1/4$	—	—	$5/3$ (min)→
$Z'_d = \frac{-53}{4},$		$\Delta_j \rightarrow$	0	$7/4\uparrow$	$5/4$	0	$\frac{-11}{2}$	$-9/4$	—	—	
				$\downarrow$							
y_4	-1	$1/3$	$-1/3$	0	②/3	1	$-1/3$	$1/3$	—	—	$1/2$(min)→
y_2	-6	$5/3$	$4/3$	1	$1/3$	0	$-2/3$	$-1/3$	—	—	5
$Z'_d = \frac{-31}{3},$		$\Delta_j \rightarrow$	$-7/3$	0	$2/3\uparrow$	$0\downarrow$	$-13/3$	$-5/3$			
y_3	-2	$1/2$	$-1/2$	0	1	$3/2$	$-1/2$	$1/2$			
y_2	-6	$3/2$	$3/2$	1	0	$-1/2$	$-1/2$	$-1/2$			
$Z'_d = -10,$		$\Delta_j \rightarrow$	-2	0	0	-1	-4	-2			

In the last row of the above table all $\Delta_j \leq 0$

$\Rightarrow$ solution of the dual is optimal and is given by

$$y_1 = 0, y_2 = \frac{3}{2}, y_3 = \frac{1}{2}, y_4 = 0$$

$$\text{Min.} Z_d = -\text{Max.} Z_d' = 10$$

Further, from the last table the solution of the primal problem is given by

$$x_1 = -\Delta_5 = 4, x_2 = -\Delta_6 = 2 \text{ and Max.} Z = \text{Min} Z_d = 10$$

EXAMPLE 4. *Using the principle of duality, solve the following LPP*

$$\textbf{Max. Z = 3}x_1 + \textbf{2}x_2$$

subject to the constraints

$$x_1 + x_2 \geq 1$$
$$x_1 + x_2 \leq 7$$
$$x_1 + 2x_2 \leq 10$$
$$x_2 \leq 3$$

and $\quad x_1, x_2 \geq 0$ [MEERUT–2016]

SOLUTION. The given problem of maximization can be written in standard form as follows:

Max. $Z = 3x_1 + 2x_2$

s.t. $\qquad -x_1 - x_2 \leq -1$

$\qquad\qquad x_1 + x_2 \leq 7$

$\qquad\qquad x_1 + 2x_2 \leq 10$

$\qquad\qquad\qquad x_2 \leq 3$

and $\qquad\qquad x_1, x_2 \geq 0$

The dual of the above primal problem is given by

Min. $Z_d = -y_1 + 7y_2 + 10y_3 + 3y_4$

s.t. $\qquad -y_1 + y_2 + y_3 \geq 3$

$\qquad -y_1 + y_2 + 2y_3 + y_4 \geq 2$

and $\qquad y_1, y_2, y_3, y_4 \geq 0$

To solve the above dual, first we change it into maximization problem and then introduce the surplus variables s_1, s_2 and artificial variable A_1, then it becomes

$$\text{Max.} Z_d' = -Z_d = y_1 - 7y_2 - 10y_3 - 3y_4 + 0s_1 + 0s_2 - MA_1$$

s.t. $\qquad -y_1 + y_2 + y_3 - s_1 + A_1 = 3$

$\qquad -y_1 + y_2 + 2y_3 + y_4 - s_2 = 2$

and $\qquad y_i, s_i \geq 0$

Now apply the simplex method in a usual manner, we have the following simplex table.

Simplex Table

B.V.	C_B	x_B	$c_j \to$ y_1 1	y_2 -7	y_3 -10	y_4 -3	s_1 0	s_2 0	A_1 $-M$	Min Ratio
A_1	$-M$	3	-1	1	1	0	-1	0	1	$3/1 = 3$
y_4	-3	2	-1	①	2	1	0	-1	0	$2/1 = 2$ (min) →
$Z'_d = -3M - 6, \Delta_j \to$			$(-M-2)$	$(M-4)$ ↑	$(M-4)$ ↓	0	$-M$	-3	0	
A_1	$-M$	1	0	0	-1	-1	-1	①	1	1 (min)
y_2	-7	2	-1	1	2	1	0	-1	0	$-$ve
$Z'_d = -M - 14, \Delta_j \to$			-6	0	$-M+4$	$-M+4$	$-M+7$ ↑	$M-7$ ↓	0	
s_2	0	1	0	0	-1	-1	-1	1	$-$	
y_2	-7	3	-1	1	1	0	-1	0	$-$	
$Z'_d = -21, \quad \Delta_j \to$			-6	0	-3	-3	-7	0	$-$	

In the last row of last table all $\Delta_j \leq 0 \Rightarrow$ solution of the dual is optimal and is given by

$$y_1 = 0, y_2 = 3, y_3 = 0, y_4 = 0 \text{ and Min } Z_d = -\text{Max } Z'_d = 21$$

Also, from the last table, the optimal solution of the given primal problem is

$$x_1 = -\Delta_5 = 7, x_2 = -\Delta_6 = 0 \text{ and Max. } Z = \text{Min } Z_d = 21$$

EXAMPLE 5. *Write the dual of the following LPP:*

$$\textbf{Max. } \textbf{Z} = \textbf{2}\textbf{x}_1 + \textbf{3}\textbf{x}_2$$

subject to the constraints

$$\textbf{2}\textbf{x}_1 + \textbf{2}\textbf{x}_2 \leq \textbf{10}$$
$$\textbf{2}\textbf{x}_1 + \textbf{x}_2 \leq \textbf{6}$$
$$\textbf{x}_1 + \textbf{2}\textbf{x}_2 \leq \textbf{6}$$

and $\qquad \textbf{x}_1, \textbf{x}_2 \geq \textbf{0}$

Solve the above primal and then find the solution of the dual.

SOUTION. Since the given primal is written in standard form so, the required dual is given by

$$\text{Min. } Z_d = 10y_1 + 6y_2 + 6y_3$$
$$\text{s.t.} \quad 2y_1 + 2y_2 + y_3 \geq 2$$
$$2y_1 + y_2 + 2y_3 \geq 3$$
and $\qquad y_1, y_2, y_3 \geq 0$

Now, we have to solve the given primal problem.

Introducing the slack variables s_1, s_2, s_3, we get

$$\text{Max. } Z = 2x_1 + 3x_2 + 0s_1 + 0s_2 + 0s_3$$
$$\text{s.t.} \quad 2x_1 + 2x_2 + s_1 = 10$$
$$2x_1 + x_2 + s_2 = 6$$

$$x_1 + 2x_2 + s_3 = 6$$

and $x_1, x_2, s_1, s_2, s_3 \geq 0$

Now apply the simplex method in a usual manner, we have the following simplex table.

Simplex Table

B.V.	C_B	X_B	x_1	x_2	s_1	s_2	s_3	Min Ratio
	$c_j \to$		2	3	0	0	0	
s_1	0	10	2	2	1	0	0	5
s_2	0	6	2	1	0	1	0	6
s_3	0	6	1	②	0	0	1	3 (min) →
$Z = C_B X_B = 0$, $\Delta_j \to$			2	3 ↑	0	0	0 ↓	
s_1	0	4	1	0	1	0	−1	4
s_2	0	3	③/2	0	0	1	−1/2	2 (min) →
x_2	3	3	1/2	1	0	0	1/2	6
$Z = C_B X_B = 9$, $\Delta_j \to$			1/2 ↑	0	0	0	−3/2 ↓	
s_1	0	2	0	0	1	−2/3	−2/3	
x_1	2	2	1	0	0	2/3	−1/3	
x_2	3	2	0	1	0	−1/3	2/3	
$Z = C_B X_B = 10$, $\Delta_j \to$			0	0	0	−1/3	− 4/3	

We observe that in the last row of last table all $\Delta_j \leq 0$

⇒ solution of the primal is optimal and is given by $x_1 = 2$, $x_2 = 2$ and Max. $Z = 10$

Solution of the dual: The solution of the dual is given by the negative Δ_3, Δ_4, Δ_5 which corresponds to s_1, s_2 and s_3 in the last simplex table.

$$\therefore \quad y_1 = -\Delta_3 = 0, y_2 = -\Delta_4 = \frac{1}{3}, y_3 = -\Delta_5 = \frac{4}{3} \text{ and Min. } Z_d = \text{Max. } Z = 10$$

7.7 SUMMARY OF THE CORRESPONDANCE BETWEEN THE PRIMAL AND ITS DUAL

The following table shows the various correspondance between the primal and dual problems.

S.No.	Primal (maximize)	Dual (minimize)
1.	Objective function Max. Z_p	Objective function : Min Z_d
2.	Coefficient matrix $A = [a_{ij}]$	Transpose A' of A
3.	Requirement vector b	Price vector C
4.	i^{th} constraints $\leq$	i^{th} variable ≥ 0
5.	i^{th} constraints $\geq$	i^{th} variable ≥ 0
6.	i^{th} constraints $=$	i^{th} variable unrestricted
7.	i^{th} variable ≥ 0	i^{th} constraints $\geq$
8.	i^{th} variable ≤ 0	i^{th} constraints $\leq$
9.	i^{th} variable unrestricted	i^{th} constraints $=$
10.	Finite optimal solution	Equal finite optimal solution
11.	Feasible and unbounded solution	Infeasible solution
12.	Infeasible solution	Feasible and unbounded solution

Exercise-7.2

1. Find the solution of the following LPP and its dual by simplex method
   Min. $Z = x_1 - x_2$
   s.t. $2x_1 + x_2 \geq 2$; $-x_1 - x_2 \geq 1$
   and $x_1, x_2 \geq 0$

2. Write the dual of the following LPP and solve it
   Max. $Z = 4x_1 + 2x_2$
   s.t. $-x_1 - x_2 \leq -3$,
   $-x_1 + x_2 \leq -2$, $x_1, x_2 \geq 0$

3. Use principle of duality, solve the following LPP
   Max. $Z = 3x_1 + 2x_2$
   s.t. $2x_1 + x_2 \leq 5$; $x_1 + x_2 \leq 3$
   and $x_1, x_2 \geq 0$

   Use principle of duality, solve the following LPP (Qus 4-7)

4. Min $Z = 2x_1 + 2x_2$
   s.t. $2x_1 + 4x_2 \geq 1$; $x_1 + 2x_2 \geq 1$
   $2x_1 + x_2 \geq 1$
   and $x_1, x_2 \geq 0$

5. Min. $Z = 3x_1 + x_2$
   s.t. $x_1 + x_2 \geq 1$
   $2x_1 + 3x_2 \geq 2$
   and $x_1, x_2 \geq 0$

6. Max. $Z = 3x_1 - 2x_2$
   s.t. $x_1 + x_2 \leq 5$, $x_1 \leq 4$, $1 \leq x_2 \leq 6$
   and $x_1, x_2 \geq 0$

7. Max. $Z = 4x_1 + 3x_2$
   s.t. $x_1 \leq 6$; $x_2 \leq 8$; $x_1 + x_2 \leq 7$, $3x_1 + x_2 \leq 15$;
   $-x_2 \leq 1$
   and $x_1, x_2 \geq 0$

8. Solve the following primal and its dual by simplex method
   Max. $Z = 40x_1 + 35x_2$
   s.t. $2x_1 + 3x_2 \leq 60$; $4x_1 + 3x_2 \leq 96$
   and $x_1, x_2 \geq 0$

9. Solve the following primal problem by simplex method and deduce from it the solution to the dual problem
   Max. $Z = 2x_1 - x_2$
   s.t. $x_1 + x_2 \leq 10$; $-2x_1 + x_2 \leq 2$

   $4x_1 + 3x_2 \geq 12$
   and $x_1, x_2 \geq 0$

10. Solve the following primal problem and its dual by simplex method
    Max. $Z = 5x_1 + 12x_2 + 4x_3$
    subject to $x_1 + 2x_2 + x_3 \leq 5$;
    $2x_1 - x_2 + 3x_3 = 2$
    and $x_1, x_2 \geq 0$

11. Apply simplex method to solve the following LPP
    Max. $Z = 30x_1 + 23x_2 + 29x_3$
    s.t. $6x_1 + 5x_2 + 3x_3 \leq 26$
    $4x_1 + 2x_2 + 5x_3 \leq 7$
    and $x_1, x_2, x_3 \geq 0$
    Hence, or otherwise find the solution to the dual of the above problem.

12. Using the principle of duality, solve the following LPP
    Max. $Z = 5x_1 - 2x_2 + 3x_3$
    s.t. $2x_1 + 2x_2 - x_3 \geq 2$
    $3x_1 - 4x_2 \leq 3$
    $x_1 + 2x_3 \leq 5$
    and $x_1, x_2, x_3 \geq 0$

13. Using duality, solve the following LPP
    Min. $Z = 3x_1 - 2x_2 + 4x_3$
    s.t. $3x_1 + 5x_2 + 4x_3 \geq 7$; $6x_1 + x_2 + 3x_3 \geq 4$;
    $7x_1 - 2x_2 - x_3 \leq 10$; $x_1 - 2x_2 + 5x_3 \geq 3$;
    $4x_1 + 7x_2 - 2x_3 \geq 2$
    and $x_1, x_2, x_3 \geq 0$

14. Using duality, solve the following primal problem
    Min. $Z = 6x + 5y + 2z$
    s.t. $x + 3y + 2z \geq 5$; $2x + 2y + z \geq 2$
    $4x - 2y + 3z \geq -1$
    and $x, y, z \geq 0$

15. Use duality theory, solve the following primal problem
    Min $Z = 10x_1 + 6x_2 + 2x_3$
    s.t. $-x_1 + x_2 + x_3 \geq 1$; $3x_1 + x_2 - x_3 \geq 2$
    and $x_1, x_2, x_3 \geq 0$

ANSWERS

1. No feasible solution for primal, unbounded solution for dual.
2. No finite optimal solution
3. $x_1 = 2$, $x_2 = 1$, Max $Z = 8$
4. $x_1 = \dfrac{1}{3}, x_2 = \dfrac{1}{3}$, Min $Z = \dfrac{4}{3}$
5. $x_1 = 0, x_2 = 1$, Min. $Z = 1$

6. $x_1 = 4, x_2 = 1$ and Max. $Z = 10$ 7. $x_1 = 4, x_2 = 3$, Max. $Z = 25$

8. **For primal:** $x_1 = 18, x_2 = 8$ and Max. $Z = 1000$; **For dual:** $y_1 = \dfrac{10}{3}, y_2 = \dfrac{25}{3}$ and Min. $Z = 1000$

9. **For primal:** $x_1 = 10, x_2 = 0$, Max. $Z = 20$; **For dual:** $y_1 = 2, y_2 = 0, y_3 = 0$, Min. $Z_d = 20$

10. **For primal:** $x_1 = \dfrac{9}{5}, x_2 = \dfrac{8}{5}, x_3 = 0$ and Max.$Z = \dfrac{141}{5}$; **For dual:** $y_1 = \dfrac{29}{5}, y_2 = \dfrac{-2}{5}$, Min.$Z_d = \dfrac{141}{5}$

11. **For Primal:** $x_1 = 0, x_2 = \dfrac{7}{2}, x_3 = 0$ and Max.$Z = \dfrac{161}{2}$; **For dual:** $y_1 = 0, y_2 = \dfrac{23}{2}$, Min.$Z_d = \dfrac{161}{2}$

12. $x_1 = \dfrac{23}{3}, x_2 = 5, x_3 = 0$, Max $Z = \dfrac{85}{3}$ 13. No feasible solution

14. $x_1 = 0, x_2 = 0, x_3 = \dfrac{5}{2}$ and Min.$Z = 5$ 15. $x_1 = \dfrac{1}{4}, x_2 = \dfrac{5}{4}, x_3 = 0$ and Min $Z = 10$

REVIEW QUESTIONS

1. Write a short note on Primal-dual relationship.
2. Define the dual of a linear programming problem.
3. Discuss the principle of duality in linear programming.
4. Write the general rule to find the solution of a primal and its dual by simplex method.
5. Write the significance of dual variable in a LP model.

MULTIPLE CHOICE QUESTIONS (CHOOSE THE MOST APPROPRIATE ONE)

1. The relation between the objective function of primal and dual problem of LPP is: [MEERUT–2013]
   (a) Max. $Z_p = -$ Min Z_d (b) Max. $Z_p = $ Min Z_d
   (c) Max $Z_p = $ Min Z_d (d) None of these

2. Dual Simplex method is applied to solve LP problems that start with: [MEERUT–2013]
   (a) feasible solution
   (b) both feasible and optimal
   (c) infeasible solution
   (d) infeasible solution and Optimal solution

3. In Dual Simplex method the key column is selected by calculating: [MEERUT–2013]
   (a) $\min\{(z_j - c_j)/x_{rj} : x_{rj} > 0\}$
   (b) $\max\{(z_j - c_j)/x_{rj} : x_{rj} < 0\}$
   (c) $\min\{(z_j - c_j)/x_{rj} : x_{rj} > 0\}$
   (d) $\max\{(z_j - c_j)/x_{rj} : x_{rj} \leq 0\}$

4. Dual of the dual is: [MEERUT–2013]
   (a) dual (b) primal
   (c) either primal or dual (d) None of the above

5. For the dual problem and primal problem, the optimum value of the objective function is: [MEERUT–2013]
   (a) zero
   (b) different
   (c) same
   (d) both (a) and (b) are true

6. If dual has an Unbounded solution, then primal has: [MEERUT–2013]
   (a) an infeasible solution
   (b) a feasible solution
   (c) an unbounded solution

   (d) None of the above

7. If the given primal is a minimization problem then its dual is:
   (a) minimize problem (b) maximize problem
   (c) (a) and (b) both (d) None of these

8. If the primal i^{th} inequality, then in dual ith variable:
   (a) $w_i \leq 0$ (b) $w_i = 0$
   (c) $w_i \geq 0$ (d) None of these

9. If a primal of LPP has finite optimal then values of objective function for primal and dual:
   (a) unequal   (b) odd solution
   (c) equal   (d) none of these

10. In LPP if both the primal and dual problem have finite optimal solution and problem object function C prime and dual then:
    (a) $Z_p \cdot 2d$ (b) $Z_p - 2d$
    (c) $Z_p = 2d$ (d) None of these

11. If the problem is in standard primal form of minimization, all the constraints involve the sign:
    (a) $\geq$ (b) $\leq$
    (c) $=$ (d) None of these

12. The requirement vector of the primal is the price vector of the:
    (a) primal   (b) dual
    (c) non-dual   (d) none of these

13. If the primal has unbounded solution then its dual has:
    (a) either no solution or an unbounded solution
    (b) no solution
    (c) an unbounded solution
    (d) (b) and (c) both

14. Standard primal problem is of maximization sign of constraints:
   (a) $\geq$
   (b) $=$
   (c) $\leq$
   (d) none of these
15. The coefficient matrix of the dual is obtained by transposing the coefficient matrix of the:
   (a) dual
   (b) primal
   (c) both (a) and (b)
   (d) none of these
16. If the primal problem has a finite feasible optimal solution then the dual has:
   (a) finite optimal F.S.
   (b) no solution
   (c) either no solution or an Unbounded solution
   (d) none of these
17. In dual simplex method, let the variable x_i leave and the variable x_j enter. Let $x_j > 0$. Then later on:
   (a) x_j can become negative
   (b) x_j will remain positive
   (c) x_j may be negative or positive
   (d) None of these
18. Requirement vector of the primal is the _____ of the dual.
   (a) row vector
   (b) price vector
   (c) column vector
   (d) none of these
19. If both the primal and dual problems have finite optimal solution and Z_p, Z_d are the optimal values of the objective functions of both problems respectively then we have:
   (a) $Z_p > Z_d$
   (b) $Z_p < Z_d$
   (c) $Z_p = Z_d$
   (d) None of these
20. If the primal problem has a unbounded feasible solution, then the dual problem has:
   (a) a finite optimal feasible solution
   (b) either no solution or an unbounded solution
   (c) no solution
   (d) none of these
21. If the primal has finite optimal solution the dual has finite optimal solution with:
   (a) equal optimal value of objective function
   (b) unbounded solution
   (c) no equal optimal value of objective function
   (d) none of these
22. The Necessary and Sufficient conditions for any LPP and its dual to have optimal solution is that both have:

(a) B.F.S.
(b) F.S.
(c) optimal Solution
(d) degenerate Solution
23. If the problem in standard primal form is of maximization, all the constraints of its dual involve the sign:
   (a) $\geq$
   (b) $\leq$
   (c) $=$
   (d) None of these
24. In standard primal problem of min., sign of constraints:
   (a) $\geq$
   (b) $\leq$
   (c) $=$
   (d) None of these
25. If the primal has Unbounded solution, then the dual has either No solution or:
   (a) an unbounded solution
   (b) bounded solution
   (c) feasible solution
   (d) basic feasible solution
26. If a primal has optimal feasible solution, then its dual has:
   (a) finite optimal solution
   (b) optimal
   (c) (a) and (b) both
   (d) none of these
27. If the ith variable of the primal is positive then ith variable of the dual is:
   (a) +ve
   (b) −ve
   (c) zero
   (d) unrestricted
28. If primal coefficient matrix A, then dual transpose of the coefficient matrix:
   (a) A'
   (b) A' or A^T
   (c) A^{-1}
   (d) none of these
29. If the problem is standard primal form is of maximization, all the constraints involve the sign:
   (a) $\geq$
   (b) $\leq$
   (c) $=$
   (d) none of these
30. The optimal values of the objective function of primal and dual problems if exists are:
   (a) one
   (b) two
   (c) zero
   (d) equal
31. If primal/dual problem has a Unbounded optimal solution, the other problem has either _____ or _____.
   (a) solution, bounded solution
   (b) no solution, bounded solution
   (c) no solution, unbounded solution
   (d) none of these

ANSWERS

1. (b)	2. (b)	3. (b)	4. (b)	5. (d)	6. (c)	7. (b)	8. (c)	9. (c)
10. (c)	11. (a)	12. (b)	13. (a)	14. (c)	15. (b)	16. (a)	17. (a)	18. (b)
19. (c)	20. (b)	21. (a)	22. (b)	23. (a)	24. (a)	25. (a)	26. (a)	27. (c)
28. (b)	29. (b)	30. (d)	31. (c)					

ARCHIVE

1. *XYZ* Company has three departments- Assembly, Painting and packing and can make three types of almirahs. An almirah of type I requires one hour of assembly, 40 minutes

of painting, and 20 minutes of packing time, respectively, Similarly, an almirah of type II needs 80 minutes, 20 minutes and one hour, respectively. The almirah of type III requires 40 minutes each of assembly, painting and packing time. The total time available at assembly, painting and packing departments is 600 hours, 400 hours and 800 hours, respectively. Determine the number of each type of almirahs that should be produced in order to maximize the profit. The unit profit for types I, II and III is ₹ 40, 80 and 60 respectively.

Suppose that the manager of *XYZ* Company is thinking of renting the production capacities of the three departments to another almirah manufacturer - *ABC* Company. *ABC* Company is interested in minizing the rental charges. On the other hand, the *XYZ* company would like to know the worth of production hours to them, in each of the departments to determine the rental rates. (a) Formulate this problem as an LP problem and solve it to determine the number of each type of almirahs that should be produced by the *XYZ* company in order to maximize its profit. (b) For LP problem in (a) formulate its dual and interpret your results.

[CA–1990; DELHI(MBA)–1999]

2. The *XYZ* company has the option of producing two products during the period of slack activity. For the next period, production has been scheduled so that the milling machine is free for 10 hours and skilled labour will have 8 hours of time available.

Product	Machine time per unit	Skilled labour time per unit	Profit contribution per unit (₹)
A	4	2	5
B	2	2	3

Solve the primal and dual LP problems and bring out the fact that the optimum solution of one can be obtained from the other. Also explain in the context of the example, what do you understand by shadow prices (or dual prices or marginal value) of resource. [JAMMU(MBA)–1996]

3. A person consumes two types of food *A* and *B* everyday to obtain 8 units of proteins, 12 units of carbohydrates, and 9 units of fats which is his daily minimum requirements. 1kg of food A contains 2, 6 and 1 units of protein, carbohydrates and fats, respectively. 1 kg of food B contain 1, 1 and 3 units of proteins, carbohydrates and fats respectively. Food A costs ₹ 8.50 per kg, while *B* costs ₹4 per kg. Determine how many kg of each food should he buy daily to minimize his cost of food and still meet the minimum requirements.

Formulate LP problem mathematically. Write its dual and solve the dual by simplex method.

[GUJRAT(MBA)–1996]

Hints and Answers

1. Let x_1, x_2 and x_3 be the number of units of the three types of almirahs, respectively to be produced. The given problem can be represented as an LP model as :

Maximize $Z = 40x_1 + 80x_2 + 60x_3$

subject to the constraints

(i) $x_1 + \left(\dfrac{4}{3}\right)x_2 + \left(\dfrac{2}{3}\right)x_3 \le 600$, (ii) $\left(\dfrac{2}{3}\right)x_1 + \left(\dfrac{1}{3}\right)x_2 + \left(\dfrac{2}{3}\right)x_3 \le 400$, (iii) $\left(\dfrac{1}{3}\right)x_1 + x_2 + \left(\dfrac{2}{3}\right)x_3 \le 800$

and $x_1, x_2, x_3 \ge 0$.

The dual objective function along with constraints which determines for *ABC* the value of the productive resources can be written as :

Minimize (total rent) $Z_y = 600y_1 + 400y_2 + 800y_3$

subject to the constraints

(i) $y_1 + \left(\dfrac{2}{3}\right)y_2 + \left(\dfrac{1}{3}\right)y_3 \ge 40$, (ii) $\dfrac{4}{3}y_1 + \dfrac{1}{3}y_2 + y_3 \ge 80$ (iii) $\left(\dfrac{2}{3}\right)y_1 + \left(\dfrac{2}{3}\right)y_2 + \left(\dfrac{2}{3}\right)y_3 \ge 60$

and $y_1, y_2, y_3 \ge 0$.

2. Primal Problem x_1, x_2 = number of units of product *A* and *B* respectively to be produced

Max $Z_x = 5x_1 + 3x_2$

subject to, (i) $4x_1 + 2x_2 \le 10$ (ii) $2x_1 + 2x_2 \le 8$ and $x_1, x_2 \ge 0$.

Ans. $x_1 = 1$, $x_2 = 3$ and Max. $Z_x = 14$

3. Primal Problem x_1, x_2 = number of units of food *A* and *B* to be consumed, respectively.

Min. $Z_x = 8.50x_1 + 4x_2$

subject to, (i) $2x_1 + x_2 \ge 8$ (ii) $6x_1 + x_2 \ge 12$ (iii) $x_1 + 3x_2 \ge 9$ and $x_1, x_2 \ge 0$.

Ans. $x_1 = 1$, $x_2 = 6$ and Min. $Z_x = \dfrac{65}{2}$.

Dual Simplex Method and Primal-Dual Algorithm

8

8.1 INTRODUCTION

It is a well known fact that simplex method deals with a basic feasible solution and simplex algorithm is terminated after achieving the optimal solution. In this algorithm, the procedure will be stopped when all $\Delta_j \leq 0$ for maximization and $\Delta_j \geq 0$ for minimization problem. But there are some situations when one or more solution variable X_B are negative and optimality condition is satisfied, then current optimal solution may not be feasible. In such type of cases it is possible to find a starting basic but not feasible solution that is dual feasible, *i.e.*, all $\Delta_j (= c_j - z_j) \leq 0$ for a maximization problem. In such cases, a type of simplex method is called dual-simplex method.

☞ REMARK
- In the dual simplex method, we always attempt to retain optimality while bringing the primal back to feasibility (*i.e.*, $x_{B_i} \geq 0$ for all i)

8.2 DUAL-SIMPLEX ALGORITHM

Dual simplex method is the technique which deals with only slack variables, while in simplex method we use surplus as well as artificial variables in some cases. Dual simplex method starts with a basic optimal solution not necessarily feasible of the primal and decrease the number of negative variable iteratively step by step maintaining the optimality criterion at each iteration. This method was developed by **C.E. Lemke**.

The steps of dual-simplex algorithm may be given as follows:

🗂 Working Procedure

STEP 1. Convert the given LPP into standard form (maximization with $\leq$ constraints)
STEP 2. Introduce slack variables whenever needed.
STEP 3. Obtain the initial basic feasible solution by putting all given variables equal to zero.

Let $X_B = (x_{B_1}, x_{B_2}, \ldots, x_{B_n})$ be the initial basic feasible solution.

We observe that if all $x_{B_i} \geq 0$ then this feasible solution is optimal but if at least one of x_{B_i} is negative, then go to the next step.

STEP 4. Construct starting simplex table as usual done in simplex table.
STEP 5. If $x_{B_r} = \min\{x_{B_i} : x_{B_i} < 0\}$ then β_r is the outgoing vector.

STEP 6. If $\dfrac{\Delta_k}{a_{rk}} = \min\limits_j \left\{ \dfrac{\Delta_j}{a_{rj}}, a_{rj} < 0 \right\}$ then α_k is the incoming vector, where $\{\alpha_1, \alpha_2, \ldots, \alpha_{n-m}\}$ is the coefficient matrix.

If all $\alpha_{rj} \geq 0, j = 1, 2, \ldots, n$ then primal solution is dual bounded (*i.e.*, infeasible).

STEP 7. To check the optimality, replace β_r by α_k by applying row operation and all basic variables reduce to non-negative values. This solution is the optimal feasible solution. If atleast one basic variable is negative, then repeat the step 5, 6 and 7 iteratively until an optimal feasible solution is achieved.

Solved Examples

EXAMPLE 1. *Apply dual simplex method, solve the following LPP*

$$\text{Min. } Z = 5x_1 + 6x_2$$

subject to the constraints

$$x_1 + x_2 \geq 2$$
$$4x_1 + x_2 \geq 4$$

and $\qquad x_1, x_2 \geq 0$

SOLUTION. Write the given LPP into standard form as follows

$$\left. \begin{array}{l} \text{Max } Z' = -5x_1 - 6x_2 \\ \text{s.t.} \quad -x_1 - x_2 \leq -2 \\ \qquad -4x_1 - x_2 \leq -4 \\ \text{and} \quad x_1, x_2 \geq 0 \end{array} \right] \qquad \qquad \dots(1)$$

We observe that all $c_j \leq 0$ and $b_i \leq 0$, so it is possible to obtain the solution by dual simplex method.

Introducing slack variables s_1 and s_2, (1) becomes

$$\text{Max } Z' = -5x_1 - 6x_2 + 0s_1 + 0s_2$$

$$\text{s.t.} \qquad -x_1 - x_2 + s_1 = -2$$

$$-4x_1 - x_2 + s_2 = -4$$

and $\qquad x_1, x_2, s_1, s_2 \geq 0$

Now, we have the following initial simplex table

Simplex Table-1

B.V.	C_B	X_B	$c_j \rightarrow$ $x_1(\alpha_1)$	-6 $x_2(\alpha_2)$	0 $s_1(\beta_1)$	0 $s_2(\beta_2)$
s_1	0	-2	-1	-1	1	0
s_2	0	-4	-4	-1	0	1

(header row: $c_j \rightarrow$ with values -5, -6, 0, 0 above $x_1(\alpha_1)$, $x_2(\alpha_2)$, $s_1(\beta_1)$, $s_2(\beta_2)$)

To find outgoing vector: We have

$$x_{B_1} = -2, x_{B_2} = -4$$

$\Rightarrow \qquad \min\{x_{B_1}, x_{B_2}\} = \min\{-2, -4\} = -4 = x_{B_2}$

$\Rightarrow \qquad\qquad r = 2$

$\Rightarrow \quad \beta_2$ is the outgoing vector.

To find incoming vector: We have

$$\frac{\Delta_k}{a_{rk}} = \min_j \left\{ \frac{\Delta_j}{a_{rj}}, a_{rj} < 0 \right\}$$

$\Rightarrow \qquad\qquad \dfrac{\Delta_k}{a_{2k}} = \min\left\{ \dfrac{\Delta_1}{a_{21}}, \dfrac{\Delta_2}{a_{22}} \right\} = \min\left\{ \dfrac{-5}{-4}, \dfrac{-6}{-1} \right\}$

$$= \frac{5}{4} = \frac{\Delta_1}{a_{21}}$$

$$\Rightarrow \qquad k = 1$$

$\Rightarrow \quad \alpha_1$ is the incoming vector.

which gives the key element $= a_{22} = -4$

Now, we have the next simplex table.

Simplex Table-2

B.V.	C_B	X_B	$x_1(\beta_2)$	$x_2(\alpha_2)$	$s_1(\beta_1)$	$s_2(\alpha_4)$
	$c_j \rightarrow$		-5	-6	0	0
s_1	0	-1	0	$-3/4$	1	$-1/4$
x_1	-5	1	1	$1/4$	0	$-1/4$
$Z' = C_B X_B = -5,\ \Delta_j \rightarrow$			0	$-19/4$	0	$-5/4$
				$\downarrow$		$\uparrow$

Here, we observe that $x_1 = 1$, $x_2 = 0$, $s_1 = -1$, $s_2 = 0$ which is not feasible and all $\Delta_j \leq 0$. Therefore, this solution is optimal but not feasible solution. Hence, this solution can be improved further as follows.

To find outgoing vector: Here,

$$x_{B_1} = -1, x_{B_2} = +1$$

Clearly, $x_{B_1} < 0$, therefore,

$$\min\{x_{B_1}, x_{B_2}\} = \min\{-1, 1\} = -1 = x_{B_1}$$

$\Rightarrow \quad \beta_1$ is the outgoing vector.

To find incoming vector:

Since,

$$\frac{\Delta_k}{a_{rk}} = \min_j \left\{ \frac{\Delta_j}{a_{rj}}, a_{rj} < 0 \right\} = \min\left\{ \frac{\Delta_2}{a_{12}}, \frac{\Delta_4}{a_{14}} \right\}$$

$$= \min\left\{ \frac{-19/4}{-3/4}, \frac{-5/4}{-1/4} \right\} = \min\left\{ \frac{19}{3}, 5 \right\} = 5 = \frac{\Delta_4}{a_{14}}$$

$\Rightarrow \quad \alpha_4 (= s_2)$ is the incoming vector.

So, we have the following simpex table.

Simplex Table-3

B.V.	C_B	X_B	x_1	x_2	s_1	s_2
	$c_j \rightarrow$		-5	-6	0	0
s_2	0	4	0	3	-4	1
x_1	-5	2	1	1	-1	0
$Z' = C_B X_B = -10,\ \Delta_j \rightarrow$			0	-1	-5	0

From the above table we observe that all $\Delta_j \leq 0$, $x_1 = 2$, $x_2 = 0$, $s_1 = 0$, $s_2 = 4$, therefore, the obtained solution is feasible and optimal and is given by

$$x_1 = 2, x_2 = 0, \text{ Min. } Z = -\text{Max. } Z' = 10$$

EXAMPLE 2. *Use dual simplex method, solve the following LPP:*

$$Max. \ Z = -3x_1 - x_2$$

subject to the constraints

$$x_1 + x_2 \geq 1$$
$$2x_1 + 3x_2 \geq 2$$

and $\qquad x_1, x_2 \geq 0$ [MEERUT–2006; GORAKHPUR–2008, 11]

SOLUTION. The given LPP can be written in standard form as follows:

$$Max. \ Z = -3x_1 - x_2$$

s.t. $\qquad -x_1 - x_2 \leq -1$
$$-2x_1 - 3x_2 \leq -2$$

and $\qquad x_1, x_2 \geq 0$

Clearly, the above problem is of maximization and all $c_j \leq 0$, so, dual simplex method is applicable.

Introducing slack variables s_1, s_2, the above LPP becomes

$$Max. \ Z = -3x_1 - x_2 + 0s_1 + 0s_2$$

s.t. $\qquad -x_1 - x_2 + s_1 = -1$
$$-2x_1 - 3x_2 + s_2 = -2$$

and $\qquad x_1, x_2, s_1, s_2 \geq 0$

The initial basic solution of this LPP is given by

$$x_1 = x_2 = 0, s_1 = -1, s_2 = -2$$

Simplex Table-1

B.V.	C_B	X_B	$c_j \rightarrow$ $x_1(\alpha_1)$	-1 $x_2(\alpha_2)$	0 $s_1(\beta_1)$	0 $s_2(\beta_2)$
s_1	0	-1	-1	-1	1	0
s_2	0	-2	-2	-3	0	1
$Z = C_B X_B = 0,$		$\Delta_j \rightarrow$	-3	-1	0	0
				$\uparrow$		$\downarrow$

To find outgoing vector:

$\because \quad x_{B_1} = -1, x_{B_2} = -2 \quad \Rightarrow \quad \min\{x_{B_1}, x_{B_2}\} = \min\{-1, -2\} = -2 = x_{B_2}$

$\Rightarrow \quad \beta_2$ is the outgoing vector.

To find incoming vector: We have

$$\frac{\Delta_k}{a_{2k}} = \min_j \left\{ \frac{\Delta_j}{a_{2j}}, a_{2j} < 0 \right\} = \min \left\{ \frac{\Delta_1}{a_{21}}, \frac{\Delta_2}{a_{22}} \right\}$$

$$= \min \left\{ \frac{-3}{-2}, \frac{-1}{-3} \right\} = \frac{1}{3} = \frac{\Delta_2}{a_{22}}$$

$\Rightarrow \quad k = 2$, *i.e.*, α_2 is the incoming vector.

So, key element $= a_{22} = -3$

Further, following the usual procedure, we have the following simplex table.

Simplex Table-2

B.V.	C_B	X_B	$c_j \rightarrow$ $x_1(\alpha_1)$	$x_2(\beta_2)$	$s_1(\beta_1)$	$s_2(\alpha_4)$
			-3	-1	0	0
s_1	0	$-1/3$	$-1/3$	0	1	$\boxed{-1/3}$
x_2	-1	$2/3$	$2/3$	1	0	$-1/3$
$Z' = -2/3$,		$\Delta_j \rightarrow$	$-7/3$	0	0	$-1/3$
					$\downarrow$	$\uparrow$

Here, we observe that, the solution given by

$$x_1 = 0, s_1 = -\frac{1}{3} \text{ is infeasible but } \Delta_j \le 0.$$

$\Rightarrow$ it can be improved further.

To find outgoing vector:

$$x_{B_1} = -\frac{1}{3}, x_{B_2} = \frac{2}{3}$$

$\Rightarrow \quad \min\{x_{B_1}, x_{B_2}\} = \min\left\{-\frac{1}{3}, \frac{2}{3}\right\} = -\frac{1}{3} = x_{B_1}$

$\Rightarrow \quad \beta_1$ is the outgoing vector.

To find incoming vector:

Since, $\dfrac{\Delta_k}{a_{rk}} = \min_{j}\left\{\dfrac{\Delta_j}{a_{rj}}, a_{rj} < 0\right\}$

$\therefore \quad \dfrac{\Delta_k}{a_{1k}} = \min_{j}\left\{\dfrac{\Delta_j}{a_{1j}}, a_{1j} < 0\right\} = \min\left\{\dfrac{\Delta_1}{a_{11}}, \dfrac{\Delta_4}{a_{14}}\right\} = \min\left\{\dfrac{-7/3}{-1/3}, \dfrac{-1/3}{-1/3}\right\} = 1 = \dfrac{\Delta_4}{a_{14}}$

$\Rightarrow \quad \alpha_4(= s_2)$ is the incoming vector

And key element $= -\dfrac{1}{3} = a_{14}$

Simplex Table-3

B.V.	C_B	X_B	$c_j \rightarrow$ $x_1(\alpha_1)$	$x_2(\beta_2)$	$s_1(\alpha_3)$	$s_2(\beta_1)$
			-3	-1	0	0
s_2	0	1	1	0	-3	1
x_2	-1	1	1	1	-1	0
$Z' = C_B X_B = -1$,		$\Delta_j \rightarrow$	-2	0	-1	0

Here, we observe that all $\Delta_j \le 0$ and $x_1 = 0, x_2 = 1, s_1 = 0, s_2 = 1$.

$\Rightarrow$ Feasible solution is optimal and is given by

$$x_1 = 0, x_2 = 1 \text{ and Max. } Z = -1$$

EXAMPLE 3. *Using dual simplex method, solve the following LPP.*

$$\text{Min. } Z = 6x_1 + 7x_2 + 3x_3 + 5x_4$$

subject to the constraints

$$5x_1 + 6x_2 - 3x_3 + 4x_4 \ge 12$$
$$x_2 + 5x_3 - 6x_4 \ge 10$$
$$2x_1 + 5x_2 + x_3 + x_4 \ge 8$$

and
$$x_1, x_2, x_3, x_4 \ge 0 \qquad \text{[MEERUT–1990, 2006]}$$

SOLUTION. First of all convert the given LPP into standard maximization form as follows.

$$\text{Max. } Z' = -6x_1 - 7x_2 - 3x_3 - 5x_4$$

s.t.
$$-5x_1 - 6x_2 + 3x_3 - 4x_4 \leq -12$$
$$-x_2 - 5x_3 + 6x_4 \leq -10$$
$$-2x_1 - 5x_2 - x_3 - x_4 \leq -8$$

and
$$x_1, x_2, x_3, x_4 \geq 0$$

Proceeding as usual, by introducing slack variables, we have

$$\text{Max. } Z' = -6x_1 - 7x_2 - 3x_3 - 5x_4 + 0s_1 + 0s_2 + 0s_3$$

s.t.
$$-5x_1 - 6x_2 + 3x_3 - 4x_4 + s_1 = -12$$
$$-x_2 - 5x_3 + 6x_4 + s_2 = -10$$
$$-2x_1 - 5x_2 - x_3 - x_4 + s_3 = -8$$

and
$$x_1, x_2, x_3, x_4, s_1, s_2, s_3 \geq 0$$

The initial basic solution is

$$x_1 = 0, x_2 = 0, x_3 = 0, x_4 = 0, s_1 = -12, s_2 = -10, s_3 = -8$$

which is not feasible.

Simplex Table

B.V.	C_B	X_B	$c_j \rightarrow$ $x_1(\alpha_1)$	$x_2(\alpha_2)$	$x_3(\alpha_3)$	$x_4(\alpha_4)$	$s_1(\beta_1)$	$s_2(\beta_2)$	$s_3(\beta_3)$
			-6	-7	-3	-5	0	0	0
s_1	0	-12	-5	$\enclose{circle}{-6}$	3	-4	1	0	0
s_2	0	-10	0	-1	-5	6	0	1	0
s_3	0	-8	-2	-5	-1	-1	0	0	1
$Z' = C_B X_B = 0,$ $\Delta_j \rightarrow$			-6	-7	-3	-5	0	0	0
				$\uparrow$			$\downarrow$		

We observe that, all $\Delta_j \leq 0$ in the above table and

$$x_{B_1} = -12, x_{B_2} = -10, x_{B_3} = -8$$

To find outgoing vector:

Since, $$x_{B_r} = \min_i \left\{ x_{B_i}, x_{B_i} < 0 \right\}$$

Therefore, $$x_{B_r} = \min\{x_{B_1}, x_{B_2}, x_{B_3}\}$$
$$= \min\{-12, -10, -8\} = x_{B_1} = -12$$

$\Rightarrow$ β_1 is the outgoing vector.

To find the incoming vector:

Since, $$\frac{\Delta_k}{\alpha_{rk}} = \min_j \left\{ \frac{\Delta_j}{a_{rj}}, a_{rj} < 0 \right\}$$

Therefore, $$\frac{\Delta_k}{a_{1k}} = \min\left\{ \frac{\Delta_1}{a_{11}}, \frac{\Delta_2}{a_{12}}, \frac{\Delta_4}{a_{14}} \right\} \qquad \{\because a_{13} > 0\}$$

$$= \min\left\{ \frac{-6}{-5}, \frac{-7}{-6}, \frac{-5}{-4} \right\} = \min\left\{ \frac{6}{5}, \frac{7}{6}, \frac{5}{4} \right\}$$

$$= \frac{7}{6} = \frac{\Delta_2}{a_{12}}$$

$\Rightarrow$ α_2 is the incoming vector.

$\therefore$ Key element $= a_{12} = -6$

Simplex Table-2

B.V.	C_B	X_B	$c_j \rightarrow$ $x_1(\alpha_1)$	-7 $x_2(\beta_1)$	-3 $x_3(\alpha_3)$	-5 $x_4(\alpha_4)$	0 s_1	0 $s_2(\beta_2)$	0 $s_3(\beta_3)$
			-6						
x_2	-7	2	$5/6$	1	$-1/2$	$2/3$	$-1/6$	0	0
s_2	0	-8	$5/6$	0	$\widehat{-11/2}$	$20/3$	$-1/6$	1	0
s_3	0	2	$13/6$	0	$-7/2$	$7/3$	$-5/6$	0	1
$Z' = C_B X_B = -14, \Delta_j \rightarrow$			$-1/6$	0	$-13/2$	$-1/3$	$-7/6$	0	0

Here, we observe that all $\Delta_j \leq 0$ but

$$x_1 = 0, x_2 = 2, x_3 = 0, x_4 = 0, s_1 = 0, s_2 = -8, s_3 = 2$$

$\Rightarrow$ solution is optimal but not feasible.

Further, $x_{B_1} = 2, x_{B_2} = -8, x_{B_3} = 2$

$\Rightarrow \qquad\qquad x_{B_2} < 0$

To find outgoing vector:

$$x_{B_r} = \min\{x_{B_1}, x_{B_2}, x_{B_3}\} = x_{B_2}$$

$\Rightarrow \qquad\qquad r = 2$

$\Rightarrow$ β_2 is the outgoing vector.

To find incoming vector:

Since, $\qquad\qquad \dfrac{\Delta_k}{a_{rk}} = \min_j \left\{ \dfrac{\Delta_j}{a_{rj}}, a_{rj} < 0 \right\}$

Therefore, $\qquad \dfrac{\Delta_k}{a_{2k}} = \min_j \left\{ \dfrac{\Delta_j}{a_{2j}}, a_{2j} < 0 \right\} = \min \left\{ \dfrac{\Delta_3}{a_{23}}, \dfrac{\Delta_5}{a_{25}} \right\} \quad (\because a_{23} < 0, a_{25} < 0)$

$$= \min \left\{ \dfrac{-15/2}{-11/2}, \dfrac{-7/6}{-1/6} \right\} = \min \left\{ \dfrac{13}{11}, 7 \right\} = \dfrac{13}{11} = \dfrac{\Delta_3}{a_{23}}$$

$\Rightarrow$ $k = 3$ i.e. α_3 is the incoming vector.

and $\qquad$ key element $= a_{23} = \dfrac{-11}{2}$

Simplex Table-3

B.V.	C_B	X_B	$c_j \rightarrow$ $x_1(\alpha_1)$	-7 $x_2(\alpha_2)(\beta_1)$	-3 $x_3(\alpha_3)(\beta_2)$	-5 $x_4(\alpha_4)$	0 s_1	0 s_2	0 $s_3(\beta_3)$
			-6						
x_2	-7	$30/11$	$25/33$	1	0	$2/33$	$-5/33$	$-1/11$	0
x_3	-3	$16/11$	$-5/33$	0	1	$-40/33$	$1/33$	$-2/11$	0
s_3	0	$8/11$	$18/33$	0	0	$-21/33$	$-8/11$	$-7/11$	1
$Z' = -258/11, \Delta_j \rightarrow$			$-38/33$	0	0	$-271/33$	$-32/33$	$-13/11$	0

Here, we observe that all $\Delta_j \leq 0$. So, solution is optimal and feasible and is given by

$$x_1 = 0, x_2 = \frac{30}{11}, x_3 = \frac{16}{11}, x_4 = 0$$

and   Min.$Z = -$Max $Z' = \dfrac{258}{11}$

EXAMPLE 4.   *Using dual simplex method, solve the following LPP*

$$\textbf{Max. } \textbf{\textit{Z}} = \textbf{--} \textbf{\textit{2x}}_\textbf{1} \textbf{--} \textbf{\textit{x}}_\textbf{3}$$

subject to the constraints

$$\textbf{\textit{x}}_\textbf{1} + \textbf{\textit{x}}_\textbf{2} - \textbf{\textit{x}}_\textbf{3} \geq \textbf{5}$$
$$\textbf{\textit{x}}_\textbf{1} - \textbf{2}\textbf{\textit{x}}_\textbf{2} + \textbf{4}\textbf{\textit{x}}_\textbf{3} \geq \textbf{8}$$

and $\textbf{\textit{x}}_\textbf{1}, \textbf{\textit{x}}_\textbf{2}, \textbf{\textit{x}}_\textbf{3} \geq \textbf{0}$

[ROORKEE(BE)–1990; BANASTHALI–1993]

SOLUTION.   Firstly, we shall write the given LPP into standard form as follows:

Max. $Z = -2x_1 + 0x_2 - x_3$

s.t. $-x_1 - x_2 + x_3 \leq -5$

$-x_1 + 2x_2 - 4x_3 \leq -8$

and $x_1, x_2, x_3 \geq 0$

Now, introducing slack variables s_1, s_2 in the above problem, we get

Max. $Z = -2x_1 + 0x_2 - x_3 + 0s_1 + 0s_2$

s.t. $-x_1 - x_2 + x_3 + s_1 = -5$

$-x_1 + 2x_2 - 4x_3 + s_2 = -8$

and $x_1, x_2, x_3, s_1, s_2 \geq 0$

Clearly, the initial basic solution is

$$x_1 = 0, x_2 = 0, x_3 = 0, s_1 = -5, s_2 = -8$$

Simplex Table-1

	$c_j \rightarrow$		-2	0	-1	0	0
B.V.	C_B	X_B	$x_1(\alpha_1)$	$x_2(\alpha_2)$	$x_3(\alpha_3)$	$s_1(\beta_1)$	$s_2(\beta_2)$
s_1	0	-5	-1	-1	1	1	0
s_2	0	-8	-1	2	-4	0	1
$Z' = C_B X_B = 0,$		$\Delta_j \rightarrow$	-2	0	-1	0	0
					$\uparrow$		$\downarrow$

To find outgoing vector: We have

$$x_{B_r} = \min\{x_{B_1}, x_{B_2}\} = \min\{-5, -8\} = -8 = x_{B_2}$$

$\Rightarrow$ $r = 2$

$\Rightarrow$ β_2 is the outgoing vector.

To find incoming vector:

Using $\dfrac{\Delta_k}{a_{rk}} = \min\limits_j \left\{ \dfrac{\Delta_j}{a_{rj}}, a_{rj} < 0 \right\}$

We get $\dfrac{\Delta_k}{a_{2k}} = \min\limits_j \left\{ \dfrac{\Delta_j}{a_{rj}}, a_{rj} < 0 \right\}$

$$= \min\left\{\frac{\Delta_1}{a_{21}}, \frac{\Delta_3}{a_{23}}\right\} = \min\left\{\frac{-2}{-1}, \frac{-1}{-4}\right\} = \min\left\{2, \frac{1}{4}\right\} = \frac{1}{4} = \frac{\Delta_3}{a_{23}}$$

$\Rightarrow \qquad k = 3$

$\Rightarrow \quad \alpha_3$ is the incoming vector.

and $\qquad$ key element $= a_{23} = -4$

Simplex Table-2

	$c_j \rightarrow$		-2	0	-1	0	0
B.V.	C_B	X_B	$x_1(\alpha_1)$	$x_2(\alpha_2)$	$x_3(\beta_2)$	$s_1(\beta_1)$	s_2
s_1	0	-7	$-5/4$	$-1/2$	0	1	$1/4$
x_3	-1	2	$1/4$	$-1/2$	1	0	$-1/4$
$Z' = C_B X_B = -2,\ \Delta_j \rightarrow$			$-7/4$	$-1/2$	0	0	$-1/4$

Here, we observe that all $\Delta_j \le 0$ and $x_1 = x_2 = 0$, $x_3 = 2$, $s_1 = -7$, $s_2 = 0$ which is optimal but not feasible.

To find outgoing vector:

$$x_{B_r} = \min\{x_{B_1}, x_{B_2}\} = \min\{-7, 2\} = -7 = x_{B_1}$$

$\Rightarrow \qquad r = 1$

$\Rightarrow \quad \beta_1$ is the outgoing vector.

To find incoming vector:

Using $\qquad \dfrac{\Delta_k}{a_{rk}} = \min\limits_j\left\{\dfrac{\Delta_j}{a_{rj}}, a_{rj} < 0\right\}$

We get $\qquad \dfrac{\Delta_k}{a_{1k}} = \min\limits_j\left\{\dfrac{\Delta_j}{a_{1j}}, a_{1j} < 0\right\} = \min\left\{\dfrac{\Delta_1}{a_{11}}, \dfrac{\Delta_2}{a_{12}}\right\}$

$$= \min\left\{\frac{-7/4}{-5/4}, \frac{-1/2}{-1/2}\right\} = \min\left\{\frac{7}{5}, 1\right\} = 1 = \frac{\Delta_2}{a_{12}}$$

$\Rightarrow \qquad k = 2$

$\Rightarrow \quad \alpha_2$ is the incoming vector and key element $= a_{12} = \dfrac{1}{2}$

Simplex Table-3

	$c_j \rightarrow$		-2	0	-1	0	0
B.V.	C_B	X_B	$x_1(\alpha_1)$	$x_2(\alpha_2)$	$x_3(\beta_2)$	s_1	s_2
x_2	0	14	$5/2$	1	0	-2	$-1/2$
x_3	-1	9	$3/2$	0	1	-1	$-1/2$
$Z' = C_B X_B = -9,\ \Delta_j \rightarrow$			$-1/3$	0	0	-1	$-1/2$

Finally, we observe that all $\Delta_j \le 0$ and $x_1 = 0$, $x_2 = 14$, $x_3 = 9$, $s_1 = s_2 = 0$.

$\Rightarrow \quad$ solution is optimal and feasible and is given by

$$x_1 = 0, x_2 = 14, x_3 = 9 \text{ and max. } Z = -9$$

Exercise-8.1

Solve the following LPP by dual simplex method.

1. Min. $Z = 2x_1 + x_2$
   s.t.
   $3x_1 + 2x_2 \geq 3$
   $4x_1 + 3x_2 \geq 6$
   $x_1 + 2x_2 \geq 3$ [MEERUT 1994; MADRAS;
   and $x_1, x_2 \geq 0$ BE(MECH), 1999]

2. Max. $Z = 2x_1 + 3x_2$
   s.t.
   $2x_1 - x_2 - x_3 \geq 3$
   $x_1 - x_2 + x_3 \geq 2$
   and $x_1, x_2, x_3 \geq 0$

3. Max. $Z = x_1 + 2x_2$
   s.t.
   $3x_1 + 2x_2 \geq 6$
   $x_1 + 6x_2 \geq 3$
   and $x_1, x_2 \geq 0$

4. Max. $Z = -3x_1 - 2x_2$
   s.t.
   $x_1 + x_2 \geq 1$
   $x_1 + x_2 \leq 7$
   $x_1 + 2x_2 \geq 10$
   $x_2 \leq 3$

and $x_1, x_2 \geq 0$

5. Min. $Z = x_1 + 2x_2$
   s.t.
   $2x_1 + x_2 \geq 4$
   $x_1 + 7x_2 \geq 7$
   and $x_1, x_2 \geq 0$

6. Min. $Z = x_1 + 2x_2 + 3x_3$
   s.t.
   $2x_1 - x_2 + x_3 \geq 4$
   $x_1 + x_2 + 2x_3 \leq 8$
   $x_2 - x_3 \geq 2$
   and $x_1, x_2, x_3 \geq 0$

7. Min. $Z = 2x_1 + 2x_2 + 4x_3$
   s.t.
   $2x_1 + 3x_2 + 5x_3 \geq 2$
   $3x_1 + x_2 + 7x_3 \leq 3$
   $x_1 + 4x_2 + 6x_3 \leq 5$
   and $x_1, x_2, x_3 \geq 0$ [PUNJAB, BE(IT), 2004]

8. Min. $Z = 3x_1 + 2x_2 + x_3 + 4x_4$
   s.t.
   $2x_1 + 4x_2 + 5x_3 + x_4 \geq 10$
   $3x_1 - x_2 + 7x_3 - 2x_4 \geq 2$
   $5x_1 + 2x_2 + x_3 + 6x_4 \geq 15$
   and $x_1, x_2, x_3, x_4 \geq 0$ [MEERUT 1995]

ANSWERS

1. $x_1 = \dfrac{3}{5}, x_2 = \dfrac{6}{5}, \text{Min}.Z = \dfrac{12}{5}$

2. $x_1 = \dfrac{5}{3}, x_2 = 0, x_3 = \dfrac{1}{3}, \text{Max}.Z = \dfrac{13}{3}$

3. $x_1 = \dfrac{15}{8}, x_2 = \dfrac{3}{16}, \text{Max } Z = \dfrac{9}{4}$

4. $x_1 = 4, x_2 = 3$ and $\text{Max}.Z = -18$

5. $x_1 = \dfrac{21}{13}, x_2 = \dfrac{10}{13}, \text{Max}.Z = \dfrac{41}{13}$

6. $x_1 = 3, x_2 = 2, x_3 = \dfrac{1}{3}, \text{Max}.Z = 7$

7. $x_1 = 0, x_2 = \dfrac{2}{3}, x_3 = 0, \text{Min}.Z = \dfrac{4}{3}$

8. $x_1 = \dfrac{65}{23}, x_2 = 0, x_3 = \dfrac{20}{23}, x_4 = 0, \text{Min}.Z = \dfrac{215}{23}$

8.3 PRIMAL-DUAL ALGORITHM

In dual simplex method, we observe that in some situations the necessity of introducing artificial variables removed. But in general, it is not always easy to find a basic solution with all $c_j - z_j \leq 0$ by dual simplex method. The removal of artificial variables are not concerned with the optimality of the given problem. At the end of phase-1 (in two phase method) or when all the artificial variables are removed in Big-M method the feasible solution may not be anywhere near optimal. So, to obtain optimality we requires a large number of iterations in phase-II. To overcome such type of difficulties we use primal-dual algorithm, which deals directly with both the primal and dual in the following manner:

STEP 1.   Start with a feasible dual solution.

STEP 2. Find infeasible solution to the primal such that following results must satisfied.

    (i) Obtain restricted primal (RP) from given primal problem by taking some variables equal to zero so that complimentary slackness theorem is satisfied and the objective function is taken as the sum of artificial variables.

    (ii) Find optimal solution of restricted primal problem by using simplex method.

    (iii) If the optimal solution of restricted primal is not optimal solution of the given primal, then again find the dual of restricted primal problem and use steps (i) and (ii).

Above process is repeated successively until the optimal solution of the given primal is obtained.

Here, it must be noted that at any stage when a feasible solution to the primal is obtained then it is also optimal to the given primal.

8.3.1 THEORY OF PRIMAL-DUAL ALGORITHM

(1) To find the initial dual solution:

Consider the primal

$$\text{Max.} Z_p = \boldsymbol{Cx} \left.\vphantom{\begin{array}{c}a\\a\\a\end{array}}\right]$$

subject to the constraints

$$\boldsymbol{Ax} = \boldsymbol{b}$$

$$\qquad \qquad \text{...(1)}$$

and

$$\boldsymbol{x} \geq 0$$

where $\boldsymbol{x} = [x_1, x_2, \ldots, x_n]'$, $\boldsymbol{C} = [c_1, c_2, \ldots, c_n]$

$\boldsymbol{A} = [a_{ij}]_{m \times n}$, $\boldsymbol{b} = [b_1, b_2, \ldots, b_m]'$

Following the usual procedure, the dual of (1) is given by

$$\text{Min.} Z_d = \boldsymbol{b'y} \left.\vphantom{\begin{array}{c}a\\a\end{array}}\right]$$

subject to the constraints

$$\qquad \qquad \text{...(2)}$$

$$\boldsymbol{A'y} \geq \boldsymbol{C'}$$

Here, $\boldsymbol{y} = [y_1, y_2, \ldots, y_m]'$ is the dual variable vector which is unrestricted in sign.

Now, to find the initial dual solution, we introduce an additional constraints to (1) such that

$$x_0 + x_1 + \ldots + x_n = b_0; \text{ where } b_0 \text{ is arbitrary large.}$$

Now (1) reduces to

$$\text{Max.} Z_p = 0 x_0 + \boldsymbol{c} \cdot \boldsymbol{x} = (0, \boldsymbol{c}) \begin{bmatrix} x_0 \\ \boldsymbol{x} \end{bmatrix} \left.\vphantom{\begin{array}{c}a\\a\\a\\a\\a\end{array}}\right]$$

s.t.

$$x_0 + x_1 + \ldots + x_n = b_0$$

$$\qquad \qquad \text{...(3)}$$

$$\boldsymbol{Ax} = \boldsymbol{b}$$

and $\quad x_0, x_1, \ldots, x_n \geq 0$

The primal (3) can be written in vector form as follows

$$\text{Max.} Z_p = [0, c] \begin{bmatrix} x_0 \\ x \end{bmatrix}$$

s.t.

$$\begin{bmatrix} 1 & 1 \\ O & A \end{bmatrix} \begin{bmatrix} x_0 \\ x \end{bmatrix} = \begin{bmatrix} b_0 \\ b \end{bmatrix}$$

and $(x_0, x) \geq 0$

...(4)

Now, modified dual of dual (4) is given by

$$\text{min.} Z_d = (b_0, b') \begin{bmatrix} y_0 \\ y \end{bmatrix}$$

s.t. $\begin{bmatrix} 1 & 0 \\ 1 & A' \end{bmatrix} \begin{bmatrix} y_0 \\ y \end{bmatrix} \geq \begin{bmatrix} 0 \\ C' \end{bmatrix}$

...(5)

Now, (5) $\Rightarrow$ $y_0 \geq 0$ and $y_0 + A'y \geq C'$ ...(6)

We observe that y_0 is non-negative and $y_1, y_2, ..., y_m$ are unrestricted in sign.

If $y_0 = 0$, then (5) reduced to the dual (2) of given primal (1).

In this case, from (6) we can easily find a solution of dual (5) which is given by

$$y = 0$$

$\therefore$ From (6) $y \geq 0$ and $y_0 \geq c_j, j = 1, 2, ..., n$

$$y_0 = \max\{0, c_j\}, j = 1, 2, ..., n$$

and $y_1 = 0 = y_2 = ... = y_m$

...(7)

Further, introducing surplus variables $s_0, s_1, ..., s_n$, then (5) becomes

$$\text{Min } Z_d = b_0 y_0 + b_1 y_1 + b_2 y_2 + ... + b_m y_m + 0 s_0 + 0 s_1 + ... + 0 s_n$$

s.t.

$$1 \cdot y_0 + 0 y_1 + 0 y_2 + ... + 0 y_m - s_0 + 0 s_1 + ... + 0 s_n = 0$$
$$1 \cdot y_0 + a_{11} y_1 + a_{21} y_2 + ... + a_{m1} y_m + 0 s_0 - s_1 + 0 s_2 + ... + 0 s_n = c_1$$
$$\vdots \qquad\qquad\qquad\qquad\qquad\qquad\qquad\qquad \vdots$$
$$1 \cdot y_0 + a_{1n} y_1 + a_{2n} y_2 + ... + a_{mn} y_m + 0 s_0 + 0 s_1 + ... - s_n = c_n$$

...(8)

and $y_0 \geq 0$, $s_0, s_j \geq 0$, $j = 1, 2, ..., n$

and $y_1, y_2, ..., y_m$ are unrestricted in sign.

Clearly, the initial solution of (8) is given by

$$s_0 = y_0, s_1 = y_0 - c_1, s_2 = y_0 - c_2, ..., s_n = y_0 - c_n$$

$\because y_0 = \max. \{0, c_j\}, j = 1, 2, ..., n$. Then we get at least one $s_0, s_1, ..., s_n$ is zero.

Therefore, solution (7) is the initial dual solution with which we start primal-dual algorithm.

And if $x_0 s_0 + x_1 s_1 + ... + x_n s_n = 0$ ...(9)

Then $Z_p = Z_d$

If along (9), $\begin{bmatrix} x_0 \\ x \end{bmatrix}$ is a feasible solution to modified primal (3) or (4), then $\begin{bmatrix} x_0 \\ x \end{bmatrix}$ will

be an optimal solution to modified form (3) and $\begin{bmatrix} y_0 \\ y \end{bmatrix}$ is an optimal solution to the modified dual (5). But if $y_0 = y$ then $Z_p = Z_d$ with condition (9).

Hence $\boldsymbol{x}$ is an optimal solution to the given primal (1) and y is an optimal solution to its dual (2). Further, if we obtain a solution to the modified dual with $y_0 \neq 0$ and a feasible solution to the modified primal such that (9) is satisfied then $x_0 = 0$ and then

$$Z_p = Z_d = b_0 y_0 + b' y$$

(2) To find Restricted Primal (RP):

Here, we add artificial variables to each of its constraints and assign the cost -1 to each artificial variable and zero price corresponding to every legistimate variable. Then the modified primal (5) so obtained is called the restricted primal and is given by

$$\text{Max.} Z^* = A_0 - A_1 - A_2 - \ldots - A_m$$
$$= 1 \cdot A_0 - 1 \cdot \boldsymbol{x_a}$$

subject to the constraints   ...(10)

$$\begin{bmatrix} 1 & 1 \\ 0 & A \end{bmatrix} \begin{bmatrix} x_0 \\ \boldsymbol{x} \end{bmatrix} + \begin{bmatrix} 1 & 0 \\ 0 & I_m \end{bmatrix} \begin{bmatrix} x_{a_0} \\ \boldsymbol{x_a} \end{bmatrix} = \begin{bmatrix} b_0 \\ \boldsymbol{b} \end{bmatrix}$$

and $[x_0, x] \geq 0$ and $[x_{a_0}, \boldsymbol{x_a}] \geq 0$

(3) To find the solution of restricted Primal:

For this, apply simplex method successively till we get $Z^* = 0$, i.e., $A_0 = A_1 = A_2 = \ldots = A_m = 0$. Here, we choose the outgoing vector and incoming vector in the following manner:

(i)   The legismate activity vectors are taken as $\alpha_j^{(1)}$ for $j = 1, 2, \ldots, n$.

(ii) A basis matrix for the constraints is taken as $\boldsymbol{B}_1$.

(iii) The price vector corresponding to the variables as $\boldsymbol{C_{B_1}^*}$.

(iv) Calculate $\Delta_j^* = c_j^* - z_j^* = c_j^* - \boldsymbol{C_{B_1}^*} \boldsymbol{B_1^{-1}} \alpha_j^{(1)}$

$$= c_j^* - \boldsymbol{C_{B_1}^*} \boldsymbol{x_j^{(1)}} = \boldsymbol{B_1^{-1}} \alpha_j^{(1)}$$

Rule to find incoming vector: The vector $\alpha_k^{(1)}$ is the incoming vector for which $s_k = 0$. So, all x_j for which s_j are strictly positive, remains zero.

Rule to find outgoing vector: Corresponding to the incoming vector $\alpha_k^{(1)}$, the outgoing vector from the basis is obtained one by one in the same manner as in phase 1.

(4) To obtain new Dual solution $\hat{s}_j$ and the improved value $\hat{Z}_d$ and $\hat{Z}^*$ of Z_d and Z^* respectively:

(i) $\hat{s}_j = s_j - \theta \Delta_j^*$ where $\theta = \min \left\{ \dfrac{s_j}{\Delta_j^*}, \Delta_j^* > 0 \right\}$

     If at least one $\hat{s}_j$ is zero then again find $\alpha_k^{(1)}$. The vector entering the primal basis at the next iteration.

(ii) The improved value of Z^* can be calculated by using the formula.

$$\hat{Z}^* = C_{B_1} \boldsymbol{X_B^{(1)}}$$

(iii) The improved value of Z_d can be calculated by using the formula.

$$\hat{Z}_d = Z_d + \theta \hat{Z}^*$$

(5) Optimality Test: The primal-dual algorithm will be stopped if any one of the following is satisfied:

(i) If we obtain $Z^* = 0$ then an optimal solution to the given primal is achieved and is given by $Z_p = Z_d$ provided $s_0 = 0 = y_0$.

(ii) If $Z^* = 0$ is obtained for $s_0 \neq 0$ then the given primal will have unbounded solution.

(iii) If $Z^* < 0$, $\Delta_j^* \leq 0 \; \forall j$ at any iteration, then primal has no solution.

8.4 STEPS OF PRIMAL-DUAL ALGORITHM

STEP 1. Find restricted primal (RP) from the given primal.

STEP 2. Find initial dual solution of the dual of restricted primal as given below

$$y_0 = \max\{0, c_j\}, c_j > 0; j = 1, 2, 3, \ldots, n$$

and $\quad y_1 = 0 = y_2 = \ldots = y_m$

$$s_0 = y_0, s_j = y_0 - c_j : j = 1, 2, 3, \ldots, n$$

STEP 3. Find incoming vector $\alpha_k^{(1)}$ corresponding to k, for which $s_k = 0$.

STEP 4. Find outgoing vector by proceeding same as in phase I of simplex method.

STEP 5. Compute Δ_j^* using the formula

$$\Delta_j^* = c_j^* - C_{B_1} x_j^{*(1)}, x_j^{*(1)} = B_1^{-1} \alpha_j^{(1)}$$

and obtain Z^* by $Z^* = C_{B_1}^* x_j^{(1)}$

and $Z_d = b_0 y_0 + b_1 y_1 + \ldots + b_m y_m$

STEP 6. Construct the simplex table as given below

	c_j^*		c_0^*	c_1^*	c_2^*	$\cdots$	c_n^*	$\cdots$	c_{n+m}^*	Min.Ratio
B_1	C_B	$X_B^{(1)}$	$x_0^{(1)}$	$x_1^{(1)}$	$x_2^{(1)}$	$\cdots$	$x_n^{(1)}$	$\cdots$	$x_{n+m}^{(1)}$	$X_B^{(1)} / x_k^{(1)}$
$x_{n+1}^{(1)}$	-1	b_0								
$x_{n+2}^{(1)}$	-1									
$\vdots$	$\vdots$									
$x_{n+m}^{(1)}$	-1									
$Z^* = C_{B_1} X_B^{(1)}$		x_j								
		Δ_j^*								
$Z^* = \sum\limits_{j=0}^{m} b_j y_j$		s_j								

STEP 7. If $Z^* = 0$ and $\Delta_j^* \leq 0 \; \forall j$, then solution is optimal, if not so, go to next step.

STEP 8. Find new dual solution by using the formula

$$\hat{s}_j = s_j - \theta \Delta_j^*$$

where $\theta = \min\left\{\dfrac{s_j}{\Delta_j^*}, \Delta_j^* > 0\right\}$

STEP 9. Repeat step (5) to (8) till we obtain the optimal solution.

 Solved Examples

EXAMPLE 1. *Using primal-dual algorithm, solve the following LPP.*

$$\text{Max. } Z = x_1 + 6x_2$$

subject to the constraints

$$x_1 + x_2 \geq 2$$
$$x_1 + 3x_2 \leq 3$$

and $\qquad x_1, x_2 \geq 0$

SOLUTION. Introducing surplus variable s_1 and slack variable s_2, we can write the given primal as

$$\text{Max. } Z = x_1 + 6x_2 + 0s_1 + 0s_2$$

s.t.

$$\left.\begin{array}{l} x_1 + x_2 - s_1 = 2 \\ x_1 + 3x_2 + s_2 = 3 \\ \text{and} \quad x_1, x_2, s_1, s_2 \geq 0 \end{array}\right] \qquad \ldots(1)$$

Now, the modified form of primal (1) is

$$\text{Max. } Z = 0 \cdot x_0 + x_1 + 6x_2 + 0s_1 + 0s_2$$

s.t.

$$\left.\begin{array}{l} x_0 + x_1 + x_2 + s_1 + s_2 = b_0 \\ 0x_0 + x_1 + x_2 - s_1 + 0s_2 = 2 \\ 0x_0 + x_1 + 3x_2 + 0s_1 + s_2 = 3 \\ \text{and} \qquad x_0, x_1, x_2, s_1, s_2 \geq 0 \end{array}\right] \qquad \ldots(2)$$

Now, the restricted primal of (1) is given by

$$\text{Max. } Z^* = -A_0 - A_1 - A_2$$

s.t.

$$\left.\begin{array}{l} x_0 + x_1 + x_2 + s_1 + s_2 = b_0 \\ 0x_0 + x_1 + x_2 - s_1 + 0s_2 + A_0 = 2 \\ 0x_0 + x_1 + 3x_2 + 0s_1 + s_2 + A_2 = 3 \\ \text{and} \quad x_0, x_1, x_2, s_1, s_2, A_0, A_1, A_2 \geq 0 \end{array}\right] \qquad \ldots(3)$$

The initial dual solution of (2) is given by

$$y_0 = \max\{0, c_j\}, c_j > 0$$
$$= \max\{0, 1, 6\} = 6$$

and
$$y_1 = y_2 = y_3 = y_4 = 0$$

So,
$$s_0 = y_0 = 6, s_1 = y_0 - c_1 = 6 - 1 = 5$$
$$s_2 = y_0 - c_2 = 6 - 6 = 0, s_3 = y_0 - c_3 = 6 - 0 = 6$$
$$s_4 = y_0 - c_4 = 6 - 0 = 6$$

and
$$Z_d = b_0 y_0 + b_1 y_1 + b_2 y_2 + b_3 y_3 + b_4 y_4 = 6b_0$$

Simplex Table-1

B_1	c_j^* $C_{B_1}^*$	$X_B^{(1)}$	0 $x_0^{(1)}$ $(\alpha_0^{(1)})$	0 $x_1^{(1)}$ $(\alpha_1^{(1)})$	0 $x_2^{(1)}$ $(\alpha_2^{(1)})$	0 $s_1^{(1)}$ $(\alpha_3^{(1)})$	0 $s_2^{(1)}$ $(\alpha_4^{(1)})$	-1 $A_0^{(1)}$ (β_1)	-1 $A_1^{(1)}$ (β_2)	-1 $A_2^{(1)}$ (β_3)	Min Ratio $X_B^{(1)}/x_2^{(1)}$
$A_0^{(1)}$	-1	b_0	1	1	1	1	1	1	0	0	$b_0/1$
$A_1^{(1)}$	-1	2	0	1	1	-1	0	0	1	0	2/1
$A_2^{(1)}$	-1	3	0	1	3	0	1	0	0	1	3/3(min)$\rightarrow$
$Z^* = C_{B_1}^* X_B^{(1)}$		x_j	0	0	0	0	0	b_0	2	3	
$=-b_0-5$		Δ_j^*	1	3	5	0	2	0	0	0	
$Z_d^* = 6b_0$		s_j	6	5	0	6	6	0	0	0	
				$\uparrow$						$\downarrow$	

In the above table, Δ_j^* is calculated by the following formula

$$\Delta_j^* = c_j^* - C_{B_1}^* x_j^{(1)}$$

and $\Delta_j^* \geq 0 \Rightarrow$ The obtained solution is not optimal.

Now, we shall improve this solution.

We have $\Delta_j = 0$ for $j = 2$ (not in basis)

$\Rightarrow \quad s_2 = 0 \quad \Rightarrow \quad \alpha_2^{(1)}$ is the incoming vector.

Also, by minimum ratio rule, β_3 is the outgoing vector and key element $a_{23} = 3$

Simplex Table-2

B_1	c_j^* $C_{B_1}^*$	$X_B^{(1)}$	0 $x_0^{(1)}$	0 $x_1^{(1)}$	0 $x_2^{(1)}$	0 $s_1^{(1)}$	0 $s_2^{(1)}$	-1 $A_0^{(1)}$ (β_1)	-1 $A_1^{(1)}$ (β_2)	Min Ratio $X_B^{(1)}/x_1^{(1)}$
$A_0^{(1)}$	-1	b_0-1	1	2/3	0	1	2/3	1	0	$(b_0-1)/(2/3)$
$A_1^{(1)}$	-1	1	0	2/3	0	-1	$-1/3$	0	1	3/2(min)$\rightarrow$
$x_2^{(1)}$	0	1	0	1/3	1	0	1/3	0	0	3
$Z^* = C_{B_1}^* X_B^{(1)} = -b_0$		x_j	0	0	1	0	0	b_0-1	1	
		Δ_j^*	1	4/3	0	0	1/3	0	0	
$Z_d^* = Z_d + \theta Z^*$		s_j	6	5	0	6	6	0	0	
$= 9b_0/2$		$\hat{s}_j$	9/4	0	0	6	19/4	0	0	
				$\uparrow$					$\downarrow$	

In the above table Δ_j^* is calculated by the following formula

$$\Delta_j^* = c_j^* - C_{B_1}^* x_j^* \text{ for all } j \text{ not in the basis.}$$

So, $\qquad \Delta_0^* = 1, \Delta_1^* = \dfrac{4}{3}, \Delta_3^* = 0, \Delta_4^* = \dfrac{1}{3}$

Clearly, $\Delta_j^* \geq 0 \implies$ solution is not optimal.

We have to again improve this solution.

The new dual solution $\hat{s}_j$ be given by $\hat{s}_j = s_j - \theta \Delta_j^*$

where, $\quad \theta = \min \left\{ \dfrac{s_j}{\Delta_j^*}, \Delta_j^* > 0 \right\}$

$$= \min \left\{ \dfrac{s_0}{\Delta_0^*}, \dfrac{s_1}{\Delta_1^*}, \dfrac{s_4}{\Delta_4^*} \right\} = \min \left\{ \dfrac{6}{1}, \dfrac{5}{4/3}, \dfrac{6}{1/3} \right\} = \min \left\{ 6, \dfrac{15}{4}, 18 \right\} = \dfrac{15}{4}$$

So, $\hat{Z}_d = Z_d + \theta Z^* = 6 b_0 + \dfrac{15}{4}(-b_0) = \dfrac{9 b_0}{4}$

and $\quad \hat{s}_0 = s_0 - \theta \Delta_0^* = 6 - \dfrac{15}{4} \times 1 = \dfrac{9}{4}$,

$$\hat{s}_1 = s_1 - \theta \Delta_1^* = 5 - \dfrac{15}{4} \times \dfrac{4}{3} = 0, \quad \hat{s}_2 = s_2 - \theta \Delta_2^* = 0 - \dfrac{15}{4} \times 0 = 0,$$

$$\hat{s}_3 = s_3 - \theta \Delta_3^* = 6 - \dfrac{15}{4} \times 0 = 6, \quad \hat{s}_4 = s_4 - \theta \Delta_4^* = 6 - \dfrac{15}{4} \times \dfrac{1}{3} = \dfrac{19}{4}$$

Clearly, $\hat{s}_2 = 0$ as $j = 2$ is not in the basis.

$\implies \quad \alpha_1^{(1)} (= x_1^{(1)})$ is the incoming vector.

Further, by minimum ratio rule, the outgoing vector is $\beta_2 (= A_1^{(1)})$

$\implies \quad$ Key element $= \dfrac{2}{3} = a_{22}$

Simplex Table-3

B_1	$C_{B_1}^*$	$X_B^{(1)}$	c_j^*						Min Ratio $X_B^{(1)} / x_0^{(1)}$
			0	0	0	0	0	−1	
			$x_0^{(1)}$	$x_1^{(1)}$ (β_2)	$x_2^{(1)}$ (β_3)	$s_1^{(1)}$	$s_2^{(1)}$	$A_0^{(1)}$ (β_1)	
$A_0^{(1)}$	−1	$b_0 - 2$	1	0	0	2	1	1	$(b_0 - 2)/1 \rightarrow$
$A_1^{(1)}$	0	3/2	0	1	0	−3/2	−1/2	0	—
$x_2^{(1)}$	0	1/2	0	0	1	1/2	1/2	0	—
$Z^* = C_{B_1}^* X_B^{(1)}$		x_j	0	3/2	1/2	0	0	$b_0 - 1$	
$= 2 - b_0$		Δ_j^*	1	0	0	2	1	0	
$\hat{Z}_d' = \hat{Z}_d + \theta Z^*$		s_j	9/4	0	0	6	19/4	0	
$= 9/2$		$\hat{s}_j$	0	0	0	3/2	5/2	0	
			$\uparrow$						

Here, Δ_j^* is calculated by the formula given by

$$\Delta_j^* = c_j^* - C_{B_1}^* X_B^{(1)}, \text{ for } j \text{ not in the basis.}$$

Therefore, $\Delta_0^* = 1, \Delta_3^* = 2, \Delta_4^* = 1$

Clearly, $\Delta_j^* > 0 \Rightarrow$ solution is not optimal. To improve it, proceed as follows.

$$\hat{s}_j = s_j - \theta \Delta_j^*$$

where

$$\theta = \min\left\{\frac{\hat{s}_j}{\Delta_j^*}, \Delta_j^* > 0\right\} = \min\left\{\frac{\hat{s}_0}{\Delta_0^*}, \frac{\hat{s}_3}{\Delta_3^*}, \frac{\hat{s}_4}{\Delta_4^*}\right\}$$

$$= \min\left\{\frac{9/4}{1}, \frac{6}{2}, \frac{19/4}{1}\right\} = \frac{9}{4} = \frac{\hat{s}_0}{\Delta_0^*}$$

So,

$$\hat{s}_0' = s_0 - \theta \Delta_0^* = \frac{9}{4} - \frac{9}{4} \times 1 = 0, \quad \hat{s}_1' = s_1 - \theta \Delta_1^* = 0 - \frac{9}{4} \times 0 = 0$$

$$\hat{s}_2' = s_2 - \theta \Delta_2^* = 0 - \frac{9}{4} \times 0 = 0, \quad \hat{s}_3' = s_3 - \theta \Delta_3^* = 6 - \frac{9}{4} \times 2 = \frac{3}{2}$$

$$\hat{s}_4' = s_4 - \theta \Delta_4^* = \frac{19}{4} - \frac{9}{4} \times 1 = \frac{5}{2}$$

which implies $\hat{s}_0' = 0$ which is corresponding to $j = 0$ not in the basis. Therefore,

$$\hat{Z}_d = Z_d + \theta \cdot Z^* = \frac{9}{4} b_0 + \frac{9}{4}(-b_0 + 2) = \frac{9}{2}$$

Also, new $\hat{s}_0 = 0 \Rightarrow \alpha_0^{(1)}(= x_0^{(1)})$ is the incoming vector and by minimum ratio rule, $\beta(= A_0^{(1)})$ is the outgoing vector.

$\Rightarrow$ Key element, $a_{11} = 1$

Now, we have the following simplex table

Simplex Table-4

B_1	$C_{B_1}^*$	$X_B^{(1)}$	c_j^* $\to$ 0 $x_0^{(1)}$ (β_1)	0 $x_1^{(1)}$ (β_2)	0 $x_2^{(1)}$ (β_3)	0 $s_1^{(1)}$	0 $s_2^{(1)}$
$x_0^{(1)}$	0	b_0-2	1	0	0	2	1
$x_1^{(1)}$	0	$3/2$	0	1	0	$-3/2$	$-1/2$
$x_2^{(1)}$	0	$1/2$	0	0	1	$1/2$	$1/2$
$Z^* = C_{B_1}^* X_B^{(1)} = 0$		x_j	b_0-2	$3/2$	$1/2$	0	0
		Δ_j^*	0	0	0	0	0
$\hat{Z}_d' = \hat{Z}_d + \theta Z^*$ $= 9/2$		$\hat{s}''$	0	0	0	$3/2$	$5/2$

Here, we observe that all $\Delta_j^* \leq 0 \Rightarrow$ solution is optimal.

Also, $Z^* = 0$ and $s_0'' = 0 = s_0$ therefore, the solution of the given primal is also optimal with $Z_p = Z_d''$

Hence, the optimal solution of the given primal is

$$x_1 = \frac{3}{2}, x_2 = \frac{1}{2} \text{ and Max.} Z_p = Z_d'' = \frac{9}{2}$$

Exercise-8.2

Solve the following LPP by primal dual algorithm:

1. Min. $Z = 3x_1 + 4x_2$
   subject to the constraints
   $2x_1 + 3x_2 \geq 8$
   $5x_1 + 2x_2 \geq 12$
   and $x_1, x_2 \geq 0$

2. Min. $Z = 2x_1 + 2x_2 + 3x_3$
   subject to the constraints
   $x_1 + x_3 \geq 1$
   $x_2 + x_3 \geq 2$
   and $x_1, x_2, x_3 \geq 0$

ANSWERS

1. $x_1 = \frac{20}{11}, x_2 = \frac{16}{11}$ and Min. $Z = \frac{104}{11}$

2. $x_1 = 0, x_2 = 1, x_3 = 1$ and Min. $Z = 5$

REVIEW QUESTIONS

1. Write the difference between regular simplex and dual simplex method.
2. Write the dual simplex algorithm.
3. Write the primal-dual simplex algorithm.
4. Write the method to choose the incoming and outgoing vector in dual simplex method.

MULTIPLE CHOICE QUESTIONS (CHOOSE THE MOST APPROPRIATE ONE)

1. Which of the following is true?
   (a) If primal has a feasible solution, then its dual will also have a feasible solution
   (b) If both primal and dual have feasible solution then both will have bounded optimal solution
   (c) If primal has no solution, then its dual will have an unbounded solution
   (d) None of these

2. In application of dual simplex method, the availability vector (RHS vector) must be:
   (a) ≥ 0
   (b) ≤ 0
   (c) no such type of restriction is required
   (d) None of these

3. If in the primal the number of constraints are m and in dual the variables are n, then:
   (a) $m \geq n$
   (b) $m \leq n$
   (c) $m = n$
   (d) None of these

4. Consider max. $Z = x_1$ subject to $x_1 + x_2 \leq 1$, $x_1 - x_2 \geq 1, x_1 \geq 0, x_2 \geq 0$. The solution of its dual is:
   (a) bounded
   (b) unbounded
   (c) degenerate
   (d) None of these

5. Let x be a non-optimal feasible solution of a LPP of maximization and y is a dual feasible solution then:
   (a) the primal objective value at x is greater than the dual objective value at y
   (b) the primal objective value at x is less than the dual objective value at y
   (c) the dual can be unbounded
   (d) None of these

6. If a slack or surplus variable s_1 is in optimal primal basis then in the optimal dual solution:
   (a) the dual variable corresponding to i^{th} primal constraints is zero
   (b) the dual variable corresponding to i^{th} primal constraints is unrestricted
   (c) both (a) and (b) are true

(d) None of these

7. If there exist feasible solutions to both the primal and its dual then:

 (a) the primal may have an unbounded solution while the dual a bounded solution

 (b) the primal may have a bounded solution while the dual an unbounded solution

 (c) both (a) and (b) are true

 (d) None of these

8. If primal: max. $Z_x = x_1 + x_2 + x_3, x_1, x_2, x_3 \geq 0$ then the solution of primal and its dual are respectively:

 (a) infeasible, feasible

 (b) feasible, infeasible

 (c) feasible, feasible

 (d) None of these

9. If the dual of the problem has infeasible solution then the value of objective function is:

 (a) bounded (b) unbounded

 (c) no solution (d) None of these

10. In LPP, if primal objective function is of maximization then for its dual the objective function is of:

 (a) maximization (b) minimization

 (c) optimization (d) None of these

11. If a slack or surplus variable s_j is in optimal primal basis, then the optimal dual solution:

 (a) the dual variable corresponding to j^{th} primal constraint is zero

 (b) the slack or surplus variable attached to j^{th} dual constraint is zero

 (c) both (a) and (b) are true

 (d) None of these

12. If the primal of LPP has bounded solution, then dual of the problem has:

 (a) optimal solution

 (b) no solution

 (c) unbounded solution

 (d) None of these

13. If the primal of LPP has no solution, then dual of the problem has:

 (a) unbounded solution

 (b) bounded solution

 (c) either no solution or is unbounded

 (d) None of these

14. If the primal has an unbounded solution then the dual has:

 (a) optimal solution (b) no solution

 (c) bounded solution (d) None of these

ANSWERS

1. (b) 2. (c) 3. (c) 4. (a) 5. (b) 6. (a) 7. (b) 8. (b) 9. (b)
10. (b) 11. (a) 12. (b) 13. (c) 14. (b)

ARCHIVE

1. Use dual simplex method to solve the LP problem Minimize $Z = 3x_1 + x_2$ subject to the constraints

 (i) $x_1 + x_2 \geq 1$, (ii) $2x_1 + 3x_2 \geq 2$ and $x_1, x_2 \geq 0$. [IAS-1990]

2. Use dual simplex method to solve the LP problem

 Minimize $Z = 10x_1 + 6x_2 + 2x_3$ subject to the constraints

 (i) $-x_1 + x_2 + x_3 \geq 1$, (ii) $3x_1 + x_2 - x_3 \geq 2$

 and $x_1, x_2, x_3 \geq 0$. [MS TIRUNVELLI-1996]

3. Use dual simplex method to solve the LP problem.

 Minimize $Z = -6x_1 - 72x_2 - 3x_3 - 5x_4$ subject to the constraints

 (i) $2x_1 + 5x_2 + x_3 + x_4 \geq 8$

 (ii) $x_2 + 5x_3 - 6x_4 \geq 10$

 (iii) $5x_1 + 6x_2 - 3x_3 + 4x_4 \geq 12$

 and $x_1, x_2, x_3, x_4 \geq 0$ [MEERUT-1990, 99]

HINTS AND ANSWERS

1. $x_1 = 0, x_2 = 1$ and Min. $Z = 1$ 2. $x_1 = 1/4, x_2 = 5/4$, $x_3 = 0$ and Min. $Z = 10$

3. $x_1 = 0, x_2 = 30/11, x_3 = 16/11, x_4 = 0$ and Max. $Z = -258/11$

□□□□

Sensitivity Analysis in LPP

9

9.1 INTRODUCTION

Consider the linear programming problem

$$\text{Max. } Z = C\mathbf{x}$$

subject to the constraints

$$A\mathbf{x} = b$$

and

$$\mathbf{x} \geq 0$$

The optimal solution of the above LPP depends upon the three parameters a_{ij}, b_i and c_j. In most of the cases, we have assumed that these parameters are constant, but there are some situations for which the values of these parameters vary time to time. Due to the variation in these parameters, optimal solution will affect. We shall discuss these changes in sensitivity analysis.

Definition. *The investigation that deals with changes in the optimal solutions due to discrete variations in the parameters a_{ij}, b_i and c_j are called sensitivity analysis.* (MEERUT-2007, 09, 12)

In this chapter, we shall discuss the effect on optimal solution due to the following changes in the given LPP.

(1) Variation in the coefficient (price vector) in the objective function, c_j.

(2) Variation in the right hand side constants (requirement vectors), b_j.

(3) Variation in the coefficient of decision variables on the left hand side of constraints, a_{ij}.

(4) Addition of a new variable to the existing list of variables in LPP.

(5) Addition of a new constraints to the original LP constraints.

☞ Remarks

- The sensitivity analysis is also known as **post-optimality analysis**, because it does not begin until the optimal solution to the given LPP is obtained.
- The sensitivity analysis is the study of knowing the effect on optimal solution of the LP model due to variations in the input coefficients or parameters.
- The sensitivity analysis provide the sensitivity range within which the LP model parameters can vary without changing the optimality of the current optimal solution.

9.2 CHANGE IN THE OBJECTIVE FUNCTION COEFFICIENTS (PRICE VECTORS), c_j

Change in the profit or cost coefficients in the objective function can occur for any basic variables or any non-basic variables. The sensitivity range for these variables is determined differently. So, there are two cases will be discussed seperately as given below:

Let X_B be the optimal basic feasible solution of an LPP, then we have

$$X_B = B^{-1} \cdot b \qquad \qquad \qquad ...(1)$$

where B is the optimal basis matrix and b is the requirement vector.

Obviously, the change in c_j does not affect X_B because (1) is independent of c_j but $\Delta_j = c_j - z_j$ depends on c_j.

Therefore, the change in c_j will affect the Δ_j, i.e., the optimality condition.

While computing Δ_j $(= c_j - z_j)$ values, following two situations may arises:

(i) The new Δ_j values satisfy, optimality condition and the solution remains unchanged, however, optimal value of the objective function may change.

(ii) The optimality condition is not satisfied. In such a case we will use simplex method to achieve optimality.

Case I.   Change in the coefficient of a non-basic variable, i.e., change in c_j when it is not in C_B

If $\Delta_j \leq 0$ for all non-basic variables in a maximization LPP, then the current optimal solution remains unchanged. Let $c_j \notin C_B$, then $\Delta_j = c_j - z_j \leq 0$ for all j not in the basis, so X_B is an optimal solution.

Suppose that Δc_j is a change in c_j which is not in the basis, then there is no change in C_B and hence there is no change in B. Therefore,

$$z_j = C_B \cdot B^{-1} \alpha_j$$

will remains unchanged.

$\Rightarrow$ X_B remains optimal.

Now, if c_j changes to $c'_j = c_j + \Delta c_j$, then by optimality, we must have

$$(c_j + \Delta c_j) - z_j \leq 0 \qquad \qquad (\because \ \Delta_j \leq 0 \ \forall \ j \text{ not in the basis})$$

$$\Rightarrow \qquad \qquad \Delta c_j \leq -(c_j - z_j)$$

$$\Rightarrow \qquad \qquad \Delta c_j \leq -\Delta_j$$

Thus, we conclude that

To retain the optimality of the current optimal solution for a change in Δc_j in c_j, we must have
$$(c_j + \Delta c_j) - z_k \leq 0$$
or $\qquad \qquad (c_j + \Delta c_j) \leq z_k$

☛ Remarks

* For any LPP of maximization, the value of c_j may be increased upto the value of z_k and decrease to $-\infty$ without affecting the optimal solution.
* In this case, there is no lower bound of Δc_j.

Case II. Change in the coefficient of a basic variable, i.e., change in c_j when it is in C_B

We know that, in a maximization LPP, the change in the coefficient say, c_j of a basic variable x_j affect the Δ_j $(= c_j - z_j)$ values corresponding to all non-basic variably in the optimal simplex table.

If $c_j \in C_B$ and we make a change in c_j, then z_j will be changed because

$$z_j = C_B B^{-1} \alpha_j$$

Since,
$$z_j = C_B B^{-1} \alpha_j = C_B \alpha_j = \sum_{i=1}^{m} C_{B_i} x_{ij}$$

Further, suppose C_{B_k} changes to $C'_{B_k} = C_{B_k} + \Delta C_{B_k}$

Then new value of z_j, *i.e.*, z_j^* is given by

$$z_j^* = \sum_{i=1}^{m} C_{B_i} x_{ij} + C'_{B_k} x_{kj} \qquad (i \neq k)$$

$$= \sum_{\substack{i=1 \\ (i \neq k)}}^{m} C_{B_i} x_{ij} + (C_{B_k} + \Delta C_{B_k}) x_{kj}$$

$$= \sum_{i=1}^{m} C_{B_i} x_{ij} + \Delta C_{B_k} x_{kj}$$

$$\Rightarrow \qquad z_j^* = z_j + \Delta C_{B_k} x_{kj}$$

$$\Rightarrow \qquad c_j - z_j^* = (c_j - z_j) - \Delta C_{B_k} x_{kj}$$

Now for optimal solution, $c_j - z_j^* \leq 0 \ \forall j$ not in the basis.

$$\Rightarrow \qquad (c_j - z_j) - \Delta C_{B_k} x_{kj} \leq 0$$

$$\Rightarrow \qquad \Delta C_{B_k} x_{kj} \leq c_j - z_j$$

So, $\Delta C_{B_k} \geq \dfrac{c_j - z_j}{x_{kj}}$, when $x_{kj} > 0$

and $\Delta C_{B_k} \leq \dfrac{c_j - z_j}{x_{kj}}$, when $x_{kj} < 0$

On combining the above two inequalities, we have

$$\max_{x_{kj} > 0} \left\{ \frac{c_j - z_j}{x_{kj}} \right\} \leq C_{B_k} \leq \min_{x_{kj} < 0} \left\{ \frac{c_j - z_j}{x_{kj}} \right\} \qquad \ldots(2)$$

for all j corresponding to which α_j is not in the optimal basis. Thus we conclude that

> If we make a change in $C_{B_k} \in C_B$, then the solution will remain optimal when the change ΔC_{B_k} in C_{B_k} satisfy the inequality (2).

New value of Objective Function

The new value of the objective function Z^* is given by

$$Z^* = C_B X_B = \sum_{\substack{i=1 \\ (i \neq k)}}^{m} C_{B_i} \cdot x_{B_i} + (C_{B_k} + \Delta C_{B_k}) x_{B_k}$$

$$= \sum_{\substack{i=1 \\ (i \neq k)}}^{m} C_{B_i} x_{B_i} + C_{B_k} \cdot x_{B_k} + \Delta C_{B_k} \cdot x_{B_k}$$

$$= \sum_{i=1}^{m} C_{B_i} x_{B_i} + \Delta C_{B_k} x_{B_k}$$

$$\Rightarrow \qquad Z^* = Z + \Delta C_{B_k} x_{B_k}$$

which conclude that the new value of the objective function is improved by $\Delta C_{B_k} x_{B_k}$, when $C_{B_k} \in C_B$ changes to $C_{B_k} + \Delta C_{B_k}$, where x_{B_k} is the basic variable corresponding to C_{B_k}.

Aliter: The sensitivity limits for the contribution per unit a basic variable can be calculated as follows:

lower limit = original value c_k – {lowest absolute value of improvement ratio or $-\infty$ (if no ratio is negative)}

upper limit = original value c_k + {lowest positive value of improvement ratio or ∞ (if no ratio is positive)}

where improvement ratio = $\dfrac{\text{per unit improvement value}}{\text{input output coefficients in the variable row}}$

$$= \frac{c_j - z_j}{a_{kj}}$$

☞ **Remarks**

- If the range of c_k is $p \le c_k \le q$ then we check the optimality at $c_k = p$ and $c_k = q$. If solution is not optimal at $c_k = p$, then range of c_k will be $p < c_k \le q$ and if solution is not optimal at $c_k = q$, then range of c_k will be $p \le c_k < q$.
- In the calculation of sensitivity analysis, the artificial variables column in the simplex table are ignored.

Solved Examples

EXAMPLE 1. **Find the optimal solution of the LPP given by**

$$\textbf{Max. } Z = 3x_1 + 5x_2$$

subject to the constraints

$$x_1 + x_2 \le 1$$
$$2x_1 + 3x_2 \le 1$$

and $x_1, x_2 \ge 0$

Obtain the variation in $c_1(= 3)$ and $c_2(= 5)$ without affecting the above optimal solution. [MEERUT 1990]

SOLUTION. Apply the simplex method in a usual manner, we have the following simplex table.

			$c_j \rightarrow$	3	5	0	0	
B.V.		C_B	X_B	x_1	x_2	s_1	s_2	Min Ratio
s_1		0	1	1	1	1	0	1/1
s_2		0	1	2	③	0	1	1/3 (min) →
$Z = C_B X_B = 0$			$\Delta_j \rightarrow$	3	5	0	0	
					↑		↓	
s_1		0	2/3	1/3	0	1	–1/3	
x_2		5	1/3	1/3	1	0	1/3	
$Z = C_B X_B = 5/3$			$\Delta_j \rightarrow$	–1/3	0	0	–5/3	

From the above table we observe that all $\Delta_j \leq 0$

$\Rightarrow$ solution is optimal.

The optimal solution is given by

$$x_1 = 0, x_2 = \frac{1}{3} \text{ and } \max Z = \frac{5}{3}$$

To find variation in $c_1 = 3$

From the final simplex table, we have that c_1 is not in the basis B. So,

$$\Delta c_1 \leq -\Delta_1$$

$$\Rightarrow \qquad \Delta c_1 \leq -\left(-\frac{1}{3}\right), i.e., \Delta c_1 \leq \frac{1}{3}$$

So, without affecting the optimality, the range of c_1 is given by

$$-\infty < c_1 < c_1 + \Delta c_1$$

$$\Rightarrow \qquad -\infty < c_1 \leq 3 + \frac{1}{3} = \frac{10}{3}$$

Hence, c_1 can vary between $-\infty$ and $\frac{10}{3}$ without affecting optimal solution.

To find the variation in $c_2 = 5$

Clearly, $c_2 \in \boldsymbol{B}$

Now, since $\quad \boldsymbol{C_B} = (0,5) = (C_{B_1}, C_{B_2})$

$$\Rightarrow \qquad C_{B_2} = 5 (= c_2)$$

$$\Rightarrow \qquad k = 2$$

We know that the range of ΔC_{B_2} is given by

$$\max_{x_{2j} > 0} \left\{ \frac{c_j - z_j}{x_{2j}} \right\} \leq \Delta C_{B_2} \leq \min_{x_{2j} < 0} \left\{ \frac{c_j - z_j}{x_{2j}} \right\}$$

Here, $x_{21} = \frac{2}{3}, x_{24} = \frac{1}{3}$. we can take x_{22} and x_{23} because they are corresponding to the basis. Clearly, $x_{2j} > 0$ for $j = 1, 4$ but there is no $x_{2j} < 0$ which implies that ΔC_{B_2} has no upper bound.

$\therefore \quad$ Range of C_{B_2} is given by

$$\max \left\{ \frac{c_1 - z_1}{x_{21}}, \frac{c_4 - z_4}{x_{24}} \right\} < \Delta C_{B_2} \leq \infty \qquad \Rightarrow \qquad \max \left\{ \frac{\Delta_1}{x_{21}}, \frac{\Delta_4}{x_{24}} \right\} \leq \Delta C_{B_2} \leq \infty$$

$$\Rightarrow \qquad \max \left\{ \frac{-1/3}{2/3}, \frac{-5/3}{1/3} \right\} \leq \Delta C_{B_2} \leq \infty \qquad \Rightarrow \qquad \max \left\{ -\frac{1}{2}, -5 \right\} \leq \Delta C_{B_2} < \infty$$

$$\Rightarrow \qquad -\frac{1}{2} \leq \Delta C_{B_2} < \infty$$

$$\Rightarrow \qquad C_{B_2} - \frac{1}{2} \leq C_{B_2} + \Delta C_{B_2} < C_{B_2} + \infty$$

$$\Rightarrow \qquad 5 - \frac{1}{2} \leq C_{B_2} + \Delta C_{B_2} < \infty \text{ or } \frac{9}{2} \leq C_{B_2} + \Delta C_{B_2} < \infty$$

Hence, we conclude that c_2 can vary between $\frac{9}{2}$ and ∞ without affecting the optimum solution.

EXAMPLE 2. *Find an optimal solution to the following LPP.*

$$\text{Max. } Z = 3x_1 + 5x_2$$

subject to the constraints

$$x_1 \leq 4$$
$$x_2 \leq 6$$
$$3x_1 + 2x_2 \leq 18$$

and $\quad x_1, x_2 \geq 0$

What happens to this solution if the objective function is changed to $Z^* = 3x_1 + x_2$ *and* $Z^* = 3x_1 + 4x_2$.

SOLUTION. Using slack variables s_1, s_2, s_3 the given LPP becomes

$$\text{Max. } Z = 3x_1 + 5x_2 + 0s_1 + 0s_2 + 0s_3$$

s.t. $\quad x_1 + x_2 + s_1 = 4$
$$x_2 + s_2 = 6$$
$$3x_1 + 2x_2 + s_3 = 18$$

and $\quad x_1, x_2, s_1, s_2, s_3 \geq 0$

The initial basic feasible solution is given by

$$x_1 = 0, x_2 = 0, s_1 = 4, s_2 = 6, s_3 = 18$$

Now, apply the simplex method in a usual manner, we have the following simplex table.

Simplex Table-1

B.V.	C_B	X_B	$c_j \to$ 3 x_1	5 x_2	0 s_1	0 s_2	0 s_3	Min Ratio X_B/x_2
s_1	0	4	1	0	1	0	0	—
s_2	0	6	0	①	0	1	0	6/11(min)→
s_3	0	18	3	2	0	0	1	18/2
$Z = C_B X_B = 0$		$\Delta_j \to$	3	5 ↑	0	0 ↓	0	
s_1	0	4	1	0	1	0	0	4/1
x_2	5	6	0	1	0	1	0	—
s_3	0	6	③	0	0	−2	1	6/3(min) →
$Z = C_B X_B = 30$		$\Delta_j \to$	3 ↑	0	0	−5	0 ↓	
s_1	0	2	0	0	1	2/3	−1/3	
x_2	5	6	0	1	0	1	0	
x_1	3	2	1	0	0	−2/3	1/3	
$Z = C_B X_B = 36$		$\Delta_j \to$	0	0	0	−3	−1	

In the last row of the above table all $\Delta_j \leq 0$

$\Rightarrow \quad$ solution is optimal and is given by

$$x_1 = 2, x_2 = 6 \text{ and Max. } Z = 36$$

Variation in c_2

Using final table

$$C_B = (0,5,3) = (C_{B_1}, C_{B_2}, C_{B_3})$$

$\Rightarrow \qquad C_{B_2} = 5 = C_2$

$\Rightarrow \qquad k = 2$

The range of ΔC_{B_2} is given by

$$\max_{x_{kj} \geq 0} \left\{ \frac{c_j - z_j}{x_{kj}} \right\} \leq \Delta C_{B_2} \leq \max_{x_{kj} < 0} \left\{ \frac{c_j - z_j}{x_{kj}} \right\}$$

$$\Rightarrow \qquad \max_{x_{2j} > 0} \left\{ \frac{\Delta_j}{x_{kj}} \right\} \leq \Delta C_{B_2} \leq \min_{x_{2j} < 0} \left\{ \frac{\Delta_j}{x_{kj}} \right\}$$

Here, $j = 4, 5$ (not in the basis)

$$x_{24} = 1 \text{ and } x_{25} = 0$$

We observe that $x_{24} > 0$ and there is no $x_{2j} < 0$, so that ΔC_{B_2} has no upper bound.

$\therefore$ The range of ΔC_{B_2} is given by

$$\frac{\Delta_4}{x_{24}} \leq \Delta C_{B_2} < \infty$$

$\Rightarrow \qquad -\dfrac{3}{1} \leq \Delta C_{B_2} < \infty \qquad\qquad \Rightarrow \qquad -3 \leq \Delta C_{B_2} < \infty$

$\Rightarrow \qquad -3 \leq \Delta c_2 < \infty \qquad\qquad\qquad\qquad (\because C_{B_2} = c_2)$

$\Rightarrow \qquad 5 - 3 \leq \Delta c_2 < 5 + \infty \qquad \Rightarrow \qquad 2 \leq c_2 < \infty$

$\Rightarrow \quad c_2$ lies between 2 and ∞ without affecting the optimal solution.

Now, if c_2 changes to 1, *i.e.*, if the function is $Z^* = 3x_1 + x_2$ then optimal solution will be changed.

To find new optimal solution

If the objective function is changed from Z to $Z^* = 3x_1 + x_2$ then C_B in the final simplex table will be $C_B = (0, 1, 3)$. So, modified final simplex table is given by:

Simplex Table-2

B.V.	C_B	X_B	x_1	x_2	s_1	s_2	s_3	Min Ratio X_B/s_2
		$c_j \rightarrow$	3	1	0	0	0	
s_1	0	2	0	0	1	(2/3)	–1/3	2/(2/3)=3 (min)→
x_2	1	6	0	1	0	1	0	6/1 = 6
x_1	3	2	1	0	0	–2/3	1/3	—
$Z^* = C_B X_B = 12$			$\Delta_j \rightarrow$ 0	0	0 ↑	1	–1 ↓	
s_2	0	3	0	0	3/2	1	–1/2	
x_2	1	3	0	1	–3/2	0	1/2	
x_1	3	4	1	0	1	0	0	
$Z^* = C_B X_B = 15$			$\Delta_j \rightarrow$ 0	0	–3/2	0	–1/2	

From the last row of the above table, we observe that all $\Delta_j \leq 0$. Hence, new optimal solution is given by

$$x_1 = 4, x_2 = 3 \text{ and max } Z^* = 15$$

Also, if the objective function changed to $Z^* = 3x_1 + 4x_2$ then max Z = max Z^*, because c_2 lies between 2 and ∞ ($c_2 = 4$).

EXAMPLE 3. *Find the limits of the variations of c_1, c_2, c_3, c_4, c_5 and c_6 respectively of the following LPP for which the optimal solution remains optimal.*

$$\text{Max. } Z = -x_2 + 3x_2 - 2x_5$$

subject to the constraints

$$x_1 + 3x_2 - x_3 + 2x_5 = 7$$
$$- 2x_2 + 4x_3 + x_4 = 12$$
$$- 4x_2 + 3x_3 + 8x_5 + x_6 = 10$$

and $\qquad x_j \geq 0 \ \forall \ j = 1, 2, ..., 6$ $\qquad$ [MEERUT-2005 BP]

SOLUTION. Apply the simplex method, we have the following simplex table

Simplex Table

B.V.	C_B	X_B	$c_j \rightarrow$ 0 x_1	−1 x_2	3 x_3	0 x_4	−2 x_5	0 x_6	Min Ratio
x_1	0	7	1	3	−1	0	2	0	—
x_4	0	12	0	−2	4	1	0	0	12/4 = 3(min)→
x_6	0	10	0	− 4	3	0	8	1	10/3
$Z = C_B X_B = 0$		$\Delta_j \rightarrow$	0	−1	3	0	−2	0	
				$\uparrow$	$\downarrow$				
x_1	0	10	1	(5/2)	0	1/4	2	0	4 (min) →
x_3	3	3	0	−1/2	1	1/4	0	0	—
x_6	0	1	0	−5/2	0	−3/4	8	1	—
$Z = C_B X_B = 9$		$\Delta_j \rightarrow$	0	1/2	0	−3/4	−2	0	
			$\downarrow$	$\uparrow$					
x_2	−1	4	2/5	1	0	1/10	4/5	0	
x_3	3	5	1/5	0	1	3/10	2/5	0	
x_6	0	11	1	0	0	−1/2	10	1	
$Z = C_B X_B = 11$		$\Delta_j \rightarrow$	−1/5	0	0	− 4/5	−12/5	0	

Clearly all $\Delta_j \leq 0 \Rightarrow$ solution is optimal and is given by

$$x_1 = 0, x_2 = 4, x_3 = 5, x_4 = 0, x_5 = 0, x_6 = 11, \text{ Max. } Z = 11$$

Now, from final table, we observe that

$$C_B = (-1, 3, 0) = (c_2, c_3, c_6)$$

Clearly, c_1, c_4 and c_5 are not in C_B. So, limits of c_1, c_4 and c_5 are given by

$$\Delta c_1 \leq -\Delta_1 \Rightarrow \Delta c_1 \leq \frac{1}{5}$$

$$\Delta c_4 \leq -\Delta_4 \Rightarrow \Delta c_4 \leq \frac{4}{5}$$

$$\Delta c_5 \leq -\Delta_5 \Rightarrow \Delta c_5 \leq \frac{12}{5}$$

for which the optimal solution remains the same.

Further, c_2, c_3 and c_6 are in C_B

Now $\qquad C_B = (-1, 3, 0) = (c_2, c_3, c_6) = (C_{B_1}, C_{B_2}, C_{B_3})$

i.e. $\qquad C_{B_1} = c_2, C_{B_2} = c_3, C_{B_3} = c_6$

The limit in $C_{B_1} = c_2$ is given by

$$\max_{x_{ij} > 0} \left\{ \frac{\Delta_j}{x_{ij}} \right\} \le \Delta C_{B_1} \le \min_{x_{ij} < 0} \left\{ \frac{\Delta_j}{x_{ij}} \right\} \text{ for } j = 1, 4, 5 \text{ (which are not in the basis)}$$

Now, $x_{11} = \dfrac{2}{5}, x_{14} = \dfrac{1}{10}, x_{15} = \dfrac{4}{5}$ and there is no $x_{ij} < 0$.

$\therefore$ limit in $C_{B_1} = c_2$ is given by

$$\max \left\{ \frac{\Delta_1}{x_{11}}, \frac{\Delta_4}{x_{14}}, \frac{\Delta_5}{x_{15}} \right\} \le \Delta c_2 < \infty \Rightarrow \max \left\{ \frac{-1/5}{2/5}, \frac{-4/5}{1/10}, \frac{-12/5}{4/5} \right\} \le \Delta c_2 < \infty$$

$$\Rightarrow \quad \max \left\{ -\frac{1}{2}, -8, -3 \right\} \le \Delta c_2 < \infty \qquad \Rightarrow \qquad -\frac{1}{2} \le \Delta c_2 < \infty$$

and limit in $C_{B_2} = c_3$ is given by

$$\max_{x_{2j} > 0} \left\{ \frac{\Delta_j}{x_{2j}} \right\} \le \Delta C_{B_3} \le \min \left\{ \frac{\Delta_j}{x_{2j}} \right\} \text{ for } j = 1, 4, 5 \text{ not in the basis.}$$

Now, $x_{21} = \dfrac{1}{5}, x_{24} = \dfrac{3}{10}, x_{25} = \dfrac{2}{5}$

and there is no $x_{2j} < 0$, therefore the limit in $C_{B_2} = c_3$ is given by

$$\max \left\{ \frac{\Delta_1}{x_{21}}, \frac{\Delta_4}{x_{24}}, \frac{\Delta_5}{x_{25}} \right\} \le \Delta c_3 < \infty$$

$$\Rightarrow \quad \max \left\{ \frac{-1/5}{1/5}, \frac{-4/5}{3/10}, \frac{-12/5}{2/5} \right\} \le \Delta c_3 < \infty$$

$$\Rightarrow \quad \max \left\{ -1, -\frac{8}{3}, -6 \right\} \le \Delta c_3 < \infty \quad \Rightarrow \quad -1 \le \Delta c_3 < \infty$$

Also, the limit in $C_{B_2} = c_6$ is given by

$$\max_{x_{3j} > 0} \left\{ \frac{\Delta_j}{x_{3j}} \right\} \le \Delta C_{B_3} \le \min_{x_{3j} < 0} \left\{ \frac{\Delta_j}{x_{3j}} \right\}, \text{ for } j = 1, 4, 5$$

Now, $x_{31} = 1, x_{34} = -\dfrac{1}{2}, x_{35} = 10$, so limit in $C_{B_3} = c_6$ is given by

$$\max \left\{ \frac{\Delta_1}{x_{31}}, \frac{\Delta_5}{x_{35}} \right\} \le \Delta c_6 \le \min \left\{ \frac{\Delta_4}{x_{34}} \right\}$$

$$\Rightarrow \quad \max \left\{ \frac{-1/5}{1}, \frac{-12/5}{10} \right\} \le \Delta c_6 \le \min \left\{ \frac{-4/5}{-1/2} \right\}$$

$$\Rightarrow \quad -\frac{1}{5} \le \Delta c_6 \le \frac{8}{5}$$

9.3 VARIATION IN THE REQUIREMENT VECTOR, b_i

Consider an LPP of maximization

$$\text{Max}. Z = \boldsymbol{Cx}$$

subject to the constraints

$$\boldsymbol{Ax} \geq \boldsymbol{b}$$

and

$$\boldsymbol{x} \geq 0$$

...(1)

Clearly, the optimality condition of (1) is given by

$$\Delta_j = c_j - z_j = c_j - \boldsymbol{C_B} \cdot \boldsymbol{B}^{-1} \cdot \alpha_j \leq 0 \quad \forall j \text{ not appearing in the basis and its solution is}$$

given by $X_B = \boldsymbol{B}^{-1} \boldsymbol{b}$. Since, b_i values are not associated with Δ_j, therefore any change in b_i does not affect the optimality condition. However, it affects the values of the basic variables and the value of the objective function because the determination of solution values $(X_B = \boldsymbol{B}^{-1} \cdot \boldsymbol{b})$, values of b is involved.

Let b_k is changed to $b_k + \Delta b_k$

Then new value of $\boldsymbol{b}$ is given by

$$\begin{aligned}
\boldsymbol{b}^* &= [b_1, b_2, \ldots, b_k + \Delta b_k, \ldots, b_m]' \\
&= [b_1 + 0, b_2 + 0, \ldots, b_k + \Delta b_k, \ldots, 0 + b_m]' \\
&= [b_1, b_2, \ldots, b_m]' + [0, 0, \ldots, 0 b_k, \ldots, b_m]' \\
&= \boldsymbol{b} + [0, 0, \ldots, \Delta b_k, \ldots, b_m]' = \boldsymbol{b} + \Delta b_k
\end{aligned}$$

Further if $\boldsymbol{x_B^*}$ is the new solution corresponding to $\boldsymbol{b}^*$, then we have

$$\boldsymbol{x_B^*} = \boldsymbol{B}^{-1} \cdot \boldsymbol{b}^* \text{ where } \boldsymbol{B} \text{ is the optimal basis.}$$

Let $\boldsymbol{B}^{-1} = \{\beta_1, \beta_2, \ldots, \beta_k, \ldots \beta_m\}$

where $\beta_i = \begin{bmatrix} \beta_{1i} \\ \beta_{2i} \\ \vdots \\ \beta_{mi} \end{bmatrix}$ for $i = 1, 2, \ldots, m$

Then

$$\begin{aligned}
x_B^* &= [\beta_1, \beta_2, \ldots, \beta_k, \ldots, \beta_m] \cdot \{\boldsymbol{b} + [0, 0, \ldots, \Delta b_k, 0, \ldots, 0]'\} \\
&= [\beta_1, \beta_2, \ldots, \beta_k, \ldots, \beta_m] \cdot \boldsymbol{b} + [\beta_1, \beta_2, \ldots, \beta_k, \ldots, \beta_m] \cdot [0, 0, \ldots, \Delta b_k, 0, \ldots, 0]' \\
&= \boldsymbol{B}^{-1} \cdot \boldsymbol{b} + \beta_k \Delta b_k \\
&= [x_{B_1}, x_{B_2}, \ldots, x_{B_k}, \ldots, x_{B_m}]' + [\beta_{1k}, \beta_{2k}, \ldots, \beta_{kk}, \ldots, \beta_{mk}]' \cdot \Delta b \\
&= [x_{B_1} + \beta_{1k} \Delta b_k, x_{B_2} + \beta_{2k} \Delta b_k, \ldots, x_{B_k} + \beta_{kk} \Delta b_k, \ldots, x_{B_m} + \beta_{mk} \Delta b_k]'
\end{aligned}$$

$$\Rightarrow \quad \boldsymbol{x_{B_i}^*} = x_{B_i} + \beta_{ik} \Delta b_k \text{ for } i = 1, 2, \ldots, m$$

Now, if x_B^* is feasible then $x_B^* \geq 0$, which implies that

$$x_{B_i} + \beta_{ik} \Delta b_k \geq 0 \quad \forall i = 1, 2, \ldots, m$$

$$\Rightarrow \qquad \Delta b_k \geq -\frac{x_{B_i}}{\beta_{ik}}, \text{ when } \beta_{ik} > 0$$

and $\qquad \Delta b_k \leq -\dfrac{x_{B_i}}{\beta_{ik}}$, when $\beta_{ik} < 0$

On combining both the above inequalities, we get

$$\max_{\beta_{ik}>0}\left\{-\frac{x_{B_i}}{\beta_{ik}}\right\} \leq \Delta b_k \leq \min_{\beta_{ik}<0}\left\{-\frac{x_{B_i}}{\beta_{ik}}\right\} \qquad \ldots(1)$$

which shows that the new solution x_B^* remains feasible when the range of Δb_k is given by (1)

☞ Remark

- If one or more values in the X_B column of the simplex table are negative, the dual simplex method can be used to get an optimal solution to the new problem by maintaining feasibility.

New Value of the Objective Function

Let $\qquad Z = C_B X_B = \displaystyle\sum_{i=1}^{m} C_{B_i} \cdot x_{B_i} \qquad \ldots(2)$

If b_k is changed to $b_k + \Delta b_k$ then new value of the objective function is given by

$$Z^* = C_B x_B^* = \sum_{i=1}^{m} C_{B_i} \cdot x_{B_i}^*$$

$$= \sum_{i=1}^{m} C_{B_i}(x_{B_i} + \beta_{ik}\Delta b_k)$$

$$= \sum_{i=1}^{m} C_{B_i} \cdot x_{B_i} + \sum_{i=1}^{m} \beta_{ik}\Delta b_k = Z + \sum_{i=1}^{m} \beta_{ik}\Delta b_k$$

Hence, we conclude that the new solution x_B^* remains optimal and feasible when the range of Δb_k is given by (1) and the objective function is increased by $\displaystyle\sum_{i=1}^{m} \beta_{ik}\Delta b_k$

☞ Remark

- The range of variation in the availability of b_i can also be obtained by using condition of feasibility of the current optimal solution, i.e., $X_B = B^{-1}b \geq 0$

🧺 Solved Examples

EXAMPLE 1. **Solve the following LPP:**
$$\textbf{Max. } Z = 3x_1 + 5x_2$$
subject to the constraints
$$x_1 \leq 4$$
$$3x_1 + 2x_2 \leq 18$$
and $\qquad x_1, x_2 \geq 0$

Find the range of b_1 and b_2 so that the solution remains optimal feasible.

SOLUTION. The given LPP of maximization is written in standard form. So, using slack variables s_1, s_2 the given problem becomes

Max. $Z = 3x_1 + 5x_2 + 0s_1 + 0s_2$

s.t. $\qquad x_1 + 0x_2 + s_1 = 4$
$$3x_1 + 2x_2 + s_2 = 18$$
and $\qquad x_1, x_2, s_1, s_2 \geq 0$

Now, apply the simplex method in a usual manner, we have the following simplex table.

Simplex Table

B.V.	C_B	X_B	x_1 (α_1)	x_2 (α_2)	s_1 (β_1)	s_2 (β_2)	Min Ratio X_B/x_2
		$c_j \rightarrow$	3	5	0	0	
s_1	0	4	1	0	1	0	—
s_2	0	18	3	②	0	1	18/2 (min) →
$Z = C_B X_B = 0$		$\Delta_j \rightarrow$	3	5 ↑	0	0 ↓	
s_1	0	$4 (x_{B_1})$	1	0	1	0	
x_1	5	$9 (x_{B_2})$	3/2	1	0	1/2	
$Z = C_B X_B = 45$		$\Delta_j \rightarrow$	−9/2	0	0	−5/2	

Clearly, all $\Delta_j \leq 0 \Rightarrow$ solution is optimal and is given by

$$x_1 = 0, \ x_2 = 9 \text{ and Max. } Z = 45$$

Here, we observe that

$$\boldsymbol{B} = (\alpha_3, \alpha_2) \text{ and } \boldsymbol{b} = [4 \quad 18]' = [b_1 \quad b_2]'$$

So, $\boldsymbol{B}^{-1} = (\beta_1, \beta_2) = \begin{bmatrix} 1 & 0 \\ 0 & 1/2 \end{bmatrix} = \begin{bmatrix} \beta_{11} & \beta_{12} \\ \beta_{21} & \beta_{22} \end{bmatrix}$

To find the variation in b_1:

We have $\max\limits_{\beta_{i1}>0} \left\{ \dfrac{-x_{B_i}}{\beta_{i1}} \right\} \leq \Delta b_1 \leq \min\limits_{\beta_{i1}<0} \left\{ \dfrac{-x_{Bi}}{\beta_{i1}} \right\}$

Here, $\beta_{11} = 1 > 0$ and no $\beta_{i1} < 0$ so that Δb_1 has no upper bound, so range of Δb_1 is given by

$$-\frac{x_{B1}}{\beta_{11}} \leq \Delta b_1 < \infty$$

$$\Rightarrow \qquad -\frac{4}{1} \leq \Delta b_1 < \infty \qquad\qquad (\because \ x_{B_1} = 4)$$

$$\Rightarrow \qquad -4 \leq \Delta b_1 < \infty$$

Thus, range in b_1 is given by

$$4 - 4 \leq b_1 < 4 + \infty$$

$$\Rightarrow \qquad 0 \leq b_1 < \infty$$

To find the variation in b_2:

We have the range of Δb_2 is

$$\max\limits_{\beta_{i2}>0} \left\{ \dfrac{-x_{B_i}}{\beta_{i2}} \right\} \leq \Delta b_2 \leq \min\limits_{\beta_{i2}<0} \left\{ \dfrac{-x_{B_i}}{\beta_{i2}} \right\}$$

Here, we have

$$\beta_{12} = 0, \ \beta_{22} = \frac{1}{2} > 0 \text{ and no } \beta_{i2} < 0$$

Therefore, Δb_2 has no upper bound.

So range of Δb_2 is given by

$$\frac{-x_{B_2}}{\beta_{22}} \leq \Delta b_2 < \infty$$

$$\Rightarrow \quad \frac{-9}{1/2} \leq \Delta b_2 < \infty \quad \Rightarrow \quad -18 \leq \Delta b_2 < \infty$$

Hence, the range of b_2 is given by

$$18 - 18 \leq b_2 < 18 + \infty, \text{ i.e., } 0 \leq b_2 < \infty$$

EXAMPLE 2. *Find the optimal solution of the problem*

$$\text{Max. } Z = 6x_1 + 8x_2$$

subject to the constraints

$$5x_1 + 10x_2 \leq 60$$

$$4x_1 + 4x_2 \leq 40$$

and $\quad x_1, x_2 \geq 0$

Also, apply sensitivity analysis to find the solution of the given LPP if

(i) RHS vector $\begin{bmatrix} 60 \\ 40 \end{bmatrix}$ *of the constraints of the LPP is changed to* $\begin{bmatrix} 40 \\ 20 \end{bmatrix}$.

(ii) the RHS vector $\begin{bmatrix} 60 \\ 40 \end{bmatrix}$ *of the constraints is changed to* $\begin{bmatrix} 20 \\ 40 \end{bmatrix}$ [MEERUT–2007]

SOLUTION.   Using slack variables s_1, s_2 the given LPP becomes

Max. $Z = 6x_1 + 8x_2 + 0s_1 + 0s_2$

s.t.

$$5x_1 + 10x_2 + s_1 = 60$$

$$4x_1 + 4x_2 + s_2 = 40$$

and $\quad x_1, x_2, s_1, s_2 \geq 0$

Apply the simplex method in a usual manner, we have the following simplex table

Simplex Table-1

B.V.	C_B	X_B	x_1	x_2	s_1	s_2	Min Ratio
		$c_j \rightarrow$	6	8	0	0	
s_1	0	60	5	⑩	1	0	60/10=6(min)→
s_2	0	40	4	4	0	1	40/4 = 10
$Z = 0$		$\Delta_j \rightarrow$	6	8↑	0	0	
x_2	8	6	1/2	1	1/10	0	12
s_2	0	16	②	0	−2/5	1	16/2 (min) →
$Z = 48$		$\Delta_j \rightarrow$	2↑	0	− 4/5	0	
x_2	8	2	0	1	1/5	−1/4	
x_1	6	8	1	0	−1/5	1/2	
$Z = C_B X_B = 64$		$\Delta_j \rightarrow$	0	0	−2/5	−1	

Clearly, all $\Delta_j \leq 0 \Rightarrow$ solution is optimal and is given by
$$x_1 = 8, x_2 = 2 \text{ and max. } Z = 64$$
and
$$B^{-1} = \begin{bmatrix} -1/5 & 1/2 \\ 1/5 & -1/4 \end{bmatrix} = \frac{1}{20}\begin{bmatrix} -4 & 10 \\ 4 & -5 \end{bmatrix}$$

(i) If $\boldsymbol{b} = \begin{bmatrix} 60 \\ 40 \end{bmatrix}$ is changed to $\boldsymbol{b'} = \begin{bmatrix} 40 \\ 20 \end{bmatrix}$, the new value of the basic variables

will become
$$X_B = B^{-1}\boldsymbol{b'}$$

i.e.,
$$X_B = \begin{bmatrix} x_1 \\ x_2 \end{bmatrix} = \frac{1}{20}\begin{bmatrix} -4 & 10 \\ 4 & -5 \end{bmatrix}\begin{bmatrix} 40 \\ 20 \end{bmatrix} = \begin{bmatrix} 2 \\ 3 \end{bmatrix}$$

$\Rightarrow \quad x_1 = 2, x_2 = 3$

$\because \ x_1$ and x_2 both are non-negative therefore, this solution is basic feasible solution. Hence, new optimal value of Z is given by
$$Z = 6 \times 2 + 8 \times 3 = 36$$

(ii) If $\boldsymbol{b} = \begin{bmatrix} 60 \\ 40 \end{bmatrix}$ is changed to $\boldsymbol{b''} = \begin{bmatrix} 20 \\ 40 \end{bmatrix}$, the new value of the basic variables in

final iteration in the above simplex table is
$$X_B = B^{-1} \cdot \boldsymbol{b''}, \text{i.e.,} \begin{bmatrix} x_1 \\ x_2 \end{bmatrix} = \frac{1}{20}\begin{bmatrix} -4 & 10 \\ 4 & -5 \end{bmatrix}\begin{bmatrix} 20 \\ 40 \end{bmatrix} = \begin{bmatrix} 16 \\ -6 \end{bmatrix}$$

$\Rightarrow \quad x_1 = 16, x_2 = -6$
$\Rightarrow \quad x_2 < 0 \Rightarrow$ solution is infeasible.
Hence, by dual simplex method, the modified simplex table (from the last table) can be written as

Simplex Table-2

B.V.	$c_j \rightarrow$		6	8	0	0
	C_B	X_B	x_1	x_2	s_1	s_2
x_1	6	16	1	0	$-1/5$	$1/2$
x_2	8	-6	0	1	$1/5$	$-1/4$
$Z = C_B X_B = 48$		$\Delta_j \rightarrow$	0	0	$-2/5$	-1
				$\downarrow$		$\uparrow$
x_1	6	4	1	2	$1/5$	0
s_2	0	24	0	-4	$-4/5$	1
$Z = C_B X_B = 24$		$\Delta_j \rightarrow$	0	-4	$-6/5$	0

Clearly in the last row of the above table all $\Delta_j \leq 0$.

$\Rightarrow \quad$ solution is optimal and feasible and is given by
$$x_1 = 4, x_2 = 0 \text{ and Max. } Z = 24$$

<u>EXAMPLE 3.</u> **For the following LPP**

$$Max.\ Z = -x_1 + 2x_2 - x_3$$

subject to the constraints

$$3x_1 + x_2 - x_3 \le 10$$ [MEERUT-1990, 93, 98, 2005, 18]
$$-x_1 + 4x_2 + x_3 \ge 6$$
$$x_2 + x_3 \le 4$$

and $$x_1, x_2, x_3 \ge 0$$

Find the seperate range of b_1, b_2, b_3 consistent with the optimal solution.

<u>SOLUTION.</u> Using slack variables s_1, s_3 and surplus variable s_2 and artificial variable A in the given LPP. Then we have

$$Max.\ Z = -x_1 + 2x_2 - x_3 + 0s_1 + 0s_2 + 0s_3 - MA$$

subject to

$$3x_1 + x_2 - x_3 + s_1 = 10$$
$$-x_1 + 4x_2 + x_3 - s_2 + A = 6$$
$$x_2 + x_3 + s_3 = 4$$

and $$x_1, x_2, x_3, s_1, s_2, s_3, A \ge 0$$

Now apply the simplex method in a usual manner, we have the following simplex table.

Simplex Table

B.V.	C_B	X_B	$c_j \to$ x_1 -1	x_2 $+2$	x_3 -1	s_1 0	s_2 0	s_3 0	A $-M$	Min Ratio
s_1	0	10	3	1	-1	1	0	0	0	10/1
A	$-M$	6	-1	4	1	0	-1	0	1	6/4 (Min) $\to$
s_3	0	4	0	①	1	0	0	1	0	4/1
$Z = C_B X_B,$ $= -6M$		$\Delta_j \to$	$-1-M$	$4M+2$	$M-1$	0	$-M$	0	0	
				$\uparrow$				$\downarrow$		
s_1	0	17/2	13/4	0	$-5/4$	1	1/4	0	—	(17/2)/(1/4)
x_2	2	3/2	$-1/4$	1	1/4	0	$-1/4$	0	—	—
s_3	0	5/2	1/4	0	3/4	0	⑴/4	1	—	(5/2)/(1/4)
$Z = 3$		$\Delta_j \to$	$-1/2$	0	$-3/2$	0	1/2	0		
							$\uparrow$	$\downarrow$		
s_1	0	6	3	0	-2	1	0	-1		
x_2	2	4	0	1	1	0	0	1		
s_2	0	10	1	0	3	0	1	4		
$Z = C_B X_B = 8,$		$\Delta_j \to$	-1	0	-3	0	0	-2		

We observe that all $\Delta_j \leq 0 \Rightarrow$ solution is optimal, feasible and is given by

$$x_1 = 0, x_2 = 4, x_3 = 0 \text{ and max. } Z = 8$$

Further, $\boldsymbol{C_B} = (0, 2, 0), \boldsymbol{B} = [s_1, x_2, s_2]$

$$X_B = (6, 4, 10) = (x_{B_1}, x_{B_2}, x_{B_3})$$

$$\boldsymbol{b} = (10, 6, 4) = (b_1, b_2, b_3)$$

and $\boldsymbol{B}^{-1} = \begin{bmatrix} 1 & 0 & -1 \\ 0 & 1 & 1 \\ 0 & 0 & 4 \end{bmatrix} = (\beta_1, \beta_2, \beta_3)$

To find the variation in b_1:

$\because$ $b_1 = 10$, then the range of Δb_1 is given by

$$\max_{\beta_{i1}>0} \left\{ \frac{-x_{B_i}}{\beta_{i1}} \right\} \leq \Delta b_1 \leq \min_{\beta_{i1}<0} \left\{ \frac{-x_{B_i}}{\beta_{i1}} \right\}$$

Also $\beta_{11} = 1 > 0$ and there is no $\beta_{i1} < 0$. Therefore, range of Δb_1 is given by

$$\frac{-x_{B_1}}{\beta_{11}} \leq \Delta b_1 < \infty$$

$\Rightarrow$ $\dfrac{-6}{1} \leq \Delta b_1 < \infty$

$\Rightarrow$ $-6 \leq \Delta b_1 < \infty$

So, variation in b_1 is given by

$$10 - 6 \leq b_1 < 10 + \infty$$

$\Rightarrow$ $4 \leq b_1 < \infty$

To find the variation in b_2:

Clearly, $b_2 = 6$. Then range of Δb_2 is given by

$$\max_{\beta_{i2}>0} \left\{ \frac{-x_{B_i}}{\beta_{i2}} \right\} \leq \Delta b_2 \leq \min_{\beta_{i2}<0} \left\{ \frac{-x_{B_i}}{\beta_{i2}} \right\}$$

Here, we have $\beta_{11} = 0, \beta_{22} = 1$ and $\beta_{32} = 0$ and there is no $\beta_{i2} < 0 \Rightarrow$ There is no upper bound of Δb_2. Hence, the range of Δb_2 is given by

$$\frac{-x_{B_2}}{\beta_{22}} \leq \Delta b_2 < \infty \quad \Rightarrow \quad \frac{-4}{1} \leq \Delta b_2 < \infty$$

$\Rightarrow$ $-4 \leq \Delta b_2 < \infty$

Thus, the variation in b_2 is given by

$$6 - 4 \leq b_2 < 6 + \infty$$

$\Rightarrow$ $2 \leq b_2 < \infty$

To find the variation in b_3:

Clearly, $b_3 = 4$. Then range of Δb_3 is given by

$$\max_{\beta_{i3}>0} \left\{ -\frac{x_{B_i}}{\beta_{i3}} \right\} \leq \Delta b_3 \leq \min_{\beta_{i3}<0} \left\{ \frac{-x_{B_i}}{\beta_{i3}} \right\}$$

Here, we have $\beta_{13} = -1, \beta_{23} = 1, \beta_{33} = 4$

Then range of Δb_3 is given by

$$\max_{\beta_{i3}>0} \left\{ \frac{-x_{B_2}}{\beta_{23}}, \frac{-x_{B_3}}{\beta_{33}} \right\} \le \Delta b_3 \le \frac{-x_{B_1}}{\beta_{13}}$$

$$\Rightarrow \qquad \max \left\{ \frac{-4}{1}, \frac{-10}{4} \right\} \le \Delta b_3 \le \frac{-6}{-1}$$

$$\Rightarrow \qquad \qquad -\frac{5}{2} \le \Delta b_3 \le 6$$

Hence, the variation in b_3 is given by

$$4 - \frac{5}{2} \le b_3 \le 4 + 6$$

$$\Rightarrow \qquad \qquad \frac{3}{2} \le b_3 \le 10$$

9.4 VARIATION IN THE ELEMENTS a_{ij} OF THE COEFFICIENT MATRIX A

Let us suppose the elements of coefficient matrix A is changed. Then we have the following two cases:

(i) Change in a coefficient when variable is a non-basic variable.

(ii) Change in a coefficient when variable is a basic variable.

CASE I. When a non-basic column $a_k \in \textbf{B}$ changed to a_k^*. Then solution will remain optimal if following condition is satisfied

$$c_k - z_k^* = c_k - \textbf{\textit{C}}_{\textbf{B}} \cdot \textbf{\textit{B}}^{-1} a_k^* \le 0$$

Otherwise the simplex method is continued, after column k of the simplex table is updated, by introducing the non-basic variable x_k into the basis.

The range for the discrete change Δa_{ij} in the coefficient of non-basic variable x_j in the constraint, i can be obtained by solving following linear inequalities

$$\max_{C_B \cdot \beta_i > 0} \left\{ \frac{\Delta_j}{C_B \cdot \beta_i} \right\} \le \Delta a_{ij} \le \min_{C_B \cdot \beta_i < 0} \left\{ \frac{\Delta_j}{C_B \cdot \beta_i} \right\}$$

Here, β_i is i^{th} column of B^{-1}.

☞ **Remark**

- If a $C_B \cdot \beta_i = 0$ then, Δa_{ij} is unrestricted in sign.

CASE II. If a basic variable $a_k \in \textbf{B}$ is changed to a_k^*. Then to maintain both feasibility and optimality of the current solution, the following conditions are satisfied.

(i) $\max_{k \ne p} \left\{ \dfrac{-x_{B_k}}{x_{B_k}\beta_{pi} - x_{Bp}\beta_{ki} > 0} \right\} \le \Delta a_{ij} \le \min_{k \ne p} \left\{ \dfrac{-x_{B_k}}{x_{B_k}\beta_{pi} - x_{Bp}\beta_{ki} < 0} \right\}$

(ii) $\max \left\{ \dfrac{\Delta_j}{\Delta_j\beta_{pi} - y_{pj}\textbf{\textit{C}}_{\textbf{B}} \cdot \beta_i > 0} \right\} \le \Delta a_{ij} \le \min \left\{ \dfrac{\Delta_j}{\Delta_j\beta_{pi} - y_{pj}\textbf{\textit{C}}_{\textbf{B}} \cdot \beta_i < 0} \right\}$

Solved Examples

EXAMPLE I. *Solve the following LPP*

$$\text{Max. } Z = -x_1 + 3x_2 - 2x_3$$

subject to the constraints

$$3x_1 - x_2 + 2x_3 \le 7$$
$$-2x_1 + 4x_2 \le 12$$
$$-4x_1 + 3x_2 + 8x_3 \le 10$$

and $\qquad x_1, x_2, x_3 \ge 0$

Hence, discuss the effect of the following changes in the optimal solution. Also,

(i) Find the range for discrete changes in the coefficients a_{13} and a_{23} consistent with the optimal solution of the given LPP

(ii) x_1-column in the problem is changed from $[3, -2, -4]'$ to $[3, 2, -4]'$.

(iii) x_3-column in the problem is changed from $[2, 0, 8]'$ to $[3, 1, 6]'$.

SOLUTION. Using slack variables s_1, s_2, s_3 we can write the given LPP as follows.

$$\text{Max. } Z = -x_1 + 3x_2 - 2x_3 + 0s_1 + 0s_2 + 0s_3$$

subject to the constraints

$$3x_1 - x_2 + 2x_3 + s_1 = 7$$
$$-2x_1 + 4x_2 + s_2 = 12$$
$$-4x_1 + 3x_2 + 8x_3 + s_3 = 10$$

and $\qquad x_1, x_2, x_3, s_1, s_2, s_3 \ge 0$

Now, apply the Big-M method in a usual manner, we have the following final simplex table (of optimal solution)

Optimal Simplex Table

	$c_j \rightarrow$		-1	3	-2	0	0	0
B.V.	C_B	X_B	x_1	x_2	x_3	s_1	s_2	s_3
x_1	-1	4	1	0	$4/5$	$2/5$	$1/10$	0
x_2	3	5	0	1	$2/4$	$1/5$	$3/10$	0
s_3	0	11	0	0	10	1	$-1/2$	1
$Z = 11$		$\Delta_j \rightarrow$	0	0	$-12/5$	$-1/5$	$-4/5$	0

From the above table, the optimal feasible solution is given by

$$x_1 = 4, x_2 = 5, x_3 = 0 \text{ and Max. } Z = 11$$

Here, we observe that

$$B^{-1} = \begin{bmatrix} 2/5 & 1/10 & 0 \\ 1/5 & 3/10 & 0 \\ 1 & -1/2 & 1 \end{bmatrix} = [\beta_1 \quad \beta_2 \quad \beta_3]$$

Then $\qquad C_B \cdot \beta_1 = -1\left(\dfrac{2}{5}\right) + 3\left(\dfrac{1}{5}\right) + 0(1) = \dfrac{1}{5}$

$$C_B \cdot \beta_2 = -1\left(\dfrac{1}{10}\right) + 3\left(\dfrac{3}{10}\right) + 0\left(\dfrac{-1}{2}\right) = \dfrac{8}{10}$$

$$C_B \cdot \beta_3 = -1(0) + 3(0) + 0(1) = 0$$

Clearly, x_1, x_2 and s_3 belong to the basis, so any discrete change in coefficients belonging to any of these column vector may affect both feasibility and optimality, while any change in the non-basic variables (x_3, s_1 and s_2) column vector may affect only optimality.

(i) Range for change in a_{13} and a_{23} in the x_3-column is given by

$$\max\left(\frac{c_3 - z_3}{\mathbf{C_B} \cdot \beta_1}\right) = \max\left\{\frac{-12/5}{1/5}\right\} \leq \Delta a_{13} \Rightarrow \Delta a_{13} \geq -12$$

and $$\max\left(\frac{c_3 - z_3}{\mathbf{C_B} \cdot \beta_2}\right) = \max\left\{\frac{-12/5}{8/10}\right\} \leq \Delta a_{23} \Rightarrow \Delta a_{23} \geq -3$$

(ii) **(a) Feasibility condition:** For $i = 2$ (constraints), $p = 1$ (column) and $k = 2, 3$ (column of $\mathbf{B}^{-1}$) we have
For $k = 2$

$$x_{B_k}\beta_{pi} - x_{B_p}\beta_{ki} = x_{B_2}\beta_{12} - x_{B_1}\beta_{22}$$

$$= 5\left(\frac{1}{10}\right) - 4\left(\frac{3}{10}\right) = -\frac{7}{10}$$

For $k = 3$, $x_{B_3}\beta_{12} - x_{B_1}\beta_{32} = 11\left(\frac{1}{10}\right) - 4\left(-\frac{1}{2}\right) = \frac{31}{10}$

Therefore, the range to maintain feasibility of the existing optimal solution is

$$-\frac{5}{31/10} \leq \Delta a_{21} \leq \frac{-5}{-7/10}$$

$$2 - \left(\frac{50}{31}\right) \leq a_{21} < 2 + \left(\frac{50}{7}\right) \quad \Rightarrow \quad \frac{-12}{31} \leq a_{21} \leq \frac{64}{7}$$

(b) Optimality condition:

$$\Delta_3\beta_{12} - y_{13}\mathbf{C_B} \cdot \beta_2 = -\frac{12}{5}\left(\frac{1}{10}\right) - \frac{4}{5}\left(\frac{8}{10}\right) = -\frac{44}{50}$$

$$\Delta_4\beta_{12} - y_{14}\mathbf{C_B} \cdot \beta_2 = -\frac{1}{5}\left(\frac{1}{10}\right) - \frac{2}{5}\left(\frac{8}{10}\right) = -\frac{17}{50}$$

$$\Delta_5\beta_{12} - y_{15}\mathbf{C_B} \cdot \beta_2 = -\frac{4}{5}\left(\frac{1}{10}\right) - \frac{1}{10}\left(\frac{8}{10}\right) = -\frac{16}{100}$$

$\therefore$ Required range is given by

$$-\infty \leq \Delta a_{21} \leq \min\left\{\frac{-12/5}{-44/50}, \frac{-1/5}{-17/50}, \frac{-4/5}{-16/100}\right\}$$

$$\Rightarrow \quad -\infty \leq \Delta a_{21} \leq \frac{10}{17}$$

$$\Rightarrow \quad -\infty \leq a_{21} \leq \frac{44}{17}$$

(iii) Let us suppose column vector $\mathbf{a_3}$ of original LPP is changed from $[2, 0, 8]'$ to $[3, 1, 6]'$. Then modified value of $c_3 - z_3^*$ for this column is

$$\mathbf{B}^{-1} \cdot a_3^* = \begin{bmatrix} 2/5 & 1/10 & 0 \\ 1/5 & 3/10 & 0 \\ 1 & -1/2 & 1 \end{bmatrix}\begin{bmatrix} 3 \\ 1 \\ 0 \end{bmatrix} = \begin{bmatrix} 13/10 \\ 9/10 \\ 17/2 \end{bmatrix}$$

$$c_3 - z_3^* = c_3 - C_B \cdot \boldsymbol{B}^{-1} \cdot a_3^* = -2 - [-1,3,0] \begin{bmatrix} 13/10 \\ 9/10 \\ 17/2 \end{bmatrix} = -\frac{34}{10}$$

9.5 ADDITION OF A NEW VARIABLE

Let us suppose an extra variable x_{n+1} be added. Then solution will remain feasible but it may no longer feasible. If a new variable x_{n+1} is added to the problem, then it will introduce an additional column say α_{n+1} to the coefficient matrix A and an extra cost c_{n+1} will introduced in C. Due to this, the optimality of the solution will be affected. To see the impact of this addition on the current optimal solution, we compute the following

$$y_{n+1} = \boldsymbol{B}^{-1} a_{n+1}$$

and $\qquad c_{n+1} - z_{n+1} = c_{n+1} - C_B y_{n+1}$

Then there are following two cases:

(i) If $c_{n+1} - z_{n+1} \le 0$ then $X_B = 0$, hence current solution is optimal.

(ii) If $c_{n+1} - z_{n+1} > 0$ then current optimal solution can be improved by introducing a new column α_{n+1} in the basis to find the new optimal solution. To improve the solution, we start with the last simplex table of original problem by introducing one extra column α_{n+1} to new variable x_{n+1}.

9.6 ADDITION OF A NEW CONSTRAINT

Let Z be the optimal value of the objective function of given LPP and Z^*, the optimal value of the objective fucntion of new problem which is obtained by adding an extra constraint to the original LPP.

If $Z^* > Z$. Then since, new optimal solution satisfies all the constraints (including additional constraints) so that it is also the optimal solution of the original problem, so, $Z^* > Z$ gives a contradiction.

Hence, $Z^* \le Z$ (in maximization case).

Here, we have the following two possibilities,

(i) If the optimal solution of the original LPP satisfies the additional constraints, then it is also the optimal solution of new problem and additional new constraint is called redundant constraint.

(ii) If the optimal solution of the original problem does not satisfy the additional constraint, then we find the optimal solution of new problem in the following manner.

New optimal solution: Let us suppose $\boldsymbol{B}$ is the optimal basis for the original LPP and $\boldsymbol{B_1}$, the optimal basis for new problem. Clearly, $\boldsymbol{B_1}$ will be a square matrix of order $(m+1)$ ($\because$ an extra constraint is added to the original problem) and $\boldsymbol{B_1}$ is a square matrix of order m.

So,

$$\boldsymbol{B_1} = \begin{bmatrix} \boldsymbol{B} & 0 \\ \alpha & \pm 1 \end{bmatrix} \qquad \qquad \dots(1)$$

Here the last column of $\boldsymbol{B_1}$ is correspond to slack, surplus and artificial vector associated with the extra constraints and r is a row vector of the coefficients in extra constraints of the variable corresponding to the vector in the optimal basis $\boldsymbol{B}$.

Now, we can easily obtained $\boldsymbol{B_1}^{-1}$ (by using partition method) such that

$$B_1^{-1} = \begin{bmatrix} \boldsymbol{B}^{-1} & 0 \\ \mp\alpha\boldsymbol{B}^{-1} & \pm 1 \end{bmatrix} \qquad \ldots(2)$$

Now, let $a_{m+1,j}$ be the coefficient of x_j in new $(m+1)^{\text{th}}$ constraints and α_j^*, the column vector of the coefficient x_j in the new problem.

Then $x_j^* = B_1^{-1}\alpha_j^* = \begin{bmatrix} \boldsymbol{B}^{-1} & 0 \\ \mp\boldsymbol{B}^{-1} & \pm 1 \end{bmatrix}\alpha_j^*$, where x_j^* is correspond to x_j for new problem.

Further, since $\alpha_j^* = \begin{bmatrix} \alpha_j \\ a_{m+1,j} \end{bmatrix}$ then we have

$$x_j^* = \begin{bmatrix} \boldsymbol{B}^{-1} & 0 \\ \mp\alpha\boldsymbol{B}^{-1} & \pm 1 \end{bmatrix}\begin{bmatrix} \alpha_j \\ a_{m+1,j} \end{bmatrix} = \begin{bmatrix} \boldsymbol{B}^{-1}\alpha_j \\ \mp\alpha\boldsymbol{B}^{-1}\alpha_j \pm a_{m+1,j} \end{bmatrix} \qquad (\because\ x_j = \boldsymbol{B}^{-1}\alpha_j)$$

and
$$z_j^* = C_{B_1}x_j^*$$

$$= (C_B, C_{B_{m+1}})\begin{bmatrix} x_j \\ \mp\alpha x_j \pm a_{m+1,j} \end{bmatrix}$$

$$= C_B x_j + C_{B_{m+1}}(\mp\alpha x_j \pm a_{m+1,j})$$

$$\Rightarrow \qquad z_j^* = z_j + C_{B_{m+1}}(\mp\alpha x_j \pm a_{m+1,j}) \qquad \ldots(3)$$

Now we have the following cases:

(i) If slack or surplus variables is introduced in the additional new constraints, then
$$C_{B_{m+1}} = 0$$
Then from (3)

$$z_j^* = z_j,\ i.e.,\ c_j - z_j^* = c_j - z_j$$

$\Rightarrow \qquad c_j - z_j^*$ remains the same (unchanged)

Finally, since the optimal solution of the original problem does not satisfy the new constraints, so that slack or surplus variables, introduced in the new constraints is negative. Therefore, in this case we can apply the dual simplex method to find an optimum solution of the new problem.

(ii) If the new constraints is a perfect inequality, *i.e.*, an artificial variable is introduced, then an additional vector is an artificial vector. Then we have the following two possibilities:

   (a) If the artificial variable in the basis solution is negative, we can assign a price zero to the artificial variable and the dual simplex method is used to remove the artificial variable from the basis.

   (b) If the artificial variable in the basis solution is positive, we can assign a cost of $-M$ to it and use simplex (standard) method for removal of the artificial variable from the basis. In this case the value of $c_j - z_j$ will be changed.

Solved Examples

EXAMPLE 1. *Solve the following LPP*

$$\text{Max. } Z = 3x_1 + 5x_2$$

subject to the constraints

$$x_1 + x_3 = 4$$
$$3x_1 + 2x_2 + x_4 = 18$$

and $\qquad x_1, x_2, x_3, x_4 \geq 0$

If a new variable x_5 is added to the above LPP with price 7 then we have the following problem

$$\text{Max } Z' = 3x_1 + 5x_2 + 7x_5$$

s.t. $\qquad x_1 + x_3 + x_5 = 4$
$$3x_1 + 2x_2 + x_4 + 2x_5 = 18$$

and $\qquad x_1, x_2, x_3, x_4, x_5 \geq 0$

Find the solution of the new problem.

SOLUTION. **Solution of the original problem:**

The given problem can be written as

$$\text{Max. } Z = 3x_1 + 5x_2 + 0x_3 + 0x_4$$

s.t.

$$x_1 + 0x_2 + x_3 + 0x_4 = 4$$
$$3x_1 + 2x_2 + 0x_3 + x_4 = 18$$

and $\qquad x_1, x_2, x_3, x_4 \geq 0$

Now, apply the simplex method in a usual manner, we have the following simplex table.

Simplex Table-1

		$c_j \rightarrow$	3	5	0	0	
B.V.	C_B	X_B	x_1 (α_1)	x_2 (α_2)	x_3 (α_3)	x_4 (α_4)	**Min Ratio**
x_3	0	4	1	0	1	0	—
x_4	0	18	3	②	0	1	18/2 (min) →
$Z = C_BX_B = 0$		$\Delta_j \rightarrow$	3	5 ↑	0	0 ↓	
x_3	0	4	1	0	1	0	
x_2	5	9	3/2	1	0	1/2	
$Z = C_BX_B = 45$		$\Delta_j \rightarrow$	–9/2	0	0	–5/2	

Clearly, all $\Delta_j \leq 0$ $\Rightarrow$ solution is optimal and is given by

$$x_1 = 0, x_2 = 9, x_3 = 4, x_4 = 0 \text{ and Max. } Z = 45$$

Now, the initial basis of the original LPP is

$$\boldsymbol{B} = (\alpha_3, \alpha_4) \text{ and } \boldsymbol{B}^{-1} = \begin{bmatrix} 1 & 0 \\ 0 & 1/2 \end{bmatrix}$$

Solution of New problem:

If one extra variable x_5 with price 7 is added to the original LPP, then new LPP is

$$\text{Max. } Z' = 3x_1 + 5x_2 + 7x_5$$

s.t.
$$x_1 + x_2 + x_5 = 4$$
$$3x_1 + 2x_2 + x_4 + 2x_5 = 18$$

and $\quad x_1, x_2, x_3, x_4, x_5 \geq 0$

Here, we have

$$\alpha_5 = \begin{bmatrix} 1 \\ 2 \end{bmatrix}, \text{ then } x_5 = B^{-1}\alpha_5 = \begin{bmatrix} 1 & 0 \\ 0 & 1/2 \end{bmatrix}\begin{bmatrix} 1 \\ 2 \end{bmatrix} = \begin{bmatrix} 1 \\ 1 \end{bmatrix}$$

Now, $\quad \Delta_5 = c_5 - z_5 = c_5 - C_B B^{-1}\alpha_5$

$$= 7 - (0,5)\begin{bmatrix} 1 & 0 \\ 0 & 1/2 \end{bmatrix}\begin{bmatrix} 1 \\ 2 \end{bmatrix}$$

$$= 7 - (0,5)\begin{bmatrix} 1 \\ 1 \end{bmatrix} = 7 - 5 = 2 > 0$$

$\Rightarrow$ solution is not optimal.

Thus, it can be improved by introducing α_5 the column corresponding to new variable in the last simplex table. So, we start with the last simplex table as follows:

Simplex Table-2

B.V.	C_B	X_B	x_1	x_2	x_3	x_4	x_5	Min Ratio
$c_j \rightarrow$			3	5	0	0	7	
x_3	0	4	1	0	1	0	①	4/1 (min) $\rightarrow$
x_2	5	9	3/2	1	0	1/2	1	9/2
$Z = C_B X_B = 45$		$\Delta_j \rightarrow$	-9/2	0	0	-5/2	2	
						$\downarrow$	$\uparrow$	
x_5	7	4	1	0	1	0	1	
x_2	5	5	1/2	1	-1	1/2	0	
$Z = C_B X_B = 53$		$\Delta_j \rightarrow$	-13/2	0	-2	-5/2	0	

Clearly, in the last row of the above table all $\Delta_j \leq 0$

$\Rightarrow$ solution is optimal and is given by

$$x_1 = 0, x_2 = 5, x_3 = 0, x_4 = 0, x_5 = 4 \text{ and max } Z' = 53$$

EXAMPLE 2. *Consider the following LPP*

$$\textbf{max. } Z = 3x_1 + 5x_2$$

subject to the constraints

$$3x_1 + 2x_2 \leq 18$$
$$x_1 + 2x_2 \leq 4$$
$$x_2 \leq 6$$

and $\quad\quad x_1, x_2 \geq 0$

Find the optimal solution of the given LPP. Also

(i) if the variable x_6 is added to the given LPP, then find an optimal solution to the new LPP. It is given that the coefficient of x_6 in the constraints of the problem are 1, 1 and 1 and its coefficients in the objective function is 2.

(ii) Discuss the effect on the optimal basic feasible solution by adding a new constraints $2x_1 + x_2 \leq 8$ to the given set of constraints.

SOLUTION. Apply the simplex method in a usual manner, the final simplex table is given as under

Simplex Table-1

B.V.	C_B	X_B	x_1	x_2	s_1	s_2	s_3
		$c_j \rightarrow$	3	5	0	0	0
x_1	3	2	1	0	1/3	0	−2/3
s_2	0	0	0	0	−2/3	1	4/3
x_2	5	6	0	1	0	0	1
$Z = C_B X_B = 36$		$\Delta_j \rightarrow$	0	0	−1	0	−3

Clearly, all $\Delta_j \leq 0 \Rightarrow$ solution is optimal and is given by

$$x_1 = 2, x_2 = 6 \text{ and max. } Z = 36$$

(i) After adding a new variable x_6 as given, the new LPP is

Max. $Z = 3x_1 + 5x_2 + 2x_6$

s.t. $\qquad 3x_1 + 2x_2 + x_6 \leq 18$

$\qquad\qquad x_1 + 2x_2 + x_6 \leq 4$

$\qquad\qquad\qquad x_1 + x_6 \leq 6$

and $\qquad\qquad x_1, x_2, x_6 \geq 0$

Clearly, the column vector associated with variable x_6 is $a_6 = (1, 1, 1)$. Then using above table, we have

$$y_6 = \mathbf{B}^{-1} a_6 = \begin{bmatrix} 1/3 & 0 & -2/3 \\ -2/3 & 1 & 4/3 \\ 0 & 0 & 1 \end{bmatrix} \begin{bmatrix} 1 \\ 1 \\ 1 \end{bmatrix} = \begin{bmatrix} -1/3 \\ 5/3 \\ 1 \end{bmatrix}$$

$\because \qquad \mathbf{C_B} = (3, 0, 5)$. Therefore,

$$c_6 - z_6 = c_6 - C_B y_6$$

$$= 2 - (3, 0, 5) \begin{bmatrix} -1/3 \\ 5/3 \\ 1 \end{bmatrix} = -2 \leq 0$$

$\Rightarrow$ Optimality of the current solution remains unaffected with the addition of x_6.

(ii) Since, the optimal basic feasible solution given in the above table does not satisfy the additional constraint $2x_1 + x_2 \leq 8$, so use additional slack variable for this constraint, the above table becomes

Simplex Table-2

B.V.	C_B	X_B	x_1	x_2	s_1	s_2	s_3	s_4
		$c_j \rightarrow$	3	5	0	0	0	0
x_1	3	2	1	0	1/3	0	−2/3	0
s_2	0	0	0	0	−2/3	1	4/3	0
x_2	5	6	0	1	0	0	1	0
s_4	0	8	0	1	0	0	0	1
$Z = C_B X_B = 36$		$\Delta_j \rightarrow$	0	0	−1	0	−3	0

We observe that the matrix B has been changed due to row 4, so coefficient in row 4 must be zero, which can be done by using the following operations.

$$R_4(\text{new}) \rightarrow R_4(\text{old}) - 2R_1 - R_3$$

Then we have the following table.

Simplex Table-3

B.V.	C_B	X_B	$c_j \rightarrow$ 3	5	0	0	0	0
			x_1	x_2	s_1	s_2	s_3	s_4
x_1	3	2	1	0	1/3	0	−2/3	0
s_2	0	0	0	0	−2/3	1	4/3	0
x_2	5	6	0	1	⓪	0	1	0
s_4	0	−2	0	0	−2/3	0	1/3	1
$Z = 36$		$\Delta_j \rightarrow$	0	0	−1 ↑	0	1/3	1 ↓

The solution in the above table is optimal but not feasible. So, we apply the dual simplex method in a usual manner, the new obtained solution is given in the following table.

Simplex Table-4

B.V.	C_B	X_B	$c_j \rightarrow$ 3	5	0	0	0	0
			x_1	x_2	s_1	s_2	s_3	s_4
x_1	3	1	1	0	0	0	−1/2	1/2
s_2	0	2	0	0	0	1	1	−1
x_2	5	6	0	1	0	0	1	0
s_1	0	3	0	0	1	0	−1/2	−3/2
$Z = C_B X_B = 33$		$\Delta_j \rightarrow$	0	0	0	0	−7/2	−3/2

We observe that all $\Delta_j \leq 0 \Rightarrow$ solution is optimal feasible and is given by

$$x_1 = 1, \, x_2 = 6 \text{ and max. } Z = 33$$

EXAMPLE 3. *Consider the following table which presents an optimal solution to some LPP*

B.V.	C_B	X_B	$c_j \rightarrow$ 2	4	1	3	2	0	0	0
			x_1	x_2	x_3	x_4	x_5	s_1	s_2	s_3
			(α_1)	(α_2)	(α_3)	(α_4)	(α_5)	(α_6)	(α_7)	(α_8)
x_1	2	3	1	0	0	−1	0	1/2	1/5	−1
x_2	4	1	0	1	0	2	1	−1	0	1/2
x_3	1	7	0	0	1	−1	−2	5	−3/10	2
$Z = C_B X_B = 17$		$\Delta_j \rightarrow$	0	0	0	−2	0	−2	−1/10	−2

(i) *If the additional constraints $2x_1 + 3x_2 - x_3 + 2x_4 - 4x_5 \leq 5$ were annexed to the system, would there be any change in the optimal solution.*

(ii) *If the additional constraints $3x_1 + x_2 + 2x_3 + x_4 + 9x_5 \leq 19$ were annexed to the system, would there be any change in the optimal solution? If yes, find the new optimal solution.*

SOLUTION.   (i) From the given table, we observe that all $\Delta_j \le 0$ and $x_1 = 3, x_2 = 1, x_3 = 7$, $x_4 = x_5 = 0$
$\Rightarrow$ solution is optimal feasible.
$\because$ New constraint is $2x_1 + 3x_2 - x_3 + 2x_4 - 4x_5 \le 5$
clearly satisfies the obtained solution
$(\because 2x_1 + 3x_2 - x_3 + 2x_4 - 4x_5 = 2(3) + 3(1) - 7 + 2(0) - 4(0) = 2 < 5)$
Hence, this is also the optimal solution of the new problem and new constraint is redundant one.

(ii) Clearly, the obtained solution $x_1 = 3, x_2 = 1, x_3 = 7, x_4 = x_5 = 0$ does not satisfy the new constraint $3x_1 + x_2 + 2x_3 + x_4 + 9x_5 \le 19$ so solution is not optimal for the new problem.
Now, we have to find the optimal solution of the new problem.
Introducing slack variable s_4 to the new constraint, we get
$$3x_1 + x_2 + 2x_3 + x_4 + 9x_5 + 0s_1 + 0s_2 + 0s_3 + s_4 = 19$$
The optimal basis of the original problem is $B = (\alpha_1, \alpha_2, \alpha_3)$
If α_9 is the column vector corresponding to slack variable s_4, then
$$\alpha_9 = \begin{bmatrix} 0 \\ 0 \\ 0 \\ 1 \end{bmatrix}$$
We assume that the cost price $c_9 = 0$ in the new objective function. Now introducing this constraint in optimal table of the original problem, we get the following simplex table.

Simplex Table-1

B.V.	C_B	X_B	$c_j \rightarrow$	2	4	1	3	2	0	0	0	0
				x_1 (α_1)	x_2 (α_2)	x_3 (α_3)	x_4 (α_4)	x_5 (α_5)	s_1 (α_6)	s_2 (α_7)	s_3 (α_8)	s_4 (α_9)
x_1	2	3		1	0	0	-1	0	$1/2$	$1/5$	-1	0
x_2	4	1		0	1	0	2	1	-1	0	$1/2$	0
x_3	1	7		0	0	1	-1	-2	5	$-3/10$	2	0
s_4	0	19		3	1	2	1	9	0	0	0	1

Here, we observed that identity matrix is disturbed (due to the inclusion of fourth row). Now identity matrix can be obtained by performing $R_4 - 3R_1$, $R_4 - R_2$ and $R_4 - 2R_3$.
Then we get the following simplex table.

Simplex Table-2

B.V.	C_B	X_B	$c_j \rightarrow$	2	4	1	3	2	0	0	0	0
				x_1 (β_1)	x_2 (β_2)	x_3 (β_3)	x_4	x_5	s_1	s_2	s_3	s_4 (β_4)
x_1	2	3		1	0	0	-1	0	$1/2$	$1/5$	-1	0
x_2	4	1		0	1	0	2	1	-1	0	5	0
x_3	1	-7		0	0	1	-1	-2	5	$-3/10$	2	0
x_4	0	-5		0	0	0	4	12	$-21/2$	0	$-3/2$	1
$Z = C_B X_B = 17$			$\Delta_j \rightarrow$	0	0	0	-2	0	-2 $\uparrow$	$-1/10$	-2	0 $\downarrow$

Here, we observe that all $\Delta_j \le 0$.

$\Rightarrow$ Solution is optimal but not feasible (as $s_4 = -5$). Therefore, we apply the dual simplex method to improve the present solution.

To find the outgoing vector:

$$\because \quad x_{B_r} = \min\{x_{B_i} : x_{B_i} < 0\}$$

$$= \min\{x_{B_1}, x_{B_2}, x_{B_3}, x_{B_4}\}$$

$$= x_{B_4} \qquad\qquad (\because \ x_{B_1}, x_{B_2}, x_{B_3} > 0)$$

$$\Rightarrow \qquad r = 4$$

$$\Rightarrow \qquad \beta_4 \text{ is the outgoing vector.}$$

To find incoming vector:

$$\because \quad \frac{\Delta_k}{a_{rk}} = \min_j\left\{\frac{\Delta_j}{a_{rj}}, a_{rj} < 0\right\} = \min_j\left\{\frac{\Delta_j}{a_{4j}}, a_{4j} < 0\right\} \text{ for } j = 4,5,6,7,8$$

$$= \min\left\{\frac{\Delta_6}{a_{46}}, \frac{\Delta_7}{a_{47}}\right\} \qquad (\because \text{ only } a_{46}, a_{47} < 0)$$

$$= \min\left\{\frac{-2}{-21/2}, \frac{-2}{-3/2}\right\} = \min\left\{\frac{4}{21}, \frac{4}{3}\right\} = \frac{4}{21} = \frac{\Delta_6}{a_{46}}$$

$$\Rightarrow \qquad k = 6$$

$$\Rightarrow \qquad \alpha_6(= s_1) \text{ is the incoming vector}$$

and key element $= a_{46} = \dfrac{-21}{2}$

Now, we have the following simplex table

Simplex Table-3

B.V.	C_B	X_B	x_1 (β_1)	x_2 (β_2)	x_3 (β_3)	x_4	x_5	s_1 (β_4)	s_2	s_3	s_4
$c_j \to$			2	4	1	3	2	0	0	0	0
x_1	2	58/21	1	0	0	−17/21	4/5	0	1/5	−15/14	1/21
x_2	4	31/21	0	1	0	34/21	−1/7	0	0	9/14	−2/21
x_3	1	97/21	0	0	1	19/21	26/7	0	−3/10	9/7	10/21
s_1	0	10/21	0	0	0	−8/21	−8/7	1	0	−3/2	−2/21
$Z = C_B X_B$ $= 337/21$		$\Delta_j \to$	0	0	0	−58/21	−16/7	0	−1/10	−12/7	−4/21

From the above table, we observe that all $\Delta_j \le 0$ and

$$x_1 = \frac{58}{21}, x_2 = \frac{31}{21}, x_3 = \frac{97}{21}, x_4 = 0, s_1 = \frac{10}{21}, s_2 = 0, s_3 = 0 \text{ and } s_4 = 0$$

$\Rightarrow$ solution is optimal and feasible

Hence, optimal solution is given by

$$x_1 = \frac{58}{21}, x_2 = \frac{31}{21}, x_3 = \frac{97}{21}, x_4 = 0$$

and $\text{Max.} Z = \dfrac{337}{21}$, which is less than the original maximum value of Z.

Exercise-9.1

1. Solve the following LPP:

Max. $Z = 15x_1 + 45x_2$

subject to the constraints

$x_1 + 16x_2 \leq 240$

$0.5x_1 + 2x_2 + x_3 = 162$

$x_2 + x_4 = 50$

and $x_1, x_2, x_3, x_4 \geq 0$

Find how much can c_1 be changed without affecting the optimality of the solution.

2. Find an optimal solution of the following LPP

Max. $Z = 5x_1 + 3x_2$

subject to the constraints

$3x_1 + 5x_2 \leq 15$

$5x_1 + 2x_2 \leq 10$

and $x_1, x_2 \geq 0$

Also, find the range of c_1 without affecting the optimality of the solution.

3. Solve the following LPP

Max. $Z = 3x_1 + 4x_2 + x_3 + 7x_4$

subject to the constraints

$8x_1 + 3x_2 + 4x_3 + x_4 \leq 7$

$2x_1 + 6x_2 + x_3 + 5x_4 \leq 3$

$x_1 + 4x_2 + 5x_3 + 2x_4 \leq 8$

and $x_1, x_2, x_3, x_4 \geq 0$

Discuss the effect of discrete changes in the requirements, i.e., b_1, b_2 and b_3 so that the solution remains optimal feasible.

4. Solve the following LPP

Max. $Z = 3x_1 + 5x_2$

subject to the constraints

$x_1 + x_3 = 4$

$3x_1 + 2x_2 + x_4 = 18$

and $x_1, x_2, x_3, x_4 \geq 0$

Would there is any change in the optimal solution if the additional constraints

(i) $x_2 \leq 10$ or (ii) $x_2 \leq 6$

is added to the above LPP. In case the optimal solution changes, find the new optimal feasible solution.

5. Solve the following LPP

Max. $Z = 2x_1 + x_2 + 3x_3$

subject to the constraints

$x_1 + x_2 + 2x_3 \leq 5$

$2x_1 + 3x_2 + 4x_3 \leq 12$

and $x_1, x_2, x_3 \geq 0$

What will happen if a new constraint $2x_1 + 2x_2 + 4x_3 \geq 14$ is added?

6. Consider the following table which presents an optimal solution to some LPP

B.V.	C_B	X_B	c_j				
			x_1	x_2	x_3	s_1	s_2
x_1	2	1	1	0	1/2	4	−1/2
x_2	3	2	0	1	1	−1	2
$Z = C_B X_B$		x_j	1	2	0	0	0
= 8		Δ_j	0	0	−3	−5	−5

For the above problem assuming that s_1 and s_2 were in that order in the initial identity matrix. Calculate the following:

(i) How much can be b_1 and b_2 be increased without effecting the optimality and feasibility of the solution.

(ii) How much c_3 can be increased before the present basic solution will no longer to be optimal.

Answers

1. $x_1 = 184, x_2 = 35, x_3 = 0, x_4 = 15, \text{Max } Z = 4335$ and $\dfrac{45}{4} < c_1 \leq \dfrac{225}{8}$

2. $x_1 = \dfrac{20}{19}, x_2 = \dfrac{45}{19}, \text{Max.} Z = \dfrac{235}{19}, \dfrac{9}{5} \leq c_1 \leq \dfrac{15}{2}$

3. $x_1 = \dfrac{16}{19}, x_2 = 0, x_3 = 0, x_4 = \dfrac{5}{19}, \text{Max.} Z = \dfrac{83}{19}$; $\dfrac{3}{5} \leq b_1 \leq 12; \dfrac{7}{4} \leq b_2 \leq \dfrac{323}{25}; \dfrac{28}{19} \leq b_3 < \infty$

4. $x_1 = 0, x_2 = 9, x_3 = 4, x_4 = 0, \text{Max. } Z = 45$

  (i) No change   (ii) $x_1 = 2, x_2 = 6, x_3 = 2, x_4 = 0, \text{Max. } Z = 36$

5. $x_1 = 3, x_2 = 2, x_3 = 0, \text{Max. } Z = 8$. When $2x_1 + 2x_2 + 4x_3 \geq 14$ is added there is no feasible solution

6. (i) $-\dfrac{1}{4} \leq \Delta b_1 \leq 2; -1 \leq \Delta b_2 \leq 2$ (ii) $-\infty < c_3 \leq 4$

REVIEW QUESTIONS

1. Write a short note on sensitivity analysis.
2. Discuss the effect of addition of a constraint in sensitivity analysis.
3. Find the limit of variation of elements a_{ik} so that optimal feasible solution of $Ax = b$, $x \geq 0$, max. $Z = Cx$ remains optimal feasible solution when:
   (i) $a_k \in B$ (ii) $a_k \notin B$
4. Discuss the effect of discrete changes in the requirement (on RHS of the inequality) for the LPP of maximization.

MULTIPLE CHOICE QUESTIONS (CHOOSE THE MOST APPROPRIATE ONE)

1. Sensitivity analysis:
   (i) is also called post optimality analysis as it is carried out after the optimal solution is obtained
   (ii) allows the decision-maker more meaningful information about changes in the LP parameters
   (iii) provides the range within which a parameter may change without affecting optimality
   (a) (i)   (b) (ii)
   (c) (ii), (iii)   (d) All of the above

2. The addition of a new variable to give LPP then problem is called:
   (a) analysis
   (b) sensitivity analysis
   (c) mono analysis
   (d) none of these

3. If $c_k \notin C_B$ changes to $c_k + \Delta c_k$ such that $\Delta c_k = -\Delta k$, there is no lower bound to:
   (a) Δc_j (b) Δc_k
   (c) Δb_j (d) None of these

4. If in a LPP, the changes in the requirement vector b_j is called:
   (a) transportation analysis
   (b) dual analysis
   (c) sensitivity analysis
   (d) none of these

5. In a LPP, the changes in the elements a_{ij} of the coefficient matrix is called:
   (a) sensitivity analysis
   (b) post analysis
   (c) both (a) and (b)
   (d) none of these

6. If $c_k \notin C_B$ changes to $c_k + \Delta c_k$ such that $\Delta c_k \leq z_k - c_k \ (= -\Delta k)$ the value of the objective function and the optimal solution of the problem remains:
   (a) changed   (b) unchanged

   (c) unbounded   (d) none of these

7. The changes in the optimal solutions due to discrete variation in the parameters a_{ij}, b_j and c_j are called:
   (a) sensitivity analysis
   (b) dual analysis
   (c) simplex analysis
   (d) none of these

8. The addition of a new constraint to given LPP. Then the problem is called:
   (a) post analysis   (b) mono analysis
   (c) analysis   (d) none of these

9. If in a LPP the variations in the price vector C is called:
   (a) simplex analysis
   (b) sensitivity analysis
   (c) transportation analysis
   (d) none of these

10. In a LPP the variation in $c_j \notin C_B$ then its changes to:
    (a) $c_k + \Delta c_k$ (b) $c_j + \Delta c_j$
    (c) c_j (d) None of these

11. If in a LPP the variations in the price vector 'C' and requirement vector 'b' is called:
    (a) post analysis
    (b) sensitivity analysis
    (c) both (a) and (b)
    (d) none of these

12. Which of the following is not correct?
    (a) After the attainment of an optimum solution of an LPP, it is desired to study the effect of changes in the different parameters of the problem on the Current Optimum Solution
    (b) An analysis of post optimal solutions is known as Post-Optimality analysis or Sensitivity analysis
    (c) Post optimality analysis study only the continuous changes in the parameters of LPP

(d) Post optimality analysis form an integral part of formulating an LPP

13. LP context, post-optimal analysis is a technique to:
    (a) determine how optimum solution to an LPP changes is response to problem inputs
    (b) allocate resources optimality
    (c) minimize cost of operations
    (d) spellout the relation between dual and its primal

14. Which of the following is not correct?
    (a) For any changes in the objective function coefficients, the optimal function values of the decision variables would change
    (b) The optimality of the current solution may be affected if right hand side of the constraints is changed
    (c) The feasibility of the current optimum solution may be affected if right hand side of the constraint is changed
    (d) When a new constraint is introduced or one of the current constraint is deleted from an LPP, the post-optimal analysis is due to structural changes

15. Which of the following is not correct?
    (a) Post-optimal analysis is normally carried out after the optimum solution is reached
    (b) Addition of a constraint may affect the current optimum solution
    (c) Addition of a new variable may disturb the feasibility of the current optimum solution
    (d) Addition of new constraints in an LPP can never improve the optimal value of the objective function

16. Which of the following is not correct?
    (a) Changes in the right hand side values of the constraints within the allowable limits would neither change the basis nor the objective function value of an LPP
    (b) Deletion of an existing variable may affect the feasibility of the current optimum Solution
    (c) When multiple changes take place in the objective function or in the RHS values of the constraints, then the cent-percent Rule may be used to determine whether they would affect the current solution
    (d) Changes in the coefficient matrix of the constraints can be analysed to determine their effect on the optimum solution

17. If the optimal solution of the original LPP

satisfies the new constraint, it is also an optimal solution of new LPP. In this case the _____ constraint is redundant.
    (a) additional   (b) subtraction
    (c) multiplying   (d) none of these

18. If the solution of LPP is optimal for $c_j = a$ and $c_j = b$ then the variation in c_j will be:
    (a) $a < c_j < b$ (b) $a < c_j \leq b$
    (c) $a \leq c_j < b$ (d) $a \leq c_j \leq b$

19. In a LPP, the variation in $c_j \in C_B$, the range of ΔC_{B_k} such that the solution remains optimal is given by $\max\limits_{y_{kj}>0}\left[\dfrac{c_j - z_j}{y_{kj}}\right] \leq \_\_ \geq \min\limits_{y_{kj}<0}\left[\dfrac{c_j - z_j}{y_{kj}}\right]$:
    (a) C_{B_k} (b) ΔC_{B_k}
    (c) Δc_j (d) None of these

20. If the addition of new constraint alters the nature of the problem, then the new problem must be solved as a:
    (a) fresh problem   (b) above Problem
    (c) no problem   (d) none of these

21. If the solution of LPP is not optimal for $c_j = a$ and $c_j = b$, then the variation in c_j will be:
    (a) $a < c_j < b$ (b) $a < c_j < b$
    (c) $a \leq c_j < b$ (d) $a \leq c_j \leq b$

22. If no $y_{kj} > 0$ there is no _____ to ΔC_{B_k} :
    (a) change   (b) lower bound
    (c) upper bound   (d) None of these

23. If the solution of LPP is not optimal for $c_j = a$, then the variation in c_j will be:
    (a) $a < c_j \leq b$ (b) $a \leq c_j < b$
    (c) $a \leq c_j \leq b$ (d) $a < c_j < b$

24. If the optimal solution of the original LPP does not satisfies the new constraint, it is also an optimal solution of the new constraint, it is also an optimal solution of the new LPP. In this case the _____ constraint is not redundant:
    (a) subtracting   (b) additional
    (c) multiplying   (d) none of these

25. Addition of a new constraint and deletion of an existing constraint simultaneously to a LPP:
    (a) disturbs feasibility only
    (b) disturbs optimality only
    (c) may disturbs both feasibility and optimality
    (d) None of these

26. Change in availability vector and addition of a new constraint simultaneously to a LPP:

(a) may disturb feasibility

(b) may disturb optimality

(c) may disturb both feasibility and optimality

(d) None of these

27. Change in availabilities and costs of a LPP simultaneouly:

(a) disturb feasibility only

(b) disturb optimality only

(c) may disturb both feasibility and optimality

(d) None of these

28. Addition of a new variable and deletion of an existing variable to LPP simultaneously:

(a) disturb feasibility only

(b) disturb optimality only

(c) may disturb both feasibility and optimality

(d) None of these

29. Let the S_F of LPP be non empty and bounded (that is both from above and below) and let an additional constraint be added to it. Then the new S_F:

(a) may become empty

(b) may become unbounded and consequently the solution may become unbounded

(c) may become empty or may become unbounded

(d) none of these

30. Let the S_F of a LPP be a non-empty and bounded (that is both from above and below) and let a constraint be deleted from it. Then the new S_F:

(a) may become empty

(b) may become unbounded and consequently the solution may become unbounded

(c) may become empty or may become unbounded

(d) none of these

31. The use of cutting plane method:

(a) yields better value of objective function

(b) reduces the Number of constraints in the given problem

(c) require use of standard LP approach between each cutting plane application

(d) all of the above

32. While solving LP problem any Non-integer variable in the solution is picked up to:

(a) enter the solution

(b) obtain the cut constraint

(c) leave the solution

(d) none of these

33. Which of the following is the consequence of adding a new cut constraint to an Optimal Simplex table:

(a) addition of a new variable to the table

(b) makes the previous Optimal Solution infeasible

(c) eliminates non-integer solution from the solution space

(d) all of the above

34. To ensure best marginal increase in the objective function value, a resource value may be increased whose shadow price is comparatively:

(a) larger

(b) smaller

(c) neither (a) nor (b)

(d) both (a) and (b)

35. In sensitivity analysis of the coefficient of the non-basic variable in cost minimization LP problem, the upper sensitivity limit is:

(a) original value + lowest positive value of improvement ratio

(b) original value – lowest absolute value of improvement ratio

(c) positive infinity

(d) negative infinity

36. In a mixed integer programming problem:

(a) all of the decision variables require integer solution

(b) few of the decision variables require integer solution

(c) different objective functions are mixed together

(d) none of these

37. In a Branch and bound minimization tree, the lower bounds on objective function value:

(a) do not decrease in value

(b) do not increase in value

(c) remain constant

(d) none of these

38. The 0-1 integer programming problem:

(a) requires the decision variables to have values between zero and one

(b) requires that the constraints all have coefficients between zero and one

(c) requires that the decision variables have coefficients between zero and one

(d) all of the above

39. Addition of an additional constraint in the existing constraints will cause a:
    (a) change in objective function coefficients (c_j)
    (b) change in coefficients a_{ij}
    (c) both (a) and (b)
    (d) none of the above

40. In the Branch and bound approach to a max. problem, a node is terminated if:
    (a) a node has an infeasible solution
    (b) a node yields a solution that is feasible but not an integer
    (c) upper bound is less than the current sub problem's lower bound
    (d) all of the above

41. The entering variable in the sensitivity analysis of objective function coefficients is always a:
    (a) decision variable (b) n o n - b a s i c variable
    (c) basic variable   (d) slack variable

42. The part of the feasible solution space eliminated by plotting a cut contains:
    (a) only non-integer solutions
    (b) only integer solutions
    (c) both (a) and (b)
    (d) none of these

43. Branch and bound method divides the feasible solution space into smaller parts by:
    (a) branching   (b) bounding
    (c) enumerating   (d) all of the above

44. Sensitivity analysis:
    (a) is also called post-optimality analysis as it is carried out after the optimal solution is obtained
    (b) allows the decision maker more meaningful information about the changes in the LP model parameter
    (c) provides the range within which a parameter may change without affecting optimality
    (d) all of the above

45. To obtain Optimality of Current Optimal

Solution for a change Δc_k in the coefficient c_k of non-basic variable X_k, we must have:
    (a) $\Delta c_k = z_k - c_k$ (b) $\Delta c_k = z_k$
    (c) $c_k + \Delta c_k = z_k$ (d) $\Delta c_k \geq z_k$

46. When an additional variable is added in LP model, the existing Optimal Solution can further be improved if:
    (a) $z_j - c_j \leq 0$ (b) $z_j - c_j \geq 0$
    (c) both (a) and (b)   (d) none of these

47. While performing sensitivity analysis, the Upper bound infinity of the value of the right hand side of a constraint means that:
    (a) the Constraint is redundant
    (b) the Shadow price for the constraint is zero
    (c) there is Slack in the constraint
    (d) none of the above

48. Rounding off solution values of decision variables in a LP Problem may not be acceptable because:
    (a) it does not satisfy constraints
    (b) it violates non-negativity conditions
    (c) objective function value is less than the objective function value of LP
    (d) none of the above

49. If the additional constraint is added in an equation and an artificial variable appear in the basis of the new problem, the new optimal solution is obtained by:
    (a) assigning Zero cost coefficient to the artificial variable if it appears in the basis at negative value
    (b) assigning $-M$ cost coefficient to the artificial variable if it appears in the basis at positive value
    (c) either (a) or (b)
    (d) none of the above

50. A Non-basic variable should be brought into the new solution mix provided its contribution rate (c_j) is:
    (a) $c_j^* = c_j + (z_j - c_j)$ (b) $c_j^* > c_j + (z_j - c_j)$
    (c) $c_j^* < c_j + (z_j - c_j)$ (d) none of these

Answers

1. (a)	2. (b)	3. (b)	4. (c)	5. (a)	6. (b)	7. (a)	8. (a)	9. (b)
10. (a)	11. (c)	12. (c)	13. (a)	14. (b)	15. (d)	16. (b)	17. (a)	18. (d)
19. (b)	20. (b)	21. (c)	22. (b)	23. (a)	24. (b)	25. (c)	26. (a)	27. (c)
28. (c)	29. (a)	30. (b)	31. (c)	32. (b)	33. (d)	34. (a)	35. (c)	36. (b)
37. (b)	38. (a)	39. (c)	40. (d)	41. (d)	42. (a)	43. (a)	44. (b)	45. (c)
46. (a)	47. (b)	48. (d)	49. (c)	50. (c)				

 **ARCHIVE**

1. Solve the following LP problem
   Maximize $Z = 5x_1 + 12x_2 + 4x_3$,
   Subject to the constraints
   (i) $x_1 + 2x_2 + x_3 \leq 5$,
   (ii) $2x_1 - x_2 + 3x_3 = 2$
   and $x_1, x_2, x_3 \geq 0$.
   (a) Discuss the effect of changing the requirement vector from $[5, 2]^T$ to $[7, 2]^T$ on the optimum solution.
   (b) Discuss the effect of changing the requirement vector from $[5, 2]^T$ to $[3, 9]^T$ on the optimum solution.
   (c) Which resource should be increased and how much to achieve the best marginal increase in the value of the objective function?   [DAYAL BAGH (M.Tech.)- 1998]

2. Consider the following LP problem.
   Maximize $Z = 3x_1 + 5x_2 + 4x_3$,
   Subject to the constraints
   (i) $2x_1 + 3x_2 \leq 8$,
   (ii) $2x_2 + 5x_3 \leq 10$,
   (iii) $3x_1 + 2x_2 + 4x_3 \leq 15$
   and $x_1, x_2, x_3 \geq 0$.
   The optimal solution is given in the following table.

			$C_j \rightarrow$	3	5	4	0	0	0
C_B	B	$b(= X_B)$	x_1	x_2	x_3	s_1	s_2	s_3
5	x_2	$\dfrac{50}{41}$	0	1	0	$\dfrac{15}{41}$	$\dfrac{8}{41}$	$\dfrac{-10}{41}$
4	x_3	$\dfrac{62}{41}$	0	0	1	$\dfrac{-6}{41}$	$\dfrac{5}{41}$	$\dfrac{4}{41}$
3	x_1	$\dfrac{29}{41}$	1	0	0	$\dfrac{-2}{41}$	$\dfrac{-12}{41}$	$\dfrac{15}{41}$
$Z = \dfrac{765}{41}$		Z_j	3	5	4	$\dfrac{45}{41}$	$\dfrac{24}{41}$	$\dfrac{11}{41}$
		$C_j - Z_j$	0	0	0	$\dfrac{-45}{41}$	$\dfrac{-24}{41}$	$\dfrac{-11}{41}$

   (a) How much C_3 and C_4 can be increased till the optimal solution given in the table remains optimal? Also find the new value of the objective function if possible.

   (b) Find the range over which b_2 can be changed maintaining the feasibility of the solution.   [DELHI -1995]

3. Given the following LP problem
   Maximize $Z = -x_1 + 2x_2 - x_3$
   Subject to the constraints
   (i) $3x_1 + x_2 - x_3 \leq 10$,
   (ii) $-x_1 + 4x_2 + x_3 \geq 6$,
   (iii) $x_2 + x_3 \leq 4$.
   and $x_1, x_2, x_3 \geq 0$.
   Determine the effect of discrete changes in $b_i (i = 1, 2, 3)$ on the optimal solution shown in the table.

			$C_j \rightarrow$	-1	2	-1	0	0	0	-M
Cost per unit C_B	Variables in Basis B	Solution values $b(= X_B)$	x_1	x_2	x_3	s_1	s_2	s_3	A_1
0	s_1	6	3	0	-2	1	0	-1	0
2	s_2	4	0	1	1	0	0	1	0
0	s_3	10	1	0	3	0	1	4	-1
$Z = 8$		Z_j	0	2	2	0	0	2	0
		$c_j - Z_j$	-1	0	-3	0	0	-2	-M

[MEERUT-1993]

4. Given the LP problem.
   Max. $Z = -x_1 + 2x_2 - x_3$
   Subject to,
   (i) $3x_1 + x_2 - x_3 \leq 10$;
   (ii) $-x_1 + 4x_2 + x_3 \geq 6$;
   (iii) $x_2 + x_3 \leq 4$
   and $x_1, x_2, x_3 \geq 0$.
   Determine the range for discrete changes in the resource values $b_2 = 10$ and $b_3 = 6$ of the LP model so as to maintain optimality of the current solution.   [MEERUT-1998]

5. Find the optimal solution to the LP problem.
   Max. $Z = 15x_1 + 45x_2$
   Subject to,
   (i) $x_1 + 16x_2 \leq 250$,
   (ii) $5x_1 + 2x_2 \leq 162$,
   (iii) $x_2 \leq 50$.
   and $x_1, x_2 \geq 0$.
   If Max. $Z = \Sigma C_j x_j, j = 1, 2$ and C_2 is kept fixed at 45, determine how much can C_1 be changed without affecting the optimal solution of the problem.   [BOMBAY-1991]

Hints and Answers

1. $x_1 = \dfrac{9}{5}, x_2 = \dfrac{8}{5}, x_3 = 0$ and Max. $Z = \dfrac{141}{5}$.

   (a) If the requirement vector changes from $[5, 2]^T$ to $[7, 2]^T$, then the solution remains feasible and optimal with new values $x_1 = \dfrac{11}{5}, x_2 = \dfrac{12}{5}, x_3 = 0$ and Max. $Z = \dfrac{199}{5}$.

   (b) If the requirement vector changes from $[5, 2]^T$ to $[3, 9]^T$, then solution becomes infeasible. To remove infeasibility, apply dual simplex method. After solving, we get the optimal solution, $x_1 = 0, x_2 = 0, x_3 = 3$, Max. $Z = 12$.

   (c) First resource can be increased indefinitely.

2. (a) $\dfrac{5}{4} \leq c_3 \leq \dfrac{23}{2}, \dfrac{594.5}{45} \leq$ Max. $Z \leq \dfrac{1230}{41}$

   $-\infty \leq c_4 \leq \dfrac{45}{41}$, max. $Z = \dfrac{765}{41}$

   (b) $\dfrac{15}{4} \leq b_2 \leq \dfrac{209}{12}$

3. $a_{12} \geq -163, a_{22} \geq -149, a_{13} \geq -15, a_{23} \geq \dfrac{34}{53}, -\infty \leq a_{33} \leq \infty, 11 \leq a_{24} < \infty$.

4. $b_2 < 10; -\dfrac{5}{2} \leq b_3 \geq 6$

5. $\left(15 - \dfrac{195}{16}\right) \leq c_1 \leq \left(15 + \dfrac{195}{16}\right)$

▢▢▢▢

Integer Programming $\mathbf{10}$

10.1 INTRODUCTION

In linear programming problem, we observe that each decision variables, slack and surplus variables, can take any real or fractional values. But there are some situations in which the fractional values of these variables has no significance. The integer programming is a linear programming in which some or all variables $x_1, x_2, ..., x_n$ are permitted to take only integral values.

Definition. *Integer linear programming are those in which some or all of the variables are restricted to integer or discrete values.*

10.2 NEED OF INTEGER LINEAR PROGRAMMING

The integer linear programming has important applications in business and industry. Actually in any situation involving the decision of the type "either-or" *i.e.* either to do the activity or not to do can be viewed as integer linear programming. Besides these, in the manufacturing decisions of trucks, of cars etc., the quantity manufactured can be a whole number. Actually all the allocation problems requiring the allocations of men, machines or vehicles etc. to activities in a programming problem will be an integer linear programming as such things can be assigned in integer quantities not in fraction.

10.3 TYPES OF INTEGER LINEAR PROGRAMMING PROBLEMS

There are following three types of integer linear programming problem:

(1) **Pure Integer Linear Programming Problem:** An integer linear programming is said to be a pure integer linear programming when all its decision variables are restricted to be integer.

(2) **Mixed Integer Linear Programming Problem:** An integer linear programming problem is said to be mixed integer programming when some, but not all of its decision variables are restricted to be integer.

(3) **Zero-one Integer Linear Programming Problem:** An integer linear programming problem in which all the decision variables are restricted to integer of 0 or 1 is called zero-one integer linear programming problem.

10.4 METHODS TO SOLVE AN INTEGER LINEAR PROGRAMMING PROBLEM

There are following two methods:

10.4.1 GOMORY'S CUTTING PLANE METHOD

Consider a linear integer programming problem. A systematic procedure for solving all pure integer linear programming problem was first developed by R.E. Gomory in 1958. Later, he extended the procedure to solve the mixed integer programme.

First find the optimal solution to the usual linear programming problem by the simplex method ignoring the integer-valued restriction. If in the optimal solution all the variables have integer values, then it is also the optimal solution of the given LPP. But if not, then modify the LPP by systematically introducing a new constraints called secondary or Gomory's constraints which essentially represents necessary conditions for integrability and eliminates some non-integer solutions without loosing any integral solution. After adding the second constraints, the problem is solved by dual simplex algorithm to get an optimal integral solution. If optimal integer valued solution is obtained, then the problem is solved to get an integer valued optimum solution. Repeat this process iteratively until one obtains the required integer valued optimal solution. In this method, our main work is to enlarge the continuous ILPP obtained by ignoring the integral valued restriction of the ILPP by introducing the Gomory's constraints (Gomory's cut) and hence the construction of such constraints is very important job in this method, and need special attention.

10.4.2 CONSTRUCTION OF GOMORY'S CONSTRAINTS

To find the Gomory's constraints, use the fact that a solution satisfy the constraints of the given IPP also satisfies any other constraints obtained by adding or subtracting two more given constraints or obtained by multiplying a constraint by a non-zero real number.

Let $X_B = (x_{B_1}, x_{B_2}, \ldots, x_{B_m})$ be the optimal solution obtained by regular simplex method by ignoring integer condition on the variables of maximization LPP. Then we get the final simplex table as given below.

B.V.	C_B	X_B	x_1 (β_1)	x_2 (β_2)	...	x_i (β_i)	...	x_m (β_m)	x_{m+1}	...	x_n
x_1	C_{B_1}	x_{B_1}	1	0	...	0	...	0	$a_{1(m+1)}$	...	a_{1n}
x_2	C_{B_2}	x_{B_2}	0	1		$\vdots$		$\vdots$	$a_{2(m+1)}$	...	a_{2n}
$\vdots$	$\vdots$	$\vdots$	$\vdots$	$\vdots$		$\vdots$		$\vdots$	$\vdots$		$\vdots$
x_i	C_{B_i}	x_{B_i}	0	0		1		$\vdots$	$a_{i(m+1)}$	...	a_{in}
$\vdots$	$\vdots$	$\vdots$	$\vdots$	$\vdots$		$\vdots$		$\vdots$	$\vdots$		$\vdots$
x_m	C_{B_m}	x_{B_m}	0	0		0		1	$a_{m(m+1)}$	...	a_{mn}
		x_j	x_{B_1}	x_{B_2}	...	x_{B_i}	...	x_{B_m}	0	...	0

Let us suppose x_i is not an integer. Then from the above table (i^{th} row) we have

$$x_{B_i} = 0 \cdot x_1 + 0 \cdot x_2 + \ldots + 1 \cdot x_i + \ldots + 0 \cdot x_m + a_{i(m+1)}x_{m+1} + \ldots + a_{in}x_n$$

$$\Rightarrow \qquad x_{B_i} = x_i + \sum_{j=m+1}^{n} a_{ij} x_j$$

$$\Rightarrow \qquad x_i = x_{B_i} - \sum_{j=m+1}^{n} a_{ij} x_j \qquad \qquad \dots(1)$$

Since, we have x_{B_i} is a non-integer, therefore, we have

$$x_{B_i} = [x_{B_i}] + f_{B_i} \quad \text{and} \quad a_{ij} = [a_{ij}] + f_{ij} \qquad \qquad \dots(2)$$

where $[\cdot]$ denote the integral part and f_{B_i} and f_{ij} are the positive fractional parts of x_{B_i} and a_{ij} respectively.

Therefore,

$$\left.\begin{array}{l} [x_{B_i}] \leq x_{B_i}, 0 \leq f_{B_i} < 1 \\ [a_{ij}] \leq a_{ij}, 0 \leq f_{ij} < 1 \end{array}\right\} \qquad \qquad \dots(3)$$

Using (2) in (1), we get

$$x_i = [x_{B_i}] + f_{B_i} - \sum_{j=m+1}^{n} ([a_{ij}] + f_{ij}) x_j$$

$$\Rightarrow \qquad x_i - [x_{B_i}] + \sum_{j=m+1}^{n} f_{ij}; x_j = f_{B_i} - \sum_{j=m+1}^{n} [a_{ij}] x_j$$

or $\quad x_i - [x_{B_i}] + \sum_{j=m+1}^{n} [a_{ij}] x_j = f_{B_i} - \sum_{j=m+1}^{n} f_{ij} x_j \qquad \dots(4)$

Since, for optimal integer solution x_i for $i = 1, 2, \dots, m$ are all integers and also for x_j for $j = m + 1, m + 2, \dots, n$ are all integers, then from (4)

$$f_{B_i} - \sum_{j=m+1}^{n} f_{ij} x_j \quad \text{must be an integer}$$

Now, since, $0 < f_{ij} < 1$ and $x_{ij} \geq 0$ then

$$\sum_{j=m+1}^{n} f_{ij} \cdot x_j > 0$$

$$\Rightarrow \qquad f_{B_i} - \sum_{j=m+1}^{n} f_{ij} \cdot x_j \leq f_{B_i}$$

But $0 \leq f_{B_i} < 1$, therefore, $f_{B_i} - \sum_{j=m+1}^{n} f_{ij} \cdot x_j < 1$

Also, $f_{B_i} - \sum_{j=m+1}^{n} f_{ij} x_j$ is an integer.

$\Rightarrow \quad$ It should be either 0 or negative integer.

So, $\qquad\qquad f_{B_i} - \sum\limits_{j=m+1}^{n} f_{ij} x_j \leq 0$

$\Rightarrow \qquad\qquad - \sum\limits_{j=m+1}^{n} f_{ij} x_j \leq -f_{B_i}$ $\qquad\qquad\qquad\qquad$ ...(5)

The inequality given above by (5) is called Gomory's constraints. Adding a non-negative slack variable s_i, Gomory's constraints become

$$- \sum\limits_{j=m+1}^{n} f_{ij} x_j + s_i = -f_{B_i} \qquad\qquad\qquad ...(6)$$

It is called Gomory's cutting plane (fractional cut)

Now, since x_j ($j = m + 1, ..., n$) are non-basic variables, so they are all zero which implies that

$\qquad\qquad s_i = -f_{B_i}$, which is infeasible.

Now, adding the Gomory's constraints equation (6) to the table-1 we get the new table-2 with one extra column of s_i (β_{m+1}).

B.V.	C_B	X_B	x_1 (β_1)	x_2 (β_2)	...	x_i (β_i)	...	x_m (β_m)	x_{m+1}	...	x_n	s_i (β_{m+1})
x_1	C_{B_1}	x_{B_1}	1	0	...	0	...	0	$a_{1(m+1)}$	...	a_{1n}	0
x_2	C_{B_2}	x_{B_2}	0	1	...	0	...	0	$a_{2(m+1)}$	...	a_{2n}	0
$\vdots$	$\vdots$	$\vdots$	$\vdots$	$\vdots$	...	$\vdots$	...	$\vdots$	$\vdots$		$\vdots$	$\vdots$
x_i	C_{B_i}	x_{B_i}	0	0	...	i	...	0	$a_{i(m+1)}$	...	a_{in}	0
$\vdots$	$\vdots$	$\vdots$	$\vdots$	$\vdots$	...	$\vdots$					$\vdots$	$\vdots$
x_m	C_{B_m}	x_{B_m}	0	0	...	0		0	$a_{m(m+1)}$	...	a_{mn}	0
s_i	0	$-f_{B_i}$	0	0	...	0		0	$f_{i(m+1)}$	...	f_{in}	1
		x_j	x_{B_1}	x_{B_2}	...	x_{B_i}	...	x_{B_m}	0	...	0	$-f_{B_i}$

Now since s_i is negative, then the optimal solution given by above table is not feasible, so dual simplex method can be applied to clear this feasibility and get new optimal solution. If this solution has integer values of all decision variables, then this is the required solution and we end the process, otherwise we construct a new Gomory's constraints as obtained earlier to find further new optimal solution. Repeat the above process until an integer solution is obtained.

Working Procedure

STEP 1. Write the given LPP into maximization form.

STEP 2. Make all $b_i's$ positive.

STEP 3. Apply the regular simplex method, by ignoring integer condition on variables and find optimal solution.

STEP 4. If the obtained optimal solution have all integer solution, then obtained solution will be the required solution. But if at least one of the variable is not an integer, go to step 5.

STEP 5. (i) If only one variable is not an integer, then from the simplex table of step 3, the row corresponding to non-integer, variable will generate Gomory's constraints.

(ii) If more than one variables are non-integer then select that variable which has largest fractional value. In case of tie, select the constraints having the lowest contribution for maximization or highest cost for minimization problem.

Alternatively, select the constraints with

$$\max\left\{\frac{f_{B_i}}{\sum\limits_{j=m+1}^{n} f_{ij}}\right\}$$

STEP 6. Add the Gomory's constraints (obtained in the above step) with non-negative slack variables s_i into the final simplex table.

STEP 7. Obtain the optimal solution of the table in the above table by dual simplex method, so that s_i is the outgoing vector.

STEP 8. If the obtained solution (obtained in step 7) has all integral values then it is the required solution of the given integer programming. If it is not an integer solution, then repeat step 6 to 8 until we get an optimal feasible integer solution.

10.4.3 PROPERTIES OF GOMORY'S ALGORITHM

The Gomory's algorithm has the following properties:

(1) Additional linear constraints never cut off that portion of the original feasible solution space which contains a feasible integer solution to the original problem.

(2) Each new additional constraints cuts-off the current non-integer optimal solution to the linear programming problem.

☛ **REMARK**

• If the given integer programming has a constraints with fractional coefficients, then transform the constraints such that the coefficients are whole numbers.

Solved Examples

EXAMPLE 1. *Find the optimum integer solution to the following LPP by Gomory technique.*

$$Max.\ Z = x_1 + 2x_2$$

subject to the constraints

$$2x_2 \leq 7$$
$$x_1 + x_2 \leq 7$$
$$2x_1 \leq 11$$

and $\qquad x_1,\ x_2 \geq 0$ *and* $x_1,\ x_2$ *are integers.* [SHIVAJI–(1985)]

SOLUTION. Clearly, the given LPP of maximization is in standard form with all $b_i's \geq 0$.

Proceeding as usual introducing slack variables s_1, s_2 and s_3, we can write

$$\text{Max. } Z = x_1 + 2x_2 + 0s_1 + 0s_2 + 0s_3$$

subject to the constraints

$$0x_1 + 2x_2 + s_1 = 7$$
$$x_1 + x_2 + s_2 = 7$$
$$2x_1 + s_3 = 11$$

and $\qquad x_1, x_2, s_1, s_2, s_3 \geq 0$

Now apply the simplex method in a usual manner, we get the following simplex table.

B.V.	C_B	X_B	x_1	x_2	s_1 (x_3)	s_2 (x_4)	s_3 (x_5)	Min Ratio
$c_j \rightarrow$			1	2	0	0	0	
s_1	0	7	0	2	1	0	0	7/2 (min)→
s_2	0	7	1	1	0	1	0	7/1
s_3	0	11	2	0	0	0	1	—
$Z = C_B X_B = 0,$	$\Delta_j \rightarrow$		1	2	0	0	0	
x_2	2	7/2	0	1	1/2	0	0	—
s_2	0	7/2	1	1	–1/2	1	0	(7/2)/1 (min)→
s_3	2	11	2	0	0	0	1	11/2
$Z = C_B X_B = 7,$	$\Delta_j \rightarrow$		1	0	–1	0	0	
x_2	2	7/2	0	1	1/2	0	0	
x_1	1	7/2	1	1	–1/2	1	0	
s_3	0	4	0	0	1	–2	1	
$Z = C_B X_B = 21/2$	$\Delta_j \rightarrow$		0	0	–1/2	–1	0	

From above table, we observe that all $\Delta_j \leq 0$ and $x_1 = x_2 = \dfrac{7}{2}$. Therefore, the solution under the above table is optimal and feasible but not integer.

Now, apply Gomory technique to find the integer solution.

$$\because \qquad x_1 = \frac{7}{2} = 3 + \frac{1}{2} \quad \text{and} \quad x_2 = \frac{7}{2} = 3 + \frac{1}{2}$$

So, $f_{B_1} = \dfrac{1}{2}, f_{B_2} = \dfrac{1}{2} \Rightarrow$ fractional part for both x_1 and x_2 are same. Select x_1-row arbitrarily.

Now, we find first Gomory's constraints as given under.

$$\because \qquad -\sum_{j=m+1}^{n} f_{ij} \cdot x_j \leq -f_{B_i}$$

Here, we have $i = 1, m = 2, n = 5, x_3 = s_1, x_4 = s_2, x_5 = s_3$, so

$$-\sum_{j=3}^{5} f_{ij} x_j \leq -f_{B_1}$$

$\Rightarrow f_{13}x_3 - f_{13}x_4 - f_{15}x_5 \leq -f_{B_1}$

Now, from x_1-row, we have

$$a_{13} = -\frac{1}{2} = -1 + \frac{1}{2} \quad \Rightarrow \quad f_{13} = \frac{1}{2}$$

$$a_{14} = 1 \Rightarrow f_{14} = 0$$

$$a_{15} = 0 \Rightarrow f_{15} = 0$$

Thus $-\frac{1}{2}s_1 \leq -\frac{1}{2}$, which is the first Gomory's constraint.

Now, introducing a non-negative slack variable s', the first Gomory's constraint becomes

$$-\frac{1}{2}s_1 + s' = -\frac{1}{2}$$

Now, adding this Gomory's constraint equation into the above table, we get the table as given below.

B.V.	C_B	X_B	$c_j \rightarrow$ 1 x_1 (β_2)	2 x_2 (β_1)	0 s_1 (x_3)	0 s_2 (x_4)	0 s_3 (β_3)	0 s' (β_4)
x_2	2	7/2	0	1	1/2	0	0	0
x_1	1	7/2	1	0	-1/2	1	0	0
s_1	0	4	0	0	1	-2	1	0
s'	0	-1/2	0	0	-1/2	0	0	1
$Z = C_B X_B = 21/2,$	$\Delta_j \rightarrow$		0	0	-1/2	-1	0	0
					$\uparrow$			

Clearly $s' = -\frac{1}{2} < 0$ then we apply dual simplex method. We can find the new optimal solution. Here, by dual simplex method, the vector $s'(\beta_4)$ will be the outgoing vector.

$\Rightarrow \qquad\qquad r = 4$

Now, by minimum ratio rule

$$\frac{\Delta_k}{a_{rk}} = \min_j \left\{ \frac{\Delta_j}{a_{rj}}, a_{rj} < 0 \right\}$$

$\Rightarrow \qquad\qquad \dfrac{\Delta_k}{a_{4k}} = \min_j \left\{ \dfrac{\Delta_j}{a_{4j}}, a_{4j} < 0 \right\} = \min \left\{ \dfrac{\Delta_3}{a_{43}} \right\} = \dfrac{\Delta_3}{a_{43}}$

$\Rightarrow \qquad\qquad k = 3$

$\Rightarrow \quad x_3 \, (s_1)$ is the incoming vector and hence the key element $= a_{43} = \dfrac{1}{2}$

Now again apply the simplex method in a usual manner, we have the following simplex table.

	$c_j \rightarrow$		1	2	0	0	0	0
B.V.	C_B	X_B	x_1 (β_2)	x_2 (β_1)	s_1 (β_4)	s_2 (x_4)	s_3 (x_5)	s' (x_6)
x_2	2	3	0	1	0	0	0	1
x_1	1	4	1	0	0	1	0	−1
s_3	0	3	0	0	0	−2	1	2
s_1	0	1	0	0	1	0	0	−2
$Z = C_B X_B = 10,$	$\Delta_j \rightarrow$		0	0	0	−1	0	−1

From the above table, we observe that all $\Delta_j \leq 0$ and $x_1 = 4$, $x_2 = 3$ and Max. $Z = 10$ which shows that this solution is optimal, feasible and integer.

EXAMPLE 2. **Solve the following integer linear programming problem**

$$\text{Max. } Z = 5x_1 + 7x_2$$

subject to the constraints

$$-2x_1 + 3x_2 \leq 6$$
$$6x_1 + x_2 \leq 30$$

and $x_1, x_2 \geq 0$ **and integer.**

Interpret graphically.

SOLUTION. Firstly, we solve the given problem by simplex method by ignoring integer condition as given under.

Introducing the slack variables s_1, s_2 the given maximization IPP with all $b_i's > 0$ becomes

$$\text{Max. } Z = 5x_1 + 7x_2 + 0s_1 + 0s_2$$
$$\text{s.t.} \quad -2x_1 + 3x_2 + s_1 = 6$$
$$6x_1 + x_2 + s_2 = 30$$
$$\text{and} \quad x_1, x_2, s_1, s_2 \geq 0$$

Apply the simplex method in a usual manner, we have the following simplex table.

	$c_j \rightarrow$		5	7	0	0	Min Ratio
B.V.	C_B	X_B	x_1	x_2	s_1	s_2	
s_1	0	6	−2	③	1	0	6/3 (min)→
s_2	0	30	6	1	0	1	30/1
$Z = C_B X_B = 0,$	$\Delta_j \rightarrow$		5	7	0	0	
				↑			
x_2	7	2	−2/3	1	1/3	0	—
s_2	0	28	20/3	0	−1/3	1	28/(20/3) →
$Z = 14,$	$\Delta_j \rightarrow$		29/3	0	−7/3	0	
			↑				
x_2	7	24/5	0	1	3/10	1/10	
x_1	5	21/5	1	0	−1/20	3/20	
$Z = C_B X_B = 273/5,$	$\Delta_j \rightarrow$		0	0	−37/20	−29/20	

We observe that all $\Delta_j \leq 0 \Rightarrow$ solution is feasible and optimal and is given by

$$x_1 = \frac{21}{5}, x_2 = \frac{24}{5}, \text{Max } Z = \frac{273}{5}, \text{ which is not an integer solution.}$$

Clearly, we can write

$$x_1 = 4 + \frac{1}{5}, x_2 = 4 + \frac{4}{5}$$

Now, to construct Gomory's constraints, select x_2-row as it is and has the larger fractional part which is $\frac{4}{5}$. The Gomory's constraints is given by

$$-\sum_{j=m+1}^{n} f_{ij}x_j \leq f_{B_i} \qquad \ldots(1)$$

Clearly, we have $i = 2$ (x_2-row), $m = 2$, $n = 4$

Then from (1), we have

$$-\sum_{j=3}^{4} f_{2j}x_j \leq -f_{B_2}$$

$$\Rightarrow \quad -f_{23}x_3 - f_{24}x_4 \leq -f_{B_2}$$

$$\Rightarrow \quad -f_{23}s_1 - f_{24}s_2 \leq -f_{B_2} \qquad \ldots(2)$$

From the above table, we have

$$\left.\begin{array}{l} x_{B_2} = \dfrac{24}{5} \text{ therefore } f_{B_2} = \dfrac{4}{5} \\[2mm] a_{23} = \dfrac{3}{10} \text{ therefore } f_{23} = \dfrac{3}{10} \\[2mm] a_{24} = \dfrac{1}{10} \text{ therefore } f_{24} = \dfrac{1}{10} \end{array}\right] \qquad \ldots(3)$$

Using (3) in (2) we get

$$-\frac{3}{10}s_1 - \frac{1}{10}s_2 \leq -\frac{4}{5} \qquad \ldots(4)$$

which is the first Gomory's constraint.

Now introducing a non-negative slack variable s', the Gomory's constraint becomes

$$-\frac{3}{10}s_1 - \frac{1}{10}s_2 + s' = -\frac{4}{5}$$

Adding this constraint, we get the following simplex table

B.V.	C_B	X_B	$c_j \rightarrow$				
			5	7	0	0	0
			x_1 (β_2)	x_2 (β_1)	s_1 (x_3)	s_2 (x_4)	s' (β_3)
x_2	7	24/5	0	1	3/10	1/10	0
x_1	5	21/5	1	0	$-1/20$	3/20	0
s'	0	$-4/5$	0	0	$\boxed{-3/10}$	$-3/10$	1
		$\Delta_j \rightarrow$	0	0	$-37/20$	$-29/20$	0
					$\uparrow$		

Since, $s' = -\dfrac{4}{5} < 0$, thus the solution obtained in above table is not feasible so we apply dual simplex method to obtain the new optimal solution.

By dual simplex method, the outgoing vector is s' i.e. β_3

$\Rightarrow$ $r = 3$

Then by minimum ratio rule

$$\frac{\Delta_k}{a_{rk}} = \min_j \left\{ \frac{\Delta_j}{a_{rj}}, a_{rj} < 0 \right\}$$

$\Rightarrow$

$$\frac{\Delta_k}{a_{3k}} = \min_j \left\{ \frac{\Delta_j}{a_{3j}}, a_{3j} < 0 \right\} = \left\{ \frac{\Delta_3}{a_{33}}, \frac{\Delta_4}{a_{34}} \right\}$$

$$= \left\{ \frac{-37/20}{-3/10}, \frac{-29/20}{-1/10} \right\} = \left\{ \frac{37}{6}, \frac{29}{2} \right\} = \frac{37}{6} = \frac{\Delta_3}{a_{33}}$$

$\Rightarrow$ $k = 3$, i.e., $x_3 (= s_1)$ is the incoming vector.

Thus, the key element $= a_{33} = \dfrac{-3}{10}$

Now, proceed the simplex method in a usual manner, we have the following simplex table

B.V.	C_B	X_B	$c_j \rightarrow$ x_1 (β_2)	5 x_2 (β_1)	7 s_1 (β_3)	0 s_2 (x_4)	0 s' (x_5)
x_2	7	4	0	1	0	0	1
x_1	5	13/3	1	0	0	1/6	−1/6
s_1	0	8/3	0	0	1	1/3	−10/3
$Z=C_B X_B=149/3$	$\Delta_j \rightarrow$		0	0	0	−5/6	−37/6

In the above table, we observe that all $\Delta_j \le 0$, thus the solution is optimal and is given by

$$x_1 = \frac{13}{3}, x_2 = 4 \quad \text{and max. } Z = \frac{149}{3}$$

Here, it is also clear that this solution is not integer $\left(\because x_1 = \dfrac{13}{3} \right)$

So, we have to construct Gomory's constraints

We have $-\sum\limits_{j=m+1}^{n} f_{ij} x_j \le f_{B_i}$

Here, $i = 1, m = 3, n = 5$ (m = No. of rows in the table)

Thus, we have

$$-\sum\limits_{j=4}^{5} f_{ij} x_j \le -f_{B_1}$$

$\Rightarrow$ $-f_{14} x_4 - f_{15} x_5 \le -f_{B_1}$ or $-f_{14} s_2 - f_{15} s' \le -f_{B_1}$ ...(1)

Using the table

$$\left.\begin{array}{l} x_{B_1} = \dfrac{13}{3} = 4 + \dfrac{1}{3} \Rightarrow f_{B_1} = \dfrac{1}{3} \\[2mm] a_{14} = \dfrac{1}{6} \Rightarrow f_{14} = \dfrac{1}{6} \\[2mm] a_{15} = -\dfrac{1}{6} \Rightarrow a_{15} = -1 + \dfrac{5}{6} \Rightarrow f_{15} = \dfrac{5}{6} \end{array}\right] \qquad \text{...(2)}$$

Putting all these values in (1) we get

$$-\dfrac{1}{6} s_2 - \dfrac{5}{6} s' \le -\dfrac{1}{3} \qquad \text{...(3)}$$

which is the required Gomory's second constraint.

Further, introducing a slack variable s'', (3) becomes $-\dfrac{1}{6} s_2 - \dfrac{5}{6} s' + s'' = -\dfrac{1}{3}$

Now, adding (3) to the above table we get the new table as given below.

B.V.	C_B	X_B	$c_j \rightarrow$ 5 x_1 (β_2)	7 x_2 (β_1)	0 s_1 (β_3)	0 s_2 (x_4)	0 s' (x_5)	0 s'' (β_4)
x_2	7	4	0	1	0	0	1	0
x_1	5	13/3	1	0	0	1/6	−1/6	0
s_1	0	8/5	0	0	1	1/3	−10/3	0
s''	0	−1/3	0	0	0	⟨−1/6⟩	−5/6	1
		$\Delta_j \rightarrow$	0	0	0	−5/6 ↑	−37/6	0

Here, $s'' = -\dfrac{1}{3} < 0 \Rightarrow$ obtained solution is optimal but not feasible. So again apply the dual simplex method.

Clearly, s'' is the outgoing vector.

$\Rightarrow \qquad \beta_4$ is the outgoing vector.

$\Rightarrow \qquad r = 4$

Now, by minimum ratio rule

$$\dfrac{\Delta_k}{a_{rk}} = \min_j \left\{ \dfrac{\Delta_j}{a_{rj}}, a_{rj} < 0 \right\}$$

$$\Rightarrow \qquad \dfrac{\Delta_k}{a_{4k}} = \min_j \left\{ \dfrac{\Delta_j}{a_{4j}}, a_{4j} < 0 \right\} \; \forall j, \text{ not in the basis}$$

$$= \min \left\{ \dfrac{\Delta_4}{a_{44}}, \dfrac{\Delta_5}{a_{45}} \right\} = \min \left\{ \dfrac{-5/6}{-1/6}, \dfrac{-37/6}{-5/6} \right\}$$

$$= \min \left\{ 5, \dfrac{37}{6} \right\} = 5 = \dfrac{\Delta_4}{a_{44}}$$

$$\Rightarrow \qquad k = 4$$

$\Rightarrow \quad x_4 (= s_2)$ is the incoming vector and key element $= a_{44} = -\dfrac{1}{6}$

Now, apply the simplex method in a usual manner, we have the following table.

B.V.	C_B	X_B	$c_j \rightarrow$ 5 x_1 (β_2)	7 x_2 (β_1)	0 s_1 (β_3)	0 s_2 (β_4)	0 s' (x_5)	0 s'' (x_6)
x_2	7	4	0	1	0	0	1	0
x_1	5	4	1	0	0	0	−1	1
s_1	0	2	0	0	1	0	−5	2
s_2	0	2	0	0	0	1	5	− 6
$Z=C_BX_B=48,$		$\Delta_j \rightarrow$	0	0	0	0	−2	−5

We observe that here all $\Delta_j \leq 0$

$\Rightarrow$ solution is optimal and is given by

$x_1 = 4, x_2 = 4$ and Max. $Z = 48$, integer solution

Graphical interpretation:

In the adjoining figure, the solution space is shown by the region $OABCO$ and the non-integer solution is obtained at $B\left(\dfrac{21}{5}, \dfrac{24}{5}\right)$. To find integer solution, we add two Gomory's constraints one by one as follows:

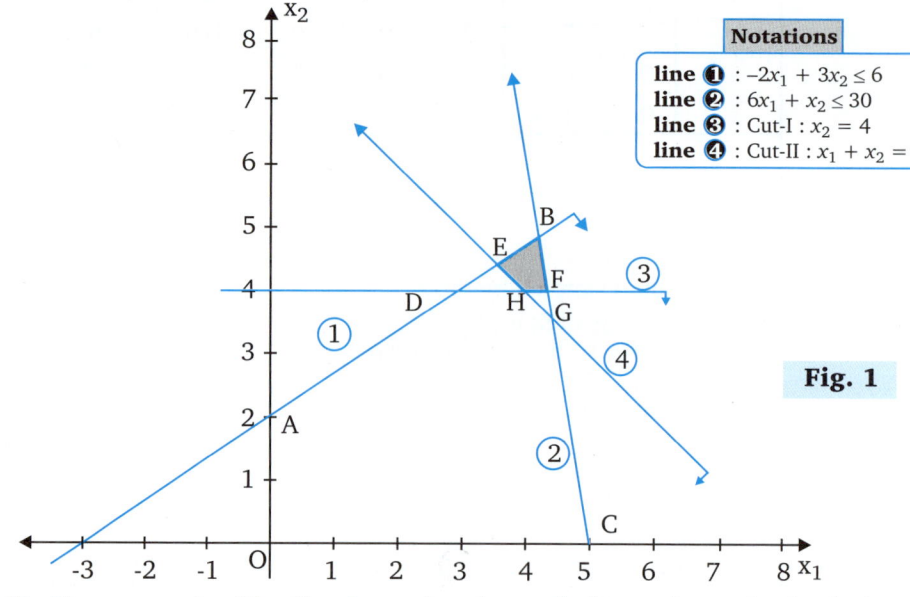

Notations

line **1** : $-2x_1 + 3x_2 \leq 6$
line **2** : $6x_1 + x_2 \leq 30$
line **3** : Cut-I : $x_2 = 4$
line **4** : Cut-II : $x_1 + x_2 = 8$

Fig. 1

(i) First constraint (Cut-I) reduces the given solution to the optimal solution at $F\left(\dfrac{13}{3}, 4\right)$ and $\text{Max } Z = \dfrac{149}{3}$

(ii) The second constraints (Cut-II) reduces to the solution thus obtained Cut-I to the optimal solution at $H(4, 4)$ and max. $Z = 48$.

In the above figure Cut-I and Cut-II reduces to the region $OABCO$ to the region $EBFHE$, shown by the shaded area.

Also, Cut-I: $-\dfrac{3}{10}s_1 - \dfrac{1}{10}s_2 \le -\dfrac{4}{5}$ $\Rightarrow$ $3s_1 + s_2 \ge 8$

$\Rightarrow$ $\quad 3(6 + 2x_1 - 3x_2) + 30 - 6x_1 - x_2 \ge 8$

$\Rightarrow$ $\quad -10x_2 + 48 \ge 8$, $i.e.$, $x_2 \le 4$

Similarly, Cut-II is $x_1 + x_2 \le 8$.

EXAMPLE 3. **Solve the following LPP by Gomory's technique.**

$$\textbf{Max. } \textbf{Z} = \textbf{3}\textbf{x}_2$$

subject to the constraints

$$\textbf{3}\textbf{x}_1 + \textbf{2}\textbf{x}_2 \le \textbf{7}$$

$$\textbf{x}_1 - \textbf{x}_2 \ge \textbf{--2}$$

and $\qquad \textbf{x}_1, \textbf{x}_2 \ge \textbf{0} \textbf{ and are integers.}$

Interpret graphically. [MEERUT–2005, 11]

SOLUTION. Write the given LPP in standard form (with all $b_i's > 0$) as follows:

Max. $Z = 0x_1 + 3x_2$

s.t.

$$3x_1 + 2x_2 \le 7$$
$$-x_1 + x_2 \le 2$$

and $\qquad x_1, x_2 \ge 0$

Now, introducing slack variables s_1 and s_2, we can write

Max $Z = 0x_1 + 3x_2 + 0s_1 + 0s_2$

s.t.

$$3x_1 + 2x_2 + s_1 = 7$$
$$-x_1 + x_2 + s_2 = 2$$ $\qquad\qquad$ ...(1)

and $\qquad x_1, x_2, s_1, s_2 \ge 0$

Apply simplex method in a usual manner, we have the following table.

B.V.	C_B	X_B	x_1	x_2	s_1	s_2	Min Ratio
		$c_j \rightarrow$	0	3	0	0	
s_1	0	7	3	2	1	0	7/2
s_2	0	2	−1	①	0	1	2/1 (min)→
$Z = 0,$		$\Delta_j \rightarrow$	0	3	0	0	
s_1	0	3	5	0	1	−2	3/5 (min)→
x_2	3	2	⊝1	1	0	1	—
$Z=C_BX_B=6,$		$\Delta_j \rightarrow$	3	0	0	−3	
x_1	0	3/5	1	0	1/5	−2/5	
x_2	3	13/5	0	1	1/5	3/5	
$Z=C_BX_B=39/5,$		$\Delta_j \rightarrow$	0	0	−3/5	−9/5	

We observe that all $\Delta_j \le 0$ $\Rightarrow$ solution is optimal and is given by

$$x_1 = \dfrac{3}{5}, x_2 = \dfrac{13}{5}, \text{Max } Z = \dfrac{39}{5}$$

Here, x_1, x_2 are not integer, so we can write

$$x_1 = 0 + \frac{3}{5} \text{ and } x_2 = 2 + \frac{3}{5} \Rightarrow f_{B_1} = f_{B_2} = \frac{3}{5}$$

Since, the fractional value of x_1 and x_2 are same, so to select x_1-row or x_2-row, we choose the row with maximum $\dfrac{f_{B_i}}{\sum\limits_{j=m+1}^{n} f_{ij}}$

Here, we have $m = 2$, $n = 4$, $f_{B_1} = f_{B_2} = \dfrac{3}{5}$

For x_1-row

$$\frac{f_{B_1}}{\sum\limits_{j=3}^{4} f_{1j}} = \frac{f_{B_1}}{f_{13} + f_{14}}$$

Since $\quad a_{13} = \dfrac{1}{5} \Rightarrow f_{13} = \dfrac{1}{5}$

$$a_{14} = -\frac{2}{5} = -1 + \frac{3}{5} \Rightarrow f_{14} = \frac{3}{5}$$

$\therefore \quad \dfrac{f_{B_1}}{\sum\limits_{j=3}^{4} f_{1j}} = \dfrac{3/5}{\dfrac{1}{5} + \dfrac{3}{5}} = \dfrac{3/5}{4/5} = \dfrac{3}{4}$

For x_2-row

$$\frac{f_{B_2}}{\sum\limits_{j=3}^{4} f_{2j}} = \frac{f_{B_2}}{f_{23} + f_{24}}$$

$\because \quad a_{23} = \dfrac{1}{5} \Rightarrow f_{23} = \dfrac{1}{5}; a_{24} = \dfrac{3}{5} \Rightarrow f_{24} = \dfrac{3}{5}$

$\therefore \quad \dfrac{f_{B_2}}{\sum\limits_{j=3}^{4} f_{2j}} = \dfrac{3/5}{\dfrac{1}{5} + \dfrac{3}{5}} = \dfrac{3}{4} \qquad \Rightarrow \qquad \max\left\{\dfrac{f_{B_i}}{\sum\limits_{j=3}^{4} f_{ij}}\right\} = \dfrac{3}{4}$

So, we may select at random any one of these.

Let us suppose we select x_1-row. We have to construct Gomory's constraints
The Gomory's constraints is given by

$$-\sum_{j=m+1}^{n} f_{ij}x_j \leq -f_{B_i}$$

Here, we have $i = 1$, $m = 2$, $n = 4$

$\therefore \qquad \sum\limits_{j=3}^{4} f_{ij}x_j \leq -f_{B_1}$

$\Rightarrow \qquad -f_{13}x_3 - f_{14}x_4 \leq -f_{B_1} \qquad \Rightarrow \qquad -f_{13}s_1 - f_{14}s_2 \leq -f_{B_1}$

$\Rightarrow \qquad -\dfrac{1}{5}s_1 - \dfrac{3}{4}s_2 \leq -\dfrac{3}{5}$

which is the first Gomory's constraint.

Further, introducing a non-negative slack variable s', then Gomory's constraints becomes

$$-\frac{1}{5}s_1 - \frac{3}{5}s_2 + s' = -\frac{3}{5}$$

Also, adding this new constraints in the last simplex table we get

B.V.	C_B	X_B	$c_j \rightarrow$	0	3	0	0	0
				x_1 (β_1)	x_2 (β_2)	s_1 (x_3)	s_2 (x_4)	s' (β_3)
x_1	0	3/5		1	0	1/5	−2/5	0
x_2	3	13/5		0	1	1/5	3/5	0
s'	0	−3/5		0	0	−1/5	−3/5	1
			$\Delta_j \rightarrow$	0	0	−3/5	−9/5	0
						$\uparrow$	$\downarrow$	

Here, $s' = -\frac{3}{5} < 0$, which is not feasible. Thus we have to apply dual simplex method to find the new optimal solution.

By dual simplex method, the vector $s' (= \beta_3)$ is the outgoing vector.

$$\Rightarrow \qquad r = 3$$

Then by minimum ratio rule

$$\frac{\Delta_k}{a_{rk}} = \min_j \left\{ \frac{\Delta_j}{a_{rj}}, a_{rj} < 0 \right\}$$

$$\Rightarrow \qquad \frac{\Delta_k}{a_{3k}} = \min_j \left\{ \frac{\Delta_j}{a_{3j}}, a_{3j} < 0 \right\} = \min \left\{ \frac{\Delta_3}{a_{33}}, \frac{\Delta_4}{a_{34}} \right\}$$

$$= \min \left\{ \frac{-3/5}{-1/5}, \frac{-9/5}{-3/5} \right\} = \min\{3, 3\} = 3 = \frac{\Delta_3}{a_{33}} \text{ or } \frac{\Delta_4}{a_{34}}$$

$$\Rightarrow \qquad k = 3 \text{ or } 4$$

Case 1 When $k = 3$

In this case $x_3 (= s_1)$ is the incoming vector and the key element $= a_{33} = -\frac{1}{5}$

Then apply the simplex method in a usual manner, we get the following simplex table.

B.V.	C_B	X_B	$c_j \rightarrow$	0	3	0	0	0
				x_1 (β_1)	x_2 (β_2)	s_1 (β_3)	s_2 (x_4)	s' (x_5)
x_1	0	0		1	0	0	−1	1
x_2	3	2		0	1	0	0	1
s_1	0	3		0	0	1	3	−5
$Z = C_B X_B = 6,$			$\Delta_j \rightarrow$	0	0	0	0	−3

We observe that all $\Delta_j \leq 0$ and $x_1 = 0$, $x_2 = 2$ and max. $Z = 6$

$\Rightarrow \qquad$ solution is optimal and integer.

Case 2 When $k = 4$

In this case $x_4 (= s_2)$ will be the incoming vector and key element $= a_{34} = -\dfrac{3}{5}$. Then again apply simplex method in a usual manner, we have the following simplex table.

B.V.	C_B	X_B	x_1 (β_1)	x_2 (β_2)	s_1 (β_3)	s_2 (x_4)	s' (x_5)
		$c_j \rightarrow$	0	3	0	0	0
x_1	0	1	1	0	1/3	0	−2/3
x_2	3	2	0	1	0	0	1
s_1	0	1	0	0	1/3	1	−5/3
$Z = C_B X_B = 6,$		$\Delta_j \rightarrow$	0	0	0	0	−3

We observe that all $\Delta_j \le 0$ and $x_1 = 0$, $x_2 = 2$ and max. $Z = 6$

$\Rightarrow$ solution is optimal and integer.

Graphical Interpretation:

In the adjoining figure, the solution space is given by $OABCO$.

The non-integer optimal solution is obtained at the point $B\left(\dfrac{3}{5}, \dfrac{13}{5}\right)$.

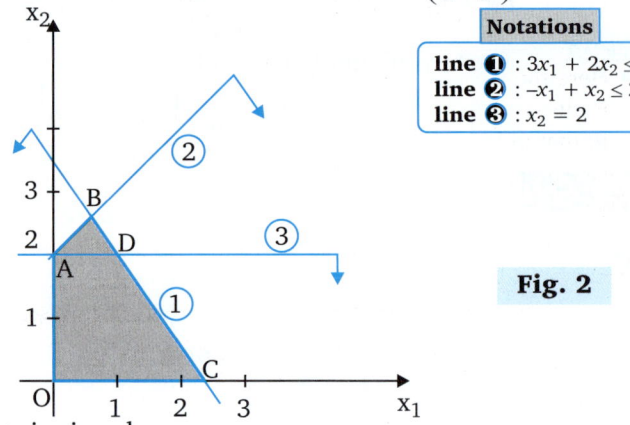

Notations

line ❶ : $3x_1 + 2x_2 \le 7$
line ❷ : $-x_1 + x_2 \le 2$
line ❸ : $x_2 = 2$

Fig. 2

The Gomory's constraints is given by

$$-\frac{1}{5}s_1 - \frac{3}{5}s_2 \le -\frac{3}{5} \quad \Rightarrow \quad s_1 + 3s_2 \ge 3$$

Since, $s_1 = 7 - 3x_1 - 2x_2$ and $s_2 = 2 + x_1 - x_2$ (From (1))

Then Gomory's constraints reduces to

$$(7 - 3x_1 - 2x_2) + 3(2 + x_1 - x_2) \ge 3$$

$\Rightarrow$ $-5x_2 + 13 \ge 3$, *i.e.*, $x_2 \le 2$

On drawing $x_2 = 2$, the region $OABCO$ reduces to the feasible region $OADCO$ and the point $A(0, 2)$ and the point $D(1, 2)$ give the required integer solution given by

(i) $x_1 = 0$, $x_2 = 2$, max. $Z = 6$
(ii) $x_1 = 1$, $x_2 = 2$, max. $Z = 6$

10.5 PROBLEM ON MIXED INTEGER LINEAR PROGRAMMING

There are some situations in which only some variables are restricted to be an integer. To solve such type of problem, we proceed as follows:

STEP 1. Solve the given problem as regular LPP by simplex method by ignoring integer condition.

STEP 2. Use Gomory's technique corresponding to the integer variables one by one as given in step 3.

STEP 3. If only one of the integer restricted variables has the fractional value then corresponding to the row in which this fractional variables lies in the optimal simplex table from the Gomory's constraints as given below:

$$- \sum_{j \in \mathbf{R}^+} y_{ij} x_j - \left(\frac{f_{B_i}}{f_{B_i} - 1} \right) \sum_{j \in \mathbf{R}^-} y_{ij} x_j \leq -f_{B_i}$$

where $x_{B_i} = [x_{B_i}] + f_{B_i}$

$[x_{B_i}] =$ largest integral part of x_{B_i}

$f_{B_i} =$ largest fractional part of x_{B_i} such that $0 \leq f_{B_i} < 1$

$\mathbf{R}^+ = [j : y_{ij} \geq 0]$ $\mathbf{R}^- = [j : y_{ij} < 0]$

and $\mathbf{R} = [\mathbf{R}^-, \mathbf{R}^+]$ set of indices corresponding to all non-basic variables.
Here, the Gomory's cutting plane is given by

$$- \sum_{j \in \mathbf{R}^+} y_{ij} x_j - \left(\frac{f_{B_i}}{f_{B_i} - 1} \right) \sum_{j \in \mathbf{R}^-} y_{ij} x_j + s_i = -f_{B_i}$$

where s_i is the slack variables.

 REMARK

- Here the value of the objective function in the optimal solution of mixed integer programming is always superior to or at least equal to that of all integer LPP and is always inferior to or euqal to that of the original LPP.

Solved Examples

EXAMPLE 1. *Solve the following LPP by Gomory's technique.*

$$\textbf{\textit{Max. Z = 4x}}_1 \textbf{\textit{ + 6x}}_2 \textbf{\textit{ + 2x}}_3$$

subject to the constraints

$$4x_1 + 4x_2 \leq 5$$
$$- x_1 + 6x_2 \leq 5$$
$$- x_1 + x_2 + x_3 \leq 5$$

and $x_1, x_2, x_3 \geq 0$ *and are integers.*

SOLUTION. The given maximization problem is in standard form. Introducing slack variables s_1, s_2, s_3 in the given LPP, we get

$$\text{Max } Z = 4x_1 + 6x_2 + 2x_3 + 0s_1 + 0s_2 + 0s_3$$
s.t.

$$
\left.
\begin{aligned}
4x_1 - 4x_2 + s_1 &= 5 \\
-x_1 + 6x_2 + s_2 &= 5 \\
-x_1 + x_2 + x_3 + s_3 &= 5 \\
\text{and} \quad x_1, x_2, x_3, s_1, s_2, s_3 &\geq 0
\end{aligned}
\right] \quad \ldots(1)
$$

Apply the simplex method in a usual manner, we get the following simplex table.

B.V.	C_B	X_B	x_1	x_2	x_3	s_1 (x_4)	s_2 (x_5)	s_3 (x_6)	Min Ratio
$c_j \rightarrow$			4	6	2	0	0	0	
s_1	0	5	4	-4	0	1	0	0	—
s_2	0	5	-1	⑥	0	0	1	0	5/6 (min)$\rightarrow$
s_3	0	5	-1	1	1	0	0	1	5/1
$Z = C_B X_B = 0,\ \Delta_j \rightarrow$			4	6	2	0	0	0	
s_1	0	25/3	10/3	0	0	1	2/3	0	5/2 (min)$\rightarrow$
x_2	6	5/6	$-1/6$	1	0	0	1/6	0	
s_3	0	25/6	$-5/6$	0	1	0	$-1/6$	1	
$Z = C_B X_B = 5,\ \Delta_j \rightarrow$			5	0	2	0	-1	0	
x_1	4	5/2	1	0	0	3/10	1/5	0	
x_2	6	5/4	0	1	0	1/20	1/5	0	
s_3	0	25/4	0	0	1	1/4	0	1	
$Z = C_B X_B = 35/2,\ \Delta_j \rightarrow$			0	0	2	$-3/2$	-2	0	
x_1	4	5/2	1	0	0	3/10	1/5	0	
x_2	6	5/4	0	1	0	1/20	1/5	0	
x_3	2	25/4	0	0	1	1/4	0	1	
$Z = C_B X_B = 30,\ \Delta_j \rightarrow$			0	0	0	-2	-2	-2	

Here, we observe that all $\Delta_j \leq 0$ and $x_1 = \dfrac{5}{2}, x_2 = \dfrac{5}{4}, x_3 = \dfrac{25}{4}$ and max. $Z = 25$.

Therefore, solution is optimal, feasible and non-integer.

Here, x_1, x_3 are constraint to be an integer.

Now, $x_1 = \dfrac{5}{2} = 2 + \dfrac{1}{2}, x_3 = \dfrac{25}{4} = 6 + \dfrac{1}{4}$

$\Rightarrow \qquad\qquad f_{B_1} = \dfrac{1}{2}, f_{B_3} = \dfrac{1}{4}$

$\Rightarrow \qquad\qquad f_{B_1} > f_{B_3}$

$\Rightarrow x_1$-row is selected to construct first Gomory's constraints

First Gomory's constraints

$\because \qquad -\sum\limits_{j=m+1}^{n} f_{ij} x_j \leq -f_{B_i}$

Here, $i = 1, m = 3, n = 6$

$\qquad -\sum\limits_{j=4}^{6} f_{ij} x_j \leq -f_{B_1} \qquad \Rightarrow \qquad -f_{14} x_4 - f_{15} x_5 - f_{16} x_6 \leq -f_{B_1}$

From x_1-row, we have

$$a_{14} = \frac{3}{10} \quad \Rightarrow \quad f_{14} = \frac{3}{10}$$

$$a_{15} = \frac{1}{5} \quad \Rightarrow \quad f_{15} = \frac{1}{5}$$

$$a_{16} = 0 \quad \Rightarrow \quad f_{16} = 0$$

and $x_4 = s_1, x_5 = s_2, x_6 = s_3$

$$\Rightarrow \quad -\frac{3}{10}s_1 - \frac{1}{5}s_2 \le -\frac{1}{2}, \text{ which is the first Gomory's constraints.}$$

Now introducing a non-negative slack variable s' to the last table of the above simplex table we get the table as given below.

B.V.	C_B	X_B	$c_j \rightarrow$ 4 x_1 (β_1)	6 x_2 (β_2)	2 x_3 (β_3)	0 s_1 (x_4)	0 s_2 (x_5)	0 s_3 (x_6)	0 s' (β_4)
x_1	4	5/2	1	0	0	3/10	1/5	0	0
x_2	6	5/4	0	1	0	1/20	1/5	0	0
x_3	2	25/4	0	0	1	1/4	0	1	0
s'	0	−1/2	0	0	0	−3/10	−1/5	0	1
$Z = C_B X_B = 30, \Delta_j \rightarrow$			0	0	0	−2	−2	−2	0
						↑			↓

Clearly, $s' = -\frac{1}{2} < 0$. Thus, we apply dual simplex method.

Here, s' $(= \beta_4)$ will be the outgoing vector.

$$\Rightarrow \qquad r = 4$$

Then by minimum ratio rule, we have

$$\frac{\Delta_k}{a_{rk}} = \min_j \left\{ \frac{\Delta_j}{a_{rj}}, a_{rj} < 0 \right\}$$

$$\Rightarrow \qquad \frac{\Delta_k}{a_{4k}} = \min_j \left\{ \frac{\Delta_j}{a_{4j}}, a_{4j} < 0 \right\} = \min \left\{ \frac{\Delta_4}{a_{44}}, \frac{\Delta_5}{a_{45}} \right\}$$

$$= \min \left\{ \frac{-2}{-3/10}, \frac{-2}{-1/5} \right\} = \min \left\{ \frac{20}{3}, 10 \right\}$$

$$= \frac{20}{3} = \frac{\Delta_4}{a_{44}}$$

$$\Rightarrow \quad k = 4, \text{ i.e., } x_4 \ (= s_1) \text{ is the incoming vector.}$$

and key element $= a_{44} = \dfrac{-3}{10}$

Now proceeding as usual, we get the following simplex table.

B.V.	C_B	X_B	$c_j \rightarrow$ x_1 (β_1)	x_2 (β_2)	x_3 (β_3)	s_1 (β_4)	s_2 (x_5)	s_3 (x_6)	s' (x_7)	Min Ratio
			4	6	2	0	0	0	0	
x_1	4	2	1	0	0	0	0	0	1	
x_2	6	7/6	0	1	0	0	1/6	0	1/6	
x_3	2	35/6	0	0	1	0	–1/6	1	5/6	
s_1	0	5/3	0	0	0	1	2/3	0	–10/3	
$Z=C_BX_B=80/3$,		$\Delta_j \rightarrow$	0	0	0	0	–2/3	–2	–20/3	

Here, we observe that all $\Delta_j \leq 0$ and $x_1 = 2, x_2 = \dfrac{7}{6}, x_3 = \dfrac{35}{6}$ and $\max.Z = \dfrac{80}{3}$.

Clearly this solution is optimal, feasible but not integer.

Now, select x_3-row to construct second Gomory's constraints

$$\because \qquad x_3 = \frac{35}{6} = 5 + \frac{5}{6} \qquad\qquad \Rightarrow \qquad f_{B_3} = \frac{5}{6}$$

Now, $-\sum\limits_{j=m+1}^{n} f_{ij} \cdot x_j \leq -f_{B_i}$

Here, $i = 3, m = 4, n = 7$

So, $-\sum\limits_{j=5}^{7} f_{3j}x_j \leq -f_{B_3}$

$$\Rightarrow \qquad -f_{35}x_5 - f_{36}x_6 - f_{37}x_7 \leq -\frac{5}{6}$$

Clearly, from x_3-row above table, we have

$$a_{35} = -\frac{1}{6} = -1 + \frac{5}{6} \qquad\qquad \Rightarrow \qquad f_{35} = \frac{5}{6}$$

$$a_{36} = 1 \qquad\qquad\qquad\qquad \Rightarrow \qquad f_{36} = 0$$

$$a_{37} = \frac{5}{6} \qquad\qquad\qquad\qquad \Rightarrow \qquad f_{37} = \frac{5}{6}$$

$x_5 = s_2, x_6 = s_3, x_7 = s'$

So, second Gomory's constraint is given by

$$-\frac{5}{6}s_2 - \frac{5}{6}s' \leq -\frac{5}{6}$$

Now introducing a non-negative slack variable s'', the above constraint becomes

$$-\frac{5}{6}s_2 - \frac{5}{6}s' + s'' = -\frac{5}{6}$$

Also by adding the constraint to the above table, we get the following table.

B.V.	C_B	X_B	x_1	x_2	x_3	s_1	s_2	s_3	s'	s''
$c_j \rightarrow$			4	6	2	0	0	0	0	0
x_1	4	2	1	0	0	0	0	0	1	0
x_2	6	7/6	0	1	0	0	1/6	0	1/6	0
x_3	2	35/6	0	0	1	0	−1/6	1	5/6	0
s_1	0	5/3	0	0	0	1	2/3	0	−10/3	0
s''	0	−5/6	0	0	0	0	−5/6	0	−5/6	1
$Z = C_B X_B = 80/3$, $\Delta_j \rightarrow$			0	0	0	0	−2/3	−2	−20/3	0
							↑			

We observe that $s'' = -\dfrac{5}{6} < 0$, then dual simplex method can be applied to get the new optimal solution.

By dual simplex method, the vector s'' ($= \beta_5$) will leave the basis.

Clearly, β_5 is the outgoing vector then $r = 5$.

Then, by minimum ratio rule

$$\frac{\Delta_k}{a_{rk}} = \min_j \left\{ \frac{\Delta_j}{a_{rj}}, a_{rj} < 0 \right\}$$

$$\Rightarrow \qquad \frac{\Delta_k}{a_{5k}} = \min_j \left\{ \frac{\Delta_j}{a_{5j}}, a_{5j} < 0 \right\} = \min \left\{ \frac{\Delta_5}{a_{55}}, \frac{\Delta_7}{a_{57}} \right\}$$

$$= \min \left\{ \frac{-2/3}{-5/6}, \frac{-20/3}{-5/6} \right\} = \min \left\{ \frac{4}{5}, 8 \right\} = \frac{4}{5} = \frac{\Delta_5}{a_{55}}$$

$$\Rightarrow \qquad k = 5$$

$$\Rightarrow \qquad x_5 \ (= s_2) \text{ is the incoming vector and key element} = a_{55} = \frac{5}{6}$$

Again apply simplex method in a usual manner, we have the following simplex table.

B.V.	C_B	X_B	x_1	x_2	x_3	s_1	s_2	s_3	s'	s''
$c_j \rightarrow$			4	6	2	0	0	0	0	0
			(β_1)	(β_2)	(β_3)	(β_4)	(β_5)	(x_6)	(x_7)	(x_8)
x_1	4	2	1	0	0	0	0	0	1	0
x_2	6	1	0	1	0	0	0	0	0	1/5
x_3	2	6	0	0	1	0	0	1	1	−1/5
s_1	0	1	0	0	0	1	0	0	− 4	4/5
s_2	0	1	0	0	0	0	1	0	1	−6/5
$Z = C_B X_B = 26$, $\Delta_j \rightarrow$			0	0	0	0	0	−2	−6	− 4/5

Here, we observe that all $\Delta_j \leq 0$ and $x_1 = 2, x_2 = 1, x_3 = 6$ and Max. $Z = 26$ which shows that the solution is optimal feasible and integer.

10.6 THE BRANCH AND BOUND TECHNIQUE

This technique is applicable to both the IPP, pure as well as mixed and involve the continuous integer programming problem ignoring the integer valued condition. The branch and bound technique was originally developed by **A.H. Land** and **A.G. Doig** and later was modified by **R.S. Dakin**.

In this method, we first obtain optimum solution in a usual manner, then divide the feasible region into smaller regions by deleting parts that contain no feasible integer solution.

This is the most general technique for the solution of an IPP in which only a few or all the variables are constraints by their upper or lower bound or both.

In a maximization problem, the value of the objective function at the LPP optimum will always be an upper bound on the optimal integer programming objective. Also, any integer feasible point is always a lower bound on the optimal LP objective value. This is the basic idea behind the development of Branch and bound technique.

 Working Procedure

To solve the given integer programming problem by Branch and bound technique, we use the following steps:

STEP 1. Solve the given integer programming problem by ignoring integer condition on the variables by graphical or simplex method.

STEP 2. Check the following points of optimality
   (i) If the variables in the optimum solution are all integers, then this is the required solution.
   (ii) If the value of the some variables in the optimum solution is not an integer then go to the next step.

STEP 3. If some variables say x_j, is not an integer then
$$[x_j] < x_j < [x_j] + 1, \text{ where } [\cdot] \text{ is the integral value.}$$
But any feasible integer value of x_j must satisfy one of the two conditions given by
$$x_j \leq [x_j] \text{ or } x_j \geq [x_j] + 1$$
Now, since the variables has no integer value between $[x_j]$ and $[x_j] + 1$ then these two conditions are mutually exclusive and when they are applied separately to the given LPP, we get two different subproblems (called nodes)
   (i) **Subproblem I:** Given LPP with $x_j \leq [x_j]$.
   (ii) **Subproblem II:** Given LPP with $x_j \geq [x_j] + 1$.

STEP 4. Solve both the subproblems and test the integrability of the optimal solution and observe that
   (i) if the optimal solution of both the subproblems are integral valued, then required solution is that which give larger value of the objective function.
   (ii) if any one subproblem has integral valued optimal solution and the other subproblem has no feasible optimal solution, then the required solution is that of the subproblem having integral valued optimal solution.
   (iii) if one subproblem has optimal integral valued and the other subproblem has non-integer valued solution then record the integral valued solution and repeat step 3 and step 4 to the other subproblem till all the integral valued solutions are recorded.

STEP 5. From all the recorded integral valued solutions, choose that integral valued solution which gives the largest value of the objective function. This is the required optimal solution of the given IPP.

- The above technique is easily extended to solve mixed integer problem. Subdivision then are generated solely by the integral variables.
- If any subdivisions needs not to be subdivided then such subdivision is comprehend.
- If the given IPP is of minimization the procedure remains the same except that upper bound are used. Therefore, the value of the first integer valued solution becomes an upper bound for the given problem and the programs are eliminated when their objective function values are greater than the current upper bound.

Solved Examples

EXAMPLE 1. *Use Branch and bound technique, solve the following LPP:*

$$Max.\ Z = 7x_1 + 9x_2$$

subject to the constraints

$$-x_1 + 3x_2 \leq 6$$
$$7x_1 + x_2 \leq 35$$
$$x_1 \leq 7$$
$$x_2 \leq 7$$

and $x_1, x_2 \geq 0$ *and are integers.*   [KANPUR–1996]

SOLUTION. Ignoring integer condition, we may solve the given problem by graphical method. Here, we have the following graph.

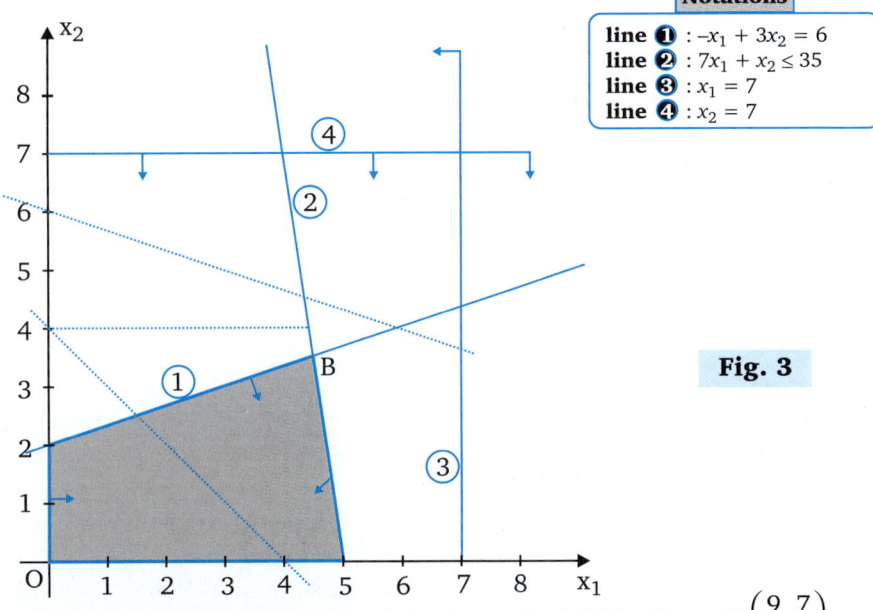

Notations

line ❶ : $-x_1 + 3x_2 = 6$
line ❷ : $7x_1 + x_2 \leq 35$
line ❸ : $x_1 = 7$
line ❹ : $x_2 = 7$

Fig. 3

The optimal solution of the given problem is obtained at the point $B\left(\dfrac{9}{2}, \dfrac{7}{2}\right)$ at B

and Max. $Z = 63$ which is taken as an upper bound of Z.

Clearly, the optimal solution is not integer valued

i.e., $x_1 = \dfrac{9}{2}, x_2 = \dfrac{7}{2}$

Let us suppose x_1 is selected for branching, then

$$x_1 = \dfrac{9}{2}$$

$\Rightarrow$ $4 < x_1 < 5$

Now, we have following two subproblems.

subproblem-1	subproblem-2
Max. $Z = 7x_1 + 9x_2$	Max. $Z = 7x_1 + 9x_2$
s.t.	s.t.
$-x_1 + 3x_2 \leq 6$	$-x_1 + 3x_2 \leq 6$
$7x_1 + x_2 \leq 35$	$7x_1 + x_2 \leq 35$
$x_1 \leq 4$	$5 \leq x_1 \leq 7$
$x_2 \leq 7$	$x_2 \leq 7$
and $x_1, x_2 \geq 0$ and are integers	and $x_1, x_2 \geq 0$ and are integers

Now, we solve these subproblem by graphical method separately.

The graph of subproblem-1 is given below.

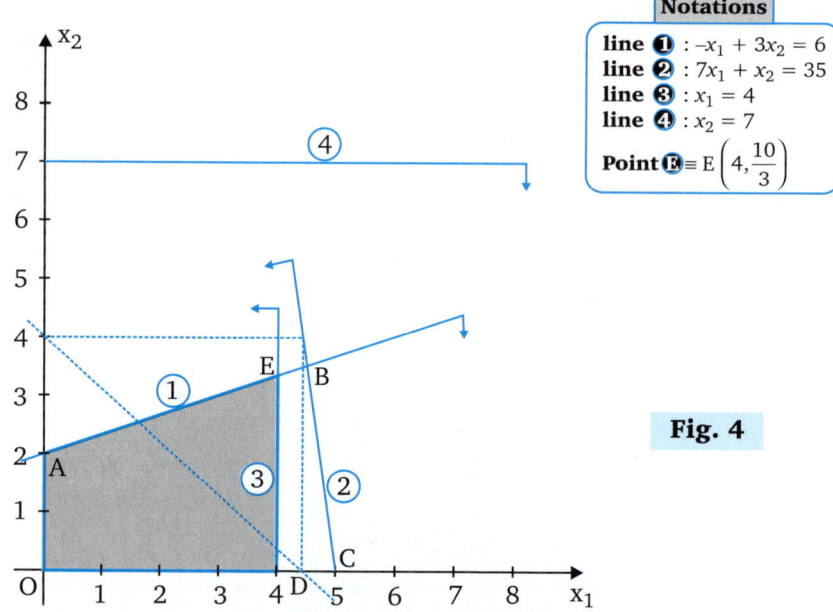

Notations

line ❶ : $-x_1 + 3x_2 = 6$
line ❷ : $7x_1 + x_2 = 35$
line ❸ : $x_1 = 4$
line ❹ : $x_2 = 7$

Point ❸ ≡ E $\left(4, \dfrac{10}{3}\right)$

Fig. 4

From the above graph, the optimal solution of subproblem-1 is given by

$$x_1 = 4, x_2 = \dfrac{10}{3}, \text{Max.} Z = 58$$

Similarly, the graph of subproblem-2 is given below

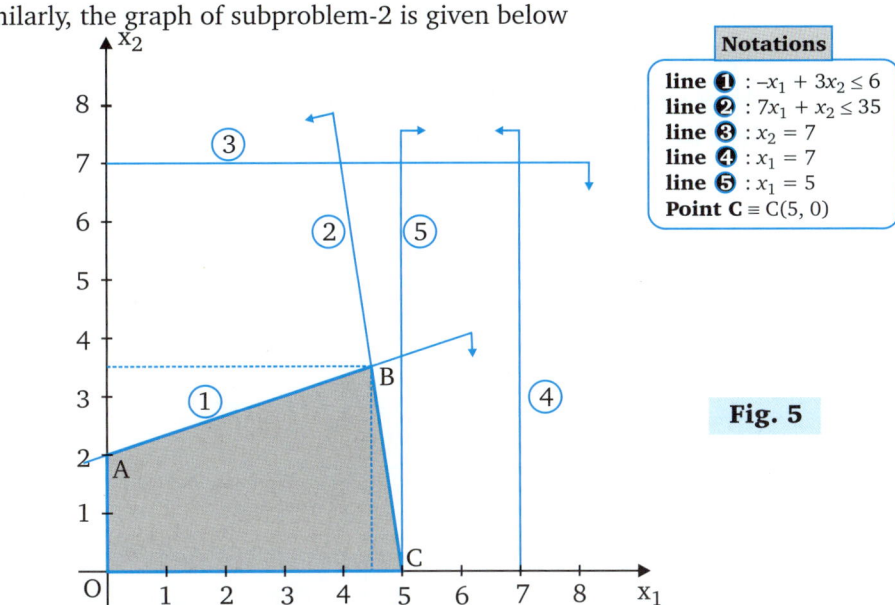

Notations

line ❶ : $-x_1 + 3x_2 \leq 6$
line ❷ : $7x_1 + x_2 \leq 35$
line ❸ : $x_2 = 7$
line ❹ : $x_1 = 7$
line ❺ : $x_1 = 5$
Point **C** ≡ C(5, 0)

Fig. 5

From the above figure the optimal solution of subproblem-2 is obtained at $B(5, 0)$ and is given by $x_1 = 5$, $x_2 = 0$ and Max. $Z = 35$

Here, we observe that subproblem-2 has integral valued solution and subproblem-1 has not integral valued solution. So, in subproblem-1, x_2 is the branching variable. Thus, subproblem-1 has two subproblems as given below.

subproblem-1(a)	subproblem-1(b)
Max. $Z = 7x_1 + 9x_2$ s.t. $\quad -x_1 + 3x_2 \leq 6$ $\quad 7x_1 + x_2 \leq 35$ $\quad\quad\quad x_1 \leq 4$ $\quad\quad\quad x_2 \leq 3$ and $x_1, x_2 \geq 0$ and are integers.	Max. $Z = 7x_1 + 9x_2$ s.t. $\quad -x_1 + 3x_2 \leq 6$ $\quad 7x_1 + x_2 \leq 35$ $\quad\quad\quad x_1 \leq 4$ $\quad\quad 4 \leq x_2 \leq 7$ and $x_1, x_2 \geq 0$ and are integers.

Now, graph of subproblem-1(a) is given as follows.

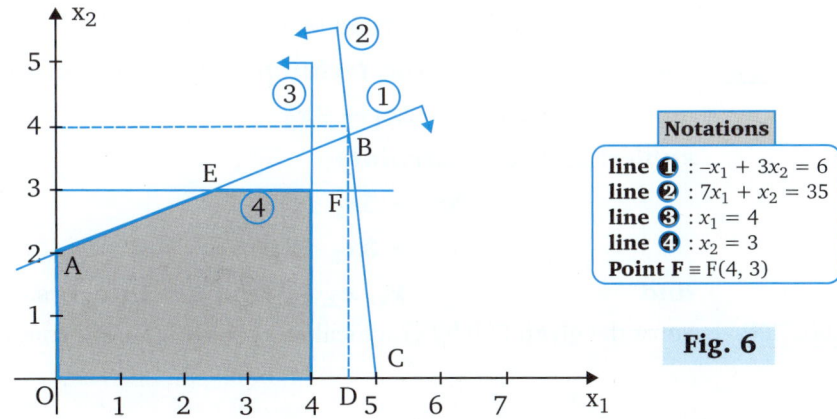

Notations

line ❶ : $-x_1 + 3x_2 = 6$
line ❷ : $7x_1 + x_2 = 35$
line ❸ : $x_1 = 4$
line ❹ : $x_2 = 3$
Point **F** ≡ F(4, 3)

Fig. 6

The optimal solution of the above problem is given at the point $F(4, 3)$ and optimal solution is given by $x_1 = 4, x_2 = 3$ and Max. $Z = 55$.

Now, we have to find the solution of subproblem 1(b).

The graph of subproblem-1(b) is given as below.

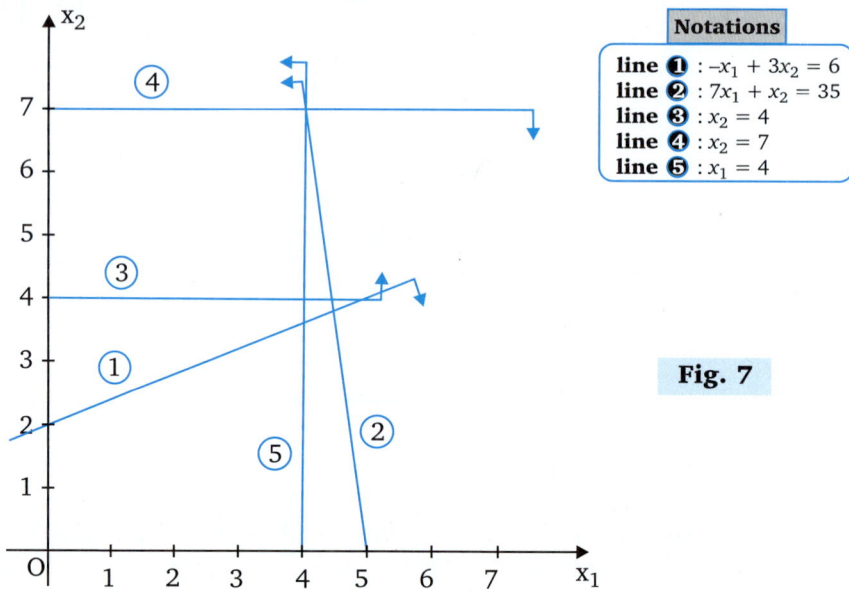

Notations

line ❶ : $-x_1 + 3x_2 = 6$
line ❷ : $7x_1 + x_2 = 35$
line ❸ : $x_2 = 4$
line ❹ : $x_2 = 7$
line ❺ : $x_1 = 4$

Fig. 7

In the above problem, no feasible region is obtained. Hence, subproblem-1(b) has no feasible solution.

Finally, we conclude that, in the subproblems we get the following integral valued solutions:

(i) $x_1 = 5, x_2 = 0$, Max. $Z = 35$

(ii) $x_1 = 4, x_2 = 3$, Max. $Z = 55$

In which the larger value of $Z = 55$

Hence, the required solution is given by

$$x_1 = 4, x_2 = 3 \text{ and Max. } Z = 55$$

EXAMPLE 2.   ***Use Branch and bound technique, solve the following LPP.***

$$\textbf{Max. } \textbf{\textit{Z}} = \textbf{2}\textbf{\textit{x}}_\textbf{1} + \textbf{3}\textbf{\textit{x}}_\textbf{2}$$

subject to the constraints

$$\textbf{6}\textbf{\textit{x}}_\textbf{1} + \textbf{5}\textbf{\textit{x}}_\textbf{2} \leq \textbf{25}$$

$$\textbf{\textit{x}}_\textbf{1} + \textbf{3}\textbf{\textit{x}}_\textbf{2} \leq \textbf{10}$$

and   ***$x_1, x_2 \geq 0$ and are integers.***   [MEERUT–1996]

SOLUTION.   Solve the given LPP by graphical method by ignoring conditions.

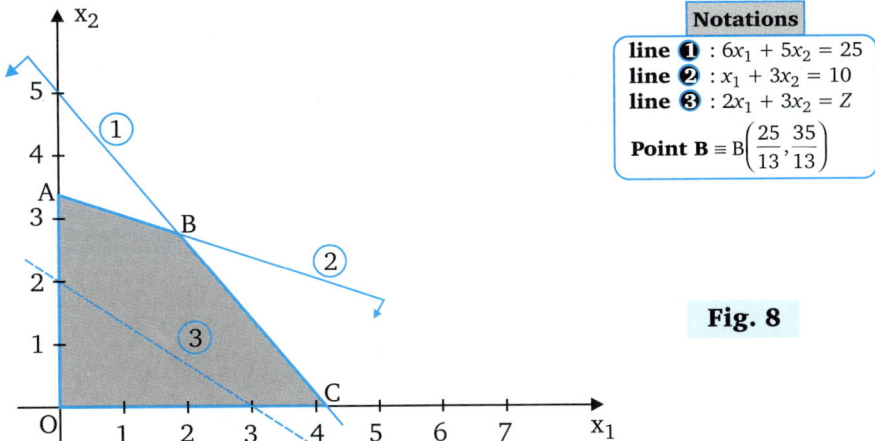

Fig. 8

Clearly, the optimal solution can be obtained at the point $B\left(\dfrac{25}{13},\dfrac{35}{13}\right)$ and is given by

$$x_1 = \frac{25}{13}, x_2 = \frac{35}{13} \text{ and Max. } Z = 12$$

We observe that both x_1 and x_2 are non-integers. Thus, select one variable arbitrarily say x_2 for branching then

$$x_2 = \frac{35}{13} \quad \Rightarrow \quad 2 < x_2 < 3$$

Now, we form the following two subproblems by adding new constraints either $x_2 \leq 2$ or $x_2 \geq 3$ to the original LPP.

subproblem-1	subproblem-2
Max $Z = 2x_1 + 3x_2$ s.t. $\quad 6x_1 + 5x_2 \leq 25$ $\quad x_1 + 3x_2 \leq 10$ $\quad\quad\quad x_2 \leq 2$ and $x_1, x_2 \geq 0$ are integers	Max $Z = 2x_1 + 3x_2$ s.t. $\quad 6x_1 + 5x_2 \leq 25$ $\quad x_1 + 3x_2 \leq 10$ $\quad\quad\quad x_2 \geq 3$ and $x_1, x_2 \geq 0$ are integers

The graph of the subproblem-1 is given as below.

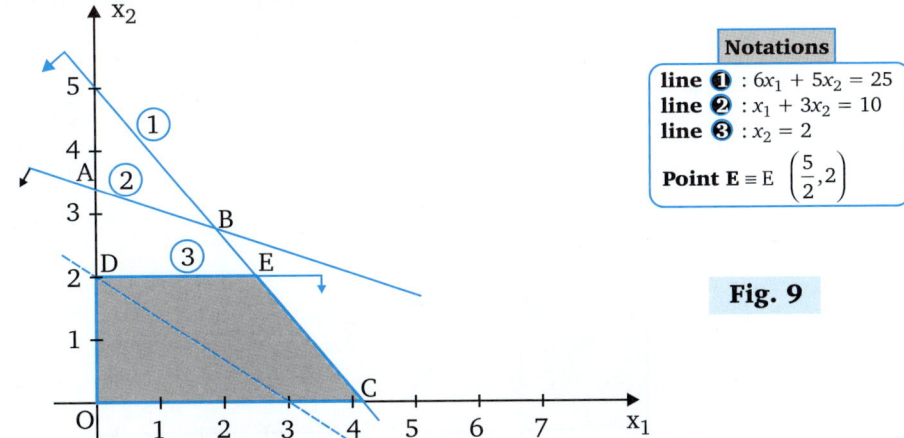

Fig. 9

Clearly, the optimal solution of the above problem is given below.

$x_1 = \dfrac{5}{2}, x_2 = 2$ and Max $Z = 11$

Now, the graph of subproblem-2 is given below.

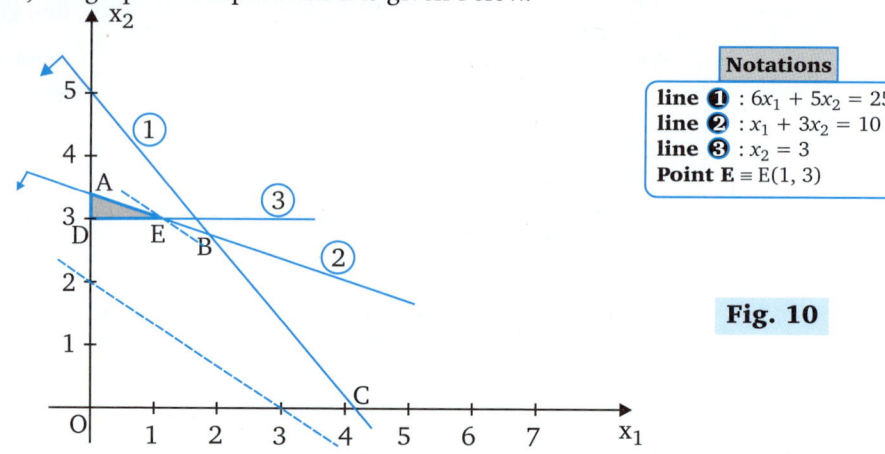

Notations

line ① : $6x_1 + 5x_2 = 25$
line ② : $x_1 + 3x_2 = 10$
line ③ : $x_2 = 3$
Point E $\equiv$ E(1, 3)

Fig. 10

From the above graph, the optimum solution of subproblem-2 is given by

$x_1 = 1, x_2 = 3$ and Max. $Z = 11$

Clearly, it is an integer valued solution and Max. $Z = 11$ is taking as lower bound of Z. (Here, no need to branch subproblem-2)

Now subproblem-1 is further divided into two subproblems as given below.

In subproblem-1 $x_1 = \dfrac{5}{2} \Rightarrow x_1 \leq \left[\dfrac{5}{2}\right]$ or $x_1 \geq \left[\dfrac{5}{2}\right] + 1$

$\Rightarrow \qquad x_1 \leq 2$ or $x_1 \geq 3$

subproblem-1(a)	subproblem-1(b)
Max. $Z = 2x_1 + 3x_2$	Max. $Z = 2x_1 + 3x_2$
subject to the constraints	subject to the constraints
$6x_1 + 5x_2 \leq 25$	$6x_1 + 5x_2 \leq 25$
$x_1 + 3x_2 \leq 10$	$x_1 + 3x_2 \leq 10$
$x_2 \leq 2$	$x_2 \leq 2$
$x_1 \leq 2$	$x_1 \geq 3$
and $x_1, x_2 \geq 0$ are integers	and $x_1, x_2 \geq 0$ are integers

The graph of subproblem-1(a) is given as below.

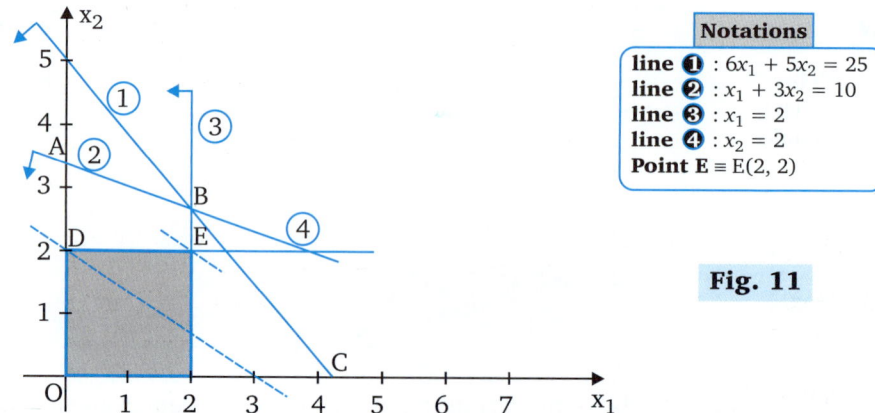

Notations

line ① : $6x_1 + 5x_2 = 25$
line ② : $x_1 + 3x_2 = 10$
line ③ : $x_1 = 2$
line ④ : $x_2 = 2$
Point E $\equiv$ E(2, 2)

Fig. 11

From the above graph, we have
$$x_1 = 2, x_2 = 2 \text{ and Max. } Z = 10$$
Similarly, the graph of subproblem-1(b) is given as below.

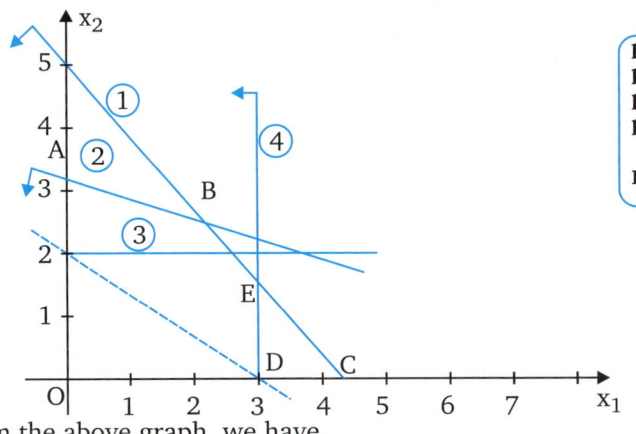

Notations

line ❶ : $6x_1 + 5x_2 = 25$
line ❷ : $x_1 + 3x_2 = 10$
line ❸ : $x_2 = 2$
line ❹ : $x_1 = 3$

Point E ≡ E$\left(3, \dfrac{7}{5}\right)$

Fig. 12

From the above graph, we have
$$x_1 = 3, x_2 = \frac{7}{5} \text{ and Max.} Z = \frac{51}{5}$$

We observe that subproblem-1(a) has integer solution with max. $Z = 10$ which is less than lower bound so that it is comprehend.

Subproblem-1(b) has one non-integer solution $x_2 = \dfrac{7}{5}$.

Now, subproblem-1(b) can be further branched with $x2$ as the branching variable but in this problem, Max. $Z = \dfrac{51}{5}$ which is less than lower bound 1, which does not assume a solution better than one already obtained. This subproblem-1(b) is comprehend.

Hence, we conclude that, the solution of subproblems are as follows.
(i) $x_1 = 1, x_2 = 3$ and max. $Z = 11$ for subproblem 2
(ii) $x_1 = 2, x_2 = 2$ and max. $Z = 10$ for subproblem 1(a)

Clearly, the larger $Z = 11$

Hence, the required solution is given by
$$x_1 = 1, x_2 = 3 \text{ and Max. } Z = 11$$

Exercise-10.1

Using Gomory's cutting plane method solve the following integer programming problems.

1. Max. $Z = x_1 + x_2$
   s.t. $3x_1 + 2x_2 \le 5$
          $x_2 \le 5$
   and $x_1, x_2 \ge 0$ are integer.

   [MADRAS(BE)–1990]

2. Max. $Z = 2x_1 + x_2$
   s.t. $2x_1 + 5x_2 \le 17$
          $3x_1 + 2x_2 \le 10$
   and $x_1, x_2 \ge 0$ are integers.

3. Max. $Z = 3x_1 + 4x_2$
   s.t.
   $x_1 + x_2 \le 4$
   $\dfrac{3}{5}x_1 + x_2 \le 3$
   and $x_1, x_2 \ge 0$ are integers.

4. Max. $Z = 4x_1 + 3x_2$
   s.t.
   $x_1 + 2x_2 \le 4$
   $2x_1 + x_2 \le 6$
   and $x_1, x_2 \ge 0$ are integers.

5. Max. $Z = x_1 + 5x_2$
s.t. $x_1 + 10x_2 \leq 20$
           $x_1 \leq 2$
and $x_1, x_2 \geq 0$ and are integers.

6. Min. $Z = 20x_1 + 22x_2 + 18x_3$
s.t.
   $4x_1 + 6x_2 + x_3 \geq 54$
   $4x_1 + 4x_2 + 6x_3 \geq 65$
   $0 \leq x_1, x_2, x_3 \leq 7$
and x_1, x_2, x_3 are integers.

7. Max. $Z = 2x_1 + 10x_2 + x_3$
s.t.
   $5x_1 + 2x_2 + x_3 \leq 15$
   $2x_1 + x_2 + 7x_3 \leq 20$
   $x_1 + 3x_2 + 2x_3 \leq 25$
and $x_1, x_2, x_3 \geq 0$ and are integers.

8. Max. $Z = 3x_1 + 2x_2 + 5x_3$
s.t.
   $5x_1 + 3x_2 + 7x_3 \leq 28$
   $4x_1 + 5x_2 + 5x_3 \leq 30$
and $x_1, x_2, x_3 \geq 0$ are integers.

9. Max. $Z = 3x_1 + 4x_2$
s.t. $3x_1 + 2x_2 \leq 8$
     $x_1 + 4x_2 \geq 10$
and $x_1, x_2 \geq 0$ are integers.

10. Max. $Z = x_1 + x_2$
s.t. $2x_1 + 5x_2 \leq 16$
     $6x_1 + 5x_2 \leq 30$
and $x_1, x_2 \geq 0$ and x_1 is an integer.

[PUNJAB(ME(Mech.))–1986]

Using branch and bound technique, solve the following LPP:

11. Max. $Z = x_1 + x_2$
s.t.
     $3x_1 + 2x_2 \leq 12$
         $x_2 \leq 2$
and $x_1, x_2 \geq 0$ are integers

12. Max. $Z = 6x_1 + 8x_2$
s.t.
  $4x_1 + 16x_2 \leq 32$
  $14x_1 + 4x_2 \leq 28$
and $x_1, x_2 \geq 0$ are integers.

13. Max. $Z = 3x_1 + 4x_2$
s.t.
  $7x_1 + 16x_2 \leq 52$
   $3x_1 - 2x_2 \leq 18$
and $x_1, x_2 \geq 0$ are integers.

14. Max. $Z = x_1 + 4x_2$
s.t.
     $x_1 + x_2 \leq 5$
  $10x_1 + 6x_2 \leq 45$
and $x_1, x_2 \geq 0$
Use x_1 as branching variable.

17. Max. $Z = 3x_1 + 2x_2$
s.t.
  $2x_1 + 2x_2 \leq 7$
       $x_1 \leq 2$
       $x_2 \leq 2$
and $x_1, x_2 \geq 0$ are integers.

18. Max. $Z = 3x_1 + 4x_2$
s.t. $3x_1 - x_2 + x_3 = 12$
   $3x_1 + 11x_2 + x_4 = 66$
and $x_1, x_2, x_3, x_4 \geq 0$ are integers.

19. Max. $Z = x_1 + x_2$
s.t.
   $4x_1 - x_2 \leq 10$
   $2x_1 + 5x_2 \leq 10$
   $4x_1 - 3x_2 \leq 6$
and $x_1, x_2 \geq 0$ are integers.

20. Max. $Z = 2x_1 + 20x_2 - 10x_3$
s.t. $2x_1 + 20x_2 + 4x_3 \leq 15$
   $6x_1 + 20x_2 + 4x_3 = 20$
and $x_1, x_2, x_3 \geq 0$ are integers.

ANSWERS

1. $x_1 = 0, x_2 = 2$, Max. $Z = 2$ **2.** $x_1 = 3, x_2 = 0$, Max. $Z = 6$ **3.** $x_1 = 3, x_2 = 1$, Max. $Z = 13$

4. $x_1 = 3, x_2 = 0$, Max. $Z = 12$ **5.** $x_1 = 2, x_2 = 1$, Max. $Z = 7$ **6.** $x_1 = 2, x_2 = 7, x_3 = 5$, Min $Z = 284$

7. $x_1 = 0, x_2 = 7, x_3 = 1$, Max. $Z = 71$ **8.** $x_1 = x_2 = 0, x_3 = 4$, Max. $Z = 20$

9. $x_1 = 0, x_2 = 4$, Max. $Z = 16$ **10.** $x_1 = 4, x_2 = \dfrac{6}{5}$, Max. $Z = \dfrac{26}{5}$

11. $x_1 = 2, x_2 = 2; x_1 = 3, x_2 = 1; x_1 = 4, x_2 = 0$ and Max. $Z = 4$

12. $x_1 = 0, x_2 = 2$, Max. $Z = 16$ **13.** $x_1 = 5, x_2 = 8$, Max. $Z = 19$

14. $x_1 = 1, x_2 = 1$, Max. $Z = 5$ **15.** $x_1 = 2, x_2 = 8$, Max. $Z = 10$ **16.** $x_1 = 3, x_2 = 2$, Max. $Z = 23$

17. $x_1 = 2, x_2 = 2$, Max. $Z = 10$ **18.** $x_1 = 5, x_2 = 4, x_3 = x_4 = 0$, Max. $Z = 31$

19. $x_1 = 2, x_2 = 1$, Max. $Z = 3$ **20.** $x_1 = 2, x_2 = 0, x_3 = 2$, Max. $Z = -16$

1. What do you mean by integer linear programming.
2. Explain all integer linear programming.
3. Discuss the need of integer linear programming.
4. Write the general form of integer linear programing.
5. Write a short note on Integer linear programming.
6. Write the short note on branch and bound technique. [MEERUT–2008, 10, 12]
7. Write the short note on Gomory's cutting method.

MULTIPLE CHOICE QUESTIONS (CHOOSE THE MOST APPROPRIATE ONE)

1. In mixed-integer programming problem: [MEERUT–2013]
   (a) all of decision variables require integer solution
   (b) few of the decision variable require integer solution
   (c) all decision variables are restricted to integer values of either 0 or 1
   (d) all of the above

2. In an I.P.P. rounding off solution values of decision variables in L.P.P. may not be acceptable, because: [MEERUT–2013]
   (a) it may violates non-negative condition
   (b) it does not satisfy the constraints
   (c) objective function value of the I.P.P. is more than the Objective function of the L.P.P.
   (d) none of the above

3. An Integer Programming Problem is a LPP with: [MEERUT–2013]
   (a) some variables take non-negative integer
   (b) all variables take non-negative integer
   (c) some/all variables take non-negative integer
   (d) none of the above

4. Which of the following is not correct? [MEERUT–2013]
   (a) an IPP that have only one constraint is called Knapsack problem
   (b) an IPP that has no constraint is known as a Knapsack problem
   (c) a travelling salesman problem can be solved using Branch and Bound method
   (d) variables in an IPP that are not integer constrained are called continuous variables

5. In the Branch and Bound approach to a maximization integer LP problem, a node is terminated if: [MEERUT–2013]
   (a) a node has an infeasible solution

(b) a node yields a solution that is feasible but not an integer
(c) upper bound is less than the current subproblem's lower bound
(d) all of the above

6. Branch and Bound algorithm is applicable to: [MEERUT–2013]
   (a) mixed I.P.P.   (b) all I.P.P.
   (c) both (a) and (b)   (d) none of the above

7. In bounded variable algorithm: [MEERUT–2013]
   (a) lower bound of a decision variable can never be converted into non-negative decision variable
   (b) upper bound of a decision variable can always be converted into non-negative decision variable
   (c) lower and Upper bounds of a decision variable are 0 and ∞ respectively in the case of unbounded variables
   (d) the lower bound constraints $l_j \le x_j$ is converted as $x'_j (= x_j - l_j) \le 0$, where x_j are slack variables

8. Lower bound constraints are handled by substituting: [MEERUT–2013]
   (a) $x_j = u_j - x'_j$ (b) $x_j = u_j + x'_j$
   (c) $x_j = l_j + x'_j$ (d) $x_j = l_j - x'_j$

9. Branch and Bound method divides the feasible solution space into smaller parts by:
   (a) enumerating   (b) bounding
   (c) branching   (d) all of the above

10. The use of Cutting plane method: [MEERUT–2013]
    (a) require the use of standard linear programming approach between each cutting plane application
    (b) yields better value of the objective function
    (c) reduces the number of constraints in the given problem
    (d) both (b) and (c)

11. The 0-1 integer programming problem:

(a) requires the decision variables to have values between Zero and One

(b) requires that the constraints all have coefficients between Zero and One

(c) requires that the decision variables have coefficients between Zero and One

(d) all of the above

12. The part of the feasible solution space eliminated by putting a cut contains:

(a) only non integer solution

(b) only integer solution

(c) both (a) and (b)

(d) none of the above

13. While solving IP problem any non-integer variable in the solution is picked upto:

(a) obtain the cut constraint

(b) enter the solution

(c) leave the solution

(d) none of the above

14. Rounding off solution values of decision variables in an LP problem may not be acceptable because:

(a) it does not satisfy constraints

(b) it violates non-negativity conditions

(c) objective function value is less than the objective function value of LP

(d) none of the above

15. Which of the following is the consequence of adding a new cut constraint to an optimal simplex table:

(a) addition of a new variable to the table

(b) makes the previous optimal solution infeasible

(c) eliminates non-integer solution from the solution space

(d) all of the above

16. A non-integer variable is chosen in the optimal simplex table of integer LP problem to:

(a) leaves the basis

(b) enter the basis

(c) to construct the Gomory cut

(d) none of the above

17. The corners of the reduced feasible region of integer LP problem contains:

(a) only integer solution

(b) optimal integer solution

(c) only non-integer solution

(d) all of the above

18. In a branch and bound minimization tree, the lower bounds on Objective function value:

(a) do not decrease in value

(b) do not increase in value

(c) remain constant

(d) None of these

19. Modifications made for the mixed integer cutting plane method are:

(a) top most rows of the simplex table contains integer variables

(b) values of the Objective function is bounded

(c) row corresponding to an integer variable serve as a source row

(d) all of the above

20. The situation of multiple solutions arises with:

(a) cutting plane method

(b) branch and bound method

(c) both (a) and (b)

(d) none of these

21. While applying cutting plane method, dual simplex is used to maintain:

(a) optimality   (b) feasibility

(c) both (a) and (b)   (d) none of the above

ANSWERS

1. (c)	**2.** (d)	**3.** (b)	**4.** (c)	**5.** (c)	**6.** (c)	**7.** (c)	**8.** (c)	**9.** (d)
10. (a)	**11.** (a)	**12.** (a)	**13.** (a)	**14.** (d)	**15.** (d)	**16.** (c)	**17.** (a)	**18.** (b)
19. (c)	**20.** (d)	**21.** (b)						

ARCHIVE

1. The owner of a readymade garments store sells two types of shirts-Zee-shirts and Button-down shirts. He make a profit of ₹ 3 and ₹ 12 per shirt on Zee-shirts and Button-down shirts, respectively. He has two tailors, A and B at his disposal to stitch the shirts. Tailors A and B can devote at the most 7 hours and 15 hours per day, respectively Both these shirts are to be stitched by both the tailors. Tailors A and B spend 2 hours and 5 hours, respectively in stiching one zee-shirt and 4 hours and 3 hours, respectively in stiching a Button-down shirt. How many shirts of both types should be stitched in order to maximize daily profit?

(a) Formulate and solve this problem as an L.P. problem.

(b) If the optimal solution is not integer-valued, use Gomory technique to derive the optimal integer solution.

[DELHI(MBA)–1995]

2. Solve the following mixed-integer programming problem :

Max. $Z = -3x_1 + x_2 + 3x_3$

Subject to the coinstrits,

$-x_1 + 2x_2 + x_3 \le 4,$

$2x_2 - \dfrac{3}{2}x_3 \le 1,$

$x_1 - 3x_2 + 2x_2 \le 3.$

and $x_1, x_2, \ge 0$, x_3 non-negative integer.

[PUNJAB(ME(MECH.))–1987]

3. A manufacturer of toys makes two types of toys, A and B. Processing of these two toys is done on two machines X and Y. The toy A requires 2 hours on machine X and 5 hours on machine Y. There are 16 hours of time per day available on machine X and 30 hours on machine Y. The profit obtained on both the toys is the same, c.c. ₹ 5 per toy. Formulate and solve this problem as an integer LP problem to determine the daily production of each of the two toys? [DELHI(MBA)–1998]

4. A company produces two products A and B. Each unit of product A requires one hour of engineering services and five hours of machine time. To produce one unit of product B, two hours of engineering and 8 hours of machine time are needed. There are 100 hours of engineering and 400 hours of machine time available. The cost of production is a non-linear function of the quality produced as given in the following table :

Product A		Product B	
Production (Units)	Unit cost (₹)	Production (Units)	Unit cost (₹)
0-50	10	0-40	7
50-100	8	40-100	3

The unit selling price of product A is ₹ 12 and of product B is ₹ 14. The company would like a production plan which gives the number of units of A and the number of units of B to be produced that will maximize profit. Formulate and solve this problem as an integer linear programming problem help the company to maximize the total revenul.

[DELHI(MBA)–1995, 99]

5. The non-inter optimal solution of a maximization LP problem is given below. Find the integer optimal solution in which x_1 is

integer.

$C_j \rightarrow$		-1	2	-1	0	0	0
Variables in Basis **B**	Solution values $b(= X_B)$	x_1	x_2	x_3	s_1	s_2	s_3
s_1	$\dfrac{15}{2}$	0	$-\dfrac{1}{2}$	$\dfrac{1}{2}$	1	$-\dfrac{1}{2}$	0
x_2	$\dfrac{25}{6}$	1	$\dfrac{5}{6}$	$\dfrac{1}{6}$	0	$\dfrac{1}{6}$	0
s_3	$\dfrac{35}{6}$	0	$\dfrac{13}{6}$	$\dfrac{17}{6}$	0	$-\dfrac{1}{6}$	1

[PUNJAB(ME(Mech.)–1987]

6. An air conditioning and refrigeration company has been awarded a contract for the air conditioning of a new computer installation. The company has to make a choice between two alternatives: (a) hire one or more refrigeration technicians for 6 hours a day, or (b) hire one or more part time refrigeration apprentice technicians for 4 hours a day. The wage rate of refrigeration technicians is ₹ 400 per day, while the corresponding rate for apprentice technicians is ₹ 160 per day. The company wants to engage the technicians on work for not more than 25 man hours per day and also limit the charges to technicians to ₹ 41800. The company estimates that the productivity of the refrigiration technician is 8 units and that of part time apprentice technician is 3 units. Formulate and solve this problem as integer LP problem to enable the company to select the optimal number of technicians and apprentics.

[DELHI(MBA)–1989, 97]

7. A firm makes two products : X and Y, and has total production capacity of 9 tonnes per day, X and Y requiring the same production capacity. The firm has a permanent contract to supply at least 2 tonnes of X and at least 3 tonnes of Y per day to another company. Each tonnes of X requires 20 machine-hours production time and each tonnes of Y requires 50 machine hours production time. The daily maximum possible number of machine-hours is 350. All the firms output can be sold and the profit made is ₹ 80 per tonnes of X and ₹ 120 per tonne of Y. It is required to determine the production schedule for maximum profit and to calculate this profit. [DELHI(MBA)–1995, 99]

HINTS AND ANSWERS

1. (a) Max. $Z = 3x_1 + 12x_2$

Subject to the constraints

$$2x_1 + 4x_2 \le 7$$

$$5x_1 + 3x_2 \leq 15$$

and $\quad\quad\quad\quad\quad\quad\quad\quad\quad x_1, x_2 \geq 0$

$x_1 = 0,\ x_2 = \dfrac{7}{4}$ and Max. $Z = 21$

(b) $x_1 = 1, x_2 = 1$ and Max. $Z = 15$

2. Integer LP problem expressed in its standard form and ignoring the integer requirements, the optional solution of the problem using the simplex method is obtained.

The optimal mixed integer solution is :

$$x_1 = 0, x_2 = \frac{8}{7}, x_3 = 1 \text{ and Max. } Z = \frac{29}{7}$$

3. x_1 and x_2 = number of units of toy A and B, respectively to be produced.

Max. $Z = 5x_1 + 5x_2$

Subject to the constraints

$$2x_1 + 5x_2 \leq 16,$$
$$6x_1 + 5x_2 \leq 30,$$

and $\quad\quad\quad\quad\quad\quad\quad\quad x_1, x_2 \geq 0$

$x_1 = 3, x_2 = 2$ and Max. $Z = 25$

4. x_1 and x_2 = number of units of product A and B respectively to be produced.

Max. $Z = 12x_1 - \{10x_1 - 2(x_1 - 50)\} + 14x_2 - \{7x_2 - 4(x_2 - 40)\}$

Subject to the constraints

$$x_1 + 2x_2 \leq 100,$$
$$5x_1 + 8x_2 \leq 400,$$

and $\quad\quad\quad\quad\quad\quad\quad\quad x_1, x_2 \geq 0$

5. $x_1 = 4,\ x_2 = \dfrac{1}{5}, x_3 = 0$ and Max. $Z = 24.8$

❑❑❑❑

Goal Programming 11

11.1 INTRODUCTION

Goal programming is an approach used for solving a multi objective optimization problem that balances trade off in conflicting objectives. It is an approach of deriving a best possible 'satisfactory' level of goal attainment. The business management has to achieve multiple goals or objectives *i.e.* the decision criteria of the management involves multiple goals. Hence, the goal programming assumes greater importance as a powerful tool to handle multiple decision criteria.

11.2 CONCEPTS OF GOAL PROGRAMMING

The concept of goal programming was introduced by **Charnes** and **Coopor** in 1961. They suggested a method for solving an infeasible linear programming arising from various resource constraints or goals. **Ijiri** developed the concept of different priority levels to the goals and different weights for goals at same priority level in 1965. **Lee (1972)** and **Ignizis (1976)** have written text books on the subject of goal programming.

In goal programming model, the decision variables of the model are to be different first. The goal related to the problem are to be listed down and ranked in order of priority. Since it is not possible to achieve every goal to the extent desired by the decision maker attempts are made to achieve each goal sequentially rather than simultaneously upto a satisfactory level rather than optimal level.

Thus a goal programming is often referred to as a lexico graphic method which consists of formulating an objective function in which the various goals are satisfied in order of their relative importance.

The goal programming technique is used in optimization of multiple objective goals by minimizing the derivation for each of the objective (goals) from the desired target that are set according to the priorities.

> Goal programming tries to minimize the deviations from the targets that are set according to their priorities. It begins with the most important goal and continues until the achievement of a less important goal.

11.3 GOAL PROGRAMMING MODEL FORMULATION

The fomulation of goal programming is similar to that of LP model. The main difference between L.P. and G.P. is that L.P. optimize (maximize or minimize) a single objective function whereas G.P. minimize the deviations between the target values of the objectives and the realized results.

The general form of goal programming model is as follows:

$$\text{minimize } Z = \sum_{i=1}^{m} w_i(d_i^- + d_i^+)$$

subject to the constraints

$$\sum_{j=1}^{n} a_{ij}x_j + d_i^- - d_i^+ = b_i, \quad i = 1, 2, ..., m$$

and $x_j, d_i^-, d_i^+ \geq 0 \forall i, j$

where

$d_i^- = $ negative deviation from i^{th} goal (under-achievement of the profit goal)

$d_i^+ = $ positive deviation from the ith goal (over-achievement of the profit goal)

Since, both under and over-achievement of a goal cannot be achieved simultaneously (either one or both of these deviational variables will be equal to zero *i.e.* $d_i^- \times d_i^+ = 0$. Hence, it either assumes a positive value or the other must be zero and *vice-versa*.

☞ REMARK
- The lower order goals are considered only after the higher goals are achieved.

11.3.1 SINGLE GOAL MODELS

Let us suppose one unit of effort applied to activity x_j might contribute an amount a_{ij} towards the ith goal.

If the target level for the ith level is fully achieved, the ith constraints is written as

$$\sum_{j=1}^{n} a_{ij}.x_j = b_i$$

where $d_i^- = $ negative deviation from i^{th} goal (amount below the target value)

$d_i^+ = $ positive deviation from the ith goal (amount above the target value)

In this case, the above stated i^{th} goal can be written as

$$\sum_{j=1}^{n} a_{ij}.x_j + d_i^- - d_i^+ = b_i, \quad i = 1, 2, ...n$$

(value of the objective) + (amount below the goal) – (amount above the goal) = Goal

☞ REMARKS
- The goal deviational variables must be non-negative.
- The deviational variables in goal programming model are equivalent to slack and surplus variables in linear programming model.
- The deviational variable d_i^+ is removed from the objective function of goal programming which over achievement is acceptable.
- The deviational variable d_i^- is removed from the objective function of the goal programming when under achivement is acceptable.
- If exact attainment of the goal is desired, then both d_i^- and d_i^+ are included in the objective function and ranked according to their pre-emptive priority factor from most important to the least *i.e.* in goal programming, the lower order goals are considered only after the higher goals are achieved.

 **Solved Examples**

EXAMPLE 1. *A manufacturing firm produces two type of products say A and B. The unit profit from product A is ₹ 100 and that from product B is ₹ 50. The goal of the firm is to earn a total profit of exactly ₹1700 in the next week. Formulate the problem.*

SOLUTION. Let x_1, x_2 be the no. of units of products A and B produced respectively. Then the LP formulation of the given problem is

$$\text{Max. } Z = 100x_1 + 50x_2$$

subject to the constraints

$$100x_1 + 5x_2 = 1700$$

and

$$x_1, x_2 \geq 0$$

Now, we have to formulate the goal programming
Clearly, the goal of the firm is to earn a profit of ₹ 1700 per week then

d_i^- = under achievement of the profit of goal of ₹ 1700

d_i^+ = over achievement of the profit of goal of ₹ 1700

So, the only constraints of the problem is

$$100x_1 + 50x_2 + d_i^- - d_i^+ = 1700$$

Hence, the required goal programming is given by

$$\text{Min. } Z = d_i^- + d_i^+$$

subject to the constraints

$$100x_1 + 50x_2 + d_i^- - d_i^+ = 1700$$

and

$$x_1, x_2, d_i^-, d_i^+ \geq 0$$

11.3.2 MULTIPLE GOAL MODELS

Multiple goal models are of the following three types.
   (i)   Multiple goal models with equal priorities (or non-priorities)
   (ii) Multiple goal models with unequal priorities.
   (iii) Multiple goal models with priorities and weights.

(i) MULTIPLE GOALS WITH EQUAL PRIORITIES

The multiple goal model with equal priorities may exist with less probability of all the three types but it is easy to deal mathematically.

To make it clear, consider the following examples.

 Solved Examples

EXAMPLE 1. *An office equipment manufacturer produces two kinds of products chairs and lamps. Production of either a chair or a lamp requires one hour of production capacity in the plant. The plant has a maximum production capacity of 50 hours per week because of the limited sales capacity, the maximum number of chairs and lamps that can be sold are 6 and 8 respectively. The gross margin from the sale of a chair is ₹ 90 and ₹ 60 for a lamp. The plant manager desires to determine the no. of units of each product that should be produced per week in consideration of the following equally ranked goals.*
Goal 1 Available production capacity should be utilized as much as possible but not exceeded.

Goal-2 *Sales of two products should be as much as possible.*

Goal-3 *Overtime should not exceed 20 percent of available production time.*

[DELHI(MBA)–1999]

Formulate this problem as a G.P. model so that the plant manager may achieve his goals as clearly as possible.

SOLUTION. Let us suppose the no. of units of the product chairs and lamps per week be x_1 and x_2 respectively.

Then as per given,

$$x_1 + x_2 + d_1^- - d_1^+ = 50 \qquad \text{(according to the first goal)}$$

The constraints according to the second goal are

$$x_1 + d_2^- = 6$$

$$x_2 + d_3^- = 8$$

Similarly, constraints according to the third goal is given by

$$d_1^+ + d_4^- - d_4^+ = 10$$

Hence, the given problem can be stated as a goal programming model as

$$\text{Min. } Z = d_1^+ + d_2^- + d_3^- + d_4^-$$

subject to the constraints

$$x_1 + x_2 + d_1^- - d_1^+ = 50$$

$$x_1 + d_2^- = 6$$

$$x_2 + d_3^- = 8$$

$$d_1^+ + d_4^- - d_4^+ = 10$$

and

$$x_1, x_2, d_1^-, d_1^+, d_2^-, d_3^-, d_4^+, d_4^- \geq 0$$

Here, we have used the following symbols

d_1^- = under utilization of product capacity of 50 hours.

d_1^+ = over utilization of product capacity of 50 hours.

d_2^- = under achievement of sales goals of chairs per week.

d_3^- = under achievement of sales goals of lamps per week.

d_4^- = over achievement of overtime hours per week.

d_4^+ = under achievement of overtime hours per week

EXAMPLE 2. *An office equipment manufacturer produces two kinds of products, computer covers and flopply boxes production of either a computer cover or a floppy box requires 1 hour of production capacity in the plant. The plant has a maximum production capacity of 10 hours per day. The gross margin from the sale of a computer cover is ₹180 and ₹140 for a floppy box.*

Formulate as a G.P. with the following equally ranked goals

(i) to earn a profit of ₹ 800 per day.

(ii) because of the limted sales capacity the maximum number of computer covers and flopy boxes that can be sold are 6 and 8 per day respectively.

SOLUTION. Let x_1 and x_2 be the number of units of the product computer covers and floppy boxes produced per day respectively.

Then linear programing problem can be stated as

$$\text{Max. } Z = 180x_1 + 140x_2$$

subject to be constraints

$$x_1 + x_2 \leq 10$$
$$x_1 \leq 6$$
$$x_2 \leq 8$$

and $\qquad x_1, x_2 \geq 0$

Now, the goals for the GP are

(i) To earn the profit of ₹ 800

(ii) The maximum no. of computer covers and floppy boxes that can be sold 6 and 8 per day respectively.

Therefore, G.P. constraints for Goal (i) is

$$80x_1 + 40x_2 + d_1^- - d_1^+ = 800$$

and G.P. constraints for goal (ii) are

$$x_1 + d_2^- = 6$$
$$x_2 + d_3^- = 8$$

where,

$d_1^- = $ under achievement of the profit of ₹ 800

$d_1^+ = $ over achievement of the profit of ₹ 800

$d_2^- = $ under achievement of sales target 6 of the computer covers

$d_3^- = $ over achievement of sales target 8 of the floppy boxes.

Hence, the required goal programming (G.P) is given as below:

$$\text{Min. } Z = d_1^- + d_2^- + d_3^-$$

subject to the constraints

$$80x_1 + 40x_2 + d_1^- - d_1^+ = 800$$
$$x_1 + x_2 \leq 10$$
$$x_1 + d_2^- = 6$$
$$x_2 + d_3^- = 8$$

and $\qquad x_1, x_2, d_1^-, d_1^+, d_2^-, d_3^- \geq 0$

(ii) Multiple Goals with Priorities

In any firm, when management has multiple goals, then they put their goals in order of their priorities *i.e.* they want the most important goal to be achieved fully or very near to the full satisfaction in comparison to the other.

To form the goal programming in such situation, we assign the priority coefficients $P_1, P_2, \ldots$ with highest priority, 2nd priority, 3rd priority, …etc., respectively. These priority coefficients have no numerical value, they simply represent the level of priority.

 Solved Examples

EXAMPLE 1. *A firm manufactures two products. Each product requires time in two production departments. Product 1 requires 20 hours in department-1 and 10 hours in department-2. Product 2 requires 10 hours in department-1 and 10 hours in department-2. Production time is limited in department-1 to 160 hours and 140 hours in department-2. Contribution to profits by two products is ₹40 and ₹80 respectively. Management has established the following goal priorities.*

Priority-1 (P_1) *To meet production goals of 2 units of each product*

Priority-2 (P_2) *To earn profit of ₹4000.*

SOLUTION.　Let x_1, x_2 be the no. of units of product-1 and product-2 produced respectively. Then we can formulate the LPP as given below.

$$\text{Max. } Z = 40x_1 + 80x_2$$

subject to the constraints

$$20x_1 + 10x_2 \le 160 \qquad \text{(department-1)}$$
$$10x_1 + 10x_2 \le 140 \qquad \text{(department-2)}$$

and 　　　　　　　　$x_1, x_2 \ge 0$

Now, to formulate a G.P., the constraints for the goal priority P_1 are

$$x_1 + d_1^- - d_1^+ = 2$$
$$\text{and} \quad x_2 + d_2^- - d_2^+ = 2$$

and the constraint for the goal priority P_2 is given by

$$40x_1 + 80x_2 + d_3^- - d_3^+ = 4000$$

clearly, to achieve first goal d_1^- and d_2^- should be minimized and to achieve second goal d_3^- should be minimized.

Hence, the formulation of G.P. is given by

$$\text{Min. } Z = P_1 d_1^- + P_1 d_2^- + P_2 d_3^-$$

subject to the constraints

$$20x_1 + 10x_2 \le 160$$
$$10x_1 + 10x_2 \le 140$$
$$x_1 + d_1^- - d_1^+ = 2$$
$$x_2 + d_2^- - d_2^+ = 2$$
$$40x_1 + 80x_2 + d_3^- - d_3^+ = 4000$$

and 　　　　$x_1, x_2, d_1^-, d_1^+, d_2^-, d_2^+, d_3^-, d_3^+ \ge 0$

(III) MULTIPLE GOALS WITH PRIORITIES AND WEIGHTS

Sometimes, there are some problem in which two or more goals have same level of priorities but the different improtance. In such type of cases, different weights are used to reflect the difference of their weights within the same level of priorities.

Solved Examples

EXAMPLE 1.　***A production manager is faced with the problem of job allocation to his two production teams. The production rate of team-1 is 8 units per hour while the production rate of team-2 is 5 units per hour. The normal working hours for each of the teams is 40 hours per week. The production manager has prioritized the following goals for the coming week.***

P_1 : *Avoid under achievement of the desired production level of 650 units.*

P_2 : *Overtime operation of team-1 is limited to 5 hours.*

P_3 : *The total overtime for both teams should be minimized.*

P_4 : *Any under utilization of regular working hours of the teams should be avoided assign different weights according to the relative productivity of the two teams.*

Formulate the problem as G.P. model.

SOLUTION. Let x_1 and x_2 be the number of hours of working of team-1 and team-2 per week respectively. Since the maximum production level is 650 units, then production volume constraint is given by

$$8x_1 + 5x_2 + d_1^- - d_1^+ = 650$$

Where $\quad d_1^- =$ under-achievement of the production target

$\qquad d_1^+ =$ over-achievement of the production target

Further, since the normal working hours of each team are 40 hours per week. So the overtime constraints for the two team is given by

$$x_1 + d_2^- - d_2^+ = 40 \qquad \text{(for team-1)}$$

and $\qquad x_2 + d_3^- - d_3^+ = 40 \qquad \text{(for team-2)}$

where, $\quad d_2^- =$ under-achievement of the normal working hours by team-1

$\qquad d_2^+ =$ over-achievement of the normal working hours by team-1

$\qquad d_3^- =$ under-achievement of the usual working hours by team-2

$\qquad d_3^+ =$ over-achievement of the usual working hours by team-2

Now, the overtime constraint for the team-1 is given by

$$d_2^+ + d_4^- - d_4^+ = 5$$

where d_4^- and d_4^+ are the under-achievement and over-achievement of overtime per week by team-1 respectively.

Since the production rate of team-1 per hour is 8 units and that of team-2 per hour is 5 units. Also the utilization of regular working hours of each team should be minimized in the ratio 8:5 of the relative productivity of team-1 and team-2. Now, we have to form the objective function.

We observe that the first goal priority P_1 of the manager is to avoid under achievement of the desired production level of 650 units then $P_1 d_1^-$ is the team of goal-1 in the objective function. Now in second priority P_2, the overtime operation of team-1 is restricted to 5 hours, then $P_2 d_4^+$ is the term of goal-2 in the objective function.

Also, in third priority P_3, the total overtime for both teams should be minimized, therefore $P_3(d_2^+ + d_3^+)$ is the objective function term for goal-3.

Similarly, in P_4, we have $8P_4 d_2^- + 5P_4 d_3^-$ is the objective function term. Hence the required goal programming is stated as

$$\text{Min. } Z = P_1 d_1^- + P_2 d_4^+ + P_3(d_2^+ + d_3^+) + P_4(8d_2^- + 5d_3^-)$$

subject to the constraints

$$8x_1 + 5x_2 + d_1^- - d_1^+ = 650$$

$$x_1 + d_2^- - d_2^+ = 40$$

$$x_2 + d_3^- - d_3^+ = 40$$

$$d_2^+ + d_4^- - d_4^+ = 5$$

and $\qquad x_1, x_2, d_1^-, d_1^+, d_2^-, d_2^+, d_3^-, d_3^+, d_4^-, d_4^+ \geq 0$

11.4 GENERAL FORM OF GOAL PROGRAMMING PROBLEM

Let us suppose a problem has m goals, p structural constraints, n decision variables and k-level of priorities.

Then a general GP can be written as

$$\text{Min. } Z = \sum_{i=1}^{m} \sum_{r=1}^{k} P_r(w_{i,r}^+ \, d_i^+ + w_{i,r}^- \, d_i^+)$$

subject to the constraints

$$\sum_{j=1}^{n} a_{ij} \cdot x_j + d_i^- - d_i^+ = b_i \qquad \text{for } i = 1, 2, ...,m$$

$$\sum_{j=1}^{n} a_{ij} x_j (\leq,=,\geq) b_i \qquad \text{for } i = m + 1, m + 2, .. \ m + p$$

and $x_j, d_i^-, d_i^+ \geq 0$ for $i = 1, 2, 3,...m, j = 1,2,3,...n$

where, P_r = The priority coefficient for r^{th} priority

$w_{i,r}^-$ = The relative weight of d_i^- variable in the rth priority level.

$w_{i,r}^+$ = The relative weight of the d_i^+ variable in the rth priority level.

☞ REMARKS

- Two type of variables are taken in the formulation namely decision variables (x_i) and the deviational variables d^- and d^+.
- Two classes of constraints can exist in a given GP model, structural constraints, which are not directly related to the goals and goal constraints which are directly related to the goals.
- In most cases, a goal constraints will have both the deviational variables even when both deviational variables do not appear simultaneously in the objective function.

11.5 METHOD OF SOLUTION OF A GOAL PROGRAMMING (GP) PROBLEM

To solve a GP problem, we have following two methods
(i) Graphical method
(ii) GP algorithm or modified simplex method

11.5.1 GRAPHICAL METHOD

We used graphical method when the GP model involve two decision variables. The method is quite similar to the graphical method of an LPP. In LPP the objective function of one goal only is to be optimized whereas in the GP models, the objective function of the deviational variables from multiple goals in respective order of priorities are minimized so that the achievement of the goals of higher order are not affected.

To solve a GP model by graphical method, we use the following procedure.

Working Procedure

STEP 1. Graph (Plot) all the structural constraints and identify the feasible region. If there is no structural constraints then the feasible region is the first quadrant (*i.e.* $x_1 \geq 0, x_2 \geq 0$). If no feasible region exists then there is no solution to the problem.

STEP 2. Draw the lines corresponding to the goal constraints by setting the deviational variables in the goal constraints equal to zero.

STEP 3. Identify the top-priority solution which is obtained by determining the points within the feasible region so obtained in step 1 that satisfy the highest priority goal.

STEP 4. Sequentially consider the remaining goals and the point that satisfy them to the greatest extent possible. Make sure that a lower priority goal is not achieved by reducing the degree of achievement of higher priority goals.

STEP 5. Repeat the step 4 until all levels of priority have been investigated.

Solved Examples

EXAMPLE 1. *An office equipment manufacturer produces two kinds of products: chairs and lamps. Production of either a chair or a lamp requires one hour of production capacity in the plant. The plant has a* *maximum capacity of 10 hours per week. The gross margin from the sale of a chair is ₹80 and ₹40 for that of a lamp. Formulate and solve the problem as a GP model with the following equally ranked goals.*

(i) to earn a profit of ₹800 per week

(ii) the maximum no. of chairs and lamps that can be sold are 6 and 8 per week respectively. [MEERUT–2004, 06]

SOLUTION: Clearly, the G.P. model of the given problem is

$$\text{Min. } Z = d_1^- + d_2^- + d_3^-$$

subject to the constraints

$$80x_1 + 40x_2 + d_1^- - d_1^+ = 800$$
$$x_1 + d_2^- = 6$$
$$x_1 + d_3^- = 8$$

and $\quad x_1, x_2, d_1^-, d_1^+, d_2^-, d_3^- \geq 0$

Now, we have to solve the above GP by graphical method.

STEP 1. Let us take $d_1^- = d_1^+ = d_2^- = d_3^- = 0$ and then plot all the constraints. The arrows (← or →) are then associated with each line to represent under achievement or over achievement.

STEP 2. Because, all the goals are equally ranked and the objective of the problem is to minimize d_1^-, d_2^- and d_3^-. So, we set $d_1^- = 0, d_2^- = 0$ and $d_3^- = 0$. Then we get an intersection point $P(6, 8)$. At $P(6, 8)$ the profit of the firm is ₹800 and min. $Z = 0$.

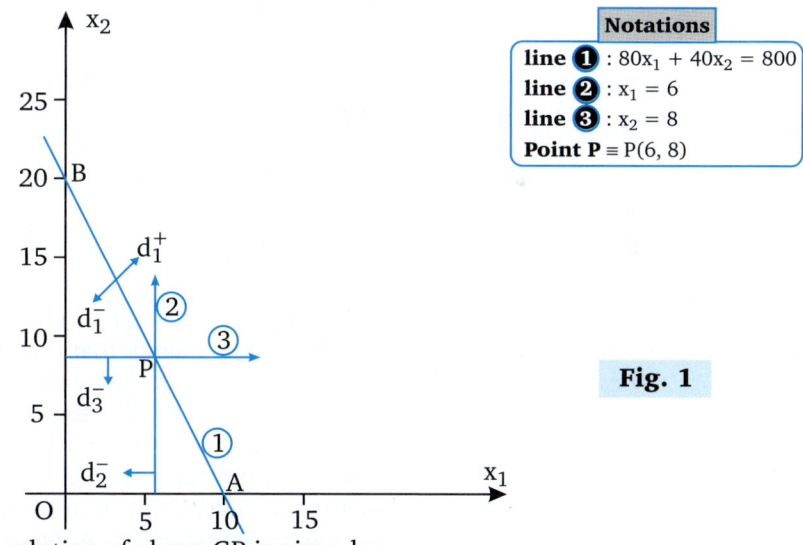

Notations
line ❶ : $80x_1 + 40x_2 = 800$
line ❷ : $x_1 = 6$
line ❸ : $x_2 = 8$
Point $P \equiv P(6, 8)$

Fig. 1

Hence, the solution of above GP is given by

$$x_1 = 6, x_2 = 8 \text{ and Min. } Z = 0.$$

EXAMPLE 2. **Solve the following G.P. by graphical method**

$$\text{Min. } Z = P_1 d_1^- + P_2 (2d_2^- + d_3^-) + P_3 d_1^+$$

subject to the constraints

$$x_1 + x_2 + d_1^- - d_1^+ = 10 \qquad \qquad \dots(1)$$

$$x_1 + d_2^- = 6 \qquad \qquad \text{...(2)}$$

$$x_2 + d_3^- = 8 \qquad \qquad \text{...(3)}$$

and $\qquad x_1, x_2, d_1^-, d_1^+, d_2^-, d_3^- \geq 0$

SOLUTION. To solve the above problem, we use the following steps

STEP 1 Taking $d_1^- = d_1^+ = d_2^- = d_3^- = 0$ and then plot the goal constraints as shown in the the following figure

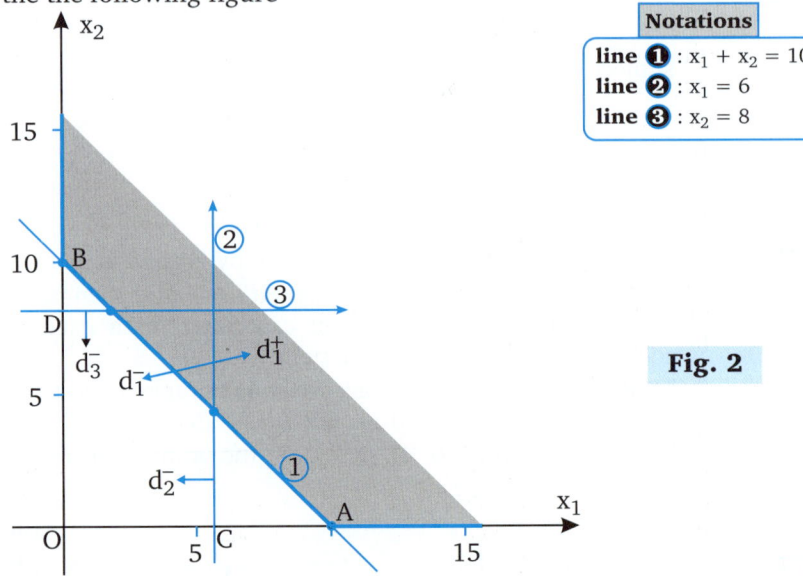

Notations

line ❶ : $x_1 + x_2 = 10$
line ❷ : $x_1 = 6$
line ❸ : $x_2 = 8$

Fig. 2

Now we have to identify the priority goal
First Priority Goal
d_1^- is the coefficient of P_1 in the objective function. Setting $d_1^- = 0$.
The feasible region is the shaded region. Any point in this shaded region satisfies the first goal as for any point of this region under achievement of production capacity *i.e.* $d_1^- = 0$.
In the second priority, the manufacturer wants to sell as many chairs and lamps as possible. The maximum no. of chairs and lamps that can be sold per week are 6 and 8 respectively. But his priority to achieve the sales goals of chairs to 6 is the first priority which is achieved by setting $d_2^- = 0$. The target of sales goal of lamps to 10 (maximum) can be obtained by setting $d_3^- = 0$. Therefore, the second priority goals are satisfied at the point $P(6, 8)$ in the feasible region of the top priority.
Similarly in the third priority d_1^+, over utilization of production capacity of 10 hours operation is to be minimized which is minimum when $d_1^+ = 0$. Clearly $d_1^+ = 0$ at all points on the line segment AB at which the first priority goal is achieved. But at any point of the line segment AB, the second priority goal can not be achieved. Hence, we can not achieve the third goal without sacrificing the second goal. So, keeping d_1^+ minimum possible, the solution of the goal programming problem is achieved at $P(6, 8)$ at which the first two top priorities goals are achieved fully, but the third priority goal is achieved as much as possible.
For $\quad d_2^- = 0, d_3^- = 0 \quad$ from (2) and (3) $\quad x_1 = 6, x_2 = 8$
$\therefore \quad$ using $d_1^- = 0, x_1 = 6, x_2 = 8$ in (1) we get $d_1^+ = 4$

Finally, we conclude that the manufacturer should produce 6 chairs and 8 lamps per week, so that his first two goals are fully achieved and the third goal is the overtime operation of the plant is minimized to 4 hours per week.

11.5.2 MODIFIED SIMPLEX METHOD TO SOLVE A GOAL PROGRAMMING MODEL

The steps of modified simplex method for a GP model are as follows:

STEP 1.   **(Formulation of Initial Table)**

Construct the initial table in the same way as that for LPP. The top most row of the table will have the coefficients of decision variables and deviational weights c_j, in the objective function. Below the entries of c_j, the coefficient of $x_j's$ and d_i^- and d_i^+ are placed in the appropriate column. Now draw a horizontal line below these entries and write the pre-empty priorities factor $P_1, P_2, \ldots$ in X_B column, starting from bottom to top i.e., first priority P_1 written at the bottom and the least priority is written at the top. In GP model, we minimize the unattained portion of goals as much as possible by minimizing the deviational variables.

STEP 2.   **(Test of Optimality)**

Compute the values z_j and $c_j - z_j$ seperately by each of the ranked goals, $P_1, P_2, \ldots$ etc., in the same way as in usual simplex method. The optimal criterion z_j or $c_j - z_j$ becomes a matrix of order $k \times n$, where k represents the no. of pre-empty priority levels and n denotes the number of total variables i.e., decision variable and deviational variables.

Further, check $c_j - z_j$ for top priority goal P_1. If all $c_j - z_j \geq 0$ in P_1 row or there is a zero in P_1 row in X_B-column then goal P_1 is achieved and then we go to step 5. But if at least one of the entries is negative and there is no zero in X_B column in P_1-row, then goal P_1 is not achieved and we go to the next step.

STEP 3.   **(To Determine Incoming and Outgoing Vector)**

Select the most negative entry in P_1 row of matrix $c_j - z_j$. The variable in the column corresponding to the most negative entry is the incoming variable. If there is a tie, then check the next lower priority level. The column corresponding to the most negative entry in the lower priority level, out of the columns in which there is a tie in P_1-row is selected as incoming variable.

When we have obtained incoming vector, then we find the outgoing vector as in usual simplex method in LPP by minimum ratio rule and the element at the intersection of the incoming vector column and the minimum ratio row is called key element.

STEP 4.   Reduce the key element to 1 as in usual simplex method and with its help using row operation, all other elements in the corresponding column are reduced to zero. So we get a new reduced matrix. From this reduced matrix, find the value of z_j or $c_j - z_j$ for each of the ranked goals $P_1, P_2, \ldots$. Now again examine $c_j - z_j$ for top priority goal P_1. If all the entries in P_1-row are positive, then P_1 is achieved. If at least one of the entries in P_1 is negative then P_1 still not achieved. Then to achieve P_1, we repeat step 3 and 4.

STEP 5.   If the first priority goal P_1 is achieved, then we proceed as above to achieve the next priority goal P_2. The goal P_2 can not be further improved from its present level if there is positive entry in P_1 row below the most negative entry in P_2 row. Continue the above process until the lowest priority goal say P_k is also fully achieved or is achieved nearest to the satisfaction. Again the goal P_k can not be further improved from its present level if there is a positive entry in the higher priority goals $P_1, P_2, \ldots, P_{k-1}$ rows below the most negative entry in P_k row.

Solved Examples

EXAMPLE 1. *An office equipment manufacturer produces two kinds of products: chairs and lamps. Production of either a chair or a lamp requires one hour of production capacity in a plant. The plant has a maximum production capacity of 10 hours per week. Because of the limited sales capacity, the maximum number of chairs and lamps that can be sold in 6 and 8 per week respectively. The gross margin from the sale of a chair is ₹ 80 and ₹ 40 for that of a lamp.*

The plant manager has set the following goals arranged in order of importance.

(i) He wants to avoid any under utilization of production capacity.

(ii) He wants to sell as many chairs and lamps as possible since the gross margin from the sale of a chair is set at double the amount of profit from a lamp, he has double as much as desire to achieve the sales goal, for chairs as for lamps.

(iii) He wants to minimize overtime operation of the plant as much as possible.

Formulate and solve this problem as a GP problem so that the plant manager makes a decision that will achieve his goals as closely as possible. [DELHI-1999]

SOLUTION. Following the usual procedure, the G.P. model of the above problem is as follows.

$$\text{Min. } Z = P_1 d_1^- + P_2 (2d_2^- + d_3^-) + P_3 d_1^+$$

subject to the constraints

$$x_1 + x_2 + d_1^- - d_1^+ = 10$$
$$x_1 + d_2^- = 6$$
$$x_2 + d_3^- = 8$$

and

$$x_1, x_2, d_1^-, d_1^+, d_2^-, d_3^- \geq 0$$

Now to solve it, we proceed as follows:

Taking $x_1 = 0$, $x_2 = 0$ and $d_1^+ = 0$ we get $d_1^- = 10$, $d_2^- = 6$, $d_3^- = 8$ which is the initial basic feasible solution

STEP 1 (Construction of initial table-1)

B.V.	C_B	X_B	$c_j \rightarrow$ O x_1	O x_2	P_1 d_1^- (x_3)	P_3 d_1^+ (x_4)	$2P_2$ d_2^- (x_5)	P_2 d_3^- (x_6)	Min. ratio X_B / x_1
d_1^-	P_1	10	1	1	1	−1	0	0	$\dfrac{10}{1}$
d_2^-	$2P_2$	6	1	0	0	0	1	0	$\dfrac{6}{1}$ (min) →
d_3^-	P_2	8	0	1	0	0	0	1	
$c_j - z_j$	P_3	0	0	0	0	1	0	0	
	P_2	20	−2	−1	0	0	0	0	
	P_1	10	−1	−1	0	1	0	0	

↑

In the above table

$$c_1 - z_1 = c_1 - C_B x_1 = 0 - (P_1, 2P_2, P_2)\begin{pmatrix} 1 \\ 1 \\ 0 \end{pmatrix} = -P_1 - 2P_2$$

$$c_2 - z_2 = c_2 - C_B x_2 = 0 - (P_1, 2P_2, P_2)\begin{pmatrix} 1 \\ 0 \\ 1 \end{pmatrix} = -P_1 - P_2$$

$$c_4 - z_4 = c_4 - C_B x_4 = P_3 - (P_1, 2P_2, P_2)\begin{pmatrix} -1 \\ 0 \\ 0 \end{pmatrix} = P_3 + P_1$$

and $Z = C_B x_B = (P_1, 2P_2, P_2)\begin{pmatrix} 10 \\ 6 \\ 8 \end{pmatrix} = 10P_1 + 20P_2$

The $c_j - z_j$ matrix is obtained by the coefficients of P_1, P_2 and P_3 in $c_j - z_j$ values.

STEP 2.

The most negative entry in P_1-row is -1 which occurs for column x_1 as well as for column x_2. So there is a tie, so we check the lower priority P_2. In P_2-row, the most negative entry is -2 which is in x_1-column thus x_1-column is selected as key element *i.e.*, x_1 is the incoming vector and by minimum ratio rule d_2^- is the outgoing vector. Hence, the key element is 1 *i.e.* a_{21}.

STEP 3.

Since the key element is already 1. So using row transformation, the elements other than the key element are reduced to zero. Then we have the following reduced table.

Reduced Table-2

B.V.	C_B	X_B	$c_j \to$ 0 x_1 (d_2^-)	0 x_2	P_1 (x_3) d_1^-	P_3 d_1^+ (x_4)	$2P_2$ x_5	P_2 d_3^- (x_6)	Min. ratio X_B / x_2
d_1^- P_1		4	0	①	1	-1	-1	0	$\dfrac{4}{1}$ (min.) $\to$
x_1	0	6	1	0	0	0	1	0	
d_3^- P_2		8	0	1	0	0	0	1	$\dfrac{8}{1}$
$c_j - z_j$ P_3	0	0	0	0	1	0	0		
P_2	8	0	-1	0	0	2	0		
P_1	4	0	-1	0	1	1	0		

$\uparrow$

In the above table

$$c_2 - z_2 = c_2 - C_B x_2 = 0 - (P_1, 0, P_2)\begin{bmatrix} 1 \\ 0 \\ 1 \end{bmatrix} = -P_1 - P_2$$

$$c_4 - z_4 = c_4 - C_B x_4 = P_3 - (P_1, 0, P_2) \begin{bmatrix} -1 \\ 0 \\ 0 \end{bmatrix} = P_3 + P_1$$

$$c_5 - z_5 = c_5 - C_B x_5 = 2P_2 - (P_1, 0, P_2) \begin{bmatrix} -1 \\ 1 \\ 0 \end{bmatrix} = 2P_2 + P_1$$

and $Z = C_B x_B = (P_1, 0, P_2) \begin{bmatrix} 4 \\ 6 \\ 8 \end{bmatrix} = 4P_1 + 8P_2$

Now from the above table we observe that in P_1-row, the most negative element is -1 which occurs in column x_2 i.e. x_2 is the incoming vector and by minimum ratio rule, d_1^- is the outgoing vector and the key element is 1 i.e. a_{12}.

Further, using row transformation as usual, the next reduced table is as given below.

Reduced Table-3

B.V.	C_B	X_B	$c_j \to$ x_1 (d_2^-)	0 x_2 (d_1^-)	P_1 (x_3)	P_3 d_1^+ (x_4)	$2P_2$ (x_5)	P_2 d_3^- (x_6)	Min. ratio X_B / d_1^+
x_2	0	4	0	1	1	-1	-1	0	$-$
x_1	0	6	1	0	0	0	1	0	$-$
d_3^-	P_2	4	0	0	-1	①	1	1	$\dfrac{4}{1} \to$
$c_j - z_j$ $\quad P_3$	0	0	0	0	1	0	0		
P_2	4	0	0	1	-1	1	0		
P_1	0	0	0	1	0	0	0		

$$\uparrow$$

In the above table, we compute $c_j - z_j$ for $j = 3, 4, 5$ (not in the basis)

$$c_3 - z_3 = c_3 - C_B x_3 = P_1 - (0, 0, P_2) \begin{bmatrix} 1 \\ 0 \\ -1 \end{bmatrix} = P_1 + P_2$$

$$c_4 - z_4 = c_4 - C_B x_4 = P_3 - (0, 0, P_2) \begin{bmatrix} -1 \\ 0 \\ 1 \end{bmatrix} = P_3 - P_2$$

$$c_5 - z_5 = c_5 - C_B x_5 = 2P_2 - (0, 0, P_2) \begin{bmatrix} -1 \\ 1 \\ 1 \end{bmatrix} = P_2$$

and $Z = C_B X_B = (0, 0, P_2) \begin{bmatrix} 4 \\ 6 \\ 4 \end{bmatrix} = 4P_2 = 0 \cdot P_3 + 4P_2 + 0 \cdot P_1$

From the above table we observe that there is zero in x_B column in P_1 row so the goal P_1 is fully achieved.

STEP 4

As goal P_1 is achieved, we move to goal P_2. In P_2 row of table 3, the most negative element is -1 which occurs in d_1^+ column, so d_1^+ is the incoming vector and by minimum ratio rule, d_3^- is the outgoing vector and the key element is 1 *i.e.* a_{34}. Now repeat step 3, we get the next reduced table 4.

Reduced Table-4

B.V.	C_B	X_B	$c_j \rightarrow$		P_1	P_3	$2P_2$	P_2	Min. ratio
			0	0					
			x_1	x_2	(x_3)	(x_4)	(x_5)	(x_6)	
			(d_2^-)	(d_1^-)		(d_3^-)			
x_2	0	8	0	1	0	0	0	1	
x_1	0	6	1	0	0	0	1	0	
d_1^+	P_3	4	0	1	-1	1	1	1	
$c_j - z_j$ P_3		4	0	0	1	0	-1	-1	
P_2		0	0	0	0	0	2	1	
P_1		0	0	0	1	0	0	0	

In the above table, we have computed $c_j - z_j$ for $j = 3, 5, 6$ (non-basic variable)

$$c_3 - z_3 = c_3 - C_B x_3 = P_1 - (0,0,P_3)\begin{bmatrix} 0 \\ 0 \\ -1 \end{bmatrix} = P_1 + P_3$$

$$c_5 - z_5 = c_5 - C_B x_5 = 2P_2 - (0,0,P_3)\begin{bmatrix} 0 \\ 1 \\ 1 \end{bmatrix} = 2P_2 - P_3$$

$$c_6 - z_6 = c_6 - C_B x_6 = P_2 - (0,0,P_3)\begin{bmatrix} 1 \\ 0 \\ 1 \end{bmatrix} = P_2 - P_3$$

and $Z = C_B X_B = (0,0,P_3)\begin{bmatrix} 8 \\ 6 \\ 4 \end{bmatrix} = 4P_3$

In the above table, we observe that in P_1 and P_2 rows all the elements are non-negative and these are zero in x_B column in P_1 and P_2 rows. Therefore, the goals P_1 and P_2 are fully achieved.

Now, we move to goal P_3 and observe that in P_3-row the most negative element is -1 which occurs in column x_5 and x_6. But in higher priority goal P_2, in P_2 row there are positive entries in column x_5 and x_6 below the most negative element.

Hence goal P_3 can not be achieved.

Again from table 4 we have

$$x_1 = 6, x_2 = 8, d_1^+ = 4, d_1^- = 0, d_2^- = 0, d_3^- = 0$$

Hence, the optimal solution of present goal programming is

$$x_1 = 6, x_2 = 8, d_1^+ = 4$$

EXAMPLE 2. *A company produces two kind of products: A and B. Production of either A or B requires 3 hours of production capacity in the plant. The plant has a maximum production capacity of 30 hours per week to manufacture these two products, has set the following goals arranged in the order of importance.*

   (i) To avoid any underutilization of production capacity.

   (ii) To limit the overtime to 5 hours.

  (iii) To minimize the overtime operations of the plant as much as possible. Formulate this problem as a goal programming and then solve it by modified simplex method.

SOLUTION. Let x_1 and x_2 be the number of units of product A and B produced respectively. As per given

      no. of hours per unit to produce product A = 3 hours

      no. of hours per unit to produce product B = 3 hours

maximum normal production capacity per week = 30 hours

upper limit for overtime hours per week = 5 hours.

Then clearly, production capacity constraints is given by

$$3x_1 + 3x_2 + d_1^- - d_1^+ = 30$$

where d_1^- and d_1^+ are the under achievement and over achievement of the production target respectively.

Also, the overtime constraints is given by

$$d_1^+ + d_2^- - d_2^+ = 5$$

where d_2^- and d_2^+ are respectively the under achievement and over achievement of overtime target.

Now, we have to define the objective function.

Let us suppose that P_1, P_2 and P_3 be the goal priorities.

Clearly, the first priority P_1 of the manager is to avoid underutilization of production capacity so that d_1^- is to be minimized, then $P_1 d_1^-$ is the objective function term for goal-1.

Now, the second goal priority P_2 of the manager is to limit the overtime hours to 5 hours so that d_2^+ is to be minimized, then $P_2 d_2^+$ is the objective function term for goal-2. The third goal priority P_3 is to be minimized overtime operation of the plant as much as possible so that d_1^+ is to be minimized. Therefore $P_3 d_1^+$ is the objective function term for goal-3.

∴   the objective function is

$$\text{Min. } Z = P_1 d_1^- + P_2 d_2^+ + P_3 d_1^+$$

Thus, the required GPP is given by

$$\text{Min. } Z = P_1 d_1^- + P_2 d_2^+ + P_3 d_1^+$$

subject to the constraints

$$3x_1 + 3x_2 + d_1^- - d_1^+ = 30$$
$$d_1^+ + d_2^- - d_2^+ = 5$$

and $$x_1, x_2, d_1^-, d_1^+, d_2^-, d_2^+ \geq 0$$

Now, we apply the modified simplex method and get the initial table as given below.

Initial Table-1

B.V.	c_j C_B	X_B	0 x_1	0 x_2	P_1 d_1^- (x_3)	P_3 d_2^+ (x_4)	0 d_2^- (x_5)	P_2 d_2^+ (x_6)	Min. Ratio
d_1^-	P_1	30	③	3	1	−1	0	0	$\dfrac{30}{3}$(min) →
d_2^-	0	5	0	0	0	1	1	−1	—
$c_j - z_j$	P_3	0	0	0	0	1	0	0	
	P_2	0	0	0	0	0	0	1	
	P_1	30	−3	−3	0	1	0	0	
			↑			↓			

In the above table, we have to compute $c_j - z_j$ for $j = 1, 2, ..., 6$

$$c_1 - z_1 = c_1 - C_B x_1 = 0 - (P_1, 0)\begin{bmatrix} 3 \\ 0 \end{bmatrix} = -3P_1$$

$$c_2 - z_2 = c_2 - C_B x_2 = 0 - (P_1, 0)\begin{pmatrix} 3 \\ 0 \end{pmatrix} = -3P_1$$

$$c_3 - z_3 = c_3 - C_B x_3 = P_1 - (P_1, 0)\begin{bmatrix} 1 \\ 0 \end{bmatrix} = P_1 - P_1 = 0$$

$$c_4 - z_4 = c_4 - C_B x_4 = P_3 - (P_1, 0)\begin{bmatrix} -1 \\ 1 \end{bmatrix} = P_3 + P_1$$

$$c_5 - z_5 = c_5 - C_B x_5 = 0 - (P_1, 0)\begin{bmatrix} 0 \\ 1 \end{bmatrix} = 0$$

$$c_6 - z_6 = c_6 - C_B x_6 = P_2 - (P_1, 0)\begin{bmatrix} 0 \\ -1 \end{bmatrix} = P_2$$

∴ The criterion matrix is obtained by entering the coefficient of P_1, P_2 and P_3 in each value of $c_j - z_j$.

Now, $Z = C_B X_B = (P_1, 0)\begin{pmatrix} 30 \\ 5 \end{pmatrix} = 30P_1 = 30P_1 + 0P_2 + 0P_3$

The coefficient of P_1, P_2 and P_3 in Z are written in column x_B in the criterion matrix $c_j - z_j$.

Further, since the most negative entry in P_1 row is −3 which occurs for column x_1 as well as the column x_2. Also in P_2 and P_3 rows, there is zero in x_B column in $c_j - z_j$ matrix so that P_1 goal is the only unattained goal, so there is no other row above P_1-row with attainment so that tie is broken randomly and the column x_1 is selected as incoming vector only by minimum ratio rule d_1^- in the outgoing vector and the key element is 3 i.e. a_{11}.

Now reduce the key element to 1 by dividing the d_1^- row by 3 and using the row operation, reduce other elements in c_1-column equal to zero so that the next reduced table is as given below.

	c_j		0	0	P_1	P_3	0	P_2
B.V.	C_B	X_B	x_1	x_2	d_1^- (x_3)	d_1^+ (x_4)	d_2^- (x_5)	d_2^+ (x_6)
x_1	0	10	1	1	1/3	−1/3	0	0
d_2^-	0	5	0	0	0	1	1	−1
$c_j - z_j$	P_3	0	0	0	0	1	0	0
	P_2	0	0	0	0	0	0	1
	P_1	0	0	0	1	0	0	0

In the above table we have computed $c_j - z_j$ for (non-basic) $j = 2, 3, 4, 6$ as follows:

$$c_2 - z_2 = c_2 - C_B x_2 = 0 - (0,0)\begin{bmatrix}1\\0\end{bmatrix} = 0 = 0P_1 + 0P_2 + 0P_3$$

$$c_3 - z_3 = c_3 - C_B x_3 = P_1 - (0,0)\begin{bmatrix}1/3\\0\end{bmatrix} = P_1 = P_1 + 0P_2 + 0P_3$$

$$c_4 - z_4 = c_4 - C_B x_4 = P_3 - (0,0)\begin{bmatrix}-1/3\\0\end{bmatrix} = P_3 = 0P_1 + 0P_2 + P_3$$

$$c_6 - z_6 = c_6 - C_B x_6 = P_2 - (0,0)\begin{bmatrix}0\\-1\end{bmatrix} = P_2 = 0P_1 + P_2 + 0P_3$$

and $Z = C_B X_B = (0,0)\begin{bmatrix}10\\5\end{bmatrix} = 0 = 0P_1 + 0P_2 + 0P_3$

Also, from the above table, we observe that in criterion matrix $c_j - z_j$, there is zero in column x_B in P_3-row, in P_2-row, in P_1-row.

Hence all the priorities of the goals are fully achieved

Hence, the optimal solution of GPP is given by

$$x_1 = 10, x_2 = 0, d_1^- = 0, d_1^+ = 0, d_2^- = 5, d_2^+ = 0 \text{ and Min. } Z = 0$$

EXAMPLE 3. *An office equipment produces two kinds of products chairs and lamps. Production of either a chair or a lamp requires one hour of production capacity in the plant. The plant has a maximum production capacity in the plant. The plant has a maximum production capacity of 50 hours per week.*

Because of the limited sales capacity, the maximum no. of chairs and lamps that can be sold are 6 and 8 per week respectively. The gross margin from the sale of a chair is ₹ 90 and ₹ 60 for a lamp. The plant manager desires to determine the no. of units of each product that should be produced per week in consideration of the following set of goals.

(i) Available production capacity should be utilized as much as possible but not exceed.

(ii) Sales of the two products should be as much as possible.

(iii) *Overtime should not exceed 20 percent of available production time. Formulate and solve this problem as a G.P. model so that the plant manager may achieve his goals as closely as possible.*

SOLUTION.   Let x_1, x_2 be the no. of chairs and lamps produced per week and

$d_1^- =$ Time in hours by which the production capacity is under utilized.

$d_1^+ =$ Time in hours by which the production capacity is over utilized.

$d_2^- =$ Number by which the sales of six chairs is under achieved.

$d_3^- =$ Number by which the sales of eight lamps is under achieved.

$d_{12}^- =$ Time in hours by which the overtime of 10 hours (20% of 50) is underachieved.

$d_{12}^+ =$ Time in hours by which the overtime of 10 hours is over achieved.

Also, there is no priority of 3 goals to be achieved.

Then the formulation of the given problem as a G.P. model is given as follows:

$$\text{Min. } Z = d_1^+ + d_2^- + d_3^- + d_{12}^+$$

subject to the constraints

$$x_1 + x_2 + d_1^- - d_1^+ = 50$$
$$x_1 + d_2^- = 6$$
$$x_2 + d_3^- = 8$$
$$d_1 + d_{12}^- - d_{12}^+ = 10$$

and $\quad x_1, x_2, d_1^-, d_1^+, d_2^-, d_3^-, d_{12}^-, d_{12}^+ \geq 0$

Further, since there is no priority in the 3 goals to be achieved, therefore it can be solved by usual simplex method. Now applying the simplex method in a usual manner, we have the following simplex table.

B.V.	C_B	X_B	$c_j \rightarrow$ 0 x_1	0 x_2	0 d_1^-	1 d_1^+	1 d_2^-	1 d_3^-	0 d_{12}^-	1 d_{12}^+	Min. Ratio X_B / x_1
d_1^-	0	50	1	1	1	−1	0	0	0	0	50
d_2^-	1	6	①	0	0	0	1	0	0	0	6(min.) →
d_3^-	1	8	0	1	0	0	0	1	0	0	—
d_{12}^-	0	10	0	0	0	1	0	0	1	−1	—
$Z = 14$	$c_j - z_j$		−1 ↑	−1	0	1	0	0	0	1	X_B / x_2
d_1^-	0	44	0	1	1	−1	−1	0	0	0	44
x_1	0	6	1	0	0	0	1	0	0	0	—
d_3^-	1	8	0	①	0	0	0	1	0	0	8(min.) →
d_{12}^-	0	10	0	0	0	1	0	0	1	−1	—
$Z = 8$	$c_j - z_j$		0	−1 ↑	0	1	1	0	0	0	
d_1^-	0	36	0	0	1	−1	−1	−1	0	0	
x_1	0	6	1	0	0	0	1	0	0	0	
x_2	0	8	0	1	0	0	0	1	0	0	
d_{12}^-	0	10	0	0	0	1	0	0	1	−1	
$Z = 0$	$c_j - z_j$		0	0	0	1	1	1	0	1	

Clearly, in the last row of the above table all $c_j - z_j \geq 0$

$\Rightarrow$ solution is optimal and is given by

$$x_1 = 6, \, x_2 = 8, \, d_1^- = 36, \, d_{12}^- = 10$$

Here, $d_1^- = 36$. Therefore, production capacity is under utilized by 36 hours. Hence there is no question of overtime.

EXAMPLE 4. *A textile company produces two types of material A and B. The average production rates for the material A and B are identical at 1000 m/hrs. By running two shifts the operational capacity of the plant is 80 hours per week. The marketing department report that maximum estimated sales for the following week is 70,000 meters of material A and 45,000 meters of material B. According to the account department the profit from one meter of material A is 2.50 and from one meter of material B is 1.50. The management of the company decide that a stable employment level is the primary goal for the firm. Thus, whenever there is a demand exceeding normal production capacity. The management simply expands production capacity by providing overtime. However management feels that overtime operation of the plant of more than 10 hours per week should be avoided because of the accelerating costs. The management has the following goals.*

Goal (1) *The first goal is to avoid any under utilization of production capacity. i.e., to maintain stable employment at normal capacity.*

Goal (2) *To limit the overtime operation of the plant to 10 hours.*

Goal (3) *To achieve the sale 70,000 meters of material A and 45000 meters of material B.*

Goal (4) *To minimize the overtime operation of the plant as much as possible. Formulate and solve the problem as a GPP to help the management for the best decision.*

SOLUTION. Let x_1, x_2 be the no. of hours per week spent for producing the material A and B respectively.

Also, d_1^- = under utilization of the production capacity of 80 hours

d_1^+ = over utilization of the production capacity of 80 hours

d_2^- = under achievement of the sales of 70000 of material A in meter

d_3^- = under achievement of the sales of 45000 of material B in meter

d_4^- = under achievement of overtime of 10 hours

d_4^+ = over achievement of overtime of 10 hours

Thus, the production capacity constraints is given by

$$x_1 + x_2 + d_1^- - d_1^+ = 80$$

Also, sales constraints are given by

$$1000x_1 + 1000d_2^- = 70000 \Rightarrow x_1 + d_2^- = 70$$

$$1000x_2 + 1000d_3^- = 45000 \Rightarrow x_2 + d_3^- = 45$$

and, overtime operation constraints is given by

$$d_1^+ + d_4^- - d_4^+ = 10$$

Now, we have to find the objective function in the following manner:

(i) The first goal priority P_1 of the plant is to avoid under utilization of production capacity i.e. d_1^- is to be minimized and $P_1 d_1^-$ is the objective function term for goal-1.

(ii) The second goal priority P_2 of the plant is to limit the overtime operation to 10 hours i.e. d_4^+ is to minimized and $P_2 d_4^+$ is the objective function term for goal-2.

(iii) The third goal priority P_3 of the plant is to achieve sales goals of material A and material B i.e., d_2^- and d_3^- are to be minimized. But the profit from one meter of material A is ₹ 2.50 and ₹ 1.50 from one meter of material B. Also the production rate for both the material A and B is the same as 1000 meter per hour, so the hourly profit of A and B is in the ratio of 2.50 : 1.50 or 5 : 3 then $P_3(5d_2^- + 3d_3^-)$ is the objective function term for goal-3.

(iv) The fourth goal priority P_4 is to minimize the overtime operation of the plant as much as possible i.e. d_1^+ is to minimized then $P_4 d_1^+$ is the objective function term for goal-4.

Keeping in mind the above fact, the required GPP is given by

$$\text{Min. } Z = P_1 d_1^- + P_2 d_4^+ + P_3(5d_2^- + 3d_3^-) + P_4 d_1^+$$

subject to the constraints

$$x_1 + x_2 + d_1^- - d_1^+ = 80$$
$$x_1 + d_2^- = 70$$
$$x_2 + d_3^- = 45$$
$$d_1^+ + d_4^- - d_4^+ = 10$$

and $x_1, x_2, d_1^-, d_1^+, d_2^-, d_3^-, d_4^+, d_4^- \geq 0$

To solve the above GPP, we proceed as follows:

Firstly we construct the initial table as given below.

Initial Table-1

B.V.	c_j		0 x_1	0 x_2	P_1 d_1^- (x_3)	P_4 d_1^+ (x_4)	$5P_3$ d_2^- (x_5)	$3P_3$ d_3^- (x_6)	0 d_4^- (x_7)	P_2 d_4^+ (x_8)	Min. Ratio X_B / x_1
	C_B	X_B									
d_1^-	P_1	80	1	1	1	−1	0	0	0	0	80/1
d_2^-	$5P_3$	70	①	0	0	0	1	0	0	0	75/1(min.)→
d_3^-	$3P_3$	45	0	1	0	0	0	1	0	0	—
d_4^-	0	10	0	0	0	1	0	0	1	−1	—
$c_j - z_j$	P_4	0	0	0	0	1	0	0	0	0	
	P_3	485	−5	−3	0	0	0	0	0	0	
	P_2	0	0	0	0	0	0	0	0	1	
	P_1	80	−1	−1	0	1	0	0	0	0	
			↑								

In the above table, we have computed $c_j - z_j$ for $j = 1, 2, 4, 8$ (not in the basis)

$$c_1 - z_1 = c_1 - C_B x_1 = 0 - (P_1, 5P_3, 3P_3, 0)\begin{bmatrix} 1 \\ 1 \\ 0 \\ 0 \end{bmatrix} = -P_1 - 5P_3$$

$$c_2 - z_2 = c_2 - C_B x_2 = 0 - (P_1, 5P_3, 3P_3, 0)\begin{bmatrix} 1 \\ 0 \\ 1 \\ 0 \end{bmatrix} = -P_1 - 3P_3$$

$$c_4 - z_4 = c_4 - C_B x_4 = P_4 - (P_1, 5P_3, 3P_3, 0)\begin{bmatrix} -1 \\ 0 \\ 0 \\ 1 \end{bmatrix} = P_4 + P_1$$

and $\quad Z = C_B x_B = (P_1, 5P_3, 3P_3, 0)\begin{bmatrix} 80 \\ 70 \\ 45 \\ 10 \end{bmatrix} = 80P_1 + 485P_3$

The above criterian matrix $c_j - z_j$ is obtained by the coefficients of P_1, P_2, P_3 and P_4 and x_B column is obtained by the coefficients of P_1, P_2, P_3 and P_4 in Z-value.

Now, in P_1-row, the most negative element is –1. Thus solution is not optimal.

Then following the usual procedure, we have the following reduced table.

Reduced Table-1

B.V.	c_j		0	0	P_1	P_4	$5P_3$	$3P_3$	0	P_2	Min. Ratio
	C_B	X_B	x_1	x_2	d_1^-	d_1^+	d_2^-	d_3^-	d_4^-	d_4^+	X_B / x_2
d_1^-	P_1	10	0	①	1	–1	–1	0	0	0	10/1(min.)→
x_1	0	70	1	0	0	0	1	0	0	0	—
d_3^-	$3P_3$	45	0	1	0	0	0	1	0	0	45/1
d_4^-	0	10	0	0	0	1	0	0	1	–1	—
$c_j - z_j$	P_4	0	0	0	0	1	0	0	0	0	
	P_3	135	0	–3	0	0	5	0	0	0	
	P_2	0	0	0	0	0	0	0	0	1	
	P_1	10	0	–1	0	1	1	0	0	0	
				↑							
x_2	0	10	0	1	1	–1	–1	0	0	0	
x_1	0	70	1	0	0	0	1	0	0	0	
d_3^-	$3P_3$	35	0	0	–1	1	1	1	0	0	
d_4^-	0	10	0	0	0	①	0	0	1	–1	

$c_j - z_j$	P_4	0	0	0	1	1	0	0	0	0
	P_3	105	0	0	3	-3	2	0	0	0
	P_2	0	0	0	0	0	0	0	0	1
	P_1	0	0	0	1	0	0	0	0	0
x_2	0	20	0	1	1	0	-1	0	1	-1
x_1	0	70	1	0	0	0	1	0	0	0
d_3^-	$3P_3$	25	0	0	-1	0	1	1	-1	1
d_1^+	P_4	10	0	0	0	1	0	0	1	-1
$c_j - z_j$	P_4	10	0	0	0	0	0	0	-1	1
	P_3	75	0	0	3	0	2	0	3	-3
	P_2	0	0	0	0	0	0	0	0	1
	P_1	0	0	0	1	0	0	0	0	0

From the above table we observe that the first two goals are fully achieved. In third priority goal P_3, the most negative entry is -3 which occurs in column d_4^+. But in P_2 (higher priority) row, the element below -3 is positive so that we can not improve P_3 and similarly P_4.

$\therefore$ solution is given by

$$x_1 = 70, \ x_2 = 20, \ d_1^+ = 10, \ d_3^- = 25, \ d_1^- = 0, \ d_2^- = 0, \ d_4^- = 0, \ d_4^+ = 0$$

and in P_1 and P_2 rows there is zero in column x_B so that the goals P_1 and P_2 are fully achieved and there is 10 hours $(d_1^+ = 10)$ over achievement of the plants and $d_3^- = 25$. So that 25000 meters under achievement in the sales goal of material B is obtained.

Hence, the required solution is given by

$$x_1 = 70, x_2 = 20$$

$\Rightarrow$ company should produce 70000 meters of material A and 20,000 meters of material B.

EXAMPLE 5. *A company manufacturers two products radios and transistors which must be processed through assembly and finishing department. Assembly has 90 hours available, finishing can handle upto 72 hours of work. Manufacturing one radio requires 6 hours in assembly and 3 hours in finishing. Each transistor requires 3 hours in assembly and 6 hours in finishing. The profit is ₹120 per radio and ₹90 per transistor. The company has established the following goals and has assigned them priorities P_1, P_2, P_3 (P_1 is most important) as follows:*

P_1 : Produce to meet a radio goal of 13

P_2 : Reach a profit goal of ₹1950

P_3 : Produce to meet a transistor goal of 5

Formulate the problem as a GPP and find the optimum solution.

SOLUTION: Let x_1, x_2 be the number of radios and transistors manufactured respectively. Also let,

d_1^- = amount by which the profit goal is under achieved

d_1^+ = amount by which the profit goal is over achieved

d_2^- = amount by which the radio goal is under achieved

d_2^+ = amount by which the radio goal is over achieved

d_3^- = amount by which the transistor goal is under achieved

d_3^+ = amount by which the transistor goal is over achieved

Then proceeding as usual, the given problem can be formulated as a GPP as follows:

$$\text{Min. } Z = P_1 d_2^- + P_2 d_1^- + P_3 d_3^-$$

subject to the constraints

$$120x_1 + 90x_2 + d_1^- - d_1^+ = 1950$$
$$x_1 + d_2^- - d_2^+ = 13$$
$$x_2 + d_3^- - d_3^+ = 5$$
$$6x_1 + 3x_2 \leq 90$$
$$3x_1 + 6x_2 \leq 72$$

and $\qquad x_1, x_2, d_1^-, d_1^+, d_2^-, d_2^+, d_3^-, d_3^+ \geq 0$

Now to solve the above GPP, we use the slack variables s_1 and s_2 such that

$$\text{Min. } Z = P_1 d_2^- + P_2 d_1^- + P_3 d_3^-$$

subject to the constraints

$$120x_1 + 90x_2 + d_1^- - d_1^+ = 1950$$
$$x_1 + d_2^- - d_2^+ = 13$$
$$x_2 + d_3^- - d_3^+ = 5$$
$$6x_1 + 3x_2 + s_1 = 90$$
$$3x_1 + 6x_2 + s_2 = 72$$

and $\qquad x_1, x_2, d_1^-, d_1^+, d_2^-, d_2^+, d_3^-, d_3^+, s_1, s_2 \geq 0$

Now we formulate the starting table as follows:

Initial Table-1

B.V.		$c_j \rightarrow$	0	0	0	0	P_2	0	P_1	0	P_3	0	Min. Ratio
	C_B	X_B	x_1	x_2	s_1	s_2	d_1^-	d_1^+	d_2^-	d_2^+	d_3^-	d_3^+	X_B / x_1
d_1^-	P_2	1950	120	90	0	0	1	−1	0	0	0	0	1950/120
d_2^-	P_1	13	①	0	0	0	0	0	1	−1	0	0	13/1 (min.)→
d_3^-	P_3	5	0	1	0	0	0	0	0	0	1	−1	—
s_1	0	90	6	3	1	0	0	0	0	0	0	0	90/6
s_2	0	72	3	6	0	1	0	0	0	0	0	0	72/3
$c_j - z_j$	P_3	5	0	1	0	0	0	0	0	0	0	1	
	P_2	1950	−120	−90	0	0	0	−1	0	0	0	0	
	P_1	13	−1	0	0	0	0	0	0	1	0	0	
			↑										

In the above table

$$c_1 - z_1 = 0 - (P_2, P_1, P_3, 0, 0) \begin{bmatrix} 120 \\ 1 \\ 0 \\ 6 \\ 3 \end{bmatrix} = -120P_2 - P_1 + 0P_3$$

$$c_2 - z_2 = 0 - (P_2, P_1, P_3, 0, 0) \begin{bmatrix} 90 \\ 0 \\ 1 \\ 3 \\ 6 \end{bmatrix} = -90P_2 + 0P_1 + 1P_3$$

$$c_3 - z_3 = 0 - (P_2, P_1, P_3, 0, 0) \begin{bmatrix} 0 \\ 0 \\ 0 \\ 1 \\ 0 \end{bmatrix} = 0P_2 + 0P_1 + 0P_3$$

$$c_4 - z_4 = 0 - (P_2, P_1, P_3, 0, 0) \begin{bmatrix} 0 \\ 0 \\ 0 \\ 0 \\ 1 \end{bmatrix} = 0P_2 + 0P_1 + 0P_3$$

$$c_5 - z_5 = P_2 - (P_2, P_1, P_3, 0, 0) \begin{bmatrix} 1 \\ 0 \\ 0 \\ 0 \\ 0 \end{bmatrix} = 0P_2 + 0P_1 + 0P_3$$

$$c_6 - z_6 = 0 - (P_2, P_1, P_3, 0, 0) \begin{bmatrix} -1 \\ 0 \\ 0 \\ 0 \\ 0 \end{bmatrix} = P_2 + 0P_1 + 0P_3$$

$$c_7 - z_7 = P_1 - (P_2, P_1, P_3, 0, 0) \begin{bmatrix} 0 \\ 1 \\ 0 \\ 0 \\ 0 \end{bmatrix} = 0P_2 + 0P_1 + 0P_3$$

$$c_8 - z_8 = 0 - (P_2, P_1, P_3, 0, 0) \begin{bmatrix} 0 \\ -1 \\ 0 \\ 0 \\ 0 \end{bmatrix} = 0P_2 + 1 \cdot P_1 + 0P_3$$

$$c_9 - z_9 = P_3 - (P_2, P_1, P_3, 0, 0) \begin{bmatrix} 0 \\ 0 \\ 1 \\ 0 \\ 0 \end{bmatrix} = 0P_2 + 0P_1 + 0P_3$$

$$\text{and} \quad c_{10} - z_{10} = 0 - (P_2, P_1, P_3, 0, 0) \begin{bmatrix} 0 \\ 0 \\ -1 \\ 0 \\ 0 \end{bmatrix} = 0P_2 + 0P_1 + 1 \cdot P_3$$

$$\text{and} \quad Z = C_B x_B = (P_2, P_1, P_3, 0, 0) \begin{bmatrix} 1950 \\ 13 \\ 5 \\ 90 \\ 72 \end{bmatrix} = 1950P_2 + 13P_1 + 5P_3$$

Now the most negative entry in P_1 is -1 in the first column
$\Rightarrow x_1$ is the incoming variable and by minimum ratio rule d_2^- is the outgoing variable
So, the key element = 1 *i.e.*, a_{21}
Now proceeding as usual, we have the following revised table

Revised Table

B.V.	C_B	X_B	$c_j \rightarrow$ 0 x_1	0 x_2	0 s_1	0 s_2	P_2 d_1^-	0 d_1^+	P_1 d_2^-	0 d_2^+	P_3 d_3^-	0 d_3^+	Min. Ratio X_B / d_2^+
d_1^-	P_2	390	0	90	0	0	1	−1	−120	120	0	0	390/120
x_1	0	13	1	0	0	0	0	0	1	−1	0	0	—
d_3^-	P_3	5	0	1	0	0	0	0	0	0	1	−1	—
s_1	0	12	0	3	1	0	0	0	−6	⑥	0	0	12/6(min.)→
s_2	0	33	0	0	0	1	0	0	−3	3	0	0	33/3
$c_j - z_j$	P_3	5	0	−1	0	0	0	0	0	0	0	1	
	P_2	390	0	−90	0	0	0	1	120	−120	0	0	
	P_1	0	0	0	0	0	0	0	1	0	0	0	
										↑			

In the above table

$$c_2 - z_2 = 0 - (P_2, 0, P_3, 0, 0) \cdot \begin{bmatrix} 90 \\ 0 \\ 1 \\ 3 \\ 6 \end{bmatrix} = -90P_2 - P_3$$

$$c_6 - z_6 = 0 - (P_2, 0, P_3, 0, 0) \begin{bmatrix} -1 \\ 0 \\ 0 \\ 0 \\ 0 \end{bmatrix} = P_2$$

$$c_7 - z_7 = P_1 - (P_2, 0, P_3, 0, 0) \begin{bmatrix} -120 \\ -1 \\ 0 \\ -6 \\ -3 \end{bmatrix} = P_1 + 120P_2$$

$$c_8 - z_8 = 0 - (P_2, 0, P_3, 0, 0) \begin{bmatrix} 120 \\ -1 \\ 0 \\ 6 \\ 3 \end{bmatrix} = -120P_2$$

$$\text{and} \quad c_{10} - z_{10} = 0 - (P_2, 0, P_3, 0, 0) \begin{bmatrix} 0 \\ 0 \\ -1 \\ 0 \\ 0 \end{bmatrix} = P_3$$

The value of $c_j - z_j$ in P_1, P_2, P_3 rows may also be found easily by making 1 at the place of key element use it to reduce all entries in P_1, P_2, P_3 rows corresponding to the column of key element to zero.
Also,

$$Z = C_B X_B = (P_2, 0, P_3, 0, 0) \begin{bmatrix} 390 \\ 13 \\ 5 \\ 12 \\ 33 \end{bmatrix} = 390P_2 + 5P_3$$

Clearly all entries in P_1-row are ≥ 0, so the priority goal P_1 is achieved. Now we have to proceed to achieve the goal P_2 without affecting the achievement of top priority goal P_1.
In the P_2-row of the above table most negative value is -120 in column

corresponding to variable d_2^+, which is taken as entering variable. Now by minimum ratio rule, x_3 in 4th row is the outgoing vector

$\Rightarrow$ $6(= a_{48})$ is the key element

Now following the usual procedure, we have the following reduced table.

Reduced Table-1

	c_j		0	0	0	0	P_2	0	P_1	0	P_3	0	Min. Ratio
B.V.	C_B	X_B	x_1	x_2	s_1	s_2	d_1^-	d_1^+	d_2^-	d_2^+	d_3^-	d_3^+	X_B / x_2
d_1^-	P_2	150	0	30	−20	0	1	−1	0	0	0	0	150/30
x_1	0	15	1	1/2	1/6	0	0	0	0	0	0	0	15/(1/2)
d_3^-	P_3	5	0	1	0	0	0	0	0	0	1	−1	5/1
d_2^+	0	2	0	①/2	1/6	0	0	0	−1	1	0	0	2/(1/2)(min)→
s_2	0	27	0	9/2	−1/2	1	0	0	0	0	0	0	27/(9/2)=6
$c_j - z_j$	P_3	5	0	−1	0	0	0	0	0	0	0	1	
	P_2	150	0	−30	20	0	0	1	0	0	0	0	
	P_1	0	0	0	0	0	0	0	1	0	0	0	
				↑									

In the above table, for non-basic variables

$$c_2 - z_2 = 0 - (P_2, 0, P_3, 0, 0)\begin{bmatrix} 30 \\ 1/2 \\ 1 \\ 1/2 \\ 9/2 \end{bmatrix} = -30P_2 - P_3$$

$$c_3 - z_3 = 0 - (P_2, 0, P_3, 0, 0)\begin{bmatrix} -20 \\ 1/6 \\ 0 \\ 1/6 \\ -1/2 \end{bmatrix} = 20P_2$$

$$c_6 - z_6 = 0 - (P_2, 0, P_3, 0, 0)\begin{bmatrix} -1 \\ 0 \\ 0 \\ 0 \\ 0 \end{bmatrix} = P_2$$

$$c_7 - z_7 = P_1 - (P_2, 0, P_3, 0, 0)\begin{bmatrix} 0 \\ 0 \\ 0 \\ -1 \\ 0 \end{bmatrix} = P_1$$

$$c_{10} - z_{10} = 0 - (P_2, 0, P_3, 0, 0) \begin{bmatrix} 0 \\ 0 \\ -1 \\ 0 \\ 0 \end{bmatrix} = P_3$$

$$\text{and} \quad Z = C_B x_B = (P_2, 0, P_3, 0, 0) \begin{bmatrix} 150 \\ 15 \\ 5 \\ 2 \\ 27 \end{bmatrix} = 150 P_2 + 5 P_3$$

Again in P_2-row, $c_2 - z_2$ is negative $\Rightarrow$ solution is not optimal from P_2 point of view. So, we take x_2 in second column corresponding to most negative entry in P_2 row as entering vector, by minimum ratio rule d_2^+ in 4th row is the outgoing vector, therefore the key element is $\frac{1}{2} (= a_{42})$.

Now following the usual procedure, we have the following reduced table.

Reduced Table-2

B.V.		$c_j \rightarrow$	0	0	0	0	P_2	0	P_1	0	P_3	0	Min. Ratio
	C_B	X_B	x_1	x_2	s_1	s_2	d_1^-	d_1^+	d_2^-	d_2^+	d_3^-	d_3^+	
d_1^-	P_2	30	0	0	−30	0	1	−1	60	−60	0	0	
x_1	0	13	1	0	0	0	0	0	1	−1	0	0	
d_3^-	P_3	1	0	0	−1/3	0	0	0	2	−2	1	−1	
x_2	0	4	0	1	1/3	0	0	0	−2	2	0	0	
s_2	0	9	0	0	−2	1	0	0	9	−9	0	0	
$c_j - z_j$	P_3	5	0	0	1/3	0	0	0	−2	2	0	1	
	P_2	150	0	0	30	0	0	1	−60	60	0	0	
	P_1	0	0	0	0	0	0	0	1	0	0	0	

Now,

$$c_3 - z_3 = 0 - (P_2, 0, P_3, 0, 0) \begin{bmatrix} -30 \\ 0 \\ -1/3 \\ 1/3 \\ -2 \end{bmatrix} = 30 P_2 + \frac{1}{3} P_3$$

$$c_6 - z_6 = 0 - (P_2, 0, P_3, 0, 0) \begin{bmatrix} -1 \\ 0 \\ 0 \\ 0 \\ 0 \end{bmatrix} = P_2$$

$$c_7 - z_7 = P_1 - (P_2, 0, P_3, 0, 0) \begin{bmatrix} 60 \\ 1 \\ 2 \\ -2 \\ 9 \end{bmatrix} = P_1 - 60P_2 - 2P_3$$

$$c_8 - z_8 = 0 - (P_2, 0, P_3, 0, 0) \begin{bmatrix} -60 \\ -1 \\ -2 \\ 2 \\ -9 \end{bmatrix} = 60P_2 + 2P_3$$

$$c_{10} - z_{10} = 0 - (P_2, 0, P_3, 0, 0) \begin{bmatrix} 0 \\ 0 \\ -1 \\ 0 \\ 0 \end{bmatrix} = P_3$$

$$\text{and} \quad Z = C_B X_B = (P_2, 0, P_3, 0, 0) \begin{bmatrix} 150 \\ 15 \\ 5 \\ 2 \\ 27 \end{bmatrix} = 150P_2 + 5P_3$$

We observe that in the above table –60 is the negative entry in P_2. But P_2 can not be improved further as there is a positive entry below this element in P_1-row (top priority). Similarly if we move to improve P_3 then it is also not possible as there is positive entry in row P_1 below the negative entry in row P_3.

$\Rightarrow$ P_2 and P_3 can not be improved further.

Therefore, the solution of the above GPP is given as below:

$$x_1 = 13, x_2 = 4 \qquad d_1^- = 30, d_3^- = 1 \qquad d_1^+ = 0 = d_2^- = d_2^+ = d_3^-$$

$\Rightarrow$ Total 13 radios and 4 transistors should be manufactured.

Also, we observe that the first priority goal P_1 is fully achieved. The second priority goal P_2 is missed by ₹30 ($\because$ Profit $= 120 \times 13 + 90 \times 4 = 1920$ and 1950–1920=30) and the last priority goal P_3 is also missed by 1 transistor ($\because$ 5–4=1).

EXAMPLE 6. *A manufacturer produces two products A and B. Each product requires time in two production departments. Product-A requires 20 hours in department-1 and 10 hours in department-2. Product-B requires 10 hrs in department-1 and 10 hours in department-2. Production time is limited in department-1 to 60 hrs and in department-2 to 40 hours. Contribution to profit for the two products is ₹40 and ₹80 respectively. Managements objectives are*

(i) to maximize profit

(ii) at least two units of each product are desired

Management considers the deviation of ₹1 from the profit goal equal to one unit deviation from the product goal.

Formulate the above problem as GPP. Also, solve it by graphical as well as modified simplex method.

SOLUTION. Let x_1 and x_2 be the no. of units of product A and B produced respectively. Also, let

$d_1^- =$ amount by which the maximum profit of ₹1000 is under achieved

$d_1^+ =$ amount by which the maximum profit of ₹1000 is over achieved

$d_2^- =$ no. of units by which the product of 2 units of product A is under achieved

$d_2^+ =$ no. of units by which the product of 2 units of product A is over achieved

$d_3^- =$ no. of units by which the product of 2 units of product B is under achieved

$d_3^+ =$ no. of units by which the product of 2 units of product B is over achieved

Then following the usual procedure, we have the following GPP

$$\text{Min. } Z = d_1^- + d_2^- + d_3^-$$

subject to the constraints

$$20x_1 + 10x_2 \le 60 \qquad \text{[time in dep.1 constraints]}$$
$$10x_1 + 10x_2 \le 40 \qquad \text{[time in dep.2 constraints]}$$
$$40x_1 + 80x_2 + d_1^- - d_1^+ = 1000 \quad \text{[max. profit goal]}$$
$$x_1 + d_2^- - d_2^+ = 2 \qquad \text{[Production goal of unit A]}$$
$$x_2 + d_3^- - d_3^+ = 2 \qquad \text{[Production goal of unit B]}$$

and $\qquad x_1, x_2, d_1^-, d_1^+, d_2^-, d_2^+, d_3^-, d_3^+ \ge 0$

Further since, priorities of the goals is not given and the deviation of ₹1 from the profit goal equals to one unit deviation from the product goal, therefore this problem can be solved as usual by graphical or simplex method.

(i) Solution by Graphical Method

Taking the deviational variables equal to zero, plot all the lines on the graph as shown in the figure and attach the arrows associated with the line to represent under and over achievement of the goals.

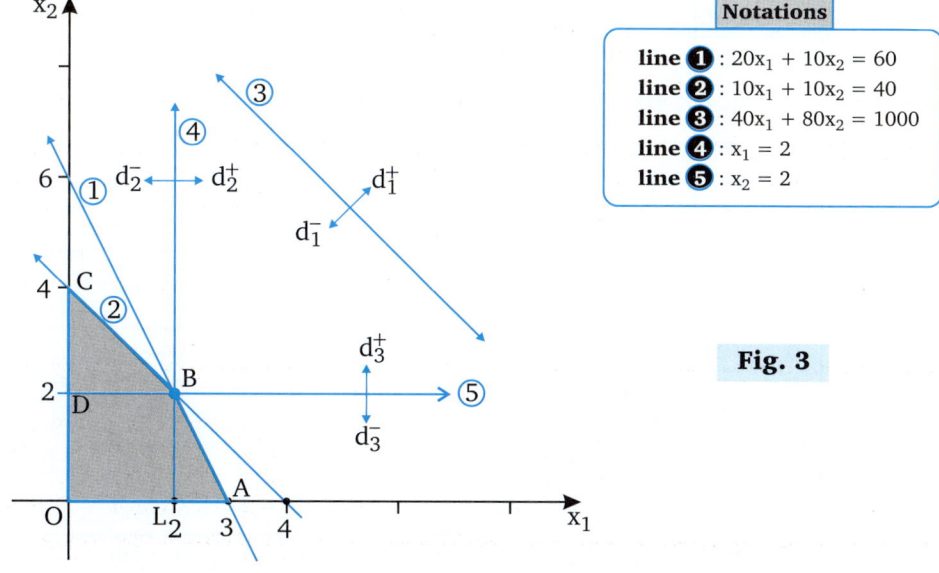

Notations

line ❶ : $20x_1 + 10x_2 = 60$
line ❷ : $10x_1 + 10x_2 = 40$
line ❸ : $40x_1 + 80x_2 = 1000$
line ❹ : $x_1 = 2$
line ❺ : $x_2 = 2$

Fig. 3

Clearly *OABCO* is the permissible region satisfying (1) and (2). The goals of the problem is to minimize d_1^-, d_2^-, d_3^-.

From the figure, it is clear that d_2^- is 0 (minimum) on the right of the line *LB* and d_3^- is 0 (minimum) above the line *BD* while d_1^- is 0 in the region not containing origin divided the line 3, which is far away from the permissible region *OABCO*. So, $d_1^- \neq 0$, and $d_1^+ = 0$

Moving towards the permissible region, we observe that point C of the permissible region is nearest to the line 3.

So, at C, $x_1 = 0, x_2 = 4$ and at C the maximum profit is missed by $d_1^- = 1000 - 0 - 80 \times 4 = $ ₹680, which can be taken as the satisfactory optimal solution.

(ii) Solution by Simplex Method

Introducing slack variables s_1 and s_2, the given problem becomes

$$\text{min. } Z = d_1^- + d_2^- + d_3^-$$

subject to the constraints

$$20x_1 + 10x_2 + s_1 = 60$$
$$10x_1 + 10x_2 + s_2 = 40$$
$$40x_1 + 80x_2 + d_1^- - d_1^+ = 1000$$
$$x_1 + d_2^- - d_2^+ = 2$$
$$x_2 + d_3^- - d_3^+ = 2$$

and $x_1, x_2, s_1, s_2, d_1^-, d_1^+, d_2^-, d_2^+, d_3^-, d_3^+ \geq 0$

Now apply the simplex method in a usual manner, we have the following computational table.

Simplex Table

B.V.	c_j		0	0	0	0	1	0	1	0	1	0	Min. Ratio
	C_B	X_B	x_1	x_2	s_1	s_2	d_1^-	d_1^+	d_2^-	d_2^+	d_3^-	d_3^+	X_B / x_2
s_1	0	60	20	10	1	0	0	0	0	0	0	0	60/10
s_2	0	40	10	10	0	1	0	0	0	0	0	0	40/10
d_1^-	1	1000	40	80	0	0	1	−1	0	0	0	0	1000/80
d_2^-	1	2	1	0	0	0	0	0	1	−1	0	0	—
d_3^-	1	2	0	①	0	0	0	0	0	0	1	−1	2/1(min.)→
Z=1004	$c_j - z_j$		−41	−81 ↑	0	0	0	1	0	1	0	1	X_B / d_3^+
s_1	0	40	20	0	1	0	0	0	0	0	−10	10	40/10
s_2	0	20	10	0	0	1	0	0	0	0	−10	⑩	20/10(min)→
d_1^-	1	840	40	0	0	0	1	−1	0	0	−80	80	840/80
d_2^-	1	2	1	0	0	0	0	0	1	−1	0	0	—
x_2	0	2	0	1	0	0	0	0	0	0	1	−1	—
Z=922	$c_j - z_j$		−41	0	0	0	0	1	0	1	80	−80 ↑	

s_1	0	20	10	0	1	-1	0	0	0	0	0	0
d_3^+	0	2	1	0	0	1/10	0	0	0	0	-1	1
d_1^-	1	680	-40	0	0	-8	1	-1	0	0	0	0
d_2^-	1	2	1	0	0	0	0	0	1	-1	0	0
x_2	0	4	1	1	0	1/10	0	0	0	0	0	0
$Z=762$	$c_j - z_j$		39	0	0	8	0	1	0	1	0	0

We observe that in the last row of the above table all $c_j - z_j \geq 0$

$\Rightarrow$ solution is optimal and is given by

$$x_1 = 0, x_2 = 4, \quad d_1^- = 680, d_2^- = 2, d_3^+ = 2, \quad d_1^+ = d_2^+ = d_3^- = 0$$

Hence, we conclude that the solution does not achieve the production target of at least two units of each product A and B and also the profit target is missed by ₹680.

EXAMPLE 7. *A manufacturing firm produces two types of product A and B. According to past experience, production of either product A or B requires an average of one hour in the plant. The plant has a normal production capacity of 400 hours a month. The marketing department of the firm report that because of the limited market, the maximum number of product A and B that can be sold in a month is 240 and 300 respectively. The net profit from the sale of products A and B are ₹800 and ₹400 respectively. The manager of the firm has set the following goals arranged in the order of importance.*

P_1 : *He wants to avoid any under-utilization of normal production capacity*

P_2 : *He wants to sale possible units of products A and B. Since the net profit from the sale of product A is twice the amount from product B, therefore the manager has twice as much desire to achieve sales for product A as for product B.*

P_3 : *He wants to minimize the overtime operation of the plant as much as possible.*

Formulate the problem as the GPP. Also solve it by graphical as well as modified simplex method.

SOLUTION. Let x_1 and x_2 be the number of products A and B respectively.

Also, let

$\qquad d_1^-$ = hours by which the production capacity is under utilized

$\qquad d_1^+$ = hours by which the production capacity is over utilized

$\qquad d_2^-$ = number by which sale of product A is under achieved

$\qquad d_3^-$ = number by which sale of product B is under achieved

Then, following the usual procedure, the required GPP is given by

$$\text{Min. } Z = P_1 d_1^- + P_2(2d_2^- + d_3^-) + P_3 d_1^+$$

subject to the constraints

$$x_1 + x_2 + d_1^- - d_1^+ = 400 \qquad \text{(production goal)}$$
$$x_1 + d_2^- = 240 \qquad \text{(sale of unit 1 goal)}$$
$$x_2 + d_3^- = 300 \qquad \text{(sale of unit 2 goal)}$$

and $\qquad x_1, x_2, d_1^-, d_2^-, d_3^- \geq 0$

Since, maximum number of sale of product A and B are given, so there is no need of over achievements d_2^+ and d_3^+.

Solution by Graphical Method

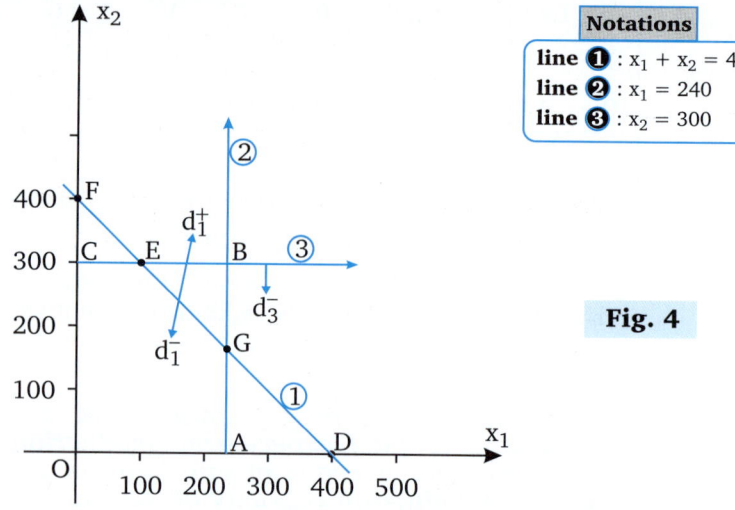

Notations

line **❶** : $x_1 + x_2 = 400$
line **❷** : $x_1 = 240$
line **❸** : $x_2 = 300$

Fig. 4

Clearly, the first priority goal $d_1^- = 0$ is fully achieved on the line DF and in the above it away from the origin. In the second priority goal, the sales goal of product A is completely achieved on the line AB and on the right of it while the sale goal of product is completely achieved on the line BC and above it. So first two goals are completely achieved at the point $B(240, 300)$. At this point the third priority is not achieved and it cannot be achieved at the expense of the first two goals.

Hence, the solution of the problem is given by

$$x_1 = 240, x_2 = 300, \ d_1^- = 0 = d_2^- = d_3^-$$
$$d_1^+ = 240 + 300 + 0 - 400 = 140$$

Solution by Modified Simplex Method

Following the usual procedure we have the following initial table.

Initial Table

B.V.			0	0	P_1	P_3	$2P_2$	P_2	Min. Ratio
	$c_j \rightarrow$								
	C_B	X_B	x_1	x_2	d_1^-	d_1^+	d_2^-	d_3^-	X_B / x_1
d_1^-	P_1	400	1	1	1	−1	0	0	400/1
d_2^-	$2P_2$	240	①	0	0	0	1	0	240/1 (min.) →
d_3^-	P_2	300	0	1	0	0	0	1	—
$c_j - z_j$	P_3	0	0	0	0	1	0	0	
	P_2	780	−2	−1	0	0	0	0	
	P_1	400	−1	−1	0	−1	0	0	
			↑						

In the above table (for non-basic variables)

$$c_1 - z_1 = 0 - (P_1, 2P_2, P_2) \begin{bmatrix} 1 \\ 1 \\ 0 \end{bmatrix} = -P_1 - 2P_2$$

$$c_2 - z_2 = 0 - (P_1, 2P_2, P_2) \begin{bmatrix} 1 \\ 0 \\ 1 \end{bmatrix} = -P_1 - P_2$$

$$c_4 - z_4 = P_3 - (P_1, 2P_2, P_2) \begin{bmatrix} 1 \\ 0 \\ 0 \end{bmatrix} = -P_1 + P_3$$

and $\qquad Z = C_B x_B = (P_1, 2P_2, P_2) \begin{bmatrix} 400 \\ 240 \\ 300 \end{bmatrix} = 400P_1 + 780P_2$

key element is $= 1 (a_{21})$

Now we have the following reduced table

Reduced Table-1

B.V.	$c_j \rightarrow$		0	0	P_1	P_3	$2P_2$	P_2	Min. Ratio
	C_B	X_B	x_1	x_2	d_1^-	d_1^+	d_2^-	d_3^-	X_B / x_2
d_1^-	P_1	160	0	①	1	−1	−1	0	160/1(min.)→
x_1	0	240	1	0	0	0	1	0	—
d_3^-	P_2	300	0	1	0	0	0	1	300/1
$c_j - z_j$	P_3	0	0	0	0	1	0	0	
	P_2	300	0	−1	0	0	2	0	
	P_1	160	0	−1	0	1	1	0	
				↑					

In the above table (for non-basic variables)

$$c_2 - z_2 = 0 - (P_1, 0, P_2) \begin{bmatrix} 1 \\ 0 \\ 1 \end{bmatrix} = -P_1 - P_2;$$

$$c_4 - z_4 = P_3 - (P_1, 0, P_2) \begin{bmatrix} -1 \\ 0 \\ 0 \end{bmatrix} = P_1 + P_3$$

$$c_5 - z_5 = 2P_2 - (P_1, 0, P_2) \begin{bmatrix} -1 \\ 1 \\ 0 \end{bmatrix} = P_1 + 2P_2$$

$$\text{and} \quad Z = C_B x_B = (P_1, 0, P_2) \begin{bmatrix} 160 \\ 240 \\ 300 \end{bmatrix} = 160P_1 + 300P_2$$

The key element $= 1(a_{12})$

Further, the next reduced table is given as under.

Reduced Table-2

B.V.			$c_j \rightarrow$	0	0	P_1	P_3	$2P_2$	P_2	Min. Ratio
	C_B	X_B	x_1	x_2	d_1^-	d_1^+	d_2^-	d_3^-	X_B / d_1^+
d_1^-	P_1	400	1	1	1	−1	0	0	−ve
d_2^-	$2P_2$	240	①	0	0	0	1	0	—
d_3^-	P_2	300	0	1	0	0	0	1	140/1(min.)$\rightarrow$
$c_j - z_j$	P_3	0	0	0	0	1	0	0	
	P_2	140	0	0	1	−1	1	0	
	P_1	0	0	0	1	0	0	0	
						↑			

In the above table (for non-basic variables)

$$c_3 - z_3 = P_1 - (0,0,P_2)\begin{bmatrix} 1 \\ 0 \\ -1 \end{bmatrix} = P_1 + P_2, \quad c_4 - z_4 = P_3 - (0,0,P_2)\begin{bmatrix} -1 \\ 0 \\ 1 \end{bmatrix} = -P_2 + P_3$$

$$c_5 - z_5 = 2P_2 - (0,0,P_2)\begin{bmatrix} -1 \\ 1 \\ 1 \end{bmatrix} = P_2 \quad \text{and} \quad Z = C_B x_B = (0,0,P_2)\begin{bmatrix} 160 \\ 240 \\ 140 \end{bmatrix} = 140P_2$$

Then key element $= 1(a_{21})$

Then next reduced table is given as below

Reduced Table-3

B.V.			$c_j \rightarrow$	0	0	P_1	P_3	$2P_2$	P_2
	C_B	X_B	x_1	x_2	d_1^-	d_1^+	d_2^-	d_3^-
x_2	0	300	0	1	0	0	0	1
x_1	0	240	1	0	0	0	1	0
d_1^+	P_3	140	0	0	−1	1	1	1
$c_j - z_j$	P_3	140	0	0	1	0	−1	−1
	P_2	0	0	0	0	0	2	1
	P_1	0	0	0	1	0	0	0
						↑		

Here, $c_3 - z_3 = P_1 - (0,0,P_3)\begin{bmatrix} 0 \\ 0 \\ -1 \end{bmatrix} = P_1 + P_3$

$$c_5 - z_5 = 2P_2 - (0,0,P_3)\begin{bmatrix} 0 \\ 1 \\ 1 \end{bmatrix} = 2P_2 - P_3$$

$$c_6 - z_6 = P_2 - (0,0,P_3)\begin{bmatrix} 1 \\ 0 \\ 1 \end{bmatrix} = P_2 - P_3$$

and $\qquad Z = C_B X_B = (0,0,P_3)\begin{bmatrix} 400 \\ 240 \\ 140 \end{bmatrix} = 140P_3$

In the last table we observe that all $c_j - z_j \geq 0$ in P_1 and P_2 rows and $(0,0)$ in X_B column. Therefore the priority goals P_1 and P_2 are fully achieved. In P_3 - row all $c_j - z_j$ are not non-negative and the most negative entries are $-1, -1$ in this row. But we cannot improve P_3 further because below $-1, -1$ in the last two columns, there are positive entries in higher priority P_2.

Also, $x_1 = 300, x_2 = 240, d_1^+ = 140, d_1^- = 0, d_2^- = d_3^- = 0$

Hence, the optimal solution is given by

$$x_1 = 300, x_2 = 240$$

i.e. first two goals are fully achieved while third priority goal is missed by 140 hours (140 hours overtime is required)

 Exercise-11.1

1. Find x_1, x_2 to minimize $Z = (d_1^-, d_2^-)$ subject to the constraints

$$x_1 + x_2 + d_1^- - d_1^+ = 20$$
$$4x_1 + 5x_2 + d_2^- - d_2^+ = 150$$

and $\qquad x_1, x_2, d_1^-, d_1^+, d_2^-, d_2^+ \geq 0$

2. Find x_1, x_2 to minimize $Z = (3d_1^+ + 2d_2^+, d_3^- d_4^-)$ subject to the constraints

$$x_1 + x_2 + d_1^- - d_1^+ = 8$$
$$x_1 + d_2^- - d_2^+ = 3$$
$$3x_1 + 5x_2 + d_3^- - d_3^+ = 65$$
$$x_1 + x_2 + d_4^- - d_4^+ = 65$$

and $x_1, x_2, d_1^-, d_1^+, d_2^-, d_2^+, d_3^-, d_3^+, d_4^-, d_4^+ \geq 0$

3. Solve the following GPP by modified simplex method

Min. $Z = P_1 d_1^- + P_2 d_4^+ + P_3 (2d_2^-$
$\qquad \qquad + d_3^-) + P_4 d_1^+$

subject to the constraints

$$x_1 + x_2 + d_1^- - d_1^+ = 10$$
$$x_1 + d_2^- = 6$$
$$x_2 + d_3^- = 8$$
$$d_1^+ + d_4^- - d_4^+ = 2$$

and $x_1, x_2, d_1^-, d_1^+, d_2^-, d_3^-, d_4^-, d_4^+ \geq 0$

ANSWERS

1. $x_1 = \dfrac{75}{2}, x_2 = 0$, Min. $Z = (0,0), d_1^- = \dfrac{35}{2}$ (or) $x_1 = 0, x_2 = 30$, Min. $Z = (0,0), d_1^- = 10$

2. $x_1 = 0, x_2 = 8$, Min. $Z = (0, 25, 27)$ \qquad 3. $x_1 = 6, x_2 = 6, d_3^- = 2, d_1^+ = 2$

REVIEW QUESTIONS

1. Write the application of goal programming.
2. What is goal programming? Distinguish it from linear programming.
3. Define the following:
   (i) Differential weights
   (ii) Deviational variables
   (iii) Priority factors
4. Identify the importance areas where GP can be used effectively.

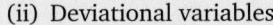 **MULTIPLE CHOICE QUESTIONS (CHOOSE THE MOST APPROPRIATE ONE)**

1. Deviation Variables in G.P. model must satisfy the following conditions: [MEERUT 2013]
   (a) $d_i^+ \times d_i^- = 0$
   (b) $d_i^+ - d_i^- = 0$
   (c) $d_i^+ + d_i^- = 0$
   (d) None of the above

2. If the targeted value of each goal in the solution value, X_B Column is zero, then it indicates: [MEERUT 2013]
   (a) multiple solution
   (b) optimum solution
   (c) infeasible solution
   (d) none of the above

3. The concept of goal programming was introduced by: [MEERUT 2013]
   (a) Chames
   (b) Cooper
   (c) Chames and Cooper
   (d) Lee and Ignizio

4. Goal Programming:
   (a) requires only that decision maker knows whether the goal is direct profit maximization or cost minimization.
   (b) allows you to have multiple goals, with or without priorities.
   (c) is an approach to achieve goal of a solution to all integer LP Problems.
   (d) None of the above

5. The GP approach attempts to achieve each objective:
   (a) sequentially   (b) simultaneously
   (c) both (a) and (b) (d) none of these

6. In GP problem, a constraint having unachieved variable is expressed as:
   (a) an equality constraint
   (b) a less than or equal to type constraints
   (c) a greater than or equal to type constraint
   (d) all of the above

7. The use of GP model preferred when:
   (a) goals are satisfied in an ordinal sequence
   (b) goals are multiple incommensurable
   (c) more than one objective is set to achieve
   (d) all of the above

8. In optimal simplex table of GP problem, two or more $c_j - z_j$ rows indicate:
   (a) unequal priority goals
   (b) equal priority goals
   (c) priority goals
   (d) unattainable goals

9. Consider a goal with constraints:$g_1(x_1, x_2, ..., x_n) + d_1^- \geq b_1(d_1^- \geq 0)$ with d_1^- in the objective function then:
   (a) the goal is to minimize under achievement
   (b) the constraint is achieve provided $d_1^- > 0$
   (c) both (a) and (b)
   (d) none of the above

10. In G.P. problem, a goal constraint having over achievement variable is expressed as a:
    (a) $\geq$ constraint   (b) $\leq$ constraint
    (c) = constraint   (d) all of the above

11. In GP, at optimality which of the following conditions indicated that a goal has been exactly satisfied:
    (a) positive deviational variable is in the solution mix with a negative value.
    (b) both positive and negative deviational variables are in the solution mix.
    (c) both positive and negative deviational variables are not in the solution mix.
    (d) none of the above

12. In simplex method of goal programming, the variable to enter the solution mix is selected with:
    (a) lowest priority row and most negative $c_j - z_j$ value in it
    (b) lowest priority row and largest positive $c_j - z_j$ value in it
    (c) higest priority row and most negative $c_j - z_j$ value in it
    (d) higest priority row and most positive $c_j - z_j$ value in it

13. The deviational variable in the basis of the initial simplex table of GP problem is:
    (a) positive deviational variable
    (b) negative deviational variable
    (c) both (a) and (b)
    (d) artificial variable

14. Consider a goal with constraint $g_1(x_1, x_2, ..., x_n) + d_1^- - d_1^+ = b_1$ and the term $3d_1^- + 2d_1^+$ in the objective function the decision maker:

    (a) prefers $g_1(x_1, x_2, ..., x_n) \geq b_1$, rather than $\leq b_1$
    (b) prefers $g_1(x_1, x_2, ..., x_n) \leq b_1$, rather than $\geq b_1$
    (c) not concerned with either $\leq$ or $\geq$
    (d) none of the above

15. In GP problem goals are assigned priorities such that:

    (a) higher priority goals must be achieved before lower priority goals
    (b) goals may not have equal priority
    (c) goals of greatest importance are given lowest priority
    (d) all of the above

16. For applying a GP approach decision maker must

    (a) set targets for each of the goals
    (b) assign pre-empvitive priority to each goal
    (c) assume that linearity exists in the use of resources to achieve goals
    (d) all of the above

17. Which of the following is a step of algorithm to formulate GP model:

    (a) Identify the goals and constraints on availability of resources (or constraints) which may restrict achievement of the goals (targets)
    (b) Determine priority to be associated with

each goal in such a way that goals with priority level P_1 are most important, those with priority level P_2 are next most important, and so on.

    (c) Define the decision variables
    (d) all are true

18. Which of the following is a step of algorithm to formula GP model:

    (a) Formulate the constraints in same manner as in LP model.
    (b) For each constraint, develop an equation by adding deviational variable d_i^- and d_i^+. These variables indicate the possible deviations below or above the target value (right hand side of each constraint).
    (c) Write the objective function in terms of minimizing a prioritized function of the deviational variables
    (d) All are true

19. Which of the following is a step of obtain graphical solution of goal prgramming:

    (a) graph all system constraints and identify the feasible solution space.
    (b) graph the straight lines corresponding to the goal constraints, labelling the deviational variables.
    (c) written the feasible solutions space identified, in determine the point or points that best satisfy the highest priority goal.
    (d) All are true

Answers

1. (a)	2. (b)	3. (c)	4. (a)	5. (a)	6. (c)	7. (d)	8. (c)	9. (c)
10. (c)	11. (c)	12. (c)	13. (a)	14. (a)	15. (a)	16. (d)	17. (d)	18. (d)
19. (d)								

ARCHIVE

1. A camera company manufactures two types of cameras. The production process for manufacturing the camera is such that two departmental operations are required. To produces their standard camera requires 2 hours of production time in department-1 and 3 hours in department-2. To produce their deluxe model requires 4 hours of production time in department-1 and 3 hours in department-2. This labour time is a same what restrictive factor since the company has a general policy of avoiding overtime, if possible. The manufacturer's profit on each

standard camera is ₹30 while the profit on the deluxe model is ₹40. The management has set the following goals arranged in the order of importance (pre-emptive factors).

P_1 : Avoid overtime operation in each department.

P_2 : Prior-sales records indicate that on the average, a minimum of 10 standard and 10 deluxe cameras can be sold weekly. Management would like to meet these sales goal. Since production time may limit producing the number of each camera and since the deluxe camera has a higher profit margin, the sales goals should be weighted by the profit contribution for the respective camera's *i.e.* ₹30 for the standard camera and ₹40 for the deluxe camera.

P_3: To maximize profit.

Formulate the problem as a GPP. Also solve it by graphical and modified simplex methods.

2. Suppose two products are to be produced in a given department of a manufacturer, quantities of two products are denoted by x_1 and x_2 respectively. A product mix is to be obtained by utilizing two limited resources: labour and raw material. Each unit of the first product requires two hours of labour and three units of raw material. Each unit of the

second product requires 4 hours of labour and 4 units of raw material. Every day 20 hours of labour and 24 units of raw material are available. The goals before the management according to priorities are as follows:

(i) The profit per day should be at least ₹36 assuming that the profit per unit of the products are ₹8 and ₹6 respectively.

(ii) Because of marketing condition as well as production substitutability, the number of units of product 1 should be double the number of units of product 2.

(iii) The labour should be fully utilized.

Formulate this problem as GPP and solve it.

HINTS AND ANSWERS

1. Min. $Z = P_1(d_1^+ + d_2^+) + P_2(3d_3^- + 4d_4^-) + P_3d_5^-$ subject to $2x_1 + 4x_2 + d_1^- - d_1^+ = 80$,

$3x_1 + 3x_2 + d_2^- - d_2^+ = 80$, $x_1 + d_3^- - d_3^+ = 10$, $x_2 + d_4^- - d_4^+ = 10$, $30x_1 + 40x_2 + d_5^- - d_5^+ = 1500$

and $x_1, x_2, d_1^-, d_1^+, d_2^-, d_2^+, d_3^-, d_3^+, d_4^-, d_4^+, d_5^-, d_5^+ \geq 0$, solution is $x_1 = \dfrac{40}{3}$, $x_2 = \dfrac{40}{3}$. First two priorities are achieved fully second priority goal is over achieved by $\dfrac{10}{3}$ unit each, third is missed by $\dfrac{1700}{3}$ rupees

2. Min. $Z = P_1d_1^- + P_2(d_2^- + d_2^+) + P_3d_3^-$ subject to $8x_1 + 6x_2 + d_1^- - d_1^+ = 36$ $x_1 - 2x_2 + d_2^- - d_2^+ = 0$,

$2x_1 + 4x_2 + d_3^- = 20$, $3x_1 + 4x_2 + d_4^- = 24$ and $x_1, x_2, d_i^-, d_i^+ \geq 0$. The optimal solution is

$x_1 = \dfrac{24}{5}, x_2 = \dfrac{12}{5}$, the total profit is $\dfrac{264}{5}$. Goal 1 is over achieved with the quantum of over achievement $d_1^+ = \dfrac{84}{5}$. Goal 2 has been achieved since $d_1^+ = d_1^- = 0$ and $x_1 = 2x_2$. Goal 3 has not been achieved. The degree of under achievement is $d_3^- = \dfrac{4}{5}$. Hence, there is an under utilization of available labour to the extent of $\dfrac{4}{5}$ hours.

Parametric Linear Programming

12.1 INTRODUCTION

In sensitivity analysis, we study the changes in the problem data that can be made without changing the optimal solution *i.e.* we did not concern about the variable that would enter the basis and the variable that would leave the basis *i.e.* in sensitivity analysis, we have consider the impact on optimal solution of LP model due to discrete changes in parameters. In this chapter we will discuss another parameter variation analysis, known as parametric analysis to obtain various feasible solution of LP model which become optimal one after the other due to continuous variations in the parameters.

12.2 PARAMETRIC PROGRAMMING

When LP model parameters changes as a linear function of a single parameter then this technique is called linear parametric programming. In this analysis we have to keep a minimum additional efforts required to take care of changes in the optimal solution due to variation in LP model parameters over a range of variation.

Definition. *The investigation which deals with the effect of simultaneous changes of all components of C or b in the optimal solution of the problem is called parametric linear programming.* [MEERUT–2007, 10]

In this chapter we shall discuss the parametric analysis only for following two parameters:

(1) systematic variation in the objective function coefficients, c_j
(2) systematic variation in resource availability, b_i

12.3 SYSTEMATIC VARIATION IN THE OBJECTIVE FUNCTION COEFFICIENTS, c_j

Let us define the parametric linear programming as follows:

$$\text{Max.} Z = (C + C'\lambda)x = C^* \cdot x$$

subject to the constraints

$$Ax = b$$

$$\text{and} \qquad x \geq 0 \qquad \qquad \qquad \dots(1)$$

We have to find the solution of (1) for each λ such that $\delta \leq \lambda < \phi$ where δ is very small and finite, ϕ finite and large and C, C', b are finite known vectors.

Let us assume that the problem is non-degenerate and has a basic feasible solution, therefore it can be solved for $\lambda = \delta$ by simplex method. Here, we may have following two cases:

(i) the problem has a finite optimal solution for $\lambda = \delta$.

(ii) the problem has no finite optimal solution for $\lambda = \delta$.

CASE (1) WHEN THE PROBLEM HAS A FINITE OPTIMAL SOLUTION FOR $\lambda = \delta$

Let us suppose that $\boldsymbol{B}$ be the optimal basis, X_B the optimal solution, $\boldsymbol{C_B^*}$ the corresponding price vector and z_j^* the value of z_j at $\lambda = \delta$.

Clearly, $X_B = \boldsymbol{B}^{-1}\boldsymbol{b}$ implies it is independent of $\boldsymbol{C}^* = \boldsymbol{C} + \boldsymbol{C'}\lambda$

$\Rightarrow$ feasibility of the solution remains unaffected when $\boldsymbol{C}^*$ is changed due to the change in λ.

Also, for the optimality, we must have $c_j - z_j^* \leq 0$ for all j not in the basis $\boldsymbol{B}$.

We observed that, the condition of optimality will be affected due to the variation in $\boldsymbol{C}^*$, because $c_j^* - z_j^*$ depends upon $\boldsymbol{C}^*$. Therefore, the change in λ, cause the disturbances in optimality. Thus, in case of λ increases through δ, the solution X_B will remain optimal if $c_j^* - z_j^* \leq 0 \ \forall j$.

Now, consider

$$c_j^* - z_j^* = c_j^* - \boldsymbol{C_B^*} \cdot \boldsymbol{B}^{-1} \cdot \alpha_j$$

$$= (c_j + c_j'\lambda) - (\boldsymbol{C_B} + \boldsymbol{C_B'}\lambda)\boldsymbol{B}^{-1}\alpha_j$$

$$= c_j - \boldsymbol{C_B} \cdot \boldsymbol{B}^{-1} \cdot \alpha_j + \lambda(c_j' - \boldsymbol{C_B'} \cdot \boldsymbol{B}^{-1}\alpha_j)$$

$$= c_j - z_j + \lambda(c_j' - z_j') \qquad \qquad \text{...(2)}$$

Clearly, X_B will remain optimal for those λ for which

$$c_j - z_j + \lambda(c_j' - z_j') \leq 0, \text{ for } j \text{ not in the basis.} \qquad \text{...(3)}$$

But if we consider $\lambda = \delta = 0$ (by shifting the origin at δ). Then from (2)

$$c_j^* - z_j^* = c_j - z_j \qquad \qquad \text{(for } j \text{ not in the basis } \boldsymbol{B})$$

$$\leq 0 \qquad \qquad (\because X_B \text{ is optimal and feasible solution)}$$

Now, we have following two cases:

(i) If $c_j' - z_j' \leq 0 \ \forall j$ not in $\boldsymbol{B}$, then from (3) $c_j^* - z_j^* \leq 0$. In this case the solution X_B will be optimal and feasible for all values of $\lambda \geq \delta$.

(ii) If $c_j' - z_j' < 0$ for at least one j, then X_B will be optimal for those λ for which $c_j^* - z_j^* < 0$.

$\Rightarrow \ (c_j - z_j) + \lambda(c_j' - z_j') \leq 0$

which can be written as

$$\lambda \geq -\frac{(c_j - z_j)}{(c_j' - z_j')} \text{ for } c_j' - z_j' > 0$$

and $\qquad \lambda \leq -\frac{(c_j - z_j)}{(c_j' - z_j')} \text{ for } c_j' - z_j' < 0$

On combining both the above inequalities, we have

$$\max_{(c_j' - z_j') < 0} \left[-\frac{c_j - z_j}{c_j' - z_j'} \right] \leq \lambda \leq \min_{c_j' - z_j' > 0} \left[-\frac{c_j - z_j}{c_j' - z_j'} \right] \qquad \text{...(4)}$$

Here, we observe that

(i) If there is no $c'_j - z'_j < 0$, then λ has no lower bound.

(ii) If there is no $c'_j - z'_j > 0$, then λ has no upper bound.

Now, let $\bar{\lambda} = \min\limits_{c'_j - z'_j > 0}\left[-\dfrac{c_j - z_j}{c'_j - z'_j}\right]$ and $\underline{\lambda} = \max\limits_{c'_j - z'_j < 0}\left[-\dfrac{c_j - z_j}{c'_j - z'_j}\right]$

Obviously, the solution X_B will remain optimal and feasible when $\underline{\lambda} \le \lambda \le \bar{\lambda}$ ...(5)

Further if $\bar{\lambda} = \infty$ then given problem has optimal solution for all $\lambda \ge \underline{\lambda}$ which implies that the problem has optimal and feasible solution at $\lambda = \phi$.

Further, if $\bar{\lambda}$ is finite, then we have to improve the range of λ beyond $\bar{\lambda}$ as given below.

If $\bar{\lambda}$ is finite, then suppose that

$$\bar{\lambda} = \min\limits_{c'_j - z'_j > 0}\left\{-\dfrac{c_j - z_j}{c'_j - z'_j}\right\} = -\dfrac{c_k - z_k}{c'_k - z'_k} \qquad ...(6)$$

Now, if at least one $a_{ik} > 0$, then to improve the range of λ, we assume that

$$\lambda = \bar{\lambda} = -\dfrac{c_k - z_k}{c'_k - z'_k}$$

$\Rightarrow \qquad (c_k - z_k) + \lambda(c'_k - z'_k) = 0$

$\Rightarrow \qquad c^*_k - z^*_k = 0 \qquad$ (Using (2))

Also, for $\lambda > \bar{\lambda}$, we have

$$\lambda > -\dfrac{c_k - z_k}{c'_k - z'_k}$$

$\Rightarrow \qquad (c_k - z_k) + \lambda(c'_k - z'_k) > 0$

$\Rightarrow \qquad c'_k - z^*_k > 0$

$\Rightarrow \qquad \Delta^*_k > 0$

$\Rightarrow$ α_k can be introduced in the basis for the better solution of the problem.

Now, if all corresponding $a_{ik} \le 0$ then for $\lambda > \bar{\lambda}$, the given problem has no optimal solution

$\Rightarrow$ We cannot improve the range of λ beyond $\bar{\lambda}$.

Therefore, in case of at least one $a_{ik} > 0$, we can improve the range of λ beyond $\bar{\lambda}$ and α_k is introduced in basis $\boldsymbol{B}$.

If $\boldsymbol{B_1}$ be the basis obtained by introducing α_k for outgoing vector α_l (selected by minimum ratio rule) and let $\boldsymbol{x_{B_1}}$ be the corresponding solution and let $(c^*_j - z^*_j)^{(1)}$ be the value of $c^*_j - z^*_j$ for new basis $\boldsymbol{B_1}$. Then

$$\left.\begin{array}{l}(c^*_j - z^*_j)^{(1)} = (c^*_j - z^*_j) - (c^*_k - z^*_k)\dfrac{a_{ij}}{a_{jk}} \text{ for } j \ne k\end{array}\right] \qquad ...(7)$$

and $(c^*_k - z^*_k)^{(1)} = 0$

Now, for $\lambda = \bar{\lambda}$, we have

$$(c^*_j - z^*_j)^{(1)} = (c^*_j - z^*_j) - (c^*_k - z^*_k)\dfrac{a_{ij}}{a_{jk}}$$

$$= c_j^* - z_j^* \qquad\qquad (\because\ c_k^* - z_k^* = 0 \text{ at } \lambda = \bar{\lambda})$$

$$\Rightarrow \qquad (c_j^* - z_j^*)^{(1)} \le 0 \qquad\qquad (\because\ c_j^* - z_j^* \le 0)$$

Thus, the new solution x_{B_1} is optimal if $\lambda = \bar{\lambda}$. Further, in the next iteration the vector, α_l can not be further introduced if

$$(c_l^* - z_l^*)^{(1)} \le 0$$

Now, $\qquad (c_l^* - z_l^*)^{(1)} = (c_l^* - z_l^*) - (c_k^* - z_k^*)\dfrac{a_{il}}{a_{lk}}$

$$= -\dfrac{c_k^* - z_k^*}{a_{lk}} \qquad\qquad (\because\ \text{when } a_{il} = 1,\ c_l^* - z_l^* = 0)$$

$$= -\dfrac{1}{a_{lk}}[(c_k - z_k) + \lambda(c_k' - z_k')] \qquad\qquad \text{(By (2))}$$

$$\le 0$$

$$\Rightarrow \qquad -(c_k - z_k) - \lambda(c_k' - z_k') \le 0 \qquad\qquad (\because\ a_{lk} > 0)$$

$$\Rightarrow \qquad (c_k - z_k) + \lambda(c_k' - z_k') \ge 0$$

$$\Rightarrow \qquad \lambda \ge -\left[\dfrac{c_k - z_k}{c_k' - z_k'}\right]$$

$$\Rightarrow \qquad \lambda \ge \bar{\lambda} \qquad\qquad \left(\because\ \bar{\lambda} = -\left[\dfrac{c_k - z_k}{c_k' - z_k'}\right]\right)$$

which is true.

Hence, we proceed in this manner, from one range to another range of λ beyond $\bar{\lambda}$ until we arrive at $\lambda = \phi$. Clearly, above process is valid because no basis is repeated for this.

Now, suppose that $\bar{\lambda}$ is replaced by $\bar{\lambda} + \varepsilon$ where $\varepsilon > 0$ is arbitrary small. Since the problem is non-degenerate so we either solve the problem for $\lambda = \bar{\lambda} + \varepsilon$ or the problem has no finite optimal solution for $\lambda = \bar{\lambda} + \varepsilon$. Therefore, we can not remain indefinitely at a value of λ such that $\bar{\lambda} = \underline{\lambda} = \lambda$ and after getting a basis, we can not we can not return to it to any basis corresponding to lower value of λ as for $\lambda > \bar{\lambda}$, we have new solution.

☞ **REMARK**

- Here $\bar{\lambda}$ and $\underline{\lambda}$ are known as characteristic value of λ and the optimal solution corresponding to $\underline{\lambda}$ and $\bar{\lambda}$ are called characteristic solution.

CASE (2) IF THE PROBLEM HAS NO FINITE OPTIMAL SOLUTION FOR $\lambda = \delta$

For a basic feasible solution, in simplex method, we get some $c_j^* - z_j^* > 0$ for which all $a_{ik} \le 0$ at any stage. Suppose $c_k^* - z_k^*$ be the corresponding value of $c_j - z_j$. Therefore, for $\lambda = \delta$, a vector a_{ik} chosen to enter the basis cannot go to the basis as all $a_{ik} \le 0$.

If a_{ik} is the vector chosen to the basis, then we have

$$c_k^* - z_k^* = (c_k - z_k) + \lambda(c_k^* - z_k^*)$$

Now we have the following possibilities:

(i) If $c_k' - z_k' \ge 0$. Then clearly we have

$$(c_k - z_k) + \lambda(c'_k - z'_k) > 0 \text{ for all } \lambda > \delta$$

which imples that

$$(c^*_k - z^*_k) > 0 \text{ for all } \lambda > \delta$$

$\Rightarrow \quad \Delta^*_k > 0$ for all $\lambda > \delta$

$\Rightarrow \quad$ Problem has no finite optimal solution for any $\lambda \geq \delta$.

(ii) If $c'_k - z'_k < 0$, then

$$(c_k - z_k) + \lambda(c'_k - z'_k) > 0$$

$\Rightarrow \qquad\qquad (c^*_k - z^*_k) > 0 \text{ for all } \lambda$

$\Rightarrow \qquad\qquad \lambda < -\left\{ \dfrac{c_k - z_k}{c'_k - z'_k} \right\} = \lambda' \text{ (say)}$

$\therefore$ Problem will have no finite optimal solution for all λ such that $\delta \leq \lambda \leq \lambda_1$

For $\lambda = \lambda'$ if $(c_j - z_j) + \lambda'_1(c'_j - z'_j) \leq 0 \; \forall j$ then the problem will have a finite optimal solution. Therefore, upper bound of λ can be taken as

$$\lambda_1 = \min_{(c'_j - z'_j) > 0} -\left\{ \dfrac{c_j - z_j}{c'_j - z'_j} \right\}$$

So, when $\lambda'_1 \leq \lambda \leq \lambda_1$, the problem has a finite optimal solution and we can proceed as in case-1.

Further, if not all $(c_j - z_j) + \lambda'_1(c'_j - z'_j) \leq 0$, then for some value of j, a vector α_j with $(c_j - z_j) + \lambda'_1(c'_j - z'_j) > 0$ can be introduced to the basis for the improvement of the solution and the criterion for the improvement will have such iteration untill we obtain

$$(c_j - z_j) + \lambda'_1(c'_j - z'_j) \leq 0 \;\; \forall j$$

or to that iteration until a vector α_p with $(c_p - z_p) + \lambda'_1(c'_p - z'_p) > 0$ and all $\alpha_{ip} \leq 0$ is obtained. Here, in the former case, we shall use the method of case-1 and in the later case if $c'_p - z'_p \geq 0$, then there will be no finite optimal solution for $\lambda \geq \lambda'_1$ and if $c'_p - z'_p < 0$ then no finite optimal solution exists for $\lambda < \lambda'_2$ where

$$\lambda'_2 = -\dfrac{c_p - z_p}{c'_p - z'_p} \text{ and } \lambda'_1 < \lambda'_2$$

and then proceed to find whether a finite solution exists for $\lambda = \lambda'_2$.

Repeat the above procedure, until a finite optimal solution is obtained for some value of λ in which we can use the case-1 or obtain the information for no finite optimal solution for λ which is case-2.

Working Procedure

STEP 1. Start with a finite optimal solution of the problem associated with the basis B for $\lambda = \delta$.

STEP 2. Find a set of critical values $\lambda_1, \lambda_2, ..., \lambda_p$ of λ and the corresponding characteristic solutions and a series of basis $B_1, B_2, ..., B_p$ (B_p may not exist) with the following properties:

  (i) Each basis B_i differ from B_{i-1} by a single vector of λ_1 is determined by a unique minimum.

  (ii) Basis B_i is a optimal basis for all values of λ satisfy

$$\lambda_i \leq \lambda \leq \lambda_{i+1} \text{ for } i = 1, 2, ..., p-1$$

  (iii) B_p is an optimal basis for every value of λ such that $\lambda \geq \lambda_p$, if B_p exists (case-1) or there exists no finite optimal solution for $\lambda \geq \lambda_p$.

Solved Examples

EXAMPLE 1. *Consider the following LPP:*

$$Max. \ Z = (4 + 2\lambda)x_1 + (6 - 2\lambda)x_2 + (2 + 2\lambda)x_3$$

subject to the constraints

$$x_1 + x_2 + x_3 \le 3$$
$$x_1 + 4x_2 + 7x_3 \le 9$$

and $\qquad\qquad x_1, x_2, x_3 \ge 0$

Find the range of λ over which the solution remains basic feasible and optimal.

SOLUTION. Introducing slack variables s_1 and s_2, the given LPP can be written as

$$Max. \ Z = (4 + 2\lambda)x_1 + (6 - 2\lambda)x_2 + (2 + 2\lambda)x_3 + 0s_1 + 0s_2$$

s.t. $\qquad x_1 + x_2 + x_3 + s_1 = 3$
$$x_1 + 4x_2 + 7x_3 + s_2 = 9$$

and $\qquad x_1, x_2, x_3, s_1, s_2 \ge 0$

Now, $C^* = (4 + 2\lambda, 6 - 2\lambda, 2 + 2\lambda, 0, 0)$

The initial basic feasible solution of the given LPP is

$$x_1 = 0, x_2 = 0, x_3 = 0, s_1 = 3 \text{ and } s_2 = 9$$

Simplex Table-1

B.V.	C_B^*	X_B	$c_j \to$ 4 x_1	6 x_2	2 x_3	0 s_1	0 s_2	Min. Ratio X_B/x_2
s_1	0	3	1	1	1	1	0	3/1
s_2	0	9	1	④	7	0	1	9/4 (min) →
$Z = C_B^* X_B = 0$		$c_j^* \to$	$4 + 2\lambda$	$6 - 2\lambda$	$2 + 2\lambda$	0	0	
$c_j^* - z_j^*$		$c_j - z_j$	4	6	2	0	0	
		$c_j' - z_j'$	2	–2	2	0	0	

In the above table

$$c_1^* - z_1^* = c_1^* - C_B^* x_1 = 4 + 2\lambda = (c_1 - z_1) + \lambda(c_1' - z_1')$$

$\Rightarrow \quad c_1 - z_1 = 4, c_1' - z_1' = 2$

Also, $c_2^* - z_2^* = c_2^* - C_B^* x_2 = 6 - 2\lambda \quad \Rightarrow \quad c_2 - z_2 = 6, c_2' - z_2' = -2$

$$c_3^* - z_3^* = c_3^* - C_B^* x_3 = 2 + 2\lambda \quad \Rightarrow \quad c_3 - z_3 = 2, c_3' - z_3' = 2$$

$$c_4^* - z_4^* = c_4^* - C_B^* s_1 = 0 \qquad \Rightarrow \quad c_4 - z_4 = 0, c_4' - z_4' = 0$$

$$c_5^* - z_5^* = c_5^* - C_B^* s_2 = 0 \qquad \Rightarrow \quad c_5 - z_5 = 0, c_5' - z_5' = 0$$

Here, we have that x_2 is the incoming vector and s_2 is the outgoing vector (by minimum ratio rule)

Now, apply simplex method in a usual manner, we have the following simplex table.

Simplex Table-2

B.V.	C_B^*	X_B	$c_j \to$ 4 x_1	6 x_2	2 x_3	0 s_1	0 s_2	Min. Ratio X_B/x_1
s_1	0	3/4	3/4	0	–3/4	1	–1/4	$(3/4)/(3/4) = 1$ (min) $\to$
x_2	$6 - 2\lambda$	9/4	1/4	1	7/4	0	1/4	$(9/4)/(1/4)$
$Z = C_B^* X_B$ = 27 / 2		$c_j^* \to$	$4 + 2\lambda$	$6 - 2\lambda$	$2 + 2\lambda$	0	0	
$c_j^* - z_j^*$	$c_j - z_j$		5/2	0	–17/2	0	–3/2	
	$c_j' - z_j'$		5/2	0	11/2	0	1/2	
			$\uparrow$		$\downarrow$			

In the above table, we have

$$c_1^* - z_1^* = \frac{5}{2} + \lambda \frac{5}{2}, c_2^* - z_2^* = 0$$

$$c_3^* - z_3^* = -\frac{17}{2} + \frac{11\lambda}{2}, c_4^* - z_4^* = 0$$

$$c_5^* - z_5^* = -\frac{3}{2} + \frac{\lambda}{2}$$

Obviously, x_1 is the incoming vector and s_1 is the outgoing vector.

Simplex Table-3

B.V.	C_B^*	X_B	$c_j \to$ 4 x_1	6 x_2	2 x_3	0 s_1	0 s_2	Min. Ratio
x_1	$4 + 2\lambda$	1	1	0	–1	4/3	–1/3	
x_2	$6 - 2\lambda$	2	0	1	2	–1/3	1/3	
$Z = C_B^* X_B$ = $16 - 2\lambda$		$c_j^* \to$	$4 + 2\lambda$	$6 - 2\lambda$	$2 + 2\lambda$	0	0	
$c_j^* - z_j^*$	$c_j - z_j$		0	0	–6	–10/3	–2/3	
	$c_j' - z_j'$		0	0	8	–10/3	4/3	

Further, we have to calculate $c_j^* - z_j^*$ for $j = 3, 4, 5$ (not in the basis) in the following manner

$$c_3^* - z_3^* = (2 + \lambda) - C_B^* x_3 = (2 + 2\lambda) - (4 + 2\lambda, 6 - 2\lambda)\begin{bmatrix} -1 \\ 2 \end{bmatrix}$$

$$= (2 + 2\lambda) - \{-(4 + 2\lambda) + 2(6 - 2\lambda)\}$$

$$= (2 + 2\lambda) - (8 - 6\lambda) = -6 + 8\lambda$$

$$= (c_3 - z_3) + \lambda(c_3' - z_3')$$

Thus, $c_3 - z_3 = -6, c_3' - z_3' = 8$

$$c_4^* - z_4^* = 0 - C_B^* x_4 = 0 - (4 + 2\lambda, 6 - 2\lambda)\begin{bmatrix} 4/3 \\ -1/3 \end{bmatrix} \qquad (\because x_4 = s_1)$$

$$= -\frac{4}{3}(4 + 2\lambda) + \frac{1}{3}(6 - 2\lambda) = \frac{16}{3} - \frac{8}{3}\lambda + \frac{6}{3} - \frac{2}{3}\lambda$$

$$= \frac{-10}{3} - \frac{10}{3}\lambda = (c_4 - z_4) + \lambda(c_4' - z_4')$$

$\Rightarrow \quad c_4 - z_4 = -\frac{10}{3}, c_4' - z_4' = -\frac{10}{3}$

Also, $\quad c_5^* - z_5^* = c_5^* - C_B^* x_5 = c_5^* - C_B^* s_2 \qquad (\because x_5 = s_2)$

$$= 0 - (4 + 2\lambda, 6 - 2\lambda)\begin{bmatrix} -1/3 \\ 1/3 \end{bmatrix}$$

$$= \frac{1}{3}(4 + 2\lambda) - \frac{1}{3}(6 - 2\lambda) = \frac{1}{3}(-2 + 4\lambda)$$

$$= -\frac{2}{3} + \frac{4}{3}\lambda = (c_5 - z_5) + \lambda(c_5' - z_5')$$

$\Rightarrow \quad c_5 - z_5 = -\frac{2}{3}, c_5' - z_5' = \frac{4}{3}$

Now, the third simplex table shows that all $c_j - z_j \le 0$.

$\Rightarrow$ solution is optimal for $\lambda = 0$ and is given by

$$x_1 = 1, x_2 = 2, x_3 = 0 \text{ and max } Z = 16$$

We know that the solution will remain optimal for those value of λ for which

$$\underline{\lambda} \le \lambda \le \bar{\lambda}$$

where

$$\underline{\lambda} = \max_{(c_j - z_j) < 0}\left\{-\frac{c_j - z_j}{c_j' - z_j'}\right\} = \left\{\frac{-10/3}{-10/3}\right\} = 1$$

and

$$\bar{\lambda} = \min_{(c_j' - z_j') > 0}\left\{-\frac{c_j - z_j}{c_j' - z_j'}\right\} = \min\left\{-\frac{c_3 - z_3}{c_3' - z_3'}, -\frac{c_5 - z_5}{c_5' - z_5'}\right\}$$

$$= \min\left\{-\frac{-6}{7}, -\frac{-2/3}{4/3}\right\} = \min\left\{\frac{6}{7}, \frac{1}{2}\right\} = \frac{1}{2} = \frac{c_k - z_k}{c_k' - z_k'}$$

$\Rightarrow \qquad k = 5$

Thus, $x_1 = 1, x_2 = 2$ and $x_3 = 0$ is an optimal solution for the given problem for all values of λ such that

$$-1 \le \lambda \le \frac{1}{2}$$

and Max. $Z = 16 - 2\lambda$.

Now, since $\bar{\lambda} = \frac{1}{2}$, which is finite and $a_{25} = \frac{1}{3} > 0$. Thus, we will try to improve the range of λ beyond $\frac{1}{2}$. When $\lambda > \frac{1}{2}$ then x_5 (s_2) can be introduced in the basis and by minimum ratio rule x_2 is the outgoing vector.

Now, introducing s_2 in place of x_2 in the basis, the next optimal simplex table is as follows.

Simplex Table-4

B.V.	C_B^*	X_B	$c_j \rightarrow$ 4 x_1	6 x_2	2 x_3	0 s_1	0 s_2
x_1	$4 + 2\lambda$	3	1	1	1	1	0
s_2	0	6	0	3	6	-1	1
$Z = C_B^* X_B$ $= 12 + 6\lambda$		$c_j^* \rightarrow$	$4 + 2\lambda$	$6 - 2\lambda$	$2 + 2\lambda$	0	0
$c_j^* - z_j^*$		$c_j - z_j$	0	2	-2	-4	0
		$c_j' - z_j'$	0	-4	0	-2	0

In the above table

$$c_2^* - z_2^* = c_2^* - \textbf{C}_B^* \textbf{x_2} = (6 - 2\lambda) - (4 + 2\lambda, 0)\begin{bmatrix}1\\3\end{bmatrix}$$

$$= 6 - 2\lambda - 4 - 2\lambda = 2 - 4\lambda = (c_2 - z_2) + \lambda(c_2' - z_2')$$

$\Rightarrow$ $c_2 - z_2 = 0$ and $c_2' - z_2' = -4$

Now, $\quad c_3^* - z_3^* = c_3^* - \textbf{C}_B^* \textbf{x_3} = (2 + 2\lambda) - (4 + 2\lambda, 0)\begin{bmatrix}1\\6\end{bmatrix}$

$$= (2 + 2\lambda) - (4 + 2\lambda) = -2$$

$$= (c_3 - z_3) + \lambda(c_3' - z_3')$$

$\Rightarrow$ $\quad c_3 - z_3 = -2, c_3' - z_3' = 0$

and $\quad c_4^* - z_4^* = c_4^* - \textbf{C}_B^* \textbf{x_4} = c_4^* - \textbf{C}_B^* s_1$

$$= 0 - (4 + 2\lambda, 0)\begin{bmatrix}1\\-1\end{bmatrix} = -4 - 2\lambda$$

$$= (c_4 - z_4) + \lambda(c_4' - z_4')$$

$\Rightarrow$ $\quad c_4 - z_4 = -4, c_4' - z_4' = -2$

For $\lambda \geq \dfrac{1}{2}, c_2^* - z_2^* \leq 0, c_3^* - z_3^* < 0, c_4^* - z_4^* \leq 0$

Therefore, the solution for $\lambda \geq \dfrac{1}{2}$ is optimal and this optimal solution is given by

$$x_1 = 3, x_2 = 0, x_3 = 0 \text{ and max. } Z = 12 + 6\lambda$$

Now, for $\lambda < -1$, then from table-3

$$c_4^* - z_4^* = \frac{10}{3}(1 + \lambda) > 0 \text{ and } a_{14} = \frac{4}{3} > 0$$

So, for $\lambda < -1$, the solution in simplex table-3 is not optimal. In this case, the corresponding vector x_4 (s_1) is the incoming vector and by minimum ratio rule x_1 will be the outgoing vector. Introducing s_1 in place of x_1, the next optimal table from table-3 is given below.

	$c_j \rightarrow$		4	6	2	0	0
B.V.	C_B^*	X_B	x_1	x_2	x_3	s_1	s_2
s_1	0	3/4	3/4	0	–3/4	1	–1/4
x_2	6 – 2λ	9/4	1/4	1	7/4	0	1/4
$Z = C_B^* X_B = \dfrac{27}{2} - \dfrac{9}{2}\lambda$	$c_j^* \rightarrow$		4 + 2λ	6 – 2λ	2 + 2λ	0	0
$c_j^* - z_j^*$	$c_j - z_j$		5/2	0	–17/2	0	–3/2
	$c_j' - z_j'$		5/2	0	11/2	0	1/2

From the above table, we observe that

$$c_1^* - z_1^* = \frac{5}{2}(1+\lambda), c_3^* - z_3^* = -\frac{17}{3} + \frac{11}{2}\lambda$$

$$c_5^* - z_5^* = -\frac{3}{2} + \frac{1}{2}\lambda$$

Clearly, for $\lambda < -1$, $c_j^* - z_j^* \leq 0 \ \forall j = 1,3,5$ (not in the basis). Hence, the solution is optimal and is given by

$$x_1 = 0, x_2 = \frac{9}{4}, x_3 = 0 \text{ and } Max.Z = \frac{27}{2} - \frac{9}{2}\lambda.$$

EXAMPLE 2. *For the following LPP:*

$$\textbf{Max. } \textbf{Z = (3 – 6λ)}\textbf{x}_1 \textbf{ + (2 – 2λ)}\textbf{x}_2 \textbf{ + (5 + 5λ)}\textbf{x}_3$$

subject to the constraints

$$\textbf{x}_1 \textbf{ + 2x}_2 \textbf{ + x}_3 \textbf{ ≤ 430}$$

$$\textbf{3x}_1 \textbf{ + 2x}_3 \textbf{ ≤ 460}$$

$$\textbf{x}_1 \textbf{ + 4x}_2 \textbf{ ≤ 420}$$

and $\textbf{x}_1\textbf{, x}_2\textbf{, x}_3 \textbf{ ≥ 0}$

Find the range of λ *for which the solution remain basic feasible and optimal.*   [MEERUT–1994, 95, 98, 2007; BHARATHIDASAN–1990]

SOLUTION. The given LPP can be written as follows:

Max. Z = (3 – 6λ)x_1 + (2 – 2λ)x_2 + (5 + 5λ)x_3 + 0s_1 + 0s_2 + 0s_3

s.t. $x_1 + 2x_2 + x_3 + s_1 = 430$

         $3x_1 + 2x_3 + s_2 = 460$

         $x_1 + 4x_2 + s_3 = 420$

and $x_i, s_i \geq 0$

Simplex Table-1

B.V.	C_B^*	X_B	x_1	x_2	x_3	s_1	s_2	s_3	Min. Ratio X_B/x_3
	$c_j \rightarrow$		3	2	5	0	0	0	
s_1	0	430	1	2	1	1	0	0	430/1
s_2	0	460	3	0	②	0	1	0	460/2 (min) →
s_3	0	420	1	4	0	0	0	1	—
$Z = C_B^* X_B$ $= 0$	$c_j^* \rightarrow$		$3-6\lambda$	$2-2\lambda$	$5+5\lambda$	0	0	0	
$c_j^* - z_j^*$	$c_j - z_j$		3	2	5	0	0	0	
	$c_j' - z_j'$		-6	-2	5 ↑	0	0 ↓	0	

In the above table, we have

$$c_1^* - z_1^* = c_1^* - \boldsymbol{C_B^* x_1} = 3 - 6\lambda - (0,0,0)\begin{bmatrix} 1 \\ 3 \\ 1 \end{bmatrix} = 3 - 6\lambda = (c_1 - z_1) + \lambda(c_1' - z_1')$$

$$\Rightarrow \qquad c_1 - z_1 = 3, c_1' - z_1' = -6$$

Now, $\quad c_2^* - z_2^* = c_2^* - \boldsymbol{C_B^* x_2} = 2 - 2\lambda - (0,0,0)\begin{bmatrix} 2 \\ 0 \\ 4 \end{bmatrix} = 2 - 2\lambda$

$$\Rightarrow \qquad c_2 - z_2 = 2, c_2' - z_2' = -2$$

Also, $\quad c_3^* - z_3^* = c_3^* - \boldsymbol{C_B^* x_3} = 5 + 5\lambda - (0,0,0)\begin{bmatrix} 1 \\ 2 \\ 0 \end{bmatrix} = 5 + 5\lambda$

$$\Rightarrow \qquad c_3 - z_3 = 5, c_3' - z_3' = 5$$

Clearly, $c_3 - z_3 = 5 > 0 \Rightarrow$ corresponding vector x_3 can be introduced in the basis and by minimum ratio rule s_2 (x_5) is the outgoing vector.

Simplex Table-2

B.V.	C_B^*	X_B	x_1	x_2	x_3	s_1 (x_4)	s_2 (x_5)	s_3 (x_6)	Min. Ratio
	$c_j \rightarrow$		3	2	5	0	0	0	
s_1	0	200	$-1/2$	②	0	1	$-1/2$	0	200/2 (min) →
x_3	$5+5\lambda$	230	3/2	0	1	0	1/2	0	—
s_3	0	420	1	4	0	0	0	1	420/4
$Z = C_B^* X_B$ $= 1150(1+\lambda)$	$c_j^* \rightarrow$		$3-6\lambda$	$2-2\lambda$	$5+5\lambda$	0	0	0	
$c_j^* - z_j^*$	$c_j - z_j$		$-9/2$	2	0	0	$-5/2$	0	
	$c_j' - z_j'$		$-27/2$	-2	0	0	$-5/2$	0	
				↑	↓				

In the above table, we have calculated $c_j^* - z_j^*$ for $j = 1, 2, 3, 5$ (not in the basis) such that

$$c_1^* - z_1^* = c_1^* - \boldsymbol{C_B^* x_1} = (3 - 6\lambda)(0, 5 + 5\lambda, 0)\begin{bmatrix} -1/2 \\ 3/2 \\ 1 \end{bmatrix} = (3 - 6\lambda) - \frac{3}{2}(5 + 5\lambda)$$

$$= \frac{-9}{2} - \frac{27}{2}\lambda = (c_1 - z_1) + \lambda(c_1' - z_1')$$

$$c_1 - z_1 = -\frac{9}{2}, c_1' - z_1' = \frac{-27}{2}$$

Also, $c_2^* - z_2^* = c_2^* - \boldsymbol{C_B^* x_2} = (2 - 2\lambda) - (0, 5 + 5\lambda, 0)\begin{bmatrix} 2 \\ 0 \\ 4 \end{bmatrix}$

$$= 2 - 2\lambda = (c_2 - z_2) + \lambda(c_2' - z_2')$$

$$\Rightarrow \quad c_2 - z_2 = 2, c_2' - z_2' = -2$$

and $c_5^* - z_5^* = c_5^* - \boldsymbol{C_B^* x_5} = 0 - (0, 5 + 5\lambda, 0)\begin{bmatrix} -1/2 \\ -1/2 \\ 0 \end{bmatrix}$

$$= -\frac{5}{2} - \frac{5}{2}\lambda = (c_5 - z_5) + \lambda(c_5' - z_5')$$

$$\Rightarrow \quad c_5 - z_5 = \frac{-5}{2} \text{ and } c_5' - z_5' = \frac{-5}{2}$$

We observe that $c_2 - z_2 = 2 > 0$ therefore, corresponding vector x_2 can be introduced in the basis and also by minimum ratio rule $s_1(x_4)$ is the outgoing vector.

Simplex Table-3

B.V.	C_B^*	X_B	$c_j \rightarrow$ 3 x_1	2 x_2	5 x_3	0 s_1 (x_4)	0 s_2 (x_5)	0 s_3 (x_6)	Min. Ratio X_B/x_4
x_2	$2 - 2\lambda$	100	$-1/4$	1	0	(1/2)	$-1/4$	0	$100/(1/2)$ (min)→
x_3	$5 + 5\lambda$	230	$3/2$	0	1	0	$1/2$	0	—
s_3	0	20	2	0	0	-2	1	1	—
$Z = C_B^* X_B$ $= 1350 + 950\lambda$	$c_j^* \rightarrow$		$3 - 6\lambda$	$2 - 2\lambda$	$5 + 5\lambda$	0	0	0	
$c_j^* - z_j^*$	$c_j - z_j$		-4	0	0	-1	-2	0	
	$c_j' - z_j'$		-14	0	0	1	-3	0	
			↓			↑			

In the above table, we have compute $c_j^* - z_j^*$ for $j = 1, 4, 5$ (not in the basis) as follows:

$$c_1^* - z_1^* = c_1^* - \boldsymbol{C_B^*}\boldsymbol{x_1} = (3 - 6\lambda) - (2 - 2\lambda, 5 + 5\lambda, 0)\begin{bmatrix} -1/4 \\ 3/2 \\ 2 \end{bmatrix}$$

$$= (3 - 6\lambda) + \frac{1}{4}(2 - 2\lambda) - \frac{3}{2}(5 + 5\lambda) = -4 - 14\lambda = (c_1 - z_1) + \lambda(c_1' - z_1')$$

$\Rightarrow \qquad c_1 - z_1 = -4$ and $c_1' - z_1' = -14$

Also, $\qquad c_4^* - z_4^* = c_4^* - \boldsymbol{C_B^*}\boldsymbol{x_4} = 0 - (2 - 2\lambda, 5 + 5\lambda, 0)\begin{bmatrix} 1/2 \\ 0 \\ -1/2 \end{bmatrix}$

$$= -1 + \lambda$$

$\Rightarrow \qquad c_4 - z_4 = -1$ and $c_4' - z_4' = 1$

and $\qquad c_5^* - z_5^* = c_5^* - \boldsymbol{C_B^*}\boldsymbol{x_5} = 0 - (2 - 2\lambda, 5 + 5\lambda, 0)\begin{bmatrix} -1/4 \\ 1/2 \\ 1 \end{bmatrix}$

$$= \frac{1}{4}(2 - 2\lambda) - \frac{1}{2}(5 + 5\lambda) = -2 - 3\lambda$$

$\Rightarrow \qquad c_5 - z_5 = -2$ and $c_5' - z_5' = -3$

Here, we observe that in simplex table-3 all $c_j - z_j < 0$ so the solution at this stage is optimal for $\lambda = 0$ and is given by

$$x_1 = 0, x_2 = 100, x_3 = 230 \text{ and max. } Z = 1350$$

This solution is optimal and feasible for those value of λ for which $\underline{\lambda} \le \lambda \le \bar{\lambda}$ where

$$\underline{\lambda} = \max_{(c_j' - z_j') < 0} \left[-\frac{(c_j - z_j)}{(c_j' - z_j')} \right] = \max \left\{ -\frac{c_1 - z_1}{c_1' - z_1'}, -\frac{c_5 - z_5}{c_5' - z_5'} \right\}$$

$$= \max \left\{ -\frac{-4}{-14}, -\frac{-2}{-3} \right\} = \max \left\{ -\frac{2}{7}, -\frac{2}{3} \right\} = -\frac{2}{7}$$

and $\qquad \bar{\lambda} = \min_{(c_j' - z_j') > 0} \left\{ -\frac{c_j - z_j}{c_j' - z_j'} \right\} = \min \left\{ -\frac{c_4 - z_4}{c_4' - z_4'} \right\} = -\frac{(-1)}{1} = 1$

$\Rightarrow$ This solution is optimal and feasible for those value of λ for which $-\dfrac{2}{7} \le \lambda \le 1$.

Now, since $\bar{\lambda} = 1$, which is finite, so we can improve the range of λ beyond $\bar{\lambda} = 1$.

Now, for $\lambda > 1$, $c_4^* - z_4^* > 0$ and $a_{14} = \dfrac{1}{2} > 0$, so that $x_1 (s_1)$ can be introduced into the basis and by minimum ratio rule, s_1 is the outgoing vector.

Simplex Table-4

	$c_j \rightarrow$		3	2	5	0	0	0
B.V.	C_B^*	X_B	x_1	x_2	x_3	s_1 (x_4)	s_2 (x_5)	s_3 (x_6)
s_1	0	200	$-1/2$	2	0	1	$-1/2$	0
x_3	$5+5\lambda$	230	$3/2$	0	1	0	$1/2$	0
s_3	0	420	1	4	0	0	0	1
$Z = C_B^* X_B$ $= 1150(1+\lambda)$		$c_j^* \rightarrow$	$3-6\lambda$	$2-2\lambda$	$5+5\lambda$	0	0	0
$c_j^* - z_j^*$		$c_j - z_j$	$-9/2$	2	0	0	$-5/2$	0
		$c_j' - z_j'$	$-27/2$	-2	0	0	$-5/2$	0

Here, we observe that all $c_j^* - z_j^* \leq 0$ for $\lambda > 1$ and no $c_j' - z_j' > 0$

$\Rightarrow$ For $\lambda > 1$, optimal solution exists

$\therefore$ We conclude that the solution is optimal and feasible for those value of λ for which

$$-\frac{27}{7} \leq \lambda \leq \infty$$

Hence, $x_1 = 0, x_2 = 100, x_3 = 230$ and max. $Z = 1350 + 950\lambda$ is the optimal and feasible solution for those value of λ for which $-\frac{2}{7} \leq \lambda \leq 1$ and $x_1 = 0, x_2 = 0$, $x_3 = 230$ and Max. $Z = 1150(1 + \lambda)$ is the optimal and feasible solution for those value of λ for which $1 < \lambda < \infty$.

EXAMPLE 3. *For the following LPP,*

$$\textbf{\textit{Min. Z}} = \lambda x_1 - \lambda x_2 - x_3 + x_4$$

subject to the constraints

$$3x_1 - 3x_2 - x_3 + x_4 \geq 5$$
$$2x_1 - 2x_2 + x_3 - x_4 \leq 3$$

and $\quad x_1, x_2, x_3, x_4 \geq 0$

Find the range of λ for which the solution remains basic feasible and optimal. [MEERUT–1993, 2006; SAMBHALPUR–1986]

SOLUTION. Introducing surplus variable s_1, slack variable s_2 and artificial variable A, the given problem becomes

Max. $Z = -\lambda x_1 + \lambda x_2 + x_3 - x_4 + 0s_1 + 0s_2 - MA$

s.t. $\quad 3x_1 - 3x_2 - x_3 + x_4 - s_1 + A = 5$
$\quad\quad\quad 2x_1 - 2x_2 + x_3 - x_4 + s_2 = 3$

and $\quad\quad x_1, x_2, x_3, x_4, s_1, s_2, A \geq 0$

Now proceeding in a usual manner, we have the following simplex table.

Simplex Table-1

B.V.	C_B^*	X_B	$c_j \rightarrow$							Min. Ratio
			0	0	1	−1	0	0	−M	X_B/x_4
			x_1	x_2	x_3	x_4	s_1 (x_5)	s_2 (x_6)	A (x_7)	
A	−M	5	③	−3	−1	1	−1	0	1	5/3
s_2	0	3	2	−2	1	−1	0	1	0	3/2 (min) →
$Z = C_B^* X_B$ $= -5M$		$c_j^* \rightarrow$	−λ	−λ	1	−1	0	0	−M	
$c_j^* - z_j^*$		$c_j - z_j$	3M	−3M	1 − M	M − 1	−M	0	0	
		$c_j' - z_j'$	−1	1	0	0	0	0	0	
			↑				↓			

In the above table, we have

We compute $c_j^* - z_j^*$ for $j = 1, 2, 3, 4, 5$ (not in the basis)

Here, $\qquad c_1^* - z_1^* = c_1^* - C_B^* x_1 = -\lambda - (-M, 0)\begin{bmatrix} 3 \\ 2 \end{bmatrix} = -\lambda + 3M$

$\Rightarrow \qquad c_1 - z_1 = 3M,\ c_1' - z_1' = -1$

Similarly, $\qquad c_2^* - z_2^* = c_2^* - C_B^* x_2 = \lambda - (-M, 0)\begin{bmatrix} -3 \\ 2 \end{bmatrix} = \lambda - 3M$

$\Rightarrow \qquad c_2 - z_2 = -3M,\ c_2' - z_2' = 1$

Also, $\qquad c_3^* - z_3^* = c_3^* - C_B^* x_3 = 1 - (-M, 0)\begin{bmatrix} -1 \\ 1 \end{bmatrix} = 1 - M$

$\Rightarrow \qquad c_3 - z_3 = 1 - M,\ c_3' - z_3' = 0$

Now, $\qquad c_4^* - z_4^* = c_4^* - C_B^* x_4 = -1 - (-M, 0)\begin{bmatrix} -1 \\ 1 \end{bmatrix} = -1 + M$

$\Rightarrow \qquad c_4 - z_4 = -1 + M,\ c_4' - z_4' = 0$

and $\qquad c_5^* - z_5^* = c_5^* - C_B^* x_5 = 0 - (-M, 0)\begin{bmatrix} -1 \\ 0 \end{bmatrix} = -M$

$\Rightarrow \qquad c_5 - z_5 = -M,\ c_5' - z_5' = 0$

We observe that not all $c_j - z_j \le 0$ and not all $c_j^* - z_j^* \le 0$. Thus the initial solution is not optimal.

Also, corresponding to $c_1 - z_1 = 3M$, the vector x_1 will enter in the basis and by minimum ratio rule we find that s_2 is the outgoing vector.

Simplex Table-2

B.V.	C_B^*	X_B	$c_j \rightarrow$ 0 x_1	0 x_2	1 x_3	−1 x_4	0 s_1 (x_5)	0 s_2 (x_6)	−M A (x_7)	Min. Ratio X_B/x_4
A	−M	1/2	0	0	−5/2	(5/2)	−1	−3/2	1	(1/2)/(5/2)(min)→
x_1	−λ	3/2	1	−1	1/2	−1/2	0	1/2	0	—
$Z = C_B^* X_B$ $= \dfrac{-M}{2} - \dfrac{3\lambda}{2}$	$c_j^* \rightarrow$		−λ	λ	1	−1	0	0	−M	
$c_j^* - z_j^*$	$c_j - z_j$		0	0	$1 - \dfrac{5}{2}M$	$\dfrac{5M}{2} - 1$	−M	−3M/2	0	
	$c_j' - z_j'$		0	0	1/2	−1/2	0	1/2	0	
						↑				

In the above table, we compute $c_j^* - z_j^*$ for $j = 2, 3, 4, 5, 6$ (not in the basis), we have

$$c_2^* - z_2^* = c_2^* - C_B^* x_2 = \lambda - (-M, -\lambda) \begin{bmatrix} 0 \\ -1 \end{bmatrix}$$

$$= \lambda - \lambda = 0$$

$$\Rightarrow \quad c_2 - z_2 = 0, c_2' - z_2' = 0$$

$$c_3^* - z_3^* = c_3^* - C_B^* x_3 = 1 - (-M, -\lambda) \begin{bmatrix} -5/2 \\ 1/2 \end{bmatrix}$$

$$= 1 + \frac{5M}{2} + \frac{\lambda}{2}$$

$$\Rightarrow \quad c_3 - z_3 = 1 - \frac{5M}{2}, \ c_3' - z_3' = \frac{1}{2}$$

Similarly we may get

$$c_4 - z_4 = -1 + \frac{5}{2}M, \ c_4' - z_4' = -\frac{1}{2}$$

$$c_5 - z_5 = -M, \ c_5' - z_5' = 0$$

$$c_6 - z_6 = -\frac{3M}{2}, \ c_6' - z_6' = \frac{1}{2}$$

Again not all $c_j - z_j \leq 0$ and not all $c_j^* - z_j^* \leq 0$ so the solution in this table is not optimal.

Now, corresponding to $c_4 - z_4 = \dfrac{5M}{2} - 1$, the vector x_4 is incoming vector and by minimum ratio rule, A is outgoing vector.

Now proceed in a usual manner, we have the third simplex table.

Simplex Table-3

B.V.	C_B^*	X_B	x_1	x_2	x_3	x_4	s_1 (x_5)	s_2 (x_6)
	$c_j \rightarrow$		0	0	1	−1	0	0
x_4	−1	1/5	0	0	−1	1	−2/5	−3/5
x_1	−λ	8/5	1	−1	0	0	−1/5	1/5
$Z = C_B^* X_B$ $= -\dfrac{1}{5} - \dfrac{8\lambda}{5}$	$c_j^* \rightarrow$		−λ	λ	1	−1	0	0
$c_j^* - z_j^*$	$c_j - z_j$		0	0	0	0	−2/5	−3/5
	$c_j' - z_j'$		0	0	0	0	−1/5	1/5

In the above table we compute $c_j^* - z_j^*$ for $j = 2, 3, 4, 5, 6$ (not in the basis).

Here, $\qquad c_2^* - z_2^* = c_2^* - C_B^* x_2 = \lambda - (-1-\lambda)\begin{bmatrix} 0 \\ -1 \end{bmatrix} = \lambda - \lambda = 0$

$$= (c_2 - z_2) + \lambda(c_2' - z_2')$$

$\Rightarrow \qquad c_2 - z_2 = 0, c_2' - z_2' = 0$

Similarly, we obtain the other values of $c_j - z_j$ and $c_j' - z_j'$ as shown in the above table.

Further, the above table shows that all $c_j - z_j \leq 0$

$\Rightarrow \quad$ solution is optimal for $\lambda = 0$

Also, this optimal solution is given by

$$x_1 = \frac{8}{5}, x_2 = x_3 = 0, x_4 = \frac{1}{5}$$

and $\qquad \min. Z = -\max. Z' = \dfrac{1}{5}$

Now, this solution will remain optimal for those values of λ for which

$$\underline{\lambda} \leq \lambda \leq \overline{\lambda}$$

where $\quad \underline{\lambda} = \max_{(c_j' - z_j') < 0} \left\{ -\frac{c_j - z_j}{c_j' - z_j'} \right\} = \max\left\{ -\frac{c_5 - z_5}{c_5' - z_5'} \right\} \qquad$ [∵ only $c_5' - z_5' < 0$]

$$= -\frac{-2/5}{-1/5} = -2$$

and $\qquad \overline{\lambda} = \min_{(c_j' - z_j') > 0} \left\{ -\frac{c_j - z_j}{c_j' - z_j'} \right\} = \min\left\{ -\frac{c_6 - z_6}{c_6' - z_6'} \right\} \qquad$ [∵ only $c_6' - z_6' > 0$]

$$= -\frac{-3/5}{1/5} = 3$$

$\Rightarrow$ The solution $x_1 = \dfrac{8}{5}, x_2 = x_3 = 0, x_4 = \dfrac{1}{5}$ will remain optimal for all values of λ for which $-2 \le \lambda \le 3$.

Further, $\bar{\lambda} = \min\left\{-\dfrac{c_6 - z_6}{c_6' - z_6'}\right\} = -\dfrac{c_k - z_k}{c_k' - z_k'}$

$\Rightarrow$ $k = 6$

$\because$ $\lambda = \bar{\lambda} = 3$, finite and $a_{26} = \dfrac{1}{5} > 0$ therefore, range of λ can be improved beyond $\bar{\lambda} = 3$.

For $\lambda > 3, c_6^* - z_6^* > 0$, so that the vector $x_6 \ (= s_2)$ is the incoming vector and by minimum ratio rule, x_1 is the outgoing vector and key element $a_{26} = \dfrac{1}{5}$.

Now, we have the following simplex table

Simplex Table-4

B.V.	C_B^*	X_B	$c_j \to$ x_1	0 x_2	1 x_3	-1 x_4	0 s_1 (x_5)	0 s_2 (x_6)
x_4	-1	5	3	-3	-1	1	-1	0
x_2	0	8	5	-5	0	0	-1	1
$Z = C_B^* X_B$ $= -5$		$c_j^* \to$	$-\lambda$	λ	1	-1	0	0
$c_j^* - z_j^*$		$c_j - z_j$	3	-3	0	0	-1	0
		$c_j' - z_j'$	-1	1	0	0	0	0

In the above table, we have compute

$c_j^* - z_j^*$ for $j = 1, 2, 3, 5$ such that

$$c_1^* - z_1^* = c_1^* - C_B^* x_1 = -\lambda - (-1,0)\begin{bmatrix} 3 \\ 5 \end{bmatrix} = -\lambda + 3 = (c_1 - z_1) + \lambda(c_1' - z_1')$$

$\Rightarrow$ $c_1 - z_1 = 3, c_1' - z_1' = -1$

also, $c_2^* - z_2^* = c_2^* - C_B^* x_2 = \lambda - (1,0)\begin{bmatrix} -3 \\ -5 \end{bmatrix} = \lambda - 3$

$\Rightarrow$ $c_2 - z_2 = -3, c_2' - z_2' = 1$

Also, $c_3^* - z_3^* = c_3^* - C_B^* x_3 = 1 - (-1,0)\begin{bmatrix} -1 \\ 0 \end{bmatrix} = 1 - 1 = 0$

$\Rightarrow$ $c_3 - z_3 = 0, c_3' - z_3' = 0$

and $c_5^* - z_5^* = c_5^* - C_B^* x_5 = 0 - (-1,0)\begin{bmatrix} -1 \\ 1 \end{bmatrix} = -1$

$$\Rightarrow \qquad c_5 - z_5 = -1, c_5' - z_5' = 0$$

Clearly, for $\lambda > 3$, $c_2^* - z_2^* > 0$ so that we shall have to introduce x_2 in the basis. But corresponding to x_2 all $a_{i2} \leq 0$ so, x_2 can not be introduced in the basis and hence no optimal solution exists for $\lambda > 3$.

12.4 SYSTEMETIC LINEAR VARIATION IN b_i

Consider the parametric linear programming

$$\text{Max.} Z = \boldsymbol{C} \boldsymbol{x}$$

subject to the constraints

$$A\boldsymbol{x} \leq \boldsymbol{b} + \lambda \boldsymbol{b}' = \boldsymbol{b}(\lambda)$$

and $$\boldsymbol{x} \geq 0$$
...(1)

where λ is a parameter, $\boldsymbol{b}$, $\boldsymbol{b}'$, $\boldsymbol{C}$ and $\boldsymbol{A}$ are known vectors. We have to find the range of λ and the family of optimal solutions in their respective range of λ from $-\infty$ to ∞.

Firstly, find the optimal solution of (1) for $\lambda = 0$ by simplex method. Let $\boldsymbol{B}$ and X_B be the optimal basis and optimal basis feasible solution for $\lambda = 0$.

Now, since $\Delta_j = c_j - z_j = c_j - \boldsymbol{C_B} \boldsymbol{B}^{-1} \alpha_j$, therefore any change in λ does not affect the optimality of the solution.

At $\lambda = 0$ since $X_B = \boldsymbol{B}^{-1} \boldsymbol{b}$ therefore, when λ changes $-\infty$ to ∞, then the new basic variable reduces to

$$\hat{\boldsymbol{x}}_{\boldsymbol{B}} = \boldsymbol{B}^{-1} \cdot \boldsymbol{b}(\lambda)$$

$$= \boldsymbol{B}^{-1}(\boldsymbol{b} + \lambda \boldsymbol{b}') = \boldsymbol{B}^{-1}\boldsymbol{b} + \lambda \boldsymbol{B}^{-1}\boldsymbol{b}'$$

$$= \boldsymbol{x}_{\boldsymbol{B}} + \lambda \boldsymbol{x}_{\boldsymbol{B}}', \ \boldsymbol{x}_{\boldsymbol{B}} = \boldsymbol{B}^{-1}\boldsymbol{b} \text{ and } \boldsymbol{x}_{\boldsymbol{B}}' = \boldsymbol{B}^{-1} \cdot \boldsymbol{b}'$$

Clearly, the solution $\hat{\boldsymbol{x}}_{\boldsymbol{B}}$ will remain feasible if $\hat{\boldsymbol{x}}_{\boldsymbol{B}} \geq 0$ so for a given solution we find the range of λ within which the solution remains optimal and feasible.

Further, the new $\hat{\boldsymbol{x}}_{\boldsymbol{B}}$ will remain basic feasible if

$$\hat{\boldsymbol{x}}_{\boldsymbol{B}} \geq 0$$

$$\Rightarrow \quad \boldsymbol{X_B} + \lambda \boldsymbol{x_B'} \geq 0 \quad i.e. \, x_{B_i} + \lambda x_{B_i}' \geq 0 \text{ for } i = 1, 2, \ldots, m.$$

Thus, $$\max_{x_{B_i} > 0} \left\{ -\frac{x_{B_i}}{x_{B_i}'} \right\} \leq \lambda \leq \min_{x_{B_i} < 0} \left\{ -\frac{x_{B_i}}{x_{B_i}'} \right\}$$ ...(2)

which shows that the solution $\hat{\boldsymbol{x}}_{\boldsymbol{B}}$ will remain basic feasible and optimal when λ satisfy the above condition (2)

Now, there are following two possibilities:

(i) If there is no $x_{B_i}' < 0$, then λ has no upper bound.

(ii) If there is no $x_{B_i}' > 0$, then λ has no lower bound.

Further, if at least one of $x_{B_i}' < 0$ then from (2) we obtain the upper bound of λ (say $\overline{\lambda}$) and if at least one of $x_{B_i}' > 0$ then again from (2), we obtain the lower bound of λ (say $\underline{\lambda}$).

Hence, $\hat{\boldsymbol{x}}_{\boldsymbol{B}}$ will remain basic feasible and optimal for $\underline{\lambda} \leq \lambda \leq \overline{\lambda}$.

Also, for $\lambda > \overline{\lambda}$ (or $\lambda < \underline{\lambda}$) the basic feasible variable X_B is negative, where $x_{B_r}' < 0$. Then applying dual simplex method we can find the new optimal solution for $\lambda > \overline{\lambda}$.

 **Solved Examples**

EXAMPLE I. *For the following LPP,*

$$\text{Max. } Z = 3x_1 + 2x_2 + 5x_3$$

subject to the constraints

$$x_1 + 2x_2 + x_3 \le 430 + 100\lambda$$
$$3x_1 + 2x_3 \le 460 - 200\lambda$$
$$x_1 + 4x_2 \le 420 + 400\lambda$$

and $\quad x_1, x_2, x_3 \ge 0$

find the range of λ for which the solution remain optimal basic and feasible. [MEERUT–1986, 94, 95]

SOLUTION. Using slack variables s_1, s_2, s_3, we can write the given problem as

$$\text{Max. } Z = 3x_1 + 2x_2 + 5x_3 + 0s_1 + 0s_2 + 0s_3$$

s.t. $\quad x_1 + 2x_2 + x_3 + s_1 = 430 + 100\lambda$
$$3x_1 + 2x_3 + s_2 = 460 - 200\lambda$$
$$x_1 + 4x_2 + s_3 = 420 + 400\lambda$$

and $\quad x_1, x_2, x_3, s_1, s_2, s_3 \ge 0$

Here, we have

$$b(\lambda) = \begin{bmatrix} 430 + 100\lambda \\ 460 - 200\lambda \\ 420 + 400\lambda \end{bmatrix} = \begin{bmatrix} 430 \\ 460 \\ 420 \end{bmatrix} + \lambda \begin{bmatrix} 100 \\ -200 \\ 400 \end{bmatrix}$$

$$\Rightarrow \qquad b = \begin{bmatrix} 430 \\ 460 \\ 420 \end{bmatrix} \text{ and } b' = \begin{bmatrix} 100 \\ -200 \\ 400 \end{bmatrix}$$

In case of $\lambda = 0$, apply the simplex method, we obtain the final simplex table given as under

Simplex Table-1

B.V.		$c_j \rightarrow$	3	2	5	0	0	0
	C_B	X_B	x_1	x_2	x_3	s_1	s_2	s_3
x_2	2	100	$-1/4$	1	0	$1/2$	$-1/4$	0
x_3	5	230	$3/2$	0	1	0	$1/2$	0
s_3	0	20	2	0	0	-2	1	1
$Z = C_B X_B = 1350$		$\Delta_j \rightarrow$	-4	0	0	-1	-2	0

We observe that

$$X_B = \begin{bmatrix} 100 \\ 230 \\ 20 \end{bmatrix} = \begin{bmatrix} x_{B_1} \\ x_{B_2} \\ x_{B_3} \end{bmatrix} = \begin{bmatrix} x_2 \\ x_3 \\ s_3 \end{bmatrix}$$

$$B = [x_2, x_3, s_3] \quad \Rightarrow \quad B^{-1} = [s_1 \quad s_2 \quad s_3] = \begin{bmatrix} 1/2 & -1/4 & 0 \\ 0 & 1/2 & 0 \\ -2 & 1 & 1 \end{bmatrix}$$

$$\Rightarrow \qquad x'_B = B^{-1}b'$$

$$= \begin{bmatrix} 1/2 & -1/4 & 0 \\ 0 & 1/2 & 0 \\ -2 & 1 & 1 \end{bmatrix} \begin{bmatrix} 100 \\ -200 \\ 400 \end{bmatrix}$$

$$= \begin{bmatrix} 100 \\ -100 \\ 0 \end{bmatrix} = \begin{bmatrix} x'_{B_1} \\ x'_{B_2} \\ x'_{B_3} \end{bmatrix}$$

So, new solution is given by

$$\hat{x}_B = X_B + \lambda x'_B = \begin{bmatrix} 100 \\ 230 \\ 20 \end{bmatrix} + \lambda \begin{bmatrix} 100 \\ -100 \\ 0 \end{bmatrix} = \begin{bmatrix} 100 + 100\lambda \\ 230 - 100\lambda \\ 20 \end{bmatrix} = \begin{bmatrix} x_2 \\ x_3 \\ s_3 \end{bmatrix}$$

$$\Rightarrow \quad x_2 = 100 + 100\lambda, \; x_3 = 230 - 100\lambda, \; s_3 = 20$$

Now,
$$\bar{\lambda} = \min_{x_{B_i} < 0} \left\{ -\frac{x_{B_i}}{x'_{B_i}} \right\} = \left\{ -\frac{x_{B_2}}{x'_{B_2}} \right\} = \frac{-230}{-100} = 23$$

and
$$\underline{\lambda} = \max_{x_{B_i} > 0} \left\{ -\frac{x_{B_i}}{x'_{B_i}} \right\} = \max \left\{ -\frac{x_{B_1}}{x'_{B_1}} \right\} = \max \left\{ -\frac{100}{100} \right\} = -1$$

Thus, the solution $x_2 = -100 + 100\lambda$, $x_3 = 230 - 100\lambda$, $s_3 = 20$ and Max. $Z = C_B \cdot \hat{x}_B = 1350 - 300\lambda$ is optimal basic feasible for
$$-1 \leq \lambda \leq 23$$
For $\lambda > 23$, x_3 is negative which correspond to x'_{B_2} so all $a_{2j} \geq 0$.

$\Rightarrow$ No optimal solution exists for $\lambda > 23$

and for $\lambda < -1$, then x_2 is negative, so the column vector corresponding to x_2 will be outgoing vector.

Since, x_2 corresponds to $\hat{x}_{B_1}$, so $r = 1$

Now,
$$\frac{\Delta_k}{a_{rk}} = \min_j \left\{ \frac{\Delta_j}{a_{rj}}, a_{rj} < 0 \right\}$$

$$\Rightarrow \qquad \frac{\Delta_k}{a_{1k}} = \min_j \left\{ \frac{\Delta_j}{a_{1j}}, a_{1j} < 0 \right\}$$

$$= \min \left\{ \frac{\Delta_1}{a_{11}}, \frac{\Delta_5}{a_{15}} \right\} = \min \left\{ \frac{-4}{-1/4}, \frac{-2}{-1/4} \right\}$$

$$= \min\{16, 8\} = 8 = \frac{\Delta_5}{a_{15}}$$

$$\Rightarrow \qquad k = 5$$

$$\Rightarrow \quad s_2 \;(= x_5) \text{ is the incoming vector and key element} = a_{15} = -\frac{1}{4}$$

Now, again apply simplex method in a usual manner, we have the next optimal table given as below.

Simplex Table-2

	$c_j \rightarrow$	3	2	5	0	0	0
B.V.	C_B X_B	x_1	x_2	x_3	s_1	s_2	s_3
s_2	0 -400	1	-4	0	-2	1	0
x_3	5 430	1	2	1	1	0	0
s_3	0 420	1	4	0	0	0	1
$Z = C_B X_B = 2150$	$\Delta_j \rightarrow$	-2	-8	0	-5	0	0

From the above table, we observe that

$$\boldsymbol{X_B} = \begin{bmatrix} -400 \\ 420 \\ 420 \end{bmatrix} = \begin{bmatrix} x_{B_1} \\ x_{B_2} \\ x_{B_3} \end{bmatrix} = \begin{bmatrix} s_2 \\ x_3 \\ s_3 \end{bmatrix}$$

$$\boldsymbol{B} = [s_2, x_3, s_3] \quad \Rightarrow \quad \boldsymbol{B}^{-1} = [s_1 \quad s_2 \quad s_3] = \begin{bmatrix} -2 & 1 & 0 \\ 1 & 0 & 0 \\ 0 & 0 & 1 \end{bmatrix}$$

So,
$$\boldsymbol{x'_B} = \boldsymbol{B}^{-1}\boldsymbol{b'} = \begin{bmatrix} -2 & 1 & 0 \\ 1 & 0 & 0 \\ 0 & 0 & 1 \end{bmatrix} \begin{bmatrix} 100 \\ -200 \\ 400 \end{bmatrix} = \begin{bmatrix} -400 \\ 100 \\ 400 \end{bmatrix} = \begin{bmatrix} x'_{B_1} \\ x'_{B_2} \\ x'_{B_3} \end{bmatrix}$$

So, the new solution is given by
$$\hat{\boldsymbol{x}}_B = \boldsymbol{x_B} + \lambda \boldsymbol{x'_B}$$

$$\Rightarrow \qquad \hat{\boldsymbol{x}}_B = \begin{bmatrix} -400 \\ 430 \\ 420 \end{bmatrix} + \lambda \begin{bmatrix} -400 \\ 100 \\ 400 \end{bmatrix} = \begin{bmatrix} -400 - 400\lambda \\ 430 + 100\lambda \\ 420 + 400\lambda \end{bmatrix} = \begin{bmatrix} s_2 \\ x_3 \\ s_3 \end{bmatrix}$$

$\Rightarrow \quad s_2 = -400 - 400\lambda, \, x_3 = 430 + 100\lambda, \, s_3 = 420 + 400\lambda$

Now,
$$\bar{\lambda} = \min_{x_{B_i} < 0} \left\{ -\frac{x_{B_i}}{x'_{B_i}} \right\} = \min \left\{ -\frac{x_{B_1}}{x'_{B_1}} \right\} = -\frac{-400}{-400} = -1$$

and
$$\underline{\lambda} = \max_{x_{B_i} > 0} \left\{ -\frac{x_{B_i}}{x'_{B_i}} \right\} = \max \left\{ -\frac{x_{B_2}}{x'_{B_2}}, -\frac{x_{B_3}}{x'_{B_3}} \right\} = \max \left\{ -\frac{430}{100}, \frac{-420}{400} \right\}$$

$$= -\frac{420}{400} = -\frac{21}{20}$$

Therefore, the solution $s_2 = -400 - 400\lambda, \, x_3 = 430 + 100\lambda, \, s_3 = 420 + 400\lambda$ is

optimal feasible for $\dfrac{-21}{20} \leq \lambda \leq -1$.

If $\lambda < -\dfrac{21}{20}$, then s_3 is negative and which corresponds to x'_{B_3} and all $a_{3j} \geq 0$.

Hence, no optimal solution exists for $\lambda < -\dfrac{21}{20}$.

$\Rightarrow$ The solution $x_1 = 0, x_2 = 100 + 100\lambda, x_3 = 230 - 100\lambda$, max. $Z = 1350 - 300\lambda$ is feasible and optimal for $-1 \leq \lambda \leq 23$ and $x_1 = 0$, $x_3 = 430 + 100\lambda$, max. $Z = 2150 + 500\lambda$ is feasible and optimal for $-\dfrac{21}{20} \leq \lambda \leq -1$.

EXAMPLE 2. *For the LPP given by*

$$\text{Max. } Z = 4x_1 + 6x_2 + 2x_3$$

subject to the constraints

$$x_1 + x_2 + x_3 \leq 3 + 3\lambda$$
$$x_1 + 4x_2 + 7x_3 \leq 9 - 3\lambda$$

and $\qquad x_1, x_2, x_3 \geq 0$

find the range of λ for which the solution remains basic optimal and feasible.

SOLUTION. Using slack variables s_1 and s_2, the given LPP can be written as

$$\text{Max. } Z = 4x_1 + 6x_2 + 2x_3 + 0s_1 + 0s_2$$
$$\text{s.t.} \qquad x_1 + x_2 + x_3 + s_1 = 3 + 3\lambda$$
$$x_1 + 4x_2 + 7x_3 + s_2 = 9 - 3\lambda$$
$$\text{and} \qquad x_1, x_2, x_3, s_1, s_2 \geq 0$$

Clearly, $\mathbf{b}(\lambda) = \begin{bmatrix} 3 + 3\lambda \\ 9 - 3\lambda \end{bmatrix} = \begin{bmatrix} 3 \\ 9 \end{bmatrix} + \lambda \begin{bmatrix} 3 \\ -3 \end{bmatrix}$

$\Rightarrow \qquad \mathbf{b} = \begin{bmatrix} 3 \\ 9 \end{bmatrix}$ and $\mathbf{b}' = \begin{bmatrix} 3 \\ -3 \end{bmatrix}$

Now, for $\lambda = 0$, the optimal solution of the problem is given in the following table.

Simplex Table-1

B.V.	C_B	X_B	$c_j \rightarrow$ 4 x_1	6 x_2	2 x_3	0 s_1 (x_4)	0 s_2 (x_5)
x_1	4	1	1	0	−1	4/3	−1/3
x_2	6	2	0	1	2	−1/3	1/3
$Z = C_B X_B = 16$		$\Delta_j \rightarrow$	0	0	−6	−10/3	−2/3

Here, clearly we have

$$X_B = \begin{bmatrix} 1 \\ 2 \end{bmatrix} = \begin{bmatrix} x_{B_1} \\ x_{B_2} \end{bmatrix} = \begin{bmatrix} x_1 \\ x_2 \end{bmatrix}$$

and $\qquad \mathbf{B} = [x_1, x_2] \quad \Rightarrow \quad \mathbf{B}^{-1} = [s_1 \quad s_2] = \begin{bmatrix} 4/3 & -1/3 \\ -1/3 & 1/3 \end{bmatrix}$

So, $\qquad \mathbf{x}'_B = \mathbf{B}^{-1} \cdot \mathbf{b}' = \begin{bmatrix} 4/3 & -1/3 \\ -1/3 & 1/3 \end{bmatrix} \begin{bmatrix} 3 \\ -3 \end{bmatrix} = \begin{bmatrix} 5 \\ 2 \end{bmatrix} \begin{bmatrix} x'_{B_1} \\ x'_{B_2} \end{bmatrix}$

$\therefore$ the new solution

$$\hat{\mathbf{x}}_B = X_B + \lambda \mathbf{x}'_B$$

$$= \begin{bmatrix} 1 \\ 2 \end{bmatrix} + \lambda \begin{bmatrix} 5 \\ -2 \end{bmatrix} = \begin{bmatrix} 1+5\lambda \\ 2-2\lambda \end{bmatrix} = \begin{bmatrix} x_1 \\ x_2 \end{bmatrix}$$

$\Rightarrow \quad x_1 = 1 + 5\lambda, x_2 = 2 - 2\lambda$

Now, since $x'_{B_1} = 5, x'_{B_2} = -2$, therefore,

$$\bar{\lambda} = \min_{x_{B_i} < 0} \left\{ -\frac{x_{B_i}}{x'_{B_i}} \right\} = \min \left\{ -\frac{x_{B_2}}{x'_{B_2}} \right\} = \frac{-2}{-2} = 1$$

and $\quad \underline{\lambda} = \max_{x_{B_i} > 0} \left\{ -\frac{x_{B_i}}{x'_{B_i}} \right\} = \max \left\{ -\frac{x_{B_2}}{x'_{B_2}} \right\} = -\frac{1}{5}$

So, solution $x_1 = 1 + 5\lambda, x_2 = 2 - 2\lambda, x_3 = 0$,

Max. $Z = C_B \hat{x}_B = 4(1+5\lambda) + 6(2-2\lambda) = 16 + 8\lambda$

is optimal basic and feasible for $-\dfrac{1}{5} \leq \lambda \leq 1$

Further, for $\lambda > 1$, x_2 is negative, so vector x_2 will be the outgoing vector.

$\because \qquad x'_{B_2} = -2 < 0$ then r = 2

Now, since

$$\frac{\Delta_k}{a_{rk}} = \min_j \left\{ \frac{\Delta_j}{a_{rj}}, a_{rj} < 0 \right\}$$

$\Rightarrow \qquad \dfrac{\Delta_k}{a_{2k}} = \min_j \left\{ \dfrac{\Delta_j}{a_{2j}}, a_{2j} < 0 \right\} = \min \left\{ \dfrac{\Delta_4}{a_{24}} \right\} = \dfrac{\Delta_4}{a_{24}}$

$\Rightarrow \qquad k = 4$

$\Rightarrow \quad x_4 (= s_1)$ is the incoming vector and key element $= a_{24} = -\dfrac{1}{3}$

Now, apply the simplex method in a usual manner, we have the final optimal simplex table.

Simplex Table-2

		$c_j \rightarrow$	4	6	2	0	0
B.V.	C_B	X_B	x_1	x_2	x_3	s_1	s_2
x_1	4	9	1	4	7	0	1
s_1	0	-6	0	-3	-6	1	-1
$Z = C_B X_B = 36$		$\Delta_j \rightarrow$	0	-10	-26	0	

We have $X_B = \begin{bmatrix} 9 \\ -6 \end{bmatrix} = \begin{bmatrix} x_{B_1} \\ x_{B_2} \end{bmatrix} = \begin{bmatrix} x_1 \\ s_1 \end{bmatrix}$

$$B = [x_1, s_1] \qquad \Rightarrow \qquad B^{-1} = [s_1, s_2] = \begin{bmatrix} 0 & 1 \\ 1 & -1 \end{bmatrix}$$

Therefore, $\hat{x}_B = B^{-1} \cdot b' = \begin{bmatrix} 0 & 1 \\ 1 & -1 \end{bmatrix} \begin{bmatrix} 3 \\ -3 \end{bmatrix} = \begin{bmatrix} -3 \\ 6 \end{bmatrix} = \begin{bmatrix} x'_{B_1} \\ x'_{B_2} \end{bmatrix}$

The new solution $\hat{x}_B$ is given by

$$\hat{x}_B = X_B + \lambda x'_B = \begin{bmatrix} 9 \\ -6 \end{bmatrix} + \lambda \begin{bmatrix} -3 \\ 6 \end{bmatrix} = \begin{bmatrix} 9-3\lambda \\ -6+6\lambda \end{bmatrix} = \begin{bmatrix} x_1 \\ s_1 \end{bmatrix}$$

$\Rightarrow \qquad x_1 = 9 - 3\lambda, \, s_1 = -6 + 6\lambda$

Since, $x'_{B_1} = -3$ and $x'_{B_2} = 6$

Now, $\qquad \overline{\lambda} = \min_{x_{B_i} < 0} \left\{ -\frac{x_{B_i}}{x'_{B_i}} \right\} = \min \left\{ -\frac{x_{B_1}}{x'_{B_1}} \right\} = -\frac{9}{-3} = 3$

and $\qquad \underline{\lambda} = \max_{x_{B_i} > 0} \left\{ -\frac{x_{B_i}}{x'_{B_i}} \right\} = \max \left\{ -\frac{x_{B_2}}{x'_{B_2}} \right\} = -\frac{-6}{6} = 1$

Thus, the solution $x_1 = 9 - 3\lambda$, $x_2 = x_3 = 0$, $s_1 = -6 + 6\lambda$, $s_2 = 0$ and max. $Z = 36 - 12\lambda$ is optimal basic and feasible for $1 \leq \lambda \leq 3$.

Now, for $\lambda > 3$, x_1 is negative which is corresponding to $x'_{B_1} = -3 < 0$ and all $a_{1j} \geq 0 \Rightarrow$ optimal solution exists for $\lambda > 3$.

Again for $\lambda < -\dfrac{1}{5}$ then $x_1 < 0 \Rightarrow x_1$ is the outgoing vector.

Clearly, x_1 correspond to x'_{B_1} so $r = 1$.

Now, $\qquad \dfrac{\Delta_k}{a_{rk}} = \min_{j} \left\{ \dfrac{\Delta_j}{a_{rj}}, a_{rj} < 0 \right\} = \min_{j} \left\{ \dfrac{\Delta_j}{a_{1j}}, a_{1j} < 0 \right\}$

$$= \min \left\{ \dfrac{\Delta_3}{a_{13}}, \dfrac{\Delta_5}{a_{15}} \right\} = \min \left\{ \dfrac{-6}{-1}, \dfrac{-2/3}{-1/3} \right\}$$

$$= \min\{6, 2\} = 2 = \dfrac{\Delta_5}{a_{15}}$$

$\Rightarrow \qquad k = 5$

$\Rightarrow x_5 \, (s_2)$ is the incoming vector and key element $a_{15} = -\dfrac{1}{3}$. Now, by applying the simplex method in a usual manner we get the following simplex table.

Simplex Table-3

B.V.	C_B	X_B	x_1 4	x_2 6	x_3 2	s_1 (x_4) 0	s_2 (x_5) 0
s_2	0	-3	-3	0	3	-4	1
x_2	6	3	1	1	1	1	0
$Z = C_B X_B = 18$		$\Delta_j \rightarrow$	-2	0	-4	-6	0

In the above table, we have

$$X_B = \begin{bmatrix} -3 \\ 3 \end{bmatrix} = \begin{bmatrix} x_{B_1} \\ x_{B_2} \end{bmatrix} = \begin{bmatrix} s_2 \\ x_2 \end{bmatrix}$$

$$B = [s_2, x_2]$$

$$\Rightarrow \qquad B^{-1} = [s_1, s_2] = \begin{bmatrix} -4 & 1 \\ 1 & 0 \end{bmatrix}$$

Therefore, $\quad x_B' = B^{-1}b' = \begin{bmatrix} -4 & 1 \\ 1 & 0 \end{bmatrix}\begin{bmatrix} 3 \\ -3 \end{bmatrix} = \begin{bmatrix} -15 \\ 3 \end{bmatrix} = \begin{bmatrix} x'_{B_1} \\ x'_{B_2} \end{bmatrix}$

Hence, the new solution is given by

$$\hat{x}_B = X_B + \lambda x_B'$$

$$= \begin{bmatrix} -3 \\ 3 \end{bmatrix} + \lambda \begin{bmatrix} -15 \\ 3 \end{bmatrix} = \begin{bmatrix} -3 - 15\lambda \\ 3 + 3\lambda \end{bmatrix} = \begin{bmatrix} s_2 \\ x_2 \end{bmatrix}$$

$$\Rightarrow \qquad x_2 = 3 + 3\lambda, s_2 = -3 - 15\lambda$$

Now, since $\quad x'_{B_1} = -15, x'_{B_2} = 3$

So, $$\bar{\lambda} = \min_{x'_{B_i} < 0}\left\{-\frac{x_{B_i}}{x'_{B_i}}\right\} = \min\left\{-\frac{x_{B_1}}{x'_{B_1}}\right\} = \min\left\{-\frac{-3}{-15}\right\} = -\frac{1}{5}$$

and $$\underline{\lambda} = \max_{x'_{B_k} > 0}\left\{-\frac{x_{B_k}}{x'_{B_k}}\right\} = \max\left\{-\frac{x_{B_2}}{x'_{B_2}}\right\} = \frac{-3}{3} = -1$$

$\Rightarrow$ the solution $x_1 = 0$, $x_2 = 3 + 3\lambda$, $x_3 = 0$, $s_1 = 0$, $s_2 = -3 - 15\lambda$ and max. $Z = 18 + 18\lambda$ is optimal basis and feasible for the range $-1 \le \lambda \le -\frac{1}{5}$.

Further, for $\lambda < -1$, $x_2 < 0$ which is corresponding to x'_{B_2} also from above table-3 all $a_{2j} \ge 0$. Hence, no optimal solution exists for $\lambda < -1$.

Hence, we conclude that the solution $x_1 = 1 + 5\lambda$, $x_2 = 2 - 2\lambda$, $x_3 = 0$, max. $Z = 16 + 8\lambda$ is optimal feasible for $-\frac{1}{5} \le \lambda \le 1$. The solution $x_1 = 9 - 3\lambda$, $x_2 = 0$, $x_3 = 0$ and max. $Z = 36 - 12\lambda$ is optimal feasible for $1 \le \lambda \le 3$ and the solution is $x_1 = 0$, $x_2 = 3 + 3\lambda$, $x_3 = 0$, max. $Z = 18 + 18\lambda$ is optimal feasible for $-1 \le \lambda \le -\frac{1}{5}$.

 Exercise-12.1

1. Solve the following parametric LPP

   Max. $Z = \lambda x_1 - x_2$

   subject to the constraints

   $$3x_1 - 3x_2 \ge 5$$
   $$2x_1 + x_2 \le 3$$

   and $\qquad x_1, x_2 \ge 0$

   for $\quad -\infty \le \lambda \le \infty$.

2. Find the critical value of λ for which the solution of the following LPP is optimal basic and feasible

   Max. $Z = (3 + 3\lambda)x_1 + 2x_2 + (5 - 6\lambda)x_3$

   s.t. $\quad x_1 + 2x_2 + x_3 \le 430$

   $$3x_1 + 2x_3 \le 460$$
   $$x_1 + 4x_2 \le 420$$

   and $\qquad x_1, x_2 \ge 0$

3. Solve the following LPP for all values of λ

   Min. $Z = 2\lambda x_1 + (1 - \lambda)x_2 - 3x_3 + \lambda x_4$
   $$+ 2x_5 - 3\lambda x_6$$

   subject to the constraints

   $$x_1 + 3x_2 - x_3 + 2x_5 = 7$$
   $$- 2x_2 + 4x_3 + x_4 = 12$$
   $$- 4x_2 + 3x_3 + 8x_5 + x_6 = 10$$

   and $\qquad x_1, x_2, \ldots, x_6 \ge 0$

4. Find the variation in optimal solution of the following parametric LPP

   Min. $Z = 4x_1 + x_2 + 2x_3$

   subject to the constraints

   $$3x_1 + x_2 + 2x_3 = 3 + 3\lambda$$
   $$4x_1 + 3x_2 + 2x_3 \ge 6 + 2\lambda$$
   $$x_1 + 2x_2 + 5x_3 \le 4 - \lambda$$

   and $x_1, x_2, x_3 \ge 0$ and $\lambda \ge 0$

5. If the LPP given by

Max. $Z = x_1 + 6x_2$

subject to the constraints

$$x_1 \le 4 + 8\lambda$$
$$3x_1 + 2x_2 \le 18 - 24\lambda$$

and $\quad x_1, x_2 \ge 0$

has the optimal solution for $l = 0$ as follows:

B.V.	C_B	X_B	$c_j \to$ x_1	6 x_2	2 s_1	0 s_2
s_1	0	4	1	0	1	1
x_2	6	9	3/2	1	0	1/2
$Z = C_B X_B$ $= 36$		$\Delta_j \to$	-8	0	0	-3

Find the range of λ for which the above solution remains basic feasible and optimal $(\lambda \ge 0)$.

6. Find the range of λ over which the solution remain basic and optimal for the following LPP

Max. $Z = \lambda x_1 - \lambda x_2 - x_3 + x_4$

subject to the constraints

$$3x_1 - 3x_2 - x_3 + x_4 \ge 5$$
$$2x_1 - 2x_2 + x_3 - x_4 \le 3$$

and $\quad x_1, x_2, x_3, x_4 \ge 0$

7. Find the range of λ (≥ 0) over which the following LPP

Max. $Z = 3x_1 + 2x_2 + 5x_3$

subject to the constraints

$$x_1 + 2x_2 + x_3 \le 40 - \lambda$$
$$3x_1 + 2x_3 \le 60 + 2\lambda$$
$$x_1 + 4x_2 \le 30 - 7\lambda$$

and $\quad x_1, x_2, x_3 \ge 0$

has optimal and feasible solution.

8. Find the variation in optimal solution of the following parametric LPP given by $(\lambda \ge 0)$

Min. $Z = (4 - \lambda)x_1 + (1 - 3\lambda)x_2 + (2 - 2\lambda)x_3$

subject to the constraints

$$3x_1 + x_2 + 2x_3 = 3$$
$$4x_1 + 3x_2 + 2x_3 \ge 6$$
$$x_1 + 2x_2 + 5x_3 \le 4$$

and $\quad x_1, x_2, x_3 \ge 0$

ANSWERS

1. $x_1 = \dfrac{8}{5}, x_2 = -\dfrac{1}{5},$ Min $Z = \dfrac{1}{5} + \dfrac{8}{5}\lambda$ for $-2 \le \lambda \le 3$ and for $\lambda = 3$, multiple solution exists

2. $x_1 = 0, x_2 = 100, x_3 = 230,$ Max $Z = 1350 - 1380\lambda$ for $0 \le \lambda \le \dfrac{1}{3}$

$x_1 = 10, x_2 = 1025, x_3 = 215,$ Max $Z = 1310 - 1260\lambda$ for $\dfrac{1}{3} \le \lambda \le \dfrac{5}{12}$

$x_1 = \dfrac{460}{3}, x_2 = \dfrac{200}{3}, x_3 = 0,$ Max $Z = \dfrac{1780}{3} + 460\lambda$ for $\lambda \ge \dfrac{5}{12}$

3. $x_1 = 0, x_2 = 4, x_3 = 5, x_4 = x_5 = 0, x_6 = 11,$ Min $Z = -11 - 37\lambda$ for $-\dfrac{1}{27} \le \lambda \le 2$

$x_1 = 0, \; x_2 = \dfrac{7}{3}, x_3 = 0, x_4 = \dfrac{50}{3}, x_5 = 0, x_6 = \dfrac{58}{3},$ Min $Z = \dfrac{7}{3} - 217\lambda$ for $2 \le \lambda \le \infty$

4. $x_1 = \dfrac{3}{2}, x_2 = \dfrac{9}{2},$ Max $Z = \dfrac{45}{2}$ and $-3 \le \theta \le \dfrac{3}{7}$

5. For $0 \le \lambda \le \dfrac{3}{8}, x_1 = \dfrac{3 + 7\lambda}{5}, x_2 = \dfrac{6 - 11\lambda}{5},$ Min $Z = \dfrac{21 - 11\lambda}{5}$

For $\dfrac{3}{8} \le \lambda \le \dfrac{2}{5}, x_1 = -3 + 11\lambda, x_2 = 6 - 15\lambda,$ Min $Z = 6 + 3\lambda$

6. $x_1 = \dfrac{8}{5}, x_2 = x_3 = 0, x_4 = \dfrac{1}{5},$ Min $Z = \dfrac{1}{5} + \dfrac{8}{5}\lambda$ for $-2 \le \lambda \le 3$ and for $\lambda > 3$ no solution exists

7. $x_1 = 0, x_2 = 5 - \lambda, x_3 = 30 + \lambda,$ Max. $Z = 160 + 3\lambda$ for $0 \le \lambda \le \dfrac{10}{3}$

$x_1 = 0, \; x_2 = \dfrac{30 - 7\lambda}{4}, x_3 = 30 + \lambda,$ Max. $Z = 165 + \dfrac{3}{2}\lambda$ for $\dfrac{10}{3} \le \lambda \le \dfrac{30}{7}$

and for $\lambda > \dfrac{30}{7}$ no feasible solution exists

8. $x_1 = \dfrac{2}{5}, x_2 = \dfrac{9}{5}, x_3 = 0,$ Min $Z = \dfrac{17}{5} - \dfrac{29}{5}\lambda$

REVIEW QUESTIONS

1. What do you understand by parametric linear programming.
2. What is the parametric linear programming. How does it different from sensitivity analysis?
3. Write the types of parametric LPP.
4. Explain the different solution process of parametric linear programming.

ARCHIVE

1. Max. $Z = (6 - \lambda)x_1 + (12 - \lambda)x_2 + (4 - \lambda)x_3$
   subject to the constraints
   $$3x_1 + 4x_2 + x_3 \le 2$$
   $$x_1 + 3x_2 + 2x_3 \le 1$$
   and $x_1, x_2, x_3 \ge 0$.
   Perform a complete parametric programming analysis and identify all the critical values of the parameter λ.   [MEERUT–1986]

2. Max. $Z = 3x_1 + 2x_2 + 5x_3$
   subject to the constraints
   $$x_1 + 2x_2 + x_3 \le 430 + \lambda$$
   $$3x_1 + 2x_3 \le 460 - 4\lambda$$
   $$x_1 + 4x_2 \le 420 - 4\lambda$$
   and $x_1, x_2, x_3 \ge 0$.
   Determine the critical value of λ for which the solution remains optimal basis feasible.
                [MEERUT–1985]

3. Max. $Z = 3x_1 + 4x_2$
   subject to the constraints
   $$2x_1 + 4x_2 \le 21 - \lambda$$
   $$x_1 + x_2 \le 6 - 2\lambda$$
   and $x_1, x_2 \ge 0$.
   Determine the optimal solution for $\lambda = 0$ and find the range of λ under which the solution remains remains optimal for $\lambda \ge 0$.
   [MADURAI–1989; ANNAMALAI–1982]

4. Max. $Z = (4 - 10\lambda)x_1 + (8 - 4\lambda)x_2$
   subject to the constraints
   $$x_1 + x_2 \le 4$$
   $$2x_1 + x_2 \le 3 - \lambda$$
   and $x_1, x_2 \ge 0$.
   Study the variations in the optimum solution with the parameter λ, where $-\infty < \lambda < \infty$.

HINTS AND ANSWERS

1. $x_1 = \dfrac{2}{5}, x_2 = \dfrac{1}{5}; 0 \le \lambda \le 3$

   $x_1 = 0, x_2 = \dfrac{1}{3}; 3 \le \lambda \le 9$

   $x_1 = 0, x_2 = 0; 9 \le \lambda \le \infty$

2. The optimal solution when $\lambda = 0$.
   No optimal solution to the problem for all $\lambda > 103$ and the optimal solution at $\lambda = 103$; $x_1 = 0$, $x_2 = 0, x_3 = 20$ and Max. $Z = 100$.

3. $x_1 = \dfrac{3}{2}, x_2 = \dfrac{9}{2}; -3 \le \lambda \le \dfrac{3}{7}$

4. $x_1 = 4, x_2 = 0, -\infty \le \lambda \le -5$
   $x_1 = 0, x_2 = 5, -5 \le \lambda \le -1$
   $x_1 = 0, x_2 = 3, -1 \le \lambda \le 2$
   No fesible solution when $\lambda > 3$.

13

The Transportation Problem

13.1 INTRODUCTION

The transportation problem are concerned with the distribution of certain product from several sources to numerous localities at a minimum cost. It is one of the types of linear programming problems. The origin of transportation problem is concerned with two contributions. First was the consequence of the study entitled "The distribution of a product from several sources to numerous locations" presented by **F.L. Hitchock** in 1941 and second was the presentation of "Optimum utilization" by **T.C. Koopmans** in 1947.

From the name itself, it is clear that the transportation problem means a problem where something is to be transferred.

Let us suppose we have a product which is to be transformed from a number of centres called 'origin' or 'sources' to a number of places called 'destinations'. The cost of transportation along different routes are different and known. The main objective is to minimize the cost of associated with such transportation from the place of supply to their destination. These special type of linear programming problems are called 'transportation problem'.

13.2 MATHEMATICAL FORMULATION OF TRANSPORTATION PROBLEM

[MEERUT–2003, 05; ROHILKHAND–2000, 05; DELHI–2010; GORAKHPUR–2011; KANPUR–2012]

To design the mathematical formulation of a transportation problem, we have the following assumptions.

(i) Product can be transported easily from sources to the destination.

(ii) Total quantity of the product available at different sources is equal to the requirements at different destinations.

(iii) The unit transportation cost of the product from all sources to the destination is well known.

(iv) The transportation cost on a given route is directly proportional to the number of units transported by that route.

(v) The objective is to minimize the total cost of transportation from all sources to the destinations and not for individual supply and destination centres.

Let us have m origins and n destinations (n may or may not be equal to m) and

a_i : quantity of product available at origin i

b_j : quantity of product required at destination j

c_{ij} : cost of transporting from origin i to destination j

x_{ij} : quantity transported from origin i to destination j

Then by second assumption

$$\sum_{i=1}^{m} a_i = \sum_{j=1}^{n} b_j \qquad \qquad \ldots(1)$$

This is the case when demand is fully met from the supply.

The problem can be stated in the form of LPP as

$$\text{Min.} \, Z = \sum_{i=1}^{m} \sum_{j=1}^{n} c_{ij} x_{ij}$$

such that
$$\sum_{j=1}^{n} x_{ij} = a_i \text{ for } i = 1, 2, \ldots, m \qquad \qquad \ldots(2)$$

$$\sum_{i=1}^{m} x_{ij} = b_j \text{ for } j = 1, 2, \ldots, n \qquad \qquad \ldots(3)$$

and $\qquad \qquad x_{ij} \geq 0 \;\; \forall i = 1,2,\ldots m; j = 1,2,\ldots,n$

The above transportation problem has two set of constraints given by (2) and (3) which will be consistent if

$$\sum_{i=1}^{m} a_i = \sum_{j=1}^{n} b_j \qquad \qquad \ldots(4)$$

which shows that the necessary and sufficient condition for a transportation problem to have a feasible solution is given by (4).

☞ REMARK

- A transportation problem that satisfies the condition

$$\sum_{i=1}^{m} a_i = \sum_{j=1}^{n} b_j$$

is called balanced transportation problem.

13.2.1 TABULAR REPRESENTATION OF TRANSPORTATION PROBLEM

Let there be m sources (origin) and n destination. Then we have the following tabular representation.

Sources \ Destinations	D_1	D_2	D_3	$\ldots$	D_n	Supply
O_1	c_{11}	c_{12}	c_{13}	$\ldots$	c_{1n}	a_1
O_2	c_{21}	c_{22}	c_{23}	$\ldots$	c_{2n}	a_2
O_3	c_{31}	c_{32}	c_{33}	$\ldots$	c_{3n}	a_3
$\vdots$	$\vdots$					$\vdots$
O_m	c_{m1}	c_{m2}	c_{m3}	$\ldots$	c_{mn}	a_m
Demand	b_1	b_2	b_3	$\ldots$	b_n	$\sum_{i=1}^{m} a_i = \sum_{j=1}^{n} b_j$

13.2.2 RELATED DEFINITIONS

(1) **Feasible solution.** A feasible solution to a transportation problem is a set of non-negative individual allocations ($x_{ij} \geq 0$) which satisfies the row and column sum restrictions (*i.e.* equation (2) and (3)). [MEERUT–2001, 07]

(2) **Basic Feasible solution.** A feasible solution of an $m \times n$ transportation problem is said to be basic if the total number of positive allocations is equal to $m + n - 1$ i.e. one less than the sum of no. of rows and columns.

[MEERUT–2001,07, 12; ROHILJHAND–1999]

(3) **Optimal solution.** A feasible solution, not necessarily basic is said to be optimum if it minimize the total transportation cost. [ROHILKHAND–1999]

(4) **Non-degenerate Basic Feasible solution.** A basic feasible solution to a transportation problem of m sources and n destination is said to be non-degenerate basic feasible if

   (i)   the total no. of non-negative allocations x_{ij} is exactly equal to $m + n - 1$.

   (ii) these allocations should be in independent position (i.e. it is always impossible to form any closed circuit by joining these allocation by horizontal and vertical lines only).

(5) **Degenerate-basic Feasible solution.** A basic feasible solution to a transportation problem of m sources and n destinations is said to be degenerate if the total number of non-negative allocations is less than $(m + n - 1)$.

13.2.3 SOME THEOREMS

THEOREM 1. **(Existence of Feasible solution)** *The necessary and sufficient condition for the existence of feasible solution of a transportation problem is*

$$\Sigma a_i = \Sigma b_j : i = 1, 2, ..., m; \ j = 1, 2, ..., n$$

[MEERUT–2007; KANPUR–2009]

PROOF. Necessary Condition:

Let us suppose, a feasible solution of the transportation problem exists. Then

$$\sum_{j=1}^{n} a_{ij} = a_i : i = 1, 2, ..., m$$

and
$$\sum_{i=1}^{m} x_{ij} = b_j, j = 1, 2, ..., n$$

Now, summing over all i and j respectively, we get

$$\sum_{i=1}^{m}\sum_{j=1}^{n} x_{ij} = \sum_{i=1}^{m} a_i \quad \text{and} \quad \sum_{j=1}^{n}\sum_{i=1}^{m} x_{ij} = \sum_{j=1}^{n} b_j$$

which shows that

$$\sum_{i=1}^{m} a_i = \sum_{j=1}^{n} b_j$$

Sufficient Condition. Let $\sum\limits_{i=1}^{m} a_i = \sum\limits_{j=1}^{n} b_j = k$ (say)

If $x_{ij} = \lambda_i b_j \ \forall \ i, j, \lambda_i \neq 0$ is any real number. Then

$$\sum_{j=1}^{n} x_{ij} = \sum_{j=1}^{n} \lambda_i b_j = \lambda_i \sum_{j=1}^{n} b_j = k\lambda_i$$

$$\Rightarrow \qquad \lambda_i = \frac{1}{k}\sum_{j=1}^{n} x_{ij} = \frac{a_i}{k}$$

Therefore, $x_{ij} = \lambda_i b_j = \dfrac{a_i b_j}{k}$ for all i and j.

$\Rightarrow \qquad x_{ij} \geq 0 \;\forall\; i$ and j $\hspace{3cm}$ ($\because\; a_i \geq 0, b_j > 0$)

Hence, a feasible solution exists.

☞ REMARK

- A balanced transportation problem has a feasible solution.

THEOREM 2. **(Existence of BFS)** *Out of $(m+n)$ equations, there are only $m + n - 1$ independent equations in a transportation problem, m and n being the number of origins and destinations and any one equation can be dropped as the redundant equation.*

PROOF. Let us consider m row equations and $(n - 1)$ columns equations of the transportation problem as

$$\sum_{j=1}^{n} x_{ij} = a_i,\, i = 1, 2, \ldots, m \qquad \ldots(1)$$

and $\qquad \displaystyle\sum_{i=1}^{m} x_{ij} = b_j,\, j = 1, 2, \ldots, n-1 \qquad \ldots(2)$

Now, adding m origin constraints given in (1), we get

$$\sum_{i=1}^{m}\sum_{j=1}^{n} x_{ij} = \sum_{i=1}^{m} a_i \qquad \ldots(3)$$

Further, adding $(n - 1)$ destination constraints given in (2), we get

$$\sum_{j=1}^{n-1}\sum_{i=1}^{m} x_{ij} = \sum_{j=1}^{n-1} b_j \qquad \ldots(4)$$

Now, subtracting (4) from (3) we get

$$\sum_{i=1}^{m}\sum_{j=1}^{n} x_{ij} - \sum_{j=1}^{n-1}\sum_{i=1}^{m} x_{ij} = \sum_{i=1}^{m} a_i - \sum_{j=1}^{n-1} b_j$$

$\Rightarrow \qquad \displaystyle\sum_{i=1}^{m}\left[\sum_{j=1}^{n} x_{ij} - \sum_{j=1}^{n-1} x_{ij}\right] = \sum_{j=1}^{n} b_j - \sum_{j=1}^{n-1} b_j \hspace{2cm}$ ($\because \Sigma a_i = \Sigma b_j$)

$\Rightarrow \qquad \displaystyle\sum_{i=1}^{m} x_{in} = b_n$, which is the nth destination constraints.

$\Rightarrow$ if $(m + n - 1)$ constraints are satisfied then the $(m + n)^{\text{th}}$ constraints will be satisfied because $\Sigma a_i = \Sigma b_j$.

Therefore, we have only $(m + n - 1)$ linearly independent equations out of $(m + n)$ equations, one is redundant.

☞ REMARK

- A basic feasible solution will contain $(m + n - 1)$ positive variables, others being zero.

THEOREM 3. **(Existence of optimal solution)** *There always exists an optimal solution to a balanced transportation problem.*

PROOF. We know that the necessary and sufficient condition for a feasible solution is

$$\sum_{i=1}^{m} a_i = \sum_{j=1}^{n} b_j \qquad \ldots(1)$$

We know that for a balanced transportation problem, we always have (1)

$\Rightarrow$ A feasible solution exists of the problem *i.e.* $x_{ij} \geq 0 \ \forall \ i, j$

Now, for the constraints of the problem each $x_{ij} \leq \min\{a_i, b_j\}$

Therefore, $0 \leq x_{ij} \leq \min\{a_i, b_j\}$

$\Rightarrow$ The feasible region of the problem is non-empty, closed and bounded.

Hence, there exists an optimal solution.

13.3 SOLUTION OF THE TRANSPORTATION PROBLEM   [MEERUT–2004]

To find the solution of a transportation problem, we have to find following two solutions.

(1) Initial basic feasible solution

(2) Optimal solution

13.4 METHODS OF FINDING INITIAL SOLUTION

The initial basic feasible solution can be obtained by using any one of the following three methods.

     (i)   North West Corner Rule

     (ii)   Least Cost Entry method or Matrix-minima method   [ROHILKHAND–2003]

     (iii) Vogel's Approximation Method (VAM) or unit cost penalty method

13.4.1 NORTH WEST CORNER METHOD   [MEERUT–2005]

In this method we apply the following steps:

STEP 1.   Start with the cell (1, 1) at the upper left (north-west) corner of the matrix and allocate it as much as possible amount equal to the minimum of the supply-demand values, *i.e.*, we allocate x_{11} to the cell (1, 1) where

$$x_{11} = \min\{a_1, b_1\}$$

where a_1 is the supply amount for the first row and b_1 is the demand for the first column.

STEP 2.   (i) If $a_1 > b_1$, then move to the cell (1, 2) and allocate x_{12} where
$$x_{12} = \min\{a_1 - x_{11}, b_2\}.$$

            (ii) If $a_1 < b_1$, then move to the cell (2, 1) and allocate it as x_{21} where
$$x_{21} = \min(a_2 - b_1 - x_{11})$$

STEP 3.   Continue this process step by step till an allocation is made in the south east corner of the cell *i.e.* until all available amount is exhausted.

☛ REMARK

- Sometime, during the process of making allocation at a particular cell, supply equal demands, then next allocation of magnitude zero can be made in a cell either in the next row or column. This is called degeneracy.

Solved Examples

EXAMPLE 1.   *Find an IBFS of the following transportation problem by north-west corner rule.*

	D_1	D_2	D_3	D_4	**Supply**
O_1	1	2	3	4	6
O_2	4	3	2	0	8
O_3	0	2	2	1	10
Demand	4	6	8	6	24

(Origins)

SOLUTION.   We have $m = 3, n = 4, a_1 = 6, a_2 = 8, a_3 = 10$
$b_1 = 4, b_2 = 6, b_3 = 8, b_4 = 6$

Also, $\Sigma a_i = \Sigma b_j = 24 \Rightarrow$ Problem is balanced.

Now to find IBFS, we proceed as follows:

STEP 1. Start with the cell (1, 1) at the north-west corner of the above table and allocate it as x_{11} such that

$x_{11} = \min\{a_1, b_1\} = \min\{6, 4\} = 4$

$\Rightarrow D_1$ column is exhausted.

STEP 2. Clearly, $a_1 > b_1$ so we move towards the cell (1, 2) and allocate it as x_{12}, where

$x_{12} = \min\{a_1 - x_{11}, b_2\} = \min\{6 - 4, 6\} = \min\{2, 6\} = 2$

i.e., allocate 2 to the cell (1, 2)

$\Rightarrow O_1$ row is exhausted.

Then we move downward to the cell (2, 3) to allocate it x_{23} such that

$x_{23} = \min\{a_2 - x_{23}, b_3\} = \min\{8 - 4, 8\} = \min\{4, 8\} = 4$

i.e., allocate 4 to the cell (2, 3).

$\Rightarrow O_2$ row is exhausted.

Then we move downward to the cell (3, 3) and allocate it as x_{33}, where

$x_{33} = \min\{a_3, b_3 - x_{23}\} = \min\{10, 8 - 4\} = \min\{10, 4\} = 4$

i.e., allocate 4 to the cell (3, 3)

$\Rightarrow D_3$ column is exhausted.

Then we move towards to the cell (3, 4) and allocate it as x_{34} where

$x_{34} = \min\{a_3 - x_{33}, b_4\} = \min\{10 - 4, 6\} = \min\{6, 6\} = 6$

i.e., allocate 6 to the cell (3, 4)

$\Rightarrow O_3$ row and D_4 column are exhausted.

Thus all the row and columns are exhausted.

The above steps of allocations can be summarized in the table given below.

④	②		
1	2	3	4
	④	④	
3	3	2	8
		④	⑥
0	2	2	1

Hence, the initial basic feasible solution of the given transportation problem is given by

Total cost $= 1 \times 4 + 2 \times 2 + 4 \times 3 + 4 \times 2 + 4 \times 2 + 6 \times 1 = ₹42$

EXAMPLE 2. *Find an IBFS of the following transportation problem by north-west corner rule.*
[MEERUT–1994, 2014]

			Supply
2	7	4	5
3	3	1	8
5	4	7	7
1	6	2	14
Demand 7	9	18	34

SOLUTION. Proceed same as in example-1, we have the following allocation table

⑤ 2	7	4
② 3	⑥ 3	1
5	③ 4	④ 7
1	6	⑭ 2

Hence, total cost $= 5 \times 2 + 2 \times 3 + 6 \times 3 + 3 \times 4 + 4 \times 7 + 14 \times 2 = ₹102$

13.4.2 LOWEST COST ENTRY METHOD OR MATRIX MINIMA METHOD

In lowest cost entry method, we use the following steps:

STEP 1. Identify the cell with lowest cost. Let it be (i, j). Then allocate x_{ij} to the cell (i, j) such that

$$x_{ij} = \min\{a_i, b_j\}$$

If lowest cost cell is not unique, we may select any one of them.

STEP 2. If $x_{ij} = a_i$ then remove the ith row from the table and then demand b_j is reduced to $(b_j - a_i)$. Then go to step-3.

If $x_{ij} = b_j$, then remove the jth column from transportation table and the supply a_i is reduced to $a_i - b_j$, then go to step-3.

If $x_{ij} \neq b_j$, then remove either i^{th} row or j^{th} column but not both.

STEP 3. Repeat the above steps with reduced transportation table thus obtained in step-2 until all the available amount is exhausted.

Solved Examples

EXAMPLE 1. *Find an initial basic feasible solution of the following transportation problem by lowest cost entry method.*
[IAS–1989]

	D_1	D_2	D_3	D_4	Supply
O_1	6	4	1	5	14
O_2	8	9	2	7	16
O_3	4	3	6	2	5
Demand	6	10	15	4	35

SOLUTION. Here, we have
$$m = 3, n = 4$$
$$a_1 = 14, a_2 = 16, a_3 = 5$$
$$b_1 = 6, b_2 = 10, b_3 = 15, b_4 = 4$$
Clearly, $\Sigma a_i = \Sigma b_j = 35 \Rightarrow$ Problem is balanced.
Since, 1 is the lowest cost (at the cell (1, 3))
$\therefore$ We allocate x_{13} such that
$$x_{13} = \min\{a_1, b_3\} = \min\{14, 15\} = 14 = a_1$$
After allocating 14 to the cell (1, 3) the O_1 row is deleted and then first reduced transportation table is given as under

	D_1	D_2	D_3	D_4	
O_2	8	9	2	7	16
O_3	4	3	6	2	5
	6	10	1	4	21

In the above table, the lowest cost is 2 which is at the (2, 3) and the cell (3, 4). Choose (3, 4) cell arbitrarily and allocate x_{34} to this cell such that
$$x_{34} = \min\{a_3, b_4\} = \min\{5, 4\} = 4$$
After allocating 4 to the cell (3, 4), the D_4 column is deleted and the second reduced table is given below.

	D_1	D_2	D_3	
O_2	8	9	2	16
O_3	4	3	6	1
	6	10	1	17

In this table, the lowest cost is 2 at the cell (2, 3). Allocate x_{23} to this cell such that
$$x_{23} = \min\{a_2, b_3\} = \min\{16, 1\} = 1$$
After allocating 1 to the cell (2, 3), the D_3 column is deleted and the third reduced table is given as below.

	D_1	D_2	
O_2	8	9	15
O_3	4	3	1
	6	10	16

In the above table, the lowest cost is 3 at the cell (3, 2). Allocate x_{32} to this cell such that
$$x_{32} = \min\{a_3, b_2\} = \min\{1, 10\} = 1$$
After allocating 1 to the cell (3, 2), the O_3 row is deleted and only O_2 row is remaining in which 8 is minimum at the cell (2, 1). So, we allocate x_{21} to the cell such that
$$x_{21} = \min\{a_2, b_1\} = \min\{15, 6\} = 6$$
and finally allocate x_{22} to the remaining element 9 which is at the cell (2, 2) such that
$$x_{22} = \min\{a_2, b_2\} = \min\{9, 9\} = 9$$
Therefore, all the rows and columns are exhausted.
Finally, the summary of all the allocations is given in the table as follow.

6	4	⑭ 1	5
⑥ 8	⑨ 9	① 2	7
4	① 3	6	④ 2

Hence, total cost $= 1 \times 14 + 6 \times 8 + 9 \times 9 + 1 \times 2 + 1 \times 3 + 4 \times 2 = ` 156$

EXAMPLE 2. *Find the initial basic feasible solution of the following transportation problem by lowest cost entry method.*

			Supply
2	7	4	5
3	3	1	8
5	4	7	7
1	6	2	14
Demand 7	9	18	34

SOLUTION. Proceeding same as in example-1, we have the following allocation table:

2	② 7	③ 4
3	3	⑧ 1
5	⑦ 4	7
⑦ 1	6	⑦ 2

Hence, the total cost $= 2 \times 7 + 3 \times 4 + 7 \times 4 + 8 \times 1 + 7 \times 1 + 7 \times 2 = ` 83$

13.4.3 VOGEL'S APPROXIMATION METHOD (UNIT COST PENALTY METHOD)

To find the IBFS by Vogel's approximation method, we use the following steps.

STEP 1. Find the smallest and next to smallest costs for each row of the transportation table and then find the difference between them for each row. Write these difference (Penalties) alongside the transportation table against the respective rows. Similar exercise will be done on case of columns.

STEP 2. Select the maximum Penalty among the rows and columns penalties and if there is a tie, choose any one arbitrarily.

STEP 3. Allocate the maximum possible amount to the cell with lowest cost in that particular row or column.
Let the largest penalty correspond to i^{th} row and let c_{ij} be the smallest cost in the ith row. Allocate the amount.
$$x_{ij} = \min\{a_i, b_j\} \text{ in the cell } (i, j)$$
and then cross out i^{th} row and j^{th} column and obtain reduced matrix.

STEP 4. Now compute the row and column penalties for the reduced table and repeat step 2 and 3.

Continue this process until all the available quantity is exhausted or all the requirements are satisfied.

 Solved Examples

EXAMPLE I. *Find the initial basic feasible solution to the following transportation problem by Vogel's approximation method.*

	D_1	D_2	D_3	D_4	Supply
F_1	3	3	4	1	100
F_2	4	2	4	2	125
F_3	1	5	3	2	75
Demand	120	80	75	25	300

SOLUTION. We have $m = 3, n = 4$

$$a_1 = 100, a_2 = 125, a_3 = 75$$
$$b_1 = 120, b_2 = 80, b_3 = 75, b_4 = 25$$

$\Rightarrow$ $\sum\limits_{i=1}^{m} a_i = 300 = \sum\limits_{j=1}^{n} b_j$

$\Rightarrow$ given transportation problem is a balanced problem.

Now, we have to find the penalties as given below

Penalties

3	3	4	1	$3 - 1 = (2)$
4	2	4	2	$2 - 2 = (0)$
1	5	3	2	$2 - 1 = (1)$

Penalties $3 - 1 = (2)$ $3 - 2 = (1)$ $4 - 3 = (1)$ $2 - 1 = (1)$

Among all the above penalties, the maximum penalty = 2, which is corresponding to D_1 column and F_1 row (We can select any one of these)

Suppose we select F_1 row. In F_1 row, the minimum cost is 1 which is at the cell $(1, 4)$, so allocate x_{14} to the cell $(1, 4)$ such that

$$x_{14} = \min\{100, 25\} = 25 \text{ (demand for } D_4)$$

Delete D_4, the reduced matrix is given by

Penalties

3	3	4	$3 - 3 = (0)$
4	2	4	$4 - 2 = (2)$
1	5	3	$3 - 1 = (2)$

Penalties $3 - 1 = (2)$ $3 - 2 = (1)$ $4 - 3 = (1)$

Here, the maximum penalty = 2, which is corresponding to F_2, F_3 rows and D_1 column, so we can select anyone of these.

Now, select F_3 row in which the least cost is 1 at the cell $(3, 1)$. Therefore allocate x_{31} to the cell $(3, 1)$ such that

$$x_{31} = \min\{75, 120\} = 75, \text{ supply for } F_3$$

Now, delete F_3 after allocating 75 to the cell $(3, 1)$

We get the following reduced table

	D_1	D_2	D_3	Supply	Penalty
F_1	3	3	4	75	$4 - 3 = (1)$
F_2	4	2	4	125	$4 - 2 = (2)$
Demand	45	80	75	200	
Penalty	$4 - 3 = (1)$	$3 - 2 = (1)$	$4 - 4 = (0)$		

Clearly, the maximum penalty $= 2$, (corresponding to the cell (2,2))

$\therefore \qquad x_{22} = \min\{125, 80\} = 80$

So, allocate 80 to the cell (2, 2) and delete D_2 column.

The third reduced table is given as under

	D_1	D_3	Supply	Penalty
F_1	3	4	75	$4 - 3 = (1)$
F_2	4	4	45	$4 - 4 = (0)$
Demand	45	75	120	
Penalty	$4 - 3 = (1)$	$4 - 4 = (0)$		

Here, the maximum penalty $= 1$, $i.e.$, corresponding to the F_1 row and D_1 column.

Therefore, we can select any of these for allocation.

Let us suppose we select F_1 row for allocation.

Clearly, in F_1 row the lowest cost is 3 which is at the cell (1, 1). Thus,

$$x_{11} = \min\{75, 45\} = 45, \text{ demand for } D_1$$

So, allocate 45 to the cell (1, 1) and delete D_1.

Then we have the following table

	D_3	Supply	Penalty
F_1	4	30	(4)
F_2	4	45	(4)
Demand	75	75	
Penalty	$4 - 4 = (0)$		

Here, the maximum penalty $= 4$ which is corresponding to F_1 and F_2 rows, so we may select any one of these.

Let us suppose F_1 is selected in which 4 is the only cost which is at the cell (1, 3).

Therefore, allocate x_{13} to the cell (1, 3) where

$$x_{13} = \min\{30, 75\} = 30, \text{ supply for } F_1$$

Then delete F_1 row and get the following reduced table

		Supply
F_2	4	45
Demand	45	45

Since, there is only single element, $i.e.$, 4, so allocate 45 to this cell.

Now, all the above allocation can be shown in a single table as given bleow.

Supply

㊺ 3	3	㉚ 4	㉕ 1	100
4	㊿80 2	㊺ 4	2	125
㊽75 1	5	3	2	75

Demand: 120 | 80 | 75 | 25 | 300

We observe that here total no. of allocation is 6 which is equal to $(m + n - 1)$
$\Rightarrow$ solution is non-degenerate.

Finally, the initial basic feasible solution is given by

$$= 1 \times 25 + 1 \times 75 + 2 \times 80 + 3 \times 45 + 4 \times 30 + 4 \times 45$$

$$= ₹\ 695, \text{ which is the required total cost}$$

EXAMPLE 2. *Solve the following transportation problem to find IBFS.*

[MEERUT–2006, 07, 08; KANPUR–2010, 11]

O \ D	1	2	3	a_i
1	2	7	4	5
2	3	3	1	8
3	5	4	7	7
4	1	6	2	14
b_j	7	9	18	34

SOLUTION. We find the penalties as follows:

Penalties

			Penalties
2	7	4	2
3	3	⑧ 1	2
5	4	7	1
1	6	2	1

Penalties: 1 | 1 | 1

Penalties

			Penalties
⑤ 2	7	4	2
5	4	7	1
1	6	2	1

Penalties: 1 | 2 | 2

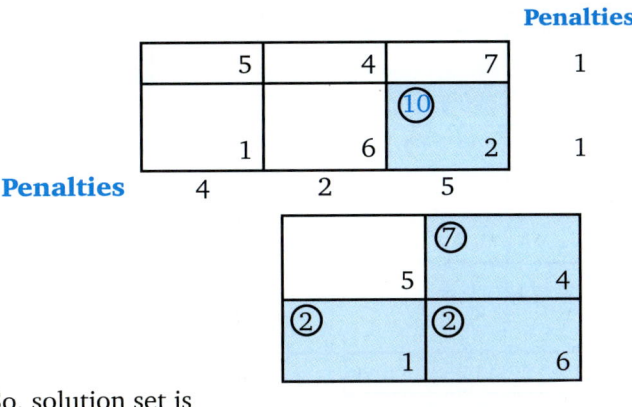

So, solution set is

5	0	0
0	0	8
0	7	0
2	2	10

Using all the above tables, the all allocations can be shown in the single table as given below

⑤ 2	7	4	5
3	3	⑧ 1	8
5	⑦ 4	⑧ 7	7
② 1	② 6	⑩ 2	14
7	9	18	

Min. cost $= 5 \times 2 + 2 \times 1 + 7 \times 4 + 6 \times 2 + 8 \times 1 + 10 \times 2$

$= 10 + 2 + 28 + 12 + 8 + 20$

$= 80$

Example 3. **Solve the following transportation problem.**

	D_1	D_2	D_3	D_4	D_5	D_6	a_i
O_1	1	2	1	4	4	2	30
O_2	3	3	2	1	4	3	50
O_3	4	2	5	9	6	2	75
O_4	3	1	7	3	4	6	20
b_j	20	40	30	10	50	25	

SOLUTION. Apply the usual procedure as above, we can find the penalties as follows:

⑳ 1	2	1	4	4	2	1
3	3	2	1	4	3	1
4	2	5	9	6	2	2
3	1	7	3	4	6	2
2	1	1	2	2	1	

2	1	4	4	2	1
3	2	1	4	3	1
㊵ 2	5	9	6	2	3
1	7	3	4	6	2
1	1	2	2	1	

1	4	4	2	1
2	1	4	3	1
5	9	6	㉕ 2	3
7	3	4	6	1
1	2	2	1	

1	4	4	3
2	1	4	1
5	9	6	1
7	3	4	1
1	2	2	

20 2	1	4	1
5	9	6	1
7	3	4	1
3	2	2	

10 1	4	3
9	6	3
3	4	1
2	2	

So, solution set is
$$
\begin{bmatrix}
20 & 0 & 10 & 0 & 0 & 0 \\
0 & 0 & 20 & 10 & 20 & 0 \\
0 & 40 & 0 & 0 & 10 & 25 \\
0 & 0 & 0 & 0 & 20 & 0
\end{bmatrix}
$$

Therefore, all allocations in a single table is given as follows:

	D_1	D_2	D_3	D_4	D_5	D_6	a_i
O_1	⑳ 1	2	⑩ 1	4	4	2	30
O_2	3	3	⑳ 2	⑩ 1	⑳ 4	3	50
O_3	4	㊴ 2	5	9	⑩ 6	㉕ 2	75
O_4	3	1	7	3	⑳ 4	6	20
b_j	20	40	30	10	50	25	

Min. cost

$= 20 \times 1 + 10 \times 1 + 20 \times 2 + 10 \times 1 + 20 \times 4 + 40 \times 2 + 10 \times 6 + 25 \times 2 + 20 \times 4$

$= 20 + 10 + 40 + 10 + 80 + 80 + 60 + 50 + 80 \quad = \quad 430$

EXAMPLE 4. **Solve the following transportation problem to find IBFS.**

	D_1	D_2	D_3	D_4	a_i
O_1	10	7	3	6	3
O_2	1	6	8	3	5
O_3	7	4	5	3	7
b_j	3	2	6	4	15

SOLUTION. First we find out penalties corresponding to each row and column by applying the usual procedure.

Penalties

					Penalties
	10	7	3	6	3
③ 1		6	8	3	2
	7	4	5	3	1
Penalties	6	2	2	3	

			a_i
7	③ 3	6	3
6	8	3	3
4	5	3	1
2	2	3	

			a_i
6	8	② 3	3
4	5	3	1
2	3	0	

② 4	③ 5	② 3

So, solution set is $\begin{bmatrix} 0 & 0 & 3 & 0 \\ 3 & 0 & 0 & 2 \\ 0 & 2 & 3 & 2 \end{bmatrix}$

Therefore, all allocations in a single table are given as below

	D_1	D_2	D_3	D_4
O_1	10	7	③ 3	6
O_2	③ 1	6	8	② 3
O_3	7	② 4	③ 5	② 3

Least cost $= 3\times1 + 2\times4 + 3\times3 + 3\times5 + 2\times3 + 2\times3 = 47$

EXAMPLE 5. *Solve the following transportation problem.*

	D_1	D_2	D_3	D_4	D_5	D_6	a_i
O_1	9	12	9	6	9	10	5
O_2	7	3	7	7	5	5	6
O_3	6	5	9	11	3	11	2
O_4	6	8	11	2	2	10	9
b_j	4	4	6	2	4	2	

SOLUTION. We use VAM here to solve the given transportation problem :

	D_1	D_2	D_3	D_4	D_5	D_6	**Penalties**
	9	12	9	6	9	10	3
	7	3	7	7	5	② 5	2
	6	5	9	11	3	11	2
	6	8	11	2	2	10	4
Penalties	1	2	2	4	1	5	

Penalties

						Penalties
9	12	9	6	9		3
7	3	7	7	5		2
6	5	9	11	3		2
6	8	11	② 2	2		4
1	2	2	4	1		

				Penalties
9	12	9	9	3
7	3	7	5	2
6	5	9	3	2
6	8	11	④ 2	4
1	2	2	1	

9	12	9	3
7	④ 3	7	4
6	5	9	1
6	8	11	2
1	2	2	

9	9	0
7	7	0
6	9	3
③ 6	11	5
1	2	

9	9	0
7	7	0
① 6	9	3
1	2	

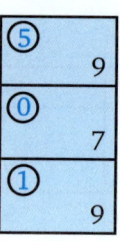

Hence, solution set is given by
$\begin{bmatrix} 0 & 0 & 5 & 0 & 0 & 0 \\ 0 & 4 & 0 & 0 & 0 & 2 \\ 1 & 0 & 1 & 0 & 0 & 0 \\ 3 & 0 & 0 & 2 & 4 & 0 \end{bmatrix}$

and Min. cost $= 1 \times 6 + 3 \times 6 + 4 \times 3 + 5 \times 9 + 1 \times 9 + 2 \times 2 + 4 \times 2 + 2 \times 5 = 112$

EXAMPLE 6. *Solve*

	O_1	O_2	O_3	O_4	Requirement
D_1	19	14	23	11	11
D_2	15	16	12	21	13
D_3	30	25	16	39	19
Available	6	10	12	15	43

SOLUTION. **Calculation of penalty by VAM Method:**

Penalty

			⑪	Penalty
19	14	23	11	3
15	16	12	21	3
30	25	16	39	9

Penalty   4   2   4   10

			④	
15	16	12	21	3
30	25	16	39	9

15   9   4   18

⑥			
15	16	12	3
30	25	16	9

15   9   4

③		
16	12	4
⑦ 25	⑫ 16	9

9   4

So, solution set is $\begin{bmatrix} 0 & 0 & 0 & 11 \\ 6 & 3 & 0 & 4 \\ 0 & 7 & 12 & 0 \end{bmatrix}$

Therefore, we have the following allocation table

	O_1	O_2	O_3	O_4	
D_1	19	14	23	⑪ 11	11
D_2	⑥ 15	③ 16	12	④ 21	13
D_3	30	⑦ 25	⑫ 16	39	19
	6	10	12	15	

Minimum cost $= 11 \times 11 + 6 \times 15 + 3 \times 16 + 4 \times 21 + 7 \times 25 + 12 \times 16 = 710$.

EXAMPLE 7. *Solve the following problem.*

	D_1	D_2	D_3	D_4	a_i
O_1	2	5	4	7	4
O_2	6	1	2	5	6
O_3	4	6	2	4	8
b_j	3	7	6	2	

SOLUTION. **Calculation of penalties:**

				Penalty
2	5	4	7	2
6	⑥ 1	2	5	1
4	6	2	4	2
Penalty 2	4	2	1	

				Penalty
2	5	4	7	2
4	6	2	② 4	2
2	1	2	3	

			Penalty
③ 2	5	4	2
4	6	2	2
2	1	2	

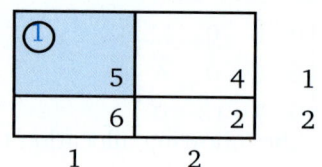

Solution set is $\begin{bmatrix} 3 & 1 & 0 & 0 \\ 0 & 6 & 0 & 0 \\ 0 & 0 & 6 & 2 \end{bmatrix}$

Therefore, we have the following allocation table

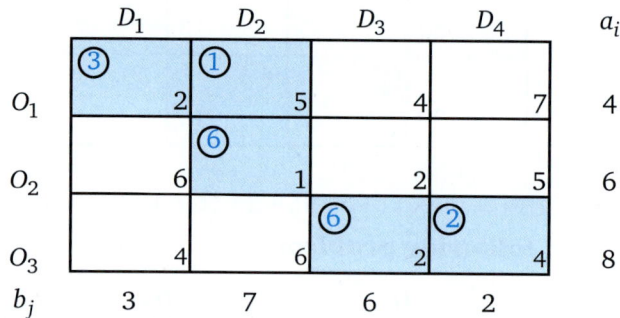

	D_1	D_2	D_3	D_4	a_i
O_1	③ 2	① 5	4	7	4
O_2	6	⑥ 1	2	5	6
O_3	4	6	⑥ 2	② 4	8
b_j	3	7	6	2	

Minimum cost $= 3 \times 2 + 1 \times 5 + 6 \times 1 + 6 \times 2 + 2 \times 4 = 37$.

EXAMPLE 8. *Solve the following transportation problem.*

	D_1	D_2	D_3	D_4	a_i
O_1	6	4	1	5	14
O_2	8	7	2	7	16
O_3	4	3	6	2	5
b_j	6	10	15	4	

SOLUTION.   **Calculation of penalty:**

                                                             Penalty

				Penalty
6	4	1	5	3
8	7	⑮ 2	7	5
4	3	6	2	1
Penalty: 2	1	1	3	

			Penalty
6	4	5	1
8	7	7	1
4	3	④ 2	1
2	1	2	

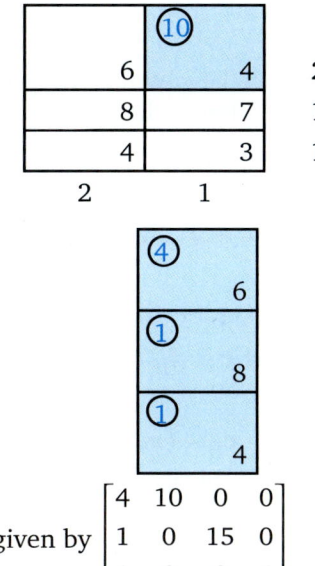

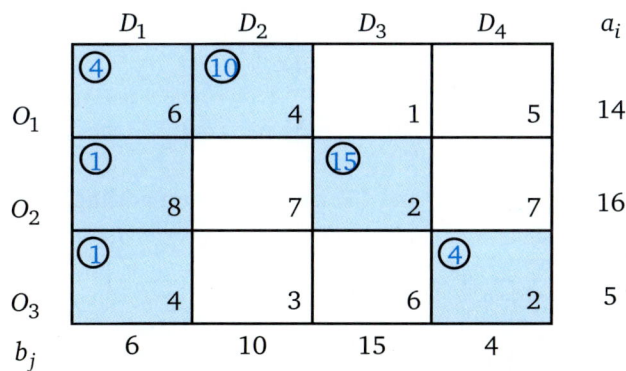

Solution set is given by $\begin{bmatrix} 4 & 10 & 0 & 0 \\ 1 & 0 & 15 & 0 \\ 1 & 0 & 0 & 4 \end{bmatrix}$

Therefore,

	D_1	D_2	D_3	D_4	a_i
O_1	④ 6	⑩ 4	1	5	14
O_2	① 8	7	⑮ 2	7	16
O_3	① 4	3	6	④ 2	5
b_j	6	10	15	4	

Minimum cost $= 4 \times 6 + 10 \times 4 + 1 \times 8 + 15 \times 2 + 1 \times 4 + 4 \times 2 = 114$.

13.5 TEST FOR OPTIMALITY

After obtaining the initial basic feasible solution of the given transportation problem by any method (discussed in previous section) we test the optimality of this solution. This test is applied to any IBFS with $(m + n - 1)$ allocations.

13.5.1 STEPPING STONE MEHTOD

In this method, we find the cell evaluation corresponding to each empty cell. For this, start with empty cell (with no allocation) and allocate 1 unit to this cell and then maintain the row and column sums by doing some necessary adjustments.

Due to such adjustments in the solution, the net change in the total cost of the transportation problem is known as cell evaluation of the cell. Then we have the following observation

(i) If the cell evaluation is positive, then the new solution increases the total transportation cost.

(ii) If the cell evaluation is negative then the new solution reduces to the transportation cost.

Thus, we conclude that if all the cell evaluation are greater than or equal to zero then we can not decrease the total cost more. Hence, the solution under test is optimal.

THEOREM I. *If a basic feasible solution of a transportation problem consist $(m + n - 1)$ independent allocations and if we have a set of arbitrary numbers u_i and v_j, $i = 1, 2, ..., m$; $j = 1, 2, ..., n$ for occupied cell (r, s) such that*

$$c_{rs} = u_r + v_s$$

Then the cell evaluation d_{ij} corresponding to each unoccupied cell (i, j) is given by

$$d_{ij} = c_{ij} - (u_i + v_j)$$

PROOF. In transportation problem of m rows and n columns we have to determine $x_{ij} \geq 0$, which

$$\text{Minimize } Z = \sum_{i=1}^{m} \sum_{j=1}^{n} c_{ij} x_{ij} \qquad \qquad ...(1)$$

subject to the restrictions

$$\sum_{j=1}^{n} x_{ij} = a_i \quad \Rightarrow \quad 0 = a_i - \sum_{j=1}^{n} x_{ij}, \ i = 1, 2, ..., m \qquad ...(2)$$

and

$$\sum_{i=1}^{m} x_{ij} = b_j \quad \Rightarrow \quad 0 = b_j - \sum_{i=1}^{m} x_{ij}, \ j = 1, 2, ..., n \qquad ...(3)$$

Now, multiplying (2) by u_i and (3) by v_j and then adding to (1), we get

$$Z = \sum_{i=1}^{m} \sum_{j=1}^{n} c_{ij} x_{ij} + \sum_{i=1}^{m} u_i \left(a_i - \sum_{j=1}^{n} x_{ij} \right) + \sum_{j=1}^{n} v_j \left(b_j - \sum_{i=1}^{m} x_{ij} \right)$$

$$\Rightarrow \qquad Z = \sum_{i=1}^{m} \sum_{j=1}^{n} [c_{ij} - (u_i + v_j)] x_{ij} + \sum_{i=1}^{m} u_i a_i + \sum_{j=1}^{n} v_j b_j \qquad ...(4)$$

Now, for zero coefficient, we must have

$$c_{rs} = u_r + v_s \qquad \qquad ...(5)$$

for each occupied cell (r, s)

Since, there are $(m + n - 1)$ occupied cells, so there are $(m + n - 1)$ equation of form (5) in $(m + n)$ unknowns u_i and v_j.

If one of these unknowns is assigned a value arbitrarily, then the cost of $(m + n - 1)$ unknowns can be solved algebraically. As any one of u_i or v_j can be assigned so we choose the u_i which has the largest number of allocations in its row and assign it the value of zero. Now, since $c_{ij} = u_i + v_j$, therefore v_j can be obtained for those columns having allocations. Further, suppose that (i, j) is the empty cell. Allocate $+1$ unit to this empty cell, then total number of allocations becomes $(m + n)$ which are independent positions, so a closed loop is formed by joining the empty cell (i, j) to the occupied cells as shown in the following figure.

	D_1	D_2	...	D_j	...	D_s	...	D_n
O_1			...		...		...	
O_2			...		...		...	
⋮			...		...		...	
O_i			...	⊕+1	...	⊕+1	...	
⋮			...		...		...	
O_r			...	⊕+1	...	⊕+1	...	
⋮			...		...		...	
O_m			...		...		...	

The loop in the above table can be shown as follows

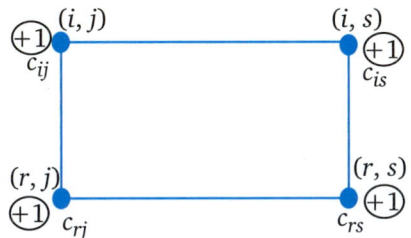

We observe that the cells (i, s), (r, s) and (r, j) are occupied cells. Then we have

$$c_{is} = u_i + v_s; \quad c_{rs} = u_r + v_s; \quad c_{rj} = u_r + v_j$$

The above loop can be formed as follows:

"Allocate $+1$ unit to the empty cell (i, j) then to maintain row and column sum unchanged we allocate -1 unit to the occupied cells (i, s) and (r, j) and $+1$ unit to the occupied cells (r, s).

Now, $\quad d_{ij}$ = cost difference between the new solution and the original solution

$$= c_{ij} - c_{is} + c_{rs} - c_{rj}$$
$$= c_{ij} - (u_i + v_s) + (u_r + v_s) - (u_r + v_j)$$
$$= c_{ij} - (u_i + v_j)$$

Here, d_{ij} gives the cell evaluation connecting empty cell (i, j) to the occupied cells by a square or rectangular shaped loop.

Similarly, we may generalise above process for an arbitrary shaped loop connecting empty cell (i, j) to the occupied cells.

Hence, the cell evaluation d_{ij} for each empty cell (i, j) is given by

$$d_{ij} = c_{ij} - (u_i + v_j)$$

13.5.2 MODIFIED DISTRIBUTION METHOD (MODI METHOD)

After determining the initial basic feasible solution of a transportation problem to check the optimality, we use the following steps:

STEP 1. Find a set of $(m + n)$ numbers u_i for $i = 1, 2, ..., m$ and v_j for $j = 1, 2, ..., n$ such that $c_{rs} = u_r + v_s$ for each occupied cell (r, s).

STEP 2. Calculate cell evaluation d_{ij} by using

$$d_{ij} = c_{ij} - (u_i + v_j)$$

for each empty cell (i, j) and enter at the upper right corner of that cell.

STEP 3. If

    (i) $d_{ij} > 0$, then the solution under test is optimal and unique.

    (ii) $d_{ij} \geq 0$, then the solution under test is optimal but not unique (an alternate optimal solution exists)

    (iii) $d_{ij} < 0$, then solution is not optimal and further improvement is needed.

STEP 4. In case at least one $d_{ij} < 0$, then select the empty cell for which the cell evaluation d_{ij} is most negative.

Allocate an unknown quantity $+\theta$ to the empty cell and construct a closed loop connecting this cell to the other occupied cells. In order to maintain rows and column sums assign $-\theta$ and $+\theta$ altenatively to the occupied cells which are used in forming the loop (at the corner point only).

STEP 5. Assign the largest possible value to θ in such a way that the value of at least one occupied cell turns to zero and in the other occupied cells, the allocations will remain non-negative. The occupied cell whose allocation becomes zero will leave the basis.

STEP 6. From step 5, we therefore get a new BFS. Return to step 3 with new BFS, repeat the process till an optimal basic feasible solution is obtained.

Here following points should be remembered:

---

(1) Every loop has an even number of cells and at least 4.

(2) Closed loops may or may not be square in shape

(3) The allocations are said to be independent position if it is not possible to increase or decrease any independent individual allocation without changing the positions of these allocations or violating the RIM conditions, a closed loop can not be formed through these allocations.

(4) Each row and column in the transportation table should have only one plus and minus sign. All cells that have a + or − sign, except the starting unoccupied cells, must be occupied cells.

 Solved Examples

EXAMPLE 1. *In the following transportation problem, find the basic feasible solution by North-West corner rule, Matrix-Minima Method and by Vogel's Method and test the optimality for each solution giving the cost of transportation for the solution. Cost of transportation in per 1000 ton in units of ₹ 1000.*

Steel mills

		S_1	S_2	S_3	S_4	Availability of coal (in 1000 tons)
	I	14	56	48	27	13
Mines	**II**	82	35	21	81	19
	III	99	31	71	63	16
Requirement in 1000 tons		7	14	21	6	48

SOLUTION.   (i) **North-West Corner Rule**

Steel mills

		S_1	S_2	S_3	S_4	Availability of coal (in 1000 tons)
Mines	I	⑦ 14	⑥ 56	48	27	13 (– 7 = 6)
	II	82	⑧ 35	⑪ 21	81	19 (– 8 = 11)
	III	99	31	⑩ 71	⑥ 63	16 (– 10 = 6)
Requirement in 1000 tons		7	14 – 6 = 8	21 – 11 = 10	6 – 6 = 0	48

Test for optimality. Find auxiliary number u_i and v_j such that $c_{ij} = u_i + v_j$ for each occupied cell

Cell (1, 1) $c_{11} = u_1 + v_1 = 14$ Let us choose $u_1 = 0$, $v_1 = 14$

Cell (1, 2) $u_1 + v_2 = 56$ $v_2 = 56$

Cell (2, 2) $u_2 + v_2 = 35$ $u_2 = -21$

Cell (2, 3) $u_2 + v_3 = 21$ $v_3 = 42$

Cell (3, 3) $u_3 + v_3 = 71$ $u_3 = 29$

Cell (3, 4) $u_3 + v_4 = 63$ $v_4 = 34$

Calculate Δ_{ij} for every non-occupied cell $\Delta_{ij} = c_{ij} - (u_i + v_j)$

Cell (1, 3) $\Delta_{13} = 48 - (u_1 + v_3) = 48 - (0 + 42) = 6$

Cell (1, 4) $\Delta_{14} = c_{14} - (u_1 + v_4) = 27 - (0 + 34) = -7$

Cell (2, 1) $\Delta_{21} = c_{21} - (u_2 + v_1) = 82 - (-21 + 14) = 89$

Cell (2, 4) $\Delta_{24} = 81 - (u_2 + v_4) = 81 - (-21 + 34) = 68$

Cell (3, 1) $\Delta_{31} = 99 - (29 + 14) = 99 - 43 = 56$

Cell (3, 2) $\Delta_{32} = 31 - (29 + 56) = -54$

Since all Δ_{ij} are not positive so the current solution is not optimal.

The cost of assignment

$= 14 \times 7 + 56 \times 6 + 35 \times 8 + 21 \times 11 + 71 \times 10 + 63 \times 6$

$= 98 + 336 + 280 + 231 + 710 + 378 = 2033$ (IBFS)

(ii) **Matrix-Minima Method**

Steel mills

		S_1	S_2	S_3	S_4	Availability
Mines	I	⑦ 14	56	48	⑥ 27	13
	II	82	35	⑲ 21	81	19
	III	99	⑭ 31	② 71	63	16
Demand		7	14	21 – 19 = 2	6 – 6 = 0	48

The initial solution is degenerate. Since only five cells are occupied which is less than $(3+4-1)=6$. So to make the number of occupied cells as 6, assign a quantity $\varepsilon \to 0$ to the remaining cell of minimum cost so that loop is not formed. Assign a quantity ε in cell (1, 3) to make the number of occupied cell 6.

Test for optimality. Find auxiliary number u_i and v_j for occupied cells such that $c_{ij} = u_i + v_j$

Cell (1, 1)	$u_1 + v_1 = 14$	Let us choose $v_1 = 0$, $u_1 = 14$
Cell (1, 3)	$u_1 + v_3 = 48$	$v_3 = 34$
Cell (1, 4)	$u_1 + v_4 = 27$	$v_4 = 13$
Cell (2, 3)	$u_2 + v_3 = 21$	$u_2 = -13$
Cell (3, 2)	$u_3 + v_2 = 31$	$v_2 = -6$
Cell (3, 3)	$u_3 + v_3 = 71$	$u_3 = 37$

Now, for each non-occupied cell calculate $\Delta_{ij} = c_{ij} - (u_i + v_j)$

Cell (1, 2) $\Delta_{12} = 56 - (14 - 6) = 48$

Cell (2, 1) $\Delta_{21} = 82 - (-13 + 0) = 95$

Cell (2, 2) $\Delta_{22} = 35 - (u_2 + v_2) = 35 - (-13 - 6) = 35 + 19 = 54$

Cell (2, 4) $\Delta_{24} = 81 - (-13 + 13) = 81 - 0 = 81$

Cell (3, 1) $\Delta_{31} = 99 - (37 + 14) = 99 - 51 = 48$

Cell (3, 4) $\Delta_{34} = 63 - (37 + 13) = 63 - 50 = 13$

Since all $\Delta_{ij} \geq 0$, therefore current solution is optimal.

Minimum cost of assignment per 1000 tons (in units of ₹ 1000)

$$= 14 \times 7 + 27 \times 6 + 21 \times 19 + 31 \times 14 + 71 \times 2$$
$$= 98 + 162 + 399 + 434 + 142 = 1235$$

(iii) Vogel's method

Steel mills

Mines		S_1	S_2	S_3	S_4	**Availability**
	I	⑦ 14	56	48	⑥ 27	$13 - 7 = 6$
	II	82	35	⑲ 21	81	19
	III	99	⑭ 31	② 71	63	16
Demand		7	14	21	6	**48**
Difference		68	4	27	36	

In this case, test for optimality is same as in matrix-minima method. So current solution is optimal.

Minimum cost of assignment (in units of ₹ 1000)

$$= 14 \times 7 + 27 \times 6 + 21 \times 19 + 31 \times 14 + 71 \times 2 = 1235$$

EXAMPLE 2. *Find the optimal basic feasible solution to the following transportation problem*

			Available
50	30	220	1
90	45	170	3
250	200	50	4
Requirement 4	2	2	8 ← **Total**

SOLUTION. Applying lowest cost entry method, we can obtain the initial basic feasible solution as follows

	1 (30)		1
2 (90)	1 (45)		3
2 (250)		2 (50)	4
4	2	2	

Here, the cost is given by

Cost $= 1 \times 3 + 2 \times 90 + 1 \times 45 + 2 \times 250 = 855$

To check the optimality of this solution, we proceed as example (1) and obtain the following table

			u_i
	(30)		−15
(90)	(45)		0
(250)		(50)	160
v_j 90	45	−110	

50	•	(220)
•	•	(170)
•	(200)	•

(Matrix for set of u_i and v_j)   (Matrix c_{ij} for empty cells)

			u_i
75	•	−125	−15
•	•	−110	0
•	205	•	160
v_j 90	45	−110	

−25	•	345
•	•	280
•	−5	•

(Matrix $(u_i + v_j)$ for empty cells)   (Matrix $c_{ij} - (u_i + u_j)$ for empty cells)

Here, all cell evaluation $\Delta_{ij} = c_{ij} - (u_i + v_j)$ are non-negative. Therefore, solution is not optimal.

Here, the most negative cell evaluation is $\Delta_{11} = -25$. Therefore, assume θ allocation in this cell and proceed as follows

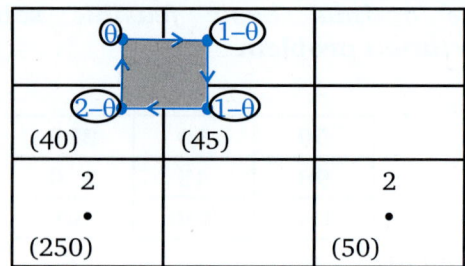

Here $\min\{1-\theta, 2-\theta\} = 0$

$\Rightarrow \quad \theta = 1$.

Therefore, the improved solution given by

1 (50)			1
1 (90)	2 (45)		3
2 (250)		2 (50)	4
4	2	2	

Now, we check the optimality of this improved solution as follows:

			u_i
(50)			50
(90)	(45)		90
(250)		(50)	250
0	-45	-200	

v_j

(Matrix for set of u_i and v_j)

•	(30)	(220)
•	•	(170)
•	(200)	•

(Matrix c_{ij} for empty cells)

			u_i
•	5	-150	50
•	•	-110	90
•	205	•	250
0	-45	-200	

v_j

(Matrix for set of u_i and v_j)

•	25	370
•	•	280
•	-5	•

(Matrix $c_{ij} - (u_i + u_j)$ for empty cells)

Here, the cell evaluation $\Delta_{32} = -5$, therefore, this solution is not optimal.

Again, to find the optimality, proceeding same as above we get the following table.

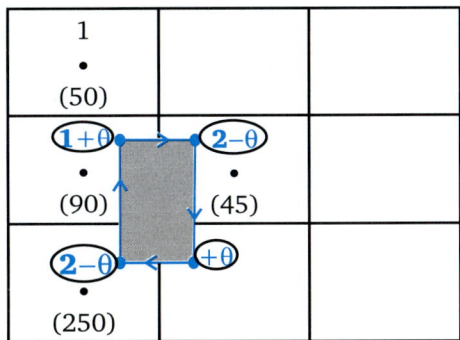

Here, $\theta = 2$. Therefore, the next improved solution is given by

1 (50)			1
3 (90)	0 (45)		3
	2 (200)	2 (50)	4
4	2	2	

By proceeding as above, we can easily see that all cell evaluation are non-negative. Hence, the solution is optimal and is given by

$$z = 1 \times 50 + 3 \times 90 + 0 \times 45 + 2 \times 200 + 2 \times 50 = 820$$

EXAMPLE 3. *The cost requirement table for the transportation problem is given as below:*

	ω_1	ω_2	ω_3	ω_4	ω_5	Available
F_1	4	3	1	2	6	40
F_2	5	2	3	4	5	30
F_3	3	5	6	3	2	20
F_4	2	4	4	5	3	10
Required	30	30	15	20	5	**100**

SOLUTION. By 'North-West Corner rule', the non-degenerate initial solution is given by following table

	ω_1	ω_2	ω_3	ω_4	ω_5	Available
F_1	㉚ 4	10 3				40
F_2		⑳ 2	⑩ 3			30
F_3			⑤ 6	⑮ 3		20
F_4				⑤ 5	⑤ 3	10
Required	30	30	15	20	5	

Now, test this solution for optimality, we get the following table:

					u_i
4	3				0
	2	3			-1
		6	3		2
			5	3	4

v_j 4   3   4   1   -1

(Matrix for set of u_i and u_j)

•	•	(1)	(2)	(6)
(5)	•	•	(4)	(5)
(3)	(5)	•	•	
(2)	(4)	(4)	•	•

(Matrix $[c_{ij}]$ for empty cells)

•	•	-3	1	7
2	•	•	4	7
-3	0	•	•	1
-6	-3	-4	•	•
Matrix $[c_{ij} - (u_i + u_j)]$ for empty cells				

Since the largest negative cell evaluation is $\Delta_{41} = -6$, allocate as much as possible to this cell (4, 1). This necessitates shifting of 5 units to this cell (4, 1) as directed by the closed loop in next table.

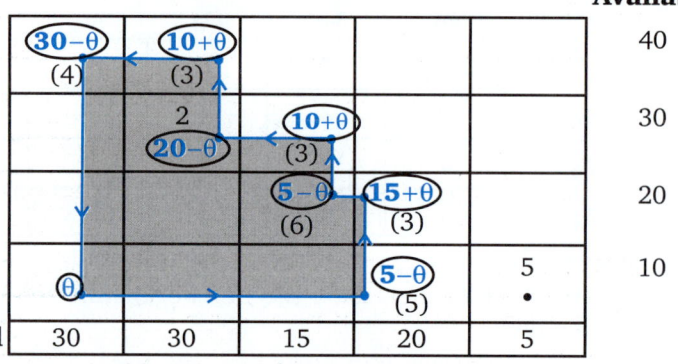

       Available

(30−θ) (4)	(10+θ) (3)				40
	2 (20−θ)	(10+θ) (3)			30
		(5−θ) (6)	(15+θ) (3)		20
(θ)			(5−θ) (5)	5 •	10

Required   30   30   15   20   5

Here maximum possible value of θ is obtained by

$$\min\{30-\theta, 20-\theta, 5-\theta, 5-\theta\} = 0,$$

i.e., $\theta = 5$ units.

Now, the revised solution is

       Availability

(25) 4	(15) 3				40
	(15) 2	(15) 3			30
		(0) 6	(20) 3		20
			(0) 5	(5) 3	10

Requirement   30   30   15   20   5

In this solution, the number of allocations becomes less than $m + n - 1$ on account of simultaneous allocation of two cells $[(3, 3), (4, 4)]$. Hence, this is a degenerate solution. Now, this degeneracy may resolve by adding Δ to one of the recently vacated cells $[(3, 3)$ or $(4, 4)]$. But in minimization problem, add Δ to recently vacated $(4, 4)$ only, because it has the lowest shipping cost of ₹ 5 per unit.

The rest of procedure will be exactly the same as explained earlier. This way, the optimum solution can be obtained.

					Available
⑤ 4		⑮ 1	⑳ 2		40
	㉚ 2				30
⑮ 3				⑤ 2	20
	⑩ 2				10
Required 30	30	15	20	5	

EXAMPLE 4. *Solve the following transportation problem*

			Supply
2	7	4	5
3	3	1	8
5	4	7	7
1	6	2	14
Requirement 7	9	18	**34**

[MEERUT–2006, 07, 08; KANPUR–2010, 11]

SOLUTION. By using Vogel's approximation method, we obtained the following initial basic feasible solution

⑤ (2)		
		⑧ (1)
	⑦ (4)	
② (1)	② (6)	⑩ (2)

Now, we check the optimality of this solution in the following manner

			u_i
(2)			1
		(1)	−1
	(4)		−2
(1)	(6)	(2)	0
v_j 1	6	2	

(Matrix for set of u_i and v_j)

•	(7)	(4)
(3)	(3)	•
(5)	•	(7)
•	•	•

(Matrix u_j for empty cells)

•	7	2
0	5	•
−1	•	0
•	•	•

•	0	1
3	−2	•
6	•	7
•	•	•

(Matrix $(u_i + v_j)$ for empty cells)   (Matrix $c_{ij} - (u_i + u_j)$ for empty cells)

Here $\Delta_{22} = -2$ is most negative value.

Hence, we can improve the solution as follows

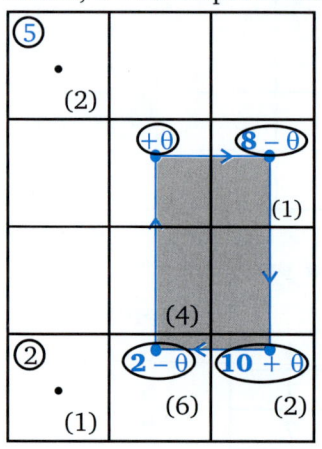 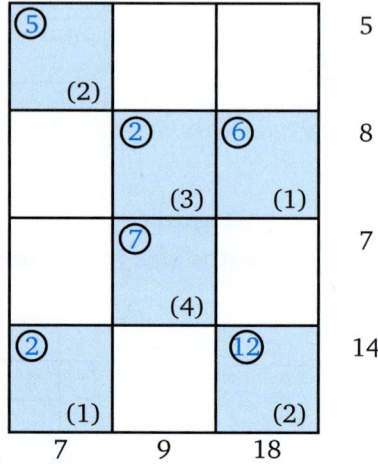

Here $\min\{8 - \theta, 2 - \theta\} = 0 \quad \Rightarrow \quad \theta = 2$

Proceeding as above, we get the following matrices :

			u_i
(2)			1
	(3)	(1)	−1
	(4)		0
(1)		(2)	0
v_j 1	4	2	

•	(7)	(4)
(3)	•	•
(5)	•	(7)
•	(6)	•

(Matrix for set of u_i and v_j)   (Matrix c_{ij} for empty cells)

•	5	3
0	•	•
1	•	2
•	4	•

•	2	1
3	•	•
4	•	5
•	2	•

(Matrix $(u_i + v_j)$ for empty cells)   (Matrix $c_{ij} - (u_i + u_j)$ for empty cells)

Since in the final table, all Δ_{ij}'s are positive, therefore, solution is optimal.

The optimal solution (minimum cost) is given by

Minimum cost = $5(2) + 2(3) + 6(1) + 7(4) + 2(1) + 12(2) = 76$.

EXAMPLE 5. *Solve the following transportation problem*

Villages

	T_1	T_2	T_3	Capacity
B_1	8	6	5	150
B_2	6	6	6	150
B_3	10	8	4	150
B_4	8	6	4	150
Demand	200	200	200	600

(Air bases label at left of B_3)

SOLUTION. **Vogel's method:**

	T_1	T_2	T_3	Capacity	Diff.
B_1	(50) -8	(50) -6	(50) -5	150	1
B_2	-6	-6	(150) -6	150	0
B_3	(150) -10	-8	-4	150	2
B_4	-8	(150) -6	-4	150	2
Demand	200	200	200	600	
Diff.	2	2	1		

Test for optimality. There are $6 = (4+3-1)$ Calculate u_i and v_j for each occupied cell such that $c_{ij} = u_i + v_j$

Cell (1, 1)	$u_1 + v_1 = -8$	Let us choose	$u_1 = 0,\ v_1 = -8$
Cell (1, 2)	$u_1 + v_2 = -6$		$v_2 = -6$
Cell (2, 3)	$u_2 + v_3 = -6$		$u_2 = -1$
Cell (3, 1)	$u_3 + v_1 = -10$		$u_3 = -2$
Cell (4, 2)	$u_4 + v_2 = -6$		$u_4 = 0$

Now, compute $\Delta_{ij} = c_{ij} - (u_i + v_j)$ for each vacant cell

Cell (2, 1) $\quad \Delta_{21} = -6 - (-1-8) = 3$

Cell (2, 2) $\quad \Delta_{22} = -6 - (-1-6) = 1$

Cell (3, 2) $\quad \Delta_{32} = -8 - (-2-6) = 0$

Cell (3, 3) $\quad \Delta_{33} = -4 - (-2-5) = 3$

Cell (4, 1) $\quad \Delta_{41} = -8 - (0-8) = 0$

Cell (4, 3) $\quad \Delta_{43} = -4 - (0-5) = 1$

Since all $\Delta_{ij} \geq 0$, therefore current assignment is optimal. The assignments are

Cell (1, 1) = 50 $\qquad$ Cell(2, 2) = 50 $\qquad$ Cell (1, 3) = 50

Cell (2, 3) = 150   Cell(3, 1) = 150   Cell (4, 2) = 150
∴ Maximum cost of assignment
$$= 8 \times 50 + 6 \times 50 + 5 \times 150 + 10 \times 150 + 6 \times 150 = 4250$$

13.6 DEGENERACY IN TRANSPORTATION PROBLEMS

[MEERUT–1994, 2001, 06]

The solution procedure for non-degenerate basic feasible solution with exactly $m+n-1$ strictly positive allocations in independent positions has been discussed so far. However, sometimes it is not possible to get such initial feasible solution to start with. Thus degeneracy occurs in the transportation problem whenever a number of occupied cells is less than $m+n-1$.

Basic feasible solution to an m-origin and n-destination transportation problem can have at most $m+n-1$ number of positive (non-zero) basic variables. If this number is exactly $m+n-1$, the BFS is said to be non-degenerate; and if less than $m+n-1$ the basic solution degenerates. It follows that whenever the number of basic cells is less than $m+n-1$, the transportation problem is a degenerate one.

Degeneracy in transportation problems can occur in two ways:
1. Basic feasible solutions may be degenerate from the initial stage onward.
2. They may become degenerate at any intermediate stage

13.6.1 RESOLUTION OF DEGENERACY

(1) Among the unoccupied cells, select one occupied cell at independent position having the least cost. If such cells are more than one, then select anyone arbitrarily.

(2) The transportation problem may also become degenerate during the solution stages. This happens when most favourable quantity is allocated to the empty cell having the largest negative cell-evaluation resulting in simultaneous vacation of two or more of currently occupied cells. To resolve degeneracy, allocate Δ to one or more of recently vacated cells so that the number of occupied cells is $m+n-1$ in the new solution.

 Solved Examples

EXAMPLE 1. *Find the optimal solution of the following transportation problem.*

Available

			Available
50	30	220	1
90	45	170	3
250	200	50	4
Required 4	2	2	**8**

SOLUTION.   Applying the Vogel's approximation method in a usual manner, the IBFS is given by

①		
50	30	220
③		
90	45	170
	②	②
250	200	50

We observe that, the total no. of allocation = 4 which is less than $m + n - 1 = 5$.

$\Rightarrow$ solution is degenerate.

To resolve this degeneracy, we allocate a very small amount ε to some suitable cell. Here, we allocate ε to the cell $(1, 2)$ getting 5 allocations at independent positions.

① 50	ⓔ 30	220
③ 90	45	170
250	② 200	② 50

Now, we check the optimality as given below.

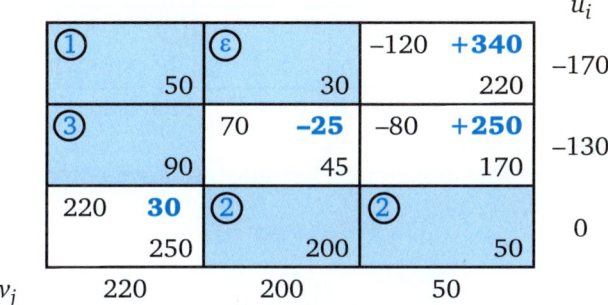

Clearly, $d_{22} = -25 < 0 \Rightarrow$ solution is not optimal.

Therefore, we take ε from cell $(1, 2)$ to the cell $(2, 2)$ and form the new table as given below

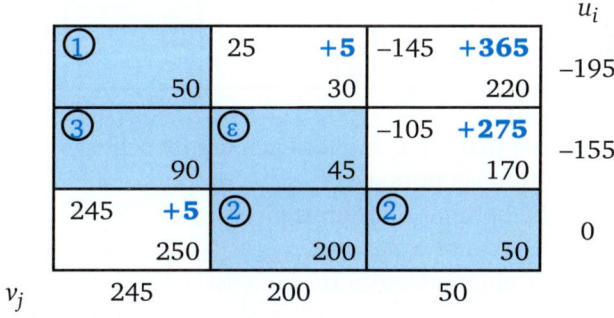

We observe that for all empty cells, all $d_{ij} > 0$

$\Rightarrow$ solution is optimal.

and is given by

$\qquad x_{11} = 1, x_{21} = 3, x_{32} = 2, x_{33} = 2$

and minimum cost $= 50 \times 1 + 90 \times 3 + 200 \times 2 + 50 \times 2 = 820$

EXAMPLE 2.   *Find the optimal basic feasible solution to the following transportation problem.*

Capacity

1	2	3	4	6
4	3	2	0	8
0	2	2	1	10

Demand 4 6 8 6 | 24 |

SOLUTION. Apply Vogel's approximation method, an IBFS is given in the following table

⑥ 1	2	3	4
4	3	② 2	⑥ 0
④ 0	2	⑥ 2	1

Clearly, total no. of allocations = 5, which is less than $m + n - 1 = 6$.

⇒ solution is degenerate

To resolve this degeneracy, we allocate a very small quantity ε to the cell (1, 1) as follows

u_i

ⓔ 1	⑥ 2	3 **0** 3	1 **+3** 4	0
0 **+4** 4	1 **+2** 3	② 2	⑥ 0	−1
④ 0	1 **+1** 2	⑥ 2	0 **+1** 1	−1

v_j 1 2 3 1

We observe that all $d_{ij} \geq 0$ for each empty cell.

⇒ solution is optimal.

Hence, the optimal solution is given by

$$x_{12} = 6, x_{23} = 2, x_{24} = 6, x_{31} = 4, x_{33} = 6$$

and the minimum transportation cost is given by

$$= 6 \times 2 + 2 \times 2 + 6 \times 0 + 4 \times 0 + 6 \times 2 = ₹ \, 28$$

EXAMPLE 3. **Solve the following transportation problem**

Available

7	4	0	5
6	8	0	15
3	9	0	9

Requirement 15 6 8 | 29 |

[MEERUT–2006]

SOLUTION. Applying north-west corner rule, the IBFS is given by

⑤		
7	4	0
⑩	⑤	
6	8	0
	①	⑧
3	9	0

Now, to check the optimality, we proceed as follows.

				u_i
⑤	9	**–5** 0	**0**	0
7	4	0		
⑩	⑤	–1	**+1**	–1
6	8	0		
7	**– 4** ①	⑧		0
3	9	0		
v_j 7	9	0		

$\because$ all d_{ij} not ≥ 0 $\Rightarrow$ solution is not optimal.

Now, the largest negative cell evaluation is $d_{12} = -5$

$\therefore$

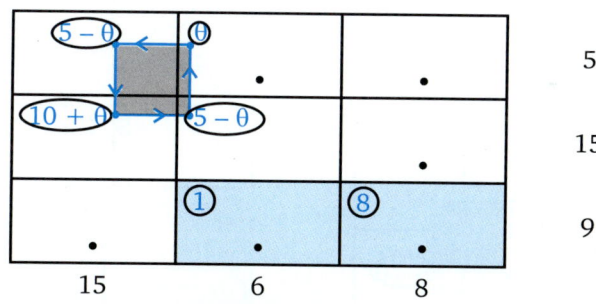

5		
15		
9		

Now, min$[5 - \theta, 5 - \theta] = 0$ $\Rightarrow$ $\theta = 5$

	⑤	
7	4	0
⑮		
6	8	0
	①	⑧
3	9	0

Here, two cell vacate, therefore, no. of allocations becomes less than $m + n - 1$ (= 5)

$\Rightarrow$ There is a degenerate solution.

Introduce a very small quantity say ε to the cell (3, 1), although least cost independent cell is (2, 3).

u_i

−2	**+9** ⑤		−5	**+5**	−5
7		4		0	
⑮	12	**−4**	3	**−3**	3
6		8		0	
⑧ε	①		⑧		0
3		9		0	

v_j 3   9   0

Clearly, all cell evaluation are not $\geq 0 \Rightarrow$ solution is not optimal

The largest negative cell $= d_{22} = -4$

Allocate as much as possible to the cell (2, 2).

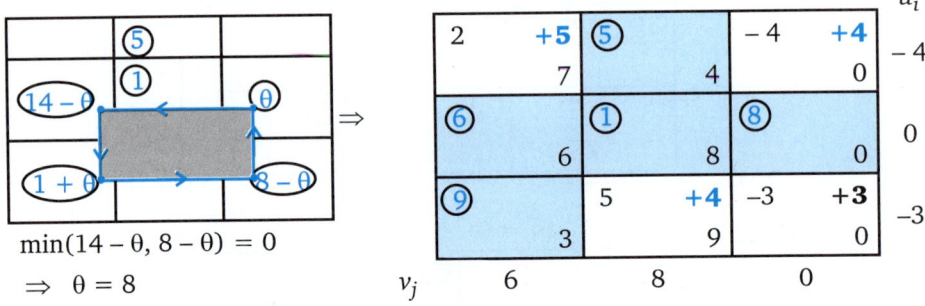

u_i

2	**+5** ⑤		−1	**+1**	−4
7		4		0	
⑭	①		3	**−3**	0
6		8		0	
①	5		**+4** ⑧		−3
3		9		0	

$\min(15 - \theta, 1 - \theta) = 0$

$\Rightarrow \theta = 1$

v_j 6   8   3

Clearly all d_{ij} are not ≥ 0.

The largest negative cell evaluation, $d_{23} = -3$.

Allocate as much as possible to the cell (2, 3)

u_i

2	**+5** ⑤		−4	**+4**	−4
7		4		0	
⑥	①		⑧		0
6		8		0	
⑨	5		**+4**	−3 **+3**	−3
3		9		0	

$\min(14 - \theta, 8 - \theta) = 0$

$\Rightarrow \theta = 8$

v_j 6   8   0

We observe that all the cell evaluation for empty cells are positive.

$\Rightarrow$ solution is optimal

and is given by

$$x_{12} = 5, x_{21} = 6, x_{22} = 1, x_{23} = 8, x_{31} = 9$$

and minimum transportation cost

$$= 4 \times 5 + 6 \times 6 + 1 \times 8 + 8 \times 0 + 3 \times 9$$

$$= ₹ \ 91$$

EXAMPLE 4. *Find the optimal solution of the following problem.*

Requirement

				Requirement
8	10	7	6	50
12	9	4	7	40
9	11	10	8	30
Availability 25	32	40	23	120

[MEERUT–2002, 03, 05]

SOLUTION. Using Vogel's approximation method in a usual manner, we have the following IBFS

(25) 8	(2) 10	7	(23) 6
12	9	(40) 4	7
9	(30) 11	10	8

Now, we have to check the optimality of this solution.

Here, we observe that, there are total 5 allocations which is less than $m + n - 1 = 6$

$\Rightarrow$ solution is degenerate

To resolve this degeneracy, allocate a small quantity ε to the cell $(2, 4)$.

				$u_i \downarrow$
(25) 8	(2) 10	3 +4 7	(23) 6	0
9 +3 12	11 −2 9	(40) 4	(ε) 7	1
9 0 9	(30) 11	4 +6 10	7 +1 8	1
$v_j \rightarrow$ 8	10	3	6	

Clearly, $d_{22} = -2 < 0$

$\Rightarrow$ solution is not optimal

Therefore, taking ε from cell $(2, 4)$ to $(2, 2)$, we form the new table as given below.

				$u_i \downarrow$
(25) 8	(2) 10	5 +2 7	(23) 6	0
7 +5 12	(ε) 9	(40) 4	5 +2 7	−1
9 0 9	(30) 11	6 +4 10	7 +1 8	1
$v_j \rightarrow$ 8	10	5	6	

Here, all $d_{ij} \geq 0 \Rightarrow$ solution is optimal and is given by
$$x_{11} = 25, x_{12} = 2, x_{14} = 23, x_{23} = 40, x_{32} = 30$$
and transportation cost
$$= 25 \times 8 + 2 \times 10 + 23 \times 6 + 40 \times 4 + 30 \times 11$$
$$= ₹\ 848$$

EXAMPLE 5. *Solve the following transportation problem*

				Available
5	3	6	5	15
10	7	12	4	11
7	5	8	4	13
Demand 8	12	13	6	39

SOLUTION. Apply Vogel's approximation method, the IBFS is given by

⑧ 5	⑦ 3	6	5
10	⑤ 7	12	⑥ 4
7	5	⑬ 8	4

Clearly, there are 5 no. of allocations which is less than $m + n - 1 = 6$
Introduce a small allocation ε as follows

$u_i \downarrow$

				$u_i \downarrow$
⑧ 5	⑦ 3	4 **+2** 6	0 **+5** 5	0
9 **+1** 10	⑤ 7	8 **+4** 12	⑥ 4	4
9 **−2** 7	7 **−2** 5	⑬ 8	⑤ε 4	4

$v_j \rightarrow$ 5   3   4   0

Clearly, all d_{ij} not $\geq 0 \Rightarrow$ solution is a degenerate solution.
Here, $d_{ij} = -2 < 0$ in both the cells, so we take ε in one of these two cells say in (3, 1) and again test the optimality we have the following table.

				$u_i \downarrow$
⑧ 5	⑦ 3	6 **0** 6	0 **+5** 5	0
9 **+1** 10	⑤ 7	10 **+2** 12	⑥ 4	4
⑤ε 7	5 **0** 5	⑬ 8	2 **+2** 4	2

$v_j \rightarrow$ 5   3   6   0

In the above table, we observe that all $d_{ij} \geq 0 \Rightarrow$ solution is optimal and is given by $x_{11} = 8, x_{12} = 7, x_{22} = 5, x_{24} = 6, x_{33} = 13$
and the minimum transportation cost
$$= 8 \times 5 + 7 \times 3 + 5 \times 7 + 6 \times 4 + 13 \times 8 = ₹\ 224$$

13.7 UNBALANCED TRANSPORTATION PROBLEM

[MEERUT–2008; GORAKHPUR–2007, 08, 11]

A transportation problem is said to be unbalanced if
$$\sum_{i=1}^{m} a_i \neq \sum_{j=1}^{n} b_j$$
An unbalanced problem having the following two forms.

(i) When $\sum\limits_{i=1}^{m} a_i < \sum\limits_{j=1}^{n} b_j$

In such type of problem, we introduce a dummy source (row) in the cost matrix with zero cost and the excess demand over the supply is entered for this dummy row.

(ii) When $\sum\limits_{i=1}^{m} a_i > \sum\limits_{j=1}^{n} b_j$

In such type of problem, we introduce a dummy demand (column) in the cost matrix with zero cost and the excess supply over the demand is entered for this dummy column.

 Solved Examples

EXAMPLE 1. *Solve the following transportation problem:*

				Supply
6	1	9	3	70
11	5	2	8	55
10	12	4	7	70
Demand 85	35	50	45	

[GORAKHPUR–2007; MEERUT–2011]

SOLUTION. Total supply $= 70 + 55 + 70 = 195$
Total demand $= 85 + 35 + 50 + 45 = 215$
$\Rightarrow \quad \Sigma a_i < \Sigma b_j$
To convert this unbalanced problem into balanced, introduce a dummy row with zero cost and the excess demand 20 (215 – 195) is entered for this dummy row.
Then we have

6	1	9	3	70
11	5	2	8	55
10	12	4	7	70
0	0	0	0	20
85	35	50	45	215

Now, apply, Vogel's approximation method, the IBFS is given in the following table

⑥⑤ 6	⑤ 1	9	3
11	㉚ 5	㉕ 2	8
10	12	㉕ 4	㊺ 7
⑳ 0	0	0	0

The IBFS is $= 65 \times 6 + 5 \times 1 + 30 \times 5 + 25 \times 2 + 25 \times 4 + 45 \times 7 + 20 \times 0 = 1010$

Now, we have to check the optimality of this solution

Clearly, the occupied cell = (1, 1), (1, 2), (2, 2), (2, 3), (3, 3), (3, 4) and (4, 1)

Taking $u_1 = 0$ we get

$$c_{11} = u_1 + v_1 \Rightarrow 6 = 0 + v_1 \Rightarrow v_1 = 6$$

Similarly, $v_2 = 1, u_2 = 4, v_3 = -2, u_3 = 6, v_4 = 1, u_4 = -6$

Further, we have to find d_{ij} for each unoccupied cell (i, j) by using $d_{ij} = c_{ij} - (u_i + v_j)$. Here, the unoccupied cells are (1, 3), (1, 4), (2, 1), (2, 4), (3, 1), (3, 2), (4, 2), (4, 3) and (4, 4). Then

$$d_{13} = c_{13} - (u_1 + v_3) = 9 - (0 - 2) = 9 + 2 = 11$$
$$d_{14} = c_{14} - (u_1 + v_4) = 3 - (0 + 1) = 3 - 1 = 2$$

Similarly, $d_{21} = 1, d_{24} = 3, d_{31} = -2, d_{32} = 5, d_{42} = 5, d_{43} = 8, d_{44} = 5$

Clearly, all d_{ij} are not $\geq 0 \Rightarrow$ solution is not optimal.

Since, $d_{31} = -2$, then allocate an unknown positive quantity θ to the cell (3, 1) and construct a closed loop connecting the cell (3, 1) and occupied cells (1, 1), (1, 2), (2, 2), (2, 3) and (3, 3). So to maintain row and column sum, we assign $-\theta$ to (1, 1) $+ \theta$ to (1, 2), $-\theta$ to (2, 2) $+\theta$ to (2, 3) and $-\theta$ to (3, 3). This loop is shown as below.

⑥⑤ − θ	⑤ + θ		
	30 − θ	㉕ + θ	
θ		㉕ − θ	

Now, min $(65 - \theta, 25 - \theta) = 0 \Rightarrow \theta = 25$. Therefore, revised table is given by

㊵ 6	㉚ 1	9	3
11	⑤ 5	㊽ 2	8
㉕ 10	12	4	㊺ 7
⑳ 0	0	0	0

Again we check the optimality of this solution

				u_i
④⓪ 6	㉚ 1	−2 +11 3 9	0 3	6
10 +1 11	⑤ 5	㊿ 2	7 +1 8	10
㉕ 10	5 +7 12	2 4	㊺ 7	10
⑳ 0	−5 +5 0	−8 +8 −3 0	+3 0	0
v_j 0	− 5	− 8	− 3	

Clearly, all $d_{ij} \geq 0 \Rightarrow$ solution is optimal and is given by

$x_{11} = 40,\ x_{12} = 30,\ x_{21} = 5,\ x_{22} = 50,\ x_{31} = 25,\ x_{34} = 45,\ x_{41} = 20$

and the minimum transportation cost is given by

$= 40 \times 6 + 30 \times 1 + 5 \times 5 + 50 \times 2 + 25 \times 10 + 45 \times 7 + 20 \times 0$

$= ₹ 960$

13.8 SOME MISCELLANEOUS SOLVED PROBLEMS

EXAMPLE I. *Solve the following transportation problem as*

		Plant			
Warehouse	**1**	**2**	**3**	**4**	**Demand**
1	6	−2	3	9	80
2	6	−2	2	6	110
3	1	−4	2	4	150
4	−1	1	−3	8	100
5	2	−1	−1	3	150
Capacity	150	200	175	100	

SOLUTION. We have

		Plant			
Warehouse	**1**	**2**	**3**	**4**	**Demand**
1	6	−2	3	9	80
2	6	−2	2	6	110
3	1	−4	2	4	150
4	−1	1	−3	8	100
5	2	−1	−1	3	150
Capacity	150	200	175	100	

To make the problem balance, we take a fictitious warehouse 6 with demand 35 and zero transportation cost throughout changing this maximization problem into minimization problem by making each entry negative.

Iteration-1

	1	2	3	4	Demand
1	⊖ −1 / 6	4 / −2	−1 / 3	(80 − θ) / 9	80
2	(110) / 6	5 / −2	−1 / 2	4 / 6	110
3	4 / 1	6 / −4	(150) 2	5 / 4	150
4	5 / −1	(80 − θ) / 1	4 / −3	(20 + θ) / 8	100
5	(40 − θ) / 2	(110 + θ) / −1	0 / −1	3 / 3	150
6	3 / 0	(10) / 0	(25) / 0	7 / 0	35
Capacity	150	200	175	100	625
v_j	−3	0	0	−7	

To find auxiliary number u_i and v_j for occupied cells

Cell (1, 4) $u_1 + v_4 = -9$ Let $v_2 = 0$, $u_1 = -9$
Cell (2, 1) $u_2 + v_1 = -6$ $u_2 = -4$
Cell (3, 3) $u_3 + v_3 = -2$ $u_3 = -2$
Cell (4, 2) $u_4 + v_2 = -1$ $u_4 = -1$
Cell (4, 4) $u_4 + v_4 = -8$ $v_4 = -7$
Cell (5, 1) $u_5 + v_1 = -2$ $v_1 = -2 - 1 = -3$
Cell (5, 2) $u_5 + v_2 = 1$ $u_5 = 1$
Cell (6, 2) $u_6 + v_2 = 0$ $v_6 = 0$
Cell (6, 3) $u_6 + v_3 = 0$ $v_3 = 0$

Since Δ_{ij} is not all positive so present B.F.S. is not optimal. So assign θ to cell (1, 1). Now, cell (5, 1) leaves the basis. Now $\theta = 40$.

Iteration-2

	1	2	3	4	Demand
1	(40) 0 / −6	4 / 2	⊖ −1 / −3	(40 − θ) 0 / −9	80
2	(110) / −6	4 / 2	0 / −2	3 / −6	110
3	5 / −1	6 / 4	(150) / −2	5 / −4	150
4	6 / 1	(40 − θ) / −1	4 / 3	(60 + θ) / −8	100
5	1 / −2	(150) / 1	0 / 1	3 / −3	150
6	4 / 0	(10 + θ) 0 / 0	(25 − θ) / 0	7 / 0	35
Capacity	150	200	175	100	625
v_j	−6	−2	−2	−9	

Calculate u_i and v_j for each occupied cell such that $c_{ij} = u_i + v_j$

Cell $(1, 1)$ $u_1 + v_1 = -6$ Let $u_1 = -6$, $v_1 = 0$

Cell $(1, 4)$ $u_1 + v_4 = -9$ $v_4 = -9$

Cell $(2, 1)$ $u_2 + v_1 = -6$ $u_2 = 0$

Cell $(3, 3)$ $u_3 + v_3 = -2$ $v_3 = 0$

Cell $(4, 2)$ $u_4 + v_2 = -1$ $v_2 = -2$

Cell $(4, 4)$ $u_4 + v_4 = -8$ $u_4 = 1$

Cell $(5, 2)$ $u_5 + v_2 = 1$ $u_5 = 3$

Cell $(6, 2)$ $u_6 + v_2 = 0$ $u_6 = 2$

Cell $(6, 3)$ $u_6 + v_3 = 0$ $v_3 = -2$

Since all Δ_{ij} are not positive, assign a quantity θ in the cell of most negative Δ_{ij}, i.e., cell $(1, 3)$. Now, $\theta = 25$. Cell $(6, 3)$ leaves.

Iteration-3

Plant

	1	2	3	4	Demand	u_i
1	(40) −6	4 2	(25) −3	15 −9	80	0
2	(110) −6	4 2	1 −2	3 −6	110	0
3	4 −1	5 4	(150) −2	4 −4	150	1
4	4 1	(15) −1	5 3	(85) −8	100	1
5	1 −2	(150) 1	1 1	3 −3	150	3
6	4 0	(35) 0	1 0	1 0	35	2
Capacity	150	200	175	100		
v_j	−6	−2	−3	−9		

Calculate u_i and v_j for each occupied cell such that $c_{ij} = u_i + v_j$

Cell $(1, 1)$ $u_1 + v_1 = -6$ Let $u_1 = -6$, $v_1 = 0$

Cell $(1, 3)$ $u_1 + v_3 = -3$ $v_3 = -3$

Cell $(1, 4)$ $u_1 + v_4 = -9$ $v_4 = -9$

Cell $(2, 1)$ $u_2 + v_1 = -6$ $u_2 = 0$

Cell $(3, 3)$ $u_3 + v_3 = -2$ $u_3 = 1$

Cell (4, 2)	$u_4 + v_2 = -1$	$v_2 = -2$
Cell (4, 4)	$u_4 + v_4 = -8$	$u_4 = 1$
Cell (5, 2)	$u_5 + v_2 = 1$	$u_5 = 3$
Cell (6, 2)	$u_6 + v_2 = 0$	$u_6 = 2$

EXAMPLE 2. *A steel company has three furnaces, five rolling mills, shipping cost of steel from furnace to the mills is shown in the following table. Obtain the optimal transportation schedule.*

Plant

	M_1	M_2	M_3	M_4	M_5	Capacity in 100 quintal
F_1	4	2	3	2	6	8
F_2	5	4	5	2	1	12
F_3	6	5	4	7	3	14
Demand in units of 100 quintal	4	4	6	8	8	34

[MEERUT–2001, 03, 07, 08; AGRA–2002]

SOLUTION.   **Iteration-1.** To make problem balance, make a fictitious column M_6 with demand 4 and cost 0 throughout. Make initial assignment by Matrix-Minima Method

Plant

	M_1	M_2	M_3	M_4	M_5	M_6	**Capacity**
F_1	3 ④ 4	2	4 3	0 2	5 ④ 6	0	8
F_2	4 5	2	6 (4+θ) 5	2	$\varepsilon - \theta$ 1	0	12
F_3	④ 6	-2 ⑥ 5	(4-θ) 4	-3 7	θ 3	-5 0	14
Demand	4	4	6	8	8	4	34
v_j	6	7	4	7	6	5	

Number of occupied cell = 7 < 9 – 1 = 8.

So assign a quantity $\varepsilon \to 0$ in the cell of minimum cost, i.e., in cell (2, 6).

Calculate u_i and v_j for each occupied cell such that $c_{ij} = u_i + v_j$

Cell (1, 2) $u_1 + v_2 = 2$ Let $v_2 = 7$, $u_3 = 0$

Cell $(1, 4)$ $u_1 + v_4 = 0$ $u_1 = -5$

Cell $(2, 4)$ $u_2 + v_4 = 2$ $u_2 = -5$

Cell $(2, 5)$ $u_2 + v_5 = 1$ $v_5 = 6$

Cell $(2, 6)$ $u_2 + v_6 = 0$ $v_6 = 5$

Cell $(3, 1)$ $u_3 + v_1 = 6$ $v_1 = 6$

Cell $(3, 4)$ $u_3 + v_4 = 4$ $v_3 = 4$

Cell $(3, 4)$ $u_3 + v_4 = 7$ $v_4 = 7$

Since all Δ_{ij} are not positive therefore assign a quantity θ in the cell of most negative Δ_{ij}, i.e., in cell $(3, 6)$.

Here $\theta = \varepsilon \to 0$.

Iteration-2

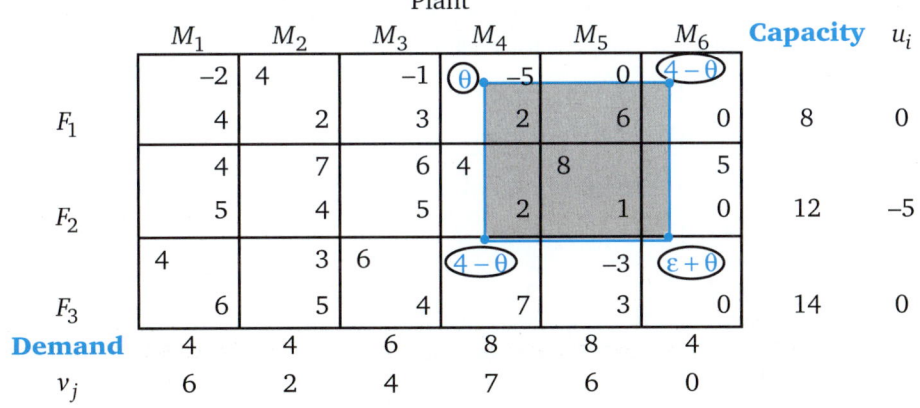

Calculate u_i and v_j for each occupied cell, as given below

Cell $(1, 2)$ $u_1 + v_2 = 2$ Let $u_3 = 0$, $v_2 = 2$

Cell $(1, 6)$ $u_1 + v_6 = 0$ $u_1 = 0$

Cell $(2, 4)$ $u_2 + v_4 = 2$ $u_2 = -5$

Cell $(3, 5)$ $u_2 + v_5 = 1$ $v_5 = 6$

Cell $(3, 1)$ $u_3 + v_1 = 6$ $v_1 = 6$

Cell $(3, 3)$ $u_3 + v_3 = 4$ $v_3 = 4$

Cell $(3, 4)$ $u_3 + v_4 = 7$ $v_4 = 7$

Cell $(3, 6)$ $u_3 + v_6 = 0$ $v_6 = 0$

Since all Δ_{ij} are not positive, assign a quantity θ in the cell of most negative Δ_{ij} such that loop is formed, i.e., in the cell $(1, 4)$. Here $\theta = 4$.

Iteration-3

	M_1	M_2	M_3	M_4	M_5	M_6	Capacity	u_i
F_1	(θ) ⁻² / 4	4 / 2	⁻¹ / 3	4 / 2	5 / 6	(ε−θ) / 0	8	0
F_2	⁻¹ / 5	2 / 4	6 / 5	4 / 2	8 / 1	/ 0	12	0
F_3	(4−θ) / 6	3 / 5	6 / 4	5 / 7	2 / 3	(4+θ) / 0	14	0
Demand	4	4	6	8	8	4		
v_j	6	2	4	2	1	0		

Number of occupied cell = 7 < 8, therefore assign a quantity $\varepsilon \to 0$ in a cell of minimum cost such that loop is not formed, *i.e.*, in cell (1, 6).

Calculate u_i and v_j for each occupied cell

Cell (1, 2)	$u_1 + v_2 = 2$	Let $v_2 = 2$, $u_1 = 0$
Cell (1, 4)	$u_1 + v_4 = 2$	$v_4 = 2$
Cell (1, 6)	$u_1 + v_6 = 0$	$v_6 = 0$
Cell (2, 4)	$u_2 + v_4 = 2$	$u_2 = 0$
Cell (2, 5)	$u_2 + v_5 = 1$	$v_5 = 1$
Cell (3, 1)	$u_3 + v_1 = 6$	$v_1 = 6$
Cell (3, 3)	$u_3 + v_3 = 4$	$v_3 = 4$
Cell (3, 6)	$u_3 + v_6 = 0$	$u_3 = 0$

Since all Δ_{ij} are not positive, assign a quantity θ in cell (1, 1).

Here, $\theta = \varepsilon \to 0$.

Iteration-4

	M_1	M_2	M_3	M_4	M_5	M_6	Capacity	u_i
F_1	(ε) / 4	(4) / 2	1 / 3	(4) / 2	5 / 6	2 / 0	8	0
F_2	1 / 5	2 / 4	3 / 5	(4) / 2	(8) / 1	2 / 0	12	0
F_3	(4) / 6	1 / 5	(6) / 4	3 / 7	0 / 3	(4) / 0	14	2
Demand	4	4	6	8	8	4		
v_j	4	2	2	2	1	−2		

Calculate u_i and v_j for each occupied cell

Cell (1, 1) $u_1 + v_1 = 4$ Let $v_1 = 4$, $u_1 = 0$

Cell (1, 2) $u_1 + v_2 = 2$ $v_2 = 2$

Cell (1, 4) $u_1 + v_4 = 2$ $v_4 = 2$

Cell (2, 4) $u_2 + v_4 = 2$ $u_2 = 0$

Cell (2, 5) $u_2 + v_5 = 1$ $v_5 = 1$

Cell (3, 1) $u_3 + v_1 = 6$ $u_3 = 2$

Cell (3, 3) $u_3 + v_3 = 4$ $v_3 = 2$

Cell (3, 6) $u_3 + v_6 = 0$ $v_6 = -2$

Since all Δ_{ij} are positive, optimal reaches.

Minimum cost $= 4 \times 2 + 4 \times 2 + 4 \times 2 + 8 \times 1 + 4 \times 6 + 6 \times 4 + 4 \times 0 = 80$.

EXAMPLE 3. *Solve the following transportation problem of minimize cost starting with degenerate bases $x_{12} = 30$, $x_{21} = 40$, $x_{32} = 20$ and $x_{43} = 60$.*

	D_1	D_2	D_3	Availability
O_1	4	5	2	30
O_2	4	1	3	40
O_3	3	6	2	20
O_4	2	3	7	60
Demand	40	50	60	

SOLUTION. **Iteration-1**

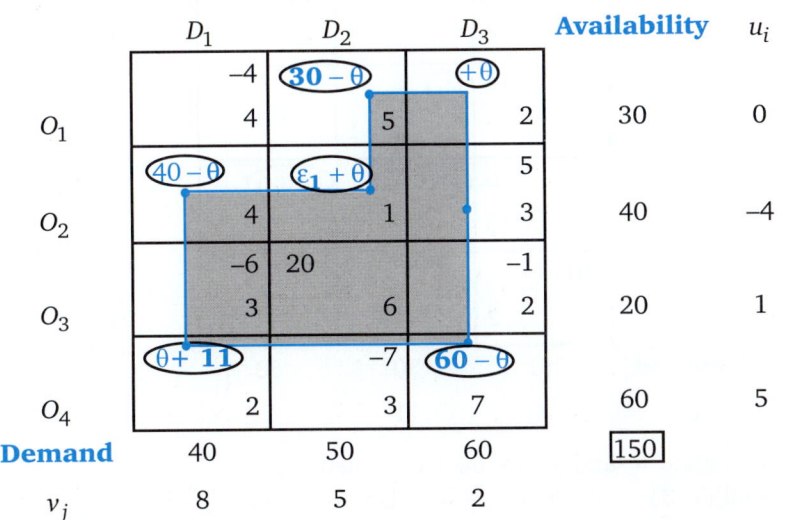

Number of occupied cell = 4 < 4 + 3 – 1 = 6. So degeneracy occurs.

To make the number of occupied cells and assign a quantity $\varepsilon \to 0$ to the two remaining cells of minimum cost, so that cell may not contain a loop.

Let ε_1 assign in cell (2, 2) and (1, 3).

Since all Δ_{ij} are not positive therefore assign a quantity θ in the cell of most negative Δ_{ij}, i.e., in cell (4, 1). Here $\theta = 30$ cell (1, 2) is leaving cell.

Iteration-2

	D_1	D_2	D_3	Availability	u_i
O_1	7 / 4	11 / 5	30 / 2	30	–5
O_2	(10 – θ) 4	(30 + θ) 1	–6 / 3	40	2
O_3	–6 / 3	(20 – θ) 6	θ / –12 / 2	20	7
O_4	(30 + θ) 2	4 / 3	(30 – θ) 7	60	0
Demand	40	50	60	150	
v_j	2	–1	7		

Since all Δ_{ij} are not positive therefore assign a quantity θ in cell (3, 3).

Here $\theta = 10$.

Iteration-3

	D_1	D_2	D_3	Availability	u_i
O_1	7 / 4	–1 / 5	(30) / 2	30	–5
O_2	12 / 4	(40) / 1	6 / 3	40	–10
O_3	6 / 3	(10 – θ) 6	(10 + θ) 2	20	5
O_4	40 / 2	θ / –9 / 3	(20 – θ) 7	60	0
Demand	40	50	60		
v_j	2	11	7		

Calculate u_i and v_j for each occupied cell

Cell (1, 3) $\qquad u_1 + v_3 = 2 \qquad$ Let $u_1 = -5$, $u_4 = 0$

Cell (2, 2) $u_2 + v_2 = 1$ $u_2 = -10$
Cell (3, 2) $u_3 + v_2 = 6$ $v_2 = 11$
Cell (3, 3) $u_3 + v_3 = 2$ $u_3 = 5$
Cell (4, 1) $u_4 + v_1 = 2$ $v_1 = 2$
Cell (4, 3) $u_4 + v_3 = 7$ $v_3 = 7$

Since all Δ_{ij} are not positive therefore assign a quantity θ in cell (4, 2).
Here $\theta = 10$. Cell (3, 2) is leaving cell.

Iteration-4

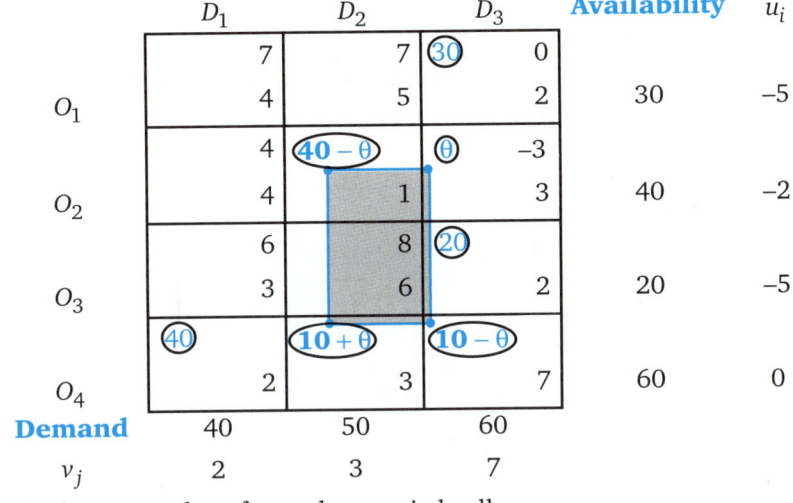

Calculate u_i and v_j for each occupied cell.

Cell (1, 3) $u_1 + v_3 = 2$ Let $u_1 = -5$, $u_4 = 0$
Cell (2, 2) $u_2 + v_2 = 1$ $u_2 = -1$
Cell (3, 3) $u_3 + v_3 = 2$ $u_3 = -5$
Cell (4, 1) $u_4 + v_1 = 2$ $v_1 = 2$
Cell (4, 2) $u_4 + v_2 = 3$ $v_2 = 3$
Cell (4, 3) $u_4 + v_3 = 7$ $v_3 = 7$

Since all Δ_{ij} are not positive therefore assign a quantity θ in cell (2, 3).

Iteration-5

	D_1	D_2	D_3	Availability	u_i
O_1	5 / 4	5 / 5	㉚ / 2	30	-3
O_2	㉚ / 4	1	⑩ / 3	40	-2
O_3	4 / 3	6 / 6	⑳ / 2	20	-3
O_4	㊵ / 2	⑳ / 3	2 / 7	60	0
Demand	40	50	60		
v_j	2	3	5		

Calculate u_i and v_j for each occupied cell

Cell (1, 3) $u_1 + v_3 = 2$ Let $v_1 = -3$, $u_4 = 0$

Cell (2, 2) $u_2 + v_2 = 1$ $u_2 = -2$

Cell (2, 3) $u_2 + v_3 = 3$ $v_3 = 5$

Cell (3, 3) $u_3 + v_3 = 2$ $u_3 = -3$

Cell (4, 2) $u_4 + v_2 = 2$ $v_1 = 2$

Cell (4, 2) $u_4 + v_2 = 3$ $v_2 = 3$

Since all Δ_{ij} are positive, hence optimal is reached.

The minimum cost of assignment

$$= 30 \times 2 + 30 \times 1 + 10 \times 3 + 20 \times 2 + 40 \times 2 + 20 \times 3$$
$$= 300$$

EXAMPLE 4. *A company has three plants and four warehouses. The supply and demand in units and the transportation costs are given. The solution of the problem is given below :*

	W_1	W_2	W_3	W_4	Supply
O_1	⑤	⑩	④	⑤	10
			10		
O_2	⑥	⑧	⑦	②	25
	20			5	
O_3	④	②	③	⑦	20
	5	10	5		
Demand	25	10	15	5	

Answer the following questions :

  (i) Is the solution feasible?

  (ii) Does the solution degenerate?

 (iii) Is this solution optimum.

 (iv) Does this problem have more than one optimum solution. If so, show all of them.

  (v) If the cost of route O_2W_3 is reduced from ₹ 7 to ₹ 6 per unit, what will be the optimum solution?   [DELHI–1987]

SOLUTION.   In the above question, North-West rule applied, so we always move to the right or down, so no loop can be formulated by drawing horizontal and vertical lines to the allocations.

  (i) Solution is feasible because we can not get more than $(m + n - 1)$ individual positive allocations, $(3 + 4 - 1) = 6$.

  (ii) We always get a non-degenerate basic feasible solution by the North-West Corner rule.

 (iii) The optimality test is applicable to a non-degenerate B.F.S. to a F.S. consisting of $(m + n - 1)$ allocations in independent positions. Since there are $m.n - (m + n - 1) - (m - 1)(n - 1)$ empty cell, so the solution is optimum.

(iv) No further possibility to obtain an optimal solution by making successive improvements to initial basic feasible solution until no further decrease in the transportation cost is possible.

(v) The solution is not feasible if replaced ₹ 7 to ₹ 6 per unit, so there is not possible to find optimum.

Exercise-13.1

1. Determine an initial basic feasible solution to the following transportation problem using VAM or Vogel's method

(i)

Destination

	D_1	D_2	D_3	D_4	D_5	Capacity
A	2	11	10	3	7	4
Origin B	1	4	7	2	1	8
C	3	9	4	8	12	9
Demand	3	3	4	5	6	21 total

[IAS–1988; MEERUT–2002, 04; AGRA–2000]

(ii)

Destination

	A_1	B_1	C_1	D_1	Supply
A	1	2	1	4	30
Origin B	3	3	2	1	50
C	4	2	5	9	20
Demand	20	40	30	10	100 total

2. Determine the optimal solution to each of the following degenerate transportation problem.

(i)

Destination

	D_1	D_2	D_3	D_4	Capacity
O_1	2	3	11	7	6
Origin O_2	1	0	6	1	1
O_3	5	5	15	10	10
Demand	7	5	3	2	17 total

(ii)

To

		D_1	D_2	D_3	D_4	a_i
	O_1	10	7	3	6	3
From	O_2	1	6	7	3	5
	O_3	7	4	5	3	7
	b_j	3	2	6	4	15 total

3. Solve the following transportation problem whose cost matrix is given below

				Capacity
5	5	4	7	5
6	4	1	2	5
5	9	1	4	6
8	3	2	4	4
6	5	3	1	6
Demand 5	8	3	10	

4. Obtain the optimal solution by using the best starting solution

Destination

					a_i
	10	20	5	7	10
	13	9	12	8	20
Origin	4	15	7	9	30
	14	7	1	0	40
	3	12	6	19	50
b_j	60	60	20	10	

5. A company has 4 warehouses and 6 stores. The capacity in the warehouse, the demand of the stores and costs (in ₹) of transporting are unit of the commodity from warehouses i to the store j are given below. How should the commodity be transported so that the total transportation cost is a minimum? Obtain the initial program by applying the

Vogel's method.

Store

		1	2	3	4	5	6	Capacity
Warehouse	1	7	5	9	5	10	7	30
	2	7	8	24	7	9	13	40
	3	4	10	5	6	10	4	20
	4	11	8	12	7	12	11	80
Demand		30	30	60	20	10	20	170

6. An oil corporation has got three refineries A, B and C and it has to send petrol to four different depots P, Q, R and S, the cost of shipping 1 gal of petrol and the available petrol at the refineries are given in the table. The demand of the depots and the capacity petrol at the refineries are also given. Find the minimum cost of shipping after obtaining an initial solution by VAM.

Depot

		P	Q	R	S	Capacity
	A	10	12	15	8	130
Refinery	B	11	11	9	10	150
	C	20	9	7	18	170
Demand		90	100	140	120	

ANSWERS

1. (i) $x_{14} = 4, x_{22} = 2, x_{25} = 6, x_{31} = 3, x_{32} = 1, x_{33} = 4, x_{34} = 1$, cost = ₹ 68

   (ii) $x_{11} = 20, x_{13} = 10, x_{22} = 20, x_{23} = 20, x_{24} = 10, x_{32} = 20$, cost = ₹ 180

2. (i) $x_{12} = 5, x_{13} = 1, x_{24} = 1, x_{31} = 7, x_{33} = 2, x_{34} = 1$, Min. cost = ₹ 102

   (Alternative solutions also exist)

   (ii) $x_{13} = 3, x_{21} = 3, x_{24} = 2, x_{32} = 2, x_{33} = 3, x_{34} = 2$, Minimum cost = ₹ 47

3. $x_{11} = 2, x_{12} = 3, x_{22} = 1, x_{24} = 4, x_{31} = 3, x_{33} = 3, x_{42} = 4, x_{54} = 6$, Min. cost = ₹ 73

4. $x_{13} = 10, x_{22} = 20, x_{31} = 30, x_{42} = 20, x_{43} = 10, x_{44} = 10, x_{51} = 30, x_{52} = 20$,

   Minimum cost = ₹ 830

5. $x_{12} = 10, x_{16} = 20, x_{21} = 30, x_{25} = 10, x_{33} = 20, x_{42} = 20, x_{43} = 40, x_{44} = 20$,

   Minimum cost = ₹ 1,370

6. $x_{11} = 90, x_{14} = 40, x_{23} = 70, x_{24} = 80, x_{32} = 100, x_{33} = 70$, Minimum cost = ₹ 3960

REVIEW QUESTIONS

1. Give a mathematical formulation of transportation problem.

2. Write the characteristic of transportation problem of linear programming.

3. Explain various method to find the initial basic feasible solution of transportation problem.

4. Write a short note on transportation problem.

5. Write the computational procedure of optimality test in transportation problem.

MULTIPLE CHOICE QUESTIONS (CHOOSE THE MOST APPROPRIATE ONE)

1. In the optimal table of a transportation problem, a zero in the North-west corner rule shows that:

   [MEERUT–2013]

   (a) An alternative optimal solution exists

   (b) The optimal solution is degenerated

   (c) Both (a) and (b) are true

   (d) None of the above

2. The dummy source or destination in a transportation problem is introduced to:

   [MEERUT–2013]

   (a) prevents solution to become degenerate

   (b) satisfy rim conditions

   (c) ensure that total cost does not exceed a limit

   (d) solve the balanced transportation problem

3. In a T.P., the solution under test will be optimal if all the cell evaluation are: [MEERUT–2013]

(a) > 0 (b) < 0

(c) ≥ 0 (d) ≤ 0

4. Transportation problem

	W_1	W_2	W_3	W_4	**Available**
F_1	19	30	50	10	7
F_2	70	30	40	60	9
F_3	40	8	70	20	18
Requirement	5	8	7	14	

has optimal solution: [MEERUT–2013]

(a) 1015 (b) 814

(c) 779 (d) 743

5. The total number of allocations in a basic feasible solution of transportation problem of $m \times n$ size is equal to: [MEERUT–2013]

(a) $m \times n$ (b) $m + n$

(c) $m + n + 1$ (d) $m + n - 1$

6. Which of the following is not correct?

(a) The transportation problem is a distribution problem

(b) A closed loop would always involve an even number of cells, subject to a minimum of 4

(c) The u_i and v_j values may be determined by initially inserting any finite number which may be positive, negative or zero to a row/column

(d) Units sents from a dummy source to various markets represent the shortfall in supply to those markets

7. The transportation problem is balanced if:

(a) total demands equals total supply irrespective of the number of sources and destinations.

(b) number of sources matches with the number of destinations.

(c) total demand and total supply are equal and the number of sources equals the number of destinations.

(d) None of the route is prohibited.

8. Which of the following is not correct?

(a) A degenerate solution may or may not be optimum.

(b) A transportation problem solution is said to be degenerate if the number of occupied cells is smaller than the number of rows plus the number of columns minus one.

(c) To remove degeneracy, an infinitesimally small quantity is placed in each of the required number of independent cells.

(d) Once non-optimum degenerate solution is obtained, the next solution is bounded to be degenerate.

9. The transportation problem deals with the transportation of:

(a) single product from a source to several destinations.

(b) a single product from several source to the several destination.

(c) a multi-product from several sources to several destinations.

(d) a single product from several sources to a destination.

10. For a transhipment problem, choose the statements which is not correct:

(a) a transhipment problem is not likely to involve a lower cost than a transportation problem, in a given situation.

(b) an 'm' source, 'n' destination transportation problem when written as a transhipment problem would have $m + n$ sources and n destinations.

(c) there is no real distinction between sources and destinations.

(d) the problem allows for the shipment of goods from one sources to another and from one destination to another.

11. Which of the following is not correct?

(a) If some $u_i + v_j - c_{ij}$ is equal to zero in the optimum solution, then the problem has multiple optimum solutions

(b) If all the cost elements c_{ij} are multiplied by a cosntant, the total cost of transportation in optimum solution shall be multiplied by the same constant

(c) The number of occupied cells involved in a closed path is always even

(d) In the transportation problem, if certain routes are prohibited then their cost elements are replaced by M (an extremely large value)

12. Which of the following is not correct?

    (a) It is possible that in some cases both, the dummy source and dummy destination, may be required to convert an unbalanced transportation problem into a balanced one

    (b) The cost elements in a dummy row/column shall always be taken equal to zero

    (c) An unbalanced transportation problem must be converted into a balanced problem before solving it

    (d) It is not necessary for the aggregate demand to be equal to the aggregate supply in a transportation problem

13. Which of the following is not correct?

    (a) A maximization transportation problem is first converted into a minimization one by subtracting each value of the given matrix from the largest value

    (b) If each cost element in a transportation problem is increased or decreased by a constant amount, it will not effect the optimum solution of the problem

    (c) Multiple optimum solutions are indicated if there are multiple zeroes for u_i and v_j values

    (d) For an optimum solutions to a transportation problem, the u_i and v_j values represent the optimum values of the dual problem

14. Which of the following is correct?

    (a) The least cost method does not provide the least cost solution to a transportation problem

    (b) The initial solution obtained by the NW corner rule would invariably be optimum

    (c) If some routes are prohibited, VAM cannot be used to find an initial solution to a transportation problem

    (d) The cost difference in the Vogel's approximation method indicate the penalties for not using the respective least cost routes

15. Which of the following is not correct?

    (a) The $u_i + v_j - c_{ij}$ value of an unoccupied cells indicates the net change in cost of re-allocating one unit through the routh involved

    (b) In time minimization problem, the cost c_{ij} is replaced by the unit time t_{ij}

    (c) Any of the $m + n - 1$ number of occupied cells would allow determining whether a given solution is optimum or not

    (d) The maximum quantity that can be re-allocated in a closed path is equal to the minimum quantity in the cells bearing negative sign

16. During an iteration while moving from one solution to the next, degeneracy may occurs when:

    (a) two or more occupied cells on the closed path with minus sign are tied for the lowest circled value

    (b) two or more occupied cells are on the closed path but neither of them represents a corner of the path

    (c) the closed path indicates a diagonal move

    (d) either of the above

17. To find a B.F.S. of a transportation problem by matrix minima method which cell we choose first:

    (a) highest cell   (b) lowest cell

    (c) zero cost   (d) none of these

18. To improve the current basic F.S. of a transportation problem if it is not optimal we allocate to the cell for which d_{ij} is:

    (a) 0

    (b) maximum and negative

    (c) minimum and negative

    (d) positive

19. A feasible solution of a T.P. is said to be optimal if it:

    (a) maximizes the total trasportation cost

    (b) minimizes the total transportation cost

    (c) balance the total transportation cost

    (d) none of these

20. An assignment problem is a special case of a $l \times n$ transportation problem in where:

(a) $l - 2n$ (b) $l = n$

(c) $l = 3n$ (d) None of these

21. In a Transportation Problem a loop may be defined as an ordered set of at least:

(a) 3 cells   (b) 4 cells

(c) 5 cells   (d) 6 cells

22. If a Transportation Problem $\Sigma a_i < \Sigma b_j$ then we introduce a dummy origin to make it a:

(a) Unbalanced T.P.   (b) Balanced T.P.

(c) L.P.P.   (d) All of these

23. For a Transportation Problem for testing the optimality we find the u_i and v_j that for all the occupied cells $c_{ij} =$

(a) $u_i - v_j$ (b) $u_i + v_j$

(c) u_i / v_j (d) $u_i \cdot v_j$

24. In transportation problem, to find BFS we start with the cell (1, 1) following some rule that rule is:

(a) N-W corner rule

(b) Lowest cost entry method

(c) Vogel's approximation method

(d) None of the above

25. A feasible solution is said to be optimal if it _____ the total transportation cost the blank space can be filled by:

(a) maximizes

(b) minimizes

(c) either maximizes or minimizes

(d) none of the above

26. To find initial BFS of transportation problem by matrix revenue method, we first choose the cell with:

(a) zero cost   (b) lowest cost

(c) highest cost   (d) none of these

27. If a B.F.S. of a $m \times n$ Transportation Problem, the number of positive allocations is at most:

(a) $m + n$ (b) $m + n - 1$

(c) $m - n$ (d) None of these

28. A solution is not a basic feasible solution in a transportation problem if after allocations:

(a) there is degeneracy

(b) there is closed loop

(c) there is no closed loop

(d) total number of allocations is one less than sum of the number of sources and destinations

29. One disadvantage of using North-west corner rule to find initial solution to the transportation problem is that:

(a) it leads to adequate initial solution

(b) it is complicated to use

(c) it does not take into account the cost of transportation

(d) all of the above

30. The dummy source or destination in a transportation problem is added to:

(a) ensure that total cost does not exceed a limit

(b) prevent solution from becoming degenerate

(c) satisfy rim conditions

(d) none of the above

31. In transportation problem, the materials are transported from 3 plants to 5 warehouses. The basic feasible solution must contain exactly which one of the following allocated cells?

(a) 8   (b) 7

(c) 5   (d) 3

32. The calculation of opportunity cost in the MODI method is analogous to a:

(a) variable in the B-columns in the Simplex method

(b) value of a variable in x_B-column of the simplex method

(c) $z_j - c_j$ value for non-basic variable columns in the simplex method

(d) none of the above

33. In a 6×6 transportation problem, degeneracy would arise, if the number of filled sports were:

(a) less than eleven   (b) equal to twelve

(c) more than twelve (d) equal to thirty six

34. The occurance of degeneracy while solving a transportation problem means that:

(a) the few allocation become negative

(b) the solution so obtained is not feasible

(c) total supply equals total demand

(d) none of the above

35. The solution of a transportation model (of dimension $m \times n$) is said to be degenerate if it has:

(a) $(m + n)$ allocation

(b) more than $(m + n - 1)$ allocation

(c) fewer than $(m + n - 1)$ allocation

(d) exactly $(m + n - 1)$ allocation

36. An unbounded cell in the transportation method is analogous to a:

(a) value in the x_B-column in the Simplex method

(b) variable not in the B-column in the Simplex method

(c) variable in the B-column in the Simplex method

(d) $z_j - c_j$ value in the Simplex method

37. The initial solution of a transportation problem can be obtained by applying any known method. However, the only condition is that:

(a) the solution may not be degenerate

(b) the RIM condition are satisfied

(c) the solution be optimal

(d) all of the above

38. An alternative optimal solution to a minimization transportation problem exists whenever opportunity cost corresponding to unused route of transportation is:

(a) negative with at least one is equal to zero

(b) positive and greater than zero

(c) positive with at least one is equal to zero

(d) none of the above

39. The solution to a transportation problem with m rows (supplies) and n columns (destinations) is feasible if number of positive allocations are:

(a) $m + n$ (b) $m \times n$

(c) $m + n - 1$ (d) $m + n + 1$

40. If we were to use opportunity cost value for an unused cell to test optimality, it should be:

(a) most positive number

(b) equal to zero

(c) most negative number

(d) any value

41. When there are 'm' rows and 'n' columns in a transportation problem, degeneracy is said to occur when the number of allocation is:

(a) less than $m - n - 1$

(b) equal to $m + n - 1$

(c) greater than $m + n - 1$

(d) less than $m + n - 1$

42. In a transportation problem, obtaining the starting BFS by VAM or any other method, a column and a row are satisfied together. This shows that:

(a) there is no feasible solution

(b) at least one basic variable is at zero level

(c) at least two basic variable is at zero level

(d) none of these

43. In the optimal table of a transportation problem, a zero in the S.W. corner shows that:

(a) the optimal solution is degenerate

(b) an alternate optimal solution exists

(c) an optimal solution does not exist

(d) None of the above

44. In a balanced transportation problem with three sources and four destinations and with availabilities 40 at each source and demand 30 at each destination, the dual variables in the optimal table corresponding to sources and destinations are respectively $-1, 2, 3$ and $0, 2, -1$, 4. Then the optimal value is:

(a) 310 (b) 320

(c) 300 (d) None of these

45. In transportation problem, one of the dual variables is assigned an arbitrary value, because:

(a) then a solution is obtained immediately

(b) one of the constraints is redundant in a T.P

(c) this facilitates construction of the loop

(d) None of these

46. In a transportation problem while obtaining the starting BFS by VAM or any other method, a column and row are satisfied together. Then treating both as satisfied (that is not writting 0 either at row or column) will give a solution:

(a) which is not basic

(b) which will be degenerate

(c) which is basic

(d) None of these

47. In a transportation problem (T.P.) the dual variables u_i and v_j are unrestricted in sign because:
(a) the TP is a minimization problem
(b) the TP is with all equality constraints
(c) in TP all decision variables are ≥ 0
(d) none of these

48. In a balanced transportation problem with m sources and n destinations, the number of linearly independent constraints is:
(a) $m + n$ (b) $m + n + 1$
(c) $m + n - 1$ (d) $m + n$

49. In a balanced transportation problem with m sources and n destinations the coefficient matrix has rank:
(a) $m + n + 1$ (b) $m + n$
(c) $m + n - 1$ (d) $m - n$

50. In an unbalanced transportation problem with m sources and n destinations the number of basic variables is:
(a) $m + n + 1$ (b) $m + n$
(c) $m + n - 1$ (d) $m + n + 2$

51. If in using VAM or any other method to obtain a starting solution the allocation made in a cell is not maximum possible, then the resulting solution:

(a) will not be BFS

(b) will be a BFS

(c) will be a basic solution but not BFS

(d) None of these

52. In Transportation Problem, if less allocation than given by loop criteria is made in the cell corresponding to entering vairable, then the resulting solution will:
(a) not be a BFS
(b) be a BFS
(c) be a basic solution but not BFS
(d) None of these

53. In Transportation Problem, if two dual variables are assigned arbitrary values, then:
(a) the method will yield correct solution
(b) the method will not yield correct solution
(c) the method may or may not yield correct solution
(d) None of these

54. In Transportation Problem, the values of dual variables are not unique in the sense that the values depend on the variable assigned arbitrarily and the values assigned:
(a) $z_{ij} - c_{ij}$ will also change
(b) $z_{ij} - c_{ij}$ are unique
(c) $z_{ij} - c_{ij}$ will not change
(d) None of these

ANSWERS

1. (a)	2. (d)	3. (b)	4. (a)	5. (d)	6. (c)	7. (a)	8. (d)	9. (b)
10. (b)	11. (c)	12. (a)	13. (c)	14. (d)	15. (c)	16. (a)	17. (b)	18. (c)
19. (b)	20. (b)	21. (b)	22. (b)	23. (b)	24. (a)	25. (b)	26. (b)	27. (b)
28. (c)	29. (c)	30. (c)	31. (a)	32. (c)	33. (a)	34. (b)	35. (c)	36. (b)
37. (c)	38. (c)	39. (c)	40. (c)	41. (d)	42. (b)	43. (b)	44. (a)	45. (b)
46. (a)	47. (b)	48. (a)	49. (c)	50. (b)	51. (a)	52. (a)	53. (b)	54. (b)

ARCHIVE

1. A company has three production facilities S_1, S_2 and S_3 with production capacity of 7, 9 and 18 units (in 100's) per week of a product, respectively. These units are to be shipped to four warehouses D_1, D_2, D_3 and D_4 with requirement of 5, 6, 7 and 14 units (in 100's) per week, respectively. The transportation cost (in ₹) per unit between factories to warehouse are given in the table below.

	D_1	D_2	D_3	D_4	Capacity
S_1	19	30	50	10	7
S_2	70	30	40	60	9
S_3	40	8	70	20	18
Demand	5	6	7	14	34

Formulate this transportation problem as an LP model to minimize the total transportation cost. [BHARTHIAR–1988; MADURAI–1989; MADRAS BE (MECH & PROD)–1990; GUHATI(MCA)–1991]

2. The following table shows all the necessary information on the availability of supply to each warehouse, the requirement of each market and unit transportation cost from each warehouse to each market.

Market

		P	Q	R	S	Supply
Ware-house	A	6	3	5	4	22
	B	5	9	2	7	15
	C	5	7	8	6	8
Demand		7	12	17	9	45

The shipping clerk has worked out the following schedule from experience : 12 units from A to Q, 1 unit from A to R, 8 units from A to S, 15 units from B to R, 7 units from C to P and 1 unit from C to R.

(a) Check and see if the clerk has the optimal schedule.

(b) Find the optional schedule and minimum total transport cost.

(c) If the clerk is approched by a carrier of route C to Q, who offers to reduce his rate in the hope of getting some business, by how much the rate should be reduced before the clerk will offer him the business. [ICWA–1987]

3. ABC limited has three production shops supplying a product to five warehouses. The cost of production varies from shop to shop and cost of transfortation from one shop to a warehouse also varies. Each shop has a specific production capacity and each warehouse has certain amount of requirement. The cost of transportation are given below :

Warehouse

		I	II	III	IV	V	Supply
Shop	A	6	4	4	7	5	100
	B	5	6	7	4	8	125
	C	3	4	6	3	4	175
Demand		60	80	85	105	70	400

The cost of manufacturing the product at different production shop is

Shop	Variable cost	Fixed cost
A	14	7,000
B	16	4,000
C	15	5,000

Find the optimum quantity to be suplied from each shop to diffferent warehouse at minimum total cost.

[DELHI(MBA)–1997; NAGPUR(MBA)–1998]

4. A manufacturer wants to ship 22 loads of his product as shown below. The matrix gives the kilometres from sources of supply to the distinations.

Destinations

		D_1	D_2	D_3	D_4	D_5	Supply
Source	S_1	5	8	6	6	3	8
	S_2	4	7	7	6	5	5
	S_3	8	4	6	6	4	9
Demand		4	4	5	4	8	22 / 25

Shipping cost is ₹ 10 per load per km. What shipping schedule should be used to minimize total transportation cost? [DELHI(MBA)–2001]

5. The following table gives the cost of transporting material from supply point A, B, C and D to demand point E, F, G, H and I.

To

		E	F	G	H	I
From	A	8	10	12	17	15
	B	15	13	18	11	9
	C	14	20	6	10	13
	D	13	19	7	5	12

The present allocation is as follows :
A to E 90, A to F 10, B to F 150, C to F 10, C to G 50, C to I 120, D to H 210, D to I 70.

(a) Check if this allowcation is optimum. If not, find an optimum schedule.

(b) If the transportation cost from A to G is reduced to 10, what will be the new optimum schedule? [IAS(MAIN)–1992]

6. A manufacturer has distribution centres at Agra, Allahabad and Kolkata. These centre have availability of 40, 20 and 40 units of his product, respectively. His retail outlets at A, B, C, D and E require 25, 10, 20, 30 and 15 units, respectively.

The transport cost per unit between each outlet is given below :

Distribution Centre	Retail outlets				
	A	B	C	D	E
Agra	55	30	40	50	40
Allahabad	35	30	100	45	60
Kolkata	40	60	95	35	30

Determine the optional distribution to minimize the cost of transportation.

[ICWA-1985; NAGPUR(MBA)-1989, DELHI(MBA)-1998]

7. A manufacturer must produce a certain product in sufficient quality to meet contracted sales in next four months. The production facilities available for this product are limited, but by different amounts in respective months. This unit cost of production also varies in each month.

The product may be produced in one month and then held for sale in a letter month, but at an estimated storage cost of ₹ 1 per unit per month. No storage cost is incurred for goods sold in the same month in which they are produced. There is no initial inventory and none is desired at the end of 4 months. Given the following table, show how much to produce in each of four months in order to minimize total cost.

Month	Contracted Sales (in units)	Maximum Production (in units)	Unit cost of Production (in units)
1	20	40	14
2	30	50	16
3	50	30	15
4	40	50	17

Formulate and solve the above problem as a transportation problem.   [DELHI(MBA)–1996]

HINTS AND ANSWERS

1. x_{ij} = number of units of the product to be be transported from factory

$i(i = 1, 2, 3)$ to warehouse $j(j = 1, 2, 3, 4)$

Min. $Z = 19x_{11}+30x_{12}+50x_{13}+10x_{14}+70x_{21}+30x_{22}+40x_{23}+60x_{24}+40x_{31}+8x_{32}+70x_{33}+20x_{34}$

Subject to the constraints

(i) Capacity constraints

$x_{11} + x_{12} + x_{13} + x_{14} = 7;\ x_{21} + x_{22} + x_{23} + x_{24} = 9;\ x_{31} + x_{32} + x_{33} + x_{34} = 18$

(ii) Requirement constraints

$x_{11} + x_{21} + x_{31} = 5;\ x_{12} + x_{22} + x_{32} = 8;\ x_{13} + x_{23} + x_{33} = 7;\ x_{14} + x_{24} + x_{34} = 14$

and $x_{ij} \geq 0$ for $i = 1, 2, 3$ and $j = 1, 2, 3, 4$

2. (a) Cell (c, s) has a negative opportunity. Thus clerk has no optimal schedule.

(b) The total minimum transportation cost is 149.

(c) ₹ 2.38

3. The optimal solution obtain by applying MODI method. The transportation cost associated with the solution is ₹ 7,605.

4. The minimum total transportation cost = ₹ 920.

5. (a) $(A, F) = 100, (B, F) = 70, (B, I) = 80, (C, E) = 90, (C, G) = 50, (C, I) = 40, (D, H) = 210,$

$(D, I) = 70$, Total cost = ₹ 6,600

(b) When transportation cost from A to G is reduced to 10, the optional schedule given in (a) remains unchanged.

6. $x_{11} = 5, x_{12} = 10, x_{13} = 20, x_{14} = 5, x_{21} = 20, x_{34} = 25, x_{35} = 15$

Total cost = 3,650.

7. The transportation problem

	1	2	3	4	Availability
1	14	15	16	17	40
2	M	15	17	18	50
3	M	M	15	16	30
4	M	M	M	17	50
Requirements	20	30	50	40	

The production schedule is

Month : 1 2 3 4
Production : 40 30 30 40
The cost is ₹2,210.

Assignment Problems

14

14.1 INTRODUCTION

An assignment problem is a special case of transportation problem. It arises because available resources have varying degrees of efficiency for performing different activities *i.e.* cost, profit or time of performing the different activities is different. This type of problem where the main motive is to allot a number of origins to the equal number of destinations at a least cost are called assignment problems.

Let us suppose there are *n* people and they are to be assigned to *n* jobs such that each person can do each job at a time with varying degree of efficiency. Let c_{ij} be the cost of assigning i^{th} person to the jth job, then the objective of assignment problem is to find an assignment (*i.e.* which job should be assigned to which particular person) that minimize the total cost of doing all the jobs.   [MEERUT–2018; ROHILKHAND–2000]

14.2 MATHEMATICAL REPRESENTATION OF ASSIGNMENT PROBLEM

[GORAKHPUR–2010; KANPUR–2012; MEERUT–1996; ROHILKHAND–2000]

The assignment problem can be stated in the form of $n \times n$ matrix. This, in general can be defined as the square transportation problem in which the no. of origins equal the number of destinations.

It can be stated mathematically as follows:

"To determine x_{ij} for i = 1, 2, ..., n and j = 1, 2, ..., n which minimize the total cost

$$Z = \sum_{i=1}^{n} \sum_{j=1}^{n} c_{ij} x_{ij}$$

subject to the constraints

$$\sum_{j=1}^{n} x_{ij} = 1 \quad (only\ one\ job\ is\ done\ by\ i^{th}\ person)$$

and

$$\sum_{i=1}^{n} x_{ij} = 1 \quad (only\ one\ person\ should\ be\ assigned\ to\ j^{th}\ job)$$

where

$$x_{ij} = \begin{cases} 1, if\ i^{th}\ person\ is\ assigned\ to\ j^{th}\ job \\ 0, otherwise \end{cases}$$

In matrix form, an assignment problem can be written as follows.

	1	2	...	j	...	n
1	c_{11}	c_{12}	...	c_{1j}	...	c_{1n}
2	c_{21}	c_{22}	...	c_{2j}	...	c_{2n}
$\vdots$	$\vdots$	$\vdots$				
i	c_{i1}	c_{i2}	...	c_{ij}	...	c_{in}
$\vdots$	$\vdots$	$\vdots$				
n	c_{n1}	c_{n2}	...	c_{nj}	...	c_{nn}

14.3 DIFFERENCE BETWEEN TRANSPORTATION AND ASSIGNMENT PROBLEM

[MEERUT–2006, 07; KANPUR–2011, 12]

The assignment problem is a variation of the transportation problem with two characteristics
  (i) The cost matrix is a square matrix
 (ii) The optimal solution for the problem would always be such that there would be only one assignment in a given row or column of the cost matrix.
In tabular form, these differences can be written as follows:

	Transportation Problem		Assignment Problem
1.	There are m sources and n destinations.	1.	There should be m persons (sources) and n jobs (destinations) (*i.e.*, $m = n$).
2.	For i^{th} source, the supply is a_i and for j^{th} destination the demand is b_j.	2.	Here, $a_i = 1$ for each person i and $b_j = 1$ for each job j.
3.	The quantity $x_{ij} \geq 0$	3.	Here, $x_{ij} = 1$ when i^{th} person is assigned to j^{th} job and $x_{ij} = 0$ when i^{th} person is not assigned to j^{th} job.
4.	It is unbalanced if $\sum\limits_{i=1}^{m} a_i \neq \sum\limits_{j=1}^{n} b_j$	4.	It is unbalanced if the number of persons is not equal to the number of jobs.

14.4 THEOREMS ON ASSIGNMENT PROBLEM

THEOREM I. **(Reduction Theorem)** *In an assignment problem, if a constant is added to or subtracted from every element of any row or column in the given cost matrix, an assignment that minimize the total cost in one matrix also minimize the total cost in the other matrix.*

[MEERUT–1996, 2005, 10; BHOPAL–2002; ROHILKHAND–1997, 99, 2001]

(OR)

If $x_{ij} = X_{ij}$ minimize $z = \sum\limits_{i=1}^{n} \sum\limits_{j=1}^{n} c_{ij} x_{ij}$ *over all $x_{ij} = 0$ or 1 such that*

$$\sum\limits_{i=1}^{n} x_{ij} = 1, \ \sum\limits_{j=1}^{n} x_{ij} = 1 \ \text{then } x_{ij} = X_{ij}$$

also minimize $z' = \sum\limits_{i=1}^{n} \sum\limits_{j=1}^{n} c'_{ij} x_{ij}$ *when $c'_{ij} = c_{ij} \pm a_i \pm b_j$ where $a_i, b_j \in R$.*

PROOF. Consider $Z' = \sum\limits_{i=1}^{n} \sum\limits_{j=1}^{n} c'_{ij} x_{ij} = \sum\limits_{i=1}^{n} \sum\limits_{j=1}^{n} (c_{ij} \pm a_i \pm b_j) x_{ij} = \sum\limits_{i=1}^{n} \sum\limits_{j=1}^{n} c_{ij} x_{ij} \pm \sum\limits_{i=1}^{n} \sum\limits_{j=1}^{n} a_i x_{ij} \pm \sum\limits_{i=1}^{n} \sum\limits_{j=1}^{n} b_j x_{ij}$

$$= Z \pm \sum_{i=1}^{n} a_i \sum_{j=1}^{n} x_{ij} \pm \sum_{j=1}^{n} b_j \sum_{i=1}^{n} x_{ij}$$

$$= Z \pm \sum_{i=1}^{n} a_i \cdot 1 \pm \sum_{j=1}^{n} b_j \cdot 1 = Z \pm \sum_{i=1}^{n} a_i \pm \sum_{j=1}^{n} b_j$$

Since $\sum_{i=1}^{n} a_i$ and $\sum_{j=1}^{n} b_j$ are independent on x_{ij}.

$\Rightarrow Z'$ is minimized whenever Z is minimized and conversely.

Hence, $x_{ij} = X_{ij}$ which minimize Z will also minimize Z'.

THEOREM 2. *In an assignment problem with cost matrix c_{ij} if all $c_{ij} \geq 0$ then a feasible solution x_{ij} which satisfies $\sum_{i=1}^{n} \sum_{j=1}^{n} c_{ij} x_{ij} = 0$ is an optimal solution for the problem.*

PROOF. Since, we have all $c_{ij} \geq 0$ and all $x_{ij} \geq 0$ therefore

$$Z = \sum_{i=1}^{n} \sum_{j=1}^{n} c_{ij} x_{ij} \geq 0$$

$\Rightarrow$ The minimum value of $Z = 0$

Hence, any feasible solution x_{ij} satisfying $\sum_{i=1}^{n} \sum_{j=1}^{n} c_{ij} \cdot x_{ij} = 0$ will be optimal.

14.5 SOLUTION OF ASSIGNMENT PROBLEMS : HUNGARIAN METHOD

[MEERUT–2000, 03, 16; UPTU(MBA)–2002]

This method was developed by Hungarian mathematician D. Konig. It works on the principle of reducing the given cost matrix to a matrix of opportunity cost, which shows the relative penalties associated with assigning a resource to an activity as opposed to making the best or least cost assignment.

In this method, we use the following procedure:

Working Procedure

STEP 1. **(Find the opportunity cost table)**

Subtract the smallest element of each row of the cost matrix $[c_{ij}]$ from all the elements of the respective rows and then modify the reduced cost matrix by subtracting the smallest element of each column from all the elements of the respective columns. These two operations on rows and columns create zeroes.

STEP 2. **(Make the assignments)**

➡ Identify rows successively from top to bottom until a row with exactly one zero element is found. Make an assignment to this single zero by making a square (□) around it and crossed off (×) all other zeroes in the corresponding column.

➡ Identify column successively from left to right with exactly one zero element that has not been assigned. Make assignment to the single zero by making a square (□) around it and crossed-off (×) all other zero elements in the corresponding rows.

➡ If a row (or column) has two or more unmarked zeroes and one cannot be chosen by inspection, then select the zero cell arbitrarily for assignment.

➡ Repeat the above steps successively until all the zeroes have been examined.

STEP 3. **(Optimality check)**

(i) If all zero elements in the matrix are either marked with square (□) or cross-off

(×) and there is exactly one assignment in each row and column then it is an optimal solution. Here the total cost associated with this solution is obtained by adding original cost figures in the occupied cells.

(ii) If a zero element in a row or column was chosen arbitrarily for assignment in the above step (i), there exist an alternate optimal solution.

(iii) If there is no assignment in a row (or column) then it implies that the total number of assignments are less than the no. of rows (or columns) in the square matrix. In such cases go to the next step.

STEP 4. **(Revision of opportunity cost matrix)**

We draw a set of horizontal and vertical lines to cover all the zeroes in the revised cost matrix obtained from step 3 by using the following procedure.

(i) For each row with no assignments mark a tick (✓).

(ii) Examine the marked rows. If any zero element occurs in those rows, mark a tick (✓) to the respective columns containing those zeroes.

(iii) Examine marked columns. If any assigned zero elements occur in those columns tick (✓) the respective rows containing those assigned zeroes.

(iv) Repeat this process until no more rows or columns can be marked.

(v) Draw a straight line through each marked column and each unmarked row.

If the no. of lines drawn (or total assignments) is equal to the number of rows (or columns), the current solution is the optimal solution otherwise go to the next step.

STEP 5. **(Find the new Revised Opportunity Cost matrix)**

(i) Choose the smallest element (say m) among the cells not covered by any line.

(ii) Subtract m from every element in the cell not covered by a line.

(iii) Add m to every element in the cell covered by two lines *i.e.* intersection of two lines.

(iv) Elements in cells covered by one line remains unchanged.

STEP 6. Repeat steps 3 to 5 until an optimal solution is obtained.

Solved Examples

EXAMPLE 1. *A department head has four subordinates and four tasks have to be performed. Subordinates differ in efficiency and tasks differ in their intrinsic difficulties. Time each man would take to perform each task is given in the effectiveness matrix. How the tasks should be allocated to each person so as to minimize the total man-hour?*

	I	II	III	IV
A	8	26	17	11
B	13	28	4	26
C	38	19	18	15
D	19	26	24	10

[KANPUR–2007; ROHILKHAND–1997, 99, 2003, 07]

SOLUTION. (i) **Row Reduced Matrix:** Subtract the smallest element of each row from all the elements of the respective rows, we get the following row-reduced matrix.

$$
\begin{array}{|cccc|}
\hline
0 & 18 & 9 & 3 \\
9 & 24 & 0 & 22 \\
23 & 4 & 3 & 0 \\
9 & 16 & 14 & 0 \\
\hline
\end{array}
$$

(ii) **Column reduced matrix:** Subtract the smallest element of each column from all the elements of respective column, we get

$$
\begin{array}{|cccc|}
\hline
0 & 14 & 9 & 3 \\
9 & 20 & 0 & 22 \\
23 & 0 & 3 & 0 \\
9 & 12 & 14 & 0 \\
\hline
\end{array}
$$

(iii) Now we have to allot zero assignments, starting from the first row. Now, we allot the 0 and cancel the corresponding zeroes in row and column

	I	II	III	IV
A	0	14	9	3
B	9	20	0	22
C	23	0	3	⧸0
D	9	12	14	0

(a) In the first row a 0 is allotted and corresponding row has been cancelled.

(b) In the second row a 0 is allotted and in the same manner corresponding row has been cancelled.

(c) But in the third row, two 0's are present. One is in the III column and other in the IV column.

We have chosen the zero in the III row as our 0 assignment. This is because if we have taken zero in the IV row as our 0 assignment, then zero in the IV row must have also cancelled.

(iv) As we have allotted the 0 assignments, we examine that 0 equals number of row. The solution becomes $A \rightarrow \text{I}$, $B \rightarrow \text{III}$, $C \rightarrow \text{II}$, $D \rightarrow \text{IV}$.

EXAMPLE 2. ***Solve the assignment problem represented by the following matrix***

	I	II	III	IV	V	VI
A	9	22	58	11	19	27
B	43	78	72	50	63	48
C	41	28	91	37	45	33
D	74	42	27	49	39	32
E	36	11	57	22	25	18
F	3	56	53	31	17	28

[KANPUR–2009; MEERUT–1990, 2012; ROHILKHAND–1996]

<u>SOLUTION.</u> **STEP-1.Row Reduced Matrix**

	I	II	III	IV	V	VI
A	0	13	49	2	10	18
B	0	35	29	7	20	5
C	13	0	63	9	17	5
D	4	15	0	22	12	5
E	25	0	46	11	14	7
F	0	53	50	28	14	25

STEP-2.Column Reduced Matrix

	I	II	III	IV	V	VI
A	0	13	49	0	0	13
B	0	35	29	5	10	0
C	13	0	63	7	7	0
D	4	15	0	20	2	0
E	25	0	46	9	4	2
F	0	53	50	26	4	20

STEP-3.Assignment Matrix

	I	II	III	IV	V	VI
A	⊠	13	49	[0]	⊠	13
B	⊠	35	29	5	10	[0]
C	13	0	63	7	7	⊠
D	4	15	[0]	20	2	⊠
E	25	[0]	46	9	4	2
F	[0]	53	50	26	4	20

Here, row 3 and column 5 have no assignments.

STEP-4. To draw the minimum number of lines, we proceed as follows:

    (i) We tick (✓) row 3 in which there is no assignment.

    (ii) We tick (✓) column 2 and 5 which have zeroes in ticked row 3.

    (iii) We tick (✓) rows 5 and 2 which have assignments in the ticked columns 2 and 6.

    (iv) Then we tick column 1 (not already ticked) which has zero in the ticked row 2.

    (v) Then, we tick row 6 which has assignment in the ticked column 1.

    (vi) Draw lines through all ticked columns 1, 2, 6. Then draw lines through unticked row 1 and 4 having zeros through which there is no line. Therefore, we get 5 lines to cover all the zeros.

	L_1	L_2				L_3	
L_4	⊠	13	49	⓪	⊠	13	
	⊠	35	29	5	10	⓪	✓(5)
	13	⊠	63	7	7	⊠	✓(1)
L_5	4	15	⓪	20	2	⊠	
	25	⓪	46	9	4	2	✓(4)
	⓪	53	50	26	4	20	✓(7)
	✓(6)	✓(2)				✓(3)	

STEP-5. Select smallest element *i.e.* 4 that do not have a line through them. Subtracting this element from all the elements that do not have a line through them and adding to every element that lies at the intersection of two lines and leaving the remaining elements unchanged. Further applying step 1 to 3, we get the following assignment matrix

	I	II	III	IV	V	VI
A	4	17	49	⓪	⊠	17
B	⓪	35	25	1	6	⊠
C	13	⊠	59	3	3	0
D	51	19	⓪	20	2	4
E	25	⓪	42	5	⊠	2
F	⊠	53	46	22	⓪	20

Hence, the optimal assignment is given by

$$A \to IV, \ B \to I, \ C \to VI, \ D \to III, \ E \to II, \ F \to V$$

with Minimum cost $= 11 + 43 + 33 + 27 + 11 + 17 = 142$.

EXAMPLE 3. *Solve the following assignment problems*

Men

		(1)	(2)	(3)	(4)	(5)
	A	11	17	8	16	20
	B	9	7	12	6	15
Job	C	13	16	15	12	16
	D	21	24	17	28	26
	E	24	10	12	11	15

SOLUTION. **(i) Row reduced matrix:** First we find out the row reduced matrix by subtracting from each element the minimum element of its row.

	(1)	(2)	(3)	(4)	(5)
A	3	9	0	8	12
B	3	1	6	0	9
C	1	4	3	0	4
D	4	7	0	11	9
E	14	0	2	1	5

(ii) **Column reduced matrix.** This can be found out by subtracting the minimum element from other elements of every column by taking row reduced matrix as the primary matrix

	(1)	(2)	(3)	(4)	(5)
A	2	9	0	8	8
B	2	1	6	0	5
C	0	4	3	0	0
D	3	7	0	11	5
E	3	0	2	1	1

(iii) **Assignment matrix**

	(1)	(2)	(3)	(4)	(5)	
A	2	9	0	8	8	✓
B	2	1	6	0	5	✓
C	0	4	3	0	0	
D	3	7	0	11	5	✓
E	3	0	2	1	1	

Since the number of 0 assignments is less than 5 *i.e.* the number of rows. Hence, the above is not the required solution.

So, now we find out the minimum element from the untouched elements of the assignment matrix or model. Subtract this element from all the untouched elements. Add this element to those elements which are at the intersection of two lines. Rest of the elements remain the same.

So, minimum element = 1.

The new assignment model becomes

	(1)	(2)	(3)	(4)	(5)	
A	1	8	0	8	7	✓
B	1	0	6	0	4	
C	0	4	4	1	0	
D	2	6	0	11	4	✓
E	3	0	3	2	0	

Number of 0 assignments ≠ 5.

So, again proceeding in the same way, we have minimum element = 1

	(1)	(2)	(3)	(4)	(5)	
A	0	7	0	7	6	
B	1	0	7	0	4	
C	0	4	5	1	0	✓
D	1	5	0	10	3	
E	3	0	4	2	0	

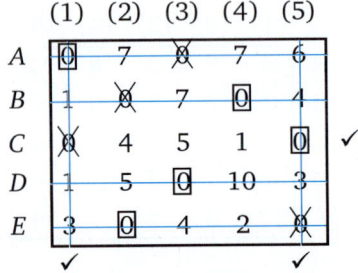

Number of $\boxed{0}$ assignments = 5.

The required solution becomes

$$A \to I, \ B \to IV, \ C \to V, \ D \to III, \ E \to II$$

EXAMPLE 4. *Five wagons are available at five stations 1, 2, 3, 4, 5. These are required at five stations I, II, III, IV and V. The mileages between various stations are given by the following matrix:*

To station

	I	II	III	IV	V
1	10	5	9	18	11
2	13	9	6	12	14
3	3	2	4	4	5
4	18	9	12	17	15
5	11	6	14	19	10

From station

How should the wagons be transported so as to minimize the total mileage covered?

SOLUTION. **Row reduced matrix**

	I	II	III	IV	V
1	5	0	4	13	6
2	7	3	0	6	8
3	1	0	2	2	3
4	9	0	3	8	6
5	5	0	8	13	4

Column reduced matrix

	I	II	III	IV	V
1	4	0	4	11	3
2	6	3	0	4	5
3	0	0	2	0	0
4	8	0	3	6	3
5	4	0	8	11	1

Assignment model

	I	II	III	IV	V
1	4	0	4	11	3
2	6	3	0	4	5
3	0	0	2	0	0
4	8	0	3	6	3
5	4	0	8	11	1

Number of $\boxed{0}$ assignments $\neq$ 5.

Again, proceeding for $\boxed{0}$ assignments, we have minimum element = 1.

New Assignment Model

The row reduced and column reduced matrix are same as the assignment model.

So, we are not writing them here again.

	I	II	III	IV	V
1	3	0	3	10	2
2	6	4	0	4	5
3	0	1	2	0	0
4	7	0	2	5	2
5	3	0	7	10	0

Again, number of 0 assignments ≠ 5.

So, the new minimum element = 2. Subtract 2 from all the untouched elements. Add this element which are at the intersection of two lines. Rest of the elements remains the same.

New Assignment Model

	I	II	III	IV	V
1	1	0	1	8	2
2	6	6	0	4	5
3	0	3	2	0	0
4	6	0	0	3	0
5	3	2	7	10	0

Again, number of 0 assignments ≠ 5.

∴   Minimum element = 3.

New Assignment Model becomes

	I	II	III	IV	V
1	1	3	4	8	3
2	3	6	0	1	5
3	0	3	3	0	0
4	3	0	0	0	0
5	0	2	4	7	0

But the above matrix is not in the reduced form. So, we proceed as follows

Row reduced Matrix

	I	II	III	IV	V
1	0	2	3	7	2
2	3	6	0	1	5
3	0	3	3	0	0
4	3	0	0	0	0
5	0	2	4	7	0

Since the column reduced matrix and the assignment model for the row reduced matrix were same, so we have given 0 assignments in the row reduced matrix itself.

∴   1 → I, 2 → III, 3 → IV, 4 → II, 5 → V

Minimum distance = 10 + 6 + 4 + 9 + 10

= 39 miles.

EXAMPLE 5. *An airline that operates seven days a week, has the time table shown below. Crews must have a minimum layover time of 1 hour between flights. Obtain the pairing of flights that minimizes layover time away from home. For any given pairing, the crew will be based at the city that results in the smaller layover. For each pair, also mention the town where crew should be based.*

(i)	Delhi-Srinagar		Srinagar-Delhi		
Flight No.	Depart	Arrive	Flight No.	Depart	Arrive
1	7:30 A.M.	9:00 A.M.	2	7:00 A.M.	10:00 A.M.
3	8:45 A.M.	9:45 A.M.	4	7:45 A.M.	10:45 A.M.
5	2:00 P.M.	3:30 P.M.	6	11:00 A.M.	2:00 P.M.
7	5:45 P.M.	7:15 P.M.	8	6:00 P.M.	9:00 P.M.
9	7:00 P.M.	8:30 P.M.	10	7:30 P.M.	10:30 P.M.

(ii)	Delhi-Calcutta		Calcutta-Delhi		
Flight No.	Depart	Arrive	Flight No.	Depart	Arrive
1	7:00 A.M.	9:00 A.M.	101	9:00 A.M.	11:00 A.M.
2	9:00 A.M.	11:00 A.M.	102	10:00 A.M.	12:00 Noon
3	1:30 P.M.	3:30 P.M.	103	3:30 P.M.	5:30 P.M.
4	7:30 P.M.	9:30 P.M.	104	8:00 P.M.	10:00 P.M.

The minimum layover time is 6 hours for part (ii).   [MEERUT-2002, 03, 09]

SOLUTION.   (i)   First, we must construct the assignment model for both the flights when their layover time is 1 hour.

Crew based at Delhi

	2	4	6	8	10
1	22	22.75	2	9	10.5
3	21.25	22	1.25	8.25	9.75
5	15.5	16.25	19.25	2.5	4
7	11.75	12.5	15.75	22.75	23.75
9	10.5	11.25	14.5	21.5	23

Crew based at Srinagar

	2	4	6	8	10
1	20.5	20.75	17.5	10.5	9
3	22.75	21	18.75	11.75	10.25
5	4	3.25	24	17	15.5
7	7.75	7	3.75	20.75	19.25
9	9	8.25	5	22	20.5

Now, we combine the above two matrices, choosing the base which gives minimum layover time. In the composite matrix, the cell marked (*) means crew based at Srinagar, otherwise it is Delhi.

	2	4	6	8	10
1	20.5 *	20.75 *	2	9	9 *
3	21.25	21 *	1.25	8.25	9.75
5	4 *	3.25 *	19.25	2.5	4
7	7.75 *	7 *	3.75 *	20.75 *	19.25 *
9	9 *	8.25 *	5 *	21.50	20.5 *

Now, we proceed as usual to solve the obtained assignment matrix

	Row reduced matrix						**Column reduced matrix**			

	2	4	6	8	10		2	4	6	8	10
1	18.5*	18.75*	0	7	7*	1	17.0*	18	0	7	5.5*
3	20	19.75*	0	12.75	11.25	3	18.5	19*	0	12.75	9.75
5	1.5*	0.75*	16.75	0	1.5	5	0*	0	16.75	0	0
7	4*	3.25*	0*	17*	15.5*	7	2.5*	2.5*	0*	17.0*	14.0*
9	4*	3.25*	0*	16.5	15.5*	9	2.5*	2.5*	0*	16.5	14.0*

Assignment matrix

	2	4	6	8	10
1	17.0*	18	[0]	7	5.5*
3	18.5	19*	0	12.75	9.75
5	[0]*	0	16.75	0	0
7	2.5*	2.5*	0*	17.0*	14.0*
9	2.5*	2.5*	0*	16.5	14.0*

Since number of [0] assignment $\neq$ 5. Hence, minimum element = 2.5.
New Assignment Model can be obtained

	2	4	6	8	10
1	14.5*	15.5*	[0]	4.5	3*
3	16.0	16.5*	0	10.25	7.25
5	0*	0*	19.25	[0]	0
7	[0]*	0*	0*	14.5*	11.5*
9	0*	[0]*	0*	14.0	11.5*

Again, number of [0] assignments $\neq$ 5.
So, minimum element = 3.
We again form new assignment model by applying usual procedure.

	2	4	6	8	10
1	14.5*	15.5*	0	1.5	0*
3	16	16.5*	0	7.25	4.25
5	3*	3*	22.25	0	0
7	3*	3*	3*	14.5*	11.5*
9	3*	3*	3*	14	11.5*

But, it is not in the reduced form. So, we first reduce it to row reduced matrix and column reduced matrix and then allot the assignments:

Assignment matrix

	2	4	6	8	10
1	14.5*	15.5*	0	1.5	[0]*
3	16	16.5*	[0]	7.25	4.25
5	3*	3*	22.25	[0]	0
7	[0]*	0*	0*	11.5*	8.5*
9	0*	[0]*	0*	11	8.5*

Since row reduced matrix, column reduced matrix and assignment model are same, so we have not mentioned them here.

So, $1 \to 10$, $3 \to 6$, $5 \to 8$, $7 \to 2$, $9 \to 4$.

Minimum layover time = 28.75 hours.

(ii) So, proceeding as usually, we first construct the tables for the layover time between two flights

Layover time in hours

When crew based at Delhi					**When crew based at Calcutta**			
	101	102	103	104		101	102	103	104

	101	102	103	104
1	24	25	6.5	11
2	22	23	28.5	9
3	17.5	18.5	24	28.5
4	11.5	12.5	18	22.5

	101	102	103	104
1	20	22	26.5	8.5
2	19	21	25.5	7.5
3	13.5	15.5	20	26
4	9	11	15.5	21.5

Now, we combine the above two tables, choosing that base which gives a lesser layover time for each pair. But before that, we multiply the above matrices by 4, to reduce the fractional complications and considering the layover time for 4 weeks.

When crew based at Delhi

	101	102	103	104
1	96	100	26	44
2	88	92	114	36
3	70	74	96	114
4	46	50	72	90

When crew based at Calcutta

	101	102	103	104
1	80	76	54	36
2	88	84	62	44
3	106	102	80	62
4	34	30	104	86

In the composite the cell marked (*) means that the crew is based at Calcutta.

	101	102	103	104
1	80 *	76	26	36 *
2	88 *	84 *	62 *	36
3	70	74	80 *	110 *
4	34 *	30 *	72	86 *

Now, we follow our usual method of solving the assignment problem.

Using general method of solving the assignment problem, we get the final assignment model as

54 *	50 *	0	10 *
52 *	48 *	26 *	0
0	4	10 *	40 *
4 *	0 *	42	56 *

Allotting the 0 assignments, we have

$$\begin{array}{cccc} 54* & 50* & \boxed{0} & 10* \\ 52* & 48* & 26* & \boxed{0} \\ \boxed{0} & 4 & 10* & 40* \\ 4* & \boxed{0}* & 42 & 56* \end{array}$$

Hence, the optimal assignment with the base of the crew is as follows:

$$1 \to 103, \quad 2 \to 104, \quad 3 \to 101, \quad 4 \to 103$$

Crew at   (Delhi)   (Delhi)   (Delhi)   (Calcutta)

14.6   THE MAXIMAL ASSIGNMENT PROBLEM

So far, we have been dealing with the assignment problems which involve minimization. But, sometimes, it may happen that instead of being minimized, the problem becomes of to be maximized. In this case, we may follow the slightly different but simple method for the solution, we can proceed as given below:

(a) We must first select the greatest element of our given assignment problem and must subtract each element from this element to obtain a new matrix. The new matrix obtained can now be solved by the usual method of assignment technique used for minimal problems.

(b) One another method is also used for the conversion of maximization to minimization. Place minus sign before each element of the given assignment matrix to obtain a new matrix. The new matrix thus can be treated as the usual assignment minimal problem.

 Solved Examples

EXAMPLE 1. *Alpha Corporation has four plants each of which can manufacture any of the four products. Production costs differ from plant to plant as do sales revenue. From the following data, obtain which product each plant should produce to maximize profit?*

Sales revenue (₹'000)					Production cost (₹'000)				
	Product					Product			
Plant ↓	1	2	3	4	Plant ↓	1	2	3	4
A	50	68	49	62	A	49	60	45	61
B	60	70	51	74	B	55	63	45	69
C	55	67	53	70	C	52	62	49	68
D	58	65	54	69	D	55	64	48	66

[ICWA–1985]

SOLUTION. In this case, we must first find out the Profit Matrix

Profit Matrix = Sales revenue – Production cost

	1	2	3	4
A	1	8	4	1
B	5	7	6	5
C	3	5	3	2
D	3	1	6	3

Since profit matrix is a maximization problem. Hence, first we convert it into minimization.

Minimization Problem

	1	2	3	4
A	7	0	4	7
B	3	1	2	3
C	5	3	5	6
D	5	7	2	5

Row Reduced Matrix

	1	2	3	4
A	7	0	4	7
B	2	0	1	2
C	2	0	2	3
D	3	5	0	3

Column Reduced Matrix

	1	2	3	4
A	5	0	4	5
B	0	0	1	0
C	0	0	2	1
D	1	5	0	1

Assignment matrix

	1	2	3	4
A	5	[0]	4	5
B	0	0	1	[0]
C	[0]	0	2	1
D	1	5	[0]	1

Since number of assignments = 4. Hence, the required solution is

$A \rightarrow 2,\ B \rightarrow 4,\ C \rightarrow 1,\ D \rightarrow 3$

Profit = $(8 + 5 + 3 + 6) \times 1000 = ₹\ 22000$.

14.7 UNBALANCED ASSIGNMENT PROBLEM

Sometimes, during a problem, the number of jobs is not equal to number of persons, that means our assignment matrix is not square. Such kind of problems are known as unbalanced assignment problem.

These type of problems are solved by the introduction of a supposed or dummy rows or columns in which each element has zero cost to form a square matrix. The new square matrix can be solved by the usual method of balanced minimal assignment problem.   [MEERUT–2004, 07, 09, 12]

Solved Examples

EXAMPLE I. *A company has 4 machines on which to do 3 jobs. Each job can be assigned to one and only one member. The cost of each job on each machine is given in the following table*   [GAUHATI(MCA)–1992]

Job↓	Machine			
	W	X	Y	Z
A	18	24	28	32
B	8	13	17	19
C	10	15	19	22

What are the job assignments that will minimize the cost?

SOLUTION. The given problem is an unbalanced problem, since the number of jobs does not equals the number of machines. So, we add a dummy row (Job) with elements of zero cost. So, we have

$$
\begin{array}{c|cccc}
 & W & X & Y & Z \\
\hline
A & 18 & 24 & 28 & 32 \\
B & 8 & 13 & 17 & 19 \\
C & 10 & 15 & 19 & 22 \\
D & 0 & 0 & 0 & 0 \\
\end{array}
$$

Row Reduced Matrix

$$
\begin{array}{c|cccc}
 & W & X & Y & Z \\
\hline
A & 0 & 6 & 10 & 14 \\
B & 0 & 5 & 9 & 11 \\
C & 0 & 5 & 9 & 12 \\
D & 0 & 0 & 0 & 0 \\
\end{array}
$$

Column Reduced Matrix

$$
\begin{array}{c|cccc}
 & W & X & Y & Z \\
\hline
A & 0 & 1 & 1 & 3 \\
B & 0 & 0 & 0 & 0 \\
C & 0 & 0 & 0 & 1 \\
D & 0 & 0 & 0 & 0 \\
\end{array}
$$

Assignment matrix

$$
\begin{array}{c|cccc}
 & W & X & Y & Z \\
\hline
A & \boxed{0} & 1 & 1 & 3 \\
B & 0 & \boxed{0} & 0 & 0 \\
C & 0 & 0 & \boxed{0} & 1 \\
D & 0 & 0 & 0 & \boxed{0} \\
\end{array}
$$

So, solution becomes $A \to W, B \to X, C \to Y, D \to Z$
But, D is only assumed job. Hence, we have
$$A \to W, B \to X, C \to Y$$

Exercise-14.1

1. Solve the following assignment problems :

(a)

Men

	I	II	III	IV	V
A	1	3	2	8	8
B	2	4	3	1	5
Tasks C	5	6	3	4	6
D	3	1	4	2	2
E	1	5	6	5	4

(b)   [MEERUT–2005, 16; ROHILKHAND–2000]

Persons

	1	2	3	4
A	10	12	19	11
Tasks B	5	10	7	8
C	12	14	13	11
D	8	15	11	9

(c)

Men

	1	2	3	4	5
I	12	8	7	15	4
II	7	9	17	14	10
Tasks III	9	6	12	6	7
IV	7	6	14	6	10
V	9	6	12	10	6

2. Find the minimum cost solution for the 5×5 assignment problem whose cost coefficients are as given below :

	1	2	3	4	5
1	–2	–4	–8	–6	–1
2	0	–9	–5	–5	–4
3	–3	–8	0	–2	–6
4	–4	–3	–1	0	–3
5	–9	–5	–8	–9	–5

3. The owner of a small machine shop has four machinists available to assign to jobs for the day. Five jobs are offered with the expected profit (in ₹) for each machinist on each job being as follows :

	A	B	C	D	E
1	6.20	7.80	5.00	10.10	8.20
Mechinist 2	7.10	8.40	6.10	7.30	5.90
3	8.70	9.20	11.10	7.10	8.10
4	4.80	6.40	8.70	7.70	8.00

Find out assignment of mechinists to jobs that will result in a maximum profit. Which job should be declined?

4. Use Hungarian method to solve the following cost minimizing assignment problem.

	1	2	3	4
I	20	22	28	15
II	16	20	12	13
III	19	23	14	25
IV	10	16	12	10

5. Solve

Jobs

	J_1	J_2	J_3	J_4	J_5
P_1	7	8	6	5	9
Persons P_2	9	6	7	6	10
P_3	8	7	9	5	6

6. Find the optimal assignment

Programs

	A	B	C
1	120	100	80
Programmers 2	70	90	110
3	110	140	120

Assign the programs to the programmers so that the total time taken is least. Here, the elements of the matrix are the time in minutes.

7. Solve the assignment problem

Location

	I	II	III	IV	V
A	15	21	6	4	9
B	3	40	21	10	7
Project C	9	6	5	8	10
D	14	8	6	9	3
E	21	16	18	7	4

8. Solve the assignment problem

Machine

		M_1	M_2	M_3
	J_1	8	7	6
Job	J_2	5	7	8
	J_3	6	8	7

ANSWERS

1. (a) $A \to I, B \to IV, C \to III, D \to II, E \to V$, Min. cost = 10.
   (b) $A \to 2, B \to 3, C \to 4, D \to I$, Minimum cost = 38.
   (c) (i) $I \to 3, II \to 1, III \to 2, IV \to 4, V \to 5$
       (ii) $I \to 3, II \to 1, III \to 4, IV \to 2, V \to 5$

2. $1 \to 3, 2 \to 2, 3 \to 5, 4 \to 4, 5 \to 1$ or $1 \to 4, 2 \to 2, 3 \to 3, 4 \to 5, 5 \to 1$, Min. cost = 36

3. $1 \to D, 2 \to B, 3 \to C, 4 \to E, 5 \to A$, Minimum cost = ₹ 37.60, Job A should be declined.

4. $1 \to 2, II \to 4, III \to 3, IV \to 1$ or $I \to 4, II \to 2, III \to 3, IV \to 1$

5. $P_1 \to J_4, P_2 \to J_2, P_3 \to J_5$; Minimum cost = 17, Jobs J_1 and J_2 left undone.

6. $1 \to C, 2 \to B, 3 \to A$, Minimum cost = 280.

7. $A \to IV, B \to I, C \to II, D \to III, E \to V$.

8. $J_1 \to M_3, J_2 \to M_2, J_3 \to M_1$ or $J_1 \to M_3, J_2 \to M_1, J_3 \to M_2$, Min. cost = 19

14.8 SOME MISCELLANEOUS PROBLEMS

EXAMPLE 1. *Solve the following minimal assignment problem.*

Men

		1	2	3	4
	I	12	30	21	15
Job	II	18	33	9	31
	III	44	25	24	21
	IV	23	30	28	14

SOLUTION. **Row Reduced Matrix.** Subtracting the smallest element of each row from every element of the corresponding row, we get the following row reduced matrix

	1	2	3	4
I	0	18	9	3
II	9	24	0	22
III	23	4	3	0
IV	9	16	14	0

Column Reduced Matrix. Subtracting smallest element of each column from every element of the corresponding column, we get the following column reduced matrix

	1	2	3	4
I	0	14	9	3
II	9	20	0	22
III	23	0	3	0
IV	9	12	14	0

Assignment Matrix. Starting with row I, we mark ☐ in the row containing only one zero and cross (×) the zeroes in the corresponding column in which ☐ lies.

	1	2	3	4
I	0	9	14	3
II	9	20	0	22
III	23	0	3	X
IV	9	12	14	0

Further, starting with column 1, we mark ☐ in the column containing only one unmarked or uncrossed zero in the above table and cross out the zeroes in the corresponding row in which the assignment ☐ is marked. Then, we get the following table.

	1	2	3	4
I	0	14	9	3
II	9	20	0	22
III	23	0	3	X
IV	9	12	14	0

In this table, every row and column have one assignment, therefore we have the following optimal assignment.

$$I \rightarrow 1, \ II \rightarrow 3, \ III \rightarrow 2, \ IV \rightarrow 4$$

EXAMPLE 2. *Solve the minimal assignment problem whose effectiveness matrix is given below*

	1	2	3	4
I	2	3	4	5
II	4	5	6	7
III	7	8	9	8
IV	3	5	8	4

SOLUTION. **Row Reduced Matrix**

	1	2	3	4
I	0	1	2	3
II	0	1	2	3
III	0	1	2	1
IV	0	2	5	1

Column Reduced Matrix

	1	2	3	4
I	0	0	0	2
II	0	0	0	2
III	0	0	0	0
IV	0	1	3	0

Assignment Matrix

	1	2	3	4
I	$\cancel{0}$	0	0	2
II	$\cancel{0}$	0	0	2
III	$\cancel{0}$	0	0	0
IV	$\boxed{0}$	1	3	$\cancel{0}$

$\Rightarrow$

	1	2	3	4
I	$\cancel{0}$	$\boxed{0}$	$\cancel{0}$	2
II	$\cancel{0}$	$\cancel{0}$	$\boxed{0}$	2
III	$\cancel{0}$	$\cancel{0}$	$\cancel{0}$	$\boxed{0}$
IV	$\boxed{0}$	1	3	$\cancel{0}$

Hence, optimal assignment is given by

$I \to 2$, $II \to 3$, $III \to 4$, $IV \to 1$ and minimum cost $= 3 + 6 + 8 + 3 = 20$.

EXAMPLE 3. *Solve the following assignment problem* [MEERUT–2009, 12]

	I	II	III	IV	V
A	1	3	2	3	6
B	2	4	3	1	5
C	5	6	3	4	6
D	3	1	4	2	2
E	1	5	6	5	4

SOLUTION. The row and column reduced matrix is given by

	I	II	III	IV	V
A	0	2	1	2	4
B	1	3	2	0	3
C	2	3	0	1	2
D	2	0	3	1	0
E	0	4	5	4	2

The assignment matrix is given by

	I	II	III	IV	V
A	$\boxed{0}$	2	1	2	4
B	1	3	2	$\boxed{0}$	3
C	2	3	$\boxed{0}$	1	2
D	2	$\boxed{0}$	3	1	$\cancel{0}$
E	$\cancel{0}$	4	5	4	2

Since row 4 and column 5 have no assignments, so we proceed as follows :

L_1						
$\boxed{0}$	2	1	2	4	$\checkmark(3)$	
1	3	2	$\boxed{0}$	3	L_2	
2	3	$\boxed{0}$	1	2	L_3	
2	$\boxed{0}$	3	1	$\cancel{0}$	L_4	
$\cancel{0}$	4	5	4	2	$\checkmark(3)$	
$\checkmark(3)$						

The smallest element that do not contain line through them is 1. Subtracting 1 from the elements that do not have a line through them, adding to every element that lies at the intersection of two lines and leaving the remaining elements unchanged and then repeating the above steps, we get the following matrix

	I	II	III	IV	V
A	Ø	1	Ø	1	3
B	2	3	2	0	3
C	3	3	0	1	2
D	3	0	3	1	Ø
E	0	3	4	3	1

Here row 1 and column 5 do not contain any assignments. Therefore, we again repeat the above steps. We draw the lines as follows

The minimum number of lines can be drawn as follows

Here, we see that even now row 1 and column 5 do not contain any assignment. So, we proceed as follows

Take the minimum element, *i.e.,* 1 and repeat the same process as above, we get

	I	II	III	IV	V
A	0	Ø	Ø	Ø	2
B	3	3	2	0	3
C	3	2	0	Ø	1
D	4	0	4	1	Ø
E	Ø	2	4	2	0

Hence, the optimal assignment is given by
$$A \rightarrow I, \ B \rightarrow IV, \ C \rightarrow III, \ D \rightarrow II, \ E \rightarrow V$$
with Minimum cost $= 1 + 1 + 3 + 1 + 4 = 10$.

EXAMPLE 4. ***Find the optimal assignment for the following problem*** [MEERUT–2007]

	I	II	III	IV
A	5	3	1	8
B	7	9	2	6
C	6	4	5	7
D	5	7	7	6

SOLUTION. **Row reduced matrix.** Find the smallest element of each row and subtract it from each element of the row, we get

	I	II	III	IV
A	4	2	0	7
B	5	7	0	4
C	2	0	1	3
D	0	2	2	1

Column reduced matrix. Find the smallest element of each column and subtract it from each element of that column, we get

	I	II	III	IV
A	4	2	0	6
B	5	7	0	3
C	2	0	1	2
D	0	2	2	0

Assignment matrix. Making an assignment to single zero, we get

	I	II	III	IV
A	4	2	[0]	6
B	5	7	⊠	3
C	2	[0]	1	2
D	[0]	2	2	⊠

Clearly row B has no assignment. So, tick that row which has no assignment corresponding to that row, tick that column which has cross zero and corresponding to that column, tick that row which has assigned zero

	I	II	III	IV	
A	4	2	0	6	✓
B	5	7	⊠	3	✓
C	2	[0]	1	2	
D	[0]	2	2	⊠	

Cross unticked rows and ticked column. Find smallest element from the remaining matrix. Subtract this smallest element from the unticked elements. Then proceeding same as usual, we get the next reduced matrix as follows :

	I	II	III	IV
A	2	0	0	4
B	3	5	0	1
C	2	0	3	2
D	0	2	4	0

Then, we have the following single zero assignment matrix.

	I	II	III	IV	
A	2	⊠	⊠	4	✓
B	3	5	[0]	1	✓
C	2	[0]	3	2	✓
D	[0]	2	4	⊠	

$$
\Rightarrow \quad
\begin{array}{c|cccc}
 & I & II & III & IV \\
\hline
A & 1 & 0 & \boxed{0} & 3 \\
B & 2 & 5 & 0 & \boxed{0} \\
C & 1 & \boxed{0} & 3 & 1 \\
D & \boxed{0} & 3 & 5 & 0 \\
\end{array}
$$

Hence, final assignment is given by
$$A \to III, \ B \to IV, \ C \to II, \ D \to I.$$

EXAMPLE 5. *Solve the following assignment problem*

$$
\begin{array}{c|cccc}
 & 1 & 2 & 3 & 4 \\
\hline
A & 10 & 12 & 19 & 11 \\
B & 5 & 10 & 7 & 8 \\
C & 12 & 14 & 13 & 11 \\
D & 8 & 15 & 19 & 9 \\
\end{array}
$$

[GORAKHPUR–2007; MEERUT–2008; KANPUR–2012]

SOLUTION. **Row reduced matrix.**

$$
\begin{array}{c|cccc}
 & 1 & 2 & 3 & 4 \\
\hline
A & 0 & 2 & 9 & 1 \\
B & 0 & 5 & 2 & 3 \\
C & 1 & 3 & 2 & 0 \\
D & 0 & 7 & 3 & 1 \\
\end{array}
$$

Column reduced matrix.

$$
\begin{array}{c|cccc}
 & 1 & 2 & 3 & 4 \\
\hline
A & 0 & 0 & 7 & 1 \\
B & 0 & 3 & 0 & 3 \\
C & 1 & 1 & 0 & 0 \\
D & 0 & 5 & 1 & 1 \\
\end{array}
$$

Assignment matrix.

$$
\begin{array}{c|cccc}
 & 1 & 2 & 3 & 4 \\
\hline
A & \cancel{0} & 0 & \boxed{7} & 1 \\
B & \cancel{0} & 3 & 0 & \boxed{3} \\
C & \boxed{1} & 1 & \cancel{0} & 0 \\
D & 0 & 5 & 1 & 1 \\
\end{array}
$$

Hence, the required assignment is given by $A \to 2, \ B \to 3, \ C \to 4, \ D \to 1$.

EXAMPLE 6. *A company is faced with the problem of assigning six different machines to five different jobs. The cost are estimated as follows*

Jobs

		1	2	3	4	5
	1	2.5	5	1	6	1
	2	2	5	1.5	7	3
Machine	3	3	6.5	2	8	3
	4	3.5	7	2	9	4.5
	5	4	7	3	9	6
	6	6	9	5	10	6

Solve the problem assuring that the objective is to minimize total cost.   [MEERUT–2006]

SOLUTION.   It is observed that the given matrix is not square, therefore we must add one fictitious job 6 (sixth column) to make it a square matrix. Therefore, we have the following matrix

	1	2	3	4	5	6
1	2.5	5	1	6	1	0
2	2	5	1.5	7	3	0
3	3	6.5	2	8	3	0
4	3.5	7	2	9	4.5	0
5	4	7	3	9	6	0
6	6	9	5	10	6	0

Rows and Column Reduced Matrix

Subtracting the smallest element of each row from every element of the corresponding row and then subtracting smallest element of each column from every element of the corresponding column, we get the following row and column reduced matrix

	1	2	3	4	5	6
1	0.5	0	0	0	0	0
2	0	0	0.5	1	2	0
3	1	1.5	1	2	2	0
4	1.5	2	1	3	3.5	0
5	2	2	2	3	5	0
6	4	4	4	4	5	0

Assignment Matrix

	1	2	3	4	5	6
1	0.5	[0]	⊠	⊠	⊠	⊠
2	[0]	0	0.5	1	2	⊠
3	1	1.5	1	2	2	[0]
4	1.5	2	1	3	3.5	⊠
5	2	2	2	3	5	⊠
6	4	4	4	4	5	⊠

Observe that the column 3, 4, 5 and rows 4, 5, 6 have no zero assignments. Now we draw the minimum number of lines

$$
\begin{array}{c}
& & & & & & L_3 \\
L_1 & \boxed{0.5} & \boxed{0} & \cancel{0} & \cancel{0} & \cancel{0} & \cancel{0} \\
L_2 & \boxed{0} & 0 & 0.5 & 1 & 2 & \cancel{0} \\
& 1 & 1.5 & 1 & 2 & 2 & \boxed{0} & \checkmark(5) \\
& 1.5 & 2 & 1 & 3 & 3.5 & \cancel{0} & \checkmark(1) \\
& 2 & 2 & 2 & 3 & 5 & \cancel{0} & \checkmark(2) \\
& 4 & 4 & 4 & 4 & 5 & \cancel{0} & \checkmark(3) \\
& & & & & & \checkmark(4)
\end{array}
$$

The smallest element among all uncovered elements in the above table is 1. Subtracting this element 1 from all uncovered elements, adding to every element that lies at the intersection of two lines. Remaining elements remain unchanged. Then reduced matrix is given below

$$
\begin{array}{c|cccccc}
& 1 & 2 & 3 & 4 & 5 & 6 \\
\hline
1 & 0.5 & 0 & 0 & 0 & 0 & 1 \\
2 & 0 & 0 & 0.5 & 1 & 2 & 1 \\
3 & 0 & 0.5 & 0 & 1 & 1 & 0 \\
4 & 0.5 & 1 & 0 & 2 & 2.5 & 0 \\
5 & 1 & 1 & 1 & 2 & 4 & 0 \\
6 & 3 & 3 & 3 & 3 & 4 & 0
\end{array}
$$

Proceeding same as above

$$
\begin{array}{c}
& & & & & & L_5 \\
L_1 & 0.5 & \cancel{0} & \cancel{0} & 0 & \cancel{0} & 1 \\
L_2 & 0 & 0 & 0.5 & 1 & 2 & 1 \\
L_3 & 0 & 0.5 & \cancel{0} & 1 & 1 & \cancel{0} \\
L_4 & 0.5 & 1 & 0 & 2 & 2.5 & \cancel{0} \\
& 1 & 1 & 1 & 2 & 4 & 0 & \checkmark(3) \\
& 3 & 3 & 3 & 3 & 4 & \cancel{0} & \checkmark(1) \\
& & & & & & \checkmark(2)
\end{array}
$$

Here, smallest element among uncovered element is 1.

$$
\Rightarrow
\begin{array}{c}
& L_2 & L_3 & L_4 & & L_5 \\
L_1 & 0.5 & 0 & 0 & 0 & 0 & 2 \\
& 0 & 0 & 0.5 & 1 & 2 & 2 & \checkmark \\
& 0 & 0.5 & 0 & 1 & 1 & 1 & \checkmark \\
& 0.5 & 1 & 0 & 2 & 2.5 & 1 & \checkmark \\
& 0 & 0 & 0 & 1 & 3 & 0 & \checkmark \\
& 2 & 2 & 2 & 2 & 3 & 0 & \checkmark \\
& \checkmark & \checkmark & \checkmark & & \checkmark
\end{array}
$$

Here, smallest element among uncovered element is 1.

	1	2	3	4	5	6
1	1.5	1	1	[0]	⊠	3
2	[0]	⊠	0.5	⊠	1	2
3	⊠	0.5	⊠	⊠	[0]	1
4	0.5	1	[0]	1	1.5	1
5	⊠	[0]	⊠	⊠	2	⊠
6	2	2	2	1	2	[0]

Hence, the optimal solution is given by

$$1 \to 4, \ 2 \to 1, \ 3 \to 5, \ 4 \to 3, \ 5 \to 2$$

with Minimum total cost = 20 *i.e.* ₹2000.

☞ REMARK

- We can also obtain other optimal solutions of the above problem as given below

$$1 \to 4, \ 2 \to 2, \ 3 \to 5, \ 4 \to 3, \ 5 \to 1$$
$$1 \to 5, \ 2 \to 1, \ 3 \to 4, \ 4 \to 3, \ 5 \to 2$$
$$1 \to 5, \ 2 \to 2, \ 3 \to 1, \ 4 \to 3, \ 5 \to 4$$
$$1 \to 5, \ 2 \to 2, \ 3 \to 4, \ 4 \to 3, \ 5 \to 1$$
$$1 \to 5, \ 2 \to 4, \ 3 \to 1, \ 4 \to 3, \ 5 \to 2$$

EXAMPLE 7. *A company has four territories open and four salesmen available for assignment. The territories are not equally rich in their sales potential. It is estimated that a typical salesman on operating in each territory would bring in the following annual sales*

Territory	I	II	III	IV
Annual sales CPs	60000	50000	40000	30000

The four salesmen are also considered to differ in ability, it is estimated that working under the same conditions, their yearly sales would be proportionally as follows

Salesman	A	B	C	D
Proportion	7	5	5	4

If the criterion is maximum expected total sales, the intuitive answer is to assign the best salesman to the richest territory, the next best salesman to the second richest and so on. Verify the answer by assignment technique.   [KANPUR–2010]

SOLUTION. Firstly, we shall construct the effectiveness matrix.

Sum of the proportion of sales of four salesmen = 7 + 5 + 5 + 4 = 21.

Now, taking the salesmen in the four territories sales are as follows :

For A: $\dfrac{7}{21} \times 6, \dfrac{7}{21} \times 5, \dfrac{7}{21} \times 4, \dfrac{7}{21} \times 3$ *i.e.* 42, 35, 28, 21 (By avoiding fractions)

For B: $\dfrac{5}{21} \times 6, \dfrac{5}{21} \times 5, \dfrac{5}{21} \times 4, \dfrac{5}{21} \times 3$ *i.e.* 30, 25, 20, 15

For C: $\dfrac{5}{21} \times 6, \dfrac{5}{21} \times 5, \dfrac{5}{21} \times 4, \dfrac{5}{21} \times 3$ *i.e.* 30, 25, 20, 15

For D: $\dfrac{4}{21} \times 6, \dfrac{4}{21} \times 5, \dfrac{4}{21} \times 4, \dfrac{4}{21} \times 3$, i.e., 24, 20, 16, 12

Therefore, the effectiveness matrix which make the total sales maximum is given by

	I	II	III	IV
A	42	35	28	21
B	30	25	20	15
C	30	25	20	15
D	24	20	16	12

Now, we convert this maximization problem into minimization problem

	I	II	III	IV
A	−42	−35	−28	−21
B	−30	−25	−20	−15
C	−30	−25	−20	−15
D	−24	−20	−16	−12

Proceeding in the usual manner, we get

L_1

0	3	6	9
0	1	2	3
0	1	2	3
0	0	0	0	L_2

Minimum element = 1.

$\Rightarrow$

L_1 L_2

0	2	5	8	✓(4)
✗	0	1	2	✓(5)
✗	✗	1	2	✓(1)
1	✗	0	✗	L_3

✓(2) ✓(3)

Minimum element = 1.

$\Rightarrow$

	I	II	III	IV
A	0	2	4	7
B	✗	0	✗	1
C	✗	✗	0	1
D	2	1	✗	0

or

	I	II	III	IV
A	0	2	4	7
B	✗	✗	0	1
C	✗	0	✗	1
D	2	1	✗	0

Hence, we get the following two optimal solutions

(i) $A \to I, B \to II, C \to III, D \to IV$ (ii) $A \to I, B \to III, C \to II, D \to IV$

Exercise-14.2

1. Solve the following assignment problems :

	I	II	III	IV
A	1	4	6	3
B	9	7	10	9
C	4	5	11	7
D	8	7	8	5

2. There are five jobs to be assigned one each to five machines and the associated cost matrix is as follows :

Machine

		1	2	3	4	5
	A	11	17	8	16	20
	B	9	7	12	6	15
Job	C	13	16	15	12	16
	D	21	24	17	28	26
	E	14	10	12	11	15

Find the optimal solution of this minimal assignment problem.

3. An automobile dealer wishes to put four repairmen to four different jobs. The repairmen have somewhat different kinds of skills and they exhibit different levels of efficiency from one job to the another. The dealer has estimated the number of man hour that would be required for each job-man combinations. This is given in the following matrix.

Job

		A	B	C	D
	1	5	3	2	8
Man	2	7	9	2	6
	3	6	4	5	7
	4	5	7	7	8

Find the optimal solution of this minimum assignment problem.

4. Solve the following assignment problem :

	I	II	III	IV	V
A	6	5	8	11	16
B	1	13	16	1	10
C	16	11	8	8	8
D	9	14	12	10	16
E	10	13	11	8	16

5. Find the optimal solution of the following minimal assignment problem

	I	II	III	IV
A	30	25	26	28
B	26	32	24	20
C	20	22	18	27
D	23	20	21	11

6. Solve the following minimal assignment problem

	1	2	3	4	5
A	8	4	2	6	1
B	0	9	5	5	4
C	3	8	9	2	6
D	4	3	1	0	3
E	9	5	8	9	5

7. Solve the following assignment problem having the following cost matrix

	1	2	3	4	5	6	7
A	35	20	60	41	27	52	44
B	51	39	42	33	65	47	58
C	25	32	53	41	50	36	43
D	32	28	40	46	33	55	49
E	43	36	45	63	57	49	42
F	27	18	31	46	35	42	34
G	48	50	72	59	43	64	58

8. A car hire company has one car at each of five depots a, b, c, d and e. A customer requires a car in each town namely A, B, C, D and E. Distances (in Kms) between depots (origins) and towns (destinations) are given in the following distance matrix.

	a	b	c	d	e
A	160	130	175	190	200
B	135	120	130	160	175
C	140	110	155	170	185
D	50	50	80	80	110
E	54	34	70	80	105

How should car be assigned to customers so as to minimize the distance travelled.

[DELHI(MBA)–1999, 2002(Similar)]

9. Solve the following cost minimizing jobs problem

	1	2	3	4	5
A	11	10	18	5	9
B	14	13	12	19	6
C	5	3	4	2	4
D	15	18	17	9	12
E	10	11	19	6	14

10. Find the optimal solution of the following problem

Machine

		1	2	3
Job	1	5	7	9
	2	14	10	12
	3	15	13	16

11. Use the Hungarian method to find which of the two jobs should be left undone when each of the four persons will do only one job in the following cost minimizing assignment problem.

Job

		J_1	J_2	J_3	J_4	J_5	J_6
Person	P_1	10	9	11	12	8	5
	P_2	12	10	9	11	9	4
	P_3	8	11	10	7	12	6
	P_4	10	7	8	10	10	5

[MEERUT–1996]

12. There are three persons P_1, P_2 and P_3 and five jobs $J_1, J_2, ..., J_5$. Each person can do only one job and a job is to be done by one person only. Using Hungarian method, find which two jobs should be left undone in the following cost minimizing problem.

	J_1	J_2	J_3	J_4	J_5
P_1	7	8	6	5	9
P_2	9	6	7	6	10
P_3	8	7	9	5	6

[MEERUT–1997, 98]

13. In a machine shop a supervisor wishes to assign five jobs among six machines. Any one of the jobs can be processed completely by any one of the machines as given below

Machines

		A	B	C	D	E	F
	1	13	13	16	23	19	9
	2	11	19	26	16	17	18
Job	3	12	11	4	9	6	10
	4	7	15	9	14	14	13
	5	9	13	12	8	14	11

The assignment of jobs to machines be on a one-to-one basis. Assign the jobs to machines so that the total cost is minimum. Find the minimum total cost. [IAS–1998]

14. Solve the following maximum assignment problem

	A	B	C	D	E
1	32	38	40	28	40
2	40	24	28	21	36
3	41	27	33	30	37
4	22	38	41	36	36
5	29	33	40	65	39

[MEERUT–2009; AGRA–1998; UPTU(MBA)–2006; DELHI(MBA)–1989]

15. A company has five jobs to be done. The following matrix shows the return in Rupees assigning i^{th} (i = 1, 2, 3, ..., 5) machine to the jth job (j = 1, 2, ..., 5). Assign the five jobs to the five machines so as to maximize the total return. [MEERUT–2010; AGRA–2002]

Machine

		1	2	3	4	5
	1	5	11	10	12	4
	2	2	4	6	3	5
Job	3	3	12	5	14	6
	4	6	14	4	11	7
	5	7	9	8	12	5

16. Four engineers are available to design four projects. Engineer 2 is not competent to design the project B. Given the following time estimates needed to each engineer to design a given project, find how should the engineers be assigned to projects so as to minimize the total design time of four projects.

[MEERUT–1996, 2009]

Project

		A	B	C	D
	1	16	14	14	12
Engineer	2	16	–	17	13
	3	11	15	21	9
	4	8	10	9	7

ANSWERS

1. $A \to I, B \to III, C \to II, D \to IV$, Minimum cost = 10.

2. $A \to 1, B \to 4, C \to 5, D \to 3, E \to 2$, Minimum cost = 60.

3. $1 \to B, 2 \to C, 3 \to D, 4 \to A$; $1 \to C, 2 \to D, 3 \to B, 4 \to A$, Min. time = 17 hrs.

4. $A \to II, B \to I, C \to V, D \to III, E \to IV$

   or $A \to II, B \to IV, C \to V, D \to I, E \to III$, Minimum cost = 34.

5. $A \to II, B \to IV, C \to I, D \to III$ or $A \to III, B \to IV, C \to I, D \to II$,

   Minimum cost = 80.

7. $A \to 2, B \to 4, C \to 6, D \to 1, E \to 7, F \to 3, G \to 5$, Minimum cost = 237 units.

8. $a \to D, b \to C, c \to B, d \to E, e \to A$, Minimum distance = 570 Kms.

9. $A \to II, B \to V, C \to III, D \to IV, E \to I$, Minimum cost = 39.

10. $1 \to 1, 2 \to 3, 3 \to 2$.

11. $P_1 \to J_5, P_2 \to J_6, P_3 \to J_4, P_4 \to J_2$, Jobs J_1 and J_3 are left undone.

12. $P_1 \to J_3, P_2 \to J_2, P_3 \to J_4$, J_1 and J_2 are left undone.

13. $1 \to F, 2 \to A, 3 \to E, 4 \to C, 5 \to D$, Minimum cost = 43.

14. $1 \to B, 2 \to A, 3 \to E, 4 \to C, 5 \to D$, Maximum sales = ₹ 221.

15. $1 \to 5, 2 \to 4, 3 \to 1, 4 \to 3, 5 \to 2$, Maximum return = 50 units.

16. $1 \to B, 2 \to D, 3 \to A, 4 \to C$, Minimum total time = 47 hours.

14.9 TRAVELLING SALESMAN PROBLEM [MEERUT–2001, 04, 09, 17]

In the travelling salesman problem, a salesman visits a certain number of cities and between every pair of cities the distance is known to him. He start from his home city, passes through each city once and returns to his home city in such a way that the route chosen by him must be shortest in distance.

A travelling slaesman wants to minimize the total distance travelled during his visit of n cities. A similar problem arises when n items say A_i, $i = 1, 2, ..., n$, are to be produced on a machine in continuation, given that c_{ij} $(i, j = 1, 2, ..., n)$ is the set up cost of the machine when item A_i is followed by A_j. Note that $c_{ij} = \infty$, when $i = j$ $i.e.$ we don't produce the item A_i again after A_j. The individual set up costs can be arranged in the form of the adjacent square matrix.

$$
\begin{array}{c|ccccc}
 & A_1 & A_1 & \cdots & A_i & \cdots & A_n \\
\hline
A_1 & \infty & c_{12} & \cdots & c_{1i} & \cdots & c_{1n} \\
A_1 & c_{21} & \infty & \cdots & c_{2i} & \cdots & c_{2n} \\
\vdots & \vdots & \vdots & & \vdots & & \vdots \\
A_1 & c_{21} & \infty & \cdots & c_{2i} & \cdots & c_{2n} \\
\vdots & \vdots & \vdots & & \vdots & & \vdots \\
A_i & c_{i2} & c_{i2} & \cdots & \infty & \cdots & c_{in} \\
\vdots & \vdots & \vdots & & \vdots & & \vdots \\
A_n & c_{n1} & c_{n2} & \cdots & c_{ni} & \cdots & \infty
\end{array}
$$

Our problem is to determine a set of n elements of this matrix, one of each row and one in each column, so as to minimize the sum of the elements determined above. Here two extra restrictions are imposed. One restriction is that we can not select the element in the leading diagonal as we have already assumed the elements of leading diagonal to be infinity. The other restriction is that we don't produce an item again until all the items are produced once.

Definition 1: *A travelling salesman problem is said to be symmetric or asymmetric according as cost matrix is symmetric or not. Thus it is symmetric if the cost from A_i to A_j is the same as that from A_j to A_i.*

Definition 2: A travelling salesman problem may be stated as follows:

Like assignment problem, if we represent" going" and "not going" of the salesman from A_i to A_j station by saying $x_{ij} = 1$ and $x_{ij} = 0$ respectively, then we wish to determine x_{ij}, $i, j = 1$, ..., n, which minimizes

$$
z = \sum_{i=1}^{n} \sum_{j=1}^{n} c_{ij}\, x_{ij}
$$

s.t.

$$
\sum_{j=1}^{n} x_{ij} = 1, \quad i = 1, 2, \ldots, n
$$

$$
\sum_{j=1}^{n} x_{ij} = 1, \quad j = 1, 2, \ldots, n
$$

$$
x_{ij} = 0 \ \text{ or } \ 1,
$$

and one extra restriction that the x_{ij} must be so chosen that no city is visited twice until the tour of all the cities is completed. Note that as we have written ∞ in the leading diagonal, x_{ij} cannot be 1 when $i = j$, because then z will not be minimum. Thus the second restriction of not going to A_i again just after A_i is automatically satisfied.

14.9.1 SOLUTION OF TRAVELLING SALESMAN PROBLEM

The problem can be solved by assignment technique.

The solution thus obtained may not be feasible for this problem. For example, if we choose the elements $c_{15}, c_{23}, c_{34}, c_{42}, c_{51}$, then the corresponding solution gives $A_1 \rightarrow A_5, A_5 \rightarrow A_1$ i.e. A_1 is followed by A_5 and A_5 by A_1. This violates our restriction. In such cases after solving the given problem by assignment technique, we use the method of enumeration by assigning the next minimum element of the matrix in place of zero. This is best explained by the following example.

Such problems occur in the field of postal deliveries, school bus routing, television relays assembly lines, production of several items by one machine.

 Solved Examples

EXAMPLE I. *Given the matrix of set up costs below, show how to sequence production so as to minimize the total set up cost per cycle.*

	A_1	A_2	A_3	A_4	A_5
A_1	∞	2	5	7	1
A_2	6	∞	3	8	2
A_3	8	7	∞	4	7
A_4	12	4	6	∞	5
A_5	1	3	2	8	∞

SOLUTION. Solving the problem by assignment technique we get the following matrix, showing a solution in terms of marked zeros.

	A_1	A_2	A_3	A_4	A_5
A_1	∞	1	3	6	[0]
A_2	4	∞	[0]	6	0
A_3	4	3	∞	[0]	3
A_4	8	[0]	1	∞	1
A_5	[0]	2	0	7	∞

The matrix does not provide the solution of the original problem as it gives $A_1 \rightarrow A_5, A_5 \rightarrow A_1$, while A_5 cannot be followed by A_1 until A_2, A_3, A_4 are produced.

Now we try to find the next best solution which also satisfies this extra restriction. The next minimum (non zero) element in the matrix is 1. We try to bring 1 into the solution. The cost 1 also occurs at three places. We shall consider all the cases one by one.

In case we bring $c_{12} = 1$ in the solution then as no other assignment can be made in the first row and second column, the resulting feasible solution will be $A_1 \to A_2, A_2 \to A_3, A_3 \to A_4, A_4 \to A_5, A_5 \to A_1$.

The selected elements for this solution are placed in the squares of the following matrix :

	A_1	A_2	A_3	A_4	A_5
A_1	∞	$\boxed{1}$	3	6	0
A_2	4	∞	$\boxed{0}$	6	0
A_3	4	3	∞	$\boxed{0}$	3
A_4	8	0	1	∞	$\boxed{1}$
A_5	$\boxed{0}$	2	0	7	∞

The cost in this reduced matrix corresponding to this feasible point is 2.

On the other hand, if we select the element $c_{43} = 1$, in the solution, then no feasible solution is available in terms of zeros or which gives the cost less than 2. Hence the best programme is

$$A_1 \to A_2 \to A_3 \to A_4 \to A_5 \to A_1.$$

The total setup cost will be $2 + 3 + 4 + 5 + 1 = 15$.

EXAMPLE 2. *Solve the travelling salesman problem given by the following data:*
$c_{12} = 20, c_{13} = 4, c_{14} = 10, c_{23} = 5, c_{34} = 6, c_{25} = 10, c_{35} = 6,$
$c_{45} = 20,$ *where* $c_{ij} = c_{ji}$ *and there is no route between cities i and j if a value for* c_{ij} *is not given above.* [MEERUT–2005, 06]

SOLUTION. The cost matrix is given as follows :

	1	2	3	4	5
1	∞	20	4	10	∞
2	20	∞	5	∞	10
3	4	5	∞	6	6
4	10	∞	6	∞	20
5	∞	10	6	20	∞

Note that in the above matrix, we have taken $c_{15} = \infty$, etc. It is taken to avoid the possibility of going from first station to 5th station as there is no such route. Note that c_{15} is not given in the data. Similarly, we have taken

$$c_{24} = \infty, c_{51} = \infty, c_{42} = \infty.$$

Now, solving the problem by assignment algorithm, we reach at the following assignment plan in terms of zero's :

	1	2	3	4	5
1	∞	12	$\boxed{0}$	$\cancel{0}$	∞
2	11	∞	$\cancel{0}$	∞	$\boxed{0}$
3	$\cancel{0}$	1	∞	$\boxed{0}$	1
4	$\boxed{0}$	∞	$\cancel{0}$	∞	9
5	∞	$\boxed{0}$	$\cancel{0}$	8	∞

In this assignment plan, the route is as follows :

$$1 \to 3 \to 4 \to 1 \to 2 \to 5 \to 2$$

This is not feasible route of the salesman problem. We bring $c_{32} = 1$ in the solution. A new assignment plan is as follows :

	1	2	3	4	5
1	∞	12	$\boxed{0}$	$\cancel{0}$	∞
2	11	∞	$\cancel{0}$	∞	$\boxed{0}$
3	$\cancel{0}$	$\boxed{1}$	∞	$\cancel{0}$	1
4	$\boxed{0}$	∞	$\cancel{0}$	∞	9
5	∞	$\cancel{0}$	$\cancel{0}$	$\boxed{8}$	∞

The optimal route is $1 \to 3 \to 2 \to 5 \to 4 \to 1$.

The corresponding total cost is $= 4 + 5 + 10 + 20 + 10 = 49$.

EXAMPLE 3. *Solve the following travelling salesman problem :*

	A	B	C	D	E
A	∞	4	7	3	4
B	4	∞	6	3	4
C	7	6	∞	7	5
D	3	3	7	∞	7
E	4	4	5	7	∞

SOLUTION. Following the general procedure of solving the assignment matrix, we have :

Row reduced matrix

	A	B	C	D	E
A	∞	1	4	0	1
B	1	∞	3	0	1
C	2	1	∞	2	0
D	0	0	4	∞	4
E	0	0	1	3	∞

Column reduced matrix

	A	B	C	D	E
A	∞	1	3	0	1
B	1	∞	2	0	1
C	2	1	∞	2	0
D	0	0	3	∞	4
E	0	0	0	3	∞

Assignment matrix

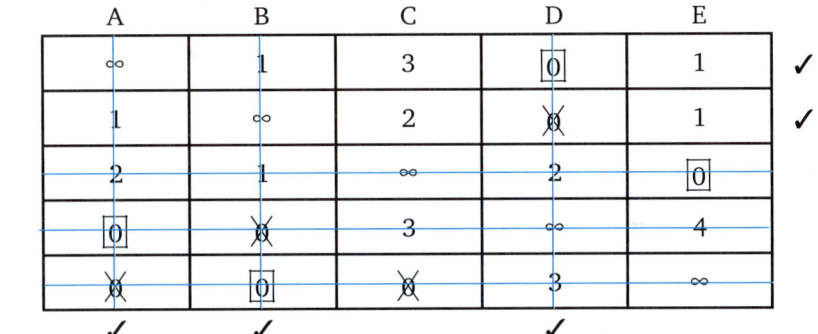

	A	B	C	D	E	
A	∞	1	3	[0]	1	✓
B	1	∞	2	⊠	1	✓
C	2	1	∞	2	[0]	
D	[0]	⊠	3	∞	4	
E	⊠	[0]	⊠	3	∞	
	✓	✓		✓		

Since the number of assignments = 4
Minimum element = 1.

	A	B	C	D	E
A	∞	0	2	0	0
B	1	∞	1	0	0
C	3	1	∞	3	0
D	1	0	3	∞	4
E	1	0	0	4	∞

Row reduced matrix

	A	B	C	D	E
A	∞	0	2	0	0
B	1	∞	1	0	0
C	3	1	∞	3	0
D	1	0	3	∞	4
E	1	0	0	4	∞

Column reduced matrix

	A	B	C	D	E
A	∞	0	2	0	0
B	0	∞	1	0	0
C	2	1	∞	3	0
D	0	0	3	∞	4
E	0	0	0	4	∞

Assignment model

	A	B	C	D	E
A	∞	0	2	⊠	⊡0
B	⊠	∞	1	⊡0	⊠
C	2	1	∞	3	⊠
D	⊡0	⊠	3	∞	4
E	0	⊠	⊡0	4	∞

Since there is no assignment in third row, so we can allot the next minimum (*i.e.,* 1) in row C.

	A	B	C	D	E
A	∞	0	2	⊠	⊡0
B	⊠	∞	1	⊡0	⊠
C	2	⊡1	∞	3	⊠
D	⊡0	⊠	3	∞	4
E	⊠	⊠	⊡0	4	∞

Required route is

$$A \to E, \ B \to D, C \to B, D \to A, E \to C, \ i.e., \ A \to E \to C \to B \to D \to A$$

Total cost = 4 + 5 + 6 + 3 + 3 = 21.

EXAMPLE 4. *Solve the following travelling sales problem :*

	A_1	A_2	A_3	A_4	A_5
A_1	∞	2	5	7	1
A_2	6	∞	3	8	2
A_3	8	7	∞	4	7
A_4	12	4	6	∞	5
A_5	1	3	2	8	∞

SOLUTION.   **Row reduced matrix**

	A_1	A_2	A_3	A_4	A_5
A_1	∞	1	4	6	0
A_2	4	∞	1	7	0
A_3	4	3	∞	0	3
A_4	8	0	2	∞	1
A_5	0	2	1	7	∞

Column reduced matrix

	A_1	A_2	A_3	A_4	A_5
A_1	∞	1	3	6	0
A_2	4	∞	0	7	0
A_3	4	3	∞	0	3
A_4	8	0	1	∞	1
A_5	0	2	0	7	∞

Assignment matrix

	A_1	A_2	A_3	A_4	A_5
A_1	∞	1	3	6	[0]
A_2	4	∞	[0]	7	⦻
A_3	4	3	∞	[0]	3
A_4	8	[0]	1	∞	1
A_5	[0]	2	⦻	7	∞

$A \to E, B \to C, C \to D, D \to B, E \to A$.

So the route obtained

$$A \to E \to A , B \to C \to D \to B .$$

But this is not a feasible route.

So, giving assignment to the next minimum to make the feasible route, we have

	A_1	A_2	A_3	A_4	A_5
A_1	∞	[1]	3	6	⦻
A_2	4	∞	[0]	7	⦻
A_3	4	3	∞	[0]	3
A_4	8	⦻	1	∞	[0]
A_5	[0]	2	⦻	7	∞

$A \to B, B \to C, C \to D, D \to E, E \to A$.

So, $\qquad A \to B \to C \to D \to E \to A$

EXAMPLE 5. ***Solve the following travelling sales problem :*** [DELHI(MBA)–1988, 99]

To item

		A	B	C	D
From item	A	∞	4	7	3
	B	4	∞	6	3
	C	7	6	∞	7
	D	3	3	7	∞

SOLUTION. **Row reduced matrix**

	A	B	C	D
A	∞	1	4	0
B	1	∞	3	0
C	1	0	∞	1
D	0	0	4	∞

Column reduced matrix

	A	B	C	D
A	∞	1	1	0
B	1	∞	0	0
C	1	0	∞	1
D	0	0	1	∞

Assignment model

	A	B	C	D
A	∞	1	1	[0]
B	1	∞	[0]	✗
C	1	[0]	∞	1
D	[0]	✗	1	∞

$$A \to D, B \to C, C \to B, D \to A.$$

So,   (1) $A \to D \to A$ (2) $B \to C \to B$.

But these are not the feasible path.

Hence, we give assignments to the next minimum element. We have,

	A	B	C	D
A	∞	1	[1]	✗
B	1	∞	✗	[0]
C	1	[0]	∞	1
D	[0]	✗	1	∞

$$A \to C, B \to D, C \to B, D \to A.$$

Hence, $A \to C \to B \to D \to A.$ Cost = 19

EXAMPLE 6. *Solve the following travelling salesman problem :*

	A	B	C	D	E	F	G
A	∞	6	12	6	4	8	1
B	6	∞	10	5	4	3	3
C	8	7	∞	11	3	11	8
D	5	4	11	∞	5	8	6
E	5	2	7	8	∞	4	7
F	6	3	11	5	4	∞	2
G	2	3	9	7	4	3	∞

SOLUTION. **Row reduced matrix**

	A	B	C	D	E	F	G
A	∞	5	11	5	3	7	0
B	3	∞	7	2	1	0	0
C	5	4	∞	8	0	8	5
D	1	0	7	∞	1	4	2
E	3	0	5	6	∞	2	5
F	4	1	9	3	2	∞	0
G	0	1	7	5	2	1	∞

Column reduced matrix

	A	B	C	D	E	F	G
A	∞	5	6	3	3	7	0
B	3	∞	2	0	1	0	0
C	5	4	∞	6	0	8	5
D	1	0	2	∞	1	4	2
E	3	0	0	4	∞	2	5
F	4	1	4	1	2	∞	0
G	0	1	2	3	2	1	∞

Assignment matrix

	A	B	C	D	E	F	G
A	∞	5	6	3	3	7	[0]
B	3	∞	2	[0]	1	0̸	0̸
C	5	4	∞	6	[0]	8	5
D	1	[0]	2	∞	1	4	2
E	3	0	[0]	4	∞	2	5
F	4	1	4	1	2	∞	0̸
G	[0]	1	2	3	2	1	∞

Since the number of [0] assignments ≠ 7. So we take minimum element = 1. Then we have the following new assignment model

New assignment model

	A	B	C	D	E	F	G
A	∞	5	5	2	2	6	[0]
B	3	∞	2	0̸	1	[0]	1
C	5	5	∞	6	[0]	8	6
D	0̸	[0]	1	∞	0̸	3	2
E	3	1	[0]	4	∞	2	6
F	3	1	3	[0]	1	∞	0̸
G	[0]	2	2	3	2	1	∞

The number of $\boxed{0}$ assignments = 7 (required) the path becomes
$A \rightarrow G \rightarrow A, B \rightarrow F \rightarrow D \rightarrow B, \ C \rightarrow E \rightarrow C$.

But it is not a feasible path. So, we must make it feasible as follows :

	A	B	C	D	E	F	G
A	∞	5	5	$\boxed{2}$	2	6	0
B	3	∞	2	0	1	0	$\boxed{1}$
C	$\boxed{5}$	5	∞	6	0	8	6
D	0	$\boxed{0}$	1	∞	0	3	2
E	3	1	$\boxed{0}$	4	∞	2	6
F	3	1	3	0	$\boxed{1}$	∞	0
G	0	2	2	3	2	$\boxed{1}$	∞

The above path has been made feasible by allotting the next minimum instead of 0 in A, B, C rows.

Thus, we get
$$A \rightarrow D \rightarrow B \rightarrow G \rightarrow F \rightarrow E \rightarrow C \rightarrow A$$
Minimum cost = $6 + 4 + 3 + 3 + 4 + 7 + 8 = 35$.

Exercise-14.3

1. Solve the travelling salesman problem.

	1	2	3	4	5
1	∞	6	12	6	4
2	6	∞	10	5	4
3	8	7	∞	11	3
4	5	4	11	∞	5
5	5	2	7	8	∞

2. Solve the travelling salesman problem.

	1	2	3	4	5	6
1	∞	20	23	27	29	34
2	21	∞	19	26	31	24
3	26	28	∞	15	36	26
4	25	16	25	∞	23	18
5	23	40	23	31	∞	10
6	27	18	12	35	16	∞

3. Solve the travelling salesman problem given in the following data.

$c_{12} = 4, c_{13} = 7, c_{14} = 3, c_{23} = 6, c_{24} = 3, c_{34} = 7$ where $c_{ij} = c_{ji}$.

4. Solve the travelling salesman problem.

	A	B	C	D	E
A	—	7	6	8	4
B	7	—	8	5	6
C	6	8	—	9	7
D	8	5	9	—	8
E	4	6	7	8	—

ANSWERS

1. $1 \rightarrow 3 \rightarrow 5 \rightarrow 2 \rightarrow 4 \rightarrow 1$, Min. cost = 27

2. $1 \rightarrow 5 \rightarrow 6 \rightarrow 3 \rightarrow 4 \rightarrow 2 \rightarrow 1$, Min. cost = 103

3. $1 \rightarrow 3 \rightarrow 2 \rightarrow 4 \rightarrow 1$, Min. cost = 19

4. $A \rightarrow C \rightarrow D \rightarrow B \rightarrow E \rightarrow A$ and $A \rightarrow E \rightarrow B \rightarrow D \rightarrow C \rightarrow A$, total cost = 30

REVIEW QUESTIONS

1. Define assignment problem.
2. Explain the Hungarian method to solve an assignment problem.
3. Explain the difference between a transportation problem and an assignment problem.
4. Write a short note on unbalanced assignment problem.
5. What is travelling salesman problem?
6. Formulate the travelling salesman problem as an assignment problem.
7. How can the travelling salesman problem be solved using assignment algorithm.

MULTIPLE CHOICE QUESTIONS (CHOOSE THE MOST APPROPRIATE ONE)

1. For solving an assignment problem, we modify cost matrix: [MEERUT–2013]
   (a) by creating zero (b) by creating infinite
   (c) by creating one (d) none of these
2. An assignment problem is a special case of $m \times n$ transportation problem in which: [MEERUT–2013]
   (a) $m = n$ (b) $2m = n$
   (c) $m = 2n$ (d) None of these
3. An assignment problem is solved by: [MEERUT–2013]
   (a) Simplex method (b) Graphical method
   (c) Vector method (d) Hungarian method
4. Due to some restriction the assignment of a particular facility to a particular job is not given, then we take that cost as: [MEERUT–2013]
   (a) 0 (b) –1
   (c) ∞ (d) None of these
5. In an unbalanced assignment problem to form a square matrix, fictitious rows or columns are added in the matrix with costs: [MEERUT–2013]
   (a) 0 (b) 1
   (c) ∞ (d) None of these
6. If the cost matrix of an assignment problem is not a square matrix, the assignment problem is called: [MEERUT–2013]
   (a) balanced (b) unbalanced
   (c) maximization (d) none of these
7. In an assignment problem with m jobs and m persons, the number of basic variables at zero level in a BFS is:
   (a) m (b) $m - 1$
   (c) $m + 1$ (d) None of these
8. The cost matrix in assignment problem is a: [MEERUT–2013]
   (a) square matrix (b) rectangle matrix
   (c) diagonal matrix (d) none of these
9. Maximization assignment problem is transformed in a minimization problem by:
   (a) deduct smallest element from all other elements of matrix

   (b) deduct all element of the row from highest element of the row
   (c) all elements of the matrix are deducted from the highest elements in the matrix
   (d) none of the above
10. Solve the following assignment problem:

    Person

	1	2	3	4
A	10	12	19	11
B	5	10	7	8
C	12	14	13	11
D	8	15	11	9

    Task

    (a) $A \to 1, B \to 2, C \to 3, D \to 4$
    (b) $A \to 2, B \to 3, C \to 4, D \to 1$
    (c) $A \to 2, B \to 3, C \to 1, D \to 4$
    (d) $A \to 3, B \to 2, C \to 1, D \to 4$
11. The assignment problem is a:
    (a) non-linear programming problem
    (b) dynamic programming problem
    (c) integer linear programming problem
    (d) integer non-linear programming problem
12. For an n × n assignment problem the number of possible solution will be:
    (a) $2n$ (b) $(n-1)!$
    (c) $n!$ (d) n^2
13. For solving an assignment problem we modify the cost matrix:
    (a) by creating zero (b) one
    (c) (a) and (b) both (d) none of these
14. The complete optimal assignment is obtained if in the reduced cost matrix of order n the number of marked '□' zero is:
    (a) less than n (b) greater than n
    (c) exactly n (d) none of these
15. In travelling salesman problem, elements of leading diagonal of the cost:
    (a) finite (b) constant
    (c) variable (d) infinite
16. If in a given problem a constant is added or subtracted to every element of a row of the cost matrix c_{ij} then an assignment which minimize the total cost for one matrix also

minimizes the total cost for other matrix then the problem is called:
(a) an assignment bvp transpite
(b) an assignment problem transpite
(c) (a) and (b) both
(d) none of these

17. In an unbalanced assignment problem to form a square matrix fictitious row or columns and are added in the matrix with costs:
(a) 0   (b) 1
(c) 2   (d) ∞

18. A salesman wants to visit n cities then the number of possible routes is:
(a) $n!$ (b) $(n-1)!$
(c) $(n+1)!$ (d) n

19. The complete optimal assignment is obtained if in the reduced cost matrix of order n the no. of assignments is:
(a) n (b) $> n$
(c) $< n$ (d) None of these

20. In travelling salesman problem, the elements of the leading diagonal of the cost are taken to be:
(a) finite   (b) infinite
(c) constant   (d) variable

21. An optimal assignment exists if the total reduced cost of the assignment is:
(a) 0   (b) 1
(c) 2   (d) 3

22. How many types travelling salesman problem?
(a) one   (b) two
(c) three   (d) four

23. If there were n workers and n jobs, there would be:
(a) $n!$ solutions   (b) $(n-1)!$ solutions
(c) $(n!)^n$ solutions   (d) n solutions

24. An assignment problem is considered as a particular case of a transportation problem, because:
(a) the number of rows equals the number of columns
(b) all $x_{ij} = 0$ or 1
(c) all rim conditions are 1
(d) all of the above

25. Maximization assignment problem is transformed into a minimization problem by:
(a) adding each entry in a column from the maximum value in that column
(b) subtracting each entry in a column from the maximum value in that column

(c) subtracting each entry in the table from the maximum value in that table
(d) any one of the above

26. While solving assignment problem an activity is assigned to a resource through a square with zero opportunity cost because the objective is to:
(a) reduce the cost of assignment to zero
(b) reduce the cost of that particular assignmen is zero
(c) minimize the cost of assignment
(d) all of the above

27. An assignment problem can be solved by:
(a) Simplex method
(b) Transportation method
(c) Both (a) and (b)
(d) None of these

28. In an assignment problem to obtain optimal assignment we draw minimum no. of lines to cover all the zero of reduced matrix then these line must through:
(a) unmarked rows
(b) unmarked columns
(c) unmarked rows and marked columns
(d) None of these

29. If an optimal assignment of an assignment problem exists then total reduced cost of the assignment will be:
(a) > 1 (b) < 1
(c) 0   (d) None of these

30. If distance between any pair of cities is independent of the direction of journey the problem is called:
(a) symmetrical   (b) asymmetrical
(c) degeneracy   (d) None of these

31. The assignment problem:
(a) requires that only one activity be assigned to each resource
(b) is a special case of transportation problem
(c) can be used to maximize resources
(d) all of the above

32. The purpose of a dummy row or column in an assignment problem is to:
(a) obtain balance between total activities and total resources
(b) prevent a solution from becoming degenerate
(c) provide the means of representing a dummy problem
(d) None of the above

ANSWERS

1. (a)	2. (a)	3. (d)	4. (c)	5. (a)	6. (b)	7. (b)	8. (a)	9. (c)
10. (b)	11. (a)	12. (c)	13. (a)	14. (c)	15. (d)	16. (a)	17. (a)	18. (b)
19. (a)	20. (b)	21. (a)	22. (b)	23. (a)	24. (d)	25. (c)	26. (c)	27. (c)
28. (c)	29. (c)	30. (a)	31. (d)	32. (a)				

ARCHIVE

1. A solicitors' firm employs typists on hourly piece-rate basis for their daily work. There are five typists and their charges and speed are different. According to an earlier understanding only one job is given one typist and the typist is paid for a full hour even if he works for a fraction of an hour. Find the least cost allocation for the following data :

Typist	Rate per hour (₹)	No. of pages typed/ hour	Job	No. of pages
A	5	12	P	199
B	6	14	Q	175
C	3	8	R	145
D	4	10	S	298
E	4	11	T	178

[DELHI(MBA)–1996; CA–1996]

2. In the modification of a plant layout of a factory four new machines M_1, M_2, M_3 and M_4 are to be installed in a machine shop. There are five vacant places A, B, C, D and E available. Because of limited space machine M_2 cannot be placed at C and M_3 cannot be placed at A. The cost of locating a machine at a place (in ₹100) is as follows :

		Location				
		A	B	C	D	E
Machine	M_1	9	11	15	10	11
	M_2	12	9	—	10	9
	M_3	—	11	14	11	7
	M_4	14	8	12	7	8

Find the optimal assignment cost.

[DELHI(MBA)–1999, 2001]

3. A construction company has requested bids for subcontracts on five different projects. Five companies have responded, their bids are represented below :

Bid Amounts (in Thousands ₹)

		I	II	III	IV	V
Bidders	1	41	72	39	52	25
	2	42	29	49	65	81
	3	27	39	60	51	40
	4	45	50	48	52	37
	5	29	40	45	26	30

Determine the minimum cost assignment of subcontracts to bidders, assuming that each bidder can recive only one contract.

[DELHI UNIV., MBA, 1988, 91]

4. The ABC Ice cream company has a distrubution depot in Greater Kailash part I for distributing ice cream in South Delhi. There are four vendors located in different parts of South Delhi (Call them A, B, C and D who have to be supplied ice cream every day. The following table displays the distance (in km) between the depot and their four vendors :

		To				
		Depot	Vendor A	Vendor B	Vendor C	Vendor D
From	Depot	∞	3.5	3	4	2
	Vendor A	3.5	∞	4	2.5	3
	Vendor B	3	4	∞	4.5	3.5
	Vendor C	4	2.5	4.5	∞	4
	Vendor D	2	3	3.5	4	∞

What route should the company can follow so that the total distance travelled is minimized?

[DELHI(MBA)–1997, 98, 2000]

5. Solve the following travelling salesman problem:

	A	B	C	D	E
A	∞	1	6	8	4
B	7	∞	8	5	6
C	6	8	∞	9	7
D	8	5	9	∞	8
E	4	6	7	8	∞

[ICWA(JUNE)–1985]

6. Solve the following travelling salesman problem :

	A	B	C	D	E
A	∞	375	600	150	190
B	375	∞	300	350	175
C	600	300	∞	350	500
D	160	350	350	∞	300
E	190	175	500	300	∞

[DELHI(MBA)–1996, 98, 99]

7. Solve the following travelling salesman problem :

	P	Q	R	S
P	∞	15	25	20
Q	22	∞	45	55
R	40	30	∞	25
S	20	26	38	∞

[DELHI(MBA)–1988, 97]

8. A city corporation has decided to carry out road repairs on main four arteries of the city. The government has agreed to make a special grant of 50 lakh towards the cost with a condition that the repairs be done at the lowest cost and quickest time. If the conditions warrant, a supplementary taken grant will also be considered favourably. The corporation has floated tenders and five contractor have sent in their bids. In order to expedite work, one road will be awarded to only one contractor.

Cost of Repairs (₹ Lakh)

Contractors/Road		R_1	R_2	R_3	R_4
	C_1	9	14	19	15
	C_2	7	17	20	19
	C_3	9	18	21	18
	C_4	10	12	18	19
	C_5	10	15	21	16

(a) Find the best way of assigning the repair work to the contractors and the costs.

(b) If it is necessary to seek supplymentary grants, what should be the amount sought?

(c) Which of the five contractors will be unsuccessful in his bid?

[ICWA, JUNE, 1987]

HINTS AND ANSWERS

1. Applying the algorithm and get a table in which entries represent the cost to be incurred due to assignment of jobs to various typists on a one-to-one basis. Then find the apportunity cost table.

   $A \to T, B \to R, C \to S, D \to P, E \to Q$ and Total cost = 399.

2. As the cost matrix is not balanced, add one dummy row with a zero cost element and apply the Hungarian method.

   $M_1 \to A, M_2 \to B, M_3 \to E, M_4 \to D, M_5 \to C$. Where M_5 is dummy. Total cost = 32.

3. $1 \to V, 2 \to II, 3 \to I, 4 \to III, 5 \to IV$; Minimum cost = 155

4. Solve the travelling salesman problem as an assignment problem and obtain an optimal solution of an assignment problem but not to the travelling salesman problem as it gives the sequence :

   Depot $\to$ Vendor $D \to$ Vendor $B \to$ Depot. This violates the condition of travelling salesman problem. Then we choose next non-zero minimum element.

   Vendor $C \to$ Vendor $B \to$ Depot $\to$ Vendor $D \to$ Vendor $A \to$ Vendor C

   Total distance to be covered in this sequence is 15 km.

5. $A \to C \to D \to B \to E \to A$; 30 km

6. $A \to D \to C \to B \to E \to A$; 165 km

7. $P \to R \to S \to Q \to P$; 98 km

8. (a) $R_1 \to C_2, R_2 \to C_4, R_3 \to C_1, R_4 \to C_5, R_5 \to C_3$

   Total cost = 54 lakh

   (b) 4 lakh

   (c) C_3

Decision Theory 15

15.1 INTRODUCTION

Decision analysis is a set of concepts and systematic procedure to improve the decision making process. A decision is simply a selection from two or more courses of action. In decision making, it is necessary to setting objectives, deciding basic policies and other major plans that will have major influence in the future profit. The concept of decision analysis can be best described by quoting R.H. Howard – "A decision is an irrevocable allocation of resources, irrevocable in the sense that it is impossible or extremely costly to change back to the situation that existed before making the decision."

Hence, the decision theory can be defined as:

'A process which results in the solution from a set of alternative courses of action that course of action which is considered to meet the objectives of the decision problem more satisfactorily than others as judged by the decision makers.'

15.2 DECISION MODELS

Decision theory can be used to find optimal strategies when a decision maker is faced with several decision alternatives and an uncertain pattern of future event. Decision models can be classified into various categories. Any decision model of any category has the following characteristics in common:

(i) **Decision Alternatives:** There is a finite number of decision alternatives available with the decision maker at each point when a decision is made. The alternative courses of action are the acts that are available to decision maker.

(ii) **State of nature:** The event identify the occurance which are outside of the decision maker's control and which determine the level of success for a given act. These events are known as "states of nature or outcomes".

☛ REMARK

- The states of nature are mutually exclusive and collectively exhaustive with respect to any decision problem.

(iii) **Pay-off:** For a given problem, a numerical value or outcome resulting from each possible combination of alternatives and state of nature is called pay-off. For each combinations of state of nature and course of action, the pay-off is calculated.

15.2.1 GENERAL FORM OF A PAY-OFF TABLE OR MATRIX

Let $N_1, N_2, ..., N_m$ be the m possible event and n alternative courses of action $A_1, A_2, ..., A_n$.

Then the pay-off corresponding to strategy A_j of the decision maker under the state of nature N_i will be denoted by $p_{ij} : i = 1, 2, ..., m; j = 1, 2, ..., n$. Then general pay-off matrix can be written as follows:

State of nature (Events)	Probability	Course of Action (strategies)			
		A_1	A_2	...	A_n
N_1	p_1	p_{11}	p_{12}	...	p_{1n}
N_2	p_2	p_{21}	p_{22}	...	p_{2n}
$\vdots$	$\vdots$	$\vdots$			
N_m	p_m	p_{m1}	p_{m2}	...	p_{mn}

☞ REMARK
- Pay-off table list the states of nature which are mutually exclusive as well as collectively exhaustive and a set of given course of action.

15.3 STEPS OF DECISION MAKING PROCESS

In the decision making process, we have the following steps:
(i) To identify and define the problem.
(ii) Listing to all possible states of nature (Future events)
(iii) Identify all the courses of action which are available to the decision maker.
(iv) Expressing the pay-off's (p_{ij}) resulting from each pair of action and state of nature.
(v) Apply an appropriate mathematical decision theory model to choose best course of action.

15.4 DECISION MAKING SITUATIONS

Here, we shall discuss the following four types of decision making environment or situations.
(i) Decision making under certainty
(ii) Decision making under risk
(iii) Decision making under uncertainty
(iv) Decision making under conflict

15.5 DECISION MAKING UNDER CERTAINTY

In decision making under certainty, the data is assumed to be known deterministically. In this type of decision making, the decision maker presumes that only one state of nature is relevent for his purpose. He identify this state of nature, takes it for granted and presumes complete knowledge as to its occurance.

The various techniques used under this categories are as follows:
(a) System of equations
(b) Linear programming
(c) Integer programming
(d) Dynamic programming
(e) Queuing models
(f) Inventory models
etc.

☞ REMARK
- In such type of decision model, only one possible state of nature (future) exists. For example, National Saving Certificate (NSC), Kisan Vikas Patra or National Saving Scheme (NSS), in which we have all the information about the future in advance.

15.6 DECISION MAKING UNDER RISK

Here, the information regarding the state of nature is probabilistic, so that the decision maker is not sure about the pay-off that will occur as a consequence of the selection of a particular course of action. Also, every course of action has several outcomes. Hence, there is more than one state of nature. Sometimes, past experience often enable the decision-maker to assign probability values of the likely possible occurance of each state of nature. After knowing the probability distribution of the state of nature, the best decision is taken by selecting that course of action that has the largest expected pay-off.

The criteria for evaluating the course of action to take a decision under risk are Expected Money Value (EMV), Exepected Opportunity Loss (EOL), Expected Profit of Perfect Information (EPPI), Expected Value of Perfect Information (EVPI) and Conditional Opportunity Loss (COL).

The objectives of decision making here is to optimize the expected pay-off, which may mean either maximization or expected profit or minimization of expected regret.

15.6.1 EXPECTED MONETARY VALUE (EMV)

For a given course of action, the expected monetary value is the weighted sum of possible pay-off for each alternatives. It can be obtained by summing the pay-off for each course of action multiplied by the probabilites associated with each state of nature.

Following working procedure can be used to calculate the EMV for various course of action.

Working Procedure

STEP 1. Identify the course of action and the state of nature.

STEP 2. Construct pay-off table listing all possible course of action and state of nature.

STEP 3. Calculate the conditional pay-off attached with each possible combination of states of nature and course of action along with the probabilities of occurance of each state of nature. Tabulate all these.

STEP 4. Calculate EMV for each course of action by multiplying the conditional pay-off by the associated probabilities and adding all these values thus obtained.

STEP 5. Identify the course of action that yield the optimal EMV.

Mathematically, we can define EMV as follows:
$$\text{EMV (Course of action } S_j) = \sum_{i=1}^{m} p_{ij} p_i$$
where m = no. of possible states of nature
p_i = probabilities of occurance of state of nature N_i
p_{ij} = pay-off associated with state of nature N_i and course of action S_{ij}

15.6.2 EXPECTED PROFIT WITH PERFECT INFORMATION (EPPI)

It is defined as the maximum monetry value that can be obtained on the basis of having perfect information about the state of nature that will occur. Now, to calculate EPPI, we use the following working procedure.

Working Procedure

STEP 1. Select the optimal course of action for each state of nature with its associate conditional pay-off from the pay-off table.

STEP 2. Multiply each conditional pay-off chosen in step 1 with the corresponding probability of occurance of that state of matter.

STEP 3. Adding all these product to get the required EPPI.

15.6.3 EXPECTED VALUE OF PERFECT INFORMATION (EVPI)

The expected value of perfect information is the expected or average return in the long run, if we have perfect information before a decision has to be made. It is defined as the maximum amount that the decision maker should be willing to pay to obtain the perfect information about the state of nature that would occur.

The expected value of perfect information (EVPI) is the expected outcome with perfect information minus the expected outcomes without perfect information (maximum EMV) *i.e.*,

$$EVPI = EPPI - maximum\ EMV\ (i.e.,\ EMV^*)$$

15.6.4 CONDITIONAL OPPORTUNITY LOSS (COL)

It is defined as the difference the highest pay-off value for a particular state of nature and the actual pay-off obtained from the particular course of action. It can be obtained for each course of action.

15.6.5 EXPECTED OPPORTUNITY LOSS (EOL)

It is another decision making criterion for decision making under risk. The EOL is defined as the difference between the highest profit for a state of nature and the actual profit obtained for the particular course of action taken *i.e.* it is the amount of pay-off that is lost by not selecting the course of action that has the greatest pay-off for the state of nature that actually occur. The course of action due to which EOL is minimum is recommended.

Mathematically, it can be defined as follows:

$$EOL\ (state\ of\ nature,\ N_i) = \sum_{i=1}^{m} l_{ij} p_i$$

where l_{ij} = opportunity loss due to state of nature N_i and course of action S_j
p_i = probability of occurance of state of nature, N_k

 Solved Examples

EXAMPLE 1. *An oil exploration company is expending its business and has to decide whether or not to drill for oil at a particular location. The results of drilling are classified into three categories – high yield worth 12 units, moderate yield with 6 units and no oil. The drilling operation costs 5 units. At a similar location 50, 30 and 20 percent of previous drillings have given high, moderate and no yield respectively. Using EMV concepts, identify the best course of action.*

SOLUTION. Let us define three states of nature as following :
N_1 = high yield
N_2 = moderate yield
N_3 = No yield

These all three are not under the control of the decision maker. Also there are following two courses of action :
S_1 = drill

S_2 = No drill

The probabilities of occurance of N_1, N_2 and N_3 are respectively given by 0.5, 0.3 and 0.2.

∴ We have the following table :

State of nature	Course of action	
	Drill (S_1)	Not Drill (S_2)
N_1	12 – 5 = 7	0
N_2	6 – 5 = 1	0
N_3	0 – 5 = – 5	0

Also, the EMV for S_1 and S_2 are given in the following table

State of nature	Probabilities	Course of action		EMV	
		S_1	S_2	S_1	S_2
N_1	0.5	7	0	0.5 × 7 = 3.5	0.5 × 0 = 0
N_2	0.3	1	0	0.3 × 1 = 0.3	0.3 × 0 = 0
N_3	0.2	– 5	0	0.2 × (–5) = –1.0	0.2 × 0 = 0
		EMV		2.8	0

⇒ EMV (S_1) = 2.8 and EMV(S_2) = 0

⇒ Optimum EMV i.e. EMV* = EMV (S_1)

⇒ Best course of action is S_1

Hence, the company should opt for drilling to have a maximum pay-off worth 2.8 units.

EXAMPLE 2. *A person who just retired from service has received a big provident fund (PF) amount. He want to invest this amount in the share market. A broker offers him the following investment alternatives and percentage return rate.*

Investment	Market conditions (Return rates)		
	Low	Medium	High
Regular shares	– 5%	10%	15%
Risky shares	– 8%	12%	20%
Property	– 10%	15%	25%

Over the past 200 days the market conditions have had medium returns rate for 100 days and high returns rate for 40 days.

On the basis of the given data, as a decision maker policy answer the following:

(i) Identify the state of nature and find the probabilities of their occurance.

(ii) Identify the course of action.

(iii) Write the pay-off matrix and find the EMV for each course of action and state the optimum investment policy for the investment.

SOLUTION. (i) Clearly, there are three state of natures namely low, medium and high market conditions, which are not under the control of decision maker. Over

200 days low, medium and high market conditions prevailed for 60, 100 and 40 (= 200 – 100 – 60) days respectively. Thus, the associated probabilities associated with state of natures– low, medium and high are 0.3, 0.5 and 0.2 respectively.

(ii) The course of actions are the investment alternatives– regular shares, risky shares or property which are in hand of the investers.

(iii) We have the following pay-off matrix.

State of nature	Probabilities	Course of Action		
		Regular shares (S_1)	Risky shares (S_2)	Property (S_3)
low	0.30	– 0.05	– 0.08	– 0.10
medium	0.50	0.10	0.12	0.15
high	0.20	0.15	0.20	0.25

Now,

$$\text{EMV}(S_1) = -0.05 \times 0.3 + 0.10 \times (0.50) + 0.15 \times (0.20)$$
$$= 0.065 = 6.5\%$$
$$\text{EMV}(S_2) = -0.08 \times 0.3 + 0.12 \times 0.5 + 0.20 \times 0.2$$
$$= 0.076 = 7.6\%$$
$$\text{EMV}(S_3) = -0.10 \times 0.3 + 0.15 \times 0.50 + 0.25 \times 0.20$$
$$= 0.095 = 9.5\%$$

Therefore, EMV* = EMV(S_3) = 9.5%

which shows that 9.5% is the highest for property, the investor should invest in property as it would be the optimum investment policy.

EXAMPLE 3. *A bread seller purchases bread packets everyday at ₹23 per packet and sells them at ₹25. The packets unsold at the end of the day are deposited of ₹22. The sales in the past have ranged from 10 to 14 packets per day and the record of sales for the past 100 days is as follows:*

Packets sold	10	11	12	13	14
No. of days	10	20	40	20	10

Find how many packets should the bread seller purchase everyday to have the maximum profit.

SOLUTION. Let $N_i : i = 1, 2, 3, 4, 5$ be the daily demand (states of nature) and $S_j : j = 1, 2, 3, 4, 5$ be the courses of action.

Now, marginal profit (MP) per packet sold

= selling price – cost price = 25 – 23 = 2

marginal loss (ML) per pack unsold = 23 – 22 = 1

Conditional profit = MP × no. of packets sold – ML × no. of packets unsold

$$= 2S - 1N = 2S - N$$

Here, S is the number of bread packets sold and N is the no. of bread packet not sold.

Now, the conditional profit values are shown in the following pay-off matrix.

State of nature (demand/ day)	Probabilities	Course of action (purchase/day) conditional profit value					Course of action (purchase/ day) expected profit in ₹				
		10	11	12	13	14	10	11	12	13	14
	(1)	(2)	(3)	(4)	(5)	(6)	(1)×(2)	(1)×(3)	(1)×(4)	(1)×(5)	(1)×(6)
10	10/100 = 0.1	20	19	18	17	16	2	1.9	1.8	1.7	1.6
11	20/100 = 0.2	20	22	21	20	19	4	4.4	4.2	4.0	3.8
12	40/100 = 0.4	20	22	24	23	22	8	8.8	9.6	9.2	8.8
13	20/100 = 0.2	20	22	24	26	25	4	4.4	4.8	5.2	5.0
14	10/100 = 0.1	20	22	24	26	28	2	2.2	2.4	2.6	2.8
EMV							20	21.7	22.8	22.7	22.0

From the above table, we observe that the highest EMV is ₹22.8 corresponding to course of action 12. Hence, the bread seller must purcahse 12 packets of bread everyday.

EXAMPLE 4. *A certain output is manufactured at ₹80 and sold at ₹140 per unit. The product is such that if it is produced but not sold during a day's time it becomes worthless. The daily sales records in the past are as follows:*

Sales/day	30	40	50	60	70
No. of days	24	24	36	24	12

(i) *Prepare a pay-off and a regret table.*

(ii) *Find the expected pay-off and regret.*

(iii) *Find the optimal act and the EVPI.*

SOLUTION. Here, the no. of items to be stocked is under control of decision maker, number of items to be stocked per day is considered as an act or 'course of action' and the daily demand of the items is uncertain and only known with prbability, it is considered as an 'event' or 'state of nature'. Each item sold within a day yields a profit of ₹(140 – 80) = ₹60, otherwise it is a dead loss of ₹80.

Now, conditional profit = MP(item sold) – ML(items not sold)

$$= \begin{cases} (140-80)D = 60D, \text{ if } D \geq S \\ (140-80)D - 80(S-D) = 140D - 80S, \text{ if } D < S \end{cases}$$

where D is the no. of item sold within a day and S is the number of items stocked.

Now, we have the following expected pay-off matrix

State of nature (S_j)	Probability $P(S_j)$	Conditional pay-off course of Action (stock)					Expected pay-off course of action (stock)				
		30	40	50	60	70	30	40	50	60	70
	(1)	(2)	(3)	(4)	(5)	(6)	(1)×(2)	(1)×(3)	(1)×(4)	(1)×(5)	(1)×(6)
30	0.2	1800	1000	200	–600	–1400	360	200	40	–120	–280
40	0.2	1800	1600	1600	800	0	360	480	320	160	0
50	0.3	1800	3000	3000	2200	1400	540	720	900	660	420
60	0.2	1800	3000	3000	3600	2800	360	480	600	720	560
70	0.1	1800	3000	3000	3600	4200	180	240	300	360	420
EMV							1800	2120	2160	1780	1120

From the above table, since the course of action 'stock 50' gives the highest EMV of ₹2160.

Hence, the optimal course of action would be to purchase or stock 50 items every day. Now, we have the following table of expected regret (opportunity loss value)

State of nature (S_j)	Prob. $P(S_j)$	Conditional opportunity loss course of action (stock)					Expected opportunity loss course of action (stock)				
		30	40	50	60	70	30	40	50	60	70
	(1)	(2)	(3)	(4)	(5)	(6)	(1)×(2)	(1)×(3)	(1)×(4)	(1)×(5)	(1)×(6)
30	0.2	0	800	1600	2400	3200	0	160	320	4800	640
40	0.2	600	0	800	1600	2400	120	0	160	320	480
50	0.3	1200	600	0	800	1600	360	180	0	240	480
60	0.2	1800	1200	600	0	800	360	240	120	0	160
70	0.1	2400	1800	1200	600	0	240	180	120	60	0
					EOL		1080	760	720	1100	1760

EXAMPLE 5. *XYZ Food products Ltd. want to determine how many kg of Shrikhand to stock on daily basis. Historical data has generated the following pattern of demand.*

Unit sold/day	180	181	182	183	184	185	186
No. of days	2	8	10	40	20	15	5

Assume that the stock levels are restricted to the range 180-186 kgs and that shrikhand left unsold at the end of the day must be disposed of due to inadequate refrigerative facility. Shrikhand cost ₹6 per kg to the company and it sells for ₹8 per kg. Then

  (i) Construct the conditional pay-off table.
  (ii) Determine the action alternatives associated with the maximum profit.
(iii) Determine EVPI.

SOLUTION.   We have
Conditional profit

$$= \text{marginal profit} \times \text{unit sold} - \text{marginal loss} \times \text{unit unsold}$$
$$= (8-6) \times \text{units sold} - 6 \times \text{units unsold}$$
$$= \begin{cases} (8-6)D = 2D, & \text{if } D \geq S \\ (8-6)D - 6(S-D) = (8D - 6S), & \text{if } D < S \end{cases}$$

where D is the no. of units demanded and S, the no. of units stocked. Now, we have the following pay-off matrix

State of nature demand/day	Prob.	Conditional pay-off course of action (stock/day)						
		180	181	182	183	184	185	186
180	0.02	360	354	348	342	336	330	324
181	0.08	360	362	356	350	344	338	332
182	0.10	360	362	364	358	352	346	340
183	0.40	360	362	364	366	360	354	348
184	0.20	360	362	364	366	368	362	356
185	0.15	360	362	364	366	368	370	364
186	0.05	360	362	364	366	368	370	372

Further, the computation of expected monetary values is given in the following table

State of nature demand/day	Prob.	Expected pay-off course of action (stock/day)						
		180	181	182	183	184	185	186
180	0.02	7.20	7.08	6.96	6.84	6.72	6.60	6.48
181	0.08	28.80	28.96	28.48	28.00	27.52	27.04	26.56
182	0.10	36.00	36.20	36.40	35.80	35.20	34.60	34.00
183	0.40	144.00	144.80	145.60	146.40	144.00	141.60	139.20
184	0.20	72.00	72.40	72.80	73.20	73.60	72.40	71.20
185	0.15	54.00	54.30	54.60	54.90	55.20	55.50	54.60
186	0.05	18.00	18.10	18.20	18.30	18.40	18.50	18.60
EMV		360.00	361.84	363.04	363.44	360.64	360.24	350.64

Clearly, in the above table the act '183 kg' yield the highest EMV of ₹363.44, therefore, optimal act for the company would be to stock 183 kg of shrikhand everyday.

15.7 DECISION MAKING UNDER UNCERTAINTY

Let the probabilities associated with occurance of different states of nature are not known *i.e.* there is no historical data available which could indicate the probability of occurance of a particular state of nature. In such type of conditions, the decision maker has no way of calculating the expected pay-off for his strategies. Such type of situations arises when new product is introduced in the market. There are following different criterion of decision making in this situation.

 (i) Optimism (maximax or minimin) criterion
 (ii) Pessimism (maxmin or minmax) criterion
 (iii) Equal probabilities (Laplace) criterion
 (iv) Coefficient of optimism (Hurwiez) criterion
 (v) Regret (salvage) criterion

15.7.1 OPTIMISM CRITERION (MAXIMAX OR MINIMIN)

In this criterion, the decision maker ensure that he should not miss the opportunity to achieve the largest possible profit or lowest possible cost.

Working Procedure

The functioning of this criterion is given in the following steps:

STEP 1. Locate the maximum (or minimum) pay-off value for profit or loss corresponding to each course of action and write it down at the location of its column.

STEP 2. Locate the maximum for profit (or minimum for cost or loss) among the pay-off obtained in step 1 and identify the corresponding course of action that will give the optimum decision.

> The maximax criterion finds the course of action or alternative strategy that maximize the maximum pay-off.

☞ REMARK
• Since, the decision maker selects an alternative with largest (or lowest) possible pay-off value, it is also called 'optimistic decision criterion'.

15.7.2 Pessimism Criterion (Minimax or Maximin Criterion)

This criterion is the decision to take the course of action which maximize the minimum possible pay-off. The decision maker ensure that he would earn no less than some specified amount. Therefore, he selects the alternative that represents the maximum of the minima (or minimum of the maxima, in case of loss) pay-off in case of profits.

Working Procedure

The functioning of this criterion is given in the following steps:

STEP 1. Locate the minimum pay-off in case of profit corresponding to each course of action or maximum in case of loss.

STEP 2. Locate the maximum for profit or minimum for the cost among those evaluated in step 1 and identify the corresponding optimum course of action.

> The pessimism criterion corresponds to identify the worst possible outcome in each course of actio, *i.e.* maximum loss or minimum outcome that would occur under each decision alternative and then selecting the best out of the worst outcome.

☞ **Remark**

- The above criterion is also called 'Waldian Criterion' because it was first suggested by 'Arabham Wald'.

15.7.3 Equal Probabilities or Laplace Criterion

When the probabilities of states of nature are not known, all the states of nature are assumed to occur with equal probabilities. The decision maker first calculate the average outcome for every course of action and then select that with the maximum number. This rule is based on the assumption that the probabilities of different states of nature are euqal.

Working Procedure

The functioning of this criterion is given in the following steps:

STEP 1. Assign equal probabilities to each state of nature.

STEP 2. Computed the expected value for each alternative by multiplying each outcome by its probability and then summing.

STEP 3. Select the best expected pay-off value *i.e.* maximum for profit and minimum for cost.

☞ **Remarks**

- This criterion is also known as criterion of insufficient reason, because, except in a few cases some information of the likelyhood of occurance of states of nature is available.
- It is also called Criterion of rationality or 'Baye's criterion'.

15.7.4 Coeffcient of Optimism or Criterion of Realism or Hurwicz criterion

This criterion is a compromise between an optimistic and pessimistic decision criterion. Hurwicz introduce the idea of a coefficient of optimism denoted by α to measure the decision-maker's degree of optimism.

Here, α lies between 0 and 1, where 0 represents a completely pessimistic attitude and 1, a completely optimistic attitude of the decision maker about the state of nature.

According to Hurwicz, the decision maker must select a course of action that maximizes.

$$H(S) = \alpha(\text{Maximum pay-off in column}) + (1 - \alpha)(\text{minimum pay-off in column})$$

Working Procedure

The functioning of this criterion is given in the following steps:

STEP 1. Identify the coefficient of optimism α and then coefficient of pessimism $(1 - \alpha)$.

STEP 2. For each course of action S_j, select the largest and the lowest pay-off value and multiply these with α and $(1 - \alpha)$ values and add to get $H(S_j)$ by using

$$H(S_j) = \alpha(\text{largest pay-off}) + (1 - \alpha)(\text{lowest pay-off})$$

15.7.5 REGRET OR SAVAGE CRITERION

Under this criterion for each conditional pay-off or cost value, a regret value equal to difference between the maximum pay-off under a state of nature and a pay-off resulting from each course of action under that strategy is calculated. This is also known as 'Opportunity loss decision criterion' or the minimax regret decision criterion. The decision maker regrets after adopting a wrong course of action resulting in an opportunity loss of pay-off that he would have liked to minimize.

 Working Procedure

The functioning of this criterion is given in the following steps:

STEP 1. Construct the pay-off from the given description of the problem.

STEP 2. Compute the conditional opportunity loss by taking the difference between the maximum pay-off of a course of action and same other courses of action resulting from a particular state of nature.

STEP 3. Find the maximum regret for each course of action. Then choose that course of aciton with the minimum of these maximum values as the optimum decision.

> Regret Pay-off = Maximum pay-off from a course of action − pay-off

 Solved Examples

EXAMPLE 1. *The conditional pay-offs (profit in ₹) of different courses of action S_1, S_2 and S_3 against the state of nature N_1, N_2, N_3 and N_4 are given in the following table.*

State of nature	Course of action		
	S_1	S_2	S_3
N_1	4000	20000	20000
N_2	– 100	5000	15000
N_3	6000	400	– 2000
N_4	18000	0	1000

Indicate the decision taken under the following approaches.
(i) Pessimistic   *(ii) Optimistic*
(iii) Equal probabilities   *(iv) Regret*   *(v) Hurwicz criterion*
The degree of optimism is given as $\alpha = 0.7$.

SOLUTION. **(i) Pessimistic criterion:** Here, the minimum pay-off in each course of action is taken and then the maximum of these minima is taken as the optimum decision.

State of nature	Course of action		
	S_1	S_2	S_3
N_1	4000	20000	20000
N_2	– 100	5000	15000
N_3	6000	400	– 2000
N_4	18000	0	1000
Minimum pay-off	– 100	0	– 2000

Now maximum $\{-100, 0, -2000\} = 0$, which is for S_2.

Hence, the optimum course of action is S_2.

(ii) **Optimistic criterion:** Here, the maximum of pay-offs from each course of action is taken and then the maximum of these maxima is taken as the optimum decisions

State of nature	Course of action		
	S_1	S_2	S_3
N_1	4000	20000	20000
N_2	– 100	5000	15000
N_3	6000	400	– 2000
N_4	18000	0	1000
Maximum pay-off	18000	20000	20000

$\Rightarrow$ max.$\{18000, 20000, 20000\} = 20000$ for S_2 and S_3.

Hence, both S_2 and S_3 are the optimum course of action.

(iii) **Equal probability (Laplace) criterion:** Here, we find the expected pay-off of each course of action as given here

$$p_1 = p_2 = p_3 = p_4 = \frac{1}{4}$$

Course of action	Expected Return (Expected pay-off)
S_1	$\frac{1}{4}(4000 - 100 + 6000 + 18000) = 6975$
S_2	$\frac{1}{4}(20000 + 5000 + 400 + 0) = 6350$
S_3	$\frac{1}{4}(20000 + 15000 - 2000 + 1000) = 85000$

Clearly, the expected return is maximum *i.e.* 85000 for S_3. Hence, S_3 is the optimum decision.

(iv) **Regret Criterion:** Using the given pay-off, we calculate the regret pay-off as shown below:

State of nature	Conditional pay-off (Course of action)			Regret (Opportunities) – Course of action		
	S_1	S_2	S_3	S_1	S_2	S_3
N_1	4000	20000	20000	16000	0	0
N_2	– 100	5000	15000	15100	10000	0
N_3	6000	400	– 2000	0	5600	8000
N_4	18000	0	1000	0	18000	17000
Maximum Regret				16000	18000	17000

Clearly, the maximum regret due to S_1, S_2 and S_3 are respectively given by 16000, 18000 and 17000.

$\therefore$ min{16000, 18000, 17000} = 16000, which is S_1.

S_1 is the optimum course of action.

(v) Hurwicz Criterion: Since, $\alpha = 0.7$ (given) $\Rightarrow 1 - \alpha = 1 - 0.7 = 0.3$

Then, select the course of action that optimize the pay-off value

$$H = \alpha(\text{best pay-off}) + (1 - \alpha)(\text{worst pay-off})$$
$$= \alpha(\text{maximum in column}) + (1 - \alpha)(\text{minimum in column})$$

	S_1	S_2	S_3
	4000	20000	20000
	– 100	5000	15000
	6000	400	– 2000
	18000	0	1000
max	18000	20000	20000
min	– 100	0	– 2000

Using the above table

$$H (S_1) = 0.7 \times 18000 + 0.3(- 100) = 12570$$
$$H (S_2) = 0.7 \times 20000 + 0.3 \times 0 = 14000$$
$$H (S_3) = 0.7 \times 20000 + 0.3(- 2000) = 13400$$

Hence, max $H(S_i)$ = max{12570, 14000, 13400} = 14000 = $H(S_2)$

Hence, S_2 provides the maximum profit i.e. ₹ 14000.

EXAMPLE 2. *A manufacturer makes a product of which the principal ingrediant is a chemical X. At this stage, the manufacturers spends ₹1000 per year on supply of X, but there is a possibility that the price may soon increase to 4 times its present figure because of a worldwide shortage of the chemical. There is another chemical Y, which the manufacturer could use in conjunction with a third chemical Z, in order to give same effect as chemical X. Chemicals Y and Z would together cost the manufacturer ₹3000 per year but their prices are unlikely to rise. What action should the manufacturer take? Apply the maximin and minimax criterion for decision making and give two sets of solutions. If the coefficient of optimum is 0.4 find the course of action that minimize the cost.* [ICWA–1988]

SOLUTION. We can summarize the above data in the following table (negative values in the table represent profits)

State of nature	Course of Action	
	S_1 (Use of Y and Z)	S_2 (Use of X)
N_1 (price of X increases)	– 3000	– 4000
N_2 (price of X does not increases)	– 3000	– 1000

(i) Maxmin criterion: We have the following table

State of nature	Course of Action	
	S_1	S_2
N_1	– 3000	– 4000
N_2	– 3000	– 1000
Column minimun	– 3000	– 4000

∴ max{–3000, – 4000} = – 3000, which is for S_1.
Hence, the manufacturer should adopt S_1.

(ii) Minimax or opportunity loss criterion:

State of nature	Course of Action	
	S_1	S_2
N_1	– 3000 – (– 3000) = 0	– 3000 – (– 4000) = 1000
N_2	– 1000 – (– 3000) = 2000	– 1000 – (– 1000) = 0
Maximun	2000	1000

∴ minimum{1000, 2000} = 1000, for S_2.
Hence, manufacturer should adopt minimum opportunity loss course of action S_2.

(iii) Hurwicz criterion: Given $\alpha = 0.4 \Rightarrow 1 - \alpha = 1 - 0.4 = 0.6$
Now, $H = \alpha$(Best pay-off) + $(1 - \alpha)$(worst pay-off)
    $= \alpha$(maximum in column) + $(1 - \alpha)$(minimum in column)

Course of action	Best pay-off	Worst pay-off	H
S_1	– 3000	– 3000	– 3000
S_2	– 1000	– 4000	– 2800

Now, since the course of action S_2 has the least cost (maximum profit)
    $= 0.4(1000) + 0.6(4000) = ₹2800$,
the manufacturer should adopt S_2.

15.8 POSTERIOR PROBABILITY AND BAYESIAN ANALYSIS

In the Bayesian decision theory we have concerned with probabilities assigned solely by the decision maker to various event. These probabilities are called prior probabilities, which are based on the current information available together with the expertise, experience and judgement of decision maker. If some extra information is required on the state of nature with a new outcome, then the prior probabilities require revision such revised probabilities in decision theory are called posterior (conditional) probabilities. These probabilities can be calculated by Baye's theorem.

15.8.1 BAYE'S THEOREM

Let $A_1, A_2, ..., A_n$ be mutually exclusive and collectively exhaustive state of nature. Their probabilities $P(A_1), P(A_2), ..., P(A_n)$ are known. There is an experiment outcome B for which the conditional probabilities $P(B|A_1), P(B|A_2), ..., P(B|A_n)$ are also known. Given the information that outcome B has occured, then the revised conditional probabilities of outcomes A_i i.e. $P(A_i|B): i = 1, 2, ..., n$ are determined as follows:

$$P(A_i \mid B) = \frac{P(A_i \text{ and } B)}{P(B)} = \frac{P(A_i \cap B)}{P(B)}$$

where
$$P(B) = P(A_1 \cap B) + P(A_2 \cap B) + ... + P(A_n \cap B)$$
$$= P(A_1)P(B \mid A_1) + P(A_2)P(B \mid A_2) + ... + P(A_n)P(B \mid A_n)$$

Then
$$P(A_i \mid B) = \frac{P(A_i)P(B \mid A_i)}{P(A_1)P(B \mid A_1) + P(A_2)P(B \mid A_2) + ... + P(A_n)P(B \mid A_n)}$$

Working Procedure

The functioning of this analysis is given in the following steps:

STEP 1. Idetify the alternative course of action, based on the set of objectives.

STEP 2. Identify the states of nature that would determine the pay-off of each course of action.

STEP 3. Obtain the numerical value to pay-off of each course of action, given the possible events.

STEP 4. Assign a numerical value (probability) to the occurance of each possible event.

STEP 5. Computed the expected value of the pay-off assigned to each course of action using the probabilities.

STEP 6. Asses the experience of loss and gain.

STEP 7. Select the alternative course of action.

> In Bayesian approach when preposterior analysis is applied to multistage decision problems. It is referred to as sequential analysis. Here, the decision maker has to make decision in a sequence, the decision coming later being dependent upon the decision choices made earlier.

Solved Examples

EXAMPLE 1. *An oil company that is engaged in oil explorations has to decide whether or not to drill for oil in Kaveri basin. There are three categories into which the drilling results are classified – high yield with 12 units, moderate with 6 units and no yield. The drilling*

Prospect	Yield		
	High	Moderate	Nil
Good	0.7	0.5	0.1
Fair	0.2	0.2	0.3
Bad	0.1	0.3	0.6

operation costs 5 units. At similar places 50, 30 and 20 percent of previous drilling have given high, moderate or no yield respectively. A seismic test is available which would indicate a good, fair or bad prospect of drilling. Past experiences indicates that the probabilities shown in the adjoining table for the result of seismic tests for the three ultimate possible yields. Find the best decision rule on the assumption that a test is made and the maximum amount that it would be wrong paying for such a test. If this is the actual cost of the test, what is the expected value of the potential well?

SOLUTION. Let N_1, N_2, N_3 be denote the state of nature High, moderate and nil respectively and two course of action S_1 for drill and S_2 for no drill. Further, suppose that good, fair or bad prospects of drilling by X, Y and Z respectively. Then we have the following initial pay-off table.

State of nature (N_i)	Probability	Course of action		Regrets		Yield prospects		
		Drill (x_1)	Do not drill (x_2)	x_1	x_2	$P(X\|N_i)$	$P(Y\|N_i)$	$P(Z\|N_i)$
N_1	0.5	7	0	0	7	0.7	0.2	0.1
N_2	0.3	1	0	0	1	0.5	0.2	0.3
N_3	0.2	– 5	0	5	0	0.1	0.3	0.6

Now, using the Baye's theorem, we have

For outcome X

$$P(X) = P(X \mid N_1)P(N_1) + P(X \mid N_2)P(N_2) + P(X \mid N_3)P(N_3)$$
$$= 0.7 \times 0.5 + 0.5 \times 0.3 + 0.1 \times 0.2 = 0.52$$

$$P(N_1 \mid X) = \frac{P(X \mid N_1)P(N_1)}{P(X)} = \frac{0.7 \times 0.5}{0.52} = 0.67$$

$$P(N_2 \mid X) = \frac{P(X \mid N_2)P(N_2)}{P(X)} = \frac{0.5 \times 0.3}{0.52} = 0.29$$

$$P(N_3 \mid X) = \frac{P(X \mid N_3)P(N_3)}{P(X)} = \frac{0.1 \times 0.2}{0.52} = 0.04$$

So, expected regret, ER $(x_1 \mid X) = 0.67 \times 0 + 0.29 \times 0 + 0.04 \times 5 = 0.20$ units
and $\quad$ ER$(x_2 \mid X) = 0.67 \times 7 + 0.29 \times 1 + 0.04 \times 0 = 4.98$ units
which shows that the expected regret is minimum for x_1 (drilling) with the value 0.20 units.

For outcome Y

$$P(Y) = P(Y \mid N_1)P(N_1) + P(Y \mid N_2)P(N_2) + P(Y \mid N_3)P(N_3)$$
$$= 0.5 \times 0.02 + 0.3 \times 0.02 + 0.2 \times 0.3 = 0.22$$
$$P(N_1 \mid Y) = \frac{0.5 \times 0.2}{0.22} = 0.46, P(N_2 \mid Y) = \frac{0.3 \times 0.2}{0.22} = 0.27 = P(N_3 \mid Y)$$

Thus ER$(x_1 \mid Y) = 0.46 \times 0 + 0.27 \times 0 + 0.27 \times 5 = 1.35$ units
and $\quad$ ER$(x_2 \mid Y) = 0.46 \times 7 + 0.27 \times 1 + 0.27 \times 0 = 3.49$ units
Hence, we can say that the expected regret (ER) is minimum for x_1 with the value of 1.35 units.

For outcome Z

$$P(Z) = P(Z \mid N_1)P(N_1) + P(Z \mid N_2)P(N_2) + P(Z \mid N_3)P(N_3)$$
$$= 0.1 \times 0.5 + 0.3 \times 0.3 + 0.6 \times 0.2 = 0.26$$

$$P(N_1 \mid Z) = \frac{0.1 \times 0.5}{0.26} = 0.19, P(N_2 \mid Z) = \frac{0.3 \times 0.3}{0.26} = 0.35$$

$$P(N_3 \mid Z) = \frac{0.6 \times 0.2}{0.26} = 0.46$$

Thus ER$(x_1 \mid Z) = 0.19 \times 0 + 0.35 \times 0 + 0.46 \times 5 = 2.30$ units
and $\quad$ ER$(x_2 \mid Z) = 0.19 \times 7 + 0.35 \times 1 + 0.46 \times 0 = 1.68$ units
which shows that expected regret is minimum for x_2 with a value of 1.68 units.
Now, the total expected regrets for X, Y and Z together are as follows:

$$ER(x_1) = 0.52 \times 0.20 + 0.22 \times 1.35 + 0.26 \times 2.30 = 0.999 \approx 1 \text{ unit.}$$
$$ER(x_2) = 0.52 \times 4.98 + 0.22 \times 3.49 + 0.26 \times 1.68 = 3.794 \text{ units}$$

We see that the expected regret is minimum for x_1 (drilling) with a value of 1 unit. Hence, the best decision for the company would be to go for drilling while paying a maximum of 1 unit for the seismic test. If this is done, the pay-off for x_1 would be $7 - 1$, $1 - 1$ and $5 - 1$ for the states N_1, N_2 and N_3 and the expected value of the potential is given by

$$EV(x_1) = 0.5 \times 6 + 0.3 \times 0 - 0.2 \times 6 = 1.8 \text{ units}$$

EXAMPLE 2. *A company is considering the introduction of a new product to its existing range. It has defined two levels of sales as 'high' and 'low' on which to base its decision and has estimated the changes that each market level will occur, together with their costs and consequent profits or losses. The information is given in the following table*

State of nature	Probability	Course of action	
		Market the product (₹)	Do not market the product (₹)
High sales	0.3	150	0
Low sales	0.7	– 40	0

The company's marketing manager suggest that a market research survey may be undertaken to provide further information on which to base the decision. On past experiences, with a certain market research organisation, the marketing manager assesses its ability to give good information in the light of subsequent actual sales achievements as follows.

Market Research	Actual sales	
	Market high	Market low
High sales forecast	0.5	0.1
Indecisive survey report	0.3	0.4
Low sales forecast	0.2	0.1

The market research survey will cost ₹20000 state whether or not there is a case for employing the market research organisation.

[DELHI(MBA)–1996, 2000]

SOLUTION. For expected monetary value, we have the following table

State of nature	Probability	Course of action		Expected profit (in ₹)	
		Market	Do not Market	Market	Do not Market
High sales	0.3	150	0	45	0
Low sales	0.7	– 40	0	– 28	0
EMV				17	0

Now, with no additional information, the company should choose course of action 'market the product'. But, if the company had the perfect information about the low sales, that it would not go ahead as the expected value is – 28000. Therefore the value of perfect information is the expected value of low sales.

Now, define the outcomes of the research survey as follows:

S_1 : high sales, S_2 : indecisive report, S_3 : low sales

N_1 : high market, N_2 : low market

Then we have the following table

Outcome	Sales prediction	
	High market (N_1)	Low market (N_2)
S_1	$P(S_1 \mid N_1) = 0.5$	$P(S_1 \mid N_2) = 0.1$
S_2	$P(S_2 \mid N_1) = 0.3$	$P(S_2 \mid N_2) = 0.4$
S_3	$P(S_3 \mid N_1) = 0.2$	$P(S_3 \mid N_2) = 0.5$

Using these information, the company can now revise the prior probabilities of outcomes to get posterior probabilities. Then revised probabilities can be calculated as follows.

State of nature	Prior probability $P(N_i)$	Conditional probability $P(S_i \mid N_i)$	Joint Probability $P(S_i \cap N_i) = P(N_i)P(S_i \mid N_i)$		
N_1	0.3	0.5	0.15	—	—
		0.3	—	0.09	—
		0.2	—	—	0.06
N_2	0.7	0.1	0.07	—	—
		0.4	—	0.28	—
		0.5	—	—	0.35
Marginal Probability			0.22	0.37	0.41

Now, the posterior probability of actual sets given the sale forecast are

Outcome (S_i)	Probability $P(S_i)$	State of nature (N_i)	Posterior Probability $P(N_i \mid S_i) = P(N_i \cap S_i) \mid P(S_i)$
S_1	0.22	N_1	0.15/0.22 = 0.681
		N_2	0.07/0.22 = 0.318
S_2	0.37	N_1	0.09/0.37 = 0.243
		N_2	0.28/0.37 = 0.756
S_3	0.41	N_1	0.06/0.41 = 0.146
		N_2	0.35/0.41 = 0.853

Finally, for each outcome the revised probabilities are now used to calculate the net expected value (EV) given the additional information supplied by that outcome as shown in the following table.

State of nature	Revised conditional profit (₹)	Sales Forecast					
		High		Indecisive		Low	
		Prob.	EV(₹)	Prob.	EV(₹)	Prob.	EV(₹)
High sales	130	0.681	88.53	0.243	31.59	0.146	18.98
Low sales	– 60	0.318	– 19.08	0.756	– 45.36	0.853	– 51.18
Expected value of sales forecast			69.45		– 13.77		– 32.20
Probability of occurance			0.22		0.37		0.41
Net Expected value (EVX Prob.)			15.279		– 5.095		13.202

Exercise-15.1

1. An electric equipment manufacturer has estimated. On the basis of the past data, the following weekly demand of distribution for the equipment:

No. demanded	Prob.	No. demanded	Prob.
0	0.14	4	0.09
1	0.27	5	0.04
2	0.27	6	0.01
3	0.18		

Each equipment costs ₹7000 and is sold for ₹10000 each. The equipment left unsold at the end of the week is disposed of ₹6000 each. What should be the weekly equipment stock so as to maximize the expected profit? Find the optimum profit per week.

2. A person has ₹20000 to invest. He was given the option to invest in three companies A, B and C. The return per year on investment depends on the performance measures that are classified into excellent, good and normal. The possible return, depending upon performance are given in the following table.

Course of action	Sales of nature		
	Excellent	Good	Normal
A	4000	2400	3000
B	6000	1600	2000
C	5000	2000	3600

What should the person decide with regard to investment on using the following criterion (i) maximin (ii) maximax (iii) equally likely (iv) regret?

3. A milk vending booth sells milk at ₹11/litre the cost price being ₹10.87/litre. The owner desires to give away unsold milk at the end of the day towards charity. The milk owner collect data based on past sales for a 100 day period.

Litre/ day	No. of days	Litre/ day	No. of days
600	1	660	25
610	1	670	15
620	2	680	10
630	6	690	7
640	10	700	3
650	20		

What should be the optimal inventory, based on the above data and the net profit for a day?

4. A newspaper distributor distributes a particular newspaper in a local market. The cost price of the newspaper is ₹1.05 and selling price is ₹1.50. An unsold copy is disposed of in the reuse market at ₹0.05. The estimates for the number of copies in demand are as follows:

Demand	15	16	17	18	19	20
Probability	0.04	0.19	0.33	0.26	0.11	0.07

How many copies of the newspaper should the distributor order so that the expected profit is maximum? Find the expected profit.

5. A flower dealer purchases flowers in the evening and delivers them to organization by 8:00 AM the next day. The daily demand for roses are as given below

Dozen of flowers	7	8	9	10
Probability	0.1	0.2	0.4	0.3

The management purchases flowers at ₹10 per dozen and sells them ₹30. All the unsold roses are donated to a nearby hospital. How many dozens of flowers should the manager order each evening to maximize the profit? Find the optimum profit.

ANSWERS

1. 3 per week, ₹4080
2. (a) A, ₹2400 (b) B, ₹6000 (c) C, ₹1000 (d) C, ₹10600
3. 660 litres, ₹833.33
4. 17 copies, ₹ 7.26
5. 9 dozen, ₹168

REVIEW QUESTIONS

1. Write a short note on 'Statistical decision theory'.
2. Write the basic steps of a decision making process.
3. Describe the meaning of EMV, EOL, EVPI.
4. What is the purpose of Bayesian analysis?
5. Write the chief characteristic of Bayesian decision-making.
6. What are the techniques used to solve decision making problems under uncertainty?
7. Write the short notes on the following:
   (i) Decision making under certainty
   (ii) Decision making under risk
   (iii) Decision making under uncertainty
8. Define the pay-off matrix.

MULTIPLE CHOICE QUESTIONS (CHOOSE THE MOST APPROPRIATE ONE)

1. The expected value of perfect information (EVPI):
   (a) equals the expected regret of the optimum decision under risk
   (b) can be determined without using probabilities
   (c) both (a) and (b) are true
   (d) none of these
2. Which of the following criteria does not apply to decision making under uncertainty:

(a) maximin   (b) maximax

(c) minimax   (d) none of these

3. The criteria which selects the action for which maximum pay-off is lowest is called:
   (a) min-min criteria (b) min-max criteria
   (c) max-min criteria (d) none of these

4. The concept of utility is a way to take into account:
   (a) oversion to risk
   (b) inclination to take risk
   (c) both (a) and (b) are true
   (d) none of these

5. Maximin return, maximax return and minimax regret are criteria that
   (a) can be used without probabilities
   (b) lead to some optimal decision
   (c) both (a) and (b) are true
   (d) none of these

6. A type of decision-making environment is:

(a) certainty   (b) uncertainty

(c) risk   (d) all are true

7. The value of the coefficient of optimism is needed while using the criterion of:
   (a) maximin   (b) realism
   (c) minimax   (d) none of these

8. The decision making criterion that should be used to achieve maximum long-term pay-off is:
   (a) maximax   (b) EOL
   (c) EMV   (d) none of these

9. The expected value of perfect information is equal to:
   (a) EPPI – min(EMV) (b) EPPI + min(EMV)
   (c) max(EOL)   (d) none of these

10. Essential characteristic of a decision model is:
    (a) state of nature   (b) pay-off
    (c) decision alternatives(d) all are ture

ANSWERS

1. (a)   2. (d)   3. (b)   4. (c)   5. (a)   6. (d)   7. (b)   8. (c)   9. (a)
10. (d)

ARCHIVE

1. An oil company is considering the purchase the mineral rights on a property of ₹100 lakh. The price includes tests to indicates whether the property has type A geological formation of type B geological formation. The company will be unable to tell the type of geological function until the purcahse is made. It is known, however that 40 percent of the land in this area has type A formation and 60 percent type B formation. If the company decides to drill on the load it will cost ₹200 lakh. If the company does drill it may hit an oil well, gas well or a dry hole. Drilling experience indicates that the probability of striking an oil well is 0.4 on type A and 0.1 on type B formation. Probability of hitting gas is 0.2 on type A and 0.3 on type B formation. The estimate discounted cash value from an oil well is ₹1000 lakh and from a gas well is ₹500 lakh. This includes everything except cost of mineral rights and cost of drilling. Using decision theory decide whether the company should purchase the mineral rights. [ICWA–1987]

2. The sensual cosmetic co. has developed a new perfume which management feels has a tremendous potential. It not only interacts with the weaver's body chemistry to create a unique fragrance but is also especially long lasting. A total of ₹10 lakh has already been spent on its development. Two marketing plans have been devised.
   (a) The first plan flows the company's usual

policy of giving small samples of the new products when other items in the company's product lines are purchased and placing advertisements in woman magzines. The plans would cost ₹5 lakh and it is believed that if might result on a high, moderate or low market response with probability of 0.2, 0.5, and 0.3 respectively. The net profit excluding development and promotion costs in these cases would be ₹20 lakh, ₹10 lakh and ₹1 lakh respectively. If it is later appears that the market response is going to below it would be possible to launch a T.V. advertise campaign. This would cost another ₹7.5 lakh. It would change the market response to high to moderate as previously described but with probability of 6.5 each.

(b) The second marketing plan is much more aggresive than the first. The emphasis would be heavily upon T.V. advertising. The total cost of this plan would be ₹15 lakh, but the market response would be either excellent or good with probabilities of 0.4 and 0.6 respectively. The profit excluding development and promotion costs would be ₹30 lakh and ₹25 lakh for two outcomes. Advise on the sequence of strategy to be followed by the company. [ICWA–1997]

ANSWERS

1. The company should purchase the mineral rights 2. Aggressive plan

INTRODUCTION

Game Theory is a special type of decision theory. It is also known as the competition strategy. A great variety of competitive situations is commonly seen in everyday life, for example, in an election, all the candidates fighting are interested to secure more votes than all the others and in military battle everyone wants to win.

Now at first, we want to define a game. "A game is an activity under a set of rules between two or more players in which each player get some gain or loss."

Definition. *Game theory is a type of decision theory in which one's choice of action is determined after assuming all possible outcomes.*

SOME BASIC DEFINITIONS

1. **n-person game.** When number of players are n where $n \geq 2$, then for $n = 2$, the game is known as 2 person game and for $n > 2$ the game is known as n person game.

2. **Zero sum and non-zero sum game.** A game is known as zero sum game in which the net gain or not profit after the game is zero means nothing comes from outside, payment is always between the players, the loss of one player is the gain of others, and a game which is not zero sum game is known as non-zero sum game.

3. **Competitive game.** A game is said to be competitive if it has the following four properties :
   (i) There should be finite number of players means $n \geq 2$.
   (ii) Each player has a finite list of his possible course of action.
   (iii) A play is said to be played when each player choose one of his course of action and no player knows the choice of action of the other player until he has decided his own.
   (iv) When each player chooses his activity, then this combination of activities gives a result according which each player gains a payment which may be –ve, +ve or zero. [MEERUT-2002, 03, UPTU MBA-04, 06, 08]

4. **Strategy.** For a given player, the strategy is given by the set of rules which specify that which of the available course of action he should make at each play. The strategy may be of two types : [MEERUT-2001, 02, 03, 04; ROHILKHAND-2002; UPTU MBA-2004]
   (a) pure strategy   (b) mix strategy
   (a) **Pure strategy.** A pure strategy is that in which one player knows what the other player is going to do. In this case, the player always choose a particular course of action.

(b) Mixed strategy. When a player does not know exactly what the other player is going to do, a probabilistic situation is obtained and such type of strategy is known as mixed strategy. Mathematically, let x_i be the probability to choose the ith activity, then we define the set

$$X = (x_1, x_2, x_3, \ldots, x_n) \text{ s.t. } x_1 + x_2 + x_3 + \ldots + x_n = 1$$

and $x_1, x_2, x_3, \ldots, x_n \geq 0$.

[MEERUT 2001, 03, 04, 12; ROHILKHAND-2002; GARHWAL-2012; UPTU MBA-2004]

5. **Two person zero sum game.** A game with only two players in which the gains of one player are the losses of another player is called a two person zero sum game. It is also called rectangular game.

[MEERUT-2001, 03, 04, 12; ROHILKHAND-2001; GARHWAL-2012; UPTU (MBA) 2005]

6. **Pay-off Matrix**

A's pay-off matrix

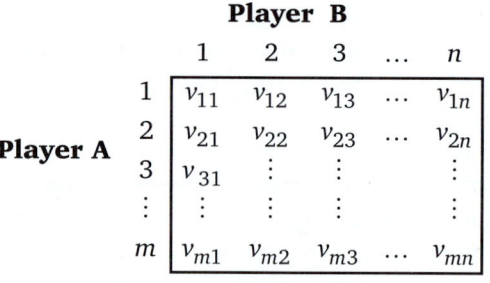

B's pay-off matrix

In A's pay-off matrix, the cell entry v_{ij} is the payment to player A when A choose the ith activity and B chooses the jth.

In a rectangular game or two person zero sum game, the player B's pay-off matrix will be the negative of A's pay-off matrix. Thus the net gain will be zero.

7. **Minimax and Maximin Principles.** If a player lists his worst possible outcomes of all his potential strategy, then he will choose the best strategy among all these outcomes. Such a principle is known as maximin principle or optimal strategy.

[AGRA-2002, MEERUT-2002, 03]

8. **Saddle Point.** A point which is minimum in its row and maximum in its column is known as the saddle point. [MEERUT 2002, 03, 04, 06, 10, 14, ROHILKHAND-2000; UPTU MBA-2004]

9. **Optimal Strategy and Value of the Game.** If in a given pay-off matrix (i, j) is the saddle point, then the player A and B are said to have the optimal strategy i and j.

The value of the $(i, j)^{\text{th}}$ cell is known as the value of the game and is denoted by v.

 Working Procedure

We can determine a saddle point by using the following steps :

STEP 1. In the given pay-off matrix, find the minimum element in each row and circle them.

STEP 2. Find the maximum element in each column and mark them by □.

STEP 3. Now find the element which have both the sign □ and O.
This point is the required saddle point.

 Solved Examples

EXAMPLE 1. *The player A's pay-off matrix is given. Find the optimal strategy by using maximin or minimax criterion.*

Player B

		I	II	III
	I	−3	−2	−3
Player A *II*		2	0	2
	III	5	−2	−4

SOLUTION. According to maximin criterion, first of all we find the minimum elements in each row and then find its maximum element. Now, we find the maximum element in column and then find its minimum.

In a game player A wants to maximum the value of the game v_{ij} by choosing one of his activity, while the player B wants to minimize the A's gain as mush as possible.

Now, in the given problem min $(-3, -2, -3) = -3$

$\qquad$ min $(2, 0, 2) = 0$

$\qquad$ min $(5, -2, -4) = -4$

and $\qquad$ max $(-3, 0, -4) = 0$

Now, $\qquad$ max $(-3, 2, 5) = 5$

$\qquad$ max $(-2, 0, -2) = 0$

$\qquad$ max $(-3, 2, -4) = 2$

$\qquad$ min $(5, 0, 2) = 0$

$\Rightarrow \qquad$ min max $v_{ij}(\overline{v}) = 0$

Here, $\qquad \underline{v} = \overline{v} = 0$

The maximin strategy used by player A = II
The minimax strategy used by player B = II
The corresponding gain $(v) = 0$.

Thus the strategy used is known as optimal strategy and the gain is known as the value of the game and the point 0 is also known as saddle point.

EXAMPLE 2. *The pay-off matrix for a two person zero sum game is given by*

Player B

		I	II	III
	I	15	2	3
Player A *II*		6	5	7
	III	−7	4	0

Find the optimal strategy and value of the game.

SOLUTION. First of all, find the minimum element in each row and maximum element in each column.

	I	II	III	Row min
I	[15]	(2)	3	2
II	6	(5)	[7]	5
III	(-7)	4	0	-7
Column max	15	5	7	

Circle the element which is minimum in each row and mark the square to the element which is maximum in each column.

The entry which have both the sign is known as saddle point.

Now the optimal strategy used by player A = II.

The optimal strategy used by player B = II.

and value of the game $v = 5$ for player A

$v = -5$ for player B.

EXAMPLE 3. *Find the saddle point and hence find the value of the game.*

Player B

		I	II	III	IV	V
	I	9	3	1	8	0
Player A	II	6	5	4	6	7
	III	2	4	4	3	8
	IV	5	6	2	2	1

SOLUTION. For finding the saddle point, find the minimum element in each row and maximum element in each column.

Table for saddle point

Player B

		I	II	III	IV	V	Row min
	I	[9]	3	1	[8]	(0)	0
	II	6	5	[(4)]	6	7	4
Player A	III	(2)	4	4	3	[8]	2
	IV	5	[6]	2	2	(1)	1
Column max		9	6	4	8	8	

The element having both the marks (O and □) = v_{23}.

So the optimal strategy used by player A = II.

The optimal strategy used by player B = III.

and value of the game is given by = 4.

So, optimal strategy = (II, III) and value of the game $v = 4$.

EXAMPLE 4. *Find the saddle point and hence find the value of the game for the given pay-off matrix.*

(a) $\begin{bmatrix} -5 & 5 & 0 & 7 \\ 2 & 6 & 1 & 8 \\ -4 & 0 & 1 & -3 \end{bmatrix}$ (b) $\begin{bmatrix} 1 & 7 & 3 & 4 \\ 5 & 6 & 4 & 5 \\ 7 & 2 & 0 & 3 \end{bmatrix}$

SOLUTION. (a) Find the minimum element in each row and maximum element in each column and then find the maximum element in the row minimum elements and find the minimum in column maximum elements.

Table for saddle point

Player B

		I	II	III	IV	Row min
	I	⊝5	5	0	7	−5
Player A	II	☐2	☐6	⊙1	☐8	1
	III	⊝4	0	1	−3	−4
Column max		2	6	1	8	

So, the required saddle point = 1.

The optimal strategy used by player A = II.

The optimal strategy used by player B = III.

The value of the game = 1.

So, optimal strategy = (II, III)

and value $v = 1$ for player A

         $= -1$ for player B.

(b) First find the min. element in each row and max. element in each column and mark them.

Table for saddle point

Player B

		I	II	III	IV	Row min
	I	⊙1	☐7	3	4	1
Player A	II	5	6	⊛4	☐5	4
	III	☐7	2	⊙0	3	0
Column max		7	7	4	5	

The point 4 having both marks (O and ☐). Therefore, it is the required saddle point.

The optimal strategy used by player A = II.

The optimal strategy used by player B = III.

So, optimal strategy = (II, III)

Value of the game $(v) = 4$.

EXAMPLE 5. *Find the range of p and q under which the given pay-off matrix gives the entry (II, II) as a saddle point.*

Player B

		I	II	III	
	I	2	4	5	[MEERUT 2007; ROHILKHAND-1999]
Player A	II	10	7	q	
	III	4	p	6	

SOLUTION. Since 7 is given to be a saddle point, so 7 will be min in II row and max in II column and after it, it will be max in all three min elements of row and min in the max elements of column. So

	I	II	III	Row min
I	②	4	5	2
II	10	⑦	q	7
III	4	p	6	4

Column max   10   7   6

Here, q should be greater than 7 and p should be less than 7.

So, the range of p and q is given by

$$p \leq 7 \quad \text{and} \quad q \geq 7$$

EXAMPLE 6. *Find the optimal strategy and value of the game whose pay-off matrix is given by*

Player B

		I	II	III	IV	V
	I	−2	0	0	5	3
Player A	II	3	2	1	2	2
	III	−4	−3	0	−2	6
	IV	5	3	−4	2	−6

SOLUTION.   **Table for saddle point**

Player B

		I	II	III	IV	V	Row min
	I	⊖2	0	0	5	3	−2
Player A	II	3	2	①	2	2	1
	III	⊖4	−3	0	−2	6	−4
	IV	5	3	−4	2	⊖6	−6

Column max   5   3   1   5   6

So, the required saddle point is given by = 1.

The optimal strategy is given by (II, III).

The value of the game = 1.

EXAMPLE 7. *Find out the saddle point and the value of the game*

Player B

		I	II	III	IV
	I	−5	2	1	20
Player A	II	5	5	4	6
	III	4	−2	0	−5

SOLUTION.   **Table for the saddle point**

Player B

		I	II	III	IV	Row min
	I	⊖5	2	1	20	−5
Player A	II	5	5	④	6	4
	III	4	−2	0	⊖5	−5

Column max   5   5   4   20

So, saddle point is given by = 4.

The optimal strategy is given by (II, III).

The value of the game = 4.

EXAMPLE 8. ***Determine the optimal strategy and value of the game***

(a)

	I	II	III
I	6	8	6
II	4	12	2

(b)

	I	II	III
I	3	0	-3
II	2	3	1
III	-4	2	-1

[MEERUT-2017]

SOLUTION. (a) **Table for saddle point**

	I	II	III	Row min.
I	⑥	8	⑥	6
II	4	⑫	②	2
Column max	6	12	6	

So, saddle point is given by 6.

Optimal strategy is given by (I, I) or (I, III).

Value of the game = 6.

(b) **Table for saddle point**

	I	II	III	Row min.
I	③	0	⊝③	-3
II	2	③	①	1
III	⊝④	2	-1	-4
Colum max	3	3	1	

So, the required saddle point is given by 1.

The optimal strategy = (II, III). Value of the game = 1.

EXAMPLE 9. ***Find the saddle point in the following case and also find the game value***

B

		I	II	III
	I	1	14	11
A	II	-9	5	-11
	III	0	-3	14

SOLUTION. Given pay-off matrix is

B

		I	II	III	Row min.
	I	1	14	11	1
A	II	-9	5	-11	-11
	III	0	-3	14	-3
Column max.		1	14	14	

To find the saddle point, find the smallest element in each row and mark it ○.
Find maximum element in each column and mark it as □. Saddle point is the

intersection of row minima and column maxima.

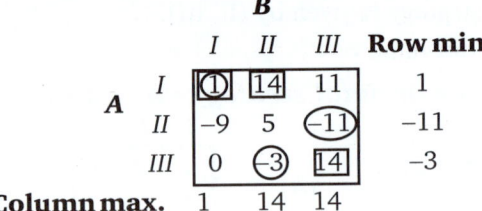

Hence, we conclude that

Strategy of A is I.

Strategy of B is I.

Value of game is 1.

EXAMPLE 10. *The player A's choice of action is given by (A_1, A_2, A_3) and B's choice of action is (B_1, B_2) only. The payment according to their activity is given as follow. Find the optimal strategy and value of the game*

Activity selected	Payment made
(A_1, B_1)	Player A pays ₹ 1 to player B
(A_2, B_1)	Player B pays ₹ 2 to player A
(A_3, B_1)	Player A pays ₹ 2 to player B
(A_1, B_2)	Player B pays ₹ 6 to player A
(A_2, B_2)	Player B pays ₹ 4 to player A
(A_3, B_2)	Player A pays ₹ 6 to player B

[GUJRAT, 1983]

SOLUTION. Firstly, make the pay-off matrix with the help of given data.

Player B

	B_1	B_2
A_1	−1	6
A_2	2	4
A_3	−2	−6

Player A

Table for the saddle point

	B_1	B_2	Row min
A_1	−1	6	−1
A_2	2	4	2
A_3	−2	−6	−6
Column max	2	6	

So, 2 is the required saddle point.

Optimal strategy is given by (A_2, B_1).

Value of the game $v = 2$.

16.2 SOLUTION OF A RECTANGULAR GAME IN TERMS OF MIXED STRATEGIES

In any game, if there is no saddle point, the two players can not use maximin, minimax strategies (pure) as their optimal strategies, then best strategies are mixed strategies. Now, two players, instead of selecting pure strategies only, may play their plays according to the predetermined set consisting probabilities corresponding to each of their pure strategies.

Thus, consider a rectangular game played by two players A (maximizing player) and B with pay off matrix $[a_{ij}]_{m \times n}$. The players A and B have m and n pure strategies respectively.

Player B

Probabilities		y_1	y_2	$\cdots$	y_j	$\cdots$	y_x
	Pure strategies	1	2	$\cdots$	j	$\cdots$	n
x_1	1	a_{11}	a_{12}	$\cdots$	a_{1j}	$\cdots$	a_{1n}
x_2	2	a_{21}	a_{22}	$\cdots$	a_{2j}	$\cdots$	a_{2n}
$\vdots$	$\vdots$	$\vdots$	$\vdots$	$\vdots$	$\vdots$	$\vdots$	$\vdots$
x_i	i	a_{i1}	a_{i2}	$\cdots$	a_{ij}	$\cdots$	a_{in}
$\vdots$	$\vdots$	$\vdots$	$\vdots$	$\cdots$	$\vdots$	$\cdots$	$\vdots$
x_m	m	a_{m1}	a_{m2}	$\cdots$	a_{mj}	$\cdots$	a_{mn}

Player A (label to left of table)

Let the mixed strategies of players A and B be respectively given by $X = (x_1, x_2, ..., x_m)$ and $Y = (y_1, y_2, ..., y_n)$.

Here, $x_1, x_2, ..., x_m$ and $y_1, y_2, ..., y_n$ are the probabilities of selecting pure strategies by A and B respectively.

Also, $\sum\limits_{i=1}^{m} x_i = 1$ and $\sum\limits_{j=1}^{n} y_j = 1$, $x_i \geq 0$, $y_j \geq 0$, $\forall\, i = 1, 2, ..., m$, $j = 1, 2, ..., n$

Further expected gain to A is

$$a_{11}x_1 + a_{21}x_2 + ... + a_{i1}x_i + ... + a_{m1}x_m = \sum\limits_{i=1}^{m} a_{i1}x_i$$

(If B uses strategy 1 with probability y_1)

$$a_{12}x_1 + a_{22}x_2 + ... + a_{i2}x_i + ... + a_{m2}x_m = \sum\limits_{i=1}^{m} a_{i2}x_i$$

(If B uses strategy 2 with probability y_2)

$$\cdots \qquad \cdots \qquad \cdots \qquad \cdots \qquad \cdots \qquad \cdots$$
$$\cdots \qquad \cdots \qquad \cdots \qquad \cdots \qquad \cdots \qquad \cdots$$

$$a_{1j}x_1 + a_{2j}x_2 + ... + a_{ij}x_i + ... + a_{mj}x_m = \sum\limits_{i=1}^{m} a_{ij}x_i$$

(If B uses strategy j with probability y_j)

$$\cdots \qquad \cdots \qquad \cdots \qquad \cdots \qquad \cdots \qquad \cdots$$
$$\cdots \qquad \cdots \qquad \cdots \qquad \cdots \qquad \cdots \qquad \cdots$$

$$a_{1n}x_1 + a_{2n}x_2 + ... + a_{in}x_i + ... + a_{mn}x_m = \sum\limits_{i=1}^{m} a_{in}x_i$$

(If B uses strategy n with probability y_n)

Then, expected gain to A is

$$E(X, Y) = \sum_{i=1}^{m} \sum_{j=1}^{n} a_{ij} x_i y_j \qquad \qquad ....(1)$$

Using maximin-minimax criterion, A selected $A_i \left(x_i \geq 0, \sum_{i=1}^{m} x_i = 1 \right)$ which will maximize

his minimum expected gain, i.e., A selects x_i which will

$$Max \left[Min \left\{ \sum_{i=1}^{m} a_{i1} x_i, \sum_{i=1}^{m} a_{i2} x_i, ..., \sum_{i=1}^{m} a_{in} x_i \right\} \right]$$

This value is known as maximin $(\underline{v})$ expected value for player A. In a similar manner, B

selects $y_j \left(y_j \geq 0, \sum_{i=1}^{n} y_j = 1 \right)$ which will minimize his maximum expected loss.

Therefore, B selects y_j which will

$$Min \left[Max \left\{ \sum_{j=1}^{n} a_{1j} y_j, \sum_{j=1}^{n} a_{2j} y_j, ..., \sum_{j=1}^{n} a_{nj} y_j \right\} \right]$$

This value is known as minimax $(\overline{v})$ expected value for B. Hence, we conclude that for

player A, the best strategy is that which maximizes the $Min_{j} \sum_{i=1}^{n} a_{ij} x_i$ and for player B best

strategy is that which minimizes the $Max \sum_{j=1}^{n} a_{ij} x_j$.

☞ REMARKS

- In case of pure strategies, $\underline{v} \leq \overline{v}$.
- Fundamental theorem of rectangular games assumes that there always exists optimum strategies such that $\underline{v} = \overline{v}$.

16.3 PROPERTIES OF OPTIMAL MIXED STRATEGIES

Some important properties of optimal mixed strategies are as follows :

1. If one of the players adheres to his optimal mixed strategy and other deviates from his optimal strategy, then deviating players can only increase his yield and can not increase in any case. At most it may be equal.

2. If one of the players adheres to his optimal strategy, then the value of the game does not change if other uses his supporting strategies only either singly or mixture.

3. If a fixed number a is added to each element of the pay off matrix, then the optimal strategies remain unchanged while the value of the game increases by a.

4. If every element of the pay off matrix is multiplied by a constant a, then the optimal strategies do not change while the value of the game becomes a times the value of the original game.

16.4 SOLUTION OF 2 x 2 GAMES WITHOUT SADDLE POINTS

THEOREM I. *For any zero sum two persons game where the optimal strategies are not pure strategies and for which A's pay-off matrix is*

$$B$$

$$
\begin{array}{c}
 & I\,(y_1) \quad II\,(y_2) \\
A \quad \begin{array}{c} I\,(x_1) \\ II\,(x_2) \end{array} & \boxed{\begin{array}{cc} a_{11} & a_{12} \\ a_{21} & a_{22} \end{array}}
\end{array}
$$

The optimal strategies (x_1, x_2) and (y_1, y_2) are given by

$$\frac{x_1}{x_2} = \frac{a_{22} - a_{21}}{a_{11} - a_{12}} \quad \text{and} \quad \frac{y_1}{y_2} = \frac{a_{22} - a_{12}}{a_{11} - a_{21}}$$

Also, the value of the game to A is given by

$$v = \frac{a_{11}a_{22} - a_{12}a_{21}}{(a_{11} + a_{22}) - (a_{12} + a_{21})}$$ [MEERUT-2002, 04, 07, 14; AGRA-2002,02]

<u>PROOF.</u> Let (x_1, x_2) and (y_1, y_2) be the mixed strategies for players A and B respectively. Then, clearly, we have

$$x_1 + x_2 = 1 \qquad \qquad \dots(1)$$
$$y_1 + y_2 = 1 \qquad \qquad \dots(2)$$

where, $x_1 \geq 0,\ x_2 \geq 0,\ y_1 \geq 0,\ y_2 \geq 0$.

Then, expected gain to A is given by $a_{11}x_1 + a_{21}x_2$, when B uses strategy I and expected gain to A is $a_{12}x_1 + a_{22}x_2$ when B uses strategy II.

Also, expected loss to B is $a_{11}y_1 + a_{12}y_2$ when A uses strategy I.

and expected loss to B is $a_{21}y_1 + a_{22}y_2$ when A uses strategy II.

Further, let v be the value of the game, then since A expects to get at least v, so

$$\left.\begin{array}{c} a_{11}x_1 + a_{21}x_2 \geq v \\ a_{12}x_1 + a_{22}x_2 \geq v \end{array}\right\} \qquad \qquad \dots(3)$$

Further, B expects to loose at most v, so

$$\left.\begin{array}{c} a_{11}y_1 + a_{12}y_2 \leq v \\ a_{21}y_1 + a_{22}y_2 \leq v \end{array}\right\} \qquad \qquad \dots(4)$$

Converting (3) and (4) into equation form, we get

$$a_{11}x_1 + a_{21}x_2 = v \qquad \qquad \dots(5)$$
$$a_{12}x_1 + a_{22}x_2 = v \qquad \qquad \dots(6)$$
$$a_{11}y_1 + a_{21}y_2 = v \qquad \qquad \dots(7)$$
$$a_{21}y_1 + a_{12}y_2 = v \qquad \qquad \dots(8)$$

From (5) and (6), we get

$$(a_{11} - a_{12})\,x_1 = (a_{22} - a_{21})\,x_2$$

$\Rightarrow$
$$\frac{x_1}{x_2} = \frac{a_{22} - a_{21}}{a_{11} - a_{12}} \qquad \qquad \dots(9)$$

Also, from (8) and (7), we get

$$\frac{y_1}{y_2} = \frac{a_{22} - a_{12}}{a_{11} - a_{21}} \qquad \qquad \dots(10)$$

Equation (9) can be written as

$$x_2 = \frac{a_{11} - a_{12}}{a_{22} - a_{21}}\, x_1$$

Putting this value in (1), we get

$$x_1 \left[1 + \frac{a_{11} - a_{12}}{a_{22} - a_{21}} \right] = 1$$

$\Rightarrow$
$$x_1 = \frac{a_{22} - a_{21}}{(a_{11} + a_{22}) - (a_{12} + a_{21})} \qquad \ldots(11)$$

and
$$x_2 = \frac{a_{11} - a_{12}}{(a_{11} + a_{22}) - (a_{12} + a_{21})} \qquad \ldots(12)$$

Similarly, from (2) and (10), we get

$$y_1 = \frac{a_{22} - a_{12}}{(a_{11} + a_{22}) - (a_{12} + a_{21})} \qquad \ldots(13)$$

and
$$y_2 = \frac{a_{11} - a_{21}}{(a_{11} + a_{22}) - (a_{12} + a_{21})} \qquad \ldots(14)$$

Putting all these values in (5), we get

$$v = \frac{a_{11}a_{22} - a_{12}a_{21}}{(a_{11} + a_{22}) - (a_{12} + a_{21})} \qquad \ldots(15)$$

Further, we have to prove that x_1, x_2, y_1, y_2 all are non-negative. Since game has no saddle point, thus the largest and second largest elements must lie on one of the diagonals. Thus, there are only the following possible ordering of the elements of the matrix.

$$a_{11} \geq a_{22} \geq a_{12} \geq a_{21}$$
$$a_{11} \geq a_{22} \geq a_{21} \geq a_{12}$$
$$a_{22} \geq a_{11} \geq a_{12} \geq a_{21}$$
$$a_{22} \geq a_{11} \geq a_{21} \geq a_{12}$$
$$a_{12} \geq a_{21} \geq a_{11} \geq a_{22}$$
$$a_{12} \geq a_{21} \geq a_{22} \geq a_{11}$$
$$a_{21} \geq a_{12} \geq a_{11} \geq a_{22}$$
$$a_{21} \geq a_{12} \geq a_{22} \geq a_{11}$$

Clearly, all ordering of the elements of the pay-off matrix x_1, x_2, y_1, y_2 given by (11), (12), (13), (14) are all non-negative.

☛ REMARK

- The above theorem is not always true for a 2×2 game with a saddle point. Thus, this formula should be applied in case of 2×2 game without saddle point.

Solved Examples

EXAMPLE 1. *In a game of matching coins with two players A wins ₹ 2 when two heads occur; wins nothing when two tails occur; and loses ₹ 1 when there are one head and one tail. Determine the pay-off matrix and best strategies for player A and B and also find the value of the game.*

[ROHILKHAND-2002, 03; MEERUT-2006]

SOLUTION. The pay-off matrix of the given problem will be

$$
\begin{array}{cc}
 & \textbf{\textit{B}} \\
 & \begin{array}{cc} H & T \end{array} \\
\textbf{\textit{A}} \quad \begin{array}{c} H \\ T \end{array} & \begin{array}{|cc|} \hline 2 & -1 \\ -1 & 0 \\ \hline \end{array}
\end{array}
$$

Now, the probability for choosing his strategy by player A

$$x_1 = \frac{a_{22} - a_{21}}{(a_{11} + a_{22}) - (a_{12} + a_{21})} = \frac{0 + 1}{(2 + 0) - (-1 - 1)} = \frac{1}{4}$$

$$x_2 = \frac{a_{11} - a_{12}}{(a_{11} + a_{22}) - (a_{12} + a_{21})} = \frac{2 + 1}{(2 + 0) - (-1 - 1)} = \frac{3}{4}$$

Probability for choosing his strategy by player B

$$y_1 = \frac{a_{22} - a_{12}}{(a_{11} + a_{22}) - (a_{12} + a_{21})} = \frac{0 - (-1)}{(2 + 0) - (-1 - 1)} = \frac{1}{4}$$

$$y_2 = \frac{a_{11} - a_{21}}{(a_{11} + a_{22}) - (a_{12} + a_{21})} = \frac{2 - (-1)}{(2 + 0) - (-1 - 1)} = \frac{3}{4}$$

Now, value of the game is given by

$$v = \frac{a_{11}a_{22} - a_{12}a_{21}}{(a_{11} + a_{22}) - (a_{12} + a_{21})}$$

$$= \frac{2 \times 0 - (-1)(-1)}{(2 + 0) - (-1 - 1)} = -\frac{1}{4}$$

Hence, Player A's optimal strategy $= \left(\dfrac{1}{4}, \dfrac{3}{4}\right)$

Player B's optimal strategy $= \left(\dfrac{1}{4}, \dfrac{3}{4}\right)$

and value of the game, $v = -\dfrac{1}{4}$.

EXAMPLE 2. *Find the optimal strategy for each of the player and the value of the game*

(a)

	B	
A	*I*	*II*
I	6	−3
II	−3	0

(b)

	B	
A	*I*	*II*
I	1	3
II	4	2

(c)

	B	
A	*I*	*II*
I	−4	6
II	2	−3

(d)

	B	
A	*I*	*II*
I	1	7
II	6	2

SOLUTION. (a) Probability for choosing his strategy by player A

$$x_1 = \frac{a_{22} - a_{21}}{(a_{11} + a_{22}) - (a_{12} + a_{21})} = \frac{0 + 3}{(6 + 0) - (-3 - 3)} = \frac{1}{4}$$

$$x_2 = \frac{a_{11} - a_{12}}{(a_{11} + a_{22}) - (a_{12} + a_{21})} = \frac{6 - (-3)}{(6 + 0) - (-3 - 3)} = \frac{3}{4}$$

Probability of choosing his strategy by player B

$$y_1 = \frac{a_{22} - a_{12}}{(a_{11} + a_{22}) - (a_{12} + a_{21})} = \frac{0 - (-3)}{(6 + 0) - (-3 - 3)} = \frac{1}{4}$$

$$y_2 = \frac{a_{11} - a_{21}}{(a_{11} + a_{22}) - (a_{12} + a_{21})} = \frac{6 - (-3)}{(6+0) - (-3-3)} = \frac{3}{4}$$

So, the optimal strategy for player A = $\left(\frac{1}{4}, \frac{3}{4}\right)$

the optimal strategy for player B = $\left(\frac{1}{4}, \frac{3}{4}\right)$.

Now, value of the game is given by

$$v = \frac{a_{11}a_{22} - a_{12}a_{21}}{(a_{11} + a_{22}) - (a_{12} + a_{21})}$$

$$= \frac{6 \times 0 - (-3)(-3)}{(6+0) - (-3-3)} = -\frac{3}{4}$$

$$\Rightarrow \quad v = -\frac{3}{4}$$

(b)

		B	
		I	II
A	I	1	3
	II	4	2

Optimal strategy used by player A = (x_1, x_2)

$$x_1 = \frac{2-4}{(1+2) - (3+4)} = \frac{-2}{3-7} = \frac{1}{2}; x_2 = \frac{1-3}{(1+2) - (3+4)} = \frac{-2}{-4} = \frac{1}{2}$$

Optimal strategy used by player B = (y_1, y_2)

$$y_1 = \frac{2-3}{(1+2) - (3+4)} = \frac{1}{4}; \quad y_2 = \frac{1-4}{(1+2) - (3+4)} = \frac{-3}{-4} = \frac{3}{4}$$

So, the optimal strategy for player A = $\left(\frac{1}{2}, \frac{1}{2}\right)$

the optimal strategy for player B = $\left(\frac{1}{4}, \frac{3}{4}\right)$.

And, value of the game is given by

$$v = \frac{2 \times 1 - 3 \times 4}{(2+1) - (3+4)} = \frac{2-12}{3-7} = \frac{5}{2} \quad \Rightarrow \quad v = \frac{5}{2}$$

(c)

		B	
		I	II
A	I	-4	6
	II	2	-3

Let the optimal strategy used by player A = (x_1, x_2)

and the optimal strategy used by player B = (y_1, y_2)

Now, $x_1 = \dfrac{-3-2}{(-4+(-3))-(6+2)} = \dfrac{-5}{-7-8} = \dfrac{1}{3}$

$x_2 = \dfrac{-4-6}{(-4-3)-(6+2)} = \dfrac{-10}{-7-8} = \dfrac{2}{3}$

$y_1 = \dfrac{-3-6}{(-4-3)-(6+2)} = \dfrac{-9}{-7-8} = \dfrac{3}{5}$

$y_2 = \dfrac{-4-2}{(-4-3)-(6+2)} = \dfrac{-6}{-15} = \dfrac{2}{5}$

Value of the game is given by, $v = \dfrac{(-4)(-3)-6\times 2}{(-4-3)-(6+2)} = 0$

So, Optimal strategy used by player A = $\left(\dfrac{1}{3}, \dfrac{2}{3}\right)$

Optimal strategy used by player B = $\left(\dfrac{3}{5}, \dfrac{2}{5}\right)$

and value of the game, $v = 0$.

(d)

B

		I	II
A	I	1	7
	II	6	2

Let the optimal strategy used by player A = (x_1, x_2)

and the optimal strategy used by player B = (y_1, y_2)

Now, $x_1 = \dfrac{2-6}{(1+2)-(7+6)} = \dfrac{-4}{3-13} = \dfrac{2}{5}$;

$x_2 = \dfrac{1-7}{(1+2)-(7+6)} = \dfrac{-6}{3-13} = \dfrac{3}{5}$

$y_1 = \dfrac{2-7}{(1+2)-(7+6)} = \dfrac{-5}{-10} = \dfrac{1}{2}$;

$y_2 = \dfrac{1-6}{(1+2)-(7+6)} = \dfrac{-5}{-10} = \dfrac{1}{2}$

and value of the game is given by

$v = \dfrac{1\times 2 - 7\times 6}{(1+2)-(7+6)} = \dfrac{2-42}{3-13} = 4$

So, Optimal strategy used by player A = $\left(\dfrac{2}{5}, \dfrac{3}{5}\right)$

Optimal strategy used by player B = $\left(\dfrac{1}{2}, \dfrac{1}{2}\right)$

and value of the game, $v = 4$.

16.5 SOLUTION OF A GAME BY LINEAR PROGRAMMING [MEERUT-2000, 02, 03, 04, 16; ROHILKHAND- 1997]

Let a rectangular game be played by two players A (max player) and B (min player) with pay-off matrix $[a_{ij}]_{m \times n}$. The mixed strategies for A and B be respectively given by $X = [x_1, x_2, ..., x_m]$ and $Y = [y_1, y_2, ..., y_n]$. We know that A selects his optimal mixed strategies which is given by

$$\underset{x_i}{Max} \left[Min \left\{ \sum_{i=1}^{m} a_{i1}x_i, \sum_{i=1}^{m} a_{i2}x_i, ..., \sum_{i=1}^{m} a_{in}x_i \right\} \right]$$

such that $x_1 + x_2 + ... + x_m = 1, \ x_i \geq 0, \ \forall \ i = 1, 2, ..., m$.

Further, B selects his optimal strategies which will be

$$\underset{y_j}{Min} \left[Max \left\{ \sum_{j=1}^{n} a_{1j}y_j, \sum_{j=1}^{n} a_{2j}y_j, ..., \sum_{j=1}^{n} a_{mj}y_j \right\} \right]$$

such that $y_1 + y_2 + ... + y_n = 1, \ y_j \geq 0, \ \forall j = 1, 2, ..., n$

Now, if

$$Min \left\{ \sum_{i=1}^{n} a_{i1}x_i, \sum_{i=1}^{m} a_{i2}x_i, ..., \sum_{i=1}^{m} a_{in}x_i \right\} = v$$

then, A expects to gain at least v. Now, since v is minimum of all expected gains, therefore, we have

$$\sum_{i=1}^{m} a_{i1}x_i \geq v, \ \sum_{i=1}^{m} a_{i2}x_i \geq v, ..., \sum_{i=1}^{m} a_{in}x_i \geq v$$

Therefore, A's problem to determine $x_1, x_2, ..., x_n$ to minimize $z = v$, subject to the constraints

$$\left. \begin{array}{l} a_{11}x_1 + a_{21}x_2 + ... + a_{m1}x_m \geq v \\ a_{12}x_1 + a_{22}x_2 + ... + a_{m2}x_m \geq v \\ ... \qquad ... \qquad ... \qquad ... \\ ... \qquad ... \qquad ... \qquad ... \\ a_{1n}x_1 + a_{2n}x_2 + ... + a_{mn}x_m \geq v \end{array} \right] \qquad ....(1)$$

$$x_1 + x_2 + ... + x_m = 1, \quad x_1, x_2, x_m \geq 0$$

For $v > 0$, it is enough to prove that all the elements of the pay-off matrix are positive. If all are not positive, then we can add a sufficient large quantity to every element of the pay-off matrix so that they all become positive. Then, we can take v as positive.

Putting $\dfrac{x_1}{v} = X_1, \dfrac{x_2}{v} = X_2, ..., \dfrac{x_m}{v} = X_m$ in (1), we get

$$a_{11}X_1 + a_{21}X_2 + ... + a_{m1}X_m \geq 1$$

$$a_{12}X_1 + a_{22}X_2 + ... + a_{m2}X_m \geq 1$$

$$... \qquad ... \qquad ... \qquad ...$$

$$... \qquad ... \qquad ... \qquad ...$$

$$a_{1n}X_1 + a_{2n}X_2 + ... + a_{mn}X_m \geq 1$$

and

$$X_1 + X_2 + ... + X_m = \frac{1}{v}, \quad X_1, X_2, ..., X_m \geq 0$$

Now, $$Max. \ v = Min \left(\frac{1}{v}\right)$$

$$= Min \left\{\frac{x_1 + x_2 + \dots + x_m}{v}\right\}$$

$$= Min \ (X_1 + X_2 + \dots + X_m)$$

Hence, the given rectangular game reduces to the following linear programming.

$$Min \ x^* = \frac{1}{v} = X_1 + X_2 + \dots + X_m$$

subject to the constraints

$$\left.\begin{array}{l} a_{11}X_1 + a_{21}X_1 + \dots + a_{m1}X_m \geq 1 \\ a_{12}X_1 + a_{22}X_2 + \dots + a_{m2}X_m \geq 1 \\ \dots \qquad \dots \qquad \dots \qquad \dots \\ \dots \qquad \dots \qquad \dots \qquad \dots \\ a_{1n}X_1 + a_{2n}X_2 + \dots + a_{mn}X_m \geq 1 \\ X_1 \geq 0, \ \ X_2 \geq 0, \dots, X_m \geq 0 \end{array}\right] \qquad \dots(2)$$

and

For B's point of view, we want to minimize v, by solving the following linear programming

$$Maximize \ y^* = \frac{1}{v} = Y_1 + Y_2 + \dots + Y_n$$

subject to the constraints

$$\left.\begin{array}{l} a_{11}Y_1 + a_{12}Y_1 + \dots + a_{1n}Y_n \geq 1 \\ a_{21}Y_1 + a_{22}Y_1 + \dots + a_{2n}Y_n \geq 1 \\ \dots \qquad \dots \qquad \dots \qquad \dots \\ \dots \qquad \dots \qquad \dots \qquad \dots \\ a_{m1}Y_1 + a_{m2}Y_2 + \dots + a_{mn}Y_n \geq 1 \\ Y_1, Y_2, \dots, Y_m \geq 0. \end{array}\right] \qquad \dots(3)$$

and

where, $y^* = \frac{1}{v}$, $Y_1 = \frac{y_1}{v}$, $Y_2 = \frac{y_2}{v}, \dots, Y_n = \frac{y_n}{v}$

☛ REMARK
- The linear programming problems given by (2) and (3) are the duals of each other. Thus, if one problem is solved, then other is solved automatically.

16.6 MINIMAX THEOREM : FUNDAMENTAL THEOREM OF GAME THEORY

THEOREM I. *Every game can be solved in terms of mixed strategies, i.e., if mixed strategies are adopted, there always exists a value of the game, i.e., $\underline{v} = v = \overline{v}$, where $\underline{v}$ and $\overline{v}$ are maximin and minimax values of v.*

[BHOPAL-2010; SAGAR-2005; ASSAM-2011]

PROOF. If $X = (x_1, x_2, \dots, x_n)$, $Y = (y_1, y_2, \dots, y_n)$ are the mixed strategies of two players where $x_1, x_2, \dots, x_m$, $y_1, y_2, \dots, y_n$ are the probabilities with which they choose their pure strategies, then A's problem is given by,

Minimize $x^* = X_1 + X_2 + \dots + X_m$

subject to the constraints

$$a_{11}X_1 + a_{21}X_2 + \dots + a_{m1}X_m \geq 1$$
$$a_{12}X_1 + a_{22}X_2 + \dots + a_{m2}X_m \geq 1$$

$$\ldots \quad \ldots \quad \ldots \quad \ldots$$
$$a_{1n}X_1 + a_{2n}X_2 + \ldots + a_{mn}X_m \geq 1$$
$$X_1, X_2, \ldots, X_m \geq 0 \quad \text{and} \quad X_i = \frac{x_i}{v}$$

Similarly, the problem for player B is given by

Maximize $y^* = Y_1 + Y_2 + \ldots + Y_n$

subject to the constraints

$$a_{11}Y_1 + a_{12}Y_2 + \ldots + a_{1n}Y_n \leq 1$$
$$a_{21}Y_1 + a_{22}Y_2 + \ldots + a_{2n}Y_n \leq 1$$
$$\ldots \quad \ldots \quad \ldots \quad \ldots$$
$$\ldots \quad \ldots \quad \ldots \quad \ldots$$
$$a_{m1}Y_1 + a_{m2}Y_2 + \ldots + a_{mn}Y_n \leq 1$$
$$Y_1, Y_2, \ldots, Y_n \geq 0 \quad \text{and} \quad Y_j = \frac{y_j}{v}$$

Clearly, B's problem is the dual of A's problem. By duality theorem, it is known that If either the primal or the dual has a finite optimal solutions, then the other has a finite optimal solution and the optimal values of two objective functions are equal. Hence,

$$Max\ y^* = Min\ x^*$$

or $$\underline{v} = v = \overline{v}.$$

 Working Procedure

If a problem has no saddle point, then convert it to a linear programming problem for player B. Then solve it by simplex method. From the final simplex table of the solution, we can read the solution for player A by the dual method.

 Solved Examples

EXAMPLE 1. ***Solve the following game by Simplex method. Find the optimal strategy and value of the game***

$$\begin{array}{c} & B \\ A & \begin{array}{|ccc|} \hline 1 & -1 & 3 \\ 3 & 5 & -3 \\ 6 & 2 & -2 \\ \hline \end{array} \end{array}$$

[MEERUT 2003, 06]

SOLUTION. First of all we find out maximin and minimax element from row and column. From this, we can find the range in which the value of the game will lie

$$\begin{array}{ccc} & & \textbf{Row min} \\ \begin{array}{ccc} 1 & \boxed{-1} & \boxed{3} \\ 3 & \boxed{5} & \boxed{-3} \\ \boxed{6} & 2 & \boxed{-2} \end{array} & & \left.\begin{array}{c} -1 \\ -3 \\ -2 \end{array}\right\} \text{max} = -1 \end{array}$$

Column max $\underbrace{6 \quad 5 \quad 3}_{\text{min} = 3}$

$\Rightarrow \qquad -1 \leq v \leq 3$

This implies that value of the game may be negative or zero. So, we add a constant such that all entry in the given pay-off matrix become +ve.

So, we add $k = 4$. Then the new pay-off matrix will be

$$\begin{vmatrix} 5 & 3 & 7 \\ 7 & 9 & 1 \\ 10 & 6 & 2 \end{vmatrix}$$

Let the strategy used by player A $= (x_1, x_2, x_3)$

The strategy used by player B $= (y_1, y_2, y_3)$

Now, B's linear programming problem will be

Min v'

s.t. $5y_1 + 3y_2 + 7y_3 \leq v'$

$7y_1 + 9y_2 + y_3 \leq v'$

$10y_1 + 6y_2 + 2y_3 \leq v'$

and $y_1 + y_2 + y_3 = 1$, where $y_1, y_2, y_3 \geq 0$

Dividing all the equations by v' and putting $\dfrac{y_1}{v'} = Y_1$, $\dfrac{y_2}{v'} = Y_2$, $\dfrac{y_3}{v'} = Y_3$

Now, the problem reduces to

$$Max\ z = \frac{1}{v'} = \frac{y_1 + y_2 + y_3}{v'} = \frac{y_1}{v'} + \frac{y_2}{v'} + \frac{y_3}{v'}$$

$$Max\ z = Y_1 + Y_2 + Y_3$$

s.t. $5Y_1 + 3Y_2 + 7Y_3 \leq 1$

$7Y_1 + 9Y_2 + Y_3 \leq 1$

$10Y_1 + 6Y_2 + 2Y_3 \leq 1$, where, $Y_1, Y_2, Y_3 \geq 0$

So, we can write $Max\ z = Y_1 + Y_2 + Y_3 + 0s_1 + 0s_2 + 0s_3$

s.t. $5Y_1 + 3Y_2 + 7Y_3 + s_1 = 1$

$7Y_1 + 9Y_2 + Y_3 + s_2 = 1$

$10Y_1 + 6Y_2 + 2Y_3 + s_3 = 1$

B.V.	C_B	X_B	$c_j \rightarrow$ Y_1 1	Y_2 1	Y_3 1	s_1 0	s_2 0	s_3 0	Min Ratio X_B / Y_j
s_1	0	1	5	3	$\boxed{7}$	1	0	0	1/7 ←
s_2	0	1	7	9	1	0	1	0	1
s_3	0	1	10	6	2	0	0	1	1/2
			1	1	1				
Y_3	1	1/7	5/7	3/7	1	1/7	0	0	1/3
s_2	0	6/7	44/7	$\boxed{60/7}$	0	–1/7	1	0	1/10 ←
s_3	0	5/7	60/7	36/7	0	–2/7	0	1	5/36
			2/7	4/7 ↑	0	–1/7	0	0	

Y_3	1	1/10	2/5	0	1	3/20	–1/20	0	
Y_2	1	1/10	11/15	1	0	–1/60	7/60	0	
s_3	0	1/5	24/5	0	0	–1/5	–3/5	1	
z		1/5	–2/15	0	0	–2/15	–1/15	0	

Since all terms are ≤ 0. So the solution obtained will be optimal.

$$Y_1 = 0, \ Y_2 = 1/10, \ Y_3 = 1/10.$$

$$\text{Max } z = c_B X_B = 1/5$$

Now, $\qquad \text{Max } z = \dfrac{1}{v'}$

So, $\qquad \dfrac{1}{v'} = \dfrac{1}{5}$

$\Rightarrow \qquad \text{Min } v' = 5$.

and $\quad y_1 / v' = Y_1, \ \dfrac{y_2}{v'} = Y_2, \ \dfrac{y_3}{v'} = Y_3$.

$$y_1 = Y_1 v' = 0$$

$$y_2 = Y_2 v' = \dfrac{1}{10} \times 5 = \dfrac{1}{2}$$

$$y_3 = Y_3 v' = \dfrac{1}{10} \times 5 = \dfrac{1}{2}.$$

Now, to find the value of the game, we subtract the previous added constant $k = 4$ from the v'.

Then, $\qquad v = v' - 4 = 5 - 4 = 1$

$\qquad\qquad v = 1$.

Now, since player A's strategy will be the dual of the strategy of player B.

So, X_1, X_2 and X_3 will be the value of Δ_4, Δ_5 and Δ_6.

So, $\quad X_1 = 2/15 \quad X_2 = 1/15 \qquad X_3 = 0$

$$X_1 = \dfrac{x_1}{v'} \qquad X_2 = \dfrac{x_2}{v'} \qquad X_3 = \dfrac{x_3}{v'}$$

$$x_1 = X_1 v' \qquad x_2 = X_2 v' \qquad X_3 = X_3 v'$$

$$x_1 = \dfrac{2}{15} \times 5, x_2 = \dfrac{1}{15} \times 5, \ x_3 = 0 \times 5$$

$$x_1 = \dfrac{2}{3}, \ x_2 = \dfrac{1}{3}, \ x_3 = 0.$$

Hence, the optimal strategy for player $A = \left(\dfrac{2}{3}, \dfrac{1}{3}, 0 \right)$

also, the optimal strategy for player $B = \left(0, \dfrac{1}{2}, \dfrac{1}{2} \right)$

and value of the game $v = 1$.

EXAMPLE 2. *Solve the following game by linear programming problem whose pay-off matrix is given by*

Player B

		I	II	III
	I	3	-1	-3
Player A	II	-3	3	-1
	III	-4	-3	3

SOLUTION. First of all we find out maximin and minimax element from row and column. From this, we can find the range in which the value of the game will lie

Row min

$$\begin{vmatrix} 3 & -1 & -3 \\ -3 & 3 & -1 \\ -4 & -3 & 3 \end{vmatrix} \quad \begin{matrix} -3 \\ -3 \\ -4 \end{matrix} \Big\} \quad \text{max} = -3$$

Column max   3   3   3

$$\underbrace{\qquad\qquad}_{\text{min} = 3}$$

$\Rightarrow \qquad -3 \le v \le 3$

This implies that value of the game may be negative or zero. So, we add a constant such that all entry in the given pay-off matrix become positive.

So, we add $k = 5$. Then the new payoff matrix will be

$$\begin{vmatrix} 8 & 4 & 2 \\ 2 & 8 & 4 \\ 1 & 2 & 8 \end{vmatrix}$$

Let the strategy used by player A $= (x_1, x_2, x_3)$

The strategy used by player B $= (y_1, y_2, y_3)$

Now, B's problem is given by

Min v'

s.t. $\qquad 8y_1 + 4y_2 + 2y_3 \le v'$

$\qquad\qquad 2y_1 + 8y_2 + 4y_3 \le v'$

$\qquad\qquad y_1 + 2y_2 + 8y_3 \le v'$

and $\qquad y_1 + y_2 + y_3 = 1$

where $\qquad y_1, y_2, y_3 \ge 0$

Dividing all the equations by v' and putting

$$\frac{y_1}{v'} = Y_1, \quad \frac{y_2}{v'} = Y_2, \quad \frac{y_3}{v'} = Y_3$$

Now, the problem reduces to

$\qquad Max \ z = Y_1 + Y_2 + Y_3$

s.t. $\quad 8Y_1 + 4Y_2 + 2Y_3 \le 1$

$\qquad\quad 2Y_1 + 8Y_2 + 4Y_3 \le 1$

$\qquad\quad Y_1 + 2Y_2 + 8Y_3 \le 1$

where, $\quad Y_1, Y_2, Y_3 \ge 0$

The above problem can be written as

$\qquad Max \ z = Y_1 + Y_2 + Y_3 + 0s_1 + 0s_2 + 0s_3$

s.t. $\qquad 8Y_1 + 4Y_2 + 2Y_3 + s_1 = 1$

$\qquad\qquad 2Y_1 + 8Y_2 + 4Y_3 + s_2 = 1$

$\qquad\qquad Y_1 + 2Y_2 + 8Y_3 + s_3 = 1$

where, $Y_1, Y_2, Y_3, s_1, s_2, s_3 \geq 0$

B.V.	C_B	X_B	$c_j \to$ Y_1	Y_2	Y_3	s_1	s_2	s_3	Min Ratio X_B / Y_j
			1	1	1	0	0	0	
s_1	0	1	8	4	2	1	0	0	1/8 ←
s_2	0	1	2	8	4	0	1	0	1/2
s_3	0	1	1	2	8	0	0	1	1
			1 ↑	1	1	0	0	0	
Y_1	1	1/8	1	1/2	1/4	1/8	0	0	1/2
s_2	0	3/4	0	7	7/2	−1/4	1	0	3/14
s_3	0	7/8	0	3/2	31/4	−1/8	0	1	7/62 ←
			0	1/2	3/4 ↑	−1/8	0	−1	
Y_1	1	3/31	1	14/31	0	4/31	0	−1/31	3/14
s_2	0	11/31	0	196/31	0	−6/31	1	−14/31	11/196 ←
Y_3	1	7/62	0	6/31	1	−1/62	0	4/31	7/12
			0	11/31 ↑	0	−7/62	0	−3/31	
Y_1	1	1/14	1	0	0	1/7	1/14	0	
Y_2	1	11/196	0	1	0	−3/98	31/196	−1/14	
Y_3	1	5/49	0	0	1	−1/98	−3/98	1/7	
z		45/196	0	0	0	−5/49	−11/196	−1/14	

Since all terms are ≤ 0. So the solution obtained will be optimal.

$$Y_1 = 1/14, \ Y_2 = 11/196, \ Y_3 = 5/49.$$

$$\text{Max } z = \frac{45}{196}$$

Now, $$\text{Max } z = \frac{1}{v'} = \frac{45}{196}$$

$\Rightarrow$ $$\text{Min } v' = \frac{196}{45}.$$

and $\dfrac{y_1}{v'} = Y_1, \ \dfrac{y_2}{v'} = Y_2, \ \dfrac{y_3}{v'} = Y_3.$

$$y_1 = Y_1 \, v', \qquad y_2 = Y_2 \, v', \qquad y_3 = Y_3 \, v'$$

$\Rightarrow$ $$y_1 = \frac{1}{14} \times \frac{196}{45} = \frac{14}{45}$$

$$y_2 = \frac{11}{196} \times \frac{196}{45} = \frac{11}{45}$$

$$y_3 = \frac{5}{49} \times \frac{196}{45} = \frac{20}{45}.$$

and value of the game $v = v' - 5 = \frac{196}{45} - 5 = -\frac{29}{45}$

$$v = -\frac{29}{45}$$

Now, by the dual of this LPP, we can find out the optimal strategy of player A.

Hence, A's optimal strategy $= \left(\frac{20}{45}, \frac{11}{45}, \frac{14}{45}\right)$

B's optimal strategy $= \left(\frac{14}{45}, \frac{11}{45}, \frac{20}{45}\right)$

and value of the game $v = -\frac{29}{45}$.

Exercise-16.1

1. Convert the given game into linear programming problem and solve them

(a) $\begin{bmatrix} 5 & 3 & 7 \\ 7 & 9 & 1 \\ 10 & 6 & 2 \end{bmatrix}$
(b) $\begin{bmatrix} 3 & -2 & 4 \\ -1 & 4 & 2 \end{bmatrix}$
(c) $\begin{bmatrix} 0 & -1 & 1 \\ 1 & 1 & -1 \\ 1 & -1 & 0 \end{bmatrix}$
(d) $\begin{bmatrix} 3 & -2 & 4 \\ -1 & 4 & 2 \\ 2 & 2 & 6 \end{bmatrix}$

ANSWERS

1. (a) $\left(\frac{2}{3}, \frac{1}{2}, 0\right)$, $\left(0, \frac{1}{2}, \frac{1}{2}\right)$ and $v = 5$.
(b) $\left(\frac{1}{2}, \frac{1}{2}\right)$, $\left(\frac{3}{5}, \frac{2}{5}, 0\right)$ and $v = 1$.

(c) $\left(\frac{1}{2}, \frac{1}{2}, 0\right)$, $\left(0, \frac{1}{2}, \frac{1}{2}\right)$ and $v = 0$.
(d) $(0, 0, 1)$, $\left(\frac{4}{5}, \frac{1}{5}, 0\right)$ and $v = 2$.

16.7 DOMINANCE PROPERTY

16.7.1 PRINCIPLE OF DOMINANCE

If one pure strategy of a player is better or superior than another one, then the inferior strategy may be simply ignored by assigning a zero probability while searching for optimal strategies. [MEERUT-1998, 2001, 02, 03, 04, 05, 15]

16.7.2 RULES OF DOMINANCE

RULE (1). If each element in one row, say ith of the pay off matrix are less than or equal to the corresponding elements of the other row (say jth row), then player A will never choose the ith strategy, i.e., ith strategy is dominated by the jth strategy.

RULE (2). If each element in one column, say rth is greater than or equal to the corresponding elements of the other column, say jth, then the player B never use the rth strategy, i.e., jth strategy dominates the rth strategy.

RULE (3). A pure strategy may be dominated if it is inferior to an average of two or more other pure strategy.

REMARKS

- The rules of dominance are especially used for evaluation of 2 person zero sum games without saddle points. Using these dominance properties, we try to reduce the size of pay-off matrix.
- In case the pay-off matrix reduces to the size $\times$ or $m \times 2$, then we use the graphical method (discussed in next article) to find the solution of this game.
- If the dominance holds strictly, then value of the optimal strategies do coincide and when the dominance does not hold strictly, then optimal strategy may not coincide.
- These rules of dominance are used when the pay-off matrix is a profit matrix for the player A and a loss matrix for player B, otherwise the rule gets reversed.

Solved Examples

EXAMPLE 1. *Solve the game whose pay-off matrix is given below*

$$
\begin{array}{c}
 & & B \\
 & & I \quad II \quad III \\
A \quad \begin{array}{c} I \\ II \\ III \end{array} & \begin{bmatrix} 2 & 3 & 1/2 \\ 3/2 & 2 & 0 \\ 1/2 & 1 & 1 \end{bmatrix}
\end{array}
$$

SOLUTION. We have

			B		Row min
		I	II	III	
	I	2	3	1/2	1/2
A	II	3/2	2	0	0
	III	1/2	1	1	1/2
Column max		2	3	1	

There is no saddle point. To find the optimal strategy, first we will convert it into 2×2 matrix with the help of dominance property.

	I	II	III
I	2	3	1/2
II	3/2	2	0
III	1/2	1	1

Here, column III is inferior to II. So, column II will dominate the column III. So, we can delete II.

Then

	I	III
I	2	1/2
II	~~3/2~~	~~0~~
III	1/2	1

Here, row I is superior to row II. So, row I dominates the II. So, we can delete row II. Now, our game reduces to 2×2 game which can be solved by arithmetic method without saddle point.

	I	III
I	2	1/2
III	1/2	1

If (x_1, x_2, x_3) and (y_1, y_2, y_3) be the optimal strategy used by player A and B respectively, then

$$x_1 = \frac{a_{22} - a_{21}}{(a_{22} - a_{21}) + (a_{11} - a_{12})} = \frac{1 - 1/2}{(1 - 1/2) + (2 - 1/2)} = \frac{1/2}{1/2 + 3/2}$$

$$x_1 = 1/4$$

Thus $x_3 = 3/4$

So, the strategy used by player A = $\left(\frac{1}{4}, 0, \frac{3}{4} \right)$

Thus, $\quad y_1 = \frac{a_{22} - a_{12}}{(a_{22} - a_{12}) + (a_{11} - a_{21})} = \frac{1 - 1/2}{(1 - 1/2) + (2 - 1/2)} = \frac{1/2}{4/2}$

$$y_1 = 1/4$$

$$y_3 = 3/4$$

So, the strategy used by player B = $\left(\frac{1}{4}, 0, \frac{3}{4} \right)$

Now, value of the game = $\dfrac{a_{22}a_{11} - a_{12}a_{21}}{(a_{22} + a_{11}) - (a_{12} + a_{21})}$

$$= \frac{1 \times 2 - \dfrac{1}{2} \times \dfrac{1}{2}}{(1 + 2) - \left(\dfrac{1}{2} + \dfrac{1}{2} \right)} = \frac{7}{8}$$

$$v = 7/8.$$

EXAMPLE 2. *Solve the following game*

$$
\begin{array}{cc}
 & B \\
 & \begin{array}{ccc} I & II & III \end{array} \\
A \begin{array}{c} I \\ II \\ III \end{array} & \begin{array}{|ccc|} \hline 1 & 7 & 2 \\ 6 & 2 & 7 \\ 5 & 2 & 6 \\ \hline \end{array}
\end{array}
$$

[OSMANIA-1985; MEERUT-1996, 2006; ROHILKHAND-2011; PATNA-2010]

SOLUTION. We have

		B		
	I	II	III	**Row min**
A I	①	⑦	2	1
II	⑥	②	⑦	2
III	5	②	6	2
Column max	6	7	7	

There is no saddle point. To find the optimal strategy, first we will convert it into 2×2 matrix with the help of dominance property.

$$
\begin{array}{cccc}
 & I & II & III \\
I & 1 & 7 & 2 \\
II & 6 & 2 & 7 \\
III & 5 & 2 & 6 \\
\end{array}
$$

Column I is inferior to III. So, we can delete the III column.

$$
\begin{array}{c|cc}
 & I & II \\
\hline
I & 1 & 7 \\
II & 6 & 2 \\
III & 5 & 2 \\
\end{array}
$$

Row III is inferior to row II. So, we can delete the III row.

Now, our problem reduces to 2×2 matrix without saddle point.

$$
\begin{array}{c|cc}
 & I & II \\
\hline
I & 1 & 7 \\
II & 6 & 2 \\
\end{array}
$$

Let (x_1, x_2, x_3) be the optimal strategy used by A.

(y_1, y_2, y_3) be the optimal strategy used by B.

$$
\frac{x_1}{x_2} = \frac{a_{22} - a_{21}}{a_{11} - a_{12}} = \frac{2-6}{1-7} = \frac{2}{3}
$$

$$
x_1 = \frac{2}{5}, \quad x_2 = \frac{3}{5}
$$

$$
\frac{y_1}{y_2} = \frac{a_{22} - a_{12}}{a_{11} - a_{21}} = \frac{2-7}{1-6} = \frac{5}{5} = 1
$$

$$
y_1 = \frac{1}{2}, \quad y_2 = \frac{1}{2}
$$

Value of the game, $v = \dfrac{a_{11}a_{22} - a_{12}a_{21}}{(a_{11} + a_{22}) - (a_{12} + a_{21})} = \dfrac{1 \times 2 - 7 \times 6}{(1+2) - (7+6)} = \dfrac{2 - 42}{3 - 13}$

$$
v = 4 .
$$

Optimal strategy used by A $= \left(\dfrac{2}{5}, \dfrac{3}{5}, 0 \right)$

Optimal strategy used by B $= \left(\dfrac{1}{2}, \dfrac{1}{2}, 0 \right)$

Value of the game $= 4$.

EXAMPLE 3.   ***Solve the following game to find the optimal strategy***

B

$$
\begin{array}{c c|cccc}
 & & I & II & III & IV \\
\hline
 & I & 3 & 2 & 4 & 0 \\
\mathbf{A} & II & 2 & 4 & 2 & 4 \\
 & III & 4 & 2 & 4 & 0 \\
 & IV & 0 & 4 & 0 & 8 \\
\end{array}
$$

[MEERUT 1985, 2005, 2007, 11,17; ROHILKHAND-1997, 2000]

SOLUTION.   We use the dominance property to solve this game.

Row I is inferior to III. So, we can delete it.

$$
\begin{array}{c|cccc}
 & I & II & III & IV \\
\hline
I & 3 & 2 & 4 & 0 \\
II & 2 & 4 & 2 & 4 \\
III & 4 & 2 & 4 & 0 \\
IV & 0 & 4 & 0 & 8 \\
\end{array}
$$

Column I is superior to III. So, we can delete it.

$$\begin{array}{c|cccc} & I & II & III & IV \\ \hline II & 2 & 4 & 2 & 4 \\ III & 4 & 2 & 4 & 0 \\ IV & 0 & 4 & 0 & 8 \end{array}$$

Since there is no row and column which are comparable. So, we use the average. Here, the average of II and III column is inferior to I. So, we can delete column-I.

$$\begin{array}{c|ccc} & II & III & IV \\ \hline II & 4 & 2 & 4 \\ III & 2 & 4 & 0 \\ IV & 4 & 0 & 8 \end{array}$$

$$\begin{array}{c|cc} & III & IV \\ \hline II & 2 & 4 \\ III & 4 & 0 \\ IV & 0 & 8 \end{array}$$

Here, row I is equal to the average of II and III. So, we can delete row-I. Now, the reduced pay off matrix is

$$\begin{array}{c|cc} & III & IV \\ \hline III & 4 & 0 \\ IV & 0 & 8 \end{array}$$

Since no saddle point is obtained. So, the optimal strategy will be the mixed strategy.

$$\frac{x_3}{x_4} = \frac{a_{22} - a_{21}}{a_{11} - a_{12}} = \frac{8 - 0}{4 - 0} = \frac{2}{1}$$

$$x_3 = \frac{2}{3}, \quad x_4 = \frac{1}{3}$$

$$\frac{y_3}{y_4} = \frac{a_{22} - a_{12}}{a_{11} - a_{21}} = \frac{8 - 0}{4 - 0} = \frac{2}{1}$$

$$y_3 = \frac{2}{3}, \quad y_4 = \frac{1}{3}$$

Finally, Optimal strategy used by A $= \left(0, 0, \dfrac{2}{3}, \dfrac{1}{3}\right)$

Optimal strategy used by B $= \left(0, 0, \dfrac{2}{3}, \dfrac{1}{3}\right)$

Value of the game = 8/3

EXAMPLE 4. ***Find the optimal strategy and value of the game***

$$\begin{array}{cc|ccc} & & & B & \\ & & I & II & III \\ \hline & I & 2 & 0 & 3 \\ A & II & 3 & -1 & 1 \\ & III & 5 & 2 & -1 \end{array}$$

SOLUTION. Since there is no saddle point, so we use the principle of dominance in the following manner.

$$
\begin{array}{c|ccc}
 & I & II & III \\
\hline
I & 2 & 0 & 3 \\
II & 3 & -1 & 1 \\
III & 5 & 2 & -1 \\
\end{array}
$$

$$
\begin{array}{c|cc}
 & II & III \\
\hline
I & 0 & 3 \\
II & -1 & -1 \\
III & 2 & -1 \\
\end{array}
$$

$$
\begin{array}{c|cc}
 & II & III \\
\hline
I & 0 & 3 \\
III & 2 & -1 \\
\end{array}
$$

Now, it is a 2×2 game without saddle point. So, the strategy will be mixed strategy.

$$\frac{x_1}{x_3} = \frac{a_{22} - a_{21}}{a_{11} - a_{12}} = \frac{-1-2}{0-3} = \frac{1}{1}$$

$$x_1 = \frac{1}{2}, \quad x_3 = \frac{1}{2}$$

So, the strategy used by A $= \left(\frac{1}{2}, 0, \frac{1}{2}\right)$

$$\frac{y_2}{y_3} = \frac{-1-3}{0-2} = \frac{-4}{-2} = \frac{2}{1}$$

$$y_2 = \frac{2}{3}, \quad y_3 = \frac{1}{3}$$

So, the strategy used by B $= \left(0, \frac{2}{3}, \frac{1}{3}\right)$

Value of the game, $v = \dfrac{a_{22}a_{11} - a_{21}a_{12}}{(a_{11} + a_{22}) - (a_{21} + a_{12})}$

$$\Rightarrow \qquad v = 1$$

EXAMPLE 5. *Given the following pay-off matrix of player A. Obtain the optimum strategies for both the players.*

$$
\begin{array}{c|ccc}
 & \multicolumn{3}{c}{B} \\
\hline
 & 2 & 8 & 2 \\
A & 8 & 3 & 8 \\
 & 5 & 3 & 7 \\
\end{array}
$$

SOLUTION. We try to find a saddle point of the game. The game does not have a saddle point. So apply the rule of dominance.

Every element of 3rd column is greater than or equal to the first column, therefore, by dominance rule, from player B's point of view, the pure strategy (3rd column) is dominated by the first. Thus, we get the reduced matrix

$$
\begin{array}{c}
\quad B \\
A \begin{bmatrix} 2 & 8 \\ 8 & 3 \\ 5 & 3 \end{bmatrix}
\end{array}
$$

Again, every element of second row is greater than the third row, therefore by dominance rule, from player A's point of view, the pure strategy (3rd row) is dominated by two, so we get 2×2 matrix

$$
\begin{array}{c}
\quad B \\
A \begin{bmatrix} 2 & 8 \\ 8 & 3 \end{bmatrix}
\end{array}
$$

Then by formulae for 2×2 game

$$x_1 = \frac{a_{22} - a_{21}}{(a_{11} + a_{22}) - (a_{12} + a_{21})} = \frac{3 - 8}{(2+3) - (8+8)} = \frac{5}{11}$$

$$x_2 = \frac{a_{11} - a_{12}}{(a_{11} + a_{22}) - (a_{12} + a_{21})} = \frac{2 - 8}{-11} = \frac{6}{11}$$

$$y_1 = \frac{a_{22} - a_{12}}{(a_{11} + a_{22}) - (a_{12} + a_{21})} = \frac{3 - 8}{-11} = \frac{5}{11}$$

$$y_2 = \frac{a_{11} - a_{21}}{(a_{11} + a_{22}) - (a_{12} + a_{21})} = \frac{2 - 8}{-11} = \frac{6}{11}$$

Hence, we conclude the following results.

Optimal strategy for player A is $\left(\dfrac{5}{11}, \dfrac{6}{11}, 0\right)$

Optimal strategy for player B is $\left(\dfrac{5}{11}, \dfrac{6}{11}, 0\right)$.

Value of the game $v = \dfrac{58}{11}$.

 Exercise-16.2

1. Solve the following games

$$
\begin{array}{c}
\quad\quad\quad B \\
\begin{array}{cc|ccc}
& & I & II & III \\
\hline
& I & 8 & 5 & 8 \\
A & II & 8 & 6 & 5 \\
& III & 7 & 4 & 5 \\
& IV & 6 & 5 & 6 \\
\end{array}
\end{array}
$$

2. Solve the following game whose pay off matrix is given

$$
\begin{array}{c}
\quad\quad\quad\quad \textbf{Player B} \\
\begin{array}{cc|ccccc}
& & I & II & III & IV & V \\
\hline
\textbf{Player A} & I & 4 & 4 & 2 & -4 & -6 \\
& II & 8 & 6 & 8 & -4 & 0 \\
& III & 10 & 2 & 4 & 10 & 12 \\
\end{array}
\end{array}
$$

3. Solve the following game with the help of dominance property

(a) $\begin{bmatrix} 1 & 7 & 2 \\ 0 & 2 & 7 \\ 5 & 2 & 6 \end{bmatrix}$ (b) $\begin{bmatrix} 30 & 40 & -80 \\ 0 & 15 & -20 \\ 90 & 20 & 50 \end{bmatrix}$

4. Solve the following games whose pay off matrices are as follows

(a)
$$
\begin{array}{c|ccc}
& y_1 & y_2 & y_3 \\
\hline
x_1 & 60 & 50 & 40 \\
x_2 & 70 & 70 & 50 \\
x_3 & 80 & 60 & 75 \\
\end{array}
$$

(b) $\begin{bmatrix} -5 & 10 & 20 \\ 5 & -10 & -10 \\ 5 & -20 & -20 \end{bmatrix}$

ANSWERS

1. $\left(0, \dfrac{1}{4}, 0, \dfrac{3}{4}\right), \left(0, \dfrac{3}{4}, \dfrac{1}{4}\right)$ and $v = 23/4$.

2. $\left(0, \dfrac{4}{9}, \dfrac{5}{9}\right), \left(0, \dfrac{7}{9}, 0, \dfrac{2}{9}, 0\right)$ and $v = 34/9$.

3. (a) $\left(\dfrac{1}{3}, 0, \dfrac{2}{3}\right), \left(\dfrac{5}{9}, \dfrac{4}{9}, 0\right)$ and $v = 11/3$ (b) $\left(\dfrac{1}{5}, 0, \dfrac{4}{5}\right), \left(0, \dfrac{13}{15}, \dfrac{2}{15}\right)$ and $v = 24$.

4. (a) $\left(0, \dfrac{3}{7}, \dfrac{4}{7}\right), \left(0, \dfrac{5}{7}, \dfrac{2}{7}\right)$ and $v = 450/7$ (b) $\left(\dfrac{1}{2}, \dfrac{1}{2}, 0\right), \left(\dfrac{2}{3}, \dfrac{1}{3}, 0\right)$ and $v = 0$.

16.8   GRAPHICAL METHOD FOR THE SOLUTION OF 2 x n AND m x 2 GAMES   [UPTU MBA-2006]

Graphical method is used to solve $2 \times n$ or $m \times 2$ games, *i.e.*, a game with mixed strategy which has only two pure strategies for one of the players. Optimal strategies for both the players assign non-zero probabilities to the same number of pure strategies. Clearly, if one player has only two strategies, the other will also use two strategies. Graphical method helps us to find which two strategies should be used. Hence, the game reduces to 2×2 which can be solved by any method discussed earlier.

(1) GRAPHICAL METHOD FOR 2 x n GAMES

Consider a $2 \times n$ game. Assume that game has no saddle point. The pay-off matrix of this game is as follows

$$
\begin{array}{c}
 & & \textbf{\textit{B}} \\
 & & \begin{array}{cccc} 1 & 2 & \dots & n \end{array} \\
\textbf{\textit{A}} \quad \begin{array}{cc} x_1 & 1 \\ x_2 & 2 \end{array} & \left[\begin{array}{cccc} a_{11} & a_{12} & \dots & a_{1n} \\ a_{21} & a_{22} & \dots & a_{2n} \end{array} \right]
\end{array}
$$

Since the player A has two strategies, then $x_1 + x_2 = 1 \Rightarrow x_2 = 1 - x_1$, $x_1 \ge 0$, $x_2 \ge 0$. Thus, for each of the pure strategies available to the player B, the expected pay-off for the player A are tabulated as follows :

Pure strategies used for player B	$E(v)$, expected pay-off to player A
1	$a_{11}x_1 + a_{21}x_2 = a_{11}x_1 + a_{21}(1 - x_1) = (a_{11} - a_{21})x_1 + a_{21}$
2	$a_{12}x_1 + a_{22}x_2 = a_{12}x_1 + a_{22}(1 - x_1) = (a_{12} - a_{22})x_1 + a_{22}$
$\vdots$	$\vdots$
n	$a_{1n}x_1 + a_{2n}x_2 = a_{1n}x_1 + a_{2n}(1 - x_1) = (a_{1n} - a_{2n})x_1 + a_{2n}$

Clearly, A's expected pay-off varies linearly with x_1. According to maximum criterion for mixed strategies game, the player A will select that value of x_1 which will maximize his minimum expected pay-offs. To find this value, we plot the following straight lines.

$$E(v) = (a_{11} - a_{21})x_1 + a_{21}$$
$$E(v) = (a_{12} - a_{22})x_1 + a_{22}$$
$$\dots \quad \dots \quad \dots$$
$$\dots \quad \dots \quad \dots$$
$$E(v) = (a_{1n} - a_{2n})x_1 + a_{2n}$$

The lowest boundary of these lines will give the maximum expected pay-off as function of

x_1. The highest point on this lowest boundary would give the maximum expected pay-off and the optimum value of x_1.

Working Procedure

To plot the above lines

STEP 1. Draw two parallel lines one unit apart and make a scale on each. These two lines represent the two strategies available to A.

STEP 2. We draw lines to represent each to B's strategies. To represent B's 1st strategy we join a_{11} on scale I to a_{21} on scale II. This line will represent the expected pay-off line $E(v) = (a_{11} - a_{21})x_1 + a_{21}$.

STEP 3. Similarly draw other pay-off lines.

STEP 4. The lowest boundary to these lines will give the minimum expected pay-off as function of x_1. The highest point on this lower boundary will give the maximum expected pay-off to A and hence optimum value of x_1.

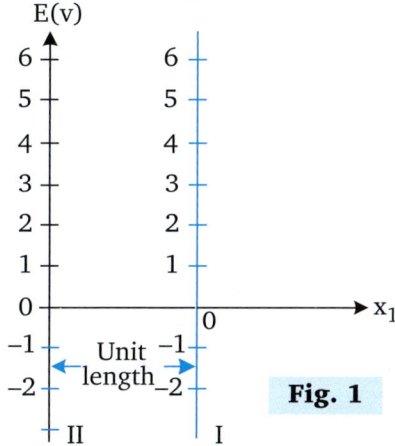

Fig. 1

In this case we determine only two strategies for player B corresponding to those two lines which pass through maximum point. In this way, the game is reduced to 2×2 game which can be solved by any method discussed earlier.

☞ REMARKS

- If more than two lines pass through this point, then any two of them having opposite signs for their slopes will be alternative optimum solutions.
- We can solve $m \times 2$ games in the same manner, except the minimax point P is the lowest point on the upper boundary.

Solved Examples

EXAMPLE 1.　**Solve the following game**

$$B$$

			I	II	III	IV
A	x_1	I	1	3	−3	7
	x_2	II	2	5	4	−6

SOLUTION. Clearly this game has no saddle point. Thus, we reduce this game into 2×2 game using graphical method.

If x_1 and x_2 are the probabilities with which the player A uses his pure strategies, then

$$x_1 + x_2 = 1, \ x_1 \geq 0, \ x_2 \geq 0$$
$$x_2 = 1 - x_1$$

Then, expected pay-off to player A for different pure strategies used by player B may be given in the following table.

Pure strategies used for player B	$E(v)$, A's expected pay-off
I	$1x_1 + 2x_2 = x_1 + 2(1 - x_1) = -x_1 + 2$
II	$3x_1 + 5x_2 = 3x_1 + 5(1 - x_1) = -2x_1 + 5$
III	$-3x_1 + 4x_2 = -3x_1 + 4(1 - x_1) = -7x_1 + 4$
IV	$7x_1 - 6x_2 = 7x_1 - 6(1 - x_1) = 13x_1 - 6$

Thus, we have to draw following four pay-off lines

$$E(v) = -x_1 + 2 \qquad\qquad \dots(1)$$
$$E(v) = -2x_1 + 5 \qquad\qquad \dots(2)$$
$$E(v) = -7x_1 + 4 \qquad\qquad \dots(3)$$
$$E(v) = 13x_1 - 6 \qquad\qquad \dots(4)$$

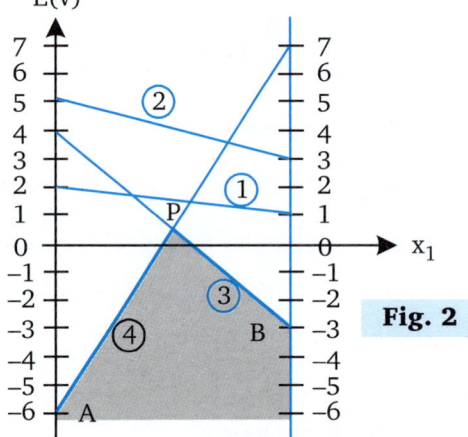

Fig. 2

From graph, lowest boundary APB to these lines give the minimum expected pay off to A. The highest point P on the lowest boundary will give the maximum expected pay-off and hence the expected value of x_1. Therefore, the best strategies for player B are III and IV pure strategies passing through point P.

Thus, the game is reduced to 2×2 game given by the following pay-off matrix.

$$
\begin{array}{c}
\qquad\qquad\qquad \textbf{\textit{B}} \\[4pt]
\begin{array}{ccc}
 & y_3 & y_4 \\
 & III & IV \\
\textbf{\textit{A}} \quad
\begin{array}{cc}
x_1 & I \\
x_2 & II
\end{array}
&
\left[\begin{array}{cc}
-3 & 7 \\
4 & -6
\end{array}\right]
\end{array}
\end{array}
$$

Using the formula of 2×2 game without saddle point, we get

$$x_1 = \frac{1}{2}, \ x_2 = \frac{1}{2}, \ y_3 = \frac{13}{20}, \ y_4 = \frac{7}{20} \ \text{ and } \ v = \frac{1}{2}$$

Hence, the solution is given by

(i) For player A, optimal mixed strategies are $\left(\dfrac{1}{2}, \dfrac{1}{2} \right)$

(ii) For player B, optimal mixed strategies are $\left(0, 0, \dfrac{13}{20}, \dfrac{7}{20} \right)$

and (iii) value of the game is $\dfrac{1}{2}$ for A and $-\dfrac{1}{2}$ for B.

EXAMPLE 2. **_Solve the following game graphically_**

$$B$$
$$\begin{array}{cc} y_1 & y_2 \\ I & II \end{array}$$

$$A \quad \begin{array}{c} I \\ II \\ III \end{array} \begin{bmatrix} 2 & 7 \\ 3 & 5 \\ 11 & 2 \end{bmatrix}$$

SOLUTION. Clearly, there is no saddle point. If y_1, y_2 are the probabilities with which the player B uses his pure strategies, then

$$\Rightarrow \qquad y_1 + y_2 = 1, \quad y_1 \geq 0, \quad y_2 \geq 0$$
$$y_2 = 1 - y_1$$

Then, B's expected pay-off for different pure strategies used by A may be given in the following table.

Pure strategies used for A	$E(v)$, B's expected pay-off
I	$2y_1 + 7y_2 = 2y_1 + 7(1 - y_1) = -5y_1 + 7 \quad$ (1)
II	$3y_1 + 5y_2 = 3y_1 + 5(1 - y_1) = -2y_1 + 5 \quad$ (2)
III	$11y_1 + 2y_2 = 11y_1 + 2(1 - y_1) = 9y_1 + 2 \quad$ (3)

Now, we draw these lines on the graph.

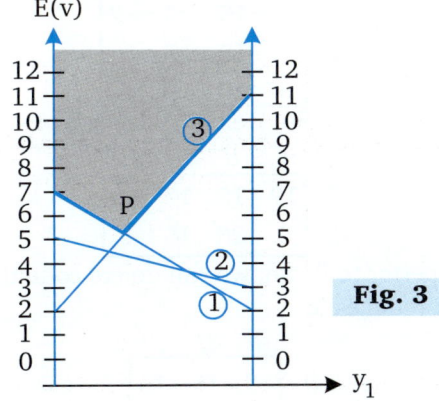

Fig. 3

Since this is a problem of $n \times 2$ type, thus minimax is the lowest point P on the upper boundary, which is the intersection of pay-off lines.

$$E(v) = -5y_1 + 7 \quad \text{and} \quad E(v) = -9y_1 + 2$$

Thus, best strategies for player A are I and III. So, the given game reduces to the following 2×2 game

B

		y_1 I	y_2 II
x_1	I	2	7
x_3	III	11	2

A

If x_1 and x_3 are the probabilities with which player A chooses strategy I and III, then by using the method of 2×2 game without saddle point, we get

$$x_1 = \frac{9}{14}, \ x_3 = \frac{5}{14}, \ y_1 = \frac{5}{14}, \ y_2 = \frac{9}{14} \quad \text{and} \quad v = \frac{73}{14}$$

Hence, optimal solution is given by

  (i) Optimal strategy for A $= \left(\dfrac{9}{14}, 0, \dfrac{5}{14} \right)$

  (ii) Optimal strategy for B $= \left(\dfrac{5}{14}, \dfrac{9}{14} \right)$

and (iii) value of the game for A $= \dfrac{73}{14}$.

EXAMPLE 3. *Solve the game whose pay-off matrix is given by*

B

		I	II	III	IV
	I	8	15	−4	−2
A	II	19	15	17	16
	III	0	20	15	5

[MEERUT-2002, 03]

SOLUTION. Since there is no saddle point. So, we will solve it with the help of dominance property as follows.

	I	II	III	IV
I	8	15	−4	−2
II	19	15	17	16
III	0	20	15	5

Row I is inferior to row II. So, we can delete row I.

	I	II	III	IV
II	19	15	17	16
III	0	20	15	5

Column III is superior to IV. So, we can delete column III.

Now, our matrix reduces to

	I	II	IV
II	19	15	16
III	0	20	5

Now, the game can not be reduced further with the help of dominance property. Then we use the graphical method

$$
\begin{array}{c|ccc}
 & y_1 & y_2 & y_3 \\
\hline
x_1 & 19 & 15 & 16 \\
1-x_1 & 0 & 20 & 5 \\
\hline
 & ① & ② & ③
\end{array}
$$

Here A is the maximum point of lowest boundary and is the intersection point of y_2 and y_3. So our matrix reduces to

$$
\begin{bmatrix}
15 & 16 \\
20 & 5
\end{bmatrix}
$$

Now,
$$\frac{x_2}{x_3} = \frac{a_{22} - a_{21}}{a_{11} - a_{12}} = \frac{5 - 20}{15 - 16} = \frac{15}{1}$$

$$x_2 = \frac{15}{16}, \ x_3 = \frac{1}{16}$$

$$\frac{y_2}{y_4} = \frac{a_{22} - a_{12}}{a_{11} - a_{21}} = \frac{5 - 16}{15 - 20} = \frac{11}{5}$$

$$y_2 = \frac{11}{16}, \ y_4 = \frac{5}{16}$$

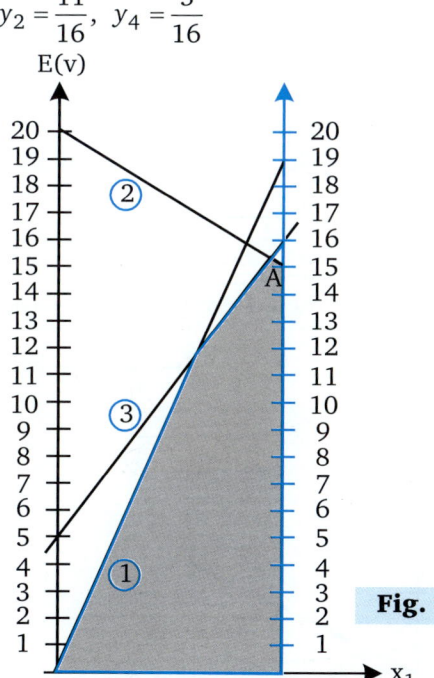

Fig. 4

and value of the game,

$$
v = \frac{a_{22}a_{11} - a_{21}a_{12}}{(a_{22} + a_{11}) - (a_{21} + a_{12})}
$$

$$
= \frac{5 \times 15 - 20 \times 16}{(5 + 15) - (20 + 16)} = \frac{245}{16}
$$

Hence,   Optimal strategy used by A $= \left(0, \dfrac{15}{16}, \dfrac{1}{16}\right)$

Optimal strategy used by B $= \left(0, \dfrac{11}{16}, 0, \dfrac{5}{16}\right)$.

and   value of the game $v = \dfrac{245}{16}$

EXAMPLE 4. **Solve the game :**

$$\begin{array}{ccc} 1 & 8 & 4 \\ 6 & 4 & 5 \\ 0 & 1 & 2 \end{array}$$

B

SOLUTION.   We have

	I	II	III	Row min
I	①	⑧	4	1
A II	⑥	④	⑤	4
III	⓪	1	2	0
Column max	6	8	5	

Since there is no saddle point. So, the optimal strategy will be mixed strategy. Now, we use the principle of dominance to solve the game.

	I	II	III
I	1	8	4
II	6	4	5
III	0	1	2

Row III is inferior to row I. So, we can delete row III.

	I	II	III
I	1	8	4
II	6	4	5

Now, we will solve it with the help of graphical method.

	y_1	y_2	y_3
x_1	1	8	4
$1-x_1$	6	4	5
	①	②	③

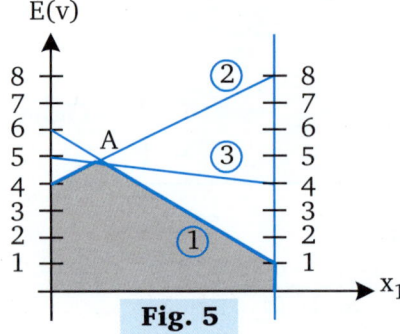

Fig. 5

A is the maximum point of lowest boundary. It is the intersection of y_1 and y_2.

So, our matrix reduces to

$$\begin{array}{c c} & \begin{array}{cc} y_1 & y_2 \end{array} \\ \begin{array}{c} x_1 \\ x_2 \end{array} & \begin{array}{|cc|} \hline 1 & 8 \\ 6 & 4 \\ \hline \end{array} \end{array}$$

$$\frac{x_1}{x_2} = \frac{a_{22} - a_{21}}{a_{11} - a_{12}} = \frac{4 - 6}{1 - 8} = \frac{2}{7}$$

$$\Rightarrow \qquad x_1 = \frac{2}{9}, \quad x_2 = \frac{7}{9}$$

$$\frac{y_1}{y_2} = \frac{a_{22} - a_{12}}{a_{11} - a_{21}} = \frac{4 - 8}{1 - 6} = \frac{4}{5}$$

$$\Rightarrow \qquad y_1 = \frac{4}{9}, \quad y_2 = \frac{5}{9}$$

So, the strategy used by A $= \left(\dfrac{2}{9}, \dfrac{7}{9}, 0 \right)$

also, the strategy used by B $= \left(\dfrac{4}{9}, \dfrac{5}{9}, 0 \right)$

and the value of the game

$$v = \frac{a_{22} a_{11} - a_{12} a_{21}}{(a_{22} + a_{11}) - (a_{12} + a_{21})} = \frac{44}{9}.$$

EXAMPLE 5. *Solve the game whose pay-off matrix to the player A is given by the table*

$$\begin{array}{c c} & B \\ & \begin{array}{ccc} I & II & III \end{array} \\ A \begin{array}{c} I \\ II \\ III \end{array} & \begin{array}{|ccc|} \hline 1 & 7 & 2 \\ 6 & 2 & 7 \\ 5 & 2 & 0 \\ \hline \end{array} \end{array}$$

SOLUTION. Using dominance rule, we observe that all elements of II row all greater than or equal to all elements of III so dominating III row by II

$$\begin{array}{c c} & B \\ & \begin{array}{ccc} I & II & III \end{array} \\ A \begin{array}{c} I \\ II \\ III \end{array} & \begin{array}{|ccc|} \hline 1 & 7 & 2 \\ 6 & 2 & 7 \\ \hline 5 & 2 & 0 \\ \hline \end{array} \end{array}$$

No row and column can be dominated now, so using graphical method

B's pure strategy	A's expected pay off $[E(v)]$
I	$E(v) = 1x_1 + 6(1 - x_1) = -5x_1 + 6$
II	$E(v) = 7x_1 + 2(1 - x_1) = 5x_1 + 2$
III	$E(v) = 2x_1 + 7(1 - x_1) = -5x_1 + 7$

Now, plotting these three lines

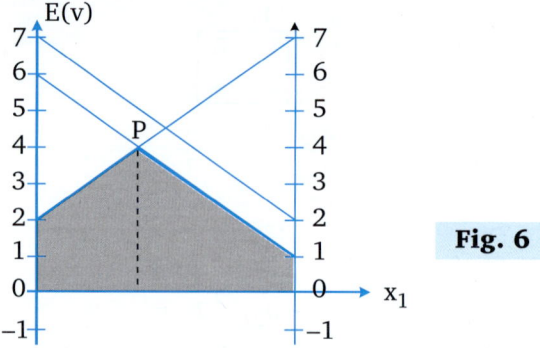

Fig. 6

Maximin point occurs at point P. The highest point P on the lowest boundary will give the largest expected gain. So the best strategies for the player B are those which pass through the point P. The reduced game to 2×2 table is

$$B$$

		I	II
A	I	1	7
	II	6	2

Now, let strategy of A is (x_1, x_2) and B is (y_1, y_2).

$$x_1 = \frac{2-6}{(1-7)+(2-6)} = \frac{-4}{-6+(-4)} = \frac{2}{5}, \quad x_2 = 1 - x_1 = 1 - \frac{2}{5} = \frac{3}{5}$$

$$y_1 = \frac{2-7}{(1-6)+(2-7)} = \frac{-5}{-10} = \frac{1}{2}, \quad y_2 = 1 - y_1 = 1 - \frac{1}{2} = \frac{1}{2}$$

Thus,   Strategy of A is $\left(\frac{2}{5}, \frac{3}{5}, 0\right)$; Strategy of B is $\left(\frac{1}{2}, \frac{1}{2}, 0\right)$.

Value of game $v = \frac{(1 \times 2) - (6 \times 7)}{-10} = 4$.

16.9 ALGEBRAIC METHOD FOR THE SOLUTION OF A GENERAL GAME

In this method, first we convert the game into inequalities and then solve them.

Let $[a_{ij}]_{m \times n}$ be the pay off matrix of a rectangular game between two persons. Let us suppose $X = (x_1, x_2, ..., x_m)$, $Y = (y_1, y_2, ..., y_n)$ be the mixed optimal strategies of player A and B respectively. Then expected gains when B used his pure strategies 1, 2, ..., n respectively are

$$\sum_{i=1}^{m} a_{i1} x_i, \ \sum_{i=1}^{m} a_{i2} x_i, \ ......, \ \sum_{i=1}^{m} a_{in} x_i$$

Let v be the value of the game, then minimum expected gain of A is v .

Thus, $\sum_{i=1}^{m} a_{i1} x_i \geq v, \ \sum_{i=1}^{m} a_{i2} x_i \geq v, \ ......, \ \sum_{i=1}^{m} a_{in} x_i \geq v$ ...(1)

In a similar way, considering B's expected losses and considering the fact that the maximum loss of B is v , we get the following system of inequalities.

$$\sum_{j=1}^{n} a_{1j} y_j \leq v, \ \sum_{j=1}^{n} a_{2j} y_j \leq v, \ ......, \ \sum_{j=1}^{n} a_{mj} y_j \leq v \qquad ...(2)$$

and

$$x_1 + x_2 + ... + x_m = 1$$
$$y_1 + y_2 + ... + y_n = 1$$ ...(3)
$$x_i \geq 0, \ x_j \geq 0, \ \forall \ i = 1, 2, ..., m \ \text{and} \ j = 1, 2, ..., n.$$

So, we have to find the values of $x_1, x_2, ..., x_m, \ y_1, y_2, ..., y_n$ such that (1), (2) and (3) are satisfied. For this, we use the following steps :

Working Procedure

STEP 1. Convert inequalities (1) and (2) as equalities. Try to solve them. If we get a solution satisfying (1), (2) and (3), then the given problem is solved completely.

STEP 2. If the system of equations obtained above is inconsistent, then we conclude that at least one of the inequalities is strict. Then, we try to solve by taking one or more inequalities as strict inequality and other as equalities until we get a solution.

(1) If for some i $(i = 1, 2, ..., m)$, $v_{i1}y_1 + v_{i2}y_2 + ... + v_{in}y_n < v$

Then, corresponding $x_i = 0$ and similarly, if for some j $(j = 1, 2, ..., n)$

$v_{1j}x_1 + v_{2j}x_2 + ... + v_{mj}x_m > v$, then corresponding $y_j = 0$.

(2) If A's optimal strategy is a mixed strategy in which exactly r pure strategies have non-zero probabilities, then the optimal strategies of other player also involves exactly r pure strategies.

Solved Examples

EXAMPLE 1. ***Find the value of the game and the optimal strategies for both players whose pay-off is given by***

B

		I	II	III
A	I	-1	2	1
	II	1	-2	2
	III	3	4	-3

SOLUTION. Clearly this game has no saddle point and can not be reduced to 2×2 game by the dominance rule. So, we solve it by algebraic method.

Let (x_1, x_2, x_3) and (y_1, y_2, y_3) be the optimal mixed strategies of the two players A and B respectively and v the value of the game.

For player A

$$-1.x_1 + 1.x_2 + 3.x_3 \geq v$$ ...(1)
$$2.x_1 - 2.x_2 + 4.x_3 \geq v$$ ...(2)
$$1.x_1 + 2.x_2 - 3.x_3 \geq v$$ ...(3)

For player B

$$-1.y_1 + 2.y_2 + y_3 \leq v$$ ...(4)
$$1.y_1 - 2.y_2 + 2.y_3 \leq v$$ ...(5)
$$3.y_1 + 4.y_2 - 3y_3 \leq v$$ ...(6)

Also,

$$x_1 + x_2 + x_3 = 1$$ ...(7)
$$y_1 + y_2 + y_3 = 1$$ ...(8)
$$x_1, x_2, x_3 \geq 0$$
$$y_1, y_2, y_3 \geq 0$$ ...(9)

Now, we have to find the values $x_1, x_2, x_3; y_1, y_2, y_3$ such that all the above relations are satisfied.

Firstly, we consider the inequalities (1) to (6) as equation :

$$-x_1 + x_2 + 3x_3 = v , \ 2x_1 - 2x_2 + 4x_3 = v , \ x_1 + 2x_2 - 3x_3 = v \left. \right\}$$
$$-y_1 + 2y_2 + y_3 = v , \ y_1 - 2y_2 + 2y_3 = v , \ 3y_1 + 4y_2 - 3y_3 = v \left. \right\} \quad ...(10)$$

On solving first three equations from (10), we get

$$x_2 = \left(\frac{2}{3}\right) v, \ x_3 = \left(\frac{3}{10}\right) v, \quad x_1 = \left(\frac{17}{30}\right) v$$

Putting all these values in (7), we get

$$v = \frac{15}{23} \quad \text{and} \quad x_1 = \frac{17}{46}, \ x_2 = \frac{10}{23}, \ x_3 = \frac{9}{46}.$$

Now on solving last three equations from (10), we get

$$y_1 = \frac{7}{23}, \ y_2 = \frac{6}{23}, \ y_3 = \frac{10}{23}$$

Hence, optimal mixed strategies for player A $\left(\frac{17}{46}, \frac{10}{23}, \frac{9}{46}\right)$

optimal mixed strategies for player B $\left(\frac{7}{23}, \frac{6}{23}, \frac{10}{23}\right)$

and value of the game $v = 15/23$.

 Exercise-16.3

1. Solve the following game graphically:

(a)

	y_1	y_2	y_3
x_1	1	3	11
$1-x_1$	8	5	2

(b)

	y_1	y_2	y_3	y_4
x_1	19	6	7	5
x_2	7	3	14	6
x_3	12	8	18	4
x_4	8	7	13	-1

(c)

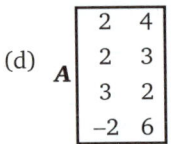

	B			
A	2	2	3	-1
	4	3	2	6

(d)

	B	
A	2	4
	2	3
	3	2
	-2	6

[MADRAS, MBA-1996]

 ANSWERS

1. (a) $\left(\frac{3}{11}, \frac{8}{11}\right), \left(0, \frac{2}{11}, \frac{9}{11}\right)$ and $v = \frac{49}{11}$ (b) $\left(\frac{3}{4}, \frac{1}{4}, 0, 0\right), \left(0, \frac{1}{4}, 0, \frac{3}{4}\right)$ and $v = \frac{21}{4}$.

(c) $\left(\frac{1}{2}, \frac{1}{2}\right), \left(0, 0, \frac{7}{8}, \frac{1}{8}\right)$ and $v = \frac{5}{2}$ (d) $\left(\frac{1}{3}, 0, \frac{2}{3}, 0\right), \left(\frac{2}{3}, \frac{1}{3}\right)$ and $v = \frac{8}{3}$.

 REVIEW QUESTIONS

1. Explain min-max and max-min principle in game theory.
2. Explain two-person zero sum game.
3. State major limitations of Game theory.
4. Define the following:
   (i) Competitive game
   (ii) pay off matrix
   (iii) pure and mixed strategies
   (iv) saddle point
   (v) Two person zero sum game
5. Write the assumptions made in the theory of game.
6. What is a game in game theory?

MULTIPLE CHOICE QUESTIONS (CHOOSE THE MOST APPROPRIATE ONE)

1. If the value of a game is zero, then the game is called:
   (a) pure game
   (b) pure strategy
   (c) fair strategy
   (d) none of these

2. Two persons zero sum game means that the:
   (a) sum of losses to one player equals the sum of gain to the other

(b) sum of losses to one player is not equal to the sum of gains to the other
(c) both (a) and (b)   (d) None of these
3. The game with saddle points are:
(a) probabilistic in nature
(b) deterministic in nature
(c) stochastic in nature
(d) none of these
4. Game theory models are classified by the:
(a) no. of players   (b) sum of all payoff
(c) no. of strategies   (d) all are true
5. When minimax and maximin criterian meets then:
(a) mixed strategies exist (b) fair game exist
(c) saddle point exits (d) None of these
6. In case there is no saddle point in a game then the game is:
(a) fair game   (b) mixed strategic game
(c) deterministc game (d) none of these
7. If there are more than two persons in a game then it is called:
(a) open game   (b) big game
(c) multiplayer game (d) none of these
8. A competitive situation is known as:
(a) competition   (b) game
(c) marketing   (d) None of these
9. A game involving n persons is known as:
(a) n-person game   (b) multiplayer game
(c) not a game   (d) None of these
10. A saddle point exists when:

(a) maximum value =maximax value
(b) minimax value = minimum value
(c) minimax value = maximum value
(d) None of these
11. Graphical method should be used to determine value of the game when size of pay-off matrix is:
(a) 2×2 (b) 3×4
(c) $2 \times n$ (d) None of these
12. In a mixed strategy game:
(a) no saddle point exits
(b) each player always selects same strategy
(c) both (a) and (b) are true
(d) None of these
13. In a pure strategy game:
(a) any strategy may be selected arbitrarily
(b) a particular strategy is selected by each player
(c) both players select their optimal strategy
(d) None of these
14. A two person game is said to be zero-sum if:
(a) gain of one player is exactly matched by a loss to the other so that their sum is equal to zero
(b) gain of one player does not match the loss to he other
(c) both the players must have an equal no. of strategies
(d) None of these

ANSWERS

| 1. (a) | 2. (a) | 3. (b) | 4. (d) | 5. (c) | 6. (b) | 7. (c) | 8. (b) | 9. (a) |
| 10. (c) | 11. (c) | 12. (a) | 13. (c) | 14. (a) | | | | |

ARCHIVE

1. For what value of λ, the game with following pay-off matrix is strictly determinable?

Player A	Player B		
	B_1	B_2	B_3
A_1	λ	6	2
A_2	–1	λ	–7
A_3	–2	4	λ

[BHARTHIAR 1989]

2. Two players A and B match coins. If the coins match, then A coins two units of value, if the coins do not match, then B coins two units of value. Determine the optimum strategies for the players and the value of the game.   [BOMBAY-1985]
3. Two competitors are competing for the market share of the similar product. The pay-off matrix in terms of their advertising plan is shown below :

Competitor A	Competitor B		
	No Advertising	Medium Advertising	Heavy Advertising
No Advert.	10	5	–2
Medium Advert.	13	12	13
Heavy Advert,	16	14	10

Suggest optimal stratigies for the two firms and the net outcome there of.
[DELHI, M.COM, 1990; MBA, 1994; HP, MBA, 1999]

4. Obtain the optimal strategies for both players and the value of the game for two-person zero-sum game whose pay-off matrix is given as follows :

Player A	Player B		
	B_1	B_2	B_3
A_1	1	3	11
A_2	8	5	2

[ICWA, JUNE 1986]

5. Obtain the optimal strategies for both persons and the value of the game for two-person zero-sum game whose pay-off matrix is as follows :

Player A	Player B	
	B_1	B_2
A_1	1	–3
A_2	3	5
A_3	–1	6
A_4	4	1
A_5	2	2
A_6	–5	0

[DIBRUGARH-1994; KARNATAKA, BE- 1994]

6. Solve the following game graphically :

Player A	Player B	
	B_1	B_2
A_1	1	2
A_2	4	5
A_3	9	–7
A_4	–3	–4
A_5	2	1

[MADURAI-1989]

7. Solve the following game approximately :

Player A	Player B		
	1	–1	–1
	–1	–1	3
	–1	2	–1

[SAMBOLPUR, 1986; MEERUT, 1988; JODHPUR 1992; DIBRUGARH, 1994]

8. Use dominance rules to reduce the size of the following pay-off matrix to (2×2) size. Find the optional strategies and value of the game.

Player A	Player B		
	B_1	B_2	B_3
A_1	3	–2	4
A_2	–1	4	2
A_3	2	4	6

[DIBRUGARH 1994; ROHILKHAND (MATHS), 1985, PUNJAB, 1995; MEERUT 2000, 2017]

HINTS AND ANSWERS

1. Ignoring the value of λ and determine the maximin and minimax values of the pay-off matrix. Saddle point is not unique, the value of the game lies between –1 and 2 *i.e.*, $-1 \le v \le 2$. For strictly determinable game, $-1 \le \lambda \le 2$.

2. The pay-off matrix for the matching player is :

Player A	Player B	
	H	T
H	2	–2
T	–2	2

The optional stratigies for two players are $H : \dfrac{1}{2}, T : \dfrac{1}{2}$ with $v = 0$.

3. Applying rules of dominance to delete first-column and then first row from the pay-off matrix.

Firm A	Firm B	
	Medium Advt. B_2	Heavy Advt. B_3
Medium Advt. A_2	12	15
Heavy Advt. A_3	14	10

The optimum strategies for A, $A_2(57\%)$ and $A_3(43\%)$ and the optimum strategies for B, $B_2(71\%)$ and $B_3(29\%)$

4. Player $A : \left(\dfrac{3}{11}, \dfrac{8}{11} \right)$, Player $B : \left(0, \dfrac{2}{11}, \dfrac{9}{11} \right)$ and $v = \dfrac{49}{11}$

5. The optimum strategies for B, $q_1 = \dfrac{4}{5}$, $q_2 = \dfrac{1}{2}$ and the optimum strategies for A,

   $p_2 = \dfrac{3}{5}, p_4 = \dfrac{2}{5}$ and $p_1 = p_3 = p_5 = p_6 = 0$. Value of the game $v = \dfrac{17}{5}$

6. $q_1 = \dfrac{12}{17}, q_2 = \dfrac{5}{17}; p_2 = \dfrac{16}{17}, p_3 = \dfrac{1}{17}$ and $p_1 = p_4 = p_5 = 0; v = \dfrac{73}{17}$

7. Player $A : A_1 = \dfrac{4}{10}, A_2 = \dfrac{3}{10}, A_3 = \dfrac{3}{10}$; Player $B : B_1 = \dfrac{4}{10}, B_2 = \dfrac{2}{10}, B_3 = \dfrac{4}{10}$

   The approximate value of the game is $-\dfrac{1}{5} \le v \le \dfrac{1}{5}$

8. Player $A : A_1 = 0, A_2 = 0, A_3 = 1$; Player $B : B_1 = \dfrac{2}{5}, B_2 = \dfrac{3}{5}, B_3 = 0$

   Value of game $v = -2$

Network Analysis (PERT/CPM) 17

17.1 INTRODUCTION

Network analysis has played an important role in engineering science, technology and management. An effective method of planning must provide a clear picture of the relationship between the activities or operations making up the project and show how delays at any particular stage will affect the remainder of the project. Therefore, the problem of networking is to find a course of action which minimizes some measure of performance. Network analysis is one of the most popular technique used for planning, scheduling monitoring and coordinating large and complex projects comparising a number of activities. It is concerned with minimizing some measure of performance of the system such as the total completion time for the project, overall cost and so on.

17.2 OBJECTIVE OF THE NETWORK ANALYSIS

The main objectives of the network analysis are as follows:
 (i) Minimization of the total cost of the project
 (ii) Minimization of the total time of the project
 (iii) Minimization of time for a given cost of the project
 (iv) Minimization of cost for a given total time of the project
 (v) Minimization of the idle resources of the project

There are many methods to study and analyse such types of projects to find some strategy so that the project completed in time with minimum cost. In this chapter we shall discuss some methods, *viz.* Gantt Chart, Critical Path Method (CPM) and Project Evaluation and Review Technique (PERT).

General Definitions

 (1) Based on graph:
   (I) Graph: A graph consists of a set of functions points called **nodes** with certain pair of nodes being joined by the line called **branches**.
   (II) End points: The two nodes joined by a branch are referred to as the end points of the branch.
   (III) Terminal branch: A branch, one of whose end points has degree one is called terminal branch.
   (IV) Degree of the node: The degree of a node is the number of branches of the graph which have this node as an end point.

(V)   **Extreme point:** A node of degree one is called an extreme point of the graph.

(VI)   **Oriented branch:** A branch of graph is said to be oriented (or directed) if there is a sense of direction attributed to the branch so that one node is considered as the point of origin and the other node as the point of destination.

(VII)   **Oriented graph:** A graph in which all the branches are oriented is called an oriented graph.

(VIII)   **Path (or chain):** A path joining nodes i and j is a sequence of branches connected these two nodes. The nodes i, j are called the extremity points of the path.

(IX)   **Loop (or cycle):** If the extremity points of the path are one and same node, then the path is called a loop.

(X)   **Connected graph:** A graph is said to be connected if there is a path connecting every pair of nodes of the graph.

(XI)   **Tree:** A tree is a connected graph which has no loops, *i.e.*, a connected graph is a tree iff the path joining any two nodes is unique.

(2) Based on Networking:

(I)   **Activity:** An activity represents some action and as such it is a time consuming part of a project. It is represented by an arrow. Each and every activity has a point of time where it begins and a point where it ends. There are following three types of activity.

    (i) Predecessor activity: An activity which must be completed before one or more other activities start is called predecessor activity.

    (ii) Successor activity: An activity which started immediately after one or more of other activities are completed is called successor activity.

    (iii) Dummy activity: An activity which does not consume either any resources and time is called dummy activity. It is added only to represent the given precedence relationship among activities of the project and is needed when

      (a) two or more parallel activities in a project have some head and tail events.

      (b) two or more activities have some of their immediate predecessor activity in common.

(II)   **Event:** The beginning and end points of an activity are called events or nodes. It is a particular instant in time and does not consume any time or resource. It is represented by a numbered circle. It may also be represented by a rectangle, square, hexagon or some other geometric shapes.

The events can be classified in the following categories,

    (i) Merge event: An event which represent the end of more than one activities is called merge event.

    (ii) Burst event: An event which represent the begining of more than one activities is called burst event.

    (iii) Merge and Burst event: An event which represent the begining of some activity and is also the end of some activity is called a merge and burst event.

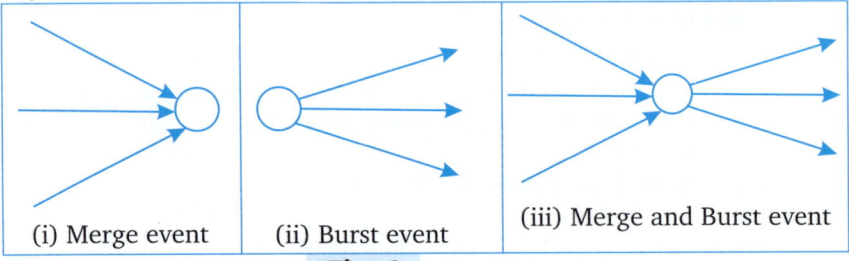

(i) Merge event   (ii) Burst event   (iii) Merge and Burst event

Fig. 1

(III) **Adjacent operations:** Two events are said to be adjacent if there is an activity connecting them.
(IV) **Adjacent activity:** The activities are adjacent if there is an event common to them.
(V) **Chain:** Chain is a sequence of adjacent activities.
(VI) **Loop:** It is a connection of an activity to an earlier event on the same path.
(VII) **Source:** It is the starting event with only succeeding but no preceeding activity.
(VIII) **Sink:** The last event showing the end of a project with only preceeding but no succeeding activity.
(IX) **Corciot:** A finite path in which the starting event connects with end points.

17.3 NETWORK

A network is a combination of several activities and events. Sometimes, in a big project many activities are done at the same time and there are many activities which can be started only at the completion of other activity. A network can be defined by a set of points of nodes that are connected by links or arrows. These links are characterized in terms of time, cost or distance involved in a traversing time. It is a linear graph consisting of a number of points or nodes, each of which is connected to one or more of the other nodes by route or edge. Hence, it is a set of operations and activities describing the time orientation of a composit project.

17.4 CHARACTERISTICS OF A NETWORK ANALYSIS

The important characteristic in a network analysis are as follows:
(1) The objective is to be finished within a specific time otherwise there is a penalty.
(2) Various activities are to be completed in a certain order and there are some activities which can start on the completion of other activities.
(3) The cost associated with any activity is proportional to its time of completion.
(4) There can be bottlenecks in the process and the resources to be allocated are limited.

17.5 ERROR IN DRAWING A NETWORK

Following are the most commonly observed errors in drawing a network.
(1) Cycling (or looping) Error: If some activity in a network form a loop or cycle, then such error is said to be looping or cycling.
For example

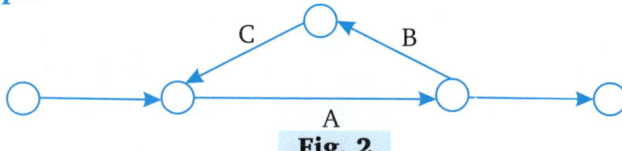

Fig. 2

In the above network, activities *A*, *B* and *C* form a loop, so the network cannot proceed.
(2) Danging Error: If some activity other than the final activity do not have any successor activity, then it is called a danging error.
For example,

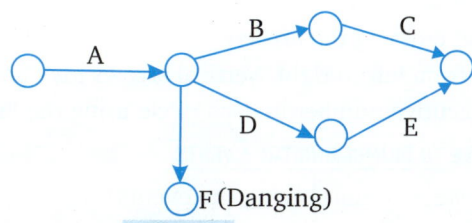

Fig. 3

☞ REMARK

- In a network all events except the first and the last of the whole project must have at least one entering and one leaving activity.

(3) Redundancy Error: In any network diagram, if dummy activity is the only activity emanating from an event, then it is called redundancy error.

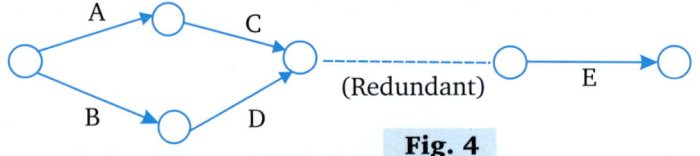

Fig. 4

17.6 CONSTRUCTION OF THE NETWORK DIAGRAM

A network can be easily drawn with the help of work breakdown schedule which represents the entire project in a systematic manner such that the interrelationship among all phases of the project are easily seen.

A network diagram is based on three main symbols, arrows ($\rightarrow$) representing various activities, circles (O) showing various events or nodes and broken arrows ($\cdots\rightarrow$) showing dummy activities for logical sequence.

The important information required for the construction of network diagram is generated by the following questions.

1. Which operation must be completed before each given operation can be started?
2. Which operation can be carried out in parallel?
3. Which operations immediately succeeds other given operations?

Working Procedure

The general procedure for the construction of a network diagram consist the following steps:

STEP 1. Prepare a list of operations (activities) their predecessor and successor operations (activities).

STEP 2. Draw a circle in the first column to represent the initial event (node).

STEP 3. List all the operations by using the circles (O) that can be started immediately, in the second column.

STEP 4. List the operations or activities that can be started immediately after the completion of the second column operation in the third column. Draw directed lines between the circles of second and third column to indicate which operation follow other operation.

STEP 5. Label each circle drawn to represent the description of the operation.

STEP 6. Proceed till all operations are completed.

STEP 7. After completing the preliminary draft including all operations, check for logical arrangements.

STEP 8. Try to avoid crossing the arrows.

STEP 9. Use arrow from left to right. Vertical arrows may be used if necessary.

STEP 10. Place consecutive number in each circle using the Fulkeron's rule. (given below)

17.6.1 FULKERSON'S RULE TO NUMBERING THE EVENTS

Following are the steps to numbering the events

STEP 1. The initial event (node) which has all outgoing vectors with no incoming arrow in numbered-1.

STEP 2. Delete all arrows coming out of event-1. Therefore, some more events (at least once) will be converted to initial events (events having no incoming arrows). Assign number 2, 3, …, to these events

STEP 3. Delete all arrows, coming out of the events, which are numbered in step-2. So, some more events (at least one) will be converted to initial events. Assign next numbers in continuation to these events.

STEP 4. Continue upto the final (end or terminal) event which has no arrow coming out of it. Assign the next number to this event.

STEP 5. Check that the numbering of event should be such that for each activity, the number of each head event in larger than the number of tail event.

Solved Examples

EXAMPLE 1. **Draw a network diagram for the following set of operations represented by separate letters.**

Operations	Post-Operations
A	precedes B, C
B	precedes D, E
C	precedes D
D	precedes F
E	precedes G
F	precedes G

[MEERUT-2016, 18]

SOLUTION. The network in this case can be easily constructed by arranging the operations in vertical columns or sequences, beginning from leaf and progressing towards the right.

Since, no operation precedes A, it will be placed in the first vertical column. After A is completed, B and C can be placed in the sequence as shown in fig. 5

D and E cannot be placed in the second vertical sequence since neither can be started until B is completed. B and C both are pre-operators of D, hence D is placed in the third sequence, E is also placed in the third sequence as it can be started as soon as B is completed. At the end of third sequence we get fig. 6.

Similarly F and G can be placed in 4th and 5th sequence respectively, as given in fig. 7.

The arrow BE in fig. 7 crosses another line. So to avoid this, if E is placed above D in third sequence, we get the fig. 8 giving the final network diagram.

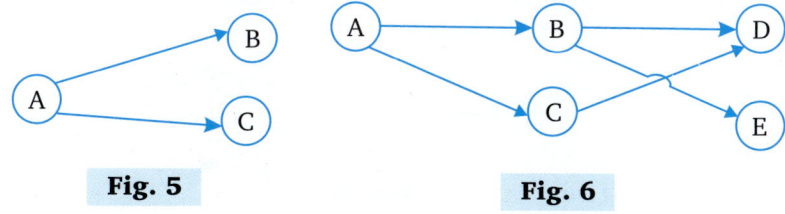

Fig. 5   Fig. 6

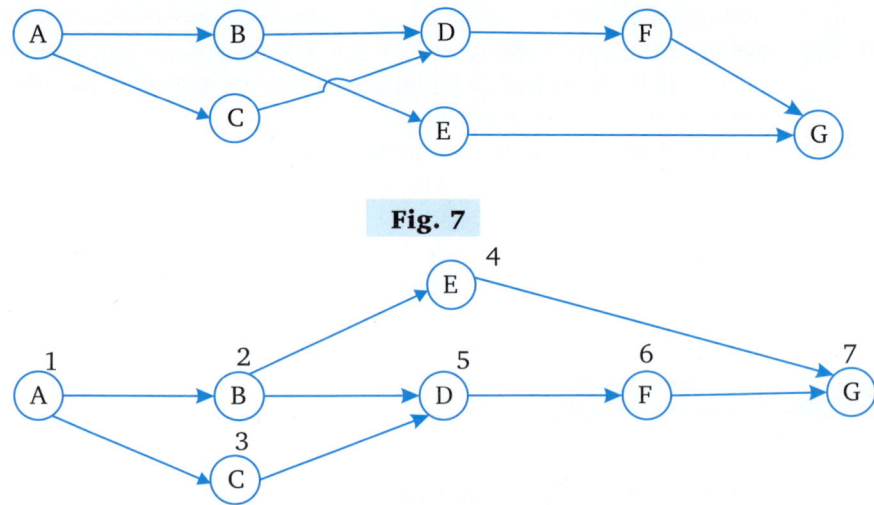

Fig. 7

Fig. 8

EXAMPLE 2. *A new type water pump is to be designed for automobile. Prepare a network diagram from the following table showing the major activities for effective control of the project and the preceding activities.*

Activity	Description	Preceding activities
A	Drawing prepared and approved	—
B	Cost analysis	A
C	Tool feasibility	A
D	Tool manufactured	C
E	Favourable cost	B, C
F	Raw material procured	E
G	Sub-assembles ordered	E
H	Sub-assembles received	G
I	Parts manufactured	D, F
J	Final assembly	H, I
K	Testing and shipment	J

[MEERUT-2017]

SOLUTION. From the above table, it is clear that

(i) The activity A has no preceding activity so it is represented by arrowed line.

$$\circ \xrightarrow{\quad A \quad} \circ$$

Fig. 9

(ii) The activities B and C can be done simultaneously and both are preceded by activity A as shown below

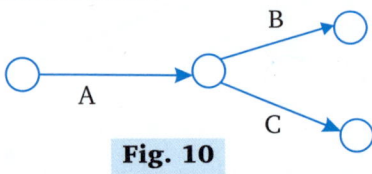

Fig. 10

(iii) Activity C precedes activity D. The activity E can be complete only after the completion of activity B and C. It may easily be seen that dependence of activity E (favourable cost) of the activity C (economic of tooling) is form a technical point of view and does not consume any resource.

So the dummy activity is shown by dotted lines as shown below:

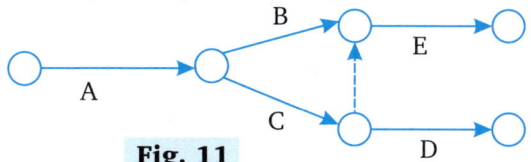

Fig. 11

Proceeding in a similar manner, we get the following graph.

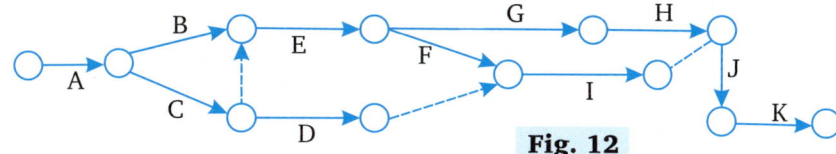

Fig. 12

Clearly, in the above draft, we have three dummy activities. So, we have to proceed to reduce the dummy activities to a minimum. Here, we get the following final network in (activity on arrow network) having only one dummy activity as shown below:

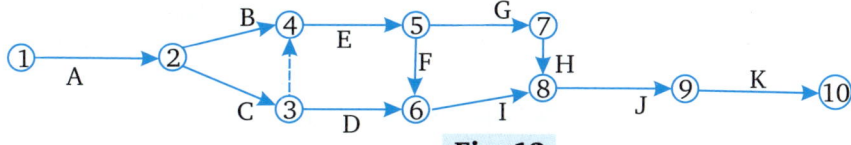

Fig. 13

Also, the activity on node diagram of the problem is shown below:

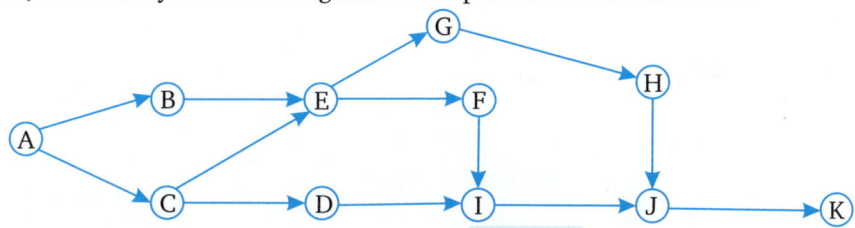

Fig. 14

EXAMPLE 3. *Develop a network based on the following information.*

Activity	Immediate predecessors
A	—
B	—
C	A
D	B
E	C, D
F	D
G	E
H	F

SOLUTION. Following the usual procedure, we have the following network diagram

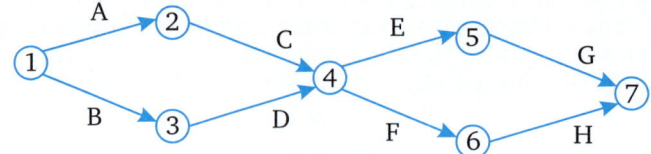

Fig. 15

Consider the activity F. According to the network, both activities C and D must be completed before we can start F but in reality, only activity D must be completed. Therefore, the network is not correct. The addition of the dummy activity and a dummy event can overcome this problem as shown below.

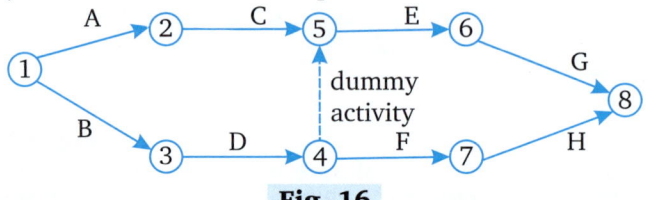

Fig. 16

Exercise-17.1

1. Construct the network diagram for the following precedence relationship

Event	A	B, C	D	E	F	G
Preceded by	—	A	B,C	C	D, E	E, F

2. Construct the network diagram for the following activity data

Activity	A	B	C	D	E	F	G	H	I	J
Preceded by	—	A	A	A	C,D	D	E	G	F,H	B,I

3. A project schedule has the following characteristics

Activity	Time	Activity	Time
1-2	4	5-6	4
1-3	1	5-7	8
2-4	1	6-8	1
3-4	1	7-8	2
3-5	6	8-10	5
4-9	5	9-10	7

4. Construct the network diagram for the following activity data

Activity	A	B	C	D	E	F	G	H
Predecessor activity	—	A	A	B	B,C	E	D,F	G

5. Draw a network diagram showing the following relationship

Activity	Predecessor activity	Activity	Predecessor activity
A	—	H	F
B	—	I	H
C	B	J	I
D	A	K	D, E, G, J
E	C	L	I
F	C	M	K, L
G	F		

ANSWERS

1.

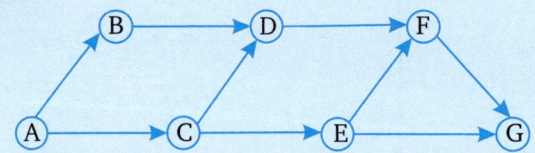

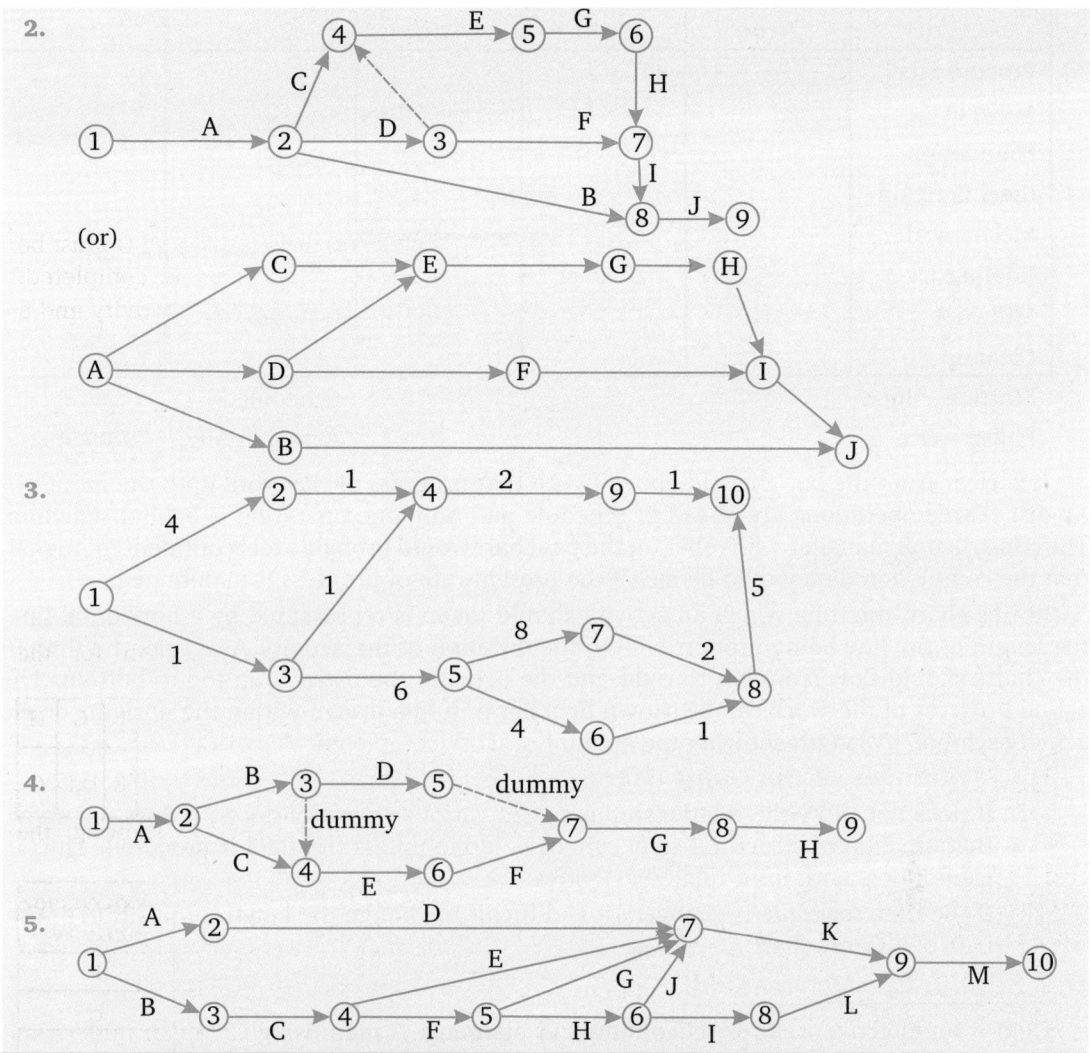

2.

(or)

3.

4.

5.

17.7 METHODS OF NETWORK ANALYSIS

In this chapter we shall discuss the problem of network analysis for the following methods.

(1) Schedule or Gantt Bar Chart

(2) Critical Path Method (CPM)

(3) Project Evaluation and Review Technique (PERT)

17.7.1 SCHEDULE CHART (GANTT BAR CHART)

A common method of studying an overall schedule is to use Gantt bar chart. Such a chart shows bar for each operation drawn to a time scale. The method of the construction of Gantt chart is explained in figure below for the construction of a small factory building. This chart displays the consecutive activities in the most suitable and significant manner.

Activity	June	July	Aug.	Sept.	Oct.
Procure Steel					
Move in					
Foundation					
Steel Erection					
Mechanical					
Roofing					
Millwork					
Cleanup					

Starting – June   Scheduled

Finish – Oct.   Actual in Aug.

It is clear from the bar chart in above figure that progress of the work upto the month of August. Three operations are ahead of schedule and only one operation is behind schedule. The construction manager who relies on the bar chart would probably feel confident in August, that the completion data would be met if the present rate of progress is maintained.

In this chart, the time which an activity should taken is represented by a horizontal line, the length of the line being proportional to the duration of the activity. As a rule, the time in the charts should flow from left to right and the activities be listed from top to bottom. The actual progress of the work can be shown by a bar or a line drawn within the uprights of the activity symbol, its length showing the amount of work completed.

Disadvantages of the Gantt chart:

(i) It does not show the relationship between various operations, *e.g.*, it does not show that erection of structural steel cannot be done unless the steel is procured. Thus it cannot deal with inter-related activities.

(ii) If the time schedule is changed, it is difficult to change the length or position of the bars of a Gantt chart.

17.7.2 CRITICAL PATH METHOD (CPM)

CPM is an effective tool for scheduling and planning. A man using CPM can understand that how the parts of a project should fit together and is able to do two things. Firstly, he can analyse graphically that how the different parts of the project fit together and test the logical of the proposed solution. Secondly, using the network diagram, he can demonstrate his schedule way to do the project and can expedite the actual doing of the work. Critical path method was developed by M.R. Walker and E.T. Dupont de nemours.

A CPM is a route between two or more operations which minimize (or maximize) some measure of performance. This can also be defined as the sequence of activities which will require greatest normal time to accomplish.

OBJECTIVES OF CPM

Following are the objectives of CPM,

(i) To find the obstacles and difficulties involved in a production process.

(ii) To assign time for each operation or activity.

(iii) To decide about the starting and finishing times of the work.

(iv) One of the main objectives of network analysis is to identify the critical path so that resources from outside the system or from the non-critical activities can be diverted to activities on critical path, if required.

 Working Procedure

In CPM, firstly the network diagram is drawn. Then after the network diagram is logically made, the required time (or some other measure of performance) to do each operation is posted above and to left of each operation circle. These times (or some other measure of performance) are then combined to develop a schedule which minimize or maximize the measure of performance for each operation. The schedule thus obtained, can be used to determine the critical path.

Let us consider a route or path (i, j) defined as a connected set of links which can be traversed to get from one stated node (i) to another node (j). Then the critical path is the path from node (i) to node (j) such that the sum of the values of the links traversed is minimum (maximum).

DETERMINATION OF CRITICAL PATH (LABELLING METHOD)

In this method, we have a set of links connecting every node in a network which does not contain any loop, is called a tree. A tree has a property that the path between any two nodes is unique.

 Working Procedure

The following are the steps in labelling methods:

STEP 1. Label origin with "distance zero". Go to step 2.

STEP 2. Look for the links whose tail ends are labelled and whose heads are unlabelled. For each such link find the sum of the labels at the tail end and its length. Tick the link for which this sum is a minimum (or maximum) and label the head with this sum. Return to the beginning of step 2.

☛ REMARKS
  • The process terminates when (a) the terminal node is labelled, (b) all nodes have been labelled, (c) the minimum of the sums formed in step 2 is finite.
  • The ticked links will form a shortest tree.
  • The main drawback of labelling method is that there may be large set of numbers at each stage.

Solved Examples

EXAMPLE 1.   *Find the critical path from A to I in the figure by labelling method.*

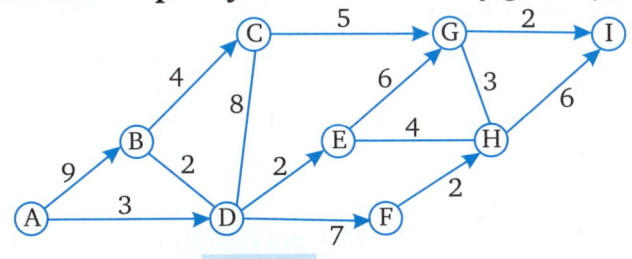

Fig. 17

SOLUTION.   Here, node 'A' is origin and I is the terminal. Now label the node 'A' as zero. The links with labelled tails and an unlabelled heads are those from A, *i.e.*, AB and the sums of their labels are

Stage 1. $AB = 0 + 9 = 9 \quad AD = 0 + 3 = 3$

The maximum of these sums is for *AB*, therefore, link *AB* is ticked and *B* is labelled with 9. Similarly, we proceed further from node *B* and have the following stages.

Stage 2. $BC = 9 + 4 = 13$,

$BD = 9 + 2 = 11$ Tick *BC* and label *C* (13)

Working through in this way, we get the following stages:

Stage 3. $CG = 13 + 5 = 18$,

$CD = 13 + 8 = 21$ Tick *CD*, label *D* (21)

Stage 4. $DE = 21 + 2 = 23$,

$DF = 21 + 7 = 28$ Tick *DF*, label *F* (28)

Stage 5. $FH = 28 + 2 = 30$ Tick *FH*, label *H* (30)

Stage 6. $HI = 30 + 6 = 36$ Tick *HI*, label *I* (36)

Hence the critical path in this case is $A \rightarrow B \rightarrow C \rightarrow D \rightarrow F \rightarrow H \rightarrow I$. with distance equal to 36.

The full working is summarized in table given below

Stage No.	Additional links considered	Sum of label lengths	Link Ticked	Node Labelled	Value of Node
1.	AB	$0 + 9 = 9$			
			AB	B	9
	AD	$0 + 3 = 3$			
2.	BC	$9 + 4 = 13$			
			BC	C	13
	BD	$9 + 2 = 11$			
3.	CG	$13 + 5 = 18$			
			CD	D	21
	CD	$13 + 8 = 21$			
4.	DE	$21 + 2 = 23$			
			DF	F	28
	DF	$21 + 7 = 28$			
5.	FH	$28 + 2 = 30$	FH	H	30
6.	HI	$30 + 6 = 36$	HI	I	36

17.8 CRITICAL PATH ANALYSIS

The objective of critical path analysis is to estimate the total project duration and to assign starting and finishing times to all activities involved in this project. This is helpful in checking actual progress against the scheduled duration of the project.

17.8.1 CRITICAL PATH

The Critical path in a network diagram is the longest continuous chain of activities through the network starting from first to the last event and is shown by thick line or double lines.

NOTATIONS

Consider the following notations associated to the activities and events :

(i, j) = Activity (i, j) with tail event i and head event j

E_i = Earliest occurance time of an event i

L_i = Latest occurance time of an event i

$ES_{i,j}$ = Earliest start time of an activity (i, j)

$LS_{i,j}$ = Latest start time of an activity (i, j)

$EF_{i,j}$ = Earliest finish time of an activity (i, j)

$LF_{i,j}$ = Latest finish time of an activity (i, j)

$t_{i,j}$ = Duration of activity (i, j)

17.8.2 DESCRIPTION OF SOME NOTATIONS

E_i : It is the earliest time at which an event i can occur without affecting the total project time.

L_i : It is the latest time at which an event i can occur without affecting the total project time.

$ES_{i,j}$ = It is the earliest possible time at which the event i, from which the activity (i, j) emanates, can start without affecting the total project time.

$LS_{i,j}$ = It is the latest possible time by which the event i from which the activity (i, j) emanates must start without affecting the total project time.

$EF_{i,j}$ = It is the earliest possible time at which an activity (i, j) can finish without affecting the total project time.

$LF_{i,j}$ = It is the latest possible time by which an activity (i, j) must finish without affecting the total project time, *i.e.*, it is the latest occurance time of an event j at which the activity (i, j) terminates.

☞ REMARKS

- For an activity (i, j), $ES_{i,j} = E_i$; $LF_{i,j} = L_j$

  $EF_{i,j} = ES_{i,j} + t_{i,j} = E_i - t_{i,j}$

  $LS_{i,j} = LF_{i,j} - t_{i,j} = L_j - t_{i,j}$

17.8.3 CALCULATION OF EARLIEST OCCURANCE AND LATEST ALLOWABLE TIMES OF EVENTS

There are following two methods for calculating the earliest occurance and latest allowable time of an event.

   (i) Forward pass method, and

   (ii) Backward pass method

17.8.4 FORWARD PASS METHOD (FOR EARLIEST EVENT TIME)

In this method, we calculate the following:

   (i) the earliest start and earliest finish time of each activity.

   (ii) the earliest occurance time of each event.

   (iii) the earliest finish time of entire project.

Working Procedure

In this method, we have the following steps:

STEP 1. Set the earliest occurance time of initial event 1 to zero, *i.e.*, $E_1 = 0$.

STEP 2. Calculate earliest start time for each activity that begins at event i (= 1), *i.e.*, $(1, j)$ is the activity that emanates from event 1 and

   (a) Calculate $ES_{i,j}$, the earliest start time of each activity $(1, j)$.

      Clearly, $ES_{i,j} = E_1 = 0$ for each activity $(1, j)$

   (b) Calculate $EF_{1,j}$, the earliest finish time of each activity $(1, j)$ that emanates from the tail event 1.

Clearly, $EF_{1,j} = ES_{1,j} + t_{1,j} = E_1 + t_{1,j} = t_{1,j}$

where $t_{1,j}$ is the duration of activity $(1, j)$.

STEP 3. Proceed to the next event say $j, j > i$.

STEP 4. Calculate the earliest occurance time for event j. This is the maximum of the earliest finish times of all activities ending into that event, *i.e.,*

$E_j = \max\{EF_{i,j}\} = \max\{E_i + t_{i,j}\}$ for all immediate predecessor activities.

STEP 5. If $j = N$ (final event) then earliest finish time for the project, *i.e.,* the earliest occurance time E_n for the final event is given by

$E_N = \max\{EF_{i,j}\} = \max \{E_{N-1} + t_{i,j}\}$ for all terminal activities.

17.8.5 BACKWARD PASS METHOD (FOR LATEST ALLOWABLE EVENT TIME)

In this method, the calculation start from the final event N and proceed through the event in decreasing order of event numbers from N and end to the initial event 1. In this method we calculate

   (i) the latest finish and latest start time of each activity.

  (ii) the latest occurance time of each event.

 (iii) the latest finish time of entire project.

 Working Procedure

The steps of procedure are given below:

STEP 1. Set the latest occurance of last event N, equal to the earliest occurance time E_N, *i.e.,*

$$L_N = E_N$$

If the target time say T, for the completion of the project is given, then take $L_N = T$ otherwise take $L_N = E_N$.

STEP 2. Calculate latest finish time of each activity (i, N) which is equal to the latest occurance time L_N of last event N, *i.e.,* $LF_{i,N} = L_N = E_N$.

STEP 3. Calculate the latest start time of each activity (L, N) which is given by

$$LS_{i,N} = LF_{i,N} = L_N - t_{i,N} \text{ for each activity } (i, N)$$

where $t_{i,N}$ is the duration of the activity (i, N)

STEP 4. Proceed backward to the event $N - 1$, next to the event N and consider all activities $(j, N - 1)$ terminating at event $N - 1$ and

   (i) Calculate $LF_{j, N-1}$, *i.e.* L_{N-1} the latest finish time of each activity $(j, N - 1)$ which is given by

$$LF_{j, N-1} = L_{N-1} = L_N - t_{N-1, N}$$

  (ii) Calculate $LS_{j, N-1}$ the latest start time of each activity $(j, N - 1)$ which is given by

$$LS_{j, N-1} = LF_{j, N-1} - t_{j, N-1} = L_{N-1} - t_{j, N-1}$$

STEP 5. Proceed backward to the event that decrease by 1, *i.e.,* proceed backward to the event no, $N - 2, N - 3, \ldots$ next to $N - 1, N - 2, \ldots$ respectively by considering all activities $(K, N - 2), (K, N - 3), \ldots$ terminating at $N - 2, N - 3, \ldots$ respectively.

Continuing in this way, the event in decreasing order of numbers, *i.e.,* through events $N, N - 1, N - 2, \ldots$ in order, read the first event-1 and calculate the latest occurance

time of start event-1 which is given by
$$L_1 = \min_s \{L_s - t_{1,s}\}$$
where s is the collection of all successor event of event-1.

☞ REMARK
- If $L_N = E_N$ then always $L_1 = E_1 = 0$.

17.9 SLACK AND FLOAT IN NETWORK ANALYSIS

There are many activities where the maximum time available to finish the activity is more than the time required to complete it, i.e., the difference between the two is known as total float available for that activity. The float of an activity has the same significance as the slack of the events. There are following three types of float.

17.9.1 TOTAL FLOAT

It is the amount of time by which an activity can be delayed if all its preceeding activities are completed at their earliest possible time and all successor activities can be delayed until their latest permissible time.

It can also be defined as the difference between the maximum time available to perform the activity and activity duration time. The total float of an activity (i, j) is denoted by $TF_{i,j}$.

If $t_{i,j}$ is the duration time of the activity (i, j), then
$$\text{total float} = TF_{i,j} = (L_j - E_i) - t_{i,j}$$
$$= (L_j - t_{i,j}) - E_i = LS_{i,j} - ES_{i,j}$$
Also,
$$TF_{i,j} = L_j - (E_i + t_{i,j}) = (F_{i,j} - EF_{i,j})$$

> The total float of an activity is equal to the difference of its latest and earliest start time or the difference of its latest and earliest finish time.

17.9.2 FREE FLOAT

The free float of an activity is defined as the amount of time by which the completion of an activity can be delayed without causing any delay in its immediate succeding activities by assuming that all activities start as early as possible. Therefore, for an activity (i, j),

Free float, $FF_{i,j}$
$$= (E_j - E_i) - t_{i,j}$$
or,
$$FF_{i,j} = E_j - (E_i + t_{i,j}) = E_j - EF_{i,j}$$

> The free float of an activity is the difference between the earliest time for the head event and the earliest completion time of the activity under consideration.

17.9.3 INDEPENDENT FLOAT

It is the amount of time by which an activity can be rescheduled without affecting the preceeding or the succeeding activities. Therefore, the independent float is the amount of time by which the start of the activity can be delayed without affecting the earliest start time of the immediate succeding activities, assuming that the preceding activity has complete at its earliest finish time.

∴ For an activity (i, j), the independent float $IF_{i,j} = (E_j - L_i) - t_{i,j}$

The value of total float for any activity can help in making following conclusions.

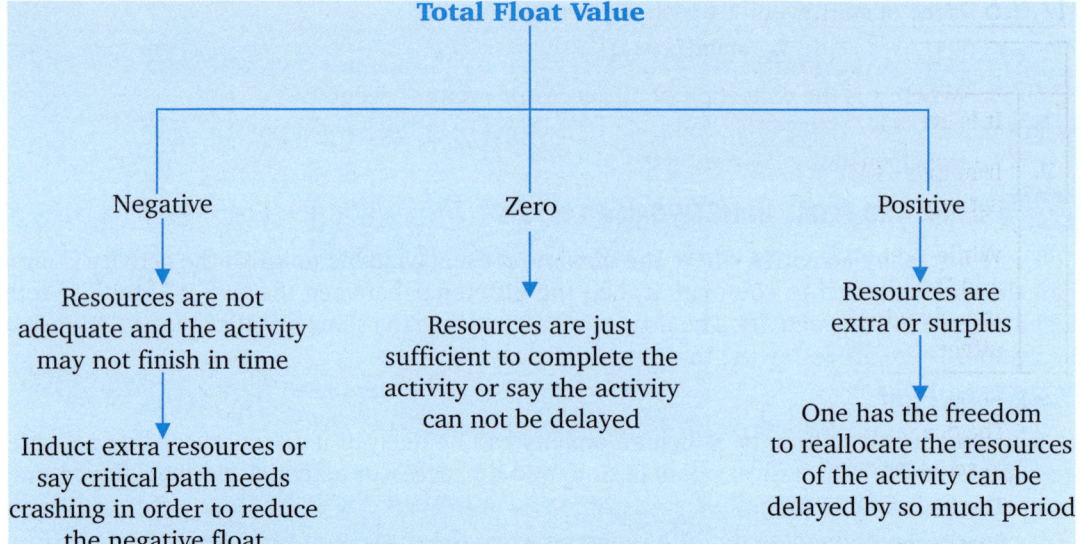

- The negative value of the independent float is considered zero.
- Latest occurance time of an event is always greater than or equal to the earliest occurance time.
- The difference between total float and free float is known as 'Interference float'.
- Total float is the extra surplus time which can be allocated to an activity or this is the period of time upto which an activity can be delayed beyond its earliest finish time without extending the overall project time. If this time is utilized by expanding the duration of any activity, the float of the preceding and the succeding activities will be reduced.

17.9.4 SLACK ON EVENTS

The slack of an event is the difference between its latest occurance time (L_i) and the earliest occurance time (E_i), i.e.,

$$\text{Event float} = L_i - E_i$$

Here, $\qquad$ head event slack $= L_j - E_j$

$$\text{tail event slack} = L_i - E_j$$

Clearly, it is a measure of how long an event can be delayed without increasing the project completion time.

Here, we observe that

(i) If $L = E$ for certain events, then such events are called critical events.

(ii) If $L \neq E$ for certain events, then slack on these events can be negative or positive.

17.9.5 SLACK ON ACTIVITY

It is the length of free time available within the estimated times of non-critical activities. The computation of activity float tells us how long an activity time may be increased without increasing the project completion time.

17.9.6 DIFFERENCE BETWEEN FLOAT AND SLACK

	Float		Slack
1.	It is activity oriented.	1.	It is event oriented.
2.	It helps to identify critical and non-critical activities.	2.	It helps to identify critical and non-critical events.
3.	While scheduling resources, independent and free floats play a dominant role where the activities can be delayed without affecting total project time.	3.	While time cost trade off is done, least slack of events is considered and the activity of that event is chosen to crash down.

☞ REMARKS

- The float of an activity has the same significance as the slack of the events. The float corresponding to activities hence to CPM and the slack corresponding to events and hence to PERT.
- The basic difference between slack and float time is that slack is used for events only whereas float is applied for activities.

17.9.7 RELATION BETWEEN FLOATS OF AN ACTIVITY

We know that

$$\text{total float, } TF_{i,j} = L_j - E_i - t_{i,j}$$

and
$$\text{free float, } FF_{i,j} = E_j - E_i - t_{i,j}$$

and
$$\text{independent float, } IF_{i,j} = E_j - L_i - t_{i,j}$$

Further, since for an event, latest occurance time can not be less than its earliest occurance time, therefore

$$L_i \geq E_i \text{ and } L_j \geq E_j$$

So,
$$TF_{i,j} = L_j - E_i - t_{i,j} \geq E_j - E_i - t_{i,j}$$

$$\Rightarrow \qquad TF_{i,j} \geq FF_{i,j} \qquad \qquad \dots(1)$$

Further,
$$FF_{i,j} = E_j - E_i - t_{i,j} = (E_j - t_{i,j}) - E_i$$
$$\geq (E_j - t_{i,j}) - L_i$$

$$\Rightarrow \qquad FF_{i,j} \geq IF_{i,j} \qquad \qquad \dots(2)$$

From (1) and (2) we conclude that

$$TF_{i,j} \geq FF_{i,j} \geq IF_{i,j}$$

$$\Rightarrow \qquad \text{Total float} \geq \text{Free float} \geq \text{Independent float}$$

17.10 CRITICAL PATH

(i) **Critical event:** An event with zero slack time is called critical event.

For critical event, $L_i = E_i$

(ii) **Critical activity:** An activity is called critical activity if a delay in its execution will cause further delay in the project completion time.

For a critical activity, total float time is zero and $E_j - E_i = L_j - L_i = t_{i,j}$

(iii) **Critical Path:** The continuous chain of critical activities from the first event to the last event in the network diagram is known as critical path.

- Critical path is the longest path starting from the first event to the last event and is shown by the thick line or double lines in the network diagram.
- The sum of the individual time durations of all critical activities on the critical path is called the total project time and it is the minimum time required to complete the project.

17.10.1 DETERMINATION OF CRITICAL PATH BASED ON TIME ESTIMATES

 Working Procedure

This procedure have the following steps:

STEP 1. Draw the network diagram of the given project and write the individual time duration of each activity above the arrow representing the activity.

STEP 2. Calculate the EST (earliest start time) of each event taken in increasing order of their numbers starting from the first to the last by forward pass method in the following manner.

Let us take EST of start event-1, *i.e.*, $E_1 = 0$

EST of event $j = E_j = \max_i \{E_i + t_{i,j}\}$

where i is the collection of tail event of all activities (i, j) of duration $t_{i,j}$ terminating in head event j. Write the EST of events in the side of corresponding event.

STEP 3. Calculate LST (latest start time) of each event taken in decreasing order of their numbers starting from the last to the first event by backward pass method in the following manner.

Let us take the LST of last event N, *i.e.*, $L_N = E_N$

LST of event $i = L_i = \min_j \{L_j - t_{i,j}\}$

where j is the collection of head events of all activities (i, j) of duration $t_{i,j}$ which emanate from tail event i. Write the LST of events in the side of the corresponding event.

STEP 4. Calculate EFT (earliest finish time) $EF_{i,j}$ and LFT (latest finish time) $LF_{i,j}$ of all activities (i, j) by the relation.

$$EF_{i,j} = E_i + t_{i,j} \text{ and } LF_{i,j} = L_i + t_{i,j}$$

and find total float, free float and independent float for each activity (i, j) by using the following formulae.

(i) total float $= (L_j - t_{i,j}) - E_i$

(ii) free float $= (E_j - E_i) - t_{i,j}$

(iii) independent float $= E_j - L_i - t_{i,j}$

STEP 5. Identify the critical activities (i, j) which satisfy all the following relations

(i) $E_i = L_i$

(ii) $E_j = L_j$

(iii) $E_j - E_i = L_j - L_i = t_{i,j}$

STEP 6. Connect the critical activities without break from the start event to end event to get the critical path. Mark this path by double line or thick line.

STEP 7. Calculate the total project time which is equal to E_N, where N is the last event. Then clearly,

$$E_N = \text{Sum of duration times of all critical activities on critical path}$$

- The slack of each node on the critical path is zero.

17.10.2 ADVANTAGES AND DISADVANTAGES OF CPM

Advantages

(i) It highlights the critical activities on which management should focus attention to reduce projection completion time.

(ii) CPM helps management in diverting resources from non-critical to critical activities.

(iii) It provides techniques of planning and scheduling a project.

(iv) It gives complete information about the significance, size, duration and performance of an activity.

(v) It helps to identify potential bottlenecks and to avoid unnecessary pressure on the paths that will not result in earlier completion of the project.

Disadvantages

(i) It operates on the assumption that there is a precise known time that each activity in the project will take, which is not true in real life situation.

(ii) It is not suitable for a situation which does not have definite start and definite finish.

(iii) CPM can not by itself solve a problem. It only facilitates a thorough examinations of the problem and alternative solution for it.

(iv) It does not incorportate statistical analysis in determining time schedule.

(v) Each time changes are introduced into the network the entire evaluation of the project has to be repeated and a new critical path has to be determined.

Solved Examples

EXAMPLE 1. *Construct a network diagram and find the critical path, EST and LST for the following data,*

Job	Description	Successor Activity	Time in minutes
S	Start	A	0
A	Processing letter	B	4
B	Dictating letter	D	3
C	Order execution	E	7
D	Supervision	E	6
E	Dispatching	F	2
F	Finish	None	0

SOLUTION. The network for the given problem is drawn in figure.

There can be two paths in the problem, namely

Path I : $A \rightarrow B \rightarrow D \rightarrow E \rightarrow F$, Path II : $A \rightarrow C \rightarrow E \rightarrow F$

In path I : Total time taken is $4 + 3 + 6 + 2 = 15$ minutes

At start A EST and LST are $(0, 0)$

At start B EST and LST are $(4, 4)$

At start D EST and LST are $(13, 13)$

and At start E EST and LST are $(15, 15)$

In path II : Total time taken is $4 + 7 + 2 = 13$ minutes

At C, EST = 4 and LST = 6

At E, EST and LST are (13, 13)

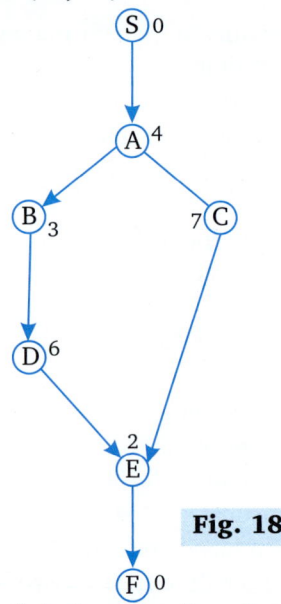

Fig. 18

Hence, the critical path is $A \rightarrow B \rightarrow D \rightarrow E \rightarrow F$ with maximum target time of 15 minutes.

EXAMPLE 2. *Calculate the various Float values for activities described in network drawn in figure given below.*

SOLUTION. The EST and LFT of various activities and their durations can be calculated easily. These can be shown in the following network diagram.

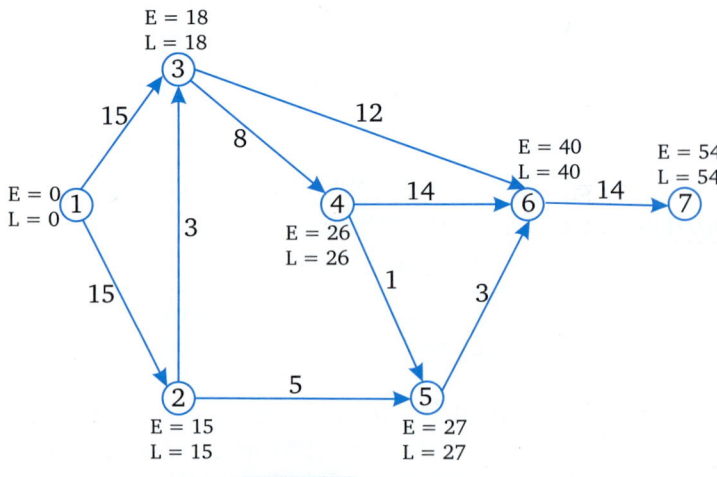

Fig. 19

The various float time estimates can be calculated in the tabular form 2 with the help of the network diagram in the above figure.

Activity	Duration	Start Times		Finish Times		Float Times		
		EST	LST	EFT	LFT	Total	Free	Independent
1-2	15	0	0	15	15	0	0	0
1-3	15	0	3	15	18	3	3	3
2-3	3	15	15	18	18	0	0	0
2-5	5	15	32	20	37	17	7	7
3-4	8	18	18	26	26	0	0	0
3-6	12	18	28	30	40	10	10	10
4-5	1	26	36	27	37	10	0	0
4-6	14	26	26	40	40	0	0	0
5-6	3	27	37	30	40	10	10	0
6-7	14	40	40	54	54	0	0	0

e.g. for activity 1–2.

$$\text{Total Float} = [\text{LFT of Head event}(2) - \text{EST of Tail event }(1)] - \text{Duration}$$
$$= [(15 - 0) - 15] = 15 - 15 = 0$$
$$\text{Free Float} = \text{Total Float} - \text{Slack Time of the Head event }(2)$$
$$= \text{Total Float} - (L - E) \text{ of event }(2)$$
$$= 0 - 0$$
$$\text{Independent Float} = \text{Free Float} - \text{Slack time of Tail event}$$
$$= 0 - 0 = 0$$

EXAMPLE 3. ***Draw a network diagram on the basis of the following data***

Activity	Duration (days)	Activity	Duration (days)
1-2	2	4-8	8
1-4	2	5-6	4
1-7	1	6-9	3
2-3	4	7-8	3
3-6	1	8-9	5
4-5	5	9-10	2

Find the critical path, total duration and slack times. [UPTU MBA-2007]

SOLUTION. Using the above data given in the table, we have the following network diagram.

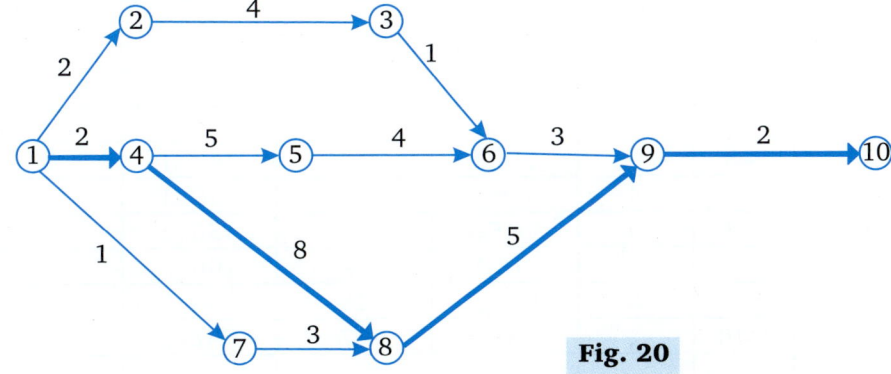

Fig. 20

Forward pass method	Backward pass method
$E_1 = 0$	$L_{10} = E_{10} = 17$
$E_2 = E_1 + t_{1,2} = 0 + 2 = 2$	$L_9 = L_{10} - t_{9,10} = 17 - 2 = 15$
$E_3 = E_2 + t_{2,3} = 2 + 4 = 6$	$L_8 = L_9 - t_{8,9} = 15 - 5 = 10$
$E_4 = E_1 + t_{1,4} = 0 + 2 = 2$	$L_7 = L_8 - t_{7,8} = 10 - 3 = 7$
$E_5 = E_4 + t_{4,5} = 2 + 5 = 7$	$L_6 = L_9 - t_{6,9} = 15 - 3 = 12$
$E_6 = E_5 + t_{5,6} = 7 + 4 = 11$	$L_5 = L_6 - t_{5,6} = 12 - 4 = 8$
$E_7 = E_1 + t_{1,7} = 0 + 1 = 1$	$L_4 = \min_{j=5,8}\{L_j - t_{4,7}\} = \min\{L_5 - t_{4,5}; L_8 - t_{4,8}\}$
	$\qquad = \min\{8 - 5, 10 - 8\} = 2$
$E_8 = \max\{E_i + t_{i,8}\}$	$L_3 = L_6 - t_{3,6} = 11 - 1 = 10$
$\quad = \max\{E_4 + t_{4,8}; E_7 + t_{7,8}\}$	
$\quad = \max\{2 + 8, 1 + 3\} = 10$	
$E_9 = \max_{i=6,8}\{E_i + t_{i,9}\}$	$L_2 = L_3 - t_{2,3} = 10 - 4 = 6$
$\quad = \max\{E_6 + t_{6,9}; E_8 + t_{8,9}\}$	
$\quad = \max\{11 + 3, 10 + 5\} = 15$	
$E_{10} = E_9 + t_{9,10} = 15 + 2 = 17$	$L_1 = \min_{j=2,4,7}\{L_j - t_{1,j}\} = \min\{L_2 - t_{1,2}; L_4 - t_{1,4}; L_7 - t_{1,7}\}$
	$\qquad = \min\{6 - 2, 2 - 2, 7 - 1\} = 0$

Now, for each activity (i, j) either start time E_i and earliest finish time $EF_{i,j} = E_i + t_{i,j}$ for tail event i, latest finish time L_j and latest start time $LS_{i,j} = L_j - t_{i,j}$ of head event j and total float are calculated in the following table.

Activity (i, j)	Duration $t_{i,j}$	Earliest time start $(ES_{i,j})$ E_i	Earliest time finish $(EF_{i,j})$ $E_i + t_{ij}$	Latest time finish $(LF_{i,j})$ L_j	Latest time start $(LS_{i,j})$ $L_j - t_{ij}$	Total float $LF_{ij} - EF_{ij}$ (or) $LS_{ij} - ES_{ij}$
(1)	(2)	(3)	(4)=(3)+(2)	(6)	(5)=(6)−(2)	(6) − (4) or (5) − (3)
—	—	—	—	$L_1 = 0$	—	—
1-2	2	0(E_1)	2	6(L_2)	4	4
1-4	2	0(E_1)	2	2(L_4)	0	0
1-7	1	0(E_1)	1	7(L_7)	6	6
2-3	4	2(E_2)	6	10(L_3)	6	4
3-6	1	6(E_3)	7	12(L_6)	11	5
4-5	5	2(E_4)	7	8(L_5)	3	1
4-8	8	2(E_4)	10	10(L_8)	2	0
5-6	4	7(E_5)	11	12(L_6)	8	1
6-9	3	11(E_6)	14	15(L_9)	12	1
7-8	3	1(E_7)	4	10(L_8)	7	6
8-9	5	10(E_8)	15	15(L_9)	10	0
9-10	2	15(E_9)	17	17(L_{10})		0
—	—	17(E_{10})	—	—	—	—

Clearly, the activity with total float zero are 1-4, 4-8, 8-9, 9-10 which are the critical activities.

Hence, the critical path is 1-4-8-9-10 shown by double lines in the above figure. Also, the slack time (in days) are as follows:

Activity	1-2	1-4	1-7	2-3	3-6	4-5	4-8	5-6	6-9	7-8	8-9	9-10
Slack	4	0	6	4	5	1	0	1	1	6	0	0

EXAMPLE 4. *All activities which together constitute a small engineering project are given in the following table. The table also shows the necessary immediate predecessors for each activity.*

Activity	A	B	C	D	E	F	G	H	I	J
Immediate Predecessor	—	A	A	A	B	C, D	D	B	E, F, G	G
Activity duration	2	3	4	5	6	3	4	7	2	3

   (i) *Construct an activity network.*

  (ii) *Find EFT for the entire project, assuming the project begins at day 0*

 (iii) *The total float for each activity*

 (iv) *The critical path*

  (v) *The latest start day for activity B*

 (vi) *The Earliest finish day for activity F*

 (vii) *The effect on the project duration if activity I were to take 3 days.*

(viii) *The effect on the project duration if activity F were to take 6 days.*

SOLUTION.   (i) Using the above data given in the table, the network diagram for the given project is as follows:

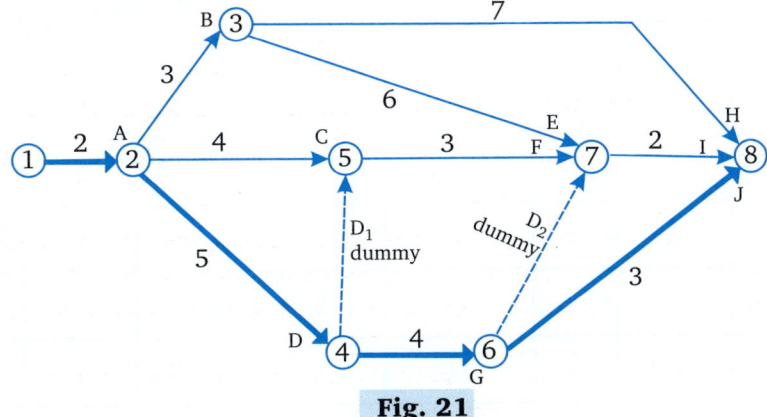

Fig. 21

(ii)

Forward pass method	Backward pass method
Taking $E_1 = 0$	Taking $L_8 = E_8 = 14$
$E_2 = E_1 + t_{1,2} = 0 + 2 = 2$	$L_7 = L_8 - t_{7,8} = 14 - 2 = 12$
$E_3 = E_2 + t_{2,3} = 2 + 3 = 5$	$L_6 = \min_{j=7,8}\{L_j - t_{i,j}\} = \min\{L_7 - t_{6,7}; L_8 - t_{6,8}\}$
	$\quad = \min\{12 - 0, 14 - 3\} = 11$
$E_4 = E_2 + t_{2,4} = 2 + 5 = 7$	$L_5 = L_7 - t_{5,7} = 12 - 3 = 9$
$E_5 = \max_{i=2,4}\{E_i + t_{i,5}\}$	$L_4 = \min_{j=5,6}\{L_j - t_{4,j}\} = \min\{L_5 - t_{4,5}; L_6 - t_{4,6}\}$
$\quad = \max\{E_2 + t_{2,5}; E_4 + t_{4,5}\}$	$\quad = \min\{9 - 0, 11 - 4\} = 7$
$\quad = \max\{2 + 4, 7 + 0\} = 7$	
$E_6 = E_4 + t_{4,6} = 7 + 4 = 11$	$L_3 = \min_{j=7,8}\{L_j - t_{3,j}\} = \min\{L_7 - t_{3,7}; L_8 - t_{3,8}\}$
	$\quad = \min\{12 - 6, 14 - 7\} = 6$
$E_7 = \max_{i=3,5}\{E_i + t_{i,7}\}$	$L_2 = \min_{j=3,4,5}\{L_j - t_{2,j}\}$
$\quad = \max\{E_3 + t_{3,7}; E_5 + t_{5,7}\}$	$\quad = \min\{L_3 - t_{2,3}; L_4 - t_{2,4}; L_5 - t_{2,5}\}$
$\quad = \max\{5 + 6, 7 + 3\} = 11$	$\quad = \min\{6 - 3, 7 - 5, 9 - 4\} = 2$
$E_8 = \max_{i=3,6,7}\{E_i + t_{i,8}\}$	$L_1 = L_2 - t_{1,2} = 2 - 0 = 0 = E_1$
$\quad = \max\{E_3 + t_{3,8}; E_6 + t_{6,8}; E_7 + t_{7,8}\}$	
$\quad = \max\{5 + 7, 11 + 3, 11 + 2\} = 14$	

Now we have the following table

Activity (i, j)	Duration $t_{i,j}$	Earliest time start E_i	Earliest time finish $EF_{i,j} = E_i + t_{i,j}$	Latest time start $LS_{i,j} = L_j - t_{i,j}$	Latest time finish L_j	Float Total $L_j - E_i - t_{i,j}$ $TF - LH$-slack	Float Free $TF - LH$-slack	Float Independent $TF -$ tail slack
(1)	(2)	(3)	(4) = (3) + (2)	(5) = (6) - (2)	(6)	(7) = (5) - (3)	(8) = (7) - (L_j - E_j)	(8) - (L_i - E_i)
—	—	—	—	—	$L_1 = 0$	—	—	—
$A(1,2)$	2	$E_1 = 0$	2	0	$L_2 = 2$	0	$0 - (L_2 - E_2)$ $= 0$	$0 - (L_1 - E_1)$ $= 0$
$B(2,3)$	3	$E_2 = 2$	5	3	$L_3 = 6$	1	$1 - (L_3 - E_3)$ $= 0$	$0 - (L_2 - E_2)$ $= 0$
$D(2,4)$	5	$E_2 = 2$	7	2	$L_4 = 7$	0	$0 - (L_4 - E_4)$ $= 0$	$0 - (L_2 - E_2)$ $= 0$
$C(2,5)$	4	$E_2 = 2$	6	5	$L_5 = 9$	3	$3 - (L_5 - E_5)$ $= 1$	$1 - (L_2 - E_2)$ $= 1$
$E(3,7)$	6	$E_3 = 5$	11	6	$L_7 = 12$	1	$1 - (L_7 - E_7)$ $= 0$	$0 - (L_3 - E_3)$ $= -1$

Cont.....

$H(3,8)$	7	$E_3 = 5$	12	7	$L_8 = 14$	2	$2-(L_8-E_8)$ $= 2$	$2-(L_3-E_3)$ $= 1$
$D_1(4,5)$	0	$E_4 = 7$	7	9	$L_5 = 9$	2	$2-(L_5-E_5)$ $= 0$	$0-(L_4-E_4)$ $= 0$
$G(4,6)$	4	$E_4 = 7$	11	7	$L_6 = 11$	0	$0-(L_6-E_6)$ $= 0$	$0-(L_4-E_4)$ $= 0$
$F(5,7)$	3	$E_5 = 7$	10	9	$L_7 = 12$	2	$2-(L_7-E_7)$ $= 1$	$1-(L_5-E_5)$ $= -1$
$D_2(6,7)$	0	$E_6=11$	11	12	$L_7 = 12$	1	$1-(L_7-E_7)$ $= 0$	$0-(L_6-E_6)$ $= 0$
$J(6,8)$	3	$E_6=11$	14	11	$L_8 = 14$	0	$0-(L_8-E_8)$ $= 0$	$0-(L_6-E_6)$ $= 0$
$I(7,8)$	2	$E_7=11$	13	12	$L_8 = 14$	1	$1-(L_8-E_8)$ $= 1$	$1-(L_7-E_7)$ $= 0$
—	—	$E_8=14$	—	—	—	—	—	—

Therefore, the earliest finish date for entire project assuming the project begins at the day 0 is $E_8 = 14$ days.

(iii) From the above table, the total float of each activity are as given below:

Activity	A	B	C	D	E	F	G	H	I	J
Total float	0	1	3	0	1	2	0	2	1	0

(iv) It is clear from the table given in (ii), the activities with total float zero are

$$A(1, 2), D(2, 4), G(4, 6) \text{ and } J(6, 8)$$

which are the required critical activities.

(v) The latest start day for activity B
$$= L_3 - t_{2,3} = 6 - 3 = 3 \text{ days}$$

(vi) Earliest finish time for activity F
$$= E_5 + t_{5,7} = 7 + 3 = 10 \text{ days}$$

(vii) If duration of activity 1, i.e., $t_{7,8} = 3$ days in place of 2 days then
$$E_7 = 11 \text{ days}$$
$$E_8 = \max\{E_3 + t_{3,8}; E_6 + t_{6,8}; E_7 + t_{7,8}\}$$
$$= \max\{5 + 7, 11 + 3, 11 + 3\}$$
$$= 14 \text{ days (No change)}$$
$$L_8 = 14 \text{ days (No change)}$$
and $\quad L_7 = L_8 - t_{7,8} = 14 - 3 = 11 \text{ days}$

Therefore, $\quad E_7 = L_7 = 11 \text{ days}$

$\Rightarrow$ There will be no effect on the project duration which will remain the same as 14 days but $E_7 = L_7$ and $E_8 = L_8$ implies that the activities $I(7, 8)$ will become critical activity as in this case total float = 0.

(viii) If duration of activity F, i.e., $t_{5,7} = 6$ days in place of 3 days then we have
$$E_7 = \max\{E_3 + t_{3,7}; E_5 + t_{5,7}\} = \max\{5 + 6, 7 + 6\} = 13$$
and $\quad E_8 = \{E_3 + t_{3,8}; E_6 + t_{6,8}; E_7 + t_{7,8}\}$
$$= \max\{5 + 7, 11 + 3, 13 + 2\} = 15$$

Hence, we conclude that in this case the project will be delayed by $15 - 14 = 1$ day.

EXAMPLE 5. *Draw the network diagram from the following activities and find the critical path and total float of the activities.*

Job	A	B	C	D	E	F	G	H	I	J	K
Jobtime (days)	13	8	10	9	11	10	8	6	7	14	18
Immediate Predecessors	—	A	B	C	B	E	D, F	E	H	G, I	J

SOLUTION. Proceeding in a usual manner, the network diagram from the given activities is as given below

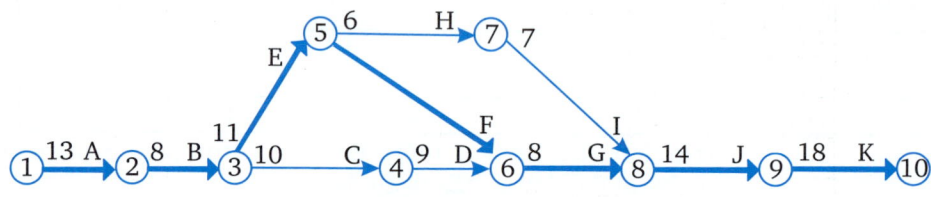

Fig. 22

Forward pass method	**Backward pass method**
Taking $E_1 = 0$	Taking $L_{10} = E_{10} = 82$
$E_2 = E_1 + t_{1,2} = 0 + 13 = 13$	$L_9 = L_{10} - t_{9,10} = 82 - 18 = 64$
$E_3 = E_2 + t_{2,3} = 13 + 8 = 21$	$L_8 = L_9 - t_{8,9} = 64 - 14 = 50$
$E_4 = E_3 + t_{3,4} = 21 + 10 = 31$	$L_7 = L_8 - t_{7,8} = 50 - 7 = 43$
$E_5 = E_3 + t_{3,5} = 21 + 11 = 32$	$L_6 = L_8 - t_{6,8} = 50 - 8 = 42$
$E_6 = \max_{i=4,5}\{E_i + t_{i,6}\}$ $= \max\{E_4 + t_{4,6}; E_4 + t_{5,6}\}$ $= \max\{31 + 9, 32 + 10\} = 42$	$L_5 = \min_{j=6,7}\{L_j - t_{5,j}\} = \min\{L_6 - t_{5,6}; L_7 - t_{5,7}\}$ $= \min\{42 - 10, 43 - 6\} = 32$
$E_7 = E_5 + t_{5,7} = 32 + 6 = 38$	$L_4 = L_6 - t_{4,6} = 42 - 9 = 33$
$E_8 = \max_{i=6,7}\{E_i + t_{i,8}\}$ $= \max\{E_6 + t_{6,8}; E_7 + t_{7,8}\}$ $= \max\{42 + 8, 38 + 7\} = 50$	$L_3 = \min_{j=4,5}\{L_j - t_{3,j}\} = \min\{L_4 - t_{3,4}; L_5 - t_{3,5}\}$ $= \min\{33 - 10, 32 - 11\} = 21$
$E_9 = E_8 + t_{8,9} = 50 + 14 = 64$	$L_2 = L_3 - t_{2,3} = 21 - 8 = 13$
$E_{10} = E_9 + t_{9,10} = 64 + 18 = 82$	$L_1 = L_2 - t_{1,2} = 13 - 13 = 0$

Now we have the following table.

Activity (i, j)	Duration $t_{i,j}$ (days)	Earliest time		Latest time		Total float
		start (E_i)	finish $(E_i + t_{i, j})$	start $(L_j - t_{i, j})$	finish (L_j)	$L_j - E_i - t_{i, j}$
(1)	(2)	(3)	(4)=(3)+(2)	(5)=(6)–(2)	(6)	(7) = (6) – (4)
—	—	—	—	—	$L_1 = 0$	—
$A(1, 2)$	13	$E_1 = 0$	13	0	$L_2 = 13$	0
$B(2, 3)$	8	$E_2 = 13$	21	13	$L_3 = 21$	0
$C(3, 4)$	10	$E_3 = 21$	31	23	$L_4 = 33$	2
$E(3, 5)$	11	$E_3 = 21$	32	21	$L_5 = 32$	0
$D(4, 6)$	9	$E_4 = 31$	40	33	$L_6 = 42$	2
$F(5, 6)$	10	$E_5 = 32$	42	32	$L_6 = 42$	0
$H(5, 7)$	6	$E_5 = 32$	38	37	$L_7 = 43$	5
$G(6, 8)$	8	$E_6 = 42$	50	42	$L_8 = 50$	0
$I(7, 8)$	7	$E_7 = 38$	45	43	$L_8 = 50$	5
$J(8, 9)$	14	$E_8 = 50$	64	50	$L_9 = 64$	0
$K(9,10)$	18	$E_9 = 64$	82	64	$L_{10} = 82$	0
—	—	$E_{10} = 82$	—	—	—	—

It is clear from the above table that floats of job A, B, E, F, G, J and K are zero, therefore the critical activities are (1, 2), (2, 3), (3, 5),(5, 6), (6, 8), (8, 9) and (9, 10). Hence, critical path is 1-2-3-5-6-8-9-10 as shown by the double line in the network diagram. Further, the total float activities are as follows:

Activity	A	B	C	D	E	F	G	H	I	J	K
Total Float	0	0	2	2	0	0	0	5	5	0	0

EXAMPLE 6. *Draw a network of the following activities and tabulate EST and LFT of each activity and the total and free float of them.*

Symbols	A	B	C	D	E	F	G	H	I	J	K	L	M
Event No.	1-2	2-3	3-4	2-4	4-5	2-5	5-8	5-6	6-8	5-7	7-8	8-9	9-10
No. of days	2	4	10	4	10	5	36	12	4	12	8	6	12

SOLUTION. The network diagram of the above activities are given below:

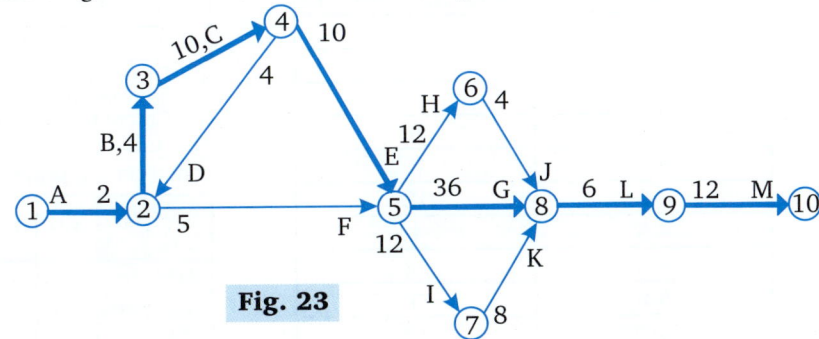

Fig. 23

Clearly, the critical path is

$$A \rightarrow B \rightarrow C \rightarrow E \rightarrow G \rightarrow L \rightarrow M$$

Now, we have the following table

Activity	Sequence code	Duration	Earliest time start	Earliest time finish	Latest time start	Latest time finish	Float total	Float free
A	1-2	2	0	2	0	2	0	0
B	2-3	4	2	6	2	6	0	0
C	3-4	10	6	16	6	16	0	0
D	2-4	4	2	6	12	16	10	10
E	4-5	10	16	26	16	26	0	0
F	2-5	5	2	7	21	26	19	19
G	5-8	36	26	62	26	62	0	0
H	5-6	12	26	38	46	58	20	20
I	6-8	4	38	42	58	62	20	20
J	5-7	12	26	38	42	54	16	16
K	7-8	8	38	46	54	62	8	8
L	8-9	6	62	68	62	68	0	0
M	9-10	12	68	82	68	80	0	0

 Exercise-17.2

1. A certain project is composed of miniactivities whose time estimates are given below.

Activity	Duration	Activity	Duration
1-2	1	4-6	2
1-3	3	5-6	4
1-4	2	6-7	6
2-5	1	5-7	3
3-5	3		

Draw the project network and find out the critical path.   [UPTU(MBA)–2008]

2. A small project consists of the following jobs where precedence relationship is given below.   [UP Tech.(MBA)–2005, 09]

Job	Duration (days)
1-2	15
1-3	15
2-3	3
2-5	5
3-4	8
3-6	12
4-5	1
4-6	14
5-6	3
6-7	14

(i) Draw an arrow diagram representing the project.

(ii) Find the critical path and the total project duration.

3. A small project consists of seven activities for which the relevant data are given below.   [UPTU(MBA)–2003]

Activity	Precedence activity	Activity duration (days)
A	—	4
B	A	7
C	—	6
D	C	5
E	B	7
F	D, E	6
G	F	5

(i) Draw the network and find the project completion time.

(ii) Calculate the total float for each activity.

4. The following table gives the activities of a construction, project and duration.

Activity (i, j)	1-2	1-3	2-3	2-4	3-4	4-5
Activity duration (days)	20	25	10	12	6	10

(i) Draw the network of the project.

(ii) Find the critical path and the minimum project time.

5. Consider the network shown in the adjoining figure.

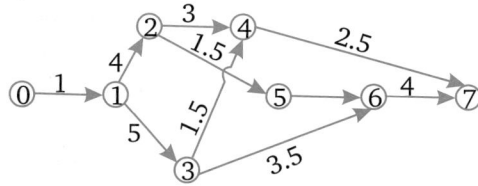

Find the total float, free float, independent float and identify the critical path.

[UPTU(B.TECH)–2007, 08, 10]

6. For a small project of 12 activities, the details are given below. Find the critical path and length of the project.

Activity	Dependance	Duration (days)
A	—	9
B	—	4
C	—	7
D	B, C	8
E	A	7
F	C	5
G	E	10
H	E	8
I	D, F, H	6
J	E	9
K	I, J	10
L	G	2

ANSWERS

1.

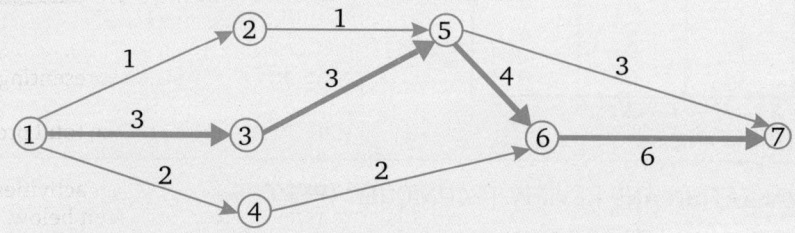

Critical path 1 — 3 — 5 — 6 — 7

2.

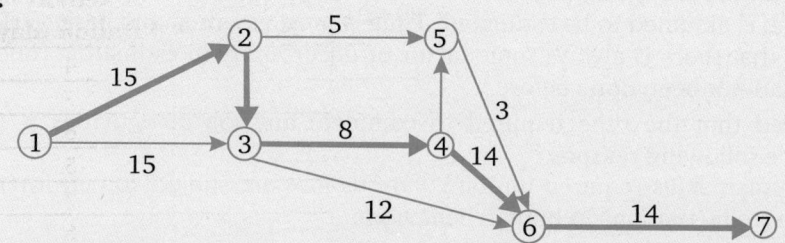

Critical path: 1 — 2 — 3 — 4 — 6 — 7. Total project duration = 54 days

3.

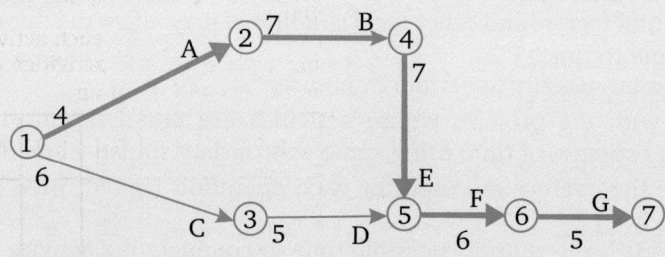

Critical path A — B — E — F — G, 1 — 2 — 4 — 5 — 6 — 7. Project completion time = 29 days

4.

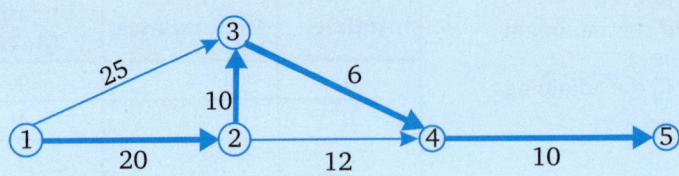

Critical path: 1 — 2 — 3 — 4 — 5

5.

Activity	Total float	Free float	Ind. float
0-1	0	0	0
1-2	2	0	0
1-3	0	0	0
2-4	3	0	– 2
2-5	2	0	– 2
3-4	3.5	0.5	0.5
3-6	0	0	0
4-7	3	3	0
5-6	2	2	0
6-7	0	0	0

6. Critical path: $A - E - H - I - K$

Total completion time = 40 days

17.11 PROJECT EVALUATION AND REVIEW TECHNIQUES (PERT)

PERT is a very popular management technique for planning and controlling the project. In PERT, the completion time is assumed to be uncertain and unknown. So, the probability of activity completion time is assumed to be estimated. Time is most essential and basic variable in PERT. It is assumed that there is always some factor of uncertainty in estimating times of any operation which had not been done before.

Clearly, we observed that the time required to complete any job or activities is non-deterministic due to the following reasons.

(1) Most of the human skills required in doing various jobs are subject to performance variability due to fluctuations in human behaviour.

(2) Uncertainty due to variation in the nature of work.

(3) Availability of resources varies from time to time.

(4) Variations in environmental factors and other local conditions may affect the duration of some activity from time to time.

(5) Research based activities are usually uncertain in nature.

The PERT technique deals with the projects whose activities are non-deterministic in nature. Here, we try to find best estimate of time using some appropriate statistical methods.

In PERT, the following three time values are used for each operation (called three time-estimates).

(i) **Optimistic time (t_0):** It is the shortest possible time to complete the activity if all goes well.

(ii) Most likely time (t_m): The model value (the most occuring duration of the activity) of the activity time distribution of the time which most often is required if the activity is replaced a no. of times.

(iii) Pessimistic time (t_p): This is the longest (maximum) time that is required to perform the activity, under extremely bad conditions.

Using the values of t_0, t_p and t_m the expected times of various activities and their standard deviation are calculated by using the following formula.

(i) Expected duration of activity: $t_e = \dfrac{1}{6}(t_0 + 4t_m + t_p)$

(ii) Standard deviation $\sigma = \dfrac{1}{6}(t_p - t_0)$ and variance $\sigma^2 = \left[\dfrac{1}{6}(t_p - t_0)\right]^2$

In PERT, for each activity we replace the three time estimates by the expected duration of the activity. Using these three time estimates, we do the forward and backward pass calculations to find the critical path.

In PERT analysis,
   (i) the duration of critical path based on the expected duration of each activity gives the expected duration of the project.
   (ii) the variance of the activities lying on the critical path are summed to find the variance of the project duration, because it is assumed that the two activities are independent.

☛ REMARK
- The activity durations are bound to follow a probability distributions called Beta (β) distribution.

17.11.1 ESTIMATE OF PROBABILITY

As we are expecting a variability in the activity duration, the total project may not be completed exactly in time. We can find the probability of actually completing the project as well as the activities in the schedule time.

The probability of completing the project by schedule time t_s is given by

$$\text{Prob.}\left(Z \le \frac{t_s - t_e}{\sigma_e}\right)$$

where,

t_e = expected time of project, i.e., sum of expected time of each activity lying on the critical path

σ_e = standard deviation of the project length

   = $\sqrt{\text{variance of the critical path}}$

   = $\sqrt{\text{sum of the variances of critical activities}}$

Here, the standard normal variate is given by

$$Z = \frac{t_s - t_e}{\sigma_e}$$

To find the probability of completing the project in some given time, we shall consider only the expected length of the critical path and its variance. The expected time of project can be calculated by adding the expected time of each activity lying on the critical path.

Solved Examples

EXAMPLE 1. *A certain project is composed of nine activities where time estimates are given below*

Activity	Expected duration		
	Optimistic	Most likely	Pessimistic
1-2	1	1	7
1-3	3	5	7
1-4	2	2	8
2-5	1	1	1
3-5	3	6	9
4-6	2	5	8
5-6	4	6	14
5-7	3	7	11
6-7	6	8	10

(i) **Draw the project network and trace all the possible paths from it. What is the expected project length?**

(ii) **Calculate the variance and standard deviations of the project length.**

(iii) **What is the probability that the project will be completed.**

   (a) **at least 4 weeks earlier than expected time**

   (b) **no more than 4 weeks later than expected time**

(iv) **If the project duration is 29 weeks, what is the probability of not meeting the due date.**

(v) **What due date has about 99% chance of completing the project?**

SOLUTION. (i) Using the above data, we have the following network diagram

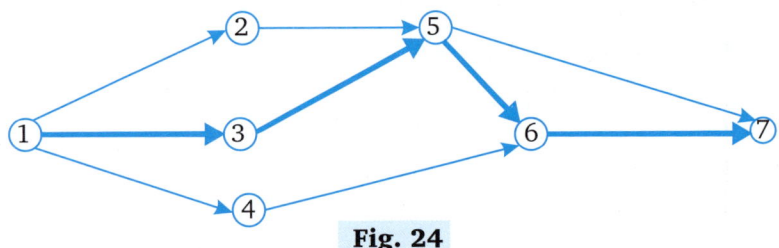

Fig. 24

All possible paths are

$1-2-5-7; 1-2-5-6-7; 1-3-5-7; 1-3-5-6-7; 1-4-6-7$

The expected activity time and variance of each activity $i-j$ can be calculated as follows:

Activity	Time estimates			Expected time $t_e = \dfrac{t_0 + 4t_m + t_p}{6}$	variance $\sigma^2 = \left(\dfrac{t_p - t_0}{6}\right)^2$
	t_0	t_m	t_p		
1-2	1	1	7	2	1
1-3	3	5	7	5	0.45
1-4	2	2	8	3	1
2-5	1	1	1	1	0
3-5	3	6	9	6	1
4-6	2	5	8	5	1
5-6	4	6	14	7	2.79
5-7	3	7	11	7	1.77
6-7	6	8	10	8	0.45

Now

Forward pass method	Backward pass method
Taking $E_1 = 0$ we have	Taking $L_7 = E_7 = 26$
$E_2 = E_1 + t_{e_{1,2}} = 0 + 2 = 2$	$L_6 = L_7 - t_{e_{6,7}} = 26 - 8 = 18$
$E_3 = E_1 + t_{e_{1,3}} = 0 + 5 = 5$	$L_5 = \min_{j=6,7} \{L_j - t_{e_{5,7}}\}$
	$\quad = \min\{L_6 - t_{e_{5,6}}, L_7 - t_{e_{5,7}}\}$
	$\quad = \min\{18 - 7, 26 - 7\} = 11$
$E_4 = E_1 + t_{e_{1,4}} = 0 + 3 = 3$	$L_4 = L_6 - t_{e_{4,6}} = 18 - 5 = 13$
$E_5 = \max_{i=2,3}\{E_i + t_{e_{i,5}}\}$	$L_3 = L_5 - t_{e_{3,5}} = 11 - 6 = 5$
$\quad = \max\{E_2 + t_{e_{2,5}}; E_3 + t_{e_{3,5}}\}$	
$\quad = \max\{2 + 1, 5 + 6\} = 11$	
$E_6 = \max_{i=4,5}\{E_i + t_{e_{i,6}}\}$	$L_2 = L_5 - t_{e_{2,5}} = 11 - 1 = 10$
$\quad = \max\{E_4 + t_{e_{4,6}}; E_5 + t_{e_{5,6}}\}$	
$\quad = \max\{3 + 5, 11 + 7\} = 18$	
$E_7 = \max_{i=5,6}\{E_i + t_{e_{i,7}}\}$	$L_1 = \min_{j=2,3,4}\{L_j - t_{e_{1,j}}\}$
$\quad = \max\{E_5 + t_{e_{5,7}}; E_6 + t_{e_{6,7}}\}$	$\quad = \min\{L_2 - t_{e_{1,2}}, L_3 - t_{e_{1,3}}, L_4 - t_{e_{1,4}}\}$
$\quad = \max\{11 + 7, 18 + 8\} = 26$	$\quad = \min\{10 - 2, 5 - 5, 13 - 3\} = 0$

Since, $E_1 = L_1, E_3 = L_3, E_5 = L_5, E_6 = L_6$ and $E_7 = L_7$ and for these activities $E_j - E_i = t_{e_{i,j}} = L_j - L_i$, so the activities 1-3, 3-5, 5-6 and 6-7 are critical activities therefore, the critical path is $1 - 3 - 5 - 6 - 7$ and expected project length $t_e = 26$ weeks.

(ii) Variance, σ^2-sum of variances for critical activities 1-3, 3-5, 5-6 and 6-7
$$= 0.45 + 1 + 2.79 + 0.45 = 4.69 \text{ weeks}$$

So, standard deviation, $\sigma_e = \sqrt{4.69} = 2.17$ weeks

(iii) (a) the probability of completing the project 4 weeks earlier than expected time, *i.e.*, time $t_s = 26 - 4 = 22$ is given by

$$\text{Prob.}\left\{Z \leq \frac{t_s - t_e}{\sigma_e}\right\} = \text{Prob.}\left\{Z \leq \frac{22 - 26}{2.17} = -1.84\right\}$$

$$= 0.5 - 0.4671 = 0.1329$$

(By Normal distribution table)

i.e., 13.29%

(b) Probability of completing the project 4 weeks later than expected time, *i.e.*, in time $t_s = 26 + 4 = 30$ weeks is given by

$$\text{Prob.}\left\{Z \leq \frac{t_s - t_e}{\sigma_e}\right\} = \text{Prob.}\left\{Z \leq \frac{30 - 26}{2.17} = 1.84\right\}$$

$$= 0.5 + 0.4671 = 0.9671$$

(By Normal distribution table)

i.e., 96.71%

(iv) If $t_s = 29$ weeks, then probability of completing the work by due date is given by

$$\text{Prob.}\left\{Z \leq \frac{t_s - t_e}{\sigma_e}\right\} = \text{Prob.}\left\{Z \leq \frac{29 - 26}{2.17} = 1.38\right\}$$

$$= 0.5 + 0.4162 = 0.9162 \quad \text{(By Normal distribution table)}$$

i.e., 91.62%

Thus, the probability of not meeting the project date of 29 weeks

$$= 1 - 0.9162 = 0.0838$$

i.e., 8.38%

(v) If t_s is the project time when the due date has about 99% chance then

$$\text{Prob}\left\{Z \leq \frac{t_s - t_e}{\sigma_e} = \frac{t_s - 26}{2.17}\right\} = 0.99$$

Then by normal distribution table

$$\frac{t_s - 26}{2.17} = 2.33$$

Hence, $t_s = 26 + 2.17 \times 2.33 = 26 + 5.06 = 31.06$ weeks ≈ 31 weeks

EXAMPLE 2. *A small project consisting of eight activities has the following characteristics.*

Activity	Preceding activity	Time estimates (in weeks)		
		most optimistic (t_0)	most likely (t_m)	most pessimistic (t_p)
A	—	2	4	12
B	—	10	12	26
C	A	8	9	10
D	A	10	15	20
E	A	7	7.5	11
F	B, C	9	9	9
G	D	3	3.5	7
H	E, F, G	5	5	5

(i) **Draw the PERT network for the project.**

(ii) **Find the critical path and the expected project length.**

(iii) **Find the expected duration and variance for each activity.**

(iv) **Prepare the activity schedule for the project.**

(v) **Calculate the variance and standard deviation of the project length.**

(vi) **If a 30 week deadline is imposed, what is the probability that the project will be finished in time.**

(vii) **If the project manager wants to be 99% sure that the project is completed on the schedule date, how many weeks before that date should he start the project work?**

SOLUTION. (i) Following the usual procedure, we have the following network diagram

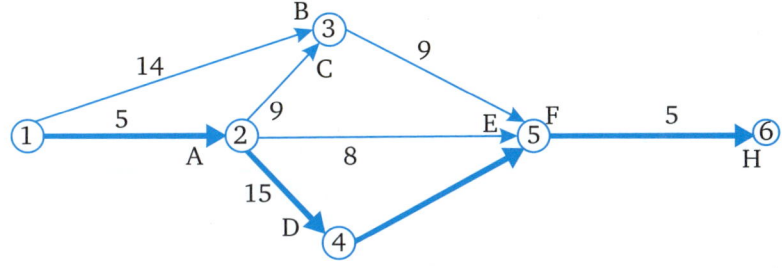

Fig. 25

(ii) We have to find the critical path and expected project length.

Forward pass method	Backward pass method
Taking $E_1 = 0$	Taking $L_6 = E_6 = 29$
$E_2 = E_1 + t_{e_{1,2}} = 0 + 5 = 5$	$L_5 = L_6 - t_{e_{5,6}} = 29 - 5 = 24$
$E_3 = \max_{i=1,2}\{E_i + t_{e_{i,3}}\}$	$L_4 = L_5 - t_{e_{4,5}} = 24 - 4 = 20$
$\quad = \max\{E_1 + t_{e_{1,3}}; E_2 + t_{e_{2,3}}\}$	
$\quad = \max\{0 + 14, 5 + 9\} = 14$	
$E_4 = E_2 + t_{e_{2,4}} = 5 + 15 = 20$	$L_3 = L_5 - t_{e_{3,5}} = 24 - 9 = 15$
$E_5 = \max_{i=2,3,4}\{E_i + t_{e_{i,5}}\}$	$L_2 = \min_{j=3,4,5}\{L_j - t_{e_{2,j}}\}$
$\quad = \max\{E_2 + t_{e_{2,5}}; E_3 + t_{e_{3,5}}; E_4 + t_{e_{4,5}}\}$	$\quad = \min\{L_3 - t_{e_{2,3}}, L_4 - t_{e_{2,4}}, L_5 - t_{e_{2,5}}\}$
$\quad = \max\{5 + 8, 14 + 9, 20 + 4\} = 24$	$\quad = \min\{15 - 9, 20 - 15, 24 - 8\} = 5$
$E_6 = E_5 + t_{e_{5,6}} = 24 + 5 = 29$	$L_1 = L_2 - t_{e_{1,2}} = 5 - 5 = 0$

Here, since $E_1 = L_1, E_2 = L_2, E_4 = L_4, E_5 = L_5$ and $E_5 = L_6$ and also for these activities $E_j - E_i = t_{e_{i,j}} = L_j - L_i$. Therefore, the critical path is $1 - 2 - 4 - 5 - 6$ i.e. $A - D - G - H$ and expected project length $= t_0 = 29$ weeks.

(iii) Now, we have to determine the expected duration time and variance of each activity as shown in the table given below.

Activity (i, j)	Time estimates			Expected activity	standard deviation	variance
	t_0	t_m	t_p	$t_e = \dfrac{t_0 + 4t_m + t_p}{6}$	$\sigma = \dfrac{t_p - t_0}{6}$	σ^2
1-2	2	4	12	5	1.67	2.79
1-3	10	12	26	14	2.67	7.13
2-3	8	9	10	9	0.33	0.11
2-4	10	15	20	15	1.67	2.79
2-5	7	7.5	11	8	0.67	0.45
3-5	9	9	9	9	0.00	0.00
4-5	3	3.5	7	4	0.67	0.45
5-6	5	5	5	5	0.00	0.00

(iv) Other activity schedule for the project are $1 - 3 - 5 - 6$ i.e. $B - F - H$ and $1 - 2 - 5 - 6$ i.e. $A - E - H$.

(v) Variance of the project length

$= $ sum of variance of critical activities A, D, G and H

$= 2.79 + 2.79 + 0.45 + 0.00 = 6.03$

and standard deviation of the project length, $\sigma_e = \sqrt{6.03} = 2.46$

(vi) The probability of the project schedule time $t_s = 30$ weeks given by

$$\text{Prob}\left\{Z \le \frac{t_s - t_e}{\sigma_e}\right\} = \text{Prob}\left\{Z \le \frac{30 - 29}{2.46}\right\}$$

$$= \text{Prob}\{Z \le 0.41\} = 0.5 + 0.1591$$

(By normal distribution table)

$$= 0.6591 \approx 0.66$$

$\Rightarrow$ there are 66% chances of completion of the project in 30 weeks

(vii) If t_s is the project schedule time (in weeks) for 99% chances of completion of the project in time then

$$\text{Prob}\left\{Z \le \frac{t_s - t_e}{\sigma_e}\right\} = 0.99$$

$\Rightarrow \qquad \dfrac{t_s - 29}{2.46} = 2.33$

$\Rightarrow \qquad t_s = 34.7$ weeks $\qquad$ (By normal distribution table)

Hence, for 99% chances for the completion of the project on the schedule date, the work should be started $34.7 - 30 = 4.7$ weeks before the start time.

EXAMPLE 3. *A project has the following activities and other characteristics*

Activity	Preceding activity	Time estimates (in weeks)		
		most optimistic (t_0)	most likely (t_m)	most pessimistic (t_p)
A	—	4	7	16
B	—	1	5	15
C	A	6	12	30
D	A	2	5	8
E	C	5	11	17
F	D	3	6	15
G	B	3	9	27
H	E, F	1	4	7
I	G	4	19	28

(i) *Draw the PERT network diagram.*

(ii) *Identify the critical path.*

(iii) *Find the expected duration and variance for each activity. What is the expected project length.*

(iv) *What is the probability that the project is completed between 35 and 40 weeks?* [DELHI(M.COM.)–1994]

SOLUTION. (i) Following the usual procedure, we get the following network diagram

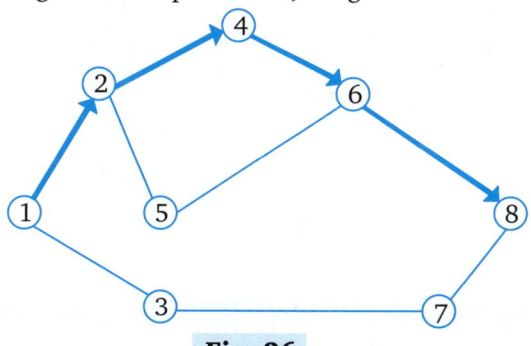

Fig. 26

(ii) The critical path is 1 – 2 – 4 – 6 – 8.

(iii) The calculation of expected times and variances are given in the following table

Activity	Time estimates			Expected activity	variance
	t_0	t_m	t_p	$t_e = \dfrac{t_0 + 4t_m + t_p}{6}$	$\sigma_i^2 = \left[\dfrac{1}{6}(t_p - t_0)\right]^2$
1-2	4	7	16	8	4
1-3	1	5	15	6	49/4
2-4	6	12	30	14	16
2-5	2	5	8	5	1
4-6	5	11	17	11	4
5-6	3	6	15	7	4
3-7	3	9	27	11	16
6-8	1	4	7	4	1
7-8	4	19	28	18	16

From the above table, we conclude that

The expected project length $t_e = 8 + 14 + 11 + 4 = 37$ weeks

and variance of the project length is the sum of its variances of each critical activities, *i.e.*,

Variance of project length, $\sigma_e^2 = 4 + 16 + 4 + 1 = 25$ weeks

(iv) The probability that the project will be completed between 35 and 40 weeks is given by

$$\text{Prob}\left\{\frac{35 - 37}{5} \le \left(Z = \frac{t_s - t_e}{\sigma_e}\right) \le \frac{40 - 37}{5}\right\}$$

$$= P(-0.8 \le Z \le 0.6)$$

$$= P(-0.8 \le Z \le 0) + P(0 \le Z \le 0.6)$$

(By normal area table)

17.12 RESOURCE LEVELLING AND TIME COST TRADE OFF ANALYSIS

It is a well known fact that the resources and the funds available to complete some project are always limited. One has to use these resources in the most efficient manner. In network analysis, the total project duration is evaluated on the basis of normal level of resources and efficient work.

17.12.1 TIME COST TRADE OFF ANALYSIS

The reduction in target time can not be done arbitrarily. It is possible only in relation with the activities lying with the critical path. Some measures of time reduction can be:

   (i) increasing manpower
  (ii) introducing shift system or overtime
 (iii) installing more machines
  (iv) use of improved methods
   (v) reducing safety margins and lowering standards
  (vi) revising the sequence of operations

All the above increase the expenditure. Though extra expenditure is incured for minimizing the duration of any activity but it may save the company from penalty due to the delay in project.

Here, we have to consider the following aspects in the analysis:

   (i) to find the total duration of a project for which the cost is minimum.
  (ii) to find the latest expenditure strategy to minimize the duration by some specified period.

17.12.2 PROJECT CRASHING

Crashing is applied to reduce the project completion time by expanding extra resources. The cost and time involved in some process can be divided in four categories, *viz.*, normal or regular costs and time and crashed cost and time.

(i) Time Crashing: The method of time reduction is called time crashing.

(ii) Cost Crashing: The method of cost reduction is called cost crashing.

17.12.3 DEFINITION OF VARIOUS TIME AND COST ESTIMATES

   (i) Normal cost and time: The minimum cost of carrying out the activity in minimum time without allocating extra resources is termed as Normal cost. Similarly, the minimum time required at the normal cost is known as Normal time.

  (ii) Crash cost and time: Crashed cost is the minimum cost of carrying on any activity using additional resources and the corresponding minimum time to complete an activity is known as Crashed time. This is the minimum time beyond which an activity time cannot be crashed even by incurring additional cost.

 (iii) Project normal cost/time: The sum of the normal costs or normal times of all activities is known as Project normal cost and Normal time. Similarly, the project crash cost/time can be defined.

  (iv) Direct Costs: These include the costs incurred for a project *i.e.* include expenditure on plant, machineries, land, building, architect's charges, engineering charges etc. These expenditures are termed direct as it is possible to allocate these costs directly to the activities.

   (v) Indirect costs: These costs are expenditures which cannot be allocated to individual activities *e.g.*, administration and establishment charges, expenditure on inventories, fixed overheads, loss of revenue etc. The longer the duration of the project, the higher is the indirect cost. These costs can be further divided into two parts (a) fixed indirect costs due to general and administrative expenses etc. which does not depend upon the progress of the project (b) variable indirect costs depending upon the duration of the project like overhead expenditure, supervision etc.

  (vi) The total cost: It is the sum of direct and indirect costs associated with a project.

The behaviour of various costs in terms of project duration can be illustrated by the following figure.

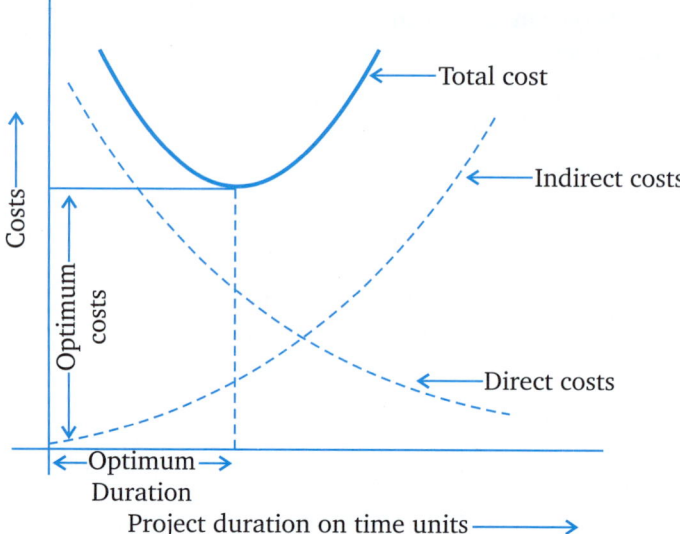

Fig. 27

(vii) Optimum Project Duration: It is observed from the study of above figure that the total cost of a project is directly related to the duration of the project. The optimal project duration is one which results in minimum overall cost.

(viii) Cost Slope: It is defined as the ratio of (crash cost – normal cost) and (normal time–crash time)

$$i.e., \quad \text{Cost slope} = \frac{(\text{Crash cost} - \text{Normal cost})}{(\text{Normal time} - \text{Crash time})}$$

It indicates the additional cost incurred per unit of time saved in reducing the duration of an activity.

17.12.4 TIME-COST OPTIMISATION/CRASHING PROCEDURE

The process of shortening the duration of a project is known as Crashing. The analysis of network and the individual activities of crashing the time upto some desired level can be classified in the following categories.

(i) The target time is to be crashed to some desired level without any cost consideration.

(ii) The target time is to be crashed to some desired level with cost consideration.

(iii) The target time is crashed to a level where increase in profits due to additional production remains more than the additional cost of time crashing.

17.12.5 OBJECTIVES OF PROJECT CRASHING

(i) To minimise the normal duration of the project.

(ii) To evolve a strategy of time crashing with minimum possible cost of crashing.

(iii) If delayed completion of some activities at any stage introduces the possibility of enhancing project duration, then time crashing by expediting one or more remaining activities on the critical path is to be explored.

(iv) To spare the resources at an early time for allocation to more profitable jobs/activities.

(v) To ensure balanced and uniform requirement of resources.

17.12.6 TIME CRASHING METHOD

It consists the following steps:

STEP 1. Identify the critical path activities and other non-critical path activities which are likely to become during the course of crashing.

STEP 2. Determine the cost slope for the activities identify in step-1.

STEP 3. Rank the activities lying on the critical path in ascending order of cost slope.

STEP 4. Crash those activities first which have the smallest cost slope to the maximum possible extent and obtain the new direct cost by cumulative addition of the cost of crashing to the normal cost.

STEP 5. **(Parallel Crashing)** During the course of crashing the activities lying on the critical path, there is always a likelihood that other paths which were non-critical before crashing may now become critical. Then the project duration can be reduced by crashing the activities of parallel critical paths simultaneously.

Hence, the lowest cost solution can be obtained by crashing the critical path activity and parallel critical path activity simultaneously, for which cost slope is lowest.

 Solved Examples

EXAMPLE 1. *The following table gives data on normal time / cost and crash time / cost for a project.*

Activity	Normal		Crash	
	Time in weeks	Cost (in ₹)	Time in weeks	Cost (in ₹)
1–2	3	300	2	400
2–3	3	30	3	30
2–4	7	420	5	580
2–5	9	720	7	810
3–5	5	250	4	300
4–5	0	0	0	0
5–6	6	320	4	410
6–7	4	400	3	470
6–8	13	780	10	900
7–8	10	1000	9	1200
		Total 4220		

Indirect cost = ₹50 per week.

(i) *What are the normal project duration and cost?*

(ii) *Crash time relevant activities systematically and determine the optimal project completion time and cost.*

SOLUTION. The network for the given problem is

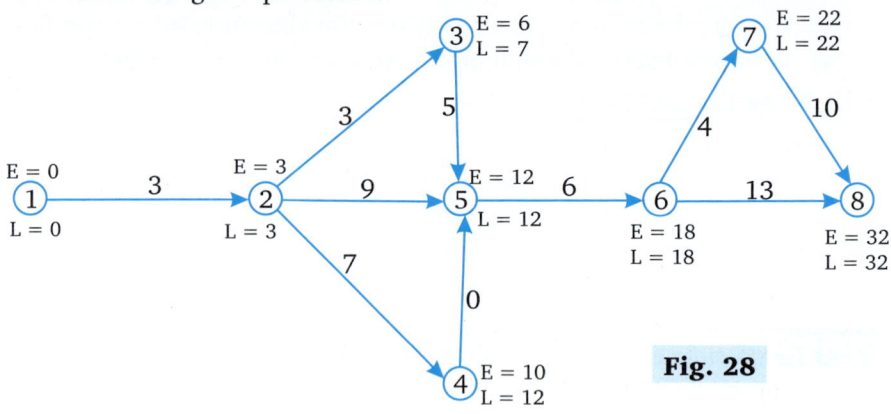

Fig. 28

The duration along with different paths are:

P_1 $1 \rightarrow 2 \rightarrow 3 \rightarrow 5 \rightarrow 6 \rightarrow 7 \rightarrow 8$ 31

P_2 $1 \rightarrow 2 \rightarrow 3 \rightarrow 5 \rightarrow 6 \rightarrow 8$ 30

P_3 $1 \rightarrow 2 \rightarrow 4 \rightarrow 5 \rightarrow 6 \rightarrow 7 \rightarrow 8$ 30

P_4 $1 \rightarrow 2 \rightarrow 4 \rightarrow 5 \rightarrow 6 \rightarrow 8$ 29

P_5 $1 \rightarrow 2 \rightarrow 5 \rightarrow 6 \rightarrow 8$ 31

P_6 $1 \rightarrow 2 \rightarrow 5 \rightarrow 6 \rightarrow 7 \rightarrow 8$ 32

(i) The critical path is $1 - 2 - 5 - 6 - 7 - 8$.

Thus the normal project duration is $(3 + 9 + 6 + 4 + 10) = 32$ weeks.

Direct cost of the project = ₹ 4220.

Indirect normal cost = $50 \times 32 = ₹1600$

∴   Normal cost of the project = ₹ 4220 + ₹ 1600 = ₹ 5820

(ii) The cost slopes for the activities on critical path are as follows:

Activities	1–2	2–5	5–6	6–7	7–8
Cost slope	$\dfrac{400 - 300}{3 - 2}$	$\dfrac{810 - 720}{9 - 7}$	$\dfrac{410 - 320}{6 - 4}$	$\dfrac{470 - 400}{4 - 3}$	$\dfrac{1200 - 1000}{10 - 9}$
	$= 100$	$= 45$	$= 45$	$= 70$	$= 200$

Here crashing cost per week is less than the indirect cost per week for the activities 2–5 and 5–6 only. Hence, crashing will be economic only if these two activities are considered for crashing.

Thus, crashing (5–6) by two weeks at a crashing cost of ₹$(45 \times 2) = ₹90$. The duration of various paths at this stage becomes.

P_1	P_2	P_3	P_4	P_5	P_6
29	28	28	27	29	30

At this stage P_6 still remains the critical path.

The problem can be solved by means of Resource allocation techniques which provides the most rational method of reducing peaks and valleys in the requirement of resources. The technique involves resources levelling and resource smoothing.

The next activity to be crashed is 2–5 with a crash cost of ₹45 per week. This activity can be crashed by two weeks. But if we crash it by two weeks then the duration of various paths becomes

P_1	P_2	P_3	P_4	P_5	P_6
29	28	28	27	27	28

and so at this stage P_1 becomes the initial path and to crash P_1 to 28 days, we have to select an activity whose cost slope is more than indirect cost.

Hence we should crash activity 2-5 by one day only as further crashing will be more expensive than the saving in indirect cost.

Hence, optimum crashed time is 29 weeks, with crashed cost equal to ₹5805.

17.13 LIMITATIONS OF PERT

1. PERT emphasis only time and not costs.
2. Time estimates to perform activities constitute a major limitations of PERT.
3. PERT does not make control automatically. It does not consider the resources required at various stages of the project. It is not universally applicable.
4. In computation of standard deviation of the critical path, independence of activities is assumed. Limitations of resources may invalidate the independence.
5. It may not be always possible to sort out completely identifiable activities and to pin point their starting and finishing points.
6. Cost-time trade-offs for deriving cost curve slopes are subjective and call for a great deal of expertise of the technology as well as genuine effort to estimate.

17.14 COMPARISON OF CPM AND PERT   [MEERUT-2010, 16, 17]

1. Both CPM / PERT techniques are based on networks.
2. Both of these uses the concepts of critical paths and slack.
3. CPM and PERT were developed under entirely different circumstances and are designed to tackle scheduling problems in various industries with different kind of problems.
4. PERT was basically developed in connection with Aero-space industries research and development projects. It is more appropriate for new ventures with rapid changes in technology and producing non-standard products involving uncertainties.
5. CPM is suitable for well known and long developed components using standard technology.
6. CPM is activity oriented where as PERT is event oriented. Here an activity represents part of the work to be completed over a period of time. Event is a point of time showing starting and finishing of the work.
7. Event or activity orientation does not affect the nature of project network which will contain same number of arrows and nodes with same type of interdependence.
8. CPM stresses more on cost considerations whereas PERT is concerned essentially with time factor.
9. CPM is a deterministic approach and does not deal with uncertainties. Here it is assumed that duration of the activities can be reduced by incurring additional resources with extra cost. The additional resource incurred in assumed to be outweighted by other advantages.

10. CPM decides that which activities are to be expedited and to what extent.
11. There are three time estimates in PERT for each of activities whereas in CPM there is single time values.
12. CPM is good for repetitive projects whereas PERT is appropriate for non-repetitive projects.
13. PERT can be analysed statistically whereas in the case of CPM statistical analysis is not possible.

 Miscellaneous Solved Examples

EXAMPLE 1. *A motorist drives into a garage, has his tank filled with petrol, has oil checked and topped up, this tyre pressure adjusted, pays for the petrol and oil and drives off. Prepare an activity dependence table and the network for the problem.*

SOLUTION. The following will be the activity dependence table and network for the problem:

Event	Description of the event	Immediately preceding events
S	Drives into the garage	–
A	Tank filled with petrol	S
B	Oil check up	A
C	Top up the oil	B
D	Adjusting the tyre pressures	S
E	Pays for the petrol and oil	D, C
F	Drives off	E

The following network can be easily constructed from the activity dependence table.

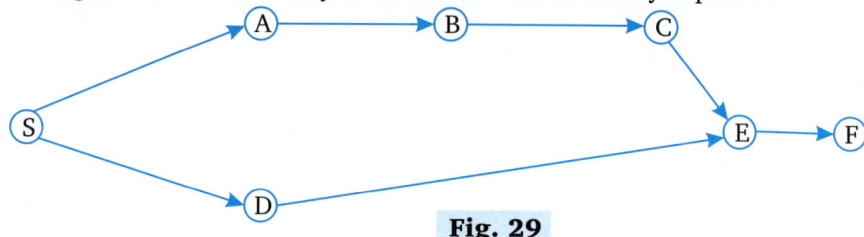

Fig. 29

EXAMPLE 2. *For the project*

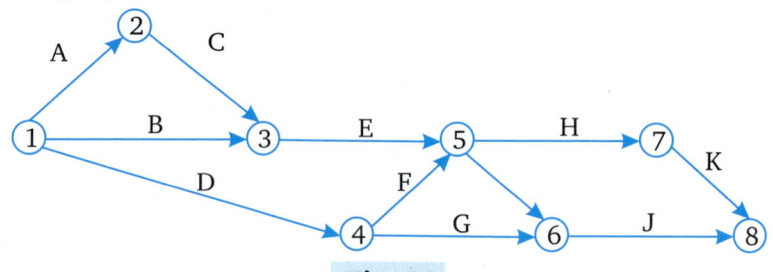

Fig. 30

Task:	A	B	C	D	E	F	G	H	I	J	K
Least time:	4	5	8	2	4	6	8	5	3	5	6
Greatest time:	8	10	12	7	10	15	16	9	7	11	13
Most likely time:	5	7	11	8	7	9	12	6	5	8	3

Find the earliest and latest expected times to each event and also the critical path in the network.

SOLUTION.　The problem can be solved by finding the expected time for each task by the formula $(t_0 + 4t_n + t_p)/6$, where t_0 is least time, t_p is greatest time and t_n is most likely time.

The calculations can be done in following tabular form:

Task	t_0	t_p	$t_0 + t_p$	t_n	$4t_n$	$t_0 + t_p + 4t_n$	Expected Time $(t_0 + t_p + 4t_n)/6$
A	4	8	12	5	20	32	5.3
B	5	10	15	7	28	43	7.2
C	8	12	20	11	44	64	10.8
D	2	7	9	3	12	21	3.3
E	4	10	14	7	28	42	7.0
F	6	15	21	9	36	57	9.6
G	8	16	24	12	48	72	12.0
H	5	9	14	6	24	38	6.3
I	3	7	10	5	20	30	5.0
J	5	11	16	8	32	48	8.0
K	6	13	19	3	12	31	5.2

The expected time for each task can be used to find EST and LFT for each event.

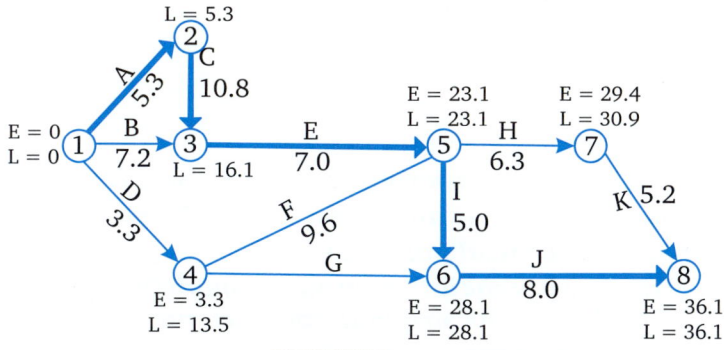

Fig. 31

Thus the critical path consisting of those events for which EST = LFT is given by $1 \rightarrow 2 \rightarrow 3 \rightarrow 5 \rightarrow 6 \rightarrow 8$ with minimum duration of 36.1 time units. The various times for each activity can be calculated in the following tabular form.

Task	Expected duration	EST	EFT	EFT	LFT
A (1–2)	5.3	0	0	5.3	5.3
B (1–3)	7.2	0	8.9	7.2	16.1
C (2–3)	10.8	5.3	5.3	16.1	16.1
D (1–4)	3.3	0	10.2	3.3	13.5
E (3–5)	7.0	16.1	16.1	23.1	23.1
F (4–5)	9.6	3.3	13.5	12.9	23.1
G (4–6)	12.0	3.3	16.1	15.3	28.1
H (5–7)	6.3	23.1	24.6	29.4	30.9
I (5–6)	5.0	23.1	23.1	28.1	28.1
J (6–8)	8.0	28.1	28.1	36.1	36.1
K (7–8)	5.0	29.4	30.9	34.6	36.1

EXAMPLE 3. *A project is represented by the network shown below and has the following data:*

Task:		A	B	C	D	E	F	G	H	I
Least time:		5	18	26	16	15	6	7	7	3
Greatest time:		10	22	40	20	25	12	12	9	5
Most likely time:		8	20	33	18	20	9	10	8	4

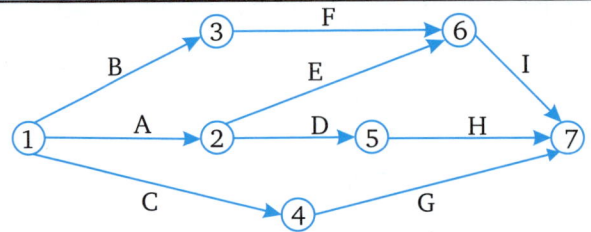

Fig. 32

Determine the following:

  (i) *Expected task time and their variance.*

 (ii) *The earliest and latest expected time to reach each node.*

(iii) *The critical path and*

(iv) *The probability of project completing at the proposed completion date if the original contract time of completing the project is 41.5 weeks.*

SOLUTION.   (i) The expected time and variance for each task can be calculated by

$$E(\text{Time}) = \frac{1}{6}(t_0 + 4t_n + t_p)$$

$$V(\text{Time}) = \left[\frac{1}{6}(t_p - t_0)\right]^2$$

where t_0 : Least time, t_p : Greatest time, t_n : Most likely time

(ii) EST and LFT for each node can be calculated on the Network by taking expected time for each task.

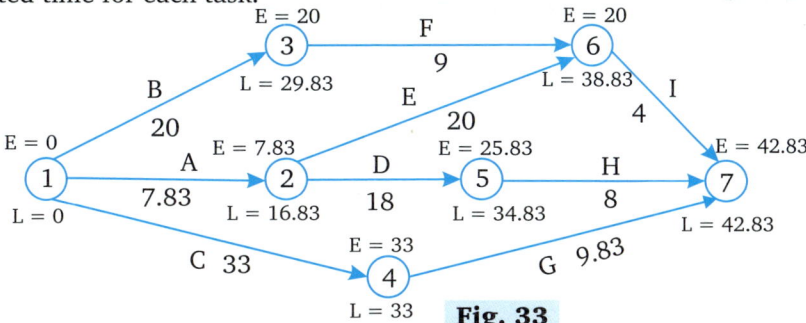

Fig. 33

(iii) The critical path is given by all those nodes for which EST = LFT, *i.e.*, $1 \to 4 \to 7$ with minimum duration of 42.83 weeks, with variance = $(6 + 0.69) = 6.69$.

(iv) Assuming the distribution of the duration to be normal with mean 42.83 weeks and variance = 6.69, the standard normal variate.

$$Z = \frac{D - 42.83}{\sqrt{(6.69)}} = \frac{41.5 - 42.83}{2.59} = -\frac{1.33}{2.59} = -0.51$$

where D is the scheduled time.

The probability $P[Z \le D]$ can be determined by the area under the standard normal curve bounded by the ordinate at $Z = -\infty$ and $Z = -0.51$. The area from the standard tables is 0.3050. Thus there are 30% chances to complete the job in less than 41.5 weeks.

Task	t_0	t_p	t_0+t_p	t_p-t_0	t_n	$4t_n$	$E(t)=(t_0+t_p+4t_n)/6$	$(t_p-t_0)^2$	$v(t)$
A	5	10	15	5	8	32	47/6 = 7.83	25	0.69
B	18	22	40	4	20	80	120/6 = 20	16	0.44
C	26	40	66	14	33	132	198/6 = 33	196	5.44
D	16	20	36	4	18	72	108/6 = 18	16	0.44
E	15	25	40	10	20	80	120/6 = 20	100	2.78
F	6	12	18	6	9	36	54/6 = 9	36	1
G	7	12	19	5	10	40	59/6 = 9.83	25	0.69
H	7	9	16	2	8	32	48/6 = 8	4	0.11
I	3	5	8	2	4	16	24/6 = 4	4	0.11

EXAMPLE 4. *A project has the following time schedule:*

Activity	Time in weeks	Activity	Time in weeks
1–2	4	5–6	4
1–3	1	5–7	8
2–4	1	6–8	1
3–4	1	7–8	2
3–5	6	8–9	1
4–9	5	8–10	8
		9–10	7

Construct Network and find: (i) T_E and T_L for each event, (ii) Float *for each activity and (iii) Critical path and its duration.*

SOLUTION.   The Network for the data is given below:

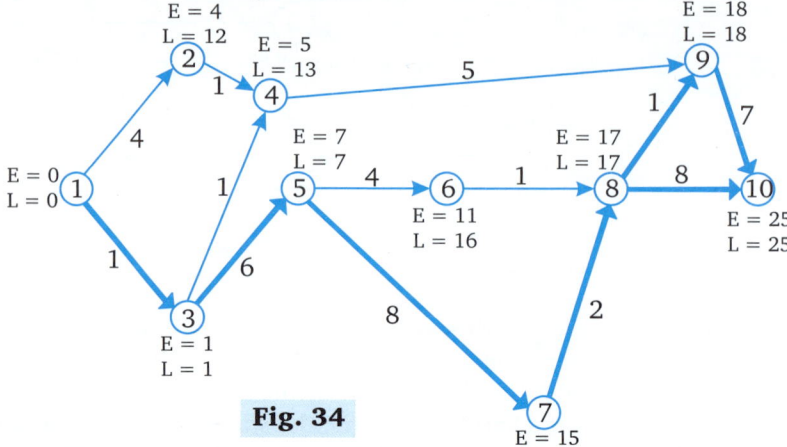

Fig. 34

(i) Calculations for T_E and T_L are shown on the network

Event No.	1	2	3	4	5	6	7	8	9	10
T_E	0	4	1	5	7	11	15	17	18	25
T_L	0	12	1	13	7	16	15	17	18	25

(ii) Using the same procedure, the floats for various activities can be calculated in following tabular form.

Activity	Duration	EST	LST	EFT	LFT	Total Float
1–2	4	0	8	4	12	8
1–3	1	0	0	1	1	0
2–4	1	4	12	5	13	8
3–4	1	1	12	2	13	11
3–5	6	1	1	7	7	0
4–9	5	5	13	10	18	8
5–6	4	7	12	11	16	5
5–7	8	7	7	15	15	0
6–8	1	11	16	12	17	5
7–8	2	15	15	17	17	0
8–9	1	17	17	18	18	0
8–10	8	17	17	25	25	0
9–10	7	18	18	25	25	0

(iii) The critical path is identified by those activities for which Float is zero. Here these are 1–3–5–7–8–9–10 or 1–3–5–7–8–10 with a duration of 25 weeks.

EXAMPLE 5. *A project has the following time schedule:*

Activity	Time in weeks	Activity	Time in weeks
1–2	2	3–7	5
1–3	2	4–6	3
1–4	1	5–8	1
2–5	4	6–9	5
3–6	8	7–8	4
		8–9	3

Construct Network and compute:

(i) Total float for each activity

(ii) Critical path and its duration

Also, find the minimum number of cranes of project must have for its activities 2–5, 3–7, 5–8 and 8–9 without delaying the project. Then is there any change required in network? If so indicate the same.

SOLUTION.　The network is drawn below

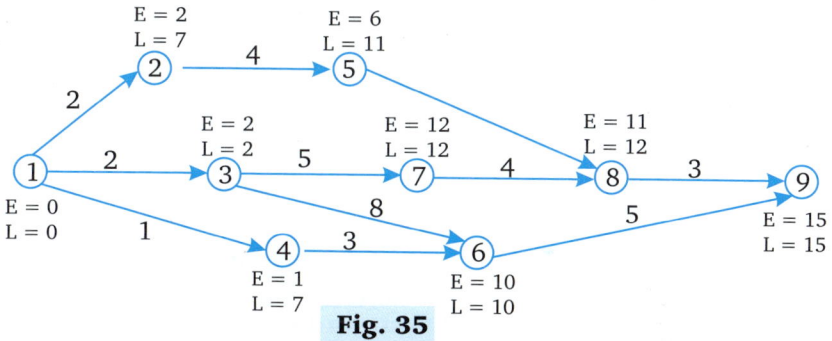

Fig. 35

EST and LFT can be calculated on the Network in usual manner.

(i) Total float for each activity can be calculated in the following tabular form.

Activity	Duration	EST = EST of tail event	EFT = EST + duration	LFT = EFT of head event	Total Float LFT – EFT
1–2	2	0	2	7	5
1–3	2	0	2	2	0
1–4	1	0	1	7	6
2–5	4	2	6	11	5
3–6	8	2	10	10	5
3–7	5	2	7	8	1
4–6	3	1	4	10	6
5–8	1	6	7	12	5
6–9	5	10	15	15	0
7–8	4	7	11	12	1
8–9	3	11	14	15	1

(ii) The critical path consists of all those activities for which Total Float is zero. In this case it is $1 \to 3 \to 6 \to 9$ with a duration of 15 months.

Now, cranes are required for the activities 2–5, 3–7, 5–8 and 8–9.

The activity 3–7	Finishes after 7 months	With one crane
The activity 2–5	Finishes after 11 months and crane is required for 4 months only.	So after finishing 3–7, the same crane can be used for 4 months.
The activity 5–8	Finishes after 12 months and for it crane is required for one month only.	So after the completion of 2–5, the same crane can be used for one month.
The activity 8–9	Finishes after 15 months and a crane is required for 3 months only.	So after the completion of 5–8 the same crane will be available for 3 months.

Therefore one crane will be sufficient to complete the task without any delay in the project.

EXAMPLE 6. *For the following project network*

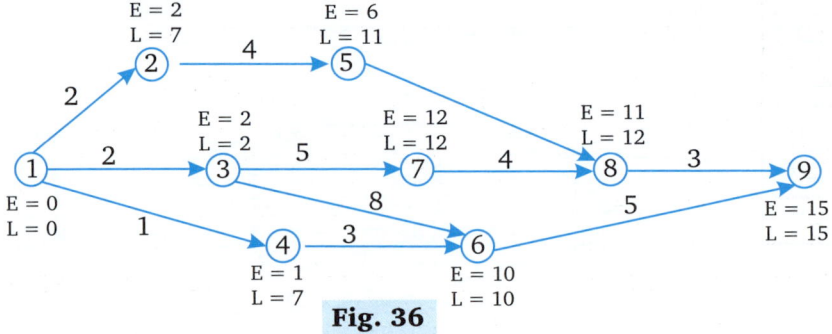

Fig. 36

 (i) Calculate for each activity, its early start, early finish, late start, late finish, total float and free float.

 (ii) Identify the critical path.

 (iii) If the project manager finds that either of the activities 2–6 or 4–5 can each be speeded up by two days at the same cost, which of the two activities should be speeded up? Explain.

 (iv) Assuming that the time estimates in days indicated in the network represent the expected duration based on three time estimates and suppose the variance along the critical path is 81 days, what is the probability that the project will be completed within 33 days? Within 44 days?

SOLUTION.   (i) The early start and latest finishing time can be calculated on the network using forward pass and backward pass technique.

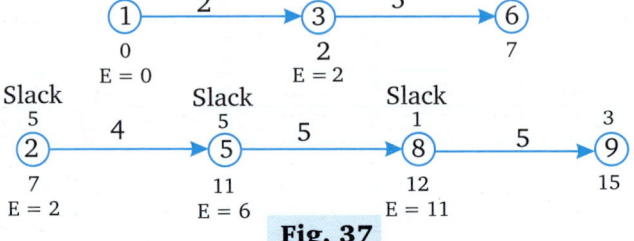

Fig. 37

The other time estimates and floats can be calculated in the following tabular form:

Activity	Duration	EST	LST = LFT duration	EFT = EST + duration	LFT	Total Float LFT – EFT	Free Float Total float – Slack of Head event
1–2	4	0	0	4	4	0	0 – 0 = 0
1–3	7	0	4	7	11	4	4 – 4 = 0
1–4	10	0	2	10	12	2	2 – 0 = 2
2–3	3	4	8	7	11	4	4 – 4 = 0
2–4	8	4	4	12	12	0	0 – 0 = 0
2–5	11	4	10	15	21	6	6 – 0 = 6
2–6	18	4	9	22	27	5	5 – 0 = 5
3–5	10	7	11	17	21	4	4 – 0 = 4
3–6	16	7	11	23	27	4	4 – 0 = 4
4–5	9	12	12	21	21	0	0 – 0 = 0
5–6	6	21	21	27	27	0	0 – 0 = 0
5–7	11	21	24	32	35	3	3 – 0 = 3
6–7	8	27	27	35	35	0	0 – 0 = 0

(ii) The critical path consists of all those nodes on the network for which EST = LFT, i.e., $1 \rightarrow 2 \rightarrow 4 \rightarrow 5 \rightarrow 6 \rightarrow 7$.

(iii) The activity 2–6 does not lie on the critical path, so if this activity is speeded up, there will be no effect on the total duration of the project where as the activity 4–5 lies on the critical path and if it is speeded up by two days the total duration of the project will becomes $35 - 2 = 33$ days.

(iv) Assuming the distribution of project duration as normal variate with mean 35 days and standard deviation $\sqrt{81} = 9$. The standard normal variate

$$Z = (X - 35)/9$$

Now probability of completing the project in 33 days

$$= p(X \leq 33) = P\left[Z \leq \frac{33 - 35}{9}\right]$$

$$= P[Z \leq -0.22] = 0.5 - 0.0871 \quad \text{(From standard normal tables)}$$

$$= 0.4129$$

Similarly, the probability of completing the project in 44 days

$$= p(X \leq 44) = P\left[Z \leq \frac{45 - 35}{9}\right]$$

$$= p(Z \leq 1) = 0.5 + 0.3413 = 0.8413$$

Exercise-17.3

1. The matrix below gives the precedence relationships and duration of a set of critical path activities and the PERT times.

Preceding activity	Duration	Following activity									Activity Times		
		A	B	C	D	E	F	G	H	I	Optimistic	Most likely	Passimistic
A	9		1		1						10	3	14
B	6						1				4	6	10
C	0					1	1				0	0	0
D	2								1	1	6	9	9
E	10								1	1	3	6	9
F	4									1	0	2	4
G	9									1	8	10	14
H	6										4	4	14
I	13										5	6	13

(a) Draw the CPM network and label the nodes.
(b) Determine the critical path time.
(c) Determine the variance of the critical path and confidence limits for the estimated duration of the total activities.

2. A critical path network is given below. The activity times are given besides each activity. Determine the critical path and the critical path time.

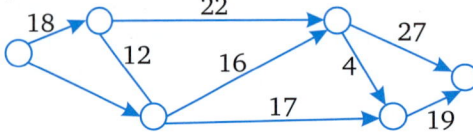

3. Draw the following network diagrams:
   (a) A project consists of 6 jobs A, B, C, D, E and F. Number the events if (i) B and C depend only on A, (ii) D depends on B but not on C, (iii) Job E depends on C and B, (iv) The project is completed when D and E are done.
   (b) A and D start at the origin; J follows F but precedes K; C follows A but precedes J; H follows D but precedes L; B follows A but precedes F; K, L and E are terminal activities; K follows G and H; E follows B and C; F is independent of C; L is independent of J.

4. The following data is derived from a problem of network

Activity	Expected time	Cost slope	Maximum time reduction
1–2	8	4	3
2–3	7	3	3
2–4	6	10	2
2–5	11	7	4
3–4	5	6	2
4–5	12	4	6
4–7	6	6	3
7–5	6	9	2

Compress the project to the least possible duration and estimate the cost of crashing.

5. Compare the EST, LST, EFT and LFT for the following network. The duration of each activity is given on the arrow.

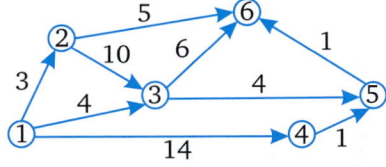

6. Determine critical path and its length, EST, EFT, LST and LFT for each activity in the project with following networks.

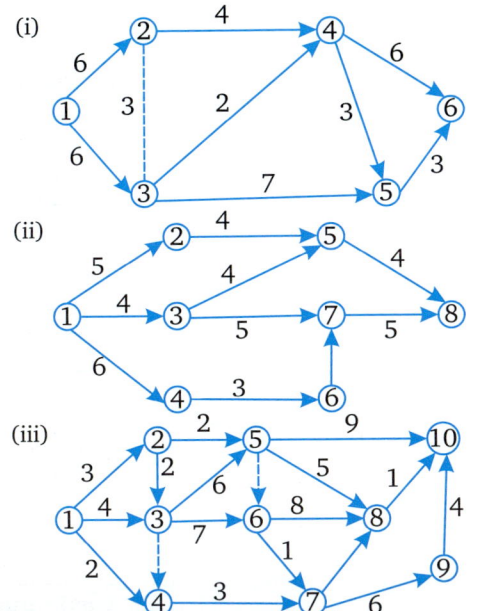

(i)

(ii)

(iii)

7. The following data pertains to some network. It is desired to compress the project to the least possible duration and estimate the extra cost.

i–j	T_n	T_c	Cost slope
1–2	3	2	700
1–3	7	4	300
2–3	5	3	100
2–4	8	6	200
3–4	4	2	400

8. A project has the following time schedule:

Activity	Time in weeks	Activity	Time in weeks
1–2	4	5–6	4
1–3	1	5–7	8
2–4	1	6–8	1
3–4	1	7–8	2
3–5	6	8–9	1
4–9	5	8–10	8
		9–10	7

Construct PERT Network and compute (i)

T_E and T_L for each event (ii) Float for each activity (iii) Critical path and its duration.

9. A project with 10 activities is tabulated below. Each activity can be relaxed and compressed with cost differential.

Activity	Normal		Crash	
	Time(days)	Cost(₹)	Time	Cost
1–2	6	100	4	120
2–3	9	200	5	280
2–4	3	80	2	110
3–4	0	0	0	0
3–5	7	150	5	180
4–6	8	250	3	375
4–7	2	120	1	170
5–8	1	100	1	100
6–8	4	180	3	200
7–8	5	130	2	220

10. A project with the following activities, duration and manpower requirements is given below:

Activity	Duration	Manpower
12	2	5
13	2	4
14	0	0
25	2	2
26	5	3
37	4	6
48	5	2
59	6	8
69	3	7
78	4	4
89	6	3

(a) Draw a network diagram of project indicating the earliest start, latest start, earliest finish, latest finish, total float, free float of each activity.

(b) There are 11 persons who can be employed for this project. Carry out the appropriate manpower levelling so that the fluctuations of work force requirement from day to day is as small as possible.

8.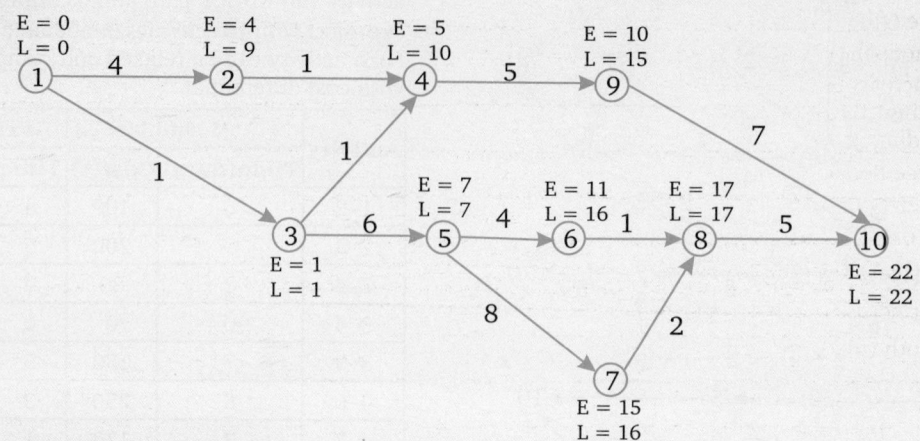

1. Define critical path, slack time, float time in the context of network model.
2. Mention the areas of application of network analysis.
3. Discuss various steps used in the application of PERT and CPM.
4. Explain the basic logic of arrow network.
5. Write the major limitations of PERT model.
6. Write the different type of float.
7. Define the term 'Critical activity'.
8. Discuss the role of statistical techniques in PERT.
9. How does PERT provides for uncertainity in activity time estimates?   [DELHI(MBA)–2000]
10. What are the basic differences between PERT and CPM.

1. The basic steps in a project are:
   (a) planning   (b) scheduling
   (c) controlling   (d) All are true
2. An activity (i, j) is said to be critical if:
   (a) $E_i = L_i$
   (b) $E_j = L_j$
   (c) $E_j - E_i = L_j - L_i = t_{ij}$
   (d) All are true
3. CPM/PERT techniques were developed first in:
   (a) USA   (b) UK
   (c) Japan   (d) None of these
4. The objective of the network analysis is to minimize:
   (a) the project duration
   (b) the project cost
   (c) both (a) and (b)
   (d) None of these
5. Of all paths through the network, the critical path has the:
   (a) maximum expected time
   (b) minimum expected time
   (c) maximum actual time
   (d) None of these
6. Network analysis have advantage in terms of project:
   (a) planning   (b) controlling
   (c) scheduling   (d) None of these
7. Estimated expected activity times in a PERT network:
   (a) makes use of three estimates
   (b) is not operated by Beta distribution
   (c) both (a) and (b) are true
   (d) None of these
8. The slack of an activity is equal to:
   (a) LS – ES   (b) LF – LS
   (c) EF – ES   (d) None of these
9. In the CPM time-cost trade off function:
   (a) the cost of normal time is 0
   (b) cost decreases linearly as time increases
   (c) both (a) and (b) are true
   (d) None of these

10. The another term commonly used for activity slack time is:
    (a) free float   (b) total float
    (c) independent float (d) all are true

11. The activity which can be delayed without affecting the execution of the immediate succeding activity is detemined by:
    (a) free float   (b) total float
    (c) independent float (d) None of these

12. Float or slack analysis is useful for:
    (a) project behind the schedule only
    (b) project ahead the schedule only
    (c) both (a) and (b) are true

(d) None of these

13. The marginal cost of crashing a network could change when the activity being crashed reaches:
    (a) its crash time
    (b) a point where another path is also critical
    (c) both (a) and (b)
    (d) None of these

14. While drawing the network diagram for each activity project, we should look what activity:
    (a) precede this activity
    (b) follow this activity
    (c) both (a) and (b)
    (d) None of these

Answers

1. (d)	2. (d)	3. (a)	4. (a)	5. (a)	6. (b)	7. (c)	8. (a)	9. (b)
10. (d)	11. (a)	12. (a)	13. (c)	14. (c)				

Archive

1. How does PERT provides for uncertainty in activity time estimates? What is the reason for using beta probability distribution?
   [DELHI(MBA)–2000]

2. Explain the significance of 'working out of float' in the network of project activities.
   [DELHI(MBA)–1998]

3. A modified association prepares an annual programme each year giving the monthly meeting dates, background on the speakers, an abstract of their talks and on alphabeting listing both by name, and medical college/hospital affiliation of all dues paying members. The programme is mailed to these members as well as to selected individuals and organisation. The activity to be performed are listed as follows:

Activity	Description	Preceding Activies	Estimated times (weeks)
A	Decide on general orientation for this years program	—	1
B	Get commitment from speakers and abstract	A	4
C	Solicit advertising to appear in the programme	A	3

Cont...

D	Mail out dues notices and wait for response	—	6
E	Prepare list of dues-paying members	D	1
F	Get copy to printer and proofread	B, C, E	2
G	Get programme printed and assemled	F	2
H	Prepare final mailing list	E	1
I	Staff envelopes and mail programmes	G, H	1

(a) Develop a network diagram to show the relationship between all activities. Specify the activities on the critical path and the project competition time.

(b) The chairman now claim that it will take 6 weeks to get commitments and abstract from the speakers. Will the delay the project completion line. How will it affect the critical path? [DELHI(MBA)–2001]

4. A multinational company wishes to launch a new fruit Yogurt in the coming season. A brief description of the activities associated with this project, their expected durations (in weeks) and their immediate predecessor(s) are given in the following table:

Activity	Predecessor	Expected time (weeks)		
		optimistic	Likely	pessimistic
A	—	2	2.5	4
B	A	3	4.7	5
C	A	2	2	3
D	C,B	1	1	2
E	D	8	12	15
F	B	4	5	7
G	F	1	1	2.5
H	F, E	4	8	10
I	G	2	3	5
J	H	3	4	5
K	D	1	1	2
L	J, F	1	1.5	3
M	L	2	2.5	3
N	M	2	4	5
O	M	1	1	1.5
P	O	0.5	0.5	1
Q	H	4	5	8
R	D, Q	1.5	2	3
S	R	1.5	1	1.5
T	S	1	2.5	3
U	T	0.5	1.5	2
V	T	0.5	1	1.5
W	U, V	1	2	4

Management of a company desire to know the realistic completion time for this project and deatailed analysis of float time.(if any)

[DELHI(MBA)–2000, 04]

5. The following table gives the activities in a construction project and other relevant information.

Activity	Immediate predecessor	Time		Direct cost (₹)	
		Normal	Crash	Normal	Crash
A	—	4	3	60	90
B	—	6	4	150	250
C	—	2	1	38	60
D	A	5	3	150	250
E	C	2	2	100	100
F	A	7	5	115	175
G	D, B, E	4	2	100	240

Indirect costs vary as follows:

Months	Cost(₹)	Months	Cost(₹)
15	600	10	100
14	500	9	75
13	400	8	50
12	250	7	35
11	175	6	25

Determine the project duration which will result in minimum total project cost.

[DELHI(MBA)–1989, OSMANIA(MBA)–1990]

HINTS AND ANSWERS

4. $A \rightarrow B \rightarrow C \rightarrow E \rightarrow N \rightarrow P \rightarrow Q \rightarrow R \rightarrow S \rightarrow T \rightarrow U \rightarrow V$, 56 days

Inventory Theory 18

18.1 INTRODUCTION

It is well known fact that the function of directing the movement of goods though the entire manufacturing from the requisitioning of raw materials to the inventory of finished goods orderly mannered. Inventory is a list of schedule of articles held on charge of a person of stock of articles and material held on behalf of the organisation.

Definition: *The inventory can be defined as a stock of goods which is kept for the future purpose.* [MEERUT–2002, 03, 16, 18; ROHILKHAND–2000]

☛ Remark
- The objective of an inventory problem is to minimize total cost or to maximize profit.

18.2 TYPES OF INVENTORY

(i) **Raw material :** The material used in manufacture of the products such as fuels, etc. is called raw material.

(ii) **Partly finished items :** The material which held between manufacturing stage is called partly finished items.

(iii) **Finished goods :** The product which are ready for sale or distribution called finished goods.

(iv) **Spare parts :** The spare parts used in the production process but do not become part of the product.

18.3 MEANING OF INVENTORY CONTROL

Inventory control is the process of deciding what and how much of various items are to be kept in stock. It also use to determine the time and quantity of various items to be procured. It is useful to reduce investment in inventories and ensuring that production process does not suffer at the same time.

18.4 NEED OF THE INVENTORY CONTROL

The need of the inventory control are due to the following facts :
(1) Inventory helps in smooth and efficient running of the business.
(2) It acts as a buffer stock when raw material are received late.
(3) It provides services to the customers immediately at a short notice.
(4) It reduces product costs.
(5) It is used to reduce financial investment in inventories.

18.5 INVENTORY COST

18.5.1 Item Cost

Item cost is defined by the price of one unit. Following are the components of item costs :

    (i) Direct material cost   (ii) Direct labour cost

    (iii) Direct expanses   (iv) Overhead cost

    (v) Profit of the manufacture

18.5.2 Set-up or Ordering cost

The cost associated with the placement of an order or purchasing or manufacturing before starting production.

$$\text{Set-up cost} = (\text{Cost per order}) \times (\text{Number of orders})$$

☛ **Remark**

- The ordering cost is generally assumed to be independent of the quantity ordered for or produced.

18.5.3 Holding cost or Inventory carrying cost

The cost of holding material inside and outside the store is known as holding cost or inventory carrying cost.   [ROHILKHAND–2001, 02]

It can be determined as follows :

    (i) Holding cost = (cost of carrying one unit of an item in the inventory for a given length of time) × (Average number of units of an item carried in the inventory for a given length of time)

    (ii) Holding cost = (cost to carry one rupee's worth of inventory per time period) × (Rupee value of units carried)

Components of Holding cost or inventory carrying cost

There are following components which constitute the holding or inventory carrying cost.

    (i) Invested capital cost

    (ii) Record-keeping and administrative cost

    (iii) Handling cost

    (iv) Storage costs

    (v) Depreciation, Deterioration and Obsolescence costs

    (vi) Taxes and insurance costs

    (vii) Purchase Price or introduction costs

    (viii) Salvage costs or selling price

☛ **Remark**

- If I is the inventory carrying charge, for holding cost expresses in rupees per year per rupee of inventory investment and C is the unit cost of the item in rupees, then the annual carrying cost for the item in rupees per year y per unit is IC. Also, I can be expressed as the percentage of the unit cost of an item to be charged as inventory carrying cost.

18.5.4 Shortage costs or Stock-out cost

The penalty costs that are incurred as a result of running out of stock *i.e.*, shortage is known as shortage or stock-out costs.   [ROHILKHAND–2001, 02]

It can be calculated as follows :

$$\text{Shortage cost} = (\text{Cost of being short one unit in the inventory planning method})$$
$$\times (\text{Average number of units short in the inventory planning period})$$

where,

Average number of units short

$$= \left\{ \frac{\left(\begin{matrix} \text{Minimum shortage in the} \\ \text{inventory planning period} \end{matrix} \right) + \left(\begin{matrix} \text{Maximum shortage in the} \\ \text{inventory planning period} \end{matrix} \right)}{2} \right\} \times \left(\begin{matrix} \text{Time for which} \\ \text{shortage occurs} \end{matrix} \right)$$

☛ Remark
- The shortage cost never appears in accounting record.

18.5.5 Cost of operating the Information Processing System

Sometimes, as stock levels change, someone must update record whether by hand or by computer, where the inventory levels are recorded daily, the operating cost is incurred in obtaining accurate physical count of inventories.

18.6 TYPES OF INVENTORY MODELS [ROHILKHAND–1999]

(i) **Deterministic model :** The inventory models, in which demand is assumed to be fixed for a subsequent period of time, are known as deterministic model.

(ii) **Probabilistic model :** The inventory models, in which the demand is a random variable having a known probabilistic distribution, are known as probabilistic models. In probabilistic model, the future demand is determined by collecting data from the past experience.

☛ Remark
- Deterministic models are those where demand and supply are known with certainty.

General Notations

1. I = The cost of carrying one rupee in inventory for a unit time.
2. C_1 = Holding cost per unit time.
3. C_2 = Shortage cost per unit time.
4. C_3 = Set-up cost per production run.
5. q = lot size per production run.
6. r = demand rate
7. R = Total demand
8. C = Average total cost per unit time
9. K = Production rate
10. z = order level or stock level
11. t = Time interval between two consecutive replenishments of inventory
12. L = Lead time

☛ Remark
- We can denote the optimal value of q, t, z by q^*, t^*, z^* respectively, for which the cost C is minimum.

Part (A) : DETERMINISTIC MODEL

18.7 ECONOMIC LOT SIZE MODEL

The evolution of a model which minimizes the total inventory cost in a particular situation is the purpose of inventory control system. The order size corresponding to it is known as economic order quantity (EOQ) or optimum size. The objective of inventory models is to determine the economic order size, optimal lead time or re-order level. The inventory problems in which the demand is assumed to be fixed and completely pre-determined are known as economic lot size problem or economic order quantity (EOQ) problem.

Definition : *The size of order which minimizes the total costs that include carrying costs, set-up costs, and shortage costs, when demand is fixed and known, is known as economic order quantity (EOQ).*

18.8 MODEL-I : ECONOMIC LOT SIZE MODEL WITH UNIFORM RATE OF DEMAND, INFINITE PRODUCTION RATE WITH NO SHORTAGE

[MEERUT–2000, 01, 03, 07, 14, 16; ROHILKHAND–2003; GARHWAL–2001; AVADH–2011]

Assumptions

(1) Demand per year is known with certainty and is at a constant rate r.
(2) Production is instantaneous *i.e.*, production rate is infinite.
(3) Lead time is zero.
(4) Stock outs are not permissible *i.e.*, shortages are not allowed.
(5) Holding cost per unit per unit time is denoted by C_1.
(6) Set-up cost per production run is denoted by C_3.

For this model, we have the following situations :

(1) Planning period is one year.
(2) Demand is deterministic.
(3) Length of time between two successive orders is t.

Formulation of the Model

Let q be the units of quantity produced per production run at interval of time t.

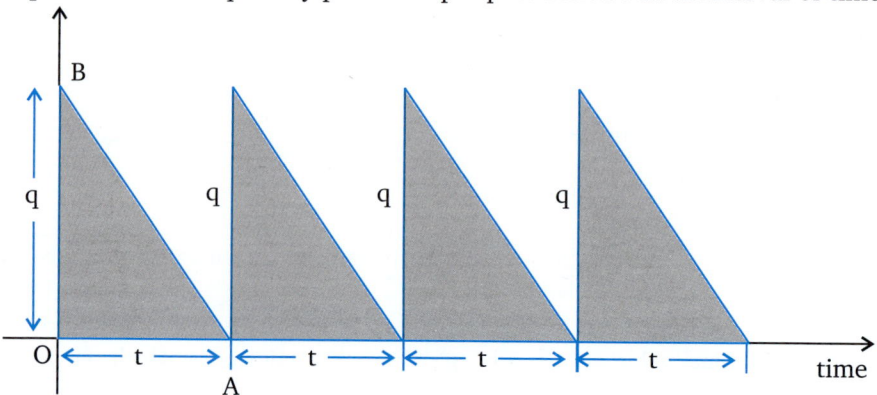

Fig. 1

We have assumed that the demand rate is r units per unit time.

Thus, total demand on one run of time interval $t = rt$

So, in the beginning, the quantity produced per production run is given by

$$q = rt \qquad ...(1)$$

Further, the cost of holding inventory $= C_1$ (Area of $\triangle OAB$)

$$= C_1 \cdot \frac{1}{2} qt$$

Therefore, total cost per production run of time t

$$= \frac{1}{2} C_1 . qt + C_3$$

where C_3 is the set-up cost.

Now, the average cost per unit time is given by

$$C(q) = \frac{1}{2} C_1 q + \frac{C_3}{t} = \frac{1}{2} C_1 q + \frac{C_3 r}{q} \qquad ...(2)$$

which is the required cost equation.

Using principle of maxima and minima, for minimum value of $C(q)$, we must have

$$\frac{dC}{dq} = \frac{1}{2} C_1 - \frac{C_3 r}{q^2} = 0$$

On solving we get

$$q = \sqrt{\left(\frac{2C_3 r}{C_1} \right)}$$

We can easily verify that, for $\quad q = \sqrt{\frac{2C_3 r}{C_1}}, \dfrac{d^2 C}{dq^2} = \dfrac{2C_3 r}{q^3}$ is positive.

Hence, minimum value of $C(q)$, is obtained for

$$q = q^* = \sqrt{\left(\frac{2C_3 r}{C_1} \right)} \qquad ...(3)$$

Equation (3) is known as Wilson Economic lot size formula or Economic lot size formula.

Further the optimum value of t is given by:

$$t = t^* = \sqrt{\left(\frac{2C_3}{C_1 r} \right)}$$

Also, the minimum cost per unit time is given by

$$C_{min.} = \frac{1}{2} C_1 \sqrt{\left(\frac{2C_3 r}{C_1} \right)} + C_3 r \sqrt{\frac{C_1}{2C_3 r}}$$

$$= \sqrt{(2C_1 C_3 r)} \qquad ...(4)$$

Observations: In this model, we observe the following:

(1) Optimum ordering interval $= \sqrt{\dfrac{2 \times \text{set up cost}}{\text{demand rate} \times \text{holding cost}}}$

(2) Optimum number of order $= \sqrt{\dfrac{\text{Total annual quantity requirement } (r)}{\text{Economic ordering quantity } (q)}}$

(3) Number of day's supply $= \dfrac{365}{\text{Optimum number of orders}}$

(4) Total inventory in one cycle of time t is given by

$$= \frac{1}{2}\, qt$$

and average inventory at any time $= \dfrac{\frac{1}{2}qt}{t} = \dfrac{q}{2}$

Hence, throughout the period, average level of inventory is $q/2$.

(5) From equation (4), if C_1 and C_3 are constants, then the minimum cost per unit time is proportional to the square root of the demand rate.

(6) If the lead time is not zero *i.e..*, L is the time gap between the time of placing an order and the time of receiving of the goods to the inventory, then we have to place the order in advance by the time L so that we may get the total quantity q as soon as the inventory falls to zero.

Second form of Model-1

Here, we want to find an economic lot size formula under the following assumptions :

[MEERUT–1994, 97, 2004, 13]

Assumptions

(1) The demand rate for product in one unit of time is λ.

(2) Lead time is zero and production rate is infinite.

(3) Price of one item of product in rupees be denoted by P.

(4) Cost of carrying one rupee to the inventory for one year is I.

(5) Set-up cost per order is C_3.

(6) Shortages are not allowed.

Formulation of the Model: To derive an economic lot size formula, we proceed as follows :
In time t, let q be the units of quantity produced per cycle.

We have already proved (in model-1) that, total inventory in one cycle $= \dfrac{1}{2}qt$

Also, average inventory is $q/2$.

Now, holding cost per item for one year $= IP$

Therefore, the total holding cost per year $= \dfrac{q}{2}IP$

Since, the no. of orders or cycles in one year $= \lambda/q$

Therefore, set-up cost in one year $= \dfrac{\lambda}{q}C_3$

Hence, total cost in one year is given by

$$C(q) = \frac{1}{2}PqI + \frac{\lambda}{q}C_3$$

Using principle of maxima and minima, for $C(q)$ to be minimum, we must have

$$\frac{dC}{dq} = 0$$

$$\Rightarrow \qquad \frac{dC}{dq} = \frac{1}{2}PI - \frac{\lambda C_3}{q^2} = 0$$

$$\Rightarrow \qquad q = \sqrt{\left(\frac{2\lambda C_3}{PI}\right)}$$

Also, we have

$$\frac{d^2C}{dq^2} = \frac{2\lambda C_3}{q^3} \text{, which is positive for } q = \sqrt{\frac{2\lambda C_3}{PI}}$$

Hence, the required economic lot size formula is given by

$$q^* = \sqrt{\frac{2\lambda C_3}{PI}}$$

☞ **Remark**

- This economic lot size formula can also be obtained from model-1 by replacing C_1 by PI and r by λ.

18.9 MODEL-2 : ECONOMIC LOT SIZE WITH DIFFERENT RATE OF DEMANDS IN DIFFERENT CYCLES

[MEERUT–2000, 17, 18]

We want to find an economic lot size formula and the minimum average cost under the same assumption as in model-1, except that demand rates are different in different production cycles.

Here, we have the following diagram of inventory level with different rates of demand in different cycles.

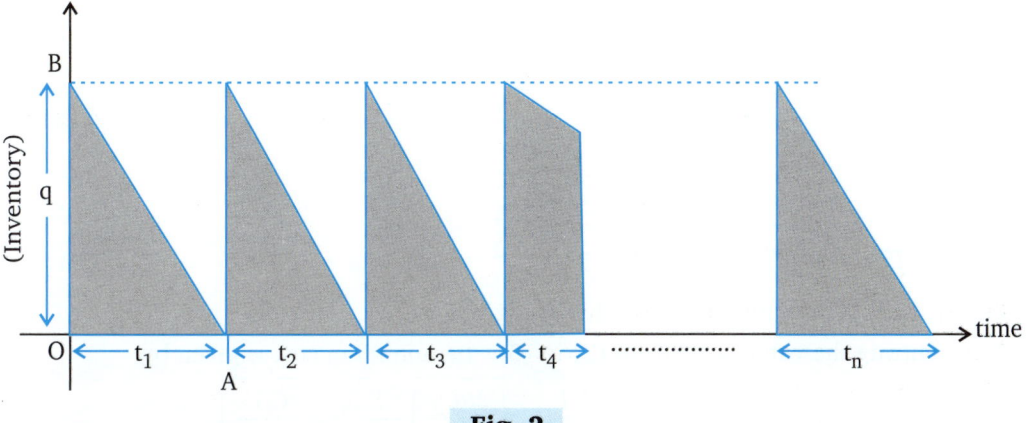

Fig. 2

Let q be the fixed quantity produced in each production cycle.

Since, R is the total demand prescribed over the time period t, the number n of production

cycles will be given by
$$n = \frac{R}{q}$$

Let $t_1, t_2, ..., t_n$ be the times of the successive production cycles such that
$$t = t_1 + t_2 + ... + t_n \qquad \qquad ...(1)$$

Clearly, the fixed quantity q, produced in the beginning of the interval t_1 with different uniform rate of demand in times $t_1, t_2, ..., t_n$.

Thus, the carrying or holding cost for the period t

$$= C_1 \text{ (sum of areas of n triangles in the above figure)}$$

$$= C_1 \left[\frac{1}{2}qt_1 + \frac{1}{2}qt_2 + ... + \frac{1}{2}qt_n \right]$$

$$= \frac{1}{2}qC_1 \left[t_1 + t_2 + ... + t_n \right]$$

$$= \frac{1}{2}qC_1 t \qquad \qquad \text{[Using (1)]}$$

Since, C_3 is the set-up cost per cycle, therefore, total set-up cost

$$= nC_3 = \frac{R}{q}C_3$$

Thus, total cost for the fixed period $t = \frac{1}{2}qC_1 t + \frac{R}{q}C_3$

Hence, the average total cost per unit time is given by

$$C(q) = \frac{1}{2}qC_1 + \frac{RC_1}{tq} \qquad \qquad ...(2)$$

which is required cost equation of this model.

For minimum value of $C(q)$ (using principle of maxima and minima), we must have

$$\frac{dC}{dq} = 0$$

$$\Rightarrow \qquad \frac{dC}{dq} = \frac{1}{2}C_1 - \frac{RC_3}{tq^2} = 0$$

$$\Rightarrow \qquad q = q^* = \sqrt{\left(\frac{2RC_3}{C_1 t} \right)} \qquad \qquad ...(3)$$

Also,
$$\frac{d^2C}{dq^2} = \frac{2RC_3}{tq^3}$$

which is positive for
$$q = \sqrt{\frac{2RC_3}{C_1 t}}$$

Hence, $C(q)$ is minimum for the value of q given by (3).

Now, putting the value of q* from (2) in (3), the minimum cost is given by

$$C_{\min} = \frac{1}{2}C_1 \sqrt{\left(\frac{2RC_3}{C_1 t} \right)} + \frac{RC_3}{t}\sqrt{\left(\frac{C_1 t}{2RC_3} \right)}$$

Hence,
$$C_{\min} = \sqrt{\left(\frac{2C_1 C_3 R}{t} \right)}$$

18.10 MODEL-3 : EOQ PRODUCTION MODEL; ECONOMIC LOT SIZE MODEL WITH UNIFORM RATE OF DEMAND, FINITE RATE OF REPLENISHMENT (PRODUCTION), HAVING NO SHORTAGE

We have to find an economic lot size formula and minimum average cost of the same assumption as in model-1 except replenishment (production) rate is finite

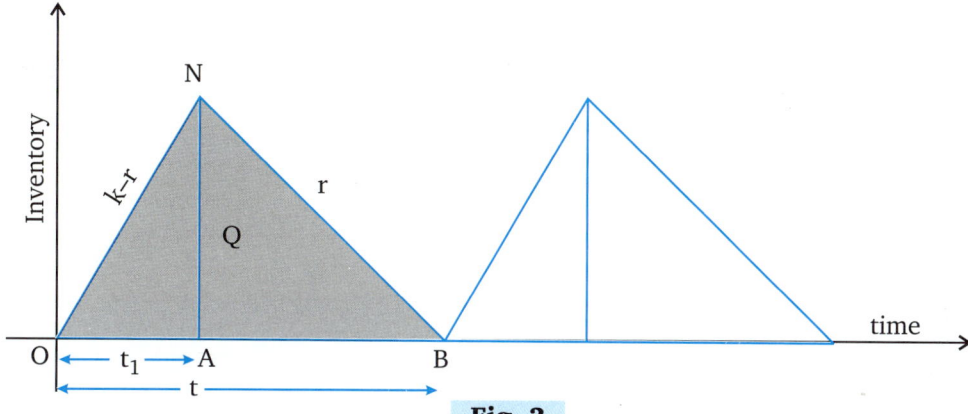

Fig. 3

Let us suppose that q be the number of items produced per production run and $k>r$ be the number of items produced per unit time.

Then, the production will continue for a time

$$t = \frac{q}{k} \qquad \qquad ...(1)$$

Also, the time of one complete production run

$$= \text{the interval between two runs}$$

$$= \frac{q}{r} = t \qquad \qquad ...(2)$$

If Q be the inventory level at the moment the production is completed (at the end of time t_1), then

$$Q = q - rt_1 = q - r.\frac{q}{k} = q\left(1 - \frac{r}{k}\right) \qquad \qquad ...(3)$$

Now, the cost of holding inventory for the period t

$$= C_1 \, (\text{Area of } \Delta \, ONB)$$

$$= C_1\left(\frac{tQ}{2}\right) = \frac{1}{2}qC_1\left(1 - \frac{r}{k}\right)t$$

Also, the set-up cost $= C_3$

which gives, the total cost per run of period t

$$= C_3 + \frac{1}{2}qC_1\left(1 - \frac{r}{k}\right)t$$

Thus, the total cost per unit time is given by

$$C(q) = \frac{C_3}{t} + \frac{1}{2}qC_1\left(1 - \frac{r}{k}\right)$$

Using (2), we get

$$C(q) = \frac{C_3 r}{q} + \frac{1}{2} q C_1 \left(1 - \frac{r}{k}\right)$$

...(4)

which is the required cost equation.

For minimum value of $C(q)$, (using principle of maxima and minima), we must have

$$\frac{dC}{dq} = 0$$

$$\Rightarrow \qquad \frac{dC}{dq} = -\frac{C_3 r}{q^2} + \frac{1}{2} C_1 \left(1 - \frac{r}{k}\right) = 0$$

$$\Rightarrow \qquad q = q^* = \sqrt{\left\{\frac{2C_3}{C_1}\left(\frac{rk}{k-r}\right)\right\}}$$

...(5)

Further, we have

$$\frac{d^2 C}{dq^2} = \frac{2C_3 r}{q^3}$$

which is always positive for $q = \sqrt{\left\{\frac{2C_3}{C_1}\left(\frac{rk}{k-r}\right)\right\}}$

Hence, we conclude that $C(q)$ given by (4) is minimum for the value of q which is given by (5).

Also, the minimum cost per unit time is given by

$$C_{min} = C_3 r \sqrt{\left[\frac{C_1}{2C_3}\left(\frac{k-r}{rk}\right)\right]} + \frac{1}{2} C_1 \left(\frac{k-r}{k}\right) \cdot \sqrt{\frac{2C_3}{C_1}\left(\frac{kr}{k-r}\right)}$$

$$= \sqrt{2C_1 C_3 r \left(1 - \frac{r}{k}\right)}$$

From (5), we also obtained, the time of one run, which is given by

$$t^* = \frac{q^*}{r} = \sqrt{\frac{2C_3 k}{C_1 r(k-r)}}$$

Observations

(1) If $k = r$, then $C_{min} = 0$ which implies that there will be no carrying cost and no set-up cost.

(2) If production rate is infinite ($k \to \infty$) then this model is exactly same as model-1.

18.11 MODEL-4: FIXED TIME MODEL

[MEERUT–1999]

Definition (order level model)

The inventory problems involving the level of the inventory in the beginning, are known as order level problems.

To derive the optimal order level and the minimum average cost

We want to derive order level model with uniform rate of demand Q to be fulfilled in constant time t, infinite rate of production and having shortage which are to be fulfilled.

Assumptions

(1) r units, per unit time is the demand rate.

(2) Production rate is infinite.

(3) Lead time is zero.

(4) C_1 is the holding cost per unit per unit time.

(5) C_2 is the shorting cost per unit per unit time.

(6) C_3 is the set-up cost per production run.

(7) Shortages, if any are allowed and backlogged.

(8) z is the order level to which the inventory is planned in the beginning of each scheduling period.

(9) Q is the total demand per production run of fixed time interval t.

Formulation of the Model

Since z be the order to which the inventory is raised in the beginning of a run of time interval t.

Therefore, total inventory is reduced to zero in time $t_1 = \dfrac{z}{r}$

Then, the shortages arise increase from O to $Q - z$ in the remaining time *i.e.,*

$$t - t_1$$

$$Q = rt$$

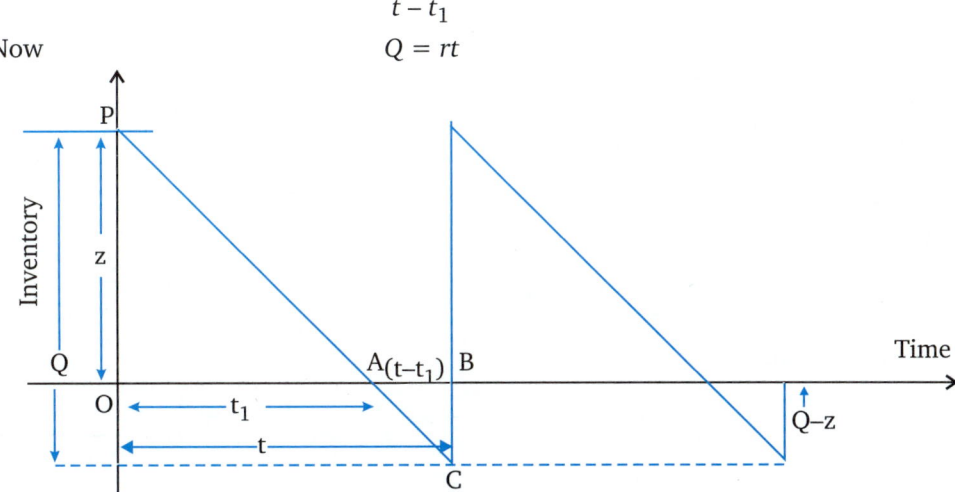

Fig. 4

Proceed as usual, the cost of holding inventory (per production run)

$$= C_1 \,(\text{Area of } \triangle OAP)$$

$$= C_1 . \frac{1}{2} z t_1 = \frac{C_1 z^2}{2r}$$

Now, the shortage cost for one run $= C_2(\text{Area of } \triangle ABC)$

$$= C_2 . \frac{1}{2}(Q - z)(t - t_1)$$

$$= \frac{1}{2}C_2(Q-z)\left(\frac{Q}{r} - \frac{z}{r}\right) = \frac{C_2}{2r}(Q-z)^2$$

Let C_3 be the set-up cost for one run, then average total cost per unit time is given by

$$C(z) = \frac{1}{t}\left[\frac{C_1}{2r}z^2 + \frac{C_2}{2r}(Q-z)^2\right] + \frac{C_3}{t}$$

$$= \frac{r}{Q}\left[\frac{C_1}{2r}z^2 + \frac{C_2}{2r}(Q-z)^2\right] + \frac{rC_3}{Q}$$

After simplification, we get

$$C(z) = \frac{C_1}{2Q}z^2 + \frac{C_2}{2Q}(Q-z)^2 + \frac{rC_3}{Q} \qquad \ldots(1)$$

For the minimum value of $C(z)$ (using principle of maxima and minima), we must have

$$\frac{dC}{dz} = 0 \implies \frac{C_1}{Q}z - \frac{C_2}{Q}(Q-z) = 0$$

On solving we get

$$z = z^* = \frac{C_2Q}{C_1+C_2} = \frac{C_2rt}{C_1+C_2} \qquad \ldots(2)$$

Equation (2) gives the optimal order level.

Further $$\frac{d^2C}{dz^2} = \frac{C_1+C_2}{Q}$$

which is always positive for $$z = \frac{C_2rt}{C_1+C_2}$$

Hence, the minimum cost per unit time is given by

$$C_{min} = \frac{C_1}{2Q}\left(\frac{C_2Q}{C_1+C_2}\right)^2 + \frac{C_2}{2Q}\left(Q - \frac{C_2Q}{C_1+C_2}\right)^2 + \frac{rC_3}{Q}$$

After simplification, we get

$$C_{min} = \frac{C_1C_2Q}{2(C_1+C_2)} + \frac{rC_3}{Q} = \frac{C_1C_2rt}{2(C_1+C_2)} + \frac{rC_3}{Q}$$

Observations

(1) If $C_1 \neq 0$, then $C_1 + C_2 > C_2$. In this case $z < Q$
which shows that the order of level to which the inventory is raised is kept less than the total demand Q to create shortages.

18.12 MODEL-5 : ECONOMIC LOT SIZE MODEL WITH UNIFORM RATE OF DEMAND, INFINITE RATE OF PRODUCTION AND WITH SHORTAGES WHICH ARE TO BE FULFILLED

Here, we want to derive an economic lot size formula which minimize the total cost.

Assumptions

(1) C_1 is the holding cost per unit per unit time.

(2) C_2 is the shortage cost per unit per unit time.

(3) C_3 is the set-up cost per run.

(4) r is the uniform demand rate.

(5) q is the order quantity per production run

(6) Production is instantaneous.

(7) Lead time is zero

(8) Shortages are allowed and backlogged.

Formulation of the model

As per given, we have that q is the order quantity per production run and let z be the order level to which the inventory is raised at the beginning of a run of time interval t.

Then , we have
$$q = rt \qquad \qquad ...(1)$$

Now, the cost of holding inventory per production run $= \dfrac{C_1 z^2}{2r}$

Also, the shortage cost for one run $= \dfrac{C_2}{2r}(q-z)^2 = \dfrac{C_2}{2r}(rt-z)^2$

Now, the total cost per run $= \dfrac{C_1}{2r}z^2 + \dfrac{C_2}{2r}(rt-z)^2 + C_3$

Hence, the average total cost per unit time

$$C(z,t) = \dfrac{C_1}{2r}\dfrac{z^2}{t} + \dfrac{C_2}{2r}\left(\dfrac{(rt-z)^2}{t}\right) + \dfrac{C_3}{t} \qquad \qquad ...(2)$$

For a function $C(z, t)$ of two variables, to be minimum, using principle of maxima and minima, we must have

$$\dfrac{\partial C}{\partial z} = 0 \quad \text{and} \quad \dfrac{\partial C}{\partial t} = 0$$

$$\dfrac{\partial C}{\partial z} = 0 \quad \Rightarrow \quad \dfrac{C_1}{r}\cdot\dfrac{z}{t} - \dfrac{C_2}{r}\dfrac{(rt-z)}{t} = 0 \qquad \qquad ...(3)$$

$$\dfrac{\partial C}{\partial t} = 0 \quad \Rightarrow \quad -\dfrac{C_1}{2r}\cdot\dfrac{z^2}{t^2} + C_2\dfrac{(rt-z)}{t} - \dfrac{C_2}{2r}\dfrac{(rt-z)^2}{t^2} - \dfrac{C_3}{t^2} = 0 \quad ...(4)$$

On solving (3), we get

$$z = \dfrac{C_2 rt}{C_1 + C_2} \qquad \qquad ...(5)$$

Now, substituting this value of z in (4), we get

$$-\dfrac{C_1}{2r}\left(\dfrac{C_2 rt}{C_1+C_2}\right)^2 + \dfrac{C_2}{t}\left(rt - \dfrac{C_2 rt}{C_1+C_2}\right) - \dfrac{C_2}{2rt^2}\left(rt - \dfrac{C_2 rt}{C_1+C_2}\right)^2 - \dfrac{C_3}{t^2} = 0$$

$$\Rightarrow \qquad C_1 C_2 r(-C_2 + 2C_1 + 2C_2 - C_1) = \dfrac{2C_3}{t^2}(C_1 + C_2)^2$$

$$\Rightarrow \qquad C_1 C_2 r(C_1 + C_2) = \dfrac{2C_3}{t^2}(C_1 + C_2)^2$$

$\Rightarrow$

$$t^2 = \frac{2C_3(C_1 + C_2)^2}{C_1 C_2 r}$$

Therefore,

$$t = t^* = \sqrt{\left(\frac{2C_3(C_1 + C_2)}{C_1 C_2 r}\right)} \qquad \ldots(6)$$

Further, we have

$$\frac{\partial^2 C}{\partial z^2} = \frac{C_1 + C_2}{rt}$$

$$\frac{\partial^2 C}{\partial t^2} = \frac{C_1 z^2}{rt^3} + C_2\left(\frac{z}{t^2}\right) - \frac{C_2}{2r}\left(\frac{2rz}{t^2} - \frac{2z^2}{t^3}\right) + \frac{2C_3}{t^3}$$

$$= \frac{C_1 z^2}{rt^3} + \frac{C_2 z^2}{rt^3} + \frac{2C_3}{t^3}$$

and

$$\frac{\partial^2 C}{\partial t \partial z} = -\frac{C_1 z}{rt^2} - \frac{C_2 z}{rt^2}$$

Now, consider $\dfrac{\partial^2 C}{\partial t^2} \cdot \dfrac{\partial^2 C}{\partial z^2} - \left(\dfrac{\partial^2 C}{\partial t \partial z}\right)^2$

$$= \left(\frac{C_1 z^2}{rt^3} + \frac{C_2 z^2}{rt^3} + \frac{2C_3}{t^3}\right)\left(\frac{C_1 + C_2}{rt}\right) - \left(-\frac{C_1 z}{rt^2} - \frac{C_2 z}{rt^2}\right)^2$$

$$= \frac{2C_3(C_1 + C_2)}{rt^4}$$

For the value of t, given by (6), we have

$$\frac{\partial^2 C}{\partial z^2} > 0, \frac{\partial^2 C}{\partial t^2} > 0$$

and

$$\frac{\partial^2 C}{\partial t^2} \cdot \frac{\partial^2 C}{\partial z^2} - \left(\frac{\partial^2 C}{\partial t \partial z}\right)^2 > 0$$

Thus by the principle of maxima and minima of two variables, we can say that
    "C is minimum for the value of t given by (6)"
Hence, the optimum order quality for the minimum cost is given by

$$q^* = rt^* = r\sqrt{\left(\frac{2C_3(C_1 + C_2)}{C_1 C_2 r}\right)}$$

$\Rightarrow$

$$q^* = \sqrt{\frac{2rC_3(C_1 + C_2)}{C_1 C_2}}$$

The above formula is known as 'Economic lot size formula'.
Now, we proceed to compute C_{min}.
Using (2) and (5), we may get

$$C(z,t) = \frac{C_1}{2rt}\left[\frac{C_2 rt}{C_1 + C_2}\right]^2 + \frac{C_2}{2rt}\left(rt - \frac{C_2 rt}{C_1 + C_2}\right)^2 + \frac{C_3}{t}$$

$$= \frac{C_1 C_2^2 rt}{2(C_1 + C_2)^2} + \frac{C_2 C_1^2 rt}{2(C_1 + C_2)^2} + \frac{C_3}{t}$$

$$= \frac{C_1 C_2 rt(C_1 + C_2)}{2(C_1 + C_2)^2} + \frac{C_3}{t} = \frac{C_1 C_2 rt}{2(C_1 + C_2)} + \frac{C_3}{t}$$

For minimum value of C, putting

$$t = t^* = \sqrt{\frac{2C_3(C_1 + C_2)}{C_1 C_2 r}}$$

in the above equation, we get

$$C_{min} = \frac{C_1 C_2 r}{2(C_1 + C_2)} \sqrt{\left[\frac{2C_3(C_1 + C_2)}{C_1 C_2 r} \right]} + C_3 \sqrt{\left[\frac{C_1 C_2 r}{2C_3(C_1 + C_2)} \right]}$$

$$= \sqrt{\left[\frac{C_1 C_2 C_3 r}{2(C_1 + C_2)} \right]} + \sqrt{\left[\frac{C_1 C_2 C_3 r}{2(C_1 + C_2)} \right]}$$

$$= 2\sqrt{\left[\frac{C_1 C_2 C_3 r}{2(C_1 + C_2)} \right]} = \sqrt{\frac{2C_1 C_2 C_3 r}{C_1 + C_2}}$$

Observations

1. If shortages are not allowed ($C_2 = \infty$), then we have $\dfrac{C_1 + C_2}{C_2} = \dfrac{C_1}{C_2} + 1 = 1$

   In this case $q^* = \sqrt{\left(\dfrac{2rC_3}{C_1} \right)}$

2. The minimum cost per unit time is reduced by allowing shortages which are to be backlogged.

18.13 MODEL-6 : ECONOMIC LOT SIZE MODEL WITH UNIFORM RATE OF DEMAND, FINITE RATE OF PRODUCTION AND HAVING SHORTAGES (WHICH ARE TO BE FULFILLED)

Assumptions

1. C_1 is the holding cost per unit per unit time.
2. C_2 is the shortage cost per unit per unit time.
3. C_3 is the set-up cost per run.
4. r is the uniform demand rate.
5. k is the production rate, which is finite.
6. Lead time is zero.
7. Shortages are allowed and backlogged.

Formulation of the Model

As per assumptions of this model, the stock is zero initially. Also, the production starts with a finite rate k units per unit time, while the demand is r units per unit time. Then, clearly we can say that the inventory increases with a rate $(k - r)$ units per unit time. Let Q be the inventory level.

Now, the inventory (stock) level at the end of continuous time t_1 is given by

$$Q = (k - r)t_1 \qquad \qquad \dots(1)$$

If the production is stopped, then inventory falls to zero with a rate r units per unit time in time say t_2. Then, further shortages arise with a rate r units per unit time and continued for a period t_3 (say).

At the end of the time t_3, let s be the shortage, then we have

$$Q = rt_2 \qquad \qquad \dots(2)$$

and

$$s = rt_3 \qquad \qquad \dots(3)$$

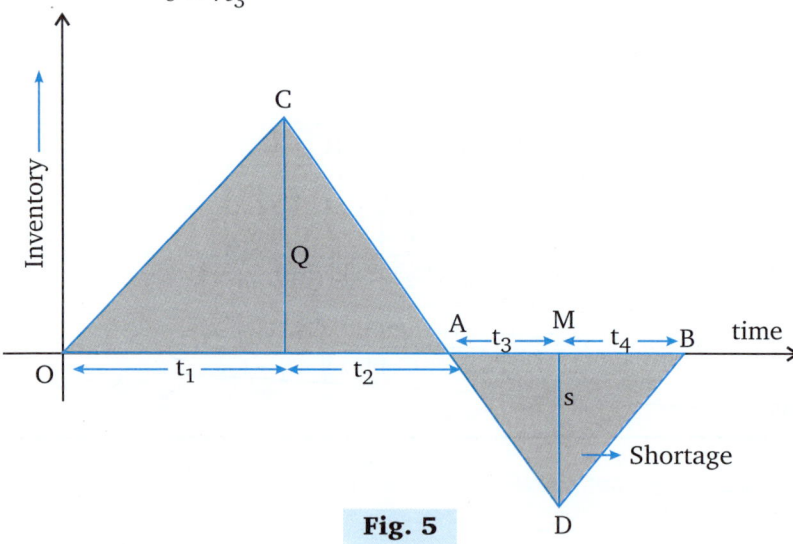

Fig. 5

Now, at the time of t_3, production starts again and all shortages are fulfilled in time t_4. Therefore, at the end of time t_4, shortages reaches to zero.

Also, assume that cycle repeat itself after the time $t_1 + t_2 + t_3 + t_4$. Further, since r is the demand rate and k is the production rate, thus the rate of reduction of shortages is $(k - r)$ units per unit time during the time t_4.

Therefore, $\qquad \qquad s = (k - r)t_4 \qquad \qquad \dots(4)$

Using (1) and (2), we have

$$Q = (k - r)t_1 = rt_2 \qquad \qquad \dots(5)$$

Also, from (3) and (4) we have

$$s = rt_3 = (k - r)t_4 \qquad \qquad \dots(6)$$

Let q units be the total demand to be fulfilled in one cycle, *i.e.*, in time $t_1 + t_2 + t_3 + t_4$. Then, we have

$$q = r(t_1 + t_2 + t_3 + t_4) \qquad \qquad \dots(7)$$

Using (5), (6) and (7), we get

$$q = r\left(\frac{rt_2}{k - r} + t_2 + t_3 + \frac{rt_3}{k - r} \right) = \frac{k(t_2 + t_3)r}{k - r} \qquad \qquad \dots(8)$$

Now, the holding cost for one cycle

$$= C_1(\text{Area of } \Delta \ OCA) = C_1 \cdot \frac{1}{2}Q(t_1 + t_2)$$

$$= \frac{C_1}{2} \cdot rt_2 \cdot \left[\frac{rt_2}{k - r} + t_2 \right] = \frac{C_1 rkt_2^2}{2(k - r)}$$

Shortage cost for one cycle $= C_2 \cdot (\text{Area of } \Delta \, ABD)$

$$= C_2 \cdot \frac{1}{2} s(t_3 + t_4)$$

$$= \frac{C_2}{2} \cdot rt_3 \left[t_3 + \frac{rt_3}{k-r} \right] = \frac{C_2 \cdot rkt_3^2}{2(k-r)}$$

Thus, the total cost per cycle

$$= \frac{C_1 rkt_2^2}{2(k-r)} + \frac{C_2 rkt_3^2}{2(k-r)} + C_3$$

$$= \frac{rk(C_1 t_2^2 + C_2 t_3^2) + 2C_3(k-r)}{2(k-r)}$$

Hence, the total average cost per cycle is given by

$$C = \frac{\left\{ \dfrac{rk(C_1 t_2^2 + C_2 t_3^2) + 2C_3(k-r)}{2(k-r)} \right\}}{(t_1 + t_2 + t_3 + t_4)}$$

Using (5) and (6), above equation reduces to

$$C(t_2, t_3) = \frac{rk(C_1 t_2^2 + C_2 t_3^2) + 2C_3(k-r)}{2(k-r)\left[\dfrac{rt_2}{k-r} + t_2 + t_3 + \dfrac{rt_3}{k-r} \right]}$$

$$= \frac{rk(C_1 t_2^2 + C_2 t_3^2) + 2C_3(k-r)}{2k(t_2 + t_3)} \qquad \dots(9)$$

Using the principle of maxima and minima, for $C(t_2, t_3)$ to be minimum, we must have

$\dfrac{\partial C}{\partial t_2} = 0$ and $\dfrac{\partial C}{\partial t_3} = 0.$

$\therefore \qquad\qquad \dfrac{\partial C}{\partial t_2} = 0$

$\Rightarrow \quad \dfrac{rk2C_1 t_2 \cdot 2k(t_2 + t_3) - \{rk(C_1 t_2^2 + C_2 t_3^2) + 2C_3(k-r)\}2k}{(2k)^2 (t_2 + t_3)^2} = 0$

$\Rightarrow \qquad C_1 t_2^2 + 2C_1 t_2 t_3 - C_2 t_3^2 - \dfrac{2C_3}{rk}(k-r) = 0 \qquad \dots(10)$

Similarly, $\qquad \dfrac{\partial C}{\partial t_3} = 0$

$\Rightarrow \qquad C_1 t_2^2 - 2C_2 t_2 t_3 - C_2 t_3^2 + \dfrac{2C_3}{rk}(k-r) = 0 \qquad \dots(11)$

Using (10) and (11), we may get

$$t_2 t_3 = \frac{2(k-r)C_3}{rk(C_1 + C_2)} \qquad \dots(12)$$

and $2C_1t_2^2 + 2C_1t_2t_3 - 2C_2t_2t_3 - 2C_2t_3^2 = 0$

$\Rightarrow$ $(C_1t_2 - C_2t_3)(t_2 + t_3) = 0$

which gives either $t_3 = -t_2$ or $t_3 = \dfrac{C_1}{C_2}t_2$

CASE 1. If $t_3 = -t_2$, then (12) gives

$$t_2^2 = -\frac{2(k-r)C_2}{rk(C_1 + C_2)} < 0 \qquad\qquad (\because\ k > r)$$

$\Rightarrow t_2$ is imaginary.

CASE 2. If $t_3 = \dfrac{C_1}{C_2}t_2$, then from (12), we have

$$t_2^2 = \frac{2C_2C_3(k-r)}{C_1rk(C_1 + C_2)}$$

Thus,

$$\left.\begin{array}{l} t_2 = t_2^* = \sqrt{\dfrac{2C_2C_3\left(1 - \dfrac{r}{k}\right)}{rC_1(C_1 + C_2)}} \\[2em] t_3 = t_3^* = \dfrac{C_1}{C_2}t_2^* = \sqrt{\dfrac{2C_1C_3\left(1 - \dfrac{r}{k}\right)}{rC_2(C_1 + C_2)}} \end{array}\right\} \qquad \ldots(13)$$

and

We can also verify that for t_2 and t_3 given by (13), $\dfrac{\partial^2 C}{\partial t_2^2} > 0, \dfrac{\partial^2 C}{\partial t_3^2} > 0$ and

$\dfrac{\partial^2 C}{\partial t_2^2} \cdot \dfrac{\partial^2 C}{\partial t_3^2} - \left(\dfrac{\partial^2 C}{\partial t_2 \partial t_3}\right)^2 > 0$. Thus, we conclude that for the values of t_2 and t_3 (given by (13)) the average cost per unit time, given by (9) is minimum. Also, from (9), the minimum cost per unit time is given by

$$C_{\min} = \sqrt{\left\{\frac{2rC_1C_2C_3\left(1 - \dfrac{r}{k}\right)}{(C_1 + C_2)}\right\}}$$

and the optimum order quantity (EOQ) is given by

$$q^* = \sqrt{\left(\frac{2rC_3}{C_1}\right)\left(\frac{C_1 + C_2}{C_2}\right)\left(\frac{1}{\left(1 - \dfrac{r}{k}\right)}\right)}$$

 Solved Examples

EXAMPLE 1. ***The storage cost of one item is ₹1 per month and set-up cost is ₹25 per run. If the production is instantaneous and the demand is 200 units per month. Find the optimal size of the batch and the best time for the replenishment of inventory.*** [MEERUT–2001, 02(BP), 09]

SOLUTION. As per given, we have

$$C_1 = 1, C_3 = 25, r = 200$$

Thus, $$q^* = \sqrt{\frac{2C_3 r}{C_1}} = \sqrt{\frac{2 \times 25 \times 200}{1}} = 100$$

and $$t^* = \frac{q^*}{r} = \frac{100}{200} = \frac{1}{2} \text{ month} = 15 \text{ days}$$

Therefore, he produce 100 units of his product at the interval of 15 days.

Also, the minimum average cost

$$= \sqrt{2C_1 C_3 r} = \sqrt{2 \times 1 \times 25 \times 200} = ₹100 \text{ per month}$$

Further, the average inventory throughout the month $= \frac{q}{2} = 100$. Thus, if he produces all 200 items in the beginning of a month, then the cost

$$= 25 + 1 \times 100 = ₹ 125$$

which is more than the cost.

EXAMPLE 2. *A manufacturer has to supply his customer with 600 units of his product per year. Shortages are not allowed and the storage cost amount to ₹0.60 per unit per year. The set-up cost per run is ₹80.00. Find the optimum run size and the minimum average yearly cost.*

[MEERUT–2001, 07, 13]

SOLUTION. As per given, we have

$$C_1 = 0.60, C_3 = 80, r = 600$$

We know that

$$q^* = \sqrt{\frac{2C_3 r}{C_1}} = \sqrt{\frac{2 \times 80 \times 600}{0.60}} = 400$$

Also, $$t^* = \frac{q^*}{r} = \frac{400}{600} = \frac{2}{3} \text{ year} = 8 \text{ months}$$

Hence, the manufacturer should produce 400 units of his product at an interval of 8 months.

Further, the minimum average cost

$$= \sqrt{2C_1 C_3 r}$$

$$= \sqrt{2 \times 0.6 \times 80 \times 600}$$

$$= ₹ 240 \text{ units/year}$$

EXAMPLE 3. *Neon lights are replaced at the rate of 100 units per day. The physical plant orders the neon lights periodically. It costs ₹100 to initiate a purchase order. A neon light kept in storage is estimated to cost ₹0.02 per day. The lead time between placing and receiving an order is 12 days. Find the optimal inventory policy for ordering the neon lights.*

[MEERUT–2002, 03(BP)]

SOLUTION. As per given: we have

$$C_3 = 100, C_1 = 0.02, r = 100$$

Putting all these values in

$$q^* = \sqrt{\frac{2C_3 r}{C_1}}$$

We get optimal order quantity $q^* = \sqrt{\dfrac{2 \times 100 \times 100}{0.02}} = 1000$ lights

Also, the length of the cycle, $t^* = \dfrac{q^*}{r} = \dfrac{1000}{100} = 10 \, \text{days}$

It is also given that

$$\text{lead time } L = 12 \text{ days}$$

and $\qquad$ length of the cycle $= 10$ days

Clearly, the lead time is greater than the length of the cycle. Thus the reordering will take place when the level of inventory is sufficient to satisfy the demand for $2 \ (= 12 - 10)$ days.

Thus, the reorder point $= 2 \times 100 = 200$ lights

Hence, we conclude that 100 lights is to be ordered when the level of inventory reaches 200 lights.

EXAMPLE 4. *A contractor has to supply 10000 bearing per day to an automobile manufacturer. He finds that when he starts a production run, he can produce 25000 bearing per day. The cost of holding a bearing in stock for one year is 2 paise and the set-up cost of a production run is ₹18. How frequently should production run be made? (Based on Model-3)* [MEERUT–1996, 2002; ROHILKHAND–1995, 98]

SOLUTION. As per given, we have $k = 25000, r = 10000$

$C_1 = 2$ paise per bearing per year $= ₹\dfrac{2}{100 \times 365} = ₹\dfrac{1}{50 \times 365}$ per bearing per day

and $C_3 = 18$

We know that lot size formula

$$q^* = \sqrt{\dfrac{2C_3}{C_1}\left(\dfrac{rk}{k-r}\right)}$$

$$= \sqrt{2 \times 18 \times 50 \times 365\left(\dfrac{10000 \times 25000}{25000 - 10000}\right)}$$

$$= 104642$$

Hence, the required time for production run

$$= \dfrac{104642}{10000} = 10.46 \ \text{days}$$

EXAMPLE 5. *A manufacturing company purchases 9000 parts of a machine for its annual requirements, ordering one month usage at a time. Each part costs ₹20. The ordering cost per order is ₹15 and the carrying charges are 15% of the average inventory per year. Suggest a more economical purchasing policy for the company.* [MEERUT–2003; DELHI–2006]

SOLUTION. As per given we have

$$r = 9000 \text{ parts per year}$$

$$C_1 = 15\% \text{ of the average inventory per year}$$

$$= 20 \times \dfrac{15}{100} = 3$$

and $\qquad C_3 = 15$

We know that, economic lot size

$$q^* = \sqrt{\frac{2C_3 r}{C_1}} = \sqrt{\frac{2 \times 15 \times 9000}{3}} = 300 \text{ units}$$

and

$$t^* = \frac{q^*}{r} = \frac{300}{9000} = \frac{1}{30}$$

Now, minimum cost per year $= \sqrt{2C_1 C_3 r} = \sqrt{2 \times 3 \times 15 \times 9000} = ₹ 900$

For the policy of ordering each month,

annual ordering cost $= 12 \times 15 = 180$

and lot size each month, $q = \dfrac{9000}{12} = 750$

$\Rightarrow$ average inventory at any time $= \dfrac{q}{2} = \dfrac{750}{2} = 375$

$\therefore$ storage cost at any time $= 375 C_1 = 375 \times 3 = 1125$

$\Rightarrow$ Total annual cost $= 1125 + 180 = ₹ 1305$

Thus, we conclude that, if the company purchase 300 parts at time intervals of 1/30 years in place of order 750 parts each month then the net saving of the company is

$$= 1305 - 900 = ₹\ 405 \text{ per year}$$

EXAMPLE 6. *An item is produced at the rate of 50 items per day. The demand occurs at the rate of 25 items per day. If the set-up cost is ₹100 per set-up and holding cost is 0.01 per unit of item per day. Find the economic lot size for one run, assuming that the shortages are not allowed.*   [MEERUT–2004, GARHWAL–2014]

SOLUTION. As per given we have

$$k = 50 \text{ items per day}$$
$$r = 25 \text{ items per day}$$
$$C_1 = ₹\ 0.01 \text{ per day}$$
$$C_3 = ₹\ 100 \text{ per run}$$

Putting all these values in economic lost size formula given by

$$q^* = \sqrt{\frac{C_3}{C_1}\left(\frac{rk}{k-r}\right)}$$

We get

$$q^* = \sqrt{\frac{2 \times 100}{0.01}\left(\frac{25 \times 50}{50-25}\right)} = 1000 \text{ items}$$

$\therefore$

$$t^* = \frac{q^*}{r} = \frac{1000}{25} = 40 \text{ days}$$

Now, minimum cost

$$= \sqrt{2C_1 C_3 r \left(1 - \frac{r}{k}\right)}$$

$$= \sqrt{2 \times 0.01 \times 100 \times 25 \left(1 - \frac{25}{50}\right)} = ₹\ 5 \text{ per day}$$

Therefore, total cost per run $= 5 \times 40 = ₹ 200$

EXAMPLE 7. *A company uses annually 24000 units of a raw material which cost ₹1.25 per unit. Placing each order costs ₹22.5 and the carrying cost is 5.4% per year of the average inventory. Find the economic lot size and the total inventory cost.* [MEERUT–1999; ASSAM–2004]

SOLUTION. As per given, we have

$$r = 24000 \text{ units per year}$$

$$C_1 = 5.4\% \text{ of the average inventory per year}$$

$$= 1.25 \times \frac{5.4}{100} = ₹\frac{27}{400} \text{ per unit per year}$$

$$C_3 = 22.5$$

Putting all these values in economic lot size formula given by

$$q^* = \sqrt{\frac{2C_3 r}{C_1}}$$

We get

$$q^* = \sqrt{\frac{2 \times 22.5 \times 24000 \times 400}{27}} = 4000 \text{ units}$$

and

$$t^* = \frac{q^*}{r} = \frac{4000}{24000} = \frac{1}{6} \text{ years} = 2 \text{ months}$$

which implies that the optimal lot size is 4000 units after every two months. Hence, total annual inventory cost

$$= \sqrt{2C_1 C_3 r} + \text{Purchasing cost per year}$$

$$= \sqrt{2 \times \frac{27}{400} \times 22.5 \times 24000} + 1.25 \times 24000 = ₹30270$$

EXAMPLE 8. *You have to supply your customers with 100 units of a certain product every Monday and only then, you obtain the product from a local suppliers at ₹60 per unit. The costs of ordering and transportation from supplier are ₹150 per order. The cost of carrying inventory is estimated at 15% per year of the cost of the product carried.*

(i) Find the total size which will minimize the cost of the system.

(ii) Find the optimal cost. [ROHILKHAND–1995]

SOLUTION. As per given, we have

$$C_1 = 15\% \text{ per year of the cost of the product carried}$$

$$= \frac{15}{100} \times 60 = ₹2 \text{ per unit per year}$$

$$= ₹\frac{9}{52} \text{ per unit per week}$$

$$C_3 = ₹150$$

$$r = 100 \text{ units per week}$$

Hence,

(i) Optimal lot size

$$q^* = \sqrt{\frac{2C_3 r}{C_1}} = \sqrt{\frac{2 \times 150 \times 100 \times 52}{9}} = 416 \text{ units}$$

and (ii) Optimal cost $= \sqrt{2C_1 C_3 r} + 60r$

$$= \left(\sqrt{2 \times \frac{9}{52} \times 150 \times 100} \right) + 60 \times 100 = ₹6072$$

EXAMPLE 9. *The demand of an item is uniform at a rate of 25 units per month. The fixed cost is ₹15 each time a production run is made. The production cost is ₹1 per item and the inventory carrying cost is ₹0.30 per item per month. If the shortage cost is ₹1.50 per item per month, determine how often to make a production run and of what size it should be.* [MEERUT–1999; ROHILKHAND–2003]

SOLUTION. Here, set-up cost is a variable. If P is the production cost of an item and q is the order quantity per production run, then set-up cost $= C_3 + Pq$

We know that, average total cost

$$C(z,t) = \frac{C_1}{2r} \cdot \frac{z^2}{t} + \frac{C_2}{2r} \frac{(rt-z)^2}{t} + \frac{C_3 + Pq}{t}$$

$$= \frac{C_1}{2r} \cdot \frac{z^2}{t} + \frac{C_2}{2r} \frac{(rt-z)^2}{t} + \frac{C_3}{t} + Pr$$

Since, pr is constant, therefore $C(z, t)$ will be minimum, when

$$t^* = \sqrt{\frac{2C_3 (C_1 + C_2)}{C_1 C_2 r}}$$

Therefore, $\quad q^* = rt^* = \sqrt{\dfrac{2r(C_1 + C_2)C_3}{C_1 C_2}}$

Now putting $\quad r = 25$ units per month,

$C_1 = ₹ 30$ per item per month
$C_2 = ₹ 1.50$ per item per month
$C_3 = ₹ 15$ per production run

Therefore, $\quad q^* = \sqrt{\dfrac{2 \times 15 \times (0.30 + 1.50) \times 25}{(0.30) \times (1.50)}} = 55$ items

and $\quad t^* = \dfrac{q^*}{r} = \dfrac{55}{25} = 2.2$ months

EXAMPLE 10. *A company uses annually 50000 units of an item each costing ₹1.20. Each order costs ₹45 and inventory carrying costs 15% of the annual average inventory value.*

(i) Find EOQ.

(ii) If the company operates 250 days a year, the procurement time is 10 days and safety stock is 500 units, find reorder level, maximum, minimum and average inventory.

SOLUTION. Here, we have

$$C_1 = \text{I.P.} = \frac{15}{100} \times 1.20 = ₹\frac{9}{50} \text{ per unit per year}$$

$$C_3 = 45$$

and total demand in one year, $R = 50{,}000$ units

$$\therefore \qquad EOQ = \sqrt{\frac{2C_3 R}{C_1}} = \sqrt{\frac{2 \times 45 \times 50000 \times 50}{9}} = 5000 \text{ units}$$

Now, since,

Number of days the company operates $= 250$

Lead time, $L = 10$ days

Safety stock, $B = 500$ units

Therefore,

Reorder level $=$ Lead time demand $+$ safety stock

$$= L \times \frac{50000}{250} + B$$

$$= 10 \times 200 + 500 = 2500 \text{ units}$$

Now maximum inventory

$$= EOQ + \text{Safety stock}$$

$$= 5000 + 500 = 5500 \text{ units}$$

Minimum inventory $=$ safety stock

$$= 500 \text{ units}$$

Hence, average inventory $= \dfrac{1}{2} \times 5000 + 500 = 3000 \text{ units}$

 ### Exercise-18.1

1. If in any model, set-up cost instead of being fixed is equal to $C_3 + bq$ where b is the set-up cost per item produced, then show that there is no change in the optimum order quantity produced due to change in the set-up cost.

2. A shopkeeper has a uniform demand of an item at the rate of 50 items per month; he buys a supplier at a cost of ₹6 per item and the cost of ordering is Rs. 10 each time. If the stock holding costs are 20% per year of stock value, how frequently should he replenish his stocks.

3. A certain item costs ₹235 per ton. The monthly requirements 5 tons and each time the stock is replenished, there is a cost of ₹ 1000. The cost of carrying inventory has been estimated at 10% of the value of the stock per year. What is the optimal order quantity?

4. Consider the inventory system with the following data in usual notation :
   $r = 100$ units per year, $I = 0.30$, $P = 0.50$ rupees per unit, $C_3 = ₹10.00$, $L = 2$ years
   Find the following
   (i) Optimal order quantity
   (ii) Re-order point
   (iii) Minimum average cost

5. A company uses annually 1200 units of a raw material costing ₹ 1.25 per unit. Placing each order costs ₹ 0.45 and the carrying costs are

15% per year unit of the average inventory. Find the economic order quantity.

6. A product is produced at the rate of 50 items per day. The demand occurs at the rate of 30 items per day. It is given that $C_3 = ₹ 100$, $C_1 = ₹ 0.05$, find the economic lot size and the associated total cost per cycle, assuming that no shortage is allowed.

7. The uniform annual demands for two bulky items are 90 tons and 160 tons respectively. The carrying costs are ₹250 and ₹200 per ton per year and set-up costs ₹50 and ₹40 per production respectively. No shortages are allowed. Space consideration restrict the average amount inventory of items to 4000 ft^3. A ton of the first item occupies 1000 ft^3 and a ton of the second item 5000 ft^3. Find the optimal lot size. [MEERUT–2007]

8. An aircraft company uses rivets at an approximate customer rate of 2500 kg per year. The rivets costs ₹30 per kgs and the company personed estimate ₹130 to place an order and the inventory carrying cost is 10% per year. How frequently should orders for rivets be placed and what quantities should be ordered. [MEERUT–2002]

9. A contractor has to supply diesel engines to a truck manufacture at the rate of 25 per day. There is a clause in the contract pandering

him ₹10 per engine per day late for missing the schedule delivery date. He finds that the cost of holding a complete engine in stock is Rs. 16 per month. The production process is such that each month (30 days), he starts a batch of engines through the shops and all these engines are available for delivery any time after the end of the month. What should his inventory level be at the beginning of each month? [MEERUT–2000]

10. For the following data, determine approximately the economic order quantities when the total value of average of inventory levels of three products is ₹1000.

Product cost	A	B	C
Holding cost	20	20	20
Cost per unit	6	7	5
Set-up cost	50	42	60
Yearly annual demand	10,0000	12000	7500

Answers

1. Lot size 100 items after every two months **3.** 71.5 tons. **4.** 365 units, 2000 units, ₹ 54.80
5. 240 units **6.** 13053, ₹11 **7.** 2 tons, 12 tons **8.** 466 kg, 5.3 orders per year **9.** 712 engines **10.** 4.7

Part (B) : PROBABILISTIC MODEL

In this system, we shall discuss some models in which demand is not known exactly. Here we shall minimize the expected costs instead actual costs. The random behaviour of demand and lead time needs to be described with probability distribution either discrete or continuous.

18.14 MODEL-I : SINGLE PERIOD MODEL WITH INSTANTANEOUS (DISCONTINUOUS) DEMAND AND TIME INDEPENDENT COSTS : NO SET-UP COST MODEL

(DELHI-2009, KANPUR-1997, GARHWAL 1993, 95, 2011; MEERUT–1996; AVADH–2012; LUCKNOW–2010)

Assumptions

(1) Reorder time is fixed *i.e.*, t is the constant interval between orders.

(2) The set-up cost is not included in the total cost.

(3) The production of the commodity is instantaneous.

(4) z is the stock level at the beginning of each period of time t.

(5) The lead time is taken to be zero.

(6) $p(r)$ = the probability of demand of r units in time t.

(7) C_1 = holding cost per unit per unit time.

(8) C_2 = shortage cost per unit per unit time.

Formulation and solution of the model

(a) When r has Discrete values

Since, the demand is instantaneous, so we have assumed that the total demand is fulfilled at the beginning of the period.

Here we have the following two cases :

Case 1 : When $r \leq z$, i.e., demand r does not exceed the stock z.

Since, the total demand r is fulfilled at the beginning of the period, then clearly $z–r$ is the inventory which is kept for time t.

Now, since the total cost is the holding cost, cost of holding $z–r$ units in stocks for unit time is given by

$$(z–r)C_1 \text{ for } r \leq z$$

So, probability of holding cost $(z-r)C_1$ is $p(r)$.

($\because$ the probability of requiring r units is $p(r)$)

Hence, the total expected cost per unit time

$$= \sum_{r=0}^{z} (z-r)C_1 p(r)$$

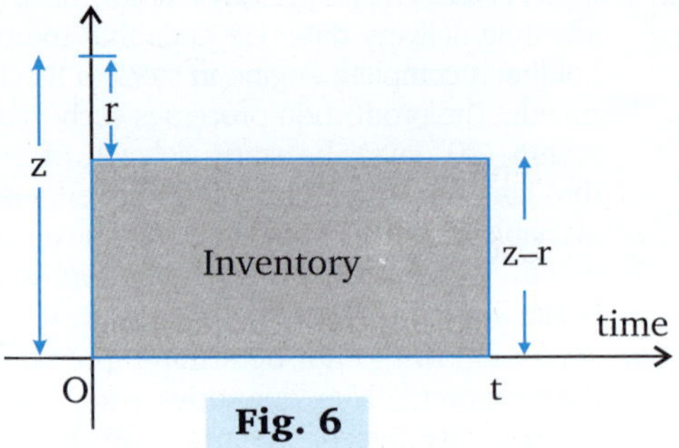

Case-2 : If $r > z$ i.e., the demand r exceeds the stock z.

In this case the total demand is not satisfied at the beginning of the period because customers demand r is more than the stock z. Therefore, $r - z$ demand remains to be satisfied.

Since, the total cost is the shortage cost. Therefore, cost of shortage of $r - z$ units for unit time

$$= (r-z)\, C_2, \text{ for } r > z$$

Then, total expected cost per unit time is given by

$$= \sum_{r=z+1}^{\infty} (r-z)C_2 p(r)$$

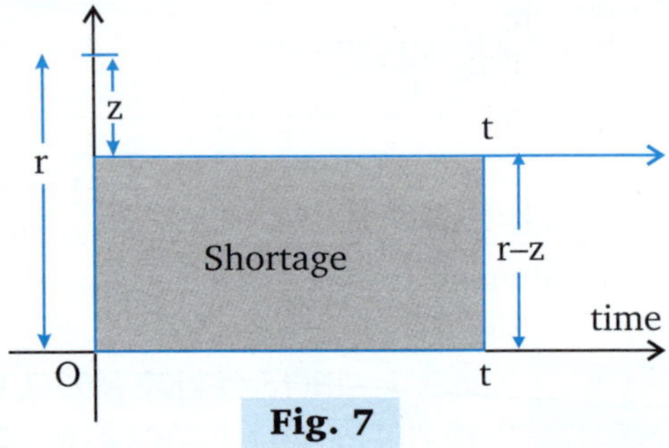

Fig. 6

Fig. 7

Total expected cost per unit time for the model is given by

$$C(z) = \sum_{r=0}^{z} (z-r)C_1 p(r) + \sum_{r=z+1}^{\infty} (r-z)C_2 p(r) \qquad \ldots(1)$$

For minima of $C(z)$, we must have

$$\Delta\, C(z-1) < 0 < \Delta C(z) \qquad \ldots(2)$$

We have

$$\Delta C(z) = C(z+1) - C(z)$$

$$= C_1 \sum_{r=0}^{z+1} \{(z+1)-r\}p(r) + C_2 \sum_{r=z+2}^{\infty} (r-z-1)\, p(r)$$

$$- C_1 \sum_{r=0}^{z} (z-r)\, p(r) - C_2 \sum_{r=z+1}^{\infty} (r-z)\, p(r)$$

Now Since $\displaystyle\sum_{r=0}^{z+1} (z+1-r)\, p(r) = \sum_{r=0}^{z} (z+1-r)\, p(r)$

and $\displaystyle\sum_{r=z+2}^{\infty} (r-z-1)\, p(r) = \sum_{r=z+1}^{\infty} (r-z-1)\, p(r)$

Therefore

$$\Delta C(z) = C_1 \sum_{r=0}^{z} p(r) - C_2 \sum_{r=z+1}^{\infty} p(r)$$

$$= C_1 \sum_{r=0}^{z} p(r) - C_2 \left[\sum_{r=0}^{\infty} p(r) - \sum_{r=0}^{z} p(r) \right]$$

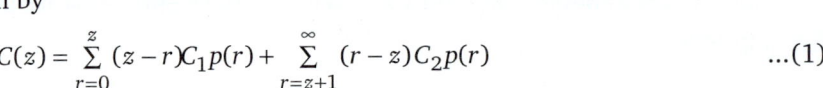

$$= (C_1 + C_2) \sum_{r=0}^{z} p(r) - C_2 \qquad (\because \Sigma p(r) = 1)$$

Similarly $\qquad \Delta C(z-1) = (C_1 + C_2) \sum_{r=0}^{z-1} p(r) - C_2$

Using all these values in (2), for minimum value of $C(z)$, we have

$$(C_1 + C_2) \sum_{r=0}^{z-1} p(r) - C_2 < 0 < (C_1 + C_2) \sum_{r=0}^{z} p(r) - C_2$$

$$\Rightarrow \qquad \sum_{r=0}^{z-1} p(r) < \frac{C_2}{C_1 + C_2} < \sum_{r=0}^{z} p(r) \qquad \ldots(3)$$

(b) When r has continuous values

In this case r is capable of being considered as a continuous variable with probability distribution function $f(r)$.

Proceeding as in part (a) (Taking $\int$ sign in place of Σ), the total expected cost is given by

$$C(z) = \int_{r=0}^{z} C_1(z-r) f(r) dr + \int_{r=z}^{\infty} C_2(r-z) f(r) dr \qquad \ldots(4)$$

For the minima of $C(z)$, we must have

$$\frac{dC}{dz} = 0$$

$$C_1 \int_{r=0}^{z} \frac{\partial}{\partial z} \{(z-r)(f(r))\}.dr + C_1 \left[(z-r) f(r) \frac{dr}{dz} \right]_{r=0}^{z}$$

$$+ C_2 \int_{r=z}^{\infty} \frac{\partial}{\partial z} \{(r-z) f(r)\} dr + C_2 \left[(r-z) f(r) \frac{dr}{dz} \right]_{r=z}^{\infty} = 0$$

$$\Rightarrow \qquad C_1 \int_{r=0}^{z} f(r) dr + C_1 . 0 - C_2 \int_{r=z}^{\infty} f(r) dr + C_2 . 0 = 0$$

$$\Rightarrow \qquad (C_1 + C_2) \int_0^z f(r) dr - C_2 = 0 \qquad (\because \int_0^\infty f(r) dr = 1)$$

$$\Rightarrow \qquad \int_0^z f(r) dr = \frac{C_2}{C_1 + C_2} \qquad \ldots(5)$$

Now for sufficiencies of minima of $C(z)$, we have

$$\frac{d^2 C(z)}{dz^2} > 0$$

$$\Rightarrow \qquad \frac{d}{dz}\left(\frac{d}{dz} C(z)\right) = (C_1 + C_2) \int_0^z \frac{\partial}{\partial z} f(r) dr + (C_1 + C_2) \left[f(r) . \frac{dr}{dz} \right]_0^z$$

$$\Rightarrow \qquad (C_1 + C_2) f(z) > 0$$

Hence, we conclude that $C(z)$ given by (4) is minimum for the value of z given by (5).

☛ Remark

- If z_0 satisfy $\sum_{r=0}^{z_0-1} p(r) = \frac{C_2}{C_1 + C_2} < \sum_{r=0}^{z_0} p(r)$ then clearly $C(z_0-1) = C(z_0)$, then optimum value of z

  is either z_0 or z_0-1. But if z_0 satisfy $\sum_{r=0}^{z_0-1} p(r) < \frac{C_2}{C_1 + C_2} = \sum_{r=0}^{z_0} p(r)$, then $C(z_0+1) = C(z_0)$.

  Then optimum value of stock z is either z_0 or z_0+1.

18.15 MODEL-2 : NO SET-UP COST MODEL; SINGLE PERIOD MODEL WITH UNIFORM DEMAND

(DELHI-2009, KANPUR-1997, GARHWAL 1993, 95, 2011; MEERUT–1995, 2004)

Assumptions

Let the demand be r units per time period. Then we have the following assumptions :

(1) Order time is fixed and known *i.e.*, t is the constant interval between orders.

(2) Set-up cost is not included in the total cost.

(3) z is the stock level at the beginning of each period of time t

(4) Demand is uniform over the period (r units per period)

(5) Lead time is zero.

(6) $p(r)$ is the probability of requiring r units in time t

(7) C_1 = holding cost per unit per unit time

(8) C_2 = shortage cost per unit per unit time (shortages are allowed and backlogged)

Formulation and solution of the model

(a) When r has discrete values

Let z be the stock level at the beginning of each period of time t. Then we have the following two cases :

Case 1 : If $r \leq z$ i.e., demand r does not exceed the stock z.

In this case, the total cost is the holding cost only *i.e.*, C_1

Now, shaded area $= C_1 . \dfrac{1}{2}[z + (z - r)]t$

$$= C_1\left(z - \dfrac{r}{2}\right)t$$

Hence, the total expected cost

$$= \sum_{r=0}^{z} C_1\left(z - \dfrac{r}{2}\right)tp(r)$$

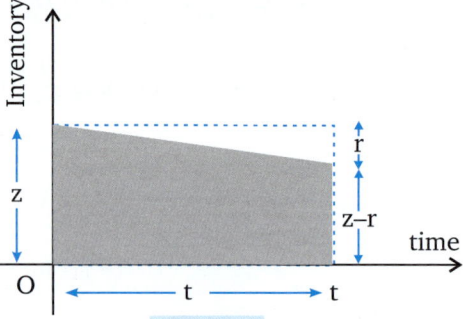

Fig. 8

Case-2 : If $r > z$. i.e., when demand r exceed the stock z.

Here, the period starts with a stock z which is supplied in time t_1 with rate r/t.

Now, for the remaining time (*i.e.*, $t-t_1$) there is a shortage which increase to $(r-z)$ by the end of the period

$$\therefore \quad z = \dfrac{r}{t}.t_1 \Rightarrow t_1 = \dfrac{zt}{r} \quad \text{and} \quad t - t_1 = \dfrac{(r-z)t}{r}$$

Now, the holding cost is $C_1 . \dfrac{zt_1}{2} = \dfrac{1}{2}C_1 \dfrac{z^2 t}{r}$

and shortage cost is given by

$$C_2 \dfrac{(r-z)(t-t_1)}{2} = C_2 \dfrac{(r-z)^2 . t}{2r}$$

So, the total expected cost is given by

$$\sum_{r=z+1}^{\infty} \dfrac{t}{2r}[C_1 . z^2 + C_2(r-z)^2]p(r)$$

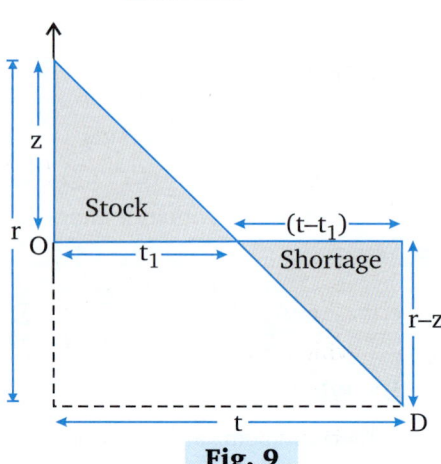

Fig. 9

$\therefore$ Total expected cost per unit time is given by

$$C(z) = C_1 \sum_{r=0}^{z}\left(z - \frac{r}{2}\right)p(r) + \sum_{r=z+1}^{\infty}\frac{1}{2r}\left[C_1.z^2 + C_2(r-z)^2\right]p(r) \qquad \ldots(1)$$

By the principle of maxima and minima, for the minimum of $C(z)$, we must have

$$\Delta C(z{-}1) < 0 < \Delta C(z) \qquad \ldots(2)$$

Now, we have

$$\Delta C(z) = C(z{+}1) - C(z)$$

$$= C_1 \sum_{r=0}^{z+1}\left(z+1 - \frac{r}{2}\right)p(r) + \sum_{r=z+2}^{\infty}\frac{1}{2r}\left[C_1.(z+1)^2 + C_2(r-z-1)^2\right]p(r)$$

$$-C_1 \sum_{r=0}^{z}\left(z - \frac{r}{2}\right)p(r) - \sum_{r=z+1}^{\infty}\frac{1}{2r}\left[C_2.z^2 + C_2(r-z)^2\right]p(r)$$

$$= C_1 \sum_{r=0}^{z}\left[\left(z+1 - \frac{r}{2}\right) - \left(z - \frac{r}{2}\right)\right]p(r) + C_1\left[z+1 - \frac{z+1}{2}\right]p(z+1)$$

$$+C_1 \sum_{r=z+1}^{\infty}\frac{1}{2r}\left[(z+1)^2 - z^2\right]p(r) - C_1\frac{1}{2(z+1)}(z+1)^2 p(z+1)$$

$$+C_2 \sum_{r=z+1}^{\infty}[(r-z-1)^2(r-z)^2]\,p(r)$$

$$= C_1 \sum_{r=0}^{z}p(r) + \frac{C_1}{2}(2z+1)\sum_{r=z+1}^{\infty}\frac{1}{r}p(r) + C_2 \sum_{r=z+1}^{\infty}\frac{1}{2r}(1 - 2r + 2z)\,p(r)$$

$$= C_1 \sum_{r=0}^{z}p(r) + C_1\left(z + \frac{1}{2}\right)\sum_{r=z+1}^{\infty}\frac{1}{r}p(r) + C_2\left(z + \frac{1}{2}\right)\sum_{r=z+1}^{\infty}\frac{1}{r}p(r) - C_2 \sum_{r=z+1}^{\infty}p(r)$$

$$= C_1 \sum_{r=0}^{z}p(r) + (C_1 + C_2)\left(z + \frac{1}{2}\right)\sum_{r=z+1}^{\infty}\frac{1}{r}p(r) - C_2\left[\sum_{r=0}^{\infty}p(r) - \sum_{r=0}^{z}p(r)\right]$$

$$= (C_1 + C_2)\left[\sum_{r=0}^{z}p(r) + \left(z + \frac{1}{2}\right)\sum_{r=z+1}^{\infty}\frac{1}{r}p(r)\right] - C_2 \quad [\because \Sigma p(r) = 1] \quad \ldots(3)$$

In a similar way, we can find

$$\Delta C(z-1) = (C_1 + C_2)\left[\sum_{r=0}^{z-1}p(r) + \left(z - \frac{1}{2}\right)\sum_{r=z}^{\infty}\frac{1}{r}p(r)\right] - C_2 \qquad \ldots(4)$$

Then, for the minimum value of z, (from (2), (3) and (4)), we have

$$(C_1 + C_2)\left[\sum_{r=0}^{z-1}p(r) + \left(z - \frac{1}{2}\right)\sum_{r=z}^{\infty}\frac{1}{r}p(r)\right] - C_2 < 0$$

$$< (C_1 + C_2)\left[\sum_{r=0}^{z}p(r) + \left(z + \frac{1}{2}\right)\sum_{r=z+1}^{\infty}\frac{1}{r}p(r)\right] - C_2$$

$$\Rightarrow \sum_{r=0}^{z-1}p(r) + \left(z - \frac{1}{2}\right)\sum_{r=z}^{\infty}\frac{1}{r}p(r) < \frac{C_2}{C_1 + C_2} < \sum_{r=0}^{z}p(r) + \left(z + \frac{1}{2}\right)\sum_{r=z+1}^{\infty}\frac{1}{r}p(r) \qquad \ldots(5)$$

which gives the required optimum value.

(b) When r has continuous values

Let r be capable of being considerable as a continuous variable with probability distribution function $f(r)$. Then proceed same as case (a) and replacing $p(r)$ by $f(r)dr$ and Σ by $\int$, we get

$$C(z) = C_1 \int_{r=0}^{z} \left(z - \frac{r}{2} \right) f(r) dr + \int_{r=z}^{\infty} \frac{1}{2r} [C_1 z^2 + C_2 (r-z)^2] f(r) dr \qquad \ldots(6)$$

Again using the principle of maxima and minima for the minimum of $C(z)$, we must have

$$\frac{dC}{dz} = 0$$

$$\Rightarrow \quad C_1 \int_0^z \frac{\partial}{\partial z} \left\{ \left(z - \frac{r}{2} \right) f(r) \right\} dr + C_1 \left[\left(z - \frac{r}{2} \right) f(r) . \frac{dr}{dz} \right]_{r=0}^{z}$$

$$+ \int_{r=z}^{\infty} \frac{1}{2r} \frac{\partial}{\partial z} \left[C_1 . z^2 + C_2 (r-z)^2 \right] f(r) dr$$

$$+ \left[\frac{1}{2r} \{ C_1 . z^2 + c_2 (r-z)^2 \} f(r) \frac{dr}{dz} \right]_{r=z}^{\infty} = 0$$

$$\Rightarrow \quad C_1 \int_0^z f(r) dr + (C_1 + C_2) \int_{r=z}^{\infty} \frac{z}{r} f(r) dr - C_2 \int_z^{\infty} f(r) dr = 0$$

$$\Rightarrow \quad C_1 \int_0^z f(r) dr + (C_1 + C_2) \int_z^{\infty} \frac{z}{r} f(r) dr - C_2 [\int_0^{\infty} f(r) dr - \int_0^z f(r) dr] = 0$$

$$\Rightarrow \quad (C_1 + C_2) \int_0^z f(r) dr + (C_1 + C_2) \int_z^{\infty} \frac{z}{r} f(r) dr - C_2 = 0$$

$$\Rightarrow \quad \int_0^z f(r) dr + \int_z^{\infty} \frac{z}{r} f(r) dr = \frac{C_2}{C_1 + C_2} \qquad \ldots(7)$$

And

$$\frac{d^2 C}{dz^2} = \frac{d}{dz} \left(\frac{d}{dz} C(z) \right)$$

$$= (C_1 + C_2) \int_0^z \left\{ \frac{\partial}{\partial z} f(r) \right\} . dr + (C_1 + C_2) \left[f(r) \frac{dr}{dz} \right]_{r=0}^{z}$$

$$= (C_1 + C_2) \left\{ \int_z^{\infty} \frac{\partial}{\partial z} \left[\frac{z}{r} f(r) \right] dr + \left[\frac{z}{r} f(r) \frac{dr}{dz} \right]_{r=z}^{\infty} \right\}$$

$$= (C_1 + C_2) f(z) + (C_1 + C_2) \int_z^{\infty} \frac{1}{r} f(r) dr - f(z) . (C_1 + C_2)$$

$$= (C_1 + C_2) \int_z^{\infty} \frac{1}{r} f(r) dr > 0$$

Hence, by the principle of maxima and minima, we conclude that $C(z)$ given by (6) is minimum for the value given by (7).

18.16 MODEL-3 : THE GENERAL SINGLE PERIOD MODEL OF PROFIT MAXIMIZATION WITH TIME INDEPENDENT COST ASSUMPTIONS

Assumptions

(1) t is the constant interval between orders.

(2) At the beginning of the period of time, t, z is the stock.

(3) lead time is zero.

(4) a is the cost of an item that is produced and b is the selling price of an item during the period t, $(b > a)$

(5) c is the selling price of an item after the end of the period, t at the beginning of which items were produced ($c < a$)

(6) d is the cost per item, in case of shortage.

Formulation and solution of the model

(a) Discrete case

Consider the following cases

Case-1 : If $r > z$ i.e., the demand exceed the stock, then the profit

$$= z(b-a) - d(r-z) = z(b-a+d) - d.r$$

Case-2 : If $r \leq z$ i.e., the demand does not exceed the stock, then the profit

$$= rb + (z - r)c - za = z(c-a) + (b-c)r$$

Thus, the net expected profit is

$$P(z) = \sum_{\substack{r=z+1 \\ (r>z)}}^{\infty} \{z(b-a+d) - dr\}p(r) + \sum_{\substack{r=0 \\ (r\leq z)}}^{z} \{z(c-a) + (b-c)r\}.p(r)$$

$$= \sum_{r=0}^{\infty} \{z(b-a+d) - dr\}p(r) - \sum_{r=0}^{z} \{z(b-a+d) - dr\}.p(r)$$

$$+ \sum_{r=0}^{z} \{z(c-a) + (b-c)r\}.p(r)\}$$

$$= (b-a+d)z \sum_{r=0}^{\infty} p(r) - d \sum_{r=0}^{\infty} r\,p(r) - (b-c+d) \sum_{r=0}^{z} (z-r)p(r)$$

$$= (b-a+d)z - d \sum_{r=0}^{\infty} r.p(r) - (b-c+d) \sum_{r=0}^{z-1} (z-r)p(r) \qquad ...(1)$$

(By using $\Sigma p(r) = 1$, $\sum_{r=0}^{z} (z-r)p(r) = \sum_{r=0}^{z-1} (z-r)p(r)$)

Now, for the maxima of $P(z)$, we have

$$\Delta P(z-1) > 0 > \Delta P(z) \qquad ...(2)$$

We know that

$\Delta P(z) = P(z+1) - P(z)$

$$= [(b-a+d)(z+1) - d \sum_{r=0}^{\infty} rp(r) - (b-c+d) \sum_{r=0}^{z} (z+1-r)p(r)]$$

$$- [(b-a+d)z - d \sum_{r=0}^{\infty} rp(r) - (b-c+d) \sum_{r=0}^{z-1} (z-r)p(r)]$$

$$= (b-a+d) - (b-c+d) \sum_{r=0}^{z} p(r) \qquad \left[\because \sum_{r=0}^{z-1} (z-r)p(r) = \sum_{r=0}^{z} (z-r)p(r)\right] \quad ...(3)$$

Similarly, we can obtain

$$\Delta P(z-1) = (b-a+d) - (b-c+d) \sum_{r=0}^{z-1} p(r) \qquad ..(4)$$

Using (3) and (4) in (2), we get

$$(b-a+d) - (b-c+d) \sum_{r=0}^{z-1} p(r) > 0 > (b-a+d) - (b-c+d) \sum_{r=0}^{z} p(r)$$

$$\Rightarrow \qquad \sum_{r=0}^{z-1} p(r) < \frac{b-a+d}{b-c+d} < \sum_{r=0}^{z} p(r) \qquad ...(5)$$

(b) If r and z are not discrete but have continuous values

Proceed same as in part (a), we get the total net expected profit for model which is given by

$$P(z) = (b - a + d)z - d\int_{r=0}^{\infty} rf(r)dr - (b - c + d)\int_{r=0}^{z}(z - r)f(r)dr \qquad ...(6)$$

Now, by the principle of maxima and minima, for the maxima of P(z), we must have

$$\frac{dP}{dz} = 0$$

$$\Rightarrow \quad (b - a + d) - d\left[\int_{r=0}^{\infty}\frac{\partial}{\partial z}\{rf(r)\}dr + \left\{rf(r)\frac{dr}{dz}\right\}_{r=0}^{\infty}\right]$$

$$- (b - c + d)\left[\int_{r=0}^{z}\frac{\partial}{\partial z}\{(z - r)f(r)\}dr + \left\{(z - r)f(r)\frac{dr}{dz}\right\}_{r=0}^{z}\right]$$

$$\Rightarrow \quad (b - a + d) - (b - c + d)\int_{0}^{z} f(r)dr = 0 \qquad ...(7)$$

$$\Rightarrow \qquad\qquad \int_{0}^{z} f(r)dr = \frac{b - a + d}{b - c + d} \qquad ...(8)$$

Also,

$$\frac{d^2p}{dz^2} = -(b - c + d)\left[\int_{0}^{z}\frac{\partial}{\partial z}f(r)dr + \left\{f(r)\frac{dr}{dz}\right\}_{0}^{z}\right]$$

$$= -(b - c + d)f(z) < 0 \qquad\qquad (\because b > c)$$

Hence, we conclude that $P(z)$ given by (6) is maximum for the value of z which satisfies (8).

☛ Remark

- In the above model, it is assumed that after one period the remaining items are either of no use or they can be sold at some less price than the cost at which they are produced.

18.17 MODEL-4: PROBABILISTIC ORDER-LEVEL SYSTEM WITH CONSTANT LEAD TIME

(a) Discrete Case: We have to find the optimum order level z so as to minimize the total expected cost under the following assumptions.

Assumptions.

1. t is the constant interval between orders.
2. z is the stock for time t.
3. x is the demand during the period t with probability $F(x)$.
4. y is the demand during the lead time L with probability $G(y)$.
5. C_1 is the holding cost per unit per unit time.
6. C_2 is the shortage cost per unit per unit time.

Formulation and Solution of the Model

Clearly, there are following three possible cases depending on the relative values of x, y and z.

Case 1. If $0 \leq y \leq z$, $0 < x \leq z - y$

In this case, the period t starts with the stock $z - y$ and end with the remaining stock $\{(z - y) - x\}$ so that the total cost is given by

C_1(inventory area of $ABCD$)

$$= C_1 \cdot \frac{1}{2}[(z-y)+(z-y-x)]t$$

$$= C_1\left(z-y-\frac{x}{2}\right)t$$

So, the total expected cost in this case is given by

$$\sum_{y=0}^{z}\sum_{x=0}^{z-y} C_1 \cdot t\left(z-y-\frac{x}{2}\right)F(x)\cdot G(y)$$

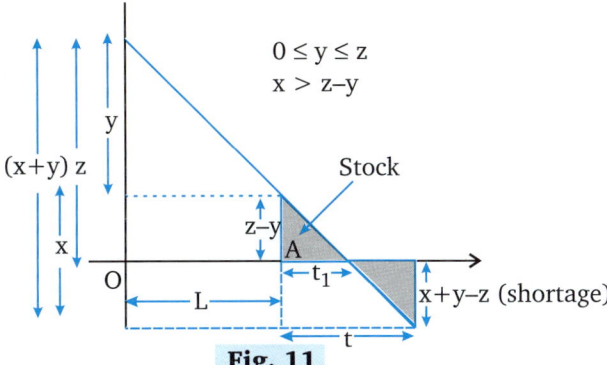

Fig. 10

Case 2. If $0 \leq y \leq z$, $x > z - y$

In this case, the period starts with a stock $(z-y)$ and end with the shortage $(x+y-z)$. Then the area representing by the inventory can be obtained as follows:

The stock $z - y$ is supplied in time t_1 with rate $\frac{x}{t}$ for remaining time $t - t_1$ there is a shortage and which increases to $x + y - z$ by the end of the time

$$\therefore \qquad z - y = \frac{x}{t} \cdot t_1$$

$$\Rightarrow \qquad t_1 = \frac{(z-y)}{x} \cdot t \text{ and } t - t_1 = \frac{x+y-z}{x} \cdot t$$

Now holding cost in this case is C_1 (Area representing stock) $= C_1 \cdot \frac{1}{2}(z-y)t_1 = \frac{C_1(z-y)^2 t}{2x}$

and shortage cost is C_2 (Area representing shortage) $= C_2 \cdot \frac{1}{2}(x+y-z)(t-t_1) = \frac{(x+y-z)^2}{2x} \cdot t$

$$\therefore \qquad \text{total expected cost} = \sum_{y=0}^{z}\sum_{x=z-y+1}^{\infty} \frac{t}{2x}[C_1(z-y)^2 + C_2(x+y-z)^2]F(x)G(y)$$

Case 3. If $y > z$, $x > 0$

Here, the period t starts with a shortage $(y-z)$ and ends with a shortage $(x+y-z)$ so that the shortage cost

$$= C_2 \times \text{(shortage area } ABCD\text{)}$$

$$= C_2 \times \frac{1}{2}[y-z+(x+y-z)] \cdot t$$

$$= C_2\left(\frac{x}{2}+y-z\right)t$$

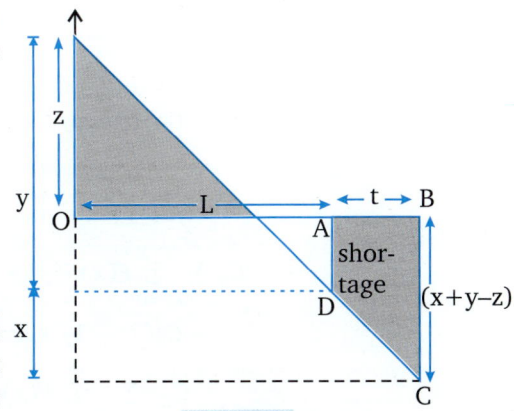

The total expected cost $C(z)$ per unit time, excluding C_3 is obtained by multiplying the cost associated with each situation by the joint probability of demands for y items during the time t and summing over the range of x and y.

Fig. 12

Therefore, total expected cost in this case is

$$= \sum_{y=z+1}^{\infty} \sum_{x=0}^{\infty} t \cdot C_2\left(\frac{x}{2} + y - z\right) F(x)G(y)$$

Hence, the total expected cost for this model is given by

$$C(z) = \sum_{y=0}^{z} \sum_{x=0}^{z-y} C_1\left(z - y - \frac{x}{2}\right) F(x)G(y)$$

$$+ \sum_{y=0}^{z} \sum_{x=z-y+1}^{\infty} \frac{1}{2x}[C_1(z-y)^2 + C_2(x+y-z)^2] \cdot F(x)G(y)$$

$$+ \sum_{y=z+1}^{\infty} \sum_{x=0}^{\infty} C_2\left(\frac{x}{2} + y - z\right) F(x)G(y) \qquad ...(1)$$

For the minima of $C(z)$, we must have

$$\Delta C(z) > 0 > \Delta C(z-1) \qquad ...(2)$$

Now, $\quad \Delta C(z) = C(z+1) - C(z)$

$$= \sum_{y=0}^{z+1} \sum_{x=0}^{z+1-y} C_1\left(z + 1 - y - \frac{x}{2}\right) F(x)G(y)$$

$$+ \sum_{y=0}^{z+1} \sum_{x=z-y+2}^{\infty} \frac{1}{2x}[C_1(z+1-y)^2 + C_2(x+y-z-1)^2]F(x)G(y)$$

$$+ \sum_{y=z+2}^{\infty} \sum_{x=0}^{\infty} C_2\left(\frac{x}{2} + y - z - 1\right) F(x)G(y) - \sum_{y=0}^{z} \sum_{x=0}^{z-y} C_1\left(z - y - \frac{x}{2}\right) F(x)G(y)$$

$$- \sum_{y=0}^{z} \sum_{x=z-y+1}^{\infty} \frac{1}{2x}[C_1(z-y)^2 + C_2(x+y-z)^2]F(x)G(y)$$

$$- \sum_{y=z+1}^{\infty} \sum_{x=0}^{\infty} C_2\left(\frac{x}{2} + y - z\right) F(x)G(y)$$

$$= \sum_{y=0}^{z} \sum_{x=0}^{z-y} C_1\left\{\left(z + 1 - y - \frac{x}{2}\right) - \left(z - y - \frac{x}{2}\right)\right\} F(x)G(y)$$

$$+ \sum_{y=0}^{z} \sum_{x=z-y+1}^{\infty} \frac{1}{2x}[C_1\{(z+1-y)^2 - (z-y)^2\}]$$

$$+ [C_2\{(x+y-z-1)^2 - (x+y-z)^2\}]F(x)G(y)$$

$$+ \sum_{y=z+1}^{\infty} \sum_{x=0}^{\infty} C_2\left\{\left(\frac{x}{2} + y - z - 1\right) - \left(\frac{x}{2} + y - z\right)\right\} F(x)G(y)$$

$$= C_1 \sum_{y=0}^{z} \sum_{x=0}^{z-y} F(x)G(y) + \sum_{y=0}^{z} \sum_{x=z-y+1}^{\infty} \frac{1}{2x}[C_1(2z - 2y) + 1]$$

$$+ C_2(2z - 2y + 2x + 1)F(x)G(y)$$

$$- C_2 \sum_{y=z+1}^{\infty} \sum_{x=0}^{\infty} F(x)G(y) \qquad ...(3)$$

Now we have

$$C_2 \sum_{y=z+1}^{\infty} \sum_{x=0}^{\infty} F(x)G(y) = C_2 \sum_{y=0}^{\infty} \sum_{x=0}^{\infty} F(x)G(y) - C_2 \sum_{y=0}^{z} \sum_{x=0}^{\infty} F(x)G(y)$$

$$= C_2 - C_2 \left[\sum_{y=0}^{z} \sum_{x=0}^{z-y} F(x)G(y) + \sum_{y=0}^{z} \sum_{x=z-y+1}^{\infty} F(x)G(y) \right]$$

(By using $\Sigma F(x) = \Sigma G(y) = 1$)

...(4)

Using (4) in (3) we get

$$\Delta C(z) = (C_1 + C_2) \sum_{y=0}^{z} \sum_{x=0}^{z-y} F(x)G(y) + \sum_{y=0}^{z} \sum_{x=z-y+1}^{\infty} \frac{1}{2x}[C_1(2z - 2y + 1)]$$

$$+ C_2(2z - 2x - 2y + 1 + 2x)] \cdot F(x)G(y) - C_2$$

$$\Rightarrow \qquad \Delta C(z) = (C_1 + C_2) \sum_{y=0}^{z} G(y) \left[\sum_{x=0}^{z-y} F(x) + \left(z - y + \frac{1}{2} \right) \right] \cdot \sum_{x=z-y+1}^{\infty} \frac{1}{x} F(x) - C_2$$

...(5)

Similarly, we may find that

$$\Delta C(z-1) = (C_1 + C_2) \sum_{y=0}^{z-1} G(y) \left[\sum_{x=0}^{z-y-1} F(x) + \left(z - y - \frac{1}{2} \right) \sum_{x=z-y}^{\infty} \frac{1}{x} F(x) \right] - C_2 \qquad ...(6)$$

Finally using (5) and (6) in (2), for minimum value of $C(z)$, we have

$$\sum_{y=0}^{z-1} G(y) \left[\sum_{x=0}^{z-y-1} F(x) + \left(z - y - \frac{1}{2} \right) \sum_{x=z-y}^{\infty} \frac{1}{x} F(x) \right] < \frac{C_2}{C_1 + C_2}$$

$$< \sum_{y=0}^{z} G(y) \left[\sum_{x=0}^{z-y} F(x) + \left(z - y + \frac{1}{2} \right) \sum_{x=z-y+1}^{\infty} \frac{1}{x} F(x) \right] \qquad ...(7)$$

Hence, optimum value of z is that which satisfies (7).

(b) When x and y are continuous

The cost equation for this model is similar to that of the discrete case (case (a)). Only $F(x)$ and $G(y)$ are replaced by probability function $f(x)dx$ and $g(y)dy$ respectively. Summation (Σ) sign is replaced by integral ($\int$) sign.

Then required cost equation is given by

$$C(z) = \int_0^z g(y)dy \int_0^{z-y} C_1 \left(z - y - \frac{x}{2} \right) f(x)dx$$

$$+ \int_0^z g(y)dy \int_{z-y}^{\infty} \left[C_1 \frac{(z-y)^2}{2x} + C_2 \frac{(x-z+y)^2}{2x} \right] f(x)dx$$

$$+ \int_z^{\infty} g(y)dy \int_0^{\infty} C_2 \left(y - z + \frac{x}{2} \right) f(x)dx$$

$$\Rightarrow \qquad \frac{dC}{dz} = (C_1 + C_2) \int_0^z g(y)dy \left[\int_0^{z-y} f(x)dx + (z-y)\int_{z-y}^{\infty} \frac{f(x)}{x}dx \right] - C_2 = 0$$

and
$$\frac{d^2 C}{dz^2} = (C_1 + C_2)\int_0^z g(y)dy \int_{z-y}^{\infty} \frac{f(x)}{x}dx > 0$$

Hence, by priniciple of maxima and minima, we conclude that the optimum order level is determined by

$$\int_0^z g(y)dy \left[\int_0^{z-y} f(x)dx + (z-y)\int_{z-y}^{\infty} \frac{f(x)}{x}dx \right] = \frac{C_2}{C_1 + C_2}$$

 Solved Examples

EXAMPLE 1. *If the demand for a certain product has a rectangular distribution between 4000 and 5000. Find the optimal order quantity if storage cost is ₹1.00 per unit and shortage cost is ₹7.00 per unit.*

SOLUTION.   We have
$$C_1 = 1, C_2 = 7$$
and for rectangular distribution
$$f(r) = \frac{1}{5000 - 4000} = \frac{1}{1000}$$
Now, the optimum order quantity is given by
$$\int_{4000}^z f(r)dr = \frac{C_2}{C_1 + C_2}$$
$$\Rightarrow \quad \frac{1}{1000}(z - 4000) = \frac{7}{1+7}$$
On solving we get $z = 4875$

EXAMPLE 2. *(Newspaper problem) A newspaper boy buys papers for 3 paise each and sells them for 7 paise each. He can not return unsold newspapers. Daily demand has the following distribution.*

No. of customers	23	24	25	26	27	28	29	30	31	32
Probability $p(r)$	0.01	0.03	0.06	0.10	0.20	0.25	0.15	0.10	0.05	0.05

If each day's demand is independent of the previous day, how many papers should he ordered each day.

[MEERUT–2005, 09, 15; GORAKHPUR–2017; GARHWAL–2012]

SOLUTION.   Here, we have
$$a = 0.03, b = 0.07, c = 0, d = 0$$
If z is the number of papers that should be ordered daily for the maximum profit, then it will satisfy

$$\sum_{r=0}^{z-1} p(r) < \frac{b-a+d}{b-c+d} < \sum_{r=0}^{z} p(r) \qquad \ldots(1)$$

Here, we have
$$\frac{b-a+d}{b-c+d} = \frac{0.07 - 0.03 + 0}{0.07 - 0 + 0} = \frac{4}{7} = 0.57$$

Now, we have the following table

z	$\sum\limits_{r=23}^{z-1} p(r)$	$\sum\limits_{r=23}^{z} p(r)$
23		0.01
24	0.01	0.04
25	0.04	0.10
26	0.10	0.20
27	0.20	0.40
28	0.40	0.65
29	0.65	0.80

From the above table, it is observed that (1) is satisfied for $z = 28$.

Hence, the boy should order 28 newspapers daily for maximum profit.

EXAMPLE 3. *A baking company sells cake by the pound. It makes a profit of 50 paise/pound on every pound sold on the day it is baked. If dispose of all cakes not sold on the day it is baked, at a loss of 12 paise/ pound. If demand is known to be rectangular between 2000 and 3000 pounds, find the optimal daily amount baked.*

[MEERUT–2004, 13; DELHI–2010; PATNA–2013]

SOLUTION. We have $b - a = ₹0.50$, $a - c = 0.12$, $d = 0$

$$f(r) = \frac{1}{3000 - 2000} = \frac{1}{1000} \text{ (for rectangular distribution)}$$

The optimal daily amount (z) baked is given by

$$\int_{2000}^{z} f(r)dr = \frac{b - a + d}{b - c + d}$$

$$\Rightarrow \frac{1}{1000}\int_{2000}^{z} dr = \frac{b - a}{(b - a) + (a - c)}$$

$$\Rightarrow \frac{z - 2000}{1000} = \frac{0.50}{0.50 + 0.12}$$

On solving for z, we get

$$z = 2806 \text{ appr.}$$

EXAMPLE 4. *A contractor of second hand motor trucks use to maintain a stock of trucks every month. The demand of the trucks occurs at a relatively constant rate but not in a constant size. The demand follow the following distribution.*

Demand(r)	0	1	2	3	4	5	6 or more
Prob. $p(r)$	0.40	0.24	0.20	0.10	0.05	0.01	0

The holding cost of an old truck in stock for one month is ₹100 and the penalty for a truck if not supplied on the demand is ₹1000. Find the optimal size of the stock for the contractor.

[MEERUT–1994, 2007; KANPUR–2013; ALLAHABAD–2009; RAJASTHAN–2007]

SOLUTION. Let z be the stock then optimal size is given by

$$\sum_{r=0}^{z-1} p(r) + \left(z - \frac{1}{2}\right)\sum_{r=z}^{\infty}\frac{1}{r}p(r) < \frac{C_2}{C_1 + C_2} < \sum_{r=0}^{z} p(r) + \left(z + \frac{1}{2}\right)\sum_{r=z+1}^{\infty}\frac{1}{r}p(r) \qquad \dots(1)$$

It is given that $C_1 = 100$, $C_2 = 1000$

$$\Rightarrow \quad \frac{C_2}{C_1 + C_2} = \frac{1000}{100 + 1000} = \frac{10}{11} = 0.909$$

Let us write $\sum_{r=0}^{z} p(r) + \left(z + \frac{1}{2}\right) \sum_{r=z+1}^{\infty} \frac{1}{r} p(r) = L(z)$

Then from (1), the optimal value of z is given by

$$L(z-1) < 0.909 < L(z) \qquad \qquad \qquad \ldots(2)$$

Now, we have the following calculation table

r	z	$z + \dfrac{1}{2}$	$p(r)$	$\displaystyle\sum_{r=0}^{z} p(r)$	$\displaystyle\sum_{r=z+1}^{\infty} \frac{1}{r} p(r)$	$\left(z + \dfrac{1}{2}\right) \displaystyle\sum_{r=z+1}^{\infty} \frac{1}{r} p(r)$	$L(z)$
(1)	(2)	(3)	(4)	(5)	(6)	(7) = (3) × (6)	(8) = (5) + (7)
0	0	0.5	0.40	0.40	0.3875	0.19375	0.59375
1	1	1.5	0.24	0.64	0.1475	0.22125	0.86125
2	2	2.5	0.20	0.84	0.0475	0.11875	0.95874

From the above table it is clear that $L(1) < 0.909 < L(2)$ value for $z = 2$, (2) is satisfied.

Hence, optimal stock level is $z = 2$.

EXAMPLE 5. *A shop owner places orders daily for goods which will be delivered 7 days later. On a certain day, the owner has 10 items in stock. Futhermore on the 6 previous days he has already placed orders for the delivery of 2, 4, 1, 10, 11 and 5 items in that order, over each of next 6 days. The inventory cost per unit, per unit time is 15% and the shortage cost per unit is 95%. The distribution of the requirement R over a 7 days period is given by*

$$f(R) = 0.02 - 0.0002R$$

Find the number of items that should be ordered on the 7th day.

[MEERUT–2006]

SOLUTION. Let z be the total no. of items required in 7 days for maximum profit.

Then $\quad \int_0^z f(R)dR = \dfrac{C_2}{C_1 + C_2}$ $\qquad \qquad \ldots(1)$

It is given that $C_1 = 15\%$ and $C_2 = 95\%$

Then from (1), we have

$$\int_0^z (0.02 - 0.0002R)dR = \frac{95}{15 + 95}$$

$$\Rightarrow \quad \left[0.02R - 0.0002\frac{R^2}{2}\right]_0^z = 0.8636$$

$$\Rightarrow \quad z^2 - 200z + 8636 = 0$$

On solving for z, we get $\quad z \approx 63$

Also, total of the orders of the six days and the previous stock

$$= [2 + 4 + 1 + 10 + 11 + 5] + 10 = 43$$

Hence, the required no. of items that should be ordered on the 7th day

$$= 63 - 43 = 20$$

EXAMPLE 6. *An ice-cream company sells one of its types of ice-cream by weight. If the product is not sold on the day it prepared, it can be sold at a loss of 50 paise per pound. But there is an unlimited market for one day old ice-cream. On the other hand, the company makes a profit of ₹3.20 on every pound of ice-cream sold on the day it is prepared. Past daily orders form a distribution with*

$$f(r) = 0.02 - 0.0002r, \ 0 \leq r \leq 100$$

How many pounds of ice-cream should the company prepare every day?

SOLUTION. Let z be the no. of ice-cream to be prepared daily for maximum profit, is given by

$$\int_0^z f(r)dr = \frac{b-a+d}{b-c+d} \qquad \qquad \ldots(1)$$

Here, $b - a = 3.20$, $a - c = 0.50$, $d = 0$, $f(r) = 0.02 - 0.0002r$
Using all these values in (1) we get

$$\int_0^z (0.02 - 0.0002r)dr = \frac{3.20}{(b-a)+(a-c)}$$

$\Rightarrow \qquad 0.02z - 0.0002z^2 = \dfrac{3.20}{3.20 + 0.50}$

$\Rightarrow \qquad 37z^2 - 7400z + 320000 = 0$

On solving for z, we get $\qquad z = 63.3$ pounds

EXAMPLE 7. *The probability distribution of monthly sales of a certain item is as follows*

Monthly sales	0	1	2	3	4	5	6
Probability	0.02	0.05	0.30	0.27	0.20	0.10	0.06

The cost of carrying inventory is ₹10 per unit per month. The current policy is to maintain a stock of four items at the beginning of each month. Assuming that the cost of shortage is proportional to both time and quantity short, obtain the inputed cost of a shortage of one item for one time unit. [MEERUT–1998, 2003]

SOLUTION. Here, we have $z = 4$, $C_1 = 10$
and

$p(0)$	$p(1)$	$p(2)$	$p(3)$	$p(4)$	$p(5)$	$p(6)$
0.02	0.05	0.30	0.27	0.20	0.10	0.06

The range of monthly sales x is given from 0 to 6.
We have to find the value of C_2. ($\because$ The problem is given in discrete case, the answer will consist of a range of values for the cost)
We know that

$$\sum_{x=0}^{z-1} p(x) + \left(z - \frac{1}{2}\right) \sum_{x=z}^{\infty} \frac{p(x)}{x} < \frac{C_2}{C_1 + C_2} < \sum_{x=0}^{z} p(x) + \left(z + \frac{1}{2}\right) \sum_{x=z+1}^{\infty} \frac{p(x)}{x}$$

The least value of C_2 can be obtained by letting

$$\frac{C_2}{C_1 + C_2} = \sum_{x=0}^{z-1} p(x) + \left(z - \frac{1}{2}\right) \sum_{x=z}^{\infty} \frac{p(x)}{x}$$

Putting all the given values in the above equation, we get

$$\frac{C_2}{10+C_2} = \sum_{x=0}^{3} p(x) + \left(4 - \frac{1}{2}\right) \sum_{x=4}^{6} \frac{p(x)}{x}$$

$$= [p(0) + p(1) + p(2) + p(3)] + \frac{7}{2}\left[\frac{p(4)}{4} + \frac{p(5)}{5} + \frac{p(6)}{6}\right]$$

$$= (0.02 + 0.05 + 0.30 + 0.27) + \frac{7}{2}\left(\frac{0.20}{4} + \frac{0.10}{5} + \frac{0.06}{6}\right)$$

$$= 0.92$$

$$\Rightarrow \qquad C_2 = \frac{9.2}{0.8} = 115, \text{ which is the least value of } C_2.$$

Similarly, the greatest value of C_2 can be obtained by

$$\frac{C_2}{10+C_2} = \sum_{x=0}^{4} p(x) + \left(4 + \frac{1}{2}\right) \sum_{x=5}^{6} \frac{p(x)}{x}$$

$$= 0.84 + \frac{9}{2} \times 0.03 = 0.975$$

$$\Rightarrow \quad \text{greatest value of } C_2 = \frac{9.75}{0.025} = 390$$

Hence, the required range of C_2 is given by $115 < C_2 < 390$.

18.18 ADVANTAGES OF INVENTORY CONTROL　　　　　　[MEERUT–2002, 03, 14, 16, 18]

Following are some advantages of inventory control.
1. It ensures a smooth running of the organisation.
2. It ensures an adequate supply of items to customers and avoid the shortages as far as possible at the minimum cost.
3. The risk of the loss due to price changes is reduced.
4. It eliminates the possibility of duplicate ordering.
5. It helps to minimize the loss.
6. It makes use of available capital in a most effective way.
7. If helps to avoid extra expenditure.
8. It utilizes the benefit of price fluctuations.

Exercise-18.2

1. A newspaper boy buys papers for ₹2.40 and sells them for ₹3.00 each. He can not return unsold newspapers. Daily demand has the following distribution

No. of customers (r)	Prob. p(r)	No. of customers (r)	Prob. p(r)
23	0.01	28	0.25
24	0.03	29	0.15
25	0.06	30	0.10
26	0.10	31	0.05
27	0.20	32	0.05

If each day's demand is independent of previous days's demand, how many papers should he order each day.　[MEERUT–2009]

2. The probability distribution of monthly sales of a certain item is as follows:

Monthly sales (r)	0	1	2	3	4	5	6
Prob. p(r)	0.01	0.06	0.25	0.35	0.20	0.03	0.10

The cost of carrying inventory is ₹30 per unit per month and cost of unit shortage is ₹70 per month. Find the optimum stock level which minimize the total expected cost.

3. A baking company sells one of its types of cakes by weight. If the product is not sold on the day it is baked, it can only be sold at a loss of 20 paise/kg. But there is unlimited market for one day old product. The cost of holding 1 kg of cake for one day is 15 paise, while the company makes a profit of one rupee/kg of cake sold on the day it is baked. If the p.d.f. of demand r is given by

$$f(r) = 0.02 - 0.0002r$$

Find how many kgs of cake company should bake daily. [MEERUT–2007]

4. A baking company sells cake by the kg. It makes a profit of ₹5.00 a kg on every kg sold on the day it is baked. It disposes off all cake not sold on the day it is baked at the loss of ₹1.20 a kg. If demand is known to be rectangular between 2000 and 3000 kg, determine the optimal daily amount baked.

[MEERUT–2008; INDORE–2011]

5. A baking company sells one of its type of cake by weight. It makes a profit of ₹2 per kg on every kg of cake sold on the day it is baked. It disposes of all cakes not sold on the day they are baked at a loss of ₹0.50. If demand is known to have p.d.f., $f(r) = 0.03 - 0.0003r$. Find the optimum amount of cake the company should bake daily. [MEERUT–2009(BP)]

Answers

1. 27   2. 4   3. 61.5   4. 2806   5. 33.33

REVIEW QUESTIONS

1. Formulate and solve a discrete stochastic model for a single product with lead time zero. The shortage and storage costs are independent of time. Set-up cost is contant.

2. Explain and solve an inventory model with instantaneous discrete random items and with no set-up cost.

3. Write a short note on Newspaper boy's problem.

4. Explain and solve that general single product model of profit maximization with time independent cost.

5. Explain with examples the probabilistic model in inventory.

6. Formulate and solve continuous probabilistic reorder point lot size model to determine optimum reorder point for a presented lot size, lead time is finite. Shortage are allowed and fully backlogged.

7. Find the optimum quantity z in a continuous simple stochastic model for a time dependent case. Shortage are allowed and backlogged fully. Set-up cost per period is constant.

8. Find the optimum value of the order quantity for a simple stochastic model when shortage are allowed and backlogged inventory carrying cost and shortage costs are dependent on units and time both. Procurement lead time is a random variable.

9. Explain the advantage and disadvantages of fixed ordered quantity system.

10. Write the objective of inventory control.

MULTIPLE CHOICE QUESTIONS (CHOOSE THE MOST APPROPRIATE ONE)

1. In the production lot size model, increasing the rate of production:
   (a) increase the optimal number of orders to place each year
   (b) does not influence the optimal number of orders
   (c) both (a) and (b) are true
   (d) None of these

2. In the EOQ model:
   (a) all demands must be satisfied
   (b) order arrive in a batch
   (c) both (a) and (b) are true
   (d) None of these

3. In the EOQ model with backlogging the optimal number of orders to backlog is:
   (a) not dependent on
   (b) directly proportional to the square root of
   (c) directly proportional to
   (d) None of these

4. The optimal number of orders per year increase when:
   (a) price increase
   (b) carry out decrease
   (c) both (a) and (b) are true
   (d) None of these

5. The stock of material kept in the stores in

anticipation of future demand is known as:

(a) inventory (b) raw material

(c) stock of material (d) None of these

6. Working class of human beings is a class of inventor is called:

(a) live stock

(b) human resource inventory

(c) human inventory

(d) None of these

7. The rent for the stores where materials are stored falls under:

(a) inventory carrying cost

(b) ordering cost

(c) stocking cost

(d) None of these

8. As the order quantity increases the cost will

reduce:

(a) insurance cost (b) ordering cost

(c) stockout cost (d) None of these

9. Losses due to deterioration, theft come under:

(a) inventory carrying charges

(b) not any cost

(c) losses due to theft

(d) None of these

10. Which of the following increases with the quantity ordered per order:

(a) holding cost (b) shortage cost

(c) ordering cost (d) None of these

11. The rent for the stores where material are kept under:

(a) holding cost (b) set-up cost

(c) ordering cost (d) None of these

Answers

1. (a) 2. (c) 3. (b) 4. (d) 5. (a) 6. (b) 7. (a) 8. (b) 9. (a)
10. (a) 11. (a)

ARCHIVE

1. A shop produces three items in lots. The demand rate for each item is constant and can be assumed to be deterministic. No back orders are to be allowed. The pertinent data for the items is given in the following table.

Items	I	II	III
Carrying cost (₹ per unit per year)	20	20	20
Set-up cost (₹ per set-up)	50	40	60
Cost per unit	6	7	5
Yearly demand (units)	10000	12000	7500

Determine approximately the economic order quantities for three items subject to the condition that the total value of average inventory levels of these items does not exceed ₹1000. [KANPUR–1993, 2007; ROHILKHAND–1993, 2014; MEERUT–2007]

2. A shopkeeper has a uniform demand of 50 items per month. He buys from a supplier at a cost of ₹6 per item and the cost of ordering is ₹10 each time. If the stock holding costs are 20% per year of stock value, how frequently should he replenish his stock? Suppose the suppliers offers a 5% discount on orders between 200 and 999 items and a 10% discount on orders exceeding or equal to 1000, can the shopkeeper reduce his cost by taking

advantage of either of these discounts.

[NAGPUR–1990, 2013]

3. A firm has experienced the probability distribution for inventory demand during the record period as given in the following table:

No. of units	30	40	50	60	70
Probability	0.2	0.2	0.3	0.2	0.1

The firm is paying carrying cost per unit year at the rate of ₹60, while stockout cost in the form of last profit A etc., estimated to be ₹400 per unit. Find out the reserve stock level that would minimize the total annual expected cost. [DELHI(MBA)–2001]

4. A newspaper boy buys papers for ₹1.30 each and sells them for ₹1.40 each. He can not return unsold newspaper. Daily demand has the following distribution:

No. of customers	23	24	25	26	27
Probability	0.01	0.03	0.06	0.10	0.20
No. of customers	28	29	30	31	32
Probability	0.25	0.15	0.10	0.05	0.05

If each day's demand is independent of the previous day's, how many newspapers he should order each day? [MEERUT–2003; BHOPAL–2006; GORAKHPUR–1999; ICWA–1990]

Hints and Answers

1. ₹1000 2. 2000 units 3. 12 units 4. 28 paper per day

19 Queuing Theory

19.1 INTRODUCTION

A group of customers/items waiting at some place to receive attention/service including those receiving the service, is known as queue or waiting line.

Some service facility waits for arrival of customers when the total capacity of system is more than the number of customers. Thus, in the absence of a perfect balance between the service facility and the customers, waiting is required either by the service facility or by the customer. The imbalance between the customers and service facility, known as congestion, which cannot be eliminated completely but efforts/techniques can be evolved to reduce the magnitude of congestion or waiting time of a new arrival in the system.

The customers arriving for service may from one queue and be serviced through only one station, they may form one queue and be serviced through several stations or they may form several queues and be served through as many stations.

In this chapter we shall discuss a mathematical theory to study the problem and try to find expected waiting time for a particular arrival and the time it will take for servicing in a given situation by calculating probability of customer's entering the system and the probability of their servicing times.

19.2 QUEUING PROCESS

A queuing process is centered around a service system (facility). The queuing system has the following characteristics:

   (i) The input (arrival pattern)
   (ii) Queue (waiting line)
   (iii) The service discipline (queue discipline)
   (iv) The service mechanism (service pattern)

All queuing situations involve the arrival of customers (input) at a service facility, where some time may be spent in waiting and then receiving the desired service. The customers arriving for service may or may not enter the system. Thus, the input pattern in the system depends on the nature of the system as well as the behaviour of the customer. The combination of these two determines the arrival pattern. After the service is completed, the customer leaves the service system (output). The departure pattern mainly depends on the service discipline.

19.2.1 INPUT PROCESS

The input process implies the mode of arrival of customers at the service facility. The number of customers emanate from finite or infinite sources. The customers arrive at the system randomly, singly or in batches. The input process is characterised by the nature of the arrivals, capacity of the system and the behaviour of the customers.

☛ REMARKS
- The size of customers arriving for servicing depends on the nature of the population, which can be finite or infinite.
- The periods between the arrivals of individual customers may be constant or scattered in some pattern.
- In general, the arrivals pattern follow a 'Poisson' distribution when the total number of arrivals during any given time interval is independent of the number of arrivals that have already occurred prior to the beginning of time interval.

19.2.2 CUSTOMER BEHAVIOUR IN A QUEUE: SOME GENERAL DEFINITIONS

(i) **Balking:** A customer may not like to join the queue seeing it very long and he may not like to wait.

(ii) **Reneging:** He may leave the queue due to impatience after joining it.

(iii) **Collusion:** Several customers may collaborate and only one of them may stand in the queue.

(iv) **Jockeying:** If there are more than one number of queues, then one customer may leave one queue to join another.

19.2.3. SERVICE FACILITY/MECHANISM

This means the arrangement of server's facility to serve the arriving customer. Service time in waiting line problems is also a statistical variable and can be studied either as the number of services completed in a given period of time or alternately the completion time of a service.

Service mechanism of any system is determined by

(i) Service facility design

(ii) Queue discipline

(1) **Service Facility Design:** The facilities at the service station can be divided in two main categories (i) Single channel and (ii) Multi-channel facilities.

(i) **Single channel :** There may be only one counter for servicing and as such only one unit can be served at a time. The next unit can be taken into service when the servicing operations on the previous unit are completed. The single channel queue can be divided in two types:

(a) Single phase   (b) Multi-phase

☛ REMARKS
- In a single phase queue, the whole service operations are completed in one stage.
- In single channel multi-phase queue, the unit taken for service has to pass through many stages before the unit goes out of the servicing channel. All the phases of service are arranged in an ordered sequence.

(ii) **Multi channel:** Due to rush of customers, management may decide to provide a number of service counters so that queue length may not become unreasonably large and the organisation may not loose customers due to long queue.

2. **Queue/Service Discipline:** Queue Discipline identifies the order in which arrivals in the system are taken into service. The Queue discipline does not always take into account the order of arrivals.

   (i) **FIFO or FCFS:** The most common discipline is First In First Out (FIFO) or First Come First Served (FCFS) discipline. In FIFO, the customers are serviced strictly in the order of their joining the system *e.g.*, queues at booking stations.

   (ii) **LCFS:** The Last Come First Served (LCFS) or Last in First Out (LIFO) System is one where the item arriving last are first to go into service.

   (iii) **SIRO:** This rule implies that arrivals are taken into service randomly, irrespective of the order of their arrivals in the system. The server chooses one of the customers to offer service at random.

☞ REMARK

- Sometimes SIRO is the only alternative to assign service, as it may not be possible to identify the order of arrivals.

19.2.4. CLASSIFICATION OF QUEUES

A queuing model is symbolically represented as $(A/B/C): (d/f)$ where,   (MEERUT 2001, 09)

A : Arrival pattern of the units, given by the probability distribution of inter-arrival time of units.

B : The probability distribution of servicing time of individuals being actually served.

C : The number of servicing channels in the system.

d : Capacity of the system i.e., the maximum number of units in the system can accommodate at any time.

f : The manner/order in which the arriving units are taken into service i.e., FIFO/LIFO/SIRO/Priority.

19.2.5 MORE DEFINITIONS RELATED TO QUEUING THEORY

   (i) **Queue length:** Number of persons in the system at any time. We may be interested in studying the distribution of queue length.

   (ii) **Waiting time:** It is the time upto which a unit has to wait before it is taken into service after arriving at the servicing station. This is studied with the help of waiting time distribution. It depends on the following :

   (a) the number of units already there in the system,

   (b) the number of servicing stations in the system,

   (c) the schedule in which units are selected for service.

(iii) **Servicing time:** It is the time taken for servicing a particular arrival.

(iv) **Average length of line:** The number of customers in the queue per unit of time.

(v) **Average idle time:** The average time for which the system remains idle.

4. NOTATIONS

Following notations will be used throughout the chapter :

X	The inter-arrival time between two successive customers (arrivals).

Y	The service time required for any customer.
w	The waiting time for any customer before it is taken into service.
v	Time spent by a customer in the system.
n	Number of customers in the system *i.e.*, in the waiting line at any time, including the number of customers being serviced.
$P_n(t)$	Probability that n customers arrive in the system in time t.
$\phi_n(t)$	Probability that n units are serviced in time t.
$U(T)$	Probability distribution of inter-arrival time P $(t \leq T)$.
$V(T)$	Probability distribution of servicing time P $(t \leq T)$.
$F(N)$	Probability distribution of queue length at any time P $(N \leq n)$.
E_n	Some state of the system at a time when there are n units in the system.
λ_n	Average number of customers arriving per unit of time when there are already n units in the system.
λ	Average number of customers arriving per unit of time.
μ_n	Average number of customers being served per unit of time when there are already n units in the system.
μ	Average number of customers being served per unit of time.
$(\lambda / \mu) = \rho$	traffic intensity.

19.3 STATES OF THE SYSTEM

(i) Steady State: When the waiting line does not depend upon time. Then it is said to be a steady state. If the average rate of arrival is less than the average rate of service, and both are constant, then system will become steady state and hence becomes independent of the initial state of the queue. Then the probability of finding a particular length of queue at any time will be same. The size of queue will fluctuate in the steady state but the statistical behaviour of the queue remains steady.

A necessary condition for the steady state is that the elapsed time since the start of the operation becomes sufficiently large *i.e.*, $(t \to \infty)$, but this condition is not sufficient as the existence of steady state also depends upon the behaviour of the system *e.g.*, if rate of arrival is greater than the rate of service then a steady state cannot be reached. Here we assume that the system acquires a steady state as $t \to \infty$, *i.e.*, the number of arrivals during a certain interval becomes independent of time.

i.e., $$\lim_{t \to \infty} P_n(t) \to P_n$$

☞ REMARK

- In the steady state system, the probability distributions of arrivals, waiting time, and servicing time does not depend on time.

(ii) Transient State: A system is said to be in "Transient state" when its operating characteristics are dependent on time. When the probability distribution of arrivals, waiting time and servicing time are dependent on time, the system is said to be in transient state.

☞ REMARK
• This state occurs at the beginning of the operation of the system.

(iii) **Explosive State:** When the waiting line increases indefinitely with time, then it is said to be explosive state. For example, arrivals in a restaurant during rush hours.

19.4 POISSON PROCESS

The Poisson process is applied to a system where the changes are independent of time i.e., the factors which affect the changes remain absolutely unchanged and the probability of occurrence of any event at any time is independent of time. An infinite sequence of independents events occurring at an instant of time form a Poisson Process, if the following conditions are satisfied:

(i) The total number of events in any time interval X does not depend on the events which has occurred before the beginning of the period, *i.e.,* the number of arrivals in non-overlapping interval are statistically independent and the process has independent and identically distributed increments.

(ii) The probability of an event occurring in a small time interval Δt is $\lambda \Delta t + O(\Delta t)^2$, where A is some constant and $O(\Delta t)^2$ means some function of Δt of order ≥ 2, *i.e.,*
$$\left| \frac{f(\Delta t)}{(\Delta t)} \right| < k \text{ as } \Delta t \to 0 \text{ for any constant } k \geq 0.$$

(iii) Two or more units cannot arrive or serviced at the same time, *i.e.,* the probability of the occurrence of more than one event in the interval t and $t + \Delta t$ is of the order $O(\Delta t)$ which is negligible.

The probability distribution of the arrival pattern can also be studied and identified through analysis of past data. It is also known as the poisson input, *i.e.,* when the arrival pattern of customers in the system follows a Poisson Process.

THEOREM I. ***Under the three conditions of a Poisson Process, the number of arrivals in a fixed time follows the Poisson law i.e., if the probability of an arrival in time interval t and $t + \Delta t$ is $\lambda \Delta t + O(\Delta t)^2$, then***

$$P_n(t) = \frac{e^{-\lambda t}(\lambda t)^n}{n!}, \quad n = 0, 1, ..., \infty \qquad ...(1)$$

PROOF. Consider the consecutive time intervals $(0, t)$ and $(t, t + \Delta t)$, where Δt is very small.

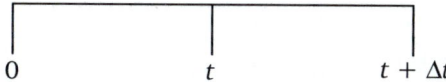

0   t $t + \Delta t$

Case (i) $n \geq 1$.

In this case, the interval $(t, t + \Delta t)$, n arrivals can take place in following three mutually exclusive cases:

(i) There are n arrivals in the interval $(0, t)$ and no arrival in the interval $(t, t + \Delta t)$.

Assumed that the number of arrivals in non-overlapping intervals are statistically independent *i.e.* total number of events in any time interval X does not depend on the events which has occurred before the beginning of the period.

If $P_n(t)$ is the probability of n arrivals from 0 to t and $P_0(\Delta t)$ is probability of no arrival in the interval $(t, t + \Delta t)$, then as the arrivals from $(0, t)$ are independent of arrivals from $(t, t + \Delta t)$, the probability of n arrivals from (0 to $t + \Delta t$) in this case, is given by

$$P_n(t + \Delta t) = P_n(t) . P_0(\Delta t) = P_n(t) [1 - \{\lambda\Delta t - o(\Delta t)\}]$$

where $P_0(\Delta t)$ = Probability that there is no arrival in the interval $(t, t + \Delta t)$

$$= 1 - [\text{Probability that there is at least one arrival in } (t, t + \Delta t)]$$

$$= 1 - [\lambda\Delta t - o(\Delta t)^2]$$

(ii) $(n - 1)$ arrivals in the interval $(0, t)$ with probability $P_{n-1}(t)$ and one arrival in the interval $(t, t + \Delta t)$ with probability $P_1(\Delta t)$. Then

$$P_n(t + \Delta t) = P_{n-1}(t) P_1(\Delta t) = P_{n-1}(t) [\lambda\Delta t + o(\Delta t)]$$

where $P_1(\Delta t)$ = Probability of one arrival in the interval $(t, t + \Delta t)$

$$= \lambda \Delta t$$

(iii) $(n - x)$ arrivals in the interval $(0, t)$ with probability $P_{n-x}(t)$ and x arrivals in the interval $(t, t + \Delta t)$ with probability $P_x(\Delta t)$.

Then,

$$P_n(t + \Delta t) = P_{n-x}(t) . P_x(\Delta t)$$

where $P_x(\Delta t) = 0$, $(\Delta t)^2 = 0$ ($\because$ two or more units cannot arrive at the same time.) Since, all these three cases are mutually exclusive. So, the probability of n arrivals in the interval (0 to $t + \Delta t$) will be given by the sum of the probabilities of these three cases *i.e.*,

$$P_n(t + \Delta t) = P_n(t).P_0(\Delta t) + P_{n-1}(t).P_1(\Delta t) + P_{n-x}(t) P_x(\Delta t)$$

$$= P_n(t) . [1 - \{\lambda\Delta t - o(\Delta t)^2\}] + P_{n-1}(t).(\lambda\Delta t) + P_{n-x}(t)\{o(\Delta t)^2\}$$

$\Rightarrow$ $P_n(t + \Delta t) = P_n(t) - \lambda P_n(t)\Delta t + P_{n-1}(t)\lambda\Delta t$ + terms of $o(\Delta t)^2$

$\therefore$ $\dfrac{P_n(t + \Delta t) - P_n(t)}{\Delta t} = -\lambda P_n(t) + \lambda P_{n-1}(t)$

Taking limit $\Delta t \to 0$, we get

$$P_n'(t) = -\lambda P_n(t) + \lambda P_{n-1}(t) \qquad \qquad ...(2)$$

Case (ii) When $n = 0$

We have $P_0(t + \Delta t)$ = Probability {no arrival in the interval $(0, t)$} $\times$

$$\text{probability \{no arrival in } (t, t + \Delta t) \}$$

$$= P_0(t) . P_0(\Delta t)$$

$$= P_0(t) . \{1 - \lambda\Delta t + o(\Delta t)^2]$$

$\therefore$ $\displaystyle\lim_{\Delta t \to 0} \dfrac{P_0(t + \Delta t) - P_n(t)}{\Delta t} = \lim_{\Delta t \to 0} \left[-\lambda P_0(t) + o(\Delta t)^2 \right]$

$\therefore$ $P_0'(t) = -\lambda P_0(t) \qquad \qquad ...(3)$

Now, $P_0(0) = 1$ and $P_n(0) = 0$, for $n > 0$.

From (3), we get $\dfrac{P_0'(t)}{P_0(t)} = -\lambda$.

On, integrating both sides w.r.t. t, we get $\log P_0(t) = -\lambda t + A$ where A, is a constant of integration.

When $t = 0$, $P_0(0) = 1$ and hence $A = 0$.

$\therefore$ $\qquad$ $\log P_0(t) = -\lambda t$ or $P_0(t) = e^{-\lambda t}$ $\qquad$ ...(4)

Again putting $n = 1$ in (2), we get

$$P_1'(t) = -\lambda P_1(t) + \lambda P_0(t)$$

i.e., $\qquad$ $P_1'(t) + \lambda P_1(t) = \lambda e^{-\lambda t}$ $\quad$ (Since $P_0(t) = e^{-\lambda t}$)

This is a linear differential equation of first order, whose solution is given by

$$e^{\lambda t}.P_1(t) = \lambda t + C.$$

When $t = 0$, $P_1(0) = 0$ and hence $C = 0$.

$\therefore$ $\qquad$ $e^{\lambda t}.P_1(t) = \lambda t$ or $P_1(t) = \lambda t e^{-\lambda t}$ $\qquad$ ...(5)

Again putting $n = 2$ in (2), we get

$$P_2'(t) = -\lambda P_2(t) + \lambda P_1(t) = -\lambda P_2(t) + \lambda^2 t e^{-\lambda t}$$

$\Rightarrow$ $\qquad$ $e^{\lambda t} P_2'(t) + \lambda e^{\lambda t} P_2(t) = \lambda^2 t$

$\Rightarrow$ $\qquad$ $\dfrac{d}{dt}\left[P_2(t).e^{\lambda t}\right] = \lambda^2 t$

$\Rightarrow$ $\qquad$ $P_2(t) = e^{-\lambda t}.\dfrac{\lambda^2 t^2}{2!} + C$; $(C = P_2(0) = 0)$

$\Rightarrow$ $\qquad$ $P_2(t) = e^{-\lambda t}.\dfrac{\lambda^2 t^2}{2!}$ and so on.

In general, $\qquad$ $P_n(t) = e^{-\lambda t}\dfrac{(\lambda t)^n}{n!}$

Hence, n arrivals in time t in the system follows a Poisson law under the given conditions.

19.5 MORE DISTRIBUTIONS

The following distributions are used for the interval of time between successive arrivals or servicing of two successive units.

(i) **Exponential Distribution :** Let x and $x + \Delta x$ be the times of two successive arrivals, then the probability that an arrival will take place in the interval $(x, x + \Delta x)$ is given by

$$f(x) = P[x < t < x + \Delta x] = \lambda e^{-\lambda x}$$

where λ is known as the parameter of the distribution and x is time between two successive arrivals.

(ii) **Regular Distribution:** If the time interval between two successive arrivals is assumed to be constant and the distribution given by

$$\left. \begin{array}{l} f(x) = 0, \ for \ x < a \\ \quad\ = 1, \ x \ge a \end{array} \right\}$$

for some constant a.

(iii) **Erlang Distribution:** If $f(x)$ is the probability that an arrival takes place in the time interval $(x, x + \Delta x)$ then for this distribution

$$f(x) = \frac{(\lambda k)^k \, x^{k-1} e^{-\lambda k}}{(k-1)!}$$

where k and λ are some positive constants with $k \geq 1$.

☛ REMARKS

- The Erlang distribution is derived by combining a number of independent exponential variates. For $k = 1$, the Erlang distribution becomes an exponential distribution and when $k = \infty$, it becomes a regular distribution.
- The inter-arrival time follows a Markov Process which states that at any instant the time until the next arrival occurs is dependent of the time that has elapsed since the occurrence of the last arrival.

THEOREM I. *If the time t between two successive arrivals, is a random variable, then it follows an exponential distribution with parameter λ, i.e., where the arrival pattern follows a Poisson process.*

$$U(t) = \lambda e^{-\lambda t}$$

PROOF. Let us denote

t_0 = instant of an arrival

$(t_0, t_0 + t)$ = no arrival

$t_0 + t + \Delta t$ = instant of another successive arrival

$$P_0(t) = \text{Probability of no arrival in Time } t$$
$$= \frac{e^{-t\lambda}(\lambda t)^0}{0!} = e^{-\lambda t},$$
$$P_0(t + \Delta t) = \text{Probability of no arrival in time } (t + \Delta t)$$
$$= e^{-\lambda(t+\Delta t)} = e^{-\lambda t} \cdot e^{-\lambda \Delta t}$$
$$= P_0(t) \, [1 - \lambda \Delta t + \text{ terms of order } (\Delta t)^2] , \text{ (expanding } e^{-\lambda \Delta t})$$

or $\quad P_0(t + \Delta t) - P_0(t) = -\lambda P_0(t) \, \Delta t + \text{ terms of } o(\Delta t)^2$

i.e., $\quad \dfrac{P_0(t + \Delta t) - P_0(t)}{\Delta t} = -\lambda e^{-\lambda t} + o(\Delta t)^2$

and $\quad \lim\limits_{\Delta t \to 0} \dfrac{P_0(t + \Delta t) - P_0(t)}{\Delta t} = -\lambda e^{-\lambda t}$ ...(1)

Now, $\quad U(t) = \lim\limits_{\Delta t \to 0} \dfrac{[\text{Probability that there is no arrival in the interval } (t, t+\Delta t)]}{\Delta t}$

$$= -dP_0(t) \, / \, dt \qquad \text{[From (1)]}$$
$$= -\lambda \, e^{-\lambda t}$$

$= \lambda e^{-\lambda t}$ (Leaving –ve sign since probability density function is always non-negative)

19.6 DISTRIBUTION OF SERVICE TIME

THEOREM I. *Under the three conditions of a Poisson process, the number of units serviced in a fixed time also follows the Poisson law i.e., if the probability of an unit being serviced in time interval t and $t + \Delta t$ is $\Delta t + o(\Delta t)^2$, then*

$$\phi_n(t) = \frac{e^{-\lambda t}(\mu t)^n}{n!}, \quad n = 0, 1, \ldots, \infty$$

PROOF. Let $(0, t)$ and $(t, t + \Delta t)$ be two consecutive time interval where Δt is very small.

Case (i) $n \geq 1$.

In the interval $(0, t + \Delta t)$, n units can be serviced in following manner:

(i) n units are serviced in the interval $(0, t)$ and no unit is serviced in the interval $(t, t + \Delta t)$. In this case $\phi_n(t + \Delta t) = \phi_n(t) . \phi_0(\Delta t)$

(ii) $(n - 1)$ units are serviced in the interval $(0, t)$ and one unit is serviced in the interval $(t, t + \Delta t)$. Hence $\phi_n(t + \Delta t) = \phi_{n-1}(t) . \phi_1(\Delta t)$

$\because$ the service rate is a Poisson process with parameter μ, therefore we have

$$\phi_0(\Delta t) = 1 - [\mu \Delta t - o(\Delta t)^2] \quad \text{and} \quad \phi_1(\Delta t) = \mu \Delta t + o(\Delta t)^2$$

Now since both these cases are mutually exclusive. So

$$\phi_n(t + \Delta t) = \phi_n(t) [1 - \mu \Delta t + o(\Delta t)^2] + \phi_{n-1}(t) [\mu \Delta t + o(\Delta t)^2]$$

i.e., $$\frac{\phi_n(t + \Delta t) - \phi_n(t)}{\Delta t} = -\mu \phi_n(t) + \mu \phi_{n-1}(t)$$

so that Limit $\Delta t \to 0$, gives

$$\phi_n'(t) = -\mu \phi_n(t) + \mu \phi_{n-1}(t)$$

Case (ii) $n = 0$.

$$\phi_0(t + \Delta t) = \phi_0(t) [1 - \mu \Delta t + o(\Delta t)^2]$$

i.e., $$\frac{\phi_0(t + \Delta t) - \phi_0(t)}{\Delta t} = -\mu \phi_0(t)$$

$\therefore$ $$\phi_0'(t) = -\mu \phi_0(t)$$

Also, $$\phi_0(0) = 1 \quad \text{and} \quad \phi_n(0) = 0 \quad \text{for} \quad n \geq 0.$$

Now, proceeding on the same lines as in theorem (1), it can be easily proved that

$$\phi_n(t) = \frac{e^{-\mu t}(\mu t)^n}{n!}$$

which shows that probability distribution of servicing process is Poisson.

☞ **REMARK**

- The distribution of inter-service time between the service of two successive units will be given by $\mu e^{-\mu t}$.

THEOREM 2. *In a system where the number of arrivals in a fixed time follows the Poisson law, λ is the mean arrival rate.*

PROOF. Let number of arrivals n at any time t be a random variable which takes the values $n = 0, 1, 2, \ldots, \infty$. So, expected number of arrivals in time t is given by,

$$E(n) = \sum_{n=0}^{\infty} n P_n(t) = \sum_{n=0}^{\infty} \frac{n e^{-\lambda t}(\lambda t)^n}{n!}$$

$$= \lambda t e^{-\lambda t} \sum_{n=1}^{\infty} \frac{(\lambda t)^{n-1}}{(n-1)!} = \lambda t e^{-\lambda t} \cdot e^{\lambda t} = \lambda t$$

$\therefore$ Mean arrival rate = (expected number of arrivals in time t)/t = $(\lambda t) t = \lambda$

Similarly, it can be shown that mean servicing rate will be μ.

19.7 CLASSIFICATION OF QUEUE MODELS

Depending upon the nature of inputs and service facilities, there can be a number of Queuing models. Queuing models can be broadly classified into three categories:

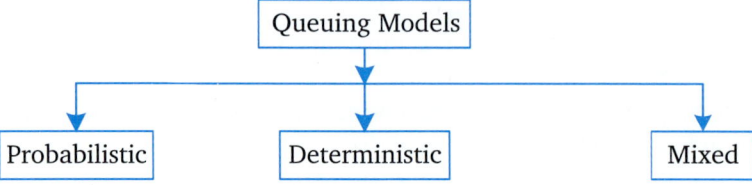

Queuing models are used to explain the descriptive behaviour of a Queuing system. These quantifies the effect of decision variables on the expected waiting times and waiting length as well as generate waiting cost and service cost information. The various systems can be evaluated through these aspects and the system which offers the minimum total cost is selected.

19.8 MODEL-I [M/M/1 : (∞/FCFS)]

[MEERUT-2000, 02, 03, 07, 12, 15, 17, 18; ROHILKHAND-2003, AGRA-2002, 03]

This is a system with Poisson input, exponential waiting time and Poisson output with single channel. Queue capacity of the system being infinite with first in first out mode.

STEP 1. To find the system of steady state equations. Let there be n units in the system including the one in service at any time *i.e.* the queue length is n at time t with probability $P_n(t)$. Since the input distribution is Poisson, the chance that a unit arrives in the system during the interval $(t, t + \Delta t)$ is

$$\lambda_n \Delta t + o(\Delta t)^2.$$

Similarly the chance that any unit leaves the system during this interval will be $\mu_n \Delta t + o(\Delta t)^2$, because the output distribution is also Poisson. The chance that more than one unit arrive or leave the system during the interval $(t, t + \Delta t)$ will be of order $o(\Delta t)^2$. Events with probabilities involving $(\Delta t)^2$ cannot happen.

If the system is in a state E_n at time $(t + \Delta t)$ with probability $P_n(t + \Delta t)$. Then there can be following three possibilities in the interval $(t, t + \Delta t)$.

(i) System is in the state E_{n-1} at time t and one unit arrives in the interval M and no unit leaves during this period.

Probability for this event = (Probability that there are $n-1$ units at time t) × Probability (one unit arrives during the time Δt) × Probability (no unit leaves in time Δt).

$$= P_{n-1}(t) \times \lambda_{n-1} \Delta t \times (1 - \mu_{n-1} \Delta t) \qquad \qquad ...(1)$$

(ii) System is in the state E_n at time t and no unit arrives and no unit leaves the system during the interval Δt. Probability of this event = Probability (there are n units at time t) $\times$ Probability (no unit arrives in time Δt) $\times$ Probability (no unit leaves in time Δt)

$$= P_n(t)\,(1 - \lambda_n \Delta t)\,(1 - \mu_n \Delta t) \qquad \dots(2)$$

(iii) The system is in the state E_{n+1} at time t and no unit arrives in times Δt, but one unit leaves the system after the completion of service.

The probability $=$ Probability (there are $n + 1$ units at time t) $\times$ Probability that no unit arrives in time Δt) $\times$ (Probability that one unit leaves in time Δt)

$$= P_{n+1}(t)\,(1 - \lambda_{n+1}\Delta t)(\mu_{n+1}\Delta t) \qquad \dots(3)$$

All these three cases are mutually exclusive. Hence the probability that the system is in state E_n at time $(t + \Delta t)$ will be equal to the sum of the probabilities in these three cases given by (1), (2) and (3). Hence

$$P_n(t + \Delta t) = P_{n-1}(t)\lambda_{n-1}\Delta t(1 - \mu_{n-1}\Delta t) + P_n(t)\,(1 - \lambda_n \Delta t)\,(1 - \mu_n \Delta t)$$
$$+ P_{n+1}(t)\,(1 - \lambda_{n+1}\Delta t)(\mu_{n+1}\Delta t)$$

$$= P_{n-1}(t)\lambda_{n-1}\Delta t + P_n(t) - P_n(t)[\lambda_n + \mu_n]\Delta t + P_{n+1}(t)\mu_{n+1}\Delta t$$

(neglecting terms of order $(\Delta t)^2$)

$\therefore \quad P_n(t + \Delta t) - P_n(t) = P_{n-1}(t)\lambda_{n-1}\Delta t - (\lambda_n + \mu_n)P_n(t)\Delta t + P_{n+1}(t)\mu_{n+1}\Delta t$

i.e., $\dfrac{P_n(t + \Delta t) - P_n(t)}{\Delta t} = P_{n-1}(t)\lambda_{n-1} - (\lambda_n + \mu_n)P_n(t) + P_{n+1}(t)\mu_{n+1}$

i.e., $\displaystyle\lim_{\Delta t \to 0} \dfrac{P_n(t + \Delta t) - P_n(t)}{\Delta t} = P_n'(t) = P_{n-1}(t)\lambda_{n-1} - (\lambda_n + \mu_n)P_n(t) + P_{n+1}(t)\mu_{n+1} \qquad \dots(4)$

Equation (4) holds only for $n > 0$. For $n = 0$, equation (1) will not hold good.

Now, $P_0(t + \Delta t) = $ [(Probability that no unit arrives in Δt) $\times$ (Probability that there was no unit at time t)] $+$ [(Probability that there was one unit at time t) $\times$ (Probability that one unit leaves but no unit arrives in the time Δt)]

$$= P_0(t)\,(1 - \lambda_0 \Delta t) + P_1(t)(1 - \lambda_1 \Delta t)\,\mu_1 \Delta t$$

$$= P_0(t) - \lambda_0 P_0(t)\Delta t + P_1(t)\mu_1 \Delta t + \text{terms of } o(\Delta t)^2$$

$\therefore \qquad P_0'(t) = -\lambda_0 P_0(t) + P_1(t)\mu_1 \qquad \dots(5)$

Now, when the system reaches the steady state then we have

$$\lim_{t \to \infty} P_n(t) = P_n \quad \text{and} \quad \lim_{t \to \infty} P_n'(t) = 0 \qquad \dots(6)$$

Also, let $\quad \lambda_n = \lambda,\ \mu_n = \mu$

Hence, under steady state of the system, (5) and (6) reduce to

$$P_{n-1}\lambda - (\lambda + \mu)P_n + P_{n+1}\mu = 0 \qquad \dots(7)$$

and $\qquad P_0' = -\lambda P_0 + \mu P_1 = 0 \qquad \dots(8)$

STEP 2. Solution of Steady State Equations

From (7), $P_{n+1}\,\mu = (\lambda + \mu)P_n - \lambda P_{n-1}$

i.e., $\qquad P_{n+1} = \left(\dfrac{\lambda}{\mu} + 1\right)P_n - \dfrac{\lambda}{\mu}P_{n-1} = (\rho + 1)P_n - \rho P_{n-1}.$

Putting $n = 1, 2, 3$, successively in the above relation, we get

$$P_2 = (\rho + 1)P_1 - \rho P_0 = (\rho + 1)\rho P_0 - \rho P_0 = \rho^2 P_0 \qquad \text{[From (8)]}$$

$$P_3 = (\rho + 1)P_2 - \rho^2 P_0 = \rho^3 P_0$$

Now, assuming the result to be true for P_n, we shall show that

$$P_{n+1} = \rho^{n+1} P_0 \qquad \qquad ...(9)$$

If, $P_r = \rho' P_0$ for all $r \leq n$

Now, we have

$$P_{n+1} = (\rho + 1)P_n - \rho P_{n-1} = (\rho + 1)\rho^n P_0 = \rho.\rho^{n-1} P_0$$

$$= (\rho + 1)\rho^n P_0 - \rho^n P_0 = \rho^n P_0(\rho + 1 - 1) = \rho^{n+1} P_0$$

But $\sum\limits_{n=0}^{\infty} P_n = 1 ; \quad \Rightarrow \quad \sum\limits_{n=0}^{\infty} \rho^n P_0 = 1.$

$\Rightarrow \quad P_0[1 + \rho + \rho^2 + ...] = 1$

$\Rightarrow \quad P_0\left(\dfrac{1}{1-\rho}\right) = 1$ [The sum is possible only when $\rho < 1$, , i.e., $(\mu > \lambda)$]

$\Rightarrow \quad P_0 = (1 - \rho) \qquad \qquad ...(10)$

Putting this value of P_0 in (9), we get

$$P_n = (1 - \rho)\rho^n \qquad \qquad ...(11)$$

which is the probability that at any time there are n units in the queue.

STEP 3. To find Probability:

We have,   Traffic intensity, $\rho = \dfrac{\text{Mean arrival rate}}{\text{Mean service rate}} = \dfrac{\lambda}{\mu}$

The unit of ρ is Erlang. Any queuing system can settle down in a steady state only when $\rho < 1$ i.e. $\lambda < \mu$.

ρ can also be written as $\dfrac{1/\mu}{1/\lambda} = \dfrac{\text{Mean service time}}{\text{Mean inter-arrival time}}$

If $\rho > 1$, then the queue length will go on increasing and will tend towards infinity with time.

To show that average number of units in a M/M/1 system is equal to $\rho/(1-\rho)$.

We have $E(n) = $ Average queue length in the system

$$= \sum\limits_{n=0}^{\infty} nP_n = \sum\limits_{n=0}^{\infty} nP_0\rho^n = P_0[\rho + 2\rho^2 + ... + n\rho^n + ...]$$

$$\text{[from (9)]}$$

$$= (1 - \rho)\rho[1 + 2\rho + 3\rho^2 + ...]$$

$$= (1 - \rho)\rho(1 - \rho)^{-2}, \text{ using Binomial Theorem}$$

$$= \dfrac{\rho}{1 - \rho} \qquad \qquad ...(12)$$

Now in a queue of length n, with one servicing channel, the waiting length i.e. the number of units waiting for service (one unit already being in service) will be $n - 1 = L$ (say).

Clearly L will also be a random variable having the same probability distribution as that of n.

STEP 4. To find the average length of the Waiting line.

We have, $\quad E(L) = \sum\limits_{n=1}^{\infty} (n-1)P_n$

(here n cannot be equal to zero because in that case waiting line length will be negative, which cannot happen)

$$= \sum_{n=1}^{\infty} nP_n - \sum_{n=1}^{\infty} P_n = E(n) - \sum_{n=1}^{\infty} P_n = \frac{\rho}{1-\rho} - \sum_{n=0}^{\infty} P_n + P_0 \quad \left(\text{Since } E(n) = \frac{\rho}{1-\rho}\right)$$

$$= \frac{\rho}{1-\rho} - 1 + (1-\rho), \qquad \left(\because \sum_{n=0}^{\infty} P_n = 1\right)$$

$$= \frac{\rho^2}{1-\rho}. \qquad \qquad \qquad \qquad \qquad \qquad \qquad \qquad \text{...(13)}$$

STEP 5. To find the average length of the waiting line with the condition that it is always greater than zero, is $1/(1-\rho)$. [MEERUT-2000; AGRA-2003]

$$E[L/L>0] = \frac{E[L]}{P[L>0]} = \frac{E[L]}{P[n>1]}$$

$$= \frac{\dfrac{\rho^2}{1-\rho}}{\sum\limits_{n=2}^{\infty} P_n} = \frac{\rho^2}{1-\rho} \times \frac{1}{(\rho^2 P_0 + \rho^3 P_0 + ...)} = \frac{\rho^2}{1-\rho} \times \frac{(1-\rho)}{\rho^2 P_0} = \frac{1}{P_0} \qquad \text{(using (9))}$$

$$= \frac{1}{1-\rho}. \qquad \qquad \qquad \qquad \qquad \qquad \qquad \qquad \text{...(14)}$$

STEP 6. To find the variance of queue length

By definition, we have

$$V(n) = E(n^2) - [E(n)]^2.$$

But $\quad E(n^2) = \sum\limits_{n=0}^{\infty} n^2 P_n = \sum\limits_{n=0}^{\infty} n^2 P_0 \rho^n \qquad \qquad (\because P_n = P_0 \rho^n)$

$$= P_0[1^2\rho + 2^2\rho^2 + 3^3\rho^3 + ...\infty \text{ terms}] = P_0\rho[1^2 + 2^2\rho + 3^2\rho^2 + ...]$$

Let $\quad S = 1 + 2^2\rho + 3^2\rho^2 + ...$

$\Rightarrow \quad \int_0^\rho S \, d\rho = [\rho + 2\rho^2 + 3\rho^3 + ...]_0^\rho = \rho[1 + 2\rho + 3\rho^2 + ...] = \rho(1-\rho)^{-2}$

Differentiating w.r.t. ρ, we get

$$S = \frac{1}{(1-\rho)^2} + \frac{2\rho}{(1-\rho)^3} = \frac{(1-\rho) + 2\rho}{(1-\rho)^3} = \frac{(1+\rho)}{(1-\rho)^3}$$

$$\therefore \qquad E(n^2) = (1-\rho)\left(\frac{\rho(1+\rho)}{(1-\rho)^3}\right) = \frac{\rho(1+\rho)}{(1-\rho)^2}.$$

Putting the values of $E(n^2)$ and $[E(n)]^2$, we get

$$V(n) = \frac{\rho(1+\rho)}{(1-\rho)^2} - \frac{\rho^2}{(1-\rho)^2} = \frac{\rho^2 + \rho - \rho^2}{(1-\rho)^2} = \frac{\rho}{(1-\rho)^2} \qquad \qquad \text{...(15)}$$

THEOREM 1. *The distribution of waiting time w of an arrival before being taken into service and of the total time v spent by an arrival in the system where v = w + servicing time, are given by*

$$(i) \quad P(w) = \begin{cases} 1-\rho & \text{for } w = 0 \\ \mu\rho(1-\rho)e^{-\mu w(1-\rho)} & \text{for } w > 0 \end{cases}$$

$$(ii) \quad P(v) = \mu(1-\rho)e^{-\mu v(1-\rho)} \qquad \text{for } v > 0$$

PROOF. (i) Let w be the waiting time of any arrival in the system before it is taken into service. This will be a random variable as its value will depend on the number of units already waiting in the system and time taken for service by each item. $P_n(w)$ denotes the probability that the waiting time of any particular unit lies between w and $(w + dw)$ when there were already n units in the queue $(n - 1)$ units waiting and one in service) when it joined the system.

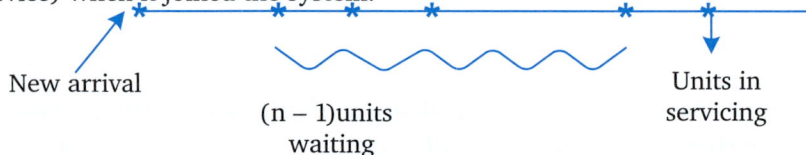

New arrival

(n – 1)units
waiting

Units in
servicing

The distribution of w can be divided into two parts :

(a) Discrete distribution, when n = 0 In this case, system is empty and the new arrival is immediately taken into service, then waiting time will be zero. Therefore, $P(w = 0) = P_0 = (1 - \rho)$.

(b) Continuous distribution, when the queue length is greater than zero. In FCFS system, any new arrival can be taken into service only when all the units waiting in the system are serviced. If there are already n units in the system, then a new arrival has to wait for time between w and $w + dw$, only when $n - 1$ units are serviced in time w and one unit in service at time w, is completely served in time dw. Since servicing distribution is also Poisson, the chance that any unit in service at the time of new arrival is serviced in time dw will be μdw and the probability that remaining $(n - 1)$ units are serviced in time w will be

$$\frac{(\mu w)^{n-1} e^{-\mu w}}{(n-1)!}.$$

So, $P_n(w)dw$ = Probability that new arrival is taken into service after a time lying between w and $w + dw$, when there were n units in the system.

= (Probability that one unit is serviced in time dw) ×

(Probability that exactly $n - 1$ units waiting, are serviced in time w).

$$= \mu dw \times \frac{(\mu w)^{n-1}e^{-\mu w}}{(n-1)!} \qquad \ldots(1)$$

Since, the queue length can vary between 1 and ∞, therefore the probability distribution of waiting time will be given by

$$P(w)dw = \sum_{n=1}^{\infty} P_n(w) \times (\text{Probability that there are n units in the system}). \, dw$$

$$= \sum_{n=1}^{\infty} P_n(w).P_n dw = \sum_{n=1}^{\infty} \mu e^{-\mu w}\frac{(\mu w)^{n-1}}{(n-1)!}(1-\rho)\rho^n dw$$

$$= \mu\rho(1-\rho)e^{-\mu w} \sum_{n=1}^{\infty} \frac{(\rho\mu w)^{n-1}}{(n-1)!} dw = \mu\,\rho(1-\rho)e^{-\mu w}e^{\rho\mu w}dw$$

$$= \mu\rho(1-\rho)e^{-\mu w(1-\rho)}dw \qquad \qquad \text{...(2)}$$

Now,

$$P(w>0) = \mu\rho(1-\rho)\int_0^\infty e^{-\mu w(1-\rho)}dw = \rho$$

Hence, Complete distribution of waiting time is given by,

$$P(w) = 1-\rho, \text{ for } w = 0 \text{ and } P(w) = \rho \text{ for } w>0$$

Here $P(w=0) + P(w>0) = (1-\rho)+\rho = 1$.

Similarly the conditional distribution of waiting time, being taken to be always greater than zero, is given by

$$P(w/w>0) = \frac{P(w)}{P(w>0)} = \frac{\mu\rho(1-\rho)e^{-\mu w(1-\rho)}}{\rho} dw$$

$$= (1-\rho)\,\mu e^{-\mu w(1-\rho)}dw \qquad \qquad \text{...(3)}$$

Clearly $\int_0^\infty P(w/w>0)dw = 1$.

(ii) Let $P_n(v)$ be the probability that an arriving unit has to spent v units of time in $M/M/1$ system, when there are already n units at the time of its arrival. Now $P_n(v)dv$ = Probability that exactly n units are served in time and the new arrival is served completely in time dv.

$$= \frac{(\mu v)^n\, e^{-\mu v}}{n!}\mu dv,\ n = 1,2,...,\infty$$

$$P_n(v)\,dv = \sum_{n=1}^{\infty} \text{ (Probability that new arrival lives in the system for } v$$

units of time and goes out before $v + dv$) $\times$ (Probability that there are n units in the system). dv

$$= \sum_{n=1}^{\infty} \frac{(\mu v)^n e^{-\mu v}}{n!}\mu P_n dv = \mu(1-\rho)\sum_{n=1}^{\infty} \frac{(\rho\mu v)^n e^{-\mu v}}{n!}.dv$$

$$= \mu(1-\rho)e^{-\mu v}\sum_{n=1}^{\infty} \frac{(\rho\mu v)^n}{n!}dv = \mu(1-\rho)e^{-\mu v}\,e^{\rho\mu v}dv$$

$$= \mu(1-\rho)\,e^{-v\mu(1-\rho)}dv. \qquad \qquad \text{....(4)}$$

Now average waiting time of an arrival in the queue is given by

$$E(w) = \text{Average waiting time}$$
$$= w \text{ (Probability that waiting time is greater than zero)}$$
$$+ 0\text{(Probability that waiting time is zero)}$$
$$= \int_0^\infty w\mu\rho(1-\rho)\,e^{-\mu w(1-\rho)}dw + \text{ zero}$$
$$= \mu\rho(1-\rho)\int_0^\infty we^{-\mu w(1-\rho)}dw = \rho\int_0^\infty \frac{t}{\mu(1-\rho)}.e^{-t}dt$$

$$\text{[Using } \mu w(1-\rho) = t\,]$$

$$= \frac{\rho}{\mu(1-\rho)}.\Gamma(2) = \frac{\rho}{\mu(1-\rho)}, \qquad (\because \Gamma(2) = 1) \qquad \text{...(5)}$$

Also the average waiting time for an arrival in the queue, when already some units are in the queue is given by

$$E(w \mid w > 0) = \frac{\int_0^\infty wP(w)dw}{P(w > 0)} = \int_0^\infty [w\mu\rho(1-\rho)e^{-\mu w(1-\rho)}dw] / \rho,$$

$$= \mu(1-\rho)\int_0^\infty we^{-\mu w(1-\rho)}dw$$

$$= \int_0^\infty \frac{t}{\mu(1-\rho)}e^{-t} dt \qquad\qquad [\text{Using } t = \mu w(1-\rho)]$$

$$= \frac{1}{\mu(1-\rho)} \qquad\qquad\qquad\qquad \ldots(6)$$

Finally, $\qquad E(v) = \int_0^\infty v.\mu(1-\rho)e^{-\mu v(1-\rho)}dv$.

$$= \mu(1-\rho)\int_0^\infty v.e^{-\mu v(1-\rho)}dv \qquad\qquad [\because t = \mu v(1-\rho)]$$

$$= \int_0^\infty \frac{t}{\mu(1-\rho)}e^{-t} = \frac{1}{[\mu(1-\rho)]} \qquad\qquad \ldots(7)$$

19.9 MODEL-2: GENERAL ERLANG QUEUING MODEL: M/M/1 : (∞/FCFS)

This model is same as model-1 with the difference that have mean arrival rate and mean service rate are not constant but depend upon n, the no. of units present in the system. Here λ_n and μ_n are mean arrival and mean service rate when there are n units in the system respectively.

Now proceed as in model-1, we have

$$P_n'(t) = -(\lambda_n + \mu_n)P_n(t) + \lambda_{n-1}P_{n-1}(t) + \mu_{n+1}P_{n+1}(t) , \text{ for } n \geq 1$$

and $\qquad P_0'(t) = -\lambda_0 P_0(t) + \mu_1 P_1(t), \quad \text{for } n \geq 0$.

In steady state system,

$$\lim_{t\to\infty} P_n(t) \to P_n \text{ and hence } \lim_{t\to\infty} P_n'(t) = 0.$$

$\therefore \qquad (\lambda_n + \mu_n)P_n = \lambda_{n-1}P_{n-1} + \mu_{n+1}P_{n+1} , \text{ for } n \geq 1 \qquad\qquad \ldots(1)$

$\qquad\qquad \lambda_0 P_0 = \mu_1 P_1 \qquad\qquad\qquad \text{for } n = 0$

i.e., $\qquad\qquad P_1 = \left(\frac{\lambda_0}{\mu_1}\right)P_0 \qquad\qquad\qquad\qquad \ldots(2)$

Putting $n = 1$ in (1), we get

$$(\lambda_1 + \mu_1)P_1 = \lambda_0 P_0 + \mu_2 P_2, \text{ but from (2)}, \mu_1 P_1 = \lambda_0 P_0$$

$$\mu_2 P_2 = \lambda_1 P_1$$

i.e., $\qquad\qquad P_2 = \frac{\lambda_1 P_1}{\mu_2} = \frac{\lambda_1}{\mu_2}.\frac{\lambda_0}{\mu_1}P_0.$

Putting $n = 2$ in (1), we get

$$(\lambda_2 + \mu_2)P_2 = \lambda_1 P_1 + \mu_3 P_3$$

and putting the values of P_1 and P_2, we get

$$P_3 = \frac{\lambda_2}{\mu_3}.\frac{\lambda_1}{\mu_2}.\frac{\lambda_0}{\mu_1}P_0 \text{ and so on.}$$

Proceeding in the same way, in general, we get

$$P_n = \frac{\lambda_{n-1}}{\mu_n} \cdot \frac{\lambda_{n-2}}{\mu_{n-1}} \cdot \frac{\lambda_{n-3}}{\mu_{n-2}} \cdots \frac{\lambda_0}{\mu_1} P_0 \qquad \dots(3)$$

If P_0 is known then all P_i's can be evaluated such that

$$\sum_{n=0}^{\infty} P_n = 1, \ i.e., \ (P_0 + P_1 + P_2 + \dots) = 1$$

$$\Rightarrow \qquad P_0 \cdot \left[1 + \frac{\lambda_0}{\mu_1} + \frac{\lambda_1}{\mu_2} \cdot \frac{\lambda_0}{\mu_1} + \dots \right] = 1$$

$$\Rightarrow \qquad P_0 . S = 1, \quad \text{where} \quad S = 1 + \frac{\lambda_0}{\mu_1} + \frac{\lambda_1}{\mu_2} \cdot \frac{\lambda_0}{\mu_1} \dots$$

Thus, $\qquad P_0 = (1 / S)$.

If S is a divergent series then $P_0 = 0$ and in that case it will be meaningless. But if S is convergent, the value of P_0 will be dependent on λ's and μ's and we can have different cases as follows:

CASE 1. If $\lambda_n = \lambda$ and $\mu_n = \mu$, *i.e.*, λ and μ are fixed and are independent of n, then the series S is convergent if $(\lambda / \mu) < 1$.

This case has already been discussed above. (In this case model-1 and model-2 both are same)

CASE 2. If $\lambda_n = \dfrac{\lambda}{n+1}$, $\mu_n = \mu$ (constant).

In this case λ_n decreases as n increases, that is arrival rate decreases with increase in queue length. Now substituting the value of λ's and μ's in the expression for S, we get

$$S = \left[1 + \frac{\lambda}{\mu} + \frac{\lambda^2}{2\mu^2} + \frac{\lambda^3}{2.3\mu^3} + \dots \right] = \left(1 + \rho + \frac{\rho^2}{2!} + \frac{\rho^3}{3!} + \dots \right),$$

$$= e^\rho, \quad \text{where} \quad \rho = \frac{\lambda}{\mu}$$

$$\therefore \qquad P_0 = e^{-\rho}$$

$$P_1 = \rho e^{-\rho}$$

$$P_2 = \frac{\rho}{2!} e^{-\rho}$$

$$\vdots \qquad \vdots$$

$$P_n = \frac{\rho^n}{n!} e^{-\rho} \qquad \dots(4)$$

$\Rightarrow P_n$ follows the Poisson distribution. So we see that in case 1 *i.e.* for fixed μ and λ the distribution is geometrical and in case 2 it is Poisson with parameter ρ.

CASE 3. $\lambda_n = \lambda$ (constant) and $\mu_n = n\mu$.

In this case, the service rate increases with increase in queue length. This is known as a problem of queue with infinite number of servicing channels. Here we assume that there are infinite number of service stations *i.e.*, service facility is available to each arrival but there may be a situation that at any time all service stations may not remain busy obviously in this case no queue will be formed *e.g.*, telephone service stations are available to all arriving calls. Then we have

$$S = 1 + \frac{\lambda}{\mu} + \frac{\lambda^2}{2\mu^2} + \frac{\lambda^3}{2.3\mu^3} + \ldots = 1 + \rho + \frac{\rho^2}{2!} + \frac{\rho^3}{3!} + \ldots = e^{\rho}$$

$\therefore \qquad P_0 = e^{-\rho}$ and so in Case 2, $P_n = e^{-\rho} \dfrac{\rho}{n!}$, \hfill ...(5)

which is again a Poisson distribution.

 Solved Examples

EXAMPLE 1. *Customers arrive at a sales counter manned by a single person according to a Poisson process with a mean rate of 20 per hour. The time required to serve a customer has an exponential distribution with a mean of 100 seconds. Find the average waiting time of a customer and queue length.* (UPTU-MBA 2008)

SOLUTION. We have

mean arrival rate λ = 20/hour,

$$\text{mean service rate } \mu = \frac{60 \times 60}{100} = 36 / \text{hour}$$

$\therefore$ Average waiting time to a customer in a queue

$$= \frac{\lambda}{\mu(\mu - \lambda)} = \frac{30}{36(36 - 20)} = 125 \text{ seconds}$$

Also, the average waiting time of a customer in the system

$$= \frac{1}{\mu - \lambda} = \frac{1}{36 - 20} \text{ hours} = 225 \text{ seconds}$$

and average queue length $= \dfrac{\rho^2}{1 - \rho} = \dfrac{\lambda^2}{\mu(\mu - \lambda)} = \dfrac{25}{36} = 0.7$ (approx.) customers.

EXAMPLE 2. *In a railway marshalling yard, goods trains arrive at a rate of 30 trains per day. Assuming that the inter-arrival time follows an exponential distribution and the service time distribution is also exponential with an average 36 minutes. Determine*
(i) The mean queue size
(ii) The probability that the queue size exceeds 10.

(MEERUT-2000, 02,16, AGRA-1999, UPTU-MBA 2006)

SOLUTION. Here, the mean arrival time $\lambda = \dfrac{30}{60 \times 24} = \dfrac{1}{46}$ trains/minute

and mean service rate $\mu = \dfrac{1}{30}$ trains/minutes

$\therefore \qquad \rho = \dfrac{\lambda}{\mu} = \dfrac{3}{4}$

(i) The mean queue size $= \dfrac{\rho}{1 - \rho} = 3$ trains

(ii) Probability [queue size ≥ 10] $= \rho^{10} = \left(\dfrac{3}{4}\right)^{10} = 0.06$

19.10 MODEL-3 : M/M/1 : (N/FCFS) (ROHILKHAND-2000, 01,17; AGRA-2000)

In this model, the capacity of the system is N and when the queue length is of size N, no new arrival can be accommodated in the system.

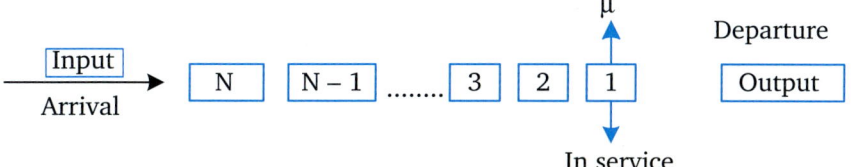

STEP-1 To find the system of steady state equation

In this model,

$$\lambda_n = \lambda, \ \mu_n = \mu, \ \text{when} \ n < N$$
$$\lambda_n = 0 \qquad \text{when} \ n \geq N$$

(Because when queue length is N, then no new arrival can be possible)

The system of difference equations of this model is

$$P_0(t + \Delta t) = P_0(t)\,[1 - \lambda \Delta t] + P_1(t)\mu \Delta t + o(\Delta t)^2, \ \text{when} \ n = 0$$

$$P_n(t + \Delta t) = P_n(t)\,[1 - (\lambda + \mu)\Delta t] + P_{n-1}(t)\lambda \Delta t + P_{n+1}(t)\mu \Delta t + o(\Delta t)^2, \ for \ 1 \leq n \leq N-1 \left.\vphantom{\begin{array}{c}1\\1\\1\end{array}}\right\}$$

$$P_N(t + \Delta t) = P_N(t)\,[1 - (\lambda_N + \mu)\Delta t] + P_{N-1}\lambda \Delta t + o(\Delta t)^2, \qquad\qquad for \ n = N$$

$$\qquad ...(1)$$

$$= P_N(t)\,[(1 - \mu \Delta t)] + P_{N-1}(t)\lambda \Delta t + o(\Delta t)^2, \ \text{since for} \ n = N, \ \lambda_N = 0.$$

On simplifying (1) and taking the limit $\Delta t \to 0$, we get

$$P_0'(t) = -\lambda P_0(t) + \mu P_1(t), \ \text{for} \ n = 0$$

$$P_n'(t) = -(\lambda + \mu)P_n(t) + \lambda P_{n-1}(t) + \mu P_{n+1}(t), \ \text{for} \ 1 \leq n \leq N \left.\vphantom{\begin{array}{c}1\\1\\1\end{array}}\right\} \qquad ...(2)$$

$$P_N'(t) = -\mu P_N(t) + \lambda P_{N-1}(t), \ \text{for} \ n = N$$

$\lim\limits_{t \to \infty} P_n(t) \to P_n$ and $P_n'(t) = 0$ then (2) becomes

$$\lambda P_0 = \mu P_1 \qquad\qquad\qquad\qquad \text{for} \ n = 0$$

$$(\lambda + \mu)P_n = \lambda P_{n-1} + \mu P_{n+1}, \qquad \text{for} \ 1 \leq n \leq N-1 \left.\vphantom{\begin{array}{c}1\\1\\1\end{array}}\right\} \qquad ...(3)$$

$$\mu P_N = \lambda P_{N-1}, \qquad\qquad\qquad \text{for} \ n = N$$

STEP-2 : To solve the steady state Equations

We have
$$P_1 = \frac{\lambda}{\mu} P_0$$

$$P_2 = \left(\frac{\lambda}{\mu}\right)^2 P_0 \qquad\qquad\qquad ...(4)$$

$$\vdots \qquad \vdots$$

$$P_N = \left(\frac{\lambda}{\mu}\right)^N P_0$$

Now, for finding P_0, we use $\sum\limits_{n=0}^{N} P_n = 1$ (Since the system is of capacity N)

i.e., $(P_0 + P_1 + ... + P_N) = 1$

i.e., $P_0(1 + \rho + \rho^2 + ... + \rho^N) = 1$

$$P_0 \frac{1 - \rho^{(N+1)}}{1 - \rho} = 1$$

$\therefore$ $$P_0 = \frac{1 - \rho}{1 - \rho^{(N+1)}}$$ ....(5)

and $$P_n = \left(\frac{1 - \rho}{1 - \rho^{(N+1)}}\right).\rho^n, \quad n = 1, 2, ..., N .$$ ....(6)

STEP3 Results and discussion
(i) Average queue length:
We know that

$$E(n) = \sum_{n=0}^{N} nP_n = \sum_{n=0}^{N} n\left(\frac{1 - \rho}{1 - \rho^{N+1}}\right)\rho^n .$$ [from (6)]

$$= \frac{(1-\rho)}{(1-\rho^{N+1})} \sum_{n=0}^{N} n\rho^n = \frac{(1-\rho)}{(1-\rho^{N+1})}[0 + \rho + 2\rho^2 + 3\rho^3 + ... + N\rho^N]$$

$$= \frac{(1-\rho)}{(1-\rho^{N+1})} \rho \cdot \left[\frac{1-\rho^N}{(1-\rho)^2} - \frac{N\rho^N}{1-\rho}\right]$$

$$= \frac{\rho}{(1-\rho^{N+1})}\left[\frac{1-\rho^N - N\rho^N + N\rho^{N+1}}{1-\rho}\right] = \frac{1 - (N+1)\rho^N + N\rho^{N+1}}{(1-\rho)(1-\rho^{N+1})} \cdot \rho$$

(ii) The average length of waiting line

$$E(L) = E(n-1) = \sum_{n=1}^{N} (n-1)P_n = \sum_{n=0}^{N} nP_n - \sum_{n=1}^{N} P_n$$

$$= E(n) - \sum_{n=0}^{N} P_n + P_0$$

$$= \frac{1 - (N+1)\rho^N + N\rho^{N+1}}{(1-\rho)(1-\rho^{N+1})} - 1 \frac{1-\rho}{1-\rho^{N+1}}$$

$$= \frac{1 - N\rho^{N-1} + (N-1)\rho^N}{(1-\rho)(1-\rho^{N+1})} \cdot \rho^2$$

(iii) The fluctuation in the queue length of the system:

$$V(n) = E(n^2) - [E(n)]^2 = \sum_{n=0}^{N} n^2 P_n - L^2$$

$$= \sum_{n=0}^{N} n^2 \frac{1-\rho}{1-\rho^{N+1}} \rho^n - L^2 = \left(\frac{1-\rho}{1-\rho^{N+1}}\right) \sum_{n=0}^{N} n^2\rho^n - L^2$$

$$= \frac{\rho(1-\rho)}{1-\rho^{N+1}}\left[1^2 + 2^2\rho + ... + N^2\rho^{N-1}\right] - L^2$$

(iv) The average waiting time in the queue:
It can be obtained by using the relation

$$E(w) = \frac{E(n)}{\lambda'} - \frac{1}{\mu} = \frac{E(L)}{\lambda'}$$ ....(7)

19.11 MODEL-4 :M/M/c : (∞ /FCFS)   (AGRA-1998, 2002; MEERUT-2014)

In this system the length of waiting line will depend on the number of channels occupied.

Case 1: If $n < c$, then there will be no problem of waiting and the rate of servicing will be nil as only n channels are busy, each serving at the rate μ .

Case 2 : In case $n = c$, all the channels will be working.

Case 3 : $n > c$, then $n - c$ persons will be in the queue and the rate of service will be $c\mu$ as all the c channels are busy.

There will be three cases in this system.

STEP-1 : To find the system of steady state equation

(i) The queue is empty at the time $(t + \Delta t)$. This will happen when the queue is empty at the time t and there is no arrival in the time interval Δt . This can also occur if one unit is in the queue at time t and it is serviced in time Δt . If there are 2 or more units in the system at time t then there will remain at least one unit in the system at time t as in time interval Δt at most one unit can leave the system after servicing. So, for n = 0 we get

$$P_0(t + \Delta t) = P_0(t)[1 - \lambda \Delta t] + P_1(t)\, \mu \Delta t (1 - \lambda \Delta t) + o(\Delta t)^2$$

(ii) When n lies between 1 and $c - 1$, then all items arriving will be immediately taken into service and n channels out of c will be busy. There can be three possibilities.

(a) n units are in the queue at time t and all are being serviced with no unit arrives or leaves the system in the time Δt . The probability of this event is

$$\underset{\substack{\text{Probability no unit}\\ \text{arrives in } \Delta t}}{P_0(t)\,(1 - \lambda \Delta t)} \quad \underset{\substack{\text{Probability no unit}\\ \text{leave. Here service ratio}\\ \text{will be } n\mu}}{(1 - n\mu)\Delta t}$$

$$= P_n(t)\,[1 - (\lambda + n\mu)\Delta t + o(\Delta t)^2\,].$$

(b) $n - 1$ units are in the queue at time t, all are being serviced, one unit arrives and no unit leaves the system. The probability of this event is

$$\underset{\substack{\text{Probability one unit}\\ \text{arrives in } \Delta t}}{P_{n-1}(t)\,.\,(\lambda\,.\,\Delta t)} \quad \underset{\substack{\text{Probability no unit}\\ \text{leave. Here service ratio}\\ \text{will be } n\mu}}{[1 - (n-1)\mu]\,\Delta t}$$

$$= P_{n-1}(t)\,[\lambda \Delta t + o(\Delta t)^2]$$

(c) There are $n + 1$ units in the queue at time t, c are being serviced, no unit arrives but one unit leaves the system in time Δt . The probability of this event is

$$\underset{\substack{\text{Probability one unit}\\ \text{arrives in } \Delta t}}{P_{n-1}(t)\,(1 - \lambda \Delta t)} \quad \underset{\substack{\text{One unit leaves in}\\ \text{time } \Delta t}}{[(n+1)\mu]\,\Delta t}$$

$$= P_{n+1}(t)\,[(n+1)\mu \Delta t + o(\Delta t)^2]$$

All these three situations are mutually exclusive. Hence, the difference equation for $1 \le n \le c - 1$, is

$$P_n(t + \Delta t) = P_n(t)[1 - (\lambda + n\mu)\Delta t] + P_{n-1}(t)\lambda \Delta t P_{n+1}(t)[(n+1)\mu \Delta t] + o(\Delta t)^2$$

(iii) Similarly for $n \ge c$, we shall have

$$P_n(t + \Delta t) = P_n(t)[1 - (\lambda + c\mu)\Delta t] + P_{n+1}(t)\lambda \Delta t + P_{n+1}(t)c\mu \Delta t + o(\Delta t)^2$$

Now as the system acquires a steady state, and limit $\Delta t \to 0$, we have

$$\lim_{t \to \infty} P_n(t) \to P_n \quad \text{and} \quad \lim_{t \to \infty} P_n'(t) \to 0 .$$

Using these, we get the following difference equations :

$$\left.\begin{array}{ll} \lambda P_0 = \mu P_1 & \text{for } n = 0 \\[4pt] \lambda P_{n-1} - (\lambda + n\mu)P_n + (n+1)\mu P_{n+1} = 0 & \text{for } 1 \le n \le c - 1 \\[4pt] \lambda P_{n-1} - (\lambda + c\mu)P_n + c\mu P_{n+1} = 0 & \text{for } n \ge c \end{array}\right\} \qquad \ldots(1)$$

STEP-2 : To solve steady state equations

Now from first equation of (1), we get $P_1 = (\lambda / \mu)P_0$

Putting $n = 1$ in the second equation, we get

$$\lambda P_0 - (\lambda + \mu)P_1 + 2\mu P_2 = 0$$

i.e., $\quad \lambda P_0 - (\lambda + \mu)(\lambda / \mu)P_0 + 2\mu P_2 = 0$

$\therefore \qquad P_2 = \dfrac{1}{2!}\dfrac{\lambda^2}{\mu^2}P_0 .$

Similarly putting $n = 2$ in second equation of (1), we have

$$P_3 = \dfrac{1}{3!}\dfrac{\lambda^3}{\mu^3}P_0$$

In general, $P_n = \dfrac{1}{n!}\dfrac{\lambda^n}{\mu^n}P_0$, for $1 \le n < c - 1$ and $\qquad P_{c-1} = \dfrac{\lambda}{(c-1)\mu}P_{c-2} .$

Putting $n = c - 1$ in the second equation of (1), we get

$$\lambda P_{c-2} - [\lambda + (c-1)\mu]P_{c-1} + c\mu P_c = 0$$

$\therefore \ \lambda P_{c-2} - [\lambda + (c-1)\mu]\dfrac{\lambda}{(c-1)\mu}P_{c-2} + c\mu P_c = 0$

i.e., $\qquad \lambda P_{c-1} = c\mu P_c \quad \text{or} \quad P_c = (\lambda / \mu c)P_{c-1}$

$\therefore \qquad P_c = \dfrac{\lambda}{\mu c}\cdot\left(\dfrac{\lambda}{\mu}\right)^{c-1}\dfrac{1}{(c-1)!}P_0 = \dfrac{1}{c!}\left(\dfrac{\lambda}{\mu}\right)^c P_0 \qquad \ldots(2)$

Putting $n = c, c+1$ in third equation of (1), we get

$$\left.\begin{array}{l} P_{c+1} = \left(\dfrac{\lambda}{c\mu}\right)P_c = \dfrac{1}{c.c!}\left(\dfrac{\lambda}{\mu}\right)^{c+1}P_0 \\[12pt] P_{c+2} = \dfrac{1}{c^2.c!}\left(\dfrac{\lambda}{\mu}\right)^{c+2}P_0 \\[12pt] P_n = \dfrac{1}{c^{n-c}.c!}\left(\dfrac{\lambda}{\mu}\right)^n P_0, \ \text{for } n \ge c \end{array}\right\} \qquad \ldots(3)$$

For determining P_0, we use the relation $\sum\limits_{n=0}^{\infty} P_n = 1$.

i.e., $\qquad 1 = P_0 + \sum\limits_{n=1}^{c-1} P_n + \sum\limits_{n=c}^{\infty} P_n$

$$= P_0 + \sum_{n=1}^{c-1} \frac{1}{n!}\left(\frac{\lambda}{\mu}\right)^n P_0 + \sum_{n=c}^{\infty} \frac{1}{c^{n-c} \, c!}\left(\frac{\lambda}{\mu}\right)^n P_0$$

$$= P_0 \left[\sum_{n=1}^{c-1} \frac{1}{n!}(c\rho)^n + \frac{c^c}{c!}\sum_{n=c}^{\infty}\left(\frac{\lambda}{c\mu}\right)^n\right]$$

$$= P_0 \left[\sum_{n=1}^{c-1} \frac{(c\rho)^n}{n!} + \frac{c^c}{c!}\sum_{n=c}^{\infty}\rho^n\right], \qquad \text{as } \frac{\lambda}{c\mu} = \rho$$

$$= P_0 \left[\sum_{n=1}^{c-1} \frac{(c\rho)^n}{n!} + \frac{c^c}{c!}\rho^c(1+\rho+\rho^2+\ldots\infty \text{ terms})\right]$$

$$= P_0 \left[\sum_{n=1}^{c-1} \frac{(c\rho)^n}{n!} + \frac{c^c}{c!}\frac{\rho^c}{1-\rho}\right]$$

$$P_0 = \frac{1}{\left[\sum\limits_{n=1}^{c-1} \frac{(c\rho)^n}{n!}\right] + \frac{(c\rho)^c}{c!(1-\rho)}} \qquad\qquad \ldots.(4)$$

and
$$P_n = \frac{1}{n!}\left(\frac{\lambda}{\mu}\right)^n P_0 \qquad \text{for } n = 0,1,2,\ldots,c-1 \qquad \ldots.(5)$$

$$= \frac{1}{c!\,c^{n-c}}\left(\frac{\lambda}{\mu}\right)^n P_0 \qquad \text{for } n = c,c+1,\ldots$$

STEP-3 : Results and Discussion

(1) Average number of units in waiting line of the system

$$= \frac{\rho}{(1-\rho)^2}P_e \qquad\qquad \ldots.(6)$$

If there are n items in queue $(n > c)$, then the items in the waiting line will be equal to $(n-c)$.

∴ Average number of units in waiting line $= E(n-c)$.

$$= \sum_{n=c}^{\infty}(n-c)P_n = \sum_{n=c}^{\infty}(n-c)\frac{c^c}{c!}\left(\frac{\lambda}{c\mu}\right)^n P_0 \qquad\qquad \text{[by (3)]}$$

$$= \sum_{n=c}^{\infty}(n-c)P_c\,\rho^{n-c} \qquad\qquad \text{[since } P_c = (1/c!)(\lambda/\mu)^c P_0]$$

$$= P_c\sum_{n=c}^{\infty}(n-c)\rho^{n-c} = P_c[0+\rho+2\rho^2+3\rho^3+\ldots]$$

$$= \frac{P_c\rho}{(1-\rho)^2}$$

(2) Average number of units in the queue $= \dfrac{P_c\rho}{(1-\rho)^2} + c\mu$.

Now, average queue length = Average number of units in waiting line

+ Number of units in service

$$= \frac{P_c \rho}{(1-\rho)^2} + c\mu .$$

(3) Average waiting time of an arrival

$$= \frac{\text{Average number of units in waiting line}}{\lambda}$$

$$= \frac{P_c \rho}{\lambda(1-\rho)^2} = \frac{\rho}{\lambda(1-\rho)^2} \times \frac{1}{c!}\left(\frac{\lambda}{\mu}\right)^c P_0$$

(4) Average time of an arrival spends in a system

$$= \frac{\text{Average number of items in the queue}}{\lambda}$$

$$= \frac{P_c \rho}{\lambda(1-\rho)} + \frac{c\rho}{\lambda} .$$

(5) Probability (that all the channels are occupied)

$$= P(n \ge c) = \left[\frac{1}{1-\rho}\right] P_c$$

(6) Probability (that some unit has to wait)

$$= P(n \ge c+1) = \frac{\rho}{1-\rho} P_c .$$

(7) The average number of units which actually wait in the system

$$= \frac{\sum\limits_{n=c+1}^{\infty} (n-c) P_n}{\sum\limits_{n=c+1}^{\infty} P_n} = \frac{1}{1-\rho}$$

(8) Average waiting time in the queue for all arrivals

$$= \frac{1}{\lambda} \sum\limits_{n=0}^{\infty} (n-c)P_n = \frac{1}{c\mu(1-\rho)^2} P_c$$

(9) Average waiting time in the queue for those who actually wait

$$= \frac{1}{c\mu - \lambda}$$

(10) Average number of items served $= \sum\limits_{n=0}^{c-1} nP_n + c \sum\limits_{n=c}^{\infty} P_n$

(11) Average number of idle channels $= c$ − Average number of items served

(12) Efficiency of M/M/c model $= \dfrac{\text{Average number of items served}}{\text{Total number of channels}}$

19.12 MODEL-5: M/M/c : (N/FCFS); N > c

This is the queuing model with Poisson arrival, Poisson service. There are $c > 1$ service channel, service rate at each channel is the same. The service discipline is first come first serve. Here the value of λ will be zero when the value of $n \geq N$ and $\mu = c\mu$ for $0 \leq n < c$ and $\mu = c\mu$ for $c \leq n \leq N$.

STEP-1: To find the system of steady state equations are their solution

we have

$$P_n' = P_{n-1}\lambda_{n-1} - (\lambda_n + \mu_n)P_n + P_{n+1}\mu_{n+1} = 0$$

or

$$P_{n+1}\mu_{n+1} = (\lambda_n + \mu_n)P_n - \lambda_{n-1}P_{n-1}$$

and

$$P_0' = -\lambda_0 P_0 + \mu_1 P_1 = 0$$

i.e.,

$$P_1 = \frac{\lambda_0}{\mu_1} P_0$$

For $n = 1$,

$$P_2\mu_2 = (\lambda_1 + \mu_1)P_1 - \lambda_0 P_0 = (\lambda_1 + \mu_1)\frac{\lambda_0}{\mu_1}P_0 - \lambda_0 P_0 = \frac{\lambda_1\lambda_0 P_0}{\mu_1}$$

i.e.,

$$P_2 = \frac{\lambda_1\lambda_0}{\mu_1\mu_2} P_0$$

For $n = 2$,

$$P_3\mu_3 = (\lambda_2 + \mu_2)P_2 - \lambda_1 P_1 = (\lambda_2 + \mu_2)\frac{\lambda_1\lambda_0}{\mu_1\mu_2}P_0 - \frac{\lambda_1\lambda_0}{\mu_1}P_0 = \frac{\lambda_1\lambda_2\lambda_0}{\mu_1\mu_2}P_0$$

i.e.,

$$P_3 = \frac{\lambda_0\lambda_1\lambda_2}{\mu_1\mu_2\mu_3} P_0$$

....

In general

$$P_n = \frac{\lambda_0\lambda_1...\lambda_{n-1}}{\mu_1\mu_2...\mu_n} P_0 = \prod_{i=0}^{n-1} \frac{\lambda_i}{\mu_{i+1}} P_0, \ n \geq 1 \qquad ....(1)$$

Also,

$$\sum_{n=0}^{\infty} P_n = 1 \quad i.e., \quad P_0 + \sum_{n=1}^{\infty} P_n = 1$$

i.e.,

$$P_0 + \sum_{n=1}^{\infty}\left[\prod_{i=0}^{\infty} \frac{\lambda_i}{\mu_{i+1}} P_0\right] = 1$$

or

$$P_0 = \left[1 - \sum_{n=1}^{\infty}\prod_{i=0}^{\infty} \frac{\lambda_i}{\mu_{i+1}}\right]^{-1} \qquad ....(2)$$

Now substituting the values of λ's and μ's in (1) and (2), we get

$$P_n = \begin{cases} \dfrac{1}{n!}\left(\dfrac{\lambda}{\mu}\right)^n P_0 ; & 0 \leq n \leq c \\[3mm] \dfrac{1}{c^{n-c}c!}\left(\dfrac{\lambda}{\mu}\right)^n P_0 ; & c \leq n \leq N \end{cases}$$

and
$$P_0 = \left[\sum_{n=0}^{c-1} \frac{1}{n!} \left(\frac{\lambda}{\mu} \right)^n + \sum_{n=c}^{N} \frac{1}{c^{n-c} c!} \left(\frac{\lambda}{\mu} \right)^n \right]^{-1}$$

In this system, a queue will be formed, only when the number of customers is more than $n-c$. Let $m = n-c$. Then average queue length of M/M/c : (N/FCFS) system is given by

$$E(m) = \sum_{n=c}^{N} (n-c)P_n = \sum_{n=c}^{N} (n-c) \frac{1}{c^{n-c} c!} \left(\frac{\lambda}{\mu} \right)^n P_0$$

$$= \frac{1}{c!} \left(\frac{\lambda}{\mu} \right)^c \sum_{n=c}^{N} \frac{(n-c)}{c^{n-c}} \left(\frac{\lambda}{\mu} \right)^{n-c} P_0 \ ,$$

$$= \frac{1}{c!} \left(\frac{\lambda}{\mu} \right)^c \sum_{n=c}^{N} (n-c) \left(\frac{\lambda}{\mu c} \right)^{n-c} P_0 \ , \text{ putting } (n-c) = x$$

$$= \frac{1}{c!} \left(\frac{\lambda}{\mu} \right)^c P_0 \sum_{x=0}^{N-c} x \, \rho^x \ , \qquad \text{putting } \rho = \frac{\lambda}{\mu c}$$

$$= \frac{(c\rho)^c}{c!} P_0 \sum_{x=0}^{N-c} x \, \rho^x$$

$$= \frac{(c\rho)^c P_0}{c!} \sum_{x=0}^{N-c} \rho.x \, \rho^{x-1} = \frac{(c\rho)^c P_0}{c!} \rho \sum_{x=0}^{N-c} \left(\frac{d}{d\rho} \rho^x \right)$$

$$= \frac{(c\rho)^c}{c!} P_0 \rho . \frac{d}{d\rho} . \frac{(1-\rho^{N-c+1})}{1-\rho}$$

$$= \frac{(c\rho)^c}{c!} P_0 \rho . \frac{(1-\rho^{N-c+1}) - (1-\rho)(N-c+1)\rho^{N-c}}{(1-\rho)^2} \qquad \text{....(3)}$$

Again $\sum_{n=c}^{N} (n-c)P_n = \sum_{n=c}^{N} nP_n - \sum_{n=c}^{N} cP_n$

and
$$E(n) = \sum_{n=0}^{N} nP_n = \sum_{n=0}^{c-1} nP_n + \sum_{n=c}^{N} nP_n$$

$$= \sum_{n=0}^{c-1} nP_n + \sum_{n=c}^{N} cP_n - \sum_{n=c}^{N} cP_n + \sum_{n=c}^{N} nP_n$$

$$= \sum_{n=0}^{c-1} nP_n + \sum_{n=c}^{N} cP_n + \sum_{n=c}^{N} (n-c)P_n$$

$$= \sum_{n=0}^{c-1} nP_n + \sum_{n=c}^{N} cP_n + E(m)$$

$$= E(m) + \sum_{n=0}^{c-1} nP_n + \left(\sum_{n=0}^{N} cP_n - \sum_{n=0}^{c-1} cP_n \right)$$

$$= E(m) + c - P_0 \sum_{n=0}^{c-1} \frac{(c-n)(\rho c)^n}{n!} \qquad \text{....(4)}$$

Solved Examples

EXAMPLE 1. *Four counters are being run on the frontier of a country to check the passports and necessary papers of the tourists. The tourists choose counter at random. If the arrivals at the frontier is Poisson at the rate λ and the service time is exponential with parameter $\lambda/2$, what is the steady stage average queue at each counter.*

[MEERUT-2002]

SOLUTION. Here, we have

$$\lambda = \lambda, \quad \mu = \lambda/2, \quad c = 4 \Rightarrow \rho = \frac{\mu}{\mu c} = \frac{1}{2}$$

$$\Rightarrow P_0 = \frac{1}{\left[\displaystyle\sum_{n=0}^{c-1} \frac{(c\rho)^n}{n!} + \frac{(c\rho)}{c!(1-\rho)}\right]} = \frac{1}{\displaystyle\sum_{n=0}^{3} \frac{2^n}{n!} + \frac{2^4}{4!\left(1-\dfrac{1}{2}\right)}}$$

$$= \frac{1}{1 + \dfrac{2}{1!} + \dfrac{2^2}{2!} + \dfrac{2^3}{3!} + \dfrac{4}{3}} = \frac{3}{23}$$

Average queue length $= \dfrac{\rho}{(1-\rho)^2} \cdot \dfrac{(c\rho)^3}{c!} \cdot P_0$

$$= \frac{\dfrac{1}{2}}{\left(1-\dfrac{1}{2}\right)^2} \cdot \frac{2^4}{4!} \cdot \frac{3}{23} = \frac{4}{23}$$

19.13 MODEL-6: M/E_k/1 : (∞/FCFS)

A system with Poisson input, Erlangian service time with k phase, single channel, infinite capacity and first come first serve discipline.

Assume that

(i) Arrival of one unit means addition of k phases in the system.

(ii) Departure of one unit implies reduction of k phase in the system.

Step-1 : To find the system of steady state equations

Let n be the number of phases in the system at any instant *i.e.*, (number of phase waiting + number of phases working). In the service channel there are k phases and average servicing time in each phase is $1/k\mu$. Thus, if m units are waiting and the unit of service channel has to complete s phases, then the total no. of phases n in the system at this instant is given by

$$n = mk + s, \text{ [due to assumptions (i) and (ii)]}$$

Let $P_n(t + \Delta t)$ be the probability of the event that there are n phases in the queue at the time $(t + \Delta t)$. This event can happen in following ways :

(i) n phases are in the queue at time t and there is no arrival in the interval Δt and no phase is completed in the time Δt. The event will have the probability equal to $P_n(t)[1 - (\lambda + k\mu)\Delta t] + o(\Delta t)^2$.

(ii) $(n + 1)$ phases are in the queue at time t and one phase is completed in time Δt. This event will have the probability equal to
$$P_{n+1}(t) . k\mu\Delta t + o(\Delta t)^2$$

(iii) There are $(n - k)$ phases in the queue at time t and k phase are added in time Δt (i.e., one unit arrives in the system). This event will have the probability
$$P_{n-k}(t) . \lambda\Delta t + o(\Delta t)^2$$

All these three cases are mutually exclusive. Hence

$$P_n(t + \Delta t) = P_n(t)[1 - (\lambda + k\mu)\Delta t] + P_{n+1}(t) . k\mu\Delta t + P_{n-k}(t)\lambda\Delta t + o(\Delta t)^2 \quad ....(1)$$

Let $\Delta t \to 0$, we get

$$P_0'(t) = -\lambda P_0(t) + k\mu P_1(t), \qquad\qquad\qquad \text{for } n = 0$$

and $\qquad\qquad P_n'(t) = -(\lambda + k\mu)P_n(t) + k\mu P_{n+1}(t) + \lambda P_{n-k}(t), \quad \text{for } n \geq k.$

In steady state, $\lim\limits_{t \to \infty} P_n(t) \to P_n$ and hence $\lim\limits_{t \to \infty} P_n'(t) \to 0$.

$$\left.\begin{array}{ll} \lambda P_0 = k\mu P_1, & \text{for } n = 0 \\ (\lambda + k\mu)P_n = k\mu P_{n+1} + \lambda P_{n-k}, & \text{for } n \geq k \end{array}\right\} \quad ....(2)$$

Putting $\rho = (\lambda / k\mu)$ in first equation of (2) and dividing the second equation by $k\mu$, we get

$$P_1 = \rho P_0$$
$$(1 + \rho)P_n = P_{n+1} + \rho P_{n-k}, \; n \geq k \qquad\qquad ....(3)$$

Step-2 : Solution of Equations

The system of equations (3) can be solved with the help of generating functions. These are known as generating functions because the various terms generate the probabilities. A generating function is defined as

$$P(z) = \sum_{n=0}^{\infty} P_n z^n$$

Now multiplying both sides of the second equation of (3) by z^n and taking sum from $n = 1$ to ∞, we get

$$(1 + \rho) \sum_{n=1}^{\infty} P_n z^n = \sum_{n=1}^{\infty} P_{n+1} z^n + \rho \sum_{n=k}^{\infty} P_{n-k} z^n$$

$$\Rightarrow \qquad (1 + \rho) \sum_{n=1}^{\infty} P_n z^n + \rho P_0 = \sum_{n=1}^{\infty} P_{n+1} z^n + \rho \sum_{n=k}^{\infty} P_{n-k} + P_1 \qquad [\text{Since } P_1 = \rho P_0]$$

$$\Rightarrow \qquad (1 + \rho) \sum_{n=0}^{\infty} P_n z^n - P_0 = \sum_{n=0}^{\infty} P_{n+1} z^n = \rho \sum_{n=k}^{\infty} P_{n-k} z^n$$

$$= \frac{1}{z} \sum_{n=0}^{\infty} P_{n+1} z^{n+1} + \rho \sum_{n=k}^{\infty} P_{n-k} z^n \quad [\text{Since } P_j = 0 \text{ if } j < 0]$$

$$= \frac{1}{z} \sum_{n=0}^{\infty} P_{n+1} z^{n+1} + \rho z^k \sum_{m=0}^{\infty} P_m z^m, \; \text{where } m = n - k$$

$$= \frac{1}{z}\left[\sum_{n=0}^{\infty} P_n z^n - P_0\right] + \rho z^k \sum_{m=0}^{\infty} P_m z^m$$

$$(1 + \rho)P(z) - P_0 = (1 / z)[P(z) - P_0] + \rho z^k P(z)$$

$$\Rightarrow \qquad z(1 + \rho) P(z) - P_0 z = P(z) - P_0 + \rho z^{k+1} P(z)$$

$$\Rightarrow \qquad P(z)[(1 + \rho)z - 1 - \rho z^{k+1}] = -P_0(1 - z)$$

$$\Rightarrow \qquad P(z)[1 - z - \rho z(1 - z^k)] = P_0(1 - z)$$

$$\therefore \qquad P(z) = \frac{P_0}{1 - \rho z \left(\frac{1 - z^k}{1 - z} \right)} = P_0 \left[1 - \rho z \left(\frac{1 - z^k}{1 - z} \right) \right]^{-1},$$

[sum of infinite G.P.]

$$\therefore \qquad P(z) = P_0 \sum_{n=0}^{\infty} (\rho z)^n \left(\frac{1 - z^k}{1 - z} \right)^n$$

$$= P_0 \sum_{n=0}^{\infty} (\rho z)^n (1 + z + z^2 + ... + z^{k-1})^n \qquad ....(4)$$

If $z = 1$, we have from (4)

$$P(1) = P_0 \sum_{n=0}^{\infty} (\rho)^n \left(1 + 1 + 1^2 + .. + 1^{k-1}, \ i.e., k \text{ terms} \right)^n$$

$$= P_0 \sum_{n=0}^{\infty} (\rho k)^n = \frac{P_0}{1 - \rho k}$$

Also from definition of $P(z)$, we have

$$P(1) = \sum_{n=0}^{\infty} P_n(1)^n = \sum_{n=0}^{\infty} P_n = 1$$

Equating these two values of $P(1)$, we get

$$\frac{P_0}{(1 - \rho k)} = 1 \quad \text{or} \quad P_0 = (1 - \rho k). \qquad ....(5)$$

Now, P_n will be the coefficient of z^n in $P(z)$, so we have

$$P(z) = P_0 \sum_{m=0}^{\infty} (\rho z)^m \left(\frac{1 - z^k}{1 - z} \right)^m = P_0 \sum_{m=0}^{\infty} (\rho z)^m (1 - z^k)^m (1 - z)^{-m}$$

But $(1 - z^k)^m = \sum_{i=0}^{\infty} (-1)^i \binom{m}{i} z^{ik}$ and $(1 - z)^{-m} = \sum_{j=0}^{\infty} (-1)^j \binom{-m}{j} z^j$

$$= \sum_{j=0}^{\infty} \binom{m + j - 1}{j} z^j$$

$$\therefore \qquad P(z) = P_0 \sum_{m=0}^{\infty} (\rho z)^m \left[\sum_{i=0}^{m} (-1)^i \binom{m}{i} z^{ik} \right] \times \sum_{j=0}^{\infty} \binom{m + j - 1}{j} z^j$$

$$= P_0 \sum_{m=0}^{\infty} \rho^m \sum_{j=0}^{\infty} \sum_{i=0}^{\infty} (-1)^i \binom{m}{i} \binom{m + j - 1}{j} z^{j+ik+m}$$

Now, let $j + ik + m = n$

$$\therefore \qquad P_n = \text{Coefficient of } z^n \text{ in } P(z) = P_0 \Sigma \rho^m (-1)^i \binom{m}{i} \binom{m + j - 1}{j}$$

$$....(6)$$

In order to obtain P_n, we give different values to $j > i$ and m so that $j + ik + m$ becomes n.

Step-3 : Results and Conclusions

(i) Average number of phases in the system

$$E(n_p) = \frac{k(k+1)\rho}{2(1-k\rho)} \qquad \qquad ...(7)$$

Multiplying (3) by n^2 taking the sum from 1 to ∞, we get

$$(1+\rho)\sum_{n=1}^{\infty} n^2 P_n = \sum_{n=1}^{\infty} n^2 P_{n+1} + \rho \sum_{n=1}^{\infty} n^2 P_{n-k}$$

$$= \sum_{n=0}^{\infty} n^2 P_{n+1} + \rho \sum_{n=k}^{\infty} n^2 P_{n-k}, \qquad \text{(as } P_j = 0, j < 0)$$

$$= \sum_{m=1}^{\infty} (m-1)^2 P_m + \rho \sum_{m=0}^{\infty} (m+k)^2 P_m$$

(putting $n+1 = m$ in first sum and $n = m+k$ in second sum)

$$= \sum_{m=1}^{\infty} (m^2 - 2m + 1)P_m + \rho \sum_{m=0}^{\infty} (m^2 + 2km + k^2)P_m$$

$$= (1+\rho)\sum_{m=1}^{\infty} m^2 P_m + \sum_{m=1}^{\infty} (-2m+1)P_m + \rho \sum_{m=0}^{\infty} (2mk+k^2)P_m$$

$\therefore \quad 2\sum_{m=1}^{\infty} m P_m - \sum_{m=1}^{\infty} P_m - P_0 + P_0 = 2k\rho \sum_{m=0}^{\infty} m P_m + \rho k^2 \sum_{m=0}^{\infty} P_m$

$\therefore \qquad 2(1-k\rho)\sum_{m=1}^{\infty} m P_m = (1+\rho k^2) - P_0 = 1 + \rho k^2 - 1 + k\rho$

$$2(1-k\rho)E(n_p) = (1+k)\,k\rho \qquad \qquad \therefore \ E(n_p) = \frac{(1+k)k\rho}{2(1-k\rho)}$$

Now k phases mean one unit in the system.

$\therefore$ Average number of units in the system $= E(n) = \dfrac{E(n_p)}{k}$

$$= \frac{(1+k)\rho}{2(1-k\rho)} \qquad \qquad ...(8)$$

$$= \frac{k+1}{2k}\frac{\lambda}{\mu-\lambda} \qquad \qquad [\text{putting } \rho = \frac{\lambda}{\mu k}]$$

(ii) Average waiting time of the phase in the system

$$= \frac{\text{Average number of phases}}{\mu} = \frac{(1+k)}{2\mu}\frac{k\rho}{1-k\rho} \qquad ...(9)$$

(iii) Average waiting time of an arrival

$$E(w) = \frac{\text{Average numbr of phases}}{k} = \frac{(1+k)\rho}{2\mu(1-k\rho)} \qquad ...(10)$$

$$= \frac{k+1}{2k}\cdot\frac{\lambda}{\mu(\mu-\lambda)}.$$

(iv) Average time an arrival spends in the system

$$E(v) = \text{Average waiting time of an arrival} + (1/\mu)$$

$$= \frac{(k+1)\rho}{2\mu(1-k\rho)} + \frac{1}{\mu}. \qquad \qquad ...(11)$$

19.14 MODEL-7: M/E_K/1 : (1/FCFS)

The capacity of the system is unity i.e. there is no queue and the system accommodates only one unit which is served through k-phases.

We have the following difference equations of the system.

STEP-1 : To find the system of the steady state equations

$$P_n(t + \Delta t) = P_n(t) \text{ [Probability that no phase is served in time } \Delta t \text{]}$$
$$+ P_{n+1}(t) \text{ [Probability that one phase is served in time } \Delta t \text{]}$$
$$= P_n(t)[1 - k\mu\Delta t] + P_{n+1}(t)k\mu\Delta t + o(\Delta t)^2, \text{ for } 1 \le n < k.$$
$$P_k(t + \Delta t) = P_k(t) \text{ [Probability that no phase serviced in time } \Delta t \text{]}$$
$$+ P_0(t) \text{ [Probability of arrival of a unit in time } \Delta t \text{]}$$
$$= P_k(t)[1 - k\mu\Delta t] + P_0(t)\lambda\Delta t + o(\Delta t)^2, \text{ for } n = k$$
$$P_0(t + \Delta t) = P_0(t) \text{ [Probability of no arrival in } \Delta t \text{]}$$
$$+ P_1(t) \text{ [Probability of one phase being served in } \Delta t \text{]}$$
$$= P_0(t)(1 - \lambda\Delta t) + P_1(t)k\mu\Delta t + o(\Delta t)^2, \text{ for } n = 0$$

Setting limit $\Delta t \to 0$, we shall have the following differential difference equations for the system

$$\left. \begin{array}{ll} P_n'(t) = -k\mu P_n(t) + k\mu P_{n+1}(t), & 1 \le n < k \\ P_k'(t) = -k\mu P_k(t) + \lambda P_0(t), n = k \\ P_0'(t) = -\lambda P_0(t) + k\mu P_1(t), & n = 0 \end{array} \right\} \quad ....(1)$$

Under steady state system, $\lim_{t\to\infty} P_n(t) \to P$ and $P_n'(t) \to 0$ for all n, and then from (1), we have

$$k\mu P_n = k\mu P_{n+1}, \ 1 \le n < k$$
$$k\mu P_k = \lambda P_0, \ n = k$$
$$\lambda P_0 = k\mu P_1, \ n = 0$$

STEP-2 : To solve the steady State Equations

$$\left. \begin{array}{ll} P_{n+1} = P_n & \text{for } 1 \le n < k \\ P_k = \left(\dfrac{\lambda}{k\mu}\right)P_0, & \text{for } n = k \\ P_1 = \left(\dfrac{\lambda}{k\mu}\right)P_0, & \text{for } n = 0 \end{array} \right\}$$

i.e.,
$$n = 1, \qquad P_2 = P_1$$
$$n = 2, \qquad P_3 = P_2$$
$$\vdots \qquad\qquad \vdots$$
$$\vdots \qquad\qquad \vdots$$
$$n = k - 1, \qquad P_k = P_{k-1}$$

i.e.,
$$P_1 = P_2 = ... = P_k$$

But from middle equation, we have $P_k = (\lambda/k\mu)P_0$. Hence

$$P_1 = P_2 = ... = P_k = (\lambda/k\mu)P_0$$

Now, $\sum\limits_{i=0}^{k} P_i = 1$ (Since capacity of the system is of one unit, *i.e.*, of k phases).

i.e.,
$$P_0 + \sum\limits_{i=1}^{k} P_i = 1 \quad \text{or} \quad P_0 + \sum\limits_{i=1}^{k}\left(\frac{\lambda}{k\mu}\right)P_0 = 1$$

or
$$\left(1+\frac{\lambda}{\mu}\right)P_0 = 1 \quad \text{or} \quad P_0 = \frac{1}{1+\rho} \qquad \qquad ....(2)$$

and hence
$$P_n = \frac{\lambda}{k\mu}\cdot\frac{1}{1+\rho} = \frac{1}{k}\cdot\frac{\rho}{1+\rho} \qquad \qquad ....(3)$$

19.15 MACHINE REPAIR PROBLEM

Whenever a machine breaks down, it will result a great loss to the organisation. Therefore, the machine repair problems are very important problems in queuing theory.

The simplest case of such problems can be a system having n machines and one mechanic for them. If any machine breaks down, the mechanic is to repair it and if during this time another machine breaks down then it can only be attended by the mechanic when first one is repaired. The time taken by the mechanic to repair any machine cannot be predicted in advanced and similarly the rate of breakdown of other machines in the system is unpredictable.

If different parts of a machine are repaired in k-phases, then the output distribution, will be the well known Erlangian distribution.

1. **Problems of Interest in connection with machine interference.**

   Here arrival rate λ will depend upon the number of machines in order at any time and the service rate μ will be constant.

   Given n machines in a system there may be many problems. Whenever, any machine in a system breaks down, it results in loss of production till it is repaired. Here, we are interested in the solution of following problems:

   (i) To find the optimum number of machines per mechanic which is most economic.

   (ii) Increasing the number of machines 'n' to eliminate the stoppages by keeping extra machines.

   (iii) There may be two alternatives, a system with n machines per mechanic and second with n' machines per mechanic, then to select the best process.

2. **M/M/1 :** Let there be n machines in the system. The arrival rate and servicing rate follow Poisson distribution with parameters λ and μ respectively. Now, if any particular machine is in working order at a time t then the probability that it will join the queue for repair in time Δt will be given by
$$\lambda\Delta t + o(\Delta t)^2, \qquad \qquad ....(1)$$
Similarly if any machine is taken for service at time t, then the probability that its service will be completed in time Δt, will be given by
$$\mu\Delta t + o(\Delta t)^2. \qquad \qquad ....(2)$$
Without loss of generality we assume that any machine on breakdown, is serviced immediately provided there is no other machine in the servicing system.

Suppose at any time t, 'b' machines are not working and are in queue for servicing. Naturally $n \geq b$. Then in a small interval of time i.e., Δt, following events are possible. The system is assumed to be in steady state.

(i) There are b machines in the queue for repair at time t and no unit arrives and none is serviced during the interval Δt. The probability of this event is
$$P_b(t)\,[1 - \lambda b\mu)\Delta t] \ + \text{terms of order } (\Delta t)^2.$$

(ii) There are $(b-1)$ machines in the queue at time t and one more machine arrives upto time $(t + \Delta t)$. The probability of this event is $P_{b-1}(t)\lambda_{b-1}\Delta t \ + \text{terms of order } (\Delta t)^2.$

(iii) There are $(b+1)$ machines at time t and one machine is served during the time $(t + \Delta t)$. The probability of this event is $P_{b+1}(t)\mu\Delta t \ + \text{terms of order } (\Delta t)^2.$

Since, all these three cases are mutually exclusive, therefore,
$$P_b(t + \Delta t) = P_b(t)\,[1 - (\lambda_b + \mu)\Delta t] + P_{b-1}(t)\lambda_{b-1}\Delta t + P_{b+1}(t)\mu\Delta t$$
$$+ \text{terms of order } (\Delta t)^2 \qquad \qquad ....(3)$$

In case b = 0, the case (ii) will not arise, and in that case
$$P_0(t + \Delta t) = P_0(t)\,[1 - (\lambda_0 + \mu)\Delta t] + P_1(t)\mu\Delta t + o(\Delta t)^2 \qquad ....(4)$$
and when $b = n$, i.e., all machines of the system are in the queue for repair, their case will not arise, and then
$$P_n(t + \Delta t) = P_n(t)\,[1 - (\lambda_n + \mu)\Delta t] + P_{n-1}(t)\lambda_{n-1} + o(\Delta t)^2 \qquad ....(5)$$
Now, $\lambda_b = (n-b)\lambda$ and $\lambda_n = 0$.
Hence, for $\Delta t \to 0$, (3) can be written as
$$P_b'(t) = -[(n-b)\lambda + \mu]P_b(t) + (n-b+1)\lambda P_{b-1}(t) + \mu P_{b+1}(t) \qquad ....(6)$$
Similarly, from (4) and (5)
$$\left.\begin{array}{l} P_0'(t) = -n\lambda P_0(t) + P_1(t)\mu \\[2mm] P_n'(t) = -\mu P_n(t) + \lambda P_{n-1}(t) \end{array}\right\} \qquad \qquad ....(7)$$

In steady state for $t \to \infty$, $P_b(t) \to P_b$ and $P_b'(t) \to 0$.
Hence, from (6) and (7), we have the following system of equations
$$\left.\begin{array}{ll} [(n-b)\lambda + \mu]P_b = (n-b+1)\lambda P_{b-1} + \mu P_{b+1} \\[2mm] n\lambda P_0 = \mu P_1, & \text{for } b = 0 \\[2mm] \mu P_n = \lambda P_{n-1}, & \text{for } b = n \end{array}\right\} \qquad ....(8)$$

Now putting $b = 1$ in the 1st equation for (8), we get
$$[(n-1)\lambda + \mu]P_1 = \lambda_n P_0 + \mu P_2$$
i.e., $\qquad (n-1)\lambda P_1 = \mu P_2$, since $\mu P_1 = \lambda_n P_0 \qquad ....(9)$

Similarly putting $b = 2$ in 1st equation of (8), we get
$$[(n-2)\lambda + \mu]P_2 = \lambda(n-1)P_1 + \mu P_3$$
i.e., $\qquad (n-2)\lambda P_2 = \mu P_3$, using (9) $\qquad\qquad ....(10)$

and so on putting $b = 3, 4, ...$, we get a recurrence formula
$$(n-b)\lambda P_b = \mu P_{b+1}, \quad \text{for } b = 0, ..., n-1 \qquad\qquad ....(11)$$
Putting $b = n-1, n-2, ..., 1$ in (11), we get
$$\lambda P_{n-1} = \mu P_n \quad \text{or} \quad P_{n-1} = (\mu / \lambda)P_n$$

$$P_{n-2} = (\mu / 2\lambda)P_{n-1} = 1/2(\mu / \lambda)^2 P_n$$

$$P_{n-3} = \frac{\mu}{3\lambda} P_{n-2} = \frac{1}{3!} \left(\frac{\mu}{\lambda} \right)^3 P_n$$

$$\vdots \qquad\qquad \vdots$$

$$P_{n-k} = \frac{1}{k!} \left(\frac{\mu}{\lambda} \right)^k P_n$$

$$\vdots \qquad\qquad \vdots$$

$$P_0 = \frac{1}{n!} \left(\frac{\mu}{\lambda} \right)^n P_n \qquad\qquad ....(12)$$

Now, $\qquad \sum\limits_{b=0}^{n} P_b = 1$, i.e., $(P_0 + P_1 + ... + P_n) = 1$.

i.e., $\qquad \left[\frac{1}{n!} \left(\frac{\mu}{\lambda} \right)^n P_n + \frac{1}{(n-1)!} \left(\frac{\mu}{\lambda} \right)^{n-1} P_n + ... + P_n \right] = 1$

$$P_n = \frac{1}{1 + \left(\dfrac{\mu}{\lambda} \right) + ... + \dfrac{1}{n!} \left(\dfrac{\mu}{\lambda} \right)^n} \qquad\qquad ....(13)$$

Here P_0 is the probability that all the machines in the system are in working order and the repairman is idle. Now we shall find expected waiting length for the problem.

Expected waiting length: If at any time t there are b machines for repair in the queue including the one in service, then the waiting length will be $b - 1$ and the average waiting length will be given by

$$L_q = E(b-1) = \sum\limits_{b=1}^{n} (b-1)P_b = \sum\limits_{b=1}^{n} bP_b - \sum\limits_{b=1}^{n} P_b$$

Now, $bP_b = 0$ for $b = 0$.

$$= \sum\limits_{b=0}^{n} bP_b - \left[\sum\limits_{b=0}^{n} P_b - P_0 \right] = \sum\limits_{b=0}^{n} bP_b - 1 + P_0 , \text{ since } \sum\limits_{b=0}^{n} P_b = 1 . \quad ....(14)$$

Now from (8), we have $(n-b)\lambda P_b = \mu P_{b+1}$

i.e., $\qquad b\lambda P_b = n\lambda P_b - \mu P_{b+1} .$

$\Rightarrow \qquad \lambda \sum\limits_{b=0}^{n} bP_b = n\lambda \sum\limits_{b=0}^{n} P_b - \mu \sum\limits_{b=0}^{n} P_{b+1} = n\lambda - \mu \sum\limits_{b=0}^{n-1} P_{b+1}$

$$= n\lambda - \mu[P_1 + P_2 + ... + P_n] = n\lambda - \mu[1 - P_0], \quad (\because \sum\limits_{b=0}^{n} P_i = 1 .)$$

$\therefore \qquad \sum\limits_{b=0}^{n} bP_b = \dfrac{n\lambda - \mu[1 - P_0]}{\lambda} = n - \left(\dfrac{\mu}{\lambda} \right)[1 - P_0]$

Putting the value of $\sum\limits_{b=0}^{n} bP_b$ in the expression for L_q, we get

$$L_q = n - (\mu / \lambda)(1 - P_0) - (1 - P_0) = n - (1 - P_0)[(\mu / \lambda) + 1] \qquad ....(15)$$

The average contribution of a machine to the waiting line will be equal to (L_q / n). This is also known as coefficient of loss for machines.

Miscellaneous Examples

EXAMPLE 1. *A TV repairman finds that the time spent on his jobs has an exponential distribution with mean 30 minutes. If he repairs sets in the order in which they came in and if the arrival of sets is approximately Poisson with an average rate of 10 per 8 hour day, what is the repairman's expected idle time each day? How many jobs are ahead for the average sets jobs brought in ?* [AGRA 1998]

SOLUTION. It is a M/M/1 : FIFO problem with $\mu = (1/30) \times 60 = 2$ sets per hour and $\lambda = 10/8 = 5/4$ sets per hour.

$$\rho = \lambda / \mu = 5 / (4 \times 2) = 5/8.$$

Now the probability that there is no unit in the queue

$$P_0 = 1 - \rho = 1 - 5/8 = 3/8.$$

Hence the expected idle time for the repairman in 8 hour day

$$= [P_0 \times 8] = 3/8 \times 8 = 3 \text{ hours}.$$

Again

$$E(n) = \frac{\lambda}{\mu - \lambda} = \frac{5/4}{2 - (5/4)} = 1\frac{2}{3} \text{ jobs}.$$

$\Rightarrow 1\frac{2}{3}$ jobs are ahead of the average set just brought in.

EXAMPLE 2. *Arrivals at a telephone booth are considered to be following Poisson law of distribution with an average time of 10 minutes between one arrival and the next. Length of a phone call is assumed to be distributed exponentially with mean 3 minutes.*

(i) What is the probability that a person arriving at the booth will have to wait?

(ii) What is the average length of queue that form from time to time?

(iii) The telephone department will install a second booth when convinced that an arrival would expect to wait at least three minutes for the phone. By how much must the flow of arrivals be increased in order to justify a second booth?

[MEERUT 2002(B.P), 03(B.P.), 04, 2018]

SOLUTION. Since the time interval between two successive arrivals is given to be 10 minutes, the mean arrival rate is (1/10) per minute.

i.e., $\lambda = (1/10)$ per minute.

Similarly $\mu = 1/3 = 0.33$ per minute.

(i) Now, the probability that an arrival does not wait on arriving $= P_0 = (1 - \rho)$.

∴ Probability that an arrival has to wait

$$= 1 - \text{Probability [an arrival does not wait on arriving]}$$

$$= 1 - P_0.$$

$$= 1 - (1 - \rho) = \rho = \frac{\lambda}{\mu} = \frac{0.10}{0.33} = 0.3$$

(ii) Average length of non-empty queue is $E[L / L > 0]$

$$= \frac{1}{1-\rho} = \frac{1}{1-0.3} = 1.43 \text{ persons.}$$

(iii) The installation of second booth will be justified if the waiting time is greater than or equal to 3. If the new arrival rate is λ', then for $\mu = 0.33$.

Putting $\lambda'/\mu = \rho$, we get

$$E(w) = \frac{\lambda'}{\mu(\mu - \lambda')} \geq 3$$

or $\qquad \lambda' = 3\mu^2 - 3\mu\lambda'$ or $\lambda' = \frac{3\mu^2}{1+3\mu}$

or $\qquad \lambda' \geq 0.16$ persons per minute.

The arrival rate should be at least 0.16 persons per minute or say one arrival in every six minutes (10 arrivals per hour) to justify the second booth.

EXAMPLE 3. *Let on average 96 patients per 24 hour day require the service of an emergency clinic. Also on average, a patient requires 10 minutes of active attention. Assume that the facility can handle only one emergency at a time. Suppose that it costs the clinic Rs. 100 per patient treated to obtain an average servicing time of 10 patients and that each minute of decrease in this average time would cost Rs. 10 per patient treated. How much would have to be budgeted by the clinic to decrease the average size of the queue from one and one third patient to half a patient?* [MEERUT-1997]

SOLUTION. Here, $\qquad \lambda = \dfrac{96}{24 \times 60} = \dfrac{1}{15}$ patients/minute

$\mu = 1/10$ patients/minutes

$\rho = \lambda/\mu = 4/6 = 2/3$.

Average number of patients in the waiting line

$$= \frac{\rho^2}{1-\rho} = \frac{(2/3)^2}{1-(2/3)} = \frac{4}{3} = 1\frac{1}{3}$$

Fraction of the time during which there are no patients $= 1 - \rho = 1/3$.

Average size of the queue $= E(L_q) = \dfrac{\rho^2}{1-\rho} = \dfrac{\lambda^2}{\mu(\mu-\lambda)}$

Here, $\qquad E(L_q) = 1/2, \lambda = 1/15$

$\therefore \qquad \dfrac{1}{2} = \dfrac{(1/15)^2}{\mu[\mu - (1/15)]}$

$\Rightarrow \qquad \mu = 2/15$ patients/minutes

$\therefore$ average rate of treatment required $= \dfrac{1}{\mu} = \dfrac{15}{2} = 7\dfrac{1}{2}$ minutes

So the decrease in the average rate of treatment $= 10 - 7\dfrac{1}{2} = 2\dfrac{1}{2}$

$\therefore$ Budget per patient $= ₹\ 100 + (5/2) \times 10 = ₹\ 125$.

$\therefore$ In order to get the required size of the queue, the budget should be increased from ₹ 100 per patient to ₹ 125 per patient.

EXAMPLE 4. *There is congestion on the platform of a railway station. The trains arrive at the rate of 30 trains per day. The waiting time for any train to hump is exponentially distributed with an average of 36 minutes. Calculate the following:*
(a) The mean queue size.
(b) The probability that queue size exceeds 9.

SOLUTION. Here $1/\lambda = (60 \times 24)/30 = 48$ minutes.

and $1/\mu = 36$ minutes, *i.e.*, $\rho = (\lambda/\mu) = 0.75$.

(a) The mean queue size,

$$E(n) = \frac{\rho}{1-\rho} = \frac{0.75}{0.25} = 3 \text{ trains.}$$

(b) Probability (queue size exceeds 9) = Probability (queue size ≥ 10).

$= 1 -$ Probability (queue size is less than 10)

$= 1 - (P_0 + P_1 + ... + P_9) = 1 - P_0(1 + \rho + ... + \rho^9)$

$$= 1 - \left[(1-\rho)\left(\frac{1-\rho^{10}}{1-\rho} \right) \right]$$

$= 1 - (1-\rho^{10}) = \rho^{10} = 0.07$ approximately.

EXAMPLE 5. *A super market has two girls ringing up sales at the counters. If the service time for each customer is exponential with mean 4 minutes, and if people arrive in a Poisson fashion at the rate of 10 an hour,*
(a) What is the probability of having to wait for service?
(b) What is the expected percentage of idle time for each girl?

[MEERUT 1999, 2004; AGRA 2001]

SOLUTION. This is an example of M/M/c model, where $c = 2$.

Here we have, $\mu = \dfrac{1}{4}$ services per minute

$$\lambda = \frac{1}{60/10} = \frac{1}{6} \text{ people per minute}$$

$c = 2.$

$$\rho = \frac{\lambda}{c\mu} = \frac{1/6}{2 \times (1/4)} = \frac{1}{3}.$$

Putting the values of c in $P_0 = \dfrac{1}{\left[\displaystyle\sum_{n=1}^{c-1} \dfrac{(c\rho)^n}{n!} + \dfrac{(c\rho)^c}{c!(1-\rho)} \right]}$, we get

$$P_0 = \frac{1}{\left[\displaystyle\sum_{n=0}^{2-1} \dfrac{(2\rho)^n}{n!} + \dfrac{(2\rho)^2}{n!(1-\rho)} \right]} = \frac{1}{(1+2\rho) + \dfrac{4\rho^2}{2(1-\rho)}}$$

$$P_0 = \frac{1}{1 + \dfrac{2}{3} + \dfrac{1}{2!}\left(\dfrac{2}{3}\right)^2 \cdot \dfrac{1}{1 - \left(\dfrac{1}{3}\right)}}, \text{ as } \rho = \frac{1}{3}$$

$$= \frac{1}{2}.$$

also, $P_1 = (\lambda / \mu) \times P_0 = (2/3) \times (1/2) = (1/3)$.

(a) Now the probability that a customer has to wait

= the probability that number of customers in the system is greater than equal to 2.

= $P[n \geq 2] = 1 - P[n < 2] = 1 - P_0 - P_1 = 1 - (1/2) - 1/3 = 0.167$.

(b) Now, we are to find the expected number of girls who are idle.

Let X denote the number of idle girls. $X = 2$ when the system is empty and both girls are free. $X = 1$ when the system contains only one unit and one of the girls is free. Hence X can take two values 2 or 1 with probability P_0 and P_1 respectively.

$$\therefore \qquad E(X) = X_1 P(X = X_1) + X_2 P(X = X_2)$$

$$= (2 \times P_0) + (1 \times P_1) = (2 \times 1/2) + 1 \times 1/3) = 4/3$$

$\therefore$ The probability of any girl being idle

$$= \frac{\text{Expected number of idle girls}}{\text{Total number of girls}} = \frac{4/3}{2} = 0.67$$

$\therefore$ Expected percentage of idle time for each girl = 67%.

EXAMPLE 6.   *A bank has two tellers working on saving accounts. The first teller handles withdrawals only. The second teller handles deposits only. It has been found that the service time distributions for both deposits and withdrawals are exponential with mean service time 3 minutes per customer. Depositors are found to arrive in a Poisson fashion with mean arrival rate 16 per hour and withdrawers arrive in a Poisson fashion with mean arrival rate 14 per hour. What would be the effect on the average waiting time for depositors and withdrawers if each teller could handle both withdrawers and depositors? What would be the effect, if this could only be accomplished by increasing the mean service time to 3.5 minutes?*

[DELHI MBA-1987; GUJRAT MBA -1988, 2000; AGRA-1998]

SOLUTION.   Mean service rate μ for both tellers = $(1/3)$ per minute = 20 per hour. Mean arrival rate of depositors $\lambda_1 = 16$ per hour.

Treating depositors and withdrawers as units of M/M/1 system with one teller do attending depositors and the other one withdrawers, we get

Expected waiting time for depositors $= E(w_1) = \dfrac{\lambda_1}{\mu(\mu - \lambda_1)}$,

$$= \frac{16}{20(20 - 16)} = \frac{1}{5} \text{ hours} = 12 \text{ minutes.}$$

Now the arrival rate of withdrawers, *i.e.*, $\lambda_2 = 14$ per hour.

$\therefore$ Expected waiting time for withdrawers $E(w_2)$,

$$= \frac{\lambda_2}{\mu(\mu - \lambda_2)} = \frac{14}{20(20 - 14)} \text{ hours} = 7 \text{ minutes.}$$

If both tellers do service for withdrawers and depositors, then the problem becomes that of two service stations with $\lambda = \lambda_1 + \lambda_2 = 16 + 14 = 30$, $\mu = 20$ per hour and $c = 2$.

Then using $P_0 = \dfrac{1}{\left[\displaystyle\sum_{n=1}^{c-1} \dfrac{(c\rho)^n}{n!} + \dfrac{(c\rho)^c}{c!(1-\rho)}\right]}$, $P_0 = 1/7$, and expected waiting time is

given by

$$E(w) = \dfrac{\rho}{\lambda(1-\rho)^2} \times \dfrac{1}{c!}\left(\dfrac{\lambda}{\mu}\right)^c P_0 \text{ , where } \rho = \dfrac{\lambda}{c\mu}$$

$$= 27/7 = 3.86 \text{ minutes.}$$

Hence when both tellers handle both withdrawals and deposits then expected waiting time is reduced. If mean service time is increased to 3.5 minutes, then

$$\mu = (1/3.5) \times 60 = (120/7) \text{ per hour.}$$

$$\dfrac{\lambda}{c\mu} = \dfrac{30}{2 \times 120} \times 7 = \dfrac{7}{8}$$

and proceeding as before,

$$P_0 = (1/15)$$

and $E(w) = 11.4$ minutes.

EXAMPLE 7. *A telephone exchange has two long distance operators. The telephone company finds that during the peak load, long distance calls arrive in a Poisson fashion at an average rate of 15 per hour. The length of service on these calls is approximately exponentially distributed with mean length 5 minutes. What is the probability that a subscriber will have to wait for this long distance call during the peak hours of the day? If subscribers will wait and serviced in turn, what is the expected waiting time?* [AGRA–2000]

SOLUTION. It is an example of the multi-channel problem with $c = 2$.

$\lambda = 15$ per hour and $\mu = (60/5) = 12$ per hour

$$\rho = \dfrac{\lambda}{c\mu} = \dfrac{15}{2 \times 12} = \dfrac{5}{8}$$

Using $P_0 = \dfrac{1}{\left[\displaystyle\sum_{n=1}^{c-1} \dfrac{(c\rho)^n}{n!} + \dfrac{(c\rho)^c}{c!(1-\rho)}\right]}$, we get

$$P_0 = \dfrac{1}{1 + \dfrac{5}{4} + \dfrac{1}{2}\cdot\dfrac{25}{16}\cdot\dfrac{1}{(1-5/8)}} = \dfrac{12}{52}$$

and $\qquad P_1 = \dfrac{\lambda}{\mu} \times P_0 = \dfrac{5}{4} \times \dfrac{12}{52} = \dfrac{15}{52}.$

(i) Now Probability that a subscriber has to wait

$$= P[n \geq 2] = 1 - P_0 - P_1 = \left(1 - \dfrac{12}{52} - \dfrac{15}{52}\right) = \dfrac{25}{52} = 0.48 .$$

(ii) expected waiting time $E(w)$

$$= \frac{\rho}{\lambda(1-\rho)^2} \times \frac{1}{c!}\left(\frac{\lambda}{\mu}\right)^c P_0.$$

$$= \frac{(5/8)}{15[1-(5.8)^2]} \times \frac{1}{2!}\left(\frac{15}{12}\right)^2 \times \frac{12}{52} \text{ hours} = 3.2 \text{ minutes}.$$

EXAMPLE 8. *An insurance company has three claim adjusters in its branch office. People with claim against the company are found to arrive in a Poisson fashion at an average rate of 20 per 8 hours day. The amount of time that an adjuster spends with a claimant is found to have an exponential distribution with mean service time 40 minutes. Claimants are processed in the order of their appearance. How many hours a week can an adjuster expect to spend with claimants?*   [ROHILKHAND 1993]

SOLUTION. This is the example of M/M/c : (∞/FIFO) model with c equal to three. Here mean arrival rate $\lambda = (20/8)$ per hour $= (5/2)$
and $\mu = (1/40)$ service per minute $= (60/40)$ per hour $= (3/2)$

Hence, $\rho = \dfrac{\lambda}{c\mu} = \dfrac{(5/2)}{3\times(3/2)} = \dfrac{5}{9}$

and using $P_0 = \dfrac{1}{\left[\displaystyle\sum_{n=1}^{c-1}\dfrac{(c\rho)^n}{n!} + \dfrac{(c\rho)^c}{c!(1-\rho)}\right]}$, we have

   $P_0 = (24/139).$

Using $P_c = \dfrac{\lambda}{\mu c}\cdot\left(\dfrac{\lambda}{\mu}\right)^{c-1}\cdot\dfrac{1}{(c-1)!}\cdot P_0$ and putting c = 1 and 2, we get

   $P_1 = (40/139)$ and $P_2 = (100/417).$

∴   Expected number of idle adjusters at any specified instant =

$3P_0$	+	$2P_1$	+	$1P_2$	$= 4/3$
(all three idle)		(two idle)		(one idle)	

Hence the probability that any adjuster is idle

$$= \frac{\text{Expected number of idle adjusters}}{\text{Total number of adjusters}} = \frac{4}{3\times 3} = \frac{4}{9}$$

∴ Probability that no adjuster is idle $= 1 - 4/9 = 5/9$

i.e., Expected weekly time that an adjuster spends with claimants

   $= 5/9 \times 40$ (treating 5 working days in a week) $= 22.2$ hours.

EXAMPLE 9. *Let there be an automobile inspection situation with three inspection stalls. Assume that cars wait in such a way that when a stall becomes vacant, the car at the head of line pulls upto it. The station can accommodate at most four cars waiting (seven in the station) at one time. The arrival pattern is Poisson with a mean of one car every minute during the peak hours. The service time is exponential with mean 6 minutes. Find the average number of customers in the system during peak hours, the average waiting* *time, and the average number per hour that cannot enter the station because of full capacity.*

SOLUTION. This is a problem of M/M/c : N/FIFO model with the following values.

λ = 1 car per minute, μ = 1/6 car per minute,

c = 3 inspection stalls, N = 7 cars.

$$\rho = \frac{\lambda}{c\mu} = \frac{1}{3 \cdot \frac{1}{6}} = 2$$

Now

$$P_0 = \left[\sum_{n=0}^{c-1} \frac{1}{n!}\left(\frac{\lambda}{\mu}\right)^n + \sum_{n=c}^{N} \frac{1}{c^{n-c}c!}\left(\frac{\lambda}{\mu}\right)^n \right]^{-1}$$

So, $$\sum_{n=c}^{N}\left[\frac{1}{c^{n-c}c!}\left(\frac{\lambda}{\mu}\right)^n \right] = \sum_{n=c}^{N} \frac{c^c}{c!}\left(\frac{\lambda}{c\mu}\right)^n = \frac{c^c}{c!}\sum_{n=c}^{N}\left(\frac{\lambda}{c\mu}\right)^n = \frac{c^c}{c!}\sum_{n=3}^{7}\left(\frac{\lambda}{c\mu}\right)^n$$

$$= \frac{c^c}{c!}\left(\frac{\lambda}{c\mu}\right)^3\left[1 + \left(\frac{\lambda}{c\mu}\right) + \left(\frac{\lambda}{c\mu}\right)^2 + \left(\frac{\lambda}{c\mu}\right)^3 + \left(\frac{\lambda}{c\mu}\right)^4 \right]$$

$$= \frac{c^c}{c!}\left(\frac{\lambda}{c\mu}\right)^3 \frac{1 - \left(\frac{\lambda}{c\mu}\right)^5}{1 - \left(\frac{\lambda}{c\mu}\right)},$$

being the sum of G.P., with common ratio $\lambda / c\mu$

$$= \frac{1}{3!}\left(\frac{\lambda}{\mu}\right)^3 \frac{1 - 2^5}{1 - 2} = \frac{1}{3!}6^3\frac{1 - 2^5}{1 - 2}$$

$\therefore$ $$P_0 = \left[\sum_{n=0}^{2} \frac{1}{n!}6^n + \frac{1}{3!}6^3\left(\frac{1-2^5}{1-2}\right) \right]^{-1}$$

$$= \left[1 + 6 + \frac{6^2}{2!} + 6^2(32 - 1) \right]^{-1} = [1141]^{-1}$$

Also, $E(m)$ = Average queue length

$$= \frac{(c\rho)^c}{c!}P_0 \cdot \rho \frac{(1 - \rho^{N-c+1}) - (1 - \rho)(N - c + 1)\rho^{N-c}}{(1-\rho)^2}$$

$$= \frac{(6)^3}{3!}P_0 \cdot 2\frac{(1 - 2^{7-3+1}) - (1-2)(7 - 3 + 1)2^{7-3}}{(1-2)^2}$$

$$= \frac{(6)^3}{3!}P_0 \cdot 2[(1 - 2^5) + 5 \cdot 2^4]$$

$$= \frac{(6)^3}{3!}P_0 \cdot 2[80 + 1 - 32] = 72P_0[49]$$

$$= \frac{72 \times 49}{1141} = 3.09 \text{ cars.}$$

Average number of cars waiting in the system $= m\, E(n)$

$$= E(m) + c - P_0 \sum_{n=0}^{c-1} \frac{(c-n)(\rho c)^n}{n!}$$

Here, $\sum_{n=0}^{c-1} \frac{(c-n)(\rho c)^n}{n!} = \left[c + (c-1)\rho c + \frac{(c-2)(\rho c)^2}{2!} \right]$

$\therefore \qquad E(n) = 3.09 + 2 - \frac{1}{1141} \left[3 + 2(6) + \frac{1(6)^2}{2!} \right]$

$$= 3.09 + 3 - \frac{1}{1141}[15 + 18] = 6.09 - \frac{33}{1141}$$

$$= 6.09 - 6.029 = 6.061 \text{ cars}$$

The average waiting time

$$E(v) = \frac{E(n)}{\lambda(1 - P_N)} = \frac{6.06}{1 - \frac{1}{1141} \cdot \frac{6^7}{3^4 \cdot 3!}} = 12.3 \text{ minutes}$$

And the expected number of cars that cannot enter the system

$$= 60\lambda\, P_N = 60 P_N$$

$$= 60 \frac{P_0 6^7}{3^4 \cdot 3!} = 30.4 \text{ car per hour.}$$

EXAMPLE 10. *Repairing a certain type of machine which breaks down in a given factory, consists of five basic steps that must be performed sequentially. The time taken to perform each of the five steps is found to have an exponential distribution with mean 5 minutes and is independent of other steps. If these machines break down in a Poisson fashion at an average rate of two per hour, and if there is only one repairman, what is the average idle time for each machine that has broken down?*

SOLUTION. The repair-time in the example will follow an Erlang distribution with $K = 5$ phases and average service time $= 5 \times 5 = 25$ minutes *i.e.*, $\mu = 1/25$.
Average inter-arrival time is 30 minutes *i.e.*, $\lambda = 1/30$.

$$\rho = \left(\frac{\lambda}{k\mu} \right) = \left(\frac{25}{30 \times 5} \right) = 0.166 \,.$$

Now expected idle time of the machine = average time spent

$$= \frac{(k+1)\rho}{2\mu(1 - k\rho)} + \frac{1}{\mu}$$

$$= \frac{(5+1) \times 0.166}{2 \times 0.04(1 - 0.83)} + 25 = 98 \text{ minutes.}$$

EXAMPLE 11. *A hospital clinic has a doctor examining every patient brought in for a general checkup. The doctor averages 4 minutes on each phase of the checkup although the distribution of time spent on each phase is approximately exponential. If each patient goes through* *four phases in the checkup and if the arrival of the patients to the doctor's office are approximately poisson at the average rate of 3 per hour, what is the average time spent by a patient waiting in the doctor's office? What is the average time spent in the examination? What is the most probable time spent in the examination?*

[MEERUT 1999, 2004 (O)]

SOLUTION. Here mean arrival rate $\lambda = 3$ patients per hour.

Service time per phase $= (1/4\mu) = 4$ minutes.

$\therefore \qquad\qquad \mu = (1/16)$ persons/minute $= (15/4)$ persons/hours.

We have, average time spent by a patient waiting in doctor's office

$$E(w) = \frac{k+1}{2k} \cdot \frac{\lambda}{\mu(\mu - \lambda)}$$

$$= \frac{4+1}{2 \times 4} \cdot \frac{3}{\dfrac{15}{4}\left(\dfrac{15}{4} - 3\right)} \text{ hours} = 40 \text{ minutes.}$$

The average time spent in examination is the mean of t when it is following the 4th member of Erlang family.

Thus average time spent $= (1/\mu) = 16$ minutes.

The most probable time spent in the examination is the modal value of the 4th member of Erlang family.

Most probable time spent $= \dfrac{k-1}{k\mu}$, k being 4.

$$= \frac{4-1}{4 \times (1/16)} = 12 \text{ minutes.}$$

EXAMPLE 12. *A colliery working one shift per day uses a large number of locomotives which break down at random intervals; on average one fails per 8 hour shift. The fitter carries out a standard maintenance schedule or each faulty loco. Each of the five main parts of this schedule takes on average (1/2) hour but the time varies widely. How much time will the fitter have for the other tasks and what is the average time a loco is out of service?*

SOLUTION. The colliery has a large number of locomotives and during 8 hours, one locomotive breaks. Hence the arrival rate (λ) of locomotives for repair will be 1/8 per hour. Now the servicing time for each part is not exactly given but we can assume that the time can follow an exponential distribution for each part of the schedule taking on average 1/2 hour for each part, the whole schedule of 5 parts will have a servicing time with Erlang distribution having average $= 5 \times 1/2 = (5/2)$ hours service time.

i.e., $\qquad \mu = (2/5)$ per hour.

The arrival rate is one per shift and the fitter takes on average 5/2 hours for the repair of a locomotive. Hence the time that filter will have for other tasks is ,

$8 - 5/2 = 5.5$ hours.

The average time for a locomotive to be out of service

$$= \frac{(k+1)\rho}{2\mu(1-k\rho)} + \frac{1}{\mu}$$

$$= \frac{(5+1)\left(\frac{1}{16}\right)}{\left(2\times\frac{2}{5}\right)\left(1-\frac{5}{16}\right)} + \frac{5}{2} = 3.18 \text{ hours, where } \rho = \frac{\lambda}{k\mu} = \frac{1}{16} \text{ and } k = 5$$

EXAMPLE 13. *A warehouse in a small state receives order for a certain item and sends them by a truck as soon as possible to the customer. The order arrives in a Poisson fashion at a mean rate of 0.9 per day. Only one item at a time can be shipped by truck from the warehouse, which is located in the central part of a state. Because the customers are located in various places in the state, the distribution of service time in days has a distribution with probability density $4te^{-2t}$. What is the expected delay between the arrival of an order and the arrival of the item to the customer? Service time here implies the time that truck takes to load, get to the customer, unload and return to the warehouse. Loading and unloading times are small as compared with travel time.*

SOLUTION. The form of the service time distribution in this case is second member of Erlangian family with $\mu = 1$. Hence

$$\mu = 1, \ \lambda = 0.9 \text{ and } k = 2.$$

Expected waiting time of an arrival,

$$= \frac{k+1}{2k} \cdot \frac{\lambda}{\mu(\mu-\lambda)} = \frac{3}{4} \times \frac{0.9}{1(1-0.9)} = 6\frac{3}{4} \text{ days,}$$

and average time spent in the system,

$$= \frac{k+1}{2k} \cdot \frac{\lambda}{\mu(\mu-\lambda)} + \frac{1}{\mu} = 6\frac{3}{4} + 1 = 7\frac{3}{4} \text{ days.}$$

EXAMPLE 14. *A firm is engaged in both shipping and receiving activities. The management is always interested in improving the efficiency by new innovations in loading and unloading procedures. The arrival distribution of trucks is found to be Poisson with arrival rate of 1.5 trucks per hour. The service time distribution is exponential with unloading rate of two trucks per hour. Find the following :*
(i) Average number of trucks in the waiting line.
(ii) The average waiting time of trucks in line.
(iii) The probability that the loading and unloading dock and workers will be idle.
(iv) What reductions are possible if lift trucks are used?

SOLUTION. This is an example of M/M/1 : (∞/FIFO) system

Hence the arrival rate $\lambda = 1.5$ per hour and the service rate $\mu = 2.0$ per hour

$$\rho = \frac{\lambda}{\mu} = \frac{1.5}{2.0} = 0.75.$$

(i) The average length of the waiting line

$$= \frac{\rho^2}{1-\rho} = \frac{(0.75)^2}{0.25} = 2.25$$

Thus, on average there will be 2.25 trucks in the waiting line.

(ii) The average waiting time of trucks in the queue,

$$= \frac{\rho}{\mu(1-\rho)} = \frac{0.75}{2(0.25)} = 1.5 \text{ hours.}$$

(iii) The probability that there is no truck in the queue i.e., the dock and workers are idle $= P_0 = 1 - \rho = 1 - 0.75 = 0.25$.

(iv) Let us suppose that when lift trucks are used then the service time becomes 12 minutes per truck i.e., $\mu = 5$ trucks per hour.

Then
$$\rho = \frac{\lambda}{\mu} = \frac{1.5}{5} = 0.3$$

Now, average length of the waiting line

$$\frac{\rho^2}{(1-\rho)} = \frac{(0.3)^2}{1-0.3} = \frac{0.75}{0.70} = 0.1286 \text{ trucks.}$$

So there is a reduction of queue length from 2.25 trucks to 0.1286 trucks. Also, the average waiting time will become

$$\frac{\rho}{(1-\rho)\mu} = \frac{(0.3)}{(0.7)\,5} = \frac{3}{35} = 0.0857.$$

So the waiting time also reduces from 1.5 hours to 0.0857 hour.

EXAMPLE 15. *The mean rate of arrival of planes at an airport during the peak period is 20 per hour but the actual number of arrivals in any hour follows a Poisson distribution. The airport can land 60 planes per hour on an average in good weather or 30 per hour in bad weather, but the actual number landed in any hour follows a Poisson distribution with the respective averages. When there is congestion, the planes are forced to flyover the field in the stack awaiting the leading of other planes that arrived earlier.*

(i) *How many planes would be flying over the field in the stack on an average in good weather and in bad weather.*

(ii) *How long a plane would be in the stack and in the process of landing in good and bad weather.*

SOLUTION. Here $\lambda = 20$ planes per hour.

The problem can be divided in two cases (i) Good weather with $\mu = 60$ planes per hour (ii) Bad weather with $\mu = 30$ planes per hours.

Case 1. In case of Good weather

Average waiting length

$$E(L) = \frac{\rho^2}{(1-\rho)} = \frac{1/9}{\left(1 - \frac{1}{3}\right)} = \frac{1}{6}.$$

Here, $\qquad\qquad\qquad\qquad \rho = \dfrac{20}{60} = \dfrac{1}{3}$

And, average waiting time of an arrival in the system will be

$$E(v) = \frac{1}{\mu(1-\rho)} = \frac{1}{60\left(1 - \dfrac{1}{3}\right)} = \frac{1}{40} \text{ hours}$$

Case 2. In case of Bad weather.

$$\rho = \frac{20}{30} = \frac{2}{3}$$

and $\qquad\qquad E(L) = \dfrac{(4/9)}{(1 - 2/3)} = \dfrac{4}{3}$

$$E(v) = \frac{1}{\mu(1-\rho)} = \frac{1}{30\left(1 - \dfrac{2}{3}\right)} = \frac{1}{10} \text{ hours.}$$

EXAMPLE 16. *At what average rate must a clerk at a supermarket work in order to ensure a probability of 0.90 that the customer will not have to wait longer than 12 minutes? It is assumed that there is only one counter, to which customers arrive in a Poisson fashion at an average rate of 15 per hour. The length of service by the clerk has an exponential distribution.*   [MEERUT 1996]

<u>SOLUTION.</u> This is an example of M/M/1 : ∞/FCFS system with parameter

$$\lambda = \frac{15}{60} = \frac{1}{4} \text{ customers/minute.}$$

Now Probability distribution of waiting time to be t in this case is

$$P(t) = \mu\rho(1-\rho)e^{-\mu t(1-\rho)}$$

Hence

Probability $\quad [t \geq 12] = \int_{12}^{\infty} \mu\rho(1-\rho)\, e^{-\mu(1-\rho)} dt$

$$= \left[\frac{\mu\rho(1-\rho)e^{-\mu(1-\rho)}}{-\mu(1-\rho)}\right]_{12}^{\infty} = \rho e^{-\mu(12)(1-\rho)}.$$

Now the probability that a customer will have to wait for more than 12 minutes

$$= 1 - \text{Probability } (t < 12) = 1 - 0.90 = 0.10$$

Hence, $\rho[e^{-12\mu(1-\rho)}] = 0.10$

But $\qquad\qquad \rho = \dfrac{\lambda}{\mu} \qquad$ i.e., $\qquad \dfrac{\lambda}{\mu}[e^{-12(\mu-\lambda)}] = 0.10$

Here, $\qquad\qquad \lambda = 1/4 \qquad \therefore \qquad \dfrac{1}{4\mu}[e^{-12(\mu - 1/4)}] = 0.10$

Solving this, we get

$$\frac{1}{\mu} = 2.48 \text{ minutes.}$$

i.e., service rate should be 0.40 per minute.

EXAMPLE 17. *Problems arrive at a computing centre in Poisson fashion at average rate of five per day. The rules of the computing centre are that any man waiting to get his problem solved must aid the man whose problem is being solved. If the time to solve a problem with one man has an exponential distribution with mean time of 1/5 day, and if the average solving time is inversely proportional to the number of people working on the problem, then approximate the expected time in the centre for a person entering the line.*

SOLUTION. This is the case of Erlang Queuing Model with $\lambda = 5$ and $\mu = 3$

Now, we have

$$P_n = e^{-\rho}\frac{\rho^n}{n!}$$

Thus the expected number of persons working at any specified instant is

$$E(n) = \sum_{n=0}^{\infty} nP_n = \sum_{n=0}^{\infty} \frac{n}{n!}e^{-\rho}.\rho^n$$

$$= e^{-\rho}\left[0 + \frac{1}{1}\rho + \frac{2}{2!}\rho^2 + \frac{3}{3!}\rho^3 + ...\right] = \rho.e^{-\rho}\left[1 + \rho + \frac{\rho^2}{2!} + ...\right]$$

$$= \rho e^{-\rho}.e^{\rho} = \rho, \text{ since } 1 + \rho + \frac{\rho}{2!} + ... = e^{\rho}$$

$$E(n) = \frac{5}{3} = 1.67 \text{ persons.}$$

Also it is given that average solving time of any problem is inversely proportional to the number of people working on the problem = 1/5 day per problem.

∴ Expected time spent by a person entering the line is

$$= \frac{1}{5} \times E(n) = \frac{1}{5} \times \frac{5}{3} = \frac{1}{3} \text{ days} = 8 \text{ hours.}$$

EXAMPLE 18. *A refinery distributes its products by trucks, loaded at the loading dock. Both company trucks and independent distributor's trucks are loaded. The independent firms complained that some times they must wait in line and thus loose money paying for a truck and driver, that is only waiting. They have asked the refinery either to put in a second loading dock or discount prices equivalent to the waiting time. The following data have been accumulated. Average arrival rate of all trucks is 2 per hour and average service rate is 3 per hour. 30% of all trucks are independent. Assuming that these rates are random according to the Poisson distributions, determine:*

(i) The probability that a truck has to wait,

(ii) The waiting time of a truck that waits, and

(iii) The expected waiting time of independent truck per day.

SOLUTION. Here $\lambda = 2$ per hour, $\mu = 3$ per hour. This is case of M/M/1 system. So

(i) Probability that a truck has to wait

$$= 1 - P_0 = 1 - (1 - \rho) = \rho = \frac{\lambda}{\mu} = \frac{2}{3}$$

(ii) The average waiting time of a truck

$$= \frac{1}{\mu - \lambda} = \frac{1}{3 - 2} = 1 \text{ per hour.}$$

(iii) There are 30% independent trucks and in full day of 8 hours, 16 trucks will arrive for loading, out of which $30\% \times 16 = 4.8$ will be for independent distributors. Thus

Expected waiting time of independent trucks per day

= (Independent trucks arriving/day) × (Expected waiting time/truck)

$$= (4.8) \frac{\rho}{\mu(1 - \rho)},$$

$$= (4.8) \frac{2/3}{3[1 - (2/3)]} = \frac{9.6}{3} = 3.2 \text{ hours/day.}$$

EXAMPLE 19. *If for a period of 2 hours in a day say 8 to 10 A.M., trains arrive at the yard every 20 minutes but the service time continues to remain 36 minutes, then calculate for this period (a) the probability that the yard is empty (b) average queue length, on the assumption that the line capacity of the yard is limited to 4 trains only.*

SOLUTION. Here $\lambda = 1/20$ trains per minute, $\mu = 1/36$ trains per minute.

Here the capacity of the system is finite i.e., N = 4.

(a) So using $P_0 = \dfrac{1 - \rho}{1 - \rho^{N+1}}$, we get

$$P_0 = \frac{1 - (36/20)}{1 - (36/20)^{4+1}} = 0.04 \qquad [\text{where } \rho = \frac{\lambda}{\mu} = \frac{36}{20}]$$

(b) Average queue length $= \left(\dfrac{1 - \rho}{1 - \rho^{N+1}} \right) \sum\limits_{n=0}^{N} n\rho^n = P_0 \sum\limits_{n=0}^{4} n\rho^n$

$$= 0.04 [\rho + 2\rho^2 + 3\rho^3 + 4\rho^4] = 2.9 \text{ or 3 trains.}$$

EXAMPLE 20. *Prove that for Erlang distribution with parameters p and k, the mode is at $(1 - 1/k)(1/\mu)$, the mean is $1/\mu$ and the variance is $1/k\mu^2$.*

SOLUTION. For Erlang distribution, the distribution of service time is given by

$$P(t) = \frac{(k\mu)^k}{\Gamma k} t^{k-1} e^{-k\mu t} = \frac{(k\mu)^k}{(k-1)!} t^{k-1} e^{-k\mu t} \qquad [\because \Gamma(k) = (k-1)!]$$

Now mode of the distribution will be that value of t for which

$$\frac{d}{dt} P(t) = 0 \text{ and } \frac{d^2}{dt^2} P(t) < 0$$

Here $\dfrac{d}{dt}P(t) = \dfrac{d}{dt}\left[\dfrac{(k\mu)^k}{(k-1)!}t^{k-1}e^{-k\mu t}\right] = \dfrac{(k\mu)^k}{(k-1)!}\left[(k-1)t^{k-2}-t^{k-1}k\mu\right]e^{-k\mu t}$

$$= \dfrac{(k\mu)^k}{(k-1)!}\left[(k-1)-k\mu t\right]e^{-k\mu t}\, t^{k-2}$$

Now for $\dfrac{d}{dt}[P(t)] = 0 \;\Rightarrow\; (k-1)-k\mu t = 0$, i.e., $\quad t = \dfrac{k-1}{k\mu}$

Also

$$\dfrac{d^2}{dt^2}P(t) = \dfrac{(k\mu)^k}{(k-1)!}\left[(k-1)(k-2)t^{k-3}-(k-1)t^{k-2}k\mu\right]e^{-k\mu t}$$

$$+\,[(k-1)t^{k-2}-t^{k-1}k\mu](-k\mu e^{-k\mu t})$$

$$= \dfrac{(k\mu)^k}{(k-1)!}\left[(k-1)(k-2)t^{k-3}-(k-1)kt^{k-2}\mu-k(k-1)t^{k-2}\mu+k^2\mu^2 t^{k-1}\right]e^{-k\mu t}$$

$$= 0 \text{ for } t = \dfrac{k-1}{k\mu}$$

Mean $= E(t)$

$$E(t) = \int_0^\infty t\cdot P(t)dt = \int_0^\infty t\cdot\dfrac{(k\mu)^k}{(k-1)!}t^{k-1}e^{-k\mu t}\, dt = \dfrac{(k\mu)^k}{(k-1)!}\int_0^\infty t^k.e^{-k\mu t}\, dt$$

[putting $k\mu t = x$, $k\mu dt = dx$ for $t=0$; $x=0, t=\infty, x=\infty$)]

$$= \dfrac{(k\mu)^k}{(k-1)!}\int_0^\infty \left(\dfrac{x}{k\mu}\right)^k e^{-x}\dfrac{dx}{k\mu} = \dfrac{1}{k!\mu}\int_0^\infty (x)^k e^{-x}dt$$

$$= \dfrac{\Gamma(k+1)}{\mu(k!)} = \dfrac{k!}{k!\mu} = \dfrac{1}{\mu}$$

$V(t) = E(t^2)-[E(t)]^2$

Here $\quad E(t^2) = \dfrac{(k\mu)^k}{(k-1)!}\int_0^\infty (t)^2 t^{(k-1)}e^{-k\mu t}dt = \dfrac{(k\mu)^k}{(k-1)!}\int_0^\infty (t)^{k+1}e^{-k\mu t}\, dt$

Now putting $k\mu t = x$, we get

$$E(t^2) = \dfrac{(k\mu)^\mu}{(k-1)!}\int_0^\infty \left(\dfrac{x}{k\mu}\right)^{k+1}e^{-x}dx = \dfrac{1}{k^2\mu^2(k-1)!}\int_0^\infty x^{k+1}e^{-x}dx$$

$$= \dfrac{1}{k\mu^2 k!}(k+1)! = \dfrac{k+1}{k}\cdot\dfrac{1}{\mu^2}$$

Hence, $V(t) = \dfrac{k+1}{k}\cdot\dfrac{1}{\mu^2}-\left(\dfrac{1}{\mu}\right)^2 = \dfrac{1}{\mu^2}\left[\dfrac{k+1}{k}-1\right] = \dfrac{1}{k\mu^2}$

EXAMPLE 21. *Customers arrive at the first class ticket counter of a theatre at a rate of 12 per hour. There is one clerk serving the customers at a rate of 30 per hour:*

(i) What is the probability that there is no customer in the counter (i.e., the system is idle)?

    *(ii) **What is the probability that there are more than 2 customers in the counter?***

    *(iii) **What is the probability that a customer is being served and nobody is waiting?***   [UPTECH-MBA 2004-05]

    *(iv) **What is the probability that there is no customer waiting to be served?***

<u>SOLUTION.</u>   Here $\lambda = 12$ per hour, $\mu = 30$ per hour.

This is an example of M/M/1 system.

  (i)   Probability that system is empty is given by

$$P_0 = 1 - \rho = 1 - \frac{\lambda}{\mu} = 1 - \frac{12}{30} = \frac{18}{30} = 0.6$$

  (ii)   Probability that there are more than two customers in the counter, i.e.,

$$P(n > 2) = 1 - P(n = 0) - P(n = 1) - P(n = 2)$$

$$= 1 - P_0 - \rho P_0 - \rho^2 P_0 = 1 - (1 + \rho + \rho^2)P_0$$

$$= 1 - (1 + 2/5 + 4/25)\, 0.6 = 1 - 0.936 = 0.064.$$

  (iii) P(a customer being served and nobody is waiting]

$$= P_1 = 0.6 \times 0.4 = 0.24.$$

  (iv) P(No customer is waiting] $= P$ (at most one customer in the counter]

$$= P_0 + P_1 = 0.6 + 0.24 = 0.84$$

EXAMPLE 22. *A car service station has two bays where service can be offered simultaneously. Due to limitation in space only four cars are accepted for servicing. The arrival pattern is Poisson with 12 cars per day. The service time for both the bays is exponentially distributed with 8 cars per day per bay. Find the average number of cars in the service station, the average number of cars waiting to be serviced and the average time a car spends in the system.*

<u>SOLUTION.</u>   This is a M/M/c : N/FIFO model with N = 4, c = 2, $\lambda = 12$ cars per day, $\mu = 8$ car per day. Now,

$$P_0 = \left[\sum_{n=0}^{c-1} \frac{1}{n!}\left(\frac{\lambda}{\mu}\right)^n + \sum_{n=c}^{N} \frac{1}{c^{n-c}c!}\left(\frac{\lambda}{\mu}\right)^n \right]^{-1} = \left[\sum_{n=0}^{1} \frac{1}{n!}\left(\frac{12}{8}\right)^n + \sum_{n=2}^{4} \frac{c^c}{c!}\left(\frac{\lambda}{c\mu}\right)^n \right]^{-1}$$

$$= \left[1 + \left(\frac{3}{2}\right) + \frac{4}{2}\left\{ \left(\frac{3}{4}\right)^2 + \left(\frac{3}{4}\right)^3 + \left(\frac{3}{4}\right)^4 \right\} \right]^{-1} = 0.1961$$

Also, $P_1 = \left(\frac{\lambda}{\mu}\right)P_0 = \frac{3}{2} \times 0.1961 = 0.2940$

$$P_2 = \frac{1}{2}\left(\frac{\lambda}{\mu}\right)^2 P_0 = \frac{1}{2!}\left(\frac{9}{4}\right)(0.1961) = 0.2260$$

$$P_3 = \frac{1}{c!c^{n-c}}\left(\frac{\lambda}{\mu}\right)^n P_0, \text{ where } n = 3$$

$$= \frac{1}{2!\,2^{3-2}}\left(\frac{3}{2}\right)^3 P_0 = \frac{1}{2\times 2}\frac{27}{8}P_0 = 0.1654$$

$$P_4 = \frac{1}{2!\,2^{4-2}}\left(\frac{3}{2}\right)^4 P_0 = \frac{1}{2\times 4}\frac{81}{16}P_0 = 0.1240$$

Now, the number of units in the system can be $n = 0, 1, 2, 3$ or 4 with probabilities P_0, P_1, P_2, P_3, P_4 respectively. Hence expected (average) number of units in the system is given by

$$E(n) = \sum_{n=0}^{4} nP_n = 0.P_0 + 1.P_1 + 2.P_2 + 3.P_3 + 4.P_4$$
$$= 0 + 1\,(0.2940) + 2(0.2260) + 3\,(0.1654) + 4\,(0.1240)$$
$$= 1.7274.$$

Now the number of units waiting in the system can be 1 and 2 when the number of units in the system are 2 and 4 respectively with probabilities P_3 and P_4.

$$\therefore \qquad E(L) = \sum_{L=1}^{2} LP_{L+c} = 1.P_3 + 2.P_4$$

$$= 1\,(0.1654) + 2\,(0.1240) = 0.4134$$

The effective arrival rate is given by

$$\lambda' = \lambda(1 - P_N) = 12(1 - P_4) = 12(1 - 0.1204) = 10.512$$

Average time a car spends in the system is given by

$$= \frac{E(n)}{\lambda'} = \frac{1.7274}{10.512} = 0.1643 \text{ days.}$$

 ### Exercise-19.1

1. Prove that if arrivals occur at random in time, then the number of arrivals occurring in a fixed time interval follows a Poisson distribution.

2. If of a period of 2 hours in a day, trains arrive at the yard every 24 minutes and the service time is 36 minutes, then calculate for this period.

   (a) The probability that the yard is empty.

   (b) Average queue length on the assumption that the line capacity of the yard is limited to 4 trains only.

3. In a car manufacturing plant, a loading crane takes exactly 10 minutes to load a car into a wagon and again come back to the position to load another car. If the arrival of cars is a Poisson stream at an average of one every 20 minutes, calculate the average waiting time of a car in a stationary state.

4. Customers arrive at one window drive-in bank according to a Poisson distribution with mean 10 per hour. Service time per customer is exponential with mean 5 minutes. The space in front of the window, including that for the serviced car can accommodate a maximum of 3 cars. Other cars can wait outside the space.

   (a) What is the probability that an arriving customer can drive directly to the space in front of the window?

   (b) What is the probability that an arriving customer will have to wait outside the indicated space?

   (c) How long an arriving customer is expected to wait before starting service?

   (d) How many space should be provided in front of the window so that all the arriving customers can wait in front of the window at least 90 percent of the time ?

5. (i) What is a multi-channel queuing problem? Deduce the difference differential equations when there are k channels. Show that, for the steady state case, the solution of the equations can be put in the form

$$P_0 = \frac{P_0 \left(\frac{\lambda}{\mu}\right)^n}{n!}, \quad 0 \le n \le k$$

and

$$= \frac{P_0 \left(\frac{\lambda}{\mu}\right)^n}{k^{(n-k)}k!}, \quad n > k$$

where,

$$P_0 = \frac{1}{\left[\sum_{n=0}^{k=1} \frac{(\lambda/\mu)^n}{n!} + \frac{(\lambda/\mu)^k}{k!}\left(1 - \frac{\lambda}{k\mu}\right)\right]}$$

For the case of a single channel, show that the mean length of waiting line is $\rho/(1-\rho)$ where ρ is the traffic intensity.

(ii) For the case of s-channels, Poisson arrivals with mean arrival rate λ and exponential service with mean service rate μ, show that the probability that an arrival has to wait is given by

$$(\mu(\lambda/\mu)^s P_0)/[(s-1)!(s\mu-\lambda)].$$

6. For the case of two channels, Poisson arrivals and exponential service, show the following:
(i) Probability that both the channels are empty is $\dfrac{2\mu - \lambda}{2\mu + \lambda}$.
(ii) Expected number in the system is $\dfrac{2\mu\lambda}{4\mu^2 + \lambda^2}$.

7. In a bank, customers arrive in a Poisson stream with mean 36 per hour. The service time per customer is negative exponential with mean .035 per hour. Assuming that the system can accommodate at most 30 customers at a time, how many tellers should be provided under each of the following two conditions ?

(a) The probability of having more than $c+3$ waiting customers, is less than 0.20, where c is the number of tellers.

(b) The expected number in the system does not exceed 3.

8. Cars arrive at a tool gate on a free way according to a Poisson distribution with mean 90 per hour. The average time for passing through the gate is 38 seconds. Drivers complain of their long waiting time. The authorities are willing to decrease the passing time through the gate to 30 seconds by introducing new automatic devices. This can be justified only if under the old system the number of waiting cars exceeds 5. In addition the percentages of idle time of the rate under the new system should not exceed 10%. Can be new device be justified?

9. A repairman is to be hired to repair machines which breakdown at an average rate of three per hour. Breakdowns are distributed in time in a manner that may be regarded as Poisson. Non-productive time on anyone machine is considered to cost the company Rs. 5 per hour. The company has narrowed the choice down to two repairmen one slow and cheap, the other fast but expensive. The slow cheap repairman demands Rs. 3 per hour, in return he will service the broken down machines exponentially at an average rate of four per hour. The fast expensive repairman demands Rs. 5 per hour and will repair machines exponentially at an average rate of six per hour. Which repairman should be hired?

10. There is a conveyor belt system on an Airport The arrival rate is one in every 15 minutes and service rate is one in every 20 minutes. Determine (a) The total number of orders that have to wait for service since start up if 20 orders were placed prior to 8.00 AM. the start up time (b) the total time from start up, until the service facility is idle.

ANSWERS

7. (a) $c = 2$ (b) $c = 2$
8. Only one condition is fulfilled, so new device not justified.
9. Fast but expensive repairman should be hourly cost = Rs. 17
10. (a) number of customers waited since the start up = 76;
(b) the facility will require 231 time units to eliminate the queue]

REVIEW QUESTIONS

1. What is a queuing problem Write the components in waiting line system.
2. Explain basic characteristic of a queuing system.
3. What do you understand by the basic queuing process?
4. Describe the different types of costs involved in a queuing system.
5. Write the major importance of queuing theory.
6. Explain the mechanism of a queuing process. Considering some illustrative situations.
7. Briefly describe the single server model.
8. Write a short note on queuing theory.

MULTIPLE CHOICE QUESTIONS (CHOOSE THE MOST APPROPRIATE ONE)

1. The most difficult aspect of performing a formed economic analysis of queuing system is :
   (a) estimate use
   (b) estimating waiting cost
   (c) both (a) and (b) are tue
   (d) none of these
2. The major goal of queue is :
   (a) minimize expected return
   (b) minimize the cost of providing service
   (c) provide models which help the manager to trade off the cost of service
   (d) none of these
3. A queue is formed when :
   (a) customers wait for service
   (b) service facilities stand idle and wait for customers
   (c) either (a) or (b)
   (d) none of these
4. A queue mode have the following characteristics :
   (a) Arrival pattern
   (b) Queue discpline
   (c) Customer's behaviour
   (d) All are true
5. As per queue discipline the following is not a negative behaviour of a customer :
   (a) balking   (b) collusion
   (c) boarding   (d) none of these

6. The system of loading and unloading of goods usually follows :
   (a) LIFO   (b) FCFS
   (c) SIRO   (d) none of these
7. A steady state exists in a queue if :
   (a) $\lambda < \mu$ (b) $\lambda > \mu$
   (c) $\lambda = \mu$ (d) none of these
8. The unit of traffic intensity is :
   (a) Erlang   (b) Poisson
   (c) Markov   (d) none of these
9. Office filing system follows :
   (a) LIFO   (b) FCFS
   (c) SIRO   (d) none of these
10. A person who leaves the queue by losing his patience to wait is said to be :
    (a) Balking   (b) Reneging
    (c) Jockeying   (d) None of these
11. The traffic intensity can be defined by :
    (a) mean arrival rate/mean service rate
    (b) μ/λ
    (c) $\lambda \times \mu$
    (d) none of these
12. The variance of queue length is :
    (a) $\dfrac{\rho}{1-\rho}$ (b) $\dfrac{\rho}{(1-\rho)^2}$
    (c) $\dfrac{\rho^2}{1-\rho}$ (d) none of these

ANSWERS

1. (a)   2. (c)   3. (c)   4. (d)   5. (c)   6. (a)   7. (c)   8. (a)   9. (a)
10. (b)   11. (a)   12. (b)

ARCHIVE

1. A Car servicing station has two boys where service can be offered simultaneously. Due to space limitations only four cars accepted for servicing. The arrival pattern is Poisson with 120 cars per day. The service time in both ways is exponentially distributed with $\mu = 96$ car per day per boy. Find the average no. of cars in service station, the average no. of cars waiting to be serviced and the average time a car spends in the system. **(IAS 1989)**

2. A mechanic repairs four machines. The mean time between service requirements in 5 hours for each machine and forms an exponential distribution. The mean repair time in one hour and also follows the same distribution pattern. Machine downtime costs ₹ 25 per hour and the mechanic costs ₹ 55 per day. Determine the following

(i) Probability that the service facility will be idle.

(ii) Probability of various number of machines (0 through 4) to be out of order and being repaired.

(iii) Expected no. of machines waiting to be repaired and being repaired.

(iv) Expected downtime cost per day.

Would it be economical to engage two mechanics, each repairing only two machines. **(DELHI MBA - 2003)**

3. A barber with one man shop takes exactly 25 minutes to complete one hair cut. If customers arrive in a Poisson pattern at an average rate of one every 40 minutes, how long on the average must a customer wait for service?

(BHOPAL-1986, 2007, MEERUT-2013)

HINTS AND ANSWERS

2. $\lambda = \dfrac{1}{5} = 0.2$ machines/hour, $\mu = 1$ machine/hour

$M = 4$ machine $\Rightarrow \rho = \dfrac{\lambda}{\mu} = 0.2$

Then using the following formulae

(i) $P_0 = \left[\sum_{n=0}^{M} \dfrac{M!}{(M-n)!} \left[\left(\dfrac{\lambda}{\mu} \right)^n \right] \right]^{-1} \Rightarrow P_0 = 0.4030$

(ii) $P_n = \dfrac{M!}{(M-n)!} \left(\dfrac{\lambda}{\mu} \right)^n . P_0$

(iii) $L_s = M - \dfrac{\mu}{\lambda}(1 - P_0) = 4 - \dfrac{1}{0.2}(1 - 0.403) = 1.015$ machines

It is not economical to engage two mechanics.

3. $\lambda = \dfrac{1}{40}, \mu = \dfrac{1}{25}$ and $k \to \infty$

Then, we have $w_q = 20.8$ minutes

$$w_s = w_q + \dfrac{1}{\mu} = 45.8 \text{ minutes.}$$

Replacement Problems 20

INTRODUCTION

The problem of replacement is experienced in systems where machines, individuals or capital assets are the main job performing units. The special feature of these units is that their level of performance/efficiency decreases with time and one has to formulate some suitable replacement policy regarding these units to keep the system upto some desired level of performance.

If in any system the efficiency of an item deteriorates with time or sometimes the item fails completely. This deterioration or breakdown affects the whole system adversely. For example, some equipment may fail due to fault in the component or it being old. Thus there may be some sort of disruption in a working environment due to the failure of machinery components, vehicle components, lighting, heating and ventilation systems. The failure occurs suddenly and unpredictably and, therefore, a system that is working efficiently can experience considerable disruption. In such situation either the item is to be replaced by new item or some remedial action, known as maintenance, is necessary to restore the efficiency of the whole system at some desired level. The cost of maintenance depends on a number of factors and a stage comes when the maintenance cost become so large that it may be profitable to replace the old item by new one. Thus, there is a need to formulate a most economic replacement policy which is in the best interest of system.

20.1.1 FAILURE MECHANISM OF ITEMS

The failure mechanism of items in any system can be divided in two main categories :

I. **Gradual Failure:** In some systems, the failure mechanism is found to be progressive, *i.e.*, with increase in the life of the item, its efficiency deteriorates with the following reasons.

   (i) the maintenance costs progressively increases with time
   (ii) the productivity of the equipment decreases and
   (iii) the resale value of the item also decreases with time, *viz.* the efficiency of the items like pistons, bearings, automobile, tyres etc.

II. **Sudden Failure:** The items ultimately fail after a period of time. The life of the equipment cannot be predicted and is some sort of random variable. The probability distribution of the life of these components can be progressive, retrogressive or random.

20.2 REPLACEMENT MODELS AND THEIR SOLUTION

The study of **Replacement** is a field of application rather than a method of analysis and mostly concerns with methods of comparing alternative replacement policies. The various types of replacement problems can be broadly classified in following categories :

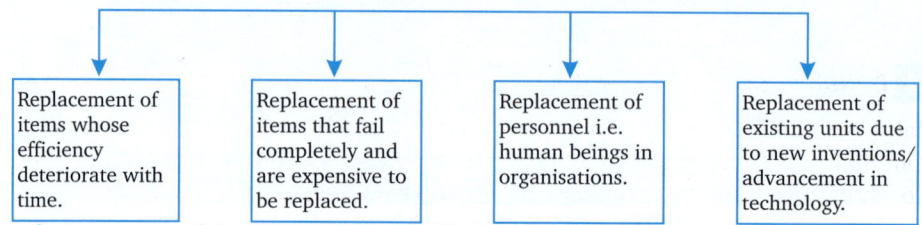

| Replacement of items whose efficiency deteriorate with time. | Replacement of items that fail completely and are expensive to be replaced. | Replacement of personnel i.e. human beings in organisations. | Replacement of existing units due to new inventions/ advancement in technology. |

(i) Replacement of items whose efficiency deteriorates with time: Here the systems can maintain the level of performance by installing a new unit at the beginning of some unit of time and decide to keep it upto some suitable period so as to minimize the maintenance and other costs.

The simplest replacement model in such cases is one where the deterioration process is predictable and is represented by an increasing maintenance cost and possibly decreasing scrap value of the equipment. Here the optimum life of the equipment is determined on the assumption that increased age reduces efficiency. The cost of new machine is like a cost of replenishment and maintenance costs corresponding to holding costs.

☛ REMARK

- These models can be solved by two methods namely (i) calculation of cost per unit of time, (ii) the present value concept which represents time value of money and enables the comparison on a one number basis.

(ii) Replacement of items that fail completely and are expensive to be replaced: Here we discuss those situations where replacement of units is done in anticipation of failure. In these models the units are assumed to have relatively constant efficiency until they fail or die. These models require knowledge of stochastic process involving probabilities of failures. Here a replacement policy is formulated to balance the wasted life of items replaced before failure against the costs incurred when items fails in service.

(iii) Replacement of human beings in organisation (staffing problem): This problem requires a separate study and involves the life distribution for the service of staff in a system.

(iv) There can also be equipment models with uncertain costs where the replacement of items may be necessary due to new researches, otherwise the system may become out of date. The construction and solution of such models will vary under different situations.

20.3 GENERAL APPROACH TO SOLVE REPLACEMENT PROBLEMS

Since the actual time of failure is unpredictable for any particular item, the likely failure pattern could be established by observation. The optimum replacement policy can be formulated by generating the probability distributions for the given situation and then using them in conjunction with relevant cost information. Here we needs following information:

(i) Objective assessment of the probability of the item failing at a particular point in time.

(ii) Assessments of the cost of replacement in terms of :

(a) actual cost of the item.

(b) direct costs of labour involved in replacement.

(c) costs of disruption in terms of lost production, lost orders etc.

20.4 CONCEPT OF PRESENT VALUE [ROHILKHAND–2002]

Many replacement problems involve huge expenditures both for purchasing as well as for maintenance of the respective units. There can be a number of alternatives and one may like to choose the one which requires minimum expenditure. The selection can be made by an approach based on the concept of present value. The present value of a number of expenditures incurred over different periods of time represents their value at the current time. It is based upon the same amount of money at the end of the time, *i.e.*, if an amount is to be spent in different years what is its worth today?

Another important consideration for present value is that generally the entrepreneur borrows money from various sources. The amount of interest paid depends on the rate of interest and period during which the borrowed money is repaid. In technical terms the fact that the amount could be invested at an interest rate r to produce the borrowed amount is known as Principal and excess amount paid is called Interest. Sum of Principal and Interest is known as Amount.

When the interest, as soon as it is due, is added to the principal of the preceding period to get a new principal and the interest of next period is calculated over this new principal then the interest is known as compound interest. Thus, if P is the principal and i the interest rate, then the amount at the end of n years with compound interest is given by

$$A = P(1+i)^n$$

or

$$P = \frac{A}{(1+i)^n} = A \times Pwf$$

where $Pwf = 1/(1+i)^n$ is known as single payment present worth factor. $v = 1/(1+i)$ is known as discount rate and is always less than one. Pwf is present value of one rupee spent after n years from now. Here P is known as the present worth of an amount A paid in n years at interest rate i. Standard tables of Pwf are available for different interest rates and time periods.

Similarly, if R denotes the uniform amount spent at the end of each year and S is the total expenditure at the end of n years, then

$$S = R[(1+i)^n - 1]/i$$

or

$$R = \frac{Si}{(1+i)^n - 1} = \frac{P(1+i)^n i}{(1+i)^n - 1}$$

or

$$P = R \cdot \frac{[(1+i)^n - 1]}{i(1+i)^n} = S(Pwf) \qquad \text{...(1)}$$

Pwf is known as uniform annual series present worth factor.

20.5 REPLACEMENT WHEN MAINTENANCE COST INCREASES WITH TIME AND THE VALUE OF MONEY REMAINS SAME

[MEERUT–2002, 03, 09, 15, 17, 18; UPTU(MBA)–2005; ROHILKHAND–2000, 03]

Here we have to find the best replacement age (time) of a machine when (i) its maintenance cost is given by a function increasing with time (ii) its scrap value is constants and (iii) money value is not considered

Symbols:

C = the capital or purchasing cost of the item.

S = the resale value or scrap value of the item (assumed constant over time.)

Case (i). *When time t is a continuous variable.* Let $C_m(t)$ be the maintenance or running cost at time t. If the item is used in the system for a period y, then the total maintenance cost or cumulative running cost incurred during period y will be

$$M(y) = \int_0^y C_m(t)\, dt \qquad \ldots(1)$$

The total cost incurred on the item during period y

= Capital cost of the item + Total maintenance cost in period y – resale value

= $C + M(y) - S$.

Hence average cost per unit of time incurred during period y on the item is given by

$$G(y) = \frac{C + M(y) - S}{y} \qquad \ldots(2)$$

We want to find that value of y for which $G(y)$ is minimum. Hence differentiating (2) w.r.t. y, we get

$$\frac{dG}{dy} = -\frac{C - S}{y^2} + \frac{C_m(y)}{y} - \frac{1}{y^2}\int_0^y C_m(t)\, dt = 0 \quad \text{(for } G \text{ to be minimum)}$$

From (1), we get

$$-\frac{C - S}{y^2} + \frac{C_m(y)}{y} - \frac{1}{y^2} M(y) = 0$$

or $$C_m(y) = \frac{C - S + M(y)}{y} = G(y) \qquad \ldots(3)$$

From (3), we conclude that replace the item when the average annual cost reaches at the minimum which will always occur at a time when the average cost becomes equal to the current maintenance cost, *i.e.*

> If time is measured continuously then the average annual cost will be minimized by replacing the machine items when the average cost to date becomes equal to the current maintenance cost.

Case (ii). *When time t is a discrete variable.* In this case the period of time is taken as one year and t can take the values 1, 2, 3, ... etc.

Then, $$M(y) = \sum_{m=1}^{y} C_m(t) = \text{Total running cost of } y \text{ years} \qquad \ldots(4)$$

∴ Total cost incurred on the item during y years

$$T(y) = C + M(y) - S = C - S + \sum_{m=1}^{y} C_m(t)$$

∴ Average annual cost incurred during y years

$$G(y) = T(y) / y = [C + M(y) - S] / y .$$ ....(5)

Now $G(y)$ will be minimum for that value of y, for which

$$G(y+1) > G(y) \quad \text{and} \quad G(y-1) > G(y)$$
$$G(y+1) - G(y) > 0 \quad \text{and} \quad G(y-1) - G(y) > 0$$

For this, we have

$$G(y+1) = \frac{\left[C - S + \sum_{m=1}^{y+1} C_m(t) \right]}{(y+1)}$$

$$\therefore \quad G(y+1) - G(y) = \frac{(C-S) + \sum_{m=1}^{y+1} C_m(t)}{(y+1)} - \frac{(C-S) + \sum_{m=1}^{y} C_m(t)}{y}$$

$$= \frac{C_m(y+1)}{y+1} + \frac{(C-S) + \sum_{m=1}^{y} C_m(t)}{y+1} - \frac{(C-S) + \sum_{m=1}^{y} C_m(t)}{y}$$

$$= \frac{C_m(y+1)}{y+1} + \left[(C-S) + \sum_{m=1}^{y} C_m(t) \right] \left[\frac{1}{y+1} - \frac{1}{y} \right]$$

$$= \frac{C_m(y+1)}{y+1} - \frac{(C-S) + \sum_{m=1}^{y} C_m(t)}{y(y+1)} = \frac{C_m(y+1)}{y+1} - \frac{G(y)}{y+1}, \text{ using (5)}$$

$$= \frac{C_m(y+1) - G(y)}{y+1}$$

Thus, $\qquad G(y+1) - G(y) > 0$, if $C_m(y+1) > G(y)$

Similarly, it can be shown that

$$G(y-1) - G(y) > 0, \text{ if } C_m(y) < G(y-1)$$ ...(6)

From equation (6) we observe that (1) do not replace, if the next years running cost is less than the previous years average total cost (2) replace at the end of y years if the next year's [i.e. of $(y + 1)^{\text{th}}$ year's] running cost is more than the average cost of yth year.

☞ REMARK
- When time is measured in discrete units, then the average annual cost will be minimized by replacing the machine when the next periods maintenance cost becomes greater than the current average cost.
- The average cost per year for each machine, assuming replacement of the optimal age (it may not be same for all machines) is computed and the machine with the lowest average cost is purchased.
- The machine should be replaced when the average cost begins to increase.

Solved Examples

EXAMPLE I. **A fleet owner finds from his past records that the costs per year of running a truck and resale values whose purchase price is ₹6000 are given as under. At what stage the replacement is due?**

Year	1	2	3	4	5	6	7	8
Running cost	1000	1200	1400	1800	2300	2800	3400	4000
Resale value	3000	1500	750	375	200	200	200	200

[ICWA–1980, MEERUT–2000, 18, DELHI–2009]

Here the scrap value decreases with time. Let $S(y)$ denote the scrap value of the truck at the end of yth year.

Here $C = 6000$. The value of $G(y)$ can be calculated from the following Table.

Table 1

Years t, y	Running cost, $C_m(t)$	Cumulative running cost $M(y)$ in ₹	Resale value $S(y)$ in ₹	$C - S(y)$ in ₹	$T(y) = C - S(y)$ $+ M(y)$ in ₹	$T(y)/y$ $= G(y)$
1	1000	1000	3000	3000	4000	4000
2	1200	2200	1500	4500	6700	3350
3	1400	3600	750	5250	8850	2950
4	1800	5400	375	5625	11025	2756
5	2300	7700	200	5800	13500	2700*
6	2800	10,500	200	5800	16300	2717
7	3400	13,900	200	5800	19700	2814
8	4000	17,900	200	5800	23700	2962

From the above table it is observed that the value of $G(y)$ in the fifth year is minimum. Hence the truck should be replaced at the end of fifth year, otherwise the average annual cost would increase.

20.6 REPLACEMENT OF THE ITEM THAT DETERIORATES WITH TIME

We know that the maintenance cost always increases with time for a machine and at a stage maintenance cost becomes so large that it becomes necessary to replace the machine with a new one. In this case, the problem of replacement is to find the best time at which the old item should be replaced.

Replacement policy for a machine (item) whose running cost (Maintenance cost) is given by a function increasing with time and its scrap value is constant and the value of money does not change with the time.

THEOREM 1. *The maintenance cost of a machine is given as a function increasing with time and its scrap value is constant.*

    (i) If time is a continuous variable, then the average annual cost of item will be minimized by replacing the machine when the average cost to date becomes equal to the current maintenance cost.

    (ii) If time is discrete variable, then the average annual cost of the machine will be minimized by replacing the machine when the next period maintenance cost becomes greater than the current average cost of that machine.

PROOF. Let C is the capital cost of a machine (item), S is the scrap value of that item which is constant here and $K_m(t)$ is the maintenance cost of the machine (item) at time t.

    (i) When time t is a continuous variable: Since, $M_m(t)$ is the maintenance cost of the machine (item) at time t and the item is to be used for n years, then the total maintenance cost of the item in n years is

$$= \int_0^n M_m(t)\, dt$$

So, the total cost of the item in n years

= capital cost of the item - scrap value of the item

+ total Maintenance cost of the item in n years.

$$= C - S + \int_0^n M_m(t)$$

The average cost of the item per year during n years is

$$A(n) = \frac{1}{n}\left[C - S + \int_0^n M_m(t)\, dt\right]$$

Now, we have to find such time for which $A_{(n)}$ will be minimum.

$$\frac{d}{dn}A(n) = 0 \implies -\frac{C-S}{n^2} - \frac{1}{n^2}\int_0^n M_m(t)dt + \frac{1}{n}M_t = 0$$

$$\implies \qquad M_t = \frac{1}{n}\left[C - S + \int_0^n M_m(t)\, dt\right]$$

which is equal to $A(n)$. Also

$$\frac{d^2}{dn^2}A(n) = \frac{2(C-S)}{n^3} + \frac{2}{n^3}\int_0^n M_m(t)dt > 0.$$

Hence, the averaged cost $A_{(n)}$ of the item per year is minimized by replacing it when the average cost to date becomes equal to the current maintenance cost.

(ii) When time t is a discrete variable: Since the time is considered in the discrete units. Then t takes the value 1, 2, 3, ...

Then the total maintenance cost of the (item) in n years is given by

$$\sum_{m=1}^{n} M_m(t)$$

So, the total cost of the item in n years is

$$C - S + \sum_{m=1}^{n} M_m(t)$$

$\therefore$ The average cost of the item per year during n years is

$$A(n) = \frac{1}{n}\left[C - S + \sum_{m=1}^{n} M_m(t)\right]. \qquad \qquad ....(1)$$

But for the value of n the Average cost $A(n)$ will be minimum if

$$\Delta A(n-1) < 0 < \Delta A(n) \qquad \qquad ....(2)$$

Now, we know that

$$\Delta A(n) = A(n+1) - A(n) \qquad \qquad ....(3)$$

On putting the value of $A(n+1)$ and $A(n)$ from (1) in (3), we get

$$\Delta A(n) = \frac{1}{n+1}\left[C - S + \sum_{m=1}^{n+1} M_m(t)\right] - \frac{1}{n}\left[C - S + \sum_{m=1}^{n} M_m(t)\right]$$

$$= \left[\frac{C-S}{n+1} + \frac{1}{n+1}\sum_{m=1}^{n+1} M_m(t)\right] - \left[\frac{C-S}{n} + \frac{1}{n}\sum_{m=1}^{n} M_m(t)\right]$$

$$= (C-S)\left(\frac{1}{n+1} - \frac{1}{n}\right) + \frac{1}{n+1}\left[\sum_{m=1}^{n} M_m(t) + M_{(n+1)}(t)\right] - \frac{1}{n}\sum_{m=1}^{n} M_m(t)$$

$$= -\frac{1}{n+1}\left[C - S + \sum_{m=1}^{n} M_m(t)\right] + \frac{1}{n+1}M_{n+1}(t)$$

∴ $\quad\quad \Delta A(n) = -\frac{A(n)}{n+1} + \frac{1}{n+1}M_{n+1}(t)$ $\quad\quad \left[\because A(n) = C - S + \sum_{m=1}^{n} M_m(t)\right]$

Now, we have

$$\Delta A(n-1) = A(n) - A(n-1)$$

$$= \left[\frac{C-S}{n} + \frac{1}{n}\sum_{m=1}^{n} M_m(t)\right] - \left[\frac{C-S}{n-1} + \frac{1}{n-1}\sum_{m=1}^{n-1} M_m(t)\right]$$

$$= -\frac{1}{n}\left[C - S + \sum_{m=1}^{n-1} M_m(t)\right] + \frac{1}{n}M_n(t)$$

Therefore, the value of $A(n)$ will be minimum for n if $\Delta A(n) > 0$ and $\Delta A(n-1) < 0$.

So, $\quad -\frac{A(n)}{n+1} + \frac{1}{n+1}M_{n+1}(t) > 0$ or $M_{n+1}(t) > A(n)$ and Similarly, $M_n(t) < A(n-1)$.

From above results it is obvious that we should not replace the Machine (item) if next year maintenance cost is less than the previous years average cost. We should replace the machine (item) if the next years maintenance cost is less than the previous years average cost.

Working Procedure

An item is continued upto the time the average cost per year of the item decreases and we are to replace it at the time when this average cost begins to increase. We know that if the replacement time of an item is n years, then we have

$$M_{n+1}(t) > A(n) \quad \text{or} \quad M_{n+1}(t) > \frac{1}{n}[C - S + M_1 + M_2 + M_3 + ... + M_n]$$

$$\frac{M_{n+1}(t)}{n+1} > \frac{1}{n(n+1)}[C - S + M_1 + M_2 + M_3 ... + M_n]$$

Now, we add $(C - S + M_1 + M_2 + M_3 + ... + M_n) / n + 1$ on both sides, we get

$$\frac{1}{n+1}[C - S + M_1 + M_2 + M_3 + ... + M_{n+1}] > \frac{1}{n(n+1)}[C - S + M_1 + M_2 + ... + M_n]$$

$$+ \frac{1}{n+1}[C - S + M_1 + M_2 + M_3 + ... + M_n]$$

⇒ $\quad \frac{1}{n+1}[C - S + M_1 + M_2 + M_3 + ... + M_{n+1}]$

$$> [C - S + M_1 + M_2 + M_3 + ... + M_n]\left[\frac{1}{n(n+1)} + \frac{1}{n+1}\right]$$

⇒ $\quad \frac{1}{n+1}[C - S + M_1 + M_2 + M_3 + ... + M_{n+1}] > \frac{1}{n}[C - S + M_1 + M_2 + M_3 + ... + M_n]$

⇒ $\quad\quad\quad\quad A(n+1) > A(n)$.

It is obvious that if the average cost of $(n + 1)$ years is greater than the average cost of n years, then we should replace the item in n years. Or we can say when the average cost begins to increase then we should replace the item.

Solved Examples

EXAMPLE 1. *The cost of a machine is ₹6100 and its scrap (resale) value is ₹100. The maintenance cost are*

Year	1	2	3	4	5	6	7	8
Maintenance cost in ₹	100	250	400	600	900	1250	1600	2000

Find the replacement age of the machine.

[MEERUT–2002, 12, 14, 17, ROHILKHAND–2000]

SOLUTION. The cost of Machine, C = ₹ 6100 and scrap value S = ₹ 100.
Now the average cost per year of the machine are as follows:

Age of replacement (in Years)	Total Running Cost ₹	Depreciation (₹) (C - S)	Total cost (₹)	Average cost per year (₹)
1	100	6000	6100	6100.00
2	350	6000	6350	3175.00
3	750	6000	6750	2250.00
4	1350	6000	7350	1837.50
5	2250	6000	8250	1650.00
6*	3500	6000	9500	1583.33*
7	5100	6000	11100	1585.71
8	7100	6000	13100	1637.50

From the above table it is clear that the average cost decreases per year upto the end of 6th year and then begins to increase. Hence, the best replacement age is 6 years. *i.e.,* we should replace the machine at the end of 6th years and the minimum average cost is ₹ 1583.33.

EXAMPLE 2. *The cost of a machine is ₹12200 and the resale value is ₹ 200. The maintenance costs of the machine are as follows:*

Year	1	2	3	4	5	6	7	8
Maintenance cost in ₹	200	500	800	1200	1800	2500	3200	4000

Find the best replacement age of the machine.

SOLUTION. The cost of the machine C = ₹ 12200 and the resale value S = ₹200.

Replacement year	Running cost (₹) $M_n(t)$	Cumulative value $\Sigma M_n(t)$	Depreciation (C - S) (₹)	Total cost (C - S + $\Sigma M_n(t)$)	Average cost (C-S+$\Sigma M_n(t)$)/y
1	200	200	12000	12200	12200.00
2	500	700	12000	12700	6350.00
3	800	1500	12000	13500	4500.00
4	1200	2700	12000	14700	3675.00
5	1800	4500	12000	16500	3300.00
6*	2500	7000	12000	19000	3166.67*
7	3200	10200	12000	22200	3171.42
8	4000	14200	12000	26200	3275.00

From the above table, it is clear that the average cost decrease per year upto the end of 6th year and then begin to increase. Hence, we should replace the machine at the end of every 6th years and minimum average cost is ₹ 3166.67.

EXAMPLE 3. *The cost of a truck is ₹ 3000. The resale value (Salvage value) and the running cost are given as follows:*

Year	1	2	3	4	5	6	7
Running cost	600	700	800	900	1000	1200	1500
Resale value	2000	1333	1000	750	500	300	300

Find the replacement age of the truck.

SOLUTION. Here $C = 3000$ is the cost of truck and S is the resale value.

Age of replacement in years	Resale value $S(₹)$	Total Running cost (₹)	Depreciation $(C - S)$ (₹)	Total cost (₹)	Average cost per year (₹)
1	2000	600	1000	1600	1600.00
2	1333	1300	1667	2967	1483.50
3	1000	2100	2000	4100	1366.67
4	750	3000	2250	5250	1312.50
5*	500	4000	2500	6500	1300.00*
6	300	5200	2700	7900	1316.67
7	3000	6700	2700	9400	1342.85

From the above table, the average cost decreases per year upto the end of 5th year and then begins to increase. Hence the best replacement age is 5 years, *i.e.,* we should replace the pick at the end of every 5th year and minimum average cost is ₹ 1300.

EXAMPLE 4. *The cost of a machine is ₹ 5000. Running cost and resale value are as given below :*

Year	1	2	3	4	5	6	7	8
Running costs (₹)	1500	1600	1800	2100	2500	2900	3400	4000
Resale value (₹)	3500	2500	1700	1200	800	500	500	500

At what year is the replacement due?   [MEERUT–1996]

SOLUTION. The cost of machine C is 5000 and average cost per year are as follows:

Age of replacement in years	Resale value S (₹)	Total Running cost (₹)	Depreciation $(C - S)$ (₹)	Total cost (₹)	Average cost per year (₹)
1	3500	1500	1500	3000	3000.00
2	2500	3100	2500	5600	2800.00
3	1700	4900	3300	8200	2733.33
4*	1200	7000	3800	10800	2700.00*
5	800	9500	4200	13700	2740.00
6	500	12400	4500	16900	2816.67
7	500	15800	4500	20300	2900.00
8	500	19800	4500	24300	3037.50

From the table, it is clear that the minimum average cost is ₹ 2700 during the 4th year. Hence, the machine should be replaced at the end of every 4th year.

EXAMPLE 5. **(a) A machine owner finds from his past records that the cost per year of maintaining a machine A whose capital cost is ₹ 6000 are as given below :**

Year	1	2	3	4	5	6	7	8
Maintenance cost	1000	1200	1400	1800	2300	2800	3400	4000
Resale value	3000	1500	750	375	200	200	200	200

What is the best replacement age of Machine A?

(b) The cost of a machine B whose capacity is 50% more than Machine A is ₹ 8000. Maintaining cost and resale value are as given below:

Year	1	2	3	4	5	6	7	8
Maintenance cost	1200	1500	1800	2450	3100	4000	5000	6100
Resale value	4000	2000	1000	500	300	300	300	300

What is the best replacement age of machine B ?

(c) A machine owner has three A type Machine. Two of these machines are two years old and one is one year old. He is considering to purchase a new machine of type B. Assuming that the loss of flexibility due to fewer machines is of no importance, and he continues to have sufficient work for three of the old machines, what should his policy be ? [MEERUT–2002, 07, ROHILKHAND–1995, 99, 2002]

SOLUTION.

(a) Here C = ₹ 6000 is the cost of Machine A. The average cost of machine A per year is as follows:

Age of replacement	Resale value S (₹)	Total Running cost (₹)	Depreciation (C - S) (₹)	Total cost (₹)	Average cost per year (₹)
1	3000	1000	3000	4000	4000.00
2	1500	2200	4500	6700	3350.00
3	750	3600	5250	8850	2950.00
4	375	5400	5625	11025	2756.00
5*	200	7700	5800	13500	2700.00*
6	200	10500	5800	16300	2717.00
7	200	14500	5800	20300	2900.00

From the table, the replacement age of machine A is 5 years and the minimum average cost is ₹ 2700.

(b) The cost of machine B is C = ₹ 8000. The average cost of machine B per year is as follows:

Age of replacement	Resale value S (₹)	Total Running cost (₹)	Depreciation $(C - S)$ (₹)	Total cost (₹)	Average cost per year (₹)
1	4000	1200	4000	5200	5200.00
2	2000	2700	6000	8700	4350.00
3	1000	4500	7000	11500	3833.33
4	500	6900	7500	14400	3600.00
5*	300	10,000	7700	17700	3540.00*
6	300	14,000	7700	21700	3616.66
7	300	19,000	7700	26700	3814.28

From the table, the minimum average cost is ₹ 3540 and the best replacement age of machine B is 5 years.

(c) It is given that the capacity of Machine B is 50% more than that of machine A. This means that two machines B are equal to the three Machines A. Since the average cost of machine B is ₹ 3540, which is equal to ₹ $\frac{2 \times 3540}{3}$ = ₹ 2360 of one machine A. Now since average cost of machine B (₹ 2360) is less than the average cost of machine A (₹ 2700). Therefore the machine A should be replaced by machine B.

Age of Replacement: Total average cost of two Machines B = ₹ 2 × 3540 = ₹ 7080.00. Now the total cost of machine A in every year are as below:

Year	1	2	3	4	5
Total cost in the year (₹)	4000	6700 – 4000 = 2700	8850 – 6700 = 2150	11025 – 8850 = 2175	13500 – 11025 = 2475

Since two of three machines are two years old and third one is one year old. Therefore, total cost for three machines A from now
= Total cost of two machines A in 3rd year
+ total cost of one machine A in the 2nd year
= ₹ 2 × 2150 + ₹ 2700 = ₹ 7000
Total cost of three machines A during the second year from now
= Total cost of the machine A in 4th year
+ Total cost of one machine A in the 3rd year
= ₹ 2 × 2175 + ₹ 2150 = ₹ 6500
and the total cost of three machine A during the third year from now
= Total cost of two machine A in 5th year
+ Total cost of one machine A in 4th year
= ₹ 2 × 2475 + ₹ 2175 = ₹ 7125
It is clear that the total average cost of two machines B (₹ 7080) is less than the total cost of three machines A in the third year from now. Hence, without waiting for their normal replacement age, three machines A should be replaced by two machines B after two years from now.

EXAMPLE 6. **(a)** *The cost of a machine A is ₹ 9000. Annual operating costs are ₹ 200 for the first year and then increase by ₹ 2000 every year. Find the best replacement age of the machine if the optimum replacement* *policy is followed. What will be the average yearly cost of owing and operating the machine? (Assume that the resale value of the machine is zero and future costs are not discounted).*

(b) *The cost of machine B is ₹ 10000. Annual operating costs are ₹ 400 for the first year, and then increase by ₹ 800 every year. You have a new machine of type A which is one year old. Should you replace it with B, and if so when? (The resale values of machine B is zero).*

(c) *Suppose you are just ready to replace machine A with another machine of the same type, when you hear that machine B will become available in a year. What should you do?*

SOLUTION.   (a)   The cost of machine A is C = 9000. Resale value (S) = 0. The maintenance costs are as follows:

Year	1	2	3	4	5
Maintenance cost	200	2200	4200	6200	8200

The average cost of machine A per year are as follows:

Replacement at the end of years	Maintenance cost (₹)	Total Running cost (₹)	Depreciation (C - S) (₹)	Total cost (₹)	Average cost per year (₹)
1	200	200	9000	9200	9200.00
2	2200	2400	9000	11400	5700.00
3*	4200	6600	9000	15600	5200.00*
4	6200	12800	9000	21800	5450.00
5	8200	21000	9000	30000	6000.00

From the table the minimum average yearly cost is ₹ 5200. Hence, the machine A should be replaced at the end of every 3rd year.

(b)   The cost of machine B is C = ₹ 10000, Resale value (S) = 0. The maintenance costs are as follows:

Year	1	2	3	4	5	6	7
Maintenance cost	400	1200	2000	2800	3600	4400	5200

The average costs per year of Machine B are follows:

Replacement at the end of year	Maintenance cost (₹)	Total Running cost (₹)	Depreciation (C - S) (₹)	Total cost (₹)	Average cost per year (₹)
1	400	400	10000	10400	10400.00
2	1200	1600	10000	11600	5800.00
3	2000	3600	10000	13600	4533.33
4	2800	6400	10000	16400	4100.00
5*	3600	10000	10000	20000	4000.00*
6	4400	14400	10000	24400	4066.66
7	5200	19600	10000	29600	4228.57

From the table, the minimum average yearly cost is ₹ 4000 and machine B should be replaced at the end of 5th year.

The minimum average cost of machine B is ₹ 4000 which is less than the minimum average cost of machine A ₹ 5200. Hence the machine A should be replaced by machine B.

Time of Replacement: When the running cost of machine A of the next year exceeds the minimum average yearly cost of machine B. That time machine A should be replaced by machine B.

Total cost of machine A per year are as follows :

Year	1	2	3	4	5
Total cost in the year (₹)	9200	11400 – 9200 = 2200	15600 –11400 = 4200	21800 – 15600 = 6200	30000 – 21800 = 8200

From the table, the running cost of machine A is ₹ 4200 in the third year which is more than the minimum average yearly cost ₹ 4000 of machine B. Therefore, the machine A is replaced by machine B after two years. But it is given that machine A is one year old. Hence, the machine A should be replaced by machine B after one year from now.

(c) It is clear from part b, the machine A should be replaced after one year from now. The machine B is also available at that time. Hence, the machine A should be replaced by machine B after one year from now.

20.7 BEST REPLACEMENT AGE OF THE ITEMS WHOSE MAINTENANCE COST INCREASE AND VALUE OF MONEY CHANGE WITH THE TIME

Let the cost of a machine be A and $C_1, C_2, C_3, ..., C_n$ be the maintenance cost in 1, 2, 3, ... n years respectively. If the depreciation ratio during a year is v then present value of maintenance costs are $C_1, C_2 v, C_3 v^2, ..., C_n v^{n-1}$ in 1, 2, 3, ... n years respectively.

The present value of the total expenditure in n years is

$$P(n) = A + C_1 + C_2 v + C_3 v^2 + ... + C_n v^{n-1}$$

$$= A + \sum_{n=1}^{n} C_n v^{n-1}$$

If the machine is replaced with a new machine after n years from now, then the $A + C_1, C_2, C_3, ..., C_n$ be the total expenditure on the new machine in years $n + 1$, $n + 2$, $n + 3$,... $n + n$ from now.

Now, the present value of all these expenditures is

$$(A + C_1) v^n + C_2 v^{n+1} + C_3 v^{n+2} + C_4 v^{n+3} + ... + C_n v^{2n-1}$$

and the present value of all these expenditure in successive n years is

$$(A + C_1) v^{2n} + C_2 v^{2n+1} + C_3 v^{2n+2} + C_4 v^{2n+3} + ... + C_n v^{2n-1} \text{ and so on.}$$

∴ The present value of the total expenditure with a replacement policy of n years is

$$C(n) = [A + C_1 + C_2 v + C_3 v^2 + ... + C_n v^{n-1}]$$

$$+ [(A + C_1) v^n + C_2 v^{n+1} + C_3 v^{n+2} + C_4 v^{n+3} + ... + C_n v^{n-1}]$$

$$+ [(A + C_1) v^{2n} + C_2 v^{(2n+1)} + C_3 v^{2n+2} + C_4 v^{2n+3} + ... + C_n v^{3n-1}] + ...$$

$$= \left[A + \sum_{m=1}^{n} C_m v^{m-1} \right] + v^n [A + C_1 + C_2 v + C_3 v^2 + C_4 v^3 + ... + C_n v^{n-1}]$$

$$+ v^{2n} [A + C_1 + C_2 v + C_3 v^2 + C_4 v^3 + ... C_n v^{n-1}] ...$$

$$= \left[A + \sum_{m=1}^{n} C_m v^{m-1} \right] + v^n \left[A + \sum_{m=1}^{n} C_m v^{m-1} \right] + v^{2n} \left[A + \sum_{m=1}^{n} C_m v^{m-1} \right] + \ldots$$

$$= \left[A + \sum_{m=1}^{n} C_m v^{m-1} \right] \left[1 + v^n + v^{2n} + \ldots \right]$$

or we can write

$$C(n) = \left[A + \sum_{m=1}^{n} C_m v^{m-1} \right] \left[\frac{1}{1 - v^n} \right] \qquad \ldots(1)$$

$$\left[\because r = v^n \quad a = 1, \text{ and } S_\infty = \frac{1}{1 - v^n} \right]$$

Here, $C(n)$ is the total amount of the money required at present of all future purchasing costs and maintenance costs on the machines with a replacement policy of n years.

Now, we have to find the value of n for which $C(n)$ will be minimum for n if

$$\Delta C(n - 1) < 0 < \Delta C(n)$$

Consider

$$\Delta C(n) = C(n + 1) - C(n)$$

$$= \left[A + \sum_{m=1}^{n+1} C_m v^{m-1} \right] \left[\frac{1}{1 - v^{n+1}} \right] - \left[A + \sum_{m=1}^{n} C_m v^{m-1} \right] \left[\frac{1}{1 - v^n} \right]$$

$$= \frac{1}{(1 - v^{n+1})(1 - v^n)} \left[A(v^{n+1} - v^n) - \left(\sum_{m=1}^{n} C_m v^{m-1} \right)(1 - v^{n+1}) + \left(\sum_{m=1}^{n+1} C_m v^{m-1} \right)(1 - v^n) \right]$$

$$= \frac{1}{(1 - v^n)(1 - v^{n+1})} \left[\sum_{m=1}^{n} C_m v^{m-1} \{(1 - v^n) - (1 - v^{n+1})\} + C_{n+1} v^n (1 - v^n) - A v^n (1 - v) \right]$$

$$= \frac{1}{(1 - v^n)(1 - v^{n+1})} \left[-\sum_{m=1}^{n+1} C_m v^{m-1} (v^n - v^{n+1}) + C_{n+1} v^n (1 - v^n) - A v^n (1 - v) \right]$$

$$= \frac{v^n}{(1 - v^n)(1 - v^{n+1})} \left[C_{n+1} (1 - v^n) - (1 - v) \left(A + \sum_{m=1}^{n+1} C_m v^{m-1} \right) \right]$$

$$= \frac{v^n (1 - v)}{(1 - v^n)(1 - v^{n+1})} \left[\frac{(1 - v^n)}{(1 - v)} C_{n+1} - \left(A + \sum_{m=1}^{n} C_m v^{m-1} \right) \right]$$

and $\Delta C(n - 1) = C(n) - C(n - 1)$

$$= \left[A + \sum_{m=1}^{n} C_m v^{m-1} \right] \left[\frac{1}{1 - v^n} \right] - \left[A + \sum_{m=1}^{n} C_m v^{m-1} \right] \left[\frac{1}{1 - v^{n-1}} \right]$$

$$= \frac{1}{(1 - v^n)(1 - v^{n-1})} \left[(1 - v^{n-1}) \left(A + \sum_{m=1}^{n} C_m v^{m-1} \right) - (1 - v^n) \left(A + \sum_{m=1}^{n+1} C_m v^{m-1} \right) \right]$$

$$= \frac{v^{n-1}(1 - v)}{(1 - v^n)(1 - v^{n-1})} \left[\frac{(1 - v^{n-1})}{(1 - v)} C_n - \left(A + \sum_{m=1}^{n-1} C_m v^{m-1} \right) \right]$$

Now $C(n)$ will be minimum for n if

$$\Delta C(n) > 0 > \Delta C(n - 1)$$

$$\frac{v^n(1-v)}{(1-v^n)(1-v^{n+1})}\left[\frac{(1-v^n)}{(1-v)}C_{n+1}-\left(A+\sum_{m=1}^{n}C_m v^{m-1}\right)\right]>0$$

$$>\frac{v^{n-1}(1-v)}{(1-v^n)(1-v^{n-1})}\left[\frac{(1-v^{n-1})}{(1-v)}C_n-\left(A+\sum_{m=1}^{n-1}C_m v^{m-1}\right)\right]$$

Since the value of v is less than one, then $\dfrac{v^{n-1}(1-v)}{(1-v^n)(1-v^{n-1})}$ and $\dfrac{v^n(1-v)}{(1-v^{n+1})(1-v^n)}$ are positive then we have

$$\frac{(1-v^n)}{(1-v)}C_{n+1}-\left(A+\sum_{m=1}^{n}C_m v^{m-1}\right)>0>\frac{(1-v^{n-1})}{(1-v)}C_n-\left(A+\sum_{m=1}^{n-1}C_m v^{m-1}\right) \qquad \dots(2)$$

We can write

$$\frac{(1-v^n)}{(1-v)}C_{n+1}-\left(A+\sum_{m=1}^{n}C_m v^{m-1}\right)>0$$

and

$$\frac{(1-v^{n-1})}{(1-v)}C_n-\left(A+\sum_{m=1}^{n}C_m v^{m-1}\right)<0$$

or

$$C_{n+1}>\frac{[1-v]\left[A+\sum_{m=1}^{n}C_m v^{m-1}\right]}{1-v^n}$$

and

$$C_n<\frac{[1-v]\left[A+\sum_{m=1}^{n}C_m v^{m-1}\right]}{1-v^{n-1}}$$

or

$$C_{n+1}>\frac{A+\sum_{m=1}^{n}C_m v^{m-1}}{1+v+v^2+\dots+v^{n-1}}=R(n)$$

and

$$C_n<\frac{A+\sum_{m=1}^{n}C_m v^{m-1}}{1+v+v^2+\dots+v^{n-2}}=R(n-1) \qquad \dots(3)$$

where, $R(n)=\dfrac{A+\sum_{m=1}^{n-1}C_m v^{m-1}}{1+v+v^2+\dots+v^{n-1}}$ and $R(n-1)=\dfrac{A+\sum_{m=1}^{n-1}C_m v^{m-1}}{1+v+v^2+\dots+v^{n-2}}$ are the

weighted average costs in n and $(n-1)$ years respectively. From (3) we observe the following result :

If the operating cost of the next year is greater than the weighted average cost of the previous year, then we should replace the machine otherwise we should not replace the machine.

☛ **REMARKS**

- If we take $v=1$ in (3), then $\because R(n)=\dfrac{A+C_1+C_2+C_n}{n}$, $\therefore C_{n+1}>R(n)$, which means we should replace the machine when the maintenance cost of the next year is greater than the average yearly cost of previous year. The resale value of the machine is not considered.

> ➡ Do not replace if the operating cost of the next year is less than the weighted average of previous costs.
> ➡ Replace if the operating costs of the next period is greater than the weighted average of previous costs.

THEOREM I. *If P(n) be the discounted cost which is investigated by taking loan, the interest rate of loan be r. The loan is repaid by fixed annual payments throughout the life of the machine. Obtain the minimum annual payment for optimum period n at which to replace.*

[MEERUT–1994, 98; ROHILKHAND–2008]

PROOF. Let x be the fixed annual payment. Then the present worth of fixed annual payment x for n years, where v is the discount rate is

$$P(n) = x + vx + v^2 x + \ldots + v^{n-1} x$$

$$= (1 + v + v^2 + \ldots + v^{n-1}) x$$

$$= \frac{(1 - v^n)}{1 - v} x \qquad \qquad \left[\because v = \frac{1}{1 - r} < 1 \right]$$

or we can write

Fixed annual payment, $x = \dfrac{(1 - v)}{(1 - v^n)} P(n)$ \hfill ...(1)

Since $1 - v$ does not depend upon n then fixed annual payment x will be minimum if $C(n)$ will be minimum where

$$C(n) = \frac{P(n)}{1 - v^n} \qquad \qquad \ldots(2)$$

i.e., $\qquad \qquad \Delta C(n) > 0 > \Delta C(n - 1)$ \hfill ....(3)

Consider, $\quad \Delta C(n) = C(n + 1) - C(n)$

$$= \frac{P(n + 1)}{1 - v^{n+1}} - \frac{P(n)}{1 - v^n} = \frac{1}{(1 - v^n)(1 - v^{n+1})} [(1 - v^n) P(n + 1) - (1 - v^{n+1}) P(n)]$$

$$= \frac{1}{(1 - v^n)(1 - v^{n+1})} [(P(n + 1) - P(n)) - v^n P(n + 1) + v^{n+1} P(n)]$$

$$= \frac{1}{(1 - v^n)(1 - v^{n+1})} [(P(n + 1) - P(n)) - v^n (P(n + 1) - v P(n))] \qquad \ldots(4)$$

But if $C_1, C_2, C_3, \ldots, C_n$ are the maintenance cost of the machine in years 1, 2, 3, ..., n respectively and the cost of machine is A. Then in n years, the present value of total expenditure is

$$P(n) = A + C_1 + C_2 v + C_3 v^2 + \ldots + C_n v^{n-1}$$

and $\qquad P(n + 1) = A + C_1 + C_2 v + C_3 v^2 + \ldots + C_n v^{n+1} + C_{n+1} v^n$

or $\qquad P(n + 1) = P(n) + C_{n+1} v^n$

On putting the values of $P(n)$ and $P(n + 1)$ in (4), we get

$$\Delta C(n) = \frac{1}{(1 - v^n)(1 - v^{n+1})} [\{P(n) + C_{n+1} v^n - P(n)\} - v^n \{P(n) + C_{n+1} v^n - v P(n)\}]$$

$$= \frac{1}{(1 - v^n)(1 - v^{n+1})} [(1 - v^n) C_{n+1} v^n - (1 - v) P(n) v^n]$$

$\therefore \qquad \Delta C(n) = \dfrac{(1 - v) v^n}{(1 - v^n)(1 - v^{n+1})} \left[\dfrac{(1 - v^n)}{(1 - v)} C_{n+1} - P(n) \right]$ \hfill ....(5)

Again consider

$$\Delta C(n + 1) = C(n) - C(n - 1) = \frac{P(n)}{1 - v^n} - \frac{P(n - 1)}{1 - v^{n-1}}$$

$$= \frac{1}{(1-v^n)(1-v^{n-1})}\left[(1-v^{n-1})P(n)-(1-v^n)P(n-1)\right]$$

$$= \frac{1}{(1-v^n)(1-v^{n-1})}\left[P(n)-P(n-1)-v^{n-1}(P(n)-vP(n-1))\right]$$

On putting the values of $P(n)$ and $P(n-1)$ in above equation, we get

$$\Delta C(n-1) = \frac{(1-v)\,v^{n-1}}{(1-v^n)(1-v^{n-1})}\left[\frac{(1-v^{n-1})}{(1-v)}C_n - P(n-1)\right] \qquad \ldots(6)$$

Now, $C(n)$ will be minimum if $\Delta C(n) > 0 > \Delta C(n-1)$

$$\therefore \quad \frac{(1-v)v^n}{(1-v^n)(1-v^{n+1})}\left[\frac{(1-v^n)}{(1-v)}C_{n+1}-P(n)\right] > 0 > \frac{(1-v)v^{n-1}}{(1-v^n)(1-v^{n-1})}\left[\frac{(1-v^{n-1})}{(1-v)}C_n - P(n-1)\right]$$

Since, $0 < v < 1$, then $1-v$ will be positive and so, $\dfrac{(1-v)v^n}{(1-v^n)(1-v^{n+1})}$ and $\dfrac{(1-v)v^{n-1}}{(1-v^n)(1-v^{n-1})}$ will be positive. So we have

$$\frac{(1-v^n)}{(1-v)}C_{n+1} - P(n) > 0 > \frac{(1-v^{n-1})}{(1-v)}C_n - P(n-1)$$

which gives the best replacement age of machine is n years for which $P(n)$ or $C(n)$ is minimum. We obtain the minimum annual payment x by $x = \dfrac{1-v}{1-v^n}.P(n)$.

 Solved Examples

EXAMPLE 1.   ***The cost pattern for two machines A and A′ when value of money is not considered, is given below***

Year	Cost at the beginning of year (in ₹)	
	Machine A	Machine A′
1	900	1400
2	600	100
3	700	700
Total	2200	2200

Obtain the cost pattern for machines A and A′ when money is worth 10% per year, and also find which machine is less costly. [MEERUT–2002]

SOLUTION.   Here, it is given that the total expenditure of Machines A and A′ in three years is ₹ 2200. Thus both the machines are equivalent if the value of money does not change with the time.

When money is worth 10% per year then discount rate, $v = \dfrac{100}{100+10} = 0.9091$.

∴ The present value in years 1, 2 and 3 of the maintenance cost of both the machines are

Year	Machine A (₹)	Machine A′ (₹)
1	900	1400
2	$600 \times v = 600 \times (0.9091) = 545.45$	$100 \times v = 100 \times (0.9091) = 90.91$
3	$700 \times v^2 = 700(0.9091)^2 = 578.52$	$700 \times v^2 = 700 \times (0.9091)^2 = 578.25$
Total (₹)	2023.97	2069.43

From the table, the present value of total expenditure for machine A is less than that for machine A′. Therefore, machine A is less costly.

EXAMPLE 2. *A manufacturer is offered two machines A and A′. Cost of machine A is ₹ 5000 and the running cost are estimated at ₹ 800 for each of the first five years, increasing by ₹ 200 per year in the sixth and subsequent years. Machine A′, which has the same capacity as machine A, costs ₹ 2500 with running cost of ₹ 1200 per year for the first six years, increasing by ₹ 200 per year in the seventh and subsequent years. If the money is worth 10% per year, which machine should be purchased? Assume that the machine will eventually be sold for scrap at negligible price.* [MEERUT–1994; ROHILKHAND–1997]

SOLUTION. Since money is worth is 10% per year then discount rate $v = \dfrac{100}{100+10} = 0.9091$, then $1 - v = \dfrac{1}{11}$.

The cost of machine A is ₹ 5000. Let n is the optimum replacement age of the machine, then we have

$$\frac{1 - v^n}{1 - v} C_{n+1} - P(n) > 0 > \frac{1 - v^{n-1}}{1 - v} C_n - P(n-1) \qquad ....(1)$$

where the value of $P(n)$ is $A + C_1 + C_2 v + C_2 v^2 + ... + C_n v^{n-1}$ then we calculate the followings:

Year n	Running cost C_n (₹)	$v^{n-1} . C_n$ (₹)	$(1 - v^{n-1}) . C_n$ (₹)	$\dfrac{(1 - v^n)}{1 - v} C_{n+1}$ (₹)	$P(n)$	$\dfrac{(1 - v^n)}{1 - v} C_{n+1} - P(n)$
1	800	800	0	800	5800	<0
2	800	727	73	1529	6527	<0
3	800	661	139	2189	7188	<0
4	800	601	199	2794	7789	<0
5	800	546	254	4169	8335	<0
6	1000	621	379	5753	8956	<0
7	1200	677	523	7502	9633	<0
8	1400	718	682	9394	10351	<0
9*	1600	746	854	11407	11097	>0
10	1800	763	1037	—	—	

where $v^{n-1} = (0.9091)^{n-1}$ for years 2, 3, ... 10.

$$v = 0.9091, \quad v^2 = 0.8264, \quad v^3 = 0.7513, \quad v^4 = 0.6830$$

$$v^5 = 0.6209, \quad v^6 = 0.5645, \quad v^7 = 0.5132, \quad v^8 = 0.4665 \text{ and } v^9 = 0.4241$$

From the table, equation (1) is satisfied for $n = 9$. Therefore, machine A should be replaced after 9 years, and if the fixed annual payment for machine A is x_1 then

$$x_1 = \frac{1 - v}{1 - v^9} . P(9) = \frac{1 - 0.9091}{1 - (0.9091)^9} \times 11097 = ₹ 1752$$

Table for Machine A′ : The cost of Machine A′ is ₹ 2500. Then we make the following table:

Year n	Running cost C_n (₹)	v^{n-1} (₹)	$v^{n-1}.C_n$ (₹)	$(1-v^{n-1}).C_{n+1}$ (₹)	$\dfrac{(1-v^n)}{1-v}C_{n+1}$ (₹)	$P(n)$	$\dfrac{(1-v^n)}{1-v}C_{n+1}$ $-P(n)$
1	1200	1	1200	0	1200	3700	< 0
2	1200	0.9091	1091	109	2288	4791	< 0
3	1200	0.8264	992	208	3278	5783	< 0
4	1200	0.7513	902	298	4180	6685	< 0
5	1200	0.6830	820	380	5005	7505	< 0
6	1200	0.6209	745	455	6710	8250	< 0
7	1400	0.5645	790	610	8369	9040	< 0
8	1600	0.5132	821	779	10560	9861	> 0
9	1800	0.4665	840	960	—	—	

From the above table, eq. (1) is satisfied for $n = 8$. Therefore, Machine A′ should be replaced after 8 years and if the fixed annual payment for machine A′ is x_2, then $x_2 = \dfrac{1-v}{1-v^8}P(8) = \dfrac{1-0.9091}{1-(0.9091)^8} \times 9861 = ₹\ 1680.$

Since, fixed annual payment at machine A′ is less than the fixed annual payment of machine A. Hence, Machine A′ should be purchased instead of machine A.

20.8 INTERVAL OF OPTIMUM REPLACEMENT

If all the items fail during a period of fixed time t', then they are replaced individually at the end of that period. If all the items, including which may be new are replaced an intervals $t = nt'$ consisting of n periods. Here, we will find the value of n for which average cost per periods of the items will be minimum. If N is the total number of items in the system in the beginning, $N(y)$ be the number of items which failed during y^{th} period, where $y = 1, 2, 3, \dots$ $(n - 1)$, C_1 is the cost per item when all the items are replaced as groups, C_2 is the cost per item for individual replacement on its failure and $C(n)$ is the total cost in the interval t, then

Total cost $\qquad C(n) = NC_1 + C_2[N(1) + N(2) + \dots + N(n-1)]$

$$= NC_1 + C_2 \sum_{y=1}^{n-1} N(y)$$

If the average cost per period of the items is $A(n)$, then

$$A(n) = \frac{1}{n} A(n)$$

Now $A(n)$ will be minimum for n if

$$\Delta A(n-1) < 0 < A(n)$$

Consider

$$\Delta A(n) = A(n+1) - A(n)$$

$$= \frac{A(n+1)}{n+1} - \frac{A(n)}{n} = \frac{C(n)+C_2N(n)}{n+1} - \frac{C(n)}{n} = \frac{C_2N(n) - \dfrac{C(n)}{n}}{n+1}$$

and $\qquad \Delta A(n-1) = A(n) - A(n-1)$

$$= \frac{C(n)}{n} - \frac{A(n-1)}{n-1} = \frac{C_2 N(n-1) - \frac{C(n-1)}{n-1}}{n}$$

Now from (1) we have

$$\frac{C_2 N(n-1) - \frac{C(n-1)}{n-1}}{n} < 0 < \frac{C_2 N(n) - \frac{C(n)}{n}}{n+1}$$

or we can write

$$C_2 N(n) - \frac{C(n)}{n-1} > 0 > C_2 N(n-1) - \frac{C(n-1)}{n-1}$$

which gives the results

$$C_2 N(n) > \frac{C(n)}{n} \quad \text{and} \quad C_2 N(n-1) < \frac{C(n-1)}{n-1}$$

Then we have the following group replacement policy:

1. Group replacement should be made at the end of nth year if the cost of individual replacement for the nth period is greater than the average cost per period by the end of n periods.

2. Group replacement should not be made at the end of nth year, if the cost of individual replacements of the end of $(n-1)^{\text{th}}$ year is not less than the average cost per period by the end of $(n-1)$ periods.

20.9 PRESENT VALUE FOR COMPARING REPLACEMENT ALTERNATIVES

The present value of all future expenditures and revenues is calculated for each alternative and the one for which the present value is minimum is preferred. Let us use the following notations :

Q	:	the annual cost.
i	:	the annual interest rate
n	:	the period in years.
P	:	the principal amount.
S	:	the scrap or salvage value.

Then the present value of total cost is given by

$P + Q$ (*Pwf* for $i\%$ interest for n years)

$$- S \text{ (} Pwf \text{ for } i\% \text{ interest for } n \text{ years)} \qquad ...(1)$$

If the annual operating costs vary for different years then the present value of these costs will be calculated on the basis of time period for which these expenditures are made *i.e.* if $Q_1, Q_2, Q_3, ..., Q_n$ are the operating *costs* for different years then the present value of the operating costs during n years will be given by

Q_1 (*Pwf* at $i\%$ interest for 1 year) + Q_2 (*Pwf* at $i\%$ interest for 2 years).

$$+ ... + Q_n \text{ (} Pwf \text{ at } i\% \text{ interest for } n \text{ years).}$$

Here the present value of total cost will be

$P + Q_1$ (*Pwf* at $i\%$ interest for 1 year) + Q_2 (*Pwf* at $i\%$ interest for 2 years)

$+ ... + Q_n$ (*Pwf* at $i\%$ interest for n year) $- S$ (*Pwf* at $i\%$ interest for n years).

Solved Examples

EXAMPLE 1. *A person is considering to purchase a machine for his factory. The related data about the alternative machine are as follows:*

	Machine A	Machine B	Machine C
Present investment (₹)	10,000	12,000	15,000
Total annual cost ₹	2,000	1,500	1,200
Life in Years	10	10	10
Salvage value ₹	500	1,000	1,200

As an advisor of the company, you have been asked to select the best machine considering 12 % normal rate of return per year.

Given Annual series Pwf @ 12% for 10 years = 5.650

Single Payment Pwf @ 12% for 10 years = 0.322

SOLUTION. Here present value of total cost of each machine or a period of 10 years is to be calculated and the machine for which the present value is least to be recommended. The calculations are given below:

	Machine A	Machine B	Machine C
Present investment	10,000	12,000	15,000
Present value of total Annual cost	2,000 × 5.650	1,500 × 5.650	1,200 × 5.650
Present value of salvage value	500 × 0.322	1,000 × 0.322	1,200 × 0.322
Present total value	₹ 21,139.00	₹ 20,153.00	₹ 21,393.60

Clearly we observed that, the machine B, having least present total value should be purchased.

EXAMPLE 2. *A company is considering the purchase of new grinder which will cost ₹ 10,000. The economic life of the machine is expected to be 6 years. The salvage value of the machine will be ₹ 2,000. The average operating and maintenance costs are estimated to be ₹ 5000 per annum.*

(a) Assuming an interest rate of 10%, determine the present value of future cost of the proposed grinder.

(b) Compare this grinder with the presently owned grinder that has an annual operating cost of ₹ 4000 per annum and expected maintenance cost of ₹ 2000 in the second year with an annual increase of ₹ 1000 thereafter.

SOLUTION. (a) Present value of annual operating cost

= 5000 (Pwf at 10% interest for 6 years) = 2,000 × 4.355

= ₹ 21775

Hence value of the salvage value = 2,000×(Pwf at 10% interest for 6 years)

= 2,000 × 0.5646 = ₹ 1129

Hence present value of total future costs

= ₹ 21,775 – ₹ 1,129 = ₹ 20,646.

(b) Here the annual operating and maintenance costs vary with time. The calculations for present value of total operational and maintenance costs are given below:

Years	Operational cost	Maintenance cost	Total Operational & maintenance cost	Pwf for single payment at 10% rate	Present value in ₹
(1)	(2)	(3)	(4)	(5)	(4 × 5)
1	4000	—	4000	.909	3636
2	4000	2000	6000	.826	4956
3	4000	3000	7000	.751	5257
4	4000	4000	8000	.683	5464
5	4000	5000	9000	.621	5589
6	4000	6000	10,000	.564	5640

Total present value ₹ 30,542.

Now from part (a), it is observed that the present value of a new grinder is ₹ 20646. Hence cost savings if a new grinder is purchased,

= ₹ 30,542 – ₹ 20,646 = ₹ 9,896.

As the cost of the new grinder is ₹ 10,000 and its cost savings are only ₹ 9,896, the management is not advised to purchase new grinder.

20.10 REPLACEMENT OF ITEMS THAT FAIL COMPLETELY AND ARE EXPENSIVE TO BE REPLACED

Consider the situations where the possibility of the failure of any item in a system increases with time. The nature of the system may be such that if the item fails then it may result in complete breakdown of the system.

When immediate replacement of the item may not be available, the break down implies loss in production, idle inventory, idle labour and many other losses, so that the failure of the item puts the organization to heavy loss. In such circumstances it is advisable to formulate a suitable replacement policy. Sometimes, the items may be replaced even when they are in working order, to decrease the probability of break down. Any of the following two courses can be followed in such situations.

(i) **Individual replacement policy.** Whenever any item fails, it should be immediately replaced.

(ii) **Group replacement policy.** Here we decide that all the items of the system should be replaced after a certain period irrespective of the fact that items have failed or not with a provision that if any item fails before this time it can be replaced individually.

20.11 MORTALITY AND MORTALITY TABLE [ROHILKHAND–2002, MEERUT–2002, 03]

These problems are the special cases of the problems in industry where the failure of any item can be treated as death and the replacement of any item on failure can be taken to be a birth in the human population.

We have the following mortality problems.

Let a large population be subjected to a given mortality curve for a very long period of time under the following assumptions.

    (i) All death's are immediately replaced by births.

and (ii) There are no other entries or exists.

Then, we have the following results:

    (i) The no. of deaths per unit time become stable.

  (ii) The age distribution ultimately becomes stable.

> The ultimate rate of death (or birth) can be obtained by dividing the size of the population by the mean age of death. Hence, the age distribution becomes unstable.

20.11.1 MORTALITY TABLE

One can never predict that when any item in a system will fail. But theory of probability can help us in solving this problem. Various methods are available for deriving the probability distribution of failures. The basic assumption we make, is that failures occur only at the end of the period. The problem is to find that value of t which minimizes total cost involved in the system.

Mortality tables for any item can be used to derive the probability distribution of life spans *e.g.* if $M(t)$ denotes the number of survivors at any time t and $M(t-1)$ is the number of survivors at time $(t-1)$, then probability that any item will fail in this time interval, will be

$$= \frac{M(t-1) - M(t)}{N} \qquad \qquad ...(1)$$

N being the number of items in the system and the conditional probability that any item survived upto age $t-1$ will die in next year, will be given by

$$= \frac{M(t-1) - M(t)}{M(t-1)} \qquad \qquad ...(2)$$

 Solved Examples

EXAMPLE 1. ***The following mortality rates have been observed for a certain type of light bulbs.***

Week	1	2	3	4	5
Percent failing at the end of week	10	25	50	80	100

There are 1000 bulbs in the beginning and it cost ₹ 1 to replace an individual bulb which has burnt out. If all bulbs were replaced simultaneously it would cost ₹ 0.25 per bulb.

It is proposed to replace all bulbs at fixed intervals, whether or not they have burnt out and to continue replacing burnt out bulbs as they fail. At what intervals should all the bulbs be replaced?

[MEERUT–2003, 05]

SOLUTION. Let P_i be the probability of failures of a bulb in i^{th} week. Then we have

$$P_1 = \frac{10}{100} = 0.10, \; P_2 = \frac{25-10}{100} = 0.15, \; P_3 = \frac{50-25}{100} = 0.25$$

$$P_4 = \frac{80-50}{100} = 0.30 \; \text{ and } \; P_5 = \frac{100-80}{100} = 0.20$$

with $P_1 + P_2 + P_3 + P_4 + P_5 = 1$. This means that a bulb cannot survive for more than five weeks.

Let $N_0 = 1000$ be the number of bulbs in the beginning and N_i be the number of burnout bulbs at the end of ith week.

Then the number of burnt out bulbs at the end of first week i.e.,
$$N_1 = N_0 P_1 = 1000 \times 0.10 = 100$$

Number of burnt out bulbs at the end of second week i.e.,
$$N_2 = N_0 P_2 + N_1 P_1 = 1000 \times 0.15 + 100 \times 0.10 = 150 + 10 = 160$$

Number of burnt out bulbs at the end of third week i.e.,
$$N_3 = N_0 P_3 + N_1 P_2 + N_2 P_1 = 1000 \times 0.25 + 100 \times 0.15 + 160 \times 0.10$$
$$= 250 + 15 + 16 = 281$$

Number of burnt out bulbs at the end of fourth week i.e.,
$$N_4 = N_0 P_4 + N_1 P_3 + N_2 P_2 + N_3 P_1$$
$$= 1000 \times 0.30 + 100 \times 0.25 + 160 \times 0.15 + 281 \times 0.10$$
$$= 300 + 25 + 24.00 + 28.1 = 377$$

Number of burnt out bulbs at the end of fifth week i.e.,
$$N_5 = N_0 P_5 + N_1 P_4 + N_2 P_3 + N_3 P_2 + N_4 P_1$$
$$= 1000 \times 0.20 + 100 \times 0.30 + 160 \times 0.25 + 281 \times 0.15 + 377 \times 0.10$$
$$= 200 + 30 + 40 + 42.15 + 37.7 = 349.85 = 350$$

and Number of burnt out bulbs at the end of sixth week i.e.,
$$N_6 = N_0 P_6 + N_1 P_5 + N_2 P_4 + N_3 P_3 + N_4 P_2 + N_5 P_1$$
$$= 1000 \times 0 + 100 \times 0.20 + 160 \times 0.15 + 281 \times 0.25$$
$$+ 377 \times 0.15 + 350 \times 0.10$$
$$= 20 + 24 + 70.25 + 56.55 + 35 = 205.8 = 206$$

Now the average life of a bulb
$$= \sum_{i=1}^{5} X_i P_i = 1 \times 0.10 + 2 \times 0.15 + 3 \times 0.25 + 4 \times 0.30 + 5 \times 0.20 = 3.35$$

The average number of burnt out bulbs every month
$$= \frac{\text{Total number of bulb in the beginning}}{\text{Average life of a bulb}}$$
$$= \frac{1000}{3.35} = 298.50 = 299$$

Average weekly cost of individual replacement
$$= 299 \times 1 = ₹ 299$$

For group replacement we make the following table

End of the week	Total cost of group replacement (₹)	Average cost per week (₹)
1.	1000 × 0.25 + 100 × 1 = 350	350.00
2.*	1000 × 0.25 + 100 × 1 + 160 × 1 = 510	255.00*
3.	1000 × 0.25 + 100 × 1 + 160 × 1 + 281 × 1 = 791	263.67
4.	1000 × 0.25 + 100 × 1 + 160 × 1 + 281 × 1 + 377 × 1 = 1168	292.00

From the table, the minimum average cost per week of group replacement is

₹ 255 after two weeks which is less than the average cost per week of individual replacement. Hence the group replacement is best after every two weeks.

EXAMPLE 2. *A large population is subject to a given mortality curve for a very long period of time. All deaths are immediately replaced by births, and there are no other entries or exits. Show that the age distribution ultimately becomes stable and that the number of deaths per unit time becomes constant.*

SOLUTION. Without any loss of generality, we may assume that no individual can survive for more than $w+1$ units of time.

Let $f(t)$ denote the number of births at time t and $p(x)$ be the probability of dying just before the age $x + 1$.

At time t, the survivors out of the $f(t-x)$ births at time $(t-x)$ are aged x. Now, $p(x)$ is a priori probability of dying just before the time $(x + 1)$. Hence the number of deaths out of the survivors aged x at time t before reaching the time $(t + 1)$, is $p(x)f(t-x)$.

So, the expected number of deaths before time $t + 1$ will be given by

$$\sum_{x=0}^{w} f(t-x)\, p(x), \quad t = w, w+1, w+2 \qquad \text{....(1)}$$

Now all deaths are replaced by births, therefore if $f(t+1)$ denotes the births at time $(t + 1)$, then

$$f(t+1) = \sum_{x=0}^{w} f(t-x) \cdot p(x) \qquad \text{....(2)}$$

This is a difference equation in t. Let the solution of this equation be given by $f(t) = A\alpha^t$, where A is some constant.

Putting the value of $f(t)$ in (2), we get

$$A\alpha^{t+1} = A \sum_{x=0}^{w} p(x)\, \alpha^{t-x}$$

$$= A[p(0)\, \alpha^t + p(1)\alpha^{t-1} + ... + p(w)\alpha^{t-w}] \qquad \text{....(3)}$$

Dividing both sides of (3) by α^{t-w}, we get

$$\alpha^{w+1} - [\alpha^w p(0) + \alpha^{w-1}p(1) + ... + p(w)] = 0 \qquad \text{....(4)}$$

(4) being an equation of degree $(w+1)$, will have $(w+1)$ roots. Let the roots be $\alpha_0, \alpha_1, ..., \alpha_w$. If we put $\alpha = 1$ in (4), then

$$\text{LH.S.} = 1 - [p(0) + p(1) + ... + p(w)]$$

$$= 1 - \sum_{x=0}^{w} p(x)$$

$$= 1 - 1,$$

($\because$ no item can survive for more than $(w + 1)$ units of time *i.e.,* $\sum_{x=0}^{w} p(x) = 1$)

$$= 0 = \text{RH.S.}$$

Hence $\alpha = 1$ is a root of (4). Let it be α_0, *i.e.,* $\alpha_0 = 1$. The most general solution of (2) will be

$$f(t) = A_0\alpha_0^t + A_1\alpha_1^t + A_2\alpha_2^t + ... + A_w\alpha_w^t$$

$$= A_0 + A_1\alpha_1^t + ... + A_w\alpha_w^t \qquad ....(5)$$

where $A_0, A_1, ..., A_w$ being arbitrary constants.

In equation (4), all the coefficients are less than 1, hence each $\alpha_i < 1$. Since $x_i < 1$, $\alpha_i^t \to 0$ as $t \to \infty \Rightarrow f(t) = A_0$, i.e., number of deaths (also number of births) at any time becomes constant.

To find the value of A_0:

Let $g(x)$ denote the probability that any individual survive for more than x time units. Then

$g(x) = 1 -$ [Probability that it dies before the age x]

$\qquad = 1 -$ [Probability that it dies before age 1 + probability that it dies before age 2 + ... + probability that it dies before age x]

$\qquad = 1 - p(0) - p(1) - .... - p(x-1)$.

Clearly, $g(0) = 1$.

Now once births and deaths have settled down to the constant A_0, the number of survivors aged x is stable at $A_0\, g\,(x)$.

Since deaths are always replaced by births, the size N of the population remains constant. Therefore,

$$N = A_0 \sum_{x=0}^{w} g(x), \quad i.e., \quad A_0 = \frac{N}{\sum_{x=0}^{w} g(x)} \qquad ...(6)$$

To show that $\sum_{x=0}^{w} g(x)$ is equal to mean age at death.

Now $\qquad \sum_{x=0}^{w} g(x) = \sum_{x=0}^{w} g(x)\, \Delta(x),\quad$ (because $\Delta(x) = x + 1 - x = 1$)

$\because \qquad \sum_{x=a}^{b} f(x)\,\Delta h(x) = f(b+1)h(b+1) - f(a)h(a) - \sum_{x=a}^{b} h(x+1)\Delta f(x)$

we get, $\quad \sum_{x=0}^{w} g(x)\,\Delta(x) = g(w+1)(w+1) - g(0).0 - \sum_{x=0}^{w}(x+1)\Delta g(x)$

But $\qquad\qquad g(w+1) = 0$,

$\qquad$ ($\because$ no individual can survive for more than $(w + 1)$ units of time) and

$\qquad\qquad\qquad \Delta g(x) = g(x+1) - g(x)$

$\qquad\qquad\qquad\qquad = [1 - p(0)... - p(x)] - [1 - p(0)... - p(x-1)] = -p(x)$

Hence, $\qquad \sum_{x=0}^{w} g(x) = \sum_{x=0}^{w} g(x)\, \Delta(x)$

$$= \sum_{x=0}^{w}(x+1)p(x) = \text{mean age at death}.$$

Putting the value of $\sum_{x=0}^{w} g(x)$ in (6), we get

$$A_0 = \frac{N}{\text{Mean age at death}} \qquad ....(7)$$

Hence the ultimate rate of death can be obtained by dividing the size of the population by the mean age at death. This implies that the age distribution ultimately becomes stable.

EXAMPLE 3. *The following failure rates have been observed for a certain type of bulbs:*

End of the month	1	2	3	4	5	6	7	8
Probability of failure to date	0.05	0.13	0.25	0.43	0.68	0.88	0.96	1.00

The cost of replacing an individual replacement is ₹ 1.25. The decision is made to replace all bulbs simultaneously at the fixed intervals. If the cost of group replacement is ₹ 0.30 per bulb and the total number of bulbs is 1000 in the beginning. What is the best interval between group replacement?

SOLUTION. It is given that the cost of group replacement is ₹ 0.30 per bulb and the cost of individual replacement is ₹ 1.25 per bulb. Let P_i be the probabilities of failure f a bulb in i^{th} month. Then we have

$$P_1 = 0.5, \quad P_2 = 0.13 - 0.05 = 0.08, \quad P_3 = 0.25 - 0.13 = 0.12$$
$$P_4 = 0.43 - 0.25 = 0.18, \quad P_5 = 0.68 - 0.43 = 0.25, \quad P_6 = 0.88 - 0.68 = 0.20$$
$$P_7 = 0.96 - 0.88 = 0.8 \quad \text{and} \quad P_8 = 1.00 - 0.96 = 0.04$$

with $P_1 + P_2 + P_3 + P_4 + P_5 + P_6 + P_7 + P_8 = 1$. This means that a bulb cannot survive for more than 8 months.

Let $N_0 = 1000$ be the number of bulbs in the beginning and M be the number of failure bulb at the end of ith month.

Then the number of failure bulbs replacement at the end of first month, i.e.,

$$N_1 = N_0 P_1 = 1000 \times 0.5 = 500$$

Number of failure bulbs replacement at the end of second moth *i.e.,*

$$N_2 = N_0 P_2 + N_1 P_1 = 1000 \times 0.08 + 50 \times 0.5 = 80 + 25 = 105$$

Number of failure bulbs replacement at the end of third month *i.e.,*

$$N_3 = N_0 P_3 + N_1 P_2 + N_2 P_1 = 1000 \times 0.12 + 50 \times 0.8 + 260 \times 0.5$$
$$= 120 + 40 + 130 = 290$$

Number of failure bulbs replacement at the end of fourth month *i.e.,*

$$N_4 = N_0 P_4 + N_1 P_3 + N_2 P_2 + N_3 P_1 = 1000 \times 0.18 + 50 \times 0.12 + 260 \times 0.8$$
$$= 125 \times 0.05 = 400$$

Similarly, we can find $N_5, N_6, N_7, N_8, \ldots$ and so on.

Now the average life of a bulb

$$= \sum_{i=1}^{8} X_i P_i = 1 \times 0.5 + 2 \times 0.08 + 3 \times 0.12$$
$$+ 4 \times 0.18 + 5 \times 0.25 + 6 \times 0.20 + 7 \times 0.08 + 3 \times 0.04$$
$$= 4.62$$

Then average number of bulbs replacement every month

$$= \frac{\text{Total number of bulbs in the beginning}}{\text{Average life}} = \frac{1000}{4.62} = 216.34 = 216$$

∴ Average monthly cost of individual replacement = 216 × 1.25 = ₹ 270.

For group replacement we make the following table:

End of the month	Total cost of group replacement (₹)	Average cost per month (₹)
1.	$1000 \times 0.30 + 500 \times 1.25 = 925$	925.00
2.	$1000 \times 0.30 + 500 \times 1.25 + 105 \times 1.25 = 1056.25$	528.12
3.*	$1000 \times 0.30 + 500 \times 1.25 + 105 \times 1.25 + 290 \times 1.25 = 1418.75$	472.91*
4.	$1000 \times 0.30 + 500 \times 1.25 + 105 \times 1.25 + 290 \times 1.25 + 4.00 \times 1.25$	479.68

From the table, the minimum average cost per month of group replacement is ₹ 472.91. Hence, the group replacement is best after every three months.

20.12 STAFFING PROBLEM

[MEERUT–2009]

The staffing problem connected with recruitment and promotion of the staff in any organisation. These problems are treated as the particular case of replacement problem, where the staff of the system is treated like a machine part. The problem of staff replacement arise due to retirements, resignations or deaths of the employee. Thus, to maintain a suitable strength of the staff in a system there is a need of some useful recruitment policy.

Solved Examples

EXAMPLE I. *Calculate the probability of a staff resignation in each year from the following survival table.*

Year No. t	0	1	2	3	4	5	6	7	8	9	10
No. of original staff in service at the end of the year	1000	940	820	580	400	280	190	130	70	30	0

SOLUTION. Let t be the number of years and $M(t)$ is number of original staff in service at the end of the tth year.

Then we have the following table:

Year t	$M(t)$	Probability that a staff will resign in t^{th} year $= \dfrac{M(t-1) - M(t)}{M}$
0	1000	-
1	940	$\dfrac{1000 - 940}{1000} = 0.06$
2	820	$\dfrac{940 - 820}{1000} = 0.12$
3	580	0.24
4	400	0.18
5	280	0.12
6	190	0.09
7	130	0.06
8	70	0.06
9	30	0.04
10	0	0.03

20.13 LIFE TESTING TECHNIQUES

We observed that for formulating some suitable replacement policy for those items which fail or in staffing-problems, the probabilities of failures with time are required. Thus we require the probability distribution of failures with time. There are various methods of deriving such distributions but most common is with the help of life curves. The method is as follows:

The group of items whose probability distribution is to be determined is installed. The number of items surviving from the initially chosen N items at the end of time interval t will be a function of t say $F_s(T)$. The probability of any item surviving at the end of time t will be given by $f(t)/N$. Now for any group the values of $f(t)/N$ obtained from past experience are plotted against t and a smooth curve passing through maximum number of points is drawn. From this curve the values of $p(t) = f(t)/N$ can be determined for various values of t and are then entered in a table, known as Life table for the group of items.

The life curves obtained by the above procedure can be classified in the following categories:

(i) **Normal:** Here the main characteristic of items is that there is only one kind of wear and the items are composed predominantly of one material only *e.g.* car tyres.

(ii) **Log Normal:** This type of curve is obtained when a series of normal curves are added together. In this case, the wear or breakdown of items occurs in various ways and items are composed of many parts *e.g.* Television sets.

(iii) **Exponential curves:** These are the curves which are frequently found in every day problems. The outcome of such curves is due to failure of items occurring randomly with time *e.g.* damages, misuse, natural calamities etc.

We can assign a particular probability distribution to each of these curves *e.g.* to an exponential curve we can fit an exponential distribution given by

$$f(x,\theta) = \frac{1}{\theta} e^{-x/\theta}, \ x > 0, \ \theta > 0 \qquad \qquad ...(1)$$

For fitting this distribution, we have to find the value of θ with the help of given set of observations. The value of θ can be estimated in a number of ways. One such method is given by Benjamin Epstein which is based on first r out of n ordered observations drawn from an experiment. The technique is known as Life testing technique of Epstein.

20.13.1 ASSUMPTIONS OF LIFE TESTING TECHNIQUES

(i) The density function of the items drawn at random for inspection is exponential given by (1).

(ii) The observations are available in an order so that $x_{1n} \le x_{2n} \le ... \le x_{rn} .. \le x_{nn}$, x_{in} $(1 \le i \le n)$ being the ith observation in a sample of n ordered observations.

(iii) The inspection is discontinued as soon as x_{rn} is obtained (*i.e.* the first r observations are made).

Under the above three assumptions, the best estimate of parameter θ in (1) with Epstein method of estimation is given by

$$\hat{\theta}_{r,n} = \frac{x_{1n} + ... + x_{rn} + (n-r)x_{rn}}{r} \qquad \qquad ....(2)$$

and the probability density function of $\hat{\theta}_{r,n}$ is given by

$$f_r(\hat{y}) = \frac{1}{(r-1)!} \left(\frac{r}{\theta}\right)^r \theta^{r-1} e^{-ry/\theta} ; \ \text{for} \ y > 0 \qquad \qquad ....(3)$$

$$= 0, \text{elsewhere.}$$

Now this can be proved that estimate $\hat{\theta}$ given by (2) is a maximum likelihood estimate which is unbiased, sufficient, efficient and consistent.

We can see that $\quad E(\hat{\theta}_{r,n}) = \theta$.

and $\qquad\qquad V(\hat{\theta}_{r,n}) = E[(\hat{\theta}_{r,n}) - \theta]^2 = \theta^2 / r$

> The main advantage of Epstein method is that we are able to estimate θ by taking only r observations where r is less than n, n being sample size in other methods.

20.14 GROUP REPLACEMENT METHOD

A group replacement policy consists of two steps. Firstly, it consists of individual replacements at the time of failure of any unit in the system and then there is group replacement of existing live units at some suitable time. Here the individual replacement at the time of failure ensures running of the system whereas group replacement after some time interval will reduce the probability of failure. Such policy requires two fold consideration namely (i) the rate of individual replacements during the period and (ii) the total cost incurred of individual and group replacement during the chosen period. The period for which the total cost incurred is minimum will be the optimum period for replacement. Thus for the formulation of group replacement policy one should know the probability of failure, loss incurred due to these failures, cost of individual replacements and cost of group replacements. The procedure for calculating rate of replacement and the total cost involved in group replacement is given below.

Let all the units in a system be replaced after a definite time t with the provision that individual replacements can be made if and when any unit fails during this time.

To find rate of replacement at time t: We have the following notations

$F(t)$	:	The number of units failing at time t.
$p(X)$	:	Probability of failure of any unit at age X.
$p(t-X)$	:	Probability of failure of any unit at age $(t - x)$.
N	:	Total number of units in the system.
$Np(X)$	:	Expected number of units to be replaced at time X
$p(X).p(t - X)$	:	Probability of failure of any unit at age X and again it fails at age $(t - X)$.

Similarly $p(X).p(b-X).p(t-b)$: Probability that any unit fails at ages X, $b-X$ and $t-b$ respectively.

Hence the total number of replacements in time t will be given by

$$F(t) = N\left[p(t) + \sum_{X=1}^{t-1} p(X)p(t-X) + \sum_{b=2}^{t-1}\sum_{X=1}^{t-1} p(X)p(b-X)p(t-b) + ... \right] \qquad ...(1)$$

The number of units failing at the end of rth period in any system can be calculated from (1).

To find cost of replacement in time t:

Let $C(t)$ be the cost of group installation after a time period t. Then

$$\text{Average cost per period} = \frac{C(t)}{t} \qquad ...(2)$$

Let C_1	:	per unit cost of replacement in a group.
C_2	:	individual replacement cost on failure.

Then $\qquad C(t) = NC_1 + C_2$ [No. of failures at the end of period one + ...

$\qquad\qquad\qquad\qquad\qquad$ + No. of failures at the end of $(t-1)^{\text{th}}$ period]

$$= NC_1 + C_2[F(1) + F(2) + ... + F(t-1)] \qquad \text{...(3)}$$

Here, $F(1)$, $F(2)$, ..., $F(t-1)$ can be calculated from (1).

For finding optimum value of t, we have to find that value of t for which (2) is minimum

i.e.,
$$\left. \begin{array}{l} \dfrac{C(t-1)}{t-1} > \dfrac{C(t)}{t} \\[2mm] \dfrac{C(t+1)}{t+1} > \dfrac{C(t)}{t} \end{array} \right\} \qquad \text{...(4)}$$

where $C(t)$ is given by (3).

Now,
$$\frac{C(t+1)}{t+1} - \frac{C(t)}{t} = \frac{N_1 + C_2 \sum\limits_{X=1}^{t} F(X)}{t+1} - \frac{C(t)}{t}$$

$$= \frac{NC_1 + C_2 \sum\limits_{X=1}^{t-1} F(X) + C_2 F(t)}{t+1} - \frac{C(t)}{t} = \frac{C(t) + C_2 F(t)}{t+1} - \frac{C(t)}{t}$$

$$= \frac{t C_2 F(t) - C(t)}{(t+1)t}$$

i.e., for (4) to be true, $t C_2 F(t) > C(t)$ or $C_2 F(t) > C(t)/t$ $\qquad \text{...(5)}$

Similarly, from $\dfrac{C(t-1)}{t-1} > \dfrac{C(t)}{t}$, we can derive the condition, that

$$C_2 F(t-1) < \frac{C(t-1)}{t-1} \qquad \text{...(6)}$$

From (2) and (6), we conclude that group replacement should be made at the end of ith period if the cost of individual replacement for i^{th} period is greater than average cost per period by the end of period t and one should not adopt a group replacement policy if the cost of individual replacement at the end of $(t-1)^{\text{th}}$ period is not less than the average cost per period through time $(t-1)$.

 Miscellaneous Solved Examples

EXAMPLE 1. *A computer contains 10,000 resistors. When anyone of the resistors fails, it is replaced. The cost of replacing of an individual resistor is ₹10. If all the resistors are replaced at the same time, the cost per resistor is ₹3.50. The percent surviving by the end of month n is as below :*

Month (n)	0	1	2	3	4	5	6
Percent surviving at the end of month	100	97	90	70	30	15	0

at what intervals should all the resistors be replaced?

[BHOPAL–2006; PATNA–2008]

SOLUTION. It is given that the cost of individual replacement is ₹ 10 and the cost of group replacement is ₹ 3.50. Let P_i be the probability of a failure of a resistor during the month n, then we have,

$$P_1 = \frac{100-97}{100} = 0.03, \quad P_4 = \frac{70-30}{100} = 0.40$$

$$P_2 = \frac{97-90}{100} = 0.07, \quad P_5 = \frac{30-15}{100} = 0.15$$

$$P_3 = \frac{90-70}{100} = 0.20, \quad P_6 = \frac{15-0}{100} = 0.15$$

such that $P_1 + P_2 + P_3 + P_4 + P_5 + P_6 = 1$, which implies that a resistor cannot survive for more than six months and assume that the fails resistors in any month are replaced at the end of that week.

Let $N_0 = 10000$ be the number of resistors in the computer in the beginning. N_i be the number of failed resistor at the end of ith month then number of failed resistors replaced at the end of first month

$\Rightarrow \qquad N_1 = N_0 P_1 = 10000 \times 0.03 = 300$

Number of failed resistors in the computer at the end of second month

$\Rightarrow \qquad N_2 = N_0 P_2 + N_1 P_2 + N_1 P_1 = 10000 \times 0.07 + 300 \times 0.03 = 709$

Number of failed resistors at the end of third month $i.e.,$

$$N_3 = N_0 P_3 + N_1 P_2 + N_2 P_1 = 10000 \times 0.20 + 300 \times 0.07 + 709 \times 0.03$$
$$= 2042$$

Number of failed resistors at the end of fourth month $i.e.,$

$$N_4 = N_0 P_4 + N_1 P_3 + N_2 P_2 + N_3 P_1$$

$$= 10000 \times 0.40 + 300 \times 0.20 + 709 \times 0.07 + 2042 \times 0.03$$

$$= 4000 + 60 + 4.9 + 61.26 = 4171$$

Number of failed resistors at the end of fifth month $i.e.,$

$$N_5 = N_0 P_5 + N_1 P_4 + N_2 P_3 + N_3 P_2 + N_4 P_1$$
$$= 10000 \times 0.15 + 300 \times 0.40 + 709 \times 0.20 + 2042 \times 0.07 + 4171 \times 0.03$$
$$= 2030$$

and number of failed resistors at the end of sixth month $i.e.,$

$$N_6 = N_0 P_6 + N_1 P_5 + N_2 P_4 + N_3 P_3 + N_4 P_2 + N_5 P_1$$
$$= 10000 \times 0.15 + 300 \times 0.15 + 709 \times 0.40 + 2042 \times 0.20$$
$$\qquad\qquad + 4171 \times 0.07 + 2030 \times 0.03$$

$$= 2590$$

Now the average life of a resistors

$$= \sum_{i=1}^{6} X_i P_i = 4.02 \text{ months}$$

Then average number of resistors replacement every month

$$= \frac{\text{Total number}}{\text{Average life}} = \frac{10000}{4.02} = 2487.56 = 2488$$

Average monthly cost of individual replacement

$$= ₹ 2488 \times 10 = ₹ 24880$$

For group replacement, we make the following table:

End of the month	Total cost of group replacement (₹)	Average cost per month (₹)
1	$10000 \times 3.5 + 300 \times 10 = 38000$	38000.00
2	$10000 \times 3.5 + 300 \times 10 + 709 \times 10 = 45090$	22545.00
3*	$10000 \times 3.50 + 300 \times 10 + 709 \times 10 + 2042 \times 10$ $= 65510$	21836.66*
4	$10000 \times 3.50 + 300 \times 10 + 709 \times 10 + 2042 \times 10$ $+ 4171 \times 10 = 107220$	26805.00
5	$10000 \times 3.50 + 300 \times 10 + 709 \times 10 + 2042 \times 10$ $+ 4.171 \times 10 + 2030 \times 10 = 127520$	25504.00

From the table, the minimum average cost per month of group replacement is ₹ 21836.66 after three months which is less than the average cost per month of individual replacement.

Hence, the group replacement is best after every three months.

EXAMPLE 2. **(a)** *Let all the item in a system be new at time zero. Each item has a probability p of failing immediately before the end of first month of life and probability $q = 1- p$ of failing immediately before the end of the second month. If all the items are replaced as they fail show that the expected number of failures f(x) at the end of the month x is given by $f(x) = \dfrac{N}{1+q}[1-(-q)^{x+1}]$, where N is the number of items in the system.* [MEERUT–1991, 96, 97, 2006]

(b) *If C_1 is the cost per item of individual replacement and C_2 is the cost per item of group replacement, then obtain the condition under which:*

(i) A group replacement policy at the end of each month is the most profitable.

(ii) A group replacement policy at the end of every other month is the most profitable.

(iii) No replacement policy is better than a policy of pure individual replacement.

SOLUTION. Let $N_0 = N$ be the number of items in the beginning and N_i be the number of replacements at the end of ith month.

If P_i is the probability of failure of an item in i^{th} month then

Number of failures at the end of first month, i.e., $N_1 = N_0 P_1 = Np = N(1-q)$

Number of failures at the end of second month, i.e., $N_2 = N_0 P_2 + N_1 P_1$

$$= Nq + N(1-q)p$$

$$= N(1-q+q^2)$$

Number of failures at the end of third month, i.e.,

$$N_3 = N_0 P_3 + N_1 P_2 + N_2 P_1$$

$$= N \times 0 + N(1-q)q + N(1-q+q^2)p$$

$$= N(1-q+q^2-q^3)$$

Proceeding in a similar way, number of failures at the end of kth month, i.e.,

$$N_k = N[1 - q + q^2 - q^3 + ... + (-1)^k q^k]$$

So, Number of failures at the end of $(k + 1)^{th}$ month, i.e., N_{k+1}

$$= N_0 P_{k+1} + N_1 P_k + N_2 P_{k-1} + ... + N_k P_1$$
$$= N[1 - q + q^2 - q^3 + ... + (-1)^{k-1} q^{k-1}].q + N[1 - q + q^2 - q^3 + ... + (-1)^k q^k].P_1$$
$$= N[1 - q + q^2 - q^3 + ... + (-1)^{k+1} q^{k+1}] \qquad [\because P_1 = 1 - q]$$

Hence by mathematical induction expected number of failures at the end of xth month is

$$N_x = N[1 - q + q^2 - q^3 + ... + (-1)^x q^x] = f(x)$$

i.e., $\quad f(x) = \dfrac{N[1 - (-q)^{x+1}]}{1 + q} \qquad [\because r = -q, n = x + 1 \text{ and } f(N) = \dfrac{a(1 - r^n)}{1 - r}]$

(b) The value of $f(x)$ fluctuate for different values of x and when $x \to \infty$, then it will reach to steady state.

$\therefore$ Expected number of failures $= \lim\limits_{x \to \infty} f(x)$.

$$= \dfrac{N}{1 + q} \qquad [\because q < 1 \therefore q^{\infty} = 0]$$

Since C_1 is the cost per item for individual replacement then the average cost per month for individual replacement will be $\left(\dfrac{N}{1 + q}\right) C_1$.

(i) Since C_2 is the cost the item for group replacement then the total average cost of group replacement at the end of first month

$$= NC_2 + N_1 C_1 = NC_2 + N(1 - q)C_1$$

If the total average cost per month of group replacement at the end of each month is less than the average monthly cost for an individual replacement. So, we have

$$= NC_2 + N(1 - q)C_1 < \dfrac{N}{1 + q} C_1$$

or $\qquad NC_2 < \dfrac{N}{1 + q} C_1 - N(1 - q)C_1$

or $\qquad C_2 < \left(\dfrac{q^2}{1 + q}\right) C_1 \qquad\qquad ....(1)$

(ii) The total average cost per month of group replacement at the end of second month

$$= \dfrac{NC_2 + N_1 C_1 + N_2 C_1}{2}$$

$$= \dfrac{NC_2}{2} + N\left[1 - q + \dfrac{q^2}{2}\right] C_1$$

Now if the total average cost per month of group replacement at the end of second month is less than the average monthly cost for an individual replacement, so we get

$$\frac{NC_2}{2} + N\left[1 - q + \frac{q^2}{2}\right]C_1 < \frac{N}{1+q}C_1$$

i.e., $$NC_2 < \frac{2N}{1+q}C_1 - 2N\left[1 - q + \frac{q^2}{2}\right]C_1 \Rightarrow C_2 < \frac{q^2(1-q)}{1+q}C_1 \quad ....(2)$$

(iii) If individual placement is better than a group replacement then we have from (1) and (2),

$$C_2 > \left(\frac{q^2}{1+q}\right)C_1 \quad \text{and} \quad C_2 > q^2\left(\frac{1-q}{1+q}\right)C_1$$

which give

$$C_1 < \frac{1+q}{q^2}C_2 \quad \text{and} \quad C_1 < \frac{1+q}{q^2(1-q)}C_2$$

Since the value of $1 - q$ is less than 1. Then both the conditions give $C_1 < \frac{1+q}{q^2}C_2$.

 Exercise-20.1

1. The cost per year of running a truck whose capital cost is ₹ 30000 are as follows. Find the best replacement age?

Year	1	2	3	4	5	6	7
Running cost (₹)	5000	6000	7000	9000	11500	14000	17000
Resale value (₹)	15000	7500	3750	1875	1000	1000	1000

2. A firm is considering when to replace its machine whose price is ₹ 12200. The Resale value of the machine is ₹ 200. The maintenance cost of the machine are as follows:

Year	1	2	3
Maintenance cost (₹)	200	500	800

Find when the new machine should be purchased?

3. The data operating cost per year and resale prices of machine A whose purchase price is ₹ 10000 are given below:

Year	1	2	3	4	5	6	7
Operating cost (₹)	1500	1900	2300	2900	3600	4500	5500
Resale value (₹)	5000	2500	1250	600	400	400	400

(a) What is the best replacement age?

(b) When machine A is two years old, machine B, which is a new model for the same usage is available. The optimum period for replacement is 4 years with an average cost of ₹ 3600. Should we change machine A with that of B ? If so when?

4. A car owner finds from his past records that the maintenance cost per year of the car whose capital cost is ₹ 8000, is as given below:

Year	1	2	3
Maintenance cost	1000	1300	1700
Resale value (₹)	4000	2000	1200

Find at which time it is profitable to replace the car.

5. Let the value of money be assumed to be 10% per year and suppose that machine A is replaced after every 3 years whereas machine B is replaced after every six year. The yearly costs of both the machines are as given below:

Year	1	2	3	4	5	6
Machine A	1000	200	400	1000	200	400
Machine B	1700	100	200	300	400	500

Find the replacement policy.

6. A machine cost ₹ 10000. Operating costs are ₹ 500 per year for the first five years. In the sixth and succeeding years operating costs increases by ₹ 100 per year. Assuming a 10% discount rate of money per year, find the optimum length of time of hold the machine before we replace it.

7. If you wish to have a return at 10% per annum on your involvement, which of the following plans would you prefer?

	Plan A (₹)	Plan B (₹)
First cost	200000	250000
Scrap value after 15 years	150000	180000
Excess of annual revenues over annual disbursement	250000	30000

8. Assuming that the present value of one rupees to be spent in a year's time is ₹ 0.9 and the capital cost of the equipment is C = ₹ 3000 and the maintenance also are given below:

Year	1	2	3	4	5	6	7
Maintenance cost (₹)	500	600	800	1000	1300	1600	2000

When should the machine be replaced?

9. A machine is priced at ₹ 60000 and running costs are estimated at ₹ 6000 for each of the first four years increasing by ₹ 2000 per year in the fifth and subsequent years. If money is worth 10% per year, when should the machine be replaced? Assume that the resale value of machine is zero.

10. A computer contains 10000 resistors. When any resistor fails, it is replaced. The cost of replacing a resistor individually is ₹ 1 only. If all the resistors are replaced at the same time, the cost of a resistor is ₹ 0.35. The percent surviving $S(t)$ by the end of month t and $P(t)$ the probability of failure during the month t are given below:

Month (t)	0	1	2	3	4	5
S(t)	100	97	90	70	30	15
P(t)		0.03	0.07	0.20	0.40	0.15

What is the optimal replacement policy?

11. A system has a large number of light bulbs all of which we must keep in working order. If a bulb fails in service, it costs ₹ 1 to replace, but if we replace all the bulbs in the same operation we can do for only ₹ 0.35 a bulb. The life distribution of the bulbs is as follows:

Week	1	2	3	4	5	6
Probabilities of the bulb failing in the week	0.09	0.16	0.24	0.36	0.12	0.03

What is the optimum period for group replacement policy ?

12. The probability P_n of failure just before weeks n are shown below. If individual replacement cost ₹ 1.25 and group replacement cost ₹ 0.50 per item, find the optimal group replacement policy.

Week	1	2	3	4	5	6
Probabilities of failure	0.01	0.03	0.05	0.07	0.10	0.15
Week	7	8	9	10	11	
Probabilities of failure	0.20	0.30	0.11	0.08	0.05	

13. The following failure rates have been observed from certain items:

End of week	1	2	3	4	5
Probability of failure to date	0.10	0.30	0.55	0.85	1.00

The cost of replacing an individual item is ₹ 1.25. The decision is made to replace all items simultaneously at fixed intervals and also to replace individual items as they fail. If the cost of group replacement is ₹ 0.50. What is the best interval of group replacements?

14. The cost of a new machine is ₹ 5000 and the maintenance cost of the nth year is given by C_n = 500 $(n - 1)$, n = 1, 2,... n. Suppose that the discount rate per year is 0.5. After how many years it will be economical to replace the machine by new one?

ANSWERS

1.	5 years	2.	After six years

3. (a) 5 groups   (b) change when machine a is 4 years old

4.	5 years	5.	6 years	6.	19 years	7.	Plan A
9.	9 years	10.	3 months	11.	3 weeks	12.	5 weeks
13.	3 weeks	14.	6 years

 ## REVIEW QUESTIONS

1. Write a short note on replacement problem.

2. (i) Write a short note on different types of replacement models.

   (ii) Discuss the importance of replacement models.

3. (i) Discuss the problem of replacement of items that fails completely.

   (ii) Write a short note on staffing problem.

4. Describe some important replacement situations and policies.   [UPTU(MBA)–2000]

5. Explain how the theory of replacement is used in the replacement of items whose mainitenance cost varies with time.
   [UPTU(MBA)–2005]

6. Write a short note on group replacement policy.

7. Write the short note on different type of replacement models.

8. Discuss the importance of replacement models.

MULTIPLE CHOICE QUESTIONS (CHOOSE THE MOST APPROPRIATE ONE)

1. The present worth factor of one rupee spent in n years time from now onwards is:
   (a) $(1 + r)^n$ (b) $(1 + r)^{-n}$
   (c) $(1 - r)^{-n}$ (d) None of these

2. The problem of replacement is faced when a working item fails:
   (a) gradually
   (b) suddenly
   (c) both (a) and (b) are true
   (d) None of these

3. The group replacement policy is suitable for similar low cost item when the failure:
   (a) occurs over a period of time
   (b) is sudden
   (c) is complete and sudden
   (d) None of these

4. When money value changes with time at 10% then profit for first year is given by:
   (a) 0.93   (b) 0.999
   (c) 1   (d) None of these

5. When the probability of failure reduces gradually, the failure mode is called:
   (a) regressive   (b) progressive
   (c) retrogressive   (d) None of these

6. When money value changes with time at 20% the discount factor for second year is:
   (a) 0.833   (b) 0.695
   (c) 1   (d) 0

7. The maintenance cost mostly depends on:
   (a) manufacturing date
   (b) running age
   (c) calender age
   (d) None of these

8. Replacement decision is very much common in the stage:
   (a) youth   (b) old
   (c) infant   (d) None of these

9. If a machine becomes old, then the failure rate expected will be:
   (a) increasing   (b) decreasing

(c) constant   (d) None of these

10. The sudden failure among items is seen as:

(a) progressive   (b) random

(c) retrogressive   (d) all are true

ANSWERS

1. (b)   2. (c)   3. (a)   4. (b)   5. (c)   6. (a)   7. (b)   8. (b)   9. (a)
10. (d)

ARCHIVE

1. An engineering company is offered a material handling equipment A. It is priced at ₹60000 including cost of installation and the cost for operation and maintenance are estimated to be ₹1000 for each of the first five years, increasing every year by ₹3000 in the sixth and subsequent years. The company expects a return of 10% on all its investment. What is the optimal replacement age period?

[PUNJAB(MBA)–1988]

Year	Maintenance cost (₹)	Resale value (₹)
1	10,000	1,30,000
2	15,000	1,20,000
3	20,000	1,15,000
4	25,000	1,05,000
5	30,000	90,000
6	40,000	75,000
7	45,000	60,000
8	50,000	50,000

Determine the time at which it is profitable to replace the truck.   [ICWA–1989]

2. A company has the option to buy one of the minicomputer minicomp and chipcomp. Mincomp costs ₹5 lakh and running and maintenance costs ₹60000 for each of the first five years, increasing by ₹20000 per year in the 6th and subsequent years. Chipcomp has the same capacity as minicomp, but costs only ₹2,50,000. However its running and maintenance costs are ₹1,20,000 per year in the first 5 years and increase by ₹20,000 per year thereafter. If the money is worth 10% per year, which computer should be purchased? What are the optimal replacement periods for each of the computers? Assume that there is no salvage value for each computer.   [CA–1990]

3. A truck owner from his past experience estimated that the maintenance cost per year of a truck whose purchase price is ₹1,50,000 and the resale value of truck will be as follows:

4. A computer has a large no. of electronic tubes. They are subject to mortality as given below:

Period	Age of failure (hrs)	Probability of failure
1	0–200	0.10
2	201–400	0.26
3	401–600	0.35
4	601–800	0.22
5	801–1000	0.07

If the tubes are group replaced and the cost of replacement is ₹15 per tube. Group replacement can be done at fixed intervals in the night shift when the computer is not normally used. Replacement of individual tubes which fail in service costs ₹60 per tube. How frequently should the tubes be replaced.

[GUJRAT(MBA)–1989]

HINTS AND ANSWERS

1. discounted factor, $d = \dfrac{1}{1+0.10} = 0.9091$

   $C = 60000$. Then proceed as usual we find that the running cost is ₹22000 is more than the weighted average cost ₹19311 for eight years. Hence, the optimal replacement age is 8 years.

2. $d = \dfrac{1}{1+0.10} = 0.0901$

   Then proceed same as in question (1). The minicomp. should be replaced after 9 years and chipcomp after 6 years.

3. Replace at the end of 4th year.

4. Growth replacement is optimal after every 400 hour.

   This will cost ₹18600 which is less than the cost associated with individual replacement of tubes.

Simulation 21

21.1 INTRODUCTION

Simulation is the most important technique used in OR for analysing a number of complex systems and estimating their characteristics. It is a technique for carrying out experiments for analysing the behaviour and evaluating the performance of a proposed system under assumed condition of reality. This technique is used when the analytical methods for analysing the characteristics of a complex system do not exists. It is a technique of manipulating the model of a system through the process of imitation. An approximate or relatively simplified experimental model of a system is used to examine the components of properties of the system, their behaviour in relation to each other and in relation to the entire system at a point of time and over a period of time under different assumed conditions. Simulation have useful applications in complex queues, inventory control, financial planning, capital budgeting etc. In this chapter, we shall discuss some of these applications.

21.2 SIMULATION

The simulation approach can be used to study almost any problem that involves uncertainty, i.e., one or more decision variables can be represented by a probability distribution like decision making under risk. This approach requires an analogous physical model to represent mathematical and logical relationship among variables of the problem under study.

Definition. *Simulation is a quantitative technique used for evaluating alternative courses of action based upon facts and assumptions with a computerized mathematical model in order to represent actual decision-making under condition of uncertainty.*

Some precise definition of simulation are given below:

(i) **Shannon's definition.** Simulation is the process of designing a model of a real time system and conducting experiments with this model for the purpose of understanding the behaviour (within the limits imposed by a criterion of set of criteria) for the operation of the system.

(ii) **Taylor's definition.** Simulation is a numerical technique for conducting experiments on a digital computer, which involves certain type of mathematical and logical relationship necessary to describe the behaviour and structure of a complex real-world system over extended period of time.

(iii) **Shubik's definition.** A simulation of a system or an organism is the operation of a model or simulator which is a representation of the system or organism. The model is amenable to manipulation which would be impossible, too expensive or unpractical to perform on the entity it portrays. The operation of the model can be studied and for it, properties concerning the behaviour of the actual system can be inferred.

(iv) Churchman's definition. Simulation is the use of system model that has the designed characteristics of reality in order to produce the essence of actual operations.

21.3 TYPES OF SIMULATION

There are so many types of simulation. Out of which, some are given below:

(i) Deterministic and Probabilistic simulation: The deterministic simulation is used when a process is very complex or consists of multiple stages with complicated interaction between them while in probabilistic simulation one or more of the independent variables is probabilistic.

(ii) Time dependent and time independent simulation: In time independent simulation, it is not important to known exactly when the event is likely to occur, while in time dependent simulation it is important to the time when the event is likely to occur.

(iii) Visual interaction simulation: Visual interaction simulation uses computer graphics displays to present the consequences of change in the value of the input variation in the model.

(iv) Corporate and financial simulation: This type of simulation is used in corporate planning in financial aspects, when risk analysis is defined.

21.4 STEPS OF SIMULATION PROCESS

A simulation study several stages of activities, which containing the following steps:

STEP 1. Define or identify the problem

The simulation process is used to solve any problem only when the assumptions required for analytical methods are not satisfied.

STEP 2. Identify the decision variables and decide the performance criterion or objective.

STEP 3. Construct a simulation model which contains certain entities, or components which are objects of interest in the system.

STEP 4. Testing and validating the model which gives results that are adequate approximations of what would occur in the real system, simulation can lead to the wrong answer. Validation achieving a sufficient level of confidence that the model does provide an adequate representation of reality.

STEP 5. Designing the experiments. It refers to controlling the conditions of study, such as the variable to include. It require to

   (i) determine factors considered fixed and variable in the model

   (ii) levels of the factors to use

   (iii) what the resulting dependent measure are going to be

   (iv) how many times the model will be replicated and length of time of replication and so on.

STEP 6. Analyse the results and draw conclusion by selecting the best course of action otherwise make required changes in model decision variables, parameters or design.

STEP 7. Documentation and implementing the findings for future use.

21.5 ADVANTAGES AND DISADVANTAGES OF SIMULATION

21.5.1 ADVANTAGES OF SIMULATION

Some important advantages of Simulation are given below:

1. This approach is relatively straight forward, flexible and can be modified to accomodate the changing environments of real situations.

2. Sometimes simulation is the only method available. It is suitable to analyse large and complex real life problems which can not be solved by ususal methods.

3. Once a model has been constructed, it may be used over and over to analyse all kind of different situations.

4. Simulation experiments are done with the model, not on the system itself. It also allows to include addtional information during analysis the most quantitative method do not permit.

5. Data for further analysis can be easily generated from a simulation model.

6. The simulation approach is easier to apply than pure analytical methods so that non-technical executives can be comprehend the problem better.

7. It provides a trial and error movement towards the optimum solution. The selection is adjusted untill it approximates the optimum solution.

8. Simulation can be used as a pre-service test to try out new policies and decision rules for operating a system before running the risk of experiments in that real system.

21.5.2 DISADVANTAGES OF SIMULATION

Some important disadvantages of Simulation are given below:

1. Sometimes, the solution methods are not as efficient as the analytical methods.

2. Sometimes, it is expensive and take a long time to develop.

3. Simulation model does not generate solution by itself but only generates a way of evaluating solutions.

4. Simulation method is descriptive rather than optimization process.

5. Each application of simulation is adhoc to a great extent.

6. Simulation does not require standard probability distributions and this allows for the inclusion of real world complications that most quantitative analysis models can not permit.

21.6 RANDOM NUMBERS AND PSEUDO-RANDOM NUMBERS

A number or a sequence of numbers, whose probability of occurance is the same as that of any other number in that sequence is called a random number and random numbers are called pseudo random numbers when they are generated using some deterministic process but qualify the predetermined statistical level of randomness.

☛ REMARKS
- A random number always lies between 0 and 1 such that $0 \leq u \leq 1$.
- The initial number used in generating pseudo random numbers is called seed.

21.7 STOCHASTIC SIMULATION

When a system contains certain factors or decision variables that can be represented by a probability distribution, the simulation model used to study this type of system is called the stochastic or probabilistic simulation model.

21.8 TECHNIQUES OF GENERATING PSEUDO-RANDOM NUMEBRS

Here, we shall discuss some of the methods of generating pseudo-random numbers.

1. **Mid-square Method:** It is one of the oldest techniques of generating pseudo-random numbers. In this method, consider an initial four digit number (seed) x_0

in [0, 1]. The seed is squared and the resulting number is supposed to have eight digits (in case of less than 8, leading zeros are inserted). Now, from this number, the middle four digits are extracted as the required number. This number again squared. Proceeding in this manner, a new pseudo number is generated each time of squaring the previously generated random number. Continue this procedure till we obtain the already generated random numbers.

Some limitations of Mid-square Method

There are some limitations of mid-square method as given below

(i) This is a time consuming method, because so many multiplications are required to access the middle digits.

(ii) This method tends to degenerate rapidly. A loop may generate and as a result the same sequence of random numbers may be repeated.

2. **Power Residual or multiplicative Congruential method:** Before discussing this method, we first define the congruent modulo system

"Two integers a and b are called Congruent modulo m (written as $a \equiv b \pmod{m}$) if and only if there exists an integer k such that $b - a = km$ (m is a positive integer), i.e., in $a \equiv b \pmod{m}$, a is the remainder on dividing b by m".

Now, we came to the main method,

In power residual or multiplicative congruential method we select the nth number x_n as the remainder of the division of the product of a constant number a and the $(n-1)^{th}$ number x_{n-1} by another constant m such that

$$x_n \equiv ax_{n-1} \pmod{m}, n = 1, 2, 3, \ldots$$

Here, seed x_0 should be chosen as randomly as possible.

The first number (i.e., for n = 1) is $x_1 \equiv ax_0 \pmod{m}$

The second number (i.e., for n = 2) is $x_2 \equiv ax_1 \pmod{m} = a^2 x_0 \pmod{m}$

The third number (i.e., for n = 3) is $x_3 = a^3 x_0 \pmod{m}$

and so on.

In general $\qquad x_k = a^k x_0 \pmod{m}$

Then, pseudo numbers $u_k \in [0, 1]$ can be obtained after dividing x_k by m, so that

$$u_k = \frac{x_k}{m}, k = 1, 2, 3, \ldots$$

3. **Mixed Congruent Method:** This method is one of the best method to generate the pseudo-random number. Here we use the following recursion formula

$$x_n = (ax_{n-1} + c)\pmod{m}, n = 1, 2, 3, \ldots$$

Then divide x_n by m to get pseudo random numbers u_n such that

$$u_n = \frac{x_n}{m}$$

Here, a, c and x_0 all are positive integers less than m. Further, m is chosen to be 2^b or 10^a according to the computer used binary or decimal. By a proper choice of c and a, all non-negative integers less than m can be generated, before any one of them is repeated. We can select c and a by using the following scheme.

(i) c must be relatively prime to m

(ii) $a \equiv 1 \pmod{p}$ for each prime factor p of m

(iii) $a \equiv 1 \pmod{4}$ if 4 is a factor of m.

Solved Examples

EXAMPLE 1. *Generate pseudo-random numbers by taking a = 2, m = 5 and $x_0 = 3$ using the multiplicative congruent method.*

SOLUTION. To generate the pseudo-random number we prepare the following table

k	x_k	$u_k = x_k/m$
1	$(2 \times 3)(\text{mod } 5) = 1$	0.2
2	$(2 \times 1)(\text{mod } 5) = 2$	0.4
3	$(2 \times 2)(\text{mod } 5) = 4$	0.8
4	$(2 \times 4)(\text{mod } 5) = 3$	0.6
5	$(2 \times 3)(\text{mod } 5) = 1$	0.2
		(Cycling start)

From the above table, it is clear that, the required random numbers generted are 0.2, 0.4, 0.8 and 0.6.

EXAMPLE 2. *Using the mid-square method, generate pseudo-random numbers starting with 0300.*

SOLUTION. We have

$$\text{seed}(x_0) = 0300$$

Then multiply 0300 by itself, i.e., by 0300 to get an eight digit number. Else, insert the requiste number of zeros at the begining to make it 8-digits number. Then retain the middle four digits cancelling the first two and last two digits. Then we get the number 900. This procedure is given as below

$$u_0 = x_0 = \underline{0300}$$
$$\underline{\times 0300}$$
$$u_1 = 00\underline{0900}00$$
$$\underline{\times 0900}$$
$$u_2 = 00\underline{8100}00$$
$$\underline{\times 8100}$$
$$u_3 = 65\underline{6100}00$$
$$\underline{\times 6100}$$
$$u_4 = 37\underline{2100}00$$
$$\underline{\times 2100}$$
$$u_5 = 04\underline{4100}00$$
$$\underline{\times 4100}$$
$$u_6 = 16\underline{8100}00 = u_2$$
$$\underline{\times 8100}$$

From the above procedure, we observe that the sequence of random numbers obtained is u_0, u_1, u_2, u_3, u_4 and u_5.

EXAMPLE 3. *Generate single decimal pseudo-random numbers by choosing c and a according to the specification given in mixed congruent method.*

SOLUTION. Let $m = p$

Then choose $c = 3$, (relatively prime to 10)

which satisfy

$$a \equiv 1 \ (\text{mod } p) \text{ for each prime factor } p \text{ of } m$$

Now, choose $a = 1$ and choose seed $x_0 = 3$.

Then we have the following table

k	x_k	$u_k = x_k/m, \ m = 10$
1	$(1 \times 3 + 3)(\text{mod } 10) = 6$	0.6
2	$(1 \times 6 + 3)(\text{mod } 10) = 9$	0.9
3	$(1 \times 9 + 3)(\text{mod } 10) = 2$	0.2
4	$(1 \times 2 + 3)(\text{mod } 10) = 5$	0.5
5	$(1 \times 5 + 3)(\text{mod } 10) = 8$	0.8
6	$(1 \times 8 + 3)(\text{mod } 10) = 1$	0.1
7	$(1 \times 1 + 3)(\text{mod } 10) = 4$	0.4
8	$(1 \times 4 + 3)(\text{mod } 10) = 7$	0.7
9	$(1 \times 7 + 3)(\text{mod } 10) = 0$	0.0
10	$(1 \times 0 + 3)(\text{mod } 10) = 3$	0.3 (seed)

Hence, the ten pseudo-random numbers generated are

0.3, 06, 0.9, 0.2, 0.5, 0.8, 0.1, 0.4, 0.7, 0.0.

EXAMPLE 4. *Generate 2-digit pseudo-random numbers using the recursion formula $x_n = (ax_{n-1} + c)(\text{mod } m)$ by taking a and c as per the restrictions given by (i), (ii) and (iii) in mixed congruent method.*

SOLUTION. Let us take $m = 100$ and $c = 7$

We have $p = 2$ or 5 as prime factors of 100. Further, 4 is a factor of 100 $(= m)$, now we must have

$$a \equiv 1 \ (\text{mod } 2), \ a \equiv 1 \ (\text{mod } 5) \text{ and } a \equiv 1 \ (\text{mod } 4)$$

$\Rightarrow \quad a \equiv 1 \ (\text{mod } 10)$

So, we can take $a = 1, 21, 41$, etc as $1 - 1, 21 - 1, 41 - 1$ are divisible by 20. Now choose $a = 1$. Further let $x_0 = 1$. Then the pseudo-random numbers $u_k = \dfrac{x_k}{m}$ are given below

$$x_1 = (ax_0 + 6)(\text{mod } 100) = (1 \times 1 + 7) \, \text{mod } 100 = 8(\text{mod } 100) = 8$$

$\Rightarrow \quad u_1 = \dfrac{8}{100} = 0.08$

$$x_2 = (ax_1 + c) \, \text{mod } 100 = (1 \times 8 + 7) \, \text{mod } 100 = 15(\text{mod } 100) = 15$$

$\Rightarrow \quad u_2 = \dfrac{15}{100} = 0.15$

$$x_3 = (ax_2 + c) \, \text{mod } 100 = (1 \times 15 + 7) \, \text{mod } 100 = 22(\text{mod } 100) = 22$$

$\Rightarrow \quad u_3 = \dfrac{22}{100} = 0.22$

Proceeding in the same manner, we have the following table:

k	x_k	u_k	k	x_k	u_k	k	x_k	u_k	k	x_k	u_k
1	8	0.08	26	83	0.83	51	58	0.58	76	30	0.30
2	15	0.15	27	90	0.90	52	65	0.65	77	40	0.40
3	22	0.22	28	97	0.97	53	72	0.72	78	47	0.47
4	29	0.29	29	04	0.04	54	79	0.79	79	54	0.54
5	36	0.36	30	11	0.11	55	86	0.86	80	61	0.61
6	43	0.43	31	18	0.18	56	93	0.93	81	68	0.68
7	50	0.50	32	25	0.25	57	00	0.00	82	75	0.75
8	57	0.57	33	32	0.32	58	07	0.07	83	82	0.82
9	64	0.64	34	39	0.39	59	14	0.14	84	89	0.89
10	71	0.71	35	46	0.46	60	21	0.21	85	96	0.96
11	78	0.78	36	53	0.53	61	28	0.28	86	03	0.03
12	85	0.85	37	60	0.60	62	35	0.35	87	10	0.10
13	92	0.92	38	67	0.67	63	42	0.42	88	17	0.17
14	99	0.99	39	74	0.74	64	49	0.49	89	24	0.24
15	06	0.06	40	81	0.81	65	56	0.56	90	31	0.31
16	13	0.13	41	88	0.88	66	63	0.63	91	38	0.38
17	20	0.20	42	95	0.95	67	70	0.70	92	45	0.45
18	27	0.27	43	02	0.02	68	77	0.77	93	52	0.52
19	34	0.34	44	09	0.09	69	84	0.84	94	59	0.59
20	41	0.41	45	16	0.16	70	91	0.91	95	66	0.66
21	48	0.48	46	23	0.23	71	98	0.98	96	73	0.73
22	55	0.55	47	30	0.30	72	05	0.05	97	80	0.80
23	62	0.62	48	37	0.37	73	12	0.12	98	87	0.87
24	69	0.69	49	44	0.44	74	19	0.19	99	94	0.94
25	76	0.76	50	51	0.51	75	26	0.26	100	01	0.01

21.9 MONTE-CARLO SIMULATION

The Monte-Carlo technique uses random numbers and is generally used to solve problems requiring decision making under uncertainty and where mathematical formulation is not possible. This technique involves repetitively conducting experiments on the model of the system under study, with a known probability distribution. If a system can not be described using a standard probability distribution such as exponential, normal or poisson, an empirical probability distribution can be constructed.

Working Procedure

It consists of the following steps:

STEP 1. Setting up probability distribution for values to be analysed.

STEP 2. Construct a cumulative probability distribution for involved random variables.

STEP 3. Generate random numbers and then assign an appropriate set of random numbers to represents value or range of values for each random variable.

STEP 4. Conducting the simulation a requisite number of times.

- 'Monte-Carlo' is the code name given by Von Neumann and S.M. Ulam to the technique of solving problems which are quite expensive for experimental solution and too much difficult for analytical treatment.

21.9.1 GENERATION OF RANDOM VARIATE

In probability theory a random sample can be generated from any known probability distribution (continuous or discrete). Therefore, in Monte-Carlo method, the sequence of random numbers becomes the basic sequences. In this section, we have to generate a random sample from the known probability distribution.

(1) CONTINUOUS RANDOM VARIABLE

Let X be a continuous random variable with the probability density function (PDF)$F(x)$ and cumulative density function (CDF)$F(x)$. Then we have

$$F(x) = P(X \le x) = \int_{-\infty}^{x} f(t)dt \qquad ...(1)$$

If U is the new random variable given by

$$U = F(x) = \int_{-\infty}^{X} f(t)dt \qquad ...(2)$$

Then $\qquad u = F(x) = \int_{-\infty}^{x} f(t)dt$ and $\dfrac{du}{dx} = f(x) \qquad ...(3)$

Using probability theory, the PDF, $g(u)$ of U is given by

$$g(u) = \frac{dx}{du} f(x) \qquad ...(4)$$

Using (3) and (4) we get

$$g(u) = \frac{f(x)}{f(x)} = 1, \quad 0 \le u < 1 \qquad \left[\because \frac{du}{dx} = F'(x) = f(x) \right]$$

which is the PDF of uniform distribution on [0, 1]

Now, $f(x)$ is monotonic $\Rightarrow x = F^{-1}(u)$ \hfill [From (3)]

Hence, we conclude that given a random number $u \in [0, 1]$, a random variate x from X can be obtained by solving

$$u = F(x) \text{ for } x$$
$$\Rightarrow \qquad x = F^{-1}(u)$$

It is called inverse transformation method.

GENERATION OF EXPONENTIAL RANDOM VARIATE

We know that, exponential distribution has important application in queuing theory for the distribution of inter arrival and service time. If the mean inter arrival time of customers in a queuing system is $\dfrac{1}{\lambda}$, then the probability density function $f(t)$ of service time T is given by

$$f(t) = \lambda e^{-\lambda t}, t \ge 0, \lambda \ge 0$$
$$F(t) = \int_{-\infty}^{t} f(t)dt = \int_{0}^{t} \lambda e^{-\lambda t} dt$$

Then

$$= 1 - e^{\lambda t}, t \ge 0$$
$$\Rightarrow \qquad t = -\frac{1}{\lambda} \log(1-u) = -\frac{1}{\lambda} \log u \qquad [u \in [0,1] \Rightarrow (1-u) \in [0,1]]$$

Hence, for a given random number u the corresponding exponential random variate t of random variable T is given by

$$t = \frac{-1}{\lambda} \log u$$

GENERATION OF UNIFORM RANDOM VARIATE

Clearly, the p.d.f., $f(x)$ of uniform distribution over $[a, b]$ is defined as follows:

$$f(x) = \begin{cases} \dfrac{1}{b-a}; & a \le x \le b \\ 0; & \text{otherwise} \end{cases}$$

The probability distribution function $F(x)$ is given by

$$F(x) = \int_0^x f(x)dx = \frac{x-a}{b-a}, a \le x \le b$$

Let us suppose, u = F(x), then $u = \dfrac{x-a}{b-a}$

i.e., $\qquad x = a + (b-a)u$ $\qquad\qquad$ ...(1)

Hence, for a given random number u, the formula given above by (1) generates a random variate x of uniformly distributed random variable X over $[a, b]$.

(2) DISCRETE RANDOM VARIABLE

In case of discrete random variable, the technique of inverse transformation can be used to generate a random sample from a discrete random variable X by using the basic sequence of random numbers.

Let X be a discrete random variable which can takes the values $x_1, x_2, ..., x_n$ with $p(X = x_i) = p_i$. Arrange x_i in ascending order ($x_1 < x_2 < x_3 < ...$) then cumulative density function of $F(x)$ is given as below:

$$F(x) = \begin{cases} 0 : x < x_1 \\ p_1 : x_1 \le x < x_2 \\ p_1 + p_2 : x_2 \le x < x_3 \\ p_1 + p_2 + p_3 : x_3 \le x < x_4 \end{cases}$$

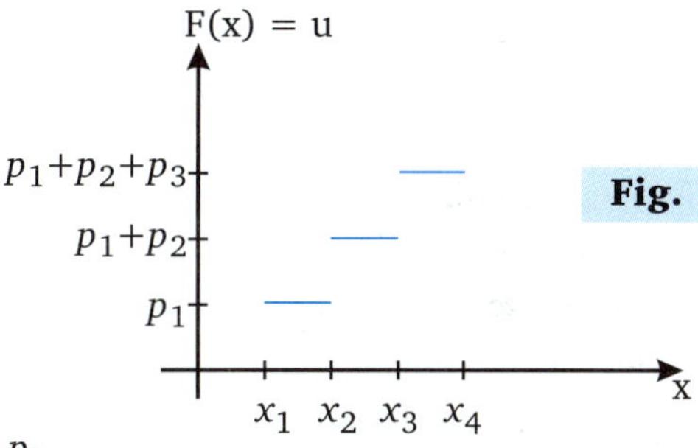

which implies

$x = x_1$ if $0 \le u < p_1$

$x = x_2$ if $p_1 \le u < p_1 + p_2$

$x = x_3$ if $p_1 + p_2 \le u < p_1 + p_2 + p_3$

Fig. 1

☞ REMARK

• The cumulative density function of $F(x)$ defined above is a step function.

21.9.2 GENERATION OF RANDOM VARIATE FROM POISSION DISTRIBUTION

We know that, for Poisson distribution

$$f_x(t) = \frac{e^{-\lambda t}(\lambda t)^x}{x!}, \quad x = 0, 1, 2, ... \qquad\qquad ...(1)$$

$\Rightarrow \qquad P(X = n) = \dfrac{e^{-\lambda}\lambda^n}{n!}, \quad n = 0, 1, 2, ...$

Here, λ is the parameter of Poission distribution X. Also, Poission distribution occurs in queue as the distribution of arrivals.

To generate a random sample X, let 1, 2, 3, ..., X, $X + 1$ be the events and

T_1 = time between start ($t = 0$) and occurance of the first event

T_2 = time between the occurance of the 1st and 2nd event

T_X = time between the occurance of the $(X-1)^{th}$ and X^{th} event

T_{X+1} = time between the occurance of the Xth and $(X+1)^{th}$ event

Let X be the number of occurance of events of during time interval $[0, t[$ then we have

$$T_1 + T_2 + \ldots T_X \leq t < T_1 + T_2 + \ldots + T_{X+1} \qquad \ldots(2)$$

where LHS of (2) is the total time of X occurances and RHS is the total time of $(X + 1)$ occurance of the event.

Let each inter-arrival time T_i be exponentially distributed random variables with parameter λ. Then X is the no. of occurances of events in $[0, t[$ is Poission distribution $P(\lambda t)$ with parameter λ. Then by using

$$t = -\frac{1}{\lambda} \log u$$

each T_i being exponentially distributed can be generated from random numbers $U_i's$ as

$$T_i = -\frac{1}{\lambda} \log U_i$$

Now, substituting this $i = 1, 2, \ldots, X, X + 1$ in (2) we get

$$-\frac{1}{\lambda} \log(U_1 U_2 \ldots U_X) \leq t < -\frac{1}{\lambda} \log(U_1 U_2 \ldots U_{X+1})$$

$$\Rightarrow \qquad U_1 U_2 \ldots U_{X+1} < e^{-\lambda t} \leq U_1 U_2 \ldots U_X$$

At t = 1, then

$$U_1 U_2 \ldots U_{X+1} < e^{-\lambda} \leq U_1 U_2 \ldots U_X$$

$$\Rightarrow \qquad U_1 U_2 \ldots U_{X+1} < A \leq U_1 U_2 \ldots U_X, \text{ where } A = e^{-\lambda} \qquad \ldots(3)$$

Above inequality (3) provides a scheme of generating random variate x from the Poisson random variable X with parameter λ.

☞ REMARK

- Here, we can not determine in advance as to how many random numbers ($u_i's$) will be required to generate one Poisson random variate x.

 Solved Examples

EXAMPLE 1. *Let a continuous random variable X have the following p.d.f.*

$$f(x) = \begin{cases} -x^3, -1 \leq x < 0 \\ \dfrac{x}{2}, 0 \leq x < 1 \\ \dfrac{1}{2}, 1 \leq x < 2 \\ 0, otherwise \end{cases}$$

Using inverse transformation method, determine a scheme for generating random variate x of X.

SOLUTION. Clearly, the cumulative distribution function $F(x)$ of X is given by

$$F(x) = \begin{cases} 0, x < -1 \\ \dfrac{1}{4}(1-x^4), -1 \le x < 0, & F(0) = \dfrac{1}{4} \\ \dfrac{1}{4}(1+x^2), 0 \le x < 1, & F(1) = \dfrac{1}{2} \\ \dfrac{x}{2}, 1 < x < 2 \end{cases} \qquad \dots (1)$$

Therefore,

$$\begin{aligned} &\text{for } -1 \le x < 0, \ \ 0 \le F(x) < \frac{1}{4} \\ &\text{for } \ \ 0 \le x < 1, \ \ \frac{1}{4} \le F(x) < \frac{1}{2} \qquad \dots (2) \\ &\text{and for } \ \ 1 \le x < 2, \ \ \frac{1}{2} \le F(x) < 1 \end{aligned}$$

Let $u = F(x)$, then from the above equations, we have the following

(i) If $0 \le u < \dfrac{1}{4}, -1 \le x < 0$ then $u = F(x) = \dfrac{1}{4}(1-x^4)$ or $x = -(1-4u)^{1/4}$

$$(\because \ x < 0)$$

(ii) If $0.25 \le u < 0.5, 0 \le x < 1$ then $\dfrac{1}{4}(1-x^2) = u$

or $x = (4u-1)^{1/2}$ (by taking positive root)

(iii) If $0.5 \le u < 1, 1 \le x < 2$ then $\dfrac{x}{2} = u$ or $x = 2u$

EXAMPLE 2. *Consider the random variable X given with the following probability distribution*

X	−1	0	1	3
$P(x) = P(X = x)$	0.2	0.3	0.4	0.1

Generate a random sample from X of size 5 using the sequence of random numbers 0.2, 0.0, 0.3, 0.5, 0.8.

SOLUTION. Clearly, CDF $F(x)$ of X is given as follows:

$$F(x) = \begin{cases} 0, & x < -1 \\ 0.2, & -1 \le x < 0 \\ 0.5, & 0 \le x < 1 \\ 0.9, & 1 \le x < 3 \\ 1, & 3 \le x < \infty \end{cases}$$

Let $u = F(x)$, then we have

$$\begin{aligned} x &= -1 \text{ for } 0 \le u < 0.2 \\ x &= 0 \text{ for } 0.2 \le u < 0.5 \\ x &= 1 \text{ for } 0.5 \le u < 0.9 \\ x &= 3 \text{ for } 0.9 \le u < 1 \end{aligned}$$

The above scheme generates random variate x of X for a given random number u. So, using the scheme, we have $x = 0$ for $u = 0.2$, $x = -1$ for $u = 0$, $x = 0$ for $u = 3$, $x = 1$ for $u = 0.5$ and $x = 1$ for $u = 0.8$.

Hence, 0, – 1, 0 ,1, 1 is a sequence of random variate x of random variable X corresponding to random numbers 0.2, 0.0, 0.3, 0.5, 0.8

EXAMPLE 3. *Generate a random sample of size five from Poisson distribution X with parameter $\lambda = 3$ by using random numbers 30, 37, 84, 71, 98, 65, 72, 19, 06, 33.*

SOLUTION. We have $A = e^{-\lambda} = e^{-3} = 0.498$, $u_1 = 0.30 < A$ and
the first random variate $x_1 = 0$
the second number $u_1 = 0.37 < A$ so the second random variate $x_2 = 0$.
For the third random variate $u_1 = 0.84$, which is not less than A

$$u_1 u_2 = 0.84 \times 0.71 = 0.5964 > A$$
$$u_1 u_2 u_3 = 0.5964 \times 0.98 = 0.5844 > A$$
$$u_1 u_2 u_3 u_4 = 0.5844 \times 0.65 = 0.03798 > A$$
$$\Rightarrow \quad u_1 u_2 u_3 u_4 < A < u_1 u_2 u_3$$

Therefore, the third random variate $x_3 = 3$
Now, to generate 4th random variate, we have $u_1 = 0.72 > A$

$$u_1 u_2 = 0.72 \times 0.19 = 1.368 < A = 0.498$$
$$\Rightarrow \quad u_1 u_2 < A < u_1$$
$$\Rightarrow \quad \text{Fourth random variate } x_4 = 1$$

For the 5th random variate we have $u_1 = 0.06 < A$
$\Rightarrow \quad 5^{th}$ random variate, $x_5 = 0$
Hence, required Poisson variates are 0, 0, 3, 1, 0.

21.10 APPLICATIONS OF SIMULATION IN OPERATIONS RESEARCH

21.10.1 APPLICATIONS OF SIMULATION IN QUEUE MODELS

Application of simulation in queue models can be understand by the following examples.

EXAMPLE 1. *In a queuing system with a single server, assume that the customers arrive such that the inter arrival time for each customer in two minutes and the service time for each customer is 3.5 minutes. Simulate the system using the 'next event increment' method for 15 minutes.*

SOLUTION. Suppose that first arrival takes place at $t = 0$. Then we have the following simulated table.

Time	Event	Total arrival so far	Total no. of customers departed	Total no. of customers in the system	Total no. of customers in the queue	Time spent between the present and preceding instant in the system	Time spent between the present and preceding instant in the queue
0.0	A	1	0	1	0	0.0	0.0
2.0	A	2	0	2	1	2.0	0.0
3.5	D	2	1	1	0	3.0	1.5
4.0	A	3	1	2	1	0.5	0.0
6.0	A	4	1	3	2	4.0	2.0
7.0	D	4	2	2	1	3.0	2.0
8.0	A	5	2	3	2	2.0	1.0
10.0	A	6	2	4	3	6.0	4.0
10.5	D	6	3	3	2	2.0	1.5
12.0	A	7	3	4	3	4.5	3.0
14.0	A, D	8	4	4	3	8.0	6.0
15.0	End	8	4	4	3	4.0	3.0
Total						39.0	24.0

From the above table, we conclude the following results.

(i) Average waiting time in the system $= \dfrac{39.0}{8} = 4.875$

(ii) Average waiting time in the queue $= \dfrac{24}{8} = 3.0$

(iii) Average no. of customers in the system $= \dfrac{39}{15} = 2.60$

(iv) Average no. of customers in the queue $= \dfrac{24}{15} = 1.60$

EXAMPLE 2. *A dentist schedules all his patients for 30 minutes appointment. Some of the patients take more or less than 30 minutes depending on the type of dental work to be done. The following summary shows the various categories of work, their probabilities and time actually needed to compute the work.*

Category of service	Time required (minutes)	Probability of category
Filling	45	0.40
Crown	60	0.15
Cleaning	15	0.15
Extraction	45	0.10
Checkup	15	0.20

Simulate the dentist's clinic for four hours and determine the average waiting time for the patients as well as the idleness of the doctor. Assume that all the patients show up at the clinic at exactly their scheduled arrival time starting at 8:00 A.M. Using the following random numbers for handling the above problem.

40   82   11   34   25   66   17   79   [CA–1990, AMIE–2005]

SOLUTION. We have the cumulative probability distribution and random number interval for service time as in the following table

Category of service	Service time required (minutes)	Probability	Cumulative Probabilities	Random number interval
Filling	45	0.40	0.40	00–39
Crown	60	0.15	0.55	40-54
Cleaning	15	0.15	0.70	55-69
Extraction	45	0.10	0.80	70-79
Checkup	15	0.20	1.00	80-99

Further we have the following arrival pattern and nature of services.

Patient no.	Scheduled arrival	Random number	Category of services	Service time (minutes)
1	8:00	40	Crown	60
2	8:30	82	Checkup	15
3	9:00	11	Filling	45
4	9:30	34	Filling	45
5	10:00	25	Filling	45
6	10:30	66	Cleaning	15
7	11:00	17	Filling	45
8	11:30	79	Extraction	45

Also, the computation of arrivals, departure and waiting of patients are given in the following table.

Time	Event (Patient no.)	Time to exist (Patient no.)	Waiting (Patient number)
8:00	1 arrive	1 (60)	—
8:30	2 arrive	1 (30)	2
9:00	1 depart, 3 arrive	2 (15)	3
9:15	2 depart	3 (45)	—
9:30	4 arrive	3 (30)	4
10:00	3 depart, 5 arrive	4 (45)	5
10:30	6 arrive	4 (15)	5, 6
10:45	4 depart	5 (45)	6
11:00	7 arrive	5 (30)	6, 7
11:30	5 depart, 8 arrive	6 (15)	7, 8
11:45	6 depart	7 (45)	8
12:00	End	7 (30)	8

The dentist was not idle during the entire simulated period.

Finally the waiting time for the patients were given in the next table

Patient	Arrival time	Service starts at	Waiting time (minutes)
1	8:00	8:00	0
2	8:30	9:00	30
3	9:00	9:15	15
4	9:30	10:00	30
5	10:00	10:45	45
6	10:30	11:30	60
7	11:00	11:45	45
8	11:30	12:30	60
Total			280

Here, the average waiting time $= \dfrac{280}{8} = 35$ minutes.

21.10.2 APPLICATION OF SIMULATION IN INVENTORY PROBLEMS

EXAMPLE 1. *A book store wishes to carry a particular book in stock. Demand is not certain and their is a lead time of 2 days for stock replenishment. The probabilities of demand are given below*

Demands (units/day)	0	1	2	3	4
Probability	0.05	0.10	0.30	0.45	0.10

Each time an order is placed, the store incures an ordering cost of ₹10 per order. The store also incures a carrying cost of ₹0.5 per book per day. The inventory carrying cost is calculated on the basis of stock at the end of each day. The manager of the book store wishes to compare two options for his inventory decision.

A : Order 5 books when present inventory plus any outstanding order falls below 8 books

B : Order 8 books when present inventory plus any outstanding order falls below 8 books

Currently (beginning of 1st day) the store has a stock of 8 books plus 6 books ordered two days ago and are expected to arrive next day. Carryout simulation run for 10 days to recommend an appropriate option. You may use random numbers in sequence. Using first no. for day one.

89   34   78   63   61   81   39   16   13   73   [AMIE–2005, ICWA–1988]

SOLUTION. We may easily obtain a probability distribution by using the daily demand distribution

Daily demand	Probability	Cumulative Probability	Random no. interval
0	0.05	0.05	00–04
1	0.10	0.15	05–14
2	0.30	0.45	15–44
3	0.45	0.90	45–89
4	0.10	1.00	90–99

It is given that stock in hand is of 8 books and stock on order is 5 books (expected next day)

Now, we have the following table for option A.

Days	Opening stock	Random no.	Resulting demand	Closing stock	Order placed	Order delivered	Avg. stock evening
1	20	0	3	17	—	—	18.5
2	17	9	7	10	15	—	13.5
3	10	1	4	6	—	—	8
4	6	1	4	2	—	—	4
5	2	5	5	0 (–3)	15	15	1
6	12	1	4	8	—	—	10
7	8	8	6	2	—	—	6
8	2	6	6	0 (– 4)	15	15	1
9	11	3	5	6	—	—	8.5
10	6	5	5	1	—	—	3.5

(Here negative entries indicates back orders)

From the above table, we conclude that

(i) Average ending cost $= \dfrac{78}{10} = 7.8$ units/day

(ii) Daily ordering cost

$\quad$ = cost of placing one order × no. of orders placed per day

$\quad$ = $50 \times 3 = ₹150$

(iii) Daily carrying cost

$\quad$ = cost of carrying one unit for one day × average ending stock

$\quad$ = $2 \times 7.8 = ₹15.60$

(iv) Total daily inventory cost

$\quad$ = daily ordering cost + daily carrying cost

$\quad$ = $150 + 15.60 = ₹165.60$

Exercise-21.1

1. In a queuing system comprising one server that follows first come first served discipline with unlimited queue length and unlimited input source, customers are arriving to avail the service. The inter arrival time distribution X and service time distribution Y are exponential with parameters $\lambda = \dfrac{1}{2}$ and $\mu = \dfrac{3}{4}$ respectively. Use random numbers for inter-arrival time and service time respectively as 20, 23, 86, 09, 92, 35, 38, 01, 24, 07, 50 and 21, 44, 27, 70, 73, 36, 59, 42, 85, 88, 51. Simulate the system for 25 units of time starting with time $t = 0$ and estimate the following characteristics of the queuing system.

   (i) Average no. of customers in the system

   (ii) Average no. of customers in the queue

   (iii) Mean waiting time of customer in the system

   (iv) Mean waiting time of a customer in the queue

   (v) The proportion of idle time of the server

2. The probability distribution of units produced (X) and the no. of vehicles available (Y) are as given below

X	P(X = x_i)	Y	P(Y = y_i)
500	0.06	5	0.16
550	0.14	6	0.36
600	0.20	7	0.20
650	0.40	8	0.16
700	0.20	9	0.12

Simulate the model for 10 days in respect of the above mentioned changes to determine the average daily output, average no. of vehicles and average no. of units store overnight.

3. In a transportation company, the finished job arriving at the checkout area is recorded for 100 trucks. Checkout takes five minutes and the checker takes care of only one truck at a time. The data is as follows:

Truck inter-arrival time (min)	Frequency	Truck inter-arrival time (min)	Frequency
1	1	6	23
2	4	7	7
3	7	8	5
4	17	9	3
5	31	10	2
		Total = 100	

After checkout, the drivers take their trucks to the next department. Using Monte-Carlo method, determine the following:

(i) Average waiting time before service

(ii) Expected longest period of waiting

Using the following random numbers

12, 81, 36, 82, 21, 74, 90, 55, 79, 70, 14, 59, 62, 57, 15, 18, 74, 11, 41, 29

4. A shopkeeper maintain a stock of Bread. On the basis of previous experience, the daily demand for the bread along with associated probabilities are given in the following table:

Daily demand (no.)	Probability
0	0.01
10	0.20
20	0.15
30	0.50
40	0.12
50	0.02

Estimate the daily average demand for the bread by simulating the demand for 10 days. Use the following random numbers

25, 39, 65, 76, 12, 05, 73, 89, 19, 49

5. A firm has three vehicles and uses whichever vehicle that is available to transport stock to its five branches. The average petrol consumption by the vehicles and the distance (to and fro) in kms of the branches are provided in the following table

Vehicles	Distance travelled (km) (per gallons)	Branch	Distance (km) (to and fro)
A	25	1	100
B	35	2	130
C	40	3	160
—	—	4	200
—	—	5	260

Assuming that a random choice of

destinations and vehicle is made, simulate 10 trips and hence estimate the average petrol consumption per trip.

6. A firm has a single channel service station with the following arrival and service time probability distribution.

Inter arrival time (min)	Prob.	Service time (min.)	Prob.
10	0.10	5	0.08
15	0.25	10	0.14
20	0.30	15	0.18
25	0.25	20	0.24
30	0.10	25	0.22
		30	0.14

The customer's arrival at the service station is a random phenomenon and the time between the arrivals varies from 10 to 30 minutes. The service time varies from 5 minutes to 30 minutes. The queuing process begins at 10:00 A.M. and proceeds for nearly 8 hours. An arrival goes to the service facility immediately, if it is free. Otherwise it will wait in a queue. The queue discipline in first come first serve. If the attendent's wages are ₹10 per hour and the customer's waiting time costs ₹15 per hour, then would it be an economical proposition to engage a second attendent? Answer using Monte-Carlo simulation technique.

ANSWERS

1. 0.37, 0.044, 1.16 units of time, 0.138 units of time, 0.6728
2. 600, 6.4, 14.5
3. 0.8 minutes, 3 minutes
4. 22 breads/day
6. Not economical

REVIEW QUESTIONS

1. Define the random and pseudo-random numbers.
2. Describe the mid-square method of generating pseudo-random numbers.
3. Write the short note on simulation.
4. State the major reasons of using simulation.
5. Write the advantages and disadvantages of simulation models. [DELHI(MBA)–1999]
6. Explain in brief the Monte-Carlo method.
7. "When it become difficult to use an optimization technique for solving a problem one has to resort to simulation technique". Discuss. [DELHI(MBA)–2003, AMIE–2005]
8. State two major reasons for using simulation. [ICWA–1985]
9. Describe the kind of problems for which Monte-Carlo will be an appropriate method of solution.
10. Explain what factors must be considered when designing a simulation experiments.

MULTIPLE CHOICE QUESTIONS (CHOOSE THE MOST APPROPRIATE ONE)

1. As simulation is not an analytical model, so result of simulation must be viewed as:
   (a) approximation
   (b) exact
   (c) simplified
   (d) none of these

2. The general purpose system simulation languages:
   (a) does not require program writing
   (b) require program writing
   (c) require predefined coding forms
   (d) none of these

3. Few causes of simulation analysis failure are:
   (a) inappropriate levels of detail
   (b) incomplete mix of essential skills
   (c) both (a) and (b) are true

   (d) none of these

4. If u is a random number then:
   (a) $0 \geq u \geq 1$ (b) $0 < u < 1$
   (c) $0 \leq u < 1$ (d) none of these

5. In Monte-Carlo method:
   (a) only random variables are used
   (b) only random numbers are used
   (c) both (a) and (b) are true
   (d) none of these

ANSWERS

1. (a)   2. (a)   3. (c)   4. (c)   5. (c)

ARCHIVE

1. A bakery keeps stock of popular brand of cake. Previous experience shows the daily demand pattern for the item with associated probabilities, as given below:

Daily demand (no.)	0	10	20	30	40	50
Probability	0.01	0.20	0.15	0.50	0.12	0.02

Use the following sequence of random numbers to simulate the demand for next 10 days. Also, examine the daily average demand for the cakes on the basis of simulated data.
[ICWA–1996]

2. A company trading in motor vehicle spare parts wishes to determine the levels of stock, it should carry for the items in its range. Demand is not certain and there is a lead time for stock replenishment. For an item A, the following information is obtained.

Demands (unit/day)	3	4	5	6	7
Prob.	0.10	0.20	0.30	0.30	0.10

Carrying cost (per unit per day)	₹2
Ordering cost (per order)	₹50
Lead time for replenishment	3 days

Stock in hand at the beginning of the simulation exercise was 20 units.

Carry out a simulation run over a period of 10 days with the objectives of evaluating the inventory rule. Order 15 units when present inventory plus any outstanding order falls below 15 units. You may use random numbers in the sequence 0, 9, 1, 1, 5, 1, 8, 6, 3, 5, 7,

1, 2, 9 using the first number for day 1. Your calculation should include the total cost of operating this inventory rule for 10 days.
[AMIE–2004]

3. The material manager of a firm wishes to determine the expected demand for a particular item in stock during the reorder lead time. This information is needed to determine how far in advance to reorder, before the stock level is reduced to zero. However both the lead time (in days) and the demand/day for the item are random variables, described by the following distribution.

Lead time (days)	Prob. of occurance	Demand/day (units)	Prob.
1	0.50	1	0.10
2	0.30	2	0.30
3	0.20	3	0.40
		4	0.20

Manually simulate the problem for 30 reorders to estimates the demand during lead time.   [DELHI(MBA)–1985, 2000]

4. A consumer sells confectionary item. Past data of demand per week (in hundred kg) with frequency is given below:

Demand/week	0	5	10	15	20	25
Frequency	2	11	8	21	5	3

Using the following sequence of random numbers, generate the demand for the next 10 weeks. Also, find the average demand per week. 35, 52, 90, 13, 23, 73, 34, 57, 35, 83, 94, 56, 67, 66, 60.   [RAJASTHAN(MBA)–2007]

HINTS AND ANSWERS

1. Expected demand = 22 units/day
2. Average ordering stock = 7.8 units/day, Daily ordering cost = ₹150; Daily carrying cost = ₹15.60
   Total daily inventory cost = ₹165.60

Dynamic Programming | 22

22.1 INTRODUCTION

Dynamic programming is a mathematical technique which is useful for solving a multistage decision problem *i.e.* for making a sequence of inter-related decision. It provides a dynamic procedure for determining the combination of decision which maximize over all effectiveness. Dynamic programming technique decompose the original problem in n-variables into n-subproblems each in one variable. We want to take such a decision at every stage so that the total effectiveness (the objective function) defined over all stage is optimal. Thus dynamic programming is a technique of recursive optimization. In this technique one has to obtain the solution in an orderly manner by starting from one stage to the next and is completed after the final stage is reached. The technique of dynamic programming was developed by **Richard Bellman** in 1950.

22.2 BELLMAN'S PRINCIPLE OF OPTIMALITY

[MEERUT–1998, 2003, 05, 06, 07, 08, 10, 14, 15; ROHILKHAND–1999; AGRA–2000]

It states that

"An optimal policy (i.e. set of decision) has the property that whatever be the initial state and initial decision, the remaining decisions must constitute an optimal policy for the state resulting from the first decision."

☞ Remark

- A problem which does not satisfy the principle of optimality cannot be solved by dynamic programming.

22.3 MULTISTAGE DECISION PROBLEM

A problem in which the decision have to be made at successive stages is called a multistage decision problem.

Though the multistage problem occur themselves sufficient frequently, it is often possible to introduce the multistage nature in the problem so that the dynamic programming may be used. It can be classified on the basis of the following properties:

(i) The outcome of a decision may be deterministic or probabilistic. In case of deterministic, given the stage of the process, the outcomes of the decision at any stage is uniquely determined and known. In probabilistic case, there is a set of possible outcomes given by a known probability distribution.

(ii) The possible decision at any stage from which we are to choose one are called 'states'. These may be finite or infinite (states are the possible situations in which the system may be at any stage).

(iii) The toal number of stages in the process may be finite or infinite and may be known and unknown.

22.4 CHARACTERSTIC OF A DYNAMIC PROGRAMMING PROBLEM [MEERUT–1998, 99]

The characteristic of a dynamic programming problem may be outlined as follows:

(1) The problem can be divided into stages with a policy decision required at each stage.

(2) Each stage has a number of states associated with it.

(3) The effect of the policy decision at each stage is to transform the current state into a state associated with the next stage.

(4) Given the current stage, an optimal policy for remaining stages is independent of the policy adopted in the previous stage.

(5) The solution procedure begins by finding the optimal policy for each state of the last stage.

(6) A functional equation is available which identify the optimal policy for each state with n-stages remaining, given the optimal policy for each state with $(n-1)$ stages left.

(7) Using the functional equation, the solution procedure moves backward stage by stage each time finding the policy when starting at the initial stage.

22.5 SOLUTION OF A MULTI-STAGE PROBLEM OF DYNAMIC PROGRAMMING WITH FINITE NUMBER OF STAGES [MEERUT–2001, 07]

The solution of problems by dynamic programming is usually done in three stages.

(i) Mathematical formulation of the problem.

(ii) The development of functional equations for the problem.

(iii) To solve functional equations for determining the optimal policy.

22.5.1 To develop a Functional Relationship: The recursive equations approach

At this stage we have to develop a recurrence relation connecting the optimal decision function for n-stage problem with the optimal decision function for the $(n-1)$-stage sub-problem, $n = 1, 2, \ldots, n$.

22.5.2 To solve the functional equation

Firstly we write the optimal decision function for one stage subproblem and then solve it.

Then we solve the optimal decision functions for 2-stage, 3-stage,... $(n - 1)$-stage and n-stage problem.

Working Procedure

STEP 1. Identify the problem decision variables and specify objective function to be optimized.

STEP 2. Divide the given problem into a number of sub-problems (stages). Identify the state variables at each stage and write down the transformation function as a function of the state variables and decision variables at the next stage.

STEP 3. Write a general recursive formula for computing the optimal policy. Also decide whether to follow the forward or the backward method (given at the end of this procedure) to solve the problem.

STEP 4. Determine the overall optimal policy or decision and its value at each stage.

22.5.3 Forward and Backward computations

If the recursive equations involved in a dynamic programming problem, are solved in the order,

$$f_1 \rightarrow f_2 \rightarrow .... \rightarrow f_n$$

then the computation involved is called forward computational procedure.
But if the recursive equations may be solved in the order $f_n \rightarrow f_{n-1} \rightarrow .... \rightarrow f_1$
Then the computation involved is called backward computational procedure.

> The solution of a recursive equation involve two types of computations according as the system is discrete or continuous. If the system is discrete, a tabulator computational scheme is followed at any stage. In each table, the no. of rows and columns are equal to the no. of corresponding feasible states values and the no. of possible decision respectively. In case of continuous system the optimal decisions at each stage are obtained by using the usual classical techniques.

22.6 TYPES OF PROBLEMS

(i) Single Additive Constraints, Multiplicatively seperable return

Consider the following problem

$$\text{Max}.Z = \prod_{j=1}^{n} f_j(y_j)$$

subject to the constraints

$$\sum_{j=1}^{n} a_j y_j = b$$

and $\qquad y_j \geq 0, a_j \geq 0$

Now introduce state variables i.e. $s_n = \sum a_j y_j = b$

$s_{j-1} = s_j - a_j y_j, \ j = 2,3,...n$

Let $F_j(s_j) = \max_{y_1, y_2, ... y_j} \prod_{1}^{j} f_j(y_j)$

Then the general recursion formula becomes

$F_j(s_j) = \max_{y_j} \left[f_i(y_i) F_{j-1}(s_j - 1) \right], \ j = n, n-1,...2$

$F_1(s_1) = f_1(y_1)$

(ii) Single Additive Constraints Additively seperable Return

Let us consider a problem in which objective function Z is an additively seperable function of n variables y_i and $f_i(y_i)$ is a function of y_i. Then we have to find $y_j, 1 \leq j \leq n$ which

Minimize $Z = \sum_{j=1}^{n} f_j(y_j)$

subject to the constraints

$$\sum_{j=1}^{n} a_j y_j \geq b$$

where a_j and b are real numbers and $a_j \geq 0, y_j \geq 0, b > 0$. In this problem each decision y_j is associated with a return function $f_j(y_j)$.

Now introduce state variables $s_0, s_1 ..., s_n$ such that

$$s_n = a_1 y_1 + a_2 y_2 + ... + a_n y_n \geq b$$
$$s_{n-1} = a_1 y_1 + a_2 y_2 + ... a_{n-1} y_{n-1} = s_n - a_n y_n$$
$$s_{n-2} = a_1 y_1 + a_2 y_2 + ... + a_{n-2} y_{n-2} = s_{n-1} - a_{n-1} y_{n-1}$$
...
$$s_1 = a_1 y_1 = s_2 - a_2 y_2$$

Also, $\quad s_{j-1} = T_j(s_j, y_j), 1 \leq j \leq n$ is the stage transformation function and $F_n(s_n)$ denote the minimum value of Z for any feasible value of s_n such that

$$F_n(s_n) = \min_{y_1, y_2, ..., y_n} [f_1(y_1) + f_2(y_2) + ... + f_n(y_n)], s_n \geq b$$

Now, first choose a particular value of y_n and minimize Z over the remaining $n-1$ variables, then

$$F_n(s_n) = \min_{y_1, y_2, ..., y_{n-1}} \left[\sum_{j=1}^{n-1} f_j(y_j) = f_n(y_n) + F_{n-1}(s_{n-1}) \right]$$

Here, values of $y_1, y_2, ..., y_{n-1}$ for which $\sum_{j=1}^{n-1} f_j(y_j)$ is minimum keeping y_n fixed, they depend upon s_{n-1} which is a function of s_n and y_n. So, minimum over all y_n for any feasible s_n would now become

$$F_n(s_n) = \min_{y_n} [f_n(y_n) + F_{n-1}(s_{n-1})]$$

If the value of $F_{n-1}(s_{n-1})$ is known for all y_n, the function to be minimized would involve only a single variable y_n and can be solve easily.

Similarly the recursion formula is

$$F_j(s_j) = \min_{y_j} \left[f_j(y_j) + F_{j-1}(s_{j-1}) \right]; 1 \leq j \leq n, \ F_1(s_1) = f_1(y_1)$$

So, starting with $F_1(s_1)$ and recursively optimizing to get $F_2(s_2), F_3(s_3), ...$, we obtain $F_n(s_n)$ for each s_n. Here, each time optimization occur over a single variable.

(iii) Single Multiplicative Constraints, Additively seperable Return

Consider the problem

$$Min. Z = f_1(y_1) + f_2(y_2) + ... + f_n(y_n)$$

subject to the constraints

$$y_1 . y_2 ... y_n \leq p, \qquad p \geq 0, \ y_i \geq 0 \quad \forall i$$

Now state variables are defined as

$$s_n = y_n y_{n-1} ... y_2 y_1 \geq p$$
$$s_{n-1} = s_n / y_n = y_{n-1} y_{n-2} ... y_2 y_1$$
.....
.....
$$s_2 = s_3 / y_3 = y_2 y_1$$
$$s_1 = s_2 / y_2 = y_1$$

Let $F(s_n)$ be the minimum value of the objective function for s_n. Then recursion formula is given by

$$f_j(s_j) = \min_{y_j}\left[f_j(y_j) + F_{j-1}(s_{j-1})\right], 2 \le j \le n$$

(iv) System Involving more than one Constraints

Dynamic programming methods can be applied to problems involving more than one constraints. In single constraint problem, there has to be single state variable for each stage while in multi constrained problem, there has to be one state variable per constraint per stage.

Solved Examples

EXAMPLE 1. *Use dynamic programming to find the maximum value of* $y_1, y_2, y_3, \ldots y_n$, *when* $y_1 + y_2 + y_3 + \ldots + y_n = c$ *and* $y_j \ge 0$ *for* $j = 1, 2,$ *3, ..., n, where c is a positive number.*

[MEERUT-1993, 98, 2002, 03, 10; IAS-1994; AGRA-2000; ROHILKHAND-1995]

SOLUTION. To solve this problem by dynamic programming, we proceed the two steps: (i) To obtain functional equations (recursive relation) (ii) solution of the functional equation (recursive relation)

Step 1. To find the Recursive relation: Let y_j be the jth parts of positive number c where $j = 1, 2, 3, \ldots, n$. Here each j corresponding to part y_j may be regarded as a stage. Now since y_j may assume any non-negative value which satisfying the constraints.

$$y_1 + y_2 + y_3 + \ldots + y_n = c$$

Therefore, alternatives at each stage are infinite. This means that y_j may be considered to be continuous. Thus, it is a problem of continuous system.

Hence the optimal decisions are obtained by using the differentiation at each stage.

Let $f_n(c)$ denote the maximum attainable product when the positive quantity c is divided into n parts. Since the quantity c is fixed so $f_n(c)$ depends upon n. The recursive relations of this problem stage by stage are as follows.

For stage 1, *i.e.,* $n = 1$, If c is divided into one part. Then we have $y_1 = c$

so, we have $f_1(c) = c$ ...(1)

For stage 2, *i.e.,* $n = 2$ if c is divided into two parts then

Let $y_1 = z$ $\therefore y_2 = c - z$

therefore $f_2(c) = \max\{y_1, y_2\} = \max_{0 \le z \le c}\{z(c - z)\}$

$\therefore$ $f_2(c) = \max_{0 \le z \le c}\{zf_1(c - z)\}$ ...(2) [using (1)]

For stage 3, *i.e.,* $n = 3$, If c is divided into three parts

Let $y_1 = z$ then we take $y_2 + y_3 = c - z$

$\therefore$ $f_3(c) = \max\{y_1, y_2, y_3\} = \max_{0 \le z \le c}\{z \cdot f_2(c - z)\}$ ...(3)

Hence, in general, for n stage problem, the recursive relation is given by

$$f_n(c) = \max_{0 \le z \le c}\{zf_{n-1}(c - z)\}$$ ...(4)

Step 2. Solution of the functional equations:

For stage 1, we have $f_1(c) = c$

For stage 2, we have $f_2(c) = \max\limits_{0 \le z \le c} \{zf_1(c-z)\}$

$$= \max\limits_{0 \le z \le c} \{z.(c-z)\} \qquad [\because f_1(c-z) = (c-z)]$$

Since, the function $z(c-z)$ has the maximum value at $z = \dfrac{c}{2}$ satisfying the restriction $0 \le z \le c$. therefore

$$f_2(c) = \left\{\frac{c}{2} \cdot \frac{c}{2}\right\} = \left\{\frac{c}{2}\right\}^2 \qquad [\text{since } c - z = \frac{c}{2}]$$

Hence, the optimal policy of two parts is $(\dfrac{c}{2}, \dfrac{c}{2})$.
For stage 3, we have

$$f_3(c) = \max\limits_{0 \le z \le c} \{zf_2(c-z)\} = \max\limits_{0 \le z \le c} \left\{z\left(\frac{c-z}{2}\right)^2\right\}$$

Since the maximum value of the function $z\left(\dfrac{c-z}{2}\right)^2$ is attained for $z = \dfrac{c}{3}$ satisfying the restriction $0 \le z \le c$. Hence,

$$f_3(c) = \frac{c}{3}(c - \frac{c}{3})^2 = (\frac{c}{3})^3$$

So, the optimal policy of three parts is $(\dfrac{c}{3}, \dfrac{c}{3}, \dfrac{c}{3})$.

Therefore, in general, we can assume for n stage problem

$$f_n(c) = \left(\frac{c}{n}\right)^n$$

Hence, the optimal policy of n parts is $\left(\dfrac{c}{n}, \dfrac{c}{n}, \dfrac{c}{n}, \dots n \text{ times}\right)$.

Using the principle of mathematical induction, we have

$$f_{n+1}(c) = \max\limits_{0 \le z \le c} \{zf_n(c-z)\} = \max\limits_{0 \le z \le c} \left\{z\left(\frac{c-z}{n}\right)^n\right\}$$

Since the function $z\left(\dfrac{c-z}{n}\right)^n$ has its maximum value at $z = \dfrac{c}{n+1}$.

So, $$f_{n+1}(c) = \left(\frac{c}{n+1}\right)^{n+1}$$

Hence, by the principle at mathematical induction, the required policy for n stage problem is $\left(\dfrac{c}{n}, \dfrac{c}{n}, \dots, \dfrac{c}{n}\right)$, and $f_n(c) = \left(\dfrac{c}{n}\right)^n$.

EXAMPLE 2. ***Obtain the minimum value of*** $x_1 + x_2 + x_3 + \dots + x_n$, ***when*** $x_1 \cdot x_2 \cdot x_3 \cdot \dots \cdot x_n = d$, $x_1, x_2, x_3, \dots, x_n \ge 0$.

[MEERUT 1981, 82, 93, 94, 2007, 08, AGRA 2007; ROHILKHAND–1995, 2005]

SOLUTION. To solve this problem by dynamic programming, we proceed the following two steps: (i) To find the functional equations (Recursive relation) (ii) Solution of functional equations (Recursive relations)

Step 1. To find the Functional equations: Let $f_n(d)$ be the minimum attainable sum $z = x_1 + x_2 + x_3 + \ldots + x_n$, when d is factorized in n factors, i.e., $x_1, x_2, x_3, \ldots, x_n$.

For stage 1, i.e., $n = 1$ we have $x_1 = d$, where d is factorized into one part only

$\therefore \qquad f_1(d) = \min z = \min\{x_1\} = d$ \qquad ...(1)

For stage 2, i.e., $n = 2$, If d is factorized into two factors.

Let $x_1 = y$ and $x_2 = \dfrac{d}{y}$

$\therefore \qquad f_2(d) = \min\{x_1 + x_2\} = \min_{0 \le y \le d} \{y \cdot \dfrac{d}{y}\}$

$\qquad = \min_{0 \le y \le d} \{y + f_1(\dfrac{d}{y})\}$ \qquad ...(2) $\quad \left[\because f_1(\dfrac{d}{y}) = \dfrac{d}{y}\right]$

For stage 3, i.e., $n = 3$, If d is factorized into three parts.

Let $x_1 = y$ and $x_2 x_3 = \dfrac{d}{y}$

$\therefore \qquad f_3(d) = \min\{x_1 + x_2 + x_3\}$

$\qquad = \min_{0 \le y \le d} \left\{y + f_2(\dfrac{d}{y})\right\}$ \qquad ...(3)

Hence, in general for n stage problem the recursive relation is given by

$$f_n(d) = \min_{0 \le y \le d} \left\{y + f_{n-1}(\dfrac{d}{y})\right\} \qquad ...(4)$$

Step 2 Solution of functional equations (recursive relation):

For stage 1, we have $f_1(d) = d$

For stage 2, we have $f_2(d) = \min_{0 \le y \le d} \left\{y + f_1(\dfrac{d}{y})\right\} = \min_{0 \le y \le d} \left\{y + \dfrac{d}{y}\right\}$

Since, the function $\left(y + \dfrac{d}{y}\right)$ has the minimum value at $y = d^{1/2}$ satisfying the restriction $0 \le z \le d$. So that $\dfrac{d}{dy}(y + \dfrac{d}{y}) = 0 \Rightarrow 1 - \dfrac{d}{y^2} = 0 \Rightarrow y = d^{1/2}$

and $\dfrac{d^2}{dy^2}\left(y + \dfrac{d}{y}\right)$ has a positive value at $y = d^{1/2}$

Therefore, $f_2(d) = d^{1/2} + \dfrac{d}{d^{1/2}} = d^{1/2} + d^{1/2} = 2d^{1/2}$ \qquad ...(5)

Hence, the optimal policy of two parts is $(d^{1/2}, d^{1/2})$.

For stage 3, we have $f_3(d) = \min_{0 \le y \le d} \left\{y + f_2\left(\dfrac{d}{y}\right)\right\} = \min_{0 \le y \le d} \left\{y + 2\left(\dfrac{d}{y}\right)^{1/2}\right\}$

Since $y + 2\left(\dfrac{d}{y}\right)^{1/2}$ has the minimum value at $y = d^{1/3}$ satisfying the restriction

$0 \le y \le d$ so that

$$\frac{d}{dy}\left\{y + 2\left(\frac{d}{y}\right)^{1/2}\right\} = 0 \ \Rightarrow \ y = d^{1/3}$$

and $\dfrac{d^2}{dy^2}\left\{y + 2\left(\dfrac{d}{y}\right)^{1/2}\right\}$ has a positive value at $y = d^{1/3}$.

$$\therefore \quad f_3(d) = d^{1/3} + 2\left(\frac{d}{d^{1/3}}\right)^{1/2} = d^{1/3} + 2d^{1/3} = 3d^{1/3} \qquad \qquad ...(6)$$

therefore, the optimal policy for three parts is $(d^{1/3}, d^{1/3}, d^{1/3})$.
Now, we can assume for n stage problem, we have

$$f_n(d) = nd^{1/n}$$

and the optimal policy for n parts is $(d^{1/n}, d^{1/n}, ..., d^{1/n})$
And, we will prove the result by induction, for this

$$f_{n+1}(d) \quad = \min_{0 \le y \le d}\left\{y + f_n\left(\frac{d}{y}\right)\right\}$$

$$= \min_{0 \le y \le d}\left\{y + n\left(\frac{d}{y}\right)^{1/n}\right\}.$$

Since, $y + n\left(\dfrac{d}{y}\right)^{1/n}$ has the minimum value at $y = d^{1/n+1}$

$$\therefore \quad f_{n+1}(d) = d^{1/n+1} + n\left(\frac{d}{d^{1/n+1}}\right)^{1/n} = (n+1)d^{1/n+1}$$

and the optimal policy for (n+1) stage is $(d^{1/n+1}, d^{1/n+1}, ..., d^{1/n+1})$.

Hence, by mathematical induction, $f_n(d) = nd^{1/n}$ and the required policy for n stage problem is $(d^{1/n}, d^{1/n}, ...d^{1/n})$.

EXAMPLE 3. ***Use dynamic programming to show that*** $-\sum\limits_{i=1}^{n} p_i \log p_i$ ***is maximum***

subjected to $\sum\limits_{i=1}^{n} p_i = 1$ ***when*** $p_1 = p_2 = p_3 = ... = p_n = 1/n$

[MEERUT-1980, 82(P), 2003 ROHILKHAND-1994, 2006, KANPUR–2000, AGRA-1996, 97, 98, 2013]

SOLUTION.   **Step 1 Functional equations (recursive relation):**
In this problem we have

$$p_1 + p_2 + p_3 + ... + p_n = 1 \text{ such that } - \sum_{i=1}^{n} p_i \log p_i \text{ is maximum.}$$

Let $f_n(1)$ be the maximum attainable value of $-\sum\limits_{i=1}^{n} p_i \log p_i$ when

$p_1 + p_2 + p_3 + \ldots + p_n = 1$

Since p_i may assume any non-negative value. So, $f_n(1)$ is a function of discrete variable.

For stage 1, *i.e.*, $n = 1$. If we divide 1 into one part p_i, then $p_1 = 1$

$\therefore \quad f_1(1) = \max\{-p_1 \log p_1\} = -1 \log 1$ ...(1)

For stage 2, *i.e.*, If we divide 1 into two parts p_1 and p_2 then

let $p_1 = z, \quad p_2 = 1 - z$

$\therefore \quad f_2(1) \quad = \max\{-p_1 \log p_1, -p_2 \log p_2\}$

$= \max\limits_{0 \leq z \leq 1} \{-z \log z - (1 - z)\log(1 - z)\}$

$= \max\limits_{0 \leq z \leq 1} \{-z \log z + f_1(1 - z)\}$ ...(2)

$[\because f_1(1 - z) = -(1 - z)\log(1 - z)]$

For stage 3, *i.e.*, If we divide 1 into three parts p_1, p_2 and p_3 then

let $p_1 = z \quad \therefore p_2 + p_3 = 1 - z$

$\therefore \quad f_3(1) \quad = \max\{-p_1 \log p_1, -p_2 \log p_2, -p_3 \log p_3\}$

$= \max\limits_{0 \leq z \leq 1} \{-z \log z + f_2(1 - z)\}$ ...(3)

Hence, in general, for n stage problem the recursive relation is given by

$f_n(1) = \max\limits_{0 \leq z \leq 1} \{-z \log z + f_{n-1}(1 - z)\}$ ...(4)

Step 2 Solution of functional equations (recursive relation):

For stage 1, we have $\quad f_1(1) \quad = -1 \log 1$

For stage 2, we have $\quad f_2(1) \quad = \max\limits_{0 \leq z \leq 1} \{-z \log z + f_1(1 - z)\}$

$= \max\limits_{0 \leq z \leq 1} \{-z \log z - (1 - z)\log(1 - z)\}$

Since the function $\{-z \log z - (1 - z)\log(1 - z)\}$ has the maximum value at $z = \dfrac{1}{2}$ satisfying the restriction $0 \leq z \leq 1$ so that

$\dfrac{d}{dz}\{-z \log z - (1 - z)\log(1 - z)\} = 0 \Rightarrow z = \dfrac{1}{2}$

and $\dfrac{d^2}{dz^2}\{-z \log z - (1 - z)\log(1 - z)\} = -\dfrac{1}{z} - \dfrac{1}{1 - z} = -4$ (negative)

therefore $f_2(1) = -\dfrac{1}{2}\log\dfrac{1}{2} - \dfrac{1}{2}\log\dfrac{1}{2} = 2\left(-\dfrac{1}{2}\log\dfrac{1}{2}\right)$ ...(5)

$\left[\because 1 - z = 1 - \dfrac{1}{2} = \dfrac{1}{2}\right]$

and the optimal policy for two parts is $\left(\dfrac{1}{2}, \dfrac{1}{2}\right)$.

For stage 3, $\quad f_3(1) \quad = \max\limits_{0 \leq z \leq 1} \{-z \log z + f_2(1 - z)\}$

$= \max\limits_{0 \leq z \leq 1} \left\{-z \log z + 2\left(-\dfrac{(1 - z)}{2}\log\dfrac{(1 - z)}{2}\right)\right\}$

Since, the function $\left\{-z\log z-(1-z)\log\dfrac{(1-z)}{2}\right\}$ is maximum at $z=\dfrac{1}{3}$ satisfying

the restriction $0\le z\le 1$ so that

$$\frac{d}{dz}\left\{-z\log z-(1-z)\log\frac{(1-z)}{2}\right\}=0 \Rightarrow -\log z+\log\frac{(1-z)}{2}-1+1=0$$

$$\Rightarrow \quad z=\frac{1}{3}$$

and $\dfrac{d^2}{dz^2}\left\{-z\log z-(1-z)\log\dfrac{(1-z)}{2}\right\}$ is negative at $z=\dfrac{1}{3}$.

$$\therefore \quad f_3(1) = -\frac{1}{3}\log\frac{1}{3}+2\left(-\frac{2/3}{2}\log\frac{2/3}{2}\right) \qquad \left[\because 1-z=1-\frac{1}{3}=\frac{2}{3}\right]$$

$$= -\frac{1}{3}\log\frac{1}{3}+2\left(-\frac{1}{3}\log\frac{1}{3}\right)=3\left(-\frac{1}{3}\log\frac{1}{3}\right) \qquad \text{...(6)}$$

and the optimal policy for three parts is $\left(\dfrac{1}{3},\dfrac{1}{3},\dfrac{1}{3}\right)$.

In general, for n stage problem, we have

$$f_n(1)=n\left(-\frac{1}{n}\log\frac{1}{n}\right) \qquad \text{...(7)}$$

Now, we will prove the result by induction.
For this, we have

$$f_{n+1}(1)=\max_{0\le z\le 1}\left\{-z\log z+f_n(1-z)\right\}$$

$$=\max_{0\le z\le 1}\left\{-z\log z+n\left(-\frac{1}{n}\log\frac{1}{n}\right)\right\}$$

Since $\left\{-z\log z+n\left(-\dfrac{1}{n}\log\dfrac{1}{n}\right)\right\}$ is maximum at $z=\dfrac{1}{n+1}$ satisfying the restriction

$0\le z\le 1$ so that

$$\frac{d}{dz}\left\{-z\log z+n\left(-\frac{1}{n}\log\frac{1}{n}\right)\right\}=0 \quad \Rightarrow z=\frac{1}{n+1}$$

and $\dfrac{d^2}{dz^2}\left\{-z\log z+n\left(-\dfrac{1}{n}\log\dfrac{1}{n}\right)\right\}$ is negative for $z=\dfrac{1}{n+1}$.

Now, $f_n(1-z) \quad =f_n\left(\dfrac{n}{1+n}\right)=n\left\{-\dfrac{\dfrac{n}{1+n}}{n}\log\dfrac{\dfrac{n}{1+n}}{n}\right\}$

$$=n\left[-\frac{1}{1+n}\log\frac{1}{1+n}\right]$$

$$\therefore \quad f_{n+1}(1) \quad =-\frac{1}{1+n}\log\frac{1}{1+n}+n\left(-\frac{1}{1+n}\log\frac{1}{1+n}\right)$$

$$=(1+n)\left(-\frac{1}{1+n}\log\frac{1}{1+n}\right)$$

and the optimal policy is $\left(\dfrac{1}{n+1}, \dfrac{1}{n+1}, \dfrac{1}{n+1}, ..., \dfrac{1}{n+1}\right)$.

Hence, the maximum value of $f_n(1)$ is $n\left(-\dfrac{1}{n}\log\dfrac{1}{n}\right)$ and the optimal policy is

$\left(\dfrac{1}{n}, \dfrac{1}{n}, \dfrac{1}{n}, ..., \dfrac{1}{n}\right)$, i.e., $p_1 = p_2 = ... p_n = \dfrac{1}{n}$.

EXAMPLE 4. *Find the functional equations for maximizing*

$$z = g_1(x_1) + g_2(x_2) + g_3(x_3) + ... + g_n(x_n)$$

subjected to $x_1 + x_2 + ... + x_n = c$, *where* $x_1, x_2, x_3, ... x_n \geq 0$. [DELHI–1995]

SOLUTION. Let $f_n(c)$ denote the maximum attainable value of

$z = g_1(x_1) + g_2(x_2) + g_3(x_3) + ... + g_n(x_n)$ when $x_1 + x_2 + ... + x_n = c$,

$x_1, x_2, x_3, ... x_n \geq 0$

Since, x_i may assume any non-negative value. So, $f_n(c)$ is a function of discrete variable and it is a continuous system problem.

For stage 1, i.e., $n = 1$, If we divide c into one part i.e., $x_1 = c$

$\therefore \qquad f_1(c) = \max\{g_1(x_1)\} = g_1(c)$ ...(1)

For stage 2, i.e. $n = 2$. If we divide c into two parts x_1 and x_2, then let $x_2 = z$, then $x_1 = c - z$.

$\therefore \qquad f_2(c) = \max\{g_1(x_1) + g_2(x_2)\}$

$\qquad = \max_{0 \leq z \leq c} \{g_2(z) + g_1(c - z)\}$

$\qquad = \max_{0 \leq z \leq c} \{g_2(z) + f_1(c - z)\}$ ...(2)

$[\because f_1(c - z) = c - z]$

For stage 3, i.e., If we divide c into three parts x_1, x_2 and x_3.

Let $x_3 = z$ then $x_1 + x_2 = c - z$

$\therefore \qquad f_3(c) = \max\{g_1(x_1) + g_2(x_2) + g_3(x_3)\}$

$\qquad = \max_{0 \leq z \leq c} \{g_3(z) + f_2(c - z)\}$ ...(3)

Hence, in general for n stage problem, we have

$\qquad f_n(c) = \max_{0 \leq z \leq c} \{g_n(z) + f_{n-1}(c - z)\}$ ...(4)

which are required functional equations.

EXAMPLE 5. *Obtain the maximum value of* $b_1x_1 + b_2x_2 + ... + b_nx_n$, *when*

$x_1 + x_2 + ... + x_n = c$, $x_1, x_2, ..., x_n \geq 0$, $i = 1, 2, ..., n$

[MEERUT 1990, 97(P), 97, 2001, 09]

SOLUTION. **Step 1 To find the Functional equations (recursive relations):**

Let $f_n(c)$ denote the maximum attainable sum of $b_1x_1 + b_2x_2 + ... + b_nx_n$ when we divide the positive quantity c into n parts. It is a n stage problem.

For stage 1, i.e., $n = 1$, If we divide c into one part then $x_1 = c$

$\therefore \qquad f_1(c) = \max\{b_1x_1\} = b_1c$ ...(1)

For stage 2, *i.e.*, $n = 2$ If we divide c into two parts x_1 and x_2 then

$$x_2 = z \qquad \therefore \quad x_1 = c - z$$

$$\therefore \qquad f_2(c) = \max\{b_1 x_1 + b_2 x_2\} = \max_{0 \le z \le c}\{b_2 z + b_1(c - z)\}$$

$$= \max_{0 \le z \le c}\{b_2 z + f_1(c - z)\} \qquad [\because f_1(c - z) = b_1(c - z)] \qquad ...(2)$$

For stage 3, *i.e.*, $n = 3$, If we divide c into three parts x_1, x_2 and x_3,

Let $\qquad x_3 = z$ then $x_1 + x_2 = c - z$

$$\therefore \qquad f_3(c) = \max\{b_1 x_1 + b_2 x_2 + b_3 x_3\} = \max_{0 \le z \le c}\{b_3 z + f_2(c - z)\} \qquad ...(3)$$

Hence, in general, for n stage problem, the recursive relation is

$$f_n(c) = \max_{0 \le z \le c}\{b_n z + f_{n-1}(c - z)\} \qquad ...(4)$$

Step 2 Solution of functional equations (recursive relations):

For stage 1, we have $f_1(c) = b_1 c$

For stage 2, we have $f_2(c) = \max_{0 \le z \le c}\{b_2 z + f_1(c - z)\}$

$$= \max_{0 \le z \le c}\{b_2 z + b_1(c - z)\} \qquad [\because f_1(c - z) = c - z]$$

$$= \max_{0 \le z \le c}\{b_1 c + (b_2 - b_1)z\}$$

It is obvious that if $b_2 > b_1$, then $\{b_1 c + (b_2 - b_1)z\}$ is maximum for z = c, satisfying the restriction $0 \le z \le c$, otherwise, it will be minimum.

$$\therefore \qquad f_2(c) = \{b_1 c + (b_2 - b_1)c\} = b_2 c$$

Therefore, the optimum policy for two parts is $x_1 = 0, x_2 = c$.

For stage 3, we have $f_3(c) = \max_{0 \le z \le c}\{b_3 z + f_2(c - z)\}$

$$= \max_{0 \le z \le c}\{b_3 z + b_2(c - z)\} = \max_{0 \le z \le c}\{b_2 c + (b_3 - b_2)z\}$$

Here, it is obvious that if $b_3 > b_2$, then $\{b_2 c + (b_3 - b_2)z\}$ is maximum for $z = c$ satisfying the restriction $0 \le z \le c$, otherwise, it will be minimum.

$$\therefore \qquad f_3(c) = \{b_2 c + (b_3 - b_2)c\} = b_3 c$$

therefore, the optimal policy for three stage problem is $x_1 = 0, x_2 = 0, x_3 = c$

Hence, in general we can say

$$f_n(c) = b_n c$$

and the optimal policy for n stage problem is

$$x_1 = 0, x_2 = 0, x_3 = 0, ..., x_n = c$$

EXAMPLE 6. *Use dynamic programming technique, obtain the minimum value of $x_1^2 + x_2^2 + x_3^2$, when $x_1 + x_2 + x_3 \ge 15$, $x_1, x_2, x_3 \ge 0$.*

[MEERUT 1984, 88, 2005, 08, AGRA–2003, 07, IAS–1995]

SOLUTION. **Step 1 To find the Functional equations:**

Since, there are three decision variables x_1, x_2 and x_3, so it is a three stage problem which can be defined as follows:

Consider $\qquad s_3 = x_1 + x_2 + x_3 \ge 15$

$$s_2 = x_1 + x_2 = s_3 - x_3$$

and $\qquad s_1 = x_1 = s_2 - x_2$

Since, it is a three stage problem. Let $f_3(s_3)$ denote the minimum attainable value of $x_1^2 + x_2^2 + x_3^2$ at the third stage, $s_3 = x_1 + x_2 + x_3$.

then,

$$f_1(s_1) = \min_{0 \le x_1 \le s_1}\{x_1^2\} = (s_2 - x_2)^2 \qquad ...(1)$$

$$f_2(s_2) = \min_{0 \le x_2 \le s_2} \{x_1^2 + x_2^2\}$$

$$= \min_{0 \le x_2 \le s_2} \{x_2^2 + f_1(s_1)\} \qquad \qquad ...(2)$$

and $\quad f_3(s_3) = \min_{0 \le x_3 \le s_3} \{x_3^2 + f_2(s_2)\} \qquad \qquad ...(3)$

Step 2 Solution of the functional equations:

From equation (1), we have

$$f_1(s_1) = (s_2 - x_2)^2$$

From equation (2), we have

$$f_2(s_2) = \min_{0 \le x_2 \le s_2} \{x_2^2 + f_1(s_1)\}$$

$$= \min_{0 \le x_2 \le s_2} \{x_2^2 + (s_2 - x_2)^2\}$$

Since, $\{x_2^2 + (s_2 - x_2)^2\}$ is minimum for $x_2 = \dfrac{s_2}{2}$ so that

$$\frac{d}{dx_2}\{x_2^2 + (s_2 - x_2)^2\} = 4x_2 - 2s_2 = 0 \Rightarrow x_2 = \frac{s_2}{2}$$

and $\quad \dfrac{d^2}{dx_2^2}\{x_2^2 + (s_2 - x_2)^2\} = 4$ at $x_2 = \dfrac{s_2}{2}$, which is positive.

$$\therefore \qquad \qquad f_2(s_2) = \left\{ \left(\frac{s_2}{2}\right)^2 + \left(s_2 - \frac{s_2}{2}\right)^2 \right\}$$

$$= \left\{ \frac{s_2^2}{4} + s_2^2 + \frac{s_2^2}{4} - s_2^2 \right\}$$

$$= \frac{s_2^2}{2}$$

Now, from equation(3), we have

$$f_3(s_3) = \min_{0 \le x_3 \le s_3} \{x_3^2 + f_2(s_2)\}$$

$$= \min_{0 \le x_3 \le s_3} \left\{ x_3^2 + \frac{(s_3 - x_3)^2}{2} \right\} \qquad \left[\because f_2(s_2) = \frac{(s_3 - x_3)^2}{2} \right]$$

Since, $\left\{ x_3^2 + \dfrac{(s_3 - x_3)^2}{2} \right\}$ is minimum for $x_3 = \dfrac{s_3}{3}$ so that

$$\frac{d}{dx_3}\left\{ x_3^2 + \frac{(s_3 - x_3)^2}{2} \right\} = 0 \Rightarrow 3x_3 - s_3 = 0 \Rightarrow x_3 = \frac{s_3}{2}$$

and $\qquad \dfrac{d^2}{dx_3^2} = 3$ at $x_3 = \dfrac{s_3}{3}$, which is positive.

$$\therefore \quad f_3(s_3) = \left\{ \left(\frac{s_3}{3}\right)^2 + \frac{1}{2}\left(s_3 - \frac{s_3}{3}\right)^2 \right\}$$

$$= \left\{ \frac{s_3^2}{9} + \frac{1}{2} \left(s_3^2 + \frac{s_3^2}{9} - \frac{2}{3} s_3^2 \right) \right\}$$

$$= \frac{s_3^2}{3}$$

But, it is given that $s_3 \geq 15$, so the minimum value of s_3 is 15.

$$\therefore \qquad x_3 = \frac{s_3}{3} = \frac{15}{3} = 5$$

$$x_2 = \frac{s_2}{2} = \frac{s_3 - x_3}{2} = \frac{15 - 5}{2} = 5$$

and $\qquad x_1 = 5$, $f_3(s_3) = \frac{s_3^2}{2} = \frac{(15)^2}{2} = 75$

Hence, the optimal policy for this problem is

$x_1 = 5$, $x_2 = 5$, $x_3 = 5$ and minimum value of $x_1^2 + x_2^2 + x_3^2$ is 75.

EXAMPLE 7. ***Use dynamic programming. Solve***

$$\text{Max } z = x_1^2 + 2x_2^2 + 4x_3, \text{ when } x_1 + 2x_2 + x_3 \leq 8, \ x_1, x_2, x_3, \geq 0$$

[MEERUT 1999(O) KANPUR 2007, AGRA 2002]

SOLUTION. **Step 1 To find the Functional equations (recursive relations):**

Since, there are three decision variables x_1, x_2 and x_3. So, It is a three stage problem.

This problem can be defined as follows:

Consider $\qquad s_3 = x_1 + 2x_2 + x_3 \leq 8$

$\qquad\qquad\qquad s_2 = x_1 + 2x_2 = s_3 - x_3$ $\qquad$ ...(1)

and $\qquad\qquad s_1 = x_1 = s_2 - 2x_2$

Since, it is a three stage problem, Let $f_i(s_i)$ denote the maximum attainable value of $z = x_1^2 + 2x_2^2 + 4x_3$ where $s_i = x_1 + x_2 + x_3 + \ldots + x_i$, $i = 1, 2, 3$.

Then

$$f_1(s_1) = \max_{0 \leq x_1 \leq s_1} \{x_1^2\} = (s_2 - 2x_2)^2 \qquad \text{...(2)}$$

$$f_2(s_2) = \max_{0 \leq 2x_2 \leq s_2} \{x_1^2 + 2x_2^2\}$$

$$= \max_{0 \leq 2x_2 \leq s_2} \{2x_2^2 + (s_2 - 2x_2)^2\}$$

$$= \max_{0 \leq x_2 \leq \frac{s_2}{2}} \{2x_2^2 + f_1(s_1)\} \qquad \text{...(3)}$$

and $\quad f_3(s_3) = \max_{0 \leq x_3 \leq s_3} \{x_1^2 + 2x_2^2 + 4x_3\}$

$$= \max_{0 \leq x_3 \leq s_3} \{4x_3 + f_2(s_2)\} \qquad \text{...(4)}$$

Step 2 Solution of functional equations:

From equation (2), we have

$$f_1(s_1) = (s_2 - 2x_2)^2 \qquad \ldots(5)$$

From equation (3), we have

$$f_2(s_2) = \max_{0 \le x_2 \le \frac{s_2}{2}} \{2x_2^2 + f_1(s_1)\}$$

$$= \max_{0 \le x_2 \le \frac{s_2}{2}} \{2x_2^2 + (s_2 - 2x_2)^2\}$$

$$= \max_{0 \le x_2 \le \frac{s_2}{2}} \{6x_2^2 - 4x_2 s_2 + s_2^2\}$$

Here, differential method is fail, so we check the max or min by the following procedure.

Since, x_2 lies within the range 0 to $\dfrac{s_2}{2}$. So the maximum value of $f_2(s_2)$

satisfying the restriction will be at $x_2 = 0$, $f_2(s_2) = s_2^2$

and at $x_2 = \dfrac{s_2}{2}$: $f_2(s_2) = 6\dfrac{s_2^2}{4} + s_2^2 - 4\dfrac{s_2^2}{2} = \dfrac{s_2^2}{2}$

It is obvious that $s_2^2 > \dfrac{s_2^2}{2}$, therefore $f_2(x_2)$ has maximum value s_2^2 at $x_2 = 0$

Again from (4), we have

$$f_3(s_3) = \max_{0 \le x_3 \le s_3} \{4x_3 + f_2(s_2)\} = \max_{0 \le x_3 \le s_3} \{4x_3 + (s_3 - x_3)^2\}$$

Since, x_2 lies within the range 0 to s_3. So, the maximum value of $f_3(s_3)$ satisfying the restriction will be

at $x_3 = 0$ $\qquad f_3(s_3) = s_3^2$ $\left.\begin{array}{l} \\ \\ \end{array}\right\}$
at $x_3 = s_3$ $\qquad f_3(s_3) = 4s_3$ $\qquad\qquad \ldots(6)$

Since $s_3 \le 8$, therefore maximum value of s_3 is 8.

From equation (6), at $x_3 = 0$, $f_3(s_3) = s_3^2 = (8)^2 = 64$

and at $x_3 = s_3 = 8$, $f_3(s_3) = 4 \times 8 = 32$.

Clearly, $64 > 32$, therefore, the maximum value of $f_3(s_3)$ is 64 at $x_3 = 0$

Now, $s_2 = s_3 - x_3 = 8 - 0 = 8$ and $x_2 = 0$

and $s_1 = s_2 - 2x_2 = 8 - 2 \times 0 = 8$ and $x_1 = s_2 - 2x_2 = 8 - 2 \times 0 = 8$

Hence, the solution is $x_1 = 8, x_2 = 0, x_3 = 0$ and Max $z = 64$

EXAMPLE 8. *Use dynamic programming, solve*

$$\textbf{Min } z = x_1^2 + 2x_2^2 + 4x_3, \textbf{ when } x_1 + 2x_2 + x_3 \ge 8, x_1, x_2, x_3 \ge 0$$

[MEERUT–1989, 94(P), 99]

SOLUTION. **Step 1 To find the Functional equation:**

Since there are three decision variable x_1, x_2 and x_3, so it is a three stage problem.

This problem can be defined as:

Consider $\qquad s_3 = x_1 + 2x_2 + x_3 \ge 8$
$\qquad\qquad\quad s_2 = x_1 + 2x_2 = s_3 - x_3$ $\left.\begin{array}{l} \\ \\ \\ \end{array}\right\}$
and $\qquad\qquad s_1 = x_1 = s_2 - 2x_2$ $\qquad\qquad\qquad \ldots(1)$

Since, It is a three stage problem. Let $f_i(s_i)$ denote the minimum attainable value of $z = x_1^2 + 2x_2^2 + 4x_3$ at i^{th} stage where $i = 1, 2, 3$.

Then

$$f_1(s_1) = \min_{0 \le x_1 \le s_1} \{x_1^2\} = (s_2 - 2x_2)^2 \qquad \ldots(2)$$

$$f_2(s_2) = \min_{0 \le x_2 \le s_2} \{x_1^2 + 2x_2^2\} = \min_{0 \le x_2 \le s_2} \{2x_2^2 + f_1(s_1)\} \qquad \ldots(3)$$

and $$f_3(s_3) = \min_{0 \le x_3 \le s_3} \{x_1^2 + 2x_2^2 + 4x_3\} = \min_{0 \le x_3 \le s_3} \{4x_3 + f_2(s_2)\} \qquad \ldots(4)$$

Step 2 Solution of functional equations:

From equation (1), we have

$$f_1(s_1) = (s_2 - 2x_2)^2$$

From equation (2), we have

$$f_2(s_2) = \min_{0 \le x_2 \le s_2} \{2x_2^2 + f_1(s_1)\}$$

$$= \min_{0 \le x_2 \le s_2} \{2x_2^2 + (s_2 - 2x_2)^2\}$$

Since, the function $\{2x_2^2 + (s_2 - 2x_2)^2\}$ will be minimum at $x_2 = \dfrac{s_2}{3}$ satisfying the restriction so that

$$\frac{d}{dx_2}\{2x_2^2 + (s_2 - 2x_2)^2\} = 0 \Rightarrow 12x_2 - 4s_2 = 0 \Rightarrow x_2 = \frac{s_2}{3}$$

and $$\frac{d^2}{dx_2^2}\{2x_2^2 + (s_2 - 2x_2)^2\} = 12 \text{ (Positive) at } x_2 = \frac{s_2}{3}$$

Therefore, the minimum value of $f_2(s_2)$ is,

$$f_2(s_2) = 2\left(\frac{s_2}{3}\right)^2 + \left(s_2 - 2 \cdot \frac{s_2}{3}\right)^2 = \frac{s_2^2}{3} \text{ at } x_2 = \frac{s_2}{3}.$$

Now, from equation (3), we have

$$f_3(s_3) = \min_{0 \le x_3 \le s_3} \{4x_3 + f_1(s_1)\}$$

$$= \min_{0 \le x_3 \le s_3} \left\{4x_3 + \frac{s_2^2}{3}\right\}$$

$$= \min_{0 \le x_3 \le s_3} \left\{4x_3 + \frac{(s_3 - x_3)^2}{3}\right\}$$

Since, the function $\left\{4x_3 + \dfrac{(s_3 - x_3)^2}{3}\right\}$ is minimum at $x_3 = s_3 - 6$. So that

$$\frac{d}{dx_3}\left\{4x_3 + \frac{(s_3 - x_3)^2}{3}\right\} = 0 \Rightarrow x_3 = s_3 - 6 \text{ and } \frac{d^2}{dx_3^2} = 0$$

therefore, the minimum value of $f_3(s_3)$ is

$$f_3(s_3) = \left\{4(s_3 - 6) + \frac{(s_3 - s_3 + 6)^2}{3}\right\} = 4s_3 - 12$$

Since, $x_3 \geq 8$. So the minimum value of s_3 is 8.

Then $f_3(s_3) = 4s_3 - 12 = 4 \times 8 - 12 = 20$

and $x_3 = s_3 - 6 = 8 - 6 = 2$ i.e., $x_3 = 2$

$\qquad s_2 = s_3 - x_3 = 8 - 2 = 6$ i.e., $s_2 = 6$

$\therefore \qquad x_2 = \dfrac{s_2}{2} = \dfrac{6}{2} = 3$ i.e., $x_2 = 3$

and $x_1 = s_2 - 2x_2 = 6 - 2 \times 2 = 2$

$\therefore \qquad x_1 = 2$

Hence, the solution is

$\qquad x_1 = 2, x_2 = 3, x_3 = 2$ and Min $z = 20$

22.7 SOLUTION OF L.P.P. BY DYNAMIC PROGRAMMING   [MEERUT–2001, 06]

A general maximization linear programming problem is

$\qquad$ Max $z = c_1 x_1 + c_2 x_2 + c_3 x_3 + \ldots + c_n x_n$

subject to the constraints

$$a_{11}x_1 + a_{12}x_2 + a_{13}x_3 + \ldots + a_{1n}x_n \leq b_1$$
$$a_{21}x_1 + a_{22}x_2 + a_{23}x_3 + \ldots + a_{2n}x_n \leq b_2$$
$$a_{31}x_1 + a_{32}x_2 + a_{33}x_3 + \ldots + a_{3n}x_n \leq b_3$$
$$\vdots \qquad \vdots \qquad \qquad \vdots \qquad \qquad \vdots$$
$$a_{m1}x_1 + a_{m2}x_2 + a_{m3}x_3 + \ldots + a_{mn}x_n \leq b_m$$
$$x_1, x_2, x_3, \ldots x_n \geq 0$$

It is n stage problem. Since we consider each activity j $(j = 1, 2, \ldots, n)$ as a individual stage, the level of activities $(x_j \geq 0)$ are the decision variables (alternatives) at stage j. Being x_j continuous, each activity has an infinite number of alternatives with in the feasible region.

We know that the allocation problems are the special case of L.P.P. which requires the allocation of available resources to the activities.

Each constraints of the L.P.P. represents the limitation of different resources and the amount of available resources are $b_1, b_2, b_3, \ldots, b_m$. Since there are m resources in this problem, therefore this problem has m state variables.

Let $(B_{1j}, B_{2j}, B_{3j}, \ldots, B_{mj})$ be the state of the system at j^{th} stage, i.e., $B_{1j}, B_{2j}, B_{3j}, \ldots, B_{mj}$ are the amounts of resources 1, 2, 3, ..., m to be allocated to stage j. Let the optimum value of objective function at j^{th} stage is $f_j(B_{1n}, B_{2n}, B_{3n}, \ldots, B_{mn})$.

Here, we shall use the backward computational procedure. The recurrence equation,

$$f_n(B_{1n}, B_{2n}, B_{3n}, \ldots, B_{mn}) = \max_{0 \leq a_{in}x_n \leq B_{in}} \{c_n x_n\} \qquad i = 1, 2, \ldots, m$$

and $\qquad f_j(B_{1j}, B_{2j}, B_{3j}, \ldots, B_{mj}) = \max_{0 \leq a_{ij}x_j \leq B_{ij}} \{c_j x_j + f_{j+1}(B_{1j} - a_{1j}x_1, B_{2j}$

$$- a_{2j}x_2, B_{3j} - a_{3j}x_3, \ldots, B_{mj} - a_{mj}x_j\}$$

$$i = 1, 2, \ldots, m, \quad j = 1, 2, 3, \ldots (n-1).$$

for $\qquad\qquad 0 \leq B_{ij} \leq b_i$ for all i and j.

For the better understanding see the following examples.

Solved Examples

EXAMPLE 1. **Solve the L.P.P.**

$$\text{Max } z = 2x_1 + 5x_2$$

subjected to $2x_1 + x_2 \le 43, \ 2x_2 \le 46, x_1, x_2 \ge 0$

by using dynamic programming technique. [MEERUT-1998, AGRA–2002]

SOLUTION. In the given problem, there are two interrelated decision variables so it is a two stage linear programming problem. Due to the two resources b_1, b_2 with available amount 43 and 46, the state vector will have two components.

Let (B_{1j}, B_{2j}) is the state of the system and let us suppose that $f_j(B_{1j}, B_{2j})$ be the optimal value of the objective function for stage 1 and 2. Using backward computational procedure, we have

$$f_2(B_{12}, B_{22}) = \max_{\substack{0 \le x_2 \le 43 \\ 0 \le 2x_2 \le 46}} \{5x_2\} = 5 \max_{\substack{0 \le x_2 \le 43 \\ 0 \le x_2 \le 23}} \{x_2\}$$

Since $\max\{x_2\}$ which satisfying $x_2 \le 43$ and $x_2 \le 23$ is minimum of $B_{12} = 43$ and $B_{22}/2 = 23$.

therefore, $\max\{x_2\} = x_2{}^* = \min\{B_{12}, B_{22}/2\}$

so, $f_2(B_{12}, B_{22}) = 5 \min\left\{B_{12}, \dfrac{B_{22}}{2}\right\}$ ...(1)

We know that the value of B_{12}, B_{22} is given by at the primary stage.

Now $f_1(B_{11}, B_{21}) = \max_{\substack{0 \le x_1 \le \frac{B_{11}}{2} \\ 0 \le x_1 \le \frac{B_{21}}{2}}} \{2x_1 + f_2(B_{11} - 2x_1, B_{21} - 0)$ ...(2)

From equation (2), we have

$$f_2(B_{12}, B_{22}) = 5 \min\left\{B_{12}, \dfrac{B_{22}}{2}\right\}$$

So, $f_2(B_{11} - x_1, B_{21} - 0) = 5 \min\left\{B_{11} - x_1, \dfrac{B_{21}}{2}\right\}$

On putting this value in equation (2), we get

$$f_1(B_{11}, B_{21}) = \max_{\substack{0 \le x_1 \le \frac{B_{11}}{2} \\ 0 \le x_1 \le \frac{B_{21}}{2}}} \left\{2x_1 + 5 \min\left(B_{11} - 2x_1, \dfrac{B_{21}}{2}\right)\right\}$$

or $f_1(B_{11}, B_{21}) = \max_{\substack{0 \le x_1 \le \frac{B_{11}}{2} \\ 0 \le x_1 \le \frac{B_{21}}{2}}} \{2x_1 + 5\min(43 - 2x_1, 23)\}$

$$= \max_{0 \le x_1 \le 21.5} \{2x_1 + 5\min(43 - 2x_1, 23)\}$$

$$f_1(B_{11}, B_{21}) = \begin{cases} 2x_1 + 5(43 - 2x_1) & 10 \le x_1 \le 23 \\ 2x_1 + 5 \times 23 & 0 \le x_1 \le 10 \end{cases}$$

$$= \begin{cases} 215 - 8x_1 & 10 \le x_1 \le 23 \\ 2x_1 + 115 & 0 \le x_1 \le 10 \end{cases}$$

Now at $x_1 = 0$ the value will be maximum and is given by $= 215 - 80 = 135$

and $\qquad x_2^* = \min\left\{ B_{12}, \dfrac{B_{22}}{2} \right\}$

Now, $\qquad B_{12} = B_{11} - 2x_1, \qquad\qquad B_{22} = B_{21} - 0$

$\qquad\qquad B_{12} = 43 - 20 = 23, \qquad B_{22} = 46$

$\qquad\qquad x_2^* = \min\{23, 23\} = 23$

Hence, the optimal solution is $x_1 = 10, x_2 = 23$ and $\text{Max } z = 135$

EXAMPLE 2. **Solve the following L.P.P. by using dynamic programming techniques**

$$\text{Max } z = 3x_1 + 5x_2$$

subjected to $\qquad x_1 \le 4$

$$x_2 \le 6$$

$$3x_1 + 2x_2 \le 18$$

and $\qquad\qquad x_1, x_2 \ge 0$ $\qquad\qquad$ [MEERUT 1998 (P), 2005, AGRA 1998, 99, 2004]

SOLUTION. In this problem there are two interrelated decision variables, therefore it is a two stage problem. Due to the three resources b_1, b_2 and b_3 with available amount 4, 6 and 18, the state vector has three components.

Let (B_{1j}, B_{2j}, B_{3j}) are the state of the system and let us suppose that $f_j(B_{1j}, B_{2j}, B_{3j})$ be the optimal value of the objective function for stage $j = 1$ and 2. Using backward computational procedure, we have

$$f_2(B_{12}, B_{22}, B_{32}) = \max_{\substack{0 \le x_2 \le B_{22} \\ 0 \le 2x_2 \le B_{32}}} \{5x_2\}$$

$$= 5 \max_{\substack{0 \le x_2 \le B_{22} \\ 0 \le x_2 \le \frac{B_{32}}{2}}} \{x_2\}$$

Since, $\max\{x_2\}$ is minimum of B_{22} and $\dfrac{B_{32}}{2}$ which satisfies $0 \le x_3 \le B_{22}$ and

$0 \le x_2 \le \dfrac{B_{32}}{2}$

therefore, $\quad \max\{x_2\} = x_2^* = \min\left\{ B_{22}, \dfrac{B_{32}}{2} \right\}$

$\therefore \quad f_2(B_{12}, B_{22}, B_{32}) = 5\min\left[B_{22}, \dfrac{B_{32}}{2} \right]$ $\qquad\qquad$ ...(1)

We know that the value of B_{12}, B_{22}, B_{32} is given by at primary stage, Now

$$f_1(B_{11}, B_{21}, B_{31}) = \max_{\substack{0 \le x_1 \le B_{11} \\ 0 \le 3x_1 \le B_{31}}} \{3x_1 + f_2(B_{11} - x_1, B_{21}, B_{31} - 3x_1)\}$$

$$= \max_{\substack{0 \le x_1 \le 4 \\ 0 \le x_1 \le 6}} \{3x_1 + f_2(B_{11} - x_1, B_{21}, B_{31} - 3x_1)\} \qquad ...(2)$$

From equation (1), we know that

$$f_2(B_{12}, B_{22}, B_{32}) = 5\min\left\{ B_{21}, \dfrac{B_{31} - 3x_1}{2} \right\}$$

so, $f_2(B_{11} - x_1, B_{21}, B_{31} - 3x_1) = 5\min\left\{B_{21}, \dfrac{B_{31} - 3x_1}{2}\right\}$

On putting this value in equation (2), we get

$$f_1(B_{11}, B_{21}, B_{31}) = \max_{\substack{0 \le x_1 \le 4 \\ 0 \le x_1 \le 6}} \left\{3x_1 + 5\min\left(B_{21}, \dfrac{B_{31} - 3x_1}{2}\right)\right\}$$

$$= \max_{\substack{0 \le x_1 \le 4 \\ 0 \le x_1 \le 6}} \left\{3x_1 + 5\min\left(6, \dfrac{18 - 3x_1}{2}\right)\right\}$$

$$= \max_{0 \le x_1 \le 4} \left\{3x_1 + 5\min\left(6, \dfrac{18 - 3x_1}{2}\right)\right\} \qquad \ldots(3)$$

Since, $\displaystyle\min_{0 \le x_1 \le 4}\left(6, \dfrac{18 - 3x_1}{2}\right) = \begin{cases} 6 & \text{if} \quad 0 \le x_1 \le 2 \\ 9 - \dfrac{3x_1}{2} & \text{if} \quad 2 \le x_1 \le 4 \end{cases}$

Therefore, from (3), we have

$$f_1(B_{11}, B_{21}, B_{31}) = \max_{0 \le x_1 \le 4} \begin{cases} 3x_1 + 30 & \text{if} \quad 0 \le x_1 \le 2 \\ 45 - \dfrac{9x_1}{2} & \text{if} \quad 2 \le x_1 \le 4 \end{cases}$$

Since, $3x_1 + 30$ is maximum at $x_1 = 2$ satisfying the restriction $0 \le x_1 \le 2$

$\therefore \quad f_1(B_{11}, B_{21}, B_{31}) = 3 \times 2 + 30 = 36$ at $x_1^* = 2$

and $\quad x_2^* = \min\left(B_{22}, \dfrac{B_{32}}{2}\right) = \min(6, 6) = 6$

Since, $B_{22} = B_{21} - 0 = 6$, $\quad B_{32} = B_{31} - 3x_1 = 12$

Hence, the optimal solution is $x_1 = 10, x_2 = 23$ and Max $z = 36$

EXAMPLE 3. *Solve the L.P.P.*

$$\textbf{Max } \textbf{\textit{z}} = \textbf{3}\textbf{\textit{x}}_\textbf{1} + \textbf{\textit{x}}_\textbf{2}$$

s.t. $\quad \textbf{2}\textbf{\textit{x}}_\textbf{1} + \textbf{\textit{x}}_\textbf{2} \le \textbf{6}, \textbf{\textit{x}}_\textbf{1} \le \textbf{2} \textbf{ and } \textbf{\textit{x}}_\textbf{2} \le \textbf{4}, \textbf{\textit{x}}_\textbf{1}, \textbf{\textit{x}}_\textbf{2}, \textbf{\textit{x}}_\textbf{3} \ge \textbf{0}$

by dynamic programming technique. [MEERUT 1994, 99 (O)]

SOLUTION. In this problem there are two interrelated decision variables, therefore it is a two stage problem. Due to the three resources $b_1 = 6, b_2 = 2$ and $b_3 = 4$, the state vector has three components. Let (B_{1j}, B_{2j}, B_{3j}) be the state of the system and let us suppose that $f_j(B_{1j}, B_{2j}, B_{3j})$ be the optimal value of the objective function for stage $j = 1$ and 2. Using Backward computational procedure, we have

$$f_2(B_{12}, B_{22}, B_{32}) = \max_{\substack{0 \le x_2 \le B_{12} \\ 0 \le x_2 \le B_{32}}} \{x_2\}$$

since, $\max\{x_2\}$ is minimum of B_{12} and B_{32} which satisfies $0 \le x_2 \le B_{12}$ and $0 \le x_2 \le B_{32}$.

Therefore, $\quad \max\{x_2\} = x_2^* = \min\{B_{12}, B_{32}\}$

$\therefore \qquad f_2(B_{12}, B_{22}, B_{32}) = \min\{B_{12}, B_{32}\} \qquad \ldots(1)$

Now, $f_1(B_{11}, B_{21}, B_{31}) = \max_{\substack{0 \le 2x_1 \le B_{11} \\ 0 \le x_1 \le B_{21}}} \{3x_1 + f_2(B_{11} - 2x_1, B_{21} - x_1, B_{31} - 0)$

$$= \max_{\substack{0 \le x_1 \le 3 \\ 0 \le x_1 \le 2}} \{3x_1 + f_2(B_{11} - 2x_1, B_{21} - x_1, B_{31}) \qquad \dots(2)$$

From equation (1), we know that

$$f_2(B_{12}, B_{22}, B_{32}) \qquad = \min\{B_{12}, B_{32}\}$$

Therefore, $f_2(B_{11} - 2x_1, B_{21} - x_1, B_{31}) = \min\{B_{11} - 2x_1, B_{31}\}$

On putting this value in equation (2).

$$f_1(B_{11}, B_{21}, B_{31}) = \max_{\substack{0 \le x_1 \le 3 \\ 0 \le x_1 \le 2}} \{3x_1 + \min(B_{11} - 2x_1, B_{31})\}$$

$$= \max_{0 \le x_1 \le 2} \{3x_1 + \min(6 - 2x_1, 4\} \qquad \dots(3)$$

$$[\because \text{ if } x_1 \le 2 \text{ then it is obvious that } x_2 \le 3\,]$$

$$= \max \begin{cases} 3x_1 + 4 & 0 \le x_1 \le 1 \\ 3x_1 + (6 - 2x_1) & 1 \le x_1 \le 2 \end{cases}$$

Since, the value of $3x_1 + 4$ will be maximum at $x_1 = 1$ and the value will be 7, and the value of $6 + x_1$ will be maximum at $x_1 = 2$ and the value will be 8. So, 8 will be the maximum value of objective function.

Therefore, $\qquad x_1^* = 2$ and max z = 8,

and $\qquad\qquad x_2^* = \min\{B_{12}, B_{32}\} = \min\{B_{21} - 2x_1, B_{31}\}$

$$= \min\{6 - 4, 4\} = 2$$

Hence, the optimum solution is given by $x_1 = 2, x_2 = 2$ and Max $z = 8$.

EXAMPLE 4. **Solve the problem:**

$$\textbf{Max } \textbf{\textit{z}} = \textbf{\textit{x}}_1 + \textbf{9}\textbf{\textit{x}}_2$$

s.t. $2\textbf{\textit{x}}_1 + \textbf{\textit{x}}_2 \le 25,\ \textbf{\textit{x}}_2 \le 11,\ \textbf{\textit{x}}_1, \textbf{\textit{x}}_2 \ge 0$

SOLUTION. There are two decision variables x_1 and x_2. So it is a two stage problem. Due to the two resources b_1 and b_2 with available amount 25 and 11, the state vector has two components.

Let (B_{1j}, B_{2j}) be the state of the system and let us suppose that $f_j(B_{1j}, B_{2j})$ be the optimal value of objective function. Using backward computational procedure, we have, $f_2(B_{12}, B_{22}) = \max_{\substack{0 \le x_2 \le B_{12} \\ 0 \le x_2 \le B_{22}}} \{9x_1\} = 9 \max_{\substack{0 \le x_2 \le 25 \\ 0 \le x_2 \le 11}} \{x_1\}$

Since, max$\{x_2\}$ is minimum of $B_{12} = 25$ and $B_{22} = 11$ which satisfies $0 \le x_2 \le 25$ and $0 \le x_2 \le 11$, i.e., therefore max$\{x_2\} = x_2^* = \min\{B_{12}, B_{22}\}$

$\therefore \qquad f_2(B_{12}, B_{22}) = 9 \min\{B_{12}, B_{22}\} \qquad \dots(1)$

Now, $\qquad f_1(B_{11}, B_{21}) = \max_{0 \le x_1 \le \frac{B_{11}}{2}} \{x_1 + f_2(B_{11} - 2x_1, B_{21} - 0)\}$

$$= \max_{0 \le x_1 \le \frac{25}{2}} \{x_1 + 9 \min(B_{11} - 2x_1, B_{21})\}$$

$\therefore \qquad f_1(B_{11}, B_{21}) = \max_{0 \le x_1 \le \frac{25}{2}} \{x_1 + 9 \min(25 - 2x_1, 11)\} \qquad \dots(2)$

Since, $\min\limits_{0 \le x_1 \le \frac{25}{2}} (25 - 2x_1, 11) = \begin{cases} 11 & 0 \le x_1 \le 7 \\ 25 - 2x_1 & 7 \le x_1 \le 25/2 \end{cases}$

Therefore, $f_1(B_{11}, B_{21}) = \max \begin{cases} x_1 + 99 & 0 \le x_1 \le 7 \\ 225 - 17x_1 & 7 \le x_1 \le 25/2 \end{cases}$

The value of $x_1 + 99$ will be maximum at $x_1 = 7$ and the value will be 106.

Therefore, $x_1^* = 7$ and $f_1(B_{11}, B_{21}) = \max z = 106$, and $x_2^* = \min\{B_{12}, B_{22}\}$
$= \min\{11, 11\} = 11$ 　　　　　$[\because B_{12} = B_{11} - 2x_1 = 11, B_{22} = B_{21} - 0 = 11]$
Hence, the optimal solution is $x_1 = 7, x_2 = 11$ and Max $z = 106$

EXAMPLE 5. **Solve the L.P.P.**

$$\text{Min } z = -2x_1 - 18x_2$$

subject to $\quad 2x_1 + x_2 \le 25, x_2 \le 15$

and $\qquad\qquad x_1, x_2 \ge 0$

by dynamic programming technique. 　　　　　　　[MEERUT 1996 (BP)]

SOLUTION. Before solving this problem, we reduces the problem into maximization as follows:

$$\text{Max. } z' = 2x_1 + 18x_2$$
$$\text{s.to} \quad 2x_1 + x_2 \le 25, x_2 \le 15$$
$$x_1, x_2 \ge 0$$

There are two interrelated decision variables in this problem, therefore, it is a two stage problem. Due to the two resources b_1 and b_2 with available amount 25 and 15, the state vector has two components.

Let (B_{1j}, B_{2j}) are the state of the system and let us suppose that $f_j(B_{1j}, B_{2j})$ be the optimal value of the objective function for stage 1 and 2. Then using backward computational procedure, we have

$$f_2(B_{12}, B_{22}) = \max\limits_{\substack{0 \le x_2 \le B_{12} \\ 0 \le x_2 \le B_{22}}} \{18x_2\} = 18 \max\limits_{\substack{0 \le x_2 \le 25 \\ 0 \le x_2 \le 15}} \{x_2\}$$

Since, $\max\{x_2\}$ is minimum of B_{12} and B_{22} which satisfies the restriction

$0 \le x_2 \le B_{12}$ and $0 \le x_2 \le B_{22}$.

$\therefore \qquad \max\{x_2\} = x_2^* = \min\{B_{12}, B_{22}\}$

Therefore, $f_2(B_{12}, B_{22}) = \min\{B_{12}, B_{22}\}$ 　　　　　　　　　　...(1)

Now, $\quad f_1(B_{11}, B_{21}) = \max\limits_{0 \le 2x_1 \le 25} \{2x_1 + f_2(B_{11} - 2x_1, B_{21} - 0)\}$

$$= \max\limits_{0 \le x_1 \le \frac{25}{2}} \{2x_1 + f_2(B_{11} - 2x_1, B_{21})\}$$

On putting this value in equation (1)

Then, $\quad f_2(B_{12}, B_{22}) = 18\min\{B_{11} - 2x_1, B_{21}\} = 18\min\{25 - 2x_1, 15\}$

$$f_1(B_{11}, B_{21}) = \max\limits_{0 \le x_1 \le \frac{25}{2}} \{2x_1 + 18\min(25 - 2x_1, 15)\}$$

$$= \max \begin{cases} 2x_1 + 18(25 - 2x_1) & 5 \le x_1 \le 25/2 \\ 2x_1 + 270 & 0 \le x_1 \le 5 \end{cases}$$

$$= \max \begin{cases} 450 - 34x_1 & 5 \le x_1 \le 25/2 \\ 2x_1 + 270 & 0 \le x_1 \le 5 \end{cases}$$

Since, the value of $450 - 34x_1$ will be maximum at $x_1 = 25/2$ and the value will be 25 and the value of $2x_1 + 270$ will be maximum at $x_1 = 5$, the value will be 280.

So, the maximum value of objective function is 280.

Therefore, $x_1^* = 5$ and $\max z' = 280$

and $\qquad x_2^* = \min\{B_{12}, B_{22}\}, \qquad\qquad B_{12} = B_{11} - 2x_1 = 25 - 10 = 15$

$\qquad\qquad x_2^* = \min\{15, 15\} = 15 \qquad\qquad B_{22} = 15$

Hence, the optimal solution is $x_1 = 5$, $x_2 = 15$ and Min $z = -\text{Max } z' = -280$.

22.8 SOLUTION OF INVENTORY PROBLEM BY A DYNAMIC PROGRAMMING TECHNIQUE

If we consider the inventory models in which demand is exactly known but different in each period. We can obtain the solution of such kind of models by dynamic programming technique.

 Solved Examples

EXAMPLE 1. *If there are n machines, which can do two jobs. If x of n machines do the first job then they produce goods worth $\phi(x) = 3x$ and if y of n machines do the second job then they produce goods worth if $\psi(y) = 2.5y$. The machines are subjected to depriciation so that after doing the first job only $a(x) = \dfrac{x}{3}$ machines remain available and after doing second job $b(y) = \dfrac{2}{3}y$ machines remain available in the begining of the second year. The process is repeated with the remaining machines. Calculate the maximum return after 3 years and obtain the optimal policy in each year.* [MEERUT–1991(S)]

SOLUTION. In this problem, we denotes the years as periods. Suppose x_j are the no. of machines devoted to job 1 in j^{th} year. M_j are the total number of machines in the begining of jth year, and $f_n(M)$ is the maximum return when there are n years left with initial number of available machines is M.

Here, we proceed the following three steps.

Step 1: Let in the third year starting with M_3 machine, x_3 and y_3 are the machines assigned to the jobs, then we have

$$f_1(M_3) = \max_{x_3, y_3} \{3x_3 + 2.5y_3\}$$

s. t. $\qquad x_3 + y_3 \le M_3, \; x_3, y_3 \ge 0$

which is a linear programming problem. If we solve this problem graphically, we have the optimal policy given by

$$x_3^* = M_3, \, y_3^* = 0$$

and $\qquad f_1(M_3) = 3M_3 + 2.5 \times 0$

$$= 3M_3 \qquad\qquad\qquad ...(1)$$

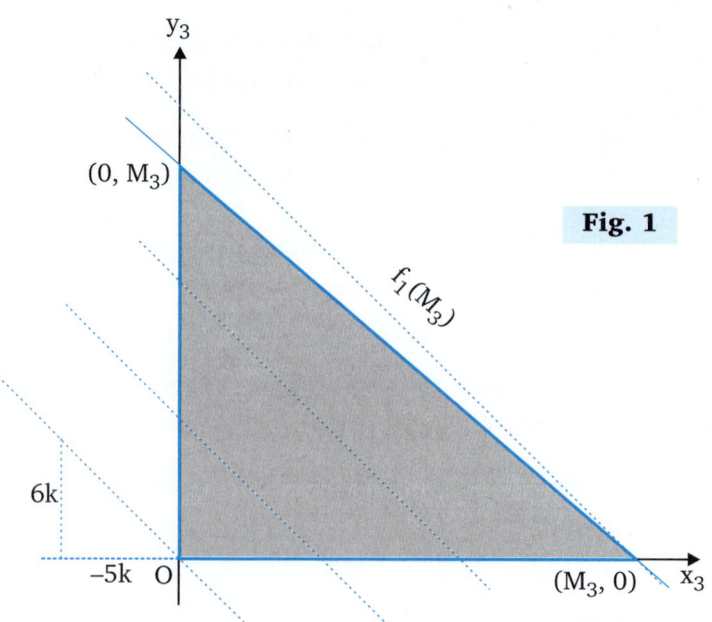

Fig. 1

Step 2: Let in the begining in the second year with machine M_2, x_2 and y_2 are the machines assigned to the jobs, then $\dfrac{x_2}{3} + \dfrac{2}{3}y_2 = M_3$ machines will remain available in the starting of the next year, then we have

$$f_2(M_2) = \max_{x_2, y_2} \left\{ 3x_2 + 2.5y_2 + f_1\left(\frac{x_2}{2} + \frac{2}{3}y_2\right) \right\}$$

s.t. $x_2 + y_2 \le M_2$ with $x_2, y_2 \ge 0$

We can also write

$f_2(M_2)$

$= \max_{x_2, y_2} \left\{ 3x_2 + 2.5y_2 \right.$

$\left. + 3\left(\dfrac{x_2}{3} + \dfrac{2}{3}y_2\right) \right\}$

$= \max_{x_2, y_2} \{ 4x_2 + 4.5y_2 \}$

which is also a linear programming problem, we can solve this problem graphically, then we have the optimal policy.

$x_2^* = 0,\ y_2^* = M_2$

and $f_2(M_2) = 4.5 M_2$

...(2)

Step 3: In this step, let in the begining of first year with machine $M_1; x_1$ and y_1 are the machines assigned to the jobs, then $\dfrac{x_1}{3} + \dfrac{2}{3}y_1 = M_2$ machines will remain available at the starting of the next year then

$$f_3(M_1) = \max_{x_1, y_1} \left\{ 3x_1 + 2.5y_1 + f_2\left(\dfrac{x_1}{3} + \dfrac{2}{3}y_1\right) \right\}$$

s.t. $\quad x_1 + y_1 \leq M_1, \; x_1, y_1 \geq 0$

We can write,

$$f_3(M_1) = \max_{x_1, y_1} \left\{ 3x_1 + 2.5y_1 + 4.5\left(\dfrac{x}{3} + \dfrac{2}{3}y_1\right) \right\} = \max_{x_1, y_1}(4.5x_1 + 5.5y_1)$$

s.t. $\quad x_1 + y_1 \leq M_1, \; x_1, y_1 \geq 0$

After solving graphically, we obtain the policy

$$x_1^* = 0, \; y_1^* = M_1 \text{ and } f_3(M_1) = 5.5M_1$$

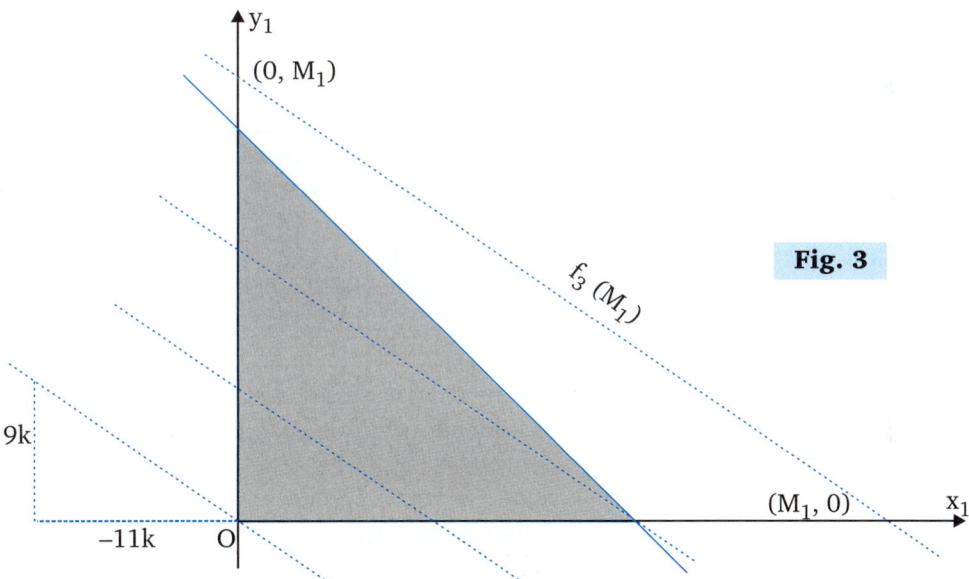

Fig. 3

But initially, we have $\qquad M_1 = n$

$$x_1^* = 0, \; y_1^* = M_1 = n, \qquad M_2 = \dfrac{x_1^*}{3} + \dfrac{2y_1^*}{3} = \dfrac{2n}{3}$$

$$x_2^* = 0, \; y_2^* = M_2 = \dfrac{2}{3}n, \qquad M_3 = \dfrac{x_2^*}{3} + \dfrac{2}{3}y_2^* = \dfrac{4}{9}n$$

$$x_3^* = M_3 = \dfrac{4n}{9}, \; y_3^* = 0$$

Hence, the optimal solution for three years is given by

Ist year: $\quad x_1^* = 0, \; y_1^* = n$

$\qquad\qquad M_1 = n$

IInd year: $\quad x_2^* = 0, \; y_2^* = \dfrac{2n}{3}$

$$M_2 = \frac{2n}{3}$$

IIIrd year: $x_3^* = \frac{4n}{9}, y_3^* = 0$

$$M_3 = \frac{4n}{9}$$

EXAMPLE 2. *A man engaged in buying and selling identical stems. He operates from a warehouse that can hold 500 items. Each month he can sell any quantity that he choose upto the stock at the begining of the month. Each month, he can buy as such as he wishes for delivery at the end of the month so long as his stock does not exceed 500 items. For the next four months he has the following error free forecasts of cost, sales price:*

Month	i	1	2	3	4
Cost	c_i	27	24	26	28
Sales price	p_i	28	25	25	27

If he currently has a stock of 200 units, what quantities should be sell and buy in the next four month? Obtain the solution using dynamic programming.   [MEERUT 1996, 99]

SOLUTION. Here, let x_i is the amount for sell during the month i, y_i is the amount to be ordered during the month i, b_i is the stock level in the begining of month i, p_i is the sale price in i^{th} month, c_i is the purchase price in the ith month and H is the warehouse capacity.

Let $f_n(b_n)$ be the maximum return when there are n months with initial stock b_n.

Using backward computational procedure, we will proceed the following steps:

Step 1: Let we are in the starting of fourth month with stock b_4 at the begining of this month. Therefore,

$$f_1(b_4) = \max_{x_4, y_4} \{p_4 x_4 - c_4 y_4\}$$

with the restrictions $0 \le x_4 \le b_4$, $y_4 \ge 0$ and $b_4 - x_4 + y_4 \le 0$

We can write $f_1(b_4) = \max_{\substack{0 \le x_4 \le b_4 \\ 0 \le y_4 \le x_4 - b_4}} \{27 - 28 y_4\}$

Since, the function $(27 x_4 - 28 y_4)$ will be maximum at $x_4 = b_4, y_4 = 0$

$\therefore$ $f_1(b_4) = 27 b_4 - 28 \times 0 = 27 b_4$

therefore, in this case, the optimal policy are $x_4^* = b_4, y_4^* = 0$ and $f_1(b_4) = 27 b_4$

Step 2: Let in the starting of third month with initial stock b_3 at the beginning of this month. Since stock level $b_4 = b_3 - x_3 + y_3$ will remain available at the begining of next month. So we have,

$$f_2(b_3) = \max_{x_3, y_3} \{p_3 x_3 - c_3 y_3 + f_1(b_3 - x_3 + y_3)\}$$

with the restrictions $0 \le x_3 \le b_3$, $y_3 \ge 0$ and $b_3 - x_3 + y_3 \le H(= 500)$

We can write $f_2(b_3) = \max_{\substack{0 \le x_3 \le b_3 \\ 0 \le y_3 \le 500 - b_3 + x_3}} \{25 x_3 - 26 y_3 + 27(b_3 - x_3 + y_3)\}$

$$= \max_{\substack{0 \le x_3 \le b_3 \\ 0 \le y_3 \le 500 - b_3 + x_3}} \{27b_3 - 2x_3 + y_3\}$$

$$= \max_{0 \le x_3 \le b_3} \{27b_3 + (500 - b_3 + x_3) - 2x_3\}$$

$$= \max_{0 \le x_3 \le b_3} \{26b_3 + 500 - x_3\}$$

which is maximum at $x_3 = 0$, therefore, the optimal policy are $x_3^* = 0$, $y_3^* = 500 - b_3 + x_3 = 500 - b_3$ and $f_2(b_3) = 26b_3 + 500$.

Step 3: Let we are in second month, three months are left with stock b_3 at the starting of this month. Since stock level $b_3 = b_2 - x_2 + y_2$ will remain available at the starting of the next month. So, we have

$$f_3(b_2) = \max_{x_2, y_2}\{p_2 x_2 - c_2 y_2 + f_2(b_2 - x_2 + y_2)\}$$

with the restrictions $0 \le x_2 \le b_2$, $y_2 \ge 0$ and $b_2 - x_2 + y_2 \le H(= 500)$

We can write $f_3(b_2) = \max_{\substack{0 \le x_2 \le b_2 \\ 0 \le y_2 \le 500 - b_2 + x_2}} \{25x_2 - 24y_2 + 26(b_2 - x_2 + y_2) + 500\}$

$$= \max_{\substack{0 \le x_2 \le b_2 \\ 0 \le y_2 \le 500 - b_2 + x_2}} \{26b_2 - x_2 + 2y_2 + 500\}$$

$$= \max_{0 \le x_2 \le b_2} \{24b_2 + x_2 + 1500\}$$

which is maximum at $x_2 = b_2$. Therefore, the optimal policy are $x_2^* = b_2$, $y_2^* = 500 - b_2 + x_2 = 500$ and $f_3(b_2) = 25b_2 + 1500$.

Step 4: Suppose, we are in the first month, four months are left with stock b_1 at the starting of this month. Since $b_2 = b_1 - x_1 + y_1$ will remain available at the starting of the next month. So, we have

$$f_4(b_1) = \max_{x_1, y_1}\{p_1 x_1 - c_1 y_1 + f_3(b_1 - x_1 + y_1)\}$$

with the restriction $0 \le x_1 \le b_1$, $y_1 \ge 0$ and $b_1 - x_1 + y_1 \le H(= 500)$

we can write $f_4(b_1) = \max_{\substack{0 \le x_1 \le b_1 \\ 0 \le y_1 \le 500 - b_1 + x_1}} \{28x_1 - 27y_1 + 25(b_1 - x_1 + y_1) + 1500\}$

$$= \max\{25b_1 + 3x_1 - 2y_1 + 1500\}$$

which is maximum at $x_1 = b_1, y_1 = 0$. Therefore, the optimal policy are $x_1^* = b_1$, $y_1^* = 0$ and $f_4(b_1) = 28b_1 + 1500$.

Since, at the starting of first month, stock is 200 units.

$\therefore \quad b_1 = 200$, so $\qquad x_1^* = 200, y_1^* = 0$

$b_2 = b_1 - x_1^* + y_1^* = 0$, $\qquad x_2^* = b_2 = 0, y_2^* = 500$

$b_3 = b_2 - x_2^* + y_2^* = 500$, $\qquad x_3^* = 0, y_3^* = 500, b_3 = 0$

$b_4 = b_3 - x_3^* + y_3^* = 500 - 0 + 0 = 500$, $\quad x_4^* = b_4 = 500, y_4^* = 0$

Hence, the optimal solution is

Month (i)	1	2	3	4
Purchase (y_i^*)	0	500	0	0
Sale (x_i) *	200	0	0	500

and the maximum return $f_4(b_1) = 28b_1 + 1500 = 28 \times 200 + 1500 = ₹\ 7100$

22.8 DIFFERENCE BETWEEN DYNAMIC AND LINEAR PROGRAMMING PROBLEM

Following table shows the basic difference between dynamic and linear programming problem

S. No.	Characteristics	Dynamic programming Problem	Linear Programming Problem
1.	Objective function	may be linear or non linear	must be linear
2.	Constraints	may be linear or non linear	must be linear
3.	Technological coefficients	must be positive	may be positive
4.	Solution Procedure	by breaking it into different stages	treated as single problem

Exercise-22.1

1. Develop the functional equations to determine $m_j (\geq 0)$ so as to Maximize $Z = \sum\limits_{i=1}^{n} m_i \left(\dfrac{p_i}{m_n} \right)^2$
   subjected to the constraints
   $$m_1 + m_2 + m_3 + \ldots + m_n = M$$

2. Solve the following problem by Dynamic programming technique.
   $$\text{Max } Z = 12x_1^2 + 27x_2^2 + 147x_3^2$$
   subject to $x_1 + x_2 + x_3 = 1, \; x_1, x_2, x_3 \geq 0$.

3. Solve the problem $\text{Max } Z = p_1 \cdot p_2 \cdot p_3 \cdots p_n$
   subjected to the constraints
   $$c_1 p_1 + c_2 p_2 + \ldots c_n p_n \leq x$$
   $$0 \leq p_i \leq 1 \text{ for } i = 1, 2, 3 \ldots n.$$

4. Using dynamic programming, show that $\sum\limits_{i=1}^{n} p_i \log p_i$ subjected to the constraints
   $\sum\limits_{i=1}^{n} p_i = 1$ is minimum, when
   $$p_1 = p_2 = p_3 = \ldots = p_n = \frac{1}{n}.$$
   [MEERUT–2000, 07; AGRA–1998]

5. Solve the problem $\text{Max } Z = x_1^2 + x_2^2 + x_3^2$ subjected to the constraints $x_1 x_2 x_3 \leq 4$, $x_1, x_2, x_3 \geq 0$, by dynamic programming technique.

6. Solve the linear programming problem
   $$\text{Max } Z = -x_1^2 - 2x_2^2 + 3x_2 + x_3$$
   subject to $x_1 + x_2 + x_3 \leq 1$ with $x_1, x_2, x_3 \geq 0$
   by dynamic programming technique.

7. Solve the following linear programming problem by Dynamic programming technique.
   $$\text{Max } Z = x_1^3 + x_2^3 + x_3^3$$
   subject to $x_1 + x_2 + x_3 \leq 6$ with $x_1, x_2, x_3 \geq 0$.

8. Using dynamic programming technique, solve

   $$\text{Min } Z = y_1^2 + y_2^2 + y_3^2$$
   subject to $y_1 + y_2 + y_3 \geq 10$ with $y_1, y_2, y_3 \geq 0$.

9. Let the function $f_n(a) = \min\limits_{R} \sum\limits_{i=1}^{n} x_i^p, \; p > 0$
   where R is defined by
   (i) $\sum\limits_{i=1}^{n} x_i \geq a, a > 0$ (ii) $x_i \geq 0 \; \forall i$
   (a) Show that $f_n(a)$ satisfies the recurence relation.
   $$f_n(a) = \min\limits_{x \leq a} \{ x^p + f_{n-1}(a - x) \}$$
   (b) If $0 < p < 1$ then show that $f_n(a) = a^p$
   (c) If $p > 1$ then show that $f_n(a) = n \left(\dfrac{a}{n} \right)^{p-1}$

10. Using dynamic programming technique, Maximize $Z = 8x_1 + 7x_2$
    subject to $2x_1 + x_2 \leq 8$
    $$5x_1 + 2x_2 \leq 15, \; x_1, x_2 \geq 0.$$

11. Solve the following L.P.P. by dynamic programming technique.
    $$\text{Max } Z = 2x_1 + 3x_2$$
    subject to $x_1 - x_2 \leq 1$
    $$x_1 + x_2 \leq 3, \; x_1, x_2 \geq 0$$

12. Solve the following by dynamic programming technique.
    (i) $\text{Max } Z = 3x_1 + 7x_2$
    subject to $x_1 + 4x_2 \leq 8$
    $$x_2 \leq 2$$
    $$x_1, x_2 \geq 0.$$
    (ii) $\text{Max } Z = 4x_1 + 3x_2$
    subject to $x_1 + 2x_2 \leq 4$

$$6x_1 + x_2 \le 6$$

$$x_1, x_2 \ge 0.$$

13. Solve the L.P.P. by dynamic programming technique

$$\text{Max } Z = 8y_1 + 7y_2$$

subject to $2y_1 + y_2 \le 8$, $5y_1 + 2y_2 \le 15$ with

$$y_1, y_2 \ge 0.$$

14. Use dynamic programming to solve

$$\text{Min } Z = 3y_1 + 5y_2$$

subject to $-3y_1 + 4y_2 \le 12$, $-2y_1 + y_2 \le 2$, $2y_1 + 3y_2 \ge 12$, $0 \le y_1 \le 4$, $y_2 \ge 2$.

15. Use dynamic programming to solve.

$$\text{Max } Z = 2x_1 + 5x_2$$

subject to $3x_1 + x_2 \le 2$, $x_2 \le 3$ and $x_1, x_2 \ge 0$.

16. Solve the following L.P.P. by dynamic programming technique.

(i) $\text{Min } Z = y_1^2 + y_2^2 + y_3^2 + ... + y_n^2$

subject to $y_1 \cdot y_2 \cdot y_3 ... y_n = c$,

$$y_1, y_2, y_3, y_4, ... y_n \ge 0.$$

[MEERUT, 2001 (BP), 02, 04]

(ii) $\text{Max } Z = x_1 \cdot x_2 \cdot x_3$

subject to $x_1 + x_2 + x_3 = 5$ with $x_1, x_2, x_3 \ge 0$.

[MEERUT-2002, 06; RAMPUR-2000; IAS-1998]

17. Solve the L.P.P.

$$\text{Max } Z = 5x_1 + 7x_2$$

subject to $x_1 + x_2 \le 4$

$$3x_1 + 8x_2 \le 24$$

$$10x_1 + 7x_2 \le 35 \text{ with } x_1, x_2 \ge 0$$

by dynamic programming technique.

[MEERUT 1994 (P)]

18. Solve the following L.P.P. by Dynamic programming technique.

$$\text{Max } Z = 4x_1 + 14x_2$$

subject to $\qquad 2x_1 + 7x_2 \le 21$

$$7x_1 + 2x_2 \le 21 \text{ and } x_1, x_2 \ge 0$$

19. A firm of manufacturers stock up every two months with certain basic material in order to carry out its production schedule. The purchase price p_n and the demand d_n, n = 1,2,...,6 are given for the next 6 bimonthly periods in the following table:

Period n	1	2	3	4	5	6
Demand d_n	8	5	3	2	7	4
Purchase price p_n	11	18	13	17	20	10

owing to limited storage space the stock must never exceed a certain value 5. The initial stock is 2 and the final stock must be nill. Use dynamic programming to ascertain the quantity to be bought at the begining of each period in such a way that the total cost will be minimum.

20. The total volume available in an air craft for 3 types of item is $13 ft^3$. The unit volume of item A is $2 ft^3$ that of item B is $3 ft^3$, and that of item C is $2 ft^3$. The cost of having a demand that occur when the system is out of stock is Rs. 600 for item A, Rs. 1200 for item B and Rs. 800 for item C. The demand for each item is Poisson distributed with mean being 5, 2 and 2 for item A, B and C respectively. How many of each item should be loaded in order to minimize to expected stock out cost?

Answers

5. $x_1 = 1, x_2 = 1, x_3 = 4$, Max. $z = 18$

6. $x_1 = 0$, $x_2 = \dfrac{1}{2}$, $x_3 = \dfrac{1}{2}$ and Max $Z = \dfrac{3}{2}$

10. $x_1 = 0$, $x_2 = \dfrac{15}{2}$ Max $Z = \dfrac{105}{2}$

11. $x_1 = 0$, $x_2 = 3$ Max $Z = 9$

12. (i) $x_1 = 8, x_2 = 0$, Max $Z = 24$

(ii) $x_1 = \dfrac{7}{10}, x_2 = \dfrac{17}{10}$ Max $Z = \dfrac{525}{10}$

13. $x_1 = 0$, $x_2 = \dfrac{75}{10}$, Max $Z = \dfrac{525}{10}$

15. $x_1 = x_2 = 3$, Max $Z = 21$

16. (i) $c^{1/n}, c^{1/n}, ..., c^{1/n}$, Min $Z = nc^{2/n}$

(ii) $x_1 = \dfrac{5}{3}, x_2 = \dfrac{5}{3}, x_3 = \dfrac{5}{3}$ and Max $Z = \dfrac{125}{27}$

17. $x_1 = \dfrac{8}{5}$, $x_2 = \dfrac{12}{5}$ and Max $Z = \dfrac{124}{5}$

18. $x_1 = 0, x_2 = 3$, Max $Z = 42$

19. $x_1 = 7$, $x_2 = 4$, $x_3 = 9$, $x_4 = 3$, $x_5 = 0$, $x_6 = 4$ and minimum purchase cost is ₹ 357.

REVIEW QUESTIONS

1. State Bellman's principle of optimality of dynamic programming.
2. Write a short note on dynamic programming.
3. Write a short note on the relation between linear programming and dynamic programming. [MEERUT 1979(S), 81, 83(P)]
4. Explain the recursive relation, using dynamic programming approach, when an N-stage objective function is to be maximized.
5. What are the essential characterstics of dynamic programming.

 ## MULTIPLE CHOICE QUESTIONS (CHOOSE THE MOST APPROPRIATE ONE)

1. The stages involved in solving an n-variable dynamic programming problems are:
   (a) $n + 1$ (b) $n + 2$
   (c) n (d) none of these
2. Bellman's principle of optimality is used to solve:
   (a) the dynamic programming problem
   (b) an assignment problem
   (c) transportation problem
   (d) none of these
3. Dynamic programming problem can be solved by using:
   (a) Hungarian algorithm
   (b) transportation algorithm
   (c) Bellmans's principle of optimality
   (d) None of these
4. A stage in a dynamic programming problem represents:
   (a) no. of decision alternatives
   (b) different time period in the planning period
   (c) both (a) and (b)
   (d) none of these
5. The return function in a dynamic programming model depends on
   (a) states   (b) stages
   (c) alternatives   (d) none of these

Answers

1. (c)   2. (a)   3. (c)   4. (b)   5. (a)

 ## ARCHIVE

1. An electronic device consists of four components each of which must function for the system to function. The system reliability can be improved by installing parallel units in one or more components. The reliability of components, R with one, two or three parallel units and the corresponding cost, C are given below. The maximum amount available for this device is 100. The problem is to determine the number of parallel units in each components.

No. of parallel units	Components							
	1		2		3		4	
	R	C	R	C	R	C	R	C
1	0.70	10	0.50	20	0.70	10	0.60	20
2	0.80	20	0.70	40	0.90	30	0.70	30
3	0.90	30	0.80	50	0.95	40	0.90	40

[PUNJAB(ME)–1987]

2. Use dynamic programming to show that $\sum_{i=1}^{n} p_i \log p_i$ subject to the constraints $\sum_{i=1}^{n} p_i = 1$, $p_i \geq 0 \ \forall \ i$ is minimum when $p_1 = p_2 = ... = p_n = 1/n$. [AMIE–2005]

3. Solve the following LPP by dynamic programming approach
   Max. $Z = 50x_1 + 100x_2$
   subject to the constraints
   $2x_1 + 3x_2 \leq 48$; $x_1 + 3x_2 \leq 42$
   $x_1 + x_2 \leq 21$ and $x_1, x_2 \geq 0$
   [SAMBHALPUR–1986]

4. In which area of organisation can dynamic programming be applied successfully?
   [DELHI(MBA)–2003]

5. Explain the recursive nature of computation in dynamic programming. [AMIE–2004]

6. State Bellman's principle of optimality.
   [AMIE–2005]

Hints and Answers

1. For maximum reliability, the device must have 1, 2, 1 and 3 units in components 1, 2, 3 and 4 respectively to attain a maximum reliability of 0.308 or 30.8 percent.

2. $f_n(1) = n \left[\dfrac{1}{n} \log \left(\dfrac{1}{n} \right) \right]$

3. $x_1 = 6, x_2 = 12$ and Max. $Z = 60$

Information Theory | 23

23.1 INTRODUCTION

Information is one of the important phenomenon in daily life. This is a basis of our various course of action and behaviour. There are numerous means for the transmission of information. It usually transmitted by human voice (as in telephone, T.V., radio, etc.) by means of letters, newspapers, books, etc.

Information theory is a new branch of probability theory. It was originated by scientisits, while studying the statistical structure of electronic communication equipments. Mathematical theory of communication was principally initiated by Claude Shannon in 1984.

Information theory can be divided into three main categories, i.e., Shannon theory, coding theory and cybematics. In Shannon theory problems and their solutions related with communication system are studied, Encoding is a method of translating information in such a way that the error of transformation of some desired information from one source to another is minimized. Cybematics deals with the communication problem enumerated in living beings and social systems.

23.2 COMMUNICATION PROCESS

The procedure by which one mind affects another is called a communication process. This may be by any means by which information is conveyed from a transmitter to receiver.

The various components of the process through which some information is transmitted from a given source to the receiver, form a communication system.

There will be following three essential parts of a simplest communication system.

(i) **Transmitter source:** It is a device that selects and transmits a sequence of symbols or it is a device which selects message at random with prescribed probability. Hence, a source of transmitted information corrsponds with the space of a random experiments.

☛ REMARK

- If the alphabets in the message are stochastically independent, then it is a zero memory source.

(ii) **Communication channel:** It is the medium through which some information is communicated. e.g., radio, T.V., newspaper, telephone, human voice, etc.

It can be classified as follows:

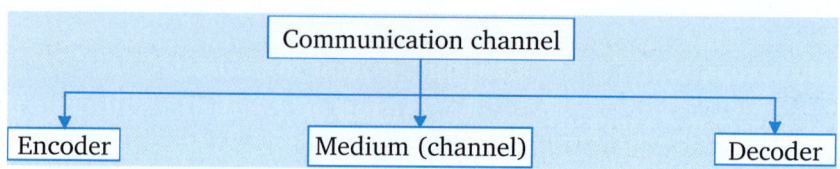

(a) **Encoder:** It is a device or procedure of translating the original information into media languages. It is used to imporove the efficiency of the medium through which the message is transformed. It acts as a sort of setup transformer.

(b) **Decoder:** Its operation is just the opposite of encoder. It is the device used to transform the encoded message in the original form which is acceptable to the receiver. It acts as a sort of step-down transformer.

(c) **Channel or mdium:** It is the medium through which coded message actually transmitted. The input to the chnnel is the encoded signal and the output is the received signal by the decoder.

(iii) **Receiver:** Receiver is the destination to which the message is transmitted. The efficiency of a communication system depends upon the quantity or quality of the information communicated by it.

SOME OTHER DEFINITIONS

(i) **Noisy or Noise:** If there are some disturbances in the communication channel then it results in distrotion of original information and the system is said to be Noisy.

For example, the disturbances in broadcasting, errors in newspaper printing etc.

(ii) **Noiseless:** A communication channel is said to be noiseless if it is not noisy.

(iii) **Codes:** The information through the channel can not travel in the same form, but each media/channel has its own language or form of transmission is known as codes.

(iv) **Computer programming:** When some information is to be processed in a computer then it is to be translated into computer language is known as computer programming.

23.3 FUNDAMENTAL THEOREM OF INFORMATION THEORY

The fundamental theorem of information theory can be stated as follows:

"It is possible to transmitt information through a noisy channel at any rate less than the channel capacity with an arbitrarily small probability of errors."

23.4 STATISTICAL NATURE OF COMMUNICATION SYSTEM

A communication channel is characterized by its input alphabets and output alphabets with a set of conditional probability that a particular output signal 'O' corresponds to a particular input signal 'I' and is written as $P(O|I)$. The probability associated with each input/output alphbets of a message can be represented by a matrix (called channel matrix).

23.4.1 CHANNEL MATRIX

If $a_1, a_2, ..., a_m$ are the inputs and $b_1, b_2, ..., b_n$ are corresponding outputs. Then if $p(b_j|a_i) = p(j|i)$ denotes the probability of getting output b_j which channel input is a_i then a channel matrix can be defined as follows:

$$\text{Output} \rightarrow$$

$$\underset{\downarrow \text{ Input}}{\begin{array}{c} a_1 \\ a_2 \\ a_3 \\ \vdots \\ a_m \end{array}} \begin{array}{cccccc} b_1 & b_2 & \cdots & b_j & \cdots & b_n \\ \begin{bmatrix} P_{1|1} & P_{2|1} & \cdots & P_{j|1} & \cdots & P_{n|1} \\ P_{1|2} & P_{2|2} & \cdots & P_{j|2} & \cdots & P_{n|2} \\ P_{1|3} & P_{2|3} & \cdots & P_{j|3} & \cdots & P_{n|3} \\ \vdots & & & & & \\ P_{1|m} & P_{2|m} & \cdots & P_{j|m} & \cdots & P_{n|m} \end{bmatrix} \end{array}$$

☛ REMARKS

- The sum of each row $\left[\sum_{j=1}^{n} P_{j|i} = 1, i = 1, 2, \ldots m \right]$ should be equal to unity.

- Each row of the channel matrix corresponds to an input of the channel and each column corresponds to a channel output.

23.4.2 MEMORYLESS CHANNEL

A memoryless channel is described by an input alphabets $A = \{x_1, x_2, \ldots, x_m\}$ and output alphabet $B = \{y_1, y_2, \ldots, y_n\}$ and a set of condiotional probability $P(y_j|x_i)$ for all i and j where $P(y_j|x_i)$ is the probability that the output symbols y_j will be reduced in the input symbols x_i in the set.

23.4.3 BINARY SYMMETRIC CHANNEL

A binary symmetric channel has just input symbols 0 and 1 with two outputs 0 and 1 and it is symmetric in the sense that

$$P(y_1|x_1) = P(y_2|x_2) = \bar{p}$$
$$P(y_1|x_2) = P(y_2|x_1) = p$$

where $\bar{p} = 1 - p$, p being the probability of error in transmission.

☛ REMARK

- A channel described by a channel matrix with one and only one non-zero element in each column as a noiseless channel, e.g. in a binary symmetric channel, the channel matrix

$$\begin{array}{c} b_j \\ \begin{array}{cc} 0 & 1 \end{array} \\ a_i \begin{array}{c} 0 \\ 1 \end{array} \begin{bmatrix} 0 & 0 \\ 1 & 1 \end{bmatrix} \end{array} \text{ is a noiseless channel.}$$

Similarly if the information passed through a communication channel, bears its effect on the future communication then it is said to be a **channel with memory** otherwise it is said to be a **memoryless channel**.

> The behaviour of the communication channel is always unpredeictable due to which the whole system become some sort of a random experiment, *i.e.*, whenever some message is communicated through a channel repeatedly then some probability can be attached to the quantity of information received by the receiver of each ocassion.

23.5 MATHEMATICAL DEFINITION OF INFORMATION

Let there be an event E which has occured. Then the information received about the event E will be given by

$$I(E) = \log \left[\frac{\text{Prob. concerning the receiver after receiving the information}}{\text{Prob. concerning the receiver before receiving the information}} \right]$$

If the channel of communication is noiseless, then the probability concerning the receiver after receiving the information will be 1, because there is no distortion of information during the process and then

$$I(E) = -\log[\text{Prob. concerning the receiver before receiving the information}]$$

For example, in the tossing of a coin, there are only two events namely Head and tail, each with probability $\frac{1}{2}$. When we toss a coin, the outcome will be either head or tail. Thus, the message that is head will have probability, concerning the receiver after receiving the message, equal to unity, i.e., it will be a noiseless process and the probability concerning the receiver before receiving the message will be equal to $\frac{1}{2}$. Hence, the information in this message is given by

$$I(E) = -\log\left(\frac{1}{2}\right)$$

23.6 AXIOMATIC APPROACH TO INFORMATION

Following are the axioms regarding the amount of information send in a message with probability p.

(i) The information depends only on the value of p lying between 0 and 1 and hence can be represented by some function of p say $g(p)$.

(ii) The message of information is a decreasing function of p. It decreases from ∞ to 0.

(iii) No information is transmitted for any event which is certain to happen, i.e., when $p = 1$, $g(p) = 0$.

(iv) If there are two independent events with corresponding probabilities p_1, p_2 then the information associated with their combined occurance is equal to the sum of the information related with their individual occurances, i.e.,

$$g(p_1 p_2) = g(p_1) + g(p_2) = g(p_2 p_1)$$

23.7 MEASURES OF INFORMATION

An amount of an information is virtually a search for statistical parameters associated with a probability scheme.

To find the formula for the amount of information, let us suppose there are n distinct models

$$\{m_1, m_2, ..., m_n\}$$

The desired amount of information $I(m_k)$ associated with the selection of a particular model m_k must be a function of the probability of choosing m_k, i.e.,

$$I(m_k) = f[p(m_k)] \qquad\qquad ...(1)$$

For simplicity, if each one of these models $m_1, m_2, ..., m_k$ is selected with an equal probabilty, then

$$p(m_1) = p(m_2) = ... = p(m_n) = \frac{1}{n}$$

$$\Rightarrow \qquad\qquad I(m_k) = f\left(\frac{1}{n}\right) \qquad\qquad \text{(From (1))} \qquad\qquad ...(2)$$

$\Rightarrow$ Amount of information is the function of n.

Further, assume that each of the models $m_1, m_2, ..., m_n$ may be ordered in one of the m distinct colours $C_1, C_2, ..., C_m$. If selection of colours is also assumed to have equal probabilities, then the amount of information $I\{C_i\}$ associated with the selection of a colour C_i among all colours is given by

$$I(C_i) = f(p(C_i)) = f\left(\frac{1}{m}\right) \qquad ...(3)$$

Now, assume that the selection of a machine is done in the following two ways:

(i) First select the machine and then the colour, the two selections being independent of each other. Therefore,

$$I(m_k \text{ and } C_i) = I(m_k) + I(C_i) = f\left(\frac{1}{n}\right) + f\left(\frac{1}{m}\right) \qquad ...(4)$$

(ii) Select the machine and its colour simultaneously at one time as one selection out of mn number of possible selection with equal probability $\frac{1}{mn}$.

So, $$I(m_k \text{ and } C_i) = f\left(\frac{1}{mn}\right) \qquad ...(5)$$

Using (4) and (5) we get

$$f\left(\frac{1}{m}\right) + f\left(\frac{1}{n}\right) = f\left(\frac{1}{mn}\right) \qquad ...(6)$$

which is a functional equation.

The functional equation (6) has the solution given by

$$f(x) = \log\left(\frac{1}{x}\right) = -\log x \qquad ...(7)$$

VERIFICATION

Using (7) we get

$$f\left(\frac{1}{n}\right) = \log n, f\left(\frac{1}{m}\right) = \log m, f\left(\frac{1}{mn}\right) = \log mn$$

Using these values in (6), we get

$\log m + \log n = \log mn$, which is always true.

Solved Examples

EXAMPLE 1. *A man is informed that when a pair of dice was rolled, the result is 7. How much inrformation is there in the message?*

SOLUTION. Here, each dice contains the no. from 1 to 6. The sum of the seven in rolling a pair of dice can be obtained in the following ways:

No. on dice I	1	2	3	4	5	6
No. on dice II	6	5	4	3	2	1

$\Rightarrow$ Favourable no. of ways of getting a sum $7 = 6$

Total no. of ways in which we can get sum $= 36$

$\therefore$ The probability concerning the receiver of getting a sum of 7 before receiving the messge $= \dfrac{6}{36} = \dfrac{1}{6}$

Hence, information in this message

$$= -\log\left(\frac{1}{6}\right) = \log 6 = \log(2)^{2.58}$$

$$= 2.58 \log 2 = 2.58 \text{ binits}$$

☞ **REMARK**

- When logarithm is taken to base 10 then unit of information is **decit** and if base is 2, the unit is **binit**.

EXAMPLE 2. *In a certain community 25% of all girls are blondes and 75% of all blondes have blue eyes. Also, 50% of all girls in the community have blue eyes. If you know that a girl has blue eyes, how much additional information do you get by being informed that she is blonde?*

SOLUTION. The probability that a girl is blonde, $(p_1) = 0.25$
The probability that a blonde has blue eyes, $(p_2) = 0.75$
The probability that a girl has blue eyes, $(p_3) = 0.50$
Then the probability that a girl is blonde and has blue eyes

$$= p_1 \times p_2$$
$$= p_3 \times \text{prob. that a blue eyed girl is blonde}$$

Hence, the probability that a blue eyed girl is a blonde is given by

$$p_x = \frac{\text{Prob. that a blonde has blue eyes}}{\text{Prob. that girl has blue eyes}} = \frac{p_1 \times p_2}{p_3}$$

Hence the additional information, we get in this case is given by

$$\log\left(\frac{1}{p_x}\right) = \log\left(\frac{p_3}{p_1 \times p_2}\right)$$

$$= \log p_3 - \log p_1 - \log p_2$$
$$= \log 0.50 - \log 0.25 - \log 0.75$$
$$= 1.42 \text{ binits}$$

EXAMPLE 3. *An alphabet consists of 8 consonants and 8 vowels. Suppose all letters of the alphabet are equally probable and there is no inter-symbol influence. If consonants are always understood correctly, but vowels are understood correctly only half of the time mistaken for other vowels, the other half of the time all vowels being involved in errors the same percentage of the time. What is the average rate of information transmission.*

SOLUTION. As per given, 50% of the letters received will be correct consonants, 25% will be the correct vowels and 25% of incorrect vowels. Now to collect the amount of information received we have the following cases:

Case-1. Correct consonants
We have

$$p_A = \text{probability before reception} = \frac{1}{16}$$
$$p_B = \text{probability after reception} = 1$$

Therefore, information/letter $= \log\left(\dfrac{1}{1/16}\right) = \log 16 = 4$ bits/letter

Case-2. Correct vowels

We have $\qquad p_A = \dfrac{1}{16}, p_B = \dfrac{1}{2}$

$\therefore \quad$ Information/letter $= \log\left(\dfrac{1/2}{1/16}\right) = \log 8 = \log 2^3$

$$= 3\log_2 2 = 3\,\text{bits/letter}$$

Case-3. Incorrect vowels

Here, we have $p_A = \dfrac{1}{16}$

and to find the probability, after reception, let E be sent and U received. We have to find the probability that an event E was sent as a consequence of receiving a U. After receiving a U, a vowel was sent and the probability of U is $\dfrac{1}{2}$ the other half of the probability is divided equally among other 7 vowels one of which is E.

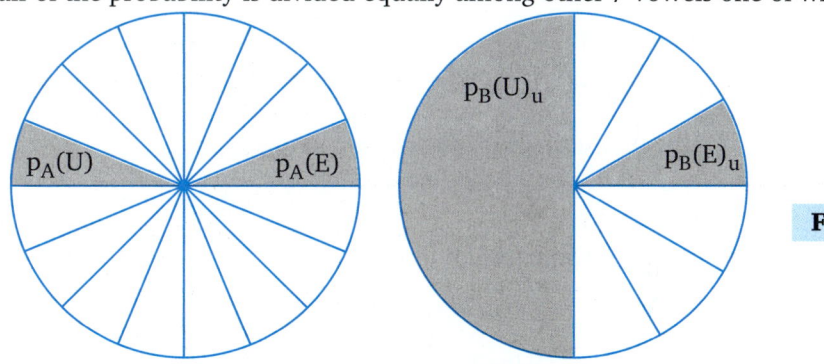

Probability before reception Probability after reception **Fig. 1**

Thus, the probability that E was sent $= \dfrac{1}{7} \times \dfrac{1}{2} = \dfrac{1}{14}$

So, $p_B = p_B(E)_u =$ probability after reception $= \dfrac{1}{14}$

Therefore, information/letter $= \log\left(\dfrac{1/14}{1/16}\right) = \log 16 - \log 14$

$$= \log 2^4 - 3 \cdot 8 = 4 - 3 \cdot 8 = 0 \cdot 2$$

So, for the average information per symbol take 50 % of case I and 25% of each of case 2 and 3. Hence, average information/symbol

$$= 0.5 \times 4 + 0.25 \times 3 + 0.35 \times 0.2$$

$$= 2.8 \text{ bits per symbol}$$

23.8 MEASURE OF UNCERTAINITY : ENTROPY

Let us suppose information-I is transmitted through some communication system then the amount of information reaching the receiver through communication channel will be some sort of random variable, *i.e.*, if the same information-I is communicated through the system n times then the amount of information reaching the receiver each time will be $m_1, m_2, ..., m_n$ with corresponding probabilitic $p(m_1), p(m_2), ..., p(m_n)$ respectively. In any particular trial, the amount of information received by the receiver can be one of the members $m_1, m_2, ..., m_n$. So, these all are mutually exclusive and information-I will constitute the whose sample space.

Therefore, we have

$$I = [m_1, m_2, ..., m_n], X = [p(m_1), p(m_2), ..., p(m_n)]$$

and

$$\sum_{i=1}^{n} p(m_i) = 1$$

The amount of information associated with the transmission of message m_k is defined by

$$I_k = -\log p(m_k) \qquad \qquad ...(1)$$

Here, I_k is called the amount of self information of the message m_k.

Now the average information per message for the source is given by

$$I = \text{Average of } I_k = -\sum_{k=1}^{n} P(m_k) \log(P(m_k)) \qquad \qquad ...(2)$$

The average information per message-I is called entropy or the communication entropy of the source. It is denoted by H.

If the same information I is communicated through a communication system n time then let $I_1, I_2, ..., I_n$ be the amounts of information reaching the receiver each time with probabilities $p_1, p_2, ..., p_n$ respectively.

Then the entropy of the source can be written as

$$H(p_1, p_2, ..., p_n) = -\sum_{i=1}^{n} p_i \log p_i$$

23.8.1 PROPERTIES OF ENTROPY FUNCTION (H)

The entropy function $H(p_1, p_2, ..., p_n)$ is given by

$$H(p_1, p_2, ..., p_n) = -\sum_{i=1}^{n} p_i \log p_i \qquad \qquad ...(1)$$

having the following properties.

(i) Continuity: *The entropy function should be continuous, i.e., $H(p_1, p_2, ..., p_n)$ should be continuous in $p_i : i = 1, 2, ..., n$. Slight changes in the probabilities of any event will not bring any significant change in the value of the function.*

PROOF. By definition of entropy function, we have

$$-H(p_1, p_2, ..., p_n)$$

$$= \sum_{i=1}^{n} p_i \log p_i$$

$$= p_1 \log p_1 + p_2 \log p_2 + ... + p_n \log p_n$$

$$= p_1 \log p_1 + p_2 \log p_2 + ... + (1 - p_1 - p_2 - ... p_{n-1}) \cdot (\log(1 - p_1 - p_2 - ... - p_{n-1}))$$

$$(\because \; \Sigma p_n = 1 \Rightarrow p_n = 1 - (p_1 + p_2 + ... + p_{n-1}))$$

Now, since $p_1, p_2, ..., p_{n-1}$ are all independent and $1 - p_1 - p_2, ..., -p_{n-1}$ being a function of $p_1, p_2, ..., p_{n-1}$ will also be continuous. Therefore, $\log(1 - p_1 - ... - p_{n-1})$ is continuous and hence, entropy function H is continuous.

(ii) **Symmetry:** *Entropy function should remain unchanged when* $p_1, p_2, \ldots, p_n$ *are interchanged with one another, i.e.,*

$$H(p_1, p_2, \ldots, p_n) = H(p_2, p_1, \ldots, p_n)$$

(iii) **External maximum property:** *Entropy function has a maximum value when all p_i's are equal.*

PROOF. By definition of entropy, we have

$$H(p_1, p_2, \ldots, p_n) = -\sum_{i=1}^{n} p_i \log p_i \text{ where } \Sigma p_i = 1$$

Also, $p_1, p_2, \ldots, p_{n-1}$ are independent and $p_n = 1 - p_1 - p_2 - \ldots - p_{n-1}$ depends on $p_1, p_2, \ldots, p_{n-1}$. Then

$$\frac{dH}{dp_i} = \frac{\partial H}{\partial p_1} \cdot \frac{\partial p_1}{\partial p_i} + \frac{\partial H}{\partial p_2} \cdot \frac{\partial p_2}{\partial p_i} + \ldots + \frac{\partial H}{\partial p_n} \cdot \frac{\partial p_n}{\partial p_i}$$

$$= \frac{d}{dp_i}(-p_i \log p_i) + \frac{d}{dp_n}(-p_n \log p_n)\frac{\partial p_n}{\partial p_i}, \quad i = 1, 2, \ldots, (n-1)$$

$$= \left(-\log p_i - \frac{p_i}{p_i}\log_2 e\right) - \left(\log p_n + \frac{p_n}{p_n}\log_2 e\right)\frac{\partial p_n}{\partial p_i}$$

$$\left[\because \frac{\partial}{\partial p_i}\log p_i = \frac{1}{p_i}\log_2 e\right]$$

$$= (-\log p_i - \log_2 e) - (\log p_n + \log_2 e)(-1)$$

$$= (\log p_n - \log p_i) = -\log\frac{p_i}{p_n}$$

Now, for the maxima of H, we must have

$$\frac{dH}{dp_i} = 0 \implies -\log\frac{p_i}{p_n} = 0, \quad i = 1, 2, \ldots, n-1$$

which is true for $p_i = p_n$

Hence, for H to be minimum we have

$$p_1 = p_2 = \ldots = p_n = \frac{1}{n}$$

$\implies$ Entropy function H is maximum when all the events are equally likely to happen.

(iv) **Additive property:**

$$H(p_1, p_2, \ldots, p_{n-1}; q_1, q_2, \ldots, q_m) = H(p_1, p_2, \ldots, p_{n-1}) + p_n \cdot H\left(\frac{q_1}{p_n}, \frac{q_2}{p_n}, \ldots, \frac{q_m}{p_n}\right)$$

where $p_n = \sum_{k=1}^{m} q_k$

PROOF. Let us assume that the event E_n with probability p_n is deivided into disjoint subsets $F_1, F_2, \ldots, F_m$ with respective probabilities $q_1, q_2, \ldots, q_m$. So,

$$p_n = q_1 + q_2 + \ldots + q_m \implies p_n = \sum_{k=1}^{m} q_k \qquad \ldots(1)$$

Now, LHS $= H(p_1, p_2, \ldots, p_{n-1}; q_1, q_2, \ldots, q_m)$

$$= -\sum_{k=1}^{n-1} p_x \log p_k - \sum_{k=1}^{m} q_k \log q_k = -\left[\sum_{k=1}^{n} p_k \log p_k - p_n \log p_n\right] - \sum_{k=1}^{m} q_k \log q_k$$

$$= H(p_1, p_2, \ldots, p_n) + \left[p_n \log p_n - \sum_{k=1}^{m} q_k \log q_k \right]$$

But $p_n \log p_n - \sum_{k=1}^{m} q_k \log q_k$

$$= p_n \left[\frac{p_n}{p_n} \log p_n \right] - p_n \sum_{k=1}^{m} \left[\frac{q_k}{p_n} \log q_k \right]$$

$$= p_n \sum_{k=1}^{m} \frac{q_k}{p_n} \log p_n - p_n \sum_{k=1}^{m} \frac{q_k}{p_n} \log q_k \qquad (\because p_n = \Sigma q_k) \text{ (From (1))}$$

$$= -p_n \sum_{k=1}^{m} \frac{q_k}{p_n} (\log q_k - \log p_n) = -p_n \sum_{k=1}^{m} \frac{q_k}{p_n} \log \frac{q_k}{p_n}$$

So, $\qquad$ LHS $= H(p_1, p_2, \ldots, p_n) + \left[-p_n \sum_{k=1}^{m} \frac{q_k}{p_n} \log \frac{q_k}{p_n} \right]$

$$= H(p_1, p_2, \ldots, p_n) + p_n \cdot H\left(\frac{q_1}{p_n}, \frac{q_2}{p_n}, \frac{q_3}{p_n} \ldots, \frac{q_m}{p_n} \right)$$

$$= \text{R.H.S.}$$

(v) Monotonicity: $H\left(\dfrac{1}{p}, \dfrac{1}{p}, \ldots, \dfrac{1}{p} \right) = f(p)$ *is a monotonically increasing function of p,*
i.e., for $p < q \Rightarrow f(p) < f(q)$.

(vi) *Let* $p_1, p_2, \ldots, p_m$ *and* $q_1, q_2, \ldots, q_m$ *be arbitrary non-negative numbers with* $\Sigma p_i = \Sigma q_i$.
Then

$$-\sum_{i=1}^{m} p_i \log p_i \leq -\sum_{i=1}^{m} p_i \log q_i$$

with equality if and only if $p_i = q_i \; \forall \; i$.

PROOF. It is a well known fact that logarithm is a convex function and we have the inequality
$\log x \leq x - 1$ with equality if and only if $x = 1$.

If we take $x = \dfrac{q_i}{p_i}$ then

$$\log\left(\frac{q_i}{p_i} \right) \leq \left(\frac{q_i}{p_i} \right) - 1 \qquad \ldots(1)$$

with equality if and only if $q_i = p_i$.
Now multiplying (1) by p_i and summing over i, we get

$$\sum_{i=1}^{m} p_i \log\left(\frac{q_i}{p_i} \right) \leq \sum_{i=1}^{m} (q_i - p_i) = 1 - 1 = 0$$

with equality if and only if $q_i = p_i \; \forall \; i$.

which prove that $\displaystyle\sum_{i=1}^{m} p_i \log q_i \leq \sum_{i=1}^{m} p_i \log p_i$

$$\Rightarrow \qquad -\sum_{i=1}^{m} p_i \log p_i \leq -\sum_{i=1}^{m} p_i \log q_i$$

with equality if and only if $p_i = q_i \; \forall \; i$

23.9 RELATIONS BETWEEN ENTROPIES

SYMBOLS

$H(X)$ = Average information per character at the source or the entropy of the source.

$H(Y)$ = Average information per character at the destination or the entropy of the receiver.

$H(X, Y)$ = Average information per pairs of transmitted and received characters or the average uncertainty (entropy of the communication system as a whole).

$H(Y|X)$ = Conditional entropy of the receiver or the measure of information about the receiving part where it is known that X is transmitted.

$H(X|Y)$ = Conditional entropy of the source or the measure of information about the source, where it is known that Y is received.

23.9.1 RELATION BETWEEN JOINT AND MARGINAL ENTROPIES

Let X and Y be two sets of messages such that

$$X = \{x_1, x_2, ..., x_m\} \text{ and } Y = \{y_1, y_2, ..., y_n\}$$

where x_i's are the channel input (message sent) and $y_j's$ are the channel output (message received)

If $p_{ij} = P(X = x_i, Y = y_j): i = 1, 2, ..., m; j = 1, 2, ..., n$ denote the probability of the joint event that message x_i is sent and message y_j is received. Now, we may define the marginal probability distribution of X and Y by

$$p_{i0} = \sum_{j=1}^{n} p_{ij} \text{ and } p_{0j} = p_{ij} \text{ for all } i, j$$

Then we have the following entropies :

(i) $H(X) = -\sum_{i=1}^{m} p_0 \log p_{i0}$

(ii) $H(Y) = -\sum_{j=1}^{n} p_{0j} \log p_{0j}$

(iii) $H(X,Y) = -\sum_{i=1}^{m} \sum_{j=1}^{n} p_{ij} \log p_{ij}$

(iv) $H(X|Y) = -\sum_{r=1}^{m} \sum_{k=1}^{n} p(x_k, y_r) \log p\{x_k | y_k\}$

(v) $H(Y|X) = -\sum_{k=1}^{n} \sum_{r=1}^{m} p(x_k, y_r) \log p\{y_r | x_k\}$

where $p(x_k | y_k) = \dfrac{p(x_k, y_k)}{p(y_k)}$

☛ REMARK

• max $H(X|Y)$ = max $H(X)$ + max $H(Y)$

THEOREM 1. $H(X,Y) \leq H(X) + H(Y)$ with equality if and only if X and Y are independent.

PROOF. We can write

$$H(X) + H(Y) = -\sum_{i=1}^{m} p_{i0} \log p_{i0} - \sum_{j=1}^{n} p_{0j} \log p_{0j}$$

$$= -\sum_{i=1}^{m}\left(\sum_{j=1}^{n}p_{ij}\right)\log p_{i0} - \sum_{j=1}^{n}\left(\sum_{i=1}^{m}p_{ij}\right)\log p_{0j}$$

$$= -\sum_{i=1}^{m}\sum_{j=1}^{n}p_{ij}\log(p_{i0}\cdot p_{0j}) = -\sum_{i=1}^{m}\sum_{j=1}^{n}p_{ij}\log q_{ij} \qquad \dots(1)$$

$$\text{where } q_{ij} = p_{i0}p_{0j}$$

Also, by definition, we have

$$H(X,Y) = -\sum_{i=1}^{m}\sum_{j=1}^{n}p_{ij}\log p_{ij} \qquad \dots(2)$$

Now, $$\sum_{i=1}^{m}\sum_{j=1}^{n}q_{ij} = \sum_{i=1}^{m}\sum_{j=1}^{n}p_{i0}p_{0j} = \left(\sum_{i=1}^{m}p_{i0}\right)\left(\sum_{j=1}^{n}p_{0j}\right) = 1$$

$$= \sum_{i=1}^{m}\sum_{j=1}^{n}p_{ij}$$

Hence, we conclude that

$$H(X, Y) \leq H(X) + H(Y)$$

with equality if and only if $q_{ij} = p_{ij} \ \forall \ i$ and j.

☞ REMARK

- The condition of equality reduced to $p_{i0}p_{0j} = p_{ij}$ meaning thereby X and Y are independent.

THEOREM 2. $H(X, Y) = H(X|Y) + H(Y) = H(Y|X) + H(X)$ where $H(X) \geq H(X|Y)$.

[DELHI–2016]

PROOF. We know that

$$H(X) = -\sum_{k=1}^{n}p\{x_k\}\log p\{x_k\} \qquad \dots(1)$$

$$H(Y) = -\sum_{r=1}^{m}p\{y_r\}\log p\{y_r\} \qquad \dots(2)$$

$$H(X,Y) = -\sum_{k=1}^{n}\sum_{r=1}^{m}p(k,r)\log p\{k,r\} \qquad \dots(3)$$

where $p(k, r)$ is the joint probability for the occurance of two events E_k and E_r simultaneously.

Further, the expression for $H(X|Y)$ and $H(Y|X)$ can also be written as

$$H(X|Y) = -\sum_{r=1}^{m}\sum_{k=1}^{n}p\{x_k, y_r\}\log p\{x_k \mid y_r\} \qquad \dots(4)$$

$$H(Y|X) = -\sum_{k=1}^{n}\sum_{r=1}^{m}p\{x_k, y_r\}\log p\{y_r \mid x_k\} \qquad \dots(5)$$

Also, we know that

$$p(x_k, y_r) = p\{x_k|y_r\}, \ p(y_r) = p\{y_r|x_k\}p(x_k)$$

$$\log p\{x_k|y_r\} = \log p\{x_k|y_r\} \qquad \dots(6)$$

$$\log p(y_r) = \log p(y_r|x_k) + \log p(x_k) \qquad \dots(7)$$

Now, using (6) and (7) directly in (1) to (4) we may easily deduce the relation given by

$$H(X, Y) = H(Y|X) + H(Y) \text{ and } H(X, Y) = H(X|Y) + H(X)$$

Now, it remains to prove that $H(X) \geq H(X|Y)$

Consider $H(X \mid Y) - H(X) = \sum\limits_{r=1}^{m} \sum\limits_{k=1}^{n} p(x_k, y_r) \log \dfrac{p(x_k)}{p(x_k \mid y_r)}$

$$\leq \sum\limits_{r=1}^{m} \sum\limits_{k=1}^{n} p(x_k, y_r) \left[\frac{p(x_k)}{p(x_k \mid y_r)} - 1 \right] \log e \qquad\qquad (\because \log x \leq (x-1))$$

But $\sum\limits_{r=1}^{m} \sum\limits_{k=1}^{n} [p\{x_k\} \cdot p\{y_r\} + p\{x_k, y_r\}] \log e = \sum\limits_{r=1}^{m} [p\{y_r\} - p\{y_r\}] \log e = 0$

Hence, $\qquad H(X \mid Y) - H(X) \leq 0$

$\Rightarrow \qquad\qquad H(X) \geq H(X \mid Y)$

Similarly, we may prove that $H(Y) \geq H(Y \mid X)$.

23.10 AXIOMS FOR AN ENTROPY FUNCTION

Here, we have the following four axioms for an entropy function:

1. The entropy function is continuous w.r.t. all the arguments.
2. Given a finite complete probability scheme $(p_1, p_2, ..., p_n)$ then

$$\max H(p_1, p_2, ..., p_n) = H\left(\frac{1}{n}, ..., \frac{1}{n} \right)$$

   $\Rightarrow$ The entropy function $H(p_1, p_2, ..., p_n)$ must take its maximum value when all events have equal probabilities $\left(p_1 = p_2 = ..., p_n = \dfrac{1}{n} \right)$

3. The average information conveyed by (X, Y) is the sum of the average information given by X and that provided by Y when X is given, i.e., $H(X \mid Y) = H(X) + H(Y \mid X)$.
4. If an impossible event is added to a scheme, the entropy of the scheme should not be effected, i.e.,

$$H(p_1, p_2, ..., p_n, 0) = H(p_1, p_2, ..., p_n)$$

23.11 UNIQUENESS THEOREM

The only function which satisfies four axioms (given above) is

$$H(p_1, p_2, ..., p_n) = \lambda \sum\limits_{i=1}^{n} p_i \log p_i$$

where λ is an arbitrary positive number and logarithm base is any number greater than 1.

PROOF. Let us consider $H\left(\dfrac{1}{n}, \dfrac{1}{n}, ..., \dfrac{1}{n} \right) = f(n)$ $\qquad\qquad\qquad\qquad$ ...(1)

Now, since $f(n) = H\left(\dfrac{1}{n}, \dfrac{1}{n}, ..., \dfrac{1}{n} \right) \leq H\left[\dfrac{1}{n+1}, \dfrac{1}{n+1}, ..., \dfrac{1}{n+1} \right] = f(n+1)$ $\qquad$ ...(2)

Therefore, $f(n) \leq f(n+1) \Rightarrow f(n)$ is a non-decreasing function of n.

By axiom (3), for any complete probability scheme consisting of the sum of m mutually exclusive schemes.

$$H(X_1, X_2, ..., X_m) = H(X_1) + H(X_2) + ... + H(X_m) = \sum\limits_{k=1}^{m} H(X_k) \qquad ...(3)$$

If each scheme consists of r equally likely events, then

$$H(X_1, X_2, ..., X_m) = m(r) = f(r^m), \ m \text{ and } r \text{ are arbitrary integers.}$$

Let us take two integers t and n such that

$$r^m < t^n < r^{m+1} \quad \Rightarrow \quad m\log r < n \log t < (m+1)\log r$$

$$\Rightarrow \quad \frac{m}{n} < \frac{\log t}{\log r} < \frac{m+1}{n} \qquad \qquad \dots(4)$$

Also, since $f(n)$ is the non-decreasing function, therefore,

$$f(r^m) \leq f(t^n) \leq f(r^{m+1})$$

$$\Rightarrow \quad mf(r) \leq nf(t) \leq (m+1)f(r)$$

$$\Rightarrow \quad \frac{m}{n} \leq \frac{f(t)}{f(r)} \leq \frac{m}{n}\,\frac{1}{} \qquad \qquad \dots(5)$$

Now, comparing (4) and (5) we get

$$\left| \frac{f(t)}{f(r)} - \frac{\log t}{\log r} \right| \leq \frac{1}{n} \qquad \qquad \dots(6)$$

Letting $n \to \infty$, then for any positive integers r and t

$$\frac{f(t)}{f(r)} - \frac{\log t}{\log r} = 0 \quad \Rightarrow \quad \frac{f(t)}{\log t} = \frac{f(r)}{\log r} = \lambda \quad \text{(say)}$$

$$\Rightarrow \quad f(t) = \lambda \log t, f(r) = \lambda \log r$$

$$\Rightarrow \quad f(n) = \lambda \log n$$

Since, $f(n)$ is non-decreasing function, therefore, λ must be positive. Hence, the uniqueness theorem is proved when all events have euqal probabilities.

☞ **REMARK**
- In a similar manner, we may prove uniqueness theorem when all possibilities are rational, but not necessarily equal.

Solved Examples

EXAMPLE 1. *Evaluate the average uncertainty associated with the probability space of events shown in the following table.*

$$\begin{bmatrix} \textbf{Event} \\ \textbf{Probability} \end{bmatrix} = \begin{bmatrix} \textbf{A} & \textbf{B} & \textbf{C} & \textbf{D} \\ \textbf{1/2} & \textbf{1/4} & \textbf{1/8} & \textbf{1/8} \end{bmatrix}$$

SOLUTION. We have

$$p_1 = \frac{1}{2}, p_2 = \frac{1}{4}, p_3 = \frac{1}{8}, p_4 = \frac{1}{8}$$

Using all these values in the equation

$$H(p_1, p_2, \dots, p_n) = -\sum_{i=1}^{n} p_i \log p_i$$

We get

$$H\left(\frac{1}{2}, \frac{1}{4}, \frac{1}{8}, \frac{1}{8}\right) = -\frac{1}{2}\log\frac{1}{2} - \frac{1}{4}\log\frac{1}{4} - \frac{1}{8}\log\frac{1}{8} - \frac{1}{8}\log\frac{1}{8}$$

$$= \frac{1}{2}\log 2 + \frac{1}{4}\log 4 + \frac{1}{8}\log 8 + \frac{1}{8}\log 8$$

$$= \frac{1}{8}(4\log 2 + 2\log 2^2 + 2\log 2^3)$$

$$= \frac{1}{8}[4 + 4 + 6]\log 2 = \frac{14}{8} \text{ bits}$$

EXAMPLE 2. *Verify the rule of additivity of entropies for events A, B and C with probabilities $\frac{1}{5}, \frac{4}{15}$ and $\frac{8}{15}$ respectively.*

SOLUTION. By definition we have

$$H(p_1, p_2, \ldots, p_n) = -\sum_{i=1}^{n} p_i \log p_i$$

Therefore,

$$H\left(\frac{1}{5}, \frac{4}{15}, \frac{8}{15}\right) = -\frac{1}{5}\log\left(\frac{1}{5}\right) - \frac{4}{15}\log\left(\frac{4}{15}\right) - \frac{8}{15}\log\left(\frac{8}{15}\right)$$

$$= -\frac{1}{15}\left[3\log\frac{1}{5} + 4\log\left(\frac{4}{15}\right) + 8\log\left(\frac{8}{15}\right)\right]$$

$$= -\frac{1}{15}[-3\log 5 + 4(\log 4 - \log 15) + 8(\log 8 - \log 15)]$$

$$= -\frac{1}{15}[-3\log 5 - 4\log(3\times 5) - 8\log(3\times 5) + 4\log(2)^2 + 8\log(2)^3]$$

$$= \frac{1}{15}[15\log 5 + 12\log 3 - 32] \qquad\qquad [\because \log_2 2 = 1]$$

$$= \log 5 + \frac{4}{5}\log 3 - \frac{32}{15}$$

Hence, we can easily verify the rule of additivity.

EXAMPLE 3. *Evaluate the entropy of the source associated with the probabilities of events shown below*

Event	A	B	C	D	E	F
Probability	1/3	1/4	1/8	1/8	1/12	1/12

SOLUTION. We have $p_1 = \frac{1}{3}, p_2 = \frac{1}{4}, p_3 = \frac{1}{8}, p_4 = \frac{1}{8}, p_5 = \frac{1}{12}$ and $p_6 = \frac{1}{12}$

$\therefore$ Required entropy

$$= H(p_1, p_2, \ldots, p_n) = -\sum_{i=1}^{6} p_i \log p_i$$

$$= -p_1 \log p_1 - p_2 \log p_2 - p_3 \log p_3 - p_4 \log p_4 - p_5 \log p_5 - p_6 \log p_6$$

$$= -\frac{1}{3}\log\frac{1}{3} - \frac{1}{4}\log\frac{1}{4} - \frac{1}{8}\log\frac{1}{8} - \frac{1}{8}\log\frac{1}{8} - \frac{1}{12}\log\frac{1}{12} - \frac{1}{12}\log\frac{1}{12}$$

$$= \frac{1}{24}[8\log 3 + 6\log 2^2 + 3\log 2^3 + 3\log 2^3 + 2\log(2^2 \times 3))$$

$$\qquad + 2\log(2^2 \times 3)]$$

$$= \frac{1}{24}[38\log 2 + 12\log 3] = \frac{1}{24}[38\log 2 + 12\log 2^{1.58}]$$

$$= \frac{1}{24}[38 + 12\times 1.58] \approx 2.37 \text{ bits}$$

EXAMPLE 4. *A transmitter has an alphbet consisting of five letters $\{x_1, x_2, x_3, x_4, x_5\}$ and the receiver has an alphabet consisting of four letters $\{y_1, y_2, y_3, y_4\}$. The joint probability for the communication are given below:*

$$
\begin{array}{c}
\begin{array}{cccc} y_1 & y_2 & y_3 & y_4 \end{array} \\
\begin{array}{c} x_1 \\ x_2 \\ x_3 \\ x_4 \\ x_5 \end{array}
\begin{bmatrix}
0.25 & 0 & 0 & 0 \\
0.10 & 0.30 & 0 & 0 \\
0 & 0.05 & 0.10 & 0 \\
0 & 0 & 0.05 & 0.10 \\
0 & 0 & 0.05 & 0
\end{bmatrix}
\end{array}
$$

Find the marginal, conditional and joint entropy for the channel (assume $0 \times \log 0 \approx 0$)   [KURUKSHETRA–2015; DELHI–1991; OSMANIA–1983]

SOLUTION. Here, we have the joint probability $p_{ij} : i = 1, 2, ..., 5, j = 1, 2, ..., 4$

$p_{10} = 0.25 + 0 = 0.25$

$p_{20} = 0.10 + 0.30 = 0.40$

$p_{30} = 0.05 + 0.10 = 0.15$

$p_{40} = 0.05 + 0.10 = 0.15$

$p_{50} = 0.05 + 0.00 = 0.05$

Similarly, $p_{01} = 0.35, p_{02} = 0.35, p_{03} = 0.20, p_{04} = 0.10$

Since, the conditional probabilities can be obtained by using the following formula

$$ p[j \,|\, i] = \frac{p_{ij}}{p_{i0}} $$

Then we get the following conditional probability matrix

$$
p[j|i] = \begin{array}{c}
\begin{array}{cccc} 1 & \quad 2 & \quad 3 & \quad 4 \end{array} \\
\begin{array}{c} 1 \\ 2 \\ 3 \\ 4 \\ 5 \end{array}
\begin{bmatrix}
1 & 0 & 0 & 0 \\
1/4 & 3/4 & 0 & 0 \\
0 & 1/3 & 2/3 & 0 \\
0 & 0 & 1/3 & 2/3 \\
0 & 0 & 1 & 0
\end{bmatrix}
\end{array}
$$

Now, the marginal entropy is given by

$$ H(X) = -\sum_{i=1}^{5} p_{i0} \log p_{i0} $$

$$ = - [(0.25) \log(0.25) + (0.40) \log(0.40) $$
$$ + 2(0.15) \log(0.15) + (0.5) \log(0.5)] $$

$$ = \frac{1}{4} \log 4 + \frac{2}{5} \log \frac{5}{2} + \frac{3}{10} \log \frac{20}{3} + \frac{1}{2} \log 2 $$

$$ = 1.3260 \, \text{bits} $$

Similarly, $$ H(Y) = -\sum_{j=1}^{4} p_{0j} \log p_{0j} $$

$$ = -[2(0.35) \log(0.35) + (0.20) \log(0.20) + (0.10) \log(0.10)] $$

$$ = \frac{7}{10} \log(2.857) + \frac{1}{5} \log 5 + \frac{1}{10} \log 10 $$

$$ = 1.8556 \, \text{bits} $$

Now, conditional entropies

$$H(Y \mid X) = -\sum_{i=1}^{5}\sum_{j=1}^{4} p_{ij} \log p(j \mid i)$$

$$= (0.25) \log 1 + (0.10) \log 4 + (0.30) \log (4/3) + (0.05) \log 3$$
$$+ (0.10) \log (3/2) + (0.05) \log 3 + (0.10) \log 3/2 + (0.05) \log 1$$

$$= \frac{1}{10}\log 4 + \frac{3}{10}\log 4 - \frac{3}{10}\log 3 + \frac{1}{20}\log 3 + \frac{1}{5}\log 3 - \frac{1}{5}\log 2$$

$$= \frac{1}{10}[4\log 4 - 2\log 2] = \frac{6}{10} = 0.60 \text{ bits}$$

Also, $H(X|Y) = H(X) + H(Y|X) - H(Y)$

$$= 1.3260 + 0.6000 - 1.8556 = 0.0704 \text{ bits}$$

Finally, the joint entropy is given by

$$H(X, Y) = H(X) + H(Y|X)$$

$$= 1.3260 + 0.6000 = 1.9260 \text{ bits}$$

EXAMPLE 5. *Find the entropy of the following probability distribution*

Event	x_1	x_2	...	x_i	...	x_{n-1}	x_n
Probabilities	$\dfrac{1}{2}$	$\dfrac{1}{2^2}$	...	$\dfrac{1}{2^i}$	...	$\dfrac{1}{2^{n-1}}$	$\dfrac{1}{2^n}$

SOLUTION. We have $p_i = \dfrac{1}{2^i} : i = 1, 2, \ldots, n-1$ and $p_n = \dfrac{1}{2^{n-1}}, \sum_{i=1}^{n} p_i = 1$

Now, $H(p_1, p_2, \ldots, p_n)$

$$= -\sum_{i=1}^{n} p_i \log p_i = -\sum_{i=1}^{n-1} p_i \log p_i - p_n \log p_n$$

$$= -\sum_{i=1}^{n-1}\left(\frac{1}{2^i}\right)\log\left(\frac{1}{2^i}\right) - \left(\frac{1}{2^{n-1}}\right)\log\left(\frac{1}{2^{n-1}}\right)$$

$$= \sum_{i=1}^{n-1}\left(\frac{1}{2^i}\right)\log(2^i) + \frac{1}{2^{n-1}}\log_2(2^{n-1})$$

$$= \sum_{i=1}^{n-1} i\left(\frac{1}{2^i}\right) + (n-1)\left(\frac{1}{2^{n-1}}\right) \qquad\qquad (\because \log_2 2 = 1)$$

$$= \left[\frac{1}{2} + \frac{2}{2^2} + \frac{3}{2^3} + \ldots + \frac{n-1}{2^{n-1}}\right] + \frac{n-1}{2^{n-1}} \ldots \qquad\qquad \ldots(1)$$

or $\dfrac{1}{2}H(p_1, p_2, \ldots, p_n) = \left[\dfrac{1}{2^2} + \dfrac{2}{2^3} + \dfrac{3}{2^4} + \ldots + \dfrac{n-1}{2^n}\right] + \dfrac{n-1}{2^n} \qquad\qquad \ldots(2)$

From (1) and (2), we get

$$H(p_1, p_2, \ldots, p_n) - \frac{1}{2}H(p_1, p_2, \ldots, p_n)$$

$$= \left\{\frac{1}{2} + \frac{1}{2^2} + \frac{1}{2^3} + \ldots + \frac{1}{2^{n-1}}\right\} + \left\{\frac{n-1}{2^{n-1}} - \frac{2(n-1)}{2^n}\right\}$$

$$\Rightarrow \quad \frac{1}{2}H(p_1, p_2, \ldots, p_n) = \left\{\frac{1}{2} + \frac{1}{2^2} + \frac{1}{2^3} + \ldots + \frac{1}{2^{n-1}}\right\} = 1 - \left(\frac{1}{2}\right)^{n-1}$$

Hence, $H(p_1, p_2, \ldots, p_n) = 2 - \left(\frac{1}{2}\right)^{n-1}$

23.12 CHANNEL CAPACITY, EFFICIENCY AND REDUNDANCY

(i) Channel capacity: We know that each transmitted symbol i in a communication process has a probability $p(j|i)$ of being communicated as j. So, a measure of mutual information in a communication system can be defined as follows:

$$I(i,j) = \log\frac{p(i|j)}{p_i} = \log\frac{p_{ij}}{p_i p_j}$$

$I(i,j)$ can be defined as the information of symbol i. Then the average mutual information obtained per symbol of pairs will be given by

$$I(X,Y) = \sum_{j=1}^{m}\sum_{i=1}^{n} p_{i,j}I(i,j) = \sum_{j=1}^{m}\sum_{i=1}^{n} p_{i,j}\log\left(\frac{p_{ij}}{p_i p_j}\right)$$

$$\Rightarrow \qquad I(X, Y) = H(X) + H(Y) - H(X, Y)$$
$$= H(X) - H(X|Y)$$

Then the capacity $\qquad C = \max I(X|Y)$

$$= \max [H(X) + H(Y) - H(X, Y)]$$
$$= \max (H(X) - H(X|Y))$$

Since, the maximum of $H(X)$ occurs when all symbols have equal probabilities, then channel capacity becomes

$$C = -\log\left(\frac{1}{n}\right) = \log n \ \text{ bits/symbol}$$

(ii) Efficiency: In one dimensional probability scheme efficiency can be defined as follows:

$$\text{Efficiency} = \frac{I(x)}{\overline{l}\log n}$$

where $\quad I(x) = $ entropy of the message in original language

$\overline{l} = $ average length of message

$n = $ no. of symbols in encoding alphabets

Similarly in a 2-dimensional channel probability scheme efficiency $= \dfrac{I(X,Y)}{\log n}$

And for noiseless channel, the efficiency is given by $\dfrac{H(X)}{\log n}$

(iii) Redundancy: The difference between the actual rate of transmission of information $I(X, Y)$ and its maximum possible value is defined as the absolute redundancy of the communication system. The ratio of absolute redundancy to the channel capacity is defined as the relative redundancy of the channel capacity. Thus,

(a) Absolute redundancy for noise free channel $= C - I(X, Y) = \log n - H(X)$

(b) Relative redundancy for noise free channel $= \dfrac{\log n - H(X)}{\log n} = 1 - \dfrac{H(X)}{\log n}$

☞ REMARK

- The efficiency of the system may also be defined as follows:
  Efficiency of the noise free channel
  $$= \frac{I(X,Y)}{\log n} = \frac{H(X)}{\log n} = (1 - \text{relative redundancy})$$

Solved Examples

EXAMPLE I. ***Find the capacity of memory less channel specified by***

$$\begin{bmatrix} 1/2 & 1/4 & 1/4 & 0 \\ 1/4 & 1/4 & 1/4 & 1/4 \\ 0 & 0 & 1 & 0 \\ 1/2 & 0 & 0 & 1/2 \end{bmatrix}$$

[ROHILKHAND–2008; SHIVAJI–1985, 2007]

SOLUTION. We know that the capacity of memory less channel is given by
$$C = \max I(x, y) = \max [H(X) + H(Y) - H(X|Y)]$$
∴ The required capacity is given by

$$C = -\sum_{i=1}^{4} p_{i1} \log p_{i1} - \sum_{i=1}^{4} p_{i2} \log p_{i2} - \sum_{i=1}^{4} p_{i3} \log p_{i3} - \sum_{i=1}^{4} p_{i4} \log p_{i4}$$

Now, using $p_{i1} = \left[\dfrac{1}{2}, \dfrac{1}{4}, \dfrac{1}{4}, 0 \right]$

$p_{i2} = \left[\dfrac{1}{4}, \dfrac{1}{4}, \dfrac{1}{4}, \dfrac{1}{4} \right]$

$p_{i3} = [0, 0, 1, 0]$

and $p_{i4} = \left[\dfrac{1}{2}, 0, 0, \dfrac{1}{2} \right]$

We get

$$C = -\left[\frac{1}{2} \log \frac{1}{2} + 2 \left(\frac{1}{4} \log \frac{1}{4} \right) + 4 \left(\frac{1}{4} \log \frac{1}{4} \right) + 1 \log 1 + 2 \left(\frac{1}{2} \log \frac{1}{2} \right) \right]$$

$$= \left(\frac{3}{2} \log 2 + 3 \log 2 \right) = \left(\frac{3}{2} + 3 \right) \log 2$$

$$\approx \frac{9}{2} \text{ bits/symbol}$$

23.13 ENCODING

We know that the main objective of information theory is to minimize the loss of information due to disturbances in the communication channel of the system. If the communicated message is lengthy, then there is high possibilities of losing information and this will be directly proportional to the length of the message. To avoid this we try to reduce the length of the message to be transmitted by transforming it in some other language. This procedure is known as coding.

Definition. *Encoding may be defined as a transformation procedure of a message from sources to receive through a noiseless channels in some code language.*

Mathematically, if $X = \{x_1, x_2,..., x_m\}$ be the set of message to be transmitted, then codes may be defined as a relationship between all possible sequences of symbols of the set X with another set $Y = \{y_1, y_2, ..., y_n\}$ of code character of alphabets.

☞ REMARKS

- Encoding is analogous to short hand in dictation which ensures quickness and accuracy in the system.
- The communication system with encoding, work according to the following diagram.

| Source |—| Encoder |—| Communication channel |—| Decoder |—| Receiver |

23.13.1 TERMINOLOGY

Here, we define some terms used in encoding.

1. **Letter, symbol or character:** It is an individual member in a set of alphabets, *i.e.*, if $\{a_1, a_2, ..., a_n\}$ is a given set of alphabets then any a_i is a letter.

2. **Message or word:** Message or word is a finite sequence of letters of alphabets. For example if $\{a_1, a_2, ..., a_n\}$ is the given set of alphabets then $a_1, a_1a_2a_3, a_1a_2a_3a_3, ...,$ are the messages.

3. **Length of word:** The no. of letters in a word is known as the length of the word. For example the length of the word $a_1a_2a_2a_4$ is 4, because it consists four alphabets.

4. **Encoding or Enciphering:** It is one to one transformation of words in original message to words in some other languages.

5. **Decoding or Deciphering:** It is the inverse process of encoding, *i.e.*, transforming the transmitted encoded message into original language.

6. **Unique Decipherability:** A code is said to be unique decipheribility if every finite sequence of code characters corresponds to at most one message.

   For example, consider the following codes.

   $$\begin{array}{ll} m_1 & 0 \\ m_2 & 10 \\ m_3 & 110 \\ m_4 & 111 \end{array}$$

   If we write 00100100011010 then it may be uniquely decoded as

   $$m_1m_1m_2m_1m_2m_1m_1m_3m_2$$

7. **Block code:** A code which establish a relationship with each of the symbol of the set X of alphabets to a fixed sequence of symbols of the set Y is called block code.

8. **Binary code:** If the set $X = \{0, 1\}$ then a block code is said to be binary code.

9. **Non-singular code:** A block code is said to be non-singular if all words of the code are distinct.

> If the original message contains n words $w_1, w_2, ..., w_n$ then encoding is one to one transformation of these words in another sequence $\{z_1, z_2, ..., z_n\}$ when w_1 corresponds to z_1, w_2 to z_2 and so on w_n corresponds to z_n.

23.13.2 USES OF ENCODING

1. It is used to increase the efficiency of transmission.

2. It is used to minimize code word length.

3. It is used to minimize the cost of transmission.

☛ REMARK

- The transmission for which the cost is minimum is considered to be efficient.

23.14 SHANNON-FANO ENCODING PROCEDURE

In this procedure, we use a sequence of binary numbers $\{0,1\}$ for encoding messages through a memoryless communication channel.

Let $X = \{x_1, x_2, ..., x_n\}$ be the list of the messages to be transmitted and $P = \{p_1, p_2, ..., p_n\}$ be their corresponding probabilities. We want to devise an encoding procedure so that a sequence of binary numbers $\{0,1\}$ of unspecified length can be associated to each message x_i. The sequence so obtained must satisfy the following conditions:

(i) No sequence of binary numbers can be obtained from any other sequence by adding additional binary terms to sequence of shorter lengths.

(ii) Binary numbers associated with each message x_i to form a sequence occur independently with equal probabilities.

Working Procedure

STEP 1. Arrange the message $x_1, x_2, ..., x_n$ in descending order in terms of their probabilities, i.e., if $p_1 > p_2 > ... > p_n$ then we have

Message: $x_1, x_2, ..., x_r, ..., x_n$

Probability: $p_1, p_2, ..., p_r, ..., p_n$

STEP 2. Divide the set of message X into two sets X_1 and X_2 of equal probabilities

Set	message	probabilities
X_1	x_1, x_2	$P(X_1) = p_1 + p_2$
X_2	$x_3, x_4, ..., x_n$	$P(X_2) = p_3 + ... + p_n$

such that $P(X_1) = P(X_2)$

STEP 3. Divide X_1 and X_2 into two subsets X_{11}, X_{12} and X_{21}, X_{22} with equal probabilities respectively.

STEP 4. Assign binary no. 0 to the first position of the coded word in each message in X_1 and binary number 1 to the first position of the coded word in each message in X_2. The same procedure must be repeated for X_1 and X_2.

STEP 5. Repeat step 3 and 4 for division and assign the binary no. until each subset contains only one message.

☛ REMARK

- If the sample space cannot be divided into two subsets of equal probabilities then above method is not optimum.

 **Solved Examples**

EXAMPLE 1. *Apply Shannon's encoding procedure to the following message ensemble*
$$X = \{x_1, x_2, x_3, x_4\}$$
$$P = \{0.4, 0.3, 0.2, 0.1\}$$
[OSMANIA–2002]

SOLUTION. We have the following table

Message	Probability	Encoded message	Length
x_1	0.4	0	1
—	—	—	—
x_2	0.3	10	2
x_3	0.2	110	3
x_4	0.1	111	3
		Average length = 1.9	

The message are first written in order of decreasing probabilities. Then partitioned the set into two subsets X_1 and X_2. Zero is assigned to each message in one subset and 1 to each of the remaining messages.

Repeat the same procedure for X_1 and X_2.

Here, the subset $X_1 = \{x_1\}$ can not be patitioned further.

But $X_2 = \{x_2, x_3, x_4\}$ can be partitioned as
$$X_{21} = \{x_2\}, X_{22} = \{x_3, x_4\}$$

Therefore, assign 0 to x_2 and 1 to each of the message x_3 and x_4.

Continue this procedure till each subset contains only one message.

∴ Entropy of the source is given by
$$H = - [0.4 \log 0.4 + 0.3 \log 0.3 + 0.2 \log 0.2 + 0.1 \log 0.1]$$
$$= 1.9 \text{ bits/message}$$

Also, the expected length is
$$\bar{l} = \Sigma p(x_i) n_i$$
$$= 0.4 \times 1 + 0.3 \times 2 + 0.2 \times 3 + 0.1 \times 3$$
$$= 1.9 \text{ bits/symbol}$$

EXAMPLE 2. *A source memory has 6 characters with the following probabilities of transmission*

A	B	C	D	E	F
1/3	1/4	1/8	1/8	1/12	1/12

Apply the Shannon-Fano encoding procedure to obtain uniquely decodable code to the above messge ensemble.

Find also the average length, efficiency and redundancy of the code you obtain.

SOLUTION. Clearly given messages are already in descending order of probabilities.

Now, divide the elements of X into X_1 and X_2 with approximately equal probabilities such that
$$X_1 = \{A, B\}, X_2 = \{C, D, E, F\}$$

$$P(X_1) = \frac{1}{3} + \frac{1}{4} = \frac{7}{12} \text{ and } P(X_2) = \frac{1}{8} + \frac{1}{8} + \frac{1}{12} + \frac{1}{12} = \frac{5}{12}$$

Further divide X_2 into two subsets X_{21} and X_{22} as given below

Subsets	Probabilities
$X_{21} = \{C,D\} = \left\{\frac{1}{8}, \frac{1}{8}\right\}$	$p(X_{21}) = \frac{1}{8} + \frac{1}{8} = \frac{1}{4}$
$X_{22} = \{E,F\} = \left\{\frac{1}{12}, \frac{1}{12}\right\}$	$p(X_{22}) = \frac{1}{12} + \frac{1}{12} = \frac{1}{6}$

Now, subsets X_{21} and X_{22} contain two elements each, therefore these can be further subdivided into two subsets as shown below

Subsets	Probabilities
$X_{211} = \{C\} = \left\{\frac{1}{8}\right\}$	$p(X_{211}) = \frac{1}{8}$
$X_{212} = \{D\} = \left\{\frac{1}{8}\right\}$	$p(X_{212}) = \frac{1}{8}$
$X_{221} = \{E\} = \left\{\frac{1}{12}\right\}$	$p(X_{221}) = \frac{1}{12}$
$X_{222} = \{F\} = \left\{\frac{1}{12}\right\}$	$p(X_{222}) = \frac{1}{12}$

Assign binary no. 0 and 1 in first position of all code words X_1 and X_2 respectively as given below.

Character	Probabilities	Partiotioning	Code word	Code word length (l)
A $\left.\right\}X_1$	1 / 3	X_{11}	00	2
B	1 / 4	X_{12}	10	2
C	1 / 8 $\left.\right\}X_{21}$	X_{211}	100	3
D $\left.\right\}X_2$	1 / 8	X_{212}	101	3
E	1 / 12 $\left.\right\}X_{22}$	X_{221}	110	3
F	1 / 12	X_{222}	111	3

(i) Entropy

$$H(X) = -\sum_{i=1}^{6} p(x_i) \log p(x_i)$$

$$= -\left[\frac{1}{3}\log\frac{1}{3} + \frac{1}{4}\log\frac{1}{4} + \frac{2}{8}\log\frac{1}{8} + \frac{2}{12}\log\frac{1}{12}\right]$$

$$= \frac{1}{3}\log 3 + \frac{1}{4}\log 4 + \frac{2}{8}\log 8 + \frac{2}{12}\log 12$$

$$= \frac{1}{3}\log 3 + \frac{1}{4}\log(2)^2 + \frac{2}{2}\log(2)^3 + \frac{2}{12}\log(2^2 \times 3)$$

$$= \left(\frac{1}{2} + \frac{4}{12} + \frac{6}{8}\right)\log 2 + \left(\frac{1}{3} + \frac{2}{12}\right)\log 3$$

$$= \frac{19}{12} + \frac{1}{2}\log 3 = 2.3752 \text{ bits}$$

(ii) Average code length of the message is given by

$$L = \sum_{i=1}^{6} I_1 p(x_i) = \frac{2}{3} + \frac{2}{4} + \frac{3}{8} + \frac{3}{8} + \frac{3}{12} + \frac{3}{12}$$

$$= \frac{29}{12} \text{ bits/symbol}$$

(iii) Efficiency of the code $= \dfrac{H(X)}{L} = \dfrac{2.3752}{29/2} = 0.9828$

(iv) Redundancy of the code

$$= 1 - \text{Efficiency}$$
$$= 1 - 0.9828 = 0.0172$$

EXAMPLE 3. *Apply the Shannon-Fano encoding procedure to the following message ensemble.*

X	x_1	x_2	x_3	x_4	x_5	x_6	x_7	x_8	x_9
P	0.49	0.14	0.14	0.07	0.07	0.04	0.02	0.02	0.01

SOLUTION. Firstly arrange the messages in descending order of their probabilities.

Now, divide the message into X_1 and X_2 of approximately equal probabilities such that $X_1 = \{x_1\}$ and $X_2 = \{x_2, x_3, ..., x_9\}$ with $P(X_1) = 0.49$ and $P(X_2) = 0.51$.

Assign 0 to each messge in the set X_1, *i.e.*, assign 0 to X_1 and 1 to each messge of X_2.

Further divide X_2 into X_{21} and X_{22} so that

$$X_{21} = \{x_2, x_3\}, X_3 = \{x_4, x_5, ..., x_9\}$$
$$P(X_{21}) = 0.28, P(X_{22}) = 0.33$$

Now each message in X_{21} will begin with code 10 and that in X_{22} will begin with 11. Further divide X_{21} and X_{22} into the subsets X_{211}, X_{212} and X_{221}, X_{222} respectively as follows.

	Probabilities	Code Word
$X_{211} = \{x_2\}$	$P(X_{211}) = 0.14$	100
$X_{212} = \{x_3\}$	$P(X_{212}) = 0.14$	101
$X_{221} = \{x_4, x_5\}$	$P(X_{221}) = 0.07 + 0.07 = 0.14$	110
$X_{222} = \{x_6, x_7, ..., x_9\}$	$P(X_{222}) = 0.09$	111

Proceeding same as above we have the following table

	Probabilities	Code Word
$X_{2211} = \{x_4\}$	$P(X_{2211}) = 0.07$	1100
$X_{2212} = \{x_5\}$	$P(X_{2212}) = 0.07$	1101
$X_{2221} = \{x_6\}$	$P(X_{2221}) = 0.04$	1110
$X_{2222} = \{x_7, x_8, x_9\}$	$P(X_{2222}) = 0.05$	1111

$X_{22221} = \{x_7\}$	0.2	11110
$X_{22222} = \{x_8, x_9\}$	0.3	11111

and

$X_{222221} = \{x_8\}$	0.02	111110
$X_{222222} = \{x_9\}$	0.01	111111

The message, their codes, the corresponding lengths are given below

Message	Code	Length
x_1	0	1
x_2	100	3
x_3	101	3
x_4	1100	4
x_5	1101	4
x_6	1110	4
x_7	11110	5
x_8	111110	6
x_9	111111	6

Finally, the average length of the message

$$L = \sum_{i=1}^{9} p(x_i) n_i$$

$$= 0.49 \times 1 + 0.14 \times 3 + 0.14 \times 3 + 0.07 \times 4 + 0.07 \times 4$$
$$+ 0.04 \times 4 + 0.02 \times 5 + 0.02 \times 6 + 0.01 \times 6$$

$$= 2.33$$

23.15 NECESSARY AND SUFFICIENT CONDITION FOR NOISELESS CODING

Noiseless coding theorem: *The necessary and sufficient condition for existence of an irreducible noiseless encoding procedure with specified word length* $n_1, n_2, ..., n_N$ *is that a set of positive integers* $\{n_1, n_2, ..., n_N\}$ *can be found such that*

$$\sum_{i=1}^{N} D^{-n_i} \leq 1, \text{ where } D \text{ is the no. of symbols in encoding apphabets}$$

PROOF. The condition is necessary

Let two messges m_1 and m_2 have the same length, *i.e.*, $n_i = n_j$. If x_1 be the number of encoded messages with only one letter can not be larger than D, then

$$x_1 \leq D \qquad \qquad ...(1)$$

Further, because of the coding restrictions the number of encoded messages of length 2 can not be larger than $(D - x_1)D$.

Therefore, $\qquad x_2 \leq (D - x_1)D = D^2 - x_1D \qquad \qquad ...(2)$

Similarly, $\qquad x_3 \leq (D - x_1)(D - x_2)D = D^3 - x_1D^2 - x_2D \qquad ...(3)$

Finally, if m is the maximum length of encoded words then

$$x_m \leq D^m - x_1 D^{m-1} - x_2 D^{m-2} - ... - x_{m-1}D \qquad ...(4)$$

Dividing both sides of (4) by D^m, we have

$$x_m D^{-m} \leq 1 - x_1 D^{-1} - x_2 D^{-2} - \ldots - x_{m-1} D^{-(m-1)}$$

$\Rightarrow \qquad x D^{-1} + x_2 D^{-2} + \ldots + x_m D^{-m} \leq 1$

$\Rightarrow \qquad \sum_{i=1}^{m} x_i D^{-i} \leq 1$ \hfill ...(5)

Now, $\qquad \sum_{i=1}^{m} x_i D^{-i} = x_1 D^{-1} + x_2 D^{-1} + \ldots + x_m D^{-m}$

$$= (D^{-1} + D^{-1} + \ldots + x_1 \text{ times})$$

$$+ (D^{-2} + D^{-2} + \ldots + x_2 \text{ times})$$

$$+ \ldots + (D^{-m} + D^{-m} + \ldots + x_m \text{ times}) \qquad ...(6)$$

Since, the expression within each bracket corresponds to a message m_1 and therefore the total no. of terms in N.

From (6), there are x_1 messages of length 1, x_2 messages of length 2, ..., x_m messages of length m.

$\therefore \qquad x_1 + x_2 + \ldots + x_m = N$

The term x_k corresponds to the encoded message of length k. These later terms can be considered as ΣD^{-n}. Therefore, we can write

$$\sum_{j=1}^{m} x_j D^{-j} = \sum_{i=1}^{N} D^{-n_i} \leq 1 \qquad \text{(Using (5))} \qquad ...(7)$$

which is the required necessary conditions.

The condition is sufficient

We have to show that the conditon

$$\sum_{j=1}^{m} x_j D^{-j} = x_1 D^{-1} + x_2 D^{-2} + \ldots + x_m D^{-m} \leq 1 \qquad ...(8)$$

is a sufficient condition.

Clearly, from (8) $\quad x_1 D^{-1} \leq 1 \qquad \Rightarrow \qquad x_1 \leq D$ \hfill ...(9)

$\qquad\qquad\qquad x_1 D^{-1} + x_2 D^{-2} \qquad \Rightarrow \qquad x_2 \leq D(D - x_1)$ \hfill ...(10)

and so on.

Since, there are conditions which have to satisfy in order to guarantee that no encoded message can be obtained from any other by the addition of a sequence of letters of the encoding alphabets.

Let D be a binary set

i.e., $\qquad\qquad\qquad\qquad D = \{a_1, a_2\}$

Then encoding theorem requires that

$$\sum_{i=1}^{N} 2^{-n_i} \leq 1 \qquad ...(11)$$

Now, the noiseless coding theorem suggest that separable codes being N words of equal length n exists if

$$\sum_{n=1}^{N} D^{-n} \leq 1$$

$\Rightarrow \quad D^{-n} + D^{-n} + \ldots + D^{-n} \ (n \text{ times}) \leq 1$

$\Rightarrow \qquad\qquad\qquad N \cdot D^{-n} \leq 1$

$\Rightarrow \qquad\qquad\qquad\qquad N \leq D^{-n}$

$\Rightarrow \quad \log N \leq n \log D$, which guarantee the existence of the desired costs.

Exercise-23.1

1. Apply Shannon-Fano entropy procedure to the following measure

X	P	X	P
x_1	1/4	x_6	1/16
x_2	1/4	x_7	1/16
x_3	1/8	x_8	1/16
x_4	1/8	x_9	1/15
x_5	1/16		

Find entropy (H) of original source and average length of encoded message.

2. Show that the entropy of the following events

is $2 - \left(\dfrac{1}{2}\right)^{n-2}$.

X	x_1	x_2	...	x_i	...	x_{n-1}	x_n
P	$\dfrac{1}{2}$	$\dfrac{1}{4}$	...	$\dfrac{1}{2^i}$	...	$\dfrac{1}{2^{n-1}}$	$\dfrac{1}{2^n}$

3. Prove that $H(p_1, p_2, ..., p_n) \leq \log n$ and equality holds iff $p_k = 1$; $k = 1, 2, 3, ..., n$.

4. If H is the entropy function, prove that
$$H(p_1, p_2, ..., p_N, q_1, q_2, ..., q_m)$$
$$= H(p_1, p_2, ..., p_n)$$
$$+ p_n H\left(\frac{q_1}{p_n}, \frac{q_2}{p_n}, ..., \frac{q_m}{p_n}\right)$$
where $p_n = \sum\limits_{k=1}^{m} q_k$

5. Show all possible sets of binary codes with prefix property for encoding the message ensemble (x_1, x_2, x_3) in words not more than three digits long.

ANSWERS

1. $H = 2.75$ bits, $\overline{l} = 2.75$

REVIEW QUESTIONS

1. Define information function. Describe its various requirements.
2. Show that the entropy function $H(X)$ achieve its maximum if all the values of X are equal probable.
3. Discuss the applications of maximum entropy principle in Operations Research.
4. State and prove Fundamental theorem of information theory.
5. Define conditional entropy.
6. Define capacity of channel, redundancy and efficiency.
7. State and prove Noiseless coding theorem.
8. Write a short note on information theory.
9. Write a short note on memoryless schemes.
10. Define entropy function and establish its formal requirements.

MULTIPLE CHOICE QUESTIONS (CHOOSE THE MOST APPROPRIATE ONE)

1. The expected value of information is given by:
   (a) $\sum\limits_{i=1}^{m} p_i I(x_i)$
   (b) $\sum q_i I(y_i)$
   (c) both (a) and (b) are true
   (d) None of these
2. Essential part of a communication system is:
   (a) source (b) receiver
   (c) both (a) and (b) (d) None of these
3. Basic requirements of logarithm entropy function is:
   (a) monotonical
   (b) grouping
   (c) both (a) and (b) are true
   (d) None of these
4. Shannon measure of expected amount of information is:
   (a) $\sum p_i \log_2 p_i$
   (b) $- \sum p_i \log_2 p_i$
   (c) both (a) and (b) are true
   (d) None of these

ANSWERS

1. (a)   **2.** (c)   **3.** (c)   **4.** (b)

ARCHIVE

1. There are 12 coins, all of equal weight except one which may be lighter or heavier. Using concepts of information theory, show that it is possible to determine which coin is heavier.

[BANGLORE–1990]

2. Apply Shannon-Fano encoding procedure to the following message ensemble

X	x_1	x_2	x_3	x_4	x_5	x_6	x_7	x_8	x_9
P(X)	0.49	0.14	0.14	0.07	0.07	0.04	0.02	0.02	0.01

3. Apply Shannon's encoding procedure to the following message ensemble.

X	A	B	C	D
P(X)	0.4	0.3	0.2	0.1

4. Verify the rule of the additivity of entropies for events A, B, C with probabilities $\frac{1}{5}, \frac{4}{15}$ and $\frac{8}{15}$ respectively.

HINTS AND ANSWERS

2. $L = 2.3$ bits/symbol, $H = 1.60 + 2 \log 5 - (1.40) \log 7$

3. $L = 1.9$ bits/symbol, $H = \log 5 - 0.3 \log 3$

Sequencing 24

24.1 INTRODUCTION

The selection of an appropriate order in which to service waiting customers or jobs is called sequencing. The total effectiveness which may be time, cost etc. is function of the order or sequence of the jobs in which they are processed. In this chapter we shall consider the problems of determining the optimal sequence of arrivals or jobs that are to be done, which minimizes the total effectiveness which may be cost or time.

24.2 GENERAL SEQUENCING PROBLEM

[MEERUT-2002, 04, 10, 16; AGRA-2001, ROHILKHAND-2000, UPTUMBA-02, 06]

Consider the problem of performing n jobs on each of m machines. We are given the order of the machines for each job, in which it should go to the machines. We also know the actual or expected time required by the jobs on each of the machines. Our problem is to find that sequence out of $(n!)^m$ sequences which minimizes the total elapsed time, *i.e.*, the time from start of the first job upto the completion of the last job.

Mathematically, if we use the notations

A_i = time estimated for the ith job on machine A, $i = 1, 2,..., n$.

(Similarly we can interpret B_i and C_i etc.)

T = the total elapsed time

then we determine a sequence of jobs, *i.e.*, a permutation of numbers 1, 2, ..., n for each machine, which minimizes the time T.

All types of sequencing problems cannot be solved. The satisfactory solutions are available only in few cases. We shall discuss here the following cases:

1. n jobs are to be processed on two machines say A and B in the order AB.
2. n jobs are to be processed on three machines A, B and C in the order ABC.
3. 2 jobs are to be processed on m machines.

24.2.1 BASIC TERMINOLOGY

1. **No. of Machines:** It means that service facilities through which a job must pass before it is completed.
2. **Processing Order:** It means the order in which various machines are required for completing the job.
3. **Processing time:** It refer to the time for each job on each machine.
4. **Total elapsed time:** The time between starting the first job and completing the last job is called total elapsed time.
5. **Idle time on a Machine:** This is the time for which a machine remains idle during the total elapsed time.

6. **No passing Rule:** It means that passing is not allowed, *i.e.*, the same order of job is maintained over each machine. If each of the *n*-jobs is to be proceed through two machines *A* and *B* in the order *AB*, then this rule state that each job will go to machine *A* first and then go to *B*.

24.2.2 BASIC ASSUMPTIONS

Some principal assumptions in this chapter are as follows:

(i) The processing times A_i's etc. are exactly known and are independent of the order of the jobs, in which they are to be processed. Such problems where times are exactly known are called *deterministic problems*.

(ii) The time taken by the jobs in going from one machine to another is negligible.

(iii) Each job, once started on a machine, is to be performed upto completion on that machine.

(iv) A job starts on the machine as soon as the job and the machine both are idle and job is next to the machine and the machine is also next to the job.

(v) There is only one machine of each type.

(vi) No machine may process more than one job at a time.

(vii) The cost of keeping the jobs in inventory (if needed) during the inprocess is same for all jobs. Also it is too small that it can be neglected.

(viii) The order of completion of jobs has no significance *i.e.*, no job is to be given priority.

(ix) Times of jobs are independent of sequence of jobs.

24.3 SEQUENCING DECISION PROBLEM FOR *n* JOBS AND TWO MACHINES

[MEERUT-2002, 04, 07; ROHILKHAND-2002]

Consider the sequencing problem of processing two jobs on two machines. For example, consider the problem of two jobs say 1 and 2 to be processed on each of the two machines *A* and *B* in the order *AB*. Processing times are as follows:

Job/Machine	A	B
1	3	5
2	5	4

There are only 2!, *i.e.*, 2 possible sequences (1, 2) and (2, 1). Corresponding to both these sequences, a total elapsed times are evaluated graphically in the below figures so called Gantt Charts. Obviously optimum sequence is (1, 2) and the total elapsed time is 12 hours.

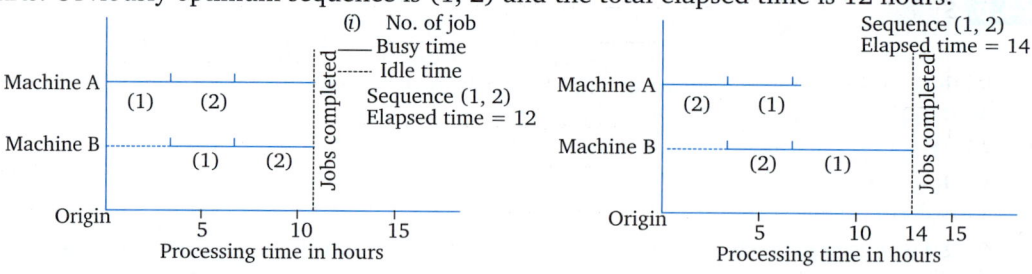

Fig. 1   **Fig. 2**

Above method is not of much practical importance even if n is small, because if there are *n* jobs and even if only two machines are involved and all jobs go over all machines in the same order, there are n ! possible sequences and it is complicated enough to evaluate each sequence.

24.4 JOHNSON'S METHOD (FOR n-JOBS 2-MACHINES)

Let us suppose n jobs $(1, 2, ..., n)$ are to be processed on two machines say A and B and A_i, B_i, $i = 1, 2, ..., n$ are the respective processing times of ith job on A and B machines respectively.

24.4.1. ASSUMPTIONS

(i) Each job is processed in order AB.

(ii) A_i = Processing time of ith job on machine A ($i = 1, 2, ..., n$)

(iii) B_i = Processing time of ith job on machine B ($i = 1, 2, ..., n$)

We want to find the sequence of jobs to be performed on two machines so that the total time (T) elapsed from the start of the first job to the completion of the last job to be minimized.

Working Procedure

STEP 1. Select the smallest processing time in the list $A_1, A_2,..., A_n$ and $B_1, B_2, ..., B_n$. If there is a tie then either of these smallest processing time may be selected or in this case consider the following cases:

  (i) Minimum of all the processing times is A_r which is also equal to B_s. Then $min(A_i, B_i) = A_r = B_s$. Then do the rth job first and s^{th} job in the end.

  (ii) If $min(A_i, B_i) = A_r$ but also $A_r = A_k$ (say) then do anyone of these jobs for which there is a tie, first.

  (iii) If there is a tie for minimum among B_i's, i.e., $Min(A_i, B_i) = B_s = B_r$ (say) then do any of these jobs in the last.

STEP 2. If the smallest processing time is A_r (i.e., in the list $A_1, ..., A_n$) then do the rth job first. On the other hand if it is B_s (i.e., in the list $B_1, B_2, ..., B_n$). Then do the sth job last.

STEP 3. Delete the times of already assigned job from both the list. If r^{th} job is assigned previously, then delete A_r and B_r both and if s^{th} job is assigned previously then delete A_s and B_s both.

STEP 4. Repeat step 1 to 3 for remaining jobs.

STEP 5. Continuing the same process until all the jobs have been ordered and get optimal sequence of jobs.

24.5 SEQUENCING DECISION PROBLEM OF n-JOB AND THREE MACHINES

[MEERUT-2000, 01, 02, 04, 06, 18 UPTU MBA-2003, 07]

Consider the problem of determining the optimal sequence of n jobs to be performed on the three machines A, B and C in the order ABC where A_i, B_i and C_i are the processing times of the ith job on the three machines respectively.

No method is yet available so far for the solution of the problem as such. However the problem of three machines can be solved by the method developed by Johnson under the following conditions:

  (i) The smallest processing time for machine A is greater than or equal to the greatest processing time for the machine B i.e., $\underset{i}{Min}(A_i) \geq \underset{i}{Max}(B_i)$

  (ii) The smallest processing time for machine C is greater than or equal to the greatest processing time for the machine B i.e., $\underset{i}{Max}(B_i) \leq \underset{i}{Min}(C_i)$

Working Procedure

STEP 1. We replace the three machines by two fictitious machines say G and H with corresponding processing times given by

$$G_i = A_i + B_i, \quad H_i = B_i + C_i, \quad i = 1, 2, ..., n$$

STEP 2. Determine the optimal sequence of jobs for these two machines G and H in the usual manner, *i.e.*, by applying the algorithm meant for problems of n jobs and two machines.

The sequences so obtained will be the optimal sequence for the original problem also.

24.6 SEQUENCING DECISION PROBLEMS FOR n-JOBS AND m-MACHINES

[MEERUT 2001; ROHILKHAND-2001; UPTU MBA-2005]

A general sequencing problem of processing n jobs through m machines say $M_1, M_2, ..., M_m$, in the order $M_1, M_2 ... M_m$, can be solved under some conditions explained below.

If $M_{ij}, i = 1, 2, ..., n, \ j = 1, 2, ..., m$ is the processing time of ith job on j^{th} machine, then calculate $\underset{i}{Min} \, M_{i1}$ and $\underset{i}{Min} \, M_{im}$ and $\underset{i}{Max} \, M_{ij}, j = 2, ..., m-1$, *i.e.*, calculate the minimum processing times for first and the last machines and maximum processing times for all the intermediate machines. The problem can be solved only if either of the following two or both the conditions are satisfied:

(i) $\underset{i}{Min} \, M_{i1} \geq \underset{i}{Max} \, M_{ij}$, for all $j = 2, 3, ..., m-1$ or

(ii) $\underset{i}{Min} \, M_{im} \geq \underset{i}{Max} \, M_{ij}$ for all $j = 2, 3, ..., m-1$.

If at least one of these two conditions is satisfied, then for two fictitious machines say G and H, their processing times, namely G_i and H_i, given by

$$G_i = M_{i1} + M_{i2} + ... + M_{i(m-1)}, \quad i = 1, 2, ..., n$$
$$H_i = M_{i2} + M_{i3} + ... + M_{im}, \quad i = 1, 2, ..., n$$

The sequence which is optimal for the problem for two machines say G and H will give the required optimal sequence for the original problem.

☛ **REMARK**

- If $M_{i2} + M_{i3} + ... + M_{i(m-1)} = c$ (constant) for all i, where c is a fixed positive quantity, then the required optimal sequence can be obtained by solving a problem involving only two extreme machines *i.e.*, solving the problem of n jobs on two machines M_1, M_m in the order $M_1 M_m$.

 Solved Examples

EXAMPLE 1. *We have five jobs, each of which has to go through the machines A and B in the order AB. Processing times are given in the table below:*

Job	Processing time in hours	
	Machines	
	A_i	B_i
1	5	2
2	1	6
3	9	7
4	3	8
5	10	4

Determine a sequence for these jobs that will minimize the total elapsed time T. [AGRA-2001; UPTU MBA-2006]

SOLUTION. The minimum time in the above table is 1 which is A_2. Hence we shall do the 2nd job first. We list the jobs as shown below

2				

Now we are left with four jobs with the processing times as shown in the given table below:

Job i	A_i	B_i
1	5	2
3	9	7
4	3	8
5	10	4

Again as the minimum time in this table 2 which is B_1, we shall do the first job in last.

2				1

Now the time for the remaining jobs are as shown in the following table:

Job i	A_i	B_i
3	9	7
4	3	8
5	10	4

Similarly using the prescribed criterion, we conclude that the optimal sequence of jobs is

2	4	3	5	1

Further the minimum elapsed time can be calculated as follows :

Job	Machine A		Idle time of A	Machine B		Idle time of B
	Time in	Time out		Time in	Time out	
2	0	1	—	1	7	1
4	1	4	—	7	15	—
3	4	13	—	15	22	—
5	13	23	—	23	27	1
1	23	28	2	28	30	1

From the above table it is clear that the total time elapsed is 30 hours and the idle time for the machine B is 3 hours. Note that the total elapsed time is equal to the sum of the idle time of B and the total processing time on machine B.

The total elapsed time can also be calculated by using Gantt Chart as follows

From the Fig. 3, it can be seen that the total elapsed time is 30 hours and the idle time of the machine B is 3 hours.

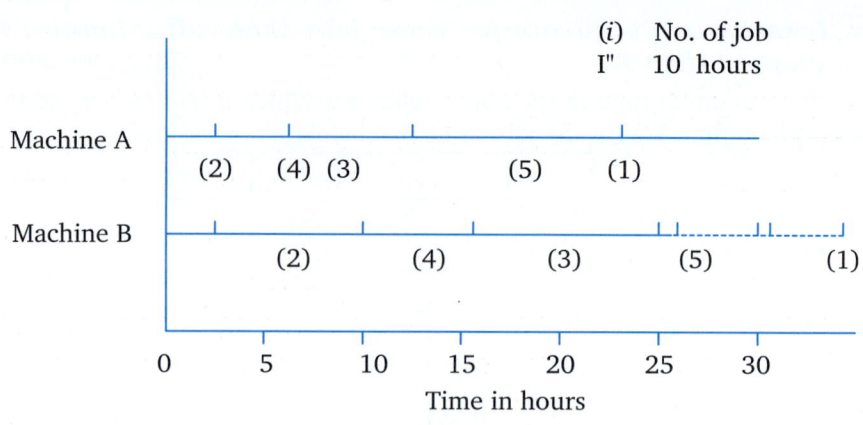

(i) No. of job
I" 10 hours

Fig. 3

☛ REMARK

• In the above problem it is to be noted that a job may be held in inventory before going to the machine. For instance 4th job will be free on machine *A* after 4th hour and will start on machine *B* after 7th hour. Therefore, it will be kept in inventory for 3 hours. So it is assumed that the storage space is available and the cost of holding the inventory for each job is either same or negligible. For short duration process problems generally it is negligible. Second general assumption is that the order of completion of jobs has no significance *i.e.,* no job claims the priority.

EXAMPLE 2. *Determine the optimal sequence of jobs which minimizes the total elapsed time based on the following information.*

Job	Processing times on the machines *A, B, C*		
	A_i	B_i	C_i
1	3	3	5
2	8	4	8
3	7	2	10
4	5	1	7
5	2	5	6

[MEERUT-1999; DELHI-1996]

SOLUTION.   We have Min. A_i = 2, Max. B_i = 5, Min. C_i = 5.

Since Max. B_i ≤ Min. C_i, the problem can be solved by the above procedure. The times for the fictitious machines G and H are given by the following table:

Job	Processing times	
	$G_i = A_i + B_i$	$H_i = B_i + C_i$
1	6	8
2	12	12
3	9	12
4	6	8
5	7	11

Note that here minimum time is 6, which is both G_1 and G_4. As there is a tie, any of the jobs first and fourth can be performed in the starting. Thus the optimal sequence may be formed in any of the following two ways:

1	4	5	3	2

4	1	5	3	2

Total elapsed time associated to the first sequence is calculated below :

Job	Machine A		B		C		Idle time
	Time in	Time out	Time in	Time out	Time in	Time out	of C
1	0	3	3	6	6	11	6
4	3	8	8	9	11	18	0
5	8	10	10	15	18	24	0
3	10	17	17	19	24	34	0
2	17	25	25	29	34	42	0

Hence the total elapsed time is 42 hours. Similarly we can show that other sequence also takes total elapsed time 42 hours.

EXAMPLE 3. *Find the optimal sequence for processing 4 jobs A, B, C, D on four Machines A_1, A_2, A_3, A_4 in the order $A_1\ A_2\ A_3\ A_4$. Processing times are as given below:*

Processing times (M_{ij}) in hours				
Job/Machine	$A_1(M_{i1})$	$A_2(M_{i2})$	$A_3(M_{i3})$	$A_4(M_{i4})$
A	15	5	4	14
B	12	2	10	12
C	13	3	6	15
D	16	0	3	19

SOLUTION.   From the above table we get for extreme machines

Min. (M_{i1}) = Min. processing time on first machine = 12
Min. (M_{i4}) = Min. processing time on last machine = 12

and for intermediate machines

Max. (M_{i2}) = Max. processing time on 2nd machine = 5
Max. (M_{i3}) = Max. processing time on 3rd machine = 10

Since Min. (M_{i1}) > Max. a_{i2} and Max. a_{i3} (both), the problem can be reduced to a problem involving only two machines G and H with processing times as

Jobs	Processing times	
	G_i	H_i
A	15 + 5 + 4 = 24	5 + 4 + 14 = 23
B	12 + 2 + 10 = 24	2 + 10 + 12 = 24
C	13 + 3 + 6 = 22	3 + 6 + 15 = 24
D	16 + 0 + 3 = 19	0 + 3 + 19 = 22

Using the algorithm for solving a sequencing problem of n jobs and 2 machines, we get the optimal sequence as given below.

D	C	B	A

Total elapsed time can be calculated as follows

Job/Machine	A_1		A_2		A_3		A_4	
	in	out	in	out	in	out	in	out
D	0	16	16	16	16	19	19	38
C	16	29	29	32	32	38	38	53
B	29	41	41	43	43	53	53	65
A	41	56	56	61	61	65	65	79

∴ Total elapsed time = 79 hours.

EXAMPLE 4. *Consider the problem of five jobs, each of which must go through the machines A, B, C in the order ABC. Processing time are:*

Job	A	B	C
1	4	5	8
2	9	6	10
3	8	2	6
4	6	3	7
5	5	4	11

Find a sequence for the five jobs that will minimize the elapsed time T.
[MEERUT 2003]

SOLUTION. We have

Min. $A_i = 4$, Max. $B_i = 6$, Min. $C_i = 6$

Since Max $B_i \le$ Min. C_i. Then we proceed as follows

Jobs	1	2	3	4	5
$G_i = A_i + B_i$	9	15	10	9	9
$H_i = B_i + C_i$	13	16	8	10	15

Then proceeding same as previous examples, we get the following optimal sequencing

5	1	4	2	3

4	1	5	2	3

1	4	5	2	3

1	5	4	2	3

4	5	1	2	3

5	4	1	2	3

EXAMPLE 5. *We have five jobs, each of which must go through the machines A, B and C in order ABC*

Job	1	2	3	4	5
Machine A	5	7	6	9	5
Machine B	2	1	4	5	3
Machine C	3	7	5	6	7

Determine a sequence for the jobs that will minimize the total elapsed time.
[MEERUT-1993; 98, 2001, 03, 2012, 2016; ROHILKHAND-2001]

SOLUTION. We have

Min. $A_i = 5$,   Max. $B_i = 5$,   Min. $C_i = 3$.

Clearly, Min $A_i \geq$ Max. B_i.

Jobs	1	2	3	4	5
$G_i = A_i + B_i$	7	8	10	14	8
$H_i = B_i + C_i$	5	8	9	11	10

The minimum time in table is 5 which is H_1. Therefore the job 1 will be done last. Thus, we get the following optimal sequence.

2	5	4	3	1

5	4	3	2	1

5	2	4	3	1

To find the elapsed time, we prepare the following table.

Job	Machine A Time in	Machine A Time out	Machine B Time in	Machine B Time out	Machine C Time in	Machine C Time out	Idle time of B	Idle time of C
2	0	7	7	8	8	15	7	8
5	7	12	12	15	15	22	4	–
4	12	21	21	26	26	32	6	4
3	21	27	27	31	32	37	1	
1	27	32	32	34	37	40	1 + 6	

Hence, the minimum elapsed time is 40 hours. We may easily verify the time for other alternative sequencing also. Idle time for machines A, B and C are respectively given by 8, 25 and 12 hours.

EXAMPLE 6. *(a) A book binder has one printing press, one binding machine and the manuscripts of a number of different books. The times required to perform the printing and binding operations for each book are known. Determine the order in which the books should be processed in order to minimize the total time required. Find also the total time required to process all the books.*

Processing time in minutes					
Book	1	2	3	4	5
Printing time	40	90	80	60	50
Binding time	50	60	20	30	40

[ROHILKHAND-1995; UPTU MBA-2005, 08]

(b) Suppose that an additional operation is added to the process described in (a) : finishing. The times required for operation are given below

Book	1	2	3	4	5
Finishing time	80	100	60	70	110

What is the order in which the books should be processed? Find also the minimum total elapsed time.

SOLUTION. (a) The optimal sequence obtained by the sequencing algorithm, meant for n jobs and 2 machines is

| 1 | 2 | 5 | 4 | 3 |

and the total minimum elapsed time is 340 minutes.

(b) Now let the three machines be P (for printing), B (for printing) and F (for finishing) then clearly min $P_i = 40$, max $B_i = 60$, min $F_i = 60$.

Since Min. $F_i \geq$ Max. B_i, the given problem can be converted into n jobs and 2 machines problem. If the new machines are G and H with processing times G_i and $G_i = P_i + B_i$, $H_i = B_i + F_i$.

Thus the new problem will become as follows:

Book	1	2	3	4	5
G_i	90	150	100	90	90
H_i	130	160	80	100	150

The optimal sequence for this problem is given below :

| 4 | 1 | 5 | 2 | 3 |

The total elapsed time can be calculated as follows:

Book	4		1		5		2		3	
	in	out	in	out	in	out	in	out	in	out
Printing	0	60	60	100	100	150	150	240	240	320
Binding	60	90	100	150	150	190	240	320	320	340
Finishing	90	160	160	240	240	350	350	450	450	510

$\Rightarrow$ Total elapsed time is 510 minutes.

EXAMPLE 7. *A readymade garments manufacturer has to process 7 items through two stages of production, i.e., cutting and sewing. The time taken for each of these items at the different stages are given below in appropriate units.*

Item		1	2	3	4	5	6	7
Processing	Cutting	5	7	3	4	6	7	12
Time	Sewing	2	6	7	5	9	5	8

(a) *Find an order in which these seven items are to be processed so as to minimise the total processing time.*   [ROHILKHAND-2003]

(b) *Suppose a third stage of production is added, say pressing and packing, with processing times as follows :*

| Processing time | 10 | 12 | 11 | 13 | 12 | 10 | 11 |

Find an order in which these seven items are to processed so as to minimize the time taken to process all the items through all the three stages.

SOLUTION. (a) The optimal sequence by the sequencing algorithm is :

| 3 | 4 | 5 | 7 | 2 | 6 | 1 |

and the total elapsed time = 46 hours.

(b) Proceed as in the previous example. The optimal sequence is

1	4	3	6	2	5	7

and the total elapsed time is 86 hours.

EXAMPLE 8. *Find an optimal sequence for the following sequencing problem of four jobs and five machines when passing is not allowed, of each processing time (in hours) is given below*

	Job	1	2	3	4
Machine	M_1	6	5	4	7
Machine	M_2	4	5	3	2
Machine	M_3	1	3	4	2
Machine	M_4	2	4	5	1
Machine	M_5	8	9	7	5

Also find the total elapsed time.   [MEERUT-2002 BP]

SOLUTION. Here, Min $M_{1i} = 4$, Max. $M_{2i} = 5$, Max. $M_{3i} = 4$, Max. $M_{4i} = 5$ and Min. $M_{5i} = 5$. Since Min. $M_{5i} \geq$ Max. M_{2i}, Max. M_{3i}, Max. M_{4i} i.e., minimum of one extreme machine is greater than or equal to the maximums of all the intermediate machines, the problem can be converted into 4 jobs and two machines problem. If the two fictitious machines are G and H, then there processing times can be calculated by

$$G_i = M_{1i} + M_{2i} + M_{3i} + M_{4i}, \quad H_i = M_{2i} + M_{3i} + M_{4i} + M_{5i}.$$

The new problem will be as follows :

Job i	1	2	3	4
Times of machine G (G_i)	13	17	16	12
Times of machine $H(H_i)$	15	21	19	10

The optimal sequence is given below

1	3	2	4

Evaluation of total elapsed time si given in the following table :

Job	1		3		2		4	
	in	out	in	out	in	out	in	out
M_1	0	6	6	10	10	15	15	22
M_2	6	10	10	13	15	20	22	24
M_3	10	11	13	17	20	23	24	26
M_4	11	13	17	22	23	27	27	28
M_5	13	21	22	29	29	38	38	43

Total minimum elapsed time = 43 hours.

24.7 SEQUENCING PROBLEM INVOLVING TWO JOBS AND m MACHINES

Here, we consider the problem in which

(i) There are only two jobs say 1 and 2 to be performed,

(ii) There are m machines say $A, B, C, ..., M$.

(iii) Exact or expected processing times of jobs on the machines are known.

(iv) The technological ordering of each job through machines is known in advance.

Then we are to determine the sequence of jobs for each machines so that the total elapsed time is minimum.

24.8 GRAPHICAL METHOD [MEERUT-2003; UPTU (MBA)-2003]

This method is applicable to the problems involving two jobs and m machines. Job 1 and Job 2 are represented by two horizontal and vertical axes respectively. On these two axes we mark the processing times of jobs on different machines in given order. Thus a horizontal line in the graph will represent the work on job 1 only while job 2 remains idle. Similarly a vertical line in the graph will represent only the work on job 2 while job 1 remains idle. A line inclined at an angle 45° with the horizontal, will represent the work on both the jobs simultaneously. Note that a job will be processed on a machine if machine is idle and is next for this job. A horizontal or vertical line will occur whenever some job is idle but the machine which is the next to this job is not idle. Also note that both the jobs cannot be processed on the same machine. Now we start with zero time (origin 0) and go on doing the jobs avoiding these shaded rectangular blocks. A best path is that which minimizes the idle time for job 1 and job 2, *i.e.*, minimizes the horizontal and vertical lines in the path. Thus we try to move along the line inclined at 45° as much as we can.

 Solved Examples

EXAMPLE I. *Use graphical method to minimize the time needed to process the following jobs on the machines shown, i.e., for each machine find the job which should be done first. Also calculate the total time needed to complete both the jobs.*

		Machines				
Job 1	Sequence	A	B	C	D	E
	Time	3	4	2	6	2
Job 2	Sequence	B	C	A	D	E
	Time	5	4	3	2	6

[AGRA-1999, 2000 MEERUT-2004]

SOLUTION. We use the following steps:

STEP 1. Mark the processing times for first and second jobs on the horizontal and vertical axis according to the technological ordering of the machines.

STEP 2. Construct the rectangular blocks by pairing the same machines as shown in Fig. 4.

STEP 3. Now mark a path from origin 0 to the point of finish by moving along the 45° line as much as possible avoiding rectangular blocks.

STEP 4. Find the total elapsed time by adding the idle time of job 1 to its processing time or adding idle time of job 2 to its processing time. These two times will be equal.

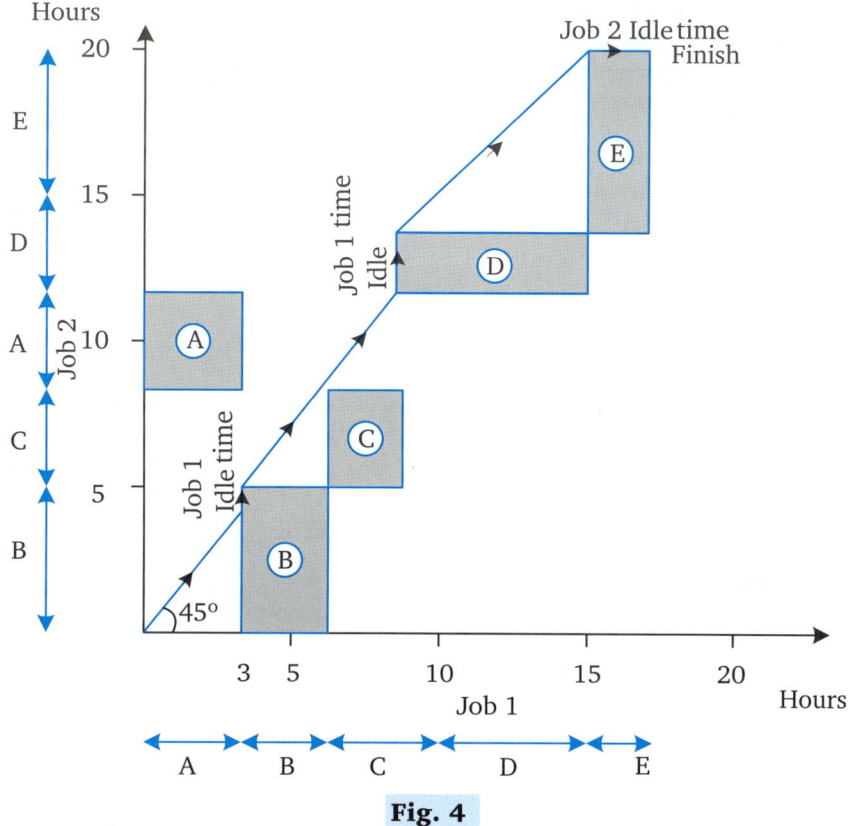

Fig. 4

The 'best' path is shown in Fig. 4 by arrows. The elapsed time is

"Processing time of Job 1 + idle time for Job 1 = 17 + 5 = 22 hours"

"Processing time of Job 2 + idle time of Job 2 = 20 + 2 = 22 hours"

Obviously in this route we have processed,

"Job 1 before 2 on machine A"

"Job 2 before 1 on machine B"

"Job 2 before 1 on machine C"

"Job 2 before 1 on machine D"

and "Job 2 before 1 on machine E"

EXAMPLE 2. *Use the graphical method to minimize the time needed to process the following jobs on the machine shown, i.e., for each machine, find the job which should be done first. Also calculate the total elapsed time to complete both jobs.*

							Total
Job-1	Machine sequence	A	B	C	D	E	17
	Time	2	3	4	6	2	
Job-2	Machine sequence	C	A	D	E	B	20
	Time	4	5	3	2	6	

[MEERUT-2002, 06, 09]

<u>**SOLUTION.**</u>   To solve the above problem by graphical method we proceed as follows:

STEP 1. Draw the set of axes at right angles to each other where x-axis represents the processing time of job-1 on different machine, while job-2 remains idle and y-axis represents processing time of job-2 while job-1 remains idle.

STEP 2. Mark the processing time for first and second job on x and y-axis respectively according to the given order of machine.

STEP 3. Construct the rectangular block by pairing the same machine as shown in the figure.

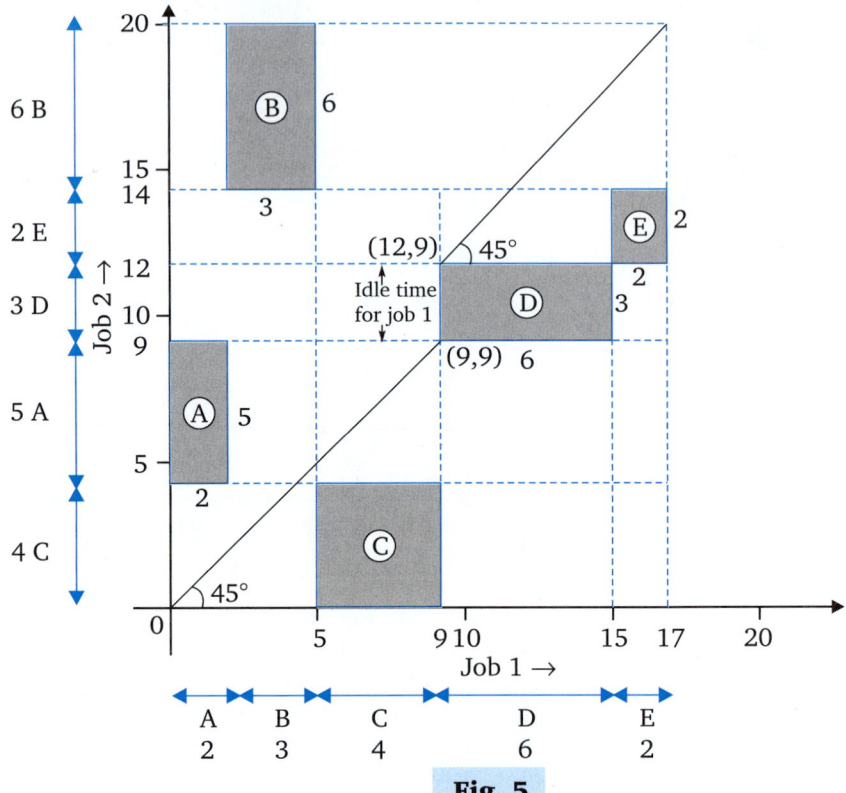

Fig. 5

STEP 4. Make a program by starting from the origin O and moving through various states of completion until we reach the point 'finish' by moving along the 45° line as much as possible avoiding rectangular blocks. Moving to the right means that job-1 is proceeding while job-2 is idle and moving upward indicate that job-2 is proceeding while job-1 is idle and moving diagonally means simultaneous work on both jobs.

STEP 5. The optimum path is one which coincides with 45° are to the maximum extent. Further both jobs can not be processed simultaneously on one machine. Graphically, it means that diagonal movement through the blocked out areas is not allowed.

STEP 6. Total elapsed time is obtained by adding the idle time for either job to the processing time for that job. The idle time for the chosen path is found to be 3 for job-1 and 0 for job-2.

Hence, total elapsed time = 20 hours.

EXAMPLE 3. *A company has five machines A, B, C, D and E. Two jobs 1 and 2 must be processed through each of these machine. The processing time (in hours) for each job on different machine are as follows:*

Job-1 Machine					
Sequence	A	B	C	D	E
Time (*h*)	2	4	5	1	2

Job-2 Machine					
Sequence	D	E	A	C	B
Time (*h*)	6	4	2	3	6

Use the graphical method to determine the total elapsed time.

SOLUTION.   Proceeding as in previous examples, we have the following steps:

STEP 1. Mark the processing times for job 1 and job 2 on the *x* and *y*-axis respectively according to the given sequential order of five machines.

STEP 2. Construct the rectangular blocks by pairing the same machines.

STEP 3. Mark a path from the origin to the end points by moving along the 45° line as much as possible.

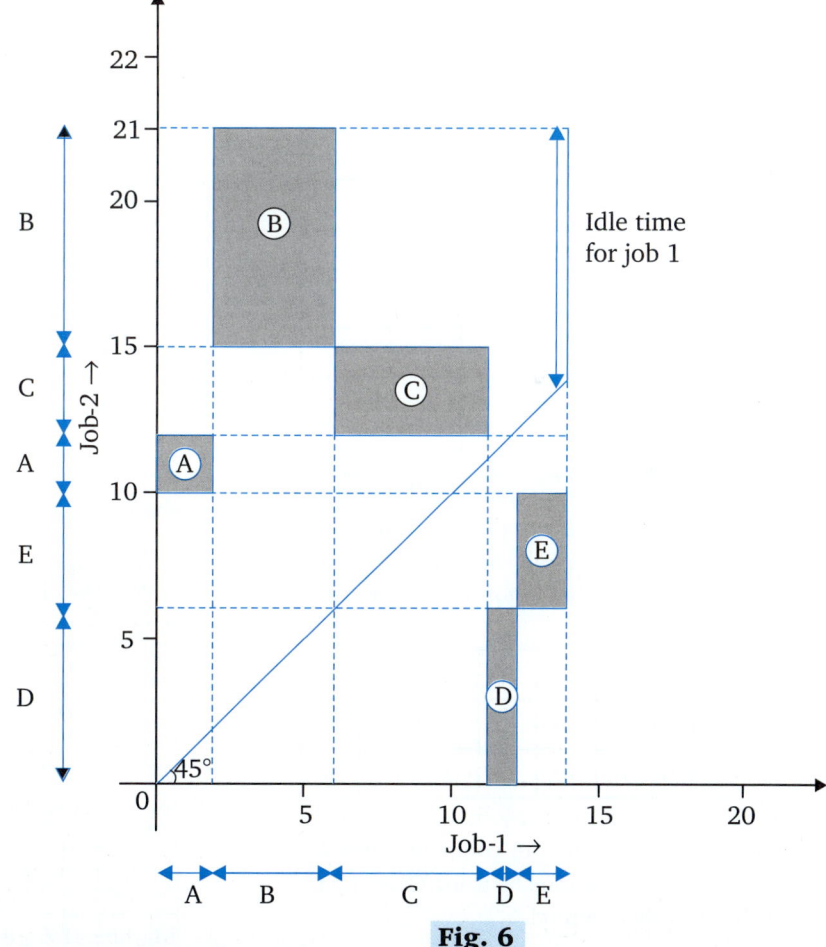

Fig. 6

From the above graph, we observe that
        Idle time for job-1 = 7 hrs.
        Processing time for job-1 = 14 hrs.
and   the total elapsed time = 14 + 7 = 21 hrs.

1. Find the sequence that minimizes the total elapsed time required to complete the following jobs.

	Job	1	2	3	4	5	6
Processing	A_i	2	5	4	3	2	1
Time	B_i	6	8	1	2	3	5

2. A book binder has one printing press, one binding machine and the manuscripts of a number of different books. The time required to perform the printing and binding operations for each book are known. We wish to determine the order in which books should be processed on the machines, in order to minimize the total time required

Book	Printing Time	Binding time
1	30	80
2	120	100
3	50	90
4	20	60
5	90	30
6	110	10

3. Six jobs go first over machine I and then over II. The order of the completion of jobs has no significance. The following table gives the machine times in hours for six jobs and the two machines:

Job no.	1	2	3	4	5	6
Machine I	5	9	4	7	8	6
Machine II	7	4	8	3	9	5

Find the sequence of the jobs that minimizes the total elapsed time to complete the jobs. Find the minimum time by using Gantt Chart or by any other method.

4. Find the sequence that minimizes the total elapsed time required to complete the following jobs.

	Processing time in hours					
No of Job	1	2	3	4	5	6
Machine A	4	8	3	6	7	5
Machine B	6	3	7	2	8	4

5. (a) We have five jobs each of which must go through the two machines in the order *AB*. Processing times are given in the table below :

Job No.	1	2	3	4	5
Machine A	10	2	18	6	20
Machine B	4	12	14	16	8

Determine a sequence for the five jobs that will minimize the total elapsed time.

(b) Find the optimal sequence of the job.

Job No.	1	2	3	4	5
Machine A	3	7	4	5	7
Machine B	6	2	7	3	4

Passing not allowed.   (ROHILKHAND-1997)

6. (a) Seven jobs each of which has to go through two machines M_1 and M_2 in order M_1M_2, take time on the machines as follows. Find in which order the jobs should be performed to minimize total time.

Job No.	1	2	3	4	5	6	7
Machine M_1	3	12	15	6	10	11	9
Machine M_2	8	10	10	6	12	1	3

[MEERUT-2002,06; AGRA-1998, 2003]

(b) Find the sequence that minimizes the total elapsed time required to complete the following jobs

Jobs	A	B	C	D	E	F	G	H	I
Machine I	2	5	4	9	6	8	7	5	4
Machine II	6	8	7	4	3	9	3	8	11

7. Find the sequence that minimizes the total elapsed time (in hours) required to complete all the following jobs on machines M_1, M_2,

M_3 in the order $M_1M_2M_3$.

Machine	Jobs				
	A	B	C	D	E
M_1	4	9	8	6	5
M_2	5	6	2	3	4
M_3	8	10	6	7	11

8. State the conditions under which the problem of processing of n jobs through three machines has been solved. Describe the corresponding algorithm.

Find the sequence that minimizes the total time required to complete the following tasks:

Task	A	B	C	D	E	F	G
Machine I	3	8	7	4	9	8	7
Machine II	4	3	2	5	1	4	3
Machine III	6	7	5	11	5	6	12

9. Find the sequence that minimizes the total time required for performing the following jobs on three machines in the order *ABC*.

Processing times in hours

Jobs	1	2	3	4	5	6
Machine A	8	3	7	2	5	1
Machine B	3	4	5	2	1	6
Machine C	8	7	6	9	10	9

10. Solve the following sequencing problems:

(i)   Processing times in hours

Jobs/Machine	M_1	M_2	M_3	M_4
A	10	3	5	14
B	12	2	6	7
C	8	4	4	12
D	15	1	7	8
E	16	5	3	10

[MEERUT 1993, 2005; ROHILKHAND-1998, 99]

(ii)   Processing times in hours

Jobs/Machine	M_1	M_2	M_3	M_4
A	13	8	7	14
B	12	6	8	19
C	9	7	5	15
D	8	5	6	15

11. Find the sequence that minimizes the total elapsed time required to complete the following tasks. Each task is processed in the order *ACB*.

Processing times

Jobs	1	2	3	4	5	6	7
Machine A	12	6	5	11	5	7	6
Machine B	7	8	9	4	7	8	3
Machine C	3	4	1	5	2	3	4

[MEERUT 2001 BP, ROHILKHAND-2001]

(iii) $A \to C \to I \to H \to B \to F \to D \to G \to E$ (iv) $A \to I \to C \to B \to H \to F \to D \to E \to G$

(v) $A \to C \to I \to B \to H \to F \to D \to G \to E$ (vi) $A \to I \to C \to H \to B \to F \to D \to E \to G$

(vii) $A \to I \to C \to B \to H \to F \to D \to G \to E$ (viii) $A \to C \to I \to H \to B \to F \to D \to E \to G$

Total min. elapsed time = 61 hours.

7. Optimal sequences :

(i) $A \to D \to E \to B \to C$ (ii) $A \to E \to D \to B \to C$ (iii) $D \to A \to E \to B \to C$

(iv) $D \to E \to A \to B \to C$ (v) $E \to D \to A \to B \to C$ (vi) $E \to A \to D \to B \to C$

Time = 51 hours

8. Optimal sequences are

| A | D | G | F | B | C | E | or | A | D | G | B | F | C | E |

Time = 59 hours.

9. Sequence is as follows

| 4 | 5 | 2 | 6 | 1 | 3 |

Min. time = 53 hours

10. (i) Sequence is as follows   (ii) Sequence is as follows

| C | A | E | D | B | | D | C | B | A |

Total time = 76 hours

11. Sequence is as follows $3 \to 5 \to 2 \to 6 \to 1 \to 4 \to 7$ or $3 \to 5 \to 6 \to 2 \to 1 \to 4 \to 7$,

Time = 59 hours.

REVIEW QUESTIONS

1. Define the problem of sequencing.
   [MEERUT-2003]

2. What is no processing rule in a sequencing problem? Explain the principal assumption made while dealing with sequencing problem.

3. Give some examples of sequencing problems in daily life.

4. Explain Johnson's procedure for determining an optimal sequence for processing n items on two machines.
   [AGRA-2002, MEERUT-2000, 01, 03, 03, 15]

5. What do you understand by the following terms in sequencing.

   (i) Job arrival pattern

   (ii) The flow pattern

   (iii) Criterion for evaluating the performance of a schedule

   (iv) No. of machines

6. Write a short note on Gantt. Chart.

7. State clearly the assumptions used in the study of sequencing problem.   [UPTU-2002]

8. Write a short note on sequencing.
   [UPTU (MBA)-2002]

MULTIPLE CHOICE QUESTIONS (CHOOSE THE MOST APPROPRIATE ONE)

1. The time between starting the first job completing the last job including the idle time is called:
   (a) the total elapsed time
   (b) idle time
   (c) the difference between idle time and total elapsed time
   (d) None of these

2. The time for which a machine does not work is known as:
   (a) the total elapsed time
   (b) idle time
   (c) wasting time
   (d) None of these

3. In a sequencing problem, the order is fixed for:
   (a) jobs
   (b) machines
   (c) both jobs and machines
   (d) None of these
4. In a sequencing algorithm:
   (a) the selection of an appropriate order for a series of jobs is to be done on a finite service facilities
   (b) all the jobs must be processed on a first come first serviced basis
   (c) all the service facilities are not of different type
   (d) None of these
5. If there are n jobs to be performed one at a time on each of m machines, the possible sequence would be:
   (a) $(n!)^m$
   (b) $(m!)^n$
   (c) n^m
   (d) None of these
6. The general assumption which is not correct in solving a sequencing problem is that:
   (a) a job once started on a machine would be performed to the point of completion uninterrupted
   (b) a machine can process more than one job at a given point of time

   (c) the time taken by different job in moving from one machine to other is negligible
   (d) None of these
7. Total elapsed time to process all jobs through two machines is given by:
   (a) $\sum\limits_{j=1}^{n} (M_{1j} + I_{1j})$
   (b) $\sum\limits_{j=1}^{n} M_{2j} + \sum\limits_{j=1}^{n} M_{1j}$
   (c) $\sum\limits_{j=1}^{n} M_{1j} + \sum\limits_{j=1}^{n} M_{2j}$
   (d) None of these
8. The minimum processing time on machine M_1 and M_2 are related as:
   (a) min. $t_{1j} = $ max. t_{2j}
   (b) min. $t_{1j} \leq $ max. t_{2j}
   (c) min. $t_{1j} \geq $ max. t_{2j}
   (d) None of these
9. If A_i, B_i and C_i are the processing time of ith job on three machines A, B and C respectively then n-jobs, three machines problem can be reduced to an n-job and 2-machine problem provided that:
   (a) min. $A_i \geq$ max. B_i and/or min. $C_i \geq$ max. B_i
   (b) min. $A_i \leq$ max. B_i and/or min. $C_i \leq$ max. B_i
   (c) min. $A_i \leq$ max. B_i and/or min. $C_i \geq$ max. B_i
   (d) None of these

Answers

1. (a) 2. (b) 3. (b) 4. (a) 5. (a) 6. (b) 7. (b) 8. (c) 9. (a)

Archive

1. A book binder has one printing press, one binding machine and manuscripts of a number of books. The time required to perform the printing and binding operations on each book are shown below. The binder wishes to determine the order in which the books should be processed, so that the total time required to process all books is minimized.

Book	1	2	3	4	5	6
Printing (hrs)	30	120	50	20	90	110
Binding time (hrs)	80	100	90	60	30	10

(ROHILKHAND-1995; UPTU MBA-2006; DELHI-1986)

2. A machine operator has to perform two operations, turning and threading on a number of different jobs. The time required to perform these operations (in minutes) for each job is known.

Job	Time of turning (minutes)	Time of threading (minutes)
1	3	8
2	12	10
3	5	9
4	2	6
5	9	3
6	11	1

Determine in order in which the jobs should be processed in order to minimize the total time required to turn cut all the jobs.

(AIME-1989)

3. A job consists of N steps. Step i takes time t_i. If these jobs are grouped somehow into station system, then twice as many units can

be produced each day. Also, two setups in parallel can also double the production rate. Critically examine the advantages of these two approaches.

4. Two jobs are to be processed on four machines A, B, C and D. The technoligical order for these jobs on machines is as follows :

Job 1	A	B	C	D
Job 2	D	B	A	C

Proceeding times are given in the following table :

	A	B	C	D
Job 1	4	6	7	3
Job 2	4	7	5	8

Find the optimal sequence of jobs on each of the machines.

HINTS AND ANSWERS

1. Optimal sequence 4 –1 – 3 – 2 – 5 – 6 elapsed time = 430 hours.
   Idle time for printing press = 10 hours and for binding machines = 40 hours.

2. Optimal sequence 4 – 1 – 3 – 2 – 5 – 6
   Elapsed time = 43 minutes. Idle time for turning operation = 1 min and for threading operation = 6 minutes

4. Idle time is 4 hrs for job –1 and zero hour for job-2.
   Elapsed time for job 1 is 20 + 4 = 24 hours.

Classical Optimization Techniques 25

25.1 INTRODUCTION

If $y = f(x)$ be a continuous function. At a point $x = x_1$, if $f(x)$ does not increase and begins to decrease, then $f(x)$ has its maximum value at $x = x_1$ and if at a point $x = x_2$, $f(x)$ does not decrease and begins to increase, then $f(x)$ has its minimum value at $x = x_2$.

If $f(x)$ is maximum at a point $x = x_1$ then $f(x)$ is an increasing function for the preceding values of x_1 and decreasing for those value of x just below x_1 or we can say derivative of the function $\left(i.e., \dfrac{dy}{dx} \right)$ will be positive before $x = x_1$ and will be negative after $x = x_1$. But $\dfrac{dy}{dx}$ is a continuous function and $\dfrac{dy}{dx}$ changes the sign from positive to negative. So, $\dfrac{dy}{dx}$ will be zero at any point.

Therefore, for a maximum value of $y = f(x)$ at a point, we have $\dfrac{dy}{dx} = 0$ and $\dfrac{dy}{dx}$ changes the sign from positive to negative. On the other hand, for a minimum value of $y = f(x)$ we have $\dfrac{dy}{dx} = 0$ and $\dfrac{dy}{dx}$ changes the sign negative to positive.

☛ REMARKS

- If $\dfrac{dy}{dx}$ changes the sign positive to negative; it means that $f(x)$ is a decreasing function of x, i.e., $\dfrac{d^2y}{dx^2} < 0$.
- If $\dfrac{dy}{dx}$ changes the sign from negative to positive, it means that $f(x)$ is an increasing function of x, i.e., $\dfrac{d^2y}{dx^2} > 0$.
- A function may have more than one maximum and minimum value.
- Any minimum value of the function $f(x)$ can be greater than any maximum value.
- Maximum and minimum values of the function occur alternately.
- Maximum and minimum values of the function are sometimes known as extreme values.
- From the definition of maxima and minima, it is clear that $\dfrac{dy}{dx} = 0$ is the necessary condition for maximum or minimum.
- $\dfrac{d^2y}{dx^2} < 0$ is sufficient condition for maximum and $\dfrac{d^2y}{dx^2} > 0$ is sufficient condition for minimum.

Working Procedure

STEP 1. Find the derivative i.e., $\dfrac{dy}{dx}$ of y of the given function.

STEP 2. Put $\dfrac{dy}{dx} = 0$ and find all the real values of x_i. (say $x_1, x_2, x_3 \ldots$).

STEP 3. Find $\dfrac{d^2 y}{dx^2}$.

STEP 4. Put $x = x_i$ in $\dfrac{d^2 y}{dx^2}$ and find the result. If result is negative then the function $f(x)$ is maximum at $x = x_i$ and max. $f(x) = f(x_i)$. On the other hand, if result is positive then the function $f(x)$ is minimum at $x = x_i$ and minimum $f(x) = f(x_i)$.

☞ REMARKS

- In a continuous function, maxima and minima values occur alternately, i.e., between two successive maxima there is one minimum and between two successive minima, there is one maximum.

- If $\dfrac{d^2 y}{dx^2}$ is equal to 0 at any point $x = x_i$ then find $\dfrac{d^3 y}{dx^3}, \dfrac{d^4 y}{dx^4}$, and find the values of these derivatives at $x = x_i$ successively and check the sign.

Solved Examples

EXAMPLE 1. *Find the value of x for which f(x)=y= x⁴+2x³– 3x²– 4x+4 is maximum or minimum and also find those value of f(x).*

SOLUTION. Here, the given function is

$$y = f(x) = x^4 + 2x^3 - 3x^2 - 4x + 4 \qquad \ldots(1)$$

So $\quad \dfrac{dy}{dx} = 4x^3 + 6x^2 - 6x - 4 = 2(x+2)(2x+1)(x-1)$

Now, put $\dfrac{dy}{dx} = 0$, we have

$$2\,(x+2)(2x+1)(x-1) = 0 \Rightarrow x = -2, -\frac{1}{2}, 1$$

Again differentiating (2) w.r.t. to x, we get

$$\dfrac{d^2 y}{dx^2} = 12x^2 + 12x - 6$$

At $\quad x = -2$, we have

$$\dfrac{d^2 y}{dx^2} = 12(-2)^2 + 12(-2) - 6 = 48 - 24 - 6 = 18 > 0$$

Since, $\dfrac{d^2 y}{dx^2} > 0$ (i.e., positive). So $f(x)$ is minimum at $x = -2$. The minimum value of $f(x)$ at $x=-2$ is given by

$$f(-2) = (-2)^4 + 2(-2)^3 - 3(-2)^2 - 4(-2) + 4 = 0$$

Now, at $x = -\dfrac{1}{2}$, we have

$$\dfrac{d^2 y}{dx^2} = 12\left(-\dfrac{1}{2}\right)^2 + 12\left(-\dfrac{1}{2}\right) - 6 = 3 - 6 - 6 = -9 < 0$$

Since, $\dfrac{d^2 y}{dx^2} < 0$ (*i.e.*, negative). So, $f(x)$ is maximum at $x = -\dfrac{1}{2}$ and maximum

value of $f(x)$ at $x = -\dfrac{1}{2}$ is

$$f\left(-\dfrac{1}{2}\right) = \left(-\dfrac{1}{2}\right)^4 + 2\left(-\dfrac{1}{2}\right)^3 - 3\left(-\dfrac{1}{2}\right)^2 - 4\left(-\dfrac{1}{2}\right) + 4 = \dfrac{81}{16}$$

Similarly, at $x = 1$, we have

$$\dfrac{d^2 y}{dx^2} = 12(1)^2 + 12(1) - 6 = 12 + 12 - 6 = 18 > 0$$

Since, $\dfrac{d^2 y}{dx^2} > 0$ (*i.e.*, positive). So $f(x)$ is minimum at $x = 1$ and minimum value

of $f(x)$ at $x = 1$ is

$$f(1) = (1)^4 + 2(1)^3 - 3(1)^2 - 4(1) + 4 = 0$$

EXAMPLE 2. *Find the maximum and minimum value of the function*

$$y = f(x) = x^3 - 12x^2 + 36x + 21$$

SOLUTION. Here, the given function is

$$y = x^3 - 12x^2 + 36x + 21$$

Now, differentiating *w.r.t.* x, we get $\dfrac{dy}{dx} = 3x^2 - 24x + 36$

Putting $\dfrac{dy}{dx} = 0$, we get $3x^2 - 24x + 36 = 0$

$\Rightarrow \qquad\qquad\qquad x^2 - 8x + 12 = 0$

$\Rightarrow \qquad\qquad\qquad (x - 2)(x - 6) = 0 \quad \text{or} \quad x = 2, 6$

Again, differentiating *w.r.t.* x, we get

$$\dfrac{d^2 y}{dx^2} = 6x - 24$$

At $x = 2$, we have

$$\dfrac{d^2 y}{dx^2} = 6(2) - 24 = -12 < 0$$

Since, $\dfrac{d^2 y}{dx^2} < 0$ so $f(x)$ is maximum at $x = 2$. The maximum value of $f(x)$ at

$x = 2$ is given by

$$f(2) = (2)^3 - 12(2)^2 + 36(2) + 21 = 53.$$

Similarly, at $x = 6$, we have

$$\dfrac{d^2 y}{dx^2} = 6 \times 6 - 24 = 36 - 24 = 12 > 0$$

Since, $\dfrac{d^2y}{dx^2} > 0$ so, $f(x)$ is minimum at $x = 6$ and minimum value of $f(x)$ at $x = 6$ is

$$f(6) = (6)^3 - 12(6)^2 + 36(6) + 21 = 21$$

EXAMPLE 3. *Investigate for maximum and minimum values, the function* $(\sin x + \cos 2x)$. [MEERUT–2005, 12]

SOLUTION. Let $\qquad\qquad y = \sin x + \cos 2x,$

$$\Rightarrow \qquad\qquad \dfrac{dy}{dx} = \cos x - 2\sin 2x = \cos x - 4\sin x \cos x$$

For stationary point

$$\dfrac{dy}{dx} = 0 \;\Rightarrow\; \cos x(1 - 4\sin x) = 0$$

or $\qquad\qquad \cos x = 0$ or $1 - 4\sin x = 0 \Rightarrow x = \dfrac{\pi}{2}$ or $\sin x = \dfrac{1}{4}$

For maxima or minima

$$\dfrac{d^2y}{dx^2} = -\sin x - 4\cos 2x = -\sin x - 4(1 - 2\sin^2 x)$$
$$= -\sin x - 4 + 8\sin^2 x$$

(i) At $x = \dfrac{\pi}{2}$,

$$\left(\dfrac{d^2y}{dx^2}\right) = -1 - 4 + 8 = 3 \qquad\qquad \text{(which is positive.)}$$

So, given function is minimum at $x = \dfrac{\pi}{2}$ and min. value of y at $x = \dfrac{\pi}{2}$ is given by

$$\sin\dfrac{\pi}{2} + \cos 2 \times \dfrac{\pi}{2} = 1 - 1 = 0.$$

(ii) At $\sin x = \dfrac{1}{4}$,

$$\left(\dfrac{d^2y}{dx^2}\right) = -\dfrac{1}{4} - 4 + 8.\dfrac{1}{16}$$
$$= \dfrac{-15}{4} \qquad\qquad \text{(which is negative)} \cdot$$

Hence, given function is maximum at $x = \sin^{-1}\dfrac{1}{4}$ and max. value at $\sin x = \dfrac{1}{4}$

is, $\dfrac{1}{4} + \left[1 - 2 \times \left(\dfrac{1}{4}\right)^2\right] = \dfrac{7}{8}$.

EXAMPLE 4. *Find the maximum value of* $(x - 1)(x - 2)(x - 3)$. [ROHILKHAND–2005]

SOLUTION. Let $\quad f(x) = (x - 1)(x - 2)(x - 3) = x^3 - 6x^2 + 11x - 6$

then $f'(x) = 3x^2 - 12x + 11$

For a maximum or minimum value of $f(x)$, we must have $f'(x) = 0$

$$\Rightarrow \qquad 3x^2 - 12x + 11 = 0$$

i.e., $\qquad x = \dfrac{12 \pm \sqrt{144 - 4 \times 3 \times 11}}{6} = 2 \pm \dfrac{1}{\sqrt{3}}$

Also $f''(x) = 6x - 12$

Now $f''[2 + (1/\sqrt{3})] = +ve$, therefore $f(x)$ has minimum value at $x = 2 + (1/\sqrt{3})$.

Again $f''[2 - (1/\sqrt{3})] = -ve$, therefore $f(x)$ has a maximum value at $x = 2 - (1/\sqrt{3}) = f[2 - (1/\sqrt{3})] = 2/3\sqrt{3}$

EXAMPLE 5. *Show that sin x(1 + cos x) is a maximum at $x = \pi/3$.*

[MEERUT–2009, 13; KANPUR–2004, 08, 10, 15]

SOLUTION. Let $\qquad f(x) = \sin x(1 + \cos x)$

$$= \sin x + \dfrac{1}{2}\sin 2x$$

Then $f'(x) = \cos x + \cos 2x$

For a maximum or a minimum value of $f(x), f'(x) = 0$

i.e., $\quad \cos x + \cos 2x = 0 \Rightarrow 2\cos^2 x + \cos x - 1 = 0$

$\Rightarrow \qquad (2\cos x - 1)(\cos x + 1) = 0$

$\therefore \qquad\qquad \cos x = 1/2, -1 \Rightarrow x = \pi/3, \pi$

Now $\qquad\quad f''(x) = -\sin x - 2\sin 2x$

$\therefore \qquad\qquad f''\left(\dfrac{\pi}{3}\right) = -\sin\left(\dfrac{\pi}{3}\right) - 2\sin\left(\dfrac{2\pi}{3}\right) = -ve$

Hence $f(x)$ is maximum at $x = \pi/3$.

EXAMPLE 6. *Find the maximum value of $(1/x)^x$.* [ROHILKHAND–2005, 06]

SOLUTION. Let $\qquad\qquad\qquad y = (1/x)^x$

$\Rightarrow \qquad\qquad\qquad \log y = x(\log 1 - \log x) = -x \log x$

$\Rightarrow \qquad\qquad\qquad \dfrac{1}{y}\dfrac{dy}{dx} = -1\log x - x(1/x) = -(1 + \log x)$

$\Rightarrow \qquad\qquad\qquad \dfrac{dy}{dx} = -y(1 + \log x) = -(1/x)^x(1 + \log x)$

For a maximum or a minimum of y, we must have $\dfrac{dy}{dx} = 0$

$\Rightarrow \qquad -(1/x)^x(1 + \log x) = 0 \Rightarrow 1 + \log x = 0 \Rightarrow x = 1/e$

$\Rightarrow \qquad\qquad\qquad \dfrac{d^2y}{dx^2} = -\dfrac{dy}{dx}(1 + \log x) - y(1/x)$

$$= -\dfrac{dy}{dx}(1 + \log x) - (1/x)^x \cdot (1/x)$$

Therefore, when $x = 1/e$,

$$\dfrac{d^2y}{dx^2} = 0 - (e)^{1/e} \cdot e = -ve$$

$\Rightarrow \qquad y$ is maximum at $x = 1/e$.

Thus the maximum value of y is given by $e^{1/e}$.

EXAMPLE 7. *Show that the semi-vertical angle of the right circular cone of given total surface (including area of the base) and maximum value is $\sin^{-1}(1/3)$.* [KANPUR-2006; MEERUT-2002, 03, 11; ROHILKHAND-2004, 09]

SOLUTION. Let x be the radius of the base, h be the height and y, the slant height of the cone. Then the total surface of the cone = constant

$$\Rightarrow \quad \pi x^2 + \pi xy = \text{constant} \qquad \ldots(1)$$

Now $\qquad V = $ volume of the cone

$$= \frac{1}{3}\pi x^2 h = \frac{1}{3}\pi x^2 (y^2 - x^2)^{1/2}$$

Since, $h = \sqrt{y^2 - x^2}$, therefore

$$V^2 = \frac{1}{9}\pi^2 x^4 (y^2 - x^2).$$

Now, V is maximum or minimum according as V^2 or $\dfrac{9V^2}{\pi^2}$ is maximum or minimum.

Let $\qquad S = \dfrac{9V^2}{\pi^2} = x^4 (y^2 - x^2).$

Then S can be regarded as a function of x because y is connected with x by (1).

We have $\qquad \dfrac{dS}{dx} = 4x^3 (y^2 - x^2) + x^4 \left\{ 2y\left(\dfrac{dy}{dx}\right) - 2x \right\} \qquad \ldots(2)$

Differentiating (1) w.r. to x, we get

$$\pi\left(2x + y + x\frac{dy}{dx}\right) = 0 \quad \text{or} \quad \frac{dy}{dx} = -\frac{2x+y}{x}$$

Substituting this value of $\dfrac{dy}{dx}$ in (2), we get

$$\frac{dS}{dx} = 4x^3 y^2 - 4x^5 + x^4 \left[-2y\frac{(2x+y)}{x} - 2x \right]$$

$$= 2x^3 y^2 - 6x^5 - 4x^4 y$$

For a maximum or a minimum of S, we must have $\dfrac{dS}{dx} = 0$

Now $\qquad \dfrac{dS}{dx} = 0$

$$\Rightarrow 2x^3(y^2 - 2xy - 3x^2) = 0 \Rightarrow 2x^3(y - 3x)(y + x) = 0$$

i.e., $\quad y = 3x$ since $x \neq 0$ and $y \neq -x$.

Again $\qquad \dfrac{d^2 S}{dx^2} = 6x^2 y^2 + 4x^3 y\frac{dy}{dx} - 30x^4 + 16x^3 y - 4x^4 \frac{dy}{dx}$

When $y = 3x, \dfrac{dy}{dx} = -5$, so when $y = 3x$, we have

$$\frac{d^2 S}{dx^2} < 0$$

Therefore S is maximum when $y = 3x$.

EXAMPLE 8. *In a submarine telegraph cable the speed of signalling varies as log $x^2 \log(1/x)$, where x is the ratio of the radius of the core to that of the* *covering. **Show that the greatest speed is attained when this ratio is**

1: $\sqrt{e}$.* $\qquad$ [MEERUT–2002, 04, 14]

SOLUTION. Let S be the speed of signalling. Then

$$S = \mu x^2 \log(1/x) = -\mu x^2 \log x \text{, where } \mu \text{ is a constant.}$$

For a maximum or a minimum of S, we have $\dfrac{dS}{dx} = 0$

i.e., $\qquad x(2\log x + 1) = 0$

$\Rightarrow \qquad\qquad x = 0 \text{ or } \log x = -1/2$

But $x = 0$ is inadmissible. Therefore

$$\log x = -1/2 \text{ or } x = e^{-1/2} = 1/\sqrt{e}$$

Now $\qquad\qquad \dfrac{d^2S}{dx^2} = -\mu(2\log x + 1) - \mu x(2/x)$

$$= -\mu(2\log x + 1) = -2\mu$$

When $x = 1/\sqrt{e}$, we have $2\log x + 1 = 0$, when $x = 1/\sqrt{e}$, we have $\dfrac{d^2S}{dx^2} = -2\mu$ which is negative.

Hence, S is maximum, when $x = 1/\sqrt{e}$.

EXAMPLE 9. ***Show that maximum rectangle that can be inscribed in a circle is square.*** $\qquad$ [MEERUT–2005, 09, 16]

SOLUTION. Let $PQRS$ be the rectangle inscribed in circle with centre O and radius a. Also, let $PQ = 2x$ and $QR = 2y$. Then

$$a^2 = x^2 + y^2 \qquad\qquad \text{...(1)}$$

Area of rectangle $PQRS$

$$A = (2x)(2y) = 4xy = 4x\sqrt{a^2 - x^2} \qquad \text{[From (1)]}$$

For maximum or minimum area, $\dfrac{dA}{dx} = 0$

$$\Rightarrow \quad 4\left\{\sqrt{a^2 - x^2} - \frac{x^2}{\sqrt{a^2 - x^2}}\right\} = 0 \Rightarrow 4\left\{\frac{a^2 - 2x^2}{\sqrt{a^2 - x^2}}\right\} = 0$$

$$\Rightarrow \qquad\qquad a^2 - 2x^2 = 0 \Rightarrow x = \frac{a}{\sqrt{2}}$$

Now $\dfrac{d^2A}{dx^2} = 4\left\{(-4x)(a^2 - x^2)^{1/2} + (a^2 - 2x^2)\left(-\frac{1}{2}\right)(a^2 - x^2)^{-3/2}(-2x)\right\}$

$$= 4\left[\frac{-4x}{\sqrt{a^2 - x^2}} + \frac{x(a^2 - 2x^2)}{(a^2 - x^2)^{3/2}}\right]$$

$$\Rightarrow \qquad\qquad \left(\frac{d^2A}{dx^2}\right)_{x = a/\sqrt{2}} = -16$$

(which is negative.)

Thus, A is max. when $x = \dfrac{a}{\sqrt{2}}$.

From (1), $\qquad y = \dfrac{a}{\sqrt{2}}$. Therefore, $x = y = \dfrac{a}{\sqrt{2}}$

Hence, area is maximum when $x = y = \dfrac{a}{\sqrt{2}}$ i.e., rectangle is square.

EXAMPLE 10. *Show that the height of the closed cylinder of given surface and greatest volume is equal to its diameter.* [MEERUT–2005]

SOLUTION. Let r be radius of base and h the height of a closed cylinder of given surface S, then

$$S = 2\pi r^2 + 2\pi rh \Rightarrow h = \frac{S - 2\pi r^2}{2\pi r} \qquad \qquad ...(1)$$

If V be volume of cylinder then

$$V = \pi r^2 h = \pi r^2\left(\frac{S - 2\pi r^2}{2\pi r}\right) = \frac{rS - 2\pi r^3}{2}$$

$$\Rightarrow \qquad \frac{dV}{dr} = \frac{S}{2} - 3\pi r^2 \qquad \qquad ...(2)$$

For max or min we have $\dfrac{dV}{dr} = 0$

$$\frac{S}{2} - 3\pi r^2 = 0 \Rightarrow S = 6\pi r^2$$

$$\Rightarrow \qquad 2\pi r^2 + 2\pi rh = 6\pi r^2$$

$$\Rightarrow \qquad h = 2r$$

From (2) $\dfrac{d^2V}{dr^2} = -6\pi r$, (–ve) for any positive value of r.

Hence V is maximum when $h = 2r$, i.e., when the height of cylinder is equal is diameter of base.

EXAMPLE 11. *Prove that a conical tent of a given capacity will required the least amount of canvas when the height is $\sqrt{2}$ times the radius of the base.* [MEERUT–2006]

SOLUTION. Let us suppose h be the height, r be the radius of the base l the slant height of the conical tent. Let V be the given capacity (i.e. volume) and S denote the area of the curved surface of the tent.

We know that $\qquad V = \dfrac{1}{3}\pi r^2 h \qquad \qquad ...(1)$

and $\qquad S = \pi lr = \pi(\sqrt{h^2 + r^2})r$

$$\Rightarrow \qquad S^2 = \pi^2 r^2 (h^2 + r^2) = u\,(\text{say}) \qquad \qquad ...(2)$$

From (1) and (2), we get

$$u = \pi^2 r^2\left[\frac{9V^2}{\pi^2 r^4} + r^2\right] = \frac{9V^2}{r^2} + \pi^2 r^4$$

$\therefore$ $\qquad$ $\dfrac{du}{dr} = -\dfrac{18V^2}{r^3} + 4\pi^2 r^3$

and $\qquad$ $\dfrac{d^2u}{dr^2} = \dfrac{54V^2}{r^4} + 12\pi^2 r^2$

Now $\qquad$ $\dfrac{du}{dr} = 0 \Rightarrow V = \left(\dfrac{2}{3\sqrt{2}}\right)\pi r^3$

for $\qquad$ $V = \left(\dfrac{2}{3\sqrt{2}}\right)\pi r^3, \dfrac{d^2u}{dr^2} > 0$

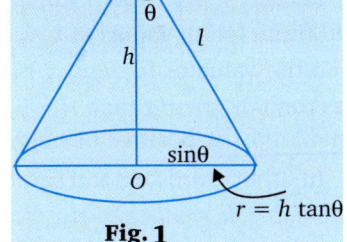

Fig. 1

i.e., u is minimum when $V = \dfrac{2}{3\sqrt{2}}\pi r^3$

i.e., when $\dfrac{2}{3\sqrt{2}}\pi r^3 = \dfrac{1}{3}\pi r^2 h$ *i.e.*, $h = r\sqrt{2}$

$\Rightarrow$ u is minimum, when $h = r\sqrt{2}$.

EXAMPLE 12. ***Show that the radius of the right circular cylinder of greatest curved surface which can be inscribed in a given cone is half that of the cone.***

SOLUTION. Let r be the radius and H, the height of the given cone

i.e., $OB = r$, $OA = H$

where O is the centre of the base circle.

Suppose x is the radius and h the height of the cylinder inscribed in the given cone.

Now triangles AOB and ADE are similar, therefore

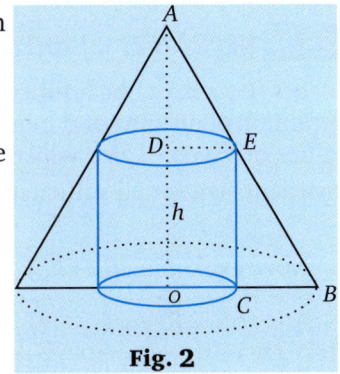

Fig. 2

$$\dfrac{AD}{AO} = \dfrac{DE}{OB} \text{ or } \dfrac{H-h}{H} = \dfrac{x}{r}$$

$\Rightarrow$ $\qquad$ $1 - \dfrac{h}{H} = \dfrac{x}{r} \Rightarrow \dfrac{h}{H} = 1 - \dfrac{x}{r}$

$\Rightarrow$ $\qquad$ $h = H\left(1 - \dfrac{x}{r}\right)$

Now the curved surface of the cylinder

$\Rightarrow$ $\qquad$ $S = 2\pi.x.h = 2\pi x H\left(1 - \dfrac{x}{r}\right) = \dfrac{2\pi H}{r}(rx - x^2)$

$\Rightarrow$ $\qquad$ $\dfrac{dS}{dx} = \dfrac{2\pi H}{r}(r - 2x)$

So $\dfrac{dS}{dx} = 0$, we get $r - 2x = 0$ or $x = \dfrac{r}{2}$

Also $\qquad$ $\dfrac{d^2S}{dx^2} = \dfrac{2\pi H}{r}(-2) < 0 \Rightarrow S$ is greatest, when $x = \dfrac{r}{2}$

i.e., when radius of the cylinder is half of that can be inscribed in a sphere of radius x.

25.2 MAXIMA AND MINIMA OF A FUNCTION OF SEVERAL INDEPENDENT VARIABLES

Let $f(x, y, z, ...)$ be a function of several independent variables x, y, z.... If f is continuous and finite for all values of $x, y, z, ...$ in the neighbourhood of $x = a, y = b, z = c, ...$ respectively, then the value of $(a, b, c, ...)$ is said to be a maximum or minimum if $f(a+h, b+k, c+l, ...)$ is less than or greater than $f(a, b, c, ...)$ for all values of $h, k, l, ...$ (where $h, k, l, ...$ are sufficiently small, may be positive or negative provided they are not all zero.)

In other words we can say, the value of $f(a, b, c, ...)$ is said to be a maximum or minimum if $f(a+h, b + k, c + l, ...) - f(a, b, c, ...)$ maintain an invariant sign (may be positive or negative) for all values of $h, k, l, ...$ positive or negative provided they are taken sufficiently small and finite.

25.2.1 STATIONARY AND EXTREME POINTS

A point $(a_1, a_2, ..., a_n)$ is called a stationary point, if all the first order partial derivatives of the function $f(x_1, x_2, ..., x_n)$ vanish at the point. A stationary point, if it is maximum or minimum is known as extreme point and the value of the function at an extreme point is known as an extreme value.

☞ **REMARK**

- A stationary point may be a maximum or minimum or neither of these two.

25.3 NECESSARY CONDITION FOR THE EXISTENCE OF MAXIMA OR MINIMA

Let $f(x, y, z, ...)$ be a function of several independent variables $x, y, z, ...$ It is clear from the definition of maxima and minima that maximum or minimum of $f(x, y, z, ..)$ will occur for those values of $x, y, z, ...$, for which the expression $f(x+h, y +k, z+l, ...) - f(x, y, z, ...)$ maintain an invariant sign for all sufficiently small and finite values of $h, k, l, ...$ positive or negative.

Now, expanding $f(x+h, y+k, z+l, ...)$ by Taylor's theorem, we have

$$f(x + h, y + k, z + l...) = f(x, y, z) + \left(h\frac{\partial f}{\partial x} + k\frac{\partial f}{\partial y} + l\frac{\partial f}{\partial z} + ... \right)$$

+ terms of second and higher order.

$$\Rightarrow \quad f(x + h, y + k, z + l...) - f(x, y, z, ...) = \left(h\frac{\partial f}{\partial x} + k\frac{\partial f}{\partial y} + l\frac{\partial f}{\partial z} + ... \right)$$

+ terms of second and higher orders.   ...(1)

Now, since $h, k, l, ...$ are sufficiently small, the first degree expression

$$\left(h\frac{\partial f}{\partial x} + k\frac{\partial f}{\partial y} + l\frac{\partial f}{\partial z} + ... \right)$$

of the equation (1) can be made to govern the sign of right hand side and hence, of the left hand side as well as. Thus, by changing the sign of the left hand side of the equation (1) will also change.

Since, left hand side is to preserve an invariable sign for maxima or minima, therefore, as a necessary condition for maximum and minimum values, we must have

$$h\frac{\partial f}{\partial x} + k\frac{\partial f}{\partial y} + l\frac{\partial f}{\partial z} + ... = 0 \qquad ...(2)$$

Now, since $h, k, l, \ldots$ are arbitrary and independent of each other, we must have

$$\frac{\partial f}{\partial x} = 0, \frac{\partial f}{\partial y} = 0, \frac{\partial f}{\partial z} = 0, \text{ etc.} \qquad \ldots(3)$$

If the number of independent variables be n, we shall get n simultaneous equations in these n variables, which will give the values $a, b, c, \ldots$ of the n variables $x, y, z, \ldots$ respectively for which $f(x, y, z, \ldots)$ will have a maximum or a minimum values.

☛ **REMARKS**

- The necessary condition for a function $f(x, y, z, \ldots)$ of the independent variables $x, y, z, \ldots$ to be maximum or minimum is given by

$$\frac{\partial f}{\partial x} = 0, \frac{\partial f}{\partial y} = 0, \frac{\partial f}{\partial z} = 0,$$

- The conditions given above is only a necessary condition for the maxima and minima of the function $f(x, y, z, \ldots)$. These conditions are not sufficient.

10.3.1 MAXIMA AND MINIMA FOR A FUNCTION OF TWO INDEPENDENT VARIABLES

(1) *To find the condition which governs the sign of a quadratic expression.*

Consider, a binary expression

$$I = ax^2 + 2hxy + by^2$$

of two variables x and y. Then I can be written as

$$I = ax^2 + 2hxy + by^2 = \frac{1}{a}[(ax + hy)^2 + (ab - h^2)y^2].$$

If $(ab - h^2)$ is positive, the sign of I will be the same as that of a.

But if $(ab - h^2)$ is negative, then, the expression within the brackets may be positive or negative and therefore we cannot say anything about the sign of expression I.

(2) *Stationary and extreme points (For the function of two independent variables):*

Let $f(x, y)$ be a function of two independent variables x and y. A point (a, b) is called a stationary point, if both the first order partial derivatives $\left(\frac{\partial f}{\partial a} \text{ and } \frac{\partial f}{\partial b}\right)$ of the function $f(x, y)$ at (a, b) vanish.

A stationary point which is either a maximum or minimum is called an extreme point.

☛ **REMARKS**

- A stationary point is not necessarily an extreme point, hence a stationary point may be a maximum or a minimum or neither of these two.
- The value of the function at extreme point is called extreme value.
- A point at which function is neither maximum nor minimum, is known as saddle point.

25.4 NECESSARY CONDITION FOR MAXIMA AND MINIMA

Let $f(x, y)$ be a function of two independent variables x and y. Then, we have the maximum or minimum of $f(x, y)$ at $x = a$ and $x = b$ if the expression $f(a + h, b + k) - f(a, b)$ is of invariable sign for all sufficiently small independent variables h and k provided both of them are not equal to zero.

We observe that,

(i) If the sign of $f(a+h, b+k) - f(a, b)$ is negative, then we have a maximum of $f(x, y)$ at $x = a, y = b$.

(ii) If the sign of $f(a+h, b+k) - f(a, b)$ is positive, then we have a minimum of $f(x, y)$ at $x = a, y = b$.

Expand $f(a+h, b+k)$ by Taylor's theorem, we have

$$f(a + h, b + k) = f(a,b) + \left(h\frac{\partial f}{\partial x} + k\frac{\partial f}{\partial y} \right)_{\substack{x=a \\ y=b}}$$

$$+ \frac{1}{2!}\left(h^2 \frac{\partial^2 f}{\partial x^2} + 2hk\frac{\partial^2 f}{\partial x\,\partial y} + k^2 \frac{\partial^2 f}{\partial y^2} \right)_{\substack{x=a \\ y=b}} + ... \qquad ...(1)$$

$$\Rightarrow \quad f(a + h, b + k) - f(a,b) = h\left(\frac{\partial f}{\partial x} \right)_{\substack{x=a \\ y=b}} + k\left(\frac{\partial f}{\partial y} \right)_{\substack{x=a \\ y=b}} + \text{term of the second and higher}$$

orders in h and k.

Now, since h and k are sufficiently small, the expression $h\left(\dfrac{\partial f}{\partial x} \right)_{\substack{x=a \\ y=b}} + k\left(\dfrac{\partial f}{\partial y} \right)_{\substack{x=a \\ y=b}}$ of the equation (1) can be made to govern the sign of right hand side and hence of the left hand side as well. Thus by changing the sign of h and k, the sign of the left hand side of the equation (1) will also change.

Since L.H.S. is to preserve an invariable sign for maximum or minimum, therefore as a necessary condition for maximum and minimum values, we must have

$$h\left(\frac{\partial f}{\partial x} \right)_{\substack{x \ a \\ y \ b}} + k\left(\frac{\partial f}{\partial y} \right)_{\substack{x \ a \\ y \ b}} = 0. \qquad ...(2)$$

If $k = 0$, we find that if $\left(\dfrac{\partial f}{\partial x} \right)_{\substack{x=a \\ y=b}} \neq 0$, the R.H.S. of (2) changes sign when h changes sign. Therefore $f(x, y)$ cannot have a maximum or minimum at $x = a, y = b$ if $\left(\dfrac{\partial f}{\partial x} \right)_{\substack{x=a \\ y=b}} \neq 0$.

Similarly, taking $h = 0$, we see that $f(x, y)$ cannot have a maximum or a minimum at $x = a$, $y = b$ if $\left(\dfrac{\partial f}{\partial y} \right)_{\substack{x=a \\ y=b}} \neq 0$.

Thus, a set of necessary conditions that $f(x, y)$ should have a maximum or minimum at $x = a, y = b$ is that

$$\left(\frac{\partial f}{\partial x} \right)_{\substack{x=a \\ y=b}} = 0 \text{ and } \left(\frac{\partial f}{\partial y} \right)_{\substack{x=a \\ y=b}} = 0.$$

25.5 SUFFICIENT CONDITION FOR MAXIMA AND MINIMA: THE LAGRANGE'S CONDITION

Let $f(x, y)$ be a function of two variables x and y.

Let
$$r = \frac{\partial^2 f}{\partial x^2}, \ s = \frac{\partial^2 f}{\partial x\,\partial y}, \ t = \frac{\partial^2 f}{\partial y^2} \text{ at } x = a \text{ and } y = b.$$

As a set of necessary conditions for a maximum or minimum at (a, b) we have

$$\frac{\partial f}{\partial x} = 0 \text{ and } \frac{\partial f}{\partial y} = 0 \text{ at } (a, b)$$

then
$$f(a + h, b + k) - f(a, b) = \frac{1}{2!}[rh^2 + 2shk + tk^2] + R \qquad ...(1)$$

Where R consists of terms of third and higher order of small quantities h and k .

Now, by taking h and k sufficiently small, the second degree terms in R.H.S. of (1) may be made to govern the sign of R.H.S. and therefore of the L.H.S. also *i.e.*, for sufficiently small values of h and k, the sign of $\frac{1}{2}(rh^2 + 2shk + tk^2) + R$ is same as that of $rh^2 + 2shk + tk^2$.

If the sign is negative, then the function is maximum at (a, b) and if the sign is positive, then the function is minimum at (a, b).

Case (i) If $(rt - s^2) > 0$.

Then, neither r nor t can be zero. Hence, we can write

$$rh^2 + 2shk + tk^2 = \frac{1}{2}[r^2h^2 + 2rshk + rtk^2] = \frac{1}{2}[(rh + sk)^2 + (rt - s^2)k^2]$$

since $rt - s^2 > 0$, therefore $(rh + sk)^2 + (rt - s^2)k^2 > 0$ for all values of h and k except when $rh + sk = 0, k = 0$ *i.e.*, at $h = 0, k = 0$, which is not possible.

Hence, in this case the expression $rh^2 + 2shk + tk^2$ will have the same sign for all values of h and k, and the sign is determined by the sign of r.

Thus, the function $f(x, y)$ will have a maximum or minimum at $x = a$ and $y = b$. If $rt - s^2 > 0$. The function $f(x, y)$ is maximum or minimum according as r is negative or positive.

Case (ii) If $(rt - s^2) < 0$.

If $rt - s^2$ is negative, we are not sure about the sign of second degree term of R.H.S. of (1) and hence there is neither a maximum nor a minimum value.

Case (iii) If $(rt - s^2) = 0$.

If $rt = s^2$, then quadratic expression $rh^2 + 2shk + tk^2$ becomes $\frac{1}{r}(hr + ks)^2$.

So that, the quadratic expression will be of the same sign as that of r or t unless

$$\frac{h}{k} = -\frac{s}{r} = \alpha \,(\text{say}) \, i.e., \, rh + sk = 0.$$

If this condition is satisfied, then the second degree expression in R.H.S. of (1) vanishes and hence, the sign of the R.H.S. of (1) depends upon third degree expression in h and k, which change sign with the change of sign of h and k and hence, the sign of L.H.S. of (1) will also change and hence, there will be neither maximum nor minimum.

Thus, the necessary condition for the existence of maxima and minima now is that the cubic terms must vanish collectively in R.H.S. of (1) when $\frac{h}{k} = -\frac{s}{r} = \alpha;$ and then the biquadratic terms of R.H.S. of (1) must collectively be of the same sign as r and t, when

$$\frac{h}{k} = -\frac{s}{r} = \alpha \quad i.e., \, hr + ks = 0$$

Hence, the case is doubtful.

Thus, if $rt - s^2 = 0$, the case is doubtful and further, investigation is needed to determine the maxima and minima of $f(x, y)$ at (a, b).

Working Procedure

To discuss the maxima and minima at $x = a, y = b$, we must find

$$r = \left(\frac{\partial^2 u}{\partial x^2}\right)_{\substack{x=a \\ y=b}}, \, s = \left(\frac{\partial^2 u}{\partial x \partial y}\right)_{\substack{x=a \\ y=b}}, \, t = \left(\frac{\partial^2 u}{\partial y^2}\right)_{\substack{x=a \\ y=b}}$$

Then, calculate $rt - s^2$.

Now following cases arise :

 (i) If $rt - s^2 > 0$, then

  (A) If r is negative then, $f(x, y)$ is maximum at $x = a, y = b$.

  (B) If r is positive then, $f(x, y)$ is minimum at $x = a, y = b$.

 (ii) If $rt - s^2 < 0$, $f(x, y)$ is neither maximum nor minimum at $x = a, y = b$.

 (iii) If $rt - s^2 = 0$, the case is doubtful, and further investigation will be required.

☛ **REMARK**

- While solving problems, we frequently used the identity, given by Lagrange.

$$\{(a^2 + b^2 + c^2)(p^2 + q^2 + r^2) - (ap + bq + cr)^2\} = \{(br - cq)^2 + (cp + ar)^2 + (aq - bp)^2\}.$$

 Solved Examples

EXAMPLE 1. **Find all maximum or minimum values of the function :**
$$f(x, y) = y^2 + x^2 y + x^4.$$

SOLUTION.   Since, we have

$$f(x, y) = y^2 + x^2 y + x^4.$$

$$\therefore \qquad \frac{\partial f}{\partial x} = 2xy + 4x^3 \text{ and } \frac{\partial f}{\partial y} = 2y + x^2.$$

For a maximum or minimum of $f(x, y)$, we must have $\dfrac{\partial f}{\partial x} = 0$ and $\dfrac{\partial f}{\partial y} = 0$

$$\therefore \qquad \frac{\partial f}{\partial x} = 0 \Rightarrow 2xy + 4x^3 = 0$$

$$\Rightarrow \qquad 2x (y + 2x^2) = 0 \qquad\qquad \text{...(1)}$$

$$\frac{\partial f}{\partial y} = 0 \Rightarrow 2y + x^2 = 0$$

Solving (1) and (2), we get $x = 0, y = 0$.

Thus $(0, 0)$ is the only point of maximum or minimum.

Now $r = \left(\dfrac{\partial^2 f}{\partial x^2}\right)_{(0,0)} = [2y + 12x^2]_{(0,0)} = 0$; $s = \left(\dfrac{\partial^2 f}{\partial x \partial y}\right)_{(0,0)} = [2x]_{(0,0)} = 0$

and $t = \left(\dfrac{\partial^2 f}{\partial y^2}\right)_{(0,0)} = [2]_{(0,0)} = 2$

$\therefore rt - s^2 = 0$ $(2) - 0^2 = 0$.

Thus, the case is doubtful and further investigation will be required.

EXAMPLE 2. **Find the maximum or minimum values of the function** $x^3 y^2 (1 - x - y)$.

SOLUTION.   Let $u = x^3 y^2 (1 - x - y)$ (ANNA–2009, JNTU–2006, 08, BHOPAL–2012)

$$\Rightarrow \quad \frac{\partial u}{\partial x} = 3x^2 y^2 (1 - x - y) - x^3 y^2 \text{ and } \frac{\partial u}{\partial y} = 2x^3 y (1 - x - y) - x^3 y^2.$$

For a maximum or minimum of u, we must have $\dfrac{\partial u}{\partial x} = 0$ and $\dfrac{\partial u}{\partial y} = 0$

$$\Rightarrow \qquad 3x^2 y^2 (1 - x - y) - x^3 y^2 = 0 \qquad\qquad \text{...(1)}$$

and $2x^3 y (1 - x - y) - x^3 y^2 = 0.$ ...(2)

Now, subtracting (2) from (1), we have
$$x^2y(1-x-y)(3y-2x)=0$$
which gives $y=\dfrac{2}{3}x$.

Putting the value of y in (1), we get $x=\dfrac{1}{2}$

So $\left(\dfrac{1}{2},\dfrac{1}{3}\right)$ be the point of maxima or minima.

Now
$$r=\frac{\partial^2 u}{\partial x^2}=6xy^2-12x^2y^2-6xy^3 \quad =-\frac{1}{9}, \text{ at } \left(\frac{1}{2},\frac{1}{3}\right)$$

$$t=\frac{\partial^2 u}{\partial y^2}=2x^3-2x^4-6x^3y \quad =-\frac{1}{8}, \text{ at } \left(\frac{1}{2},\frac{1}{3}\right)$$

$$s=\frac{\partial^2 u}{\partial x \partial y}=6x^2y-8x^3y-9x^2y^2 \quad =-\frac{1}{12} \text{ at } \left(\frac{1}{2},\frac{1}{3}\right).$$

Now, $rt-s^2 = $ positive.

Also, r is negative, hence the function u has a maximum at $x=\dfrac{1}{2}, y=\dfrac{1}{3}$.
The maximum value is

$$=\left(\frac{1}{2}\right)^3 \left(\frac{1}{3}\right)^2 \left(1-\frac{1}{2}-\frac{1}{3}\right)=\frac{1}{432}.$$

EXAMPLE 3. *Discuss the maximum or minimum values of u, where*
$$u=2a^2xy-3ax^2y-ay^3+x^3y+xy^3.$$

SOLUTION. We have
$$u=2a^2xy-3ax^2y-ay^3+x^3y+xy^3$$
which gives
$$\frac{\partial u}{\partial x}=2a^2y-6axy+3x^2y+y^3$$
and
$$\frac{\partial u}{\partial y}=2a^2x-3ax^2-3ay^2+x^3+3xy^2$$

For a maximum and minima of u, we have
$$\frac{\partial u}{\partial x}=0, \frac{\partial u}{\partial y}=0$$
which gives,
$$y(2a^2-6ax+3x^2+y^2)=0 \qquad \text{...(1)}$$
and $2a^2x-3ax^2-3ay^2+x^3+3xy^2=0 \qquad \text{...(2)}$

Equation (1) and (2) gives the following values of x and y :

$$x=0, y=0; \; x=a, y=0;$$
$$x=2a, y=0; \; x=\frac{3}{2}a, \; y=\pm\frac{1}{2}a;$$
$$x=a, y=a, \; x=\frac{1}{2}a, \; y=\frac{1}{2}a;$$
$$x=a, y=-a; \; x=\frac{1}{2}a, \; y=-\frac{1}{2}a.$$

Then, we get the following pairs of values of x and y which make the function u stationary.

$$(0,0), (a,0), (2a,0), \left(\frac{3}{2}a, \frac{1}{2}a\right), \left(\frac{3}{2}a, -\frac{1}{2}a\right), (a,a), \left(\frac{1}{2}a, \frac{1}{2}a\right), (a,-a), \left(\frac{1}{2}a, -\frac{1}{2}a\right).$$

Also

$$r = \frac{\partial^2 u}{\partial x^2} = -6ay + 6xy,$$

$$s = \frac{\partial^2 u}{\partial x \, \partial y} = 2a^2 - 6ax + 3x^2 + 3y^2,$$

and

$$t = \frac{\partial^2 u}{\partial y^2} = -6ay + 6xy.$$

For (0, 0).

$$r = 0, s = 2a^2, t = 0$$

$\Rightarrow$ $rt - s^2$, is negative.

Therefore, we have neither maximum nor a minimum of u at (0, 0).

Similarly, we can easily shown that u has neither a maximum nor a minimum at $(a, 0), (2a, 0), (a, a), (a, -a)$.

For $\left(\dfrac{3a}{2}, \dfrac{a}{2}\right)$. $r = \dfrac{3}{2}a^2, s = \dfrac{1}{2}a^2, t = \dfrac{3}{2}a^2,$

$\Rightarrow$ $rt - s^2$ is positive.

Here, since r is positive, therefore u has minimum at $\left(\dfrac{3a}{2}, \dfrac{a}{2}\right)$.

Similarly, we can check the maxima and minima at all other points.

☛ REMARK

- The point $\left(\dfrac{x_1 + x_2 + x_3}{3}, \dfrac{y_1 + y_2 + y_3}{3}\right)$ is the centroid of the given triangle.

EXAMPLE 4. ***Show that the minimum value of*** $u = xy + \left(\dfrac{a^3}{x}\right) + \left(\dfrac{a^3}{y}\right)$ ***is*** $3a^2$***.***

SOLUTION.   We have $u = xy + \left(\dfrac{a^3}{x}\right) + \left(\dfrac{a^3}{y}\right)$

$\Rightarrow$ $\dfrac{\partial u}{\partial x} = y - \dfrac{a^3}{x^2}$ and $\dfrac{\partial u}{\partial y} = x - \dfrac{a^3}{y^2}.$

For a maximum or minimum of u, we have $\dfrac{\partial u}{\partial x} = 0$ and $\dfrac{\partial u}{\partial y} = 0$

Now, $\dfrac{\partial u}{\partial x} = 0 \Rightarrow y - \dfrac{a^3}{x^2} = 0$ ...(1)

and $\dfrac{\partial u}{\partial y} = 0 \Rightarrow x - \dfrac{a^3}{y^2} = 0.$ ...(2)

Solving (1) and (2), we get, $x = a, y = a$

Now $r = \dfrac{\partial^2 u}{\partial x^2} = \dfrac{2a^3}{x^3}, s = \dfrac{\partial^2 u}{\partial x \, \partial y} = 1$ and $t = \dfrac{\partial^2 u}{\partial y^2} = \dfrac{2a^3}{y^3}.$

At $x = y = a$, we have

$$r = 2, s = 1, t = 2$$

$\Rightarrow rt - s^2 = 3 > 0.$

Thus, at (a, a), $rt - s^2 > 0$ and $r > 0$. Therefore u is minimum at $x = a$, $y = a$.

The minimum value of

$$u = a.a + \left(\frac{a^3}{a}\right) + \left(\frac{a^3}{a}\right) = 3a^2.$$

EXAMPLE 5. *Determine the points where a function $x^3 + y^3 - 3axy$ has maximum or minimum.*

SOLUTION. Here, we have $u = x^3 + y^3 - 3axy$

$$\Rightarrow \qquad \frac{\partial u}{\partial x} = 3x^2 - 3ay \text{ and } \frac{\partial u}{\partial y} = 3y^2 - 3ax.$$

For a maximum or minimum of u, we must have $\dfrac{\partial u}{\partial x} = 0$ and $\dfrac{\partial u}{\partial y} = 0$

which gives, $\qquad x^2 - ay = 0$ ...(1)

and $\qquad\qquad\quad y^2 - ax = 0$ ...(2)

Solving (1) and (2), we get

$$x = 0, y = 0; x = a, y = a.$$

Thus $(0, 0)$ and (a, a) are the stationary points of u.

Now $r = \dfrac{\partial^2 u}{\partial x^2} = 6x$, $s = \dfrac{\partial^2 u}{\partial x\,\partial y} = -3a$, $t = \dfrac{\partial^2 u}{\partial y^2} = 6y$.

For $x = 0, y = 0$

$$r = 0, s = -3a \text{ and } t = 0$$

$\therefore \qquad rt - s^2 = -9a^2 < 0$, for all values of a.

$\Rightarrow u$ is neither maximum nor minimum at $x = 0, y = 0$.

For $x = a, y = a$

$$r = 6a, s = -3a \text{ and } t = 6a$$

$\Rightarrow \qquad rt - s^2 = 27a^2 > 0$, for all values of a.

Also $r = 6a$, which is positive if $a > 0$.

Thus   (i) u is maximum at $x = a, y = a$ if $a < 0$

and   (ii) u is minimum at $x = a, y = a$ if $a > 0$.

EXAMPLE 6. *Discuss the maxima and minima of the function $u = \sin x \sin y \sin (x+y)$.*

[UPTU–2009]

SOLUTION. Here, we have $u = \sin x \sin y \sin (x + y)$

$$\Rightarrow \qquad \frac{\partial u}{\partial x} = \sin y [\sin x \cos(x + y) + \cos x \sin(x + y)]$$

and $\qquad \dfrac{\partial u}{\partial y} = \sin x [\sin y \cos(x + y) + \cos y \sin (x + y)].$

For a maxima and minima of u, we must have

$$\frac{\partial u}{\partial x} = 0 \text{ and } \frac{\partial u}{\partial y} = 0.$$

$\Rightarrow \sin y [\sin x \cos (x + y) + \cos x \sin (x+y)] = 0$

and $\sin x [\sin y \cos (x + y)+\cos y \sin (x+y)]= 0.$

Equation (1) and (2) gives

$$\tan (x + y) = - \tan x \qquad \qquad \qquad \ldots(1)$$

$\Rightarrow \qquad \qquad \tan x = \tan y$

and $\tan (x + y) = - \tan y \qquad \qquad \qquad \ldots(2)$

$\Rightarrow \qquad \qquad x = y$

From (1) and (2), we have

$$\tan 2x = - \tan x = \tan (\pi - x) \Rightarrow 2x = \pi - x$$

$\Rightarrow \qquad \qquad 3x = \pi \quad \Rightarrow x = \dfrac{\pi}{3} = y.$

Moreover, $\dfrac{\partial u}{\partial x} = 0$, gives $\sin y = 0 \Rightarrow y = 0$

and $\dfrac{\partial u}{\partial y} = 0$, gives $\sin x = 0 \Rightarrow x = 0.$

Thus, we get the following pair of values, which makes the function u stationary

$(0,0), \left(\dfrac{\pi}{3}, \dfrac{\pi}{3}\right).$

Now $\qquad \qquad r = \dfrac{\partial^2 u}{\partial x^2} = 2\sin y \cos(2x + y),$

$$s = \dfrac{\partial^2 u}{\partial x\, \partial y} = \sin 2(x + y),$$

and $\qquad \qquad t = \dfrac{\partial^2 u}{\partial y^2} = 2\sin x \cos(2y + x).$

For (0, 0).

$$r = 0, s = 0, t = 0 \Rightarrow \quad rt - s^2 = 0.$$

$\therefore$ this case is doubtful and need further investigation.

For $\left(\dfrac{\pi}{3}, \dfrac{\pi}{3}\right).$

$$r = 2\sin \dfrac{1}{3}\pi . \cos \pi = -\sqrt{3},$$

$$s = \sin \left(\dfrac{4\pi}{3}\right) = -\sin \dfrac{\pi}{3} = -\dfrac{\sqrt{3}}{2},$$

and $\qquad \qquad t = 2\sin \dfrac{1}{3}\pi \cos \pi = -\sqrt{3}.$

$\therefore \qquad \qquad rt - s^2 = \dfrac{9}{4} = $ positive.

Also $\qquad \qquad r = -\sqrt{3}.$

Hence, u has a maximum value at $\left(\dfrac{\pi}{3}, \dfrac{\pi}{3}\right).$

EXAMPLE 7. *Find a point within a triangle such that the sum of the squares of its distances from the vertices is a minimum.*

SOLUTION. Let us suppose $[(x_r, y_r) : r = 1, 2, 3]$ be the vertices of the triangle and (x, y) be any point inside the triangle.

Now, let us define a function

$$u = \sum_{r=1}^{3} [(x - x_r)^2 + (y - y_r)^2].$$

Then, we have

$$\frac{\partial u}{\partial x} = \sum 2(x - x_r) = 2[(x - x_1) + (x - x_2) + (x - x_3)]$$

and

$$\frac{\partial u}{\partial y} = \sum 2(y - y_r) = 2[(y - y_1) + (y - y_2) + (y - y_3)].$$

For a maximum or minimum of u, we must have

$$\frac{\partial u}{\partial x} = 0 \Rightarrow (x - x_1) + (x - x_2) + (x - x_3) = 0$$

$$\Rightarrow \qquad x = \frac{x_1 + x_2 + x_3}{3}$$

and

$$\frac{\partial u}{\partial y} = 0 \Rightarrow (y - y_1) + (y - y_2) + (y - y_3) = 0$$

$$\Rightarrow \qquad y = \frac{y_1 + y_2 + y_3}{3}.$$

Thus, we have

$$\left(\frac{x_1 + x_2 + x_3}{3}, \frac{y_1 + y_2 + y_3}{3} \right)$$

is the only point at which u have a maximum or minimum.

Now $r = \dfrac{\partial^2 u}{\partial x^2} = 6, s = \dfrac{\partial^2 u}{\partial x \partial y} = 0, t = \dfrac{\partial^2 u}{\partial y^2} = 6.$

At $\left[\dfrac{x_1 + x_2 + x_3}{3}, \dfrac{y_1 + y_2 + y_3}{3} \right] \Rightarrow \quad r = 6, s = 0, t = 6$

$\Rightarrow \qquad rt - s^2 = 36 > 0.$

Also, since $\qquad r > 0.$

Therefore u have a minimum value at $\left[\dfrac{x_1 + x_2 + x_3}{3}, \dfrac{y_1 + y_2 + y_3}{3} \right].$

Hence, the point

$$\left(\frac{x_1 + x_2 + x_3}{3}, \frac{y_1 + y_2 + y_3}{3} \right)$$

is the required point at which u is minimum.

EXAMPLE 8. *Show that distance l of any point (x, y, z) on the plane $2x + 3y - z = 12$ from the origin is given by*

$$l = \sqrt{[x^2 + y^2 + (2x + 3y - 12)^2]}.$$

Hence, find the point on the plane that is nearest to the origin.

SOLUTION. If l is the distance from $(0, 0, 0)$ of any point (x, y, z) then $l = \sqrt{(x^2 + y^2 + z^2)}$. If the point (x, y, z) lies on the plane $2x + 3y - z = 12$, then

$$l = \sqrt{[x^2 + y^2 + (2x + 3y - 12)^2]}$$

$[\because z = 2x + 3y - 12,$ from the equation of the plane$]$

$\therefore \qquad l^2 = x^2 + y^2 + (2x + 3y - 12)^2$

$$= 5x^2 + 10y^2 + 12xy - 48x + 72y + 144 = u \,(\text{say}).$$

Now l is maximum or minimum according as l^2 i.e., u is maximum or minimum. For a maximum or minimum of u, we get

$$\frac{\partial u}{\partial x} = 10x + 12y - 48 = 0 \quad \text{and} \quad \frac{\partial u}{\partial y} = 20y + 12x - 72 = 0$$

Solving these equations, we get $x = \dfrac{12}{7}$ and $y = \dfrac{18}{7}$.

Also $\qquad r = \dfrac{\partial^2 u}{\partial x^2} = 10, \; s = \dfrac{\partial^2 u}{\partial x \, \partial y} = 12 \; \text{and} \; t = \dfrac{\partial^2 u}{\partial y^2} = 20.$

Therefore $rt - s^2 = 10 \times 20 - (12)^2 = +$ ve, since $rt - s^2 > 0$ and $r > 0$, then u is minimum and hence l is minimum.

When $x = \dfrac{12}{7}$ and $y = \dfrac{18}{7}$. Putting these values of x and y in the equation of the plane, we get

$$z = 2 \cdot \left(\frac{12}{7}\right) + 3 \cdot \left(\frac{18}{7}\right) - 12 = -\frac{6}{7}.$$

Hence, the required point is $\left(\dfrac{12}{7}, \dfrac{18}{7}, -\dfrac{6}{7}\right)$.

EXAMPLE 9. *Find the points on $z^2 = xy + 1$ nearest to the origin.*

SOLUTION. Let l be the distance from the origin $(0, 0, 0)$ of any point (x, y, z) on the surface $z^2 = xy + 1$ $\qquad \qquad \qquad \qquad \qquad \qquad \qquad \qquad \qquad \qquad$ ...(1)

Then $\qquad l = \sqrt{x^2 + y^2 + z^2} \quad = \sqrt{(x^2 + y^2 + xy + 1)}$ $\quad$ [Using equation (1)]

Since l is always greater than zero, therefore l is maximum or minimum according as l^2, i.e., u is maximum or minimum, where $u = l^2$.

For a maximum or minimum of u, we must have

$$\frac{\partial u}{\partial x} = 2x + y = 0 \qquad \qquad \qquad \qquad \qquad \qquad \text{...(2)}$$

and $\qquad \dfrac{\partial u}{\partial y} = 2y + x = 0.$ $\qquad \qquad \qquad \qquad \qquad \qquad \qquad$ ...(3)

Solving the equation (2) and (3), we get $x = 0, y = 0$

Also $\qquad r = \dfrac{\partial^2 u}{\partial x^2} = 2, \; s = \dfrac{\partial^2 u}{\partial x \, \partial y} = 1, \quad t = \dfrac{\partial^2 u}{\partial y^2} = 2.$

$\therefore \qquad rt - s^2 = 2.\,2 - 1 = 3 > 0.$

Since at $x = 0, y = 0$, then $rt - s^2 > 0$ and $r > 0$.

Therefore u is minimum at $x = 0, y = 0$. Hence l is minimum, when $x = 0, y = 0$.

Putting $x = 0, y = 0$ in the equation (1), we get $z^2 = 1$ i.e., $z = \pm 1$.

Hence, the required points are $(0, 0, 1)$ and $(0, 0, -1)$.

Exercise-25.1

1. Discuss the maxima and minima of the function
$$f(x,y) = x^2 + y^2 + \frac{2}{x} + \frac{2}{y}.$$

2. Find the values of x and y for which the expression
$$(a_1x + b_1y + c_1)^2 + (a_2x + b_2y + c_2)^2$$
$$+ ... + (a_nx + b_ny + c_n)^2$$
is minimum.

3. Examine for maximum and minimum values of the function $f(x,y) = x^2 - 3xy + y^2 + 2x$.

4. Examine the function $f(x,y) = x^2y - y^2x - x + y$ for maxima and minima.

5. Discuss the maxima and minima of the function
$$f(x,y) = 2\sin\frac{1}{2}(x+y)\cos\frac{1}{2}(x-y) + \cos(x+y).$$

6. Find the maximum and minimum values of $u = 6xy + (47 - x - y)(4x + 3y)$.

7. Examine for extreme values
   (i) $x^2 + y^2 + 6x + 12$ [GBTU–2012]
   (ii) $x^3 + y^3 - 63(x+y) + 12xy$ [UKTU–2011]

ANSWERS

1. $f(x,y)$ is minimum at $(1, 1)$. 2. $f(x,y)$ is minimum for the value of x and y which are obtained by
$$\Sigma(a_1^2)x + (a_1b_1)y + a_1c_1 = 0 \text{ and } \Sigma(a_1b_1)x + (b_1^2)y + b_1c_1 = 0.$$

3. Stationary point is $x = \frac{4}{5}, y = \frac{6}{5}$. The function $f(x, y)$ is neither maximum nor minimum at $\left(\frac{4}{5}, \frac{6}{5}\right)$. 4. At $(1, 1)$ and $(-1, -1)$ function is neither maximum nor minimum.

5. $x = y = 2n\pi \pm \pi/2$; neither maximum nor minimum ; $x = y = n\pi + (-1)^n \pi/6$; f is maximum.

6. Maximum value is 3384.

7. (i) At $x = -3$, $y = 0$, minimum (ii) max at $(-7, -7)$ min. at $(3, 3)$ neither max nor min. at $(5, -1)$ and $(-1, 5)$.

25.6 MAXIMA AND MINIMA OF THE FUNCTION OF THREE INDEPENDENT VARIABLES

(1) *To find the condition, which governs the sign of the quadratic equation of three independent variables.*

Let I be the expression of three independent variables x, y and z given by
$$I = ax^2 + by^2 + cz^2 + 2fyz + 2gzx + 2hxy$$

I can be written as

$$I = \frac{1}{a}\left[a^2x^2 + aby^2 + acz^2 + 2afyz + 2agzx + 2ahxy\right] (a \neq 0)$$

$$= \frac{1}{a}\left[a^2x^2 + 2ax(gz + hy) + aby^2 + acz^2 + 2afyz\right]$$

$$= \frac{1}{a}\left[(ax + hy + gz)^2 + aby^2 + acz^2 + 2afyz - (gz + hy)^2\right]$$

$$= \frac{1}{a}\left[(ax + hy + gz)^2 + (ab - h^2)y^2 + 2yz(af - gh) + (ac - g^2)z^2\right]$$

Here, we observe that I be of the same sign as provided the expression within the square brackets is positive which will of course be so if $ab-h^2$ and $\{(ab-h^2)(ac-g^2)-(af-gh)^2\}$ are positive *i.e.*, if

$$ab-h^2 \text{ and } a[abc + 2fgh - af^2 - bg^2 - ch^2] \text{ are both positive.}$$

Hence, *I* will be positive if

$$a, \begin{vmatrix} a & h \\ h & b \end{vmatrix}, \begin{vmatrix} a & h & g \\ h & b & f \\ g & f & c \end{vmatrix}$$

be all positive and will be negative if these three expression are alternately negative and positive.

25.7 MAXIMA AND MINIMA FOR A FUNCTION OF THREE INDEPENDENT VARIABLES : THE LAGRANGE'S CONDITION

Let $f(x,y,z)$ be a given function of three independent variables x, y and z.

Let A, B, C, F, G, H stand for $\dfrac{\partial^2 f}{\partial x^2}, \dfrac{\partial^2 f}{\partial y^2}, \dfrac{\partial^2 f}{\partial z^2}, \dfrac{\partial^2 f}{\partial y \partial z}, \dfrac{\partial^2 f}{\partial z \partial x}, \dfrac{\partial^2 f}{\partial x \partial y}$ respectively.

Let a set of the values of x, y, z obtained by solving the equations

$$\frac{\partial f}{\partial x} = \frac{\partial f}{\partial y} = \frac{\partial f}{\partial z} = 0 \text{ be } a, b, c.$$

By Taylor's theorem, we have

$$f(a+h, b+k, c+l), -f(a,b,c) = \frac{1}{2!}\left[Ah^2 + Bk^2 + Cl^2 + 2Fkl + 2Glh + 2Hhk \right] + R \qquad ...(1)$$

where, remainder term R consist of third and higher order of same quantity (*i.e.*, h, k, l).

Now, by taking h, k, l sufficiently small, the second term of R.H.S. of (1) can be made to govern the sign of R.H.S. and therefore of L.H.S. also.

If for all such values of h, k and l, these terms be of permanent sign, then we shall have a maximum or minimum of $f(x,y,z)$ according as that sign is negative or positive.

Hence, the function will be minimum if the expression $A, \begin{vmatrix} A & H \\ H & B \end{vmatrix}, \begin{vmatrix} A & H & G \\ H & B & F \\ G & F & C \end{vmatrix}$ be all positive.

The function will have a maximum value, if the above three quantities are alternately negative and positive. If these conditions are not satisfied, we have neither a maximum nor a minimum.

Working Procedure

Let $f(x, y, z)$ be a function of three independent variables x, y and z. Find the values of triads (a,b,c) of the value x, y and z by putting $\dfrac{\partial f}{\partial x} = 0, \dfrac{\partial f}{\partial y} = 0, \dfrac{\partial f}{\partial z} = 0$ .The values of triads (a,b,c) will give the stationary values of $f(x, y, z)$.

Now, to discuss maximum and minimum values, at (a, b, c) we find the following six partial derivatives of second order

$$A = \frac{\partial^2 f}{\partial x^2}, B = \frac{\partial^2 f}{\partial y^2}, C = \frac{\partial^2 f}{\partial z^2}, F = \frac{\partial^2 f}{\partial y \partial z}, G = \frac{\partial^2 f}{\partial z \partial x}, and H = \frac{\partial^2 f}{\partial x \partial y}$$

Now, we have the following cases :

Case (i) The function $f(x,y,z)$ will be minimum at (a,b,c) if the expressions

$A, \begin{vmatrix} A & H \\ H & B \end{vmatrix}, \begin{vmatrix} A & H & G \\ H & B & F \\ G & F & C \end{vmatrix}$ be all positive at (a, b, c).

Case (ii) The function $f(x, y, z)$ will be maximum at (a, b, c) if the expressions

$$A, \begin{vmatrix} A & H \\ H & B \end{vmatrix}, \begin{vmatrix} A & H & G \\ H & B & F \\ G & F & C \end{vmatrix}$$

be alternately negative and positive.

Case (iii) If the expression, using in case (i) and (ii) neither be all positive nor having alternately negative and positive sign at (a,b,c). Then $f(x, y, z)$ is neither maximum nor minimum at (a,b,c).

☞ **REMARK**

- To find the maximum and minimum of the function at stationary point, it is sufficient to find the value of a second order partial derivative of function with respect to any of the independent variables. Then, the value of the function is maximum or minimum according as the value of this second order partial derivative at the stationary point under consideration is negative or positive.

 Solved Examples

EXAMPLE 1. *Find the maximum value of u, where* $u = \dfrac{xyz}{(a+x)(x+y)(y+z)(z+b)}$.

SOLUTION. We have $u = \dfrac{xyz}{(a+x)(x+y)(y+z)(z+b)}$

Taking, log of both the sides, we have

$$\log u = \log x + \log y + \log z - \log(a+x) - \log(x+y) - \log(y+z) - \log(z+b)$$

Differentiating w.r.t. x, we have

$$\frac{1}{u} \frac{\partial u}{\partial x} = \frac{1}{x} - \frac{1}{a+x} - \frac{1}{x+y} = \frac{ay - x^2}{x(a+x)(x+y)}$$

$\Rightarrow \qquad \dfrac{\partial u}{\partial x} = \dfrac{(ay - x^2)u}{x(a+x)(x+y)}$

Similarly $\qquad \dfrac{\partial u}{\partial y} = \dfrac{(xz - y^2)u}{y(x+y)(y+z)}$ and $\dfrac{\partial u}{\partial z} = \dfrac{(by - z^2)u}{z(y+z)(z+b)}$

For, a maxima and minima of u, we must have

$$\frac{\partial u}{\partial x} = 0 \Rightarrow ay - x^2 = 0 \; ; \; \frac{\partial u}{\partial y} = 0 \Rightarrow xz - y^2 = 0$$

and $\qquad \dfrac{\partial u}{\partial z} = 0 \Rightarrow by - z^2 = 0$

Here, we observe that $x^2 = ay$, $y^2 = xz$, $z^2 = by$ which implies that a, x, y, z and b are in G.P. Let r be the common ratio of this G.P.

Then $\qquad ar^4 = b$ or $r = \left(\dfrac{b}{a}\right)^{1/4}$

Also $\qquad x = ar, y = ar^2, z = ar^3$.

Hence, we have

$$u = \frac{ar.ar^2.ar^3}{a(1+r)ar(1+r)ar^2(1+r)ar^3(1+r)}$$

$$= \frac{1}{a(1+r)^4} = \frac{1}{a\left[1+\left(\dfrac{b}{a}\right)^{1/4}\right]^4}$$

$$= \frac{1}{\left(a^{1/4}+b^{1/4}\right)^4}$$

which gives a stationary value of u. Now, to decide whether this value of u is a maximum or a minimum, we proceed to find the second order partial derivative of u.

Here $\dfrac{\partial^2 u}{\partial x^2} = \dfrac{-2ux}{x(a+x)(x+y)} + (ay-x^2)\dfrac{\partial}{\partial x}\left[\dfrac{u}{x(a+x)(x+y)}\right]$

When $x=ar, y=ar^2, z=ar^3$, we have

$$A = \frac{\partial^2 u}{\partial x^2} = -\frac{2u}{a^2 r(1+r)^2} < 0$$

Hence, the above stationary value of u is maximum.

EXAMPLE 2. *Find the maxima and minima value of the function*
$$u = \sin x \sin y \sin z$$
where x,y and z are the vertex angles of a triangle.

SOLUTION. Here, we have

$\qquad u = \sin x \sin y \sin z$; where $x+y+z = \pi$ ...(1)

$\therefore \qquad u = \sin x \sin y \sin [\pi-(x+y)]$

$\qquad\quad = \sin x \sin y \sin(x+y)$

$\therefore \qquad \dfrac{\partial u}{\partial x} = \cos x \sin y \sin(x+y) + \sin x \sin y \cos(x+y)$

$\qquad\qquad = \sin y \sin(2x+y).$ ...(2)

Similarly $\dfrac{\partial u}{\partial y} = \sin x \sin(2y+x)$ ...(3)

For a maxima and minima, we must have

$$\frac{\partial u}{\partial x} = 0, \frac{\partial u}{\partial y} = 0$$

So, $\qquad \dfrac{\partial u}{\partial x} = 0 \Rightarrow \qquad \sin y \sin(2x+y)=0$

$\Rightarrow \qquad \sin y=0$ or $\sin(2x+y)=0 \Rightarrow y=0$ or $\sin(x+x+y)=0$

$\Rightarrow \qquad\qquad y=0$

or $\sin x \cos(x+y)+\cos x \sin(x+y)=0$

$\Rightarrow \tan(x+y) = -\tan x$

$\Rightarrow \tan(x+y) = \tan(-x) = \tan(\pi - x)$ ...(4)

$\Rightarrow \qquad x+y=\pi - x$

$\Rightarrow \qquad 2x+y=\pi$ ...(5)

Similarly, from (3), $\qquad x=0$

or $\tan(x+y) = -\tan y$ ...(6)

Now, by (4) and (6), we have

$\tan x = \tan y \Rightarrow x = y.$

Hence, by (5), we have

$3y = \pi \Rightarrow y = \pi/3$ and $x = \pi/3$

Therefore, the stationary points are $\left(\dfrac{\pi}{3}, \dfrac{\pi}{3}\right)$ and $(0, 0)$.

For (0,0): $u = 0$.

For $\left(\dfrac{\pi}{3}, \dfrac{\pi}{3}\right)$

$$r = \frac{\partial^2 u}{\partial x^2} = 2\sin y \cos(2x + y) = 2\sin\frac{\pi}{3}\cos\left(\frac{2\pi}{3} + \frac{\pi}{3}\right)$$

$$= -\sqrt{3} < 0$$

and $$s = \frac{\partial^2 u}{\partial x \partial y} = \sin(2x + 2y)$$

$$= \sin\left(\frac{2\pi}{3} + \frac{2\pi}{3}\right) = \sin\left(\frac{4\pi}{3}\right) = -\frac{\sqrt{3}}{2} < 0$$

$$t = \frac{\partial^2 u}{\partial y^2} = 2\sin x \cos(x + 2y) = 2\sin\frac{\pi}{3}\cos\pi = -\sqrt{3} < 0$$

Now $rt - s^2 = \left(-\sqrt{3}\right)\left(-\sqrt{3}\right) - \left(\dfrac{\sqrt{3}}{2}\right)^2 = \dfrac{9}{4} > 0$. Thus, $rt - s^2 > 0$ and $r < 0$.

Hence, the function u will be maximum at $\left(\dfrac{\pi}{3}, \dfrac{\pi}{3}\right)$.

 **Exercise-25.2**

1. Prove that the function $u = x^2 + y^2 + x - 2z$ $- xy$ is minimum at $\left(-\dfrac{2}{3}, -\dfrac{1}{3}, 1\right)$.

2. Find the maximum and minimum values of $u = y^2 + 2z^2 - 5x^4 + 4x^5$.

3. Find the maximum or minimum values of the function u, where $u = axy^2z^3 - x^2y^2z^3 - xy^3z^3 - xy^2z^4$

4. Find the maximum value of

$(ax + by + cz)\, e^{-\left(\alpha^2 x^2 + \beta^2 y^2 + \gamma^2 z^2\right)}$.

5. A rectangle box is placed on x-y plane. The one end of the box is at the origin. If the vertex opposite to the origin be on the plane $6x + 4y + 3z = 24$, then find the maximum value of this box.

6. In a plane triangle xyz, find the maximum value of $\sin x \sin y \sin z$.

7. A rectangular box, open at the top is to have a given capacity. Show that the domain of the box requiring least material for its construction $x = y = (2v)^{1/3}$, where $v = xyz$.

ANSWERS

2. Minimum at $(1,0,0)$, neither maximum nor minimum at $(0,0,0)$.

3. Maximum at $\left(\dfrac{a}{7}, \dfrac{2a}{7}, \dfrac{3a}{7}\right)$, max. value $= \dfrac{108a^7}{7^7}$

4. Maximum at $\left(\dfrac{a}{2\alpha^2 k}, \dfrac{b}{2\beta^2 k}, \dfrac{c}{2\gamma^2 k}\right)$ where $k = \sqrt{\left\{\dfrac{1}{2}\left(\dfrac{a^2}{\alpha^2} + \dfrac{b^2}{\beta^2} + \dfrac{c^2}{\gamma^2}\right)\right\}}$,

Maximum value $= \sqrt{\left\{\dfrac{1}{2e}\left(\dfrac{a^2}{\alpha^2}+\dfrac{b^2}{\beta^2}+\dfrac{c^2}{\gamma^2}\right)\right\}}$

5. Maximum at $\left(\dfrac{4}{3},2\right)$. maximum value $= \dfrac{64}{9}$ cube units. Neither maximum nor minimum at (0,0).

6. Maximum at $\left(\dfrac{\pi}{3},\dfrac{\pi}{3},\dfrac{\pi}{3}\right)$, maximum value$= \dfrac{3\sqrt{3}}{8}$

25.8 LAGRANGE'S METHOD OF UNDETERMINED MULTIPLIERS

Let $u=f(x_1, x_2, ..., x_n)$ be a function of n variables $x_1, x_2, ..., x_n$.

Let us suppose these variables $x_1,x_2,...,x_n$ are connected by k equations

$$g_1(x_1, x_2,..., x_n)= 0$$
$$g_2(x_1, x_2,..., x_n)= 0$$
$$\vdots \; ... \quad ... \quad ... \quad ... \; \vdots$$
$$g_k(x_1, x_2, ..., x_n)= 0$$

so, that there are $n-k$ independent variables out of these n variables. For the maxima and minima of u, we find

$$du = \frac{\partial u}{\partial x_1}dx_1 + \frac{\partial u}{\partial x_2}dx_2 +...+ \frac{\partial u}{\partial x_n}dx_n = 0 \qquad ...(1)$$

Also
$$dg_1 = \frac{\partial g_1}{\partial x_1}dx_1 + \frac{\partial g_1}{\partial x_2}dx_2 +...+ \frac{\partial g_1}{\partial x_n}dx_n = 0 \qquad ...(2)$$

$$dg_2 = \frac{\partial g_2}{\partial x_1}dx_1 + \frac{\partial g_2}{\partial x_2}dx_2 +...+ \frac{\partial g_2}{\partial x_n}dx_n = 0 \qquad ...(3)$$

$$\vdots \qquad \vdots \qquad\qquad \vdots \qquad \vdots$$

$$dg_k = \frac{\partial g_k}{\partial x_1}dx_1 + \frac{\partial g_k}{\partial x_2}dx_2 +...+ \frac{\partial g_k}{\partial x_n}dx_n = 0 \qquad ...(k+1)$$

Multiplying equation (1),(2),(3)...(k+l) by $1,l_1,l_2,...,k$ respectively and adding, we get the result, which can be written as

$$P_1dx_1+P_2dx_2+P_3dx_3+...+P_ndx_n=0 \qquad ...(4)$$

where
$$P_k= \frac{\partial u}{\partial x_k} + l_1\frac{\partial g_1}{\partial x_k} + l_2\frac{\partial g_2}{\partial x_k} +...+ l_k\frac{\partial g_k}{\partial x_k}$$

Now we have at our choice k multiple *viz* $l_1,l_2,...,l_k$ and can be chosen such that

$$P_1=0, P_2=0, ..., P_k=0$$

Then, the equation (4) reduces to

$$P_{k+1}dx_{k+1}+P_{k+2}dx_{k+2}+P_{k+3}dx_{k+3}+...+P_ndx_n=0 \qquad ...(5)$$

Now, let us suppose that out of n variables, the $(n-k)$ variables $x_{k+1}, x_{k+2}, ...,x_n$ are independent.

Then, since $n-k$ quantities $dx_{k+1}, dx_{k+2},..., dx_n$ are independent so their coefficients must be separately zero. Hence, we have

$$P_{k+1}=0, P_{k+2}=0, ..., P_n=0$$

Thus, we have $k+n$ equations

$$P_1=0, \; P_2=0, \; ..., \; P_n=0$$

and $$g_1=0, \; g_2=0, \; ..., \; g_k=0.$$

Hence, we get $(n+k)$ equations which determine the k multipliers $l_1, l_2, ..., l_k$ and get the possible value of u.

☛ REMARKS

- The Lagrange's method of undetermined multipliers is very convenient to apply. It gives the maximum and minimum values of the function without actually determining the values of the multipliers $l_1, l_2, ..., l_k$.
- It does not determine the nature of stationary point, which is the only drawback of this method.

25.8.1 APPLICATIONS OF THE METHOD OF UNDETERMINED MULTIPLIERS

The Lagrange's method of undetermined multipliers can be applied to determine the extreme values of the given functions, it does not determine the nature of stationary point. Now, it is more convenient to find out the extreme values of a function F with the help of new function, given by

$$V=g+l_1f_1+l_2f_2+...+l_mf_m$$

and use the following method. Here, we give the method for four variables x, y, u, v connected by the following two relations.

Let $F=g(x, y, u, v)$ be subjected to the conditions

$$f_1(x,y,u,v)=0 \qquad \qquad ...(1)$$

and $$f_2(x,y,u,v)=0. \qquad \qquad ...(2)$$

For the maxima and minima of F, we have

$$dF = \frac{\partial g}{\partial x}dx + \frac{\partial g}{\partial y}dy + \frac{\partial g}{\partial u}du + \frac{\partial g}{\partial v}dv = 0 \qquad \qquad ...(3)$$

Now, from (1) and (2), we have

$$df_1 = \frac{\partial f_1}{\partial x}dx + \frac{\partial f_1}{\partial y}dy + \frac{\partial f_1}{\partial u}du + \frac{\partial f_1}{\partial v}dv = 0 \qquad \qquad ...(4)$$

and $$df_2 = \frac{\partial f_2}{\partial x}dx + \frac{\partial f_2}{\partial y}dy + \frac{\partial f_2}{\partial u}du + \frac{\partial f_2}{\partial v}dv = 0 \qquad \qquad ...(5)$$

Multiplying (4) by l_1, (5) by l_2 and adding their sum to (3), we get

$$\left(\frac{\partial g}{\partial x} + l_1\frac{\partial f_1}{\partial x} + l_2\frac{\partial f_2}{\partial x} \right)dx + \left(\frac{\partial g}{\partial y} + l_1\frac{\partial f_1}{\partial y} + l_2\frac{\partial f_2}{\partial y} \right)dy$$

$$+ \left(\frac{\partial g}{\partial u} + l_1\frac{\partial f_1}{\partial u} + l_2\frac{\partial f_2}{\partial u} \right)du + \left(\frac{\partial g}{\partial v} + l_1\frac{\partial f_1}{\partial v} + l_2\frac{\partial f_2}{\partial v} \right)dv = 0 \qquad ...(6)$$

Here, we have l_1 and l_2 are arbitrary, therefore we can choose them to satisfy the two linear equations

$$\frac{\partial g}{\partial x} + l_1\frac{\partial f_1}{\partial x} + l_2\frac{\partial f_2}{\partial x} = 0 \qquad \qquad ...(7)$$

and $$\frac{\partial g}{\partial y} + l_1\frac{\partial f_1}{\partial y} + l_2\frac{\partial f_2}{\partial y} = 0 \qquad \qquad ...(8)$$

Using (7) and (8), equation (6) reduces to

$$\left(\frac{\partial g}{\partial u} + l_1\frac{\partial f_1}{\partial u} + l_2\frac{\partial f_2}{\partial u}\right)du + \left(\frac{\partial g}{\partial v} + l_1\frac{\partial f_1}{\partial v} + l_2\frac{\partial f_2}{\partial v}\right)dv = 0$$

Since, the given function contains four variables (namely x, y, u and v) and we are given two equations of conditions, therefore, only two of the variables are independent and it is immaterial which two of the four variables are regarded as independent. Let them be u and v then du and dv are also independent, therefore, their coefficients must be zero separately. Thus

$$\frac{\partial g}{\partial u} + l_1\frac{\partial f_1}{\partial u} + l_2\frac{\partial f_2}{\partial u} = 0 \qquad\qquad …(9)$$

$$\frac{\partial g}{\partial v} + l_1\frac{\partial f_1}{\partial v} + l_2\frac{\partial f_2}{\partial v} = 0 \qquad\qquad …(10)$$

Now, we have six equations namely (1),(2),(7),(8),(9) and (10) to determine the two multipliers l_1, l_2 and values of the four variables x, y, u and v for which maximum and minimum values of F are possible.

Now, defined a new function $V(x, y, u, v)$ such that

$$V(x,y,u,v) = g(x, y, u, v) + l_1 f_1(x, y, u, v) + l_2 f_2(x, y, u, v).$$

Assuming that x, y, u, v are now all independent variables. Hence, for the maxima and minima of V, we must have

$$\frac{\partial V}{\partial x} = \frac{\partial g}{\partial x} + l_1\frac{\partial f_1}{\partial x} + l_2\frac{\partial f_2}{\partial x} = 0 \qquad\qquad …(11)$$

$$\frac{\partial V}{\partial y} = \frac{\partial g}{\partial y} + l_1\frac{\partial f_1}{\partial y} + l_2\frac{\partial f_2}{\partial y} = 0 \qquad\qquad …(12)$$

$$\frac{\partial V}{\partial u} = \frac{\partial g}{\partial u} + l_1\frac{\partial f_1}{\partial u} + l_2\frac{\partial f_2}{\partial u} = 0 \qquad\qquad …(13)$$

and $$\frac{\partial V}{\partial v} = \frac{\partial g}{\partial v} + l_1\frac{\partial f_1}{\partial v} + l_2\frac{\partial f_2}{\partial v} = 0 \qquad\qquad …(14)$$

Equations (11), (12), (13) and (14) are exactly the same as the equations (7), (8), (9) and (10). Hence, the maxima and minima of $V(x, y, u, v)$ are same as those of $F(x, y, u, v)$ assuming that $V(x, y, u, v)$ the variables x, y, u, v are now all independent.

Now, we proceed to find whether the values of F obtained with the help of above equations are maximum or minimum. For this, adopt the procedure, which is discussed ahead.

From (3), we get

$$d^2F = \left(\frac{\partial}{\partial x}dx + \frac{\partial}{\partial y}dy + \frac{\partial}{\partial u}du + \frac{\partial}{\partial y}dy\right)^2 g + \left(\frac{\partial g}{\partial x}d^2x + \frac{\partial g}{\partial y}d^2y + \frac{\partial g}{\partial u}d^2u + \frac{\partial g}{\partial y}d^2v\right) … \quad …(15)$$

Also

$$d^2f_1 = \left(\frac{\partial}{\partial x}dx + \frac{\partial}{\partial y}dy + \frac{\partial}{\partial u}du + \frac{\partial}{\partial v}dv\right)^2 f_1 + \frac{\partial f_1}{\partial x}d^2x + \frac{\partial f_1}{\partial y}d^2y + \frac{\partial f_1}{\partial u}d^2u + \frac{\partial f_1}{\partial v}d^2v = 0$$
$$…(16)$$

and $$d^2f_2 = \left(\frac{\partial}{\partial x}dx + \frac{\partial}{\partial y}dy + \frac{\partial}{\partial u}du + \frac{\partial}{\partial v}dv\right)^2 f_2 + \frac{\partial f_2}{\partial x}d^2x + \frac{\partial f_2}{\partial y}d^2y + \frac{\partial f_2}{\partial u}d^2u + \frac{\partial f_2}{\partial v}d^2v = 0$$
$$…(17)$$

Multiplying (16) by l_1 and (17) by l_2 and adding their sum to (15) and using the result (11), (12),(13) and (14), we have

$$d^2F = \left(\frac{\partial}{\partial x}dx + \frac{\partial}{\partial y}dy + \frac{\partial}{\partial u}du + \frac{\partial}{\partial v}dv\right)^2 (g + l_1 f_1 + l_2 f_2)$$

$$= \left(\frac{\partial}{\partial x}dx + \frac{\partial}{\partial y}dy + \frac{\partial}{\partial u}du + \frac{\partial}{\partial v}dv\right)^2 V = d^2V.$$

Hence d^2F is equal to d^2V, where d^2V is obtained by assuming all the variables x, y, u and v as independent. Therefore, it is clear that d^2F and d^2V have the same sign. Hence, F will be minimum or maximum according as V is minimum or maximum.

☞ REMARK

- This method has the advantage over the Lagrange's methods that it enables us to decide whether the values are maximum or minimum.

Solved Examples

EXAMPLE 1. **Find the maxima and minima of $x^2+y^2+z^2$ subject to the conditions :**
$ax^2+by^2+cz^2 = 1$ and $lx+my+nz = 0$ [UKTU–2011]

SOLUTION. Here, we have
$$u = x^2 + y^2 + z^2 \qquad \qquad ...(1)$$
$$ax^2+by^2+cz^2 = 1 \qquad \qquad ...(2)$$
and $\quad lx+my+nz = 0 \qquad \qquad ...(3)$

For the maxima and minima of u, we must have
$$du = 0 \Rightarrow 2xdx + 2ydy + 2zdz = 0$$
$$\Rightarrow \qquad xdx + ydy + zdz = 0 \qquad \qquad ...(4)$$

From (2) and (3), we get
$$ax\,dx + by\,dy + cz\,dz = 0 \qquad \qquad ...(5)$$
$$ldx + mdy + ndz = 0 \qquad \qquad ...(6)$$

Now, multiplying (4) by 1, (5) by l_1 and (6) by l_2 and adding, we get
$$(x\,dx + y\,dy + z\,dz) + l_1(ax\,dx + by\,dy + cz\,dz) + l_2(l\,dx + m\,dy + n\,dz) = 0$$
$$\Rightarrow (x + al_1 x + ll_2)dx + (y + bl_1 y + ml_2)\,dy + (z + cl_1 z + nl_2)dz = 0$$

Now equating the coefficient of dx, dy, dz to zero, we get
$$x + l_1 ax + l_2 l = 0 \qquad \qquad ...(7)$$
$$y + bl_1 y + ml_2 = 0 \qquad \qquad ...(8)$$
and $\quad z + cl_1 z + nl_2 = 0 \qquad \qquad ...(9)$

Multiplying the equations (7), (8) and (9) by x, y and z respectively, and adding we get
$$x^2+y^2+z^2 + l_1(ax^2+by^2+cz^2) + l_2(lx+my+nz) = 0$$
or $\quad u + l_1.1 + l_2.0 = 0 \qquad$ [By using (1), (2) and (3)]
$$\Rightarrow \qquad l_1 = -u$$

Substituting for l_1 in the equations (7), (8) and (9), we get
$$x = \frac{l_2 l}{au - 1}, y = \frac{l_2 m}{bu - 1}, z = \frac{l_2 n}{cu - 1} \qquad \qquad ...(10)$$

Now from (10) and (3), we get

$$\frac{l_2 l^2}{au-1} + \frac{l_2 m^2}{bu-1} + \frac{l_2 n^2}{cu-1} = 0$$

or $\quad \dfrac{l^2}{au-1} + \dfrac{m^2}{bu-1} + \dfrac{n^2}{cu-1} = 0 \qquad \qquad \text{...(11)}$

which gives the maximum and minimum of $u = x^2 + y^2 + z^2$.

- Equation (11) is a quadratic in u. So it gives two stationary values of u.
- Geometrically, the surface $ax^2 + by^2 + cz^2 = 1$ represents an ellipsoid whose centre is origin, and $lx + my + nz = 0$ represents a plane passing through the origin. The points (x, y, z) satisfying both the conditions (2) and (3) lies on the conic in which (2) and (3) intersect. $x^2 + y^2 + z^2$ gives the square of the distance (x, y, z) from the origin, which is also the centre of the conic of intersection. The maximum value of this distance is the major axis of this conic, and the minimum value of this distance is the minor axis of this conic. Hence, equation (11) gives the squares of the lengths of the semi-axis of the conic of intersection.

EXAMPLE 2. **_Find the maxima and minima of $x^2 + y^2 + z^2$, where $ax^2 + by^2 + cz^2 + 2fyz + 2gzx + 2hxy = 1$._**

SOLUTION. Let $\qquad \qquad \qquad u = x^2 + y^2 + z^2 \qquad \qquad \qquad \text{...(1)}$

where the relation between the variables x, y and z is

$$ax^2 + by^2 + cz^2 + 2fyz + 2gzx + 2hxy = 1. \qquad \qquad \text{...(2)}$$

For a maximum or minima of u, we must have $du = 0$

$\Rightarrow \qquad \qquad x\,dx + y\,dy + z\,dz = 0. \qquad \qquad \qquad \text{...(3)}$

From (2), we have

$2ax\,dx + 2by\,dy + 2cz\,dz + 2fy\,dz + 2fz\,dy + 2gz\,dx + 2gx\,dz + 2hx\,dy + 2hy\,dx = 0$

$\Rightarrow (ax + hy + gz)dx + (hx + by + fz)dy + (gx + fy + cz)dz = 0. \qquad \text{...(4)}$

Now, multiplying (3) by 1 and (4) by l_1, adding, and then equating the coefficient of dx, dy, dz to zero, we have

$$x + l_1(ax + hy + gz) = 0. \qquad \qquad \text{...(5)}$$

$$y + l_1(hx + by + fz) = 0. \qquad \qquad \text{...(6)}$$

$$z + l_1(gx + fy + cz) = 0. \qquad \qquad \text{...(7)}$$

Multiplying (5) by x, (6) by y, (7) by z and adding, we get

$$x^2 + y^2 + z^2 + l_1(ax^2 + by^2 + cz^2 + 2fyz + 2gzx + 2hxy) = 0$$

$\Rightarrow \qquad \qquad \qquad u + l_1.1 = 0 \qquad \qquad \text{[From (1) and (2)]}$

$\therefore \qquad \qquad \qquad l_1 = -u.$

Hence, from (5), we have

$$x - u(ax + hy + gz) = 0$$

$\Rightarrow \qquad \qquad \left(a - \dfrac{1}{u}\right)x + hy + gz = 0 \qquad \qquad \text{...(8)}$

Similarly from (6) and (7), we get

$$hx + \left(b - \dfrac{1}{u}\right)y + fz = 0 \qquad \qquad \text{...(9)}$$

and $\qquad \qquad gx + fy + \left(c - \dfrac{1}{u}\right)z = 0 \qquad \qquad \text{...(10)}$

Eliminating x, y, z from (8), (9) and (10), we get

$$\begin{vmatrix} \left(a - \dfrac{1}{u}\right) & h & g \\ h & \left(b - \dfrac{1}{u}\right) & f \\ g & f & \left(c - \dfrac{1}{u}\right) \end{vmatrix} = 0 \qquad \text{...(11)}$$

Hence, the maximum or minimum values of u are the roots of the equation (11).

EXAMPLE 3. ***Find the maxima and minima of*** $u = x^2 + y^2$ ***subject to the condition***

$$ax^2 + 2hxy + by^2 = 1.$$

SOLUTION. Here, we have $u = x^2 + y^2$ ...(1)

where the relation between the variables x and y is

$$ax^2 + 2hxy + by^2 = 1. \qquad \text{...(2)}$$

For the maxima and minima of u, we must have

$$du = 0 \implies 2x\,dx + 2y\,dy = 0$$

$$\implies \qquad x\,dx + y\,dy = 0. \qquad \text{...(3)}$$

Now, from (2), we get

$$2ax\,dx + 2hx\,dy + 2hy\,dx + 2by\,dy = 0$$

$$\implies \qquad (ax + hy)dx + (hx + by)dy = 0 \qquad \text{...(4)}$$

Now, multiplying (3) by 1, (4) by l_1, adding and then equating the coefficients of dx, dy to zero, we have

$$x + l_1(ax + hy) = 0 \qquad \text{...(5)}$$

and $\qquad y + l_1(hx + by) = 0 \qquad \text{..(6)}$

Multiplying (5) by x, (6) by y and adding, we get

$$x^2 + y^2 + l_1(ax^2 + 2hxy + by^2) = 0$$

$$\implies \qquad u + l_1.1 = 0 \qquad \text{[Using (1) and (2)]}$$

$$\implies \qquad u = -l_1$$

Therefore, from (5), we have

$$x - u(ax + hy) = 0$$

$$\implies \qquad \left(a - \dfrac{1}{u}\right)x + hy = 0 \qquad \text{...(7)}$$

Similarly from (6), we have

$$hx + \left(b - \dfrac{1}{u}\right)y = 0 \qquad \text{...(8)}$$

Eliminating x and y from (7) and (8), we get

$$\begin{vmatrix} a - \dfrac{1}{u} & h \\ h & b - \dfrac{1}{u} \end{vmatrix} = 0 \qquad \text{...(9)}$$

Hence, the maximum or minimum values of u are the roots of the equation (9).

EXAMPLE 4. *Find the maximum value of $u=x^m y^n z^p$ subject to the condition $x+y+z=a$.*   [ANNA–2009]

SOLUTION.   Here, we have $u = x^m y^n z^p$ ...(1)

and x, y, z connected by the relation given by $x+y+z = a$ ...(2)

Taking log of both the sides of (1), we get

$$\log u = m \log x + n \log y + p \log z.$$

On differentiating, we get

$$\frac{1}{u} du = \frac{m}{x} dx + \frac{n}{y} dy + \frac{p}{z} dz$$

For the maxima and minima of u, we must have $du=0$

$\Rightarrow$ $\dfrac{m}{x} dx + \dfrac{n}{y} dy + \dfrac{p}{z} dz = 0$ ...(3)

Now, differentiating (2), we get

$$dx + dy + dz = 0.$$ ...(4)

Now, multiplying (3) by 1 and (4) by l, and equating the coefficient of dx, dy, dz to zero (after adding), we get

$$\frac{m}{x} + l = 0, \quad \frac{n}{y} + l = 0 \text{ and } \frac{p}{z} + l = 0$$

which implies

$$x = -\frac{m}{l}, y = -\frac{n}{l}, z = -\frac{p}{l}$$

Putting the values of x, y and z in (2), we get $l = -\left(\dfrac{m+n+p}{a}\right)$ therefore, we can

say that, u is stationary when

$$x = \frac{am}{m+n+p}, y = \frac{an}{m+n+p}, z = \frac{ap}{m+n+p}$$

Now, we find the nature of this stationary value of u.

Let us regard x and y as independent variable and z is a function of x and y given by (2) [It is justify, because the variables x, y and z are connected by the relation (2), any two of them may be regarded as independent].

Now from (1), we get

$$\log u = m \log x + n \log y + p \log z$$

$\therefore$ $\dfrac{1}{u} \dfrac{\partial u}{\partial x} = \dfrac{m}{x} + \dfrac{p}{z} \dfrac{\partial z}{\partial x}$ ...(5)

Now, differentiating (2) partially $w.r.t\ x$ (treating y as constant), we get

$$1 + \frac{\partial z}{\partial x} = 0 \quad \Rightarrow \quad \frac{\partial z}{\partial x} = -1$$

Put this value in (5), we get

$$\frac{1}{u} \frac{\partial u}{\partial x} = \frac{m}{x} - \frac{p}{z}$$

$\Rightarrow$ $\dfrac{1}{u} \dfrac{\partial^2 u}{\partial x^2} - \dfrac{1}{u^2}\left(\dfrac{\partial u}{\partial x}\right)^2 = -\dfrac{m}{x^2} + \dfrac{p}{z^2}\dfrac{\partial z}{\partial x} = -\dfrac{m}{x^2} - \dfrac{p}{z^2}$

At stationary point, $\dfrac{\partial u}{\partial x} = 0$

Therefore, $\dfrac{1}{u}\dfrac{\partial^2 u}{\partial x^2} = \dfrac{-m}{x^2} - \dfrac{p}{z^2}$

$\Rightarrow \qquad \dfrac{\partial^2 u}{\partial x^2} = u\left[-\dfrac{m}{x^2} - \dfrac{p}{z^2} \right] = -x^m y^n z^p \left[-\dfrac{m}{x^2} - \dfrac{p}{z^2} \right]$

which is negative for the obtained values of x, y and z.

Hence, at the stationary point, u is maximum and maximum value is

$$= \left(\dfrac{am}{m+n+p} \right)^m \left(\dfrac{an}{m+n+p} \right)^n \left(\dfrac{ap}{m+n+p} \right)^p$$

EXAMPLE 5. *In a plane triangle ABC, find the maximum value of u = cos A cosB cosC.*

[VTU–2010; ANNA–2006]

SOLUTION. Here, we have $\qquad u = \cos A \cos B \cos C$ ...(1)

Since, we know that the sum of the angles of a triangle is always 180°.

∴ The variables A, B and C are connected by the relation

$\qquad A + B + C = \pi$ ...(2)

From (1), we get

$\qquad \log u = \log \cos A + \log \cos B + \log \cos C$

$\Rightarrow \qquad \dfrac{1}{u} du = -\tan A\, dA - \tan B\, dB - \tan C\, dC.$

For the maxima and minima of u, we must have $du = 0$

$\Rightarrow \tan A\, dA + \tan B\, dB + \tan C\, dC = 0$ ...(3)

Also from (2),

$\qquad dA + dB + dC = 0$ ...(4)

Now, multiply (3) by 1, (4) by l, adding, and equating the coefficients of dA, dB and dC to zero, we get

$\qquad \tan A + l = 0; \quad \tan B + l = 0; \quad \tan C + l = 0$

$\Rightarrow \qquad l = -\tan A = -\tan B = -\tan C$

$\Rightarrow \qquad A = B = C.$

Now from (2), $\quad A = B = C = \dfrac{\pi}{3}$ i.e., the triangle is equilateral.

Now to show that the stationary value of u given by $A = B = C = \pi/3$ is maximum.

Let C be a function of A and B, regarding A and B as independent variables. From (1),

$\log u = \log \cos A + \log \cos B + \log \cos C$

$\Rightarrow \qquad \dfrac{1}{u}\dfrac{\partial u}{\partial A} = -\tan A - \tan C \dfrac{\partial C}{\partial A}$

Now, differentiating (2), partially w.r.t. A, we get

$\qquad 1 + \dfrac{dC}{dA} = 0 \quad \Rightarrow \quad \dfrac{\partial C}{\partial A} = -1$

∴ $\qquad \dfrac{1}{u}\dfrac{\partial u}{\partial A} = -\tan A + \tan C$

$$\Rightarrow \quad \frac{1}{u}\frac{\partial^2 u}{\partial^2 A} - \frac{1}{u^2}\left(\frac{\partial u}{\partial A}\right)^2 = -\sec^2 A + \sec^2 C.\frac{\partial C}{\partial A} = -(\sec^2 A + \sec^2 C)$$

At stationary point $\dfrac{\partial u}{\partial A} = 0$

$$\because \qquad\qquad \frac{\partial^2 u}{\partial^2 A} = -u\left(\sec^2 A + \sec^2 C\right) = -\text{ve for } A=B=C=\pi/3.$$

Hence, u is maximum at $A=B=C=\dfrac{\pi}{3}$ and the maximum value is given by

$$u = \left(\cos\frac{\pi}{3}\right)^3 = \left(\frac{1}{2}\right)^3 = \frac{1}{8}.$$

 Exercise-25.3

1. Find the maximum and minimum values of
$$\frac{x^2}{a^4} + \frac{y^2}{b^4} + \frac{z^2}{c^4}$$
where $lx+my+nz=0$ and $\dfrac{x^2}{a^2}+\dfrac{y^2}{b^2}+\dfrac{z^2}{c^2}=1$.

2. Find the maximum and minimum values of
$$f = a^2x^2+b^2y^2+c^2z^2$$
where $x^2+y^2+z^2=1$ and $lx+my+nz=0$.

3. Show that the maximum and minimum values of $u=x^2+y^2+z^2$ subject to the conditions
$$px+qy+rz = 0 \text{ and } \frac{x^2}{a^2}+\frac{y^2}{b^2}+\frac{z^2}{c^2}=1$$
are given by $\dfrac{a^2p^2}{u-a^2}+\dfrac{b^2q^2}{u-b^2}$.

4. Find the minimum value of $u=x+y+z$ subject to the condition $\dfrac{a}{x}+\dfrac{b}{y}+\dfrac{c}{z}=1$.

5. Find the minimum value of $u=x^2+y^2+z^2$, subject to the condition $ax+by+cz=p$.
(UKTU–2012, UPTU–2009)

6. Find the minimum value of $x+y+z$ where $xyz=c^3$.

7. Find the extreme values of $x^p y^q z^r$ subject to the condition $\dfrac{a}{x}+\dfrac{b}{y}+\dfrac{c}{z}=1$.

8. Show that the maximum and minimum values of the radii vectors of the sections of the surface
$$(x^2+y^2+z^2)^2 = \frac{x^2}{a^2}+\frac{y^2}{b^2}+\frac{z^2}{c^2}$$

by the plane $\lambda x+\mu y+\nu z=0$
are given by $\dfrac{a^2\lambda^2}{1-a^2r^2}+\dfrac{b^2\mu^2}{1-b^2r^2}+\dfrac{c^2\nu^2}{1-c^2r^2}=0$

9. Find the stationary points of the function $u=ax^p+by^q+cz^r$ subject to the condition
$$x^l+y^m+z^n=k.$$

10. If two variables x and y are connected by the relation $ax^2+by^2=ab$, show that the maximum and minimum values of the function $u=x^2+y^2+xy$ will be the roots of the equation $4(u-a)(u-b)=ab$.

11. Prove that of all rectangular parallelopipeds of the same volume, the cube has the least surface. [KURUKSHETRA–2006, UPTU–2004]

12. Prove that if $x+y+z=1$, $ayz+bzx+cxy$ has an extreme value equal to
$$\frac{abc}{2bc+2ca+2ab-a^2-b^2-c^2}$$
Also, prove if a, b, c are all positive and c lies between $a + b - 2\sqrt{ab}$ and $a + b + 2\sqrt{ab}$ this value is true maximum and that if a, b, c are all negative and c lies between
$$a + b \pm 2\sqrt{ab}. \text{ It is true minimum.}$$

13. Find the maximum value of u, when
$$u=\sin x \sin y \sin z$$
and x,y,z are the angles of a triangle.

14. Find the triangle of maximum area inscribed in a circle.

ANSWERS

1. The maximum and minimum values of the given function is given by the equation

$$\frac{l^2a^4}{a^2u-1} + \frac{m^2b^4}{b^2u-1} + \frac{n^2c^4}{c^2u-1} = 0$$

2. The maximum and minimum values of the given function is given by the equation

$$\frac{l^2}{u-a^2} + \frac{m^2}{u-b^2} + \frac{m^2}{u-c^2} = 0$$

4. Stationary points are $x = \sqrt{a}\left(\sqrt{a} + \sqrt{b} + \sqrt{c}\right), y = \sqrt{b}\left(\sqrt{a} + \sqrt{b} + \sqrt{c}\right), z = \sqrt{c}\left(\sqrt{a} + \sqrt{b} + \sqrt{c}\right)$, minimum value is $\left(\sqrt{a} + \sqrt{b} + \sqrt{c}\right)^2$.

5. Minimum value is $\dfrac{p^2}{\left(a^2 + b^2 + c^2\right)}$ **6.** u is minimum at the point $x = y = z = c$. Value is $= 3c^4$.

7. u is stationary when $\dfrac{px}{a} = \dfrac{qy}{b} = \dfrac{rc}{c} = p + q + r$, Minimum value is $\dfrac{a^p b^q c^r}{p^p q^q r^r}\left(p + q + r\right)^{p+q+r}$.

9. Stationary points are given by $\dfrac{x^{p-1}}{l/pa} = \dfrac{y^{q-m}}{m/qb} = \dfrac{z^{r-n}}{n/rc}$ **13.** u is maximum, when $x = y = z = \pi/3$.

Maximum value is $\dfrac{3\sqrt{3}}{8}$.

14. Equilateral.

25.9 LAGRANGIAN MULTIPLIERS METHOD IN NON-LINEAR PROGRAMMING

In non-linear programming problem, if the objective function is continuous and differentiable and having all equally constraints, then we have the following necessary and sufficient conditions for the optimality of objective function by Lagrange's method.

Necessary Condition:

Case-I : Two decision variables and one equally constraint

Let us consider the following non-linear programming problem.

Optimize (max. or min.) $Z = f(x_1, x_2)$

subject to the constraints

$$g(x_1, x_2) = b, b \text{ is a constant}$$

and $x_1, x_2 \geq 0$

We can write the above problem as follows :

Optimize $Z = f(x_1, x_2)$

subject to the constraints

$$h(x_1, x_2) = g(x_1, x_2) - b = 0$$

and $x_1, x_2 \geq 0$

For an unknown function λ, let us define a function

$$L(x_1, x_2, \lambda) = f(x_1, x_2) - \lambda h(x_1, x_2) \qquad \ldots(1)$$

The function $L(x_1, x_2, \lambda)$ defined above in (1) is called the Lagrangian function and λ is called the Lagrangian multiplier.

Further, suppose that L, f and h are differentiable functions w.r.t. x_1, x_2 and λ.

Therefore, the necessary conditions for maxima and minima of f, subject to the condition

$h(x_1, x_2) = 0$ are given as below :

$$\frac{\partial L}{\partial x_1} = \frac{\partial f}{\partial x_1} - \lambda \frac{\partial h}{\partial x_1} = 0$$

$$\frac{\partial L}{\partial x_2} = \frac{\partial f}{\partial x_2} - \lambda \frac{\partial h}{\partial x_2} = 0$$

and
$$\frac{\partial L}{\partial \lambda} = -h = 0$$

Hence, we conclude that the necessary and sufficient conditions for maxima and minima of $f = f(x_1, x_2)$ subject to $h(x_1, x_2) = g(x_1, x_2) - b = 0$ are that

$$\frac{\partial f}{\partial x_1} = \lambda \frac{\partial h}{\partial x_1}, \frac{\partial f}{\partial x_2} = \lambda \frac{\partial h}{\partial x_2} \text{ and } h(x_1, x_2) = 0$$

Case II. n-decision variables and one equality constraint

Consider the non-linear programming of n variables.

$$\text{Optimize } Z = f(x_1, x_2, ..., x_n) = f(X) \text{ where } X = (x_1, x_2,.., x_n) \left.\begin{matrix} \\ \\ \\ \\ \end{matrix}\right]$$

...(1)

subject to the constraints
$$g(x_1, x_2,..., x_n) = g(X) = b, b \text{ is a constant}$$

and
$$x_1, x_2, ..., x_n \geq 0$$

The NLPP (1) can also be written as

$$\text{Optimize } Z = f(x_1, x_2,..., x_n)$$

subject to the constraints
$$h(x_1, x_2,..., x_n) = g(x_1, x_2,..., x_n) - b$$

i.e.,
$$h(X) = g(X) - b$$

and
$$x_1, x_2, ..., x_n \geq 0$$

Let λ be an unknown quantity (called Lagrangian multiplier). Then consider the Lagrangian function.

$$L(x_1, x_2, ..., x_n; \lambda) = f(x_1, x_2, ..., x_n) - \lambda h(x_1, x_2,..., x_n)$$

i.e.,
$$L(X, \lambda) = f(X) - \lambda h(X)$$

Since, L, f and h are all differentiable, the necessary conditions for maxima and minima of $f(X)$ subject to $h(X) = 0$ are given by

$$\frac{\partial L}{\partial x_j} = \frac{\partial f}{\partial x_j} - \lambda \frac{\partial h}{\partial x_j} = 0 \ \forall \ j = 1, 2, ..., n$$

and
$$\frac{\partial L}{\partial \lambda} = -h = 0$$

Hence, we conclude that 'the necessary condition for maximum or minimum of $f(x_1, x_2, ..., x_n)$ subject of the constraints

$$h(x_1, x_2, .., x_n) = g(x_1, x_2, ..., x_n) - b = 0$$

are that
$$\frac{\partial f}{\partial x_j} = \lambda \frac{\partial h}{\partial x_j} \forall j = 1, 2, ..., n \text{ and } h(x_1, x_2, ..., x_n) = 0$$

Case-III : n-decision variables and two equality constraints

Consider the non-linear programming of n variables given by

$$\text{Optimize } Z = f(x_1, x_2, ..., x_n)$$

subject to the constraints

$$g_1(x_1, x_2, ..., x_n) = b_1,$$
$$g_2(x_1, x_2, ..., x_n) = b_2,$$

and $\qquad x_1, x_2, ..., x_n \geq 0, \quad b_1$ and b_2 are constants

The above problem can also be written as

$$\text{Optimize } Z = f(x_1, x_2, ..., x_n)$$

subject to the constraints

$$h_1(x_1, x_2, ..., x_n) = g_1(x_1, x_2, ..., x_n) - b_1 = 0$$
$$h_2(x_1, x_2, ..., x_n) = g_2(x_1, x_2, ..., x_n) - b_2 = 0$$

and $\qquad x_1, x_2, ..., x_n \geq 0$

Now, taking the Lagrangian multipliers λ_1 and λ_2, the Lagrangian function is given by

$$L(x_1, x_2, ..., x_n; \lambda_1, \lambda_2) = f(x_1, x_2, ..., x_n) - \lambda_1 h_1(x_1, x_2, ..., x_n)$$
$$- \lambda_2 h_2(x_1, x_2, ..., x_n)$$

which can also be written as

$$L(X, \lambda_1, \lambda_2) = f(X) - \lambda_1 h_1(X) - \lambda_2 h_2(X)$$
$$X = (x_1, x_2, ..., x_n)$$

Further since L, f and h_1, h_2 are all differentiable partially, $w.r.t.$ $x_1, x_2, ..., x_n$ and λ_1, λ_2, therefore, the necessary conditions for maximum or minimum of $f(X)$ subject to the constraints $h_i(X) = 0$, $i = 1, 2$ are given by

$$\frac{\partial L}{\partial x_j} = \frac{\partial f}{\partial x_j} - \lambda_1 \frac{\partial h_1}{\partial x_j} - \lambda_2 \frac{\partial h_2}{\partial x_j} \quad \forall j = 1, 2, ..., n$$

$$\frac{\partial L}{\partial \lambda_1} = -h_1(X) = 0 \quad \text{and} \quad \frac{\partial L}{\partial \lambda_2} = -h_2(X) = 0$$

Hence, we conclude that the necessary and sufficient conditions for maximum or minimum of $f = f(x_1, x_2, ..., x_n)$ subject to the constraints

$$h_1 = h_1(x_1, x_2, ..., x_n) = g_1(x_1, x_2, ..., x_n) - b_1 = 0$$

and $\qquad h_2 = h_2(x_1, x_2, ..., x_n) = g_2(x_1, x_2, ..., x_n) - b_2$ are that

$$\frac{\partial f}{\partial x_j} = \lambda_1 \frac{\partial h_1}{\partial x_j} + \lambda_2 \frac{\partial h_2}{\partial x_j} \quad \forall j = 1, 2, ..., n$$

$$h_1(x_1, x_2, ..., x_n) = 0 \text{ and } h_2(x_1, x_2, ..., x_n) = 0$$

Case-IV : n-decision variables and m equality constraints ($m < n$)

Consider the non-linear programming problem given by

$$\text{Optimize } Z = f(x_1, x_2, ..., x_n)$$

subject to the constraints

$$g_i(x_1, x_2, ..., x_n) = b_i$$

and $\qquad x_i \geq 0 ; \quad i = 1, 2, ..., m \ (< n)$

The above problem can also be written in the following form

$$\text{Optimize } Z = f(x_1, x_2, ..., x_n)$$

subject to the constraint

$$h_i(x_1, x_2, ..., x_n) = g_i(x_1, x_2, ..., x_n) - b_i = 0$$

and

$$x_i \geq 0, \quad i = 1, 2, ..., m \ (<n)$$

Now, take the Lagrangian multiplier $\lambda = (\lambda_1, \lambda_2, ..., \lambda_m)$, the Lagrangian function $L(x_1, x_2, ..., x_n ; \lambda)$ is given by

$$L(X, \lambda) = f(X) - \sum_{i=1}^{m} \lambda_1 h_i(X) \qquad X = (x_1, x_2, ..., x_n)$$

Assuming that L, f and h_i are all differentiable partially w.r.t., x_1, x_2, ..., x_n and $\lambda_1, \lambda_2, ..., \lambda_m$, the necessary conditions for maximum or minimum of $f(X)$ subject to $h_i(X) = 0$ $(i = 1, 2, ..., n)$ are given by

$$\frac{\partial L}{\partial x_j} = \frac{\partial f}{\partial x_j} - \sum \lambda \frac{\partial}{\partial x_j} = 0 \ \forall \ j = 1, 2, ..., n$$

and

$$\frac{\partial L}{\partial \lambda_i} = -h_i(X) = 0; \quad i = 1, 2, ..., m (<n)$$

Hence, we conclude that the necessary condition for maximum or minimum of $f(X)$ subject to $h_i(X) = g_i(X) - b_i = 0; i = 1, 2, ..., n$ are that

$$\frac{\partial f}{\partial x_j} = \sum_{i=1}^{m} \lambda_i \frac{\partial h_i}{\partial x_j} \forall \ j = 1, 2, ..., n$$

and

$$h_i(X) = 0, \quad i = 1, 2, ..., m \ (<n)$$

25.10 SUFFICIENT CONDITIONS FOR MAXIMUM OR MINIMUM OF THE OBJECTIVE FUNCTION

Consider the following non-linear programming problem of n decision variables and m equality constraints

$$\text{Optimize } Z = f(x_1, x_2, ..., x_n)$$

subject to the constraints

$$h_i (x_1, x_2, ..., x_n) = 0 \ ; \quad i = 1, 2, ..., m \ (<n)$$

Now introduce m Lagrangian multipliers $\lambda = (\lambda_1, \lambda_2, ..., \lambda_m)$ the Lagrangian function is given by

$$L(x_1, x_2, ..., x_n; \lambda) = f(x_1, x_2, ..., x_n) - \sum_{i=1}^{m} \lambda_i h_i(x_1, x_2, ...x_n) \quad m < n$$

or

$$L(X, \lambda) = f(X) - \sum_{i=1}^{m} \lambda_i h_i(X) \quad m < n$$

where

$$X = (x_1, x_2, ..., x_n) \in R^n$$

We know that the necessary conditions for stationary points of $f(X)$ may be maximum or minimum are given by

$$\frac{\partial L(X, \lambda)}{\partial x_j} = \frac{\partial f(X)}{\partial x_j} - \sum_{i=1}^{m} \lambda_i \frac{\partial h_i(X)}{\partial x_j} = 0 \quad \text{for } j = 1, 2, ..., n$$

and

$$\frac{\partial L(X, \lambda)}{\partial \lambda_i} = -\frac{\partial h_i(X)}{\partial \lambda_i} = 0 \qquad \text{for } i = 1, 2, ..., m \ (m < n)$$

The above equations will give the stationary point of $f(X)$ at which the function $f(X)$ may be maximum or minimum.

Now, let us define
$$U = \left[\frac{\partial h_i(X)}{\partial x_j}\right]_{m \times n} = \begin{bmatrix} \dfrac{\partial h_1}{\partial x_1} & \dfrac{\partial h_1}{\partial x_2}, & \cdots, & \dfrac{\partial h_1}{\partial x_n} \\[2mm] \dfrac{\partial h_2}{\partial x_1} & \dfrac{\partial h_2}{\partial x_2}, & \cdots, & \dfrac{\partial h_2}{\partial x_n} \\ \vdots & & & \\ \dfrac{\partial h_m}{\partial x_1} & \dfrac{\partial h_m}{\partial x_2}, & \cdots, & \dfrac{\partial h_m}{\partial x_n} \end{bmatrix}_{m \times n}$$

and
$$V = \left[\frac{\partial^2 L(X,\lambda)}{\partial x_i x_j}\right]_{n \times n} = \begin{bmatrix} \dfrac{\partial^2 L}{\partial x_1^2} & \dfrac{\partial^2 L}{\partial x_1 \partial x_2}, & \cdots, & \dfrac{\partial^2 L}{\partial x_1 \partial x_n} \\[2mm] \dfrac{\partial^2 L}{\partial x_2 \partial x_1} & \dfrac{\partial^2 L}{\partial x_2^2}, & \cdots, & \dfrac{\partial^2 L}{\partial x_2 \partial x_n} \\ \vdots & \vdots & & \vdots \\ \dfrac{\partial^2 L}{\partial x_n \partial x_1} & \dfrac{\partial^2 L}{\partial x_n \partial x_2}, & \cdots, & \dfrac{\partial^2 L}{\partial x_n^2} \end{bmatrix}_{n \times n}$$

and
$$O = \begin{bmatrix} 0 & 0 & \cdots & 0 \\ 0 & 0 & \cdots & 0 \\ \vdots & & & \\ 0 & 0 & \cdots & 0 \end{bmatrix}_{m \times m}$$

is a null matrix.

Then the square matrix H_B of order $(m+n) \times (m+n)$ (called bordered Hessian matrix) is given as below

$$H_B = \left[\begin{array}{c|c} O & U \\ \hline U^T & V \end{array}\right]_{(m+n) \times (m+n)}$$

Then the required sufficient conditions for maximum and minimum stationary points are as follows :

"If (x_0, λ_0) be a stationary point for the function $L(x_0, \lambda)$ and H_{0B} is the value of the corresponding bordered Hessian matrix H_B at this stationary point, then

(i) The point x_0 gives the maximum value of the objective function if starting with principal minor of order $(2m+1)$, the last $(n-m)$ principal minor of H_{0B} are of alternating signs starting with $(-1)^{m+n}$ sign.

(ii) The point x_0 gives the maximum value of the objective function, starting with the principal minor of order $(2m+1)$, the last $(n-m)$ principal minors of H_{0B} are of the sign of $(-1)^m$.

25.10.1 SOME PARTICULAR CASES :

Case I : If $n = 2$, $m = 1$: In this case, the order of H_B is 3×3 ($\because n + m = 2 + 1 = 3$)

Since $2m + 1 = 3$, $(-1)^{m+n} = (-1)^3 = -1$, $n - m = 1$

$$(-1)^m = (-1)^1 = -1$$

Hence, the extreme point x_0 gives maximum value of the objective function if

$$\Delta_3 = |H_B| < 0$$

and minimum value of the obejctive function if

$$\Delta_3 = |H_B| > 0$$

Case II : If $m = 1$, $n = 3$, then order of H_B is 4×4.

Therefore $n + m = 4$

Since $2m + 1 = 3$, $(-1)^{n+m}=1$, $(-1)^m = -1$ and $n - m = 2$

Therefore, the extreme point x_0 gives the maximum value of the objective function of $\Delta_3>0$ and $\Delta_4 < 0$ and minimum if $\Delta_3 < 0$ and $\Delta_4 < 0$.

Case III : If $n = 3$ and $m = 2$, then order of H_B is 5×5.

Since $n + m = 5$, therefore

$$2m + 1 = 5, (-1)^{n+m} = -1$$
$$(-1)^m = 1, n - m = 1$$

Thus, the extreme point x_0 gives maximum value of the objective function if $\Delta_5 = |H_B| < 0$ and minimum value of the objective function if $\Delta_5 = |H_B| > 0$

☞ REMARKS

- A stationary point may be an extreme point without satisfying the above conditions.
- If $f(X)$ is a real valued continuous and differentiable function of $X = (x_1, x_2, ..., x_n)$, then Hessian matrix of $f(X)$ is given by

$$H_B(X) = \begin{vmatrix} \dfrac{\partial^2 f}{\partial x_1^2} & \dfrac{\partial^2 f}{\partial x_1 \partial x_2} & \cdots & \dfrac{\partial^2 f}{\partial x_1 \partial x_n} \\[2ex] \dfrac{\partial^2 f}{\partial x_2.\partial x_1} & \dfrac{\partial^2 f}{\partial x_2^2} & \cdots & \dfrac{\partial^2 f}{\partial x_2 \partial x_n} \\[2ex] \vdots & & & \\[1ex] \dfrac{\partial^2 f}{\partial x_n \partial x_1} & \dfrac{\partial^2 f}{\partial x_n \partial x_2} & \cdots & \dfrac{\partial^2 f}{\partial x_n^2} \end{vmatrix}$$

- If all the leading principal minors of Hessian matrix $H(X)$ of $f(X)$ are positive in sign, then the function $f(X)$ is convex, while if the signs of leading principal minors of Hessian matrix $H(X)$ of $f(X)$ are alternatively negative and positive, then $f(X)$ is concave.

> The necessary conditions for maximum or minimum of the objective function in non-linear programming problem with equality constraints also become the sufficient conditions for a maximum of the objective function if it is concave and for a minimum of the objective function if it is convex.

 Solved Examples

<u>EXAMPLE 1.</u> ***Determine the set of necessary conditions for the non-linear programming problem given below***

$$\text{\textit{Maximize }} Z = x_1^2 + 3x_2^2 + 5x_3^2$$

subject to the constraints

$$x_1 + x_2 + 3x_3 = 2$$
$$5x_1 + 2x_2 + x_3 = 5$$

and $x_1, x_2, x_3 \geq 0$

SOLUTION. We have $f(X) = Z = x_1^2 + 3x_2^2 + 5x_3^2$, $X = (x_1, x_2, x_3)$

and $\quad g_1(X) = x_1 + x_2 + 3x_3 = 2$

$\qquad g_2(X) = 5x_1 + 2x_2 + x_3 = 5$

Define $h_1(X)$ and $h_2(X)$ such that

$\qquad h_1(X) = g_1(X) - 2 = 0$ and $h_2(X) = g_2(X) - 5 = 0$

Let λ_1, λ_2 be the multipliers then Lagrangian function $L(X, \lambda)$ is given by

$\qquad L(X, \lambda) = f(X) - \lambda_1 h_1(X) - \lambda_2 h_2(X)$

$$= (x_1^2 + 3x_2^2 + 5x_3^2) - \lambda_1(x_1 + x_2 + 3x_3 - 2) - \lambda_2(5x_1 + 2x_2 + x_3 - 5)$$

Therefore, the required necessary conditions for optimality can be obtained as follows :

$$\frac{\partial L}{\partial x_1} = 2x_1 - \lambda_1 - 5\lambda_2 = 0; \qquad \frac{\partial L}{\partial x_2} = 6x_2 - \lambda_1 - 2\lambda_2 = 0$$

$$\frac{\partial L}{\partial x_3} = 10x_3 - 3\lambda_1 - \lambda_3 = 0; \qquad \frac{\partial L}{\partial \lambda_1} = -(x_1 + x_2 + 3x_3 - 2) = 0$$

and $\qquad \dfrac{\partial L}{\partial \lambda_2} = -(5x_1 + 2x_2 + x_3 - 5) = 0$

EXAMPLE 2. *Solve the following non-linear programming problem by the method of Lagrangian multipliers.*

$$\text{Maximize } Z = 6x_1 + 8x_2 - x_1^2 - x_2^2$$

subject to the constraints

$$4x_1 + 3x_2 = 16; \qquad 3x_1 + 5x_2 = 15$$

and $\qquad x_1, x_2 \geq 0$

SOLUTION. We have

$$f(X) = Z = 6x_1 + 8x_2 - x_1^2 - x_2^2$$

$$h_1(X) = 4x_1 + 3x_2 - 16 = 0$$

$$h_2(X) = 3x_1 + 5x_2 - 15 = 0$$

Taking λ_1, λ_2 as the multipliers, the Lagrangian function is given by

$$L(X, \lambda) = 6x_1 + 8x_2 - x_1^2 - x_2^2 - \lambda_1(4x_1 + 3x_2 - 16) - \lambda_2(3x_1 + 5x_2 - 15)$$

For the optimality, the necessary conditions are given by

$$\frac{\partial L}{\partial x_1} = 6 - 2x_1 - 4\lambda_1 - 3\lambda_2 = 0 \qquad \qquad \dots(1)$$

$$\frac{\partial L}{\partial x_2} = 8 - 2x_2 - 3\lambda_1 - 5\lambda_2 = 0 \qquad \qquad \dots(2)$$

$$\frac{\partial L}{\partial \lambda_1} = -(4x_1 + 3x_2 - 16) = 0 \qquad \qquad \dots(3)$$

and $\qquad \dfrac{\partial L}{\partial \lambda_2} = -(3x_1 + 5x_2 - 15) = 0 \qquad \qquad \dots(4)$

From the above equations, we can easily find

$$x_1 = \frac{35}{11}, \ x_2 = \frac{12}{11}, \ \lambda_1 = -\frac{212}{121}, \ \lambda_2 = \frac{268}{121}$$

Therefore, the stationary point is given by

$$x_0 = (x_1, x_2) = \left(\frac{35}{11}, \frac{12}{11} \right)$$

$$\lambda_1 = -\frac{212}{121}, \lambda_2 = \frac{268}{121}$$

Here, $n = 2$, $m = 2$, so Hessian matrix of the objective function $f(X)$ at x_0 is given by

$$H(X) = \begin{bmatrix} \dfrac{\partial^2 f}{\partial x_1^2} & \dfrac{\partial^2 f}{\partial x_1 \partial x_2} \\ \dfrac{\partial^2 f}{\partial x_2 \partial x_1} & \dfrac{\partial^2 f}{\partial x_2^2} \end{bmatrix} = \begin{bmatrix} -2 & 0 \\ 0 & -2 \end{bmatrix}$$

and principal minors of $H(x_0)$ are

$$D_1 = -2, D_2 = \begin{vmatrix} -2 & 0 \\ 0 & -2 \end{vmatrix} = 4$$

$\Rightarrow$ signs are alternatively negative and positive

$\Rightarrow$ $f(X)$ is negative definite

$\Rightarrow$ $f(X)$ is concave

Hence $f(X)$ is minimum at the point $\left(\dfrac{35}{11}, \dfrac{12}{11} \right)$

and maximum value of $f(X)$ is given by

$$\frac{1997}{121} = 16.5$$

EXAMPLE 3.   ***Solve the following non-linear programming problem by using the Lagrangian multiplier method***

$$\textbf{Min.}\, Z = 2x_1^2 + x_2^2 + 3x_3^2 + 10x_1 + 8x_2 + 6x_3 - 100$$

subject to the constraints

$$x_1 + x_2 + x_3 = 20$$

and $x_1, x_2, x_3 \geq 0$

SOLUTION.   Let λ be the Lagrangian multiplier, then define the Lagrangian function such that

$$L(X, \lambda) = f(X) - \lambda h(X)$$

$$= (2x_1^2 + x_2^2 + 3x_3^2 + 10x_1 + 8x_2 + 6x_3 - 100) - \lambda(x_1 + x_2 + x_3 - 20)$$

Clearly, the necessary conditions for $f(X) = Z$ to be maximum or minimum are given by

$$\frac{\partial L}{\partial x_1} = 0 \Rightarrow 4x_1 + 10 - \lambda = 0 \qquad \qquad \dots(1)$$

$$\frac{\partial L}{\partial x_2} = 0 \Rightarrow 2x_2 + 8 - \lambda = 0 \qquad \qquad \dots(2)$$

$$\frac{\partial L}{\partial x_3} = 0 \Rightarrow \frac{\partial L}{\partial x_3} = 0 \Rightarrow 6x_3 + 6 - \lambda = 0 \qquad \qquad \dots(3)$$

and $$\frac{\partial L}{\partial \lambda} = 0 \Rightarrow -(x_1 + x_2 + x_3 - 20) = 0 \qquad \qquad \dots(4)$$

On solving the above equations (1) to (3), we can easily find that

$$x_1 = \frac{-10+\lambda}{4}, \ x_2 = \frac{-8+\lambda}{2} \text{ and } x_3 = \frac{-6+\lambda}{6}$$

Using all these values in (4), we get $\lambda = 30$

Hence, $x_1 = 5, x_2 = 11, x_3 = 4$

$\therefore$ The stationary point is $(x_1, x_2, x_3) = (5, 11, 4)$

Now, we have to check the maximum or minimum of $f(X)$ at $(5, 11, 4)$

We proceed as fallow, we have

$$H_B = \begin{bmatrix} 0 & \dfrac{\partial h}{\partial x_1} & \dfrac{\partial h}{\partial x_2} & \dfrac{\partial h}{\partial x_3} \\ \dfrac{\partial h}{\partial x_1} & \dfrac{\partial^2 L}{\partial x_1^2} & \dfrac{\partial^2 L}{\partial x_1 \partial x_2} & \dfrac{\partial^2 L}{\partial x_1 \partial x_3} \\ \dfrac{\partial h}{\partial x_2} & \dfrac{\partial^2 L}{\partial x_2 \partial x_1} & \dfrac{\partial^2 L}{\partial x_2^2} & \dfrac{\partial^2 L}{\partial x_2 \partial x_3} \\ \dfrac{\partial h}{\partial x_3} & \dfrac{\partial^2 L}{\partial x_3 \partial x_1} & \dfrac{\partial^2 L}{\partial x_3 \partial x_2} & \dfrac{\partial^2 L}{\partial x_3^2} \end{bmatrix} = \begin{bmatrix} 0 & 1 & 1 & 1 \\ 1 & 4 & 0 & 0 \\ 1 & 0 & 2 & 0 \\ 1 & 0 & 0 & 6 \end{bmatrix}$$

Here, we have

$n = 3, m = 1 \Rightarrow 2m + 1 = 3$ and $n - m = 3 - 1 = 2$

So, we have to consider last two principal minors given by

$$D_3 = \begin{vmatrix} 0 & 1 & 1 \\ 1 & 4 & 0 \\ 1 & 0 & 2 \end{vmatrix} = -6 \text{ and } D_4 = \begin{vmatrix} 0 & 1 & 1 & 1 \\ 1 & 4 & 0 & 0 \\ 1 & 0 & 2 & 0 \\ 1 & 0 & 0 & 6 \end{vmatrix} = -44$$

Since signs of D_3 and D_4 are of sign $(-1)^m = (-1)^1 = -1$

$\Rightarrow$ $f(X)$ is minimum at the point $(5, 11, 4)$ and the minimum value of $f(X)$ is given by $Z = 281$.

EXAMPLE 4. *Using Lagrangian multiplier method solve the following non-linear programming problem.*

$$\textbf{Min} . \, \boldsymbol{Z = x_1^2 + x_2^2 + x_3^2}$$

subject to the constraints

$$\boldsymbol{4x_1 + x_2^2 + 2x_3 = 14}$$

and $\boldsymbol{x_1, x_2, x_3 \geq 0}$

SOLUTION. We have $f(X) = Z = x_1^2 + x_2^2 + x_3^2$

$$h(X) = 4x_1 + x_2^2 + 2x_3 - 14 = 0$$

If λ is the multiplier, then Lagrangian function is given by

$$L(X, \lambda) = f(X) - \lambda h(X)$$
$$= x_1^2 + x_2^2 + x_3^2 - \lambda(4x_1 + x_2^2 + 2x_3 - 14)$$

For the minimum of $f(X)$, the necessary conditions are

$$\frac{\partial L}{\partial x_1} = 0 \Rightarrow 2x_1 - 4\lambda = 0$$

...(1)

$$\frac{\partial L}{\partial x_2} = 0 \Rightarrow 2x_2 - 2\lambda x_2 = 0 \qquad ...(2)$$

$$\frac{\partial L}{\partial x_3} = 0 \Rightarrow 2x_3 - 2\lambda = 0 \qquad ...(3)$$

and $\qquad \frac{\partial L}{\partial \lambda} = 0 \Rightarrow -(4x_1 + x_2^2 + 2x_3 - 14) = 0 \qquad ...(4)$

From (2), we have

$$x_2(1 - \lambda) = 0 \Rightarrow x_2 = 0 \text{ or } \lambda = 1$$

From (1) and (3), we have

$$x_1 = 2\lambda \text{ and } x_3 = \lambda$$

So, when $x_2 = 0$, then from (4), $\lambda = 1.4$. Thus $x_1 = 2.8$ and $x_3 = 1.4$ and when $\lambda = 1$ then from (1) and (3), we get $x_2 = 2, x_3 = 1$

Then from (4)

$$x_2 = 2$$

Therefore, we have the following stationary points:

$$x_0 = (x_1, x_2, x_3) = (2.8, 0, 1.4), \lambda = 1.4$$

and $\qquad x_0 = (x_1, x_2, x_3) = (2, 2, 1), \lambda = 1$

Now, for the sufficient condition of maxima or minima of $f(X)$ at these stationary points, we proceed as follows.

Let

$$H_B = \begin{bmatrix} 0 & \dfrac{\partial h}{\partial x_1} & \dfrac{\partial h}{\partial x_2} & \dfrac{\partial h}{\partial x_3} \\ \dfrac{\partial h}{\partial x_1} & \dfrac{\partial^2 L}{\partial x_1^2} & \dfrac{\partial^2 L}{\partial x_1 \partial x_2} & \dfrac{\partial^2 L}{\partial x_1 \partial x_3} \\ \dfrac{\partial h}{\partial x_2} & \dfrac{\partial^2 L}{\partial x_2 \partial x_1} & \dfrac{\partial^2 L}{\partial x_2^2} & \dfrac{\partial^2 L}{\partial x_2 \partial x_3} \\ \dfrac{\partial h}{\partial x_3} & \dfrac{\partial^2 L}{\partial x_3 \partial x_1} & \dfrac{\partial^2 L}{\partial x_3 \partial x_2} & \dfrac{\partial^2 L}{\partial x_3^2} \end{bmatrix} = \begin{bmatrix} 0 & 4 & 2x_2 & 2 \\ 4 & 2 & 0 & 0 \\ 2x_2 & 0 & 2 & 0 \\ 2 & 0 & 0 & 2 \end{bmatrix}$$

Now,

$$H_B \text{ at } x_0 = (2.8, 0, 1.4) = \begin{bmatrix} 0 & 4 & 0 & 2 \\ 4 & 2 & 0 & 0 \\ 0 & 0 & 2 & 0 \\ 2 & 0 & 0 & 2 \end{bmatrix}$$

Here, $n = 3, m = 1 \Rightarrow n - m = 3 - 1 = 2$ and $2m + 1 = 3$

Now, we have to check the sign of two principal minors D_3 and D_4

$$\therefore \qquad D_3 = \begin{vmatrix} 0 & 4 & 0 \\ 4 & 2 & 0 \\ 0 & 0 & 2 \end{vmatrix} = -32 \text{ and } D_4 = \begin{vmatrix} 0 & 4 & 0 & 2 \\ 4 & 2 & 0 & 0 \\ 0 & 0 & 2 & 0 \\ 2 & 0 & 0 & 2 \end{vmatrix} = -80$$

$\Rightarrow$ The sign of D_3 and D_4 are same as the sign of $(-1)^m = (-1)^1 = -1$

$\Rightarrow f(X)$ has local minimum at the point (2.8, 0, 1.4) and minimum value of $Z = 9.8$

Similarly

$$H_B \text{ at } x_0 = (2,2,1) = \begin{bmatrix} 0 & 4 & 4 & 2 \\ 4 & 2 & 0 & 0 \\ 4 & 0 & 2 & 0 \\ 2 & 0 & 0 & 2 \end{bmatrix}$$

and $\qquad D_3 = -64 < 0$ and $D_4 = -144 < 0$

$\Rightarrow$ $f(X)$ has local minimum at the point (2, 2, 1) and min.$Z = 9$

Finally since, $\qquad 9 < 9.8$

Hence, minimum value of Z is 9 at (2, 2, 1)

25.11   SOLUTION OF NON-LINEAR PROGRAMMING PROBLEMS WHEN CONSTRAINTS ARE NOT EQUALITY CONSTRAINTS: KUHN-TUCKER CONDITIONS

25.11.1 KUHN-TUCKER NECESSARY CONDITIONS

Consider the non-linear programming problem

$$\text{Max.} Z = f(X), \quad X = (x_1, x_2, ..., x_n)$$

subject to the constraints

$$g_i(X) \le b_i \quad i = 1, 2, ..., m(<n) ; \ b_i's \text{ are constants}$$

Define $\qquad h_i(X) = g(X) - b_i \le 0$

Now, introducing the slack variables $s_1, s_2, ..., s_m$, then above inequality constraints reduces to

$$h_i(X) + s_i^2 = 0 \qquad i = 1, 2, ..., m$$

Here, addition of s_i^2 ensure the quantity added to $h_i(x)$ to be non-negative

Then the given problem becomes

$$\text{Max.} Z = f(X), X = (x_1, x_2, ..., x_n)$$

subject to the constraints

$$h_i(X) + s_i^2 = 0 \quad \forall i = 1, 2, ..., m$$

and $\qquad X \ge 0$

which is clearly a non-linear programming problem in $(n + m)$ variables $x_j, s_i \ i = 1, 2, .., m$ $j = 1, 2, .., n$ with m equality constraints.

Now introduce the Lagrangian multipliers $\lambda_1, \lambda_2, ..., \lambda$, then Lagrangian function is defined as follows:

$$L(X, S, \lambda) = f(S) - \sum_{i=1}^{m} \lambda_i [h_i(X) + s_i^2]$$

$$X = (x_1, x_2, ..., x_n) \ S = (s_1, s_2, ... s_m) \text{ and } \lambda = (\lambda_1, \lambda_2, ..., \lambda_m)$$

If L, f and h_i are all differentiable partially w.r.t. $x_1, x_2, ..., x_n; \lambda_1, \lambda_2, ..., \lambda_m; s_1, s_2, ..., s_m$ then the necessary conditions for the stationary points are given by

$$\frac{\partial L}{\partial x_j} = \frac{\partial f}{\partial x_j} - \sum_{i=1}^{m} \lambda_i \frac{\partial h_i}{\partial x_j} = 0 \ \forall \ j = 1, 2, ..., n \qquad \qquad ...(1)$$

$$\frac{\partial L}{\partial \lambda_i} = -[h_i(X) + s_i^2] = 0; \quad i = 1, 2, ..., m \qquad ...(2)$$

and
$$\frac{\partial L}{\partial s_i} = -2s_i\lambda_i = 0; \qquad i = 1, 2, ..., m \qquad ...(3)$$

From (3), we have either $s_i = 0$ or $\lambda_i = 0$

Now, if $s_i = 0$, then (2) implies

$$h_i(X) = 0 \Rightarrow \lambda_i = 0 \text{ or } s_i = 0$$

$$\Rightarrow \qquad \lambda_i = 0 \text{ or } h_i(X) = 0$$

$$\therefore \qquad \lambda_i h_i(X) = 0 \qquad ...(4)$$

Since, $\qquad s_i^2 \geq 0$; therefore (2) implies $h_i(x) \leq 0$

When $h_i(X) < 0$ then (4) implies $\lambda_i = 0$ and when $\lambda_i > 0$, $h_i(X) = 0$

But from (4), $h_i(X) = 0$ therefore λ_i is unrestricted in sign

If $\lambda_i \neq 0$, then clearly $s_i = 0$, then from (2)

$$h_i(X) = g_i(X) - b_i = 0 \Rightarrow g_i(X) = b_i$$

Here λ_i represent the rate of change of f w.r.t. b_i

Therefore,
$$\frac{\partial f}{\partial b_i} = \lambda_i$$

Now, as the RHS of $h_i(X) \leq 0$ increases about zero, the solution space becomes less constraints $\Rightarrow f$ can not decreases

$$\Rightarrow \qquad \frac{\partial f}{\partial b_i} = \lambda_i \nleq 0$$

$$\Rightarrow \qquad \lambda_i \geq 0$$

Therefore, when $h(X) < 0$, $\lambda = 0$ and when $\lambda > 0$, $h(X) = 0$

Hence, we conclude that the Kuhn-Tucker necessary conditions for X to be a point of maximum for $f(X)$ subject to $h_i(X) = g_i(X) - b_i \leq 0$ are given by

$$\frac{\partial f}{\partial x_j} - \sum_{i=1}^{m} \lambda_i \frac{\partial h_i}{\partial x_j} = 0 \qquad \forall \ j = 1, 2, ..., n$$

and $\lambda_i h_i(X) = 0, h_i(X) \leq 0, \lambda_i \geq 0, i = 1, 2, ..., m$

and the Kuhn-Tucker conditions for the point X to be a point of minimum for $f(X)$ subject to $h_i(X) = g_i(X) - b_i \geq 0$ are given by

$$\frac{\partial f}{\partial x_j} - \sum_{i=1}^{m} \lambda_i \frac{\partial h_i}{\partial x_j} = 0 \qquad \forall \ j = 1, 2, ..., n$$

and $\lambda_i h_i(X) = 0, h_i(X) \geq 0$ and $\lambda_i \geq 0, i = 1, 2, ..., m$.

☛ REMARKS

- In case of maximization, NLPP convert all the constraints $m \leq$ type and in case of minimization, convert all the constraints in $\geq$ type.
- The Lagrangian multiplier, λ_i corresponding to $h_i(X) = 0$ must be unrestricted in sign.
- If the given problem is of minimization (having the constraints of the form $g_i(X) \geq 0$) then $\lambda_i \leq 0$ and if the problem is of maximization (having the constraints of the form $g_i(X) \leq 0$) then $\lambda_i \geq 0$.

25.11.2 KUHN-TUCKER SUFFICIENT CONDITIONS

"The Kuhn-Tucker necessary conditions for a non-linear programming problem given by

Max. $f(X)$ subject to the constraints

$$h_i(X) = g_i(X) - b_i \leq 0 \qquad i = 1, 2, ..., m \text{ and } X \geq 0$$

will also be the sufficient condition for maximum of $f(X)$ if,

(i) $f(X)$ is concave and

(ii) $h_i(X)$ are convex i.e., $-h_i(X)$ is also concave.

The Lagrangian function of the problem can be written as

$$L(X, s, \lambda) = f(X) - \sum_{i=1}^{m} \lambda_i \left[h_i(X) + s_i^2 \right]$$

where $X = (x_1, x_2, ..., x_n)$, $S = (s_1, s_2, ..., s_m)$ and $\lambda = (\lambda_1, \lambda_2, ... \lambda_m)$ and s_i are slack variables used to convert inequality constraints to equality and λ_i are the Lagrangian multipliers.

Using the slack variables s_i, we get

$$h_i(X) + s_i^2 = 0, \ i = 1, 2, ... m$$

and from the necessary conditions, we have $\lambda_i h_i(X) = 0, \ i = 1, 2, ..., m$

Therefore,

$$\lambda_i s_i^2 = -\lambda_i h_i(X) = 0, \ i = 1, 2, ..., m$$

Now, since $h_i(X)$ are convex functions of X and $\lambda_i \geq 0$

therefore, $\lambda_i h_i(X)$ is convex, i.e., $-\lambda_i h_i(X)$ is concave.

$\Rightarrow \quad -\sum_{i=1}^{m} \lambda_i h_i(X)$ is concave function of X.

$\Rightarrow \quad f(X) - \sum_{i=1}^{m} \lambda_i h_i(X)$ is concave function of X.

$\Rightarrow \quad f(X) - \sum_{i=1}^{m} \lambda_i \left[h_i(X) + s_i^2 \right]$ is concave function of X. $\qquad [\because \lambda_i s_i^2 = 0]$

$\Rightarrow \quad L(X, S, \lambda)$ is concave.

Thus, we conclude that the necessary conditions for maximum of $f(X)$ at an extreme point implies that $L(X, S, \lambda)$ also have the same extreme point. Since $L(X, S, \lambda)$ is concave, its derivative must be zero at one point, which give the absolute maximum value of $f(X)$.

Also, the Kuhn-Tucker necessary conditions for the minimization of non-linear programming problem given by

Min. $f(X)$.

subject to the constraints

$$h_i(X) = g_i(X) - b_i \geq 0, \ i = 1, 2, ..., m$$

and $X \geq 0$

are also the sufficient conditions for minimum of $f(X)$ if

(i) $f(X)$ is convex, and

(ii) $h_i(X)$ are also convex i.e., $-h_i(X)$ is concave.

If in the minimizing non-linear programming problem, the constraints are taken of the type $h_i(X) \leq 0$ then the Kuhn-Tucker necessary conditions for the optimality of the objective function will be given by

$$\frac{\partial f}{\partial x_j} - \sum_{i=1}^{m} \lambda_i \frac{\partial h_i(X)}{\partial x_j} = 0 \qquad \forall j = 1, 2, ..., n$$

$$\lambda_i h_i(X) = 0, \ h_i(X) \leq 0 \text{ and } \lambda_i \leq 0 \forall i = 1, 2, ..., m$$

and these conditions will be sufficient if $f(X)$ and all $h_i(X)$ are convex function of X.

 Solved Examples

EXAMPLE 1. *Find the value of x_1 and x_2 which*

$$\text{Maximize} \quad Z = 12x_1 + 21x_2 + 2x_1x_2 - 2x_1^2 - 2x_2^2$$

subject to the constraints

$$x_2 \leq 8$$
$$x_1 + x_2 \leq 10$$

and $\quad x_1, x_2 \geq 0$

SOLUTION. We have
$$f(X) = f(x_1, x_2) = 12x_1 + 21x_2 + 2x_1x_2 - 2x_1^2 - 2x_2^2$$
$$h_1(X) = h_1(x_1, x_2) = x_2 - 8 \leq 0$$
$$h_2(X) = h_2(x_1, x_2) = x_1 + x_2 - 10 \leq 0$$

Then the Lagrangian function L(X, S, λ) is given by

$$L(X, S, \lambda) = f(x) - \lambda_1 \left[h_1(x) + s_1^2 \right] - \lambda_2 \left[h_2(x) + s_2^2 \right]$$

Then we have the following Kuhn-Tucker necessary conditions

(i) $\quad \dfrac{\partial f}{\partial x_j} - \sum\limits_{i=1}^{2} \lambda_i \dfrac{h_i(X)}{\partial x_j} = 0$

$\Rightarrow \qquad 12 + 2x_2 - 4x_1 - \lambda_2 = 0$

and $\quad 21 + 2x_1 - 4x_2 - \lambda_1 - \lambda_2 = 0$

(ii) $\quad \lambda_i h_i(X) = 0$, $i = 1, 2$

$\Rightarrow \quad \lambda_1(x_2 - 8) = 0$ and $\lambda_2(x_1 + x_2 - 10) = 0$

(iii) $\quad h_i(X) \leq 0$

$\Rightarrow \qquad x_2 - 8 \leq 0$ and $x_1 + x_2 - 10 \leq 0$

(iv) $\quad \lambda_i \geq 0 \quad i = 1, 2$

Then we have the following cases:

Case 1: **If $\lambda_1 = 0$, $\lambda_2 = 0$** then from (i) we have

$\qquad 12 + 2x_2 - 4x_1 = 0$ and $21 + 2x_1 - 4x_2 = 0$

On solving we get $x_1 = \dfrac{15}{2}, x_2 = 9$

But x_1 and x_2 do not satisfy (iii) and therefore it may be discarded.

Case 2: **If $\lambda_1 \neq 0$, $\lambda_2 = 0$** then from (i) and (ii), we have

$$x_1 + x_2 = 10$$
$$2x_2 - 4x_1 = -12$$

$$2x_1 - 4x_2 = -12 + \lambda_1$$

On solving the above equations, we get $x_1 = 2$, $x_2 = 8$, $\lambda_1 = -16$

which violates the condition (iv) and therefore it may also be discarded.

Case 3: **If $\lambda_1 \neq 0, \lambda_2 \neq 0$ then from (ii)**

$$x_2 - 8 = 0 \quad \Rightarrow \quad x_2 = 8$$
$$x_1 + x_2 - 10 = 0 \quad \Rightarrow \quad x_1 = 2$$

Putting these values in (i) we get $\lambda_1 = -27$ and $\lambda_2 = 20$

which violates the condition (iv) and therefore may be discarded.

Case 4: **If $\lambda_1 = 0, \lambda_2 \neq 0$ then from (i) and (ii) we have**

$$2x_2 - 4x_1 = -12 + \lambda_2$$
$$2x_1 - 4x_2 = -21 + \lambda_2$$
$$x_1 + x_2 = 10$$

On solving the above equations, we get

$$x_1 = \frac{17}{4}, \; x_2 = \frac{23}{4}, \; \lambda_2 = \frac{13}{4}$$

which does not violate any of the Kuhn-Tucker conditions and therefore accepted.
Hence, the optimum solution of the given NLPP is

$$x_1 = \frac{17}{4}, \; x_2 = \frac{23}{4}, \; \lambda_1 = 0 \text{ and } \lambda_2 = \frac{13}{4}$$

and $\qquad Max.Z = \dfrac{1734}{16}$

EXAMPLE 2. *Solve the following non-linear programming problem.*

$$\text{Min } Z = (x_1 - 2)^2 + (x_2 - 1)^2$$

subject to the constraints

$$x_1^2 - x_2 \leq 0, \; x_1 + x_2 \leq 2$$

and $\qquad x_1, \; x_2 \geq 0$

SOLUTION. We have $\qquad Z = f(X) = f(x_1, x_2) = (x_1 - 2)^2 + (x_2 - 1)^2$

$$h_1(X) = -x_1^2 + x_2 \geq 0, \; h_2(X) = 2 - x_1 - x_2 \geq 0$$

and $\qquad X = (x_1, x_2) \geq 0$

Since it is a problem of minimization, therefore the signs in inequality constraints are taken as $\geq$.

Now, the Hessian matrix

$$H_B = \begin{bmatrix} \dfrac{\partial^2 f}{\partial x_1^2} & \dfrac{\partial^2 f}{\partial x_1 \partial x_2} \\ \dfrac{\partial^2 f}{\partial x_2 \partial x_1} & \dfrac{\partial^2 f}{\partial x_2^2} \end{bmatrix} = \begin{bmatrix} 2 & 0 \\ 0 & 2 \end{bmatrix}$$

whose principal minors are $D_1 = 2$, $D_2 = \begin{vmatrix} 2 & 0 \\ 0 & 2 \end{vmatrix} = 4$, clearly both are positive.

$\Rightarrow \quad f(X)$ is a convex function.

Also, $h_1(X)$ and $h_2(X)$ are convex functions of X.

Therefore the Kuhn-Tucker necessary conditions for minimum of $f(X)$ will also be the sufficient conditions.

Here, the Kuhn-Tucker necessary conditions for the minimum of $f(X)$ are given as below:

$$\frac{\partial f}{\partial x_j} = \sum_{i=1}^{2} \lambda_i \frac{\partial h_i}{\partial x_j}, \ j = 1, 2$$

$$\lambda_1 h_1(X) = 0, \ \lambda_2 h_2(X) = 0$$

$$h_1(X) \geq 0, \ h_2(X) \geq 0 \text{ and } \lambda_1, \lambda_2 \geq 0.$$

which implies

$$2(x_1 - 2) = -2x_1\lambda_1 - \lambda_2 \qquad \text{...(1)}$$

$$2(x_2 - 1) = \lambda_1 - \lambda_2 \qquad \text{...(2)}$$

$$\lambda_1(-x_1^2 + x_2) = 0 \qquad \text{...(3)}$$

$$\lambda_2(2 - x_1 - x_2) = 0 \qquad \text{...(4)}$$

$$-x_1^2 + x_2 \geq 0 \qquad \text{...(5)}$$

$$2 - x_1 - x_2 \geq 0 \qquad \text{...(6)}$$

and $\qquad \lambda_1, \lambda_2 \geq 0 \qquad \text{...(7)}$

Now, we have the following cases:

Case 1: If $\lambda_1 = \lambda_2 = 0$ Here, from (1) and (2), we get $x_1 = 2, x_2 = 1$ These values do not satisfy (5) and (6) and so are discarded.

Case 2: If $\lambda_1 = 0, \lambda_2 \neq 0$ In this case from (1) and (2)

$$2(x_1 - 2) = -\lambda_2 \text{ and } 2(x_2 - 1) = -\lambda_2 \Rightarrow x_1 - x_2 - 1 = 0 \qquad \text{...(8)}$$

and from (4), $\qquad -x_1 - x_2 + 2 = 0 \qquad \text{...(9)}$

On solving (8) and (9), we get $x_1 = \dfrac{3}{2}, x_2 = \dfrac{1}{2}$

which does not satisfy (5) and hence discarded.

Case 3: If $\lambda_1 \neq 0, \lambda_2 = 0$ Here, from (1) and (2) we have

$$x_1 - 2 = -x_1\lambda_1 \text{ and } 2(x_2 - 1) = \lambda_1$$

$\Rightarrow \quad -x_1 + 2x_1x_2 - 2 = 0 \qquad \text{...(10)}$

Again from (3) $\quad -x_1^2 + x_2 = 0 \qquad \text{...(11)}$

From (10) and (11) we get

$$2x_1^3 - x_1 - 2 = 0$$

$\Rightarrow \qquad x_1 = 1.52 \text{ and } x_2 = 2.31$

These values does not satisfy (6) and hence discarded.

Case 4: If $\lambda_1 \neq 0$ and $\lambda_2 \neq 0$ Here, from (3) and (4) we have

$$-x_1^2 + x_2 = 0 \text{ and } 2 - x_1 - x_2 = 0$$

$\Rightarrow \qquad x_2 = x_1^2 \text{ and } x_1^2 + x_1 - 2 = 0$

$\Rightarrow \qquad x_2 = x_1^2 \text{ and } (x_1 - 1)(x_1 + 2) = 0$

$\Rightarrow$ $\qquad x_1 = 1, x_2 = 1$

Putting the values of x_1 and x_2 in (1) and (2), we get,

$\qquad 2\lambda_1 + \lambda_2 = 2$ and $\lambda_1 - \lambda_2 = 0$

On solving, we get,

$$\lambda_1 = \frac{2}{3} \text{ and } \lambda_2 = \frac{2}{3} \geq 0$$

Hence, the optimal solution is given by

$$x_1 = 1, \ x_2 = 1, \ Min. Z = 1$$

EXAMPLE 3. *Solve the following non-linear programming problem:*

$$\mathbf{Max.Z = f(x_1, x_2) = 3.6x_1 - 0.4x_1^2 + 1.6x_2 - 0.2x_2^2}$$

subject to the constraints

$$\mathbf{2x_1 + x_2 \leq 10 \text{ and } x_1, x_2 \geq 0}$$

SOLUTION. We have,

$$f(X) = f(x_1, x_2) = 3.6x_1 - 0.4x_1^2 + 1.6x_2 - 0.2x_2^2$$

$$h(X) = 2x_1 + x_2 - 10 \leq 0$$

$$X = (x_1, x_2) \geq 0$$

The Hessian matrix is given by

$$H_B = \begin{bmatrix} \dfrac{\partial^2 f}{\partial x_1^2} & \dfrac{\partial^2 f}{\partial x_1 \partial x_2} \\[3mm] \dfrac{\partial^2 f}{\partial x_2 \partial x_1} & \dfrac{\partial^2 f}{\partial x_2^2} \end{bmatrix} = \begin{bmatrix} -0.8 & 0 \\ 0 & -0.4 \end{bmatrix}$$

whose principal minors are given by

$$D_1 = -0.8 < 0, \quad D_2 = \begin{vmatrix} -0.8 & 0 \\ 0 & -0.4 \end{vmatrix} = 0.32 > 0$$

which are clearly have alternate signs.

$\Rightarrow$ $f(X)$ is a concave function.

and $h(X) = 2x_1 + x_2 - 10$ is a convex function

Therefore, the Kuhn-Tucker necessary conditions for maximum of $f(X)$ are given by

$$\frac{\partial f}{\partial x_1} = \lambda \frac{\partial h}{\partial x_1}, \frac{\partial f}{\partial x_2} = \lambda \frac{\partial h}{\partial x_2}$$

$$\lambda h(X) = 0, \quad h(X) \leq 0 \text{ and } \lambda \geq 0$$

which implies

$$3.6 - 0.8 x_1 = 2\lambda \qquad \qquad \text{...(1)}$$

$$1.6 - 0.4 x_2 = \lambda \qquad \qquad \text{...(2)}$$

$$\lambda(2x_1 + x_2 - 10) = 0 \qquad \qquad \text{...(3)}$$

$$2x_1 + x_2 - 10 \leq 0 \qquad \qquad \text{...(4)}$$

and $\qquad \qquad \lambda \geq 0$ $\qquad \qquad \text{...(5)}$

Now, from (3) we have either $\lambda = 0$ or $2x_1 + x_2 - 10 = 0$

If $\lambda = 0$ then from (1) and (2), we have,

$$3.6 - 0.8x_1 = 0 \Rightarrow x_1 = 4.5$$

and $\qquad 1.6 - 0.4\,x_2 = 0 \Rightarrow x_2 = 4$

These values does not satisfy (4) therefore $\lambda \neq 0$ which implies $2x_1 + x_2 - 10 = 0$

$$\qquad\qquad\qquad\qquad\qquad ...(5)$$

Now, from (1) and (2), we have

$$x_1 = \frac{1.8 - \lambda}{0.4}, \quad x_2 = \frac{1.6 - \lambda}{0.4}$$

Then, from (5) we get $\lambda = 0.4$

$\Rightarrow \qquad\qquad x_1 = 3.5$ and $x_2 = 3$

Clearly, these values of x_1 and x_2 also satisfy (4)

Also, $\qquad\qquad \lambda = 0.4 > 0$

$\Rightarrow$ (5) is also satisfied

Hence, the optimum solution of the given problem is

$\qquad\qquad x_1 = 3.5, \ x_2 = 3$ and Max. $Z = 10.7$

EXAMPLE 4. **Solve the following non-linear programming problem**

$$\textbf{Max.}\, Z = -x_1^2 - x_2^2 - x_3^2 + 4x_1 + 6x_2$$

subject to the constraints

$$x_1 + x_2 \leq 2;\ 2x_1 + 3x_2 \leq 12 \text{ and } x_1, x_2 \geq 0$$

SOLUTION. We have, $\quad Z = f(X) = f(x_1, x_2, x_3) = -x_1^2 - x_2^2 - x_3^2 + 4x_1 + 6x_2$

$$h_1(X) = x_1 + x_2 - 2 \leq 0$$

$$h_2(X) = 2x_1 + 3x_2 - 12 \leq 0$$

Now, the Hessian matrix is given by

$$H_B = \begin{vmatrix} \dfrac{\partial^2 f}{\partial x_1^2} & \dfrac{\partial^2 f}{\partial x_1\,\partial x_2} & \dfrac{\partial^2 f}{\partial x_1\,\partial x_3} \\[2mm] \dfrac{\partial^2 f}{\partial x_2\,\partial x_1} & \dfrac{\partial^2 f}{\partial x_2^2} & \dfrac{\partial^2 f}{\partial x_2\,\partial x_3} \\[2mm] \dfrac{\partial^2 f}{\partial x_3\,\partial x_1} & \dfrac{\partial^2 f}{\partial x_3\,\partial x_2} & \dfrac{\partial^2 f}{\partial x_3^2} \end{vmatrix} = \begin{bmatrix} -2 & 0 & 0 \\ 0 & -2 & 0 \\ 0 & 0 & -2 \end{bmatrix}$$

whose principle minors are $D_1 = -2 < 0,\ D_2 = \begin{vmatrix} -2 & 0 \\ 0 & -2 \end{vmatrix} = 4 > 0$

and $\qquad D_3 = \begin{vmatrix} -2 & 0 & 0 \\ 0 & -2 & 0 \\ 0 & 0 & -2 \end{vmatrix} = -8 < 0$ which are clearly of alternate sign.

Therefore, $f(X)$ is a concave function.

Also, $h_1(X),\, h_2(X)$ are convex function of $X = (x_1, x_2)$.

$\Rightarrow$ Kuhn-Tucker necessary conditions for max. $f(X)$ will also be the sufficient conditions and are given by

$$\frac{\partial f}{\partial x_j} = \sum_{i=1}^{2} \lambda_i \frac{\partial h_i(X)}{\partial x_j}, \quad j = 1, 2, 3$$

$$\lambda_1 h_1(X) = 0, \lambda_2 h_2(X) = 0, h_1(X) \le 0, h_2(X) \le 0 \text{ and } \lambda_1, \lambda_2 \ge 0$$

which implies

$$-2x_1 + 4 = \lambda_1 + 2\lambda_2 \qquad \qquad \text{...(1)}$$

$$-2x_2 + 6 = \lambda_1 + 3\lambda_2 \qquad \qquad \text{...(2)}$$

$$-2x_3 = 0 \qquad \qquad \text{...(3)}$$

$$\lambda_1(x_1 + x_2 - 2) = 0 \qquad \qquad \text{...(4)}$$

$$\lambda_2(2x_1 + 3x_2 - 12) = 0 \qquad \qquad \text{...(5)}$$

$$x_1 + x_2 - 2 \le 0 \qquad \qquad \text{...(6)}$$

$$2x_1 + 3x_2 - 12 \le 0 \qquad \qquad \text{...(7)}$$

and $\qquad \qquad \lambda_1, \lambda_2 \ge 0 \qquad \qquad \text{...(8)}$

Now, we have the following cases:

Case 1: **If $\lambda_1 = \lambda_2 = 0$** Here, from (1) and (2) we have

$$-2x_1 + 4 = 0 \text{ and } -2x_2 + 6 = 0$$

$\Rightarrow \qquad \qquad x_1 = 2 \text{ and } x_2 = 3$

These values do not satisfy (6) and (7) and hence discarded.

Case 2: **If $\lambda_1 \ne 0, \lambda_2 = 0$** In this case, from (1) and (2), we have

$$-2x_1 + 4 = \lambda_1 \text{ and } -2x_2 + 6 = \lambda_1$$

$\Rightarrow \qquad \qquad x_1 - x_2 + 1 = 0 \qquad \qquad \text{...(9)}$

If $\lambda_1 \ne 0$ then from (4),

$$x_1 + x_2 - 2 = 0 \qquad \qquad \text{...(10)}$$

On solving (9) and (10) we get

$$x_1 = \frac{1}{2}, x_2 = \frac{3}{2}$$

Clearly, these values satisfy (6) and (7)

Also, for these values of x_1 and x_2, from (1) and (2), we get

$$\lambda_1 = 3 > 0, \lambda_2 = 0$$

For this solution, $x_1 = \frac{1}{2}, x_2 = \frac{3}{2}$ and $\max . Z = \frac{17}{2}$

Case 3: **If $\lambda_1 = 0, \lambda_2 \ne 0$** Here, from (1) and (2), we get

$$-2x_1 + 4 = 2\lambda_2 \text{ and } -2x_2 + 6 = 3\lambda_2$$

$\Rightarrow \qquad \qquad 3x_1 - 2x_2 = 0 \qquad \qquad \text{...(11)}$

If $\lambda_2 \ne 0$, then from (5), we get $2x_1 + 3x_2 - 12 = 0 \qquad \qquad \text{...(12)}$

On solving (11) and (12), we get

$$x_1 = \frac{24}{13}, x_2 = \frac{36}{13}$$

Also, from (3), $\qquad \qquad x_3 = 0$

These values does not satisfy (6) and hence discarded.

Case 4: **If $\lambda_1 \ne 0, \lambda_2 \ne 0$** From (4) and (5) we get

$$x_1 + x_2 - 2 = 0 \text{ and } 2x_1 + 3x_2 - 12 = 0$$

$\Rightarrow \qquad \qquad x_1 = -6 \text{ and } x_2 = 8$

Since, $x_1 = -6 < 0 \Rightarrow$ solution is discarded.

Hence, the optimal solution of the given problem is

$$x_1 = \frac{1}{2}, x_2 = \frac{3}{2}, x_3 = 0 \text{ and Max.}Z = \frac{17}{2}$$

EXAMPLE 5. ***Solve the following non-linear programming problem***

$$\mathbf{Max.Z} = \mathbf{f(X)} = \mathbf{f(x_1, x_2)}$$
$$= \mathbf{(200x_1 - 2x_1^2) + (500x_2 - 3x_2^2)}$$

subject to the constraints

$$\mathbf{2x_1 + x_2 \leq 140;\ 2x_1 + 3x_2 \leq 180 \text{ and } x_1, x_2 \geq 0}$$

SOLUTION. We have,

$$f(X) = (200x_1 - 2x_1^2) + (500x_2 - 3x_2^2)$$
$$h_1(X) = 2x_1 + x_2 - 140 \leq 0$$
$$h_2(X) = 2x_1 + 3x_2 - 180 \leq 0$$
$$X = (x_1, x_2) \geq 0$$

Here, the Hessian matrix is given by

$$H_B = \begin{bmatrix} \dfrac{\partial^2 f}{\partial x_1^2} & \dfrac{\partial^2 f}{\partial x_1 \partial x_2} \\[3mm] \dfrac{\partial^2 f}{\partial x_2 \partial x_1} & \dfrac{\partial^2 f}{\partial x_2^2} \end{bmatrix} = \begin{bmatrix} -4 & 0 \\ 0 & -6 \end{bmatrix}$$

whose principal minors are $D_1 = -4, D_2 = \begin{vmatrix} -4 & 0 \\ 0 & -6 \end{vmatrix} = 24 > 0$ are of alternate sign.

$\Rightarrow f(X)$ is a concave function.

Further, $h_1(x)$ and $h_2(x)$ are convex functions.

$\Rightarrow$ The kuhn-Tucker necessary conditions for max.$f(X)$ will also be the sufficient conditions.

The Kuhn-Tucker necessary conditions for maximum of $f(X)$ are given by

$$\frac{\partial f}{\partial x_1} = \sum_{i=1}^{2} \lambda_i \frac{\partial h_i(X)}{\partial x_i}; \frac{\partial f}{\partial x_2} = \sum_{i=1}^{2} \lambda_i \frac{\partial h_i(X)}{\partial x_2}$$

$$\lambda_1 h_1(X) = 0, \lambda_2 h_2(X) = 0, h_1(X) \leq 0, h_2(X) \leq 0 \text{ and } \lambda_1, \lambda_2 \geq 0$$

which implies

$$200 - 4x_1 = 2\lambda_1 + 2\lambda_2 \qquad \qquad ...(1)$$
$$500 - 6x_2 = \lambda_1 + 3\lambda_2 \qquad \qquad ...(2)$$
$$\lambda_1(2x_1 + x_2 - 140) = 0 \qquad \qquad ...(3)$$
$$\lambda_2(2x_1 + 3x_2 - 180) = 0 \qquad \qquad ...(4)$$
$$2x_1 + x_2 - 140 \leq 0 \qquad \qquad ...(5)$$
$$2x_1 + 3x_2 - 180 \leq 0 \qquad \qquad ...(6)$$
and $\qquad \lambda_1, \lambda_2 \geq 0 \qquad \qquad ...(7)$

Now, we have the following cases:

Case 1: **If $\lambda_1 = 0$, $\lambda_2 = 0$** Here, from (1) and (2), we get

$$200 - 4x_1 = 0 \text{ and } 500 - 6x_2 = 0$$

$\Rightarrow$ $\qquad$ $x_1 = 50 \text{ and } x_2 = \dfrac{250}{3}$

which does not satisfy (5) and (6) and hence discarded.

Case 2: **If $\lambda_1 \neq 0$, $\lambda_2 = 0$** In this case, from (1) and (2), we get

$$200 - 4x_1 = 2\lambda_1 \text{ and } 500 - 6x_2 = \lambda_1$$

$\Rightarrow$ $\qquad$ $x_1 - 3x_2 + 200 = 0$ $\qquad$ ...(8)

If $\lambda_1 \neq 0$, then from (3), $2x_1 + x_2 - 140 = 0$ $\qquad$ ...(9)

On solving (8) and (9) we get $x_1 = \dfrac{220}{7}$, $x_2 = \dfrac{540}{7}$

which does not satisfy (6) and hence discarded.

Case 3: **If $\lambda_1 = 0$, $\lambda_2 \neq 0$** In this case, from (1) and (2), we get

$$200 - 4x_1 = 2\lambda_2 \text{ and } 500 - 6x_2 = 3\lambda_2$$

$\Rightarrow$ $\qquad$ $3x_1 - 3x_2 + 100 = 0$ $\qquad$ ...(10)

If $\lambda_2 \neq 0$, then from (4),

$$2x_1 + 3x_2 - 180 = 0 \qquad \text{...(11)}$$

On solving (10) and (11), we get

$$x_1 = 16, x_2 = \dfrac{148}{3}$$

$\Rightarrow$ $\qquad$ $\lambda_2 = 68 > 0$

Clearly, these values of x_1 and x_2 satisfy (5)

$\Rightarrow$ $\qquad$ $x_1 = 16$, $x_2 = \dfrac{148}{3}$ is stationary point at which

$$\text{Max.} f(X) = \dfrac{60160}{3}$$

Case 4: **If $\lambda_1 \neq 0$, $\lambda_2 \neq 0$** In this case from (3) and (4), we get

$$2x_1 + x_2 - 140 = 0 \text{ and } 2x_1 + 3x_2 - 180 = 0$$

$\Rightarrow$ $\qquad$ $x_1 = 60$, $x_2 = 20$

Then, from (1) and (2), we get

$$\lambda_1 + \lambda_2 = -20 \text{ and } \lambda_1 + 3\lambda_2 = 380$$

On solving we get $\lambda_1 = -220 < 0$ and $\lambda_2 = 200 > 0$

For these values (7), is not satisfied and hence discarded.

Hence, the optimal solution is given by

$$x_1 = 16, x_2 = \dfrac{148}{3} \quad \text{and} \quad \text{Max.} Z = \dfrac{60160}{3}$$

Exercise-25.4

Solve the following non-linear programming problem using the method of Lagrangian multipliers.

1. Minimize $Z = 3e^{2x_1+1} + 2e^{x_2+5}$

subject to the constraints
$$x_1 + x_2 = 7$$
and $\quad x_1, x_2 \geq 0$

2. Min. $Z = 2x_1^2 - 24x_1 + 2x_2^2 - 8x_2 + 2x_3^2$
$$-12x_3 + 200$$
subject to the constraints
$$x_1 + x_2 + x_3 = 11$$
and $\qquad x_1, x_2, x_3 \geq 0$

3. Min. $Z = x_1^2 + x_2^2 + x_3^2$

subject to the constraints
$$x_1 + x_2 + 3x_3 = 2$$
$$5x_1 + 2x_2 + x_3 = 5$$
and $\qquad x_1, x_2, x_3 \geq 0$

4. Min. $Z = 6x_1^2 + 5x_2^2$
subject to the constraints
$$x_1 + 5x_2 = 3$$
and $\quad x_1, x_2 \geq 0$

5. Max. $Z = 4x_1 + 6x_2 - 2x_1^2 - 2x_1x_2 - 2x_2^2$
subject to the constraints
$$x_1 + 2x_2 = 2$$
and $\quad x_1, x_2 \geq 0$

6. Max. $Z = x_1^2 + 4x_1x_2 + x_2^2$

subject to the constraints
$$x_1^2 + x_2^2 = 1$$
and $\quad x_1, x_2 \geq 0$

7. Min. $Z = 4x_1^2 + 2x_2^2 + x_3^2 - 4x_1x_2$
subject to the constraints
$$x_1 + x_2 + x_3 = 15$$
$$2x_1 - x_2 + 2x_3 = 20$$
and $\qquad x_1, x_2, x_3 \geq 0$

8. Max. $Z = 7x_1 - 0.3x_1^2 + 8x_2 - 0.4x_2^2$
subject to the constraints
$$4x_1 + 5x_2 = 100$$
and $\qquad x_1, x_2 \geq 0$

Solve the following non-linear programming problem, by using Kuhn-Tucker conditions.

9. Min. $Z = \left(x_1 - \frac{9}{4}\right)^2 + (x_2 - 2)^2$

subject to the constraints
$$x_2 - x_1^2 \geq 0$$
$$x_1 + x_2 \leq 6$$
and $\qquad x_1, x_2 \geq 0$

10. Min. $Z = (x_1 - 1)^2 + (x_2 - 5)^2$
subject to the constraints
$$-x_1^2 + x_2 \leq 4$$
$$-(x_1 - 2)^2 + x_2 \leq 4$$
and $\qquad x_1, x_2 \geq 0$

11. Min. $Z = -\log x_1 - \log x_2$
subject to the constraints
$$x_1 + x_2 \leq 2$$
and $\quad x_1, x_2 \geq 0$

12. Max. $Z = 2x_1^2 + 12x_1x_2 - 7x_2^2$
subject to the constraints
$$2x_1 + 5x_2 \leq 98$$
and $\qquad x_1, x_2 \geq 0$

13. Max. $Z = 3x_1 + x_2$
subject to the constraints
$$x_1^2 + x_2^2 \leq 5$$
$$x_1 - x_2 \leq 1$$
and $\quad x_1, x_2 \geq 0$

14. Max. $Z = 10x_1 - x_1^2 + 10x_2 - x_2^2$
subject to the constraints
$$x_1 + x_2 \leq 14$$
$$-x_1 + x_2 \leq 6$$
and $\qquad x_1, x_2 \geq 0$

15. Max. $Z = 7x_1^2 - 6x_1 + 5x_2^2$
subject to the constraints
$$x_1 + 2x_2 \leq 10$$
$$x_1 - 3x_2 \leq 9$$
and $\qquad x_1, x_2 \geq 0$

16. Max $Z = 2x_1 - x_1^2 + x_2$
subject to the constraints
$$2x_1 + 3x_2 \leq 6$$
$$2x_1 + x_2 \leq 4$$
and $\qquad x_1, x_2 \geq 0$

ANSWERS

1. $x_1 = \dfrac{1}{3}(11 - \log 3)$, $x_2 = \dfrac{1}{3}(10 + \log 3)$
2. $x_1 = 6$, $x_2 = 2$, $x_3 = 3$, Min.$Z = 102$

3. $x_1 = 0.81$, $x_2 = 0.35$, $x_3 = 0.928$, Min. $Z = 1.625$
4. $x_1 = \dfrac{3}{31}$, $x_2 = \dfrac{18}{31}$, Min.$Z = \dfrac{54}{31}$

5. $x_1 = \dfrac{1}{3}$, $x_2 = \dfrac{5}{6}$, Max.$Z = \dfrac{25}{6}$
6. $x_1 = \dfrac{1}{\sqrt{2}}$, $x_2 = \dfrac{1}{\sqrt{2}}$, Max.$Z = 3$

7. $x_1 = \dfrac{11}{3}$, $x_2 = \dfrac{10}{3}$, $x_3 = 8$, Min.$Z = \dfrac{820}{9}$
8. $x_1 = 12.06$, $x_2 = 10.35$, Max. $Z = 80.73$

9. $x_1 = \dfrac{3}{2}$, $x_2 = \dfrac{9}{4}$, Min.$Z = \dfrac{\sqrt{10}}{16}$
10. $x_1 = 1$, $x_2 = 5$, Min. $Z = 10$

11. $x_1 = 1$, $x_2 = 1$, Min.$Z = 0$
12. $x_1 = 44$, $x_2 = 2$, Max.$Z = 4900$

13. $x_1 = 1.43$, $x_2 = 0.48$, Max.$Z = 4.77$
14. $x_1 = 5$, $x_2 = 5$, Max.$Z = 50$

15. $x_1 = \dfrac{48}{5}$, $x_2 = \dfrac{1}{5}$, Max.$Z = 587.72$
16. $x_1 = \dfrac{2}{3}$, $x_2 = \dfrac{14}{9}$, Max.$Z = \dfrac{22}{9}$

REVIEW QUESTIONS

1. State and prove Kuhn-Tucker necessary and sufficient conditions in non-linear programming problem.   [AGRA 2001, 03, 15; IAS 1986, 88]
2. Explain Kuhn-Tucker conditions.
3. Explain the concept of maxima and minima.
4. Explain the Lagrangian method of undetermined multipliers.
5. Interpret the Lagrangian multipliers functions.

MULTIPLE CHOICE QUESTIONS (CHOOSE THE MOST APPROPRIATE ONE)

1. A function is said to have its minimum value at the point $x = x_0$ if :
   (a) $f(x_0) > f(x_0 + h)$
   (b) $f(x_0) < f(x_0 + h)$
   (c) $f(x_0) = f(x_0 + h)$
   (d) none of these
2. The sufficient conditions for the function $f(x_1, x_2)$ to have maxima at the point (a, b) are:
   (a) $rt - \lambda^2 > 0$, $r < 0$
   (b) $rt - \lambda^2 > 0$, $r > 0$
   (c) $rt - \lambda^2 < 0$,
   (d) None of these
3. The point such that the sum of the squares of its distance from n given points is :
   (a) centroid
   (b) centre of the mean position of the given points
   (c) both (a) and (b)
   (d) none of these
4. The triangle of maximum area which can be inserted in a circle is :
   (a) equilateral
   (b) isosceles
   (c) right angled
   (d) None of these
5. If the perimeter of a triangle is constant, then its area is maximum when triangle is :
   (a) equilateral
   (b) isosceles
   (c) both (a) and (b) are true
   (d) none of these
6. The function
   $f(x, y) = x^3 + y^2 - 63(x+y) + 12xy$ has :
   (a) four stationary points
   (b) maximum at $(-7, -7)$
   (c) minimum at $(3, 3)$
   (d) all are true
7. The maximum value of
   $u = axy^2z^3 - x^2y^2z^3 - xy^3z^3 - xy^2z^4$ is :
   (a) $108\,a^7$
   (b) $\dfrac{107\,a^7}{7^7}$
   (c) $\dfrac{a^7}{7^7}$
   (d) none of these

ANSWERS

1. (b)   2. (a)   3. (c)   4. (a)   5. (a)   6. (d)   7. (b)

ARCHIVE

1. The efficiency E of a small manufacturing concern depends on the workers w and is given $10E = -\left(\dfrac{w^3}{40}\right) + 30w - 392$. Find the strength of the workers which give maximum efficiency.   [CA–1986]

2. The total profit y in rupees of a drug company from the manufacture and sale of x drug bottles is given by $y = -\left(\dfrac{x^2}{400}\right) + 2x - 80$

   (i) How many drug bottles must the company sell to achieve the maximum profit.

   (ii) What is the profit per drug bottle when this maximum is achieved.   [CA–1985]

3. Find the second order Taylor's series expansion of the function
   $$f(x_1, x_2) = x_1^2 x_2 + 5x_1 e^{x_2}$$
   about the point $x_0 = \begin{bmatrix} 1 \\ 0 \end{bmatrix}$.   [AMIE–2005]

4. Show that the demand curve $p = \dfrac{a}{x+b} - c$ and $p = (a - bx)^2$ are each downward sloping and convex from below.
    [DELHI(B.Com)–1985, 2006]

5. A firm has revenue function is given by $R = 80D$, where R is the gross revenue and D is the quantity sold and a production cost function is given by $C = 15000 + 60\left(\dfrac{D}{900}\right)^2$ Find the total profit function and the no. of units to be sold to get the maximum profit.
    [ICWA–1997]

6. A manufacturer can sell x items per week at a price $P = 20 - 0.001x$ rupees each. It cost $Y = 5x + 200$ rupees to produce x items. Find the no. of items the manufacturer has to produce per week for maximum profit.   [GUWAHATI–1989]

7. An indifference map is defined by the relation $(x + h)\sqrt{y + k} = a$, where h and k are fixed positive numbers and a is a positive parameter. By expressing y as a function of x and by finding derivatives, show that each indifference curve is downward sloping from below.   [GURUNANAK(MA-Eco)–1997]

8. A firm produces x units of output per week at a total cost of $\left(\dfrac{x^3}{3}\right) - x^2 + 5x + 3$ rupees.
   Find the output level at which:
   (i) the marginal cost (MC) and the average variable cost (AC) attain their respective minimum value and
   (ii) MC = AC   [CALCUTTA–1995]

HINTS AND ANSWERS

1. $E = -\dfrac{w^3}{400} + 3w - 392 \Rightarrow \dfrac{dE}{dw} = -\dfrac{3w^2}{400} + 3$ Now, $\dfrac{dE}{dw} = 0 \Rightarrow w = 20.$ Also, $\dfrac{d^2E}{dw^2} < 0$ at $w = 20$
   $\therefore$ Total workers = 20

2. $y = -\dfrac{x^2}{400} + 2x - 80 \Rightarrow \dfrac{dy}{dx} = -\dfrac{x}{200} + 2$ Now, $\dfrac{dy}{dx} = 0 \Rightarrow x = 400, \dfrac{d^2y}{dx^2} < 0$ at $x = 400$
   $\Rightarrow$ Company must sell $x = 400$ drug bottles to achieve the maximum profit and is equal to ₹320

3. $5 + [5,6]\begin{bmatrix} x_1 - 1 \\ x_2 \end{bmatrix}' + \dfrac{1}{2!}\begin{bmatrix} x_1 - 1 \\ x_2 \end{bmatrix}\begin{bmatrix} 2x_2 & 2x_1 + 5e^{x_2} \\ 2x_1 + 5e^{x_2} & 5x_1 e^{x_2} \end{bmatrix}\begin{bmatrix} x_1 - 1 \\ x_2 \end{bmatrix}$

4. $\dfrac{dp}{dx} = -\dfrac{a}{(x+b)^2} < 0$ and $(x+b)^2 > 0 \Rightarrow$ curve is downward sloping.
   Also, $\dfrac{d^2p}{dx^2} = \dfrac{2a}{(x+b)^2} > 0 \Rightarrow$ curve is convex below.

5. Profit = Revenue – cost = $80D - 15000 - 60\left(\dfrac{D}{900}\right)^2$, Maximum profit at $D = 5400$

6. Profit = Revenue – cost = $(20 - 0.001x) - (5x + 200)$; Maximum profit at $x = 7500$

7. $x + H = \dfrac{a}{\sqrt{y+k}} < 0$ or $x = \dfrac{a}{\sqrt{y+k}} - h$, $\dfrac{dy}{dx} = -\dfrac{a}{2(y+k)^{3/2}} < 0 \Rightarrow$ curve are sloping downward
   $\dfrac{d^2y}{dx^2} = \dfrac{3a}{2(y+k)^{5/2}} > 0 \Rightarrow$ curve are convex from below.

Non-linear Programming: Search Techniques — 26

In this chapter we shall examine the fundamental concepts and useful methods for finding minima of constrained and unconstrained functions. The study of unconstrained minimization techniques provide the basic understanding necessary for the study of constrained minimization method. The unconstrained minimization method can be used to solve certain complex engineering analysis problems.

The general form of unconstrained minimization problem can be written as follows:

Find
$$X = \begin{bmatrix} x_1 \\ x_2 \\ \vdots \\ x_n \end{bmatrix} \text{ which minimizes } f(X).$$

GENERAL DEFINITIONS

(1) **Rate of convergence:** An optimization method is said to have convergence of order p if

$$\frac{\left\| X_{i+1} - X^* \right\|}{\left\| X_i - X^* \right\|^p} \le k, \quad k \ge 0, p \ge 1$$

(2) **Linearly convergent method:** If the rate of convergence of a method is 1, *i.e.*, $p = 1, 0 \le k \le 1$ then this method is said to be linearly convergent. (slow convergence)

(3) **Quadratically convergent method:** If the rate of convergence of a method is 2, *i.e.*, $p = 2$, then this method is said to be quadratically convergence. (fast convergence)

(4) **Super linear convergence method:** An optimization method is said to have super linear convergence if

$$\lim_{i \to \infty} \frac{\left\| X_{i+1} - X^* \right\|}{\left\| X_i - X^* \right\|^p} \to 0 \text{ (Fast convergence)}$$

☞ REMARK
- The different iterative methods have different rate of convergence.

26.2 UNCONSTRAINED MINIMIZATION METHODS

The unconstrained minimization methods can be classified into following two categories:

   1. Direct search method   **2.** Descent method (Indirect search method)

In direct search method, there is a requirement of only the objective function values but not the partial derivatives of the function for finding the minimum. On the other hand the descent method requires in addition to the function values, the first and second order partial derivatives of the objective function.

☞ **REMARKS**
- The direct search methods are also known as non-gradient methods or Zeroth order methods.
- The descent methods are also known as gradient methods.
- Among the gradient methods those requiring only first order derivative are called 'First order method', those requiring both first and second order partial derivatives are known as 'Second order method'.

> Direct search methods are more suitable for simple problem involving a relatively small number of variables and generally less efficient than the descent methods.

26.3 SOME DIRECT METHOD

26.3.1 UNIVARIATE METHOD

In univariate method, we change only one variable at a time and search is carried out in the axial directions covering all the direction one by one to get a sequence of improved approximation to the minimum point. In this method we start at a base point x_i in the ith iteration, we fix the value of $(n - 1)$ variables and vary the remaining variable. Then problem becomes a one dimensional minimization problem ($\because$ only one variable is change), which can be solved easily.

Now, change any one of the $(n - 1)$ variables that were fixed in previous iteration and continue this process by take each coordinates. Finally, n directions are searched sequentially. Continue this process until no further improvement is possible in the objective function in any of the n directions of a cycle.

Working Procedure

STEP 1. Choose a fixed point $X_1(x_1, y_1)$ and a search direction $u_1 = (1, 0)^T$.

STEP 2. Take a step of size λ_1 in this direction and find the optimum point $\lambda_1 = \lambda_1^*$ for which $f(X_1 + \lambda_1^* u_1)$ is minimum.

Set a new point $X_2 = X_1 + \lambda_1^* u_1$.

STEP 3. Repeat the above steps by taking the optimum step length λ_2^* in the direction $u_2 = (0, 1)^T$ from the point X_2, to arrive at the next point X_3 given by

$$X_3 = X_2 + \lambda_2^* u_2$$

It complete the first iteration.

STEP 4. Repeat the process for the completion of the next cycle of iterations until no significant change is achieved in the value of the objective function, *i.e.*, until the variation in the value of the function is negligible.

☞ **REMARKS**
- In the above procedure, we consider the function of two variables x_1 and x_2 only, although, this procedure can be extended for function of more than two variables as well.
- Univariate method will not converge rapidly to the optimum solution as it has a tendency to oscillate with steadly decreasing progress toward the optimum.
- The univariate method can be applied to find the minimum of any function that possesses continuous derivatives.

For n-dimensional case, the search directions are given as below:

$$u_i^T = \begin{cases} (1,0,0,\ldots,0), \text{for } i = 1, n+1, 2n+1,\ldots \\ (0,1,0,\ldots,0), \text{for } i = 2, n+2, 2n+2,\ldots \\ (0,0,1,\ldots,0), \text{for } i = 3, n+3, 2n+3,\ldots \\ \vdots \\ (0,0,0,\ldots,1), \text{for } i = n, 2n, 3n,\ldots \end{cases}$$

 Solved Examples

EXAMPLE I. *Minimize* $f(X) = f(x_1, x_2) = 2x_1^2 + 3x_2^2 - x_1 x_2$ *using univariate method by taking* $X_1 = \begin{bmatrix} 1 \\ 1 \end{bmatrix}$ *as a starting point.*

SOLUTION. Let $\qquad f(x_1, x_2) = 2x_1^2 + 3x_2^2 - x_1 x_2$ $\qquad\qquad$ ...(1)

Iteration-1. Consider the unit vectors $u_1 = (1, 0)^T$ and $u_2 = (0, 1)^T$ in the directions of x_1 and x_2 respectively.

$\because \qquad\qquad X_1 = \begin{pmatrix} 1 \\ 1 \end{pmatrix}$

Then $\qquad f(x_1) = 2 \cdot 1^2 + 3 \cdot 1^2 - 1 \cdot 1 = 4$

Now take a step size λ_1 in the x_1-direction to get X_2 such that

$$X_2 = X_1 + \lambda_1 \begin{pmatrix} 1 \\ 0 \end{pmatrix} = \begin{pmatrix} 1 \\ 1 \end{pmatrix} + \begin{pmatrix} \lambda_1 \\ 0 \end{pmatrix} = \begin{pmatrix} 1+\lambda \\ 1 \end{pmatrix}$$

Now, to obtain the optimum size λ_1, we find

$$\text{Min.} f \begin{pmatrix} 1+\lambda_1 \\ 1 \end{pmatrix} = 2(1+\lambda_1)^2 + 3 - (1+\lambda_1)$$

Using the principle of maxima and minima, for the minimum of f, we have

$$\frac{df}{d\lambda_1} = 0 \implies 4(1+\lambda_1) - 1 = 0 \implies \lambda_1 = -\frac{3}{4}$$

and $\qquad\qquad \dfrac{d^2 f}{d\lambda_1^2} = 4 > 0$

$\implies$ f is minimum for $\lambda_1 = -\dfrac{3}{4}$

So, new point $X_2 = \begin{pmatrix} 1+\lambda_1 \\ 1 \end{pmatrix} = \begin{pmatrix} 1-3/4 \\ 1 \end{pmatrix} = \begin{pmatrix} 1/4 \\ 1 \end{pmatrix}$

and $\qquad f(X_2) = 2\left(\dfrac{1}{4}\right)^2 + 3(1)^2 - \dfrac{1}{4}(1) = \dfrac{23}{8} = 2.875$

Further, to find the optimum step size λ_2 in x_2-direction

Minimize $\qquad f(X_2 + \lambda_2(0,1)^T) = f\left(\begin{pmatrix} 1/4 \\ 1 \end{pmatrix} + \begin{pmatrix} 0 \\ \lambda_2 \end{pmatrix}\right)$

$$= f\left(\begin{pmatrix} 1/4 \\ 1+\lambda_2 \end{pmatrix}\right) = \dfrac{1}{8} + 3(1+\lambda_2)^2 - \dfrac{1}{4}(1+\lambda_2)$$

Again, for minimum of f,

$$\frac{df}{d\lambda_2} = 0 \implies 6(1+\lambda_2) - \frac{1}{4} = 0 \implies \lambda_2 = -\frac{23}{24} \quad \text{and} \quad \frac{d^2 f}{d\lambda_2^2} = 6 > 0$$

$\implies$ f is minimum for $\lambda_2 = -\dfrac{23}{24}$.

$\therefore$ New point $X_3 = X_2 + \lambda_2(0,1)^T = \begin{pmatrix} 1/4 \\ 1+\lambda_2 \end{pmatrix} = \begin{pmatrix} 1/4 \\ 1/24 \end{pmatrix}$

and $\qquad f(X_3) = 2\left(\dfrac{1}{4}\right)^2 + 3\left(\dfrac{1}{24}\right)^2 - \left(\dfrac{1}{4}\right)\left(\dfrac{1}{24}\right) = 0.11979$

Iteration-2. Consider X_3 as the new base point. Now, to find optimum step size λ_3 in x_3-direction, we

Minimize $f(X_3 + \lambda_3(1, 0)^T)$

$$= f\left(\begin{pmatrix} 1/4 \\ 1/24 \end{pmatrix} + \begin{pmatrix} \lambda_3 \\ 0 \end{pmatrix}\right) = f\begin{pmatrix} 1/4 + \lambda_3 \\ 1/24 \end{pmatrix}$$

$$= 2\left(\frac{1}{4}+\lambda_3\right)^2 + 3\left(\frac{1}{24}\right)^2 - \frac{1}{24}\left(\frac{1}{4}+\lambda_3\right) \qquad \text{(By (1))}$$

Again, for minimum of f, $\dfrac{df}{d\lambda_3} = 0$

$\implies \qquad 4\left(\dfrac{1}{4}+\lambda_3\right) - \dfrac{1}{24} = 0$

$\implies \qquad \lambda_3 = -\dfrac{23}{96}$

and $\qquad \dfrac{d^2 f}{d\lambda_3^2} = 4 > 0$

$\implies$ f is minimum for $\lambda_3 = -\dfrac{23}{96}$

So, new point $X_4 = \begin{pmatrix} (1/4)+(-23/96) \\ 1/24 \end{pmatrix} = \begin{pmatrix} 1/96 \\ 1/24 \end{pmatrix}$

and $\qquad f(X_4) = 2\left(\dfrac{1}{96}\right)^2 + 3\left(\dfrac{1}{24}\right)^2 - \dfrac{1}{96}\left(\dfrac{1}{24}\right) = \dfrac{23}{4608} = 0.00499$

Now considering X_4 as the new base point, to find optimum step size λ_4 in x_2-direction.

Minimize $f(X_4 + \lambda_4(0, 1)^T)$

$$= f\left(\begin{pmatrix} 1/96 \\ 1/24 \end{pmatrix} + \begin{pmatrix} 0 \\ \lambda_4 \end{pmatrix}\right) = f\begin{pmatrix} \dfrac{1}{96} \\ \dfrac{1}{24}+\lambda_4 \end{pmatrix}$$

$$= 2\left(\frac{1}{96}\right)^2 + 3\left(\frac{1}{24}+\lambda_4\right)^2 - \frac{1}{96}\left(\frac{1}{24}+\lambda_4\right)$$

Again for minimum of f, $\dfrac{df}{d\lambda_4} = 0$

$\Rightarrow \qquad 6\left(\dfrac{1}{24} + \lambda_4\right) - \dfrac{1}{96} = 0$

$\Rightarrow \qquad \lambda_4 = -\dfrac{23}{576}$

and $\qquad \dfrac{d^2 f}{d\lambda_4^2} = 6 > 0$

$\Rightarrow \quad f$ is minimum for $\lambda_4 = -\dfrac{23}{576}$

$\therefore \quad$ the new point $X_5 = \begin{pmatrix} 1/96 \\ \left(\dfrac{1}{24}\right) - \left(\dfrac{23}{576}\right) \end{pmatrix} = \begin{pmatrix} 1/96 \\ 1/576 \end{pmatrix}$

and $\qquad f(X_5) = 2\left(\dfrac{1}{96}\right)^2 + 3\left(\dfrac{1}{576}\right)^2 - \left(\dfrac{1}{96}\right)\left(\dfrac{1}{576}\right)$ $\hfill$ (By (1))

$\qquad\qquad = 0.00020797$

Hence, at this stage

$\qquad$ Minimum of $f(X) = 0.00020797$ when $x_1 = \dfrac{1}{96}$ and $x_2 = \dfrac{1}{576}$

EXAMPLE 2. *Using univariate search method,*

$\qquad\qquad$ *Minimize $f(X) = f(x_1, x_2) = x_1^2 - x_1 x_2 + 3x_2^2$*

by taking starting point as $X_1 = (1, 1)^T$.

SOLUTION. We have $f(X) = f(x_1, x_2) = x_1^2 - x_1 x_2 + 3x_2^2$ $\hfill$ ...(1)

Iteration-1.

STEP-1. Consider the unit vectors $u_1 = (1, 0)^T$ and $u_2 = (0, 1)^T$ in x_1 and x_2-directions respectively. Given $X_1 = \begin{pmatrix} 1 \\ 1 \end{pmatrix}$

$\therefore \qquad f(X_1) = 1^2 - 1 \cdot 1 + 3 \cdot 1^2 = 3$

Now, we take a step size λ_1 in the x_1-direction to arrive at the value of X_1. Now proceed same as in example-1, we have to minimize $f(1 + \lambda_1, 1)$

From (1)

$\qquad f(1 + \lambda, 1) = (1 + \lambda_1)^2 - (1 + \lambda_1) + 3$

For the minimum of f,

$\qquad \dfrac{df}{d\lambda_1} = 0 \qquad \Rightarrow \qquad 2(1 + \lambda_1) - 1 = 0 \qquad \Rightarrow \quad \lambda_1 = -1/2$

and $\qquad \dfrac{d^2 f}{d\lambda_1^2} = 2 > 0$

$\Rightarrow \qquad f$ is minimum when $\lambda_1 = -\dfrac{1}{2}$

So, the new point is given by

$$X_2 = \begin{pmatrix} 1+\lambda_1 \\ 1 \end{pmatrix} = \begin{pmatrix} 1-1/2 \\ 1 \end{pmatrix} = \begin{pmatrix} \frac{1}{2}, 1 \end{pmatrix}$$

STEP-2. Now find step size λ_2 along x_2 by minimizing $f\left(\frac{1}{2}, 1+\lambda_2\right)$

Now, $f\left(\frac{1}{2}, 1+\lambda_2\right) = \frac{1}{4} - \frac{1}{2}(1+\lambda_2) + 3(1+\lambda_2)^2$ (By (1))

Now for minimum of f

$$\frac{df}{d\lambda_2} = 0 \implies -\frac{1}{2} + 6(1+\lambda_2) = 0 \implies \lambda_2 = -\frac{11}{12}$$

and $\quad \dfrac{d^2 f}{d\lambda_2^2} = 6 > 0$

$\implies \quad f$ is minimum at $\lambda_2 = -\dfrac{11}{12}$

So, the new point is

$$X_3 = \begin{pmatrix} \frac{1}{2}, 1+\lambda_2 \end{pmatrix} = \begin{pmatrix} \frac{1}{2}, \frac{1}{12} \end{pmatrix}$$

Now, repeat the above process with starting point X_3. Go on repeating the process till the quantities $|\lambda_k|$ are less than some prefixed tolerance.

EXAMPLE 3. *Minimize* $f(X) = f(x_1, x_2) = x_1 - x_2 + 2x_1^2 + 2x_1 x_2 + x_2^2$ *with*   *the*

starting point $X_1 = [0, 0]^T$ *by Univariate method.*

SOLUTION. We have

$$f(X) = f(x_1, x_2) = x_1 - x_2 + 2x_1^2 + 2x_1 x_2 + x_2^2 \qquad \qquad ...(1)$$

We can minimize $f(X)$ by using the following steps:

STEP 1. Consider the unit vector $u_1 = (1, 0)^T$ and $u_2 = (0, 1)^T$ in x_1 and x_2 direction respectively.

Given $\quad X_1 = \begin{pmatrix} 0 \\ 0 \end{pmatrix}$

$\implies \quad f(X_1) = 0 - 0 + 2 \cdot 0^2 + 2 \cdot 0 \cdot 0 + 0^2 = 0$

We take a step size λ_1 in x_1-direction to get the value of X_2

$$\therefore \quad X_2 = X_1 + \lambda_1 \begin{pmatrix} 1 \\ 0 \end{pmatrix} = \begin{pmatrix} 0 \\ 0 \end{pmatrix} + \begin{pmatrix} \lambda_1 \\ 0 \end{pmatrix} = \begin{pmatrix} \lambda_1 \\ 0 \end{pmatrix}$$

(We can easily verify that $-u_1$ is the correct direction for minimizing f from X_1)

So, $\quad X_2 = X_1 - \lambda_1 \begin{pmatrix} 1 \\ 0 \end{pmatrix} = \begin{pmatrix} -\lambda_1 \\ 0 \end{pmatrix}$

To obtain the optimum size λ, we minimize $f\begin{pmatrix} -\lambda_1 \\ 0 \end{pmatrix}$

$$= (-\lambda_1) - 0 + 2(-\lambda_1)^2 + 0 + 0 = 2\lambda_1^2 - \lambda_1$$

Now, for the minimum of f,

$$\frac{df}{d\lambda_1} = 0 \implies 4\lambda_1 - 1 = 0$$

$$\implies \quad \lambda_1 = 1/4$$

and $\dfrac{d^2 f}{d\lambda_1^2} = 4 > 0$

$\Rightarrow$ f is minimum at $\lambda_1 = 1/4$

Now, set the new point

$$X_3 = X_1 - \lambda_1 u_1 = \begin{pmatrix} 0 \\ 0 \end{pmatrix} - \frac{1}{4}\begin{pmatrix} 1 \\ 0 \end{pmatrix} = \begin{pmatrix} -1/4 \\ 0 \end{pmatrix}$$

and $f(X_3) = f\left(-\dfrac{1}{4}, 0\right) = -\dfrac{1}{8}$ (By (1))

STEP 2.   Now considering X_3 as the new base point in $u_2 = (0, 1)^T$ direction. Then we proceed same as in example-1.

We have to minimize $f(X_3 + \lambda_2 u_2)$

Now, $f(X_3 + \lambda_2 u_2) = f(X_3 + \lambda_2(0, 1)^T)$

                  $= f(-0.25, \lambda_2)$

                  $= -0.25 - \lambda_2 + 2(0.25)^2 - 2(0.25)(\lambda_2) + \lambda_2^2$ (By (1))

                  $= \lambda_2^2 - 1.5\lambda_2 - 0.125$

Again by the principle of maxima and minima

$$\frac{df}{d\lambda_2} = 0 \Rightarrow 2\lambda_2 - 1.5 = 0, \ i.e., \lambda_2 = 0.75$$

and $\dfrac{d^2 f}{d\lambda_2^2} = 2 > 0$

$\Rightarrow$ f is minimum at $\lambda_2 = 0.75$

Now, set the new point

$$X_4 = X_3 + \lambda_2 u_2 = \begin{pmatrix} -0.25 \\ 0 \end{pmatrix} + 0.75\begin{pmatrix} 0 \\ 1 \end{pmatrix} = \begin{pmatrix} -0.25 \\ 0.75 \end{pmatrix}$$

Then by (1), $f(X_4) = -0.6875$

Continuing the above process until the optimum solution

$$X^* = \begin{pmatrix} -1.0 \\ 1.5 \end{pmatrix} \text{ with } f(X^*) = -1.25 \text{ is found.}$$

26.3.2 PATTERN SEARCH METHOD

In univariate method, we search for the minimum of the given functions along directions parallel to the coordinate axes (axial directions) only. It can be observed that this method may not converge in some cases, *i.e.*, the rate of convergence in univariate method to arrive at the minimal point is very slow. This problem can be avoided by changing the directions of search in a favourable manner instead of retaining them always parallel to the coordinate axes.

    (i) Pattern Directions: The line joining the alternate points of the search in the general direction of the minimum are known as pattern directions.

☞ REMARKS

- If the objective function is a quadratic in two variables, all pattern directions pass through the minimum.
- The above property is not valid for multivariable functions even when they are quadratic.

(ii) Pattern search method: The methods that use pattern directions as search directions are called pattern search methods.

There are following two pattern search methods.

   (i)   Hooke and Jeeve's method
   (ii) Powel's method

(1) HOOKE AND JEEVE'S METHOD

It is a sequential technique in which each step consist of the following two types of move

(i) Exploratory move

In exploratory moves, given a specified step size, which may be different for each coordinate direction and change during the search, the explanation proceeds from an initial point by specified step size in each coordinate direction. If the function value does not increase, the step is considered successful, otherwise the step is retracted and replaced by a step in the opposite direction which in turn is retained depending upon whether it succeeds or fails. Finally, the exploratory move is completed when all coordinates have been investigated. Here, the resulting point is known as base point.

(ii) Pattern move

A pattern move consists of a single step from the present base point along the line from the previous to the current base point.

The general procedure of Hooke and Jeeve's method is given as below:

 Working Procedure

STEP 1. Start with a fixed point say $X_1 = \begin{bmatrix} x_1 \\ x_2 \\ \vdots \\ x_n \end{bmatrix}$ and prescribed step size is considered to be constants (*i.e.*, $\Delta x_1 = \Delta x_2 = \dots = \Delta x_n = \text{constant}$)

STEP 2. Made the search in each direction $u_i : i = 1, 2, \dots, n$ where u_i is the unit vector along the direction x_i whose i^{th} component is 1 and all other components are zero. (The search is made first in the positive direction and then if necessary in the negative direction of each axis to arrive at a temporary base point)

STEP 3. Define the temporary base point Y_k, obtained from X_k by slightly changing (perturbed) the ith components of X_k ($Y_{k_0} = X_k$) as given below

$$Y_{ki} = \begin{cases} Y_{k,i-1} + \Delta x_i u_i; \text{if } f^+ = f(Y_{k,i-1} + \Delta x_i u_i) < f = f(Y_{k,i-1}) \\ Y_{k,i-1} - \Delta x_i u_i; \text{if } f^- = f(Y_{k,i-1} - \Delta x_i u_i) < f = f(Y_{k,i-1}) \\ Y_{k,i-1}; \text{if } f(Y_{k,i-1}) = f \le \min(f^+, f^-) \end{cases}$$

STEP 4. Continue this process of finding a new base point for $i = 1, 2, 3, \dots, n$ until all directions are covered.

STEP 5. If the point $Y_{k, n}$ remains same as X_k reduce the step length Δx_i (say by a factor of 2) set $i = 1$ and go to step 3.

But if $Y_{k, n}$ is different from X_k, obtain the new base point as

$$X_{k + 1} = Y_{k, n}$$

and go to the next step.

STEP 6. Establish a pattern direction u as
$$u = X_{k+1} - X_k$$
and find a point $Y_{k+1,0} = X_{k+1} + \lambda u$
where λ is the step length (for simplicity, it can be taken as 1)

STEP 7. Repeat the steps until a desired accuracy is achieved or the change in the value of the function satisfies the given conditions.

 **Solved Examples**

EXAMPLE 1. *Using Hooke and Jeeve's method, minimize the function*
$$f(X) = 2x_1^2 + x_2^2 + 2x_1x_2 + x_1 - x_2$$
taking the starting base point as $X_1 = (0, 0)^T$ and $\Delta x_1 = \Delta x_2 = 0.8$

SOLUTION. We have
$$f(X) = f(x_1, x_2) = 2x_1^2 + x_2^2 + 2x_1x_2 + x_1 - x_2 \qquad \ldots(1)$$

Iteration-1.

STEP 1. Take the starting base point $X_1 = \begin{bmatrix} 0 \\ 0 \end{bmatrix}$ and step length $\Delta x_1 = 0.8, \Delta x_2 = 0.8$

$\therefore \qquad f(X_1) = f(0, 0) = 0$ (By (1))

Now, moving in direction u_1 from the base point X_1. So, obtain the temporary base point as given below:
$$f^+(X_1 + \Delta x_1 \cdot u_1) = f^+[(0,0) + 0.8(1,0)] = f^+(0.8, 0) = 2.08 \nleq f(X_1)$$
$$f^-(X_1 - \Delta x_1 \cdot u_1) = f^+(-0.8, 0) = 0.48 \nleq f(X_1)$$
Now, since f^+, f^- are not less than $f(X_1)$, movement in u_1-direction is not beneficial and hence discarded.

So, take $Y_{11} = X_1$ and $f(Y_{11}) = f(X_1) = 0$.

STEP 2. Now resuming movement in u_2-direction as follows:
$$f^+(X_1 + \Delta x_2 \cdot u_2) = f^+[(0,0) + 0.8(0,1)] = f^+(0, 0.8) = -0.16 < f(X_1)$$
Since, $Y_{12} = X_1 + \Delta x_2 u_2 = (0, 0.8)$ is different from X_1, the new base point is taken as
$$Y_{12} = X_2 = (0, 0.8) \text{ and } f(X_2) = -0.16 \qquad \text{(By (1))}$$

STEP 3. In step 1 and 2, the movements have been made in both the directions (u_1 and u_2) so third movement is to be made in the first pattern direction.
$$S_{p1} = X_2 - X_1 = (0, 0.8) - (0, 0) = (0, 0.8)$$
through step length λ from X_2 so that $f(X_2 + \lambda S_{p1})$ is minimum.
Hence, $f(X_2 + \lambda S_{p1}) = f[X_2 + \lambda(X_2 - X_1)] = f\{0, 0.8(1 + \lambda)\}$
$$= [(0.8)^2(1 + \lambda)^2 - 0.8(1 + \lambda)]$$
Using the principle of extrema, for the minima of f, we must have
$$\frac{df}{d\lambda} = 0 \implies 2 \times 0.64(1 + \lambda) - 0.8 = 0 \implies \lambda = -\frac{3}{8}$$
and $\quad \dfrac{d^2f}{d\lambda^2} = 1.28 > 0$

$\implies f$ is minimum for $\lambda = -\dfrac{3}{8} = \lambda^* \text{ (say)}$

So, we obtain the new base point

$$Y_{20} = X_3 = X_2 + \lambda^*(X_2 - X_1)$$

$$= (0, 0.8) - \frac{3}{8}(0, 0.8) = (0, 0.5)$$

and $\quad f(X_3) = f(0, 0.5) = -0.25$

Iteration-2.

STEP 1. Let us move in the direction of u_1 with base point $X_3 = (0, 0.5)$

Now, $f^+(X_3 + \Delta x_1 u_1) = f^+[(0, 0.5) + (0.8, 0)] = f^+(0.8, 0.5) = 2.63$ (By (1))

$$\nleq f(X_3)$$

and $\quad f^-(X_3 - \Delta x_1 u_1) = f^-(-0.8, 0.5) = -0.57 < f(X_3) = -0.25$

Therefore, the new temporary base point is given by

$$Y_{21} = (X_3 - \Delta x_1 u_1) = (-0.8, 0.5), f(Y_{21}) = -0.57$$

STEP 2. Now moving in the direction of u_2 from the base point $Y_{21} = (-0.8, 0.5)$ to Y_{22}. Then we have

$$f^+(Y_{21} + \Delta x_2 u_2) = f^+(-0.8, 1.3) = -1.21 < f(Y_{21})$$

So, the new base point is given by

$$Y_{22} = X_4 = Y_{21} + \Delta x_2 u_2 = (-0.8, 1.3)$$

and $\quad f(X_4) = -1.21$ (By (1))

STEP 3. Now, we move along the second pattern search directions $S(p, 2) = X_4 - X_3$ starting from X_4 through optimal step length λ so that

$$f(X_4 + \lambda S_{p2}) = f[(-0.8, 1.3) + \lambda(-0.8, 0.8)]$$

$$= f[-0.8(1 + \lambda), 1.3 + \lambda(0.8)]$$

is minimum.

Now, for the minima of $f(\lambda)$ we must have

$$\frac{df}{d\lambda} = 0 \qquad \Rightarrow \qquad 1.28\lambda - 0.32 = 0$$

$$\Rightarrow \qquad \lambda = 0.25$$

and $\quad \dfrac{d^2 f}{d\lambda^2} = 1.28 > 0$

$\Rightarrow$ f is minimum at $\lambda = 0.25$

Therefore, we get the points $X_5 = Y_{30} = X_4 + \lambda S(p, 2) = (-1, 1.5)$

and $f(X_5) = -1.25$ (By (1))

Iteration-3.

STEP 1. First moving in the u_1-direction from $X_5 = (-1, 1.5)$ we have

$$f^+(X_5 + \Delta x_1 u_1) = f^+[(-1, 1.5) + 0.8(1, 0)]$$

$$= f^+(-0.2, 1.5) = 0.03 \nleq f(X_5)$$

and $\quad f^-(X_5 - \Delta x_1 u_1) = f^-[(-1, 1.5) - 0.8(1, 0)]$

$$= f^-(-1.8, 1.5) = 0.03 \nleq f(X_5)$$

Therefore, $\qquad Y_{31} = X_5$

STEP 2. Now making movement in u_2-direction from $Y_{31} = X_5$. Then

$$f^+(X_5 + \Delta x_2 u_2) = f^+[(-1, 1.5) + 0.8(0, 1)]$$

$$= f^+(-1, 2.3) = -0.61 \nless f(X_5)$$

and $f^-(X_5 - \Delta x_2 u_2) = f^-[(-1, 1.5) - 0.8(0, 1)]$

$$= f^-(-1, 0.7) = -0.61 \nless f(X_5)$$

Thus, we conclude that the movement in any of the axial directions produces no change in the value of the function f at X_5. Therefore, X_5 is an optimal solution. Hence, minimum of $f = -1.25$ when $x_1 = -1$ and $x_2 = 1.5$

(2) POWELL'S METHOD

This is most successful direct search algorithm, which is an extension of the basic pattern search method. This method is based upon the model of quadratic function and thus have a theoritical basis.

We have the following two reasons for choosing a quadratic model.

(1) It is the simplest type of non-linear function to minimize and so any general technique must work well in a quadratic, if it is to have any success with a general function.

(2) Near the optimum, all non linear functions can be approximated by a quadratic.

Before discussing the Powell's algorithm, let us have the following definitions:

(1) Conjugate Directions: Let $\mathbf{A} = [A]$ be an $n \times n$ symmetric matrix. A set of n vectors $\{s_i\}$ is said to be conjugate if

$$s_i^T A s_j = 0 \quad \forall i \neq j, \quad i = 1, 2, \ldots, n; j = 1, 2, \ldots, n$$

☞ REMARK

• The orthogonal directions are the special case of conjugate directions.

(2) Quadratically Convergent method: If a minimization method, using exact arithmetic can find the minimum point in n steps while minimizing a quadratic function in n-variables, the method is known as quadratically convergent method.

IMPORTANT RESULTS

(1) Given a quadratic function of n variables and two hyperplanes 1 and 2 of dimension $k < n$, let the constrained stationary points of the quadratic function in the hyperplane be X_1 and X_2 respectively, then the line joining X_1 and X_2 is conjugate to any line parallel to the hyperplanes.

(2) If a quadratic function given by

$$f(X) = \frac{1}{2} X^T A X + B^T X + C$$

is minimized sequentially, once along each directions of a set of n mutually conjugate directions, the minimum of the function f will be found at or before the nth step irrespective of the starting point.

POWELL'S ALGORITHM

Let f be a function of n variables which is to be minimized and $u_1, u_2, \ldots, u_n$ be the axial directions.

In this method, we start from a fixed point, carried out the search sequentially in the direction $u_1, u_2, \ldots, u_n$ in the first cycle, along $S(p,1), u_2, u_3, \ldots, u_n, S(p,1)$ in the second cycle and $S(p,2), u_3, u_4, \ldots, u_{n-1}, u_n, S(p,1), S(p,2)$ in the third cycle and so on. Proceed until the minimum point is reached. Here, $S(p,i)'s$ are the pattern search direction which are defined by

$$S(p,i) = X(i) - X(i - n)$$

and n is the number of decision variables.

☞ **REMARKS**

- The Powell's method is converge in almost two cycles of iterations, *i.e.*, it has quadratic convergence.
- This is also known as Powell's conjugate direction method.
- Since, the no. of cycles n is valid only for quadratic function, it will take generally greater than n cycles for non-quadratic functions.

 Solved Examples

EXAMPLE I. *Using Powell's method, minimize*
$$f(X) = 2x_1^2 + x_2^2 + 2x_1x_2 + x_1 - x_2$$
starting with $X_1 = (0, 0)$.

SOLUTION. We have

$$f(X) = f(x_1, x_2) = 2x_1^2 + x_2^2 + 2x_1x_2 + x_1 - x_2 \qquad \text{...(1)}$$

and base point (initial or starting point), $X_1 = (0, 0)$

Cycle-1. Here, the search will be made in the direction u_2, u_1, u_2.

Let λ be the step size in the direction u_2 from X_1 to reach X_2 such that

$$f(X_2) = f(X_1 + \lambda_1 u_2) = f\left(\begin{pmatrix} 0 \\ 0 \end{pmatrix} + \lambda_1 \begin{pmatrix} 0 \\ 1 \end{pmatrix} \right) = f\begin{pmatrix} 0 \\ \lambda_1 \end{pmatrix}$$

$$= \lambda_1^2 - \lambda_1 \text{ is minimum}$$

Now, for the minima of f, we must have

$$\frac{df}{d\lambda_1} = 0 \Rightarrow 2\lambda_1 - 1 = 0 \Rightarrow \lambda_1 = \frac{1}{2}$$

and

$$\frac{d^2 f}{d\lambda_1^2} = 2 > 0$$

$\Rightarrow$ f is minimum at $\lambda_1 = \frac{1}{2}$

Therefore, $\qquad X_2 = X_1 + \lambda_1 u_2 = (0,0) + \frac{1}{2}(0,1) = \left(0, \frac{1}{2}\right)$

$$f(X_2) = f\left(0, \frac{1}{2}\right) = -0.25 < f(X_1)$$

Now, moving from X_2 in the direction of $u_1 = (1, 0)$ through step length λ_2 arriving at the point $X_3 = X_2 + \lambda_2 u_1$

We have to find λ_2 so that $f(X_3) = f\left(\lambda_2, \frac{1}{2}\right) = 2\lambda_2^2 + 2\lambda_2 - 0.25$ is minimum.

Now, using the principle of maxima and minima for the minima of f, we must have

$$\frac{df}{d\lambda_2} = 0 \Rightarrow 4\lambda_2 + 2 = 0 \Rightarrow \lambda_2 = -\frac{1}{2}$$

and $\qquad \dfrac{d^2 f}{d\lambda_2^2} = 4 > 0$

$\Rightarrow \quad f$ is minimum at $\lambda_2 = -\dfrac{1}{2}$

Therefore, $X_3 = X_2 + \lambda_2 u_1 = \left(0, \dfrac{1}{2}\right) - \dfrac{1}{2}(1, 0) = \left(-\dfrac{1}{2}, \dfrac{1}{2}\right)$

and $f(X_3) = -0.75 < f(X_2) = -0.25$

Now, move from X_3 to X_4 taking step length of size λ_3 in the direction of $u_2 = (0, 1)$

$\therefore \qquad X_4 = X_3 + \lambda_3(0, 1) = \left(-\dfrac{1}{2}, \dfrac{1}{2}\right) + (0, \lambda_3) = \left(-\dfrac{1}{2}, \dfrac{1}{2} + \lambda_3\right)$

Now, we have to find the value of λ_3 such that the function

$$f(X_4) = f\left(-\dfrac{1}{2}, \dfrac{1}{2} + \lambda_3\right) = \lambda_3^2 - \lambda_3 - 0.75 \text{ is minimum.}$$

For minima of f, we must have

$$\dfrac{df}{d\lambda_3} = 0 \Rightarrow 2\lambda_3 - 1 = 0 \Rightarrow \lambda_3 = \dfrac{1}{2}$$

and $\qquad \dfrac{d^2 f}{d\lambda_3^2} = 2 > 0$

$\Rightarrow \quad f$ is minimum at $\lambda_3 = \dfrac{1}{2}$

$\therefore \quad X_4 = \left(-\dfrac{1}{2}, 1\right)$ and $f(X_4) = -1 < f(X_3) = -0.75$

Cycle-2. Let us take $i = 4$

Here, search will be done in the direction $S(p, 1)$, u_2, $S(p, 2)$

Now, pattern search direction

$$S(p, 1) = X_i - X_{i-n} = X_4 - X_2 = \left(-\dfrac{1}{2}, 1\right) - \left(0, \dfrac{1}{2}\right) = \left(-\dfrac{1}{2}, \dfrac{1}{2}\right)$$

From X_4, we move to X_5 by taking step length λ_4 in the direction of $S(p, 1) = \left(-\dfrac{1}{2}, \dfrac{1}{2}\right)$

$$X_5 = X_4 + \lambda_4 S(p, 1) = \left(-\dfrac{1}{2}, 1\right) + \lambda_4 \left(-\dfrac{1}{2}, \dfrac{1}{2}\right)$$

$$= \left(-\dfrac{1}{2}(1 + \lambda_4), 1 + \dfrac{1}{2}\lambda_4\right)$$

Now, we have to find the value of λ_4 so that the function

$$f(X_5) = 0.25\lambda_4^2 - 0.5\lambda_4 - 1 \text{ is minimum.}$$

Now, for the minima of f, we must have

$$\dfrac{df}{d\lambda_4} = 0 \Rightarrow 0.5\lambda_4 - 0.5 = 0 \Rightarrow \lambda_4 = 1$$

and $\qquad \dfrac{d^2 f}{d\lambda_4^2} = 0.5 > 0$

$\Rightarrow$ f is minimum at $\lambda_4 = 1$

So, $X_5 = X_4 + \lambda_4 S(p,1) = \left(-\dfrac{1}{2}, 1\right) + \left(-\dfrac{1}{2}\right)\left(\dfrac{1}{2}, \dfrac{1}{2}\right) = \left(-1, \dfrac{3}{2}\right)$

and $f(X_5) = -1.25 < f(X_4)$

Now, from X_5, we move to X_4 by taking step length λ_5 in the direction of $u_2 = (0, 1)$.

$\therefore$ $X_6 = X_5 + \lambda_5 u_2 = \left(-1, \dfrac{3}{2}\right) + \lambda_5(0,1) = \left(-1, \dfrac{3}{2} + \lambda_5\right)$

We have to find the value of λ_5 for which the function f is minimum.

For the minima of f, we must have

$$\frac{df}{d\lambda_5} = 0 \;\Rightarrow\; 2\lambda_5 = 0 \;\Rightarrow\; \lambda_5 = 0$$

and $$\frac{d^2 f}{d\lambda_5^2} = 2 > 0$$

$\therefore$ $$X_6 = \left(-1, \frac{3}{2}\right) = X_5$$

Finally, $\lambda_5 = 0$ shows that f can not be minimized in the direction of u_2 and there is no other direction to move.

Hence, Min $f = -1.25$ when $x_1 = -1, x_2 = \dfrac{3}{2}$

26.4 SOME MORE UNCONSTRAINED DIRECT SEARCH METHODS

26.4.1 RANDOM SEARCH METHOD

In these methods, we use the random numbers to find the minimum point. There are following two types of random search method:

   (i) Random Jumping method

   (ii) Random Walk method

(I) RANDOM JUMPING METHOD

In random jumping method, we find the bounds l_i and u_i for each decision variables x_i such that

$$l_i \le x_i \le u_i \quad \forall\; i$$

(l_i is called the lower bound and u_i is upper bound)

Also, we generate a set of n numbers $(r_1, r_2, ..., r_n)$ which are uniformly distributed between 0 and 1.

Then to find a point X, we use the following formula

$$X = \begin{bmatrix} x_1 \\ x_2 \\ \vdots \\ x_n \end{bmatrix} = \begin{Bmatrix} l_1 + r_1(u_1 - l_1) \\ l_2 + r_2(u_2 - l_2) \\ \vdots \\ l_n + r_n(u_n - l_n) \end{Bmatrix}$$

and the value of the function at the point X, i.e., $f(X)$

Repeat the above process to generate a large number of random variable X_i and then evaluate $f(X_i)$ for each i.

Finally, take the smallest value of $f(X_i)$.

(II) RANDOM WALK METHOD

This method is based on generating a sequence of improved approximation to the minimum each derived from the previous approximation.

Working Procedure

In this method we proceed as follows:

STEP 1. If X_i is the approximation obtained in the $(i-1)^{\text{th}}$ stage (minimum). Then
$$X_{i+1} = X_i + \lambda u_i \qquad \ldots(1)$$
where λ is the step size and u_i is the unit vector.

STEP 2. Find $f_1 = f(X_1)$

STEP 3. Using $u = \dfrac{1}{(r_1^2 + r_2^2 + \ldots + r_n^2)^{1/2}} \begin{Bmatrix} r_1 \\ r_2 \\ \vdots \\ r_n \end{Bmatrix}$

generate a set of n random numbers $r_1, r_2, \ldots, r_n$ in the interval $[-1, 1]$.

Here, $R = (r_1^2 + r_2^2 + \ldots + r_n^2)^{1/2}$

Discard the random numbers if $R > 1$ and accept if $R \le 1$.

STEP 4. Find the new vector and $X = X_1 + \lambda u$ and $f = f(X)$.

STEP 5. Find the value of f and f_1. Observe that
➡ If $f < f_1$, set the new values as $X_1 = Y, f_1 = f$ and set the next value of i.
➡ If $f \ge f_1$ go to the next step.

STEP 6. If $i \le N$, set $i = i + 1$ and go to step 3, otherwise go to next step.

STEP 7. Deduce the step length λ as $\dfrac{\lambda}{2}$. If it is less than or equal to ε go to next step, otherwise go to step 3.

STEP 8. Terminate the procedure by taking
$$X_{\text{opt}} = X_1 \text{ and } f_{\text{opt}} = f_1$$

☞ **REMARKS**
- Random search methods can work even if the objective function is discontinuous and non-differentiable at some of the points.
- These methods can be used to find the global minimum when the objective function has several relative minimum.

26.4.2 GRID SEARCH METHOD

Let l_i and u_i be the lower and upper bounds on the ith decision variables respectively. Then divide (l_i, u_i) into $k_i - 1$ equal parts so that $x_1^{(1)}, x_1^{(2)}, \ldots, x_1^{(k)}$ denote the grid point along x_i. Thus there are $k_1 \cdot k_2 \cdot \ldots \cdot k_n$ grid points in the design space.

> This method involves setting up a suitable grid in the design space, evaluating the objective function at all the grid points and finding the grid point corresponding to the lowest function value.

☞ **REMARKS**
- Clearly grid with $k_i = 3$ and 4 are shown in a 2-dimensional space.
- Grid method can be used to find a good starting point for one of the more efficient methods.

26.5 SOME DESCENT (OR INDIRECT) METHODS

In direct search method, we require only objective function values to find the solution. But we observe that, even the best direct methods can require an excessive number of function evaluations to locate the solution. This combined with the quite natural desire to seek stationary points, motivates us to consider methods that are based on gradient. The methods will be iterative, since the elements of the gradient will in general be non-linear functions of decision variables.

SOME IMPORTANT DEFINITIONS

(1) Gradient of a function: Let f be a function. Then gradient of f denoted by ∇f is an n-components vector given by

$$\nabla f = \begin{bmatrix} \partial f / \partial x_1 \\ \partial f / \partial x_2 \\ \vdots \\ \partial f / \partial x_n \end{bmatrix}$$

(2) Direction of steepest ascent: We known that, if we move along the gradient direction from any point to n-dimensional space, the function value increases at the fastest rate. Thus, the gradient direction is known as the direction of steepest ascent.

(3) Direction of steepest descent: Since the gradient vector represents the direction of steepest ascent, the negative of the gradient vector denotes the direction of steepest descent.

☞ **REMARKS**

- A function decreases most rapidly in the negative direction of the gradient.
- Indirect search methods use the derivatives along with determining the value of the function at search point. These methods, therefore are also called 'gradient methods'.

26.6 STEEPEST DESCENT OR CAUCHY METHOD

In steepest descent method we start with an initial point X_1 (base point) and iteratively move along the steepest descent direction until the optimum point is found.

Here, we use the following working procedure.

 Working Procedure

Let X_1 be the initial point and $f(x_1, x_2)$ be the function to be minimize, then proceed as follows:

STEP 1. Calculate $f(X_1)$ and search direction $S_1 = -\nabla f(X_1)$.

STEP 2. Find the optimum step length λ_1 in this direction to arrive at the point $X_2 = X_1 + S_1\lambda_1$ so that
$$f(X_2) < f(X_1)$$

Proceeding in this way until one of the following conditions is satisfied and then terminate the process.

   (i) $\left|\dfrac{\partial f}{\partial x_i}\right| < \varepsilon, \varepsilon > 0$ is a small quantity.

  (ii) Change in decision vectors in the consecutive iteration is very small, *i.e.,* $|X_{i+1} - X_i| < \varepsilon$.

 (iii) The relative change in the value of f at two consecutive steps is very small, *i.e.,*
$$\left|\frac{f(X_{i+1}) - f(X_i)}{f(X_i)}\right| < \varepsilon$$

☞ **REMARKS**

- This method is the best unconstrained minimization method.

> The direction of steepest ascent generally varies from point to point and if we make infinitely small moves along the direction of the steepest ascent, the path will be a curved line.

 Solved Examples

EXAMPLE 1. *Using steepest descent method, minimize the function*
$$f(x_1, x_2) = 2x_1^2 + x_2^2 + 2x_1 x_2 + x_1 - x_2$$

starting from the point $X_1 = (0, 0)$ **($\varepsilon = 0.01$)**

SOLUTION. We have

$$f(x_1, x_2) = 2x_1^2 + x_2^2 + 2x_1 x_2 + x_1 - x_2 \qquad \text{...(1)}$$

Differentiating (1) partially w.r.t. x_1 and x_2 we get

$$\frac{\partial f}{\partial x_1} = 4x_1 + 2x_2 + 1$$

$$\frac{\partial f}{\partial x_2} = 2x_1 + 2x_2 - 1$$

Therefore, $\nabla f(x) = \left(\frac{\partial f}{\partial x_1}, \frac{\partial f}{\partial x_2} \right) = (4x_1 + 2x_2 + 1, 2x_1 + 2x_2 - 1)$ ...(2)

Since it is given that $X_1 = (x_1, x_2) = (0, 0)$. Then from (2)

$$-\nabla f(X_1) = (-1, 1)$$

Iteration-1. We have to find the value of λ which minimize the function

$$f[X_1 - \lambda_1 \nabla f(X_1)] = f[(0, 0) + \lambda_1(-1, 1)] = f(-\lambda_1, \lambda_1)$$

$$= 2\lambda_1^2 + \lambda_1^2 - 2\lambda_1^2 - \lambda_1 - \lambda_1 = \lambda_1^2 - 2\lambda_1$$

Using the principle of maxima and minima for the minimum of f, we must have

$$\frac{\partial f}{\partial \lambda_1} = 0 \quad \Rightarrow \quad 2\lambda_1 - 2 = 0 \quad \Rightarrow \quad \lambda_1 = 1$$

and $\qquad \dfrac{\partial^2 f}{\partial \lambda_1^2} = 2 > 0$

$\Rightarrow \quad f$ is minimum at $\lambda_1 = 1$

So, the new point X_2 is given by

$$X_2 = X_1 - \lambda_1 \nabla f(X_1) = (-1, 1)$$

Then from (2) $\qquad \nabla f(X_2) = (-1, -1)$

$\Rightarrow \qquad\qquad |\nabla f(X_2)| \nleq \varepsilon$

$\Rightarrow \quad X_2$ is not an optimal point.

Iteration-2. We have to find λ_2 which minimize the function

$$f(X_3) = f[X_2 - \lambda_2 \nabla f(X_2)]$$

$$= f[(-1, -1) - \lambda_2(-1, -1)] = f(-1 + \lambda_2, 1 + \lambda_2)$$

$$= 5\lambda_2^2 - 2\lambda_2 - 1 \qquad \text{(Using (1))}$$

i.e., we have to find λ_2 which minimizes $f(X_3)$

For the minimum of f, we must have

$$\frac{\partial f}{\partial \lambda_2} = 0 \quad \Rightarrow \quad 10\lambda_2 - 2 = 0 \quad \Rightarrow \quad \lambda_2 = \frac{1}{5}$$

and $\dfrac{\partial^2 f}{\partial \lambda_2^2} = 10 > 0$

$\Rightarrow \quad f$ is minimum at $\lambda_2 = \dfrac{1}{5}$

$\therefore \quad X_3 = X_2 - \lambda_2 \nabla f(X_2) = \left(-1 + \frac{1}{5}, 1 + \frac{1}{5} \right) = (-0.8, 1.2)$

$\Rightarrow \quad \nabla f(X_3) = (0.2, -0.2)$ (By (2))

$\Rightarrow \quad |\nabla f(X_3)| \nmid 0.01$

$\Rightarrow \quad X_3$ is not an optimal point.

Iteration-3. We have to find λ_3 which minimize the function

$$f(X_4) = f[X_3 - \lambda_3 \nabla f(X_3)] = f[(-0.8, 1.2) - \lambda_3(0.2, -0.2)]$$

$$= f(-0.8 - 0.2\lambda_3, 1.2 + 0.2\lambda_3)$$

$$= 0.04\lambda_3^2 - 0.08\lambda_3 - 1.2$$ (By (1))

Now, for the minimum of f, we must have

$$\frac{\partial f}{\partial \lambda_3} = 0 \quad \Rightarrow \quad 0.08\lambda_3 - 0.08 = 0 \quad \Rightarrow \quad \lambda_3 = 1$$

and $\quad \dfrac{\partial^2 f}{\partial \lambda_3^2} = 0.08 > 0$

$\Rightarrow \quad f$ is minimum at $\lambda_3 = 1$

Therefore, $X_4 = X_3 - \lambda_3 \nabla f(X_3) = (-1, 1.4)$

and $\quad f(X_4) = -1.24$ (By (1))

$\Rightarrow \quad \nabla f(X_4) = (-0.2, -0.2)$ (By (2))

$\Rightarrow \quad |\nabla f(X_4)| \nmid 0.01$

$\Rightarrow \quad X_4$ is not an optimal point.

Iteration-4. We have to find λ_4 which minimize the function

$$f(X_5) = f[X_4 - \lambda_4 \nabla f(X_4)]$$

Here, $\quad X_5 = X_4 - \lambda_4 \nabla f(X_4)$

$$= (-1, 1.4) - 0.2(-0.2, -0.2) = (-0.96, 1.44)$$

$\Rightarrow \quad f(X_5) = 0.2\lambda_4^2 - 0.08\lambda_4 - 1.24$ (By (1))

Now, for the minimum of f, we must have

$$\frac{\partial f}{\partial \lambda_4} = 0 \quad \Rightarrow \quad 0.4\lambda_4 - 0.08 = 0 \quad \Rightarrow \quad \lambda_4 = 0.2$$

and $\quad \dfrac{\partial^2 f}{\partial \lambda_4^2} = 0.4 > 0$

$\Rightarrow \quad f$ is minimum at $\lambda_4 = 0.2$

Since, $f(X_5) = 0.2\lambda_4^2 - 0.08\lambda_4 - 1.24$

$\Rightarrow \quad \nabla f(X_5) = (0.04, -0.04)$

$\Rightarrow \quad |\nabla f(X_5)| \nmid 0.01$

$\Rightarrow \quad X_5$ is not an optimal point.

Iteration-5. Proceed same as above, we have to minimize

$$f(X_6) = f[X_5 - \lambda_5 \nabla f(X_5)] = 0.0016\lambda_5^2 - 0.0032\lambda_5 - 1.248$$

Now, for the minimum of f, we must have

$$\frac{\partial f}{\partial \lambda_5} = 0 \quad \Rightarrow \quad 0.0032\lambda_5 - 0.0032 = 0 \quad \Rightarrow \quad \lambda_5 = 1$$

and $\dfrac{\partial^2 f}{\partial \lambda_5^2} = 0.0032 > 0$

$\Rightarrow \quad f$ is minimum at $\lambda_5 = 1$

Therefore, $X_6 = X_5 - \lambda_5 \nabla f(X_5) = (-1, 1.48)$

$\Rightarrow \quad \nabla f(X_6) = (-0.04, 0.04)$

$\Rightarrow \quad |\nabla f(X_6)| \nleq 0.01$

$\Rightarrow \quad X_6$ is not an optimal point.

Iteration-6. We have to minimize

$$f(X_7) = f[X_6 - \lambda_6 \nabla f(X_6)] = 0.0016\lambda_6^2 - 1.2496$$

For the minimum of f, we must have

$$\dfrac{\partial f}{\partial \lambda_6} = 0 \quad \Rightarrow \quad 0.0032\lambda_6 = 0 \quad \Rightarrow \quad \lambda_6 = 0$$

and $\dfrac{\partial^2 f}{\partial \lambda_6^2} = 0.0032 > 0$

$\Rightarrow \quad f$ is minimum at $\lambda_6 = 0$

$\therefore \quad X_7 = f[X_6 - \lambda_6 \nabla f(X_6)] = (-1, 1.48)$

Also, $\lambda_6 = 0$ implies that, further improvement in f is not possible.

Hence, $\min f = f(X_6) = f(X_7) = -1.2496 \approx -1.25$ when $x_1 = -1$ and $x_2 = 1.48$

26.7 FLETCHER-REEVES METHOD

We know that any minimization method that make use of the conjugate directions is quadratically convergent, which ensure that the method will minimize a quadratic function in n steps or less. Since, any general function can be approximated by a quadratic near the optimum point, any quadratically convergent method is expected to find the optimum point in a finite number of iteration. The construction of conjugate directions and development of the Fletcher-Reeves method are given below.

Working Procedure

The iterative procedure of Fletcher-Reeves method consist the following steps:

STEP 1. Take any arbitrary small point X_1.

STEP 2. Obtain the first search direction using $S_1 = -\nabla f(X_1) = -\nabla f_1$

STEP 3. Find the next point X_2 such that $X_2 = X_1 + \lambda_1^* S_1$

where λ_1^* is the optimum step length in S_1-direction. Set $i = 2$ and go to the next step.

STEP 4. Obtain $\nabla f(X_i)$ and set

$$S_i = -\nabla f_i + \dfrac{|\nabla f_i|^2}{|\nabla f_{i-1}|^2} S_{i-1}$$

STEP 5. Find the optimum step length λ_i^* in S_i-direction and find the new point

$$X_{i+1} = X_i + \lambda_i^* S_i$$

STEP 6. Test the optimality of the point X_{i+1}.

If X_{i+1} is optimum, stop the process otherwise set $i = i + 1$ and go to step 4.

☞ REMARKS

- The Fletcher-Reeves method is superior to the steepest descent method and pattern search method.
- The Fletcher-Reeves method was originally proposed by Hestenes and Stiefel as a method of solving the system of linear equation derived from the stationary conditions of a quadratic.

Solved Examples

EXAMPLE 1. *Using Fletcher-Reeves method, minimize the function*

$$f(x_1, x_2) = 2x_1^2 + x_2^2 + 2x_1x_2 + x_1 - x_2$$

starting from the point $X_1 = (x_1, x_2) = \begin{bmatrix} 0 \\ 0 \end{bmatrix}$.

SOUTION. We have

$$f(x_1, x_2) = 2x_1^2 + x_2^2 + 2x_1x_2 + x_1 - x_2 \qquad ...(1)$$

$$X_1 = \begin{bmatrix} 0 \\ 0 \end{bmatrix}$$

Iteration-1. Now $\dfrac{\partial f}{\partial x_1} = 1 + 4x_1 + 2x_2$

and $\dfrac{\partial f}{\partial x_2} = -1 + 2x_1 + 2x_2$

$\therefore \quad \nabla f = (1 + 4x_1 + 2x_2, -1 + 2x_1 + 2x_2)$

Now, $\quad \nabla f(X_1) = \nabla f(0,0) = \begin{bmatrix} 1 \\ -1 \end{bmatrix}$

$\Rightarrow \qquad S_1 = -\nabla f_1 = \begin{bmatrix} -1 \\ 1 \end{bmatrix}$

Now, we have to find optimal step length λ_1^* along S_1

For this we have to minimize $f(X_1 + \lambda_1 S_1)$ w.r.t. λ_1

Here, we have $f(X_1 + \lambda_1 S_1) = f(-\lambda_1, \lambda_1) = \lambda_1^2 - 2\lambda_1 \qquad$ (By (1))

Now for the minimum of f, we must have

$$\dfrac{df}{d\lambda_1} = 0 \quad \Rightarrow \quad 2\lambda_1 - 2 = 0 \quad \Rightarrow \quad \lambda_1^* = 1$$

and $\dfrac{d^2 f}{d\lambda_1^2} = 2 > 0$

$\Rightarrow \quad f$ is minimum at $\lambda_1^* = 1$

Therefore, $X_2 = X_1 + \lambda_1^* S_1 = \begin{bmatrix} 0 \\ 0 \end{bmatrix} + 1 \begin{bmatrix} -1 \\ 1 \end{bmatrix} = \begin{bmatrix} -1 \\ 1 \end{bmatrix}$

Iteration-2.

$\because \qquad \nabla f_2 = \nabla f(X_2) = \begin{bmatrix} -1 \\ 1 \end{bmatrix}$

Now, $\qquad S_2 = -\nabla f_2 + \dfrac{|\nabla f_2|^2}{|\nabla f_1|^2} S_1$

Here, $|\nabla f_1|^2 = 2$ and $|\nabla f_2|^2 = 2$

$$\therefore \qquad S_2 = -\begin{bmatrix} -1 \\ 1 \end{bmatrix} + \left(\frac{2}{2}\right)\begin{bmatrix} -1 \\ 1 \end{bmatrix} = \begin{bmatrix} 0 \\ 2 \end{bmatrix}$$

Now, we have to minimize $f(X_2 + \lambda_2 S_2)$

Here, we have
$$f = f(-1, 1 + 2\lambda_2)$$
$$= -1 - (1 + 2\lambda_2) + 2 - 2(1 + 2\lambda_2) + (1 + 2\lambda_2)^2$$
$$= 4\lambda_2^2 - 2\lambda_2 - 1$$

For the minimum of f, we must have

$$\frac{df}{d\lambda_2} = 0 \quad \Rightarrow \quad 8\lambda_2 - 2 = 0 \quad \Rightarrow \quad \lambda_2^* = \frac{1}{4}$$

and $\dfrac{d^2 f}{d\lambda_2^2} = 8 > 0$

$\Rightarrow \quad f$ is minimum at $\lambda_2^* = \dfrac{1}{4}$

Therefore, $X_3 = X_2 + \lambda_2^* S_2 = \begin{bmatrix} -1 \\ 1 \end{bmatrix} + \dfrac{1}{4}\begin{bmatrix} 0 \\ 2 \end{bmatrix} = \begin{bmatrix} -1 \\ 1.5 \end{bmatrix}$

Iteration-3.

$$\nabla f_3 = \nabla f(X_3) = \begin{bmatrix} 0 \\ 0 \end{bmatrix}$$

$$|\nabla f_2|^2 = 2 \text{ and } |\nabla f_3|^2 = 0$$

Therefore, $S_3 = -\nabla f_3 + \dfrac{(|\nabla f_3|^2)}{|\nabla f_2|^2} \cdot S_2 = -\begin{bmatrix} 0 \\ 0 \end{bmatrix} + \begin{bmatrix} 0 \\ 2 \end{bmatrix}\begin{bmatrix} 0 \\ 0 \end{bmatrix} = \begin{bmatrix} 0 \\ 0 \end{bmatrix}$

$\Rightarrow$ There is no search direction to reduce f further.

Hence, X_3 is the optimal point and minimum of $f(x)$ can be obtained from (1) by putting $x_1 = -1$ and $x_2 = 1.5$.

 Exercise-26.1

1. Using univariate method minimize the function
$$f(X) = f(x_1, x_2) = x_1^2 + x_2^2 - 2x_1 - 4x_2$$
starting with $X_1 = \begin{bmatrix} 0 \\ 0 \end{bmatrix}$.

2. Using Hooke and Jeeve's method, minimize the function
$$f(X) = 3x_1^2 + x_2^2 - 2x_1 x_2 - 4x_1 - 3x_2$$
taking the starting base point as $X_1 = (0, 0)$ and $\Delta x_1 = \Delta x_2 = 1$.

3. Using Powel's method minimize the function
$$f(X) = f(x_1, x_2) = 4x_1^2 + 3x_2^2 - 5x_1 x_2 - 8x_1$$
taking $X_1 = (0, 0)$ as starting point.

4. Using Steepest-descent method, minimize the function
$$f(x_1, x_2) = 2x_1^2 + x_2^2$$
starting from the point $X_1 = (1, 2)$.

ANSWERS

1. $x_1 = 1, x_2 = 2$
2. $x_1 = 1.75, x_2 = 3.25, f(X) = -8.375$
3. $x_1 = 48/23, x_2 = 40/23$
4. $f(X) = 1.0336$

26.8 CONSTRAINED OPTIMIZATION PROBLEMS

In previous section, we have studied some search unconstrained optimization technique. This section deals with the techniques that are applicable to the solution of the constrained optimization problems. The general form of constrained non-linear programming problem is given as below

$$\text{Min. } f(X)$$

subject to the constraints

$$\left.\begin{array}{ll} g_i(X) \le 0 ; & i = 1, 2, ..., n \\ h_k(X) = 0; & k = 1, 2, ..., p \end{array}\right]$$

....(1)

In this section we shall discuss some methods to solve the constrained non-linear programming problems given by (1).

26.9 CHARACTERISTICS OF A CONSTRAINED PROBLEM

Following are the main characteristics of constrained non-linear programming problem.

(i) The constraints may have no effect on the optimum point.

(ii) The optimum solution occurs on a constraints boundary.

(iii) The negative of the gradient must be expressible as a positive linear combinations of the gradient of the active constraints.

(iv) If the objective function has two or more unconstrained local minima, then the constrained problem may have multiple minima.

(v) If the objective function has a single unconstrained minimum, then the constraints may introduce multiple local minima.

26.10 CONSTRAINED MINIMIZATION METHODS

There are many methods available for the solution of a constrained non-linear programming problems which are classified into following two categories :

(i) Direct methods

(ii) Indirect methods

26.11 DIRECT METHODS

In solving non-linear programming problem, if the function to be minimized is not differentiable but requires computational work, then direct methods are easy to use.

Let us discuss some direct methods one by one.

26.11.1 RANDOM SEARCH METHOD

In this method, we can use methods of unconstrained minimization with minor changes, to solve constrained optimization problem.

In this method we have the following procedure.

Working Procedure

Here we use the following steps:

STEP 1. Set a trial design vector using one random number for each design variable.

STEP 2. Verify, whether the constraints are satisfied by the trial design vector (of step-1)

(if not, continue generating new trial vectors)

STEP 3. If all the constraints are satisfied retain the current trial vector as the best design if it gives a reduced objective function value compared to the previous best available design. Otherwise go to the step-1 for new trial design vector.

At the end of generating a specified maximum number of trials design vectors is

taken as the solution of the constrained optimization problem.

☛ **REMARK**

- The random search methods are not efficient compared to the other methods.

26.11.2 COMPLEX METHOD

In this method we assume that an initial feasible point X_1, which satisfies all the constraints is available. This method is an extension upon the simplex method of unconstrained minimization to solve constrained optimization problem.

It deals with the constrained optimization problem of the type given below:

$$\text{Minimize } Z = f(X) \qquad \qquad ...(1)$$

subject to the constraints

$$g_j(X) \leq 0, \ j = 1, 2, ..., m \qquad \qquad ...(2)$$

$$X = (x_1, x_2, ..., x_n)^T$$

and x_i (l) and x_i(u) are the lower and upper bounds of x_i respectively so that

$$x_i(l) \leq x_i \leq x_i \ (u) \ \ \forall \ i = 1, 2, ..., n \qquad \qquad ....(3)$$

The condition given in (3) are called "side constraints".

GENERAL PROCEDURE

Given the set of points, the objective function is evaluated at each point and the point corresponding to the highest value is rejected. A new point is generated by reflecting the rejected point a certain distance through the centroid of the remaining point. At the new point, the performance function and the constraints are evaluated with the following alternatives :

(i) The new point is feasible and its function value is not the highest of the set of points. Then select the point that does not correspond to the highest and continue with a reflection.

(ii) The new value is feasible and its function value is the highest of the current set of points. Rather than reflecting back again, retract the point by half the distance to the previously calculated centroid.

(iii) The new point is infeasible. Retract the point by half the distance to the previously calculated centroid.

☞ **REMARK**

- The search is terminated when the pattern of points has shrunk so that the points are sufficiently close together and when the differences between the function values at the points becomes small enough.

- For a minimization problem in n variables, if we consider k points where $k \leq n + 1$, then the figure formed on joining them is called '**complex**'.
- If we consider the minimization problems in two variables, then in 2-dimension, it will have $4(n=2)$ vertices. In any dimension n, we will have $2n$ as an even number. So, when $n = 2$, k should be 4 and hence take $k = 2n = 4$. These four points will be the vertices of the complex.

 Working Procedure

To solve the problems of two variables, we use the following steps :

STEP 1.   Let one point X_1 be given, obtain remaining $(2n-4) = 3$ points X_2, X_3 and X_4 are at a time by using random number r_{ij}, $0 < r_{ij} < 1$.

STEP 2.   Calculate $x_{ij} = x_i(l) + r_{i,j} [(x_i(u) - x_i (l)]$

when $x_{ij} = i^{th}$ component of the function X_j

$$i = 1, 2 \text{ and } j = 2, 3, 4$$

The obtained points X_2, X_3 and X_4 satisfies the constraints (3) but may not satisfy all the constraints given by (2).

If, X_4 is not satisfying all the constraints in (2), then obtained a new point $X_4^{(1)}$ by moving X_4 halfway towards the centroid

$$X_0 = \frac{1}{3}(X_1 + X_2 + X_3)$$

of the remaining points X_1, X_2 and X_3

i.e.,

$$X_4^{(1)} = \frac{X_0 + X_j}{2}$$

Now check whether $X_4^{(1)}$ satisfies all the constraints in (2) or not. If not obtain, then another point $X_4^{(2)}$ by moving $X_4^{(1)}$ halfway towards the centroid X_0

Proceed in the same manner until a feasible point X_4 satisfying (2), is obtained.

$\Rightarrow$ Finally four points X_1, X_2, X_3, X_4 all satisfying the constraints given by (2). These points are the vertices of the starting complex.

STEP 3.   Calculate value of the function at X_1, X_2, X_3 and X_4 i.e., $f(X_1), f(X_2), f(X_3)$ and $f(X_4)$.

Find the largest and smallest value among them. Find a new point by using (process of reflection)

$$X_r = (1+\alpha)X_0 - \alpha X_h$$

where $\alpha \leq r$ and X_0 is the centroid of all vertices except X_h

i.e.,

$$X_0 = \frac{(X_1 + X_2 + X_3)}{3}$$

STEP 4. Now we check the feasibility of X_r.

(i) If X_r is feasible and $f(X_r) < f(X_h)$, then X_h is replaced by X_r and carry on step-3.

(ii) If $f(X_r) \leq f(X_h)$, then a new trial point X_r is found by taking a new value of

$$\alpha = \left(\frac{\text{previous value of } \alpha}{2} \right)$$

and tested further until the condition $f(X_r) < f(X_h)$ is satisfied.

Proceeding in this way we will make the value of α smaller and smaller.

(iii) If the value of X_r does not satisfy $f(X_r) < f(X_h)$ in any way, neglect the whole reflection process and start a new reflection process by taking X_h which gives the second largest value of the function.

The procedure terminate when the distance between any two vertices among X_1, X_2, X_3 and X_4 becomes smaller than the prescribed value of ε.

26.11.3 SOME FACTS ABOUT COMPLEX METHOD

(1) If the feasible region is non convex, there is no guarantee that the centroid of all feasible points is also feasible. If the centroid is not feasible then we can not apply the above procedure to find the new point.

(2) Complex method can not be used to solve problem having equality constraints.

(3) This method becomes inefficient rapidly as the number of variables increases.

 Solved Examples

EXAMPLE 1. *Using complex method, minimize the function*

$$f(X) = f(x_1, x_2) = (x_1-1)^2 + (x_2-1.5)^2 - 0.25$$

subject to the constraints

$$x_1 + x_2 \leq 4, \quad 0 \leq x_1 \leq 2 \quad \text{and} \quad 1 \leq x_2 \leq 3$$

Starting with $X_1 = \begin{pmatrix} 0.7 \\ 1.1 \end{pmatrix}$

SOLUTION. We have

$$f(X) = f(x_1, x_2) = (x_1 - 1)^2 + (x_2 - 1.5)^2 - 0.25 \qquad \ldots(1)$$

$$g(X) = x_1 + x_2 - 4 \leq 0$$

and side constraints are given by

$$0 \leq x_1 \leq 2 \quad \text{and} \quad 1 \leq x_2 \leq 3$$

Let X_1, X_2, X_3 and X_4 be four points such that $X_1 = \begin{pmatrix} 0.7 \\ 1.1 \end{pmatrix}$

and choose the random numbers

$$r_{1,2} = 0.4, \ r_{1,3} = 0.6, \ r_{1,4} = 0.8$$

$$r_{2,2} = 0.5, \ r_{2,3} = 0.7, \ r_{2,4} = 0.9$$

Now, we have to find the value of X_2, X_3 and X_4 by using the following formula

$$x_{i,j} = x_i(l) + r_{i,j}[x_i(u) - x_i(l)] \quad i = 1, 2, \ j = 1, 2, 3, 4$$

where $x_{i,j}$ is the ith component of vector X_j.

Clearly $x_1(l) = 0$ and $x_1(u) = 2$

and $\qquad\qquad x_2\ (l) = 1$ and $x_2(u) = 3$

Therefore, we have

$$x_{1,2} = x_1(l) + r_{1,2}\,[\,x_1(u) - x_1(l)] = 0.8$$

$$x_{1,3} = x_1(l) + r_{1,3}\,[\,x_1(u) - x_1(l)] = 1.2$$

$$x_{1,4} = x_1(l) + r_{1,4}\,[\,x_1(u) - x_1(l)] = 1.6$$

$$x_{2,2} = x_2(l) + r_{2,2}\,[\,x_2(u) - x_2(l)] = 2.0$$

$$x_{2,3} = x_2(l) + r_{2,3}\,[\,x_2(u) - x_2(l)] = 2.4$$

$$x_{2,4} = x_2(l) + r_{2,4}\,[\,x_2(u) - x_2(l)] = 2.8$$

Therefore, the vertices of the first complex are given by

$$X_1 = \begin{pmatrix} 0.7 \\ 1.1 \end{pmatrix},\ X_2 = \begin{pmatrix} 0.8 \\ 2 \end{pmatrix},\ X_3 = \begin{pmatrix} 1.2 \\ 2.4 \end{pmatrix} \text{ and } X_4 = \begin{pmatrix} 1.6 \\ 2.8 \end{pmatrix}$$

So $\qquad\qquad g(X_1) = x_1 + x_2 = 0.7 + 1.1 = 1.8 \le 4$

Similarly $\qquad g(X_2) = 2.8 \le 4$

$$g(X_3) = 3.6 \le 4$$

and $\qquad\qquad g(X_4) = 4.8 > 4$

$\Rightarrow\ g(X)$ is not satisfies at X_4

So, we have to replace it by some point in the feasible region as follows.

The centroid of the vertices be given by

$$X_0 = \frac{X_1 + X_2 + X_3}{3} = \begin{pmatrix} 0.9 \\ 1.83 \end{pmatrix}$$

New value of X_4 i.e., $\quad X_4^{(1)} = \frac{X_0 + X_4}{2} = \begin{pmatrix} 1.25 \\ 2.315 \end{pmatrix}$

and $\qquad\qquad g\left(X_4^{(1)}\right) = 3.565 \le 4$

$\Rightarrow\ X_4^{(1)}$ lies in the feasible region and hence initial complex has the vertices given by

$$X_1 = \begin{pmatrix} 0.7 \\ 1.1 \end{pmatrix},\ X_2 = \begin{pmatrix} 0.8 \\ 2 \end{pmatrix},\ X_3 = \begin{pmatrix} 1.2 \\ 2.4 \end{pmatrix} \text{ and } X_4 = X_4^{(1)} = \begin{pmatrix} 1.25 \\ 2.315 \end{pmatrix}$$

$\Rightarrow\qquad\qquad f(X_1) = 0,\ f(X_2) = 0.04,\ f(X_3) = 0.60,\quad f\left(X_4^{(1)}\right) = 0.4767,$

[Using (1)]

Here, $f(X_3) = 0.60$ gives the maximum value and $f(X_1) = 0$ is the minimum value.

Let us take $X_3 = X_h$ with $f(X_h) = 0.6$ and $X_1 = X(l)$ and $f(X(l)) = 0$

Then, new centroid X_0 is given by

$$X_0 = \frac{X_1 + X_2 + X_4^{(1)}}{3} = \begin{pmatrix} 0.917 \\ 1.805 \end{pmatrix}$$

$\Rightarrow \qquad f(X_0) = 0.15 \qquad\qquad\qquad$ [Using (1)]

Now, $f(X_0) < f(X_h) \Rightarrow f(X)$ is a decreasing function from $X_h (= X_3)$ towards X_0.

To find X_r, we use the following formula

$$X_r = (1+\alpha)X_0 - \alpha X_h$$

Here $\alpha = 1$ so $X_r = 2X_0 - X_h \begin{pmatrix} 0.634 \\ 1.21 \end{pmatrix}$ and $f(X_r) = -0.034944$

Since X_r is feasible and $f(X_r) < f(X_h)$, we proceed further for the different values of α as given below:

(i) Taking $\alpha = 0.1$; then, we have

$$X_r^{(1)} = 1.1X_0 - 0.1X_h = (1.1)\begin{pmatrix} 0.917 \\ 1.805 \end{pmatrix} - 0.1\begin{pmatrix} 1.2 \\ 2.4 \end{pmatrix} = \begin{pmatrix} 0.8887 \\ 1.7455 \end{pmatrix}$$

and $f\left(X_r^{(1)}\right) = -0.1773$

(ii) Taking $\alpha = 0.2$; then, we have

$$X_r^{(2)} = 1.2X_0 - 0.2X_h = (1.2)\begin{pmatrix} 0.917 \\ 1.805 \end{pmatrix} - 0.2\begin{pmatrix} 1.2 \\ 2.4 \end{pmatrix} = \begin{pmatrix} 0.8604 \\ 1.686 \end{pmatrix}$$

and $f\left(X_r^{(2)}\right) = -0.1959$

(iii) Taking $\alpha = 0.3$; then, we have

$$X_r^{(3)} = 1.3X_0 - 0.3X_h = (1.3)\begin{pmatrix} 0.917 \\ 1.805 \end{pmatrix} - 0.3\begin{pmatrix} 1.2 \\ 2.4 \end{pmatrix} = \begin{pmatrix} 0.8321 \\ 1.6265 \end{pmatrix}$$

and $f\left(X_r^{(3)}\right) = -0.2058$

(iv) Taking $\alpha = 0.4$; then, we have

$$X_r^{(4)} = 1.4X_0 - 0.4X_h = (1.4)\begin{pmatrix} 0.917 \\ 1.805 \end{pmatrix} - 0.4\begin{pmatrix} 1.2 \\ 2.4 \end{pmatrix} = \begin{pmatrix} 0.8038 \\ 1.567 \end{pmatrix}$$

and $f\left(X_r^{(4)}\right) = -0.2070$

(v) Taking $\alpha = 0.5$; then, we have

$$X_r^{(5)} = 1.5X_0 - 0.5X_h = (1.5)\begin{pmatrix} 0.917 \\ 1.805 \end{pmatrix} - 0.5\begin{pmatrix} 1.2 \\ 2.4 \end{pmatrix} = \begin{pmatrix} 0.7755 \\ 1.5075 \end{pmatrix}$$

and $f\left(X_r^{(5)}\right) = -0.1995$

We observe that, the values of f continue to decrease upto $X_r^{(4)}$ replace $X_h = X_3$ with the maximum value of $X_r^{(4)}$ to obtain the new complex with vertices given by

$$X_1 = \begin{pmatrix} 0.7 \\ 1.1 \end{pmatrix} \text{ and } f(X_1) = 0$$

$$X_2 = \begin{pmatrix} 0.8 \\ 1.1 \end{pmatrix} \text{ and } f(X_2) = 0.04$$

$$X_3 = X_r^{(4)} = \begin{pmatrix} 0.8038 \\ 1.567 \end{pmatrix} \text{ and } f(X_3) = f\left(X_r^{(4)}\right) = -0.207016$$

and $\quad X_4 = \begin{pmatrix} 1.25 \\ 2.315 \end{pmatrix}$ and $f(X_4) = f\left(X_4^{(1)}\right) = 0.476725$

Therefore,

$f(X_4) = f\left(X_4^{(1)}\right) = 0.476725$ gives the maximum values

and $f(X_3) = f\left(X_r^{(4)}\right) = -0.207016$ gives the minimum value.

Thus $\quad X_h = X_4 = \begin{pmatrix} 1.25 \\ 2.315 \end{pmatrix}$ and $f(X_4) = 0.476725$

and $\quad X(l) = X_3 = X_r^{(4)} = \begin{pmatrix} 0.8038 \\ 1.567 \end{pmatrix}$ and $f(X_3) = f(X(l)) = -0.207016$

The centroid $X_0 = \dfrac{(X_1 + X_2 + X_3)}{3} = \begin{pmatrix} 0.7679 \\ 1.557 \end{pmatrix}$ with $f(X_0) = -0.19303$

$\Rightarrow \quad |f(X(l)) - f(X_0)| = 0.014$

If the desired accuracy is $\varepsilon = 0.01$, then the above solution is accepted and hence min. $f(X) = -0.2070$ at X_3 i.e., when $X_1 = 0.8038$ and $X_2 = 1.5557$.

26.11.4 METHOD OF FEASIBLE DIRECTIONS: ZOUTENDIJK'S METHOD

In the methods of feasible directions, we select a starting point satisfying all the given constraints and move to a better point by using the following iterative scheme

$$X_{i+1} = X_i + \lambda S_i$$

Here, $\quad X_i$ = starting point (base point) for the ith iteration

S_i = direction of movement

λ = step length (distance of movement)

X_{i+1} = Final point obtained at the end of ith iteration

To find the search direction S_i, we have the following point keep in mind

Prop. (i) a small move in that direction violates no constraints

Prop.(ii) the value of the objective function can be reduced in that direction.

The new point X_{i+1} is taken as the starting point for the next iteration and we will repeat the whole procedure until a point is obtained such that no direction satisfying both (i) and (ii). Such a point denotes the constrained local minimum of the given problem.

☞ REMARKS

- A direction satisfying the prop (i) is called **feasible**.
- A direction satisfying both the above properties ((i) and (ii)) is called a **usable feasible direction**.

26.11.5 ZOUTENDIJK'S METHOD

In this method, the usable feasible direction is taken as the negative of the gradient direction if the initial point of the iteration lies in the interior of the feasible region.

But, if the initial point lies on the boundary of the feasible region, some constraints will be active.

Working Procedure

Let us consider the problem given by

$$\text{Minimize } f(X) \qquad \qquad \text{...(1)}$$

subject to the constraints

$$g_j(X) \le 0, \; j = 1, 2, ..., m \qquad \qquad \text{...(2)}$$
$$X = (x_1, x_2, ..., x_m)^T$$

To solve the above problem, we use the following steps:

STEP 1. Choose the starting point X_i ($i = 1$, for starting point) which satisfies all the constraints in (2).

STEP 2. Move to a better point X_{i+1} using the following formula

$$X_{i+1} = X_i + \lambda_i S_i$$

(Better point is the point where value of the function in (1) is less than it has at X_i)

where

X_i = starting point (base point) for i^{th} iteration

S_i = direction to move

λ_i = step length (in the direction S_i)

X_{i+1} = Point obtained at the end of ith iteration

STEP 3. Find the search direction such that

(i) even a small movement in that direction violates none of the constraints in (2).

(ii) the value of the function $f(X)$ in (1) decreases.

STEP 4. Test for the convergence. If $\left| \dfrac{f(X_i) - f(X_{i+1})}{f(X_i)} \right| \le \varepsilon_1$ and $|| X_i - X_{i+1} || \le \varepsilon_2$ terminate

the iteration by taking $X_{\text{opt}} = X_{i+1}$.

☛ REMARKS

- Care should be taken to choose λ_i such that the new point X_{i+1} is a feasible region.
- For feasible direction:

$$\left[\frac{d}{d\lambda} [g_j(X_i + \lambda_i S_i] \right]_{\lambda_i=0} = S_i^T \nabla g_j(X_i) \le 0$$

- For usable direction:

$$\left[\frac{d}{d\lambda} [f(X_i + \lambda_i S_i] \right]_{\lambda_i=0} = S_i^T \nabla f(X_i) < 0$$

and

$$\left[\frac{d}{d\lambda} [g_j(X_i + \lambda_i S_i)] \right]_{\lambda_i=0} = S_i^T \nabla g_j(X_i) \le 0$$

 Solved Examples

EXAMPLE 1. *Using Zoutendijk's method minimize the function*

$$f(X) = x_1^2 + x_2^2 - 2x_1 - 3x_2 + 3$$

subject to the constraints

$$g_1(X) = x_1 + x_2 \le 4, \text{ taking the starting point } x_1 = \begin{pmatrix} 0 \\ 0 \end{pmatrix}$$

SOLUTION.   We have

$$f(X) = f(x_1, x_2) = x_1^2 + x_2^2 - 2x_1 - 3x_2 + 3 \qquad \qquad \ldots(1)$$

$$X_1 = \begin{pmatrix} 0 \\ 0 \end{pmatrix}$$

Let $\qquad g_1(X) = x_1 + x_2 - 4 \qquad \qquad \ldots(2)$

Then $\qquad f(X_1) = 3$ and $g(X_1) = -4 < 0$ [Using (1) and (2)]

Since $g_1(X) < 0$ so search direction S_1 is given by

$$S_1 = -\nabla f(X_1) = -\begin{bmatrix} \partial f / \partial x_1 \\ \partial f / \partial x_2 \end{bmatrix}_{X_1 = (0,0)} = \begin{pmatrix} 2 \\ 3 \end{pmatrix}$$

$$\left[\because \frac{\partial f}{\partial x_1} = 2x_1 - 2 \text{ and } \frac{\partial f}{\partial x_2} = 2x_2 - 3 \right]$$

Therefore, $\qquad S_1 = \begin{pmatrix} 2/3 \\ 1 \end{pmatrix}$

Now, to find a new point X_1, we take a step of length λ_1 in the direction of $-\nabla f(X_1)$ to arrive at

$$X_2 = X_1 + \lambda_1 S_1 = \left(\frac{2}{3}\lambda_1, \lambda_1 \right)$$

$$f(X_2) = f\left(\left(\frac{2}{3}\right)\lambda_1, \lambda_1 \right) = \left(\frac{13}{9}\right)\lambda_1^2 - \frac{13}{3}\lambda_1 + 3$$

To minimize f, we must have

$$\frac{df}{d\lambda_1} = 0 \Rightarrow \frac{26}{9}\lambda_1 - \frac{13}{3} = 0 \Rightarrow \lambda_1 = 1.5$$

and $\qquad \dfrac{d^2 f}{d\lambda_1^2} = \dfrac{26}{9} > 0$

$\Rightarrow f$ is minimum when $\lambda_1 = 1.5$

Therefore, $\qquad X_2 = (0, 0) + 1.5 \, (2/3, 1) = (1, 1.5)$

and $\qquad g(X_2) = -1.5 < 0$

Therefore, the new search direction S_2 is given by

$$S_2 = -\nabla f(X_2)$$

$$= \begin{bmatrix} \partial f / \partial x_1 \\ \partial f / \partial x_2 \end{bmatrix}_{X_1 = (1, 1.5)} = \begin{pmatrix} 0 \\ 0 \end{pmatrix}$$

$\Rightarrow$ There is no search direction available to obtain minimum f.

Hence, minimum of $f = -0.25$ at $X = X_2 = (1, 1.5)$ *i.e.*, $x_1 = 1, x_2 = 1.5$

EXAMPLE 2. *Using Zoutendijk's method, minimize the function*
$$f(X) = f(x_1, x_2) = x_1^2 + x_2^2 - 4x_1 - 4x_2 + 8$$
subject to the constraints
$g_1(x_1, x_2) = x_1 + 2x_2 - 4 \leq 0$, *with the starting point* $X_1 = \begin{pmatrix} 0 \\ 0 \end{pmatrix}$.
Take $\varepsilon_1 = 0.001$, $\varepsilon_2 = 0.001$ *and* $\varepsilon_3 = 0.01$

SOLUTION. We have
$$f(X) = f(x_1, x_2) = x_1^2 + x_2^2 - 4x_1 - 4x_2 + 8 \qquad \qquad ...(1)$$
and $g_1(x_1, x_2) = x_1 + 2x_2 - 4$ $\qquad \qquad ...(2)$

Iteration-1 :

Since $g_1(X_i) < 0$, therefore search direction given by
$$S_1 = -\nabla f(X_1)$$

$\Rightarrow \qquad S_1 = -\begin{bmatrix} \partial f / \partial x_1 \\ \partial f / \partial x_2 \end{bmatrix}_{x_1} = \begin{pmatrix} 4 \\ 4 \end{pmatrix}$

$(\because$ from (1) $\dfrac{\partial f}{\partial x_1} = 2x_1 - 4$ and $\dfrac{\partial f}{\partial x_2} = 2x_2 - 4$)

$\Rightarrow$ normalized vector $S_1 = \begin{bmatrix} 1 \\ 1 \end{bmatrix}$

Now to find the new point X_2, we have to find a suitable step length along S_1. For this we have to minimize
$$f(X_1 + \lambda S_1) \text{ w.r.t. } \lambda$$
We have $f(X_1 + \lambda S_1) = f(0 + \lambda, 0 + \lambda) = f(\lambda, \lambda) = 2\lambda^2 - 8\lambda + 8$ (by (1))
Now, for the minima of f, we must have
$$\frac{df}{d\lambda} = 0 \Rightarrow 4\lambda - 8 = 0 \Rightarrow \lambda = 2$$
and $\qquad \dfrac{d^2 f}{d\lambda^2} = 4 > 0$

$\Rightarrow f$ is minimum at $\lambda = 2$
So, we have a new point given by $X_2 = \begin{bmatrix} 2 \\ 2 \end{bmatrix}$ and $g_1(X_2) = 2$
Since, $g_1' = g_1$ (at $\lambda = 0$) = -4 and $g_1'' = g_1$ (at $\lambda = 2$) = 2
so, new step length given by
$$\bar{\lambda} = -\frac{g_1'}{g_1'' - g_1'} \lambda = \frac{4}{3}$$

$\Rightarrow g_1$ at $\lambda = \bar{\lambda} = 0$ and hence $X_2 = \begin{bmatrix} 4/3 \\ 4/3 \end{bmatrix} = \begin{bmatrix} 1.333 \\ 1.333 \end{bmatrix}$
Also from (1), $f(X_2) = \dfrac{8}{9}$
Now, we check the convergence.

Here, $\left|\dfrac{f(X_1)-f(X_2)}{f(X_1)}\right| = \left|\dfrac{8-8/9}{8}\right| = \dfrac{8}{9} > \varepsilon_2$

and $\qquad ||X_1 - X_2|| = \left[\left(0-\dfrac{4}{3}\right)^2 + \left(0-\dfrac{4}{3}\right)^2\right]^{1/2} = 1.887 > \varepsilon_2$

$\Rightarrow$ Convergence criterion is not satisfied.

Iteration-2

Since $g_1 = 0$ at $X = X_2$, so we have to find a usable feasible direction.

Here direction finding problem can be stated as

$$\text{Minimize } f = -\alpha$$

subject to $t_1 + 2t_2 + \alpha + y_1 = 3$

$$-\dfrac{4}{3}t_1 - \dfrac{4}{3}t_2 + \alpha + y_2 = -\dfrac{8}{3}$$

$$t_1 + y_3 = 2$$

$$t_2 + y_4 = 2$$

$$t_1, t_2, \alpha \geq 0$$

which is a linear programming problem and having the solution

$$t_1^* = 2,\ t_2^* = \dfrac{3}{10},\ \alpha^* = \dfrac{4}{10},\ y_4^* = \dfrac{17}{10},\ y_1^* = y_2^* = y_3^* = 0$$

$$-f_{\min} = -\alpha^* = -\dfrac{4}{10}$$

$\Rightarrow$ usable feasible direction is given by

$$S = \begin{bmatrix} t_1^* - 1 \\ t_2^* - 1 \end{bmatrix} = \begin{bmatrix} 1.0 \\ -0.7 \end{bmatrix}$$

Now, we have to move along the direction $S_2 = \begin{bmatrix} 1.0 \\ -0.7 \end{bmatrix}$

from $\qquad\qquad X_2 = \begin{bmatrix} 1.333 \\ 1.333 \end{bmatrix}$

$\therefore \qquad\qquad f(X_2 + \lambda S_2) = f(1.333 + \lambda,\ 1.333 - 0.7\lambda)$

$$= 1.49\lambda^2 - 0.4\lambda + 0.889$$

Using the principle of maxima and minima, for the minimum of f, we must have

$$\dfrac{df}{d\lambda} = 0 \Rightarrow 2(1.49)\lambda - 0.4 = 0$$

$\Rightarrow \qquad\qquad \lambda = 0.134$

and $\qquad\qquad \dfrac{d^2 f}{d\lambda^2} = 2.98 > 0$

$\Rightarrow f$ is minimum at $\lambda = 0.134$

Hence, new point is given by

$$X_3 = X_2 + \lambda S_2 = \begin{bmatrix} 1.333 \\ 1.333 \end{bmatrix} + 0.134\begin{bmatrix} 1.0 \\ -0.7 \end{bmatrix} = \begin{bmatrix} 1.467 \\ 1.239 \end{bmatrix}$$

Clearly $\quad g_1(X_3) = -0.005$

$\Rightarrow X_3$ lies in the interior of the feasible domain.

Proceeding same as above , after some iteration

we get $\quad X^* = \begin{bmatrix} 1.6 \\ 1.2 \end{bmatrix}$ and $f_{min} = 0.8$.

26.11.6 ROSEN GRADIENT PROJECTION METHOD

Rosen gradient projection method is an efficient method of solving constrained non-linear programming problem. It does not require the solution of an auxiliary linear optimization problem to find the usable feasible direction. This method can be used for general non-linear programming problems where even the constraints can be non-linear along with the non-linear objective function.

However, it is more effectively used to solve the problem where all constraints are linear.

In this method, we uses the projection of the negative of the objective function gradient onto the constraints that are currently active.

Consider a problem given by

$$\text{Minimize } f(X)$$

subject to the constraints

$$g_j(X) = \sum_{i=1}^{n} a_{ij} x_i - b_j \le 0, \, j = 1,2,...,m \qquad \qquad ...(1)$$

where $X = (x_1, x_2, ..., x_n)$

Before finding the solution of (1), let us define the following terms

  (i) **Feasible point :** The set of all X such that $g_j(X), j = 1, 2, ..., m$ is called the set of feasible solution of the above equation (1). It is denoted by S_F

  i.e., $\qquad\qquad S_F = \left[X : g_j(X) \le 0, \, j = 1,2,...,m \right]$

  Then a point $X \in S_F$ is called a feasible point.

 (ii) **Active Constraints :** A constraints $\{g_j\,(X) \le 0\}$ is called active at a feasible point X if $g_i(X) = 0$.

(iii) **Projection matrix :** Let $j_1, j_2, ..., j_p$ be the indices of active constraints at a feasible point X, then the matrix of order $n \times p$ of the gradient of these active constraints.

$$N_p = [\nabla g_{j,1}, \nabla g_{j,2}, ... \nabla g_{j,p}]$$

where

$$\nabla g_j(X) = \begin{bmatrix} a_{1j} \\ a_{2j} \\ \vdots \\ a_{nj} \end{bmatrix}, \, j = j_1, j_2, ..., j_p$$

and the projection matrix P is given by

$$P = I - N \,[N^T N]^{-1} \, N^T$$

where I is the identity matrix.

Now we shall discuss the procedure, by which we solve the given constrained problem by Rosen's gradient projection method.

Working Procedure

To solve a non-linear programming problem using Rosen's gradient projection method, we use the following steps :

STEP 1. Take an initial point X_1. Check the feasibility of X_1 i.e., X_1 has to be feasible if $g_j(X_1) \leq 0, j = 1, 2, ..., m$. We have following two cases:

Case-1 : If $g_j(X_1) < 0$ for $j = 1, 2, ..., m$ then find the normalized search direction using
$$S_i = \frac{-\nabla f(X_i)}{||\nabla f(X_i)||}$$

Case-2 : If $g_j(X_1) = 0$ for $j = j_1, j_2, ..., j_p$, then find the normalized search direction using
$$S_i = \frac{-P_i \nabla f(X_i)}{|P_i \nabla f(X_i)||}$$

where $P_1 = I - N_1[N_1^T N_1]^{-1} N_1^T$

and $N_p = \left[\nabla g_{j1}(X_i) \nabla g_{j2}(X_i).... \nabla g_{jp}(X_i) \right]$

STEP 2. Here, we also have following two cases :

Case-1 : If $S_j = 0$, find the vector λ at X_i such that
$$\lambda = -\left[N_p^T N_p \right]^{-1} N_p^T \nabla f(X_i)$$

Here, if each component of λ is non-negative then X_i is the optimum value and then stop the iterations.

If some of the components of λ are negative then identify the component λ_q with the most negative value and form a new matrix N_p as
$$[\nabla g_{j1}(X_i) \nabla g_{j2}(X_i)...\nabla g_{j(q-1)}(X_i) \nabla g_{j(q+1)}(X_i)]$$
and go to step-1.

Case-2 : If $S_j \neq 0$, then evaluate the maximum step length λ_m where
$$\lambda_m = \min \{\lambda_k\}, \ \lambda_k > 0 \text{ and } k \text{ is any integer between 1 to } m$$
$$\text{other than } j_1, j_2, ..., j_p$$

Then calculate $\dfrac{df}{d\lambda}$ at $\lambda = \lambda_m$

If at $\lambda = \lambda_m, \dfrac{df}{d\lambda} = 0$ or negative, take the step length $\lambda_i = \lambda_m$.

If at $\lambda = \lambda_m, \dfrac{df}{d\lambda}$ is positive, then find the minimum step length $\lambda = \lambda_i$ by putting

$\dfrac{df}{d\lambda} = 0$

Then optimum point becomes $X_{i+1} = X_i + \lambda_i S_i$

Further,

(i) If $\lambda = \lambda_m$ or $\lambda_m \leq \lambda_i^*$ some new constraints become active at X_{i+1}. Then generate the new matrix N_e which include the gradient of all the active constraints at the point X_{i+1}.

Fix the new iteration number as $i = i+1$ and go to case-2 of step-1.

(ii) If $\lambda_i = \lambda_i^*$ and $\lambda_i^* < \lambda_m$, no new constraints will be active at X_{i+1}. Then the matrix N_p remains unchanged.

Fix the new iteration number as $i = i +1$ and go to case-1 of step-1.

Solved Examples

EXAMPLE 1. *Use the Rosen's gradient projection method to solve the following problem*

$$\textit{Minimize} \ \ f(X) = x_1^2 + x_2^2 - 2x_1 - 4x_2$$

subject to the constraints

$$g_1(x_1, x_2) = x_1 + 4x_2 - 5 \le 0$$
$$g_2(x_1, x_2) = 2x_1 + 3x_2 - 6 \le 0$$
$$g_3(x_1, x_2) = -x_1 \le 0$$
$$g_4(x_1, x_2) = -x_2 \le 0$$

starting from the point $X = \begin{bmatrix} 1 \\ 1 \end{bmatrix}$

SOLUTION. We have

$$f(X) = x_1^2 + x_2^2 - 2x_1 - 4x_2 \qquad \qquad \text{...(1)}$$

$$\left.\begin{array}{l} g_1(x_1, x_2) = x_1 + 4x_2 - 5 \\ g_2(x_1, x_2) = 2x_1 + 3x_2 - 6 \\ g_3(x_1, x_2) = -x_1 \\ \text{and } \ g_4(x_1, x_2) = -x_2 \end{array}\right\} \qquad \text{...(2)}$$

From (2)

$$g_1(1, 1) = 1 + 4 - 5 = 0 \text{ for } j = 1$$

$\Rightarrow$ only first constraint is active, $p = 1, j = 1$

$\therefore$ normalized search direction is given by

$$S_1 = \frac{-P_i \nabla f(X_i)}{||P_i \nabla f(X_i)||}$$

where

$$P_i = I - N_P [N_P^T N_P]^{-1} N_P^T$$

and

$$N_P = [\nabla g_{j1}(X_i) \nabla g_{j2}(X_i)....\nabla g_{jp}(X_i)]$$

$\Rightarrow$

$$N_1 = [\nabla g_1(X_1)] = \begin{bmatrix} 1 \\ 4 \end{bmatrix}$$

$$P_1 = I - N_1 [N_1^T N_1]^{-1} . N_1^T$$

$$= \begin{bmatrix} 1 & 0 \\ 0 & 1 \end{bmatrix} - \begin{bmatrix} 1 \\ 4 \end{bmatrix} \left[\begin{pmatrix} 1 \\ 4 \end{pmatrix} [1 \ 4] \right]^{-1} [1 \ 4]$$

$$= \begin{bmatrix} 1 & 0 \\ 0 & 1 \end{bmatrix} - \frac{1}{17} \begin{bmatrix} 1 \\ 4 \end{bmatrix} [1 \ 4] = \begin{bmatrix} 1 & 0 \\ 0 & 1 \end{bmatrix} - \frac{1}{17} \begin{bmatrix} 1 & 4 \\ 4 & 16 \end{bmatrix}$$

$$= \frac{1}{17} \begin{bmatrix} 16 & -4 \\ -4 & 1 \end{bmatrix}$$

and therefore, $\nabla f(X_1) = \begin{bmatrix} 2x_1 - 2 \\ 2x_2 - 4 \end{bmatrix}_{(1,1)} = \begin{bmatrix} 0 \\ -2 \end{bmatrix}$

Thus, $$S_1 = \frac{-P_1 \nabla f(X_1)}{||P_1 \nabla f(X_1)||}$$

$$= \dfrac{-\dfrac{1}{17}\begin{bmatrix} 16 & -4 \\ -4 & 1 \end{bmatrix}\begin{bmatrix} 0 \\ -2 \end{bmatrix}}{\left\| -\dfrac{1}{17}\begin{bmatrix} 16 & -4 \\ -4 & 1 \end{bmatrix}\begin{bmatrix} 0 \\ -2 \end{bmatrix} \right\|} = \begin{bmatrix} -0.9701 \\ 0.2425 \end{bmatrix}$$

$\Rightarrow$ $S_1 \neq 0$

Therefore, we find $\lambda_m = \min\{\lambda_k\}$, $\lambda_k > 0$ and k is any integer among 2, 3 and 4.

Set $X = \begin{bmatrix} x_1 \\ x_2 \end{bmatrix} = X_1 + \lambda S_1 = \begin{bmatrix} 1 - 0.9701\lambda \\ 1 + 0.2425\lambda \end{bmatrix}$

Now,

for $j = 2$, $g_2(X) = (2 - 1.9402\lambda) + (3 + 0.7275) - 6 = 0$ at $\lambda = \lambda_2 = -0.8245$

For $j = 3$, $g_3(X) = -(1 - 0.9701\lambda) = 0$ at $\lambda = \lambda_3 = 1.03$

For $j = 4$, $g_4(X) = -(1 + 0.2425\lambda) = 0$ at $\lambda = \lambda_4 = -4.124$

Since λ_3 is the minimum positive value, thus we take

$$\lambda_m = \lambda_3 = 1.03$$

Further,

$$f(X) = f(\lambda) = (1 - 0.9701\,\lambda)^2 + (1 + 0.2425\,\lambda)^2 - 2(1 - 0.9701\,\lambda)$$
$$- 4(1 + 0.2425\lambda) - 0.9998\,\lambda^2 - 0.4850\,\lambda - 4 \quad \text{[By (1)]}$$

Now, $\dfrac{df}{d\lambda} = -1.9996\lambda - 0.4850$

$\therefore$ At $\lambda = \lambda_m$, $\dfrac{df}{d\lambda} = 1.9996(1.03) - 0.4850 = 1.5746 > 0$

Therefore, we find minimum step length λ_1^* by putting

$$\dfrac{df}{d\lambda} = 0$$

which gives $\lambda_1 = \lambda_1^* = \dfrac{0.4850}{1.9996} = 0.2425$

Then minimum point will be given by

$$X_2 = X_1 + \lambda_1 S_1$$
$$= \begin{bmatrix} 1 \\ 1 \end{bmatrix} + 0.2425 \begin{bmatrix} -0.9701 \\ 0.2425 \end{bmatrix} = \begin{bmatrix} 0.764701 \\ 1.0588 \end{bmatrix}$$

Here $\lambda_1 = \lambda_1^*$ and $\lambda_1^* < \lambda_m \Rightarrow$ No new constraints will be active at X_2 and hence the matrix N_p remains unchanged.

Further, fix the iteration number as $i = 2$.

Now, since $g_2(X_2) = 0$, we set $p = 1$, $j = 1$, then normalized search direction S_2 is given by $S_2 = \dfrac{-P_2 \nabla f(X_2)}{\|P_2 \nabla f(X_2)\|}$

Here, we have

$$N_2 = \begin{bmatrix} 1 \\ 4 \end{bmatrix}, \quad P_2 = \dfrac{1}{17}\begin{bmatrix} 16 & -4 \\ -4 & 1 \end{bmatrix}$$

$$\nabla f(X_2) = \begin{bmatrix} 2x_1 - 2 \\ 2x_2 - 4 \end{bmatrix}_{X_2} = \begin{bmatrix} -0.4706 \\ -1.8824 \end{bmatrix}$$

Therefore

$$P_2\nabla f(X_2) = \frac{1}{17}\begin{bmatrix} 16 & -4 \\ -4 & 1 \end{bmatrix}\begin{bmatrix} -0.4706 \\ -1.8824 \end{bmatrix} = \begin{bmatrix} 0 \\ 0 \end{bmatrix}$$

$\Rightarrow \qquad S_2 = 0$

Thus we compute the step length λ at X_2 such that

$$\lambda = -[N_P^T N_P]^{-1} N_P^T \nabla f(X_i) = -\frac{1}{17}[1 \;\; 4]\begin{bmatrix} -0.4706 \\ -1.8824 \end{bmatrix}$$

$$= 0.4707 > 0$$

So, the non-negative value of λ indicates that we have obtained the optimum value and so stop further iteration.

Hence,

$$X_2 = \begin{bmatrix} 0.7647 \\ 1.0588 \end{bmatrix} \text{ and } f_{\min} = -4.059$$

26.12 INDIRECT METHODS

In using the technique of indirect-methods for solving non-linear programming problem, the function to be minimized must be differentiable.

In this section, we will discuss the following indirect methods.

26.12.1 TRANSFORMATION TECHNIQUES

Consider the constraints $g_i(X)$ of a non-linear programming problem, which have some simple forms in independent variables, then it may be possible to make use of the transformation of these variables in such a way that constraints are satisfied automatically.

Hence, it is possible to convert a constrained optimization problem into an unconstrained one by transforming the variables.

Here, we are giving some typical transformations of independent variables as given below

(i) Let l_i and u_i are the lower and upper bound of x_i such that $l_i \le x_i \le u_i$

Then these can be satisfied by transforming the variables x_i as given below

$$x_i = l_i + (u_i - l_i)\sin^2 y_i \qquad \qquad ...(1)$$

where y_i is the new variable, which may take any value.

(ii) If $x_i \in (0,1)$, then use any one of the following transformation

(a) $\qquad x_i = \sin^2 y_i$

(b) $\qquad x_i = \cos^2 y_i$

(c) $\qquad x_i = \dfrac{e^{y_i}}{e^{y_i} + e^{-y_i}}$ $\qquad\qquad ...(2)$

(d) $\qquad x_i = \dfrac{y_i^2}{1 + y_i^2}$

(iii) If decision (design) variable is restricted to assume only positive values then use any one of the following transformations

(a) $\qquad x_i = |y_i|$

(b) $\qquad x_i = y_i^2$ $\qquad\qquad ...(3)$

(c) $\qquad x_i = e^{y_i}$

(iv) If $x_i \in (-1, 1)$, then we can use any one of the following transformations :

(a) $\qquad x_i = \sin y_i$

(b) $\qquad x_i = \cos y_i$

(c) $\qquad x_i = \dfrac{2y_i}{1 + y_i^2}$ $\qquad$ ...(4)

☞ REMARKS

- To use the transformed transformation, the constrained function $g_i(X)$ should be simple.
- If it is not possible to eliminate all the constraints by changing the variables, then its better not to use the transformation method.

Solved Examples

EXAMPLE 1. *A courier service does not accept rectangular packets of more than 42 cm length. If the maximum [length + 2(width + height)] is 72 cms. Compute the maximum volume of the rectangular packet.*

SOLUTION. Let us suppose that x_1, x_2 and x_3 be the length, width and height of a rectangular packets respectively. According to given question, we formulate the problem as given below

$$\text{Max. } f(X) = x_1 x_2 x_3 \qquad ...(1)$$

subject to the constraints

$$x_1 + 2x_2 + 2x_3 \leq 72 \qquad ...(2)$$

$$x_1 \leq 42 \qquad ...(3)$$

and $\qquad x_1, x_2, x_3 \geq 0 \qquad ...(4)$

Let us use the following transformation

$$y_1 = x_1, y_2 = x_2 \text{ and } y_3 = x_1 + 2x_2 + 2x_3$$

$\Rightarrow \qquad x_3 = \dfrac{1}{2}(y_3 - y_1 - 2y_2) \qquad ...(5)$

Then constraints (1), (2) and (3) can be written as

$$\left. \begin{array}{l} 0 \leq y_1 < 42 \\ 0 \leq y_2 \leq 36 \\ 0 \leq y_3 \leq 72 \end{array} \right] \qquad ...(6)$$

Clearly the upper bound for y_i's in (6) can be easily obtained say for y_2 taking $x_1 = x_2 = 0$ in (2), we get

$$2x_2 \leq 72 \qquad \text{so } x_2 = y_2 \leq 36.$$

Now, constraints in (6) are automatically satisfied, if we define z_1, z_2 and z_3 as given below

$$y_i = l_i + (u_i - l_i) \sin^2 z_i \qquad ...(7)$$

Then from (6) and (7), we get

$$y_1 = l_1 + (u_1 - l_1) \sin^2 z_1 = 42 \sin^2 z_1 \qquad \text{...(8)}$$
$$y_2 = l_2 + (u_2 - l_2) \sin^2 z_2 = 36 \sin^2 z_2 \qquad \text{...(9)}$$
$$y_3 = l_3 + (u_3 - l_3) \sin^2 z_3 = 72 \sin^2 z_3 \qquad \text{...(10)}$$

Then original problem reduces to

$$\text{Max. } f = \frac{1}{2} y_1 y_2 (y_3 - y_1 - 2 y_2)$$

$$= \frac{1}{2} (42 \sin^2 z_1)(36 \sin^2 z_2).(72 \sin^2 z_3 - 42 \sin^2 z_1 - 72 \sin^2 z_2)$$

subject to the constraints

$$0 \le \sin^2 z_i \le 1, \ i = 1, 2, 3 \qquad \text{...(11)}$$

Now, using the principle of maxima and minima, for the minimum of f, we must have

$$\frac{\partial f}{\partial z_1} = 0 \Rightarrow \sin z_1 \cos z_1 \sin^2 z_2 \left(\sin^2 z_3 - \frac{7}{6} \sin^2 z_1 - \sin^2 z_2 \right) = 0 \qquad \text{...(12)}$$

$$\frac{\partial f}{\partial z_2} = 0 \Rightarrow \sin^2 z_1 \sin z_2 \cos z_2 \left(\sin^2 z_3 - \frac{7}{12} \sin^2 z_1 - 2 \sin^2 z_2 \right) = 0 \qquad \text{...(13)}$$

and

$$\frac{\partial f}{\partial z_3} = 0 \Rightarrow \sin^2 z_1 \sin^2 z_2 \sin z_3 \cos z_3 = 0 \qquad \text{...(14)}$$

Therefore, we have

$$\sin^2 z_1 = 0 \Rightarrow z_1 = 0 \Rightarrow y_1 = 0 \Rightarrow f = 0 \qquad \text{(Not acceptable)}$$
$$\sin^2 z_2 = 0 \Rightarrow z_2 = 0 \Rightarrow y_2 = 0 \Rightarrow f = 0 \qquad \text{(Not acceptable)}$$
$$\sin z_3 = 0 \Rightarrow z_3 = 0 \Rightarrow y_3 = 0 \Rightarrow f = 0 \qquad \text{(Not acceptable)}$$

$$\cos z_3 = 0 \Rightarrow z_3 = \frac{\pi}{2}$$

Then from (12)

$$\sin^2 z_3 - \frac{7}{6} \sin^2 z_1 - \sin^2 z_2 = 0$$
$$\Rightarrow \qquad 7 \sin^2 z_1 + 6 \sin^2 z_2 = 6 \qquad \text{...(15)}$$

and from (13)

$$\sin^2 z_3 - \frac{7}{12} \sin^2 z_1 - 2 \sin^2 z_2 = 0 \ \Rightarrow \ 7 \sin^2 z_1 + 24 \sin^2 z_2 = 12 \qquad \text{...(16)}$$

On solving (15) and (16), we get

$$\sin^2 z_2 = \frac{1}{3}$$

and

$$7 \sin^2 z_1 + 6 \left(\frac{1}{3} \right) = 6 \Rightarrow \sin^2 z_1 = \frac{4}{7}$$

Hence,

$$y_1 = 42 \left(\frac{4}{7} \right) = 24$$

$$y_2 = 36\left(\frac{1}{3}\right) = 12$$

$$y_3 = 72$$

which implies that $x_1 = 24,\ x_2 = 12, x_3 = 12$

and $\qquad\qquad$ max. $f = 3456$ cm^3.

26.12.2 PENALTY FUNCTION METHODS

Penalty function methods transforms the basic optimization problem into alternative formulation. In these methods, we transform the given problem into a sequence of problems each with no constraints.

To understand it, consider the following problem

$$\left.\begin{array}{l}\text{Minimum } f(X) \\[6pt] \text{subject to the constraints} \\[4pt] \qquad g_j(X) \le 0,\ \ j = 1,2,...,m \\[6pt] \text{and} \qquad h_i(X) = 0,\ i = 1,2,...,p\end{array}\right\} \qquad\qquad ...(1)$$

We know that penalty function method transform (1) into a sequence of problems each with no constraints. Here, penalty function method create the effect of constraints by bringing in the modification in the objective function (as done in Big-M method of artificial variable technique).

Therefore, to solve (1), let us introduce an auxiliary unconstrained function given by

$$F(X, r_k) = f(X) + P(X, r_k) \qquad\qquad ...(2)$$

where $P(X, r_k)$ is a function of constraints $g_j(X)$ and $h_i(X)$ and r_k is a positive parameter such that

$$\lim_{r_k \to 0} \min F(X, r_k) = \min f(X)$$

Now, we will formulate the penalty function.

There are following two types of penalty function method.

(i) **Interior Penalty function method :** In interior penalty function method, the initial point as well as each of the subsequent points generated, lie inside the acceptable region of the design space.

In this method, the form of the penalty function $P(X, r_k)$, which is a function of constraints functions $g_j(X)$ and $h_i(X)$ and a positive parameter r_k is given by

$$P(X, r_k) = -r_k \sum_{j=1}^{m} \frac{1}{g_j(X)} + \frac{1}{\sqrt{r_k}} \sum_{i=1}^{p} h_i^2(X)$$

Here, the minimum of the auxiliary function $F(X, r_k)$ as defined by (2) approaches the minimum of the objective function $f(X)$ from point inside the feasible region as $r_k \to 0$.

☛ REMARKS

- Interior penalty function method is also known as barrier method.
- Once the unconstrained minimization of $P(X, r_k)$ is started from any feasible point X_i the subsequent points generated will always lie within the feasible domain.
- The value of the function P will always be greater than f, since $g_j(X)$ is negative for all feasible point X.

(ii) **Exterior Penalty function method :** In the exterior penalty function method, the function P is generally taken as follows

$$P(X, r_k) = \frac{1}{r_k} \sum_{j=1}^{m} \left[\max(0, g_j(X)) \right]^2 + \frac{1}{r_k} \sum_{i=1}^{p} h_i^2(X)$$

and minimize the auxiliary function $F(X, r_k)$ for a sequence of decreasing values of r_k i.e., the minimum of $F(X, r_k)$ approaches to minimum of $f(X)$ from the point of infeasible region as $r_k \to 0$.

The value max. $\{0, g_j(X)\}$ can be defined as follows

$$\langle g_j(X) \rangle = \max.\{0, g_j(X)\} = \begin{cases} g_j(X), & \text{if } g_j(X) > 0 \text{ (constraint is violent)} \\ 0, & \text{if } g_j(X) \leq 0 \text{ (constraint is satisfied)} \end{cases}$$

☛ REMARKS

- Usually the function $P(X, r_k)$ possesses a minimum as a function of X in the infeasible region.
- From above, we observe that there will be a penalty for violating the constraints and the amount of penalty will increase at a faster rate than the amount of violation of a constant that is why it is called exterior penalty function method.

Solved Examples

EXAMPLE 1. *Using interior penalty method, solve*

$$\textbf{\textit{Minimize f(X) = 2x}}$$

subject to the constraints

$$\textbf{\textit{x}} \geq \textbf{3}$$

SOLUTION. We have to find

$$\text{Min } f(X) = 2x$$

subject to $g_i(x) = 3 - x \leq 0$

Clearly this is the problem of one variable.

Using the formula of unconstrained interior penalty function method, the auxiliary unconstrained problem can be written as

$$\text{Min } F(X, r_k) = 2x - r_k \left(\frac{1}{3-x} \right)$$

such that $\quad \lim_{r_k \to 0} \min F(X, r_k) = \min f(X)$

Using the principle of maxima and minima, for the minimization of F, we must have

$$\frac{\partial F}{\partial x} = 0 \Rightarrow 2 - \frac{r_k}{(3-x)^2} = 0 \Rightarrow x = 3 + \sqrt{\frac{r_1}{2}}$$

and $\qquad \dfrac{\partial^2 F}{\partial x^2} = \dfrac{-2r_2}{(3-x)^3} > 0 \text{ at } x = 3 + \sqrt{\dfrac{r_1}{2}}$

$\Rightarrow F$ is minimum at $x = 3 + \sqrt{\dfrac{r_1}{2}}$

Now as $r_k \to 0$ at $x = 3$

$$\lim_{r_k \to 0} \min F(X, r_k) = \min f(X) = 6 \text{ when } x = 3$$

EXAMPLE 2. *Using the exterior penalty function method, minimize the function $f(X)$ given by*

$$f(X) = x_1^2 + 2x_2^2$$

subject to the constraints

$$2x_1 + 5x_2 \leq 10$$

SOLUTION. We have

$$f(X) = x_1^2 + 2x_2^2 \qquad \qquad ...(1)$$

and $\qquad g_i(x) = 2x_1 - 5x_2 - 10 \leq 0 \qquad \qquad ...(2)$

Using the formula of exterior penalty function method the auxiliary function $F(X, r_k)$ can be written as

$$\text{Min.} \, F(X, r_k) = x_1^2 + 2x_2^2 + \frac{1}{r_k}[\max(0, 2x_1 + 5x_2 - 10)]^2$$

Using the principle of maxima and minima, for the minimization of $F\left(r = \dfrac{1}{r_k}\right)$, we must have

$$\frac{\partial F}{\partial x_1} = 0 \Rightarrow 2x_1 + \frac{4}{r_k}(2x_1 + 5x_2 - 10) = 0$$

$$\Rightarrow \qquad (2 + 8r)x_1 + 20rx_2 - 40r = 0 \qquad \qquad ...(3)$$

and $\qquad \dfrac{\partial F}{\partial x_2} = 0 \Rightarrow 4x_2 + 10r(2x_1 + 5x_2 - 10) = 0$

$$\Rightarrow \qquad 20rx_1 + (4 + 50r)x_2 - 100r = 0 \qquad \qquad ...(4)$$

On solving (3) and (4), we get

$$\frac{x_1}{160r} = \frac{x_2}{200r} = \frac{1}{8 + 132r}$$

$$\Rightarrow \qquad x_1 = \frac{40r}{2 + 33r} = \frac{40}{2r_k + 33}$$

and $\qquad x_2 = \dfrac{50r}{2 + 33r} = \dfrac{50}{2r_k + 33}$

As $r_k \to 0$, we get

$$x_1 = \frac{40}{33} \text{ and } x_2 = \frac{50}{33}$$

and Minimum of $f(X) = \left(\dfrac{40}{33}\right)^2 + 2\left(\dfrac{50}{33}\right)^2 = \dfrac{200}{33}$

EXAMPLE 3. *Using penalty function method minimize the function*

$$f(X) = f(x_1, x_2) = (x_1 + 2)^3 + 3x_2 + 1$$

subject to the constraints

$$x_1 \geq 2, \, x_2 \geq 0$$

SOLUTION. We can write the given problem as

$$f(X) = (x_1 + 2)^2 + 3x_2 + 1$$

subject to $2 - x_1 \leq 0, \, -x_2 \leq 0$

Therefore, the auxiliary unconstrained problem becomes

Minimize $F(X, r_k) = (x_1 + 2)^3 + 3x_2 + 1 - r_k \left(\dfrac{1}{2 - x_1} - \dfrac{1}{x_2} \right)$

Now, for the minimization of F, we must have

$$\frac{\partial F}{\partial x_1} = 0 \Rightarrow 3(x_1 + 2)^2 - \frac{r_k}{(2 - x_1)^2} = 0$$

$\Rightarrow \qquad\qquad x_1 = \left(4 + \sqrt{\dfrac{r_k}{3}} \right)^{1/2}$

and $\qquad\qquad \dfrac{\partial F}{\partial x_2} = 0 \Rightarrow 3 - \sqrt{\dfrac{r_k}{x_2^2}} = 0$

$\Rightarrow \qquad\qquad x_2 = \sqrt{\dfrac{r_k}{3}}$

Therefore, $x_1 = \left(4 + \sqrt{\dfrac{r_k}{3}} \right)^{1/2}$ and $x_2 = \sqrt{\dfrac{r_k}{3}}$ are the possible feasible values of x_1 and x_2.

Now as $r_k \to 0$, we get
$$x_1 = 4, x_2 = 0$$
and min. $f(X) = 65$ and $x_1 = 4, x_2 = 0$

EXAMPLE 4. *Using penalty function method,*
$$\textbf{\textit{Minimize }} f(x_1, x_2) = \frac{1}{3}(x_1 + 1)^3 + x_2$$

subject to the constraints
$$g_1(x_1, x_2) = 1 - x_1 \leq 0$$
and $\qquad g_2(x_1, x_2) = -x_2 \leq 0$

SOLUTION. Proceeding same as above, we get the auxiliary unconstrained function $F(X, r_k)$ as given below

$$F(X, r_k) = \frac{1}{3}(x_1 + 1)^3 + x_2 + r_k \left[\max(0, 1 - x_1) \right]^2 + r_k [\max(0, -x_2)]^2$$

For the minimum of F, we must have

$$\frac{\partial F}{\partial x_1} = 0 \Rightarrow (x_1 + 1)^2 - 2r_k [\max(0, 1 - x_1)] = 0$$

and $\qquad \dfrac{\partial F}{\partial x_2} = 0 \Rightarrow 1 - 2r [\max(0, -x_2)] = 0$

The above equations can be written as
$$\min\left[(x_1 + 1)^2, (x_1 + 1)^2 - 2r_k(1 - x_1) \right] = 0 \qquad\qquad \ldots(1)$$
and $\qquad \min[1, 1 + 2r_k x_2] = 0 \qquad\qquad \ldots(2)$

In equation (1), if $(x_1 + 1)^2 = 0 \Rightarrow x_1 = -1$, which violates the first constraints and if
$$(1 + 1)^2 - 2r_k(1 - x_1) = 0 \Rightarrow x_1 = -1 - r_k + \sqrt{r_k^2 + 4r_k}$$
while in equation (2), the only possibility is that

$$1 + 2r_k x_2 = 0 \Rightarrow x_2 = -\frac{1}{2r_k}$$

Therefore, the solution of the unconstrained minimization problem is given by

$$x_1^*(r_k) = -1 - r_k + r_k \left(1 + \frac{4}{r_k}\right)^{1/2} \qquad \text{...(3)}$$

and

$$x_2^*(r_k) = -\frac{1}{2r_k} \qquad \text{...(4)}$$

Hence, the optimum points are

$$x_1^* = \lim_{r_k \to \infty} x_1^*(r_k) = 1 \text{ and } x_2^* = \lim_{r_k \to \infty} x_2^*(r_k) = 0$$

and

$$\text{Min.} f(X) = \frac{8}{3} \qquad \text{(see remark)}$$

☛ REMARK

- If the function $F(X, r_k)$ is minimized for an increasing sequence of values of r_k, the unconstrained minima x_k^* converges to the optimum solution of the constrained problem $r_k \to \infty$.

Exercise-26.2

1. Minimize $f(X) = x_1^2 + x_2^2 - 6x_1 - 8x_2 + 10$
   subject to the constraints
   $$4x_1^2 + x_2^2 \le 16, \ 3x_1 + 5x_2 \le 15$$
   Using
   (i) interior penalty function method.
   (ii) exterior penalty function method.

2. Using complex method
   Minimize $f(X) = (x_1 - 1)^2 + (x_2 - 2)^2$
   subject to the constraints
   $x_1 + x_2 \le 4, x_1 - x_2 \le 2, x_1 \ge 0, x_2 \ge 0$ with
   $$X_1 = \begin{bmatrix} 0.5 \\ 1.5 \end{bmatrix}$$

3. Find the dimensions of a rectangular prism type box that has the largest volume when the sum of its length, width and height is limited to a maximum value of 60 cm and its length is restricted to a maximum value of 36 cms.

4. Consider the problem
   Minimize $f = x_1^2 + x_2^2 - 6x_1 - 8x_2 + 15$
   subject to the constraints
   $$4x_1^2 + x_2^2 \ge 16, 3x_1 + 5x_2 \le 15$$
   Normalize the constraints and find a suitable value of r_1 for use in the interior penalty function method at the starting point
   $$X_1 = \begin{bmatrix} 0 \\ 0 \end{bmatrix}.$$

ANSWERS

1. $x_1 = 3, x_2 = 4$

2. $x_1 = 1, x_2 = 2$

3. $x_1 = 20, x_2 = 20, x_3 = 20, \text{max.} f = 8000 \text{ cm}^3$

4. $\frac{1}{4}x_1^2 + \frac{1}{16}x_2^2 - 1 \le 0, \frac{x_1}{5} + \frac{x_2}{3} - 1 \le 0, r_1 = 1.5$

REVIEW QUESTIONS

1. Define each of the following:
   (i) Gradient of a function
   (ii) Conjugate direction
   (iii) Pattern direction

2. Write the characteristics of a direct search method.

3. Give the importance of the study of unconstrained minimization methods.

4. Write the necessary and sufficient conditions for the unconstrained minimum of a function.

5. State convergence criterion used in direct search method.

6. How are the search directions generated in Fletcher-Jeeves method?

7. Why Powell's method called pattern search method?

8. What is a descent method?

9. Explain the complex method.

10. Describe the following methods
   (i) Zoutendijk method
   (ii) Cutting plane method
   (iii) Rosen's method

11. Write the difference between interior and exterior penalty function method.

12. How is the direction finding problem solved in Zoutendijk's method.

13. Write a short note on
   (i) Complex method
   (ii) Penalty method

14. Describe in details, the procedure used in the following methods
   (i) Complex method
   (ii) Zoutendijk's method
   (iii) Rosen's gradient search method
   (iv) Penalty functions method

15. Discuss the transformation of variable techniques in detail.

16. Write a short note on Random search method.

MULTIPLE CHOICE QUESTIONS (CHOOSE THE MOST APPROPRIATE ONE)

1. Which one of the following is a method of feasible direction in solving NLPP using search technique :
   (a) Zoutendijk method (b) Univariate method
   (c) Powel's method   (d) None of these

2. The search technique that uses a gradient of the function to be minimized in solving an NLPP is :
   (a) Zoutendijk method (b) Univariate method
   (c) Powel's method   (d) None of these

3. Which of the following is not a direct method:
   (a) Random search method
   (b) Complex method
   (c) Zoutendijk's method
   (d) None of these

4. Which of the following is not indirect method:
   (a) Transformation of variable technique
   (b) Penalty function method
   (c) Complex method
   (d) None of these

ANSWERS

1. (a)   2. (a)   3. (d)   4. (c)

ARCHIVE

1. Use the fibonacei's method o determine the maximum of $f(x) = x(5-x)$ given that $f(x)$ is a unimodal function in the interval $[0, 8]$, where in the maximum lies.

2. Minimize the function

$$f(x) = 0.65 - \frac{0.75}{1+x^2} - 0.65x \tan^{-1}\frac{1}{x}$$

Using the golden sectoin method in the interval $[0, 8]$ with $n = 6$.

3. Minimize

$$f(x) = f(x_1, x_2) = x_1^2 + 3x_2^2 + x_1 - 17x_2 - x_1 x_2$$

using the univariate method taking $X_1 = \begin{pmatrix} 0 \\ 0 \end{pmatrix}$ as the base point.

4. Show that the Newton's method, finds the minimum of a quadratic function in one iteration.

5. Minimize the function

$$f(x_1, x_2) = x_1^2 + x_2^2 - 4x_1 - 4x_2 + 8$$

subject to $g(x_1, x_2) = x_1 + 2x_2 - 4 \le 0$

with the starting point $X_1 = \begin{bmatrix} 0 \\ 0 \end{bmatrix}$

6. Minimize $f(x_1, x_2) = x_1^2 + x_2^2 - 2x_1 - 4x_2$

subject to the constraints

$g_1(x_1, x_2) = x_1 + 4x_2 - 5 \leq 0$

$g_2(x_1, x_2) = 2x_1 + 3x_2 - 6 \leq 0$

$g_3(x_1, x_2) = -x_1 \leq 0$

$g_4(x_1, x_2) = -x_2 \leq 0$

Starting from the base point $X_1 = \begin{bmatrix} 1 \\ 1 \end{bmatrix}$

HINTS AND ANSWERS

1. $x = 2.66667, f(x) = 6.22222$ **2.** $x = 0.4055, f(x) = -0.30658$ **3.** $x_1 = 1, x_2 = 3, f(x) = -25$

5. $X = \begin{bmatrix} 1.6 \\ 1.2 \end{bmatrix}, f_{\min} = 0.8$ **6.** $X_{\text{opt}} = \begin{bmatrix} 0.7647 \\ 1.0588 \end{bmatrix}, f_{\text{opt}} = -4.059$

Non-linear Programming: Formulation and Graphical Solutions

27

27.1 INTRODUCTION

It is a well known fact that the term non-linear programming problem (NLPP) refers to the problem in which the objective function becomes non-linear or one or more of the constraints have non-linear relationship or both. In this chapter we shall discuss the formulation of non-linear programming problem and their graphical solution.

27.2 GENERAL NON-LINEAR PROGRAMMING PROBLEMS (GNLPP)

The general form of a non-linear programming problem is defined as follows,

$$\text{find} \quad x_1, x_2, ..., x_n$$

which optimize the objective function

$$Z = f(x_1, x_2, ..., x_n)$$

subject to the constraints

$$g_1(x_1, x_2, ..., x_n) \; (\leq, = \text{ or } \geq) \; b_1$$
$$g_2(x_1, x_2, ..., x_n) \; (\leq, = \text{ or } \geq) \; b_2$$
$$\vdots \qquad \vdots \qquad \vdots$$
$$\vdots \qquad \vdots \qquad \vdots$$
$$g_m(x_1, x_2, ..., x_n) \; (\leq, = \text{ or } \geq) \; b_m$$

and non-negative restrictions

$$x_j \geq 0 \quad \forall \, j = 1, 2, .., n.$$

where, either $f(x_1, x_2, ..., x_n)$ or some $g_i(x_1, x_2, ..., x_n); i = 1, 2, ..., m$ or both are non-linear.

In matrix notations, the above NLPP can be written as below

$$\text{Determine } X^T = (x_1, x_2, ..., x_n)^T$$

which optimize the objective function $Z = f(X)$

subject to the constraints

$$g_i(X) \; (\leq, = \text{ or } \geq) \quad b_i \; ; i = 1, 2, ..., m$$

and $\qquad\qquad x_i \geq 0$

27.3 CANONICAL FORM OF NON-LINEAR PROGRAMMING PROBLEM

The canonical form of a non-linear programming problem can be written as follows,

$$\text{Max. } Z = C(x_1, x_2, ..., x_n)$$

subject to the constraints

$$a(x_1, x_2, ..., x_n) \leq 0; \quad i = 1, 2, ..., m$$

and
$$x_j \geq 0, \ j = 1, 2, ..., n$$

where at least one of the function $C(x_1, x_2, ..., x_n)$ and $a_i(x_1, x_2, ..., x_n)$ or both is non-linear.

☛ Remark

• Both the function $C(x)$ and $a(x)$ defined above should be non-linear.

27.4 MATHEMATICAL FORMULATION OF NON-LINEAR PROGRAMMING PROBLEMS

Following examples give the basic idea of the formation of general non-linear programming problem.

 Solved Examples

EXAMPLE 1. *A company manufactures two products A and B. It takes 30 minutes to process one unit of product A and 15 minutes for product B and the maximum machine time available is 35 hours per week. Product A and B require 2 kgs and 3 kgs of raw material per unit respectively. The available quantity of raw material is 180 kgs per week. The product A and B which have unlimited market potential sell for ₹ 200 and ₹ 500 per unit respectively. If the manufacturing costs for product A and B are $2x_1^2$ and $3x_2^2$ respectively, formulate the non-linear programming problem, where*

$$x_1 = quantity \ of \ product \ A \ to \ be \ produced$$
$$x_2 = quantity \ of \ product \ B \ to \ be \ produced$$

SOLUTION. According to the question,

The processing time of products A and B are 30 minutes *i.e.*, 0.5 hour and 15 minutes, *i.e.*, 0.25 hour respectively. So,

total processing time for x_1 and x_2 units of two products produces per week is $0.5 x_1 + 0.25 x_2$ which should be less than or equal to the total available time of 35 hour Thus, one constraint is

$$0.5 x_1 + 0.25 x_2 \leq 35$$

$$\Rightarrow \qquad 2 x_1 + x_2 \leq 140$$

Further, since the products requires 2 kgs and 3 kgs of raw material per unit respectively, therefore the total material required per week is $2x_1 + 3x_2$ which should be less than or equal to the total quantity of raw material available per week which is 180 kg.

Therefore, second constraint is given by

$$2 x_1 + 3x_2 \leq 180$$

If Z is the total profit of the company per week, then

$$Z = (200 x_1 - 2x_1^2) + (500 x_2 - 3x_2^2)$$

Hence, the non-linear programming problem can be written as,

$$\text{Maximize } Z = (200\, x_1 - 2x_1{}^2) + (500\, x_2 - 3x_2{}^2)$$

subject to the constraints

$$2x_1 + x_2 \le 140$$
$$2x_1 + 3x_2 \le 180$$

and
$$x_1, x_2 \ge 0$$

EXAMPLE 2. *An engineering company has received a rush order for a maximum no. of two types of items that can be produced and transported during a two week period. The profit in thousand rupees on this order is related to the number of each type of item manufactured by company and is given by*

$$12x_1 + 10x_2 - x_1^2 - x_2^2 + 61$$

where x_1 is the no. of units (in thousands) of type -I item and x_2 is the number of units (in thousands) of type-II item. Because of other commitment over the next two weeks, the company has available only 60 hours in the shifting and packing department. It is assumed that every thousand unit of type-I and II items will requires 20 hours and 30 hours respectively in the shifting and packing departments. Given the above information, formulate the non-linear programming problem.

SOLUTION. Proceed same as in example-1, we have the following non-linear programming problem

$$\text{Maximize.} \quad Z = 12x_1 + 10x_2 - x_1^2 - x_2^2 + 61$$

subject to the constraints

$$20\, x_1 + 30\, x_2 \le 60$$

and
$$x_1, x_2 \ge 0$$

EXAMPLE 3. *(Production Allocation Problem)*

A manufacturing company produces a product consisting of two raw material say A and B. The production function is estimated as

$$Z = f(x_1, x_2) = 3.6\, x_1 - 0.4\, x_1^2 + 1.6\, x_2 - 0.2\, x_2^2$$

where Z represents the quantity (in tones) of the product produced and x_1, x_2 design the input amount of raw material A and B. The company has ₹ 50000 to spend on these two raw materials. The unit price of A is ₹10,000 and of B is ₹ 5000. Formulate the problem.

SOLUTION. Let x_1, x_2 be the input amounts of the material A and B respectively. Then the company will have to spend ₹ 10,000 x_1 + 5000 x_2 on the two materials. Since company has ₹ 50000 to spend on the two raw materials A and B, therefore, we have

$$10000\, x_1 + 5000\, x_2 \le 50000$$

$\Rightarrow$ $\qquad$ $2x_1 + x_2 \leq 10$

Hence, the NLPP is given as follows

Maximize $Z = f(x_1, x_2) = 3.6x_1 - 0.4x_1^2 + 1.6x_2 - 0.2x_2^2$

subject to the constraint

$$2x_1 + x_2 \leq 10$$

and $\qquad\qquad$ $x_1, x_2 \geq 0$

EXAMPLE 4. *A manufacturing company produces two products Radios and T.V. sets. Sales price relationship for these two products are given below*

Product	Quantity demanded	Unit price
Radios	$1500 - 5p_1$	p_1
T.V.	$3800 - 10\,p_2$	p_2

The total cost functions for these two products are given by $200x_1 + 0.1x_1^2$ and $300x_2 + 0.1x_2^2$ respectively. The production takes place on two assembly lines. Radio sets are assembled on assembly line-1 and T.V. sets are assembled line-II. Because of the limitations of the line capacity, the daily production is limited to no more than 80 radio sets and 60 T.V. sets. The product of both types of products requires electronic components. The production of each of these sets requires 5 units and 6 units of electronic equipments component respectively. The electronic components are supplied by another manufacturer and the supply is limited to 600 units per day. The production of one unit of radio set requires 1 man-day of labour, whereas 2 men-days of labour are required for a T.V. set. How many units of radios and T.V. sets should the company produce in order to maximize the total profit. Formulate the problem as a NLPP. (Agra-2002)

SOLUTION. Let us assume that the company manufacture x_1 and x_2 units of radio and T.V. sets respectively.

So, $\qquad\qquad$ $x_1 = 1500 - 5p_1$ $\quad$ and $x_2 = 3800 - 10\,p_2$

$\Rightarrow$ $\qquad\qquad$ $p_1 = 300 - 0.2\,x_1$ and $p_2 = 380 - 0.1\,x_2$

Now, total revenue received by the company is given by

$\qquad$ $R = p_1 x_1 + p_2 x_2 = (300 - 0.2\,x_1)x_1 + (380 - 0.1\,x_2)x_2$

and total cost of production

$\qquad\qquad$ = cost of production of radio sets + cost of production of TV sets

$\qquad\qquad$ $= (200x_1 + 0.1x_1^2) + (300x_2 + 0.1x_2^2)$

Now, if Z is the total profit of the company then

$\qquad$ Z = total revenue – total cost of production

$\qquad\qquad$ $= [(300 - 0.2x_1)x_1 + (380 - 0.1x_2)x_2] - [(200x_1 + 0.1x_1^2) + (300x_2 + 0.1x_2^2)]$

$$= 100x_1 + 80x_2 - 0.3x_1^2 - 0.2x_2^2$$

Further, since the capacity of total production of the assembly line is limited to no more than 80 radio sets and 60 T.V. sets, thus the constraints on x_1 and x_2 are given by

$$x_1 \le 80 \text{ and } x_2 \le 60$$

Also, since 5 and 6 units of electronic equipments respectively are required for the production of radio sets and T.V. sets whereas the supply of these equipments is limited to 600 units per day.

So, total no. of electronic equipments required for production

$$= 5x_1 + 6x_2 \le 600$$

and, since the production of one unit of radio set and one unit of T.V. set require 1 man-day and 2 man-day labour respectively whereas the labour supply is limited to 160 man-days.

$\Rightarrow$ Total man hour of labour required $= 1. x_1 + 2x_2 \le 160$

Also $x_1 \ge 0$, $x_2 \ge 0$ (because negative no. of sets cannot be produced)

Hence, the required NLPP is given as follows :

$$\text{Max}. Z = 100x_1 + 80x_2 - 0.3x_1^2 - 0.2x_2^2$$

subject to the constraints

$$5x_1 + 6x_2 \le 600$$
$$x_1 + 2x_1 \le 160$$
$$x_1 \le 80$$
$$x_2 \le 60$$

and

$$x_1, x_2 \ge 0$$

27.5 GRAPHICAL SOLUTION OF A NON-LINEAR PROGRAMMING PROBLEM

In a linear programming problem, the optimal solution was usually obtained at one of the extremities of the convex region generated by the constraints and the objective function of the problem. But it is not necessary to find the solution at a corner or edge of the feasible region of non-linear programming problem.

The following examples will make the method clear.

 Solved Examples

EXAMPLE 1. ***Solve the following NLPP by graphical method***

$$\text{Maximize } Z = x_1 + 2x_2$$

subject to the constraints

$$x_1^2 + x_2^2 \le 1$$
$$2x_1 + x_2 \le 2$$

and $$x_1, x_2 \ge 0$$

SOLUTION. Clearly the constraints of the given non-linear programming problem are given by

$$x_1^2 + x_2^2 \le 1$$
$$2x_1 + x_2 \le 2$$

and $$x_1, x_2 \geq 0$$

Considering these constraints as equalities, we get

$$x_1^2 + x_2^2 = 1 \quad \text{(Circle)}$$

and $$2x_1 + x_2 = 2 \quad \text{(Straight line)}$$

Also $$Z = x_1 + 2x_2 = 0 \Rightarrow \frac{x_1}{x_2} = \frac{-2}{1}$$

Drawing these in the plane, we get the following figure

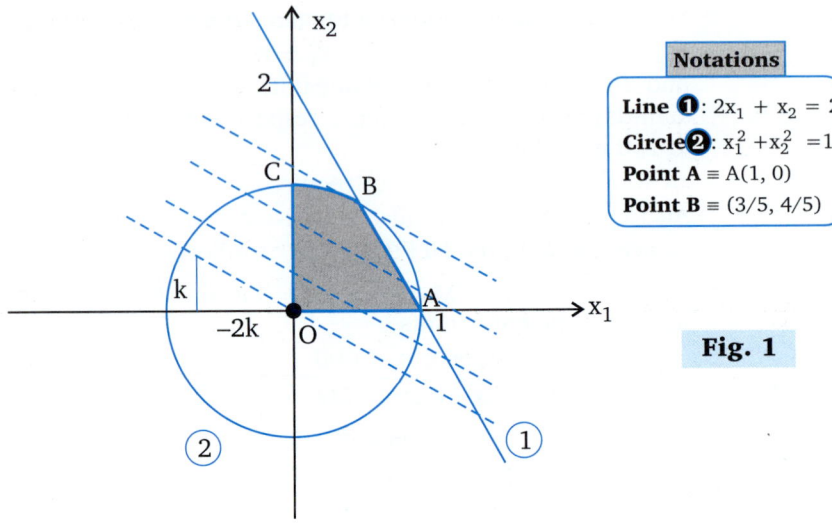

Notations

Line **❶**: $2x_1 + x_2 = 2$
Circle **❷**: $x_1^2 + x_2^2 = 1$
Point A ≡ A(1, 0)
Point B ≡ (3/5, 4/5)

Fig. 1

In the above figure, drawing the line $Z = x_1 + 2x_2 = 0$ through the origin and draw parallel to this line till we reach the extremities B of the permissible region B is the most distance point of the permissible region through with the line parallel to $Z = 0$ passes.

Here, B is the point of intersection of the circle

$$x_1^2 + x_2^2 = 1 \text{ and the line } 2x_1 + x_2 = 0.$$

Therefore, at B

$$x_1 = \frac{3}{5} = 0.6$$

$$x_2 = \frac{4}{5} = 0.8$$

and $$Z = \frac{11}{5} = 2.2$$

Hence the optimal solution is given by

$$x_1 = 0.6, x_2 = 0.8 \text{ and Max. } Z = 2.2$$

EXAMPLE 2. *Solve the following non-linear programming problem graphically. Also verify the Kuhn-Tucker necessary condition for maximum of the function*

$$\textbf{Max. } Z = 8x_1 - x_1^2 + 8x_2 - x_2^2$$

subject to the constraints

$$x_1 + x_2 \leq 12$$

$$x_1 - x_2 \geq 4$$

and $\qquad x_1, x_2 \geq 0$ [AGRA-2009]

SOLUTION. Here, the constraints of the given NLPP are

$$x_1 + x_2 \leq 12$$

$$x_1 - x_2 \geq 4$$

$$x_1, x_2 \geq 0$$

Convert the above inequations into equalities and drawing these lines in the plane, we get the permissible region DABD.

Also, the objective function

$$Z = 8x_1 - x_1^2 + 8x_2 - x_2^2 \text{ is a circle with centre } (4, 4)$$

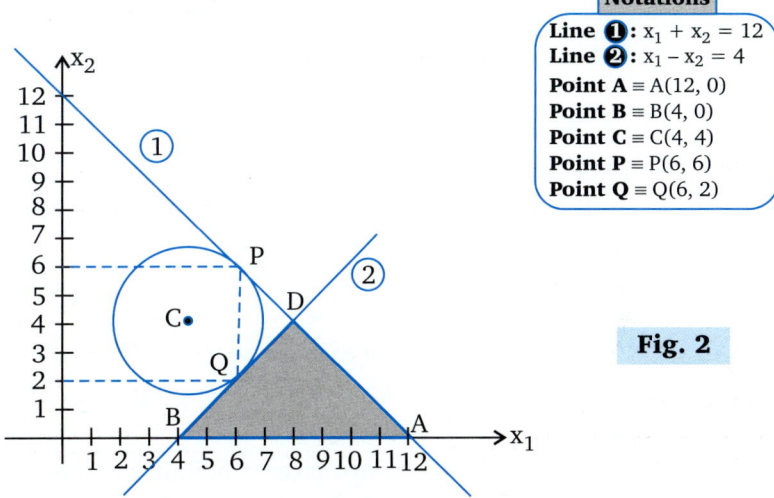

Notations

Line ①: $x_1 + x_2 = 12$
Line ②: $x_1 - x_2 = 4$
Point A $\equiv A(12, 0)$
Point B $\equiv B(4, 0)$
Point C $\equiv C(4, 4)$
Point P $\equiv P(6, 6)$
Point Q $\equiv Q(6, 2)$

Fig. 2

Now the point giving the maximum value of Z in the point at which the feasible region is tangent to the circle given by the objective function

$$Z = 8x_1 - x_1^2 + 8x_2 - x_2^2 \qquad \qquad ...(1)$$

The centre of this circle $C = C (4, 4)$

Differentiating (1) w.r.t. x_1, we get

$$0 = 8 - 2x_1 + 8\frac{dx_2}{dx_1} - 2x_2\frac{dx_2}{dx_1}$$

$$\Rightarrow \qquad \frac{dx_2}{dx_1} = \frac{2x_1 - 8}{8 - 2x_2} = \frac{x_1 - 4}{4 - x_2} = m_1 \text{ (say)}$$

Now, for the line $x_1 + x_2 = 12$, we have

$$\frac{dx_2}{dx_1} = -1 = m_2 \quad \text{(say)}$$

Clearly the circle touch this line, when

$$m_1 = m_2$$

$$\Rightarrow \qquad \frac{x_1 - 4}{4 - x_2} = -1 \Rightarrow x_2 = x_1$$

Putting this value in $x_1 + x_2 = 12$, we get

$$x_1 = x_2 = 6$$

$\Rightarrow$ the circle touches the line $x_1 + x_2 = 12$ at the point $P(6, 6)$

But the point P is not in the feasible region DABD

Again for the line $x_1 - x_2 = 4 \Rightarrow \dfrac{dx_2}{dx_1} = 1 = m_3 \quad \text{(say)}$

The circle touches this line at the point where

$$m_1 = m_3 \Rightarrow \frac{x_1 - 4}{4 - x_2} = 1 \Rightarrow x_2 = 8 - x_1$$

Using this in the line $x_1 - x_2 = 4$, we get

$$x_1 = 6 \Rightarrow x_2 = 2$$

$\Rightarrow$ the circle touches the line $x_1 - x_2 = 4$ at the point $Q\ (6, 2)$ which is the point in the feasible region DABD.

Also, for $x_1 = 6,\ x_2 = 2,\ Z = 24$, which is the required solution of given NLPP.

Kuhn-Tucker conditions:

We have

$$f(x) = 8x_1 - x_1^2 + 8x_2 - x_2^2$$

$$h_1(x) = x_1 + x_2 - 12 \leq 0$$

$$h_2(x) = 4 - x_1 + x_2 \leq 0$$

and $\qquad\qquad x_1, x_2 \geq 0$

We know that the Kuhn-Tucker condition for max. $f(x)$ are

$$\frac{\partial f(x)}{\partial x_j} = \sum_{i=1}^{2} \lambda_i \frac{\partial h_1(x)}{\partial x_j} \qquad j = 1, 2$$

$\lambda_1 h_1(x) = 0,\ \lambda_2 h_2(x) = 0,\ h_1(x) \leq 0,\ h_2(x) \leq 0$ and $\lambda_1, \lambda_2 \geq 0$

$$8 - 2 x_1 = \lambda_1 - \lambda_2 \qquad \qquad \text{...(2)}$$

$$8 - 2 x_2 = \lambda_1 + \lambda_2 \qquad \qquad \text{...(3)}$$

$$\lambda_1(x_1 + x_2 - 12) = 0 \qquad \qquad \text{...(4)}$$

$$\lambda_2(4 - x_1 + x_2) = 0 \qquad \qquad \text{...(5)}$$

$$x_1 + x_2 - 12 \leq 0 \qquad \qquad \text{...(6)}$$

$$4 - x_1 + x_2 \leq 0 \qquad \qquad \text{...(7)}$$

and $\qquad\qquad\qquad \lambda_1, \lambda_2 \geq 0 \qquad \qquad \text{...(8)}$

Clearly the point $(6, 2)$ satisfies conditions (6) and (7).

For these values from (2) and (3), we have

$$\lambda_1 - \lambda_2 = -4$$

and

$$\lambda_1 + \lambda_2 = 4$$

On solving we get, $\lambda_1 = 0, \lambda_2 = 4$

$\Rightarrow$ Condition (4), (5) and (8) are also satisfied.

Hence, the optimal solution $x_1 = 6, x_2 = 2$ obtained by graphical method satisfies all the Kuhn-Tucker condition for maximum of $f(x)$.

EXAMPLE 3. *Using graphical method, solve the following NLPP*

$$\textbf{Min.Z} = x_1^2 + x_2^2$$

subject to the constraints

$$x_1 + x_2 \geq 4$$

$$2x_1 + x_2 \geq 5$$

and

$$x_1, x_2 \geq 0$$

SOLUTION. Convert the above inquation into equalities , we get

$$x_1 + x_2 = 4 \text{ and } 2x_1 + x_2 = 5$$

Plot the above lines on the plane.

Since, $x_1, x_2 \geq 0$, the feasible region will lie in the first quadrant only.

Clearly, the constraints $x_1 + x_2 \geq 4$ is satisfied by all points lying in the region shaded by the vertical lines

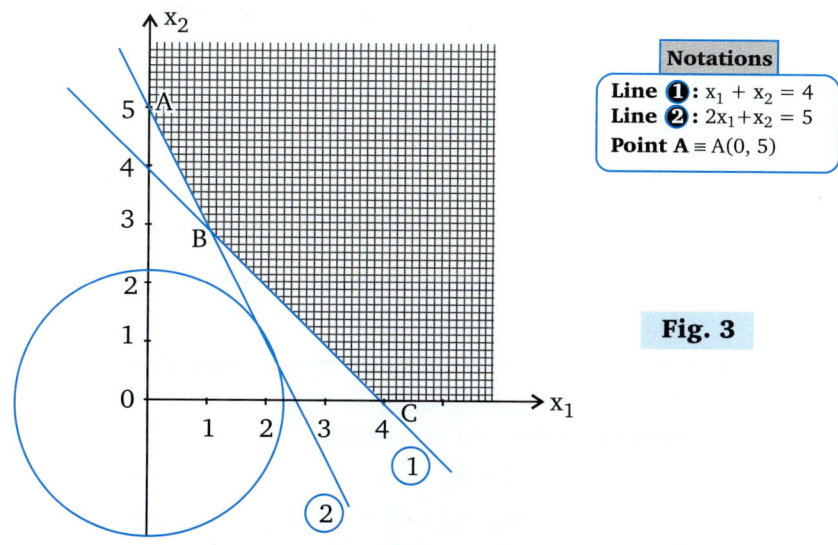

Notations

Line ❶: $x_1 + x_2 = 4$
Line ❷: $2x_1 + x_2 = 5$
Point A $\equiv$ A(0, 5)

Fig. 3

While the constraints $2x_1 + x_2 \geq 5$ is satisfied by all the points lying in the region shaded by horizontal lines only.

The shaded region by both the vertical and horizontal lines is unbounded convex feasible region x_2ABCx_1.

But we want to find a point (x_1, x_2) which gives a minimum value of $x_1^2 + x_2^2$ and lies in the convex region.

The desired point will be a point of the region at which a side of the convex region is tangent to the circle.

Now, the gradient of the circle $x_1^2 + x_2^2 = k$ (let $z = k$) is given by

$$2x_1 + 2x_2 \frac{dx_2}{dx_1} = 0 \Rightarrow \frac{dx_2}{dx_1} = -\frac{x_1}{x_2} \qquad \qquad ...(1)$$

Also, gradient of the line $x_1 + x_2 = 4$ is -1 and the gradient of the line $2x_1 + x_2 = 5$ is -2.

If the line $x_1 + x_2 = 4$ is the tangent to the circle $x_1^2 + x_2^2 = k$, then

$$\frac{dx_2}{dx_1} = -\frac{x_1}{x_2} = -1 \Rightarrow x_1 = x_2 \qquad \qquad ...(2)$$

and if the line $2x_1 + x_2 = 5$ is the tangent to the circle $x_1^2 + x_2^2 = k$, then

$$\frac{dx_2}{dx_1} = -\frac{x_1}{x_2} = -2 \Rightarrow x_1 = 2x_2 \qquad \qquad ...(3)$$

$\Rightarrow$ The point at which the line $x_1 + x_2 = 4$ is tangent to the circle is obtained by solving the equations $x_1 + x_2 = 4$ and $x_1 = x_2$ which gives $x_1 = 2, x_2 = 2$.

In a similar way, the point at which the line $2x_1 + x_2 = 5$ touches the circle is obtained by solving the equations

$2x_1 + x_2 = 5$ and $x_1 = 2x_2$ to give us $x_1 = 2, x_2 = 1$.

$\Rightarrow$ The line $x_1 + x_2 = 4$ touches the circle $x_1^2 + x_2^2 = k$ at the point $(2, 2)$ and the line $2x_1 + x_2 = 5$ touches the circle $x_1^2 + x_2^2 = k$ at the point $(2, 1)$.

But the point $(2, 1)$ lies outside the convex region and hence it is not the desired point. Hence, the required point $= (2, 2)$

and   Min. $Z = 2^2 + 2^2 = 8$

EXAMPLE 4. **Solve the following NLPP by graphical method:**

$$\textbf{Max. } Z = 2x_1 + 3x_2$$
subject to the constraints
$$x_1^2 + x_2^2 \leq 20$$
$$x_1 \cdot x_2 \leq 8$$
and $$x_1, x_2 \geq 0$$

SOLUTION.   In the given problem, clearly the objective function is linear and constraints are non-linear.

Apply the usual method to plot the given constraints on the graph.

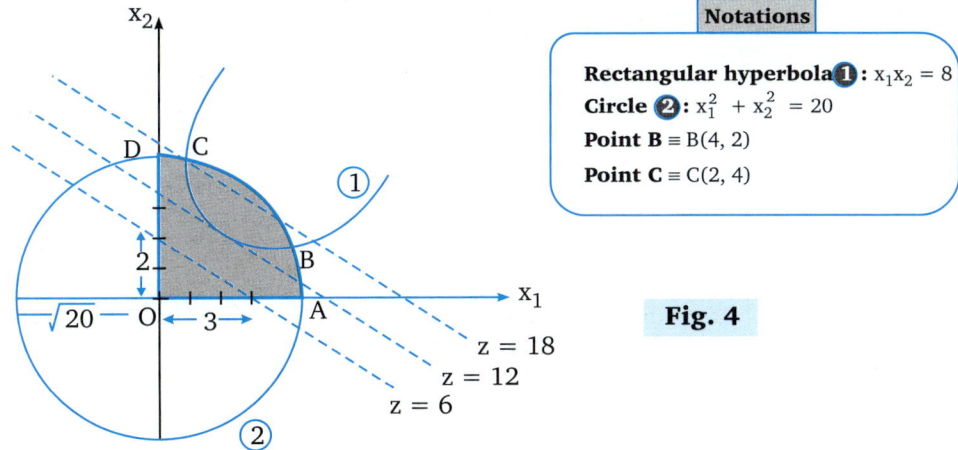

Notations

Rectangular hyperbola ①: $x_1 x_2 = 8$
Circle ②: $x_1^2 + x_2^2 = 20$
Point B $\equiv$ B(4, 2)
Point C $\equiv$ C(2, 4)

Fig. 4

Clearly the constraints $x_1^2 + x_2^2 = 20$ represents a circle with radius $\sqrt{20}$, centre (h, k) = (0, 0) and $x_1 x_2 = 8$ represents a rectangular hyperbola, whose asymptotes are represented by x_1 and x_2 axis.

On solving

$$x_1^2 + x_2^2 = 20$$

and

$$x_1 x_2 = 8$$

we get

$$(x_1, x_2) = (4, 2) \text{ and } (2, 4)$$

These solution points which also satisfy both the constraints may be obtained within the shaded non-convex region OABCDO, which is the feasible region.

Now, we are looking for such a point (x_1, x_2) within the region OABCDO where the value of the given objective function $Z = 2x_1 + 3x_2$ is maximum and this point lies in the convex part of the region. We can find such point by iso-profit method.

In this method, we saw parallel objective function $2x_1 + 3x_2 = k$ lines for different value of k and stop the process when a line touches the extreme boundary point of the feasible region for some value of k. Starting with $k = 6$ and so on, we find the iso-profit line at $k = 16$ touches the extreme boundary point C(2, 4) where Z is maximum.

Hence, the optimum solution is given by

$$x_1 = 2, x_2 = 4 \text{ and Max. } Z = 16$$

EXAMPLE 5. *Use graphical method to minimize the distance of the origin from the concave region bounded by the following constraints*

$$x_1 + x_2 \geq 4; \qquad 2x_1 + x_2 \geq 5$$

and $$x_1, x_2 \geq 0$$

Also, verify that Kuhn-Tucker necessary conditions holds at the point of minimum distance.

SOLUTION. Since the problem is of minimizing distance of solution point from origin, we have to minimize the radius of the circle touching the convex region bounded by the given constraints. Therefore, non-linear programming becomes

$$\text{Min.}\, Z(= r^2) = x_1^2 + x_2^2$$

subject to the constraints

$$x_1 + x_2 \geq 4 \qquad\qquad \text{...(1)}$$
$$2x_1 + x_2 \geq 5 \qquad\qquad \text{...(2)}$$

and $\qquad\qquad x_1, x_2 \geq 0$

Plot the above constraints on the plane graph by following the usual procedure

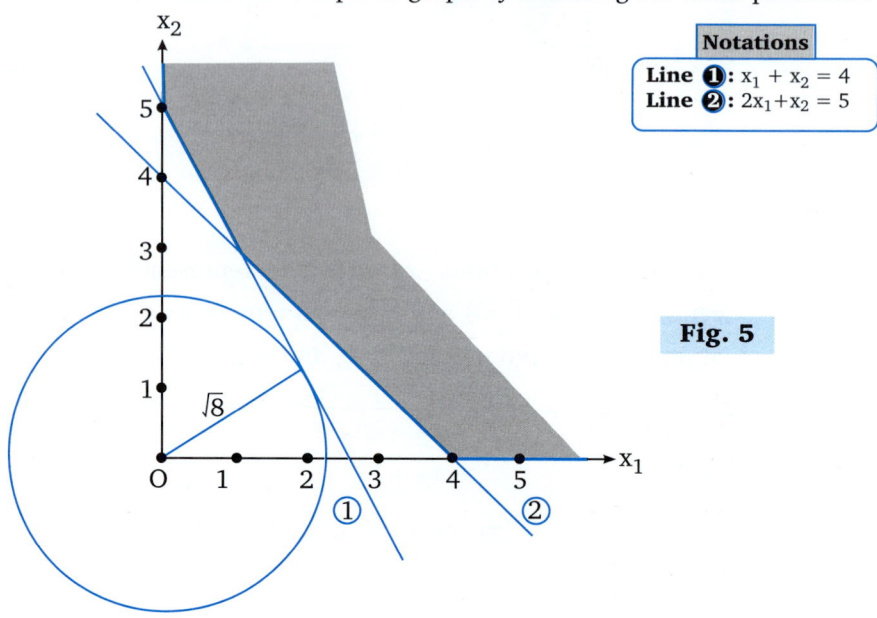

Notations

Line ①: $x_1 + x_2 = 4$
Line ②: $2x_1 + x_2 = 5$

Fig. 5

Here, the circle which represents the objective function will touch one of the sides of the convex space, convex region should be tangent to the circle.

Differentiate the equation of the circle $x_1^2 + x_2^2 = k$

We get

$$2x_1 dx_1 + 2x_2 dx_2 = 0 \Rightarrow \frac{dx_2}{dx_1} = -\frac{x_1}{x_2} \qquad\qquad \text{...(3)}$$

Also from (1) and (2), we get

$$2dx_1 + dx_2 = 0 \Rightarrow \frac{dx_2}{dx_1} = -2 \qquad\qquad \text{...(4)}$$

and $\qquad\qquad dx_1 + dx_2 = 0 \Rightarrow \frac{dx_2}{dx_1} = -1 \qquad\qquad \text{...(5)}$

Solving (3) and (4) with constraints (2), we get

$$-\frac{x_1}{x_2} = \frac{dx_2}{dx_1} = -2$$

$\Rightarrow \qquad\qquad x_1 = 2x_2$

$\Rightarrow \qquad\qquad (x_1, x_2) = (2, 1)$

Clearly this point does not satisfy the constraint $x_1 + x_2 \geq 4$.

Now solving (3) and (5) with constraint (1), we get

$$-\frac{x_1}{x_2} = \frac{dx_2}{dx_1} = -1 \Rightarrow x_1 = x_2$$

$$\Rightarrow \qquad (x_1, x_2) = (2, 2)$$

This point also satisfies the constraints and hence the optimal point for the given problem.

Therefore, the minimum distance of solution from the origin is the radius

$$r = \sqrt{x_1^2 + x_2^2} = \sqrt{8}.$$

Kuhn-Tucker Conditions

Minimize $f(x) = x_1^2 + x_2^2$

subject to the constraints

$$h_1(x) = x_1 + x_2 - 4$$

$$h_2(x) = 2x_1 + x_2 - 5$$

and $\qquad x \geq 0$

Then the Kuhn-Tucker conditions for the minimization of non-linear programming are given by

(i) $f_j(x) - \Sigma \lambda_i h_{i,j}(x) = 0$

(ii) $\lambda_i h_i(x) = 0$

(iii) $h_i(x) \geq 0$

(iv) $\lambda_i \geq 0$

where $f_j(x) = \dfrac{\partial f(x)}{\partial x_j}$, $h_{i,j} = \dfrac{\partial h_i(x)}{\partial x_j}$ $(j = 1, 2)$

Thus, we have

(i) $2x_1 - \lambda_1 - 2\lambda_2 = 0$ (ii) $2x_2 - \lambda_1 - \lambda_2 = 0$ (iii) $\lambda_1(x_1 + x_2 - 4) = 0$

(iv) $\lambda_2(2x_1 + x_2 - 5) = 0$ (v) $x_1 + x_2 - 4 \geq 0$ (vi) $2x_1 + x_2 - 5 \geq 0$

and $\lambda_1, \lambda_2 \geq 0$

Now, since the optimal solution obtained above is $x_1 = x_2 = 2$

Putting these values in constraints (i) and (ii), we get

$$\lambda_1 + 2\lambda_2 = 4 \text{ and } \lambda_1 + \lambda_2 = 4$$

On solving, we get

$$\lambda_1 = 4, \lambda_2 = 0$$

The solution $x_1 = 2$, $x_2 = 2$ and $\lambda_1 = 4$ and $\lambda_2 = 0$ satisfies all the equations (i) to (vi) and hence satisfies the Kuhn-Tucker condition.

EXAMPLE 6. *Solve graphically, the following non-linear programming problem.*

$$\textbf{Max. } Z = 10x_1 - x_1^2 + 10x_2 - x_2^2$$

subject to the constraints

$$x_1 + x_2 \leq 12$$

$$x_1 - x_2 \leq 6$$

and $\qquad x_1, x_2 \geq 0$

SOLUTION. Here, the objective function is non-linear and constraints are linear.

Firstly plot the constraints on the graph following the usual procedure.

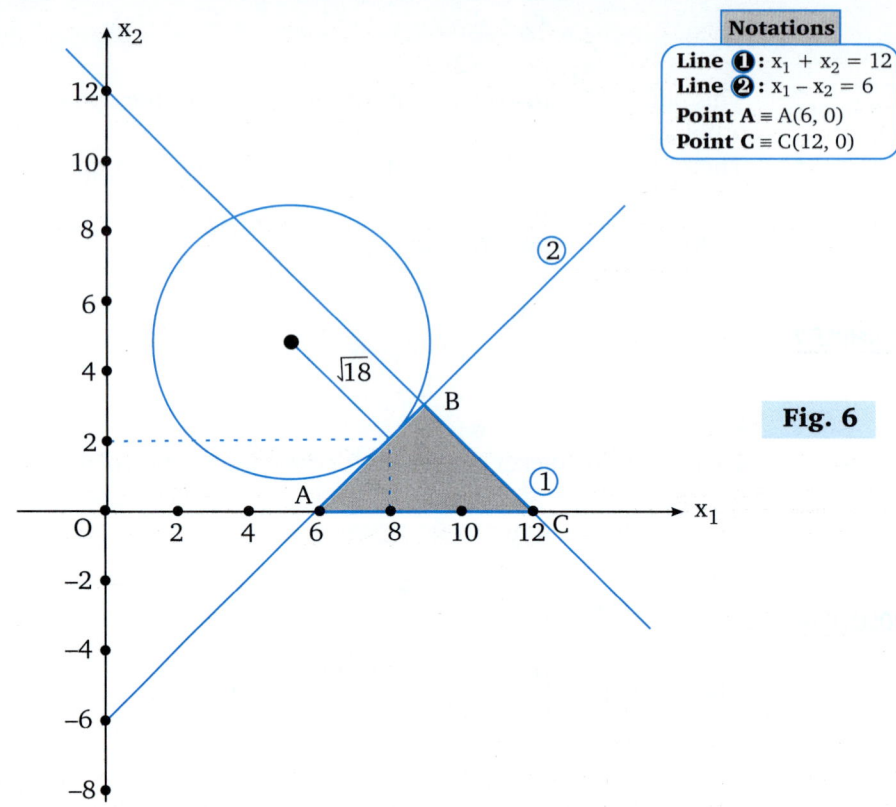

> **Notations**
>
> **Line ①**: $x_1 + x_2 = 12$
> **Line ②**: $x_1 - x_2 = 6$
> **Point A** $\equiv A(6, 0)$
> **Point C** $\equiv C(12, 0)$

Fig. 6

Since the objective function is a circle, the optimum solution point (x_1, x_2) should be a part at which the side of the convex region is tangent to the circle.

Now

$$Z = 10x_1 - x_1^2 + 10x_2 - x_2^2$$

$$\Rightarrow \qquad 10 - 2x_1 + 10\frac{dx_2}{dx_1} - 2x_2\frac{dx_2}{dx_1} = 0$$

$$\Rightarrow \qquad \frac{dx_2}{dx_1} = \frac{2x_1 - 10}{10 - 2x_2} \qquad \qquad \text{...(1)}$$

$$x_1 + x_2 = 12 \Rightarrow 1 + \frac{dx_2}{dx_1} = 0 \Rightarrow \frac{dx_2}{dx_1} = -1 \qquad \qquad \text{...(2)}$$

and $\quad x_1 - x_2 = 6 \Rightarrow 1 - \frac{dx_2}{dx_1} = 0 \Rightarrow \frac{dx_2}{dx_1} = 1 \qquad \qquad \text{...(3)}$

If the line $x_1 + x_2 = 12$ is tangent to the circle, then put $\dfrac{dx_2}{dx_1} = -1$, from (2) in (1), we get

$$\frac{dx_2}{dx_1} = \frac{2x_1 - 10}{10 - 2x_2} = -1$$

$$\Rightarrow \qquad \qquad x_2 = x_1$$

Using $x_1 = x_2$ in the equation $x_1 + x_2 = 12$, we get
$$(x_1, x_2) = (6, 6)$$
which does not satisfies the constraint $x_1 - x_2 \geq 6$ and hence not feasible.

Now (3) $\quad \Rightarrow \quad \dfrac{dx_2}{dx_1} = \dfrac{2x_1 - 10}{10 - 2x_2} = 1$

$\Rightarrow \qquad x_1 + x_2 = 10$ $\hspace{4cm}$ ...(4)

Solving (4) with $x_1 - x_2 = 6$, we get $x_1 = 8, x_2 = 2$.

$\Rightarrow (x_1, x_2) = (8, 2)$, which satisfies all the constraints also.

Hence, optimal solution is given by
$$x_1 = 8, x_2 = 2 \text{ and Max. } Z = 32.$$

EXAMPLE 7. ***Solve graphically the following non-linear programming problem.***

$$\textbf{\textit{Minimize }} Z = x_1^2 + x_2^2$$

subject to the constraints

$$x_1 + x_2 \geq 8$$
$$x_1 + 2x_2 \geq 10$$
$$2x_1 + x_2 \geq 10$$

and $\hspace{3cm} x_1, x_2 \geq 0$

SOLUTION. In the given problem, all constraints are linear but objective function is not linear (Circle). Plot the given constraints on the graph following the usual procedure

If r is the radius of the circle $Z = (r^2) = x_1^2 + x_2^2$

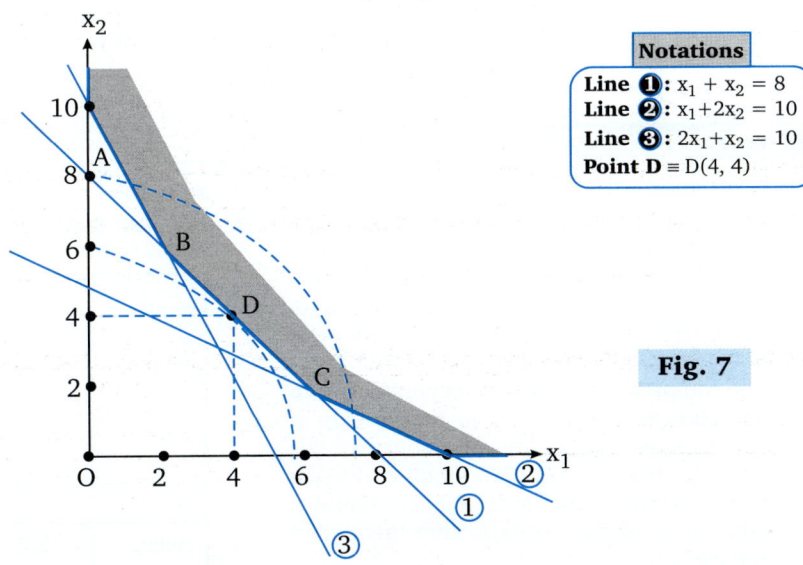

Notations
Line ❶: $x_1 + x_2 = 8$
Line ❷: $x_1 + 2x_2 = 10$
Line ❸: $2x_1 + x_2 = 10$
Point D $\equiv D(4, 4)$

Fig. 7

Then the objective function is to determine the minimum value of r so that the circle with centre $(0, 0)$ and radius r touches the solution space.

The solution point $D(4, 4)$ lies on the line $x_1 + x_2 = 8$ and line is tangent to the circle at D.

Now, since the circle touches one of the sides of the convex region, one of the side of the convex solution space would be tangent to the circle and therefore the solution can also be obtained by differentiating $Z = x_1^2 + x_2^2$ w.r.t. x_1 i.e.,

$$2x_1 + 2x_2 \frac{dx_2}{dx_1} = 0 \Rightarrow \frac{dx_2}{dx_1} = -\frac{x_1}{x_2} \qquad \ldots(1)$$

Also, differentiating the constraints equation, we get

$$dx_1 + dx_2 = 0 \Rightarrow \frac{dx_2}{dx_1} = -1 \qquad \ldots(2)$$

$$dx_1 + 2dx_2 = 0 \Rightarrow \frac{dx_2}{dx_1} = -\frac{1}{2} \qquad \ldots(3)$$

and

$$2dx_1 + dx_2 = 0 \Rightarrow \frac{dx_2}{dx_1} = -2 \qquad \ldots(4)$$

Here, three alternative solution which can now be obtained are

(i) From (1) and (2) with $x_1 + x_2 = 8$, we have

$$\frac{dx_2}{dx_1} = -\frac{x_1}{x_2} = -1 \Rightarrow x_1 = x_2 \Rightarrow (x_1, x_2) = (4, 4)$$

This solution satisfies all the constraints so it is feasible.

(ii) Using (1) and (3) with constraint $x_1 + 2x_2 = 10$, we have

$$\frac{dx_2}{dx_1} = -\frac{x_1}{x_2} = -\frac{1}{2} \Rightarrow (x_1, x_2) = (2, 4)$$

which does not satisfies the constraints.

(iii) Taking (1) and (4) with the constraints $2x_1 - x_2 = 10$, we have

$$\frac{dx_2}{dx_1} = -\frac{x_1}{x_2} = -2 \Rightarrow x_2 = \frac{x_1}{2} \Rightarrow (x_1, x_2) = (4, 2)$$

which also, does not satisfies the constraints.

Hence, the optimal solution is given by

$$x_1 = 4, x_2 = 4 \text{ and Min. } Z = 32$$

 Exercise-27.1

1. (One-Potato, two potato problem) : A frozen food company processes potatoes into packages of French fries hash browns and flakes (for meshed potatoes). At the beginning of the manufacturing process, the raw potatoes are sorted by length and quality and then allocated to the separate product lines.

The company can purchase its potatoes from two sources, which differ in their yields of various sizes and quality. Each source yields different fractions of the product. French fries, has brown and flakes. Suppose that it is possible at different costs to alter these yield some what.

Let f_1, f_2 and f_3 be the fractional yield per unit of weight of source-1 potatoes made into the three products.

Similarly, let g_1, g_2 and g_3 be the yield for source-2.

Suppose that each f_i and g_i can vary within $\pm 10\%$ of the yield shown below :

Product	Source-1	Source-2	Purchase limitation
French fries	0.2	0.3	1.8
Mash browns	0.2	0.1	1.2
Flakes	0.3	0.3	2.4
Relative profit	5.0	6.0	

Let $C_1(f_1, f_2, f_3)$ and $C_2(g_1, g_2, g_3)$ be the expense associated with obtaining these yields.

The problem is to determine how many

potatoes should the company purchase from each source? Formulate the problem as NLPP.

2. A manufacturing concern operates its two available machines to polish it metal products. The two machines are equally efficient although their maintenance costs are different. The daily maintenance and operation cost of the machine is given in rupees as the non-linear function

$$f(x_1, x_2) = 100 - 1.2x_1 - 1.5x_2 + 0.3x_1^2 + 0.5x_2^2$$

where x_1 and x_2 are the numbers of hour of operation of machine-I and machine-II respectively.

The past records of the firm indicate that the combined operating hours of two machines should be at least 35 hours, a day in order to perform a satisfactory jobs. However, the production manager wishes to operate machine-I at least 6 hours more than machine-II because of the higher repair cost of the later. Find the optimal hours of operating the two machines and the minimum daily cost. Formulate the problem as NLPP.

Solve the following NLPP by graphical method

3. Maximize. $Z = 100x_1 - x_1^2 + 100x_2 - x_2^2$

subject to the constraints
$$x_1 + x_2 \geq 80$$

$$x_1 + 2x_2 \leq 100$$
and $$x_1, x_2 \geq 0$$

4. Minimize $f(Z) = (x_1 - 1)^2 + (x_2 - 2)^2$
subject to the constraints
$$0 \leq x_1 \leq 2, 0 \leq x_1 \leq 1$$

5. Min. $Z = 4(x_1 - 6)^2 + 6(x_2 - 2)^2$
subject to the constraints
$$0.5\, x_1 + x_2 \leq 4;\ 3x_1 + x_2 \leq 15;\ x_1 + x_2 \geq 1$$
and $x_1, x_2 \geq 0$

6. Min. $Z = (x_1 - 2)^2 + (x_2 - 1)^2$
subject to the constraints
$$-x_1^2 + x_2 \geq 0;\ -x_1 - x_2 + 2 \geq 0 \text{ and } x_1, x_2 \geq 0$$

Solve the following problems graphically and show that Kuhn-tucker necessary conditions for a maxima do not hold. What do you conclude?

7. Max. $Z = x_1$
subject o the constraints
$$(1 - x_1)^2 - x_2 \geq 0$$
$$x_1, x_2 \geq 0$$

8. Max. $Z = x_1$
subject to the constraints
$$(3 - x_1)^3 - (x_2 - 2) \geq 0$$
$$(3 - x_1)^3 - (x_2 - 2) \geq 0$$
and $x_1, x_2 \geq 0$

Answers

1. Max. $Z = 5(f_1, f_2, f_3) + 6(g_1, g_2, g_3)$ subject to the constraints
$$0.2\,f_1 + 0.3\,g_1 \leq 1.8\ ;\ 0.2\,f_2 + 0.1\,g_2 \leq 1.2;$$
$$0.3\,f_3 + 0.3\,g_3 \leq 2.4\ ;\ f_1, f_2, f_3$$

2. Min. $Z = 100 - 1.2\,x_1 - 1.5\,x_2 + 0.3x_1^2 + 0.5x_2^2$
subject to the constraints $6x_1 + x_2 \geq 35; x_1 \geq 0; x_2 \geq 0$

3. $x_1 = 4, x_2 = 2, x_3 = 1$, Min $Z = -35$

4. $x_1 = 0, x_2 = 1$, Min $Z = 2$ 5. $x_1 = \dfrac{129}{29}, x_2 = \dfrac{48}{29}$, Min. $Z = \dfrac{7800}{841}$

6. $x_1 = 1, x_2 = 1$, Min $Z = 1$

7. $x_1 = 0, x_2 = 0$, Max. $Z = 1$, Constraint qualification is not satisfied.

8. $x_1 = 3, x_2 = 2$, Max. $Z = 3$, Constraint qualification is not satisfied.

REVIEW QUESTIONS

1. Give a formulation of the general mathematical programming problem and obtain the linear programming as a special case of the same.

2. What is non-linear programming problem?

3. Define non-linear function and linear constraints?

4. Describe elementary solution techniques of non-linear programming problem.

ARCHIVE

1. A company manufacture two products A and B. It takes 30 minutes to process one unit of product A and 15 minutes for each unit of B and the maximum machine time available is 35 hours per week. Products A and B require 2 Kgs and 3 Kgs of raw material per unit respectively. The available quantity of raw material is considered to be 180 Kgs per week. The products A and B which have unlimited market potential sell for ₹200 and ₹500 per unit respectively. If the manufacturing costs for products A and B are $2x^2$ and $3y^2$ respectively, find how much of each product should be produced per week where x = quantity of product A to be produced and y = quantity of product B to be produced.

[DELHI(MBA)–1986, 2002]

2. The total profit of restaurant was found to depend mostly on the amount of money spent on advertising and the quality of the preparation of the food (measured in terms of the salaries paid to the chefs). Infact the manager of the restaurant found that if he pays his chefs x ₹/hour and spends ₹y a week on advertising, the restaurant's weekly profit (in ₹) will be

$$Z = 412x + 806y + x^2 - y^2 - xy$$

What hourly wages should the manager pay his chefs and how much should he spend on advertising so as to maximize the restaurant's profit? [DELHI(MBA)–2004]

Hints and Answers

1. Max. $Z = (200 - 2x^2) + (500 - 2y^2)$

   subject to $0.5x + 0.25y \leq 35$

   $2x + 3y \leq 180$ and $x \geq 0; y \geq 0$

Non-linear Programming Techniques: Quadratic and Separable Programming

28

28.1 INTRODUCTION

In linear programming problem, because of linearity, we are able to develop a very efficient algorithm called the simplex method but in non-linear programming problem no such general algorithm exists for the solution. However for problems with certain suitable structures, efficient algorithm have been developed. To solve non-linear programming problem, we use quadratic and separable programming for which specific computational method have been designed. In a mathematical programming, when the objective function is quadratic and constraints are linear, we apply quadratic programming techniques. Separable programming is also a simplex based method and is applicable if the objective function and constraints are separable function.

28.2 QUADRATIC PROGRAMMING

An non-linear programming problem in which the objective function is quadratic and the constraints are linear is called quadratic programming problem.

The general structure of quadratic programming problem is given as below:

$$\text{Optimize } Z = \sum_{j=1}^{n} C_j x_j + \frac{1}{2} \sum_{j=1}^{n} \sum_{k=1}^{n} x_j d_{jk} x_k$$

subject to the constraints

$$\sum_{j=1}^{n} a_{ij} x_j \leq b_i$$

and $\qquad x_j \geq 0, \quad i = 1, 2, \ldots, m; j = 1, 2, \ldots, n$

where $\qquad\qquad d_{jk} = d_{kj} \quad \forall j, k \text{ and } b_i \geq 0$

In matrix form the above quadratic programming problem (QPP) can be written as

$$\text{Optimize } Z = \mathbf{C}X + \frac{1}{2} X^T \mathbf{D} X$$

subject to the constraints

$$\mathbf{A}X \leq \mathbf{b} \qquad \text{and} \qquad X \geq 0$$

where $\qquad\qquad X = (x_1, x_2, \ldots, x_n)^T; \qquad \mathbf{C} = (C_1, C_2, \ldots, C_n)$

$$\mathbf{b} = (b_1, b_2, \ldots, b_m)^T$$

$\mathbf{D} = [d_{jk}]$ is an $n \times n$ symmetric matrix.

$\mathbf{A} = [a_{ij}]$ is $m \times n$ matrix.

☞ **REMARKS**
- The matrix D defined above is symmetric and positive definite.
- Objective function of the quadratic programming problem is strictly convex in X for minimization and concave in X for maximization.
- If the matrix D is a null matrix, the QPP reduced to the standard LPP.

28.3 KUHN-TUCKER CONDITIONS FOR QUADRATIC PROGRAMMING PROBLEMS

Consider the following QPP

$$\text{Max } f(X) = Z = \sum_{j=1}^{n} C_j x_j + \frac{1}{2} \sum_{j=1}^{n} d_{jk} x_j x_k$$

subject to the constraints

$$\sum_{j=1}^{n} a_{ij} x_j \le b_i \quad i = 1, 2, ..., m; j = 1, 2, ..., n$$

and $\qquad x_j \ge 0$

where $\qquad d_{jk} = d_{kj}$ and $b_i \ge 0 \quad \forall\ i$

Introducing slack variables s_i^2 and r_j^2, then above problem becomes

$$\max Z = \sum_{j=1}^{n} C_j x_j + \frac{1}{2} \sum_{j=1}^{n} \sum_{k=1}^{n} d_{jk} x_j x_k$$

subject to the constraints

$$a_i' X + s_i^2 = b_i \quad i = 1, 2, ..., m$$

$$-x_j + r_j^2 = 0 \quad j = 1, 2, ..., n$$

Now, define the Lagrangian function such that

$$L(x, s, \mu, \lambda, r) = f(X) - \sum_{i=1}^{m} \lambda_i (a_i' X + s_i^2 - b_i) - \sum_{j=1}^{n} \mu_j (-x_j + r_j^2)$$

Forming the necessary condition, we obtain

$$\frac{\partial L}{\partial x_j} = \frac{\partial f}{\partial x_j} - \sum_{i=1}^{m} \lambda_i a_{ij} + \mu_j = 0, \quad j = 1, 2, ..., n \qquad ...(1)$$

$$\lambda_i (a_i X - b_i) = 0 \qquad ...(2)$$

$$\mu_i x_i = 0 \qquad ...(3)$$

$$AX \le b \qquad ...(4)$$

and $\qquad X, \lambda, \mu \ge 0$

Equation (1) can be rewritten as

$$\frac{\partial L}{\partial x_j} = \left[C_j + \frac{1}{2} \left(2 \sum_{k=1}^{n} C_{jk} x_k \right) \right] - \sum_{i=1}^{m} \lambda_i a_{ij} + \mu_j = 0, \quad j = 1, 2, ..., n$$

Letting $s_i^2 = q_i \ge 0$ then above equation becomes,

$$-\mu_j + C_j + \sum_{k=1}^{n} d_{jk} x_k - \sum_{i=1}^{m} \lambda_i a_{ij} = 0, \quad j = 1, 2, ..., n \qquad ...(5)$$

$$AX + Iq = b, \quad X \ge 0, q \ge 0, \lambda, \mu \ge 0$$

and $\qquad \lambda_i q_i = 0, \ i = 1, 2, ..., m, \ \mu_j x_j = 0, \ j = 1, 2, ..., n$

It must be noted that except for the final condition $\lambda_i q_i = 0 = \mu_j x_j$, the remaining equations are linear in X, λ, μ and q. The problem thus becomes equivalent to finding the solution to a set of linear equations which also satisfies the additional conditions $\lambda_i q_i = 0 = \mu_j x_j$. Since, $f(X)$ is strictly concave and the solution space is convex the feasible solution satisfying all these conditions must give the optimum solution directly.

28.4 WOLFE'S MODIFIED SIMPLEX METHOD

Consider the following quadratic programming problem

$$\text{Max. } Z = f(X) = \sum_{j=1}^{n} C_j x_j + \frac{1}{2} \sum_{j,k=1}^{n} x_j d_{jk} x_k$$

subject to the constraints

$$\sum_{j=1}^{n} a_{ij} x_j \le b_i$$

and
$$x_j \ge 0, \ i = 1, 2, ..., m; j = 1, 2, ..., n$$
where
$$d_{jk} = d_{kj} \text{ and } b_i \ge 0$$

Wolfe suggested a solution procedure for the above QPP using the ordinary simplex method with one slight modification as given in the following working procedure.

Working Procedure

STEP 1. Write the given QPP into standard maximization form (i.e., if it is a problem of minimization, then first convert it into maximization form).

STEP 2. Convert the inequality constraints into equations by introducing the slack variable s_i^2 in the ith constraints ($i = 1, 2, ..., m$) and the slack variables s_{m+j}^2 in the ith non-negative constraints ($j = 1, 2, ..., n$).

STEP 3. Construct the Lagrangian function

$$L(X, S, \lambda) = f(X) - \sum_{i=1}^{m} \lambda_i \left[\sum_{j=1}^{n} a_{ij} x_j - b_i + s_i^2 \right] - \sum_{j=1}^{n} \lambda_{m+j}(-x_j + s_{m+j}^2)$$

STEP 4. Construct the Kuhn-Tucker conditions by differentiating $L(X, S, \lambda)$ partially w.r.t. x, s and λ and put the first order partial derivatives equal to zero.

STEP 5. Introduce non-negative artificial variables $A_j : j = 1, 2, ..., n$ in the Kuhn-Tucker conditions and get the following equations

$$C_j + \sum_{k=1}^{n} d_{jk} x_k - \sum_{j=1}^{m} \lambda_i a_{ij} + \lambda_{m+j} = 0, \quad j = 1, 2, ..., m$$

STEP 6. Construct the objective function Max. $Z = -A_1 - A_2 - ... - A_n$.

STEP 7. Apply the simplex method to obtain initial basic feasible solution of the following LPP

$$\text{Max. } Z = -A_1 - A_2 - ... - A_n$$
subject to the constraints

$$\sum_{k=1}^{n} d_{jk} x_k - \sum_{i=1}^{m} \lambda_i a_{ij} + \lambda_{m+j} + A_j = -C_j$$

$$\sum_{j=1}^{n} a_{ij} x_j + x_{n+i} = b_i$$

and $A_j, \lambda_i, \lambda_{m+j}, x_j \ge 0$

where $x_{n+1} = s_i^2$, $i = 1, 2, ..., n$ and satisfying the complementary slackness conditions given by

$$\sum_{j=1}^{n} \lambda_{m+j} x_j + \sum_{j=1}^{m} x_{n+i} \lambda_i = 0$$

STEP 8. Use Phase-I of the artificial variable techniques (two phase method) to find the optimum solution to the LPP of step 7, which satisfy the complementary slackness conditions.

☞ REMARK
- The solution of the above system is obtained by using Phase-I of simplex method. Since, our aim is to obtain a feasible solution, the solution does not require the consideration of Phase-II. The only necessary thing is to maintain the conditions $\lambda_i s_i = 0 = \mu_j x_j$ all the time. This implies that if λ_i is in the basic solution with positive values, then x_i can not be basic with positive value. Similarly μ_j and x_j can not be positive simultaneously.

> Phase-I will end in the usual manner with the sum of all artificial variables equal to zero only if the feasible solution to the problem exists.

Solved Examples

EXAMPLE 1. *Solve the following QPP using Wolfe's modified simplex method.*

$$\text{Max. } Z = 4x_1 + 6x_2 - 2x_1^2 - 2x_1x_2 - 2x_2^2$$

subject to the constraints

$$x_1 + 2x_2 \leq 2$$

and $$x_1, x_2 \geq 0$$

[IAS–1994; GUWAHATI(MCA)–1992; DELHI(MBA)–2012]

SOLUTION. Using the slack variables s_1^2, s_2^2, s_3^2, the modified problem can be written as

$$\text{Max. } Z = 4x_1 + 6x_2 - 2x_1^2 - 2x_1x_2 - 2x_2^2$$

subject to the constraints

$$x_1 + 2x_2 + s_1^2 = 2$$

$$-x_1 + s_2^2 = 0$$

$$-x_2 + s_3^2 = 0$$

Let us define the Lagrangian function as given below

$$L(x_1, x_2; s_1, s_2, s_3; \lambda_1, \lambda_2, \lambda_3) = (4x_1 + 6x_2 - 2x_1^2 - 2x_1x_2 - 2x_2^2)$$

$$- \lambda_1(x_1 + 2x_2 + s_1^2 - 2) - \lambda_2(-x_1 + s_2^2)$$

$$- \lambda_3(-x_2 + s_3^2)$$

$$= 0$$

For the necessary condition of maxima, we must have

$$\frac{\partial L}{\partial x_1} = 0 \Rightarrow \qquad 4 - 4x_1 - 2x_2 - \lambda_1 + \lambda_2 = 0$$

$$\frac{\partial L}{\partial x_2} = 0 \Rightarrow \qquad 6 - 2x_1 - 4x_2 - 2\lambda_1 + \lambda_3 = 0$$

$$\frac{\partial L}{\partial s_1} = 0 \Rightarrow \quad -2\lambda_1 s_1 = 0; \quad \frac{\partial L}{\partial s_2} = 0 \quad \Rightarrow \quad -2\lambda_2 s_2 = 0$$

$$\frac{\partial L}{\partial s_3} = 0 \Rightarrow \quad -2\lambda_3 s_3 = 0; \quad \frac{\partial L}{\partial \lambda_1} = 0 \quad \Rightarrow \quad x_1 + 2x_2 + s_1^2 - 2 = 0$$

$$\frac{\partial L}{\partial \lambda_2} = 0 \Rightarrow \quad -x_1 + s_2^2 = 0$$

and $\quad \dfrac{\partial L}{\partial \lambda_3} = 0 \Rightarrow \quad -x_2 + s_3^2 = 0$

After some simplification, we have the following system of equations

$$\left. \begin{array}{r} 4x_1 + 2x_2 + \lambda_1 - \lambda_2 = 4 \\ 2x_1 + 4x_2 - 2\lambda_1 - \lambda_3 = 6 \\ x_1 + 2x_2 + s_1^2 = 2 \end{array} \right] \qquad \ldots(1)$$

and $\quad \left. \begin{array}{l} \lambda_1 s_1^2 + \lambda_2 s_2^2 + \lambda_3 s_3^2 = 0 \Rightarrow \lambda_1 s_1^2 + \lambda_2 x_1 + \lambda_3 x_2 = 0 \\ x_1, x_2, s_1^2, \lambda_i \geq 0, \ \ i = 1, 2, 3 \end{array} \right] \qquad \ldots(2)$

Now, to obtain the solution to the above simultaneous equations (1), introduce the artificial variables $A_1, A_2 \ (\geq 0)$ in the first two constraints of (1). Then given problem can be written as

$$\text{Max. } Z = -A_1 - A_2$$

subject to the constraints

$$4x_1 + 2x_2 + \lambda_1 - \lambda_2 + A_1 = 4$$
$$2x_1 + 4x_2 + 2\lambda_1 - \lambda_3 + A_2 = 6$$
$$x_1 + 2x_2 + x_3 = 2$$

and $\quad x_1, x_2, x_3 \geq 0, A_1, A_2, \lambda_i \geq 0$

We note that here we have replaced s_1^2 by x_3 and satisfied complementary slackness conditions $\Sigma \lambda_i x_i = 0$. Now, apply the procedure of phase-1 of two phase method we have the following simplex table

Simplex Table-1

	C_j		0	0	0	0	0	0	-1	-1	Min. Ratio
B.V.	C_B	X_B	x_1	x_2	x_3	λ_1	λ_2	λ_3	A_1	A_2	X_B/x_1
A_1	-1	4	④	2	0	1	-1	0	1	0	1 (min) →
A_2	-1	6	2	4	0	2	0	-1	0	1	3
x_3	0	2	1	2	1	0	0	0	0	0	2
	Δ_j		6 ↑	6	0	3	-1	-1	0	0	

In the above table, clearly we have x_1 and x_2 are the variables with the most positive entry. Let x_1 enter in the basis and A_1 will leave. Then next simplex table is given as below.

Simplex Table-2

	C_j		0	0	0	0	0	0	−1	Min. Ratio
B.V.	C_B	X_B	x_1	x_2	x_3	λ_1	λ_2	λ_3	A_2	$X_B\|x_2$
x_1	0	1	1	1/2	0	1/4	−1/4	0	0	2
A_2	−1	4	0	3	0	3/2	1/2	−1	1	4/3
x_3	0	1	0	③/2	1	−1/4	1/4	0	0	2/3 (min) →
	Δ_j		0	3 ↑	0	3/2	1/2	−1	0	

In the above table, x_2 is the variable with the most positive entry (*i.e.*, 3) in the Δ_j-row so x_2 enter in the basis and x_3 will leave. Then we have the following simplex table.

Simplex Table-3

	C_j		0	0	0	0	0	0	−1	Min. Ratio
B.V.	C_B	X_B	x_1	x_2	x_3	λ_1	λ_2	λ_3	A_2	$X_B\|\lambda_1$
x_1	0	2/3	1	0	−1/3	1/3	−1/3	0	0	2
A_2	−1	2	0	0	−2	②	0	−1	1	1 (min) →
x_2	0	2/3	0	1	2/3	−1/6	1/6	0	0	—
	Δ_j		0	0	−2	2 ↑	0	−1	0	

Here, λ_1 is the variable with the most positive entry 2 in Δ_j-row so λ_1 enter in the basis and A_2 will leave. Then we have the following simplex table.

Simplex Table-4

	C_j		0	0	0	0	0	0
B.V.	C_B	X_B	x_1	x_2	x_3	A_1	A_2	A_3
x_1	0	1/3	1	0	0	0	−1/3	1/6
λ_1	0	1	0	0	−1	1	0	−1/6
x_2	0	5/6	0	1	1/2	0	1/6	−1/6
	Δ_j		0	0	0	0	0	0

In the above table, all $\Delta_j \leq 0 \Rightarrow$ solution is optimal and is given by
$$x_1 = 1/3, x_2 = 5/6 \text{ and}$$

$$\text{Max.} Z = 4\left(\frac{1}{3}\right) + 6\left(\frac{5}{6}\right) - 2\left(\frac{1}{3}\right)^2 - 2\left(\frac{1}{3}\right)\left(\frac{5}{6}\right) - 2\left(\frac{5}{6}\right)^2 = \frac{25}{6}$$

<u>**EXAMPLE 2.**</u> *Solve the following QPP using Wolfe's method:*

$$\textbf{Max. } \textbf{Z} = 2x_1 + x_2 - x_1^2$$

subject to the constraints

$$2x_1 + 3x_2 \leq 6$$

$$2x_1 + x_2 \leq 4$$

and $x_1, x_2 \geq 0$ [ANDHRA(BE)–1996; MADRAS–1990; MEERUT(M.PHIL)–2012]

SOLUTION. Considering non-negativity constraints conditions $x_1 \geq 0$, $x_2 \geq 0$ as inequality constraints and adding the slack variables s_1^2, s_2^2, s_3^2 and s_4^2 to all the inequalities, the above problem can be written as

$$\text{Max. } Z = 2x_1 + x_2 - x_1^2$$

subject to the constraints

$$2x_1 + 3x_2 + s_1^2 = 6$$

$$2x_1 + x_2 + s_2^2 = 4$$

$$-x_1 + s_3^2 = 0$$

$$-x_2 + s_4^2 = 0$$

Let us define the Lagrangian function as given below

$$L(X,S,\lambda) = (2x_1 + x_2 - x_1^2) - \lambda_1(2x_1 + 3x_2 + s_1^2 - 6) - \lambda_2(2x_1 + x_2 + s_2^2 - 4)$$
$$- \lambda_3(-x_1 + s_3^2) - \lambda_4(-x_2 + s_4^2)$$

Now, the necessary condition for the maximum of L (and hence for Z) we must have

$$\frac{\partial L}{\partial x_1} = 0 \quad \Rightarrow \quad -2 - 2x_1 - 2\lambda_1 - 2\lambda_2 + \lambda_3 = 0$$

$$\frac{\partial L}{\partial x_2} = 0 \quad \Rightarrow \quad 1 - 3\lambda_1 - \lambda_2 + \lambda_4 = 0$$

$$\frac{\partial L}{\partial s_1} = 0 \quad \Rightarrow \quad -2\lambda_1 s_1 = 0, \quad \frac{\partial L}{\partial s_2} = 0 \quad \Rightarrow \quad -2\lambda_2 s_2 = 0$$

$$\frac{\partial L}{\partial s_3} = 0 \quad \Rightarrow \quad -2\lambda_3 s_3 = 0, \quad \frac{\partial L}{\partial s_4} = 0 \quad \Rightarrow \quad -2\lambda_4 s_4 = 0$$

$$\frac{\partial L}{\partial \lambda_1} = 0 \quad \Rightarrow \quad 2x_1 + 3x_2 + s_1^2 - 6 = 0$$

$$\frac{\partial L}{\partial \lambda_2} = 0 \quad \Rightarrow \quad 2x_1 + x_2 + s_2^2 - 4 = 0$$

$$\frac{\partial L}{\partial \lambda_3} = 0 \quad \Rightarrow \quad -x_1 + s_3^2 = 0$$

$$\frac{\partial L}{\partial \lambda_4} = 0 \quad \Rightarrow \quad -x_2 + s_4^2 = 0$$

On simplification, above equations can be written as

$$2x_1 + 2\lambda_1 + 2\lambda_2 - \lambda_3 = 2$$

$$3\lambda_1 + \lambda_2 - \lambda_4 = 1$$

$$2x_1 + 3x_2 + s_1^2 = 6$$

$$2x_1 + x_2 + s_2^2 = 4$$

$$\lambda_1 s_1 = \lambda_2 s_2 = 0$$

$$\lambda_3 x_1 = \lambda_4 x_2 = 0$$

and $\quad x_1, x_2; \lambda_1, \lambda_2, \lambda_3, \lambda_4; s_1, s_2 \geq 0$

Now, introducing the artificial variables A_1 and A_2, the modified QPP becomes

$$\text{Max. } Z = -A_1 - A_2$$

subject to the constraints

$$2x_1 + 2\lambda_1 + 2\lambda_2 - \lambda_3 + A_1 = 2$$
$$3\lambda_1 + \lambda_2 - \lambda_4 + A_2 = 1$$
$$2x_1 + 3x_2 + s_1^2 = 6$$
$$2x_1 + x_2 + s_2^2 = 4$$
$$\lambda_1 s_1 = \lambda_2 s_2 = 0$$
$$\lambda_3 x_1 = \lambda_4 x_2 = 0$$
$$x_i, s_j, \lambda_k, A_l \geq 0$$

Now, apply the phase-1 of two phase method, we have the following simplex table

Simplex Table-1

C_j			0	0	0	0	0	0	0	0	-1	-1	Min. Ratio
B.V.	C_B	X_B	x_1	x_2	λ_1	λ_2	λ_3	λ_4	s_1	s_2	A_1	A_2	
A_1	1	2	②	0	2	2	-1	0	0	0	1	0	1 (min) →
A_2	1	1	0	0	3	1	0	-1	0	0	0	1	—
s_1	0	6	2	3	0	0	0	0	1	0	0	0	3
s_2	0	4	2	1	0	0	0	0	0	1	0	0	2
	Δ_j		2	0	5	3	-1	-1	0	0	0	0	
			↑										

In the above table, the largest value in Δ_j-row is 5, but we can not enter λ_1 (or λ_2) in the basis because complementry slackness conditions $\lambda_1 s_1 = \lambda_2 s_2 = 0$. Since, $\lambda_3 = 0$, x_1 can be entered into the basis with A_1 as leaving variable.

Then, new simplex table is given as under

Simplex Table-2

C_j			0	0	0	0	0	0	0	0	-1	Min. Ratio
B.V.	C_B	X_B	x_1	x_2	λ_1	λ_2	λ_3	λ_4	s_1	s_2	A_2	
x_1	0	1	1	0	1	1	1/2	0	0	0	0	—
A_2	1	1	0	0	3	1	0	-1	0	0	1	—
s_1	0	4	0	③	-2	-2	1	0	1	0	0	4/3 (min)→
s_2	0	2	0	1	-2	-2	1	0	0	1	0	2
	Δ_j		0	0	3	1	0	-1	0	0	0	
				↑								

In the above table, again we can not enter λ_1, λ_2, and λ_3 in the basis because s_1, s_2 and x_1 are already in the basis. Therefore, enter x_2 into the basis with s_1 as the leaving variable because $\lambda_4 = 0$. The next simplex table is given as follows:

Simplex Table-3

	C_j		0	0	0	0	0	0	0	0	-1	Min. Ratio
B.V.	C_B	X_B	x_1	x_2	λ_1	λ_2	λ_3	λ_4	s_1	s_2	A_2	
x_1	0	1	1	0	1	1	$-1/2$	0	0	0	0	1
A_2	1	1	0	0	③	1	0	-1	0	0	1	1/3 (min)→
x_2	0	4/3	0	1	$-2/4$	$-2/3$	1/3	0	1/3	0	0	
s_2	0	2/3	0	0	$-4/3$	$-4/3$	2/3	0	$-1/3$	1	0	
	Δ_j		0	0	3	1	0	-1	0	0	-1	
					↑							

In the above table, since $s_1 = 0$, λ_1 can be entered into the basis and A_2 will leave the basis. Then we have the following simplex table

Simplex Table-4

	C_j		0	0	0	0	0	0	0	0
B.V.	C_B	X_B	x_1	x_2	λ_1	λ_2	λ_3	λ_4	s_1	s_2
x_1	0	2/3	1	0	0	2/3	$-1/2$	1/3	0	0
λ_1	0	1/3	0	0	1	1/3	0	$-1/3$	0	0
x_2	0	14/9	0	1	0	$-4/9$	1/3	$-2/9$	1/3	0
s_2	0	10/9	0	0	0	$-8/9$	2/3	$-4/9$	$-1/3$	1
	Δ_j		0	0	0	0	0	0	0	0

Here, clearly all $\Delta_j \leq 0 \Rightarrow$ solution is optimal and is given by

$$x_1 = \frac{2}{3}, x_2 = \frac{14}{9}, \lambda_1 = \frac{1}{3}, \lambda_2 = \lambda_3 = \lambda_4 = 0, s_1 = 0, s_2 = \frac{10}{9}$$

This solution also satisfy the complementary slackness conditions

$$\lambda_1 s_1 = \lambda_2 s_2 = 0, \lambda_3 x_1 = \lambda_4 x_2 = 0$$

Also, $Z^* = 0$, so current solution is also feasible. Hence, the maximum value of the objective function of the given QPP is given by

$$\text{Max.} Z = 2\left(\frac{2}{3}\right) + \left(\frac{14}{9}\right) - \left(\frac{2}{3}\right)^2 = \frac{22}{9}$$

EXAMPLE 3. *Solve the following quadratic programming problem by using Wolfe's modified simplex method.*

$$\text{Max. } Z = 2x_1 + 3x_2 - 2x_1^2$$

subject to the constraints

$$x_1 + 4x_2 \leq 4$$
$$x_1 + x_2 \leq 2$$

and $\qquad x_1, x_2 \geq 0$

SOLUTION. Using the slack variables s_1^2, s_2^2, s_3^2 and s_4^2 in the given QPP, we get the following modified form.

$$\text{Max. } Z = 2x_1 + 3x_2 - 2x_1^2$$

subject to the constraints

$$x_1 + 4x_2 + s_1^2 = 4$$

$$x_1 + x_2 + s_2^2 = 2$$

$$-x_1 + s_3^2 = 0$$

$$-x_2 + s_4^2 = 0$$

Now, the Lagrangian function is given by

$$L(x_1, x_2, s_1, s_2, s_3, s_4, \lambda_1, \lambda_2, \lambda_3, \lambda_4)$$

$$= (2x_1 + 3x_2 - 2x_1^2) - \lambda_1(x_1 + 4x_2 + s_1^2 - 4)$$

$$- \lambda_2(x_1 + x_2 + s_2^2 - 2) - \lambda_3(-x_1 + s_3^2) - \lambda_4(-x_2 + s_4^2)$$

Now, the necessary conditions of the minimum of L (and hence Z) we must have

$$\frac{\partial L}{\partial x_1} = 0 \quad \Rightarrow \quad 2 - 4x_1 - \lambda_1 - \lambda_2 + \lambda_3 = 0$$

$$\frac{\partial L}{\partial x_2} = 0 \quad \Rightarrow \quad 3 - 4\lambda_1 - \lambda_2 + \lambda_4 = 0$$

$$\frac{\partial L}{\partial s_1} = 0 \quad \Rightarrow \quad -2\lambda_1 s_1 = 0; \frac{\partial L}{\partial s_2} = 0 \quad \Rightarrow \quad -2\lambda_2 s_2 = 0$$

$$\frac{\partial L}{\partial s_3} = 0 \quad \Rightarrow \quad -2\lambda_3 s_3 = 0; \frac{\partial L}{\partial s_4} = 0 \quad \Rightarrow \quad -2\lambda_4 s_4 = 0$$

$$\frac{\partial L}{\partial \lambda_1} = 0 \quad \Rightarrow \quad x_1 + 4x_2 + s_1^2 - 4 = 0$$

$$\frac{\partial L}{\partial \lambda_2} = 0 \quad \Rightarrow \quad x_1 + x_2 + s_2^2 - 2 = 0$$

$$\frac{\partial L}{\partial \lambda_3} = 0 \quad \Rightarrow \quad -x_1 + s_3^2 = 0$$

and $$\frac{\partial L}{\partial \lambda_4} = 0 \quad \Rightarrow \quad -x_2 + s_4^2 = 0$$

After some simplification, we get

$$\left.\begin{array}{l} 4x_1 + \lambda_1 + \lambda_2 - \lambda_3 = 2 \\ 4\lambda_1 + \lambda_2 - \lambda_4 = 3 \end{array}\right] \qquad \text{...(1)}$$

and

$$\left.\begin{array}{l} x_1 + 4x_2 + s_1^2 = 4 \\ x_1 + x_2 + s_2^2 = 2 \\ \lambda_1 s_1^2 + \lambda_2 s_2^2 + x_1 \lambda_3 + x_2 \lambda_4 = 0 \end{array}\right] \qquad \text{...(2)}$$

$x_1, x_2, s_1, s_2, s_3, s_4, \lambda_1, \lambda_2, \lambda_3, \lambda_4 \geq 0$

Now, proceed same as in previous questions, introduce the artificial variables A_1, A_2 (≥ 0) in the equations of (1) we get the following form

$$\text{Max. } Z = -A_1 - A_2$$

subject to the constraints

$$4x_1 + \lambda_1 + \lambda_2 - \lambda_3 + A_1 = 2$$
$$4\lambda_1 + \lambda_2 - \lambda_4 + A_2 = 3$$
$$x_1 + 4x_2 + x_3 = 4$$
$$x_1 + x_2 + x_4 = 2$$

and $x_i, s_i, A_i, \lambda_i \geq 0$

Here, we have replaced s_1^2 by x_3 and s_2^2 by x_4 and satisfied the complementry slackness conditions $\Sigma\lambda_i x_i = 0$. Now, the optimum solution of the above problem shall be obtained using phase-1 of two phase method. So, we have the following initial simplex table.

Simplex Table-1

B.V.	C_B	X_B	C_j 0 x_1	0 x_2	0 x_3	0 x_4	0 λ_1	0 λ_2	0 λ_3	0 λ_4	-1 A_1	-1 A_2	Min. Ratio $X_B \mid x_1$
A_1	-1	2	④	0	0	0	1	1	-1	0	1	0	1/2(min) →
A_2	-1	3	0	0	0	0	4	1	0	-1	0	1	—
x_3	0	4	1	4	1	0	0	0	0	0	0	0	4
x_4	0	2	1	1	0	1	0	0	0	0	0	0	2
		Δ_j	4 ↑	0	0	0	5	2	-1	-1	0	0	

In the above table we observe that λ_1 is the variable with the most positive value (i.e., 5) and so it should enter the basis, but it will not because x_3 is in the basis (complementary slackness condition $\lambda_1 x_3 = 0$). The next most positive entry is 4 for the x_1 column in Δ_j-row. Hence, x_1 enter in the basis and A_1 will leave. Then we have the following simplex table.

Simplex Table-2

B.V.	C_B	X_B	C_j 0 x_1	0 x_2	0 x_3	0 x_4	0 λ_1	0 λ_2	0 λ_3	0 λ_4	-1 A_2	Min. Ratio $X_B \mid x_2$
x_1	0	1/2	1	0	0	0	1/4	1/4	$-1/4$	0	0	—
A_2	-1	3	0	0	0	0	4	1	0	-1	1	—
x_3	0	7/2	0	④	1	0	$-1/4$	$-1/4$	1/4	0	0	7/8(min)→
x_4	0	3/2	0	1	0	1	$-1/4$	$-1/4$	1/4	0	0	3/2
		Δ_j	0	0 ↑	0	0	4	1	0	-1	0	

Here, we see that either λ_1 or λ_2 will enter in the basis, but since x_3 and x_4 are still in the basis they can not enter the basis because of the complementary slackness conditions $\lambda_1 x_3 = 0$ and $\lambda_2 x_4 = 0$. Here, x_2 is eligible to enter in the basis because λ_4 is not in the basis ($\lambda_4 x_2 = 0$). Then we have the following simplex table.

Simplex Table-3

C_j			0	0	0	0	0	0	0	0	-1	Min. Ratio
B.V.	C_B	X_B	x_1	x_2	x_3	x_4	λ_1	λ_2	λ_3	λ_4	A_2	X_B/λ_2
x_1	0	1/2	1	0	0	0	1/4	$-1/4$	$-1/4$	0	0	2
A_2	-1	3	0	0	0	0	④	1	0	-1	1	3/8(min)→
x_2	0	7/8	0	1	1/4	0	$-1/16$	$-1/16$	1/16	0	0	—
x_4	0	5/8	0	0	$-1/4$	1	$-3/16$	$-3/16$	3/16	0	0	—
	Δ_j		0	0	0	0	4	1	0	-1	0	

In the above table, we observe that either λ_1 or λ_2 will enter in the basis but x_4 is still in the basis so λ_2 can not enter because of the complementary slackness conditions $\lambda_2 x_4 = 0$. So, λ_1 enter in the basis.

Hence, we have the following simplex table

Simplex Table-4

C_j			0	0	0	0	0	0	0	0
B.V.	C_B	X_B	x_1	x_2	x_3	x_4	λ_1	λ_2	λ_3	λ_4
x_1	0	5/16	1	0	0	0	0	$-5/16$	$-1/4$	1/16
λ_2	0	3/4	0	0	0	0	1	1/4	0	$-1/4$
x_2	0	59/64	0	1	1/4	0	0	$-3/64$	1/16	$-1/64$
x_4	0	49/64	0	0	$-1/4$	1	0	$-9/64$	3/16	$-3/64$
	Δ_j		0	0	0	0	0	0	0	0

In the above table, we observe that all $\Delta_j \leq 0$.

$\Rightarrow$ Solution is optimum and is given by

$$x_1 = \frac{5}{16}, x_2 = \frac{59}{64} \text{ and Max. } Z = 3.19$$

EXAMPLE 4. **Solve the following QPP by using Wolfe's method**

$$\textbf{Max. } Z = 6x_1 + 3x_2 - 4x_1x_2 - 2x_1^2 - 3x_2^2$$

subject to the constraints

$$x_1 + x_2 \leq 1$$
$$2x_1 + 3x_2 \leq 4$$

and $x_1, x_2 \geq 0$

SOLUTION.   Using the slack variables s_1^2, s_2^2, s_3^2 and s_4^2 we have the modified form of the given problem as follows:

$$\text{Max. } Z = 6x_1 + 3x_2 - 4x_1x_2 - 2x_1^2 - 3x_2^2$$

s.t. $x_1 + x_2 + s_1^2 = 1$

$$2x_1 + 3x_2 + s_2^2 = 4$$

$$-x_1 + s_3^2 = 0$$

$$-x_2 + s_4^2 = 0$$

Now, construct the Lagrangian function as given below

$$L(x_1, x_2; s_1, s_2, s_3, s_4; \lambda_1, \lambda_2, \lambda_3, \lambda_4)$$

$$= (6x_1 + 3x_2 - 4x_1 x_2 - 2x_1^2 - 3x_2^2)$$

$$- \lambda_1(x_1 + x_2 + s_1^2 - 1) - \lambda_2(2x_1 + 3x_2 + s_2^2 - 4)$$

$$- \lambda_3(-x_1 + s_3^2) - \lambda_4(-x_2 + s_4^2)$$

For the maxima of L (and hence Z) we must have

$$\frac{\partial L}{\partial x_1} = 0 \quad \Rightarrow \quad 6 - 4x_2 - 4x_1 - \lambda_1 - 2\lambda_2 + \lambda_3 = 0$$

$$\frac{\partial L}{\partial x_2} = 0 \quad \Rightarrow \quad 3 - 4x_1 - 6x_2 - \lambda_1 - 3\lambda_2 + \lambda_4 = 0$$

$$\frac{\partial L}{\partial s_1} = 0 \quad \Rightarrow \quad -2\lambda_1 s_1 = 0; \frac{\partial L}{\partial s_2} = 0 \quad \Rightarrow \quad -2\lambda_2 s_2 = 0$$

$$\frac{\partial L}{\partial s_3} = 0 \quad \Rightarrow \quad -2\lambda_3 s_3 = 0; \frac{\partial L}{\partial s_4} = 0 \quad \Rightarrow \quad -2\lambda_4 s_4 = 0$$

$$\frac{\partial L}{\partial \lambda_1} = 0 \quad \Rightarrow \quad x_1 + x_2 + s_1^2 - 1 = 0$$

$$\frac{\partial L}{\partial \lambda_2} = 0 \quad \Rightarrow \quad 2x_1 + 3x_2 + s_2^2 - 4 = 0$$

$$\frac{\partial L}{\partial \lambda_3} = 0 \quad \Rightarrow \quad -x_1 + s_3^2 = 0$$

and $$\frac{\partial L}{\partial \lambda_4} = 0 \quad \Rightarrow \quad -x_2 + s_4^2 = 0$$

After some simplification, we get

$$\left. \begin{array}{l} 4x_1 + 4x_2 + \lambda_1 + 2\lambda_2 - \lambda_3 = 6 \\ 4x_1 + 6x_2 + \lambda_1 + 3\lambda_2 - \lambda_4 = 3 \\ x_1 + x_2 + s_1^2 = 1 \end{array} \right] \qquad \ldots(1)$$

and

$$\left. \begin{array}{l} 2x_1 + 3x_2 + s_2^2 = 4 \\ \lambda_1 s_1^2 + \lambda_2 s_2^2 + x_1 \lambda_3 + x_2 \lambda_4 = 0 \\ x_1, x_2, s_1^2, s_2^2, \lambda_i \geq 0 \end{array} \right] \qquad \ldots(2)$$

Proceed same as in previous questions, let us introduce artificial variables A_1 and A_2 (≥ 0) in the first two constraints of (1), we have the following problem

$$\text{Max } Z = -A_1 - A_2$$

subject to the constraints

$$4x_1 + 4x_2 + \lambda_1 + 2\lambda_2 - \lambda_3 + A_1 = 6$$

$$4x_1 + 6x_2 + \lambda_1 + 3\lambda_2 - \lambda_4 + A_2 = 3$$

$$x_1 + x_2 + x_3 = 1$$

$$2x_1 + 3x_2 + x_4 = 4$$

and $\quad x_1, x_2, x_3, x_4, A_1, A_2, \lambda_i \geq 0 \ \forall \ i = 1, 2, 3, 4$

Here, we have replaced s_1^2 by x_3 and s_2^2 by x_4 and satisfied the complementary slackness conditions $\Sigma \lambda_i x_i = 0$. The optimum solution to the above LPP will now be obtained by phase-1 of two phase method. Thus, we have the following starting table,

Simplex Table-1

B.V.	C_j C_B	X_B	0 x_1	0 x_2	0 x_3	0 x_4	0 λ_1	0 λ_2	0 λ_3	0 λ_4	-1 A_1	-1 A_2	Min. Ratio $X_B \mid x_2$
A_1	-1	6	4	4	0	0	1	2	-1	0	1	0	3/2
A_2	-1	3	4	⑥	0	0	1	3	0	-1	0	1	1/2(min) →
x_3	0	1	1	1	1	0	0	0	0	0	0	0	1
x_4	0	4	2	3	0	1	0	0	0	0	0	0	4/3
	Δ_j		8	10 ↑	0	0	2	5	-1	-1	0	0	

In the above table, we observe that x_2 is the variable with the most positive entry 10 in the Δ_j-row and hence will enter the basis and A_2 will leave. Then we have the following simplex table.

Simplex Table-2

B.V.	C_j C_B	X_B	0 x_1	0 x_2	0 x_3	0 x_4	0 λ_1	0 λ_2	0 λ_3	0 λ_4	-1 A_1	Min. Ratio $X_B \mid x_1$
A_1	-1	4	4/3	0	0	0	1/3	0	-1	2/3	1	3
x_2	0	1/2	②/③	1	0	0	1/6	1/2	0	$-1/6$	0	3/4(min) →
x_3	0	1/2	1/3	0	1	0	$-1/6$	$-1/2$	0	1/6	0	3/2
x_4	0	5/2	0	0	0	1	$-1/2$	$-3/2$	0	1/2	0	—
	Δ_j		4/3 ↑	0	0	0	1/3	0	-1	2/3	0	

In the above table, we see that x_1 is the variable with the most positive entry 4/3 in the Δ_j-row and hence will enter the basis and x_2 will leave. Then we have the following simplex table.

Simplex Table-3

B.V.	C_j C_B	X_B	0 x_1	0 x_2	0 x_3	0 x_4	0 λ_1	0 λ_2	0 λ_3	0 λ_4	-1 A_1	Min. Ratio
A_1	-1	3	0	-2	0	0	0	-1	-1	1	1	3
x_1	0	3/4	1	3/2	0	0	1/4	3/4	0	$-1/4$	0	—
x_3	0	1/4	0	$-1/2$	1	0	$-1/4$	$-3/4$	0	①/④	0	4/3(min)→
x_4	0	5/2	0	0	0	1	$-1/2$	$-3/2$	0	1/2	0	5
	Δ_j		0	-2	0	0	0	-1	-1	1 ↑	0	

Here, λ_4 is the variable with the most positive entry 1 in Δ_j-row and hence will enter the basis and x_3 will leave. Then the next simplex table is given as below:

Simplex Table-4

	C_j		0	0	0	0	0	0	0	0	-1	Min. Ratio
B.V.	C_B	X_B	x_1	x_2	x_3	x_4	λ_1	λ_2	λ_3	λ_4	A_1	$X_B \mid \lambda_1$
A_1	-1	6	4	4	0	0	①	2	-1	0	1	$6 \rightarrow$
x_1	0	1	1	1	1	0	0	0	0	0	0	—
λ_4	0	1	0	-2	4	0	-1	-3	0	1	0	—
x_4	0	2	0	1	-2	1	0	0	0	0	0	—
	Δ_j		0	4	0	0	1	2	-1	0	0	
							↑					

In the above table, either x_2 or λ_1 will enter the basis but λ_4 is still in the basis, so x_2 can not enter in the basis because of the complementary slackness condition $\lambda_2 x_4 = 0$. Hence, λ_1 enter in the basis and A_1 leaves. Therefore, we have the following simplex table.

Simplex Table-5

	C_j		0	0	0	0	0	0	0	0
B.V.	C_B	X_B	x_1	x_2	x_3	x_4	λ_1	λ_2	λ_3	λ_4
λ_1	0	6	4	4	0	0	1	2	-1	0
x_1	0	1	1	1	1	0	0	0	0	0
λ_4	0	7	4	2	4	0	0	-1	-1	1
x_4	0	2	0	1	-2	1	0	0	0	0
	Δ_j		0	0	0	0	0	0	0	0

In the above table, we observe that all $\Delta_j \leq 0$.

$\Rightarrow$ solution is optimum and is given by

$$x_1 = 1, x_2 = 0$$

and $\qquad$ Max. $Z = 6(1) + 3(0) - 4(1)(0) - 2(1)^2 - 3(0)^2 = 4$

28.5 BEALE'S METHOD

In Wolfe's method, we use the Kuhn-Tucker conditions, but in this method instead of Kuhn-Tucker conditions, we used the results based on calculus.

Consider the quadratic programming problem given by

$$\text{Max. } Z = f(X) = C^T \cdot X + \frac{1}{2} X^T Q X$$

subject to the constraints

$$AX (\leq, \geq \text{ or } =) b$$

and $\qquad X \geq 0$, where $X = (x_1, x_2, ..., x_n) \in R^n$

A is a $m \times n$, b is $m \times 1$, C is $n \times 1$ and Q is an $n \times n$ symmetric matrix.

To solve the above QPP by Beale's method, we use the following procedure,

Working Procedure

STEP 1. Convert the given QPP into the standard form of maximization.

STEP 2. Set the given QPP in standard form using slack and surplus variables.

STEP 3. Choose arbitrarily any m variables as the basic variables such that remaining $(n - m)$ variables becomes non-basic.

STEP 4. Write each basic variable X_B in terms of non-basic variables X_{NB} and u_i if any using the given constraints.

STEP 5. Write the objective function $f(X)$ in terms of non-basic variables X_{NB} (and u_i if any)

STEP 6. Obtain the partial derivaties of $f(X)$ w.r.t. the non-basic variables at the point $X_{NB} = 0$ (or $u = 0$). Then we have the following three cases:

CASE 1 If $\left(\dfrac{\partial f}{\partial X_{NB_k}}\right)_{\substack{X_{NB}=0 \\ u=0}} = 0$ for each $k = 1, 2, ..., n - m$ and $\left(\dfrac{\partial f}{\partial u_i}\right)_{\substack{X_{NB}=0 \\ u=0}} = 0$ then the current basic solution is optimal. Then go to step 9.

CASE 2 If $\left(\dfrac{\partial f}{\partial X_{NB_k}}\right)_{\substack{X_{NB}=0 \\ u=0}} > 0$ for at least one k, then select the most positive one. Then the corresponding non-basic variables will enter the basis. Then go to step 7.

CASE 3 If $\left(\dfrac{\partial f}{\partial X_{NB_k}}\right)_{\substack{X_{NB}=0 \\ u=0}} = 0$ for each $k = 1, 2, ..., n - m$ and $\left(\dfrac{\partial f}{\partial u_i}\right)_{\substack{X_{NB_k}=0 \\ u=0}} \neq 0$ for some $i = r$, then introduce a new non-basic variable $u_j = \dfrac{1}{2}\left(\dfrac{\partial f}{\partial u_r}\right)$ and treat u_r as a basic variable and go to step 4.

STEP 7. Consider $X_{NB_i} = x_k$ to be the entering variable identified in the above step 6. Find the minimum ratio, i.e., $\min\left\{\dfrac{\alpha_{h_0}}{|\alpha_{hk}|}, \dfrac{\gamma_{k_0}}{|\gamma_{kk}|}\right\}$ for all basic variables x_h, where α_{h_0} is the constant term and α_{hk} is the coefficient of x_k in the expression of basic variables x_k when expressed in terms of the non-basic ones and γ_{k_0} is the constant term and γ_{kk} is the coefficient of x_k in $\dfrac{\partial f}{\partial x_k}$. Here, we have the following cases:

CASE 1 If the minimum ratio occur for some $\dfrac{\alpha_{h_0}}{|\alpha_{hk}|}$, the corresponding basic variables x_h will be the outgoing vector, i.e., x_h will leave the basis.

CASE 2 If the minimum ratio occurs for some $\dfrac{\gamma_{k_0}}{|\gamma_{kk}|}$, there exist criterion corresponding to a non-basic variables. Then we introduce an additional non-basic variable (called free variable) defined by

$$u_i = \frac{1}{2}\frac{\partial f}{\partial x_k}, u_i \text{ is unrestricted}$$

STEP 8. Go to step 4 and repeat the same procedure until an optimum basic solution is obtained.

STEP 9. Find the optimum value of X_B and $f(X)$ by setting $X_{NB} = 0$ in the expression obtained in steps 4 and 5.

☞ **REMARKS**

- The free variable u_j is introduced in the set of constraints only for computational purpose and its value is zero at the next feasible solution.
- Since u_j is unrestricted in sign, so while evaluating $\dfrac{\partial f}{\partial u_j}$, both increase and decrease must be checked.

> At any iteration, if a free variable becomes a basic variable and is non-zero then drop the new constraints containing it. This should be done because it is a free variable and therefore will neither be chosen to leave the basis nor will appear in the selection of leaving variable.

 Solved Examples

EXAMPLE 1. *Solve the following quadratic programming problem by Beale's method*

$$\textbf{Max. } Z = 2x_1 + 3x_2 - x_1^2$$

subject to the constraints

$$x_1 + 2x_2 \leq 4$$

and $\quad x_1, x_2 \geq 0$

SOLUTION. Using the slack variable s_1, the above QPP can be written as

$$\text{Max. } Z = 2x_1 + 3x_2 - x_1^2$$

subject to the constraints

$$x_1 + 2x_2 + s_1 = 4$$

and $\quad x_1, x_2, s_1 \geq 0$

Treat s_1 as a basic variable and x_1, x_2 as non-basic.

i.e., $\quad X_B = (s_1)$ and $X_{NB} = (x_1, x_2)$

Then we have

$$s_1 = 4 - x_1 - 2x_2 \qquad \qquad \dots(1)$$

and $\quad f(X) = 2x_1 + 3x_2 - x_1^2 \qquad \qquad \dots(2)$

Differentiating (2) partially w.r.t. x_1 and x_2 we get

$$\left(\frac{\partial f}{\partial x_1} \right)_{X_{NB}=0} = (2 - 2x_1)_{x_1, x_2=0} = 2$$

and $\quad \left(\dfrac{\partial f}{\partial x_2} \right)_{X_{NB}=0} = (3)_{x_1, x_2=0} = 3$

Hence, the most positive is α_2, therefore x_2 will enter in the basis.

Now, $\qquad \beta_1 = \min \left\{ \dfrac{\alpha_{30}}{|\alpha_{32}|}, \dfrac{\gamma_{20}}{|\gamma_{22}|} \right\} = \min \left\{ \dfrac{4}{|-2|}, \dfrac{3}{0} \right\} = 2$

$\Rightarrow \quad s_1$ is the outgoing vector and hence leave the basis.

Thus, we have $\quad X_B = (x_2), X_{NB} = (x_1, s_1)$

Then $\qquad x_2 = \dfrac{1}{2}(4 - x_1 - s_1) = 2 - \dfrac{1}{2}x_1 - \dfrac{1}{2}s_1$

$\Rightarrow \qquad f = 2x_1 + 3\left(2 - \frac{1}{2}x_1 - \frac{1}{2}x_3\right) - x_1^2$

$$= 2x_1 + 6 - \frac{3}{2}x_1 - \frac{3}{2}s_1 - x_1^2$$

$$= 6 + \frac{1}{2}x_1 - x_1^2 - \frac{3}{2}s_1$$

$\Rightarrow \qquad \left(\frac{\partial f}{\partial x_1}\right)_{X_{NB}=0} = \left(\frac{1}{2} - 2x_1\right)_{x_1, x_2 = 0} = \frac{1}{2}$

and $\qquad \left(\frac{\partial f}{\partial x_2}\right)_{X_{NB}=0} = \left(-\frac{3}{2}\right)_{x_1, x_2 = 0} = -\frac{3}{2}$

Here, x_1 enter in the basis.

Now, $\qquad \beta_2 = \min\left\{\frac{\alpha_{20}}{|\alpha_{21}|}, \frac{\gamma_{10}}{|\gamma_{11}|}\right\} = \min\left\{\frac{2}{\left|-\frac{1}{2}\right|}, \frac{1/2}{(-2)}\right\} = \frac{1}{4}$

which is corresponds to $\dfrac{\gamma_{10}}{|\gamma_{11}|}$ so x_2 does not enter in the basis.

Thus, we have to introduce a non-basic variable u_1 such that

$$u_1 = \frac{1}{2}\frac{\partial f}{\partial x_1} = \frac{1}{2}\left(\frac{1}{2} - 2x_1\right) = \frac{1}{4} - x_1 \;\Rightarrow\; x_1 = \frac{1}{4} - u_1$$

$$X_B = (x_1, x_2), \; X_{NB} = (u_1, s_1)$$

$$x_2 = 2 - \frac{1}{2}x_1 - \frac{1}{2}s_1 = 2 - \frac{1}{2}\left(\frac{1}{4} - u_1\right) - \frac{1}{2}s_1$$

$$= 2 - \frac{1}{8} + \frac{1}{2}u_1 - \frac{1}{2}s_1 = \frac{15}{8} + \frac{1}{2}u_1 - \frac{1}{2}s_1$$

and $\qquad f = 6 + x_1\left(\frac{1}{2} - x_1\right) - \frac{3}{2}s_1 = 6 + \left(\frac{1}{4} - u_1\right)\left(\frac{1}{2} - \frac{1}{4} + u_1\right) - \frac{3}{2}s_1$

$$= 6 + \left(\frac{1}{4} - u_1\right)\left(\frac{1}{4} + u_1\right) - \frac{3}{2}s_1 = 6 + \frac{1}{16} - u_1^2 - \frac{3}{2}s_1$$

$$= \frac{97}{16} - u_1^2 - \frac{3}{2}s_1$$

Differentiating the f defined above partially w.r.t. s_1 and u_1 we get

$$\left(\frac{\partial f}{\partial s_1}\right)_{X_{NB}=0} = \left(-\frac{3}{2}\right)_{x_1, x_2 = 0} = \frac{-3}{2}$$

and $\left(\dfrac{\partial f}{\partial u_1}\right)_{X_{NB}=0} = (-2u_1)_{x_1, x_2 = 0} = 0$

which gives the optimal solution and is given by

$$x_1 = \frac{1}{4}, x_2 = \frac{15}{8} \text{ and Max } Z = \frac{97}{16}.$$

EXAMPLE 2. *Solve the following quadratic programming problem by Beale's method.*

$$\text{Max. } Z = 2x_1 + 3x_2 - 2x_2^2$$

subject to the constraints

$$x_1 + 4x_2 \le 4$$
$$x_1 + x_2 \le 2$$

and $\qquad x_1, x_2 \ge 0$ $\qquad$ [BHARATHIAR–1992; SAMBHALPUR–2006; RAJ.–2013]

SOLUTION. Introducing slack variables s_1 and s_2 the given QPP can be written as

$$\text{Max. } Z = 2x_1 + 3x_2 - 2x_2^2$$

subject to the constraints

$$x_1 + 4x_2 + s_1 = 4 \qquad \qquad \qquad \dots(1)$$
$$x_1 + x_2 + s_2 = 4 \qquad \qquad \qquad \dots(2)$$

and $\qquad x_1, x_2, s_1, s_2 \ge 0$

Making s_1, s_2 as basic variables and x_1, x_2 non-basic, then

$$X_B = (s_1, s_2) \text{ and } X_{NB} = (s_1, s_2)$$

Now, from (1), $\qquad s_1 = 4 + 1(-x_1) + 4(-x_2)$

and from (2), $\qquad s_2 = 2 + 1(-x_1) + 2(-x_2)$

We choose the initial basic feasible solution $x_1 = x_2 = 0$, $s_1 = 4$ and $s_2 = 2$.

Then, the initial value of the objective function is $Z = 0$.

also $\qquad \qquad X_B = (s_1, s_2) = (4, 2)$

and $\qquad \qquad X_{NB} = (x_1, x_2) = (0, 0)$

Expressing Z in terms of non-basic variables x_1 and x_2, we get

$$f = Z = 2x_1 + 3x_2 - 2x_2^2$$

$$\Rightarrow \qquad \qquad \frac{\partial f}{\partial x_1} = 2, \frac{\partial f}{\partial x_2} = 3 - 4x_2$$

$$\Rightarrow \qquad \qquad \left(\frac{\partial f}{\partial x_1}\right)_{X_{NB}=0} = (2)_{x_1, x_2=0} \text{ and } \left(\frac{\partial f}{\partial x_2}\right)_{X_{NB}=0} = (3 - 4x_2)_{x_1, x_2=0} = 3$$

Here, the most positive is α_2.

$\Rightarrow$ x_2 will enter in the basis.

Now, critical value β_1 of x_2 is given by

$$\beta_1 = \min\left\{\frac{4}{4}, \frac{2}{1}\right\} = 1$$

The partial derivative $\dfrac{\partial f}{\partial x_2}$ becomes zero at $x_2 = \dfrac{3}{4}$ $(x_1 = 0)$ therefore,

$$\beta_2 = \frac{|\alpha_2|}{2\gamma_{22}} = \frac{|3|}{2(2)} = \frac{3}{4}$$

The new value of the entering variable x_2 is given by

$$x_2 = \min\{\beta_1, \beta_2\} = \left\{1, \frac{3}{4}\right\} = \frac{3}{4}$$

$\Rightarrow$ x_2 does not enter in the basis.

Therefore, introduce a non-basic variable u_1 such that

$$u_1 = \frac{\partial f}{\partial x_2} = 3 - 4x_2 \Rightarrow 4x_2 + u_1 = 3$$

$\therefore$ New $X_B = (s_1, s_2, u_1)$

and $X_{NB} = (x_1, x_2)$

Now, introduce x_2 into the basis and remove u_1 from the basis as in the following table :

Simplex Table-1

B.V.	X_B	x_1	x_2	s_1	s_2	u_1
s_1	1	1	0	1	0	1
s_2	5/4	1	0	0	1	1/4
x_2	3/4	0	1	0	0	$-1/4$

Here, $X_B = (s_1, s_2, x_2) = (1, 5/4, 3/4)$ and $X_{NB} = (x_1, u_1) = (0, 0)$

Now, expressing basic variables into non basic variables as given below :

$$x_2 = \frac{3}{4} - \frac{1}{4}u_1, \quad s_1 = 1 - x_1 - u_1, \quad s_2 = \frac{5}{4} - x_1 - \frac{1}{4}u_1$$

Eliminating the basic variable x_2 from the objective function and expressing it in terms of non-basic variables x_1 and u_1, we get

$$f = 2x_1 + 3\left(\frac{3}{4} - \frac{u_1}{4}\right) - 2\left(\frac{3}{4} - \frac{u_1}{4}\right)^2 = \frac{9}{8} + 2x_1 - \frac{u_1^2}{8}$$

$\Rightarrow$ $\dfrac{\partial f}{\partial x_1} = 2, \quad \dfrac{\partial f}{\partial u_1} = -\dfrac{u_1}{4}$

$\Rightarrow$ $\left(\dfrac{\partial f}{\partial x_1}\right)_{X_{NB}=0} = (2)_{\substack{x_1=0 \\ u_1=0}} = 2$ and $\left(\dfrac{\partial f}{\partial u_1}\right)_{X_{NB}=0} = \left(-\dfrac{u_1}{4}\right)_{(0,0)}$

Here, $\alpha_1 = 2$, $\alpha_2 = 0$. Choosing x_1 to enter in the basis and using the above table, we get

(i) The largest value of x_1 without deriving any basic variables x_1, s_2 and x_2 to zero.

Since $x_2 = \dfrac{3}{4} - \dfrac{1}{4}u_1, \quad s_1 = 1 - x_1 - u_1, \quad s_2 = \dfrac{5}{4} - x_1 - \dfrac{1}{4}u_1$

$\Rightarrow$ $\beta = \min\left\{\dfrac{1}{1}, \dfrac{(5/4)}{1}\right\} = 1$

(ii) Since $\dfrac{\partial f}{\partial x_1} \neq 0 \Rightarrow \beta_2 = 2$

Therefore,

$$x_1 = \min\{\beta_1, \beta_2\} = 1$$

This value of x_1 corresponds to β_1. Thus the new optimal solution is given in the following table:

Simplex Table-2

B.V.	X_B	x_1	x_2	s_1	s_2	u_1
x_1	1	1	0	1	0	1
s_2	1/4	0	0	-1	1	$-3/4$
x_2	3/4	0	1	0	0	$-1/4$

In the above table,
$$X_B = (x_1, s_2, x_2) = (1, 1/4, 3/4)$$
and
$$X_{NB} = (s_1, u_1) = (0, 0)$$
Now, expressing basic variables x_1, x_2, s_2 in terms of non-basic variables s_1, u_1, we get
$$x_1 = 1 - s_1 - u_1, \quad s_2 = \frac{1}{4} + s_1 + \frac{3}{4}u_1, \quad x_2 = \frac{3}{4} + \frac{1}{4}u_1$$
Also, expressing objective function Z in terms of non-basic variables s_1 and u_1, we get
$$f = \frac{9}{8} + 2(1 - s_1 - u_1) - \frac{1}{8}u_1^2 = \frac{25}{8} - 2s_1 - 2u_1 - \frac{1}{8}u_1^2$$

$$\frac{\partial f}{\partial s_1} = -2 \Rightarrow \left(\frac{\partial f}{\partial s_1}\right)_{X_{NB}=0} = (-2)_{(s_1, u_1)=(0,0)} = -2 < 0$$

and $\quad \dfrac{\partial f}{\partial u_1} = -2 - \dfrac{1}{4}u_1 \Rightarrow \left(\dfrac{\partial f}{\partial u_1}\right)_{X_{NB}=0} = \left(-2 - \dfrac{1}{4}u_1\right)_{(0,0)} = -2 < 0$

$\Rightarrow$ solution is optimal and is given by
$$x_1 = 1, \ x_2 = \frac{3}{4} \text{ and Max. } Z = \frac{25}{8}$$

EXAMPLE 3. *Solve the following QPP by using Beale's method*
$$\textbf{Min. } Z = 6 - 6x_1 + 2x_1^2 - 2x_1x_2 + 2x_2^2$$
subject to the constraints
$$x_1 + x_2 \leq 2$$
and $\qquad x_1, x_2 \geq 0$

SOLUTION. Convert the given minimization QPP as maximization, we get
$$\text{Max}(-Z) = f = -6 + 6x_1 - 2x_1^2 + 2x_1x_2 - 2x_2^2$$
Now, using slack variable $s_1 (\geq 0)$ the given QPP can be written as follows :
$$\text{Max}(-Z) = f = -6 + 6x_1 - 2x_1^2 + 2x_1x_2 - 2x_2^2$$

subject to the constraints
$$x_1 + x_2 + s_1 = 2$$
$$x_1, x_2, s_1 \geq 0$$
Choosing s_1 as basic variable and x_1, x_2 as non-basic.

Therefore, $\qquad X_B = (s_1)$ and $X_{NB} = (x_1, x_2)$
Now expressing basic variables in f in terms of non-basic variables, we get
$$s_1 = 2 - x_1 - x_2$$
and $\qquad f = -6 + 6x_1 - 2x_1^2 + 2x_1x_2 - 2x_2^2$

$$\Rightarrow \left(\frac{\partial f}{\partial x_1}\right)_{X_{NB}=0} = (6 - 4x_1 + 2x_2)_{\substack{x_1=0 \\ x_2=0}} = 6 \text{ and } \left(\frac{\partial f}{\partial x_2}\right)_{X_{NB}=0} = (-2x_1 + 4x_2)_{\substack{x_1=0 \\ x_2=0}} = 0$$

Here α_1 is most positive, so x_1 will enter the basis.

Now, $\min\left\{\dfrac{\alpha_{30}}{|\alpha_{31}|},\dfrac{\gamma_{10}}{|\gamma_{11}|}\right\} = \min\left\{\dfrac{2}{|-1|},\dfrac{6}{|-4|}\right\} = \dfrac{3}{2}$

which corresponds to $\dfrac{\gamma_{10}}{|\gamma_{11}|}$

$\Rightarrow$ s_1 does not leave the basis.

Therefore, we introduce a new non-basic variable u_1, such that

$$u_1 = \frac{1}{2}\frac{\partial f}{\partial x_1} = 3 - 2x_1 + x_2$$

$\therefore$ $\qquad X_B = (s_1, x_1),\ X_{NB} = (x_2, u_1)$

$\Rightarrow$ $\qquad x_1 = \dfrac{3}{2} - \dfrac{1}{2}u_1 + \dfrac{1}{2}x_2$

$$s_1 = 2 - x_1 - x_2 = 2 - \left(\frac{3}{2} - \frac{1}{2}u_1 + \frac{1}{2}x_2\right) - x_2$$

$$= \frac{1}{2} + \frac{1}{2}u_1 - \frac{3}{2}x_2$$

and $\qquad f = -6 + 6x_1 - 2x_1^2 + 2x_1 x_2 - 2x_2^2 = -6 + 2x_1(3 - x_1 + x_2) - 2x_2^2$

$$= -6 + 2\left(\frac{3}{2} - \frac{1}{2}u_1 + \frac{1}{2}x_2\right)\left(3 - \frac{3}{2} + \frac{1}{2}u_1 - \frac{1}{2}x_2 + x_2\right) - 2x_2^2$$

$$= -6 + (3 - u_1 + x_2)\left(\frac{3}{2} + \frac{1}{2}u_1 + \frac{1}{2}x_2\right) - 2x_2^2$$

$$= -6 + \frac{1}{2}(3 + x_2 - u_1)(3 + x_2 + u_1) - 2x_2^2$$

$$= -6 + \frac{1}{2}\left\{(3 + x_2)^2\right\} - 2x_2^2$$

$$= -6 + \frac{1}{2}\left(9 + x_2 - 6x_2 - u_1^2\right) - 2x_2^2$$

$$= -\frac{3}{2} - \frac{u_1^2}{2} - 3x_2 - \frac{3}{2}x_2^2$$

$\Rightarrow$ $\qquad \left(\dfrac{\partial f}{\partial x_2}\right)_{X_{NB}=0} = (3 - 3x_2)_{\substack{x_1=0 \\ u_1=0}} = 3$

and $\qquad \left(\dfrac{\partial f}{\partial u_1}\right)_{X_{NB}=0} = (-u_1)_{\substack{x_1=0 \\ u_1=0}} = 0$

$\Rightarrow$ x_2 enter in the basis.

Now, $\min\left\{\dfrac{\alpha_{30}}{|\alpha_{32}|},\dfrac{\alpha_{10}}{|\alpha_{12}|},\dfrac{\gamma_{20}}{|\gamma_{22}|}\right\} = \min\left\{\dfrac{1/2}{|-3/2|},\dfrac{3/2}{|1/2|},\dfrac{3}{|-3|}\right\} = \dfrac{1}{3}$

$\Rightarrow$ s_1 will leave the basis and thus now

$\qquad X_B = (x_1, x_2)$ and $X_{NB} = (u_1, s_1)$

Now, $\qquad x_2 = \dfrac{1}{3} + \dfrac{1}{3}u_1 - \dfrac{2}{3}s_1$

$$x_1 = \frac{3}{2} - \frac{1}{2}u_1 + \frac{1}{2}x_2 = \frac{3}{2} - \frac{1}{2}u_1 + \frac{1}{2}\left(\frac{1}{3} + \frac{1}{3}u_1 - \frac{2}{3}s_1\right)$$

$$= \frac{3}{2} - \frac{1}{2}u_1 + \frac{1}{6} + \frac{1}{6}u_1 - \frac{1}{3}s_1$$

$$= \frac{5}{3} - \frac{1}{3}u_1 + \frac{1}{3}s_1$$

and

$$f = -\frac{3}{2} - \frac{u_1^2}{2} + \frac{3}{2}x_2(2 - x_2)$$

$$= -\frac{3}{2} - \frac{u_1^2}{2} + \frac{3}{2}\left(\frac{1}{3} + \frac{1}{3}u_1 - \frac{2}{3}s_1\right)\left(2 - \frac{1}{3} - \frac{1}{3}u_1 + \frac{2}{3}s_1\right)$$

$$= -\frac{3}{2} - \frac{u_1^2}{2} + \frac{1}{6}(1 + u_1 - 2s_1)(5 - u_1 + 2s_1)$$

$$= -\frac{3}{2} - \frac{u_1^2}{2} + \frac{1}{6}\left(5 - u_1 + 2s_1 + 5u_1 - u_1^2 + 2s_1 u_1 - 10s_1 + 2s_1 u_1 - 4s_1^2\right)$$

$$= -\frac{3}{2} - \frac{u_1^2}{2} + \frac{5}{6} + \frac{4u_1}{6} - \frac{8}{6}s_1 - \frac{u_1^2}{6} + \frac{4}{6}s_1 u_1 - \frac{4}{6}s_1^2$$

$$= -\frac{4}{6} - \frac{4}{6}u_1^2 + \frac{4}{6}u_1 - \frac{4}{3}s_1 - \frac{4}{6}s_1^2 + \frac{4}{6}s_1 u_1$$

$$= -\frac{2}{3} + \frac{2}{3}u_1 - \frac{2}{3}u_1^2 - \frac{4}{3}s_1 + \frac{2}{3}s_1 u_1 - \frac{2}{3}s_1^2$$

which implies

$$\left(\frac{\partial f}{\partial s_1}\right)_{X_{NB}=0} = \left(-\frac{4}{3} - \frac{2}{3}u_1 - \frac{4}{3}s_1\right)_{\substack{x_1=0 \\ u_1=0}} = -\frac{4}{3}$$

$$\left(\frac{\partial f}{\partial u_1}\right)_{X_{NB}=0} = \left(\frac{2}{3} - \frac{4}{3}u_1 + \frac{2}{3}s_1\right)_{\substack{s_1=0 \\ u_1=0}} = \frac{2}{3}$$

Since $\dfrac{\partial f}{\partial u_1} \neq 0$, this solution can be improved further.

Let x_1 does not enter in the basis, so we introduce another non-basic free variable u_2 such that

$$u_2 = \frac{1}{2}\frac{\partial f}{\partial u_1} = \frac{1}{3} - \frac{1}{3}u_1 + \frac{1}{3}s_1$$

Now $X_B = (x_1, x_2, u_1)$ and $X_{NB} = (s_1, u_2)$

$$u_1 = \frac{1}{2} - \frac{3}{2}u_2 + \frac{1}{2}s_1$$

$$x_2 = \frac{1}{3} + \frac{1}{3}u_1 - \frac{2}{3}s_1 = \frac{1}{3} + \frac{1}{3}\left(\frac{1}{2} - \frac{3}{2}u_2 + \frac{1}{2}s_1\right) - \frac{2}{3}s_1$$

$$= \frac{1}{2} - \frac{1}{2}u_2 - \frac{1}{2}s_1$$

$$x_1 = \frac{5}{3} - \frac{1}{3}u_1 - \frac{1}{3}s_1 = \frac{5}{3} - \frac{1}{3}\left(\frac{1}{2} - \frac{3}{2}u_2 + \frac{1}{2}s_1\right) - \frac{1}{3}s_1$$

$$= \frac{5}{3} - \frac{1}{6} + \frac{1}{2}u_2 - \frac{1}{6}s_1 - \frac{1}{3}s_1$$

$$= \frac{3}{2} + \frac{1}{2}u_2 - \frac{1}{2}s_1$$

and $\quad f = -\frac{2}{3} + \frac{2}{3}u_1 - \frac{4}{3}s_1 + \frac{2}{3}s_1 u_1 - \frac{2}{3}u_1^2 - \frac{2}{3}s_1^2$

$$= -\frac{2}{3} + \frac{2}{3}u_1(1 + s_1 - u_1) - \frac{4}{3}s_1 - \frac{2}{3}s_1^2$$

$$= -\frac{2}{3} + \frac{2}{3}\left(\frac{1}{2} - \frac{3}{2}u_2 + \frac{1}{2}s_1\right)\left(1 + s_1 - \frac{1}{2} + \frac{3}{2}u_2 - \frac{1}{2}s_1\right) - \frac{4}{3}s_1 - \frac{2}{3}s_1^2$$

$$= -\frac{2}{3} + \frac{1}{6}(1 - 3u_2 + s_1)(2 + 2s_1 - 1 + 3u_2 - s_1) - \frac{4}{3}s_1 - \frac{2}{3}s_1^2$$

$$= -\frac{2}{3} + \frac{1}{6}(1 - 3u_2 + s_1)(1 + s_1 + 3u_2) - \frac{4}{3}s_1 - \frac{2}{3}s_1^2$$

$$= -\frac{2}{3} + \frac{1}{6}\{(1 + s_1)^2 - 9u_2^2\} - \frac{4}{3}s_1 - \frac{2}{3}s_1^2$$

$$= -\frac{2}{3} + \frac{1}{6}[1 + s_1^2 + 2s_1 - 9u_2^2] - \frac{4}{3}s_1 - \frac{2}{3}s_1^2$$

$$= -\frac{2}{3} + \frac{1}{6} + \frac{s_1^2}{6} + \frac{1}{3}s_1 - \frac{3}{2}u_2^2 - \frac{4}{3}s_1 - \frac{2}{3}s_1^2$$

$$= -\frac{1}{2} - \frac{1}{2}s_1^2 - s_1 - \frac{3}{2}u_2^2$$

which implies

$$\left(\frac{\partial f}{\partial s_1}\right)_{X_{NB}=0} = (-s_1 - 1)_{\substack{s_1=0 \\ u_2=0}} = -1 \text{ and } \left(\frac{\partial f}{\partial u_2}\right)_{X_{NB}=0} = \left(-\frac{3}{2} \cdot 2u_2\right)_{\substack{s_1=0 \\ u_2=0}} = 0$$

which gives an optimum solution. Now ignoring the free variable u_1, the optimal solution is given by

$$x_1 = \frac{3}{2}, \ x_2 = \frac{1}{2}, \ Z = -f = -\left(-\frac{1}{2}\right) = \frac{1}{2}$$

EXAMPLE 4. *Solve the following QPP by Beale's method*

$$\textbf{\textit{Min. }} Z = -4x_1 + x_2^2 - 2x_1 x_2 + 2x_2^2$$

subject to the constraints

$$2x_1 + x_2 \geq 6$$
$$x_1 - 4x_2 \geq 0$$

and $\qquad x_1, x_2 \geq 0$

SOLUTION. Introducing the surplus variables s_1 and s_2 and convert the problem of minimization into maximization, we can write

$$\text{Max. } Z = 4x_1 - x_1^2 + 2x_1 x_2 - 2x_2^2$$

s.t. $\quad 2x_1 + x_2 - s_1 = 6 \Rightarrow s_1 = -6 + 2x_1 + x_2$

$$x_1 - 4x_2 - s_2 = 0 \Rightarrow s_2 = x_1 - 4x_2$$

and $\qquad x_1, x_2, s_1, s_2 \geq 0$.

Let s_1, s_2 be the basic and x_1, x_2 be the non-basic variables.

Now
$$\alpha_1 = \left(\frac{\partial f}{\partial x_1}\right)_{X_{NB}=0} = (4 - 2x_1 + 2x_2)_{\substack{x_1=0 \\ x_2=0}} = 4 > 0$$

$$\alpha_2 = \left(\frac{\partial f}{\partial x_2}\right)_{X_{NB}=0} = (2x_1 + 4x_2)_{\substack{x_1=0 \\ x_2=0}} = 0$$

$$X_B = (s_1, s_2) = (-6, 0), \ X_{NB} = (x_1, x_2) = (0, 0)$$

Here
$$\alpha_1 = 4, \ \alpha_2 = 0$$

Clearly both are positive, so choose
x_1(most positive value of α_1) to enter into the basis

Then critical value β_1 of x_1 is given by

$$\beta_1 = \min.\left\{-\frac{6}{|2|}, \frac{0}{|1|}\right\} = -3$$

$\Rightarrow s_1$ will leave the basis. Expressing the new basic variables x_1, s_2 and f in terms of new non-basic variables x_2 and s_1 such that

$$x_1 = 3 - \frac{1}{2}x_2 + \frac{1}{2}s_1, \ s_2 = 3 - \frac{3}{2}x_2 + \frac{1}{2}s_1$$

$$f = 4\left(3 - \frac{1}{2}x_2 + \frac{1}{2}s_1\right) - \left(3 - \frac{1}{2}x_2 + \frac{1}{2}s_1\right)^2 + 2\left(3 - \frac{1}{2}x_2 + \frac{1}{2}s_1\right)x_2 - 2x_2^2$$

$$= 9 + x_2 - s_1 + \frac{3}{2} + \frac{3}{2}x_2 s_1 - \frac{13}{4}x_2^2 - \frac{1}{4}s_1^2$$

which implies
$$\left(\frac{\partial f}{\partial x_2}\right)_{X_{NB}=0} = \left(1 + \frac{3}{2}s_1 - \frac{13}{2}x_2\right)_{\substack{x_2=0 \\ s_1=0}} = 1$$

$\Rightarrow \qquad \left(\frac{\partial f}{\partial s_1}\right)_{X_{NB}=0} = \left(-1 + \frac{3}{2}x_2 - \frac{1}{2}s_1\right)_{\substack{x_2=0 \\ s_1=0}} = -1$

$\Rightarrow x_2$ will enter the basis.

Now compute $\min\left\{\dfrac{3}{|-1/2|}, \dfrac{3}{|-3/2|}\right\} = 2$

Since, the minimum ratio corresponds to β_2, we introduce a non-basic free variable defined by

$$u_1 = \frac{1}{2}\frac{\partial f}{\partial x_2} = \frac{1}{2} + \frac{3}{4}s_1 - \frac{13}{4}x_2$$

Therefore, $\qquad X_B = (x_1, s_2, x_2)$ and $X_{NB} = (s_1, u_1)$

Expressing basic variables and f in terms of non-basic variables, we have

$$x_1 = \frac{38}{13} - \frac{3}{26}s_1 + \frac{2}{13}u_1, \ x_2 = \frac{2}{13} + \frac{3}{13}s_1 - \frac{4}{13}u_1, \ s_2 = \frac{30}{13} - \frac{27}{26}s_1 + \frac{18}{13}u_1$$

$$f = 9 + \frac{1}{13}(2 + 3s_1 - 4u_1) - s_1 + \frac{3}{26}s_1(2 + 3s_1 - 4u_1) - \frac{1}{52}(2 + 3s_1 - u_1)^2 - \frac{1}{4}s_1^2$$

which implies

$$\left(\frac{\partial f}{\partial s_1}\right)_{X_{NB}=0} = \left(\frac{3}{13} - 1 + \frac{3}{26}(2 - u_1) + \frac{18}{26}s_1 - \frac{6}{52}(2 + 3s_1 - 4u_1) - \frac{1}{2}s_1\right)_{\substack{s_1=0 \\ u_1=0}}$$

$$= -\frac{9}{13} < 0$$

and $\left(\dfrac{\partial f}{\partial x_1}\right)_{X_{NB}=0} = \left(-\dfrac{4}{13} - \dfrac{12}{26}s_1 + \dfrac{8}{52}(2 + 3s_1 - 4u_1)\right)_{\substack{s_1=0 \\ u_1=0}} = 0$

Clearly both the above values ≤ 0. So the optimal value of Z is obtained by setting $u_1 = 0$, $s_1 = 0$ in the current objective function given by

$$Z = 9 + \frac{2}{13} - \frac{2}{52} = \frac{474}{52}$$

Hence, the optimum solution of the given QPP is

$$x_1 = \frac{38}{13}, \ x_2 = \frac{2}{13} \text{ and Min. } Z = \frac{474}{52}$$

28.6 SEPARABLE PROGRAMMING

We know that a separable programming is an indirect method to solve non-linear programming problem. It is useful for solving those non-linear programming problem in which the objective function and constraints are separable (A function of n variables $f(x_1, x_2, ..., x_n)$ is said to be separable if it can be written as the sum of n functions $f_1(x_1), f_2(x_2), ..., f_n(x_n)$).

If function is not separable, then it can be made separable by using simplified approximations.

28.7 SOME GENERAL DEFINITIONS RELATED TO SEPARABLE PROGRAMMING

(I) SEPARABLE PROGRAMMING PROBLEM

In a non-linear programming problem if the objective function can be expressed as a linear combinations of several different single-variable functions, of which some or all are non-linear. Such type of NLPP is called separable programming problem.

(II) SEPARABLE CONVEX PROGRAMMING

A separable programming in which separate functions are convex, also the non-linear function $f(x)$ is convex in case of minimization and concave in case of maximization is called separable convex programming.

28.8 PIECEWISE LINEAR APPROXIMATION OF NON-LINEAR FUNCTIONS

Consider

$$\text{Optimize } Z = \sum_{j=1}^{n} f_j(x_j)$$

subject to the constraints

$$\sum_{j=1}^{n} a_{ij}x_j = b_i, \ i = 1, 2, ..., m$$

and $\qquad x_j \geq 0$

where $f_j(x_j)$ is the jth separable function which is to be approximated over a defined interval.

Let $(a_k, b_k) \ \forall \ k = 1, 2, ..., p$ be the kth breaking point joining a linear segment which approximated $f(x)$. Further let us define the weight function w_k such that

$$\sum_{k=1}^{p} w_k = 1$$

Further, add an additional constraint, (if necessary) w_{k+1} such that w_k and w_{k+1} are equated to zero to find the weighted average of breaking point. Therefore, $f(x)$ is approximated as given below

$$f(x) = \sum_{k=1}^{p} b_k w_k, \quad x = \sum_{k=1}^{p} a_k w_k$$

provided

$$0 \leq w_1 \leq y_1$$

$$0 \leq w_2 \leq y_1 + y_2$$

$$0 \leq w_3 \leq y_2 + y_3$$

$$::$$

$$0 \leq w_{k-1} \leq y_{k-2} + y_{k-1}$$

$$0 \leq w_k \leq y_{k-1}$$

and

$$\sum_{k=1}^{p} w_k = 1, \quad \sum_{k=1}^{p-1} y_k = 1$$

$$y_k = 0 \ \text{or} \ 1 \ \forall \ k$$

The variables for approximation are now w_k and y_k.

28.9 MIXED INTEGER APPROXIMATION OF SEPARABLE PROBLEMS

We can approximate a single variable non-linear separable function by a piecewise linear function using mixed integer programming.

Let us suppose the no. of breaking point for j^{th} variable x_j be equal to p_j and g_{ik} be the kth breaking values. If w_{jk} be the weight associated with the kth breaking point of jth variable x_j. Then we have the following mixed integer programming

$$\text{Optimize} \ . \ Z = \sum_{j=1}^{n} \sum_{k=1}^{p_j} f_j(a_{jk}) w_{jk}$$

subject to the constraints

$$Z = \sum_{j=1}^{n} \sum_{k=1}^{p_j} g_{ij} \, | \, a_{jk} \, | \, w_{jk} \leq b_i, \ i = 1, 2, ..., m$$

$$0 \ \leq \ w_{j1} \leq y_{j1}$$

$$\vdots \qquad \vdots \ \vdots \qquad \vdots \ \vdots \qquad \vdots$$

$$0 = w_{jk} \leq y_{jk-1} + y_{jk} \ , \ k = 2, 3, ..., p_j^{-1}$$

and
$$\sum_{k=1}^{P_j} w_{jk} = 1, \quad \sum_{k=1}^{P_j-1} y_{jk} = 1,$$
$$y_{jk} = 0 \text{ or } 1 \ \forall \ j \text{ and } k.$$

28.10 VALIDITY OF MIXED INTEGER APPROXIMATION

The mixed integer approximation is valid under the following two conditions:
(i) For each j, no more than two w_{jk} should appear in the basis
(ii) Two w_{jk} can be positive only if they are adjacent

☞ REMARKS
- The restricted basic method gives only a local optimum, while the mixed integer programming method gives the global optimum.
- The solution obtained above may not be feasible for the original non-linear programming.

28.11 METHOD OF SOLUTION OF SEPARABLE PROGRAMMING PROBLEMS

To solve the given separable programming, use the following procedure:

 Working Procedure

We use the following steps :
STEP 1. Convert the minimization problem into maximization.
STEP 2. Examine the concavity (convexity) conditions of the function $f_j(x_j)$ and $g_{ij}\, x_j$. If condition is satisfied, go to the next step otherwise stop.
STEP 3. Divide the given interval $0 \le x_j \le t_j$ into a no. of breaking points a_{jk} ($k = 1, 2, ..., p_j$) so that
$$a_{j1} = 0, a_{j1} < a_{j2} ... < a_{jp_j}$$
STEP 4. For each point a_{jk} obtain piecewise linear approximation $f_j(x_j)$ and $g_{ij}(x_j) \ \forall \ i$ and j.
STEP 5. Solve the resulting LPP by two-phase method by assuming w_{i1} ($i = 1, 2, ..., m$) as artificial variable.
STEP 6. Find the optimum solution of the original non-linear programming problem by using
$$x_j^* = \sum_{k=1}^{P_j} a_{jk} w_{jk}, \qquad j = 1, 2, ..., n$$

☞ REMARKS
- The solution space of the approximate problem may have additional extreme points which do not exist in the solution space of the original problem.
- Separable programming gives the approximate solution of the problem.
- To find the better approximation, have the greater no. of breaking points.

 Solved Examples

EXAMPLE 1. *Solve the following NLPP by using separable programming algorithm.*
$$\textbf{Max. } Z = x_1 + x_2^4$$
subject to the constraints
$$3x_1 + 2x_2^2 \le 9$$
and $\qquad x_1, x_2 \ge 0$

SOLUTION. Here, we have the following separable functions.

$$f_1(x_1) = x_1, \qquad f_2(x_2) = x_2^2$$
$$g_{11}(x_1) = 3x_1, \qquad g_{12}(x_2) = 2x_2^2$$

Clearly the function $f_1(x_1)$ and $g_{11}(x_1)$ satisfy concavity (convexity) conditions.

From the constraints of the given problem, we have

$$x_1 \leq 3 \text{ and } x_2 \leq \sqrt{9/2} = 2.13 < 3$$

So, 3 is the upper limit for x_1 and x_2 both.

$\therefore$ Divide the interval [0, 3] into four breaking points a_{jk} of equal intervals such that

$$a_{j1} = 0, \quad a_{j1} < a_{j2} < a_{j3} < a_{j4} = 3$$

$\therefore$ The piecewise linear approximation for $f_2(x_2)$ and $g_{12}(x_2)$ are given in the following table:

k	a_{jk}	$f_2(x_2 = a_{jk})$	$g_{12}(x_2 = a_{jk})$
1	0	0	0
2	1	1	2
3	2	16	8
4	3	81	18

So, $\qquad f_2(x_2) = w_{21}f_2(a_{21}) + w_{22}f_2(a_{22}) + w_{23}f_2(a_{23}) + w_{24}f_2(a_{24})$

$$= w_{21}(0) + w_{22}(1) + w_{23}(16) + w_{24}(81)$$

$$= w_{22} + 16 w_{23} + 81 w_{24}$$

and $\qquad g_{12}(x_2) = w_{21}g_{12}(a_{21}) + w_{22}g_{12}(a_{22}) + w_{23}f_{12}(a_{23}) + w_{24}g_{12}(a_{24})$

$$= w_{21}(0) + w_{22}(2) + w_{23}(8) + w_{24}(18)$$

$$= 2w_{22} + 8 w_{23} + 18 w_{24}$$

Therefore, the approximated linear programming problem is given below

$$\text{Max. } f(x) = x_1 + w_{22} + 16 w_{23} + 81w_{24}$$

subject to the constraints

$$3x_1 + 2w_{22} + 8w_{23} + 18 w_{24} \leq 9$$

$$w_{21} + w_{22} + w_{23} + w_{24} = 1$$

and $\qquad x_1, w_{21}, w_{22}, w_{23}, w_{24} \geq 0$

with the following two conditions:

(i) For each j, no more than two w_{jk} are positive

and (ii) if two w_{jk} are positive, they must correspond to adjacent point.

Now, treating w_{21} as the artificial variable, solve the given LPP by two-phase simplex method as follows :

The simplex tables of phase-2 are given as below :

Simplex Table-1

B.V.	C_j		1	1	16	81	0	0	Min. Ratio
	C_B	X_B	x_1	w_{22}	w_{23}	w_{24}	s_1	w_{21}	
s_1	0	9	3	2	8	18	1	0	9/8
w_{21}	0	1	0	1	①	1	0	1	1/1(min.) →
	$\Delta_j = C_j - Z_j$		1	1	16 ↑	81	—	—	

From the above table we observed that w_{24} should enter the basis. Since w_{21} is artificial basic variable, it must be dropped before w_{21} enter the basis. By minimum ratio rule, s_1 is the leaving variable. So w_{24} can not enter the basis. So, consider the next best entering variable w_{23}.
Again the artificial variable w_{21} must be dropped.
Then, we have the following table.

Simplex Table-2

B.V.	C_j		1	1	16	81	0	Min. Ratio
	C_B	X_B	x_1	w_{22}	w_{23}	w_{24}	s_1	
s_1	0	1	3	–6	0	⑩	1	1/10 (min.) →
w_{23}	16	1	0	1	1	1	0	1/1
	Δ_j		1	–15	–	65 ↑	–	

Clearly, in the above table w_{24} is the entering variable.
Also, s_1 is the leaving variable. Then we have the following table.

Simplex Table-3

B.V.	C_j		1	1	16	81	0	Min. Ratio
	C_B	X_B	x_1	w_{22}	w_{23}	w_{24}	s_1	
w_{24}	81	1/6	3/10	–6/10	0	1	1/10	
w_{23}	16	9/10	–3/10	16/10	1	0	–1/10	
	Δ_j		–37/2	24	–	–	13/2	

From the above table, we observe that w_{22} should enter in the basis which is not possible because w_{24} can't be dropped from the current solution. So procedure terminated.
Hence, the optimum solution of the approximate LP is given by

$$w_{23} = \frac{9}{10}, \ w_{24} = \frac{1}{10} \quad \text{Max.} f(x) = 22.5$$

Hence, the optimum solution of the original NLPP is given by

$$x_j = \sum_{k=1}^{4} a_{jk} w_{jk}, \ j = 1, 2$$

$$\Rightarrow \quad x_2 = a_{21} \omega_{21} + a_{22} \omega_{22} + a_{23} \omega_{23} + a_{24} w_{24}$$

$$= 0.0 + 1.0 + 2.\frac{9}{10} + 3\left(\frac{1}{10}\right) = 2.1$$

$$x_1 = 0$$

and Max. $f(x) = 22.5$

Exercise-28.1

Use Wolfe's method, solve the following quadratic programming problem:

1. Min. $Z = 6 - 6x_1 + 2x_1^2 - 2x_1x_2 + 2x_2^2$
   subject to the constraints
   $$x_1 + x_2 \leq 2$$
   and $x_1, x_2 \geq 0$

2. Min. $Z = x_1^2 + x_2^2 + x_3^2$
   subject to the constraints
   $$x_1 + x_2 + 3x_3 = 2$$
   $$5x_1 + 2x_2 + x_3 = 5$$
   and $x_1, x_2, x_3 \geq 0$

3. Max. $Z = 8x_1 + 10x_2 - x_1^2 - x_2^2$
   subject to the constraints
   $$3x_1 + 2x_2 \leq 6$$
   and $x_1, x_2 \geq 0$

Use the Beale's method solve the following QPP:

4. Max. $Z = 10x_1 + 25x_2 - 10x_1^2 - x_2^2 - 4x_1x_2$
   subject to the constraints
   $$x_1 + 2x_2 \leq 10$$
   $$x_1 + x_2 \leq 9$$

and $x_1, x_2 \geq 0$

5. Max. $Z = 2x_1x_2 - 5x_1 - 13x_2 + 3x_2^2 - 10$
   subject to the constraints
   $$x_1 + x_2 \leq 1$$
   $$4x_1 + x_2 \geq 2$$
   and $x_1, x_2 \geq 0$

6. Max. $Z = 4x_1 + 6x_2 - x_1^2 - 3x_2^2$
   subject to the constraints
   $$x_1 + x_2 \leq 4$$
   and $x_1, x_2 \geq 0$

Solve the following NLPP using separable programming algorithm.

7. Max. $Z = 3x_1 + 2x_2$
   subject to the constraints
   $$4x_1^2 + x_2^2 \leq 16$$
   and $x_1, x_2 \geq 0$

8. Max. $Z = 16 - 2(x_1 - 3)^2 - (x_2 - 7)^2$
   subject to the constraints
   $$x_1^2 + x_2 \leq 16$$
   and $x_1, x_2 \geq 0$

ANSWERS

1. $x_1 = \dfrac{3}{2}, x_2 = \dfrac{1}{2}, \text{Max } Z = \dfrac{1}{2}$

2. $x_1 = \dfrac{81}{100}, x_2 = \dfrac{7}{20}, x_3 = \dfrac{7}{20}, \text{Max. } Z = \dfrac{17}{20}$

3. $x_1 = \dfrac{4}{13}, x_2 = \dfrac{33}{13}, \text{Max.} Z = \dfrac{267}{13}$

4. $x_1 = 0, x_2 = 5, \text{Max. } Z = 100$

5. $x_1 = 1, x_2 = 0, \text{Max.} Z = -15$

6. $x_1 = 2, x_2 = 1, \text{Max. } Z = 7$

7. $x_1 = 1, x_2 = \dfrac{24}{7}, \text{Max.} f = \dfrac{69}{7}$

8. $x_1 = 3, x_2 = 7, \text{Max.} f = 16$

 REVIEW QUESTIONS

1. Define Concavity and Convexity of a function.
2. Write a short note on Quadratic programming problem.
3. Write a short note on Separable programming problem.
4. Write the procedure of Wolfe's method.
5. Write the procedure of Beale's method.
6. Explain the Kuhn-Tucker conditions in details.

 MULTIPLE CHOICE QUESTIONS (CHOOSE THE MOST APPROPRIATE ONE)

1. An approximate solution of a non-linear programming problem is obtained by using a:
   (a) separable programming
   (b) quadratic programming
   (c) both (a) and (b)
   (d) none of these

2. While solving an NLPP, it is reduced to an LPP if it is:
   (a) QPP
   (b) SPP
   (c) both (a) and (b)
   (d) none of these

3. Every NLPP can be solved by using a/an:
   (a) SPP
   (b) QPP
   (c) both (a) and (b)
   (d) none of these

ARCHIVE

1. Is it correct to say that in a quadratic programming problem the objective function and the constraints both should be quadratic? If not give your own comments.   [AMIE–2005]

2. Show that the non-linear non-convex programming problem given by

Minimize $Z = a_0 + b_{01}x_1 + \left(\sum_{j=2}^{5} b_{0j}x_j \right) x_1$

subject to the constraints

$0 \le a_{i1}x_1 + \left(\sum_{j=2}^{5} a_{ij}x_j \right) x_1 \le b_i : i = 1, 2, \dots, 5$

$l_i \le x_i \le u_i : i = 1, 2, \dots, 5$

can be transformed into a concave LP problem by setting $y_i = x_i \cdot x_i$, $(i = 1, 2, \dots, 5)$ and $y_1 = x_1$ where $a_0, b_{0j}, a_{ij}, b_i, l_i, u_i$ are real constants.

3. Solve: Min $Z = x_1^2 + x_2^2 + 5$

subject to the constraints

$3x_1^4 + x_2 \le 243$

$x_1 + 2x_2^2 \le 32$

and $x_1, x_2 \ge 0$ [AMIE–2000]

4. Show that if $f_{0j}(x_j)$ is strictly convex and $f_{ij}(x_j)$ is concave for $i = 1, 2, \dots, m$, then we can discard the additional restriction in the approximated separable non-linear programming problem given by

Min $Z = \sum_{j=1}^{n} f_{0j}(x_j)$

subject to the constraints $\sum_{j=1}^{n} f_{0j}(x_j) \ge b_i :$

$i = 1, 2, \dots, m.$

APPENDIX
(Selected Tables)

Table 1. Logarithms

	0	1	2	3	4	5	6	7	8	9	Mean Differences								
											1	2	3	4	5	6	7	8	9
10	0000	0043	0086	0128	0170	0212	0253	0294	0334	0374	4	8	12	17	21	25	29	33	37
11	0414	0453	0492	0531	0569	0607	0645	0682	0719	0755	4	8	11	15	19	23	26	30	34
12	0792	0828	0864	0899	0934	0969	1004	1038	1072	1106	3	7	10	14	17	21	24	28	31
13	1139	1173	1206	1239	1271	1303	1335	1367	1399	1430	3	6	10	13	16	19	23	26	29
14	1461	1492	1523	1553	1584	1614	1644	1673	1703	1732	3	6	9	12	15	18	21	24	27
15	1761	1790	1818	1847	1875	1903	1931	1959	1987	2014	3	6	8	11	14	17	20	22	25
16	2041	2068	2095	2122	2148	2175	2201	2227	2253	2279	3	5	8	11	13	16	18	21	24
17	2304	2330	2355	2380	2405	2430	2455	2480	2504	2529	2	5	7	10	12	15	17	20	22
18	2553	2577	2601	2625	2648	2672	2695	2718	2742	2765	2	5	7	9	12	14	16	19	21
19	2788	2810	2833	2856	2878	2900	2923	2945	2967	2989	2	4	7	9	11	13	16	18	20
20	3010	3032	3054	3075	3096	3118	3139	3160	3181	3201	2	4	6	8	11	13	15	17	19
21	3222	3243	3263	3284	3304	3324	3345	3365	3385	3404	2	4	6	8	10	12	14	16	18
22	3424	3444	3464	3483	3502	3522	3541	3560	3579	3598	2	4	6	8	10	12	14	15	17
23	3617	3636	3655	3674	3692	3711	3729	3747	3766	3784	2	4	6	7	9	11	13	15	17
24	3802	3820	3838	3856	3874	3892	3909	3927	3945	3962	2	4	5	7	9	11	12	14	16
25	3979	3997	4014	4031	4048	4065	4082	4099	4116	4133	2	3	5	7	9	10	12	14	15
26	4150	4166	4183	4200	4216	4232	4249	4265	4281	4298	2	3	5	7	8	10	11	13	15
27	4314	4330	4346	4362	4378	4393	4409	4425	4440	4456	2	3	5	6	8	9	11	13	14
28	4472	4487	4502	4518	4533	4548	4564	4579	4594	4609	2	3	5	6	8	9	11	12	14
29	4624	4639	4654	4669	4683	4698	4713	4728	4742	4757	1	3	4	6	7	9	10	12	13
30	4771	4786	4800	4814	4829	4843	4857	4871	4886	4900	1	3	4	6	7	9	10	11	13
31	4914	4928	4942	4955	4969	4983	4997	5011	5024	5038	1	3	4	6	7	8	10	11	12
32	5051	5065	5079	5092	5105	5119	5132	5145	5159	5172	1	3	4	5	7	8	9	11	12
33	5185	5198	5211	5224	5237	5250	5263	5276	5289	5302	1	3	4	5	6	8	9	10	12
34	5315	5328	5340	5353	5366	5378	5391	5403	5416	5428	1	3	4	5	6	8	9	10	11
35	5441	5453	5465	5478	5490	5502	5514	5527	5539	5551	1	2	4	5	6	7	9	10	11
36	5563	5575	5587	5599	5611	5623	5635	5647	5658	5670	1	2	4	5	6	7	8	10	11
37	5682	5694	5705	5717	5729	5740	5752	5763	5775	5786	1	2	3	5	6	7	8	9	10
38	5798	5809	5821	5832	5843	5855	5866	5877	5888	5899	1	2	3	5	6	7	8	9	10
39	5911	5922	5933	5944	5955	5966	5977	5988	5999	6010	1	2	3	4	5	7	8	9	10
40	6021	6031	6042	6053	6064	6075	6085	6096	6107	6117	1	2	3	4	5	6	8	9	10
41	6128	6138	6149	6160	6170	6180	6191	6201	6212	6222	1	2	3	4	5	6	7	8	9
42	6232	6243	6253	6263	6274	6284	6294	6304	6314	6325	1	2	3	4	5	6	7	8	9
43	6335	6345	6355	6365	6375	6385	6395	6405	6415	6425	1	2	3	4	5	6	7	8	9
44	6435	6444	6454	6464	6474	6484	6493	6503	6513	6522	1	2	3	4	5	6	7	8	9
45	6532	6542	6551	6561	6571	6580	6590	6599	6609	6618	1	2	3	4	5	6	7	8	9
46	6628	6637	6646	6656	6665	6675	6684	6693	6702	6712	1	2	3	4	5	6	7	7	8
47	6721	6730	6739	6749	6758	6767	6776	6785	6794	6803	1	2	3	4	5	5	6	7	8
48	6812	6821	6830	6839	6848	6857	6866	6875	6884	6893	1	2	3	4	4	5	6	7	8
49	6902	6911	6920	6928	6937	6946	6955	6964	6972	6981	1	2	3	4	4	5	6	7	8
50	6990	6998	7007	7016	7024	7033	7042	7050	7059	7067	1	2	3	3	4	5	6	7	8
51	7076	7084	7093	7101	7110	7118	7126	7135	7143	7152	1	2	3	3	4	5	6	7	8
52	7160	7168	7177	7185	7193	7202	7210	7218	7226	7235	1	2	2	3	4	5	6	7	7
53	7243	7251	7259	7267	7275	7284	7292	7300	7308	7316	1	2	2	3	4	5	6	6	7
54	7324	7332	7340	7348	7356	7364	7372	7380	7388	7396	1	2	2	3	4	5	6	6	7

Table 1. Logarithms (Contd.)

	0	1	2	3	4	5	6	7	8	9	Mean Differences								
											1	2	3	4	5	6	7	8	9
55	7404	7412	7419	7427	7435	7443	7451	7459	7466	7474	1	2	2	3	4	5	5	6	7
56	7482	7490	7497	7505	7513	7520	7528	7536	7543	7551	1	2	2	3	4	5	5	6	7
57	7559	7566	7574	7582	7589	7597	7604	7612	7619	7627	1	2	2	3	4	5	5	6	7
58	7634	7642	7649	7657	7664	7672	7679	7686	7694	7701	1	1	2	3	4	4	5	6	7
59	7709	7716	7723	7731	7738	7745	7752	7760	7767	7774	1	1	2	3	4	4	5	6	7
60	7782	7789	7796	7803	7810	7818	7825	7832	7839	7846	1	1	2	3	4	4	5	6	6
61	7853	7860	7868	7875	7882	7889	7896	7903	7910	7917	1	1	2	3	4	4	5	6	6
62	7924	7931	7938	7945	7952	7959	7966	7973	7980	7987	1	1	2	3	3	4	5	6	6
63	7993	8000	8007	8014	8021	8028	8035	8041	8048	8055	1	1	2	3	3	4	5	5	6
64	8062	8069	8075	8082	8089	8096	8102	8109	8116	8122	1	1	2	3	3	4	5	5	6
65	8129	8136	8142	8149	8156	8162	8169	8176	8182	8189	1	1	2	3	3	4	5	5	6
66	8195	8202	8209	8215	8222	8228	8235	8241	8248	8254	1	1	2	3	3	4	5	5	6
67	8261	8267	8274	8280	8287	8293	8299	8306	8312	8319	1	1	2	3	3	4	5	5	6
68	8325	8331	8338	8344	8351	8357	8363	8370	8376	8382	1	1	2	3	3	4	4	5	6
69	8388	8395	8401	8407	8414	8420	8426	8432	8439	8445	1	1	2	2	3	4	4	5	6
70	8451	8457	8463	8470	8476	8482	8488	8494	8500	8506	1	1	2	2	3	4	4	5	6
71	8513	8519	8525	8531	8537	8543	8549	8555	8561	8567	1	1	2	2	3	4	4	5	5
72	8573	8579	8585	8591	8597	8603	8609	8615	8621	8627	1	1	2	2	3	4	4	5	5
73	8633	8639	8645	8651	8657	8663	8669	8675	8681	8686	1	1	2	2	3	4	4	5	5
74	8692	8698	8704	8710	8716	8722	8727	8733	8739	8745	1	1	2	2	3	4	4	5	5
75	8751	8756	8762	8768	8774	8779	8785	8791	8797	8802	1	1	2	2	3	3	4	5	5
76	8808	8814	8820	8825	8831	8837	8842	8848	8854	8859	1	1	2	2	3	3	4	5	5
77	8865	8871	8876	8882	8887	8893	8899	8904	8910	8915	1	1	2	2	3	3	4	4	5
78	8921	8927	8932	8938	8943	8949	8954	8960	8965	8971	1	1	2	2	3	3	4	4	5
79	8976	8982	8987	8993	8998	9004	9009	9015	9020	9025	1	1	2	2	3	3	4	4	5
80	9031	9036	9042	9047	9053	9058	9063	9069	9074	9079	1	1	2	2	3	3	4	4	5
81	9085	9090	9096	9101	9106	9112	9117	9122	9128	9133	1	1	2	2	3	3	4	4	5
82	9138	9143	9149	9154	9159	9165	9170	9175	9180	9186	1	1	2	2	3	3	4	4	5
83	9191	9196	9201	9206	9212	9217	9222	9227	9232	9238	1	1	2	2	3	3	4	4	5
84	9243	9248	9253	9258	9263	9269	9274	9279	9284	9289	1	1	2	2	3	3	4	4	5
85	9294	9299	9304	9309	9315	9320	9325	9330	9335	9340	1	1	2	2	3	3	4	4	5
86	9345	9350	9355	9360	9365	9370	9375	9380	9385	9390	1	1	2	2	3	3	4	4	5
87	9395	9400	9405	9410	9415	9420	9425	9430	9435	9440	0	1	1	2	2	3	3	4	4
88	9445	9450	9455	9460	9465	9469	9474	9479	9484	9489	0	1	1	2	2	3	3	4	4
89	9494	9499	9504	9509	9513	9518	9523	9528	9533	9538	0	1	1	2	2	3	3	4	4
90	9542	9547	9552	9557	9562	9566	9571	9576	9581	9586	0	1	1	2	2	3	3	4	4
91	9590	9595	9600	9605	9609	9614	9619	9624	9628	9633	0	1	1	2	2	3	3	4	4
92	9638	9643	9647	9652	9657	9661	9666	9671	9675	9680	0	1	1	2	2	3	3	4	4
93	9685	9689	9694	9699	9703	9708	9713	9717	9722	9727	0	1	1	2	2	3	3	4	4
94	9731	9736	9741	9745	9750	9754	9759	9763	9768	9773	0	1	1	2	2	3	3	4	4
95	9777	9782	9786	9791	9795	9800	9805	9809	9814	9818	0	1	1	2	2	3	3	4	4
96	9823	9827	9832	9836	9841	9845	9850	9854	9859	9863	0-	1	1	2	2	3	3	4	4
97	9868	9872	9877	9881	9886	9890	9894	9899	9903	9908	0	1	1	2	2	3	3	4	4
98	9912	9917	9921	9926	9930	9934	9939	9943	9948	9952	0	1	1	2	2	3	3	4	4
99	9956	9961	9965	9969	9974	9978	9983	9987	9991	9996	0	1	1	2	2	3	3	3	4

Table 2. Antilogarithms

	0	1	2	3	4	5	6	7	8	9	Mean Differences								
											1	2	3	4	5	6	7	8	9
.00	1000	1002	1005	1007	1009	1012	1014	1016	1019	1021	0	0	1	1	1	1	2	2	2
.01	1023	1026	1028	1030	1033	1035	1038	1040	1042	1045	0	0	1	1	1	1	2	2	2
.02	1047	1050	1052	1054	1057	1059	1062	1064	1067	1069	0	0	1	1	1	1	2	2	2
.03	1072	1074	1076	1079	1081	1084	1086	1089	1091	1094	0	0	1	1	1	1	2	2	2
.04	1096	1099	1102	1104	1107	1109	1112	1114	1117	1119	0	1	1	1	1	2	2	2	2
.05	1122	1125	1127	1130	1132	1135	1138	1140	1143	1146	0	1	1	1	1	2	2	2	2
.06	1148	1151	1153	1156	1159	1161	1164	1167	1169	1172	0	1	1	1	1	2	2	2	2
.07	1175	1178	1180	1183	1186	1189	1191	1194	1197	1199	0	1	1	1	1	2	2	2	2
.08	1202	1205	1208	1211	1213	1216	1219	1222	1225	1227	0	1	1	1	1	2	2	2	3
.09	1230	1233	1236	1239	1242	1245	1247	1250	1253	1256	0	1	1	1	1	2	2	2	3
.10	1259	1262	1265	1268	1271	1274	1276	1279	1282	1285	0	1	1	1	1	2	2	2	3
.11	1288	1291	1294	1297	1300	1303	1306	1309	1312	1315	0	1	1	1	2	2	2	2	3
.12	1318	1321	1324	1327	1330	1334	1337	1340	1343	1346	0	1	1	1	2	2	2	3	3
.13	1349	1352	1355	1358	1361	1365	1368	1371	1374	1377	0	1	1	1	2	2	2	3	3
.14	1380	1384	1387	1390	1393	1396	1400	1403	1406	1409	0	1	1	1	2	2	2	3	3
.15	1413	1416	1419	1422	1426	1429	1432	1435	1439	1442	0	1	1	1	2	2	2	3	3
.16	1445	1449	1452	1455	1459	1462	1466	1469	1472	1476	0	1	1	1	2	2	2	3	3
.17	1479	1483	1486	1489	1493	1496	1500	1503	1507	1510	0	1	1	1	2	2	2	3	3
.18	1514	1517	1521	1524	1528	1531	1535	1538	1542	1545	0	1	1	1	2	2	2	3	3
.19	1549	1552	1556	1560	1563	1567	1570	1574	1578	1581	0	1	1	1	2	2	3	3	3
.20	1585	1589	1592	1596	1600	1603	1607	1611	1614	1618	0	1	1	1	2	2	3	3	3
.21	1622	1626	1629	1633	1637	1641	1644	1648	1652	1656	0	1	1	2	2	2	3	3	3
.22	1660	1663	1667	1671	1675	1679	1683	1687	1690	1694	0	1	1	2	2	2	3	3	3
.23	1698	1702	1706	1710	1714	1718	1722	1726	1730	1734	0	1	1	2	2	2	3	3	4
.24	1738	1742	1746	1750	1754	1758	1762	1766	1770	1774	0	1	1	2	2	2	3	3	4
.25	1778	1782	1786	1791	1795	1799	1803	1807	1811	1816	0	1	1	2	2	2	3	3	4
.26	1820	1824	1828	1832	1837	1841	1845	1849	1854	1858	0	1	1	2	2	3	3	3	4
.27	1862	1866	1871	1875	1879	1884	1888	1892	1897	1901	0	1	1	2	2	3	3	3	4
.28	1905	1910	1914	1919	1923	1928	1932	1936	1941	1945	0	1	1	2	2	3	3	4	4
.29	1950	1954	1959	1963	1968	1972	1977	1982	1986	1991	0	1	1	2	2	3	3	4	4
.30	1995	2000	2004	2009	2014	2018	2023	2028	2032	2037	0	1	1	2	2	3	3	4	4
.31	2042	2046	2051	2056	2061	2065	2070	2075	2080	2084	0	1	1	2	2	3	3	4	4
.32	2089	2094	2099	2104	2109	2113	2118	2123	2128	2133	0	1	1	2	2	3	3	4	4
.33	2138	2143	2148	2153	2158	2163	2168	2173	2178	2183	0	1	1	2	2	3	3	4	4
.34	2188	2193	2198	2203	2208	2213	2218	2223	2228	2234	1	1	2	2	3	3	4	4	5
.35	2239	2244	2249	2254	2259	2265	2270	2275	2280	2286	1	1	2	2	3	3	4	4	5
.36	2291	2296	2301	2307	2312	2317	2323	2328	2333	2339	1	1	2	2	3	3	4	4	5
.37	2344	2350	2355	2360	2366	2371	2377	2382	2388	2393	1	1	2	2	3	3	4	4	5
.38	2399	2404	2410	2415	2421	2427	2432	2438	2443	2449	1	1	2	2	3	3	4	4	5
.39	2455	2460	2466	2472	2477	2483	2489	2495	2500	2506	1	1	2	2	3	3	4	5	5
.40	2512	2518	2523	2529	2535	2541	2547	2553	2559	2564	1	1	2	2	3	4	4	5	5
.41	2570	2576	2582	2588	2594	2600	2606	2612	2618	2624	1	1	2	2	3	4	4	5	5
.42	2630	2636	2642	2649	2655	2661	2667	2673	2679	2685	1	1	2	2	3	4	4	5	6
.43	2692	2698	2704	2710	2716	2723	2729	2735	2742	2748	1	1	2	3	3	4	4	5	6
.44	2754	2761	2767	2773	2780	2786	2793	2799	2805	2812	1	1	2	3	3	4	4	5	6
.45	2818	2825	2831	2838	2844	2851	2858	2864	2871	2877	1	1	2	3	3	4	5	5	6
.46	2884	2891	2897	2904	2911	2917	2924	2931	2938	2944	1	1	2	3	3	4	5	5	6
.47	2951	2958	2965	2972	2979	2985	2992	2999	3006	3013	1	1	2	3	3	4	5	5	6
.48	3020	3027	3034	3041	3048	3055	3062	3069	3076	3083	1	1	2	3	4	4	5	6	6
.49	3090	3097	3105	3112	3119	3126	3133	3141	3148	3155	1	1	2	3	4	4	5	6	6

Table 2. Antilogarithms (Contd.)

	0	1	2	3	4	5	6	7	8	9	1	2	3	4	5	6	7	8	9
.50	3162	3170	3177	3184	3192	3199	3206	3214	3221	3228	1	1	2	3	4	4	5	6	7
.51	3236	3243	3251	3258	3266	3273	3281	3289	3296	3304	1	2	2	3	4	5	5	6	7
.52	3311	3319	3327	3334	3342	3350	3357	3365	3373	3381	1	2	2	3	4	5	5	6	7
.53	3388	3396	3404	3412	3420	3428	3436	3443	3451	3459	1	2	2	3	4	5	6	6	7
.54	3467	3475	3483	3491	3499	3508	3516	3524	3532	3540	1	2	2	3	4	5	6	6	7
.55	3548	3556	3565	3573	3581	3589	3597	3606	3614	3622	1	2	2	3	4	5	6	7	7
.56	3631	3639	3648	3656	3664	3673	3681	3690	3698	3707	1	2	3	3	4	5	6	7	8
.57	3715	3724	3733	3741	3750	3758	3767	3776	3784	3793	1	2	3	3	4	5	6	7	8
.58	3802	3811	3819	3828	3837	3846	3855	3864	3873	3882	1	2	3	4	4	5	6	7	8
.59	3890	3899	3908	3917	3926	3936	3945	3954	3963	3972	1	2	3	4	5	5	6	7	8
.60	3981	3990	3999	4009	4018	4027	4036	4046	4055	4064	1	2	3	4	5	6	6	7	8
.61	4074	4083	4093	4102	4111	4121	4130	4140	4150	4159	1	2	3	4	5	6	7	8	9
.62	4169	4178	4188	4198	4207	4217	4227	4236	4256	4256	1	2	3	4	5	6	7	8	9
.63	4266	4276	4285	4295	4305	4315	4325	4335	4345	4355	1	2	3	4	5	6	7	8	9
.64	4365	4375	4385	4395	4406	4416	4426	4436	4446	4457	1	2	3	4	5	6	7	8	9
.65	4467	4477	4487	4498	4508	4519	4529	4539	4550	4560	1	2	3	4	5	6	7	8	9
.66	4571	4581	4592	4603	4613	4624	4634	4645	4656	4667	1	2	3	4	5	6	7	9	10
.67	4677	4688	4699	4710	4721	4732	4742	4753	4764	4775	1	2	3	4	5	7	8	9	10
.68	4786	4797	4808	4819	4831	4842	4853	4864	4875	4887	1	2	3	4	6	7	8	9	10
.69	4898	4909	4920	4932	4943	4955	4966	4977	4989	5000	1	2	3	5	6	7	8	9	10
.70	5012	5023	5035	5047	5058	5070	5082	5093	5105	5117	1	2	4	5	6	7	8	9	11
.71	5129	5140	5152	5164	5176	5188	5200	5212	5224	5236	1	2	4	5	6	7	8	10	11
.72	5248	5260	5272	5284	5297	5309	5321	5333	5346	5358	1	2	4	5	6	7	9	10	11
.73	5370	5383	5395	5408	5420	5433	5445	5458	5470	5483	1	3	4	5	6	8	9	10	11
.74	5495	5508	5521	5534	5546	5559	5572	5585	5598	5610	1	3	4	5	6	8	9	10	12
.75	5623	5636	5649	5662	5675	5689	5702	5715	5728	5741	1	3	4	5	7	8	9	10	12
.76	5754	5768	5781	5794	5808	5821	5834	5848	5861	5875	1	3	4	5	7	8	9	11	12
.77	5888	5902	5916	5929	5943	5957	5970	5984	5998	6012	1	3	4	5	7	8	10	11	12
.78	6026	6039	6053	6067	6081	6095	6109	6124	6138	6152	1	3	4	6	7	8	10	11	13
.79	6166	6180	6194	6209	6223	6237	6252	6266	6281	6295	1	3	4	6	7	9	10	11	13
.80	6310	6324	6339	6353	6368	6383	6397	6412	6427	6442	1	3	4	6	7	9	10	12	13
.81	6457	6471	6486	6501	6516	6531	6546	6561	6577	6592	2	3	5	6	8	9	11	12	14
.82	6607	6622	6637	6653	6668	6683	6699	6714	6730	6745	2	3	5	6	8	9	11	12	14
.83	6761	6776	6792	6808	6823	6839	6855	6871	6887	6902	2	3	5	6	8	9	11	13	14
.84	6918	6934	6950	6966	6982	6998	7015	7031	7047	7063	2	3	5	6	8	10	11	13	15
.85	7079	7096	7112	7129	7145	7161	7178	7194	7211	7228	2	3	5	7	8	10	12	13	15
.86	7244	7261	7278	7295	7311	7328	7345	7362	7379	7396	2	3	5	7	8	10	12	13	15
.87	7413	7430	7447	7464	7482	7499	7516	7534	7551	7568	2	3	5	7	9	10	12	14	16
.88	7586	7603	7621	7638	7656	7674	7691	7709	7727	7745	2	4	5	7	9	11	12	14	16
.89	7762	7780	7798	7816	7834	7852	7870	7889	7907	7925	2	4	5	7	9	11	13	14	16
.90	7943	7962	7980	7998	8017	8035	8054	8072	8091	8110	2	4	6	7	9	11	13	15	17
.91	8128	8147	8166	8185	8204	8222	8241	8260	8279	8299	2	4	6	8	9	11	13	15	17
.92	8318	8337	8356	8375	8395	8414	8433	8453	8472	8492	2	4	6	8	10	12	14	15	17
.93	8511	8531	8551	8570	8590	8610	8630	8650	8670	8690	2	4	6	8	10	12	14	16	18
.94	8710	8730	8750	8770	8790	8810	8831	8851	8872	8892	2	4	6	8	10	12	14	16	18
.95	8913	8933	8954	8974	8995	9016	9036	9057	9078	9099	2	4	6	8	10	12	15	17	19
.96	9120	9141	9162	9183	9204	9226	9247	9268	9290	9311	2	4	6	8	11	13	15	17	19
.97	9333	9354	9376	9397	9419	9441	9462	9484	9506	9528	2	4	7	9	11	13	15	17	20
.98	9550	9572	9594	9616	9638	9661	9683	9705	9727	9750	2	4	7	9	11	13	16	18	20
.99	9772	9795	9817	9840	9863	9886	9908	9931	9954	9977	2	5	7	9	11	14	16	18	20

TABLE 3. POISSON DISTRIBUTION

x/λ	0.1	0.2	0.3	0.4	0.5	0.6	0.7	0.8	0.9	1.0
0	.9048	.8187	.7408	.6703	.6065	.5488	.4966	.4493	.4066	.3679
1	.0905	.1637	.2222	.2681	.3033	.3293	.3476	.3595	.3659	.3679
2	.0045	.0164	.0333	.0536	.0758	.0988	.1217	.1438	.1647	.1839
3	.0002	.0011	.0033	.0072	.0126	.0198	.0284	.0383	.0494	.0613
4	.0000	.0001	.0002	.0007	.0016	.0030	.0050	0.0077	.0111	.0153
5	.0000	.0000	.0000	.0001	.0002	.0004	.0007	.0012	.0020	.0031
6	.0000	.0000	.0000	.0000	.0000	.0000	.0001	.0002	.0003	.0005
7	.0000	.0000	.0000	.0000	.0000	.0000	.0000	.0000	.0000	.0001

x/λ	1.1	1.2	1.3	1.4	1.5	1.6	1.7	1.8	1.9	2.0
0	.3329	.3012	.2725	.2466	.2231	.2019	.1827	.1653	.1496	.1353
1	.3662	.3014	.3543	.3452	.3347	.3230	.3106	.2957	.2842	.2707
2	.2014	.2169	.2303	.2417	.2510	.2584	.2640	.2678	.2700	.2707
3	.0738	.0867	.0998	.1128	.1255	.1378	.1496	.1607	.1710	.1804
4	.0203	.0260	.0324	.0395	.0471	.0551	.0636	.0723	.0812	.0902
5	.0045	.0062	.0084	.0111	.0141	.0176	.0216	.0260	.0309	.0361
6	.0008	.0012	.0018	.0026	.0035	.0047	.0061	.0078	.0098	.0120
7	.0001	.0002	.0003	.0005	.0008	.0011	.0015	.0020	:0027	.0034
8	.0000	.0000	.0000	.0001	.0001	.0002	.0003	.0005	.0006	.0009
9	.0000	.0000	.0000	.0000	.0000	.0000	.0001	.1001	.0001	.0002

x/λ	2.1	2.2	2.3	2.4	2.5	2.6	2.7	2.8	2.9	3.0
0	.1225	.1108	.1003	.0907	.0821	.0743	.0672	.0608	.0550	.0498
1	.2572	.2438	.2306	.2177	.2052	.1931	.1815	.1703	.1596	.1494
2	.2700	.2681	.2652	.2613	.2505	.2510	.2450	.2384	.2314	.2240
3	.1890	.1966	.2033	.2090	.2138	.2176	.2205	.2225	.2237	.2240
4	.0992	.1082	.1169	.1254	.1336	.1414	.1488	.1557	.1622	.1680
5	.0417	.0476	.0538	.0602	.668	.0735	.0804	.0872	.0940	.1008
6	.0146	.0174	.0206	.0241	.0278	.0319	.0362	.0407	.0455	.0504
7	.0044	.0055	.0068	.0083	.0099	.0118	.0139	.0163	.0188	0.216
8	.0011	.0015	.0019	.0025	.0031	.0038	.0047	.0057	.0068	.0081
9	.0003	.0004	.0005	.0007	.0009	.0011	.0014	.0068	.0022	.0027
10	.0001	.0001	.0001	.0002	.0002	.0003	.0004	.0005	.0006	.0008
11	.0000	.0000	.0000	.0000	.0000	.0001	.0001	.0001	.0002	.0002
12	.0000	.0000	.0000	.0000	.0000	.0000	.0000	.0000	.0000	.0001

TABLE 3. POISSON DISTRIBUTION (CONTD.)

x/λ	3.1	3.2	3.3	3.4	3.5	3.6	3.7	3.8	3.9	4.0
0	.0450	.0408	.0369	.0334	.0302	.0273	.0247	.0224	.0224	.0183
1	.1397	.1304	.1217	.1135	.1057	.0984	.0915	.0850	.0789	.0733
2	.2165	.2087	.2008	.1929	.1850	.1771	.1692	.1615	.1539	.1465
3	.2237	.2226	.2209	.2186	.2158	.2125	.2087	.2046	.2001	.1954
4	.0734	.1781	.1823	.1858	.1888	.1912	.1931	.1944	.1951	.1954
5	.1075	.1140	.1203	.1264	.1322	.1377	.1429	.1477	.1522	.1563
6	.0555	.0608	.0662	.0716	.0771	.0826	.0881	.0936	.0989	.1042
7	.0246	.0278	.0312	.0348	.0385	.0425	.0466	.008	.0551	.1595
8	.0095	.0111	.0129	.0148	.0159	.0191	.0215	.0241	.0269	.0298
9	.0033	.0040	.0047	.0056	.0066	.0076	.0089	.0102	.0116	.0132
10	.0010	.0013	.0016	.0019	.0023	.0028	.0033	.0039	.0045	.0053
11	.0003	.0004	.0005	.0006	.0007	.0009	.0011	.0013	.0016	.0019
13	.0000	.0000	.0000	.0000	.0001	.0001	.0001	.0001	.0002	.0002
14	.0000	.0000	.0000	.0000	.0000	.0000	.0000	.0000	.0000	.0001

x/λ	4.1	4.2	4.3	4.4	4.5	4.6	4.7	4.8	4.9	5.0
0	.0166	.0150	.0136	.0123	.0111	.0101	.0091	.0082	.0074	.0067
1	.0679	.0630	.0583	.0540	.0500	.0462	.0427	.0395	.0365	.0337
2	.1393	.1323	.1254	.1188	.1125	.1063	.1005	.0948	.0894	.0842
3	.1904	.1852'	.1798	.1743	.1687	.1631	.1574	.1517	.1460	.1404
4	.1951	.1944	.1933	.1917	.1898	.1875	.1849	.1820	.1789	.1775
5	.1600	.1633	.1662	.1687	.1708	.1725	.1738	.1747	.1753	.1755
6	.1093	.1143	.1191	.1237	.1281	.1323	.1362	.1398	.1432	1462
7	.0640	.0686	.0732	.0778	.0824	.0869	.0914	.0959	.1002	.1044
8	.0328	.0360	.0393	.0428	.0463	.0500	.0537	.0537	.0614	.0653
9	.0150	.0168	.0188	.0209	.0232	.0255	.0280	.0307	.0334	.0363
10	.0061	.0071	.0081	.0092	.0104	.0118	.0132	.0147	.0164	.0181
11	.0023	.0027	.0032	.0037	.0043	.0049	.0056	.0064	.0073	.0082
12	.0008	.0009	.0011	.0014	.0016	.0019	.0022	.0026	.0030	.0034
13	.0002	.0003	.0004	.0005	.0006	.0007	.0008	.0009	.0011	.0013
14	.0001	.0001	.0001	.0001	.0002	.0002	.0003	.0003	.0004	.0005
15	.0000	.0000	.0000	.0000	.0001	.0001	.0001	.0001	.0001	.0002

TABLE 3. POISSON DISTRIBUTION (CONTD.)

x/λ	5.1	5.2	5.3	5.4	5.5	5.6	5.7	5.8	5.9	6.0
0	.0061	.0055	.0050	.0045	.0041	.0037	.0033	.0030	.0027	.0025
1	.0311	.0287	.0265	.0244	.0225	.0207	.0191	.0176	.0162	.0149
2	.0793	.0746	.0701	.0659	.0618	.0580	.0544	.0509	.0477	.0446
3	.1348	.1293	.1239	.1185	.1133	.1082	.1033	.0985	.0938	.0892
4	.1719	.1681	.1641	.1600	.1558	.1515	.1472	.1428	.1383	.1339
5	.1753	.1748	.1740	.1728	.1714	.1697	.1678	.1656	.1632	.1606
6	.1490	.1515	.1537	.1555	.1571	.1584	.1594	.1601	.1605	.1606
7	.1086	.1125	.1163	.1200	.1234	.1267	.1298	.1326	.1353	.1377
8	.0692	.0731	.0711	.0810	.0849	.0887	.0925	.0962	.0998	.1033
9	.0362	.0423	.0454	.0486	.0519	.0552	.0586	.0620	.0654	.0688
10	.0200	.0220	.0241	.0262	.0285	.0309	.0334	.0359	.0386	.0413
11	.0093	.0104	.0116	.0129	.0143	.0157	.0173	.0190	.0207	.0225
12	.0039	.0045	.0051	.0058	.0065	.0073	.0082	.0092	.0102	.0113
13	.0015	.0104	.0021	.0024	.0028	.0032	.0036	.0041	.0046	.0052
14	.0006	.0007	.0008	.0009	.0011	.0013	.0015	.0017	.0019	.0022
15	.0002	.0002	.0003	.0003	.0004	.0005	.0006	.0007	.0008	.0009
16	.0001	.0001	.0001	.0001	.0001	.0002	.0002	.0002	.0003	.0003
17	.0000	.0000	.0000	.0000	.0000	.0000	.0001	.0001	.0001	.0001

x/λ	6.1	6.2	6.3	6.4	6.5	6.6	6.7	6.8	6.9	7.0
0	.0022	.0020	.0018	.0017	.0015	.0014	.0012	.0011	.0010	.0000
1	.0137	.0126	.0116	.0106	.0098	.0090	.0082	.0076	.0070	.0064
2	.0417	.0390	.0364	.0340	.0318	.0296	.0276	.0258	.0240	.0223
3	.0848	.0806	.0765	.0726	.0688	.0652	.0617	.0584	.0552	.0521
4	.1294	.1249	.1203	.1162	.1118	.1076	.1034	.0992	.0952	.0912
5	.1579	.1549	.1519	.1487	.1454	.1420	.1385	.1349	.1314	.1277
6	.1605	.1601	.1595	.1586	.1575	.1562	.1546	.1529	.1511	.1490
7	.1399	.1418	.1435	.1450	.1462	.1472	.1480	.1486	.1489	.1490
8	.I 066	.1099	.1130	.1160	.1188	.1215	.1240	.1263	.1284	.1304
9	.0723	.0757	.0791	.0825	.0858	.0891	.0923	.0954	.0985	.1014
10	.0441	.0469	.0498	.5285	.0558	.0588	.0618	.0679	.0679	.0710
11	.0245	.0265	.0285	.0307	.0330	.0353	.0377	.0401	.0426	.0452
12	.0124	.0137	.0150	.0164	.0179	.0194	.0210	.0227	:0245	.0264
13	.0058	.0065	.0073	.0081	.0089	.0098	.0108	.0119	.0130	.0142
14	.0025	.0029	.0033	.0037	.0041	.0046	.0052	.0058	,0064	.0071
15	.0010	.0012	.0014	.0016	.0018	.0020	.0023	.0026	.0029	.0033
16	.0004	.0005	.0005	.0006	.0007	.0008	.0010	.0011	.0013	.0014
17	.0001	.0002	.0002	.0002	.0003	.0003	.0004	.0004	.0005	.0006
18	.0000	.0001	.0001	.0001	.0001	. 0001	.0001	.0002	.0002	.0002
19	.0000	.0000	.0000	.0000	.0000	.0000	.0000	.0001	.0001	.0001

TABLE **4.** AREA UNDER THE STANDARD NORMAL
DISTRIBUTION FOR NEGATIVE VALUES OF Z

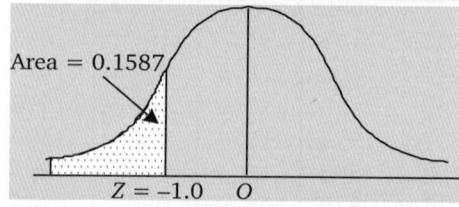

Z to First Decimal	Second Decimal									
	.00	.01	.02	.03	.04	.05	.06	.07	.08	.09
−3.0	.0014	.0013	.0013	.0012	.0012	.0011	.0011	.0011	.0010	.0010
−2.9	.0019	.0018	.0018	.0017	.0016	.0016	.0015	.0015	.0014	.0014
−2.8	.0026	.0025	.0024	.0023	.0023	.0022	.0021	.0021	.0020	.0019
−2.7	.0035	.0034	.0033	.0032	.0031	.0030	.0029	.0028	.0027	.0026
−2.6	.0047	.0045	.0044	.0043	.0041	.0040	.0039	.0038	.0037	.0036
−2.5	.0062	.0060	.0059	.0057	.0055	.0054	.0052	.0051	.0049	.0048
−2.4	.0082	.0080	.0078	.0075	.0073	.0071	.0069	.0068	.0066	.0064
−2.3	.0107	.0104	.0102	.0099	.0096	.0094	.0091	.0089	.0087	.0084
−2.2	.0139	.0136	.0132	.0129	.0126	.0122	.0119	.0116	.0113	.0110
− 2.1	.0179	.0174	.0170	.0166	.0162	.0158	.0154	.0150	.0146	.0143
−2.0	.0228	.0222	.0217	.0212	.0207	.0202	.0197	.0192	.0188	.0183
− 1.9	.0287	.0281	.0274	.0268	.0262	.0256	.0250	.0244	.0238	.0233
− 1.8	.0359	.0352	.0344	.0336	.0329	.0322	.0314	.0307	.0300	.01.94
− 1.7	.0446	.0436	.0427	.0418	.0409	.0401	.0392	.0384	.0375	.0367
− 1.6	.0548	.0537	.0526	.0516	.0505	.0495	.0485	.0475	.0465	.0455
− 1.5	.0668	.0655	.0643	.0630	.0618	.0606	.0594	.0582	.0570	.0559
− 1.4	.0808	.7938	.0778	.0764	.0749	.0735	.0722	.0708	.0694	.0681
−1.3	.0968	.0951	.0934	.0918	.0901	.0855	.0869	.0853	.0838	.0823
− 1.2	.1151	.1131	.1112	.1093	.1075	.1056	.1038	.1020	.1003	.0985
−1.1	.1357	.1335	.1314	.1292	.1271	.1251	.1230	.1210	.1190	.1170
−1.0	.1587	.1562	.1529	.1515	.1492	.1469	.1446	.1423	.1401	.1379
−0.9	.1841	.1814	.1785	.1762	.1736	.1711	.1685	.1660	.1635	.1611
−0.8	.2119	.2090	.2061	.2033	.2005	.1977	.1949	.1922	.1894	.1867
−0.7	.2420	.2389	.2358	.2327	.2297	.2266	.2236	.2206	.2177	.2143
−0.6	.2743	.2709	.2676	.2643	.2611	.2578	.2546	.2514	.2483	.2451
−0.5	.3085	.3050	.3015	.2981	.2946	.2912	.2877	.2843	.2810	.2776
−0.4	.3446	.3409	.3372	.3336	.3300	.3264	.3228	.3192	.3156	.3121
−0.3	.3821	.3783	.3745	.3707	.3669	.3632	.3594	.3557	.3520	.3483
−0.2	.4207	.4168	.4129	.4090	.4052	4013	.3974	.3936	.3897	.3859
− 0.1	.4602	.4562	.4522	.4483	.4443	.4404	.4364	.4325	.4286	.4247
−0.0	.5000	.4960	.4920	.4880	.4840	.4801	.4761	.4721	.4681	.4641

TABLE 5. AREA UNDER THE STANDARD NORMAL DISTRIBUTION FROM EXTREME LEFT TO POSITIVE VALUES OF Z

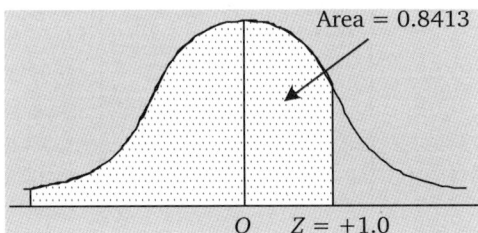

Area = 0.8413

O   Z = +1.0

Example. To find the area under the curve from extreme left $Z = -\infty$ to a point $Z = 1.0$ to the right of the mean, look up the value opposite 1.0 in the table; 0.8413 of the area under the curve lies between the extreme left ($Z = -\infty$) to a z value.

Z to First Decimal	Second Decimal									
	.00	.01	.02	.03	.04	.05	.06	.07	.08	.09
0.0	.5000	.5040	.5080	.5120	.5160	.5199	.5239	.5279	.5319	.5359
0.1	.5398	.5438	.5478	.5517	.5557	.5596	.5636	.5675	.5714	.5753
0.2	.5793	.5832	.5871	.5910	.5948	.5987	.6026	.6064	.6103	.6141
0.3	.6179	.6217	.6255	.6293	.6331	.6368	.6406	.6443	.6480	.6517
0.4	.6554	.6591	.6628	.6664	.6700	.6736	.6772	.6808	.6841	.6879
0.5	.6915	.6950	.6985	.9019	.7054	.7088	.7123	.7157	.7190	.7224
0.6	.7257	.7291	.7324	.7357	.7389	.7422	.7454	.7406	.7517	.7549
0.7	.7580	.7611	.7642	.7673	.7703	.7734	.7764	.7794	.7823	.7852
0.8	.7881	.7910	.7939	.7967	.7995	.8023	.8051	.8078	.8106	.8133
0.9	.8159	.8186	.8212	.8238	.8264	.8289	.8315	.8340	.8365	.8389
1.0	.8413	.8438	.8461	.8485	.8508	.8531	.8554	.8577	.8599	.8621
1.1	.8643	.8665	.8686	.8708	.8729	.8749	.8770	.8790	.8910	.8830
1.2	.8849	.8869	.8888	.8907	.8925	.8944	.8962	.8980	.8997	.9015
1.3	.9032	.9049	.9066	.9082	.9099	.9115	.9131	.9147	.9162	.9177
1.4	.9192	.9207	.9222	.9236	.9251	.9265	.9278	.9292	.9306	.9319
1.5	.9332	.9345	.9357	.9370	.9382	.9394	.9406	.9418	.9430	.9441
1.6	.9452	.9463	.9474	.9485	.9495	.9505	.9515	.9525	.9535	.9545
1.7	.9554	.9564	.9573	.9582	.9591	.9599	.9608	.9616	.9625	.9633
1.8	.9641	.9649	.9656	.9664	.9671	.9678	.9686	.9693	.9700	.9706
1.9	.9713	.9719	.9726	.9732	.9783	.9744	.9750	.9756	.9762	.9767
2.0	.9772	.9778	.9783	.9788	.9793	.9798	.9803	.9808	.9812	.9817
2.1	.9821	.9826	.9830	.9834	.9838	.9842	.9846	.9850	.9854	.9857
2.2	.9861	.9865	.9868	.9871	.9874	.9878	.9881	.9884	.9887	.9890
2.3	.9893	.9896	.9898	.9901	.9904	.9906	.9909	.9911	.9913	.9916
2.4	.9918	.9920	.9922	.9924	.9926	.9928	.9930	.9932	.9934	.9936
2.5	.9938	.9940	.9941	.9943	.9944	.9946	.9932	.9949	.9951	.9952
2.6	.9953	.9955	.9956	.9957	.9958	.9960	.9961	.9962	.9963	.9964
2.7	.9965	.9966	.9967	.9968	.9969	.9970	.9971	.9972	.9973	.9974
2.8	.9974	.9975	.9976	.9977	.9977	.9979	.9979	.9979	.9980	.9981
2.9	.9981	.9982	.9982	.9983	.9984	.9984	.9985	.9985	.9986	.9986
3.0	.9986	.9987	.9987	.9988	.9988	.9988	.9989	.9989	.9990	.9990

TABLE 6. PROPOSITIONS OF TOTAL AREA UNDER THE NORMAL CURVE FROM ∞ TO t, WHERE $t / (x - m\mu)\sigma$

t	$\psi(t)$	t	$\psi(t)$	t	$\psi(t)$	t	$\psi(t)$
0.00	0.5000	0.65	0.7422	1.30	0.9032	1.95	0.9744
0.01	0.5040	0.66	0.7454	1.31	0.9049	1.96	0.9750
0.02	0.5080	0.67	0.7486	1.32	0.9066	1.97	0.9756
0.03	0.5120	0.68	0.7517	1.33	0.9082	1.98	0.9761
0.04	0.5160	0.69	0.7549	1.34	0.9099	1.99	0.9767
0.05	0.5199	0.70	0.7580	1.35	0.9115	2.00	0.9772
0.06	0.5239	0.71	0.7611	1.36	0.9131	2.02	0.9783
0.07	0.5279	0.72	0.7642	1.37	0.9147	2.04	0.9793
0.08	0.5319	0.73	0.7673	1.38	0.9162	2.06	0.9803
0.09	0.5359	0.74	0.7703	1.39	0.9177	2.08	0.9812
0.10	0.5398	0.75	0.7734	1.40	0.9192	2.10	0.9821
0.11	0.5438	0.76	0.7764	1.41	0.9207	2.12	0.9830
0.12	0.5478	0.77	0.7794	1.42	0.9222	2.14	0.9838
0.13	0.5517	0.78	0.7823	1.43	0.9236	2.16	0.9846
0.14	0.5557	0.79	0.7852	1.44	0.9251	2.18	0.9854
0.15	0.5596	0.80	0.7881	1.45	0.9265	2.20	0.9861
0.16	0.5636	0.81	0.7910	1.46	0.9279	2.22	0.9868
.0.17	0.5675	0.82	0.7939	1.47	0.9292	2.24	0.9875
0.18	0.5714	0.83	0.7967	1.48	0.9306	2.26	0.9881
0.19	0.5753	0.84	0.7995	1.49	0.9319	2.28	0.9887
0.20	0.5793	0.85	0.8023	1.50	0.9332	2.30	0.9893
0.21	0.5832	0.86	0.8051	1.51	0.9345	2.32	0.9898
0.22	0.5871	0.87	0.8078	1.52	0.9357	2.34	0.9904
0.23	0.5910	0.88	0.8106	1.53	0.9370	2.36	0.9909
0.24	0.5948	0.89	0.8133	1.54	0.9382	2.38	0.9913
0.25	0.5987	0.90	0.8159	1.55	0.9394	2.40	0.9918
0.26	0.6026	0.91	0.8186	1.56	0.9406	2.42	0.9922
0.27	0.6064	0.92	0.8212	1.57	0.9418	2.44	0.9927
0.28	0.6103	0.93	0.8238	1.58	0.9429	2.46	0.9931
0.29	0.6141	0.94	0.8264	1.59	0.9441	2.48	0.9934
0.30	0.6179	0.95	0.8289	1.60	0.9252	2.50	0.9938
0.31	0.6217	0.96	0.8315	1.61	0.9463	2.52	0.9941
0.32	0.6255	0.97	0.8340	1.62	0.9474	2.54	0.9945
0.33	0.6293	0.98	0.8365	1.63	0.9484	2.56	0.9948
0.34	0.6331	0.99	0.8389	1.64	0.9495	2.58	0.9951
0.35	0.6368	1.00	0.8413	1.65	0.9505	2.60	0.9953
0.36	0.6406	0.01	0.8438	1.66	0.9515	2.62	0.9956
0.37	0.6443	1.02	0.8461	1.67	0.9525	2.64	0.9959
0.38	0.6480	1.03	0.8485	1.68	0.9535	2.66	0.9961
0.39	0.6517	1.04	0.8508	1.69	0.9545	2.68	0.9963

TABLE 7. PROPOSITIONS OF TOTAL AREA UNDER THE NORMAL CURVE FROM ∞ TO t, WHERE t / (x − mµ)σ (CONTD.)

t	$\psi(t)$	t	$\psi(t)$	t	$\psi(t)$	t	$\psi(t)$
0.40	0.6554	1.05	0.8531	1.70	0.9554	2.70	0.9965
0.41	0.6591	1.06	0.8554	1.71	0.9564	2.72	0.9967
0.42	0.6628	1.07	0.8577	1.72	0.9573	2.74	0.9969
0.43	0.6664	1.08	0.8599	1.73	0.9582	2.76	0.9971
0.44	0.6700	0.09	0.8621	1.74	0.9591	2.78	0.9973
0.45	0.6736	1.10	0.8643	1.75	0.5999	2.80	0.9974
0.46	0.6772	1.11	0.8665	1.76	0.9608	2.82	0.9976
0.47	0.6808	1.12	0.8686	1.77	0.9616	2.84	0.9977
0.48	0.6844	1.13	0.8708	1.78	0.9625	2.86	0.9979
0.49	0.6879	1.14	0.8729	1.79	0.9633	2.88	0.9980
0.50	0.6915	1.15	0.8749	1.80	0.9641	2.90	0.9981
0.51	0.6950	1.16	0.8770	1.81	0.9649	2.92	0.9982
0.52	0.6985	1.17	0.8190	1.82	0.9656	2.94	0.9984
0.53	0.7019	1.18	0.8810	1.83	0.9664	2.96	0.9985
0.54	0.7054	1.19	0.8830	1.84	0.9671	2.98	0.99116
0.55	0.7088	1.20	0.8849	1.85	0.9678	3.00	0.99865
0.56	0.7123	1.21	0.8869	1.86	0.9686	3.20	0.99931
0.57	0.7190	1.22	0.8888	1.87	0.9693	3.40	0.99966
0.58	0.7190	1.23	0.8907	1.88	0.9699	3.60	0.999841
0.59	0.7224	1.24	0.8925	1.89	0.9706	3.80	0.999928
0.60	0.7257	1.25	0.8944	1.90	0.9713	4.00	0.999968
0.61	0.7291	1.26	0.8962	1.91	0.9719	4.50	0.999997
0.62	0.7324	1.27	0.8980	1.92	0.9726	5.00	0.999997
0.63	0.7357	1.28	0.8997	1.93	0.9732		
0.64	0.7389	1.29	0.9015	1.94	0.9738		

TABLE 8. RANDOM NUMBERS

39 65 76 45 45	19 90 69 64 61	20 26 36 31 62	58 24 97 14 97	95 06 70 99 00
73 71 23 70 90	65 97 60 12 11	31 56 34 19 19	47 83 75 51 33	30 62 38 20 44
72 20 47 33 84	61 67 47 97 19	98 40 07 17 66	23 05 09 51 80	59 78 11 52 69
75 17 25 69 17	17 95 21 78 48	24 33 45 77 48	69 81 84 09 29	93 22 70 45 80
37 48 79 88 74	63 52 06 34 30	01 31 60 10 27	35 07 79 71 53	28 99 52 01 41
02 89 08 16 94	85 53 83 29 95	56 27 09 24 43	21 78 55 09 82	72 61 88 73 61
87 18 15 70 07	37 49 79 12 38	48 13 93 15 96	41 92 45 71 51	09 18 25 58 94
98 83 71 70 15	89 09 39 59 24	00 06 41 41 20	14 36 59 25 47	54 45 17 24 89
10 08 58 07 04	76 62 60 48 68	58 76 17 14 86	58 53 11 52 21	66 04 18 72 87
47 90 56 37 31	71 82 13 50 41	27 55 10 24 92	28 04 67 53 44	95 23 00 84 47
93 05 31 03 07	34 18 04 52 35	74 13 39 55 22	68 95 23 92 35	36 63 70 35 31
21 89 11 47 99	11 20 99 45 18	76 51 94 84 86	13 79 93 37 55	98 16 04 41 67
95 18 94 36 97	23 37 83 28 71	79 57 95 13 91	09 61 87 25 21	56 20 11 32 44
97 08 31 55 73	10 65 81 92 59	77 31 61 95 46	20 44 90 32 65	23 99 76 75 63
69 26 88 86 13	59 71 74 17 32	48 38 75 93 29	73 37 32 04 05	60 82 29 20 25
41 27 10 25 03	87 63 93 95 17	81 83 83 04 49	77 45 85 50 51	79 88 01 97 30
91 94 50 63 62	08 61 74 51 63	92 79 43 83 79	29 18 94 51 23	14 85 11 47 23
80 06 54 18 47	08 52 85 08 40	48 40 35 94 22	72 65 71 08 86	50 03 42 99 36
76 72 77 63 99	89 85 84 46 06	64 71 06 21 66	89 37 20 70 01	61 65 70 22 12
59 40 24 13 75	42 29 82 23 19	07 94 76 10 08	81 30 15 89 14	81 83 17 16 33
63 62 06 34 41	79 53 36 02 95	94 61 09 43 62	20 21 14 68 86	84 95 48 46 45
78 47 23 53 90	79 93 96 38 63	34 85 92 05 09	85 43 10 72 73	14 93 87 81 40
87 68 62 15 43	97 48 72 66 48	53 16 71 13 81	59 97 50 99 92	24 62 20 42 30
47 60 92 10 77	26 97 05 73 51	88 46 38 00 58	72 63 49 29 31	75 70 16 08 24
56 88 87 59 41	06 87 37 78 48	65 88 69 58 39	88 02 84 27 82	85 81 56 39 38
22 17 68 65 84	86 02 22 57 51	68 69 80 95 44	11 29 01 95 80	49 34 35 86 47
19 36 27 59 46	39 77 32 77 09	79 57 92 36 59	89 74 39 82 15	05 50 94 34 74
16 77 23 02 77	28 06 24 35 93	22 45 44 84 11	87 80 61 65 31	09 71 91 74 25
78 43 66 07 61	97 66 63 99 61	80 45 67 93 82	59 73 19 85 23	53 33 65 97 21
03 28 28 26 08	69 30 16 09 05	53 58 47 70 93	66 56 45 65 79	45 56 20 19 47
04 31 17 21 56	33 63 99 19 87	26 72 39 27 67	53 77 57 68 93	60 61 97 22 61
61 06 98 03 91	87 14 77 43 96	43 00 65 98 50	45 60 33 01 07	98 99 46 50 47
23 58 35 26 00	99 53 93 61 28	52 70 05 48 34	56 65 05 61 86	90 92 10 79 80
15 39 25 70 99	93 86 52 77 65	15 35 59 05 28	22 87 26 07 47	86 96 98 29 06
58 71 96 30 24	18 46 23 34 27	85 13 99 24 44	49 18 09 79 49	74 16 32 23 02
93 22 53 64 39	07 10 63 76 35	87 03 04 79 88	08 33 33 85 51	55 34 57 72 69
78 76 58 54 74	92 38 70 97 92	52 06 79 79 45	82 63 18 27 44	69 66 92 19 09
61 81 31 96 82	00 57 25 60 56	46 72 60 18 77	55 66 12 62 11	09 99 55 64 57
42 88 07 10 05	24 98 65 08 21	47 21 61 88 32	27 80 30 21 60	10 92 35 36 12
77 94 30 05 33	28 10 99 00 27	12 73 73 99 12	39 99 57 94 82	96 88 87 17 91

TABLE 9. PRESENT VALUES

Year	1%	2%	3%	4%	5%	6%	7%	8%	9%	10%
1	0.990	0980	0.971	0.962	0.952	0.943	0.935	0.926	0.917	0.909
2	0.980	0.961	0.943	0.925	0.907	0.890	0.873	0.857	0.842	0.826
3	0.971	0.942	0.915	0.889	0.864	0.840	0.816	0.794	0.772	0.751
4	0.961	0.924	0.888	0.855	0.823	0792	0.763	0.735	0.708	0.683
5	0.951	0.906	0.863	0.822	0.784	0.747	0.713	0.681	0.650	0.621
6	0.942	0.888	0.837	0.790	0.746	0.705	0.666	0.630	0.596	0.564
7	0.933	0.871	0.813	0.760	0.711	0.665	0.623	0.583	0.547	0.513
8	0.923	0.853	0.789	0.731	0.677	0.627	0.582	0.540	0.502	0.467
9	0.914	0.837	0.766	0.703	0.645	0.592	0.544	0.500	0.460	0.424
10	0.905	0.820	0.744	0.676	0.614	0.558	0.508	0.463	0.422	0.386
11	0.896	0.804	0.722	0.650	0.585	0.527	0.475	0.429	0.388	0.350
12	0.887	0.789	0.701	0.625	0.557	0.497	0.444	0.397	0.356	0.319
13	0.879	0.773	0.681	0.601	0.530	0.469	0.415	0.368	0.326	0.290
14	0.870	0.758	0.661	0.577	0.505	0.442	0.388	0.340	0.299	0.263
15	0.861	0.743	0.642	0.555	0.481	0.417	0.362	0.315	0.275	0.239
16	0.853	0.728	0.623	0.534	0.458	0.394	0.339	0.292	0.252	0.218
17	0.844	0.714	0.605	0.513	0.436	0.371	0.317	0.270	0.231	0.198
18	0.836	0.700	0.587	0.494	0.416	0.350	0.296	0.250	0.212	0.180
19	0.828	0.686	0.570	0.475	0.396	0.331	0.277	0.232	0.194	0.164
20	0.820	0.673	0.554	0.456	0.377	0.312	0.258	0.215	0.178	0.149
21	0.811	0.660	0.538	0.439	0.359	0.294	0.242	0.199	0.164	0.135
22	0.803	0.647	0.522	0.422	0.342	0.278	0.226	0.184	0.150	0.123
23	0.795	0.634	0.507	0.406	0.326	0.262	0.211	0.170	0.138	0.112
24	0.788	0.622	0.492	0.390	0.310	0.247	0.197	0.158	0.126	0.102
25	0.780	0.610	0.478	0.375	0.295	0.233	0.184	0.146	0.116	0.092
30	0.742	0.552	0.412	0.308	0.231	0.174	0.131	0.099	0.075	0.057
35	0.706	0.500	0.355	0.253	0.181	0.130	0.094	0.068	0.049	0.036
40	0.672	0.453	0.307	0.208	0.142	0.097	0.67	0.046	0.032	0.022
45	0.639	0.410	0.264	0.171	0.111	0.073	0.48	0.031	0.021	0.014
50	0.608	0.372	0.228	0.141	0.087	0.054	0.34	0.021	0.013	0.09

TABLE 10. VALUES OF e^x AND e^{-x}

x	e^x	e^{-x}	x	e^x	e^{-x}
0.00	1.000	1.000	3.00	20.086	0.049
0.10	1.105	0.904	3.10	22.198	0.045
0.20	1.221	0.818	3.20	24.533	0.040
0.30	1.349	0.740	3.30	27.113	0.036
0.40	1.491	0.670	3.40	29.964	0.033
0.50	1.648	0.606	3.50	33.115	0.030
0.60	1.822	0.548	3.60	36.598	0.027
0.70	2.015	0.496	3.70	40.447	0.024
0.80	2.225	0.449	3.80	44.701	0.022
0.90	2.459	0.406	3.90	49.402	0.020
1.00	2.718	0.367	4.00	54.598	0.018
1.10	3.004	0.332	4.10	60.340	0.016
1.20	3.320	0.301	4.20	66.686	0.014
1.30	3.669	0.272	4.30	73.700	0.013
1.40	4.055	0.246	4.40	81.451	0.012
1.50	4.481	0.223	4.50	90.017	0.011
1.60	4.953	0.201	4.60	99.484	0.010
1.70	5.473	0.182	4.70	109.95	0.009
1.80	6.049	0.165	4.80	121.51	0.008
1.90	6.685	0.149	4.90	134.29	0.007
2.00	7.389	0.135	5.00	148.41	0.006
2.10	8.166	0.122	5.10	164.02	0.006
2.20	9.025	0.110	5.20	181.27	0.005
2.30	9.974	0.100	5.30	200.34	0.004
2.40	11.023	0.090	5.40	221.41	0.004
2.50	12.182	0.082	5.50	244.69	0.004
2.60	13.464	0.074	5.60	270.43	0.003
2.70	14.880	0.067	5.70	298.87	0.003
2.80	16.445	0.060	5.80	330.30	0.003
2.90	18.174	0.055	5.90	365.04	0.002
3.00	20.086	0.049	6.00	403.43	0.002

Bibliography

1.	Ashour, S.	Sequencing theory, Springer-Verlag, Berlin (1972)
2.	Bellman, R.F.	Applied dynamic programming, Princeton University Press, Berlin (1988)
3.	Dantzig, G.B.	Linear Programming and extensions, Princeton University Press, Berlin (1988)
4.	Gopal Krishnan, P. and M. Sundaresan	Material Management : An Integrated approach, PHI, India (1982)
5.	Goel, B.S. and Mittal S.K.	Operation Research, Pragati Prakashan, Meerut (1998)
6.	Hadley, G.	Non-linear and Dynamic Programming; Addison Wesley Reading mass, Massachusetts (1964)
7.	Howard, R.	Dynamic Programming and Markov Process, Wiley, New York (1960)
8.	Jianhua, W.	The theory of games, Oxford University, Press (1988)
9.	Kamboo, N.S.	Mathematical programming techniques, East-West Press, India (1984)
10.	Pundir, S.K.	Linear Programming, CBS, New Delhi (2019)
11.	Pundir, S.K.	Linear Programming with Game theory, CBS, New Delhi (2019)
12.	Rao, S.S.	Engineering Optimization, New Age International, Delhi (2013)
13.	Sharma, J.K.	Operation Research, MacMillan Company, New Delhi (2013)
14.	Sharma, S.D.	Operation Research, KedarNath RamNath, Meerut (2009)
15.	Smith, D.E.	Quantitative Business Analysis, John Willey and sons (1977)

16.	Taha, H.A.	Operation Research – An Introduction, Maxwell Macmillan, New York (1989)
17.	Vajda, S.	Theory of linear and non-linear programming, Longman Green and Co. London (1974)
18.	Vohra, N.D.	Quantitative techniques in Management, TMH, New Delhi (1990)
19.	Wagner, A.M.	Principles of Operations Research, PHI, India (1975)
20.	Yadav, S.R. and Malik A.K.	Operation Research, Oxford University Press, Delhi (2014)
21.	Zoints, S.	Linear and Integer Programming, Prantice-Hall, Eaglewood N. J. (1969)
22.	Zangwill, W.I.	Non-linear Programming – A unified approach, Eaglewood Cliffs, N.J. : Prantice Hall, 1969

INDEX

Other Useful Books
by the Same Author

Measure Theory

Classical Algebra

Complex Analysis

Functional Analysis

Numerical Analysis

Linear Programming

Mathematical Analysis

Applied Discrete Structures

Linear Programming with Game theory

Handbook of Mathematics for Biosciences

A Competitive Approach to Linear Algebra

A Competitive Approach to Modern Algebra

Handbook of Mathematics for Pharmacy students

Fundamental Mathematics for Computer Applications

Integral Transform Methods in Science and Engineering

Numerical and Statistical Computing in Science and Engineering